KU-075-436

Contents

Standard Handbook of Fastening and Joining

Other McGraw-Hill Handbooks of Interest

Avallone and Baumeister • Marks' Standard Handbook for Mechanical Engineers
Brady and Clauser • Materials Handbook
Brater and King • Handbook of Hydraulics
Callender • Time-Saver Standards for Architectural Design Data
Chow • Handbook of Applied Hydrology
Crocker and King • Piping Handbook
Davis and Sorenson • Handbook of Applied Hydraulics
DeChiara and Callender • Time-Saver Standards for Building Types
Fink and Beaty • Standard Handbook for Electrical Engineers
Guyer • Handbook of Applied Thermal Design
Harris • Shock and Vibration Handbook
Hicks • Standard Handbook of Engineering Calculations
Higgins • Maintenance Engineering Handbook
Ireson and Coombs • Handbook of Reliability Engineering and Management
Juran and Gryna • Juran's Quality Control Handbook
Karassi, Krutzsch, Fraser, and Messina • Pump Handbook
Maynard • Handbook of Business Administration
McPartland • McGraw-Hill's National Electrical Code® Handbook
Merritt • Building Design and Construction Handbook
Merritt • Standard Handbook for Civil Engineers
Parmley • Field Engineers' Manual
Parmley • HVAC Field Manual
Parmley • Mechanical Components Handbook
Perry • Engineering Manual
Rohsenow, Hartnett, and Ganić • Handbook of Heat Transfer Applications
Rohsenow, Hartnett, and Ganić • Handbook of Heat Transfer Fundamentals
Rothbart • Mechanical Design and Systems Handbook
Smeaton • Motor Application and Maintenance Handbook
Smeaton • Switchgear and Control Handbook
Tuma • Engineering Mathematics Handbook
Tuma • Handbook of Physical Calculations
Tuma • Handbook of Structural and Mechanical Matrices

#39229122

Standard Handbook of Fastening and Joining

ROBERT O. PARMLEY, P.E. Editor in Chief

Consulting Engineer, Morgan & Parmley, Ltd.
Ladysmith, Wisconsin

D
621.88
PAR

Third Edition

McGraw-Hill
New York San Francisco Washington, D.C. Auckland Bogotá
Caracas Lisbon London Madrid Mexico City Milan
Montreal New Delhi San Juan Singapore
Sydney Tokyo Toronto

Library of Congress Cataloging-in-Publication Data

Standard handbook of fastening and joining / Robert O. Parmley, editor-in-chief. — 3rd ed.
p. cm.
Includes index.
ISBN 0-07-048589-5 (alk. paper)
1. Fasteners—Handbooks, manuals, etc. 2. Joints (Engineering)--Handbooks, manuals, etc. I. Parmley, Robert O.
TJ1320.S74 1996
621.8′8—dc20 96-27689
CIP

McGraw-Hill
A Division of The McGraw-Hill Companies

Copyright © 1997, 1989, 1977 by The McGraw-Hill Companies, Inc. All rights reserved. Printed in the United States of America. Except as permitted under the United States Copyright Act of 1976, no part of this publication may be reproduced or distributed in any form or by any means, or stored in a data base or retrieval system, without the prior written permission of the publisher.

1 2 3 4 5 6 7 8 9 0 DOC/DOC 9 0 1 0 9 8 7 6

ISBN 0-07-048589-5

UNIVERSITY OF STRATHCLYDE
27 JAN 1997
UNIVERSITY LIBRARY

The sponsoring editor for this book was Robert Esposito, the editing supervisor was Paul R. Sobel, and the production supervisor was Donald F. Schmidt. It was set in Gael by Pro-Image Corporation.

Printed and bound by R. R. Donnelley & Sons Company.

McGraw-Hill books are available at special quantity discounts to use as premiums and sales promotions, or for use in corporate training programs. For more information, please write to the Director of Special Sales, McGraw-Hill, 11 West 19th Street, New York, NY 10011. Or contact your local bookstore.

Information contained in this work has been obtained by The McGraw-Hill Companies, Inc. ("McGraw-Hill") from sources believed to be reliable. However, neither McGraw-Hill nor its editor in chief and authors guarantees the accuracy or completeness of any information published herein and neither McGraw-Hill nor its editor in chief and authors shall be responsible for any errors, omissions, or damages arising out of use of this information. This work is published with the understanding that McGraw-Hill and its editor in chief and authors are supplying information but are not attempting to render engineering or other professional services. If such services are required, the assistance of an appropriate professional should be sought.

This book is printed on recycled, acid-free paper containing a minimum of 50% recycled de-inked fiber.

To the memory of my dad

Contributors

CHRISTOPHER T. ANDERSON *Product Manager, Motoman Corp., West Carrolton, OH*

MARK BOYER *Harris Calorific, Division of Emerson Electric, Cleveland, OH*

RONALD B. BORG *Consultant, Peterborough, Ontario, Canada*

HARRY S. BRENNER, P.E. *President, Almay Research & Testing Corp., Los Angeles, CA*

JOSEPH F. BRIGGS *Manager, Service Engineering, Aeroquip Corp., Jackson, MI*

ALGER B. COLTHORP, P.E. *Chief Engineer, Insituform Technologies Inc., Chesterfield, MO*

LAWRENCE B. CURTIS *Fisher Gauge, Ltd., Peterborough, Ontario, Canada*

EINAR A. ERIKSEN *Manager of Engineering Services, Waldes Truarc, Inc., Long Island City, NY*

PATRICK J. GOLDEN, P.E. *Sales Engineer, Elk River Concrete Products, Minneapolis, MN*

GIRARD S. HAVILAND *Consultant, Naples, ME*

DR. LESLIE A. HORVE, P.E. *V.P. of Technology, Chicago Rawhide Mfg. Co., Elgin, IL*

MERLIN T. LEBAKKEN, P.E. *President, Power System Engineering, Inc., Madison, WI*

JOSEPH F. LOPES *V.P. of Engineering, Penn Engineering & Mfg. Corp., Danboro, PA*

MILFORD RIVET & MACHINE TECHNICAL STAFF *Milford Rivet & Machine Co., Milford, CT*

WILLIAM O'KEEFE *Senior Technical Editor, Power Magazine, New York, NY*

ROBERT O. PARMLEY, P.E. *President, Morgan & Parmley, Ltd., Ladysmith, WI*

RICHARD S. SABO *Manager, Publicity and Educational Services, The Lincoln Electric Co., Cleveland, OH*

GERALD L. SCHNEBERGER, Ph.D., P.E. *President, Training Resources, Inc., Flint, MI*

STEPHEN M. WARD *Chief Engineer, Spirolor Kaydon Corp., St. Louis, MO*

Special Consultants

WILL BARBEAU[2]
Barbeau Associates, Inc.
Barrington, RI

MICHAEL K. BONNER[1]
Technology Services Inc.
Port Washington, NY

GERALD L. BRADSHAW,[1] **SR., C.E.T.**
Ladysmith, WI

JERRY BRISTOL[2]
Loctite Corporation
Newington, CT

CHARLES HERBRUCK, SR.[1]
Herbruck-Mills, Inc.
Cleveland, OH

LEONARD KIRSCH[1 & 2]
Kirsch Communications
Merrick, NY

ROBIN McFARLIN[2]
CEM Company, Inc.
Danielson, CT

GEORGE H. MORGAN, P.E.[1]
Ladysmith, WI

TED SCHAUMBURG[2]
Drive-Loc, Inc.
Sycamore, IL

DAVID P. HAMMER
McKinney Advertising & Public Relations
Philadelphia, PA

[1]First edition only.
[2]Second edition only.

Preface

Harnessing natural forces and constructing devices to improve the quality of life has greatly depended upon our ability to fasten and join components together to form useful things. The story of fastening technology parallels the history of civilization, with both developing in concert; each relying upon the other for advancement. Certainly the interchangeability of standardized fasteners is one of the secrets behind the phenomenal advances of Western technology.

A humorist once said that, "if the wheel was the greatest invention of antiquity, the second greatest was the axle." However, it must be pointed out that neither would have achieved much success if someone had not connected them together. It is this principal of fastening, joining, connecting, attaching, securing and locking of various independent components into useful assemblies, designed to perform specific tasks, which has been key to advancing our technology. Remember that the sum of the individual parts of a machine or system is less than the value of the total assembled mechanism.

My thirty-five plus years of compiling data on fastening and joining technology has resulted in massive files and mounds of material. In a way it is unfortunate that the handbook page limit restricts the coverage because there are areas that should be included. However, if a page barrier was not established, I fear we would never reach publication. In any event, we have endeavored to be wise in our selection of material.

This third edition has been significantly updated and expanded and has grown to twenty-five separate sections. While much of the data from previous editions remains in tact because of its time-tested value, over a quarter of the original material has been either revised or expanded. A significant sub-section has been added to Pipe Fastening entitled, Pipe Relining Technology. This recently perfected technology is the successful practice of rehabilitating existing, deteriorated pipelines by in-place relining. The Lumber and Timber section has been expanded to include typical log connections. Both Standard Pins and Locking Components sections have additional material. Five completely new sections have been added. They are: Shafts and Couplings, Seals and Packings, Self-Clinching Fasteners, Robotic Assembly and Innovative Connections. The later is included to stimulate creative fastening using standard components.

The United States has been struggling seriously with metric conversion since the early 1970s, which has produced only marginal results. However, recent legislation may accelerate its use. The 1988 Omnibus Trade and Competitiveness Act and the subsequent Executive Order by President Bush in 1991 mandated federal agencies to use metric measurement in government contracting and procurement. As a result of these governmental requirements, the General Services Administration began requiring metrication in federal construction projects commencing on January 1, 1994. Additionally, the Federal Highway Administration set a target date of October 1, 1996 to implement the use of metrics in all federally funded highway projects which includes bridges.

The Concrete Reinforcing Steel Institute (CRSI) and the Steel Manufacturers Association (SMA) support a "soft" metric conversion, rather than a "hard" transition. Certainly the established inch-pound sizes could continue to be produced and their dimensions be converted to metric equivalents. This logical compromise would avoid confusion, eliminate the need for new tooling and not force producers and fabricators to maintain extra inventory to supply both the private and government markets.

In an effort to meet the evolving metrication because of an emerging increase of global trade and the governmental mandates, this third edition has continued to sprinkle the text with metric fastening and joining where feasible. We have continued to include a full section on metrication and conversion in this edition. However, the dominant measurement has remained in typical English units. I believe that a "total" conversion to metrics in this edition would be premature at this point in time and would serve no practical purpose.

As with previous editions, our present effort is greatly indebted to the many sources that contributed to producing this handbook. Many technical societies, manufacturing organizations, construction firms, individual engineers and technical research facilities were extremely cooperative in supplying their respective technical expertise. While these individuals and organizations are not listed in this Preface, we have continued our policy to note each source of specific material on the page where it appears throughout the handbook to insure proper credit.

Again, I want to acknowledge the clerical assistance provided by Lana and Ethne throughout the years. A special thanks to Wayne Parmley for producing a top quality book cover design for this third edition.

As long as assembly is a key factor in civilization's development, fastening and joining technology will play a major role. It has a great heritage and an unlimited future affecting all trades, skills and disciplines.

Finally, it is my desire that our original goal of providing a reliable, accurate and practical handbook of standard fastening and joining technology has continued into this latest effort. I hope this edition will serve you well.

Robert O. Parmley, P.E., CMfgE, CSI

Preface to Second Edition

As alluded to in the Preface to the first edition, a second edition of this handbook required a wider coverage of relevant material, to include some peripheral areas as well as an update on advancing fastener technology.

Well, here we are almost 15 years later with an ever expanding collection of technical material which must be culled to fit within the limited number of pages allocated to this second edition. Hopefully, we have been true to the original concept of the first edition and wisely chosen from the current data so that this present effort will be of value.

Almost half of the original text has been revised and updated. Additionally, six new sections have been included. They are: Expansion Joints, Concrete Fastening, Injected Metal Assembly, Sheet Metal Assembly, Retaining Compounds, and Rope Splicing and Tying.

Fasteners have literally held the world together throughout recorded history. The tale of fastening and joining technology parallels the history of humanity's technical advancement, both evolving in harmony and each dependent upon the other. From the ancient mechanic who first connected a wheel onto an axle to the 250,000 fasteners used on the Saturn V rocket, humanity has continued to develop this technology into a highly specialized branch of engineering. According to recognized authorities, the U.S. fastener industry produces an estimated 200 billion fasteners a year in many thousands of different styles, designs, and sizes.

In recent years, more manufacturing companies are seriously looking at the feasibility of integrating robots into their assembly process. Typical or standard assembly methods have historically been one of the areas of highest direct labor costs, in some cases, accounting for more than 50 percent of the total cost of manufacturing. Yet there has been a relatively low use of automation in this area. Products have not been adequately adapted for robotic assembly; existing assemblies generally require human dexterity, visual skills, and judgment to deal with components that an industrial robot may find difficult or sometimes impossible to handle; robotic machinery has not been typically versatile enough to be cost-effective for limited or small production runs; and robots have rarely been either speedy or sophisticated enough for precision assembly. However, I believe that robotic assembly is in its infancy and within a decade or two the previously mentioned problems will be solved so that automation will become the major method used for manufacturing assembly. Because robotic assembly is not presently an industry norm or considered standard, we have not included any technical presentation in this edition. I predict, however, that a section on robotic assembly will be a major segment in our third edition.

As in the first edition, the reader will find ample tables, data, and standards with applicable illustrations in this expanded second effort. Each section was prepared and/or revised by well-qualified engineers with proven expertise.

The sources of technical data have been properly credited at their respective locations throughout the handbook. The fine cooperation from the many technical institutions, societies, manufacturing firms, consultants, special advisors, and contributors was a repeat of that experienced during preparation of the first edition. I personally thank them and consider each one a partner in this work.

In the same vein, I again want to thank the contributors to the first edition, especially those whose material is also included in this current edition.

Let me add a special note recognizing Wayne's illustrating contributions in various segments of the text and his original conception for the cover, nor can I forget Ethne's untiring efforts in typing portions of the manuscript and related correspondence.

One final note: Recognizing the important and growing role of women in the fields of engineering and industry, every effort has been made in this second edition to present material in gender-neutral language. However, for a few terms such as "manhole," "yoeman," and the names of some knots no generally accepted gender-neutral equivalents exist, and occasionally, to avoid the awkwardness of the "he/she" or "her/his" types of grammatical construction, the older use of the pronouns "he" and "his" to refer to people in general have been allowed to stand. In all such cases, the words should be taken in a purely generic sense, intended to apply to women as well as to men.

Robert O. Parmley, P.E., CMfgE

Preface to First Edition

Fastening and joining is defined as an act of bringing together, connecting, uniting, becoming one, or becoming a unit. Therefore, it is with this technology that *Standard Handbook of Fastening and Joining* is concerned.

From almost the dawn of time, man used his ingenuity to fasten and join components of similar or different materials. In reality, this may well have been his first technology, whether it was used in the fabrication of tools or in connection with other basic skills necessary to survival.

Archaeologists have excavated remains of old civilizations which reveal the uses of this art: statues with separate eyes glued to their sockets, iron pins used as hinge pivots, Roman ships caulked with tars, egg whites used to laminate gold leaf to wood, wooden pegs used as dowels, primitive attempts at threaded bolts, and ancient ways of forging iron.

The Renaissance advanced the technology and widened its scope of application, particularly in the mechanical areas.

With the Industrial Revolution came further-refined techniques. But it was the scientific revolution in the second quarter of this century and the development of new metals, plastics, and synthetic materials, plus the increase of world-wide trade, that created a need for standards to be established, especially in the area of mechanical fastening. The technology of fastening and joining has come of age and has taken its rightful place as a catalyst in modern engineering.

It should be mentioned that modern engineering incorporates quick disassembly features into many designs. This is done for two reasons. First, easy disassembly offers fewer maintenance problems when replacement of a component is required. Second, it assures greater value at the end of a unit's life when basic materials are separated. Reusable parts and fasteners can be recycled into the manufacturing process, thus relieving some of the strain placed on our environment and our natural resources.

This handbook was conceived in an attempt to categorize, define, and list standards and to present the latest engineering data on fastening and joining. The editor, the contributors, the consultants, and the publisher feel that an all-inclusive work on this technology is long overdue.

Fastening technology's multitude of new designs have been developed in such astronomical quantities in recent years that it is an impossible task to mention all fasteners. Therefore, we have illustrated only the "standard" fastening and joining techniques that have proven themselves.

It is hoped that this handbook will be successful enough to warrant revision and updating. At that time, a wider coverage should be considered to include peripheral areas, new designs, and updated technology.

This handbook is intended to be helpful to all engineering disciplines. Since fastening and joining is an all-encompassing term, we have designed this work to be of value to each general area of engineering. There are areas devoted specifically to mechanical, civil, electrical, structural, and manufacturing functions and to their particular fastening problems.

Each section has been prepared by a well-qualified engineer with the expertise to provide a practical presentation. Much research and editing and many long hours of rewriting went with this effort to achieve a useful and long-lasting reference for the entire engineering profession.

Throughout this handbook, the reader will find applicable tables, data, and standards as developed by various institutes, societies, and respected manufacturing firms. Our largest source for this wealth of information was The American National Standards Institute as published by The American Society of Mechanical Engineers, 345 East 47th Street, New York, NY 10017. We have included extracts of these standards; copies of the complete works or standards may be obtained from the publisher at the address noted above.

The Industrial Fasteners Institute, 1505 East Ohio Bldg., 1717 E. 9th Street, Cleveland, OH 44114, has supplied our project with valuable data; their research on metric conversion has been particularly useful in the preparation of Section 16. In anticipation of the proposed national conversion, we have attempted to include in the text metric measurements where convenient. Section 16 has all the applicable conversion tables as standardized by the International Organization for Standardization (ISO).

The cooperation that I have experienced from the contributors has been fantastic. Their spirit of dedication to our project has kept our ship on course through many troubled waters. I take my hat off to these fellow engineers for a job well done.

The publishers have given me free rein in developing the outline, in selecting the contributors, and in finalizing the text. Their expert guidance throughout the project was most welcome and needed.

A special thanks to the many technical organizations, societies, and manufacturing firms who supplied much tabular data and many illustrations in their specific fields. Their contributions are noted throughout the handbook.

A list of special consultants appear elsewhere, but a particular mention should be made recognizing their yeoman work. Without their tireless effort this handbook would not have reached completion. One of these consultants, G. L. Bradshaw, Sr., not only assisted in the illustrating and preparation of artwork, but provided much encouragement by his enthusiasm for our project and his tireless performance of many peripheral duties.

A well-seasoned engineer with a well-rounded background who is willing to assist and advise a handbook editor is invaluable. George H. Morgan, P.E., performed this function. His wise counsel has strengthened the fiber of the material and ensured the practical format we originally sought.

Rupert LeGrand, Editor of *The New American Machinist's Handbook;* Robert Abbott, Editor of *Product Engineering;* and Robert Kelly, Editor of *Assembly Engineering* were all very helpful in steering me to prospective contributors and to various sources of useful technical material.

My wife deserves much credit for this handbook's completion. Her encouragement and untiring effort in typing letters, memos, and the manuscript were a constant marvel to me. I would like to express my gratitude to her for helping me make a dream come true.

Robert O. Parmley

Standard Handbook of Fastening and Joining

Section 1

Threaded Fasteners — Descriptions and Standards

Part 1 Standard Threaded Fasteners

HARRY S. BRENNER, P.E.

President, Almay Research and Testing Corp.,
Los Angeles, California

INTRODUCTION

The general acceptance and wide-ranging use of threaded fasteners is often taken for granted in our current high-technology society. In reality, though, the modern threaded-fastening system represents a complex and often critical design element which has been under continual development and refinement through innovative production, application experience, and a degree of standardization. With more emphasis being placed on reliability and long-life service of structures and equipment, the engineering community is recognizing that the so-called "common" fastener is really not that common after all. Certainly, for all the significant designs being created, the fact remains that these designs must still be put together and function properly for the purpose intended.

1-1 Advantages of Threaded Fasteners There are a number of recognized techniques for joining and assembling structures and materials. As a prime joining system, standard threaded fasteners offer certain distinct advantages. They are commercially available in a wide range of styles, materials, and sizes often permitting fastening where other systems are not effective or efficient. They are capable of joining same or dissimilar materials in uniform as well as in unusual joint configurations. They can be easily installed in factories or in the field with both standard manual or power installation tools, with a maximum of safety. And in particular, where subsequent maintenance and/or repair may be required, threaded fasteners provide the same ease in removal and replacement.

Standard threaded fasteners are highly sophisticated mechanical products subject to rigid requirements of standardization and interchangeability. Basically, the standard threaded fastening system consists of an external (male) threaded element such as a bolt or screw, and a mating internal (female) threaded element such as a nut, insert, or tapped hole. Interestingly enough, the two components of the system, which are so dependent on each other, are fabricated by different production methods, and often by companies specializing in the manufacture of only one of the major components. The ability to assemble and use male and female threaded elements, irrespective of the manufacturing source, emphasizes an additional advantage and consideration of fasteners as a joining technique: low unit cost of the fastener itself, and competitive installed cost of the fastener assembly.

Threaded fasteners may be used for one function or a combination of various functions. Primarily, fasteners are intended as structural, load-carrying elements. However, they may also be selected for their nonmagnetic properties, for corrosive or other environmental exposure conditions, or even for decorative appearance. For these reasons, standard threaded fasteners are available in various combinations of materials, strength levels, and finishes. Implicit in the accurate design application of standard fasteners is an understanding of the basic functions and practical limitations of the fastener system.

1-2 Screw Threads Fundamental to the threaded fastener is the screw thread, which has often been described as an inclined plane wrapped around a cylinder. While there are early records of fasteners and devices employing a screw thread principle, it was characteristic that such "nuts and bolts" were generally not interchangeable and usually required specific matching or fitting to permit their use. The engineering effort to establish working standards for screw threads did not fully materialize until the middle of the nineteenth century. In England, Sir Joseph Whitworth proposed a screw thread system featuring a constant thread angle of 55 degrees, which became known as the Whitworth Thread. About the same time in the United States, William Sellers developed a screw thread system based on a standard thread angle of 60 degrees, which formed the basis for the American National Thread. The eventual adoption of each system was instrumental in contributing to the success of the industrial revolution, and each system remained dominant as a respective national standard for about a century.

In 1948, culminating many years of intensive effort, representatives of Great Britain, Canada, and the United States signed a Declaration of Accord establishing the Unified Screw Thread system. The standards developed defined the thread form based on a 60-degree angle, tolerances and size limits, and diameter-pitch combinations for use by the three countries, and most English (inch) measurement system nations. The elements of the Unified Screw Thread design form are shown in Fig. 1-1. In the United

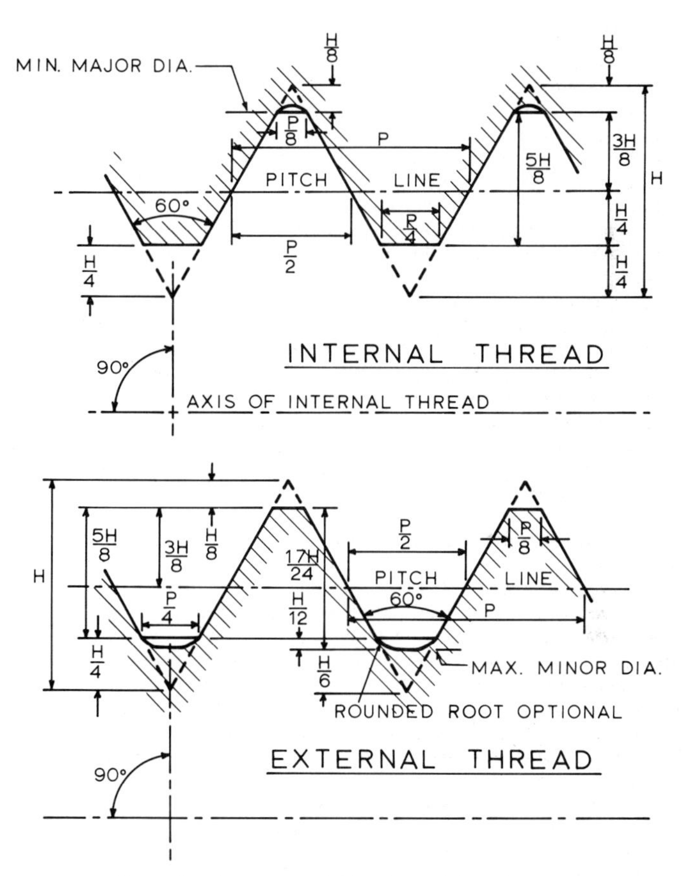

FIG. 1-1 Elements of Unified Screw Thread form.

States, the basic documents outlining complete information on screw threads are the National Bureau of Standards Handbook H-28, *Screw-Thread Standards for Federal Services*, and American National Standard ANSI B1.1-1974, *Unified Inch Screw Threads*.

Here, specific note should be made of the present activity and interest in adopting the metric system as the future prime or preferred measurement system in the United States. As part of the overall metrication study, considerable effort is under way both in this country and through the International Standards Organization (ISO) to explore and possibly develop a uniform screw thread system and related product standards which will be accepted on a truly international basis. Should such metric standards evolve and be adopted, it is anticipated that general conversion would be phased into both production and design after adequate notice and complete engineering data are made available to the engineering community.

1-3 Prime Fastener Functions Prime functions of a threaded fastener system (bolt and nut assembly) are to join and hold structures together, and to satisfactorily carry the applied structural loads under the specific design environment. Diversity of equipment, products, and structures using fasteners points up the fact that design environments are not always uniform, and can even vary for different applications on the same structure. The potential design environments for a fastener system are quite broad and may involve static loadings such as tension, shear, or bending, or dynamic loadings such as fatigue, vibration, or shock. Should elevated-temperature exposure be a design consideration, creep, relaxation, or stress rupture-strength properties may be of concern. Where the fastener system may be subjected to a corrosive environment, not only is corrosion protection a factor, but also the stress-corrosion properties of the fastener may be significant in avoiding premature service failure. Occasionally, a criterion other than strength may predominate, as in the case where fasteners may be required for their nonmagnetic properties. Practically, initial design analysis of the application should establish which are the more important or significant criteria required of the fastener for long-life joint integrity.

1-4 Product and Design Specification Distinction Even though design environments vary, because of their very nature, threaded fasteners have readily lent themselves to standardization. The standards and specifications approach has permitted logical groupings of commonly used fasteners by strength level, material, finish, and function to meet the range of objectives imposed by actual service. Additionally, the definition of standards has contributed to the assurance of general availability and interchangeability at competitive cost. While much deserved attention has been devoted to the development of product specifications and standards, it is important to recognize that there is a clear distinction between the specifications for the product, and the design or application specification for the same fastener system.

The *product specification* is essentially a manufacturing control document and normally identifies the specific material, finish, method of manufacture (where critical), heat treatment or strength level, and other related processes or requirements vital to proper fabrication. A key feature of most product specifications is the quality assurance provision outlining test sampling to confirm that the dimensional and performance objectives for the finished fastener have in fact been met. In essence, then, the product specification is essentially a contract to assure that the producer can furnish the finished fastener to high quality standards. However, the fastener as manufactured and supplied to the customer does not know where or how it will be used.

As differentiated from the product specification, the *design specification* normally outlines the conditions for use of the fastener. The design specification is intended to ensure structural and performance integrity of the assembled joint, with the threaded fastener system considered as a critical part of the joint. In the past, design criteria have been established and applied by major industry groups based on prior problems and successful solutions developed within the industry. Such criteria often reflect structural philosophy, including allowable working loads or stress levels and factors of safety, optimum or recommended fastener series for particular applications, installation and/or tightening requirements, and often special considerations requiring attention to safety, possible servicing, repair, replacement, etc. Prominent examples of design specifications in practice are the various building codes supporting design of structures and related requirements for fastening systems.

While industrywide application criteria are gaining greater recognition, it is significant that not every major industry group has adopted or established uniform design standards covering threaded fasteners. In such instances, primary design solution has generally been identified as a fundamental company prerogative. As a result, design standards representing particular experience falling within an industry are often noted in individual company manuals or specifications. These application specifications are important and constitute the basis for extensive and usually practical design use of fastening systems.

PRODUCT SPECIFICATION CONSIDERATIONS

1-5 Product Specifications Standard finished fasteners supplied to appropriate product specifications and standards normally are designated to meet certain minimum specified performance requirements. Such requirements take into account the dramatic transformation of raw material shapes into final product form and basically include dimensional tolerances, metallurgical control, and mechanical performance properties. Each characteristic should be considered important in its own right and also essential to the overall quality of the finished fastener and its subsequent satisfactory functional use. Some of the more significant product criteria usually referenced in specifications are as follows:

Dimensional. Dimensions and permissible tolerances are detailed on standard or product drawings. Adherence to specified tolerance limits is the key to dimensional interchangeability of standard uniform series of fasteners. While in-process inspections are often made during manufacture, product acceptance should be predicated on dimensional characteristics of the finished fastener. Because of the large numbers of fasteners normally associated with production lots, statistical sampling is often permitted in accordance with acceptable quality levels (AQLs) established for critical and noncritical dimensional characteristics. A commonly referenced source for sampling plans is MIL-STD-105.

Threads. Inspection of fastener threads represents a special dimensional gaging requirement. Individual thread characteristics of concern to the manufacturer may include major-, minor-, and pitch-diameter tolerance, angle error, lead error, drunken threads, etc. However, the concern of the user is that the finished threads are functional within the established class limits; i.e., the cumulative individual variations are still permissible as meeting the acceptable tolerance range. At present, there are several prominent thread-gaging systems, including functional and "go" and "not go" (or "lo") types of gages. Details for thread gaging and proper use of gages are summarized in American National Standard ANSI B1.2-1974, *Gages and Gaging for Unified Screw Threads.* Applicable product specifications should be consulted to confirm preferred or referee thread-gaging systems for specific types of fastener and class of thread fit.

Metallurgical. Metallurgical requirements, as distinguished from mechanical properties, are intended to control quality, uniformity, and often method of manufacture of the fastener. For example, bolts specified for fabrication by heading produce a unique or characteristic grain-flow pattern which can be readily distinguished from similar-configuration bolts which have been machined. Metallurgical examination of grain-flow patterns can detect acceptable as well as poor heading practice. Marginally headed fasteners pose a problem in that potential service failure can occur from undesirable stress concentrations or internal defects. Conversely, threads formed by the rolling process are beneficial in improving static and fatigue strength, and have work-effect patterns that can be observed in metallurgical study. Other inspections such as checking for discontinuities, internal defects, and grain size indicative of proper heat treatment, and study for possible detrimental effects such as excessive decarburization, can be made only by metallurgical examination.

Macro (low magnification) examination is intended for evaluation of heading and grain-flow properties. Micro (high magnification) examination is intended for evaluation of grain size, work effect, decarburization, and evidence of possible defects such as cracks, laps, seams, etc. Where metallurgical properties must be confirmed, product specifications will normally identify or reference proper etchants and location and tolerance limits for observed discontinuities. In particular, many government specifications define permissible and nonpermissible discontinuities as illustrated in Fig. 1-2.

Macrophotographs of a steel Grade 8 hexagon-head cap screw and a steel socket-head cap screw showing representative grain flow patterns are illustrated in Fig. 1-3. Microphotographs of rolled threads in steel and corrosion-resistant steel bolts showing work effect at the root radius are illustrated in Fig. 1-4. An illustration of thread laps and cracks found by metallurgical examination is shown in Fig. 1-5, while decarburization detected in a steel fastener is shown in Fig. 1-6.

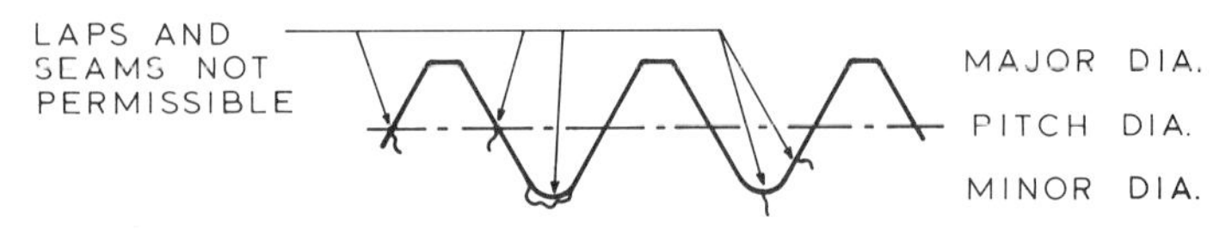

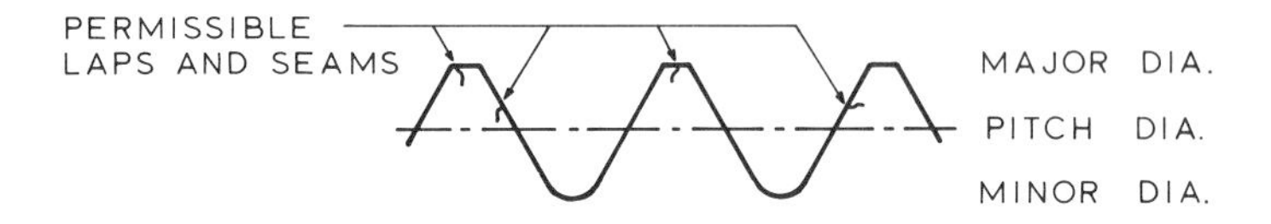

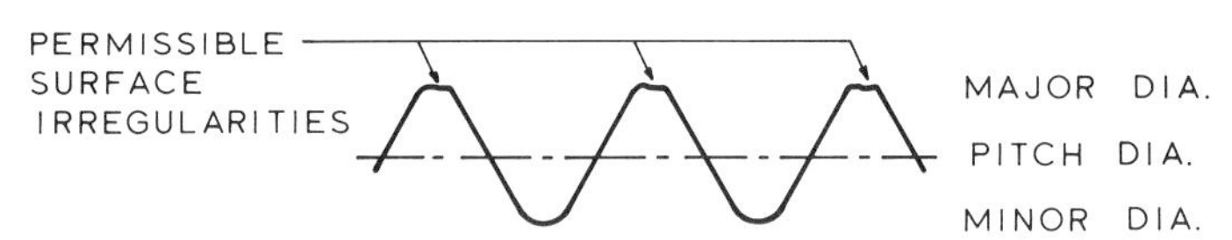

FIG. 1-2 Screw thread permissible and nonpermissible discontinuity locations.

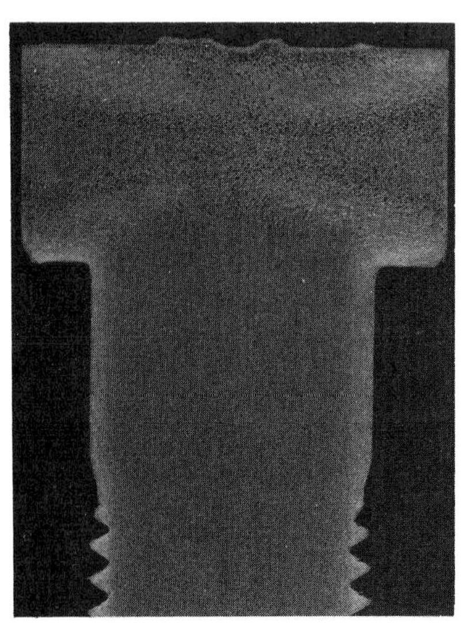

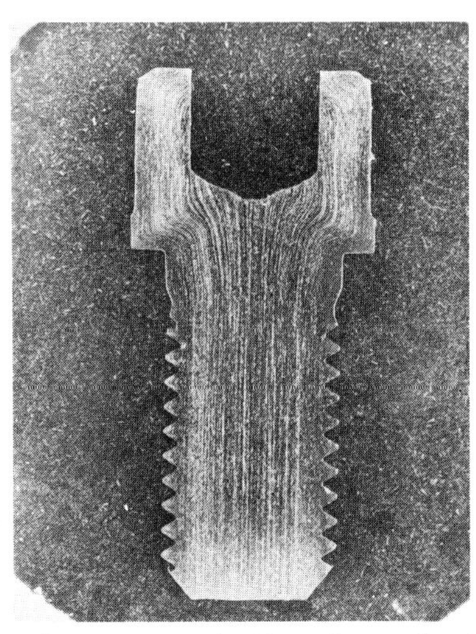

FIG. 1-3 Macrophotographs illustrating bolt grain-flow patterns. A—Macrophotograph of Grade 8 hex head bolt showing grain flow pattern. B—Macrophotograph of socket head cap screw showing grain flow pattern. (*Almay Research and Testing Corp.*)

Surface Inspection. Surface defects in finished fasteners can occur from heat treating and/or quenching, or from surface seams present in the original material used for the fabrication of the fastener, as illustrated in Fig. 1-7. To detect presence of such indications, magnetic-particle inspection can be used for ferrous materials, and fluorescent-penetrant inspection can be employed for fasteners manufactured from non-

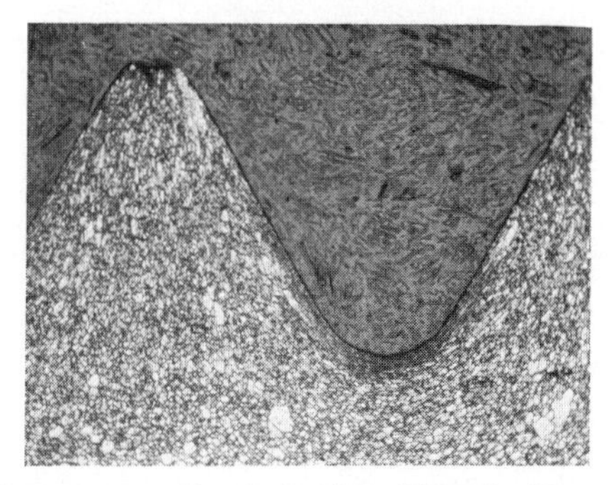

FIG. 1-4 Microphotographs illustrating work effect at root radius, indicative of thread-rolling process. A—Rolled thread in alloy steel bolt. Note flow pattern and work effect at root radius. B—Rolled thread in corrosion resistant steel bolt. (*Almay Research and Testing Corp.*)

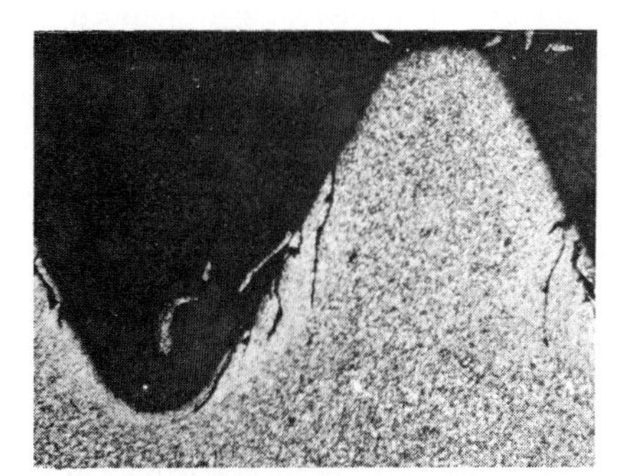

FIG. 1-5 Laps and seams in screw thread found by metallurgical examination. (*Almay Research and Testing Corp.*)

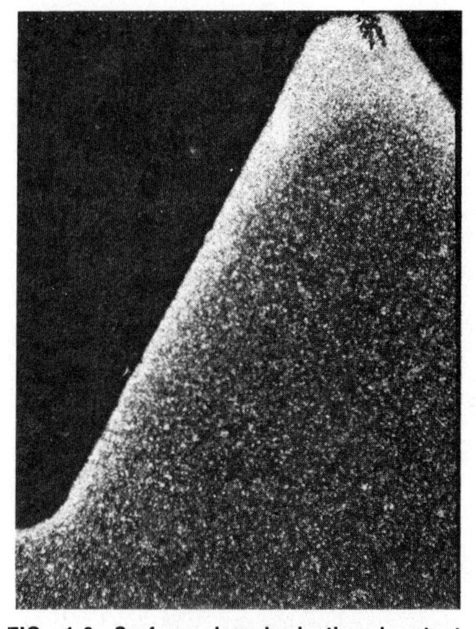

FIG. 1-6 Surface decarburization in steel fastener found by metallurgical examination. (*Almay Research and Testing Corp.*)

ferrous materials. The presence of surface indications will usually warrant further metallurgical examination to identify the nature, depth, and severity of the "defect." It should be understood that all surface indications are not necessarily cause for rejection, but rather that the surface inspection technique is a valuable tool in isolating potential metallurgical problems which could seriously affect the performance of highly stressed fasteners.

Tensile Strength. The foremost mechanical property associated with standard threaded fasteners is tensile strength. It is the guide to the user of the strength class and of the ability of the fastener to carry or transfer load to specified minimum values. Processing of parent materials to develop required strength levels may involve heat treatment, as in the case of some aluminum alloys, carbon and alloy steels, and martensitic corrosion-resistant steels. Other materials, notably austenitic stainless steels, respond to cold working (or drawing) to achieve increase in strength properties during manufacture.

Tensile strength is determined from the formula

$$S_t = P/A_s$$

where S_t = tensile strength, psi
P = tensile load, lb
A_s = tensile stress area, sq in.

For any specified tensile strength level then, the minimum tensile load requirement for a fastener can be calculated in terms of $P = S_t A_s$. From this basic relationship, a significant consideration is the definition of the tensile stress area A_s for the fastener thread. Since the screw thread is formed on a helix angle, the stress area does not lend itself to absolute definition. In fact, over the years, there have been several different screw thread stress areas used in calculating minimum tensile-load values based on objective tensile-strength criteria. At present, the commonly accepted formula for tensile stress area for static tensile-strength properties of standard fasteners is referenced in NBS Handbook H-28 as

$$A_s = 3.1416\left(\frac{E}{2} - \frac{3H}{16}\right)^2$$

or

$$A_s = 0.7854(D - 0.9743/n)^2$$

where E = basic pitch diameter
D = basic major diameter
n = threads per inch

For $3H/16$, refer to Handbook H-28.

By contrast, it should be noted that for high-strength aircraft fasteners, the aerospace industry has predicated tensile stress area on the pitch diameter area, as outlined in NAS 1348. For other mechanical properties, such as stress rupture and fatigue strength, the sectional area at the minor diameter of the thread is often employed. Comparison of the thread tensile-stress areas for the three conditions discussed is listed in Table 1-1. An awareness of the basis of the tensile stress area defined for a specific fastener series therefore becomes of value in using and understanding the strength range capability of the fasteners.

Several other product specification considerations are important in evaluating tensile strength properties of finished fasteners. For standard bolts and screws, ultimate failure is designed to occur in the threads because of the smaller sectional area. Many specifications do not permit bolt-head failures during acceptance testing, since they may be indicative of metallurgical problems or poor manufacturing quality.

Also, test experience has shown that thread engagement, hardness, and position of the mating nut have an influence on the tensile strength of the bolt. For standard series bolts, test requirements in general call for six threads exposed when positioning the mating test nut. Because of shorter thread lengths, aerospace bolts are usually tested with two threads exposed. Interestingly enough, tests conducted on grade 8 hexagon-head bolts with two threads and with six threads exposed, as illustrated in Fig. 1-8, indicated a difference of approximately 15,000 psi in the actual tensile strength of the same lot of fasteners. The higher breaking loads were observed with the nut positioned at two threads from the run-out. Strain hardening and notch sensitivity of the exposed threads apparently are factors contributing to this phenomenon. In

FIG. 1-7 Surface seam in headed bolt blank. Surface indications are detected by either magnetic particle or fluorescent particle inspection. (*Almay Research and Testing Corp.*)

TABLE 1-1 Representative Tensile Stress Areas for Fine- and Coarse-Thread Bolts

Nominal thread size	Tensile stress area,° sq in.	Sectional area at minor dia.,† sq in.	NAS 1348 pitch dia. stress area,‡ sq in.
	FINE-THREAD SERIES		
1/4-28	0.0364	0.0326	0.0388
5/16-24	0.0580	0.0524	0.0614
3/8-24	0.0878	0.0809	0.0950
7/16-20	0.1187	0.1090	0.1288
1/2-20	0.1599	0.1486	0.1717
9/16-18	0.203	0.189	0.2176
5/8-18	0.256	0.240	0.2724
3/4-16	0.373	0.351	0.3952
7/8-14	0.509	0.480	0.5392
1-12	0.663	0.625	0.7027
	COARSE-THREAD SERIES		
1/4-20	0.0318	0.0269	
5/16-18	0.0524	0.0454	
3/8-16	0.0775	0.0678	
7/16-14	0.1063	0.0933	
1/2-13	0.1419	0.1257	
9/16-12	0.182	0.162	
5/8-11	0.226	0.202	
3/4-10	0.334	0.302	
7/8-9	0.462	0.419	
1-8	0.606	0.551	

° Tensile stress area per NBS Handbook H28, where

$$A_s = 3.1416\left(\frac{E}{2} - \frac{3H}{16}\right)^2$$

† Minor-diameter area per NBS Handbook H28 at $D - 2h_b$.
‡ National Aerospace Standard for externally threaded fasteners in 160 thru 260 ksi range with threads rolled after heat treatment.

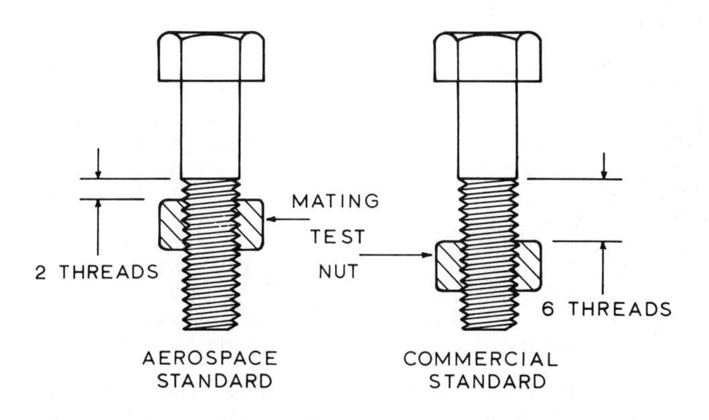

FIG. 1-8 Effect of positioning of mating test nut for routine static tensile-strength test.

view of the sensitivity of this condition, applicable specification requirements for type and positioning of the mating nut should be observed to accurately evaluate tensile strength properties.

A separate problem is associated with large-diameter bolts. Where testing-machine capacity was not available or adequate, some early product specifications permitted machining of reduced-gage-section coupons from the bolts in order to establish tensile strength characteristics. Typical machined test coupons are illustrated in Fig. 1-9. However, caution should be exercised in evaluating the test results from machined coupons since mechanical properties do not always correlate with the results obtained from full-scale bolt testing. Contributions of work effect in rolled threads and in the head by upsetting or forging are often lost in a standard machined specimen. This is particularly noticed in corrosion-resistant steel and similar materials where fastener mechanical properties have been achieved by cold work. For this reason, major em-

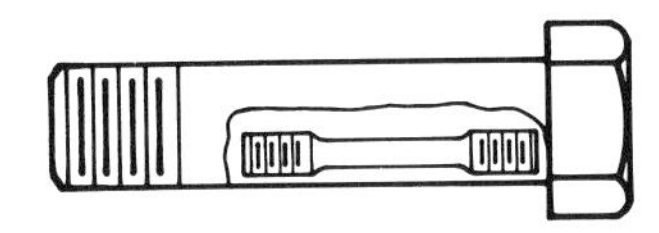

REPRESENTATIVE STANDARD ROUND 2 IN. GAGE LENGTH TENSION TEST SPECIMEN MACHINED FROM LARGE DIAMETER BOLT

MINIMUM RADIUS RECOMMENDED 3/8 IN., BUT NOT LESS THAN 1/8 IN. PREMITTED

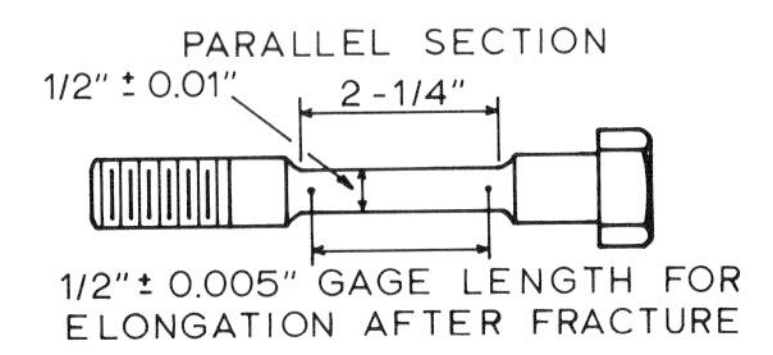

TYPICAL REDUCED GAGE SECTION TENSION TEST SPECIMEN MACHINED FROM FINISHED BOLT

FIG. 1-9 Typical standards for machining reduced-gage-section material test coupons from large-diameter bolts.

phasis should be placed on evaluation of full-scale bolts wherever possible to assure confirmation of the actual properties of the fasteners which will be used in service.

Proof Load. Whereas testing to destruction is a confirmation of ultimate tensile strength, there is often an equal concern for another criterion known as *proof load* for certain classes of standard fasteners. This property, in a sense, represents the usable strength range of the fastener for many design functions. Critically, when subjected to proof-load exposure, the fastener should not exhibit permanent set as determined by length measurement prior to and after loading. As with other critical fastener characteristics, evaluation of proof-load properties, when specified, should be performed on the actual finished fasteners.

Occasionally, proof load and yield strength are interpreted as being the same. As mentioned earlier, a threaded fastener is a complex component affected by changes of cross section, notches, and other regions of stress concentration, as well as by manufacturing processes. To date there is no common or verified standard for universally determining actual yield strength of a threaded fastener. The 0.2 percent offset yield-strength criterion identified for parent material coupons should not be considered as a blanket correlation with comparable properties for threaded fasteners fabricated from the same material.

Wedge Tensile Strength. As a measure of ductility and head integrity, some product specifications call for tensile strength evaluation using a wedge positioned under the bolt head as shown in Fig. 1-10. Wedge angle specified for standard bolts up to 1-inch diameter is usually 10 degrees, and for bolts larger than 1-inch diameter the angle is normally 6 degrees. In the case of quenched and tempered bolts which are fully threaded to the head, wedge angle is held at 6 degrees for sizes up to $^3/_4$-inch, and at 4 degrees for bolts larger than $^3/_4$-inch diameter. The use of the wedge introduces artificial bending in the bolt, forcing stress concentrations at the bolt head and in the threads. When such requirements are specified, failures under ultimate load are limited to the bolt threads or bolt shank area.

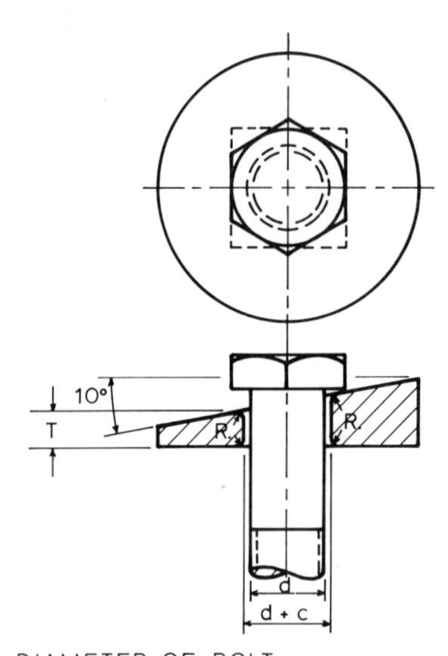

d = DIAMETER OF BOLT
c = CLEARANCE OF WEDGE HOLE
R = RADIUS
T = THICKNESS OF WEDGE AT SHORT SIDE HOLE (EQUAL TO 1/2 DIA. OF BOLT)

FIG. 1-10 Typical 10-degree wedge under bolt head for performing wedge tensile-strength evaluation.

Shear Strength. Shear strength of standard threaded fasteners as a prime mechanical property has not generally been referenced in industrial or commercial specifications, even though these fasteners are often used in shear-design applications. However, attention is currently being devoted to possible requirements in this area which may be reflected in future specifications. Aerospace bolts, on the other hand, are covered by rigid requirements for shear strength as well as for tensile strength. As previously indicated, normal heat treatment is undertaken primarily to achieve tensile strength response. For many of the original aerospace steel bolt series heat treated to 125 ksi, shear strength was estimated at 60 percent of the tensile strength and this ratio was accepted as a standard for this class of fastener. The continuing development of higher-tensile-strength aircraft bolts, however, has shown that the estimated shear-to-tensile ratio is not a constant and can vary from about 54 to 65 percent. For this reason, where bolts may be subject to shear-loading service, the developed shear properties of the manufactured fastener assume greater significance, particularly in the higher strength grades. Product specifications or specific customer requirements should therefore be reviewed and observed where shear strength is established as a criterion in addition to minimum tensile-strength objectives.

Hardness. Brinell or Rockwell hardness testing is a common technique for estimating the tensile strength properties for steel fasteners. It is recognized as a valuable tool indicative of tensile properties, but should not be construed as a direct measure of actual tensile strength of finished fasteners. Again, it is emphasized that true tensile strength can be determined only from full-scale fastener evaluation. There are instances, though, where production bolts are too short to permit a tensile strength test, and in such instances, lot acceptance may be predicated on the results of hardness testing. As a manufacturing in-process control check, hardness testing is often considered as nondestructive. On a finished and plated fastener, however, hardness readings may not be valid unless the plating is removed and a flat and uniform surface prepared for the hardness indentation.

Microhardness. Microhardness testing is normally used in conjunction with metallurgical evaluation, primarily for determination of surface decarburization or carburization. Microhardness-type instruments have the ability to make measurements 0.001 inch apart. While visual metallurgical examination can detect obvious decarburization, specification criteria for partial decarburization is normally based on a differential of three points *RC* between carburization zone and uniform parent material. Where particular limits are noted for decarburization depth, the microhardness traverse provides a means for accurate measurement. The same technique, of course, can similarly be employed for measurement of case-hardening depth. Such tests are required on prepared specimens.

Protective Coatings. There are several important reasons for applying protective coatings or finishes to standard threaded fasteners. Foremost, coatings are employed as a protective finish or barrier for the fastener base metal. In addition, they may be specified for wear resistance, for unique or special decorative appearance, as a base for subsequent painting, and as a thread lubricant. Quite often, these functions overlap and selection of a coating takes into account the multiple advantages available, as well as the economics of the finish applied. The range of commercial finishes includes anodizing and chemical surface treatment for aluminum alloys, oxide treatment for brass, passivation for corrosion-resistant steels, and a wide spectrum of protective coatings for steels such as galvanizing, electroplated zinc, electroplated cadmium, conversion coatings (phosphate and oil), electroplated nickel and chromium, and mechanical plating. From a product-specification viewpoint, the main concern is that the specific finish has been properly and fully applied to provide the required protection for the parent fastener in storage and in service.

For electrodeposited platings in particular, perhaps the foremost criteria are thickness of plating, adhesion to the base metal, and porosity. Each characteristic is critical in contributing to the overall performance of the plating. A breakdown of any individual property could result in a defective coating, even if the remaining conditions are otherwise within specification.

Evaluation of the adequacy of a coating or finish may be by means of a humidity test, a salt-spray exposure test, or by measurement of coating thickness as in the case of electrodeposited platings. Thickness measurements can be made nondestructively by means of eddy-current instruments or by magnetic measuring devices capable of identifying thickness of nonmagnetic coatings on ferrous materials. Unfortunately, electroplated coatings are not consistently uniform in thickness and have a tendency to build up on corners or at sharp changes in section. Precise measurement of thickness may therefore require sectioning and microscopic analysis.

The severest test condition for overall performance of plating or finish is the controlled salt-spray exposure, which is intended as an accelerated-type test. The reliability of associating number of hours of test-chamber exposure with actual years of service life is questionable. However, the salt-spray test has been widely used to screen poor or defective plating, as evidenced by the appearance of white corrosion products, red rust, pitting, flaking, or other attack. Hours of salt-spray exposure are generally related to the type of finish and/or thickness of coating specified for the fastener.

Hydrogen Embrittlement. Fasteners fabricated from steels and alloy steels heat-treated above 160 ksi tensile strength are particularly susceptible to hydrogen embrittlement, unless preventive care is exercised during manufacture. Hydrogen embrittlement is a unique and special phenomenon wherein atomic hydrogen is infused or absorbed into high-strength steel. Under stress, which implies service application, hydrogen contamination can induce premature brittle failure without warning.

Sources of hydrogen during manufacture can be attributed to material pickling, alkaline or acid cleaning, and to final electroplating operations, with electroplating possibly considered the severest contributor to critical contamination. For high-heat-treat fasteners which have been zinc- or cadmium-plated by an electrodeposition process, the recommended control is to bake immediately after plating at 375°F to free any surface-trapped hydrogen to prevent diffusion into the steel.

Actually, there is no visual or other accepted metallurgical examination which can satisfactorily detect presence of hydrogen embrittlement. The technique employed is to expose the fastener to a stress-loading condition from 24 to 200 hours. If hydrogen contamination is present, brittle fracture will generally be observed within the test time period in the threads or at the head-to-shank junction, which are locations of high stress concentration. Loading may be either direct dead weight, or induced by torquing. Normally, such loading approximates 75 percent of the ultimate tensile load, or the load corresponding to the estimated yield strength of the fastener.

Fatigue Strength. For the most part, classes of bolts identified as *fatigue-rated fasteners* are limited, examples being engine connecting-rod bolts and aerospace-series fasteners. Fatigue is that special condition where the component is subjected to repetitive cyclic loading or stressing. Fatigue life is sensitive to part design and inherent stress concentrations, and is dependent on the stress intensity, with cycle life increasing as the stress level is lowered. The nature of fatigue is such that under repetitive or dynamic loading, failure can occur at strength levels well below the rated static strength. Unlike a ductile static failure which is often identified by elongation or yielding, failure by fatigue gives no advance warning and is usually catastrophic.

Good design practice dictates that attention be given to joints subjected to fluctuating or repetitive load conditions. However, to minimize potential service fatigue problems, good design also requires that the threaded fastener system be of high quality with care taken during manufacture to develop optimum fatigue strength properties. Extensive study has confirmed that the three main areas of stress concentration on a bolt are at the head-to-shank juncture, at the thread runout, and at the first thread engagement with the mating nut. With an appreciation of this analysis, it is recognized that fatigue strength is a property which must be built into a threaded fastener by virtue of design configuration and manufacturing processes. For example, fatigue performance of bolts with rolled threads is markedly improved over identical bolts with cut threads. Carrying this illustration one step further, threads rolled after heat treatment have shown better fatigue strength than bolts with threads rolled before heat treatment. As in the case of aerospace-series fasteners, separate cold working of the head-to-shank fillet radius is significant in improving head fatigue strength and overall fatigue performance. Special head designs for better stress distribution and particular surveillance of material quality and metallurgical characteristics to avoid laps, cracks, decarburization, etc., also contribute to basic fatigue resistance of finished fasteners.

The majority of commercial and industrial standard threaded fasteners are not designated as fatigue-rated, but still have been acceptably used under structural fatigue applications. Because of the continuing emphasis of fatigue as a design concern, there is an awareness of the performance advantages of bolts specifically designed and manufactured to improve fatigue strength properties over existing standards. In developing a fatigue-rated fastener system, test data is plotted in the form of an *S-N* curve, where *S* is stress level and *N* is number of cycles, as typically shown in Fig. 1-11. Both the shape of the curve and the endurance limit are significant criteria in judging the quality of the manufactured fastener. For quality assurance evaluation, though, it is not practical to perform routine product tests to the endurance limit. Since there is usually a definite relationship for an overall *S-N* performance curve, where specification fatigue tests are required, testing is normally undertaken at a higher stress level to permit accelerated evaluation.

Product specifications requiring fatigue tests of finished fasteners should be reviewed for specific details pertaining to test method. Procedures originally developed within the aerospace industry and widely observed are referenced in MIL-STD-1312, Test 11, and in Specification NAS 1069.

Environmental Temperature Strength. The design of equipment and structures is not limited to ambient-temperature service conditions. Practical design areas may involve exposure to elevated temperatures, as in engines and high-pressure steam systems, or to cryogenic (subzero) temperatures, as in refrigeration systems. The current interest in structural projects in the Arctic has particularly heightened concern over low-temperature performance of threaded fasteners.

The primary distinction for fasteners intended for environmental temperature exposure is the selection of special materials designed for this function. Fastener product specifications carefully screen and define materials based on both room-temperature and service-temperature mechanical properties. Generally, high-temperature strength characteristics are lower than the properties developed for the same fastener at room temperature, while low-temperature strength is higher than obtained at room temperature.

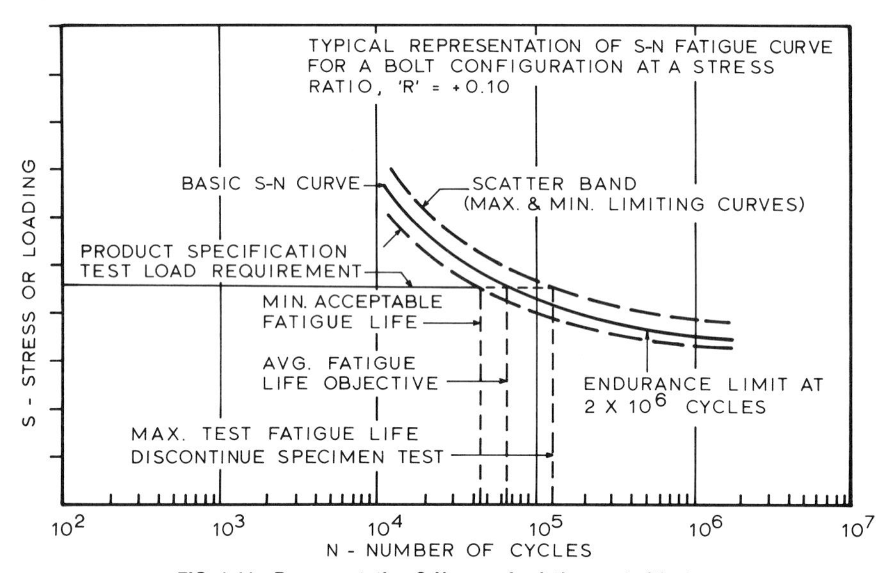

FIG. 1-11 Representative *S-N* curve for fatigue-rated fastener.

From the wealth of data available on parent materials, basic relationships of room-temperature and service-temperature strength properties are fairly well established. For the manufactured fastener, some specifications permit evaluation at room temperature and essentially extrapolate expected properties at operating temperature (elevated or cryogenic). For fasteners deemed critical, however, confirmation of fastener strength at temperature may be just as important as room-temperature strength. Such tests are normally short-time tests, and because of the expense and special test-support facilities needed to simulate temperature environment, product specifications often outline such requirements on the agreement of the purchaser and manufacturer.

Stress-Rupture Strength. A special condition relating to elevated-temperature strength properties of structural bolts is stress-rupture strength. As differentiated from short-time elevated-temperature strength, stress-rupture strength is the design characteristic which governs extended-life structural service at operating temperature. This criterion is a direct function of temperature, time, and applied stress. Fasteners exposed to long-time elevated temperatures may be found in conventional and nuclear power plants, jet engines, superheated steam systems, etc.

Actually, there are two criteria involved in evaluation of high-temperature material properties; namely, creep strength and rupture strength. Under long-time stressing at elevated temperature, metals will continue to elongate or deform at some constant rate, with the creep strength defined as the stress which produces a particular rate of elongation, or *creep rate*. Rupture strength, as the identification implies, is the stress the material can sustain at a particular temperature for a specified period of time without failure, and is usually lower than the short-time tensile strength for the same temperature condition. Most product specifications, especially for aerospace fasteners, are more concerned with stress-rupture-strength properties since they are influenced by material selection and by manufacturing controls such as cold work, heat treatment, and grain size.

Depending on design application, long-time elevated-temperature service can range from a few hours up to several thousands of hours. Aside from research studies intended to develop design parameters, it is not practical to conduct extensive long-time tests for routine quality-assurance acceptance. Product specification requirements for high-temperature fasteners, therefore, normally reference stress-rupture-test strength requirements at from 23 to 100 hours, taking into account prior correlation of longer-time strength properties at that temperature.

Magnetic Permeability. Certain design applications, especially in electronic or guidance systems, require that structural components be nonmagnetic. Certain classes of materials are obviously nonmagnetic, such as aluminum alloys, brass, and titanium alloys. Although austenitic stainless steels as a class are generally nonmagnetic, by virtue of cold heading and cold working, some residual magnetism can be introduced into finished threaded fasteners. While such small amounts of induced magnetism might be considered negligible, in the vicinity of magnetic-sensing instruments excessive residual magnetism could affect the performance of such instruments.

As a check for residual magnetism in austenitic stainless steel fasteners, mainly in the aerospace and defense industries, magnetic permeability is inspected by a "go, no-go" type instrument conforming to specification MIL-I-17214. For most parts, performance requirements are established on the basis that magnetic permeability shall be less than 2.0 (air = 1.0) for a field strength $H = 200$ oersteds.

1-6 Test Methods The above summary of the various requirements pertaining to manufacturing and performance controls for standard threaded fasteners is indicative of the importance assigned to these properties in product specifications. It is to be recognized that if it is important to specify performance objectives in the first place, it is just as vital to assure that the support quality-assurance testing standards are equally accurate and capable of discriminating between acceptable and rejectable fasteners. Occasionally, the significance of product testing is glossed over or taken for granted. Experience has shown that test results can vary for the same fastener, depending on *how* the test was run. As reviewed earlier, the differential in bolt tensile strength depending on position of the mating test nut is cited as only one example.

Most product specifications outline testing requirements or reference test procedures to be followed, as in ASTM A370, Federal Test Method Standard 151, and related ASTM testing procedures. More recently, a comprehensive series of test methods specifically directed to standardize all aspects of routine and advanced test evaluation for military fastener series has been issued as MIL-STD-1312. It is understood that a similar effort is being explored to undertake development of comparable test procedures for commercial and industrial fasteners. In the interim, especially where unique or special performance requirements may be specified for standard threaded fasteners not covered by ASTM procedures, the use of MIL-STD-1312 as a reference test guide may be of value.

1-7 Specification Sources Product specifications for standard threaded fasteners (bolts and nuts) have been generated by a number of major national organizations. Certain of the fastener standards have been promulgated to support applications in a specific industry. For the most part, though, the nationally recognized specifications cover threaded fasteners commonly used by a broad base of American industry. For general reference, sources for some of the prominent classes of specifications in present use are as shown in the following list.

Type specification	*Source*
Federal specifications	Naval Publications and Forms Center, 5801 Tabor Avenue, Philadelphia, PA 19120
Handbook H28, National Bureau of Standards	Superintendent of Documents, Government Printing Office, Washington, DC 20025
Military specifications	Naval Publications and Forms Center, 5801 Tabor Avenue, Philadelphia, PA 19120
ANSI standards	American National Standards Institute, 1430 Broadway, New York, NY 10018; or American Society of Mechanical Engineers, 345 E. 47th Street, New York, NY 10017
ASTM specifications	American Society for Testing and Materials, 1916 Race Street, Philadelphia, PA 19103
SAE specifications	Society of Automotive Engineers, 400 Commonwealth Drive, Warrendale, PA 15096
IFI documents	Industrial Fasteners Institute, 1505 E. Ohio Building, Cleveland, OH 44114
NAS specifications (National Aerospace Standards Committee)	National Standards Association, 1321 Fourteenth St. N.W., Washington, DC 20005
AMS specifications (Aeronautical Materials Specifications)	Society of Automotive Engineers, 400 Commonwealth Drive, Warrendale, PA 15096
AAR specifications	Association of American Railroads, 59 E. Van Buren Street, Chicago, IL 60605
API specifications	American Petroleum Institute, 300 Corrigan Tower Building, Dallas, TX 75201
EEI specifications	Edison Electric Institute, 750 Third Avenue, New York, NY 10017
IOS recommendations (International Organization for Standardization)	American National Standards Institute, 1430 Broadway, New York, NY 10018

DESIGN CONSIDERATIONS

Structural design, whether it be for a massive heavy-duty structure or a miniature subsystem, involves the same basic challenge: joining and assembling the components in an economical manner with maximum joint integrity. Underlying successful design is the consideration and analysis of the joint on an overall basis, with appreciation that the fastener may be only one vital part of the complete structural system. Obviously, it is important to start with fasteners which are of good quality and which will perform under the expected design conditions. The previous sections have touched on features which influence manufactured quality of threaded fasteners and have highlighted the importance of product specification controls in assuring and confirming that these performance features are in fact built into the finished fasteners. Care exercised in implementing a sound quality-assurance system may be well worth the effort to avoid service failures which may otherwise be attributed to questionable quality fasteners.

The realm of design takes into account where and how fasteners are to be used. The approach to joint design varies markedly from industry to industry. In part, this reflects the difference in types of materials and structures being joined, unique or unusual environmental service conditions, the factor of safety normally required, anticipated service life, special considerations for tooling and assembly, and whether the joint is intended to be permanent or removable. If there is no single set of design standards which can be set down for use of all standard threaded fasteners, experience nevertheless suggests certain guidelines which can contribute to sound design practice.

That valid design principles are significant is borne out by some industry estimates which have assigned as much as 50 to 60 percent of the cost of the structure to the requirements of joint design, analysis, and actual physical fastening. A working knowledge and understanding of the essentials of the threaded fastening system and its function therefore become a valuable tool in original design and in maintenance.

1-8 Thread Selection The Unified Screw Thread form includes standards for both coarse thread and fine thread series. Since standard bolts in the popular nominal sizes up to 1-inch diameter are available in either thread series, what are the determining factors for selection? There are several differences in performance, suggesting some study prior to specifying a thread choice.

The fine-thread series, by virtue of finer pitch and resultant smaller thread depth, produces a tensile stress area which is greater than the area for a corresponding coarse-thread fastener. The net effect is that the fine-thread series is characteristically stronger than the coarse-thread series. A stronger fastener develops higher working and clamp loads, which are certainly prime design objectives. In addition, because fine threads have a small lead angle, they are advantageous where fine adjustments may be required. Fine threads are easier to tap in harder materials, but are generally limited to shorter thread engagements than coarse threads. Also, the fine thread is preferred for thin-walled materials where tapping may be necessary.

Coarse threads offer the advantage of easier assembly with less likelihood of cross threading. The additional clearance allowed for coarse threads permits a greater latitude for plating or other finishes, and at the same time is not as seriously affected during assembly by presence of contaminants or burrs. Where seizing or corrosion are encountered in service, coarse threads make for easier removal than comparable fine-thread fasteners.

The coarse-thread series is probably the most widely used for construction and general applications, although specialized industries such as the automotive and aircraft groups tend to rely more heavily on fine-thread-series fasteners. It is interesting to note, however, that one of the concepts being advanced in the study of the new proposed metric fasteners is a single thread series for all general commercial and industrial applications.

1-9 Corrosion For many applications, the problems of corrosion pose an overriding concern in design. It is important to understand that there are several distinct types of corrosion including galvanic corrosion, concentration-cell corrosion, stress corrosion, fretting corrosion, pitting, and oxidation. Probably the most common form of corrosion is rust associated with steel structures and steel fasteners, although the effects of corrosive attack can be observed in many other structural materials.

In the case of galvanic corrosion, the combination of two dissimilar metals with an electrolyte is all that is needed to form a battery. The use of dissimilar metals in structural design is not uncommon, particularly where fasteners are a different material from the structure being joined. The necessary ingredient to induce corrosion, the electrolyte, may be present in the form of ocean salt spray, rain, dew, snow, high humidity, or even air pollution. The anode-cathode effect of the battery couple actually results in loss of electrons and material where the metals are apart in the electromotive series.

Concentration-cell corrosion and pitting are similar types of corrosion in that only one metal and an electrolyte are sufficient to set up an attack system. As corrosion progresses, a differential in concentration of oxygen at the metal surface and in the electrolyte produces a highly effective localized battery with resultant corrosion and metal attack.

Stress corrosion represents a particular condition where cracks are induced and propagated under combined effects of stress and corrosion environments. Structures or components with high stress concentrations (such as threaded fasteners) are susceptible to this type of attack when under load. The initial corrosion may occur at a point of high stress contributing to crack initiation which can be either intergranular or transgranular. Continued exposure to the corrosion environment will propagate the crack, resulting in serious, if not catastrophic, failure.

Other corrosion systems can be equally severe. Alkalis, acids, marine atmosphere, industrial pollution, road salts, in addition to water products, all exert their influence. Common to all is the fact that corrosion which can adversely affect operating life or function is encountered normally after the structure is put into service. Corrosion protection at design inception should be a major objective of good joint design.

The first step is to define or identify the specific anticipated corrosion exposure in order to control or minimize its consequences in service. As noted, part of the corrosion problem arises when fasteners of one material are used to join a structure of a different material. Initially, an attempt should be made to select fastener materials which are as compatible as practical with the structure being joined; i.e., electromotive potential should be minimized. A corollary consideration is a protective coating or finish for the fastener to provide protection for the fastener base metal and/or the joint material. Additional protection may be afforded by supplemental coatings or finishes for the entire joint, since holes drilled through a structure are bare until treated. The range of protective coatings includes primers, paints, insulation materials, inhibitors, plating, and greases. The choice depends on the severity of the application and the esthetics of the product requirement. As part of the design, care should be taken to prevent accumulation of corrosion elements by providing adequate draining, and also to avoid areas of high stress concentration subjected to possible corrosive environments.

1-10 Washers Invariably, when standard fasteners are discussed, they are referred to as a *bolt-nut* system. Standard flat washers are very rarely thought of as part of the system, and yet they often play an important role in fastening.

Washers are used to distribute bearing stresses under bolt heads and nuts, especially when joining softer structural materials. They can be used as a corrosion barrier or insulation as part of a corrosion-protection system in separating otherwise highly anodic-cathodic materials. Washers used under the fastener element being tightened (bolt head or nut) provide a good surface for uniform torque control, as in the case of the washers specified for use with the A325 and A490 series structural fasteners. And in some instances, washers can be used to accommodate or adjust for proper grip length of bolts in an installed joint.

Flat washers are available in a wide variety of materials and finishes compatible with mating threaded fasteners. There are many higher-strength structural materials which do not particularly warrant the use of a washer to satisfy joing structural strength, and there is some concern that the addition of a washer constitutes an added element, added expense, and added weight. However, judicious use of washers for the reasons noted may enhance overall joint efficiency and design and should be considered on that basis.

1-11 Joint Design Mechanical fastening of structures, equipment, or components requires attention to details of joint design and proposed assembly techniques. There are really no standards for uniform joint configurations, although there has been no limit to the versatility of methods for fastening and connecting structures. The more popular types of joints include the lap joint, the flanged joint, and the butt joint. Actually, there are any number of variations and combinations of these joint configurations which can be assembled by through fasteners (bolt and nut) or by securing a bolt in a tapped hole.

It has been demonstrated that modern threaded fasteners are effective as either permanent or removable connections. Still, special considerations are necessary for both conditions. Joints employing permanent fasteners can be readily and more accurately assembled under shop conditions. The same joints assembled under field conditions may require additional working tolerances and may be limited by available installation tooling. Similarly, for removable fastener joints, wrench clearance, spacing, and accessibility must all be considered. Factory assembly is often geared to the use of special tooling to expedite and facilitate fastener installation. Under field service repair or maintenance, the limiting factors may be available manual tools and working clearances.

Joint design also includes study and analysis of the working forces acting on the joint. The types of loading may include tension, shear, bending, or fatigue. Even though the same fastener assembly could conceivably be used for each application, several guidelines for structural design are applicable.

1-12 Tensile Loading A tensile joint is where the applied load is in line with the fastener axis. In theory, if the tensile preload induced in the bolt system meets or exceeds the applied tensile loading on the joint, the fastener will not sense any increased tensile load and the joint will remain rigid. The key here is developing the proper tensile preload in the threaded fastener system. The action of assembling and tightening a bolt and nut produces an elongation and resultant tensile prestress in the fastener, while at the same time equal compressive stresses are introduced into the material being joined. This model of the system is assumed valid so long as the applicable stresses are within the elastic limit in accordance with Hooke's law.

For most design involving static tensile loading, the assumptions are true enough. In practice, loads tend to fluctuate and tightened fasteners may actually sense tensile loadings in excess of the induced preload. An increase in the tensile load on the fastener is accompanied by a decrease in the compressive load in the joint and resultant loss of clamping load effect. Where joints are designed to be rigid, excessive loss of clamp load can pose a serious structural problem.

Several significant considerations are developed from the tensile-joint analysis. The bolt proof load or yield strength may be a more important design criterion than the ultimate tensile strength rating for the fastener, based on load capability of the fastener within the elastic limit of the material. Obviously, however, the higher the tensile strength of a fastener system, the higher the yield strength with a higher clamp-load capability. The higher-strength fastener series having greater clamp-load properties has permitted reduction in the number of fasteners required, or use of smaller-diameter fasteners, to achieve the same design objective.

The foregoing discussion is applicable to rigid joints. The case where gaskets are used, as in flexible joints, presents a different problem and requires a different analysis. For the gasket to be effective, it is important for all fasteners in the joint to be uniformly loaded to the design stress condition. Any increase in tensile loading in a gasketed joint adds to the bolt preload, usually requiring either a higher-strength or larger-diameter fastener than would be required for the same clamp-load objective in a rigid joint.

While most strength criteria are predicated on the properties of the male threaded element (bolt or screw), it should be recognized that the mating nut plays a significant role in developing the full rated strength of the fastener system. The nut should be capable of sustaining the minimum ultimate tensile strength of the bolt. Generally, matching nuts require a minimum height equal to the basic diameter of the fastener in order to develop the rated tensile strength. The same minimum thread length requirement pertains when bolts are inserted into tapped threads in high-strength-steel structures. Where tapping is necessary in softer materials, thread engagement varying from $1\frac{1}{2}$ to 2 times the nominal fastener diameter may be needed to develop full bolt strength.

1-13 Shear Loading A shear joint is one in which the applied loading is at right angles to the fastener axis, or across the bolt shank. Shear joints fall into two separate categories. The first is the friction connection, which depends on high clamp load induced in the mating threaded fasteners. As long as the mating faying surfaces are free of dirt, lubricant, paint, etc., the high clamp load contributes to high frictional forces in the joint. The frictional forces developed are normally well within the allowable design shear-stress ratings, particularly in the construction industry. Should joint shear load exceed the frictional stress, joint slipping would occur to the full tolerance of the clearance hole, resulting in shear forces acting on the bolt and bearing forces acting on the joint material. The second type of shear joint, therefore, is based on bolt shear strength and plate bearing strength properties.

For friction-type joints, it is evident that clean and flat working surfaces are necessary for high joint efficiency. Also, close control must be maintained of torque on the fastener connections to assure that the design preload is developed in the fasteners. For bearing-type joints, care should be taken to specify bolts which are long enough to permit the full bolt shank to act through the shear planes. Occasionally, improper or short-length bolts are used with a portion of the bolt threads resisting shear loading, as illustrated in Fig. 1-12. The results of test studies have confirmed that substantial loss of bolt shear strength is caused by smaller net sectional area in the screw threads. Some design requirements reflect a reduction of almost 20 percent in bolt shear strength when threads are in bearing. To take full advantage of strength properties, preferred design suggests that the full shank body be positioned in the shear planes, also as illustrated in Fig. 1-12. The airframe industry, for example, maintains a strict prohibition against the use of any threads in bearing in a shear-joint connection.

Another consideration for bearing-type shear joints is whether the bearing strength of the joint material, or the shear strength of the fastener, is critical. With a bolt installed in a hole acting against the plate, the bearing stress is calculated on an area equal to the thickness of the plate times the nominal diameter of the bolt or fastener. In a joint where the plates are relatively thin, the plates may fail, making the joint bearing-critical. Using the same diameter fastener in a joint where the plates are relatively heavy, the full shear strength of the fasteners can be realized, making the joint shear-critical. These conditions are illustrated in Fig. 1-13.

One of the major aspects of joint design is working within the bearing-strength rating for the joint material. Mechanical property data established for most of the common structural materials include bearing strength as a function of edge distance. Edge distance is usually defined as the distance from the center line of the bolt hole to the edge of the plate, and is expressed as a function of bolt diameter. For example, an edge distance of 2.0 for a 1-inch-diameter bolt would mean that the center line of the mating hole was 2 inches from the end of the plate. Edge distances used in design range from 1.5 to 2.5, depending on the structure and material being employed. Since bearing strength properties vary with edge distance, some caution may be warranted where repairs are undertaken. To save a structure with a defective hole, or possibly to repair

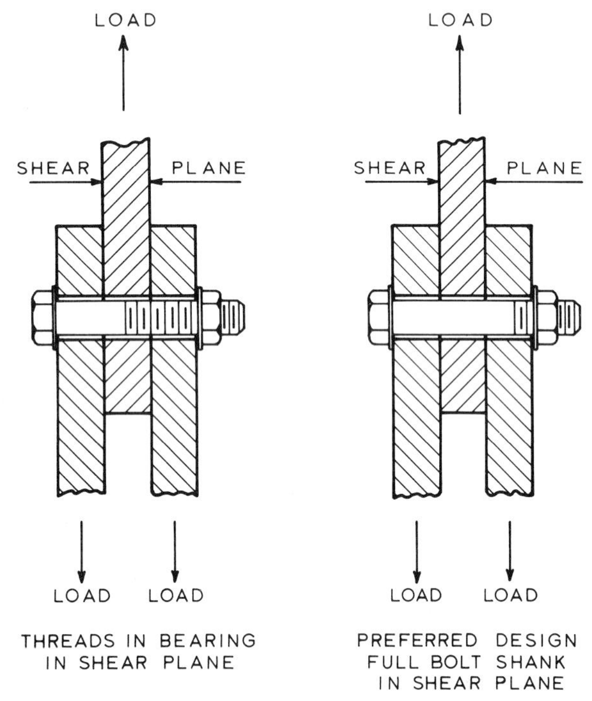

FIG. 1-12 Effect of bolt threads in bearing under shear joint loading.

a damaged joint, there is a tendency to open up the hole to permit the use of the next larger size fastener. This action will change the edge-distance relationship. So long as the joint is not critical in bearing, it may be a perfectly acceptable approach to saving the original joint.

1-14 Bending Loading Unfortunately, structural joints are not always loaded in pure tension or in pure shear. Many structural applications are subjected to bending forces which effectively result in combined tension and shear loads acting simultaneously on the fastener. To establish design limits for fasteners under bending-load conditions, interaction curves are usually generated for the fastener or fastener series. A typical interaction curve is illustrated in Fig. 1-14. Such a curve is developed from actual tests where a number of bolts are evaluated at different combinations of applied tensile and shear loadings. From the curve, it is possible to calculate maximum allowable design stresses for tension and for shear at any degree of bending. For illustrative purposes, the bolt stresses at a bending condition of 45 degrees are shown. Interaction curves are basically empirical and are tailored to specific bolt series taking into account strength level, material, and head design.

Even with valid interaction curves for design support, additional caution must be taken when working with joints subjected to bending. Because of the notch factor, screw threads are sensitive to applied bending stresses. Where severe or critical bending is anticipated in a joint, optimum-quality fasteners and maximum safety factors may well be advised.

1-15 Fatigue Loading As noted earlier, fatigue is the condition associated with repetitive or dynamic loading sensed by a structure or component. Many types of equipment are subjected to continuous dynamic stresses during their service life. Although extensive testing and experience have contributed to the ability to control

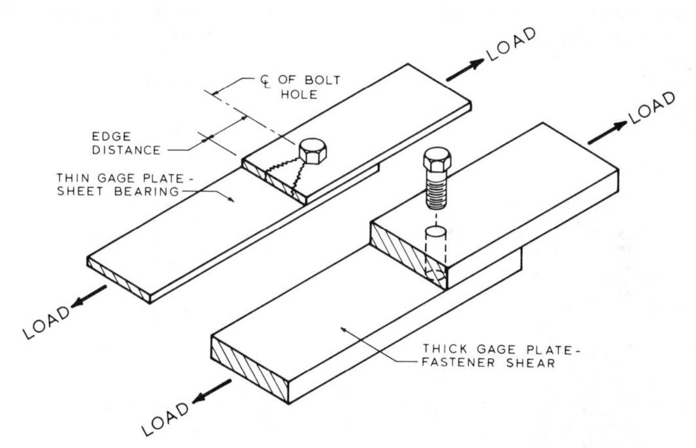

FIG. 1-13 Typical joints showing effect of sheet gage in design.

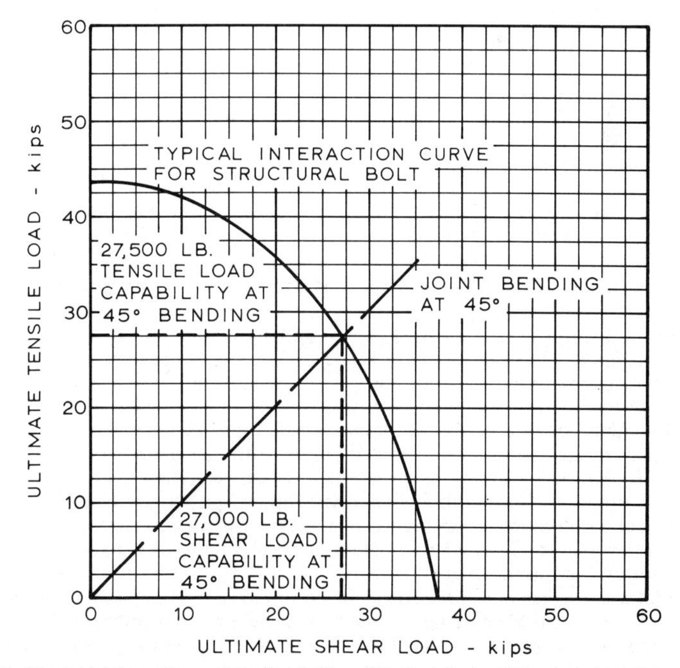

FIG. 1-14 Typical interaction curve for bolts subjected to bending (combined tension and shear) loading.

or minimize fatigue effects, the majority of service failures are still fatigue-related. This fact alone suggests the importance of good joint design.

There are two basic categories of joints subjected to fatigue loading: shear-fatigue-loaded joints and tension-fatigue-loaded joints. In the case of shear-loaded joints, fatigue failure normally occurs in the plate or sheet material. Applicable dynamic stresses, hole preparation, hole clearances, and fastener preload are just some of the factors which affect shear-joint-fatigue life.

As for tension-fatigue-loaded joints, the basic principles outlined for static-tensile-loaded joints are equally applicable. The prime factor is the adequate preloading of the fastener to meet or exceed anticipated dynamic or cyclic loading on the joint. A properly tightened and preloaded bolt will realize only minimal external tensile forces imposed by repetitive tensile loadings. The stresses expected during the service life of the equipment should be defined for accurate design. This may involve detailed experimental stress analysis and simulated service operation, and may be easier said than done.

As a result of the *S-N* curve relationship, substantially lower allowable design stresses have to be accommodated in fatigue joints than in comparable static-loaded joints. This is particularly true where long-life service is required. Effective use of fasteners specifically designed for fatigue strength and particular attention to preload control are perhaps two of the more important steps which can be taken to overcome fatigue effects in structural joints.

1-16 Vibration Fatigue loading is presumed to be relatively high repetitive or cyclic stresses in relation to the strength of the threaded fastener, or the joint. On the other hand, vibratory forces are considered to be relatively low, but may be associated with various ranges of operating frequencies. Critical combinations of frequency, amplitude, and loading can force a structure into resonance with catastrophic results. However, for most structural joints the combination of these conditions is not sufficient to produce resonance. Yet, vibration is present in equipment subjected to dynamic forces and its implications are quite serious for the fastener system.

Extensive or continued vibration affects threaded fasteners in a joint, without incurring structural damage to the joint material. It was indicated that the thread is essentially an inclined plane. Vibration or shock loading has a tendency to loosen bolts and nuts. The mechanism is complex, but the results are serious. Under the worst condition, it is possible for a nut to literally walk off a bolt thread. In other cases, vibration can be sufficient to slightly loosen a bolted connection, with resultant loss of preload. In such a case, with relaxation of preload or clamp load, the threaded fastener may sense higher fatigue stresses than were originally intended for the application.

Again, initial preload control of the threaded fastener assembly is vital to resisting vibration, shock, and impact loading. Where vibration is unusually severe, additional safety features may be necessary for proper joint integrity. Such devices include self-locking nuts, safety wire, cotter pins, and adhesive locking systems. As with fatigue, vibration analysis is complicated and costly, but sometimes there is no alternative to assure operating performance of structural joints.

TORQUING AND PRELOAD CONSIDERATIONS

The continued emphasis on achieving and maintaining proper preload control in threaded fasteners points up one dramatic fact. In the last analysis, the person who has the greatest influence on the structural integrity of an assembled joint is the man with the wrench in his hand. If a threaded fastener is torqued too high, there is a danger of failure on installation by stripping the threads or breaking the bolt, or making the fastener yield excessively. If the bolt is torqued too low, a low preload will be induced in the fastener assembly, possibly inviting fatigue and/or vibration failure. For every bolt system, there is an optimum preload objective, which is obtained by proper torquing of the bolt-nut combination.

The amount of tightening normally required to elongate a bolt and induce a desired tensile preload or clamp load in the fastener system is generally identified as the torque-tension relationship. This relationship is not always fully appreciated and is often misused. Unless the torque-tension relationship is accurately applied, there is an obvious danger of losing all of the advantages gained by good joint design.

1-17 Bolt Properties Threaded fasteners such as bolts and screws respond differently to application of straight tensile loading than to tightening by torquing. A bolt subjected to a static tensile-strength test will develop a load-elongation curve as illustrated in Fig. 1-15. The same bolt torqued with a nut to induce a tension preload will generate another load-elongation curve, also shown in Fig. 1-15. A comparison of the two curves indicates that the bolt which has been torqued does not develop as high a tensile strength as the bolt which was statically loaded. The additional torsional component introduced by the torquing action on the fastener accounts for this phenomenon. Apparent from Fig. 1-15 is the essentially straight-line relationship between increasing tensile load and bolt elongation up to the approximate yield point of the

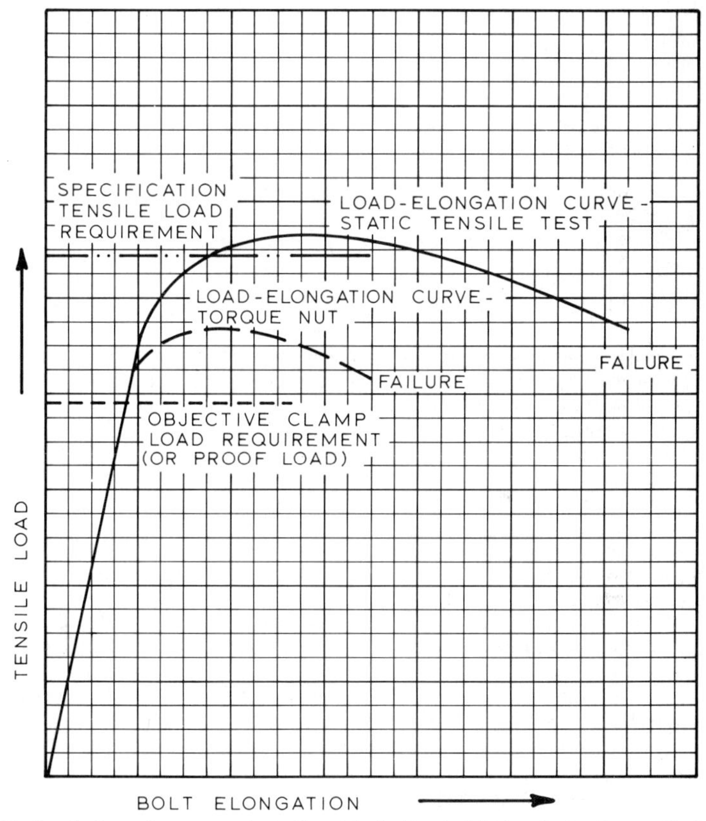

FIG. 1-15 Load-elongation curves for bolts subjected to straight tensile loading and to torquing.

fastener. While the test curves vary, in practice, a bolt torqued within its yield strength will be capable of developing the full rated tensile strength when subjected to additional tensile loading in excess of the preload value.

For this reason, many design requirements are predicated on torquing fasteners that develop a tensile preload of 70 to 80 percent of the rated static tensile strength. Torquing in excess of this objective can result in excessive bolt elongation and failure.

1-18 Torque-Tension Characteristics From the straight-line relationship of the load-elongation curve, it is evident that strict controls are imperative for developing clamp loads within design objectives. The problem of torquing to achieve tensile preload involves the critical area of friction coefficients of the fastener system. Investigators have fairly well established that as much as 90 percent of applied torque is used in overcoming friction in threaded fasteners, with high points of friction located

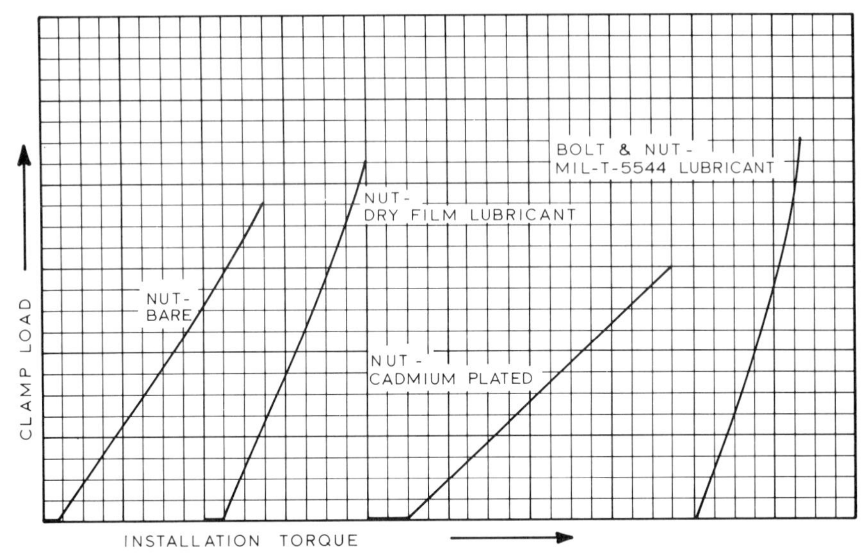

FIG. 1-16 Representative torque-tension curves for ¼-inch diameter alloy steel high strength cadmium-plated bolts showing influence of different surface finish conditions on mating nuts.

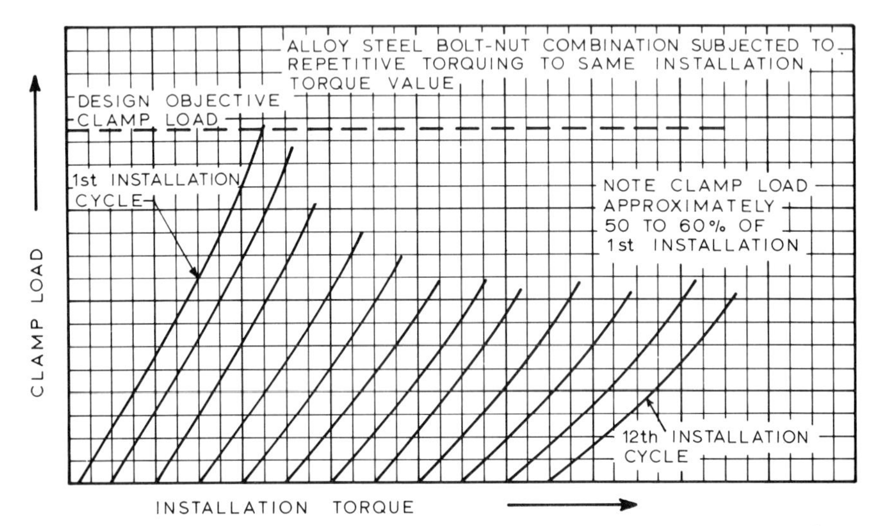

FIG. 1-17 Torque-tension curves showing effect of repetitive installation of same bolt-nut combination.

under the bolt head, under the nut, and in the actual fastener threads. This extensive work has indicated that the following formula can be used as an estimate of tension preload caused by torquing:

$$T = K\,D\,P$$

where T = installation torque, lb-in.
K = torque coefficient
D = nominal bolt diameter, inches
P = clamp load objective, lb

From the formula it can be seen that the prime variable is the torque coefficient K. The torque coefficient varies with different finishes, platings, and lubricant coatings normally found with standard fasteners. It was noted earlier that platings and protective coatings are used for protection of the base metal as well as for their lubrication qualities, which are sometimes vital in torque control. For example, both zinc and cadmium are popular platings. There is a noticeable difference in friction characteristics between these platings, making it necessary to rely on different installation-torque values.

Recent studies of high-strength-steel, cadmium-plated bolts using mating nuts with different surface finish conditions indicate the wide spread in both torque and preload properties, as illustrated in Fig. 1-16. In order to maintain torque control, some fasteners are finished with supplemental zinc, oil, or molybdenum disulfide coatings applied over basic cadmium or zinc platings. Supplemental lubricants have a direct effect on torque-tension characteristics, making the use of standard tables questionable for all installation conditions.

Another factor often overlooked is that the installation-torque value normally specified for a bolt-nut combination is essentially valid only for the initial assembly of the fastener. Continued use of the same bolt-nut assembly tends to alter and change the coefficient of friction properties of the nut, resulting in lower preload in the bolt after as few as five installations. Figure 1-17 illustrates this effect with data from other tests showing that preload loss for the same fastener combination can range as much as 30 to 60 percent after 10 installations.

The major emphasis in maintaining control of preload is on accurate torque control. For critical applications, the use of a calibrated torque wrench is a necessity. Even with a torque wrench, a wide spread of preload values can be observed. To compare the effects of different torque coefficients, tightening-torque values for standard bolt series have been calculated and listed in Table 1-2. A difference of 25 percent in tight-

TABLE 1-2 Calculated Tightening-Torque Values for Standard Bolt Series to Develop Rated Clamp Load Objectives

	SAE Grade 2 bolts			SAE Grade 5 bolts			SAE Grade 8 bolts		
Nominal thread size	Clamp load, lb°	Tightening torque		Clamp load, lb°	Tightening torque		Clamp load, lb°	Tightening torque	
		$K = .15$	$K = .20$		$K = .15$	$K = .20$		$K = .15$	$K = .20$
Coarse-Thread Series									
1/4-20	1,300	49†	65†	2,000	75†	100†	2,850	107†	143†
5/16-18	2,150	101†	134†	3,350	157†	210†	4,700	220†	305†
3/8-16	3,200	15	20	4,950	23	31	6,950	32.5	44
7/16-14	4,400	24	30	6,800	37	50	9,600	53	70
1/2-13	5,850	36.5	49	9,050	57	75	12,800	80	107
9/16-12	7,500	53	70	11,600	82	109	16,400	115	154
5/8-11	9,300	73	97	14,500	113	151	20,300	159	211
3/4-10	13,800	129	173	21,300	200	266	30,100	282	376
Fine-Thread Series									
1/4-28	1,500	55†	75†	2,300	85†	115†	3,250	120†	163†
5/16-24	2,400	112†	150†	3,700	173†	230†	5,200	245†	325†
3/8-24	3,600	17	22.5	5,600	26	35	7,900	37	50
7/16-20	4,900	27	36	7,550	42	55	10,700	59	78
1/2-20	6,600	41	55	10,200	64	85	14,400	90	120
9/16-18	8,400	59	79	13,000	92	122	18,300	129	172
5/8-18	10,600	83	110	16,300	128	170	23,000	180	240
3/4-16	15,400	144	193	23,800	223	298	33,600	315	420

° Clamp load objective predicated on developing 75 percent of the proof loads specified for the respective bolt series.

† Torque values for 1/4- and 5/16-inch sizes are in lb-in. All other torque values are in lb-ft. Torque values calculated from formula $T = KDL$, where T = tightening torque, lb-in
K = torque coefficient
D = nominal bolt diameter, inches
L = clamp load objective, lb

TABLE 1-3 Suggested Tightening-Torque Values for Nonferrous Threaded Fasteners
This table is offered as the suggested maximum torquing values for threaded products and is only a guide. Actual tests were conducted on dry, or near-dry, products. Mating parts were wiped clean of chips and foreign matter. A lubricated bolt requires less torque to attain the same clamping force as a nonlubricated bolt.
All values shown on chart except for nylon represent a safe working torque; in the case of nylon only, the figures represent breaking torque.

Bolt size	18–8 stainless steel	Brass	Silicon bronze	Aluminum 2024-T4	316 stainless steel	Monel	Nylon
	VALUES IN POUND-INCHES						
2-56	2.5	2.0	2.3	1.4	2.6	2.5	0.44
2-64	3.0	2.5	2.8	1.7	3.2	3.1	
3-48	3.9	3.2	3.6	2.1	4.0	4.0	
3-56	4.4	3.6	4.1	2.4	4.6	4.5	
4-40	5.2	4.3	4.8	2.9	5.5	5.3	1.19
4-48	6.6	5.4	6.1	3.6	6.9	6.7	
5-40	7.7	6.3	7.1	4.2	8.1	7.8	
5-44	9.4	7.7	8.7	5.1	9.8	9.6	
6-32	9.6	7.9	8.9	5.3	10.1	9.8	2.14
6-40	12.1	9.9	11.2	6.6	12.7	12.3	
8-32	19.8	16.2	18.4	10.8	20.7	20.2	4.3
8-36	22.0	18.0	20.4	12.0	23.0	22.4	
10-24	22.8	18.6	21.2	13.8	23.8	25.9	6.61
10-32	31.7	25.9	29.3	19.2	33.1	34.9	8.2
1/4"-20	75.2	61.5	68.8	45.6	78.8	85.3	16.0
1/4"-28	94.0	77.0	87.0	57.0	99.0	106.0	20.8
5/16"-18	132	107	123	80	138	149	34.9
5/16"-24	142	116	131	86	147	160	
3/8"-16	236	192	219	143	247	266	
3/8"-24	259	212	240	157	271	294	
7/16"-14	376	317	349	228	393	427	
7/16"-20	400	327	371	242	418	451	
1/2"-13	517	422	480	313	542	584	
1/2"-20	541	443	502	328	565	613	
9/16"-12	682	558	632	413	713	774	
9/16"-18	752	615	697	456	787	855	
5/8"-11	1110	907	1030	715	1160	1330	
5/8"-18	1244	1016	1154	798	1301	1482	
3/4"-10	1530	1249	1416	980	1582	1832	
3/4"-16	1490	1220	1382	958	1558	1790	
7/8"-9	2328	1905	2140	1495	2430	2775	
7/8"-14	2318	1895	2130	1490	2420	2755	
1"-8	3440	2815	3185	2205	3595	4130	
1"-14	3110	2545	2885	1995	3250	3730	
	VALUES IN POUND-FEET						
1 1/8"-7	413	337	383	265	432	499	
1 1/8"-12	390	318	361	251	408	470	
1 1/4"-7	523	428	485	336	546	627	
1 1/4"-12	480	394	447	308	504	575	
1 1/2"-6	888	727	822	570	930	1064	
1 1/2"-12	703	575	651	450	732	840	

Reprinted by permission of ITT Harper

TABLE 1-4 Suggested Locknut Tightening-Torque Values for Use with Standard Series Bolts

Locknut size and threads per inch	Steel hex locknuts						Steel hex flange locknuts					
	Grade B locknuts			Grade C locknuts			Grade F locknuts			Grade G locknuts		
	Clamp load, lb*	Locknut tightening torque‡		Clamp load, lb†	Locknut tightening torque‡		Clamp load, lb*	Locknut tightening torque‡		Clamp load, lb†	Locknut tightening torque‡	
		Max	Min		Max	Min		Max	Min		Max	Min
Coarse-Thread Series												
1/4-20	2,000	85	60	2,850	125	85	2,000	95	65	2,850	150	100
5/16-18	3,350	150	110	4,700	190	130	3,350	180	120	4,700	240	155
3/8-16	4,950	20	14.5	6,950	28	20	4,950	26	16	6,950	32	21
7/16-14	6,800	32	23	9,600	43	31	6,800	42	28	9,600	51	34
1/2-13	9,050	50	37	12,800	62.5	45	9,050	57	38	12,800	85	55
9/16-12	11,600	70	50	16,400	95	70	11,600	85	55	16,400	120	80
5/8-11	14,500	95	70	20,300	122.5	90	14,500	112	75	20,300	143	95
3/4-10	21,300	165	125	30,100	210	155	21,300	152	102	30,100	240	160
7/8-9	29,500	250	185	41,600	312.5	225						
1-8	38,700	375	275	54,600	462.5	350						
Fine-Thread Series												
1/4-28	2,300	90	65	3,250	125	85	2,300	115	75	3,250	160	105
5/16-24	3,700	160	120	5,200	200	140	3,700	200	130	5,200	230	155
3/8-24	5,600	22	16	7,900	29	21	5,600	25	17	7,900	33	22
7/16-20	7,550	34	24	10,700	43	31	7,550	45	30	10,700	60	40
1/2-20	10,200	52.5	37.5	14,400	70	50	10,200	66	44	14,400	89	59
9/16-18	13,000	77.5	57.5	18,300	95	70	13,000	94	62	18,300	132	88
5/8-18	16,300	97.5	72.5	23,000	125	90	16,300	120	80	23,000	175	115
3/4-16	23,800	165	120	33,600	210	155	23,800	192	128	33,600	270	170
7/8-14	32,400	270	200	45,800	312.5	225						
1-14	43,300	400	300	61,100	500	362.5						

* Clamp loads for Grade B and Grade F locknuts equal 75% of the proof loads specified for SAE J429 Grade 5 and ASTM A449 bolts.
† Clamp loads for Grade C and Grade G locknuts equal 75% of the proof loads specified for SAE J429 Grade 8 and ASTM A354 Grade BD bolts.
‡ Torque values for 1/4- and 5/16-in. sizes are in lb-in. All other torque values are in lb-ft.
SOURCE: The Industrial Fasteners Institute.

ening torque requirements can be noted based on *K* factors of .20 and .15. The torque coefficient of .20 is an estimate for unlubricated and dry fasteners, while the coefficient of .15 approximates a plated finish. The use of additional lubricants such as oil, MIL-T-5544 graphite grease, waxes, etc., can vary the *K* factor from 0.07 to .30. For specific design applications with exacting preload control requirements, it may be necessary to test and evaluate the specific bolts and nuts intended for the design application to define necessary installation-torque values.

Suggested tightening-torque values for nonferrous threaded fasteners are noted in Table 1-3. Suggested locknut tightening-torque values for use with standard series bolts are noted in Table 1-4. As with all previous discussion concerning torque-tension properties, these values are intended as guidelines. It is recommended that they be substantiated for particular design joints.

1-19 Turn-of-the-Nut Method Under field service conditions, when torque wrenches are not available, an alternate technique has been employed for tightening to develop preload. This system requires the nut to be seated to a finger-tight condition. From this position, the nut is subjected to an additional turn to induce preload. The amount of the turn required will vary with bolt diameter, thread series, bolt length, etc., and will usually be specified for the fastener system. The technique should be considered as a less exact substitute for calibrated equipment. The system requires sensitive determination of the snug-tight position of the nut. It is not considered applicable where self-locking nuts are used since the finger-tight condition is difficult to determine.

Part 2 Threaded Fastener Standards

Edited by

ROBERT O. PARMLEY, P.E.

Consulting Engineer, Ladysmith, Wisconsin

Threaded fasteners have been standardized by design, dimension, material, and mechanical quality.

This segment of Sec. 1 will, therefore, separate threaded fasteners into major categories and present them in tabular form for convenient reference.

1-20 Mechanical Requirements General mechanical requirements for bolts, screws, studs, and nuts are shown in an easy-reference listing. General screw thread

TABLE 1-5 Mechanical Requirements For Bolts, Screws, Studs, and Nuts

GRADE[1]	GENERAL DESCRIPTION	MECHANICAL REQUIREMENTS							
		BOLTS, SCREWS AND STUDS						NUTS	
		FULL SIZE BOLTS, SCREWS, STUDS		MACHINED TEST SPECIMENS OF BOLTS, SCREWS, STUDS			HARDNESS ROCKWELL	PROOF LOAD STRESS	HARDNESS ROCKWELL
		YIELD[2] STRENGTH Min psi	TENSILE STRENGTH Min psi	YIELD[2] STRENGTH Min psi	TENSILE STRENGTH Min psi	ELONGATION[3] %Min	Min	psi	Min
303-A	Austenitic Stainless Steel-Sol. Annealed	30,000	75,000	30,000	75,000	20	B75	75,000	B75
304-A	Austenitic Stainless Steel-Sol. Annealed	30,000	75,000	30,000	75,000	20	B75	75,000	B75
304	Austenitic Stainless Steel-Cold Worked	50,000	90,000	45,000	85,000	20	B85	90,000	B85
304-SH	Austenitic Stainless Steel-Strain Hardened	See Note 6	See Note 6	See Note 6	See Note 6	15	C25	See Note 6	C20
305-A	Austenitic Stainless Steel-Sol. Annealed	30,000	75,000	30,000	75,000	20	B70	75,000	B70
305	Austenitic Stainless Steel-Cold Worked	50,000	90,000	45,000	85,000	20	B85	90,000	B85
305-SH	Austenitic Stainless Steel-Strain Hardened	See Note 6	See Note 6	See Note 6	See Note 6	15	C25	See Note 6	C20
316-A	Austenitic Stainless Steel-Sol. Annealed	30,000	75,000	30,000	75,000	20	B70	75,000	B70
316	Austenitic Stainless Steel-Cold Worked	50,000	90,000	45,000	85,000	20	B85	90,000	B85
316-SH	Austenitic Stainless Steel-Strain Hardened	See Note 6	See Note 6	See Note 6	See Note 6	15	C25	See Note 6	C20
XM7-A	Austenitic Stainless Steel-Sol. Annealed	30,000	75,000	30,000	75,000	20	B70	75,000	B70
XM7	Austenitic Stainless Steel-Cold Worked	50,000	90,000	45,000	85,000	20	B85	90,000	B85
384-A	Austenitic Stainless Steel-Sol. Annealed	30,000	75,000	30,000	75,000	20	B70	75,000	B70
384	Austenitic Stainless Steel-Cold Worked	50,000	90,000	45,000	85,000	20	B85	90,000	B85

SOURCE: Industrial Fasteners Institute. (See IFI-104 Standard for footnotes.)

TABLE 1-6 Mechanical Requirements For Bolts, Screws, Studs, and Nuts

GRADE[1]	GENERAL DESCRIPTION	MECHANICAL REQUIREMENTS							
		BOLTS, SCREWS AND STUDS						NUTS	
		FULL SIZE BOLTS, SCREWS, STUDS		MACHINED TEST SPECIMENS OF BOLTS, SCREWS, STUDS			HARDNESS ROCKWELL	PROOF LOAD STRESS	HARDNESS ROCKWELL
		YIELD[2] STRENGTH Min psi	TENSILE STRENGTH Min psi	YIELD[2] STRENGTH Min psi	TENSILE STRENGTH Min psi	ELONG-ATION[3] % Min	Min	psi	Min
410-H	Martensitic Stainless Steel-Hardened and Tempered	95,000	125,000	95,000	125,000	20	C22	125,000	C22
410-HT	Martensitic Stainless Steel-Hardened and Tempered	135,000	180,000	135,000	180,000	12	C36	180,000	C36
416-H	Martensitic Stainless Steel-Hardened and Tempered	95,000	125,000	95,000	125,000	20	C22	125,000	C22
416-HT	Martensitic Stainless Steel-Hardened and Tempered	135,000	180,000	135,000	180,000	12	C36	180,000	C36
430	Ferritic Stainless Steel	40,000	70,000	40,000	70,000	20	B75	70,000	B75
464-HF	Naval Brass	15,000	52,000	14,000	50,000	25	B56	52,000	B56
464	Naval Brass	27,000	60,000	25,000	57,000	25	B65	60,000	B65
462	Naval Brass	27,000	52,000	24,000	50,000	20	B65	52,000	B65
642	Aluminum Bronze	35,000	72,000	35,000	72,000	15	B75	72,000	B75
630	Aluminum Bronze	50,000	105,000	50,000	105,000	10	B90	105,000	B90
614	Aluminum Bronze	40,000	75,000	40,000	75,000	30	B70	75,000	B70
510	Phosphor Bronze	35,000	60,000	35,000	60,000	15	B60	60,000	B60
675	Manganese Bronze	22,000	55,000	22,000	55,000	20	B60	55,000	B60
655-HF	Silicon Bronze	20,000	52,000	18,500	50,000	20	B60	52,000	B60
655	Silicon Bronze	38,000	70,000	36,000	68,000	15	B75	70,000	B75
651	Silicon Bronze	45,000	75,000	42,500	72,000	8	B75	75,000	B75
661	Silicon Bronze	38,000	70,000	38,000	70,000	15	B75	70,000	B75
NICU-A-HF	Nickel-Copper Alloy A	25,000	70,000	25,000	70,000	20	B70	70,000	B70
NICU-A	Nickel-Copper Alloy A	40,000	80,000	40,000	80,000	20	B80	80,000	B80
NICU-B	Nickel-Copper Alloy B	40,000	80,000	40,000	80,000	20	B80	80,000	B80
NICU-K[7]	Nickel-Copper-Aluminum Alloy	90,000	130,000	90,000	130,000	20	C24	130,000	C24
24T4	Aluminum Alloy	40,000	55,000	40,000	55,000	14	B70	55,000	B70
61T6	Aluminum Alloy	35,000	42,000	35,000	42,000	12	B50	42,000	B50

SOURCE: Industrial Fasteners Institute. (See IFI-104 Standard for footnotes.)

symbols and tensile stresses are listed. Standard grade markings for steel bolts and screws are also tabulated.

1-21 Bolts Bolts are externally threaded fasteners whose shank portions are threaded with standard threads and whose unthreaded segments are machined or ground in one or more places or over its entire length to a diameter approximately that of its thread-root diameter. Standard designs commence on page 1-33.

Bolts may have heads of various designs, such as square, hexagon, round, countersunk, elliptic, or oval. Eyebolts and bent bolts are also a major design and are illustrated.

1-22 Nuts A nut is a block of metal, usually of a square or hexagon shape, which has a hole drilled through its center and is thus internally threaded to mate with a standard bolt. The tables following will show the standardized designs, and also depict hexagon locknuts, slotted or castle nuts, and wing nuts. Standard designs start on page 1-52.

1-23 Screws Screws, as tabulated in this section, include machine, cap, set, thumb, socket, lag, miniature, and self-drilling designs. (See page 1-62.)

1-24 Metric Sizing Metric threaded fasteners and thread data are tabulated on pages 1-86 through 1-88.

TABLE 1-7 General Screw Thread Symbols

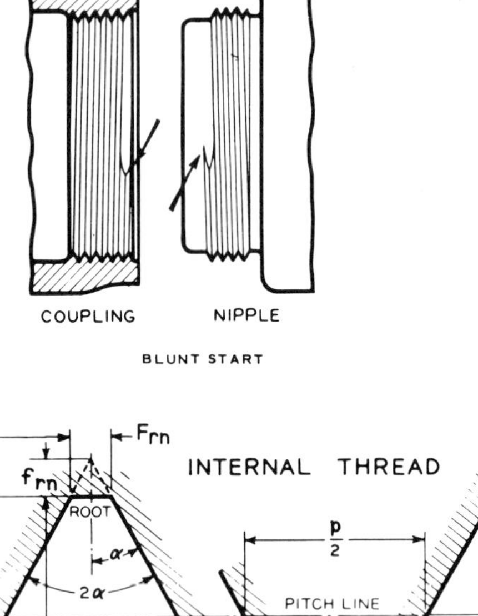

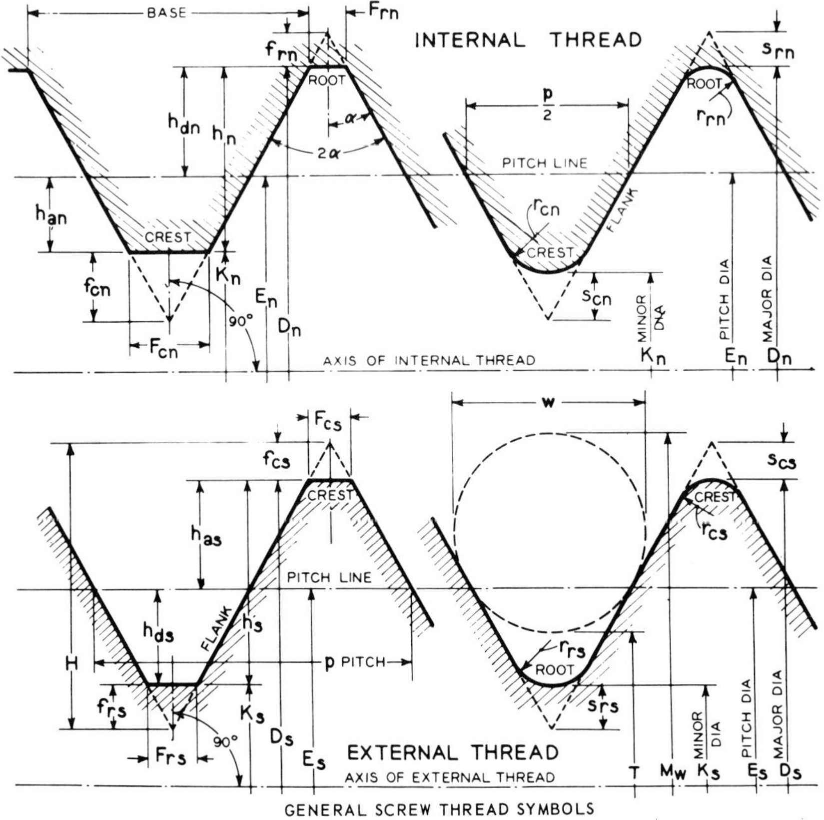

NOTE: These diagrams are not intended to show standard thread forms but illustrate only the applications of symbols.

SOURCE: ANSI B1.7-1965 (R1972), as published by the American Society of Mechanical Engineers.

TABLE 1-8 Tensile Stress Areas and Threads per Inch

PRODUCT SIZE DIA. in.	COARSE THREAD		FINE THREAD	
	STRESS AREA, sq in.	THREADS per in.	STRESS AREA, sq in.	THREADS per in.
6	0.00909	32	0.01015	40
8	0.0140	32	0.01474	36
10	0.0175	24	0.0200	32
12	0.0242	24	0.0258	28
1/4	0.0318	20	0.0364	28
5/16	0.0524	18	0.0580	24
3/8	0.0775	16	0.0878	24
7/16	0.1063	14	0.1187	20
1/2	0.1419	13	0.1599	20
9/16	0.1820	12	0.2030	18
5/8	0.2260	11	0.2560	18
3/4	0.3340	10	0.3730	16
7/8	0.4620	9	0.5090	14
1	0.6060	8	0.6630	12
1-1/8	0.7630	7	0.8560	12
1-1/4	0.9690	7	1.0730	12
1-3/8	1.1550	6	1.3150	12
1-1/2	1.4050	6	1.5810	12

Tensile stress areas are computed using the following formula:

$$A_s = 0.7854 \left[D - \frac{0.9743}{n}\right]^2$$

Where A_s = tensile stress area in square inches

D = nominal size (basic major diameter) in inches

n = number of threads per inch

SOURCE: Industrial Fasteners Institute.

TABLE 1-9 ASTM and SAE Grade Markings for Steel Bolts and Screws

Grade Marking	Specification	Material
NO MARK	SAE–Grade 1	Low or Medium Carbon Steel
	ASTM–A 307	Low Carbon Steel
	SAE–Grade 2	Low or Medium Carbon Steel
	SAE–Grade 5	Medium Carbon Steel, Quenched and Tempered
	ASTM–A 449	
	SAE–Grade 5.2	Low Carbon Martensite Steel, Quenched and Tempered
A 325	ASTM–A 325 Type 1	Medium Carbon Steel, Quenched and Tempered
A 325	ASTM–A 325 Type 2	Low Carbon Martensite Steel, Quenched and Tempered
A 325	ASTM–A 325 Type 3	Atmospheric Corrosion (Weathering) Steel, Quenched and Tempered
BB	ASTM–A 354 Grade BB	Low Alloy Steel, Quenched and Tempered
BC	ASTM–A 354 Grade BC	Low Alloy Steel, Quenched and Tempered
	SAE–Grade 7	Medium Carbon Alloy Steel, Quenched and Tempered, Roll Threaded After Heat Treatment
	SAE–Grade 8	Medium Carbon Alloy Steel, Quenched and Tempered
	ASTM–A 354 Grade BD	Alloy Steel, Quenched and Tempered
A 490	ASTM–A 490	Alloy Steel, Quenched and Tempered

ASTM Standards:

A 307 – Low Carbon Steel Externally and Internally Threaded Standard Fasteners.
A 325 – High Strength Steel Bolts for Structural Steel Joints, Including Suitable Nuts and Plain Hardened Washers.
A 449 – Quenched and Tempered Steel Bolts and Studs.
A 354 – Quenched and Tempered Alloy Steel Bolts and Studs with Suitable Nuts.
A 490 – Quenched and Tempered Alloy Steel Bolts for Structural Steel Joints.

SAE Standard:

J 429 – Mechanical and Quality Requirements for Externally Threaded Fasteners.

SOURCE: ANSI B18.2.1-1972, as published by the American Society of Mechanical Engineers.

TABLE 1-10 Dimensions of Square Bolts

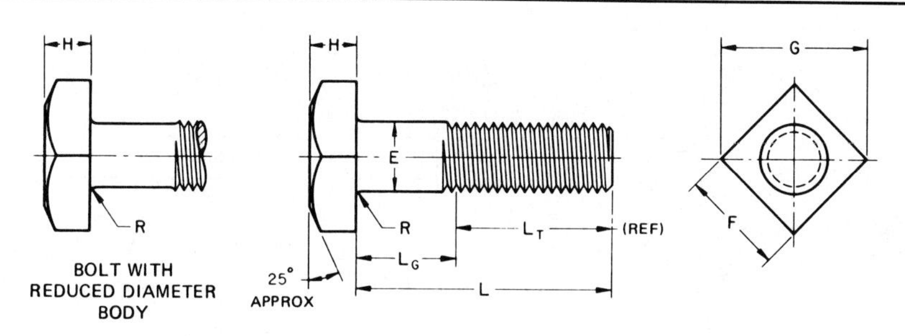

Nominal size of basic product dia., in.		E, body dia., in. (max)	F, width across flats, in.			G, width across corners, in.		H, height, in.			R, radius of fillet, in.		L_T, thread length for bolt lengths	
			Basic	Max	Min	Max	Min	Basic	Max	Min	Max	Min	6 in. and shorter (basic)	Over 6 in. (basic)
$\frac{1}{4}$	0.2500	0.260	$\frac{3}{8}$	0.375	0.362	0.530	0.498	$\frac{11}{64}$	0.188	0.156	0.03	0.01	0.750	1.000
$\frac{5}{16}$	0.3125	0.324	$\frac{1}{2}$	0.500	0.484	0.707	0.665	$\frac{13}{64}$	0.220	0.186	0.03	0.01	0.875	1.125
$\frac{3}{8}$	0.3750	0.388	$\frac{9}{16}$	0.562	0.544	0.795	0.747	$\frac{1}{4}$	0.268	0.232	0.03	0.01	1.000	1.250
$\frac{7}{16}$	0.4375	0.452	$\frac{5}{8}$	0.625	0.603	0.884	0.828	$\frac{19}{64}$	0.316	0.278	0.03	0.01	1.125	1.375
$\frac{1}{2}$	0.5000	0.515	$\frac{3}{4}$	0.750	0.725	1.061	0.995	$\frac{21}{64}$	0.348	0.308	0.03	0.01	1.250	1.500
$\frac{5}{8}$	0.6250	0.642	$\frac{15}{16}$	0.938	0.906	1.326	1.244	$\frac{27}{64}$	0.444	0.400	0.06	0.02	1.500	1.750
$\frac{3}{4}$	0.7500	0.768	$1\frac{1}{8}$	1.125	1.088	1.591	1.494	$\frac{1}{2}$	0.524	0.476	0.06	0.02	1.750	2.000
$\frac{7}{8}$	0.8750	0.895	$1\frac{5}{16}$	1.312	1.269	1.856	1.742	$\frac{19}{32}$	0.620	0.568	0.06	0.02	2.000	2.250
1	1.0000	1.022	$1\frac{1}{2}$	1.500	1.450	2.121	1.991	$\frac{21}{32}$	0.684	0.628	0.09	0.03	2.250	2.500
$1\frac{1}{8}$	1.1250	1.149	$1\frac{11}{16}$	1.688	1.631	2.386	2.239	$\frac{3}{4}$	0.780	0.720	0.09	0.03	2.500	2.750
$1\frac{1}{4}$	1.2500	1.277	$1\frac{7}{8}$	1.875	1.812	2.652	2.489	$\frac{27}{32}$	0.876	0.812	0.09	0.03	2.750	3.000
$1\frac{3}{8}$	1.3750	1.404	$2\frac{1}{16}$	2.062	1.994	2.917	2.738	$\frac{29}{32}$	0.940	0.872	0.09	0.03	3.000	3.250
$1\frac{1}{2}$	1.5000	1.531	$2\frac{1}{4}$	2.250	2.175	3.182	2.986	1	1.036	0.964	0.09	0.03	3.250	3.500

SOURCE: ANSI B18.2.1-1972, as published by the American Society of Mechanical Engineers.

TABLE 1-11 Dimensions of Hex Bolts

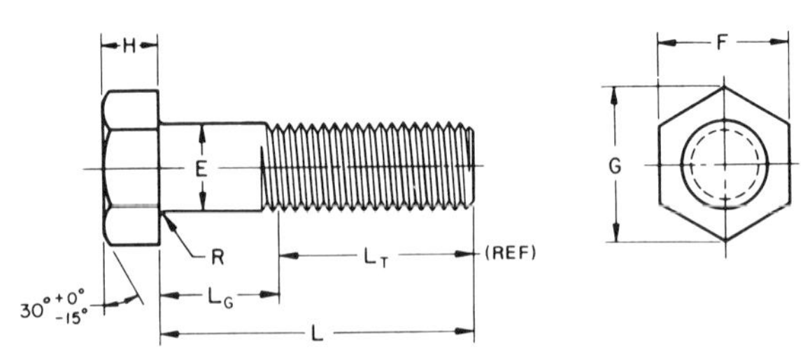

Dimensions of Hex Bolts

Nominal size of basic product dia., in.		E, body dia., in. (max)	F, width across flats, in.			G, width across corners, in.		H, height, in.			R, radius of fillet, in.		L_T, thread length for bolt lengths	
			Basic	Max	Min	Max	Min	Basic	Max	Min	Max	Min	6 in. and shorter (basic)	Over 6 in. (basic)
1/4	0.2500	0.260	7/16	0.438	0.425	0.505	0.484	11/64	0.188	0.150	0.03	0.01	0.750	1.000
5/16	0.3125	0.324	1/2	0.500	0.484	0.577	0.552	7/32	0.235	0.195	0.03	0.01	0.875	1.125
3/8	0.3750	0.388	9/16	0.562	0.544	0.650	0.620	1/4	0.268	0.226	0.03	0.01	1.000	1.250
7/16	0.4375	0.452	5/8	0.625	0.603	0.722	0.687	19/64	0.316	0.272	0.03	0.01	1.125	1.375
1/2	0.5000	0.515	3/4	0.750	0.725	0.866	0.826	11/32	0.364	0.302	0.03	0.01	1.250	1.500
5/8	0.6250	0.642	15/16	0.938	0.906	1.083	1.033	27/64	0.444	0.378	0.06	0.02	1.500	1.750
3/4	0.7500	0.768	1 1/8	1.125	1.088	1.299	1.240	1/2	0.524	0.455	0.06	0.02	1.750	2.000
7/8	0.8750	0.895	1 5/16	1.312	1.269	1.516	1.447	37/64	0.604	0.531	0.06	0.02	2.000	2.250
1	1.0000	1.022	1 1/2	1.500	1.450	1.732	1.653	43/64	0.700	0.591	0.09	0.03	2.250	2.500
1 1/8	1.1250	1.149	1 11/16	1.688	1.631	1.949	1.859	3/4	0.780	0.658	0.09	0.03	2.500	2.750
1 1/4	1.2500	1.277	1 7/8	1.875	1.812	2.165	2.066	27/32	0.876	0.749	0.09	0.03	2.750	3.000
1 3/8	1.3750	1.404	2 1/16	2.062	1.994	2.382	2.273	29/32	0.940	0.810	0.09	0.03	3.000	3.250
1 1/2	1.5000	1.531	2 1/4	2.250	2.175	2.598	2.480	1	1.036	0.902	0.09	0.03	3.250	3.500
1 3/4	1.7500	1.785	2 5/8	2.625	2.538	3.031	2.893	1 5/32	1.196	1.054	0.12	0.04	3.750	4.000
2	2.0000	2.039	3	3.000	2.900	3.464	3.306	1 11/32	1.388	1.175	0.12	0.04	4.250	4.500
2 1/4	2.2500	2.305	3 3/8	3.375	3.262	3.897	3.719	1 1/2	1.548	1.327	0.19	0.06	4.750	5.000
2 1/2	2.5000	2.559	3 3/4	3.750	3.625	4.330	4.133	1 21/32	1.708	1.479	0.19	0.06	5.250	5.500
2 3/4	2.7500	2.827	4 1/8	4.125	3.988	4.763	4.546	1 13/16	1.869	1.632	0.19	0.06	5.750	6.000
3	3.0000	3.081	4 1/2	4.500	4.350	5.196	4.959	2	2.060	1.815	0.19	0.06	6.250	6.500
3 1/4	3.2500	3.335	4 7/8	4.875	4.712	5.629	5.372	2 3/16	2.251	1.936	0.19	0.06	6.750	7.000
3 1/2	3.5000	3.589	5 1/4	5.250	5.075	6.062	5.786	2 5/16	2.380	2.057	0.19	0.06	7.250	7.500
3 3/4	3.7500	3.858	5 5/8	5.625	5.437	6.495	6.198	2 1/2	2.572	2.241	0.19	0.06	7.750	8.000
4	4.0000	4.111	6	6.000	5.800	6.928	6.612	2 11/16	2.764	2.424	0.19	0.06	8.250	8.500

SOURCE: ANSI B18.2.1-1972, as published by the American Society of Mechanical Engineers.

TABLE 1-12 Dimensions of Heavy Hex Structural Bolts

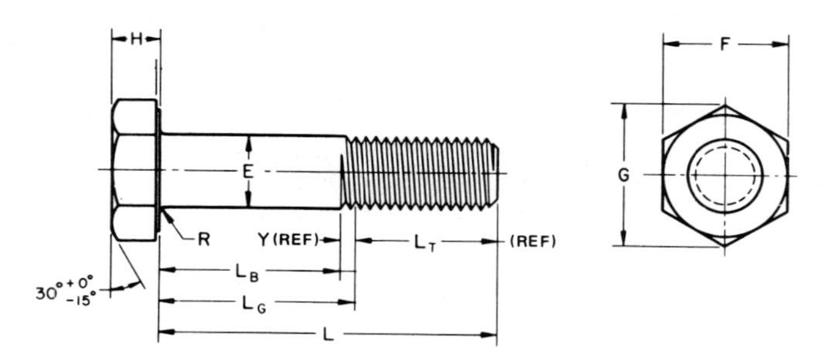

Nominal size or basic product dia., in.	*E*, body dia., in.		*F*, width across flats, in.			*G*, width across corners, in.		*H*, height, in.			*R*, radius of fillet, in.		L_T, thread length, in.	*Y*, transition thread length, in.	Runout of bearing surface, *FIR*, in.
	Max	Min	Basic	Max	Min	Max	Min	Basic	Max	Min	Max	Min	(basic)	(max)	(max)
1/2 0.5000	0.515	0.482	7/8	0.875	0.850	1.010	0.969	5/16	0.323	0.302	0.031	0.009	1.00	0.19	0.016
5/8 0.6250	0.642	0.605	1 1/16	1.062	1.031	1.227	1.175	25/64	0.403	0.378	0.062	0.021	1.25	0.22	0.019
3/4 0.7500	0.768	0.729	1 1/4	1.250	1.212	1.443	1.383	15/32	0.483	0.455	0.062	0.021	1.38	0.25	0.022
7/8 0.8750	0.895	0.852	1 7/16	1.438	1.394	1.660	1.589	35/64	0.563	0.531	0.062	0.031	1.50	0.28	0.025
1 1.0000	1.022	0.976	1 5/8	1.625	1.575	1.876	1.796	39/64	0.627	0.591	0.093	0.062	1.75	0.31	0.028
1 1/8 1.1250	1.149	1.098	1 13/16	1.812	1.756	2.093	2.002	11/16	0.718	0.658	0.093	0.062	2.00	0.34	0.032
1 1/4 1.2500	1.277	1.223	2	2.000	1.938	2.309	2.209	25/32	0.813	0.749	0.093	0.062	2.00	0.38	0.035
1 3/8 1.3750	1.404	1.345	2 3/16	2.188	2.119	2.526	2.416	27/32	0.878	0.810	0.093	0.062	2.25	0.44	0.038
1 1/2 1.5000	1.531	1.470	2 3/8	2.375	2.300	2.742	2.622	15/16	0.974	0.902	0.093	0.062	2.25	0.44	0.041

SOURCE: ANSI B18.2.1-1972, as published by the American Society of Mechanical Engineers.

TABLE 1-13 Dimensions of Round-Head Bolts

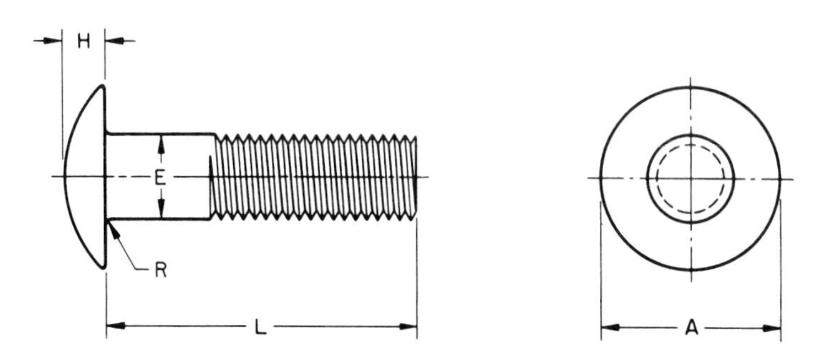

Nominal Size[1] or Basic Bolt Diameter		E Body Diameter		A Head Diameter		H Head Height		R Fillet Radius
		Max	Min	Max	Min	Max	Min	Max
No. 10	0.1900	0.199	0.182	0.469	0.438	0.114	0.094	0.031
1/4	0.2500	0.260	0.237	0.594	0.563	0.145	0.125	0.031
5/16	0.3125	0.324	0.298	0.719	0.688	0.176	0.156	0.031
3/8	0.3750	0.388	0.360	0.844	0.782	0.208	0.188	0.031
7/16	0.4375	0.452	0.421	0.969	0.907	0.239	0.219	0.031
1/2	0.5000	0.515	0.483	1.094	1.032	0.270	0.250	0.031
5/8	0.6250	0.642	0.605	1.344	1.219	0.344	0.313	0.062
3/4	0.7500	0.768	0.729	1.594	1.469	0.406	0.375	0.062
7/8	0.8750	0.895	0.852	1.844	1.719	0.469	0.438	0.062
1	1.0000	1.022	0.976	2.094	1.969	0.531	0.500	0.062

[1] Where specifying nominal size in decimals, zeros preceding decimal and in the fourth decimal place shall be omitted.

SOURCE: ANSI B18.5-1971, as published by the American Society of Mechanical Engineers.

TABLE 1-14 Dimensions of Round-Head Square-Neck Bolts

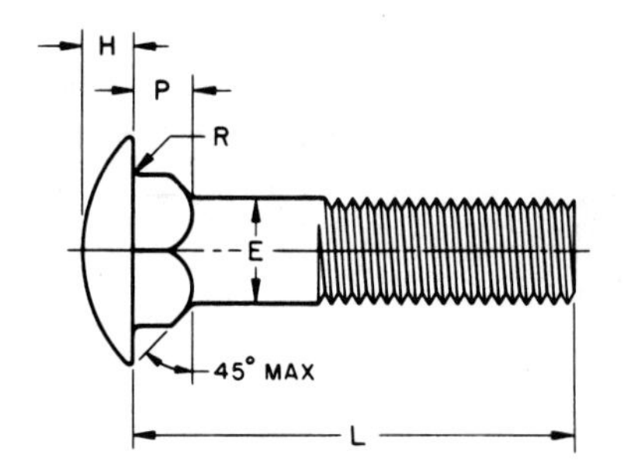

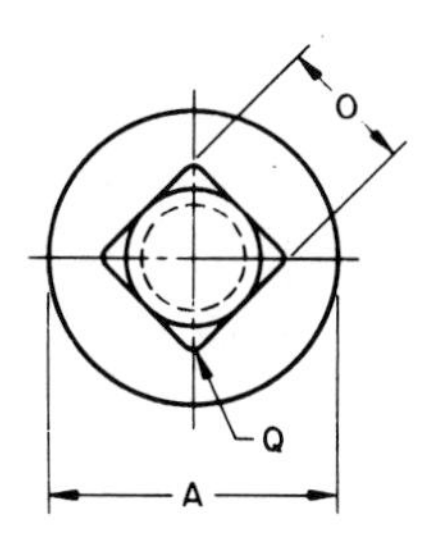

Nominal Size[1] or Basic Bolt Diameter	E		A		H		O		P		Q	R
	Body Diameter		Head Diameter		Head Height		Square Width		Square Depth		Corner Radius on Square	Fillet Radius
	Max	Min	Max	Min	Max	Min	Max	Min	Max	Min	Max	Max
No. 10 0.1900	0.199	0.182	0.469	0.438	0.114	0.094	0.199	0.185	0.125	0.094	0.031	0.031
1/4 0.2500	0.260	0.237	0.594	0.563	0.145	0.125	0.260	0.245	0.156	0.125	0.031	0.031
5/16 0.3125	0.324	0.298	0.719	0.688	0.176	0.156	0.324	0.307	0.187	0.156	0.031	0.031
3/8 0.3750	0.388	0.360	0.844	0.782	0.208	0.188	0.388	0.368	0.219	0.188	0.047	0.031
7/16 0.4375	0.452	0.421	0.969	0.907	0.239	0.219	0.452	0.431	0.250	0.219	0.047	0.031
1/2 0.5000	0.515	0.483	1.094	1.032	0.270	0.250	0.515	0.492	0.281	0.250	0.047	0.031
5/8 0.6250	0.642	0.605	1.344	1.219	0.344	0.313	0.642	0.616	0.344	0.313	0.078	0.062
3/4 0.7500	0.768	0.729	1.594	1.469	0.406	0.375	0.768	0.741	0.406	0.375	0.078	0.062
7/8 0.8750	0.895	0.852	1.844	1.719	0.469	0.438	0.895	0.865	0.469	0.438	0.094	0.062
1 1.0000	1.022	0.976	2.094	1.969	0.531	0.500	1.022	0.990	0.531	0.500	0.094	0.062

[1] Where specifying nominal size in decimals, zeros preceding decimal and in the fourth decimal place shall be omitted.

SOURCE: ANSI B18.5-1971, as published by the American Society of Mechanical Engineers.

TABLE 1-15 Dimensions of Round-Head Fin-Neck Bolts

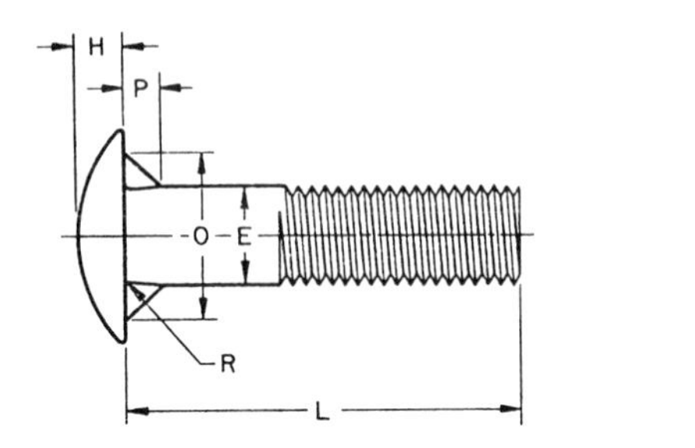

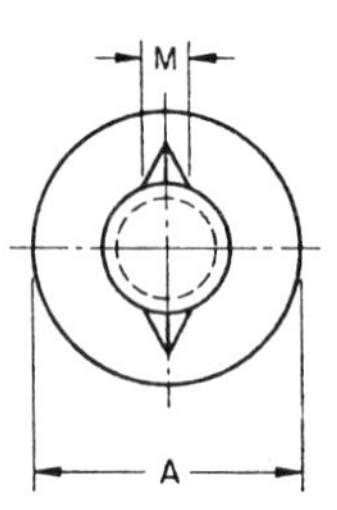

Nominal size or basic bolt diameter, in.		*E*, body diameter, in.		*A*, head diameter, in.		*H*, head height, in.		*M*, fin thickness, in.		*O*, distance across fins, in.		*P*, fin depth, in.		*R*, fillet radius, in. (max)
		Max	Min	Max	Min	Max	Min	Max	Min	Max	Min	Max	Min	
No. 10	0.1900	0.199	0.182	0.469	0.438	0.114	0.094	0.098	0.078	0.395	0.375	0.088	0.078	0.031
1/4	0.2500	0.260	0.237	0.594	0.563	0.145	0.125	0.114	0.094	0.458	0.438	0.104	0.094	0.031
5/16	0.3125	0.324	0.298	0.719	0.688	0.176	0.156	0.145	0.125	0.551	0.531	0.135	0.125	0.031
3/8	0.3750	0.388	0.360	0.844	0.782	0.208	0.188	0.161	0.141	0.645	0.625	0.151	0.141	0.031
7/16	0.4375	0.452	0.421	0.969	0.907	0.239	0.219	0.192	0.172	0.739	0.719	0.182	0.172	0.031
1/2	0.5000	0.515	0.483	1.094	1.032	0.270	0.250	0.208	0.188	0.833	0.813	0.198	0.188	0.031

When nominal size is specified in decimals, zeroes preceding decimal and in the fourth decimal place shall be omitted.

SOURCE: ANSI B18.5-1971, as published by the American Society of Mechanical Engineers.

TABLE 1-16 Dimensions of Round-Head Ribbed-Neck Bolts

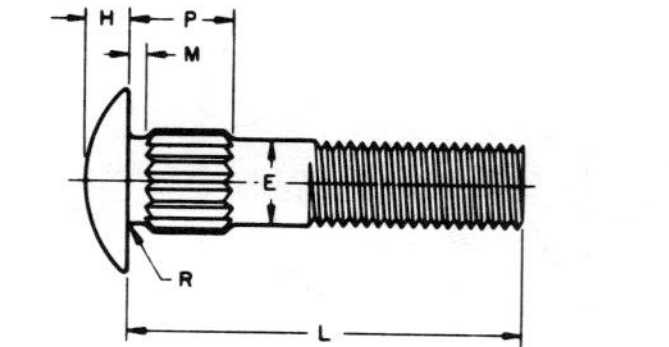

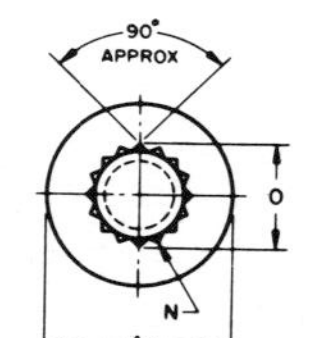

Nominal Size[1] or Basic Bolt Diameter	E Body Diameter		A Head Diameter		H Head Height		M Head to Ribs For Lengths of		N Number of Ribs	O Diameter Over Ribs	P Depth Over Ribs For Lengths of			R Fillet Radius
							7/8 and Shorter	1 in. and Longer			7/8 and Shorter	1 in. and 1 1/8	1 1/4 and Longer	
	Max	Min	Max	Min	Max	Min	±0.031*		Approx	Min	±0.031			Max
No. 10 0.1900	0.199	0.182	0.469	0.438	0.114	0.094	0.031*	0.063	9	0.210	0.250	0.407	0.594	0.031
1/4 0.2500	0.260	0.237	0.594	0.563	0.145	0.125	0.031*	0.063	10	0.274	0.250	0.407	0.594	0.031
5/16 0.3125	0.324	0.298	0.719	0.688	0.176	0.156	0.031*	0.063	12	0.340	0.250	0.407	0.594	0.031
3/8 0.3750	0.388	0.360	0.844	0.782	0.208	0.188	0.031*	0.063	12	0.405	0.250	0.407	0.594	0.031
7/16 0.4375	0.452	0.421	0.969	0.907	0.239	0.219	0.031*	0.063	14	0.470	0.250	0.407	0.594	0.031
1/2 0.5000	0.515	0.483	1.094	1.032	0.270	0.250	0.031*	0.063	16	0.534	0.250	0.407	0.594	0.031
5/8 0.6250	0.642	0.605	1.344	1.219	0.344	0.313	0.094	0.094	19	0.660	0.313	0.438	0.625	0.062
3/4 0.7500	0.768	0.729	1.594	1.469	0.406	0.375	0.094	0.094	22	0.785	0.313	0.438	0.625	0.062

[1] Where specifying nominal size in decimals, zeros preceding decimal and in the fourth decimal place shall be omitted.

* Tolerance on the No. 10 through 1/2 in. sizes for nominal lengths 7/8 in. and shorter shall be +0.031 and −0.000.

SOURCE: ANSI B18.5-1971, as published by the American Society of Mechanical Engineers.

TABLE 1-17 Dimensions of Round, Countersunk, and Square-Neck Plow Bolts

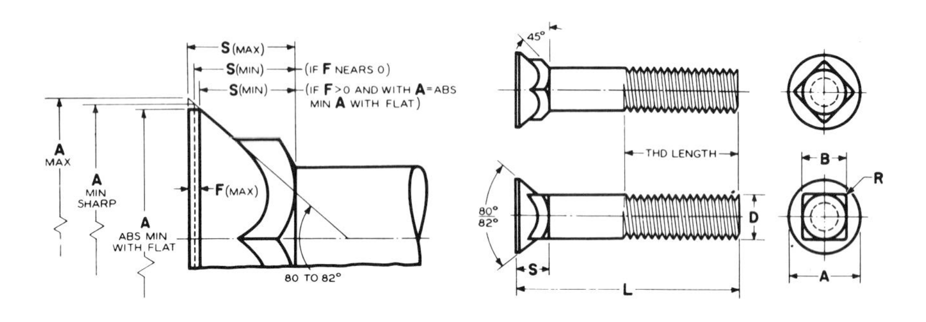

D	A			F	S		B		R
Nominal Diameter of Bolt	Diameter of Head			Feed Thickness	Depth of Square and Head		Width of Square		Radius on Corners of Square
	Max	Min Sharp	Abs Min With Flat	Max	Max	Min	Max	Min (Basic)	Max
5/16	0.605	0.578	0.563	0.025	0.269	0.243	0.325	0.313	1/32
3/8	0.708	0.671	0.656	0.031	0.312	0.281	0.387	0.375	3/64
7/16	0.826	0.781	0.766	0.036	0.364	0.328	0.450	0.438	3/64
1/2	0.945	0.890	0.875	0.042	0.417	0.375	0.515	0.500	3/64
*9/16	1.045	1.000	0.969	0.045	0.461	0.416	0.578	0.563	5/64
5/8	1.147	1.094	1.063	0.050	0.506	0.456	0.640	0.625	5/64
3/4	1.303	1.250	1.219	0.050	0.541	0.491	0.765	0.750	5/64
7/8	1.512	1.469	1.406	0.063	0.626	0.563	0.906	0.875	3/32
1	1.700	1.656	1.594	0.063	0.690	0.627	1.031	1.000	3/32
				REPAIR°					
5/16	0.556	0.531	0.516	0.020	0.232	0.212	0.325	0.313	1/32
3/8	0.659	0.624	0.609	0.025	0.272	0.247	0.387	0.375	3/64
7/16	0.779	0.734	0.719	0.030	0.324	0.294	0.450	0.438	3/64
1/2	0.898	0.843	0.828	0.035	0.375	0.340	0.515	0.500	3/64
9/16	1.002	.953	0.922	0.040	0,423	0.383	0.578	0.563	5/64
5/8	1.096	1.047	1.016	0.040	0.458	0.418	0.640	0.625	5/64
3/4	1.252	1.203	1.172	0.040	0.493	0.453	0.765	0.750	5/64
7/8	1.465	1.422	1.359	0.050	0.573	0.523	0.906	0.875	3/32
1	1.653	1.609	1.547	0.050	0.637	0.587	1.031	1.000	3/32

All dimensions given in inches, unless otherwise specified.

See Introductory notes for threads, points, material, length and length tolerance, and thread length and thread length tolerance.

°The letter "R" shall be shown on top of the repair head, to distinguish it from the regular head bolt.

*This size is not recommended.

If the method of manufacture permits, it is recommended that the same radius be maintained on each of all four corners of the square.

SOURCE: ANSI B18.9-1958, as published by the American Society of Mechanical Engineers.

TABLE 1-18 Dimensions of Countersunk Bolts and Slotted Countersunk Bolts

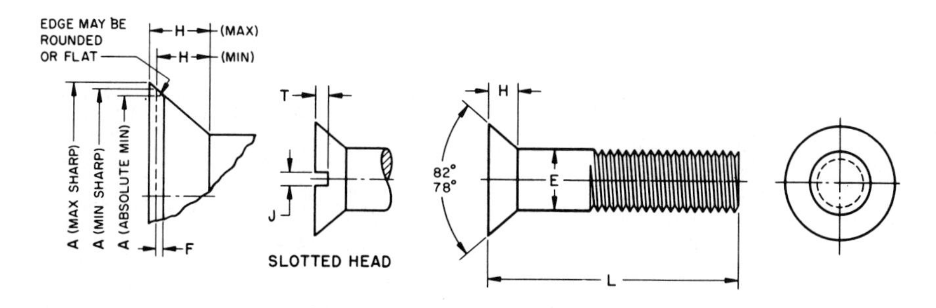

Nominal Size[2] or Basic Bolt Diameter	E		A			F[5]	H		J[1]		T[1]	
	Body Diameter		Head Diameter			Flat on Min Dia Head	Head Height		Slot Width		Slot Depth	
	Max	Min	Max, Edge Sharp[3]	Min, Edge Sharp[4]	Absolute Min, Edge Rounded or Flat	Max	Max	Min	Max	Min	Max	Min
1/4 0.2500	0.260	0.237	0.493	0.477	0.445	0.018	0.150	0.131	0.075	0.064	0.068	0.045
5/16 0.3125	0.324	0.298	0.618	0.598	0.558	0.023	0.189	0.164	0.084	0.072	0.086	0.057
3/8 0.3750	0.388	0.360	0.740	0.715	0.668	0.027	0.225	0.196	0.094	0.081	0.103	0.068
7/16 0.4375	0.452	0.421	0.803	0.778	0.726	0.030	0.226	0.196	0.094	0.081	0.103	0.068
1/2 0.5000	0.515	0.483	0.935	0.905	0.845	0.035	0.269	0.233	0.106	0.091	0.103	0.068
5/8 0.6250	0.642	0.605	1.169	1.132	1.066	0.038	0.336	0.292	0.133	0.116	0.137	0.091
3/4 0.7500	0.768	0.729	1.402	1.357	1.285	0.041	0.403	0.349	0.149	0.131	0.171	0.115
7/8 0.8750	0.895	0.852	1.637	1.584	1.511	0.042	0.470	0.408	0.167	0.147	0.206	0.138
1 1.0000	1.022	0.976	1.869	1.810	1.735	0.043	0.537	0.466	0.188	0.166	0.240	0.162
1 1/8 1.1250	1.149	1.098	2.104	2.037	1.962	0.043	0.604	0.525	0.196	0.178	0.257	0.173
1 1/4 1.2500	1.277	1.223	2.337	2.262	2.187	0.043	0.671	0.582	0.211	0.193	0.291	0.197
1 3/8 1.3750	1.404	1.345	2.571	2.489	2.414	0.043	0.738	0.641	0.226	0.208	0.326	0.220
1 1/2 1.5000	1.531	1.470	2.804	2.715	2.640	0.043	0.805	0.698	0.258	0.240	0.360	0.244

1 Head shall be unslotted, unless otherwise specified. Slot dimensions are same as Slotted Flat Countersunk Head Cap Screws in American National Standard, ANSI B18.6.2.

2 Where specifying nominal size in decimals, zeros preceding decimal and in the fourth decimal place shall be omitted.

3 Maximum head height calculated on maximum sharp head diameter, basic bolt diameter and 78° head angle.

4 Minimum head height calculated on minimum sharp head diameter, basic bolt diameter and 82° head angle.

5 Flat on minimum diameter head calculated on minimum sharp and absolute minimum head diameters and 82° head angle.

SOURCE: ANSI B18.5-1971, as published by the American Society of Mechanical Engineers.

TABLE 1-19 Dimensions of 114-Degree Countersunk Neck Bolts

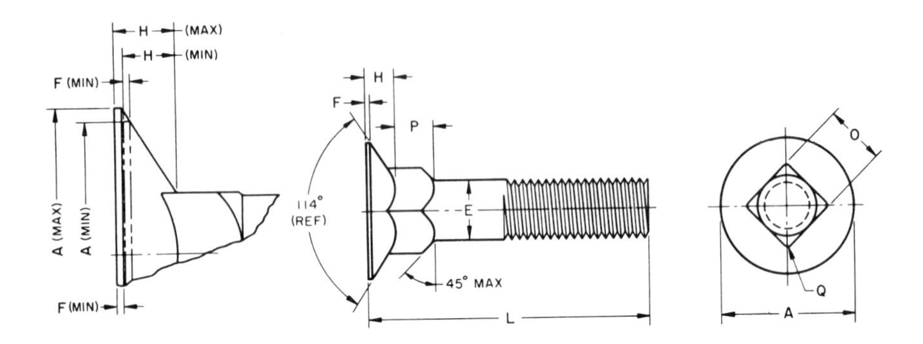

Nominal Size[1] or Basic Bolt Diameter	E		A		F	H		O		P		Q
	Body Diameter		Head Diameter		Flat on Head	Head Height		Square Width		Square Depth		Corner Radius on Square
	Max	Min	Max	Min	Min	Max	Min	Max	Min	Max	Min	Max
No. 10 0.1900	0.199	0.182	0.548	0.500	0.015	0.131	0.112	0.199	0.185	0.125	0.094	0.031
1/4 0.2500	0.260	0.237	0.682	0.625	0.018	0.154	0.135	0.260	0.245	0.156	0.125	0.031
5/16 0.3125	0.324	0.298	0.821	0.750	0.023	0.184	0.159	0.324	0.307	0.219	0.188	0.031
3/8 0.3750	0.388	0.360	0.960	0.875	0.027	0.212	0.183	0.388	0.368	0.250	0.219	0.047
7/16 0.4375	0.452	0.421	1.093	1.000	0.030	0.235	0.205	0.452	0.431	0.281	0.250	0.047
1/2 0.5000	0.515	0.483	1.233	1.125	0.035	0.265	0.229	0.515	0.492	0.312	0.281	0.047
5/8 0.6250	0.642	0.605	1.495	1.375	0.038	0.316	0.272	0.642	0.616	0.406	0.375	0.078
3/4 0.7500	0.768	0.729	1.754	1.625	0.041	0.368	0.314	0.768	0.741	0.500	0.469	0.078

[1] Where specifying nominal size in decimals, zeros preceding decimal and in the fourth decimal place shall be omitted.

SOURCE: ANSI B18.5-1971, as published by the American Society of Mechanical Engineers.

TABLE 1-20 Dimensions of Flat Countersunk Head Elevator Bolts

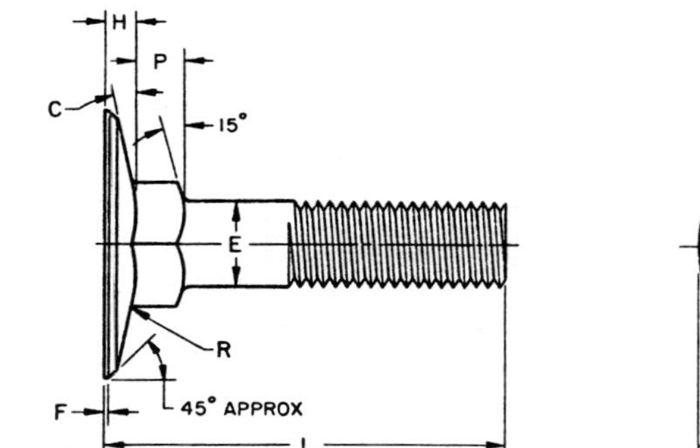

Nominal size* or basic bolt diameter, in.	*E*, body diameter, in.		*A*, head diameter, in.			*C*, head angle, degrees (ref.)	*F*, flat on min. dia. head, in. (max)	*H*, head height, in.		*O*, square width, in.		*P*, square depth, in.		*Q*, corner radius on square, in. (max)	*R*, fillet radius, in. (max)
	Max	Min	Max., edge sharp	Min., edge sharp	Min., edge flat			Max	Min	Max	Min	Max	Min		
No. 10 0.1900	0.199	0.182	0.790	0.750	0.740	9	0.025	0.082	0.062	0.210	0.185	0.125	0.094	0.031	0.031
1/4 0.2500	0.260	0.237	1.008	0.969	0.938	9	0.035	0.098	0.078	0.280	0.245	0.219	0.188	0.031	0.031
5/16 0.3125	0.324	0.298	1.227	1.188	1.157	9	0.035	0.114	0.094	0.342	0.307	0.250	0.219	0.031	0.031
3/8 0.3750	0.388	0.360	1.352	1.312	1.272	11	0.040	0.145	0.125	0.405	0.368	0.250	0.219	0.047	0.031
7/16 0.4375	0.452	0.421	1.477	1.438	1.397	13	0.040	0.176	0.156	0.468	0.431	0.281	0.250	0.047	0.031
1/2 0.5000	0.515	0.483	1.602	1.562	1.522	12	0.040	0.176	0.156	0.530	0.492	0.281	0.250	0.047	0.031

* When nominal size is specified in decimals, zeroes preceding decimal and in the fourth decimal place shall be omitted.
SOURCE: ANSI B18.5-1971, as published by the American Society of Mechanical Engineers.

TABLE 1-21 Dimensions of Step Bolts

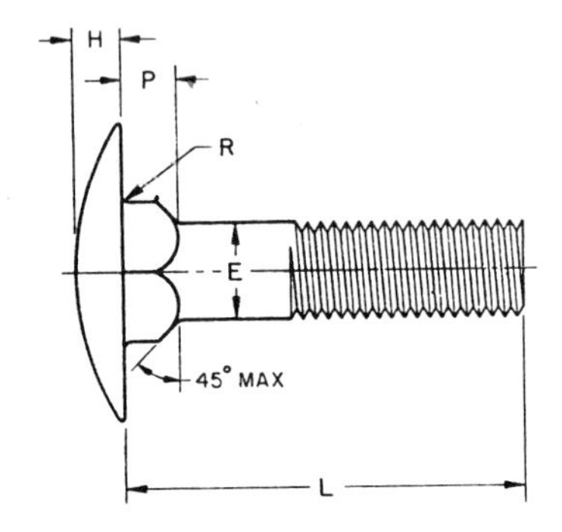

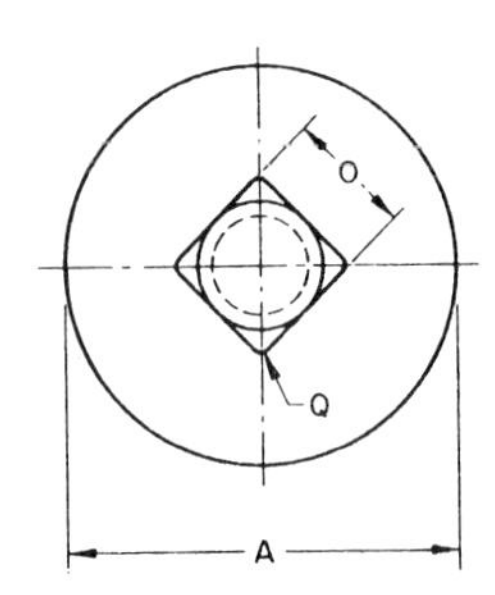

Nominal size° or basic bolt diameter, in.	*E*, body diameter, in.		*A*, head diameter, in.		*H*, head height, in.		*O*, square width, in.		*P*, square depth, in.		*Q*, corner radius on square, (max)	*R*, fillet radius, in. (max)
	Max	Min	Max	Min	Max	Min	Max	Min	Max	Min		
No. 10 0.1900	0.199	0.182	0.656	0.625	0.114	0.094	0.199	0.185	0.125	0.094	0.031	0.031
1/4 0.2500	0.260	0.237	0.844	0.813	0.145	0.125	0.260	0.245	0.156	0.125	0.031	0.031
5/16 0.3125	0.324	0.298	1.031	1.000	0.176	0.156	0.324	0.307	0.187	0.156	0.031	0.031
3/8 0.3750	0.388	0.360	1.219	1.188	0.208	0.188	0.388	0.268	0.219	0.188	0.047	0.031
7/16 0.4375	0.452	0.421	1.406	1.375	0.239	0.219	0.452	0.431	0.250	0.219	0.047	0.031
1/2 0.5000	0.515	0.483	1.594	1.563	0.270	0.250	0.515	0.492	0.281	0.250	0.047	0.031

° When nominal size is specified in decimals, zeroes preceding decimal and in the fourth decimal place shall be omitted.

SOURCE: ANSI B18.5-1971, as published by the American Society of Mechanical Engineers.

TABLE 1-22 Dimensions of T-Head Bolts

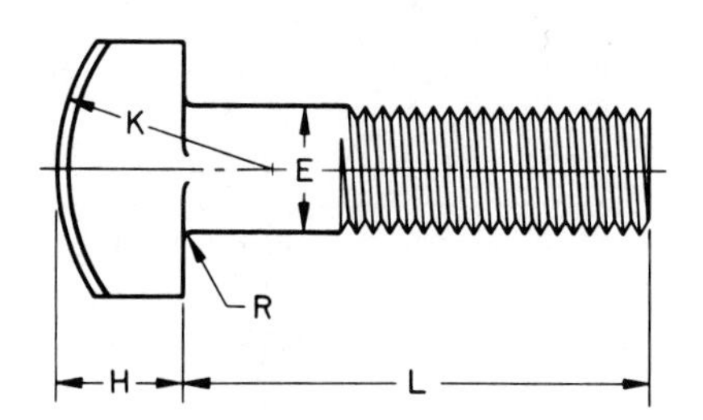

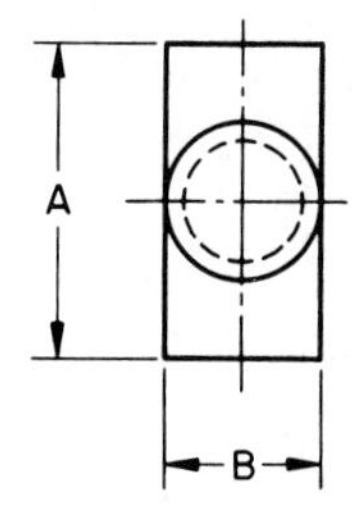

Dimensions of T-Head Bolts

Nominal Size[1] or Basic Bolt Diameter		E		A		B		H		K	R
		Body Diameter		Head Length		Head Width		Head Height		Head Radius	Fillet Radius
		Max	Min	Max	Min	Max	Min	Max	Min	Basic	Max
1/4	0.2500	0.260	0.237	0.500	0.488	0.280	0.245	0.204	0.172	0.438	0.031
5/16	0.3125	0.324	0.298	0.625	0.609	0.342	0.307	0.267	0.233	0.500	0.031
3/8	0.3750	0.388	0.360	0.750	0.731	0.405	0.368	0.331	0.295	0.625	0.031
7/16	0.4375	0.452	0.421	0.875	0.853	0.468	0.431	0.394	0.356	0.875	0.031
1/2	0.5000	0.515	0.483	1.000	0.975	0.530	0.492	0.458	0.418	0.875	0.031
5/8	0.6250	0.642	0.605	1.250	1.218	0.675	0.616	0.585	0.541	1.062	0.062
3/4	0.7500	0.768	0.729	1.500	1.462	0.800	0.741	0.649	0.601	1.250	0.062
7/8	0.8750	0.895	0.852	1.750	1.706	0.938	0.865	0.776	0.724	1.375	0.062
1	1.0000	1.022	0.976	2.000	1.950	1.063	0.990	0.903	0.847	1.500	0.062

[1] Where specifying nominal size in decimals, zeros preceding decimal and in the fourth decimal place shall be omitted.

SOURCE: ANSI B18.5-1971, as published by the American Society of Mechanical Engineers.

TABLE 1-23 Dimensions of Askew-Head Bolts

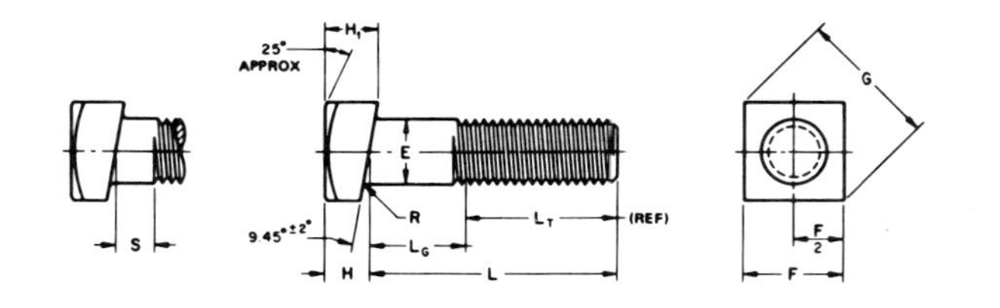

Nominal Size or Basic Bolt Diameter	E		F			G		H_1		H	R		S	L_T	
	Body Dia		Width Across Flats			Width Across Corners		Height		Mid Height	Radius of Fillet		Unthreaded Length	Thread Length For Bolt Lengths: 6 in. and Shorter	Over 6 in.
	Max	Min	Basic	Max	Min	Max	Min	Max	Min	Ref	Max	Min	Max	Basic	Basic
3/8 0.3750	0.388	0.360	9/16	0.562	0.544	0.795	0.747	0.317	0.277	0.250	0.03	0.01	0.250	1.000	1.250
1/2 0.5000	0.515	0.482	3/4	0.750	0.725	1.061	0.995	0.411	0.371	0.328	0.03	0.01	0.312	1.250	1.500
5/8 0.6250	0.642	0.605	15/16	0.938	0.906	1.326	1.244	0.520	0.480	0.422	0.06	0.02	0.344	1.500	1.750
3/4 0.7500	0.768	0.729	1 1/8	1.125	1.088	1.591	1.494	0.614	0.574	0.500	0.06	0.02	0.406	1.750	2.000
7/8 0.8750	0.895	0.852	1 5/16	1.312	1.269	1.856	1.742	0.723	0.683	0.594	0.06	0.02	0.438	2.000	2.250
1 1.0000	1.022	0.976	1 1/2	1.500	1.450	2.121	1.991	0.801	0.761	0.656	0.09	0.03	0.500	2.250	2.500

SOURCE: ANSI B18.2.1-1972, as published by the American Society of Mechanical Engineers.

TABLE 1-24 Elliptic-Neck Track Bolts

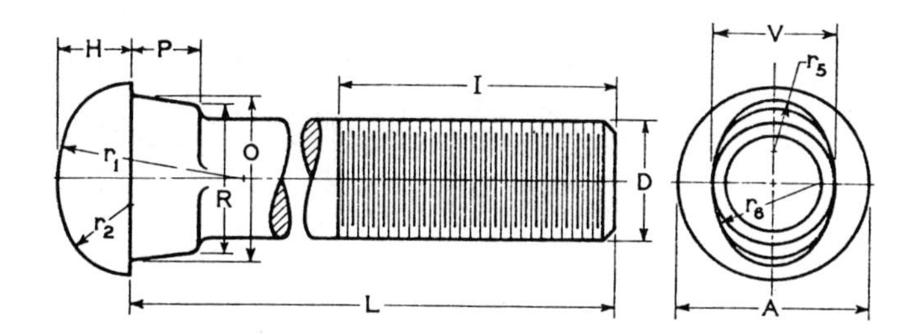

Nominal diameter over thread*	Head				Neck						Length under head	Minimum thread length	Threads per inch
D	A	H	r_1	r_2	O	R	P	r_5	r_6	V	L	I	
3/4	1 9/32	15/32	1 5/32	7/16	63/64	61/64	7/16	9/32	41/64	Same as body diameter of bolt	Under 7 inch by steps of 1/4 inch From 7 to 10 inch by steps of 1/2 inch	1 3/4	10
7/8	1 31/64	35/64	1 25/64	33/64	1 3/16	1 5/32	1/2	5/16	51/64			2	9
1	1 11/16	5/8	1 5/8	19/32	1 3/8	1 11/32	9/16	3/8	15/16			2 1/4	8
1 1/8	1 57/64	45/64	1 55/64	43/64	1 1/2	1 15/32	5/8	7/16	1			2 1/2	7
ADDITIONAL SIZES NOW IN USE BUT NOT RECOMMENDED FOR NEW DESIGN													
13/16	1 25/64	33/64	1 9/32	31/64	1 1/8	1 3/32	15/32	19/64	25/32	Same as above	Same as above	1 3/4	10
15/16	1 19/32	19/32	1 33/64	9/16	1 7/32	1 3/16	17/32	23/64	25/32			2	9
1 1/16	1 51/64	43/64	1 3/4	41/64	1 7/16	1 13/32	19/32	25/64	31/32			2 1/4	8

All dimensions are given in inches.

Tolerances: Length L plus or minus 1/8 in.
Neck (O and R) plus or minus 1/32 in.
Head (A and H) plus or minus 1/16 in.

Screw threads: The screw threads on track bolts shall conform to the American Standard coarse-thread series with fit as specified. They may be formed by cutting or rolling.

Either a free or wrench-turn fit shall be furnished at the option of the purchaser in accordance with the following provisions:

1. *Free fit*—Threads for free fit shall conform as closely as practicable to the American Standards limits for coarse-thread series Class 2 for bolts and nuts. The maximum allowable torque shall be 5 lb pull applied to the end of a 24-in. wrench.

2. *Wrench-turn fit*—The nut shall have a free fit for at least two threads in starting on the bolts. When engaged for the thickness of the nut plus two threads, the nut shall show the following minimum and maximum resistance, and for the remainder of the screw length shall not exceed the maximum resistance to turning as expressed in pounds pull applied to the end of a 24-in. wrench:

	Min lb	*Max lb*
Low-carbon nuts	5	45
Medium-carbon nuts	5	55

* In ordering bolts, specify the nominal diameter D, over the threads and not the body diameter.

SOURCE: ANSI B18.10-1963, as published by the American Society of Mechanical Engineers.

TABLE 1-25 Oval-Neck Track Bolts

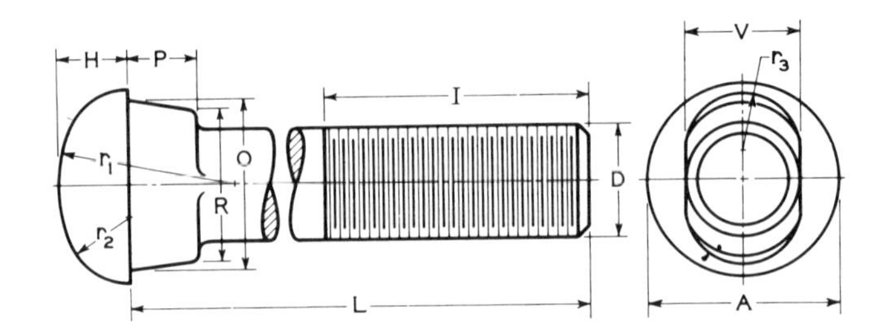

Nominal diameter over thread*	Head				Neck					Length under head,	Minimum thread length,	Threads per inch
D	A	H	r_1	r_2	O	R	P	r_3	V	L	I	
1/2	7/8	5/16	11/16	9/32	5/8	19/32	5/16			Under 7 inch	1 1/8	13
5/8	1 5/64	25/64	59/64	23/64	13/16	25/32	3/8	1/2 body	Same as	by steps	1 1/4	11
3/4	1 9/32	15/32	1 5/32	7/16	1 1/16	1 1/32	7/16	diameter	body	of 1/4 inch	1 3/4	10
7/8	1 31/64	35/64	1 25/64	33/64	1 7/32	1 3/16	1/2	of bolt	diam-	From 7 to	2	9
1	1 11/16	5/8	1 5/8	19/32	1 3/8	1 11/32	9/16		eter	10 inch	2 1/4	9
1 1/8	1 57/64	45/64	1 55/64	43/64	1 17/32	1 1/2	5/8		of bolt	by steps	2 1/2	7
										of 1/2 inch		
ADDITIONAL SIZES NOW IN USE BUT NOT RECOMMENDED FOR NEW DESIGN												
13/16	1 9/32	15/32	1 5/32	7/16	1 1/16	1 1/32	7/16	Same	Same	Same	1 7/8	10
15/16	1 31/64	35/64	1 25/64	33/64	1 7/32	1 3/16	1/2	as	as	as	2 1/8	9
1 1/16	1 11/16	5/8	1 5/8	19/32	1 3/8	1 11/32	9/16	above	above	above	2 3/8	8

All dimensions are given in inches.
Tolerances: Length L plus or minus 1/8 in.
Neck (O and R) plus or minus 1/32 in.
Head (A and H) plus or minus 1/16 in.

Screw threads: The screw threads on track bolts shall conform to the American Standard coarse-thread series with fit as specified. They may be formed by cutting or rolling.

Either a free or wrench-turn fit shall be furnished at the option of the purchaser in accordance with the following provisions:

1. *Free fit*—Threads for free fit shall conform as closely as practicable to the American Standards limits for coarse-thread series Class 2 for bolts and nuts. The maximum allowable torque shall be 5-lb pull applied to the end of a 24-in. wrench.
2. *Wrench turn fit*—The nut shall have a free fit for at least two threads in starting on the bolts. When engaged for the thickness of the nut plus two threads, the nut shall show the following minimum and maximum resistance, and for the remainder of the screw length shall not exceed the maximum resistance to turning as expressed in pounds pull applied to the end of a 24-in. wrench.

	Min lb	*Max lb*
Low-carbon nuts	5	45
Medium-carbon nuts	5	55

* In ordering bolts specify the nominal diameter D over the threads and not the body diameter. Standard sizes for industrial use are 1/2, 5/8, 3/4, 7/8, and 1 in. with free fit ASA heavy square nuts.

SOURCE: ANSI B18.10-1963, as published by the American Society of Mechanical Engineers.

TABLE 1-26 Type 1 Regular-Pattern Eyebolt

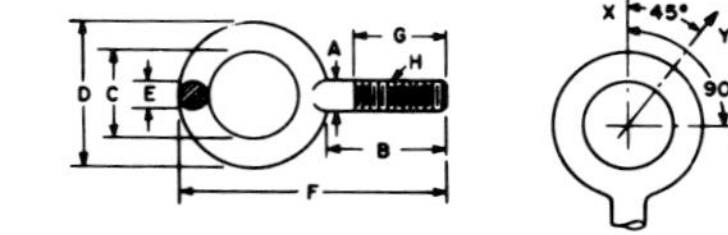

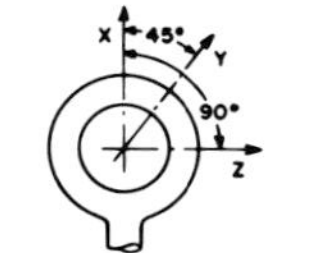

Nominal Size		*A* Shank Dia	*B* Shank Length	*C* Eye ID	*D* Nominal Eye OD	*E* Eye Sect. Dia	*F* Overall Length	*G* Min. Length Full Thread	*H* Thread Size UNC-2A	W_x Safe Working Load, lb at 0° (1)	W_y Safe Working Load, lb at 45° (1)	W_z Safe Working Load, lb at 90° (1)
1/4*	0.25	0.25 0.28	1.00 1.06	0.69 0.81	1.19	0.19 0.25	2.06 2.38	0.75	1/4–20 or .250–20	400	80	60
5/16*	0.31	0.31 0.34	1.12 1.19	0.81 0.94	1.44	0.25 0.31	2.44 2.75	0.81	5/16–18 or .3125–18	800	160	120
3/8*	0.38	0.38 0.41	1.25 1.38	0.94 1.06	1.69	0.31 0.38	2.81 3.19	0.88	3/8–16 or .375–16	1400	280	210
7/16	0.44	0.44 0.47	1.38 1.50	1.00 1.12	1.81	0.34 0.41	3.06 3.44	1.00	7/16–14 or .4375–14	2000	400	300
1/2*	0.50	0.50 0.53	1.50 1.62	1.12 1.25	2.12	0.44 0.50	3.50 3.88	1.12	1/2–13 or .500–13	2600	520	390
9/16	0.56	0.56 0.59	1.62 1.75	1.19 1.31	2.25	0.47 0.53	3.75 4.12	1.25	9/16–12 or .5625–12	3000	600	450
5/8*	0.62	0.62 0.66	1.75 1.88	1.31 1.44	2.56	0.56 0.62	4.19 4.56	1.38	5/8–11 or .625 –11	4000	800	600
3/4*	0.75	0.75 0.78	2.00 2.12	1.44 1.56	2.81	0.62 0.69	4.69 5.06	1.62	3/4–10 or .750–10	6000	1200	900
7/8*	0.88	0.88 0.91	2.25 2.38	1.56 1.69	3.19	0.75 0.81	5.31 5.69	1.81	7/8–9 or .875–9	6600	1300	1000
1*	1.00	1.00 1.06	2.50 2.62	1.69 1.81	3.56	0.88 0.94	5.94 6.31	2.06	1–8 or 1.000–8	8000	1600	1200
1 1/8	1.12	1.12 1.19	2.75 2.88	1.94 2.06	4.06	1.00 1.06	6.69 7.06	2.31	1 1/8–7 or 1.125–7	10000	2000	1500
1 1/4*	1.25	1.25 1.34	3.00 3.12	2.12 2.25	4.44	1.09 1.16	7.31 7.69	2.50	1 1/4–7 or 1.250–7	15000	3000	2250
1 1/2*	1.50	1.50 1.59	3.50 3.62	2.44 2.56	5.19	1.31 1.38	8.56 8.94	3.00	1 1/2–6 or 1.500–6	18000	3600	2700
1 3/4	1.75	1.75 1.84	3.75 3.88	2.75 3.00	6.00	1.50 1.62	9.50 10.12	3.19	1 3/4–5 or 1.750–5	22000	4400	3300
2*	2.00	2.00 2.09	4.00 4.25	3.06 3.44	6.88	1.75 1.88	10.56 11.44	3.38	2–4 1/2 or 2.000–4.50	26000	5200	3900
2 1/2*	2.50	2.50 2.62	5.00 5.25	3.81 4.19	8.50	2.19 2.31	13.19 14.06	4.25	2 1/2–4 or 2.500–4	32000	6400	4800

*Preferred

(1) These safe working loads are based on the following percentages of minimum proof loads shown in ASTM A 489; 0° –66.6 per cent, 45° – 13.3 per cent, 90° – 10.0 per cent.

All dimensions in inches.

SOURCE: ANSI B18.15-1969, as published by the American Society of Mechanical Engineers.

TABLE 1-27 Type 2 Shoulder-Pattern Eyebolt

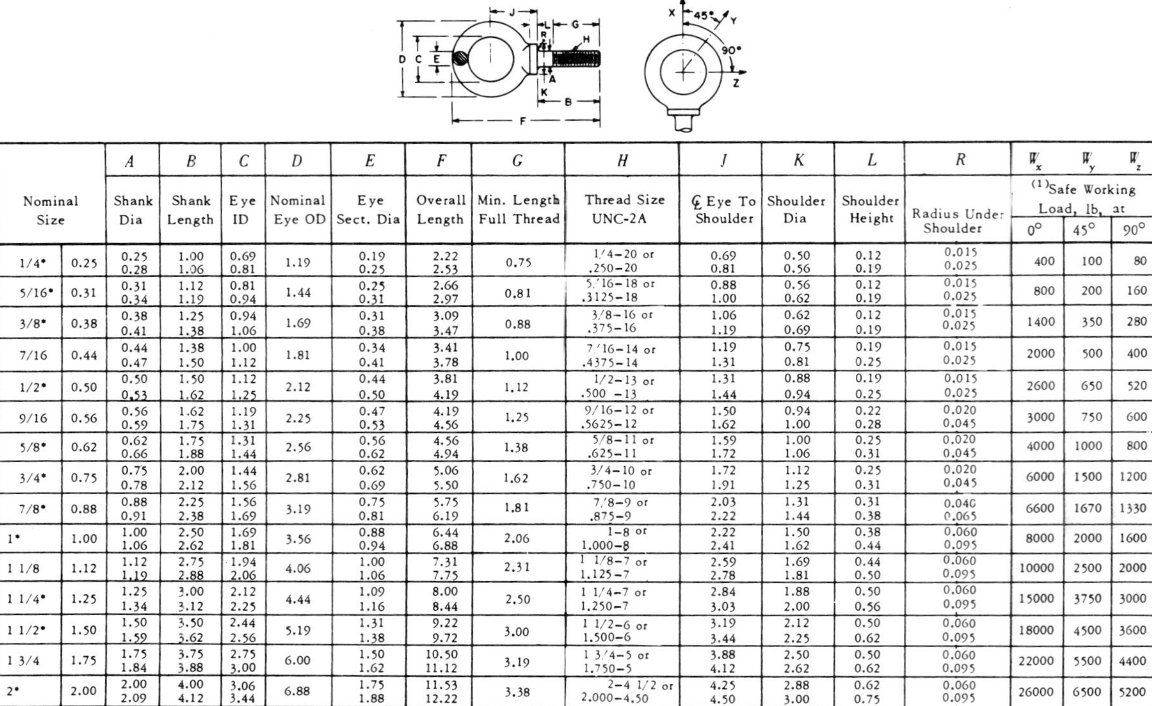

Nominal Size		A Shank Dia	B Shank Length	C Eye ID	D Nominal Eye OD	E Eye Sect. Dia	F Overall Length	G Min. Length Full Thread	H Thread Size UNC-2A	J ℄ Eye To Shoulder	K Shoulder Dia	L Shoulder Height	R Radius Under Shoulder	W_x [1]Safe Working Load, lb, at 0°	W_y 45°	W_z 90°
1/4*	0.25	0.25 0.28	1.00 1.06	0.69 0.81	1.19	0.19 0.25	2.22 2.53	0.75	1/4–20 or .250–20	0.69 0.81	0.50 0.56	0.12 0.19	0.015 0.025	400	100	80
5/16*	0.31	0.31 0.34	1.12 1.19	0.81 0.94	1.44	0.25 0.31	2.66 2.97	0.81	5/16–18 or .3125–18	0.88 1.00	0.56 0.62	0.12 0.19	0.015 0.025	800	200	160
3/8*	0.38	0.38 0.41	1.25 1.38	0.94 1.06	1.69	0.31 0.38	3.09 3.47	0.88	3/8–16 or .375–16	1.06 1.19	0.62 0.69	0.12 0.19	0.015 0.025	1400	350	280
7/16	0.44	0.44 0.47	1.38 1.50	1.00 1.12	1.81	0.34 0.41	3.41 3.78	1.00	7/16–14 or .4375–14	1.19 1.31	0.75 0.81	0.19 0.25	0.015 0.025	2000	500	400
1/2*	0.50	0.50 0.53	1.50 1.62	1.12 1.25	2.12	0.44 0.50	3.81 4.19	1.12	1/2–13 or .500 –13	1.31 1.44	0.88 0.94	0.19 0.25	0.015 0.025	2600	650	520
9/16	0.56	0.56 0.59	1.62 1.75	1.19 1.31	2.25	0.47 0.53	4.19 4.56	1.25	9/16–12 or .5625–12	1.50 1.62	0.94 1.00	0.22 0.28	0.020 0.045	3000	750	600
5/8*	0.62	0.62 0.66	1.75 1.88	1.31 1.44	2.56	0.56 0.62	4.56 4.94	1.38	5/8–11 or .625–11	1.59 1.72	1.00 1.06	0.25 0.31	0.020 0.045	4000	1000	800
3/4*	0.75	0.75 0.78	2.00 2.12	1.44 1.56	2.81	0.62 0.69	5.06 5.50	1.62	3/4–10 or .750–10	1.72 1.91	1.12 1.25	0.25 0.31	0.020 0.045	6000	1500	1200
7/8*	0.88	0.88 0.91	2.25 2.38	1.56 1.69	3.19	0.75 0.81	5.75 6.19	1.81	7/8–9 or .875–9	2.03 2.22	1.31 1.44	0.31 0.38	0.040 0.065	6600	1670	1330
1*	1.00	1.00 1.06	2.50 2.62	1.69 1.81	3.56	0.88 0.94	6.44 6.88	2.06	1–8 or 1.000–8	2.22 2.41	1.50 1.62	0.38 0.44	0.060 0.095	8000	2000	1600
1 1/8	1.12	1.12 1.19	2.75 2.88	1.94 2.06	4.06	1.00 1.06	7.31 7.75	2.31	1 1/8–7 or 1.125–7	2.59 2.78	1.69 1.81	0.44 0.50	0.060 0.095	10000	2500	2000
1 1/4*	1.25	1.25 1.34	3.00 3.12	2.12 2.25	4.44	1.09 1.16	8.00 8.44	2.50	1 1/4–7 or 1.250–7	2.84 3.03	1.88 2.00	0.50 0.56	0.060 0.095	15000	3750	3000
1 1/2*	1.50	1.50 1.59	3.50 3.62	2.44 2.56	5.19	1.31 1.38	9.22 9.72	3.00	1 1/2–6 or 1.500–6	3.19 3.44	2.12 2.25	0.50 0.62	0.060 0.095	18000	4500	3600
1 3/4	1.75	1.75 1.84	3.75 3.88	2.75 3.00	6.00	1.50 1.62	10.50 11.12	3.19	1 3/4–5 or 1.750–5	3.88 4.12	2.50 2.62	0.50 0.62	0.060 0.095	22000	5500	4400
2*	2.00	2.00 2.09	4.00 4.12	3.06 3.44	6.88	1.75 1.88	11.53 12.22	3.38	2–4 1/2 or 2.000–4.50	4.25 4.50	2.88 3.00	0.62 0.75	0.060 0.095	26000	6500	5200

*Preferred.

[1] Applies if shoulder is properly seated. Otherwise safe working load of Type 1 applies. These safe working loads are based on the following percentages of minimum proof loads shown in ASTM A 489; 0° – 66.6 per cent, 45° – 16.7 per cent, 90° – 13.3 per cent.

All dimensions in inches.

SOURCE: ANSI B18.15-1969, as published by the American Society of Mechanical Engineers.

TABLE 1-28 Bent Bolts

J-BOLT
(D x L x C x E x T x M x N)

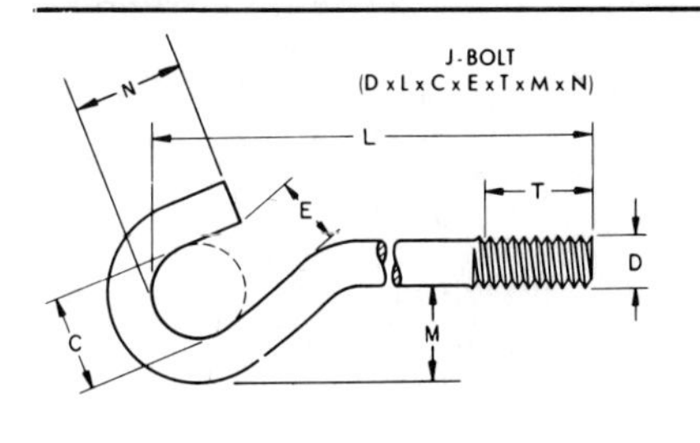

EYE BOLT, CLOSED
(D x L x C x T)

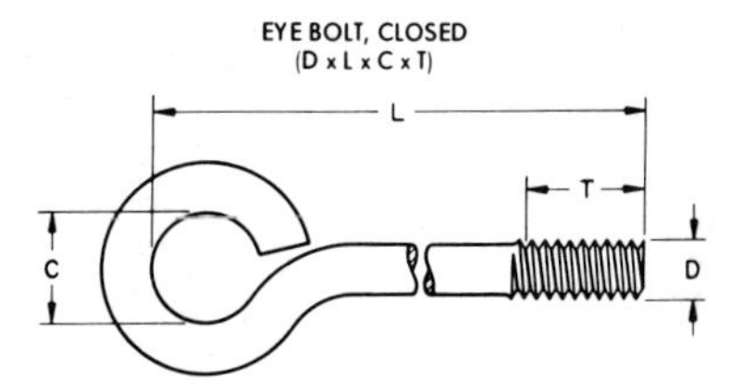

HOOK BOLT, ROUND BEND
(D x L x C x A x T)

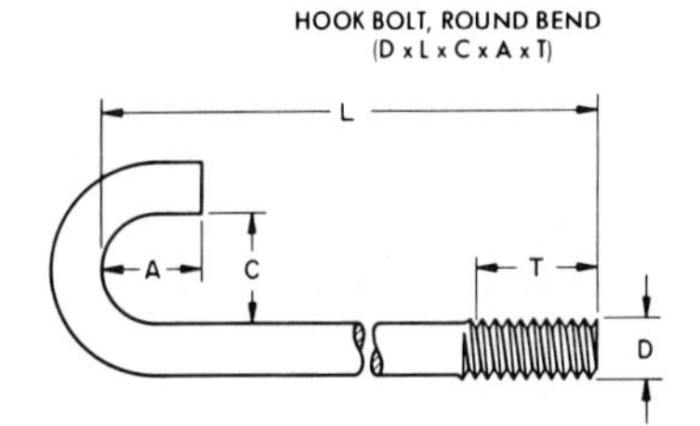

EYE BOLT, OPEN
(D x L x C x E x T)

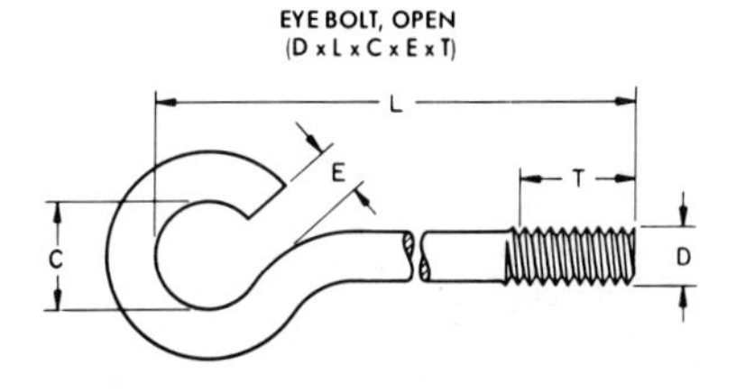

HOOK BOLT, SQUARE BEND
(D x L x C x A x T)

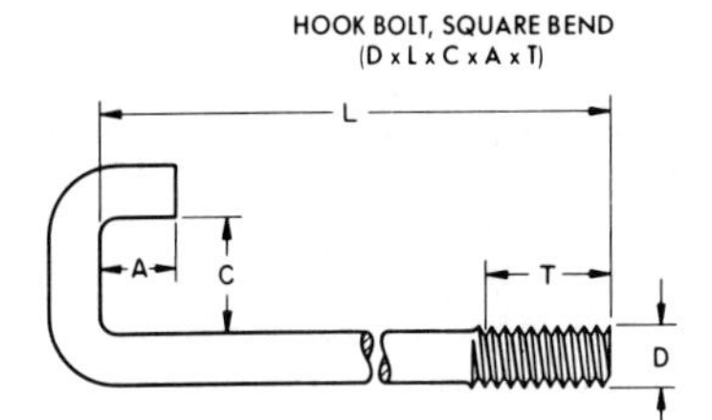

U-BOLT, ROUND BEND
(D x L x C x T)

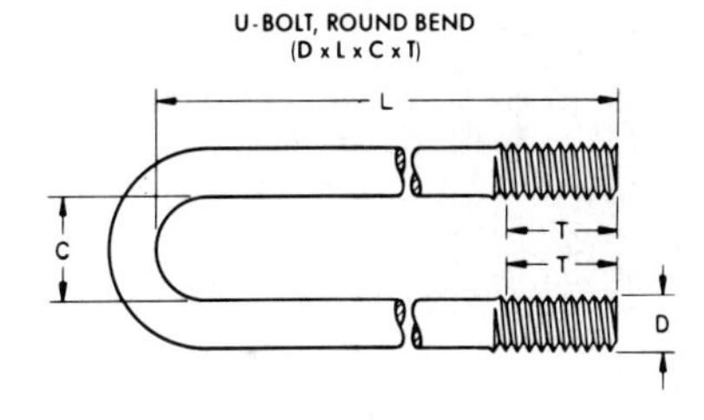

HOOK BOLT, RIGHT ANGLE BEND
(D x L x C x T)

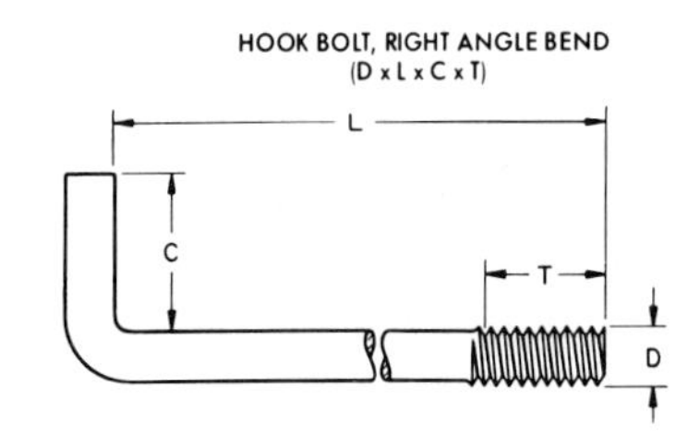

U-BOLT, SQUARE BEND
(D x L x C x T)

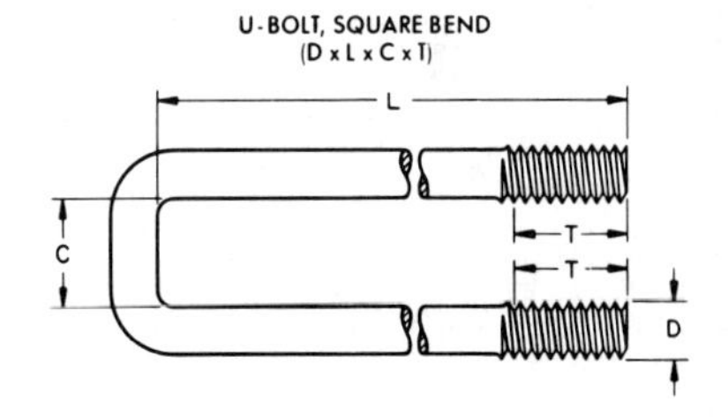

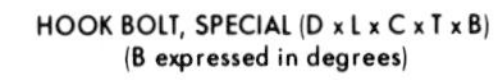
HOOK BOLT, SPECIAL (D x L x C x T x B)
(B expressed in degrees)

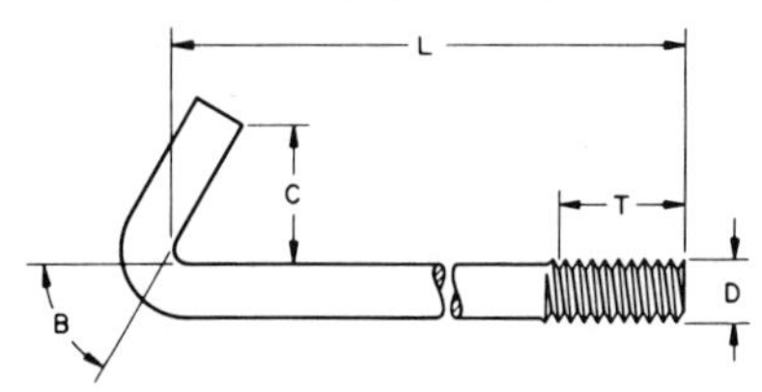

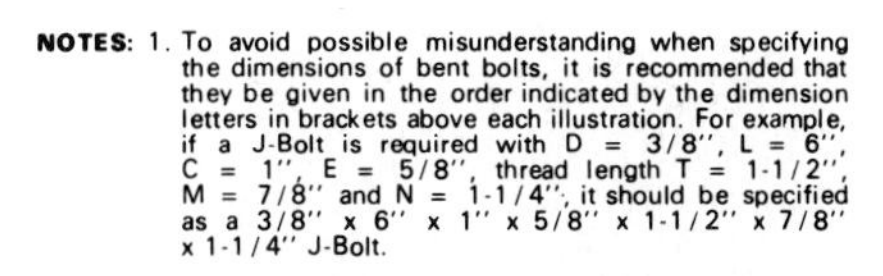
NOTES:

1. To avoid possible misunderstanding when specifying the dimensions of bent bolts, it is recommended that they be given in the order indicated by the dimension letters in brackets above each illustration. For example, if a J-Bolt is required with D = 3/8″, L = 6″, C = 1″, E = 5/8″, thread length T = 1-1/2″, M = 7/8″ and N = 1-1/4″, it should be specified as a 3/8″ x 6″ x 1″ x 5/8″ x 1-1/2″ x 7/8″ x 1-1/4″ J-Bolt.
2. Thread length is the distance from the extreme end of bolt to (and including) the last complete (full form) thread.
3. Threads may be cut or rolled; blank portion of rolled thread bolts may be slightly undersize, the nominal size of the bolt referring to the basic major diameter of the threaded portion.

SOURCE: Industrial Fasteners Institute.

TABLE 1-29 Dimensions of Square Nuts

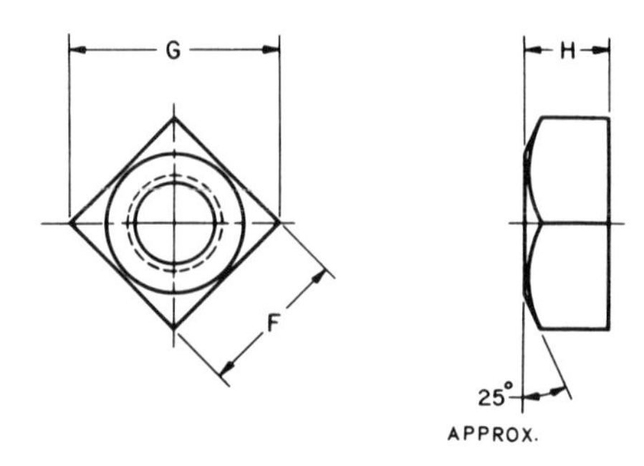

Nominal Size or Basic Major Dia of Thread		F Width Across Flats			G Width Across Corners		H Thickness		
		Basic	Max	Min	Max	Min	Basic	Max	Min
1/4	0.2500	7/16	0.438	0.425	0.619	0.584	7/32	0.235	0.203
5/16	0.3125	9/16	0.562	0.547	0.795	0.751	17/64	0.283	0.249
3/8	0.3750	5/8	0.625	0.606	0.884	0.832	21/64	0.346	0.310
7/16	0.4375	3/4	0.750	0.728	1.061	1.000	3/8	0.394	0.356
1/2	0.5000	13/16	0.812	0.788	1.149	1.082	7/16	0.458	0.418
5/8	0.6250	1	1.000	0.969	1.414	1.330	35/64	0.569	0.525
3/4	0.7500	1 1/8	1.125	1.088	1.591	1.494	21/32	0.680	0.632
7/8	0.8750	1 5/16	1.312	1.269	1.856	1.742	49/64	0.792	0.740
1	1.0000	1 1/2	1.500	1.450	2.121	1.991	7/8	0.903	0.847
1 1/8	1.1250	1 11/16	1.688	1.631	2.386	2.239	1	1.030	0.970
1 1/4	1.2500	1 7/8	1.875	1.812	2.652	2.489	1 3/32	1.126	1.062
1 3/8	1.3750	2 1/16	2.062	1.994	2.917	2.738	1 13/64	1.237	1.169
1 1/2	1.5000	2 1/4	2.250	2.175	3.182	2.986	1 5/16	1.348	1.276

SOURCE: ANSI B18.2.2-1972, as published by the American Society of Mechanical Engineers.

TABLE 1-30 Track Bolt Nuts

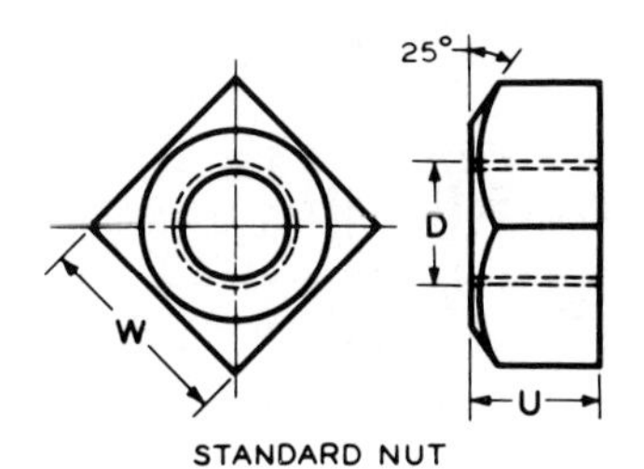

STANDARD NUT

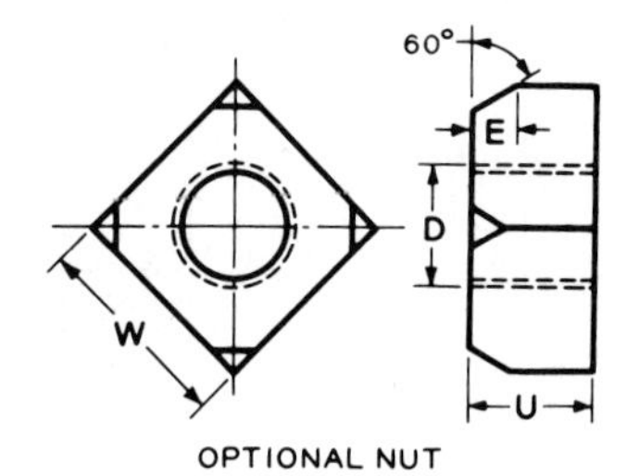

OPTIONAL NUT

Nominal Diameter	Width Across Flats W			Thickness U						Chamfer (Optional Nut only) Note (c)
				Recommended for Medium Carbon Nuts			Recommended for Low Carbon Nuts			
D	Nom	Max	Min	Nom	Max	Min	Nom	Max	Min	E
3/4	1- 1/4	1.2500	1.212	3/4	0.774	0.710	7/8	0.901	0.833	1/4
7/8	1- 7/16	1.4375	1.394	7/8	0.901	0.833	1	1.028	0.956	1/4
1	1- 5/8	1.6250	1.575	1	1.028	0.956	1-1/8	1.155	1.079	3/8
1-1/8	1-13/16	1.8125	1.756	1-1/8	1.155	1.079	1-1/4	1.282	1.187	1/2
ADDITIONAL SIZES – NOW IN USE BUT NOT RECOMMENDED FOR NEW DESIGN										
13/16	1- 1/4	1.2500	1.212	----	----	----	7/8	0.901	0.833	1/4
15/16	1- 1/2	1.5000	1.450	----	----	----	1-1/8	1.155	1.079	3/8
1	1- 1/2	1.5000	1.450	----	----	----	1-1/8	1.155	1.079	3/8
1- 1/16	1- 5/8	1.6250	1.575	----	----	----	1-1/8	1.155	1.079	3/8
1- 1/8	1-11/16	1.6875	1.631	----	----	----	1-1/4	1.282	1.187	1/2

Notes: (a) All dimensions are given in inches.
(b) 60 deg chamfer is optional when specified. (Dimensions for medium carbon nut are same as American Standard Heavy Nut, ASA B18.2 for sizes shown.)
(c) This dimension is not specified in ASA B18.2, which specifies diameter of top circle instead.

SOURCE: ANSI B18.10-1963, as published by the American Society of Mechanical Engineers.

TABLE 1-31 Dimensions of Hex Locknuts

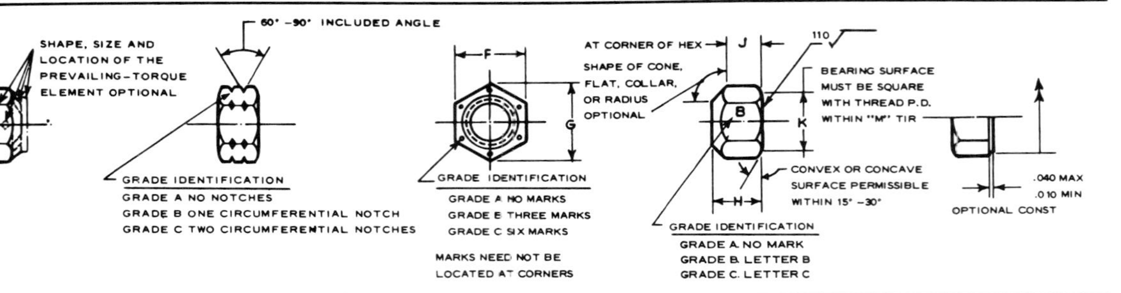

NOMINAL SIZE OR BASIC MAJOR DIA OF THREAD	WIDTH ACROSS FLATS F			WIDTH ACROSS CORNERS G		THICKNESS: INSERT TYPE LOCKNUT H	THICKNESS: ALL METAL TYPE LOCKNUT H	THICKNESS: ALL TYPES OF LOCKNUTS H	HEIGHT OF HEX J	DIA OF BEARING SURFACE K		ANGULARITY OF BEARING SURFACE TIR M
	Basic	Max	Min	Max	Min	Max	Max	Min	Min	Max	Min	Max
No. 4 0.1120	1/4	0.251	0.241	0.289	0.275	0.163	0.163	0.087	0.066	0.251	0.238	.008
6 0.1380	5/16	.313	.302	.361	.344	.188	.171	.102	.075	.313	.297	.008
8 0.1640	11/32	.345	.332	.397	.378	.239	.191	.117	.083	.345	.328	.009
10 0.1900	3/8	.376	.362	.433	.413	.249	.241	.117	.083	.376	.357	.009
12 0.2160	7/16	.438	.423	.505	.482	.328	.241	.148	.103	.438	.416	.010
1/4 0.2500	7/16	.4385	.428	.505	.488	.328	.288	.212	.145	.438	.416	.010
5/16 0.3125	1/2	.5020	.489	.577	.557	.359	.336	.258	.166	.502	.475	.011
3/8 0.3750	9/16	.5645	.551	.650	.628	.469	.415	.320	.198	.564	.534	.012
7/16 0.4375	11/16	.6895	.675	.794	.768	.524	.463	.365	.223	.689	.653	.013
1/2 0.5000	3/4	.7520	.736	.866	.840	.609	.573	.427	.262	.752	.712	.014
9/16 0.5625	7/8	.8770	.861	1.010	.982	.656	.621	.473	.286	.877	.830	.015
5/8 0.6250	15/16	.9395	.922	1.083	1.051	.765	.731	.535	.329	.939	.890	.016
3/4 0.7500	1-1/8	1.1270	1.088	1.299	1.240	.890	.827	.617	.382	1.127	1.069	.018
7/8 0.8750	1-5/16	1.3145	1.269	1.516	1.447	.999	.922	.724	.450	1.314	1.246	.020
1 1.0000	1-1/2	1.5020	1.450	1.732	1.653	1.124	1.018	.831	.513	1.502	1.425	.022
1-1/8 1.1250	1-11/16	1.6895	1.631	1.949	1.859	1.281	1.176	.939	.576	1.689	1.603	.025
1-1/4 1.2500	1-7/8	1.8770	1.812	2.165	2.066	1.422	1.272	1.030	.628	1.877	1.781	.028
1-3/8 1.3750	2-1/16	2.0645	1.994	2.382	2.273	1.609	1.399	1.138	.681	2.064	1.959	.031
1-1/2 1.5000	2-1/4	2.2520	2.175	2.598	2.480	1.671	1.526	1.245	.757	2.252	2.138	.034

1. All dimensions are in inches.
2. Except as noted dimensions apply to all grades of locknuts.
3. Tapped holes shall be countersunk on the bearing face. The maximum countersink diameter shall be the thread basic (nominal) major diameter plus .030 in. for 3/8 in. nuts or smaller, and 1.08 times the basic major diameter for nuts larger than 3/8 in. No part of the threaded portion shall project beyond the bearing surface.
4. Axis of tapped hole shall be concentric with axis of locknut body within a tolerance of 1.5 percent (3 percent TIR) of the maximum width across flats.

SOURCE: Industrial Fasteners Institute.

TABLE 1-32 Dimensions of Hex Nuts and Hex Jam Nuts

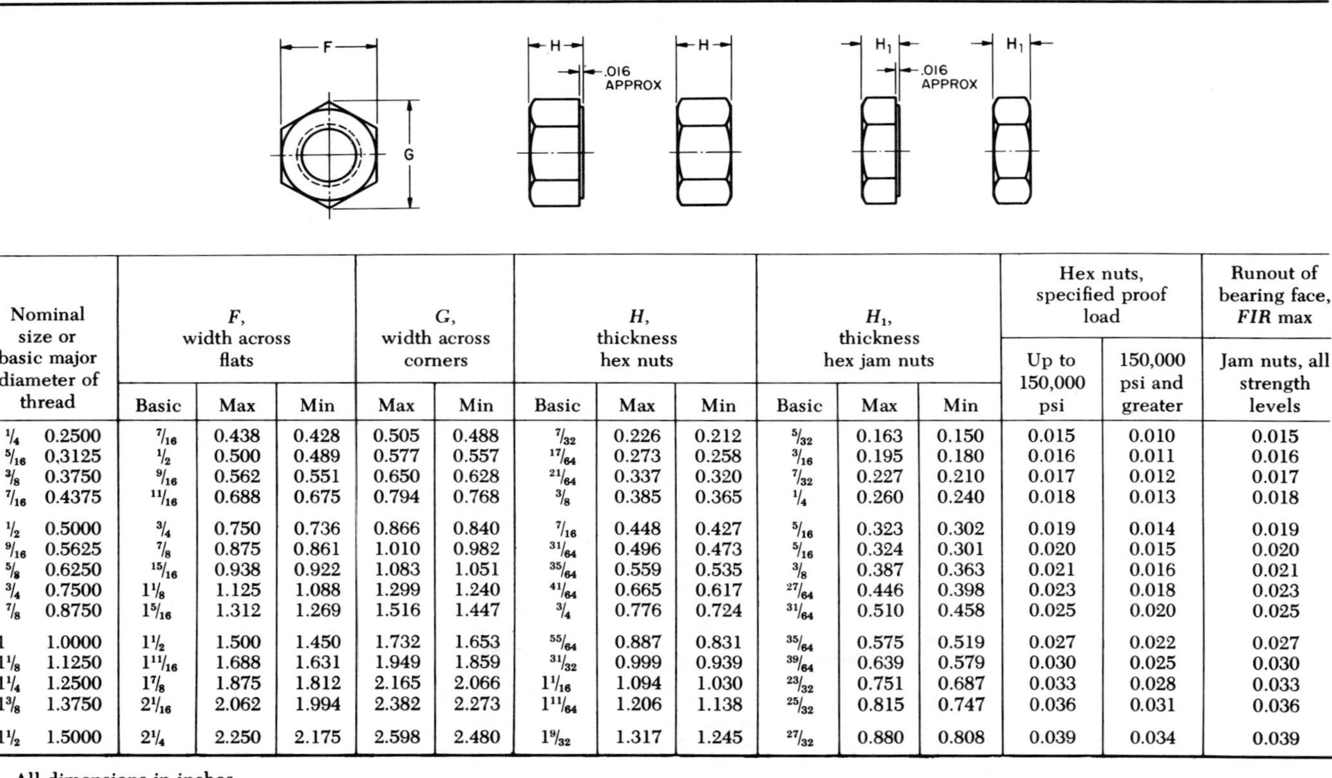

Nominal size or basic major diameter of thread	*F*, width across flats			*G*, width across corners		*H*, thickness hex nuts			H_1, thickness hex jam nuts			Hex nuts, specified proof load		Runout of bearing face, *FIR* max
	Basic	Max	Min	Max	Min	Basic	Max	Min	Basic	Max	Min	Up to 150,000 psi	150,000 psi and greater	Jam nuts, all strength levels
1/4 0.2500	7/16	0.438	0.428	0.505	0.488	7/32	0.226	0.212	5/32	0.163	0.150	0.015	0.010	0.015
5/16 0,3125	1/2	0.500	0.489	0.577	0.557	17/64	0.273	0.258	3/16	0.195	0.180	0.016	0.011	0.016
3/8 0.3750	9/16	0.562	0.551	0.650	0.628	21/64	0.337	0.320	7/32	0.227	0.210	0.017	0.012	0.017
7/16 0.4375	11/16	0.688	0.675	0.794	0.768	3/8	0.385	0.365	1/4	0.260	0.240	0.018	0.013	0.018
1/2 0.5000	3/4	0.750	0.736	0.866	0.840	7/16	0.448	0.427	5/16	0.323	0.302	0.019	0.014	0.019
9/16 0.5625	7/8	0.875	0.861	1.010	0.982	31/64	0.496	0.473	5/16	0.324	0.301	0.020	0.015	0.020
5/8 0.6250	15/16	0.938	0.922	1.083	1.051	35/64	0.559	0.535	3/8	0.387	0.363	0.021	0.016	0.021
3/4 0.7500	1 1/8	1.125	1.088	1.299	1.240	41/64	0.665	0.617	27/64	0.446	0.398	0.023	0.018	0.023
7/8 0.8750	1 5/16	1.312	1.269	1.516	1.447	3/4	0.776	0.724	31/64	0.510	0.458	0.025	0.020	0.025
1 1.0000	1 1/2	1.500	1.450	1.732	1.653	55/64	0.887	0.831	35/64	0.575	0.519	0.027	0.022	0.027
1 1/8 1.1250	1 11/16	1.688	1.631	1.949	1.859	31/32	0.999	0.939	39/64	0.639	0.579	0.030	0.025	0.030
1 1/4 1.2500	1 7/8	1.875	1.812	2.165	2.066	1 1/16	1.094	1.030	23/32	0.751	0.687	0.033	0.028	0.033
1 3/8 1.3750	2 1/16	2.062	1.994	2.382	2.273	1 11/64	1.206	1.138	25/32	0.815	0.747	0.036	0.031	0.036
1 1/2 1.5000	2 1/4	2.250	2.175	2.598	2.480	1 9/32	1.317	1.245	27/32	0.880	0.808	0.039	0.034	0.039

All dimensions in inches.
SOURCE: ANSI B18.2.2-1972, as published by the American Society of Mechanical Engineers.

TABLE 1-33 Dimensions of Heavy Hex Flat Nuts and Heavy Hex Flat Jam Nuts

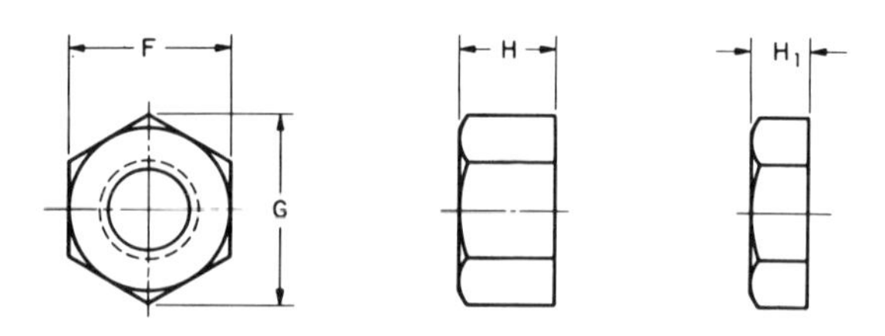

Nominal size or basic major diameter of thread		*F*, width across flats			*G*, width across corners		*H*, thickness, heavy hex flat nuts			H_1, thickness, heavy hex flat jam nuts		
		Basic	Max	Min	Max	Min	Basic	Max	Min	Basic	Max	Min
1 1/8	1.1250	1 13/16	1.812	1.756	2.093	2.002	1 1/8	1.155	1.079	5/8	0.655	0.579
1 1/4	1.2500	2	2.000	2.119	2.526	2.209	1 1/4	1.282	1.187	3/4	0.782	0.687
1 3/8	1.3750	2 3/16	2.188	2.119	2.526	2.416	1 3/8	1.409	1.310	13/16	0.846	0.747
1 1/2	1.5000	2 3/8	2.375	2.300	2.742	2.622	1 1/2	1.536	1.433	7/8	0.911	0.808
1 3/4	1.7500	2 3/4	2.750	2.662	3.175	3.035	1 3/4	1.790	1.679	1	1.040	0.929
2	2.0000	3 1/8	3.125	3.025	3.608	3.449	2	2.044	1.925	1 1/8	1.169	1.050
2 1/4	2.2500	3 1/2	3.500	3.388	4.041	3.862	2 1/4	2.298	2.155	1 1/4	1.298	1.155
2 1/2	2.5000	3 7/8	3.875	3.750	4.474	4.275	2 1/2	2.552	2.401	1 1/2	1.552	1.401
2 3/4	2.7500	4 1/4	4.250	4.112	4.907	4.688	2 3/4	2.806	2.647	1 5/8	1.681	1.522
3	3.0000	4 5/8	4.625	4.475	5.340	5.102	3	3.060	2.893	1 3/4	1.810	1.643
3 1/4	3.2500	5	5.000	4.838	5.774	5.515	3 1/4	3.314	3.124	1 7/8	1.939	1.748
3 1/2	3.5000	5 3/8	5.375	5.200	6.207	5.928	3 1/2	3.568	3.370	2	2.068	1.870
3 3/4	3.7500	5 3/4	5.750	5.562	6.640	6.341	3 3/4	3.822	3.616	2 1/8	2.197	1.990
4	4.0000	6 1/8	6.125	5.925	7.073	6.755	4	4.076	3.862	2 1/4	2.326	2.112

All dimensions in inches.
SOURCE: ANSI B18.2.2-1972, as published by the American Society of Mechanical Engineers.

TABLE 1-34 Dimensions of Hex Slotted Nuts

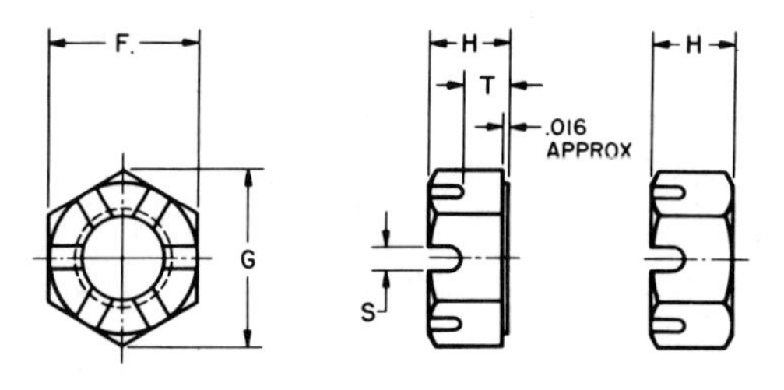

Nominal size or basic major dia. of thread	F, width across flats			G, width across corners		H, thickness			T, unslotted thickness		S, width of slot		Runout of bearing surface, *FIR* max
	Basic	Max	Min	Max	Min	Basic	Max	Min	Max	Min	Max	Min	
1/4 0.2500	7/16	0.438	0.428	0.505	0.488	7/32	0.226	0.212	0.14	0.12	0.10	0.07	0.015
5/16 0.3125	1/2	0.500	0.489	0.577	0.557	17/64	0.273	0.258	0.18	0.16	0.12	0.09	0.016
3/8 0.3750	9/16	0.562	0.551	0.650	0.628	21/64	0.337	0.320	0.21	0.19	0.15	0.12	0.017
7/16 0.4375	11/16	0.688	0.675	0.794	0.768	3/8	0.385	0.365	0.23	0.21	0.15	0.12	0.018
1/2 0.5000	3/4	0.750	0.736	0.866	0.840	7/16	0.448	0.427	0.29	0.27	0.18	0.15	0.019
9/16 0.5625	7/8	0.875	0.861	1.010	0.982	31/64	0.496	0.473	0.31	0.29	0.18	0.15	0.020
5/8 0.6250	15/16	0.938	0.922	1.083	1.051	35/64	0.559	0.535	0.34	0.32	0.24	0.18	0.021
3/4 0.7500	1 1/8	1.125	1.088	1.299	1.240	41/64	0.665	0.617	0.40	0.38	0.24	0.18	0.023
7/8 0.8750	1 5/16	1.312	1.269	1.516	1.447	3/4	0.776	0.724	0.52	0.49	0.24	0.18	0.025
1 1.0000	1 1/2	1.500	1.450	1.732	1.653	55/64	0.887	0.831	0.59	0.56	0.30	0.24	0.027
1 1/8 1.1250	1 11/16	1.688	1.631	1.949	1.859	31/32	0.999	0.939	0.64	0.61	0.33	0.24	0.030
1 1/4 1.2500	1 7/8	1.875	1.812	2.165	2.066	1 1/16	1.094	1.030	0.70	0.67	0.40	0.31	0.033
1 3/8 1.3750	2 1/16	2.062	1.994	2.382	2.273	1 11/64	1.206	1.138	0.82	0.78	0.40	0.31	0.036
1 1/2 1.5000	2 1/4	2.250	2.175	2.598	2.480	1 9/32	1.317	1.245	0.86	0.82	0.46	0.37	0.039

All dimensions in inches.
SOURCE: ANSI B18.2.2-1972, as published by the American Society of Mechanical Engineers.

TABLE 1-35 Dimensions of Hex Thick Slotted Nuts

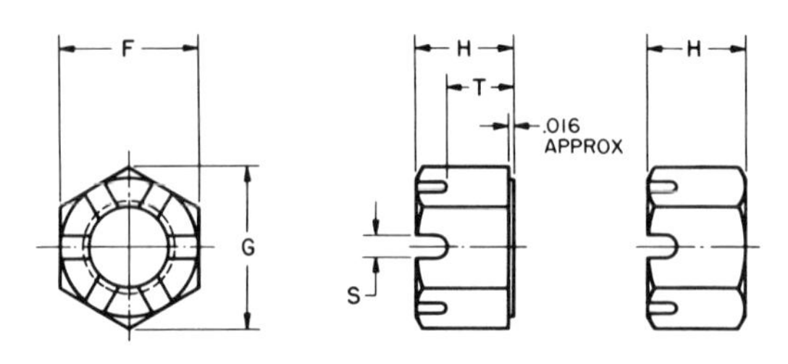

Nominal size or basic major dia. of thread	*F*, width across flats			*G*, width across corners		*H*, thickness			*T*, unslotted thickness		*S*, width of slot		Runout of bearing surface, *FIR* max
	Basic	Max	Min	Max	Min	Basic	Max	Min	Max	Min	Max	Min	
1/4 0.2500	7/16	0.438	0.428	0.505	0.488	9/32	0.288	0.274	0.20	0.18	0.10	0.07	0.015
5/16 0.3125	1/2	0.500	0.489	0.577	0.557	21/64	0.336	0.320	0.24	0.22	0.12	0.09	0.016
3/8 0.3750	9/16	0.562	0.551	0.650	0.628	13/32	0.415	0.398	0.29	0.27	0.15	0.12	0.017
7/16 0.4375	11/16	0.688	0.675	0.794	0.768	29/64	0.463	0.444	0.31	0.29	0.15	0.12	0.018
1/2 0.5000	3/4	0.750	0.736	0.866	0.840	9/16	0.573	0.552	0.42	0.40	0.18	0.15	0.019
9/16 0.5625	7/8	0.875	0.861	1.010	0.982	39/64	0.621	0.598	0.43	0.41	0.18	0.15	0.020
5/8 06250	15/16	0.938	0.922	1.083	1.051	23/32	0.731	0.706	0.51	0.49	0.24	0.18	0.021
3/4 0.7500	1 1/8	1.125	1.088	1.299	1.240	13/16	0.827	0.798	0.57	0.55	0.24	0.18	0.023
7/8 0.8750	1 5/16	1.312	1.269	1.516	1.447	29/32	0.922	0.890	0.67	0.64	0.24	0.18	0.025
1 1.0000	1 1/2	1.500	1.450	1.732	1.653	1	1.018	0.982	0.73	0.70	0.30	0.24	0.027
1 1/8 1.1250	1 11/16	1.688	1.631	1.949	1.859	1 5/32	1.176	1.136	0.83	0.80	0.33	0.24	0.030
1 1/4 1.2500	1 7/8	1.875	1.812	2.165	2.066	1 1/4	1.272	1.228	0.89	0.86	0.40	0.31	0.033
1 3/8 1.3750	2 1/16	2.062	1.994	2.382	2.273	1 3/8	1.399	1.351	1.02	0.98	0.40	0.31	0.036
1 1/2 1.5000	2 1/4	2.250	2.175	2.598	2.480	1 1/2	1.526	1.474	1.08	1.04	0.46	0.37	0.039

All dimensions in inches.
SOURCE: ANSI B18.2.2-1972, as published by the American Society of Mechanical Engineers.

TABLE 1-36 Dimensions of Hex Castle Nuts

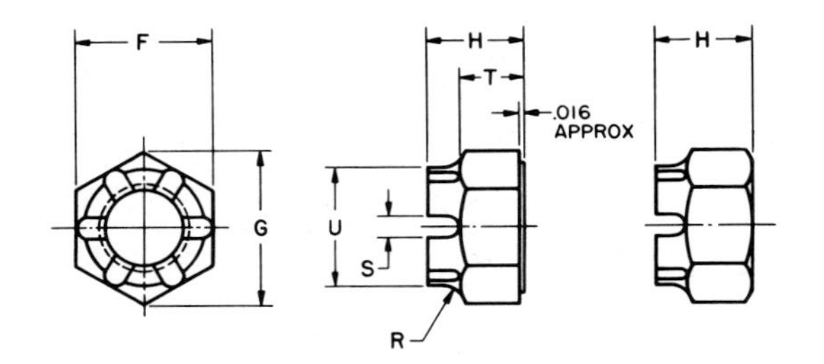

Nominal size or basic major dia. of thread		F, width across flats			G, width across corners		H, thickness			T, unslotted thickness and height of flats			S, width of slot		R, radius of fillet, ±0.010	U, dia. of cylindrical part, min	Runout of bearing surface, *FIR* max
		Basic	Max	Min	Max	Min	Basic	Max	Min	Nom	Max	Min	Max	Min			
1/4	0.2500	7/16	0.438	0.428	0.505	0.488	9/32	0.288	0.274	3/16	0.20	0.18	0.10	0.07	0.094	0.371	0.015
5/16	0.3125	1/2	0.500	0.489	0.577	0.557	21/64	0.336	0.320	15/64	0.24	0.22	0.12	0.09	0.094	0.425	0.016
3/8	0.3750	9/16	0.562	0.551	0.650	0.628	13/32	0.415	0.398	9/32	0.29	0.27	0.15	0.12	0.094	0.478	0.017
7/16	0.4375	11/16	0.688	0.675	0.794	0.768	29/64	0.463	0.444	19/64	0.31	0.29	0.15	0.12	0.094	0.582	0.018
1/2	0.5000	3/4	0.750	0.736	0.866	0.840	9/16	0.573	0.552	13/32	0.42	0.40	0.18	0.15	0.125	0.637	0.019
9/16	0.5625	7/8	0.875	0.861	1.010	0.982	39/64	0.621	0.598	27/64	0.43	0.41	0.18	0.15	0.156	0.744	0.020
5/8	0.6250	15/16	0.938	0.922	1.083	1.051	23/32	0.731	0.706	1/2	0.51	0.49	0.24	0.18	0.156	0.797	0.021
3/4	0.7500	1 1/8	1.125	1.088	1.299	1.240	13/16	0.827	0.798	9/16	0.57	0.55	0.24	0.18	0.188	0.941	0.023
7/8	0.8750	1 5/16	1.312	1.269	1.516	1.447	29/32	0.922	0.890	21/32	0.67	0.64	0.24	0.18	0.188	1.097	0.025
1	1.0000	1 1/2	1.500	1.450	1.732	1.653	1	1.018	0.982	23/32	0.73	0.70	0.30	0.24	0.188	1.254	0.027
1 1/8	1.1250	1 11/16	1.688	1.631	1.949	1.859	1 5/32	1.176	1.136	13/16	0.83	0.80	0.33	0.24	0.250	1.411	0.030
1 1/4	1.2500	1 7/8	1.875	1.812	2.165	2.066	1 1/4	1.272	1.228	7/8	0.89	0.86	0.40	0.31	0.250	1,570	0.033
1 3/8	1.3750	2 1/16	2.062	1.994	2.382	2.273	1 3/8	1.399	1.351	1	1.02	0.98	0.40	0.31	0.250	1.726	0.036
1 1/2	1.5000	2 1/4	2.250	2.175	2.598	2.480	1 1/2	1.526	1.474	1 1/16	1.08	1.04	0.46	0.37	0.250	1.881	0.039

All dimensions in inches.
SOURCE: ANSI B18.2.2-1972, as published by the American Society of Mechanical Engineers.

TABLE 1-37 Dimensions of Type A Wing Nuts

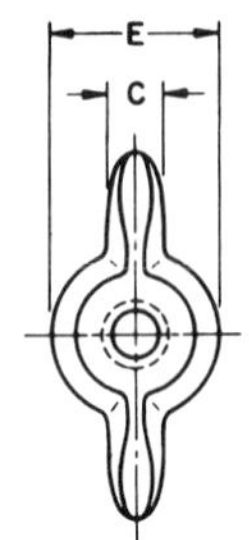

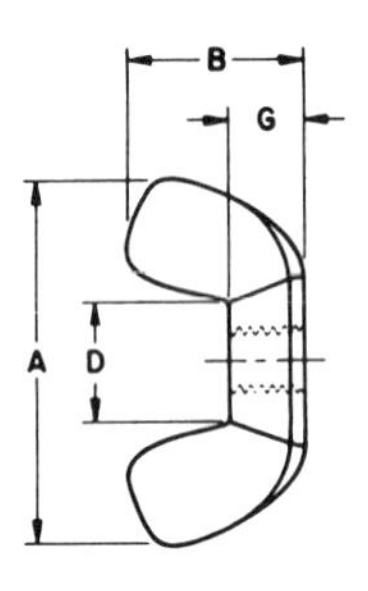

Nominal Size or Basic Major Diameter of Thread[1]	Threads per Inch	Series	Nut Blank Size (Ref)	A		B		C		D		E		G	
				Wing Spread		Wing Height		Wing Thickness		Between Wings		Boss Diameter		Boss Height	
				MAX	MIN	MAX	MIN	MAX	MIN	MAX	MIN	MAX	MIN	MAX	MIN
3 (0.0990)	48 & 56	Heavy	AA	0.72	0.59	0.41	0.28	0.11	0.07	0.21	0.17	0.33	0.29	0.14	0.10
4 (0.1120)	40 & 38	Heavy	AA	0.72	0.59	0.41	0.28	0.11	0.07	0.21	0.17	0.33	0.29	0.14	0.10
5 (0.1250)	40 & 44	**Light**	**AA**	**0.72**	**0.59**	**0.41**	**0.28**	**0.11**	**0.07**	**0.21**	**0.17**	**0.33**	**0.29**	**0.14**	**0.10**
		Heavy	A	0.91	0.78	0.47	0.34	0.14	0.10	0.27	0.22	0.43	0.39	0.18	0.14
6 (0.1380)	32 & 40	Light	AA	0.72	0.59	0.41	0.28	0.11	0.07	0.21	0.17	0.33	0.29	0.14	0.10
		Heavy	**A**	**0.91**	**0.78**	**0.47**	**0.34**	**0.14**	**0.10**	**0.27**	**0.22**	**0.43**	**0.39**	**0.18**	**0.14**
8 (0.1640)	32 & 36	**Light**	**A**	**0.91**	**0.78**	**0.47**	**0.34**	**0.14**	**0.10**	**0.27**	**0.22**	**0.43**	**0.39**	**0.18**	**0.14**
		Heavy	B	1.10	0.97	0.57	0.43	0.18	0.14	0.33	0.26	0.50	0.45	0.22	0.17
10 (0.1900)	24 & 32	**Light**	**A**	**0.91**	**0.78**	**0.47**	**0.34**	**0.14**	**0.10**	**0.27**	**0.22**	**0.43**	**0.39**	**0.18**	**0.14**
		Heavy	B	1.10	0.97	0.57	0.43	0.18	0.14	0.33	0.26	0.50	0.45	0.22	0.17
12 (0.2160)	24 & 28	**Light**	**B**	**1.10**	**0.97**	**0.57**	**0.43**	**0.18**	**0.14**	**0.33**	**0.26**	**0.50**	**0.45**	**0.22**	**0.17**
		Heavy	C	1.25	1.12	0.66	0.53	0.21	0.17	0.39	0.32	0.58	0.51	0.25	0.20
1/4 (0.2500)	20 & 28	**Light**	**B**	**1.10**	**0.97**	**0.57**	**0.43**	**0.18**	**0.14**	**0.39**	**0.26**	**0.50**	**0.45**	**0.22**	**0.17**
		Regular	C	1.25	1.12	0.66	0.53	0.21	0.17	0.39	0.32	0.58	0.51	0.25	0.20
		Heavy	D	1.44	1.31	0.79	0.65	0.24	0.20	0.48	0.42	0.70	0.64	0.30	0.26
5/16 (0.3125)	18 & 24	**Light**	**C**	**1.25**	**1.12**	**0.66**	**0.53**	**0.21**	**0.17**	**0.39**	**0.32**	**0.58**	**0.51**	**0.25**	**0.20**
		Regular	D	1.44	1.31	0.79	0.65	0.24	0.20	0.48	0.42	0.70	0.64	0.30	0.26
		Heavy	E	1.94	1.81	1.00	0.87	0.33	0.26	0.65	0.54	0.93	0.86	0.39	0.35
3/8 (0.3750)	16 & 24	**Light**	**D**	**1.44**	**1.31**	**0.79**	**0.65**	**0.24**	**0.20**	**0.48**	**0.42**	**0.70**	**0.64**	**0.30**	**0.26**
		Regular	E	1.94	1.81	1.00	0.87	0.33	0.26	0.65	0.54	0.93	0.86	0.39	0.35
7/16 (0.4375)	14 & 20	**Light**	**E**	**1.94**	**1.81**	**1.00**	**0.87**	**0.33**	**0.26**	**0.65**	**0.54**	**0.93**	**0.86**	**0.39**	**0.35**
		Heavy	F	2.76	2.62	1.44	1.31	0.40	0.34	0.90	0.80	1.19	1.13	0.55	0.51
1/2 (0.5000)	13 & 20	**Light**	**E**	**1.94**	**1.81**	**1.00**	**0.87**	**0.33**	**0.26**	**0.65**	**0.54**	**0.93**	**0.86**	**0.39**	**0.35**
		Heavy	F	2.76	2.62	1.44	1.31	0.40	0.34	0.90	0.80	1.19	1.13	0.55	0.51
9/16 (0.5825)	12 & 18	Heavy	F	2.76	2.62	1.44	1.31	0.40	0.34	0.90	0.80	1.19	1.13	0.55	0.51
5/8 (0.6250)	11 & 18	Heavy	F	2.76	2.62	1.44	1.31	0.40	0.34	0.90	0.80	1.19	1.13	0.55	0.51
3/4 (0.7500)	10 & 16	Heavy	F	2.76	2.62	1.44	1.31	0.40	0.34	0.90	0.80	1.19	1.13	0.55	0.51

[1] Where specifying nominal size in decimals, zeros in the fourth decimal place shall be omitted.

SOURCE: ANSI B18.17-1968, as published by the American Society of Mechanical Engineers.

TABLE 1-38 Dimensions of Type D, Style 2 and 3 Wing Nuts

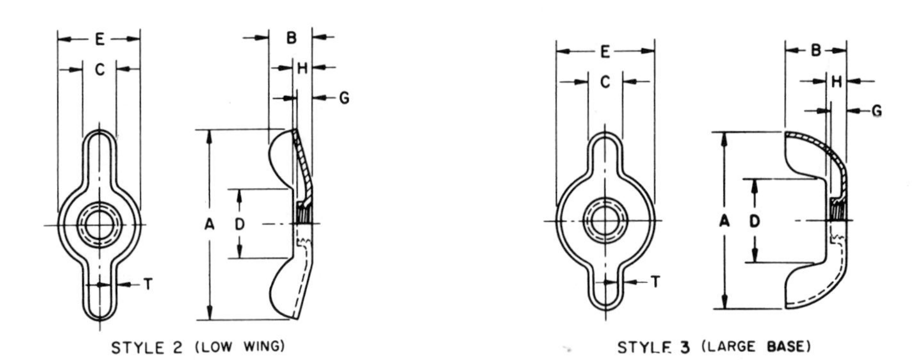

STYLE 2 (LOW WING) STYLE 3 (LARGE BASE)

Dimensions of Type D, Style 2 Wing Nuts

Nominal Size or Basic Major Diameter of Thread[1]	Threads per Inch	Series	Nut Blank Size (Ref)	A Wing Spread		B Wing Height		C Wing Thickness		D Between Wings	E Boss Diameter		G Boss Height	H Wall Height	T Stock Thickness	
				MAX	MIN	MAX	MIN	MAX	MIN	MIN	MAX	MIN	MIN	MIN	MAX	MIN
5 (0.1250)	40	Regular	A	1.03	0.97	0.25	0.19	0.19	0.13	0.30	0.40	0.34	0.07	0.09	0.04	0.03
6 (0.1380)	32	Regular	A	1.03	0.97	0.25	0.19	0.19	0.13	0.30	0.40	0.34	0.08	0.09	0.04	0.03
8 (0.1640)	32	Regular	A	1.03	0.97	0.25	0.19	0.19	0.13	0.30	0.40	0.34	0.08	0.09	0.04	0.03
10 (0.1900)	24 & 32	Regular	B	1.40	1.34	0.34	0.28	0.25	0.18	0.32	0.53	0.47	0.09	0.16	0.05	0.04
		Heavy	C	1.21	1.16	0.28	0.26	0.31	0.25	0.60	0.61	0.55	0.09	0.13	0.05	0.04
12 (0.2160)	24	Regular	C	1.21	1.16	0.28	0.26	0.31	0.25	0.60	0.61	0.55	0.11	0.13	0.05	0.04
1/4 (0.2500)	20	Regular	C	1.21	1.16	0.28	0.26	0.31	0.25	0.60	0.61	0.55	0.11	0.13	0.05	0.04

[1] Where specifying nominal size in decimals, zeros in the fourth decimal place shall be omitted.

Dimensions of Type D, Style 3 Wing Nuts

Nominal Size or Basic Major Diameter of Thread[1]	Threads per Inch	Series	Nut Blank Size (Ref)	A Wing Spread		B Wing Height		C Wing Thickness		D Between Wings	E Boss Diameter		G Boss Height	H Wall Height	T Stock Thickness	
				MAX	MIN	MAX	MIN	MAX	MIN	MIN	MAX	MIN	MIN	MIN	MAX	MIN
10 (0.1900)	24 & 32	Light	A	1.31	1.25	0.48	0.42	0.29	0.23	0.47	0.65	0.59	0.08	0.12	0.04	0.03
		Regular	C	1.40	1.34	0.53	0.47	0.25	0.19	0.50	0.75	0.69	0.08	0.14	0.04	0.03
12 (0.2160)	24	Regular	B	1.28	1.22	0.40	0.34	0.23	0.17	0.59	0.73	0.67	0.11	0.12	0.04	0.03
1/4 (0.2500)	20	Light	B	1.28	1.22	0.40	0.34	0.23	0.17	0.59	0.73	0.67	0.11	0.12	0.04	0.03
		Regular	E	1.78	1.72	0.66	0.60	0.31	0.25	0.70	1.03	0.97	0.14	0.17	0.06	0.04
		Heavy	D	1.47	1.40	0.50	0.44	0.37	0.31	0.66	1.03	0.97	0.14	0.14	0.08	0.06
5/16 (0.3125)	18	Regular	E	1.78	1.72	0.66	0.60	0.31	0.25	0.70	1.03	0.97	0.14	0.17	0.06	0.04
		Heavy	D	1.47	1.40	0.50	0.44	0.37	0.31	0.66	1.03	0.97	0.14	0.14	0.08	0.06

[1] Where specifying nominal size in decimals, zeros in the fourth decimal place shall be omitted.

SOURCE: ANSI B18.17-1968, as published by the American Society of Mechanical Engineers.

TABLE 1-39 Dimensions of Hex Cap Screws (Finished Hex Bolts)

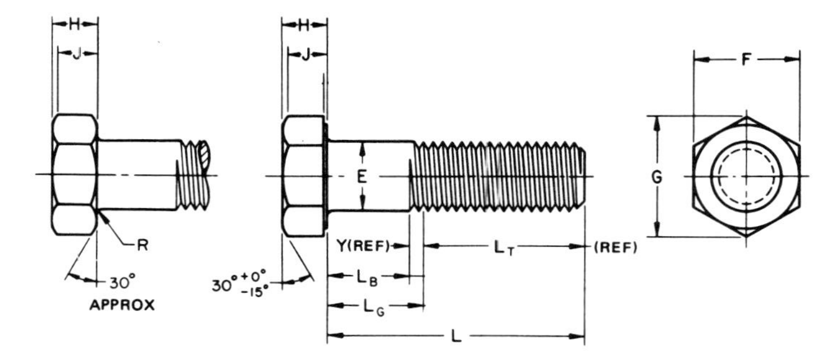

Nominal size or basic product dia.	*E*, Body dia.		*F*, Width across flats			*G*, Width across corners		*H*, Height			*J*, Wrench-ing height	L_T Thread length for screw lengths		*Y*, Transition thread length for screw lengths		Runout of bearing surface, *FIR* max
	Max	Min	Basic	Max	Min	Max	Min	Basic	Max	Min	min	6 in. and shorter, basic	Over 6 in., basic	6 in. and shorter, max	Over 6 in., max	
1/4 0.2500	0.2500	0.2450	7/16	0.438	0.428	0.505	0.488	5/32	0.163	0.150	0.106	0.750	1.000	0.400	0.650	0.010
5/16 0.3125	0.3125	0.3065	1/2	0.500	0.489	0.577	0.557	13/64	0.211	0.195	0.140	0.875	1.125	0.417	0.667	0.011
3/8 0.3750	0.3750	0.3690	9/16	0.562	0.551	0.650	0.628	15/64	0.243	0.226	0.160	1.000	1.250	0.438	0.688	0.012
7/16 0.4375	0.4375	0.4305	5/8	0.625	0.612	0.722	0.698	9/32	0.291	0.272	0.195	1.125	1.375	0.464	0.714	0.013
1/2 0.5000	0.5000	0.4930	3/4	0.750	0.736	0.866	0.840	5/16	0.323	0.302	0.215	1.250	1.500	0.481	0.731	0.014
9/16 0.5625	0.5625	0.5545	13/16	0.812	0.798	0.938	0.910	23/64	0.371	0.348	0.250	1.375	1.625	0.750	0.750	0.015
5/3 0.6250	0.6250	0.6170	15/16	0.938	0.922	1.083	1.051	25/64	0.403	0.378	0.269	1.500	1.750	0.773	0.773	0.17
3/4 0.7500	0.7500	0.7410	1 1/8	1.125	1.100	1.299	1.254	15/32	0.483	0.455	0.324	1.750	2.000	0.800	0.800	0.020
7/8 0.8750	0.8750	0.8660	1 5/16	1.312	1.285	1.516	1.465	35/64	0.563	0.531	0.378	2.000	2.250	0.833	0.833	0.023

1	1.0000	1.0000	0.9900	1 1/2	1.500	1.469	1.732	1.675	39/64	0.627	0.591	0.416	2.250	2.500	0.875	0.875	0.026
1 1/8	1.1250	1.1250	1.1140	1 11/16	1.688	1.631	1.949	1.859	11/16	0.718	0.658	0.461	2.500	2.750	0.929	0.929	0.029
1 1/4	1.2500	1.2500	1.2390	1 7/8	1.875	1.812	2.165	2.066	25/32	0.813	0.749	0.530	2.750	3.000	0.929	0.929	0.033
1 3/8	1.3750	1.3750	1.3630	2 1/16	2.062	1.994	2.382	2.273	27/32	0.878	0.810	0.569	3.000	3.250	1.000	1.000	0.036
1 1/2	1.5000	1.5000	1.4880	2 1/4	2.250	2.175	2.598	2.480	15/16	0.974	0.902	0.640	3.250	3.500	1.000	1.000	0.039
1 3/4	1.7500	1.7500	1.7380	2 5/8	2.625	2.538	3.031	2.893	1 3/32	1.134	1.054	0.748	3.750	4.000	1.100	1.100	0.046
2	2.0000	2.0000	1.9880	3	3.000	2.900	3.464	3.306	1 7/32	1.263	1.175	0.825	4.250	4.500	1.167	1.167	0.052
2 1/4	2.2500	2.2500	2.2380	3 3/8	3.375	3.262	3.897	3.719	1 3/8	1.423	1.327	0.933	4.750	5.000	1.167	1.167	0.059
2 1/2	2.5000	2.5000	2.4880	3 3/4	3.750	3.625	4.330	4.133	1 17/32	1.583	1.479	1.042	5.250	5.500	1.250	1.250	0.065
2 3/4	2.7500	2.7500	2.7380	4 1/8	4.125	3.988	4.763	4.546	1 11/16	1.744	1.632	1.151	5.750	6.000	1.250	1.250	0.072
3	3.0000	3.0000	2.9880	4 1/2	4.500	4.350	5.196	4.959	1 7/8	1.935	1.815	1.290	6.250	6.500	1.250	1.250	0.079

All dimensions in inches.
SOURCE: ANSI B18.2.1-1972, as published by the American Society of Mechanical Engineers.

TABLE 1-40 Dimensions of Square Head Set Screws

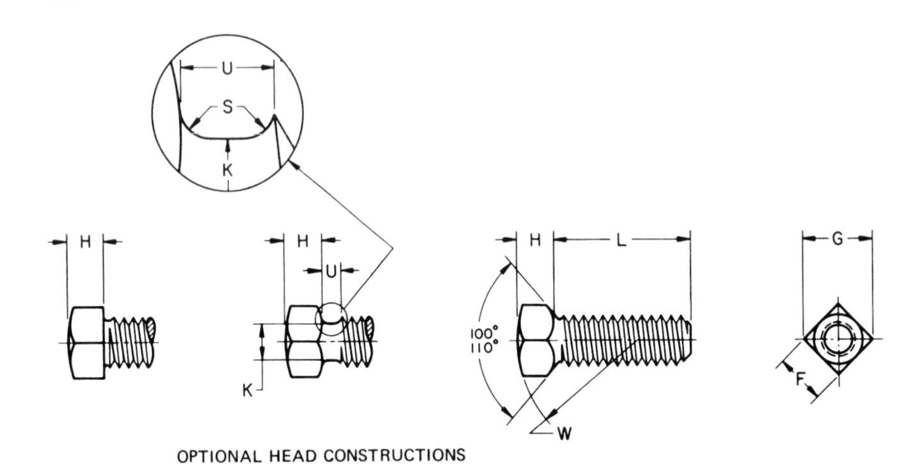

Dimensions of Square Head Set Screws

Nominal Size[1] or Basic Screw Diameter		F		G		H		K		S	U	W
		Width Across Flats		Width Across Corners		Head Height		Neck Relief Diameter		Neck Relief Fillet Radius	Neck Relief Width	Head Radius
		Max	Min	Max	Min	Max	Min	Max	Min	Max	Min	Min
10	0.1900	0.188	0.180	0.265	0.247	0.148	0.134	0.145	0.140	0.027	0.083	0.48
1/4	0.2500	0.250	0.241	0.354	0.331	0.196	0.178	0.185	0.170	0.032	0.100	0.62
5/16	0.3125	0.312	0.302	0.442	0.415	0.245	0.224	0.240	0.225	0.036	0.111	0.78
3/8	0.3750	0.375	0.362	0.530	0.497	0.293	0.270	0.294	0.279	0.041	0.125	0.94
7/16	0.4375	0.438	0.423	0.619	0.581	0.341	0.315	0.345	0.330	0.046	0.143	1.09
1/2	0.5000	0.500	0.484	0.707	0.665	0.389	0.361	0.400	0.385	0.050	0.154	1.25
9/16	0.5625	0.562	0.545	0.795	0.748	0.437	0.407	0.454	0.439	0.054	0.167	1.41
5/8	0.6250	0.625	0.606	0.884	0.833	0.485	0.452	0.507	0.492	0.059	0.182	1.56
3/4	0.7500	0.750	0.729	1.060	1.001	0.582	0.544	0.620	0.605	0.065	0.200	1.88
7/8	0.8750	0.875	0.852	1.237	1.170	0.678	0.635	0.731	0.716	0.072	0.222	2.19
1	1.0000	1.000	0.974	1.414	1.337	0.774	0.726	0.838	0.823	0.081	0.250	2.50
1 1/8	1.1250	1.125	1.096	1.591	1.505	0.870	0.817	0.939	0.914	0.092	0.283	2.81
1 1/4	1.2500	1.250	1.219	1.768	1.674	0.966	0.908	1.064	1.039	0.092	0.283	3.12
1 3/8	1.3750	1.375	1.342	1.945	1.843	1.063	1.000	1.159	1.134	0.109	0.333	3.44
1 1/2	1.5000	1.500	1.464	2.121	2.010	1.159	1.091	1.284	1.259	0.109	0.333	3.75

[1] Where specifying nominal size in decimals, zeros preceding decimal and in the fourth decimal place shall be omitted.

SOURCE: ANSI B18.6.2-1972, as published by the American Society of Mechanical Engineers.

TABLE 1-41 Dimensions of Square Head Set Screws

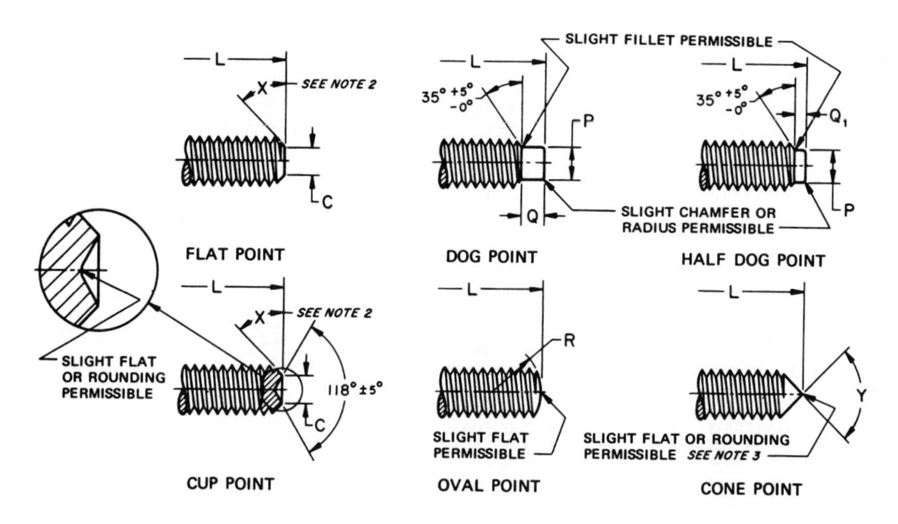

Dimensions of Square Head Set Screws

Nominal Size[1] or Basic Screw Diameter	C		P		Q		Q_1		R	Y
	Cup and Flat Point Diameters		Dog and Half Dog Point Diameters		Point Length				Oval Point Radius	Cone Point Angle 90° ±2° For These Nominal Lengths or Longer; 118° ±2° For Shorter Screws
					Dog		Half Dog		+0.031 -0.000	
	Max	Min	Max	Min	Max	Min	Max	Min		
10 0.1900	0.102	0.088	0.127	0.120	0.095	0.085	0.050	0.040	0.142	1/4
1/4 0.2500	0.132	0.118	0.156	0.149	0.130	0.120	0.068	0.058	0.188	5/16
5/16 0.3125	0.172	0.156	0.203	0.195	0.161	0.151	0.083	0.073	0.234	3/8
3/8 0.3750	0.212	0.194	0.250	0.241	0.193	0.183	0.099	0.089	0.281	7/16
7/16 0.4375	0.252	0.232	0.297	0.287	0.224	0.214	0.114	0.104	0.328	1/2
1/2 0.5000	0.291	0.270	0.344	0.334	0.255	0.245	0.130	0.120	0.375	9/16
9/16 0.5625	0.332	0.309	0.391	0.379	0.287	0.275	0.146	0.134	0.422	5/8
5/8 0.6250	0.371	0.347	0.469	0.456	0.321	0.305	0.164	0.148	0.469	3/4
3/4 0.7500	0.450	0.425	0.562	0.549	0.383	0.367	0.196	0.180	0.562	7/8
7/8 0.8750	0.530	0.502	0.656	0.642	0.446	0.430	0.227	0.211	0.656	1
1 1.0000	0.609	0.579	0.750	0.734	0.510	0.490	0.260	0.240	0.750	1 1/8
1 1/8 1.1250	0.689	0.655	0.844	0.826	0.572	0.552	0.291	0.271	0.844	1 1/4
1 1/4 1.2500	0.767	0.733	0.938	0.920	0.635	0.615	0.323	0.303	0.938	1 1/2
1 3/8 1.3750	0.848	0.808	1.031	1.011	0.698	0.678	0.354	0.334	1.031	1 5/8
1 1/2 1.5000	0.926	0.886	1.125	1.105	0.760	0.740	0.385	0.365	1.125	1 3/4

[1] Where specifying nominal size in decimals, zeros preceding decimal and in the fourth decimal place shall be omitted.
[2] Point angle X shall be 45° plus 5°, minus 0°, for screws of nominal lengths equal to or longer than those listed in Column Y, and 30° minimum for screws of shorter nominal lengths.
[3] The extent of rounding or flat at apex of cone point shall not exceed an amount equivalent to 10 per cent of the basic screw diameter.

SOURCE: ANSI B18.6.2-1972, as published by the American Society of Mechanical Engineers.

TABLE 1-42 Dimensions of Slotted Fillister Cap Screws

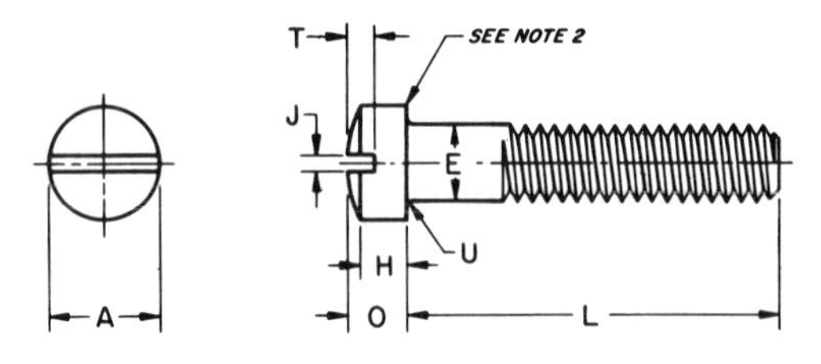

Nominal Size[1] or Basic Screw Diameter	E Body Diameter		A Head Diameter		H Head Side Height		O Total Head Height		J Slot Width		T Slot Depth		U Fillet Radius	
	Max	Min	Max	Min	Max	Min	Max	Min	Max	Min	Max	Min	Max	Min
1/4 0.2500	0.2500	0.2450	0.375	0.363	0.172	0.157	0.216	0.194	0.075	0.064	0.097	0.077	0.031	0.016
5/16 0.3125	0.3125	0.3070	0.437	0.424	0.203	0.186	0.253	0.230	0.084	0.072	0.115	0.090	0.031	0.016
3/8 0.3750	0.3750	0.3690	0.562	0.547	0.250	0.229	0.314	0.284	0.094	0.081	0.142	0.112	0.031	0.016
7/16 0.4375	0.4375	0.4310	0.625	0.608	0.297	0.274	0.368	0.336	0.094	0.081	0.168	0.133	0.047	0.016
1/2 0.5000	0.5000	0.4930	0.750	0.731	0.328	0.301	0.413	0.376	0.106	0.091	0.193	0.153	0.047	0.016
9/16 0.5625	0.5625	0.5550	0.812	0.792	0.375	0.346	0.467	0.427	0.118	0.102	0.213	0.168	0.047	0.016
5/8 0.6250	0.6250	0.6170	0.875	0.853	0.422	0.391	0.521	0.478	0.133	0.116	0.239	0.189	0.062	0.031
3/4 0.7500	0.7500	0.7420	1.000	0.976	0.500	0.466	0.612	0.566	0.149	0.131	0.283	0.223	0.062	0.031
7/8 0.8750	0.8750	0.8660	1.125	1.098	0.594	0.556	0.720	0.668	0.167	0.147	0.334	0.264	0.062	0.031
1 1.0000	1.0000	0.9900	1.312	1.282	0.656	0.612	0.803	0.743	0.188	0.166	0.371	0.291	0.062	0.031

[1] Where specifying nominal size in decimals, zeros preceding decimal and in the fourth decimal place shall be omitted.

[2] A slight rounding of the edges at periphery of head shall be permissible provided the diameter of the bearing circle is equal to no less than 90 per cent of the specified minimum head diameter.

SOURCE: ANSI B18.6.2-1972, as published by the American Society of Mechanical Engineers.

TABLE 1-43 Dimensions of Type A Regular and Heavy Thumb Screws

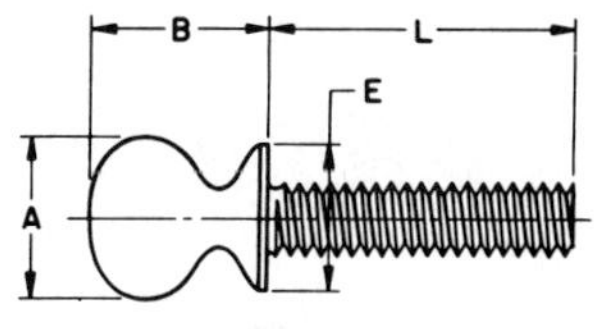
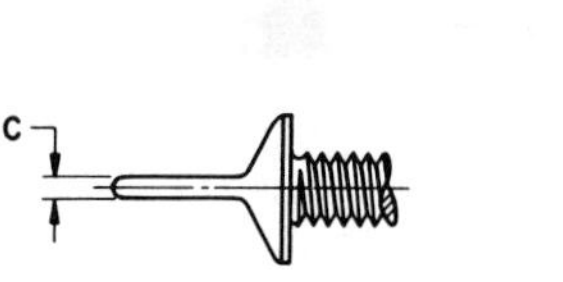
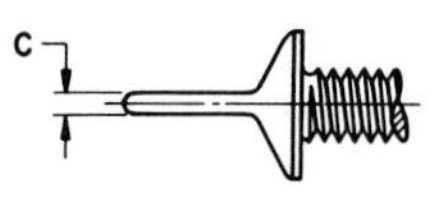

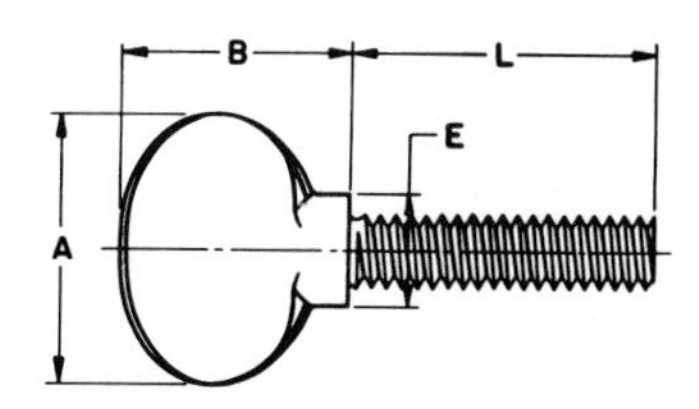
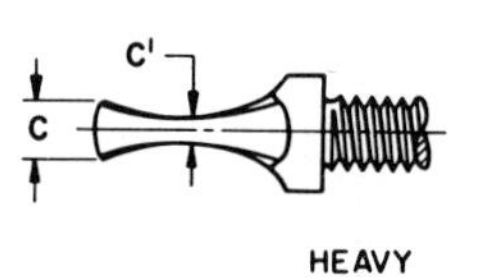

Dimensions of Type A, Regular Thumb Screws

Nominal Size or Basic Screw Diameter[2]	Threads per Inch	A Head Width		B Head Height		C Head Thickness		E Shoulder Diameter		L Practical Screw Lengths	
		MAX	MIN	MAX	MIN	MAX	MIN	MAX	MIN	MAX	MIN
6 (0.1380)	32	0.31	0.29	0.33	0.31	0.05	0.04	0.25	0.23	0.75	0.25
8 (0.1640)	32	0.36	0.34	0.38	0.36	0.06	0.05	0.31	0.29	0.75	0.38
10 (0.1900)	24 & 32	0.42	0.40	0.48	0.46	0.06	0.05	0.35	0.32	1.00	0.38
12 (0.2160)	24	0.48	0.46	0.54	0.52	0.06	0.05	0.40	0.38	1.00	0.38
1/4 (0.2500)	20	0.55	0.52	0.64	0.61	0.07	0.05	0.47	0.44	1.50	0.50
5/16 (0.3125)	18	0.70	0.67	0.78	0.75	0.09	0.07	0.59	0.56	1.50	0.50
3/8 (0.3750)	16	0.83	0.80	0.95	0.92	0.11	0.09	0.76	0.71	2.00	0.75

Dimensions of Type A, Heavy Thumb Screws

Nominal Size or Basic Screw Diameter[2]	Threads per Inch	A Head Width		B Head Height		C Head Thickness		C[1] Head Thickness		E Shoulder Diameter		L Practical Screw Lengths	
		MAX	MIN	MAX	MIN	MAX	MIN	MAX	MIN	MAX	MIN	MAX	MIN
10 (0.1900)	24	0.89	0.83	0.84	0.72	0.18	0.16	0.10	0.08	0.33	0.31	2.00	0.50
1/4 (0.2500)	20	1.05	0.99	0.94	0.81	0.24	0.22	0.10	0.08	0.40	0.38	3.00	0.50
5/16 (0.3125)	18	1.21	1.15	1.00	0.88	0.27	0.25	0.11	0.09	0.46	0.44	4.00	0.50
3/8 (0.3750)	16	1.41	1.34	1.16	1.03	0.30	0.28	0.11	0.09	0.55	0.53	4.00	0.50
7/16 (0.4375)	14	1.59	1.53	1.22	1.09	0.36	0.34	0.13	0.11	0.71	0.69	2.50	1.00
1/2 (0.5000)	13	1.81	1.72	1.28	1.16	0.40	0.38	0.14	0.12	0.83	0.81	3.00	1.00

SOURCE: ANSI B18.17-1968, as published by the American Society of Mechanical Engineers.

TABLE 1-44 Dimensions of Hexagon and Spline Socket-Head Cap Screws (1960 Series)

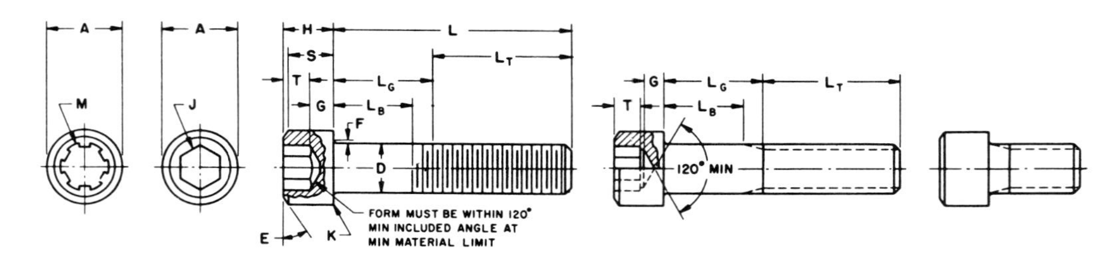

Nominal size or basic screw diameter		*D*, body diameter		*A*, head diameter		*H*, head height		*S*, head side height, min	*M*, spline socket size, nom	*J*, hexagon socket size, nom		*T*, key engagement, min	*G*, wall thickness, min	*F*, fillet extension		*K*, chamfer or radius, max
		Max	Min	Max	Min	Max	Min							Max	Min	
0	0.0600	0.0600	0.0568	0.096	0.091	0.060	0.057	0.054	0.060		0.050	0.025	0.020	0.007	0.003	0.003
1	0.0730	0.0730	0.0695	0.118	0.112	0.073	0.070	0.066	0.072	1/16	0.062	0.031	0.025	0.007	0.003	0.003
2	0.0860	0.0860	0.0822	0.140	0.134	0.086	0.083	0.077	0.096	5/64	0.078	0.038	0.029	0.008	0.004	0.003
3	0.0990	0.0990	0.0949	0.161	0.154	0.099	0.095	0.089	0.096	5/64	0.078	0.044	0.034	0.008	0.004	0.003
4	0.1120	0.1120	0.1075	0.183	0.176	0.112	0.108	0.101	0.111	3/32	0.094	0.051	0.038	0.009	0.005	0.005
5	0.1250	0.1250	0.1202	0.205	0.198	0.125	0.121	0.112	0.111	3/32	0.094	0.057	0.043	0.010	0.006	0.005
6	0.1380	0.1380	0.1329	0.226	0.218	0.138	0.134	0.124	0.133	7/64	0.109	0.064	0.047	0.010	0.006	0.005
8	0.1640	0.1640	0.1585	0.270	0.262	0.164	0.159	0.148	0.168	9/64	0.141	0.077	0.056	0.012	0.007	0.005
10	0.1900	0.1900	0.1840	0.312	0.303	0.190	0.185	0.171	0.183	5/32	0.156	0.090	0.065	0.014	0.009	0.005
1/4	0.2500	0.2500	0.2435	0.375	0.365	0.250	0.244	0.225	0.216	3/16	0.188	0.120	0.095	0.014	0.009	0.008
5/16	0.3125	0.3125	0.3053	0.469	0.457	0.312	0.306	0.281	0.291	1/4	0.250	0.151	0.119	0.017	0.012	0.008
3/8	0.3750	0.3750	0.3678	0.562	0.550	0.375	0.368	0.337	0.372	5/16	0.312	0.182	0.143	0.020	0.015	0.008
7/16	0.4375	0.4375	0.4294	0.656	0.642	0.438	0.430	0.394	0.454	3/8	0.375	0.213	0.166	0.023	0.018	0.010
1/2	0.5000	0.5000	0.4919	0.750	0.735	0.500	0.492	0.450	0.454	3/8	0.375	0.245	0.190	0.026	0.020	0.010
5/8	0.6250	0.6250	0.6163	0.938	0.921	0.625	0.616	0.562	0.595	1/2	0.500	0.307	0.238	0.032	0.024	0.010
3/4	0.7500	0.7500	0.7406	1.125	1.107	0.750	0.740	0.675	0.620	5/8	0.625	0.370	0.285	0.039	0.030	0.010

7/8	0.8750	0.8750	0.8647	1.312	1.293	0.875	0.864	0.787	0.698	3/4	0.750	0.432	0.333	0.044	0.034	0.015
1	1.0000	1.0000	0.9886	1.500	1.479	1.000	0.988	0.900	0.790	3/4	0.750	0.495	0.380	0.050	0.040	0.015
1 1/8	1.1250	1.1250	1.1086	1.688	1.665	1.125	1.111	1.012		7/8	0.875	0.557	0.428	0.055	0.045	0.015
1 1/4	1.2500	1.2500	1.2336	1.875	1.852	1.250	1.236	1.125		7/8	0.875	0.620	0.475	0.060	0.050	0.015
1 3/8	1.3750	1.3750	1.3568	2.062	2.038	1.375	1.360	1.237		1	1.000	0.682	0.523	0.065	0.055	0.015
1 1/2	1.5000	1.5000	1.4818	2.250	2.224	1.500	1.485	1.350		1	1.000	0.745	0.570	0.070	0.060	0.015
1 3/4	1.7500	1.7500	1.7295	2.625	2.597	1.750	1.734	1.575		1 1/4	1.250	0.870	0.665	0.080	0.070	0.015
2	2.0000	2.0000	1.9780	3.000	2.970	2.000	1.983	1.800		1 1/2	1.500	0.995	0.760	0.090	0.075	0.015
2 1/4	2.2500	2.2500	2.2280	3.375	3.344	2.250	2.232	2.025		1 3/4	1.750	1.120	0.855	0.100	0.085	0.031
2 1/2	2.5000	2.5000	2.4762	3.750	3.717	2.500	2.481	2.250		1 3/4	1.750	1.245	0.950	0.110	0.095	0.031
2 3/4	2.7500	2.7500	2.7262	4.125	4.090	2.750	2.730	2.475		2	2.000	1.370	1.045	0.120	0.105	0.031
3	3.0000	3.0000	2.9762	4.500	4.464	3.000	2.979	2.700		2 1/4	2.250	1.495	1.140	0.130	0.115	0.031
3 1/4	3.2500	3.2500	3.2262	4.875	4.837	3.250	3.228	2.925		2 1/4	2.250	1.620	1.235	0.140	0.125	0.031
3 1/2	3.5000	3.5000	3.4762	5.250	5.211	3.500	3.478	3.150		2 3/4	2.750	1.745	1.330	0.150	0.135	0.031
3 3/4	3.7500	3.7500	3.7262	5.625	5.584	3.750	3.727	3.375		2 3/4	2.750	1.870	1.425	0.160	0.145	0.031
4	4.0000	4.0000	3.9762	6.000	5.958	4.000	3.976	3.600		3	3.000	1.995	1.520	0.170	0.155	0.031

All dimensions in inches.
SOURCE: ANSI B18.3-1969, as published by the American Society of Mechanical Engineers.

TABLE 1-45 Dimensions of Hexagon and Spline Socket Set Screws

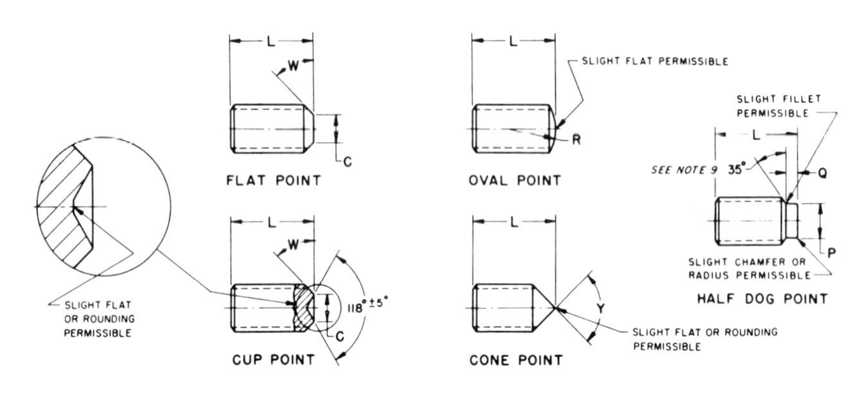

Nominal size or basic screw diameter		Half dog point				B, Shortest optimum nominal length to which column T_H applies			B_1, Shortest optimum nominal length to which column T_S applies		
		P, diameter		Q, length		Cup and flat points	90° cone and oval points	Half dog point	Cup and flat points	90° cone and oval points	Half dog point
		Max	Min	Max	Min						
0	0.0600	0.040	0.037	0.017	0.013	7/16	1/8	7/64	1/16	1/8	7/64
1	0.0730	0.049	0.045	0.021	0.017	1/8	9/64	1/8	3/32	9/64	1/8
2	0.0860	0.057	0.053	0.024	0.020	1/8	9/64	9/64	3/32	9/64	9/64
3	0.0990	0.066	0.062	0.027	0.023	9/64	5/32	5/32	3/32	5/32	5/32
4	0.1120	0.075	0.070	0.030	0.026	9/64	11/64	5/32	3/32	11/64	5/32
5	0.1250	0.083	0.078	0.033	0.027	3/16	3/16	11/64	1/8	3/16	11/64
6	0.1380	0.092	0.087	0.038	0.032	11/64	13/64	3/16	1/8	13/64	3/16
8	0.1640	0.109	0.103	0.043	0.037	3/16	7/32	13/64	3/16	7/32	13/64
10	0.1900	0.127	0.120	0.049	0.041	3/16	1/4	15/64	3/16	1/4	15/64
1/4	0.2500	0.156	0.149	0.067	0.059	1/4	5/16	19/64	1/4	5/16	19/64
5/16	0.3125	0.203	0.195	0.082	0.074	5/16	25/64	23/64	5/16	25/64	23/64
3/8	0.3750	0.250	0.241	0.099	0.089	3/8	7/16	7/16	3/8	7/16	7/16
7/16	0.4375	0.297	0.287	0.114	0.104	7/16	35/64	31/64	7/16	35/64	31/64
1/2	0.5000	0.344	0.334	0.130	0.120	1/2	39/64	35/64	1/2	39/64	35/64
5/8	0.6250	0.469	0.456	0.164	0.148	5/8	49/64	43/64	5/8	49/64	43/64
3/4	0.7500	0.562	0.549	0.196	0.180	3/4	29/32	51/64	3/4	29/32	51/64
7/8	0.8750	0.656	0.642	0.227	0.211	7/8	1 1/8	63/64	7/8	1 1/8	63/64
1	1.0000	0.750	0.734	0.260	0.240	1	1 17/64	1 1/8			
1 1/8	1.1250	0.844	0.826	0.291	0.271	1 1/8	1 25/64	1 3/16			
1 1/4	1.2500	0.938	0.920	0.323	0.303	1 1/4	1 1/2	1 5/16			
1 3/8	1.3750	1.031	1.011	0.354	0.334	1 3/8	1 21/32	1 7/16			
1 1/2	1.5000	1.125	1.105	0.385	0.365	1 1/2	1 51/64	1 9/16			
1 3/4	1.7500	1.312	1.289	0.448	0.428	1 3/4	2 7/32	1 61/64			
2	2.0000	1.500	1.474	0.510	0.490	2	2 25/64	2 5/64			

All dimensions in inches.

SOURCE: ANSI B18.3-1969, as published by the American Society of Mechanical Engineers.

TABLE 1-46 Optional Types of Cup Points

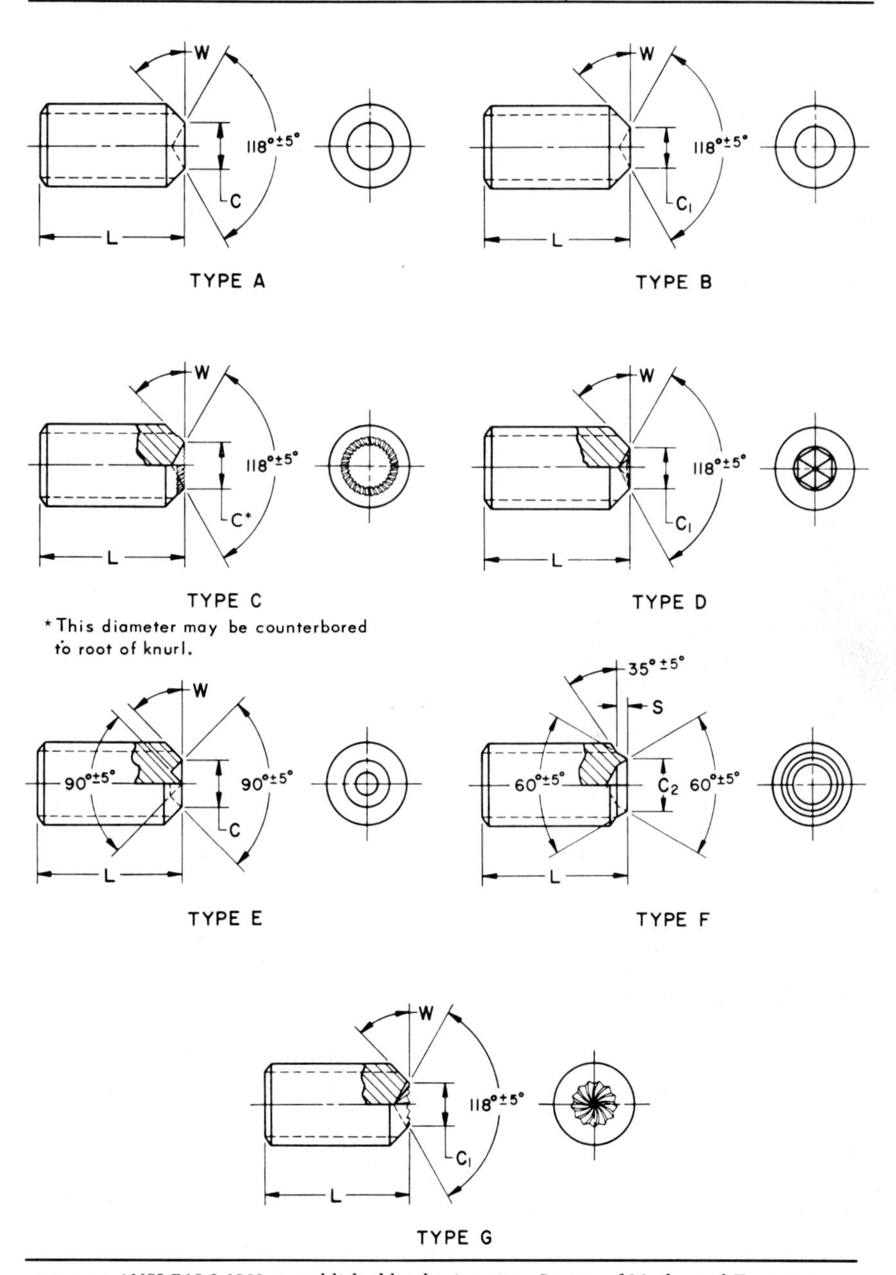

SOURCE: ANSI B18.3-1969, as published by the American Society of Mechanical Engineers.

TABLE 1-47 Dimensions of Spline Sockets

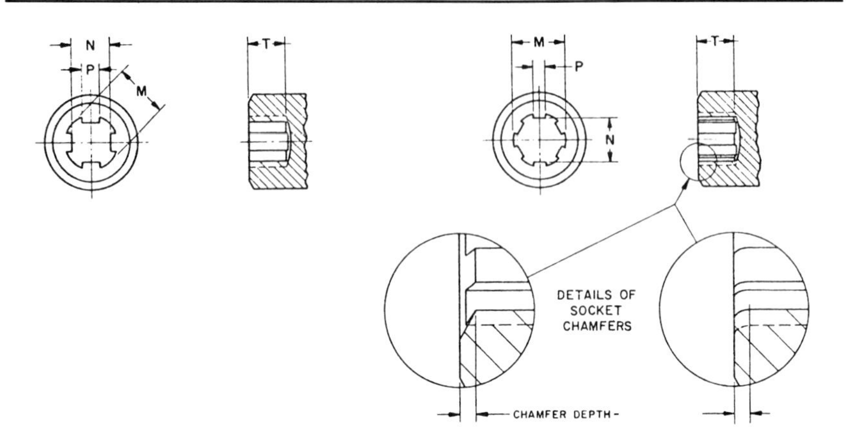

Nominal socket and key size	Number of teeth	*M*, socket major diameter		*N*, socket minor diameter		*P*, width of tooth	
		Max	Min	Max	Min	Max	Min
0.033	4	0.0350	0.0340	0.0260	0.0255	0.0120	0.0115
0.048	6	0.050	0.049	0.041	0.040	0.011	0.010
0.060	6	0.062	0.061	0.051	0.050	0.014	0.013
0.072	6	0.074	0.073	0.064	0.063	0.016	0.015
0.096	6	0.098	0.097	0.082	0.080	0.022	0.021
0.111	6	0.115	0.113	0.098	0.096	0.025	0.023
0.133	6	0.137	0.135	0.118	0.116	0.030	0.028
0.145	6	0.149	0.147	0.128	0.126	0.032	0.030
0.168	6	0.173	0.171	0.150	0.147	0.036	0.033
0.183	6	0.188	0.186	0.163	0.161	0.039	0.037
0.216	6	0.221	0.219	0.190	0.188	0.050	0.048
0.251	6	0.256	0.254	0.221	0.219	0.060	0.058
0.291	6	0.298	0.296	0.254	0.252	0.068	0.066
0.372	6	0.380	0.377	0.319	0.316	0.092	0.089
0.454	6	0.463	0.460	0.386	0.383	0.112	0.109
0.595	6	0.604	0.601	0.509	0.506	0.138	0.134
0.620	6	0.631	0.627	0.535	0.531	0.149	0.145
0.698	6	0.709	0.705	0.604	0.600	0.168	0.164
0.790	6	0.801	0.797	0.685	0.681	0.189	0.185

All dimensions in inches.
SOURCE: ANSI B18.3-1969, as published by the American Society of Mechanical Engineers.

TABLE 1-48 Dimensions of Hexagon Keys and Bits

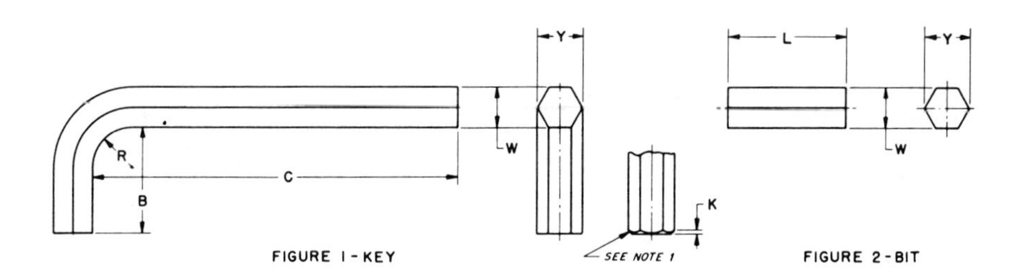

FIGURE 1 - KEY

FIGURE 2 - BIT

Dimensions of Hexagon Keys and Bits

Nominal Key or Bit and Socket Size		W Hexagon Width Across Flats		Y Hexagon Width Across Corners		B Length of Short Arm		C Length of Long Arm				R Radius of Bend	L Length of Bit	K Chamfer
								Short Series		Long Series				
		Max	Min	Max	Min	Max	Min	Max	Min	Max	Min	Min	±0.062	Max
	0.028	0.0280	0.0275	0.0314	0.0300	0.312	0.125	1.312	1.125	2.688	2.500	0.062	...	0.003
	0.035	0.0350	0.0345	0.0393	0.0378	0.438	0.250	1.312	1.125	2.766	2.578	0.062	...	0.004
	0.050	0.0500	0.0490	0.0560	0.0540	0.625	0.438	1.750	1.562	2.938	2.750	0.062	...	0.006
1/16	0.062	0.0625	0.0615	0.0701	0.0680	0.656	0.469	1.844	1.656	3.094	2.906	0.062	...	0.008
5/64	0.078	0.0781	0.0771	0.0880	0.0859	0.703	0.516	1.969	1.781	3.281	3.094	0.078	...	0.008
3/32	0.094	0.0937	0.0927	0.1058	0.1035	0.750	0.562	2.094	1.906	3.469	3.281	0.094	...	0.009
7/64	0.109	0.1094	0.1079	0.1238	0.1210	0.797	0.609	2.219	2.031	3.656	3.469	0.109	...	0.014
1/8	0.125	0.1250	0.1235	0.1418	0.1390	0.844	0.656	2.344	2.156	3.844	3.656	0.125	...	0.015
9/64	0.141	0.1406	0.1391	0.1593	0.1566	0.891	0.703	2.469	2.281	4.031	3.844	0.141	...	0.016
5/32	0.156	0.1562	0.1547	0.1774	0.1745	0.938	0.750	2.594	2.406	4.219	4.031	0.156	...	0.016
3/16	0.188	0.1875	0.1860	0.2135	0.2105	1.031	0.844	2.844	2.656	4.594	4.406	0.188	...	0.022
7/32	0.219	0.2187	0.2172	0.2490	0.2460	1.125	0.938	3.094	2.906	4.969	4.781	0.219	...	0.024
1/4	0.250	0.2500	0.2485	0.2845	0.2815	1.219	1.031	3.344	3.156	5.344	5.156	0.250	...	0.030
5/16	0.312	0.3125	0.3110	0.3570	0.3531	1.344	1.156	3.844	3.656	6.094	5.906	0.312	...	0.032
3/8	0.375	0.3750	0.3735	0.4285	0.4238	1.469	1.281	4.344	4.156	6.844	6.656	0.375	...	0.044
7/16	0.438	0.4375	0.4355	0.5005	0.4944	1.594	1.406	4.844	4.656	7.594	7.406	0.438	...	0.047
1/2	0.500	0.5000	0.4975	0.5715	0.5650	1.719	1.531	5.344	5.156	8.344	8.156	0.500	...	0.050
9/16	0.562	0.5625	0.5600	0.6420	0.6356	1.844	1.656	5.844	5.656	9.094	8.906	0.562	...	0.053
5/8	0.625	0.6250	0.6225	0.7146	0.7080	1.969	1.781	6.344	6.156	9.844	9.656	0.625	...	0.055
3/4	0.750	0.7500	0.7470	0.8580	0.8512	2.219	2.031	7.344	7.156	11.344	11.156	0.750	...	0.070
7/8	0.875	0.8750	0.8720	1.0020	0.9931	2.469	2.281	8.344	8.156	12.844	12.656	0.875	...	0.076
1	1.000	1.0000	0.9970	1.1470	1.1350	2.719	2.531	9.344	9.156	14.344	14.156	1.000	...	0.081
1 1/4	1.250	1.2500	1.2430	...	...	3.250	2.750	11.500	11.000	...	...	1.250	3.750	0.092
1 1/2	1.500	1.5000	1.4930	...	...	3.750	3.250	13.500	13.000	...	...	1.500	4.500	0.104
1 3/4	1.750	1.7500	1.7430	...	...	4.250	3.750	15.500	15.000	...	...	1.750	5.250	0.115
2	2.000	2.0000	1.9930	...	...	4.750	4.250	17.500	17.000	...	...	2.000	6.000	0.126
2 1/4	2.250	2.2500	2.2430	...	...	5.250	4.750	19.500	19.000	...	...	2.250	6.750	0.137
2 3/4	2.750	2.7500	2.7420	...	...	6.250	5.750	23.500	23.000	...	...	2.750	8.250	0.159
3	3.000	3.0000	2.9920	...	...	6.750	6.250	25.500	25.000	...	...	3.000	9.000	0.171

SOURCE: ANSI B18.3-1969, as published by the American Society of Mechanical Engineers.

TABLE 1-49 Dimensions of Spline Keys

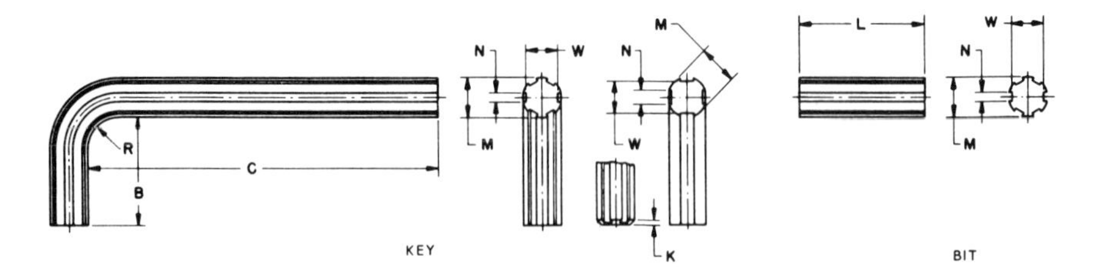

Dimensions of Spline Keys

Nominal Key and Socket Size	M		W		Number of Splines	N		B		C				R	K
	Major Diameter		Minor Diameter			Width of Space		Length of Short Arm		Length of Long Arm				Radius of Bend	Chamfer
										Short Series		Long Series			
	Max	Min	Max	Min		Max	Min	Max	Min	Max	Min	Max	Min	Min	Max
0.033	0.0330	0.0320	0.0250	0.0240	4	0.0140	0.0130	0.312	0.125	1.312	1.125	...	...	0.062	0.003
0.048	0.0480	0.0470	0.0390	0.0380	6	0.0130	0.0120	0.438	0.250	1.312	1.125	...	...	0.062	0.004
0.060	0.0600	0.0590	0.0490	0.0480	6	0.0160	0.0150	0.625	0.438	1.750	1.562	...	...	0.062	0.006
0.072	0.0720	0.0710	0.0620	0.0610	6	0.0190	0.0180	0.656	0.469	1.844	1.656	...	...	0.062	0.008
0.096	0.0960	0.0950	0.0790	0.0775	6	0.0240	0.0230	0.703	0.516	1.969	1.781	...	...	0.078	0.008
0.111	0.1110	0.1100	0.0940	0.0925	6	0.0280	0.0270	0.750	0.562	2.094	1.906	...	...	0.094	0.009
0.133	0.1330	0.1310	0.1140	0.1120	6	0.0340	0.0320	0.797	0.609	2.219	2.031	3.656	3.469	0.125	0.014
0.145	0.1450	0.1435	0.1240	0.1225	6	0.0355	0.0340	0.844	0.656	2.344	2.156	3.844	3.656	0.125	0.015
0.168	0.1680	0.1660	0.1440	0.1420	6	0.0410	0.0390	0.891	0.703	2.469	2.281	4.031	3.844	0.156	0.016
0.183	0.1830	0.1815	0.1580	0.1565	6	0.0440	0.0425	0.938	0.750	2.594	2.406	4.219	4.031	0.156	0.016
0.216	0.2160	0.2145	0.1840	0.1825	6	0.0550	0.0535	1.031	0.844	2.844	2.656	4.594	4.406	0.188	0.022
0.251	0.2510	0.2495	0.2140	0.2125	6	0.0655	0.0640	1.125	0.938	3.094	2.906	4.969	4.781	0.219	0.024
0.291	0.2910	0.2895	0.2460	0.2445	6	0.0775	0.0760	1.219	1.031	3.344	3.156	5.344	5.156	0.250	0.030
0.372	0.3720	0.3705	0.3100	0.3085	6	0.0975	0.0960	1.344	1.156	3.844	3.656	6.094	5.906	0.312	0.032
0.454	0.4540	0.4525	0.3770	0.3755	6	0.1185	0.1170	1.469	1.281	4.344	4.156	6.844	6.656	0.375	0.044
0.595	0.5950	0.5935	0.5000	0.4975	6	0.1460	0.1445	1.719	1.531	5.344	5.156	8.344	8.156	0.500	0.050
0.620	0.6200	0.6175	0.5240	0.5215	6	0.1615	0.1590	1.844	1.656	5.844	5.656	9.094	8.906	0.500	0.053
0.698	0.6980	0.6955	0.5930	0.5905	6	0.1805	0.1780	1.844	1.656	5.844	5.656	...	...	0.562	0.055
0.790	0.7900	0.7875	0.6740	0.6715	6	0.1975	0.1950	1.969	1.781	6.344	6.156	...	...	0.625	0.070

SOURCE: ANSI B18.3-1969, as published by the American Society of Mechanical Engineers.

TABLE 1-50 Dimensions of Hexagon and Spline Socket Button-head Cap Screws

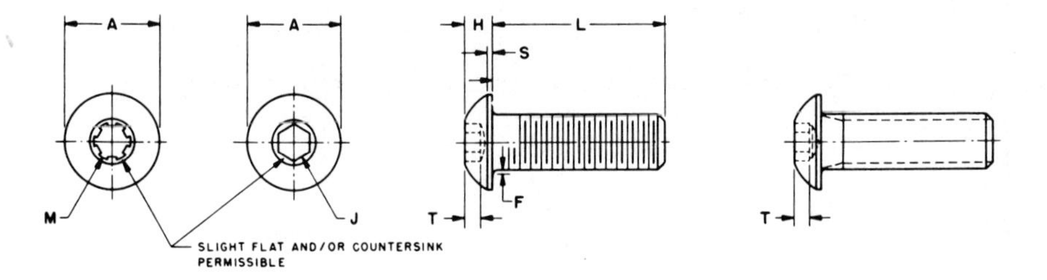

Nominal size or basic screw diameter		*A*, head diameter		*H*, head height		*S*, head side height (ref)	*M*, spline socket size (nom)	*J*, hexagon socket size, (nom)	*T*, key engage-ment (min)	*F*, fillet extension		*L*, maximum standard length (nom)
		Max	Min	Max	Min					Max	Min	
0	0.0600	0.114	0.104	0.032	0.026	0.010	0.048	0.035	0.020	0.010	0.005	1/2
1	0.0730	0.139	0.129	0.039	0.033	0.010	0.060	0.050	0.028	0.010	0.005	1/2
2	0.0860	0.164	0.154	0.046	0.038	0.010	0.060	0.050	0.028	0.010	0.005	1/2
3	0.0990	0.188	0.176	0.052	0.044	0.010	0.072	1/16 0.062	0.035	0.010	0.005	1/2
4	0.1120	0.213	0.201	0.059	0.051	0.015	0.072	1/16 0.062	0.035	0.010	0.005	1/2
5	0.1250	0.238	0.226	0.066	0.058	0.015	0.096	5/64 0.078	0.044	0.010	0.005	1/2
6	0.1380	0.262	0.250	0.073	0.063	0.015	0.096	5/64 0.078	0.044	0.010	0.005	5/8
8	0.1640	0.312	0.298	0.087	0.077	0.015	0.111	3/32 0.094	0.052	0.015	0.010	3/4
10	0.1900	0.361	0.347	0.101	0.091	0.020	0.145	1/8 0.125	0.070	0.015	0.010	1
1/4	0.2500	0.437	0.419	0.132	0.122	0.031	0.183	5/32 0.156	0.087	0.020	0.015	1
5/16	0.3125	0.547	0.527	0.166	0.152	0.031	0.216	3/16 0.188	0.105	0.020	0.015	1
3/8	0.3750	0.656	0.636	0.199	0.185	0.031	0.251	7/32 0.219	0.122	0.020	0.015	1 1/4
1/2	0.5000	0.875	0.851	0.265	0.245	0.046	0.372	5/16 0.312	0.175	0.030	0.020	2
5/8	0.6250	1.000	0.970	0.331	0.311	0.062	0.454	3/8 0.375	0.210	0.030	0.020	2

All dimensions in inches.
SOURCE: ANSI B18.3-1969, as published by the American Society of Mechanical Engineers.

TABLE 1-51 Dimensions of Hexagon Socket-head Shoulder Screws

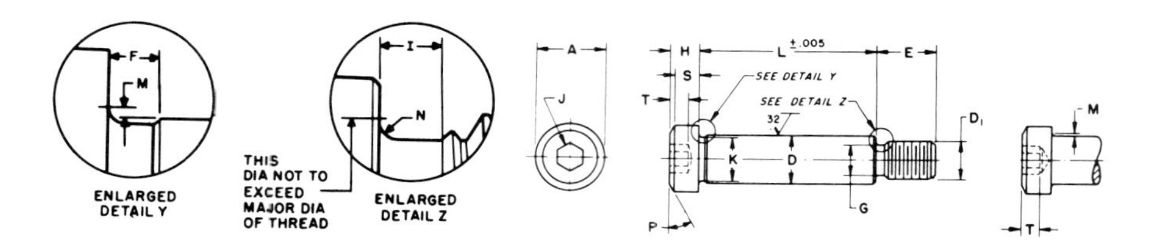

Nominal size or basic shoulder diameter	*D*, shoulder diameter		*A*, head diameter		*H*, head height		*S*, head side height (min)	*J*, hexagon socket size (nom)	*T*, key angagement (min)	*M*, head fillet extension above *D*		*I*, thread neck width	*F*, shoulder neck width (max)	*N*, thread neck fillet	
	Max	Min	Max	Min	Max	Min				Max	Min			Max	Min
1/4 0.250	0.2480	0.2460	0.375	0.357	0.188	0.177	0.157	1/8 0.125	0.094	0.014	0.009	0.062	0.093	0.023	0.017
5/16 0.312	0.3105	0.3085	0.438	0.419	0.219	0.209	0.183	5/32 0.156	0.117	0.017	0.012	0.075	0.093	0.028	0.022
3/8 0.375	0.3730	0.3710	0.562	0.543	0.250	0.240	0.209	3/16 0.188	0.141	0.020	0.015	0.083	0.093	0.031	0.025
1/2 0.500	0.4980	0.4960	0.750	0.729	0.312	0.302	0.262	1/4 0.250	0.188	0.026	0.020	0.093	0.093	0.035	0.029
5/8 0.625	0.6230	0.6210	0.875	0.853	0.375	0.365	0.315	5/16 0.312	0.234	0.032	0.024	0.115	0.093	0.042	0.036
3/4 0.750	0.7480	0.7460	1.000	0.977	0.500	0.490	0.421	3/8 0.375	0.281	0.039	0.030	0.136	0.093	0.051	0.045
1 1.000	0.9980	0.9960	1.312	1.287	0.625	0.610	0.527	1/2 0.500	0.375	0.050	0.040	0.150	0.125	0.055	0.049
1 1/4 1.250	1.2480	1.2460	1.750	1.723	0.750	0.735	0.633	5/8 0.625	0.469	0.060	0.050	0.166	0.125	0.062	0.056

All dimensions in inches.
SOURCE: ANSI B18.3-1969, as published by the American Society of Mechanical Engineers.

TABLE 1-52 Dimensions of Hexagon and Spline Socket Flat Countersunk-head Cap Screws

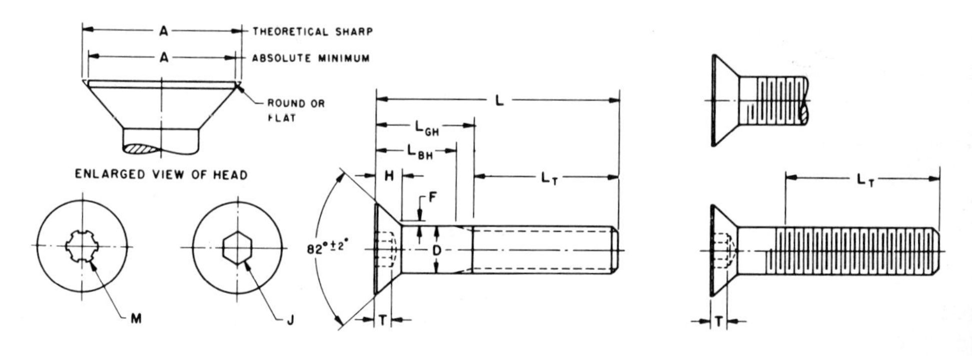

Nominal size or basic screw diameter		D, body diameter		A, head diameter		H, head height		M, spline socket size	J, hexagon socket size (nom)	T, key engagement (min)	F, fillet extension above D max (max)
		Max	Min	Theoretical sharp max	Abs. min	Reference	Flushness tolerance				
0	0.0600	0.0600	0.0568	0.138	0.117	0.044	0.006	0.048	0.035	0.025	0.006
1	0.0730	0.0730	0.0695	0.168	0.143	0.054	0.007	0.060	0.050	0.031	0.008
2	0.0860	0.0860	0.0822	0.197	0.168	0.064	0.008	0.060	0.050	0.038	0.010
3	0.0990	0.0990	0.0949	0.226	0.193	0.073	0.010	0.072	1/16 0.062	0.044	0.010
4	0.1120	0.1120	0.1075	0.255	0.218	0.083	0.011	0.072	1/16 0.062	0.055	0.012
5	0.1250	0.1250	0.1202	0.281	0.240	0.090	0.012	0.096	5/64 0.078	0.061	0.014
6	0.1380	0.1380	0.1329	0.307	0.263	0.097	0.013	0.096	5/64 0.078	0.066	0.015
8	0.1640	0.1640	0.1585	0.359	0.311	0.112	0.014	0.111	3/32 0.094	0.076	0.015
10	0.1900	0.1900	0.1840	0.411	0.359	0.127	0.015	0.145	1/8 0.125	0.087	0.015
1/4	0.2500	0.2500	0.2435	0.531	0.480	0.161	0.016	0.183	5/32 0.156	0.111	0.015
5/16	0.3125	0.3125	0.3053	0.656	0.600	0.198	0.017	0.216	3/16 0.188	0.135	0.015
3/8	0.3750	0.3750	0.3678	0.781	0.720	0.234	0.018	0.251	7/32 0.219	0.159	0.015
7/16	0.4375	0.4375	0.4294	0.844	0.781	0.234	0.018	0.291	1/4 0.250	0.159	0.015
1/2	0.5000	0.5000	0.4919	0.938	0.872	0.251	0.018	0.372	5/16 0.312	0.172	0.015
5/8	0.6250	0.6250	0.6163	1.188	1.112	0.324	0.022	0.454	3/8 0.375	0.220	0.015
3/4	0.7500	0.7500	0.7406	1.438	1.355	0.396	0.024	0.454	1/2 0.500	0.220	0.015
7/8	0.8750	0.8750	0.8647	1.688	1.604	0.468	0.025		9/16 0.562	0.248	0.015
1	1.0000	1.0000	0.9886	1.938	1.841	0.540	0.028		5/8 0.625	0.297	0.015
1 1/8	1.1250	1.1250	1.1086	2.188	2.079	0.611	0.031		3/4 0.750	0.325	0.031
1 1/4	1.2500	1.2500	1.2336	2.438	2.316	0.683	0.035		7/8 0.875	0.358	0.031
1 3/8	1.3750	1.3750	1.3568	2.688	2.553	0.755	0.038		7/8 0.875	0.402	0.031
1 1/2	1.5000	1.5000	1.4818	2.938	2.791	0.827	0.042		1 1.000	0.435	0.031

All dimensions in inches.
SOURCE: ANSI B18.3-1969, as published by the American Society of Mechanical Engineers.

TABLE 1-53 Dimensions of Round, Countersunk, and Reverse-key Cap Screws

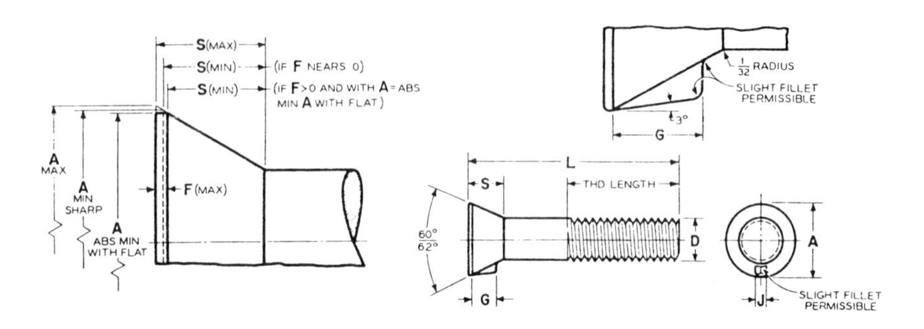

DIMENSIONS
(Round, Countersunk, Reverse Key)

REGULAR

D, Nominal diameter of bolt	*A*, diameter of head			*F*, feed thickness, (max)	*S*, head height		*J*, width of key		*G*, Key length	
	Max	Min sharp	Abs min with flat		Max	Min	Max	Min	Min	Max
5/16	0.592	0.578	0.563	0.025	0.233	0.208	0.156	0.151	0.185	0.200
3/8	0.661	0.640	0.625	0.031	0.239	0.208	0.156	0.151	0.187	0.202
7/16	0.776	0.749	0.734	0.036	0.282	0.246	0.156	0.151	0.227	0.242
1/2	0.892	0.859	0.844	0.042	0.328	0.286	0.156	0.151	0.267	0.282
°9/16	1.021	0.984	0.969	0.045	0.383	0.338	0.156	0.151	0.321	0.336
5/8	1.121	1.078	1.063	0.050	0.414	0.364	0.156	0.151	0.348	0.363
3/4	1.277	1.234	1.219	0.050	0.440	0.390	0.156	0.151	0.375	0.390
REPAIR†										
5/16	0.554	0.546	0.531	0.020	0.201	0.181	0.156	0.151	0.158	0.173
3/8	0.623	0.609	0.594	0.025	0.207	0.182	0.156	0.151	0.158	0.173
7/16	0.738	0.718	0.703	0.030	0.250	0.220	0.156	0.151	0.199	0.214
1/2	0.853	0.828	0.813	0.035	0.295	0.260	0.156	0.151	0.239	0.254
9/16	0.984	0.953	0.938	0.040	0.352	0.312	0.156	0.151	0.293	0.308
5/8	1.077	1.046	1.031	0.040	0.378	0.338	0.156	0.151	0.321	0.336
3/4	1.234	1.203	1.188	0.040	0.404	0.364	0.156	0.151	0.347	0.362

All dimensions given in inches, unless otherwise specified.
° This size is not recommended.
† The letter *R* shall be shown on top of the repair head, to distinguish it from the regular head bolt.
SOURCE: ANSI B18.3-1969, as published by the American Society of Mechanical Engineers.

TABLE 1-54 Dimensions of Square Lag Screws

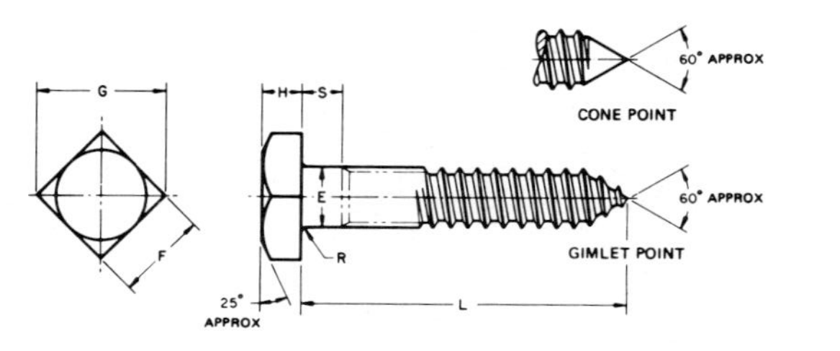

Nominal Size or Basic Product Dia		E		F			G		H			S	R	
		Body or Shoulder Dia		Width Across Flats			Width Across Corners		Height			Shoulder Length	Radius of Fillet	
		Max	Min	Basic	Max	Min	Max	Min	Basic	Max	Min	Min	Max	Min
No. 10	0.1900	0.199	0.178	9/32	0.281	0.271	0.398	0.372	1/8	0.140	0.110	0.094	0.03	0.01
1/4	0.2500	0.260	0.237	3/8	0.375	0.362	0.530	0.498	11/64	0.188	0.156	0.094	0.03	0.01
5/16	0.3125	0.324	0.298	1/2	0.500	0.484	0.707	0.665	13/64	0.220	0.186	0.125	0.03	0.01
3/8	0.3750	0.388	0.360	9/16	0.562	0.544	0.795	0.747	1/4	0.268	0.232	0.125	0.03	0.01
7/16	0.4375	0.452	0.421	5/8	0.625	0.603	0.884	0.828	19/64	0.316	0.278	0.156	0.03	0.01
1/2	0.5000	0.515	0.482	3/4	0.750	0.725	1.061	0.995	21/64	0.348	0.308	0.156	0.03	0.01
5/8	0.6250	0.642	0.605	15/16	0.938	0.906	1.326	1.244	27/64	0.444	0.400	0.312	0.06	0.02
3/4	0.7500	0.768	0.729	1 1/8	1.125	1.088	1.591	1.494	1/2	0.524	0.476	0.375	0.06	0.02
7/8	0.8750	0.895	0.852	1 5/16	1.312	1.269	1.856	1.742	19/32	0.620	0.568	0.375	0.06	0.02
1	1.0000	1.022	0.976	1 1/2	1.500	1.450	2.121	1.991	21/32	0.684	0.628	0.625	0.09	0.03
1 1/8	1.1250	1.149	1.098	1 11/16	1.688	1.631	2.386	2.239	3/4	0.780	0.720	0.625	0.09	0.03
1 1/4	1.2500	1.277	1.223	1 7/8	1.875	1.812	2.652	2.489	27/32	0.876	0.812	0.625	0.09	0.03

SOURCE: ANSI B18.2.1-1972, as published by the American Society of Mechanical Engineers.

TABLE 1-55 Dimensions of Hex Lag Screws

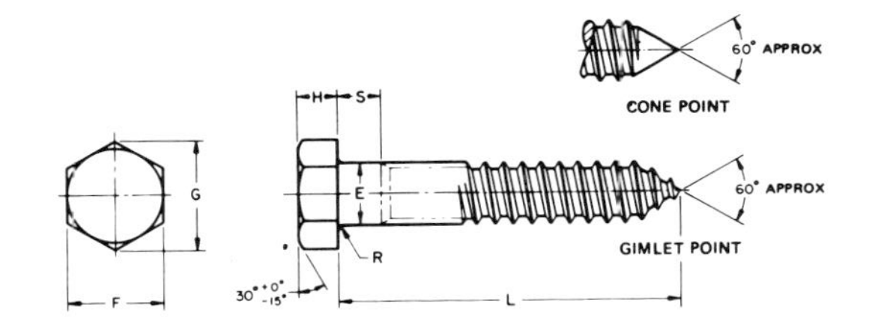

Nominal Size or Basic Product Dia		E		F			G		H			S	R	
		Body or Shoulder Dia		Width Across Flats			Width Across Corners		Height			Shoulder Length	Radius of Fillet	
		Max	Min	Basic	Max	Min	Max	Min	Basic	Max	Min	Min	Max	Min
No. 10	0.1900	0.199	0.178	9/32	0.281	0.271	0.323	0.309	1/8	0.140	0.110	0.094	0.03	0.01
1/4	0.2500	0.260	0.237	7/16	0.438	0.425	0.505	0.484	11/64	0.188	0.150	0.094	0.03	0.01
5/16	0.3125	0.324	0.298	1/2	0.500	0.484	0.577	0.552	7/32	0.235	0.195	0.125	0.03	0.01
3/8	0.3750	0.388	0.360	9/16	0.562	0.544	0.650	0.620	1/4	0.268	0.226	0.125	0.03	0.01
7/16	0.4375	0.452	0.421	5/8	0.625	0.603	0.722	0.687	19/64	0.316	0.272	0.156	0.03	0.01
1/2	0.5000	0.515	0.482	3/4	0.750	0.725	0.866	0.826	11/32	0.364	0.302	0.156	0.03	0.01
5/8	0.6250	0.642	0.605	15/16	0.938	0.906	1.083	1.033	27/64	0.444	0.378	0.312	0.06	0.02
3/4	0.7500	0.768	0.729	1 1/8	1.125	1.088	1.299	1.240	1/2	0.524	0.455	0.375	0.06	0.02
7/8	0.8750	0.895	0.852	1 5/16	1.312	1.269	1.516	1.447	37/64	0.604	0.531	0.375	0.06	0.02
1	1.0000	1.022	0.976	1 1/2	1.500	1.450	1.732	1.653	43/64	0.700	0.591	0.625	0.09	0.03
1 1/8	1.1250	1.149	1.098	1 11/16	1.688	1.631	1.949	1.859	3/4	0.780	0.658	0.625	0.09	0.03
1 1/4	1.2500	1.277	1.223	1 7/8	1.875	1.812	2.165	2.066	27/32	0.876	0.749	0.625	0.09	0.03

SOURCE: ANSI B18.2.1-1972, as published by the American Society of Mechanical Engineers.

TABLE 1-56 Dimensions of Binding Head Screws (Slotted Head—Miniature)

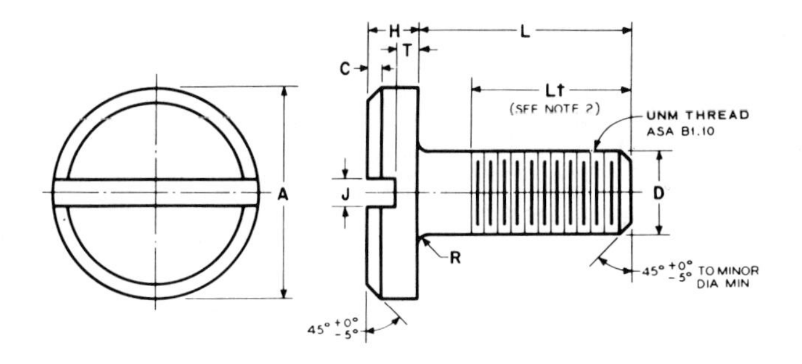

BINDING HEAD

Size designation°	Threads per inch	*D*, basic major dia. (max)	Binding head dimensions										
			A, head dia.		*H*, head height		*J*, slot width		*T*,† slot depth		*C*, chamfer (max)	*R*, radius	
			Max	Min	Max	Min	Max	Min	Max	Min		Max	Min
40 UNM	**254**	**0.0157**	**0.041**	**0.039**	**0.010**	**0.008**	**0.006**	**0.004**	**0.005**	**0.003**	**0.002**	**0.004**	**0.002**
45 UNM	254	0.0177	0.045	0.043	0.011	0.009	0.006	0.004	0.006	0.004	0.002	0.004	0.002
50 UNM	**203**	**0.0197**	**0.051**	**0.049**	**0.012**	**0.010**	**0.008**	**0.005**	**0.006**	**0.004**	**0.003**	**0.004**	**0.002**
55 UNM	203	0.0217	0.056	0.054	0.014	0.012	0.008	0.005	0.007	0.005	0.003	0.004	0.002
60 UNM	**169**	**0.0236**	**0.062**	**0.058**	**0.016**	**0.013**	**0.010**	**0.007**	**0.008**	**0.006**	**0.004**	**0.006**	**0.003**
70 UNM	145	0.0276	0.072	0.068	0.018	0.015	0.010	0.007	0.009	0.007	0.004	0.006	0.003
80 UNM	**127**	**0.0315**	**0.082**	**0.078**	**0.020**	**0.017**	**0.012**	**0.008**	**0.010**	**0.007**	**0.005**	**0.008**	**0.004**
90 UNM	113	0.0354	0.092	0.088	0.022	0.019	0.012	0.008	0.011	0.008	0.005	0.008	0.004
100 UNM	**102**	**0.0394**	**0.103**	**0.097**	**0.025**	**0.021**	**0.016**	**0.012**	**0.012**	**0.009**	**0.006**	**0.010**	**0.005**
110 UNM	102	0.0433	0.113	0.107	0.028	0.024	0.016	0.012	0.014	0.011	0.006	0.010	0.005
120 UNM	**102**	**0.0472**	**0.124**	**0.116**	**0.032**	**0.028**	**0.020**	**0.015**	**0.016**	**0.012**	**0.008**	**0.012**	**0.006**
140 UNM	85	0.0551	0.144	0.136	0.036	0.032	0.020	0.015	0.018	0.014	0.008	0.012	0.006

° Bold face type indicates preferred sizes.
† *T* measured from bearing surface.

MATERIAL: Corrosion resistant steels, ASTM designation A276
Class 303, Cond. A
Class 416, Cond. A, heat treat to approx. 120,000–150,000 psi (Rockwell C28–34)
Class 420, Cond. A, heat treat to approx. 220,000–240,000 psi (Rockwell C50–53)
Brass, temper half-hard ASTM designation B16
Nickel silver, temper hard ASTM designation B151 alloy C

MACHINE FINISH: Machined surface roughness of heads shall be approximately 63 microinches arithmetical average determined by visual comparison.

APPLIED COATINGS: Corrosion-resistant steel, passivate.
Brass, bare, black oxide, or nickel flash.
Nickel silver, none.

NOTES:

1. The diameter of the unthreaded body shall not be more than the maximum major diameter nor less than the minimum pitch diameter of the thread.
2. For screw lengths four times the major diameter or less, thread length L_t shall extend to within two threads of the head bearing surface. Screws of greater length shall have complete threads for a minimum of four major diameters.
3. Screws shall be free of all projecting burrs, observed at 3X magnification.
4. All dimensions are in inches.

SOURCE: ANSI B18.11-1961, as published by the American Society of Mechanical Engineers.

TABLE 1-57 Dimensions of Fillister-Head Screws (Slotted Head—Miniature)

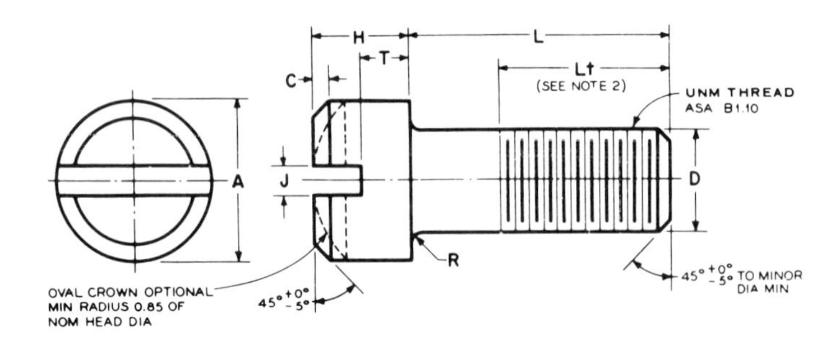

Size Desig-nation (a)	Thds per Inch	D Basic Major Dia	Fillister Head Dimensions									
			A Head Dia		H Head Hgt		J Slot Width		T (b) Slot Depth		C Chamfer	R (c) Radius
		Max	Max	Min	Max	Min	Max	Min	Max	Min	Max	Max
30 UNM	**318**	**0.0118**	**0.021**	**0.019**	**0.012**	**0.010**	**0.004**	**0.003**	**0.006**	**0.004**	**0.002**	**0.002**
35 UNM	282	0.0138	0.023	0.021	0.014	0.012	0.004	0.003	0.007	0.005	0.002	0.002
40 UNM	**254**	**0.0157**	**0.025**	**0.023**	**0.016**	**0.013**	**0.005**	**0.003**	**0.008**	**0.006**	**0.002**	**0.002**
45 UNM	254	0.0177	0.029	0.027	0.018	0.015	0.005	0.003	0.009	0.007	0.002	0.002
50 UNM	**203**	**0.0197**	**0.033**	**0.031**	**0.020**	**0.017**	**0.006**	**0.004**	**0.010**	**0.007**	**0.003**	**0.002**
55 UNM	203	0.0217	0.037	0.035	0.022	0.019	0.006	0.004	0.011	0.008	0.003	0.002
60 UNM	**169**	**0.0236**	**0.041**	**0.039**	**0.025**	**0.021**	**0.008**	**0.005**	**0.012**	**0.009**	**0.004**	**0.003**
70 UNM	145	0.0276	0.045	0.043	0.028	0.024	0.008	0.005	0.014	0.011	0.004	0.003
80 UNM	**127**	**0.0315**	**0.051**	**0.049**	**0.032**	**0.028**	**0.010**	**0.007**	**0.016**	**0.012**	**0.005**	**0.004**
90 UNM	113	0.0354	0.056	0.054	0.036	0.032	0.010	0.007	0.018	0.014	0.005	0.004
100 UNM	**102**	**0.0394**	**0.062**	**0.058**	**0.040**	**0.035**	**0.012**	**0.008**	**0.020**	**0.016**	**0.006**	**0.005**
110 UNM	102	0.0433	0.072	0.068	0.045	0.040	0.012	0.008	0.022	0.018	0.006	0.005
120 UNM	**102**	**0.0472**	**0.082**	**0.078**	**0.050**	**0.045**	**0.016**	**0.012**	**0.025**	**0.020**	**0.008**	**0.006**
140 UNM	85	0.0551	0.092	0.088	0.055	0.050	0.016	0.012	0.028	0.023	0.008	0.006

(a) Bold face type indicates preferred size (b) "T" measured from bearing surface (c) Relative to max. major dia

MATERIAL: CORROSION RESISTANT STEELS: ASTM Designation A276
CLASS 303, COND A
CLASS 416, COND A, HEAT TREAT TO APPROX 120,000–150,000 PSI (ROCKWELL C28–34)
CLASS 420, COND A, HEAT TREAT TO APPROX 220,000–240,000 PSI (ROCKWELL C50–53)
BRASS: TEMPER HALF HARD ASTM Designation B16
NICKEL SILVER: TEMPER HARD ASTM Designation B151, Alloy C

MACHINE FINISH: Machined surface roughness of heads shall be approximately 63MU in. arithmetical average determined by visual comparison.

APPLIED COATINGS: CORROSION RESISTANT STEEL: Passivate
BRASS: Bare, Black Oxide or Nickel Flash.
NICKEL SILVER: None

NOTE:
1. The diameter of the unthreaded body shall not be more than the maximum major diameter nor less than the minimum pitch diameter of the thread.
2. For screw lengths four times the major diameter or less, thread length (Lt) shall extend to within two threads of the head bearing surface. Screws of greater length shall have complete threads for a minimum of four major diameters.
3. Screws shall be free of all projecting burrs, observed at 3X magnification.
4. All dimensions are in inches.

SOURCE: ANSI B18.11-1961, as published by the American Society of Mechanical Engineers.

TABLE 1-58 Dimensions of Threads and Points for Types BSD and CSD Self-drilling Tapping Screws

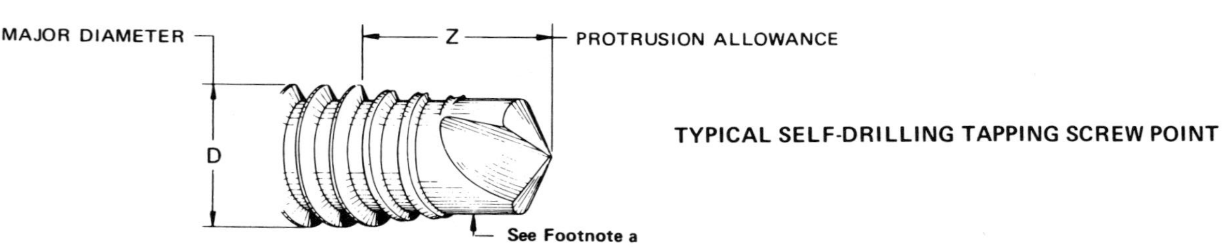

TYPICAL SELF-DRILLING TAPPING SCREW POINT

Nominal Size[b] or Basic Screw Diameter	TYPE BSD							TYPE CSD					TYPES BSD AND CSD							
	Threads Per Inch	D Major Diameter		d Minor Diameter		Z^c Protrusion Allowance (Ref)		Threads Per Inch	D Major Diameter		Z^c Protrusion Allowance (Ref)		L Minimum Practical Nominal Screw Lengths (Ref)							
													Style 2 Points				Style 3 Points			
													Formed		Milled		Formed		Milled	
		Max	Min	Max	Min	Style 2 Point	Style 3 Point		Max	Min	Style 2 Point	Style 3 Point	90° Heads	Csk Heads	90° Heads	Csk Heads	90° Heads	Csk Heads	90° Heads	Csk Heads
4 0.1120	24	0.114	0.110	0.086	0.082	0.163	—	40	0.1120	0.1072	0.130	—	5/16	3/8	3/8	7/16	—	—	—	—
6 0.1380	20	0.139	0.135	0.104	0.099	0.190	0.220	32	0.1380	0.1326	0.152	0.172	5/16	3/8	3/8	7/16	3/8	7/16	7/16	1/2
8 0.1640	18	0.166	0.161	0.122	0.116	0.211	0.251	32	0.1640	0.1586	0.162	0.202	3/8	7/16	7/16	1/2	7/16	1/2	1/2	9/16
10 0.1900	16	0.189	0.183	0.141	0.135	0.235	0.300	24	0.1900	0.1834	0.193	0.258	7/16	1/2	15/32	19/32	1/2	9/16	9/16	21/32
12 0.2160	14	0.215	0.209	0.164	0.157	0.283	0.353	24	0.2160	0.2094	0.223	0.293	1/2	5/8	17/32	21/32	1/2	5/8	21/32	25/32
1/4 0.2500	14	0.246	0.240	0.192	0.185	0.318	0.393	20	0.2500	0.2428	0.275	0.350	1/2	5/8	17/32	11/16	1/2	5/8	11/16	27/32

[a] Drill portion of points may be milled and/or cold formed and details of point taper and flute design shall be optional with the manufacturer provided the screws meet the performance requirements specified in this standard and are capable of drilling the maximum panel thicknesses shown in Table 5 prior to thread pick-up.

[b] Where specifying nominal size in decimals, zeros preceding decimal and in fourth decimal place shall be omitted.

[c] Protrusion allowance Z is the distance, measured parallel to the axis of screw, from the extreme end of the point to the first full form thread beyond the point and encompasses the length of drill point and the tapered incomplete threads. It is intended for use in calculating the maximum effective design grip length Y on the screw in accordance with the following:

$$Y = L\ min - Z$$

SOURCE: Industrial Fasteners Institute.

TABLE 1-59 Thread and Point Dimensions of High-performance Thread-rolling Screws

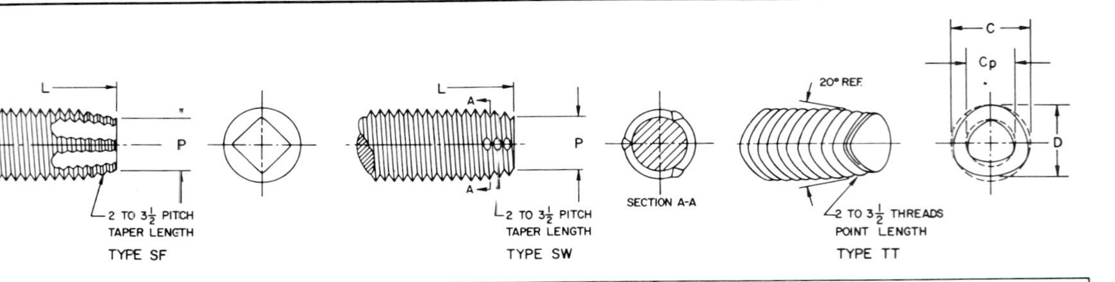

NOM SCREW SIZE AND THREADS** PER INCH	SCREW DIA	TYPES SF AND SW		TYPE SW		TYPE SF		TYPE TT				
				P*		P		C		D		CP
		MAJOR DIAMETER		POINT DIAMETER		POINT DIAMETER		DIAMETER OF CIRCUMSCRIBING CIRCLE		MEASUREMENT ACROSS CENTER		CIRCUM-SCRIBING CIRCLE (POINT)
	Basic	Max	Min	Max	Min	Max	Min	Max	Min	Max	Min	Max
No. 2-56	0.0860	0.0860	0.0813	0.0645	0.0595	–	–	0.0875	0.0835	0.0840	0.0800	0.070
3-48	.0990	.0990	.0938	.0726	.0676	–	–	.1010	.0970	.0970	.0930	.081
4-40	.1120	.1120	.1061	.0810	.0760	0.086	0.077	.1145	.1105	.1095	.1055	.090
5-40	.1250	.1250	.1191	.0930	.0880	.099	.090	.1275	.1235	.1225	.1185	.103
6-32	.1380	.1380	.1312	.0960	.0910	.106	.095	.1410	.1350	.1350	.1290	.111
8-32	.1640	.1640	.1571	.1165	.1115	.132	.121	.1670	.1610	.1610	.1550	.137
10-24	.1900	.1900	.1818	.1295	.1245	.147	.133	.1940	.1880	.1850	.1790	.153
1/4-20	.2500	.2500	.2408	.1800	.1750	.198	.181	.2550	.2490	.2460	.2400	.206
5/16-18	.3125	.3125	.3026	.2300	.2250	.255	.236	.3180	.3120	.3080	.3020	.264
3/8-16	.3750	.3750	.3643	.2870	.2820	.310	.289	.3810	.3750	.3710	.3650	.320
7/16-14	.4375	.4375	.4258	.3370	.3320	.361	.340	.4445	.4385	.4315	.4255	.375
1/2-13	.5000	.5000	.4876	.3920	.3870	.416	.395	.5075	.5015	.4935	.4875	.433

* The tabulated values apply to screw blanks before roll threading.

**Fine thread series screws are also available.

SOURCE: Industrial Fasteners Institute.

TABLE 1-60 Self-drilling Tapping-Screw Selection Chart

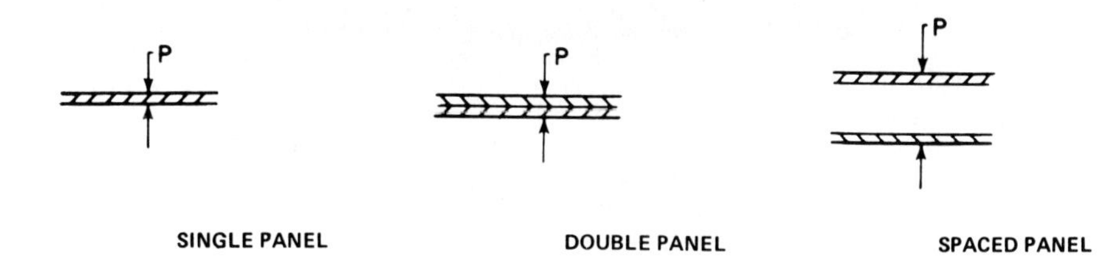

SINGLE PANEL DOUBLE PANEL SPACED PANEL

TYPICAL PANEL CONFIGURATIONS

SELF-DRILLING TAPPING SCREW SELECTION CHART

SCREW TYPE	POINT STYLE	NOMINAL SCREW SIZE	P[a] RECOMMENDED PANEL THICKNESS, in.
BSD and CSD	2	4	0.080 Max
		6	0.090 Max
		8	0.100 Max
		10	0.110 Max
		12	0.140 Max
		1/4	0.175 Max
	3	6	0.090–0.110
		8	0.100–0.140
		10	0.110–0.175
		12	0.110–0.210
		1/4	0.110–0.210

SOURCE: Industrial Fasteners Institute.

TABLE 1-61 Metric Thread Data

Metric thread data, M profile, internal and external

Thread Designation Dia. × Pitch, mm	Tap Drill, mm	Pitch Dia. 6H, Internal, mm	Pitch Dia. 6g, External, mm
M1.6 × 0.35	1.25	1.373	1.291
M2 × 0.4	1.60	1.740	1.654
M2.5 × 0.45	2.05	2.208	2.117
M3 × 0.5	2.50	2.675	2.580
M3.5 × 0.6	2.90	3.110	3.004
M4 × 0.7	3.30	3.545	3.433
M5 × 0.8	4.20	4.480	4.361
M6 × 1	5.00	5.350	5.212
M8 × 1.25	6.70	7.188	7.042
M8 × 1	7.00	7.350	7.212
M10 × 1.5	8.50	9.026	8.862
M10 × 1.25	8.70	9.188	9.042
M10 × 0.75	—	9.513	9.391
M12 × 1.75	10.20	10.863	10.679
M12 × 1.5	—	11.026	10.854
M12 × 1.25	10.80	11.188	11.028
M12 × 1	—	11.350	11.206
M14 × 2	12.00	12.701	12.503
M14 × 1.5	12.50	13.026	12.854
M15 × 1	—	14.350	14.206
M16 × 2	14.00	14.701	14.503
M16 × 1.5	14.50	15.026	14.854
M17 × 1	—	16.350	16.206
M18 × 1.5	16.50	17.026	16.854
M20 × 2.5	17.50	18.376	18.164
M20 × 1.5	18.50	19.026	18.854
M20 × 1	—	19.350	19.206
M22 × 2.5	19.50	20.376	20.164
M22 × 1.5	20.50	21.026	20.854
M24 × 3	21.00	22.051	21.803
M24 × 2	22.00	22.701	22.493
M25 × 1.5	—	24.026	23.854

SOURCE: McGraw-Hill Machining and Metalworking Handbook. copyright 1994 by R. Walsh. Reproduced with permission of The McGraw-Hill Companies.

TABLE 1-62 Dimensions for Metric Hex-cap Screws

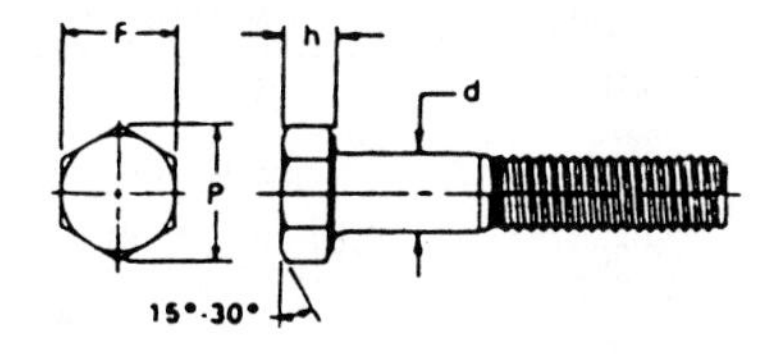

Diameter & thread pitch	d	F	P	h
M5 - 0.8	5.00	8.00	9.24	3.65
M6 - 1	6.00	10.00	11.55	4.15
M8 - 1.25	8.00	13.00	15.01	5.50
M10 - 1.50	10.00	16.00	18.48	6.63
M12 - 1.75	12.00	18.00	20.78	7.76
M14 - 2	14.00	21.00	24.25	9.09
M16 - 2	16.00	24.00	27.71	10.32
M20 - 2.5	20.00	30.00	34.64	12.88
M24 - 3	24.00	36.00	41.57	15.44
M30 - 3.5	30.00	46.00	53.12	19.48
M36 - 4	36.00	55.00	63.51	23.38

Note: Tabulated dimensions are n millimeters and are maximum values.

SOURCE: McGraw-Hill Machining and Metalworking Handbook. Copyright 1994 by R. Walsh. Reproduced with permission of The McGraw-Hill Companies.

TABLE 1-63 Dimensions for Metric Standard Nuts, Hexagonal

Size	Flats	Points	Thickness
M1.6×0.35	3.20	3.70	1.30
M2×0.4	4.00	4.62	1.60
M2.5×0.45	5.00	5.77	2.00
M3×0.5	5.50	6.35	2.40
M3.5×0.6	6.00	6.93	2.80
M4×0.7	7.00	8.08	3.20
M5×0.8	8.00	9.24	4.70
M6×1	10.00	11.55	5.20
M8×1.25	13.00	15.01	6.80
M10×1.5	16.00	18.48	8.40
M12×1.75	18.00	20.78	10.80
M14×2	21.00	24.25	12.80
M16×2	24.00	27.71	14.80
M20×2.5	30.00	34.64	18.00
M24×3	36.00	41.57	21.50
M30×3.5	46.00	53.12	25.60
M36×4	55.00	63.51	31.00

Note: Tabulated dimensions are in millimeters. 1 mm = 0.03937"

SOURCE: McGraw-Hill Machining and Metalworking Handbook. Copyright 1994 by R. Walsh. Reproduced with permission of The McGraw-Hill Companies.

Section 2

Standard Pins

ROBERT O. PARMLEY, P.E.

Consulting Engineer, Ladysmith, Wisconsin

2-1 Scope This section describes, in text and illustrations, the major types and designs of all standard pins in present-day use, as well as their design functions and applications.

As a fastener, the pin has evolved from a solid, straight cylindrical form into a multi-shaped family of modern-design cousins which serve an ever-widening realm of employment. It should be noted here that the square pin or key is described in sec. 10, "Locking Components."

Basically, a pin is a machine element or fastening component which secures the position of two or more parts of a machine, mechanism, or assembly relative to each other. The choice of type and size of a given pin application must be based on a sound balance of stress and strain of the pin itself and the forces on the parts to be connected, as well as their material compatibility.

Reference is made here to the technical paper entitled "The Pin" by M. J. Schilhansl, published by ASME, no. 58-SA-23. This work has established and captured the guidelines for strength calculations of basic pin stress and strain forces. It is suggested by the author that the reader obtain a copy of this text if an in-depth analysis of pin strength characteristics is desired. Space and format of this section do not allow the duplication of this landmark paper; however, application segments and data have been incorporated with permission from the publisher and author.

Since World War II and the dawn of the space age, many new pin designs have flowed from the drawing boards of design engineers. Older basic types have also become more prominent in all facets of assembly.

A satellite field of assembly equipment has developed and we will explore this area later in the section.

Pins used for heavy structural connections are described in Section 9, Structural Steel Connections. These structural pins meet the specifications of AISC and AASHO.

This section will isolate each major pin type and describe its design and areas of usage. Tables, charts, and illustrations, together with related engineering data, are displayed. In some cases, typical as well as unusual applications are shown to demonstrate versatility and stimulate ideas.

2-2 Straight Cylindrical Pins The solid or straight cylindrical pin is undoubtedly the original basic pin. Usually cut from bar stock, it is square cut on each end or chamfered as shown in Fig. 2-1.

Nominal sizes for standard straight pins are listed in Table 2-1.

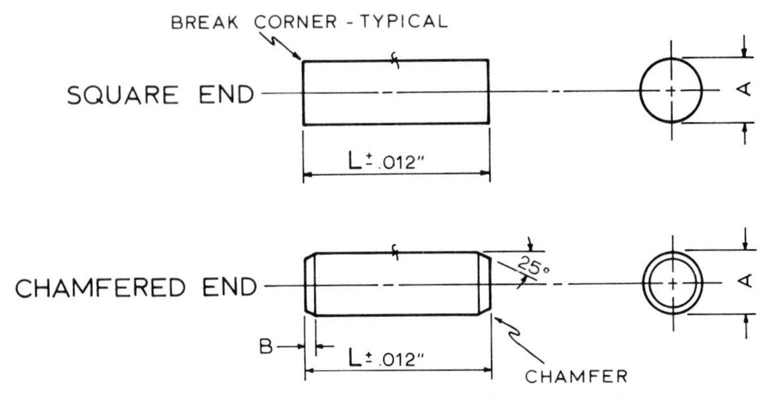

FIG. 2-1 Straight cylindrical pins. Basic solid pins, usually cut and machined from bar stock.

TABLE 2-1 Dimensions of Straight Pins

Nominal Diameter	Diameter A		Chamfer B
	Max	Min	
0.062	0.0625	0.0605	0.015
0.094	0.0937	0.0917	0.015
0.109	0.1094	0.1074	0.015
0.125	0.1250	0.1230	0.015
0.156	0.1562	0.1542	0.015
0.188	0.1875	0.1855	0.015
0.219	0.2187	0.2167	0.015
0.250	0.2500	0.2480	0.015
0.312	0.3125	0.3095	0.030
0.375	0.3750	0.3720	0.030
0.438	0.4375	0.4345	0.030
0.500	0.500	0.4970	0.030

All dimensions are given in inches.

These pins must be straight and free from burrs or any other defects that will affect their serviceability.

SOURCE: ANSI B5.20-1958, as published by American Society of Mechanical Engineers.

Straight cylindrical pins are often used to transmit torque in round shafts drilled to receive the pin in which both parts are equally strong in shear, because they are fabricated from the same material. When this condition exists, it should be noted that when the pin diameter equals 40 percent of the shaft diameter, the shearing stress in the pin equals the shearing stress in the shaft.

Thus:

$$T = 2\pi r^2 RS \qquad (2\text{-}1)$$

where:

T = torque delivered by pin, in.-lb
r = radius of pin, in.
R = radius of shaft, in.
S = shearing stress in pin, lb per sq in.

DOWEL PINS

2-3 Dowel Pins Dowel pins are an outgrowth of the straight cylindrical pin and are most often used in machine and tool fabrication including jigs and fixtures (see section 2-47, Jigs and Fixtures).

A nominal-size reamer is employed to develop the holes into which these pins are tap-, press-, or drive-fit by typical insertion methods.

Standard-size pins are for original assembly, whereas oversize pins are for reassembly or replacement.

If a dowel pin is driven into a blind hole and no provision was made for the release of air, the worker installing the pin may be exposed to physical harm and the associated components damaged. One simple method of avoiding this dangerous situation is to provide a small flat surface along the full length of the pin to permit an escape avenue for air as the dowel is driven home. There are specially fabricated dowel pins that have full-length helical grooves to relieve air pressure during insertion. Dowel pins that are repeatedly removed have their exposed ends drilled and tapped to receive a removal tool. These pins are known as "pull-out dowel pins." Their use avoids the often dangerous and time consuming "do-it-yourself" method of removal.

2-4 Hardened Dowel Pins Hardened dowel pins are bullet-nosed on the lead or entry end (Fig. 2-2).

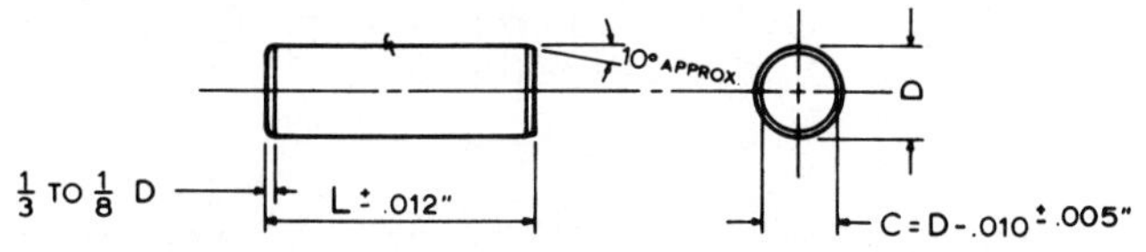

FIG. 2-2 Hardened dowel pin. Basic geometry of typical hardened dowel pin. *(Adapted from ANSI B5.20-1958, as published by American Society of Mechanical Engineers.)*

2-5 Soft Dowel Pins Soft dowel pins generally are chamfered on both ends as shown in Fig. 2-3.

Dowel pins are fabricated from commercial bar-stock sizes. Table 2-2 lists nominal sizes under 1-inch diameter for soft dowel pins.

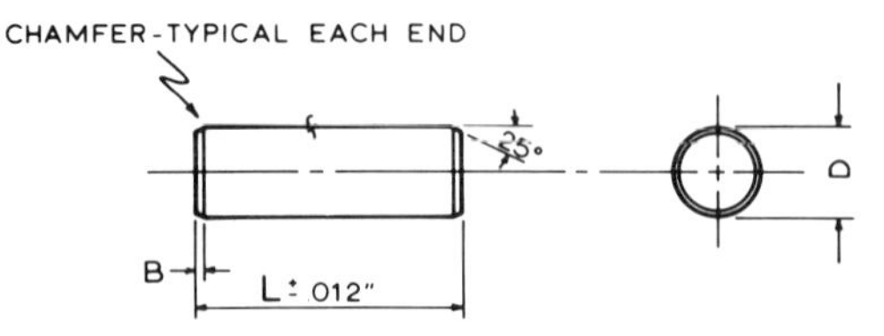

FIG. 2-3 Soft dowel pin. Commercial standard design. *(Adapted from ANSI B5.20-1958, as published by American Society of Mechanical Engineers.)*

TABLE 2-2 Dimensions of Ground Dowel Pins

Nominal Diameter	Diameter A		Chamfer B
	Max	Min	
0.062	0.0600	0.0595	0.015
0.094	0.0912	0.0907	0.015
0.109	0.1068	0.1063	0.015
0.125	0.1223	0.1218	0.015
0.156	0.1535	0.1530	0.015
0.188	0.1847	0.1842	0.015
0.219	0.2159	0.2154	0.015
0.250	0.2470	0.2465	0.015
0.312	0.3094	0.3089	0.030
0.375	0.3717	0.3712	0.030
0.438	0.4341	0.4336	0.030
0.500	0.4964	0.4959	0.030
0.625	0.6211	0.6206	0.045
0.750	0.7458	0.7453	0.045
0.875	0.8705	0.8700	0.060
1.000	0.9952	0.9947	0.060

All dimensions are given in inches.

Maximum diameters are graduated from 0.0005 on ⅟₁₆ in. pins to 0.0028 on 1-in. pins under the minimum commercial bar stock sizes.

SOURCE: ANSI B5.20-1958, as published by the American Society of Mechanical Engineers.

TABLE 2-3 Tolerances for General Fits

CLASS NO.	DESCRIPTION	FORMULA
1	LOOSE FIT	$+ 0.0025\sqrt[3]{d}$
2	FREE FIT	$+ 0.0013\sqrt[3]{d}$
3	MEDIUM FIT	$+ 0.0008\sqrt[3]{d}$
4	SNUG FIT	$+ 0.0006\sqrt[3]{d}$

NOTE: SEE MANUFACTURER'S RECOMMENDATIONS FOR WRINGING, TIGHT, MEDIUM AND HEAVY FORCE OR SHRINK FITS.

Various standards have been developed for different degrees of fit. Table 2-3 lists the accepted hole tolerances for each general type of fit.

TAPERED PINS

2-6 Standard Taper Pins Standard taper pins are generally used to transmit small torques or to position components. They are sized to be inserted by drive fit. Table 2-4 shows their dimensions.

Taper-pin diameters at the large end are shown in Table 2-5 and their standard sizes are detailed in Table 2-6.

CLEVIS PINS

2-7 Clevis Pins Clevis pins employ cotter pins (see sec. 2-41) for securing their position. Figure 2-4 illustrates the design.

Refer to Table 2-7 for standard clevis pin dimensions.

SPIRALLY COILED PINS

2-8 Spirally Coiled Pins The spirally coiled or wrapped pin is shown in Fig. 2-5. The shape of the cross section is very close to the geometrical configuration of the Spiral of Archimedes. The departure from this shape is made to minimize the area of surface along which there is no contact between the hole wall and the pin, and to prevent the end of the outer coil from moving in the peripheral direction over the adjacent coil.*

Spirally coiled pins are among the most versatile fasteners available today because of their flexibility both before and after insertion. All spring pins must have adequate flexibility to permit being driven into a hole with a diameter less than the expanded diameter of the pin. Spiral-wrapped pins, as spirally coiled pins are often called, are not only designed to be flexible during insertion, but also to remain a flexible element once inserted into the hole. Being flexible, they absorb shock and dynamic loads, reduce the possibility of hole damage, and have a dampening effect on vibration transmitted from one element to another. They are true spring pins before, during, and after insertion; it is this factor that sets them apart from other pin designs.

These spring pins can replace other pin types and designs in a multitude of assemblies and, under certain circumstances, they can replace even rivets, nuts, and bolts. This spring-pin design may also be used in pivots, shafts, retainers, stops, and locators.

A spring pin should not merely be designed and selected for a specific hole size; the material into which the hole has been drilled or cast should also be taken into consideration. Obviously, a spring pin that enlarges the hole into which it is inserted or damages the hole because of its inflexibility, when subjected to loads, defeats its own purpose.

2-9 Standard-duty Pins Standard-duty pins are designed to yield the optimum combination of shear strength and dynamic-shock-absorbing quality. Their superior flexibility keeps insertion forces at a practical minimum. These pins will not enlarge the hole by excessive radial tension when used in most common nonheated materials.

2-10 Heavy-duty Pins Heavy-duty pins are used where higher shear strength is a prerequisite and shock-absorbing quality is of a secondary importance. Although heavy-duty pins are still a flexible element when inserted, the external forces required to flex the pin are much greater. Accordingly, the dampening effect of this kind of pin is limited when it is subjected to light loads.

2-11 Light-duty Pins Light-duty pins are designed for those materials and applications where even the radial tension of standard-duty pins is excessive. Examples of such materials are some plastics, glass, and ceramics. Light-duty pins have also found substantial use in applications requiring moderate strength, since they provide substantial installation savings because of increased insertion speed.

Table 2-8 shows the general data for stock spirally coiled pins in the three duty ratings. Table 2-9 lists the metric sizes now manufactured and carried as stock items.

* M. H. Schilhansl, "The Pin," ASME publication no. 58-SA-23.

TABLE 2-4 Tapered Pin Dimensions

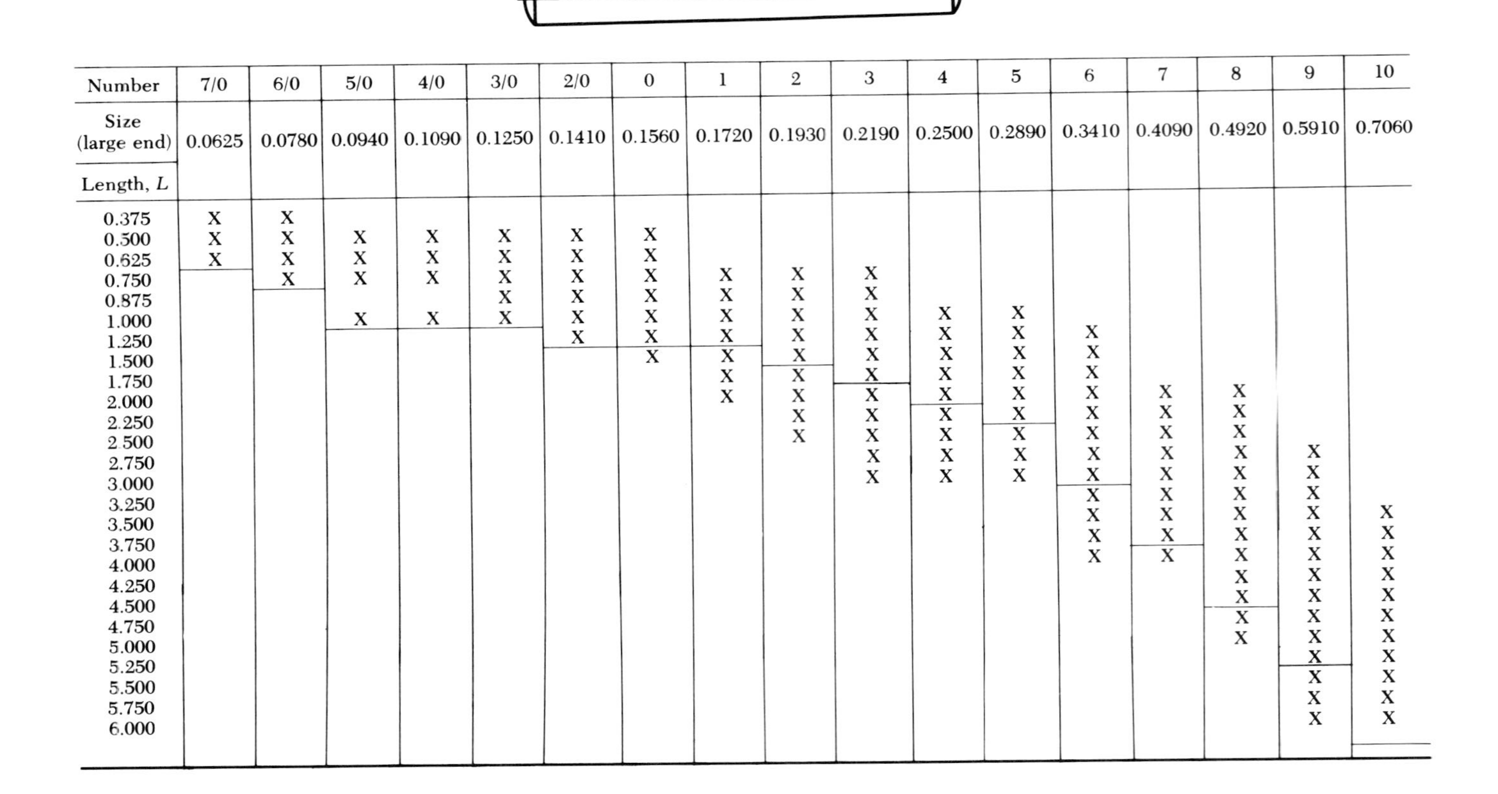

Number	7/0	6/0	5/0	4/0	3/0	2/0	0	1	2	3	4	5	6	7	8	9	10
Size (large end)	0.0625	0.0780	0.0940	0.1090	0.1250	0.1410	0.1560	0.1720	0.1930	0.2190	0.2500	0.2890	0.3410	0.4090	0.4920	0.5910	0.7060
Length, L																	
0.375	X	X															
0.500	X	X	X	X	X	X	X										
0.625	X	X	X	X	X	X	X										
0.750		X	X	X	X	X	X	X	X	X							
0.875					X	X	X	X	X	X							
1.000			X	X	X	X	X	X	X	X	X	X					
1.250						X	X	X	X	X	X	X	X				
1.500							X	X	X	X	X	X	X				
1.750								X	X	X	X	X	X				
2.000								X	X	X	X	X	X	X	X		
2.250									X	X	X	X	X	X	X		
2.500									X	X	X	X	X	X	X		
2.750										X	X	X	X	X	X	X	
3.000										X	X	X	X	X	X	X	
3.250													X	X	X	X	
3.500													X	X	X	X	X
3.750													X	X	X	X	X
4.000													X	X	X	X	X
4.250															X	X	X
4.500															X	X	X
4.750															X	X	X
5.000															X	X	X
5.250																X	X
5.500																X	X
5.750																X	X
6.000																X	X

All dimensions are given in inches. Standard reamers are available for pins given above the line.
Pins nos. 11 (size 0.8600), 12 (size 1.032), 13 (size 1.241), and 14 (1.523) are special sizes—hence their lengths are special.
To find small diameter of pin, multiply the length by 0.02083 and subtract the result from the large diameter.

Types	*Commercial*	*Precision*
Sizes	7/0 to 14	7/0 to 10
Tolerance on diameter	(+0.0013, −0.0007)	(+0.0013, −0.0007)
Taper	1/4 in./ft	1/4 in./ft
Length tolerance	(±0.030)	(±0.030)
Concavity tolerance	None	0.0005 up to 1 in. long 0.001 1 1/16 to 2 in. long 0.002 2 1/16 in. and longer

SOURCE: ANSI B5.20-1958, as published by The American Society of Mechanical Engineers.

TABLE 2-5 Taper-Pin Diameters at Large End

NO.	SIZE	NO.	SIZE	NO.	SIZE
7/0	0.0625	1	0.1720	8	0.4920
6/0	0.0780	2	0.1930	9	0.5910
5/0	0.0940	3	0.2190	10	0.7060
4/0	0.1090	4	0.2500	11	0.8600
3/0	0.1250	5	0.2890	12	1.0320
2/0	0.1410	6	0.3410	13	1.2410
0	0.1560	7	0.4090	14	1.5230

NOTE: TO FIND SMALL END DIAMETER, MULTIPLY LENGTH BY 0.02083 AND SUBTRACT PRODUCT FROM LARGE END DIAMETER.

SOURCE: Adapted from ANSI B5.20-1958, as published by American Society of Mechanical Engineers.

TABLE 2-6 Standard Taper Pins

NO.	LENGTH OF TAPER PIN, IN.																									
	3/8	1/2	5/8	3/4	7/8	1	1 1/4	1 1/2	1 3/4	2	2 1/4	2 1/2	2 3/4	3	3 1/4	3 1/2	3 3/4	4	4 1/4	4 1/2	4 3/4	5	5 1/4	5 1/2	5 3/4	6
7/0	■	■	■																							
6/0	■	■	■	■																						
5/0		■	■	■		■																				
4/0		■	■	■		■																				
3/0		■	■	■	■	■																				
2/0		■	■	■	■	■	■																			
0		■	■	■	■	■	■	■																		
1				■	■	■	■	■	■	■																
2				■	■	■	■	■	■	■	■	■														
3				■	■	■	■	■	■	■	■	■	■	■												
4						■	■	■	■	■	■	■	■	■												
5						■	■	■	■	■	■	■	■	■												
6							■	■	■	■	■	■	■	■	■	■	■	■								
7										■	■	■	■	■	■	■	■	■	■	■	■					
8										■	■	■	■	■	■	■	■	■	■	■	■	■				
9													■	■	■	■	■	■	■	■	■	■	■	■	■	■
10																■	■	■	■	■	■	■	■	■	■	■

SHADED ENTRIES INDICATE STANDARD LENGTHS OF PINS MADE IN THE SIZES INDICATED.

SOURCE: Adapted from ANSI B5.20-1958, as published by American Society of Mechanical Engineers.

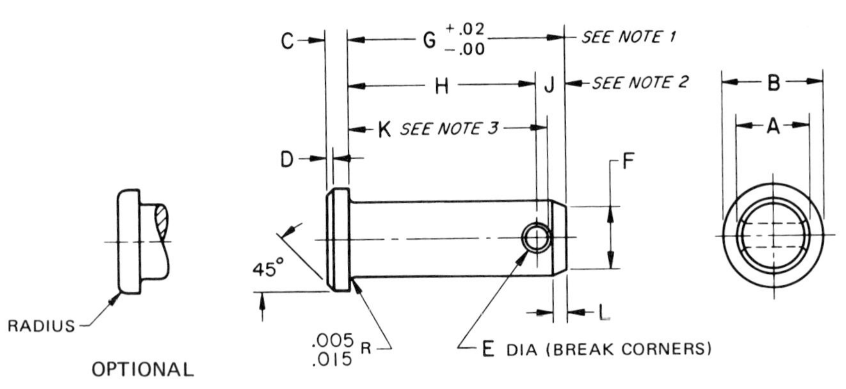

FIG. 2-4 Clevis pin. Standard design as approved by the American National Standards Institute. See Table 2-7 for dimensions. (*Source: ANSI B18.8.1-1972, as published by American Society of Mechanical Engineers.*)

TABLE 2-7 Dimensions of Clevis Pins

Nominal Size[1] or Basic Pin Diameter		A		B		C		D	E		F		G[2]	H		J[3]	K[4]		L		Recommended Cotter Pin Nominal Size	
		Shank Diameter		Head Diameter		Head Height		Head Chamfer	Hole Diameter		Point Diameter		Pin Length	Head to Center of Hole		End to Center Ref	Head to Edge of Hole Ref		Point Length			
		Max	Min	Max	Min	Max	Min	±0.01	Max	Min	Max	Min	Basic	Max	Min	Basic	Max	Min	Max	Min		
3/16	0.188	0.186	0.181	0.32	0.30	0.07	0.05	0.02	0.088	0.073	0.15	0.14	0.58	0.504	0.484	0.09	0.548	0.520	0.055	0.035	1/16	0.062
1/4	0.250	0.248	0.243	0.38	0.36	0.10	0.08	0.03	0.088	0.073	0.21	0.20	0.77	0.692	0.672	0.09	0.736	0.708	0.055	0.035	1/16	0.062
5/16	0.312	0.311	0.306	0.44	0.42	0.10	0.08	0.03	0.119	0.104	0.26	0.25	0.94	0.832	0.812	0.12	0.892	0.864	0.071	0.049	3/32	0.093
3/8	0.375	0.373	0.368	0.51	0.49	0.13	0.11	0.03	0.119	0.104	0.33	0.32	1.06	0.958	0.938	0.12	1.018	0.990	0.071	0.049	3/32	0.093
7/16	0.438	0.436	0.431	0.57	0.55	0.16	0.14	0.04	0.119	0.104	0.39	0.38	1.19	1.082	1.062	0.12	1.142	1.114	0.071	0.049	3/32	0.093
1/2	0.500	0.496	0.491	0.63	0.61	0.16	0.14	0.04	0.151	0.136	0.44	0.43	1.36	1.223	1.203	0.15	1.298	1.271	0.089	0.063	1/8	0.125
5/8	0.625	0.621	0.616	0.82	0.80	0.21	0.19	0.06	0.151	0.136	0.56	0.55	1.61	1.473	1.453	0.15	1.548	1.521	0.089	0.063	1/8	0.125
3/4	0.750	0.746	0.741	0.94	0.92	0.26	0.24	0.07	0.182	0.167	0.68	0.67	1.91	1.739	1.719	0.18	1.830	1.802	0.110	0.076	5/32	0.156
7/8	0.875	0.871	0.866	1.04	1.02	0.32	0.30	0.09	0.182	0.167	0.80	0.79	2.16	1.989	1.969	0.18	2.080	2.052	0.110	0.076	5/32	0.156
1	1.000	0.996	0.991	1.19	1.17	0.35	0.33	0.10	0.182	0.167	0.93	0.92	2.41	2.239	2.219	0.18	2.330	2.302	0.110	0.076	5/32	0.156

[1] Where specifying nominal size in decimals, zeros preceding decimal shall be omitted.

[2] Lengths tabulated are intended for use with standard clevises, without spacers. When required, it is recommended that other pin lengths be limited wherever possible to nominal lengths in 0.06 in. increments.

[3] Basic "J" dimension (distance from centerline of hole to end of pin) is specified for calculating hole location from underside of head on pins of lengths not tabulated.

[4] Reference dimension provided for convenience in design layout and is not subject to inspection.

Note: See Fig. 2-4 for clevis pin designs.

SOURCE: ANSI B18.8.1-1972, as published by American Society of Mechanical Engineers.

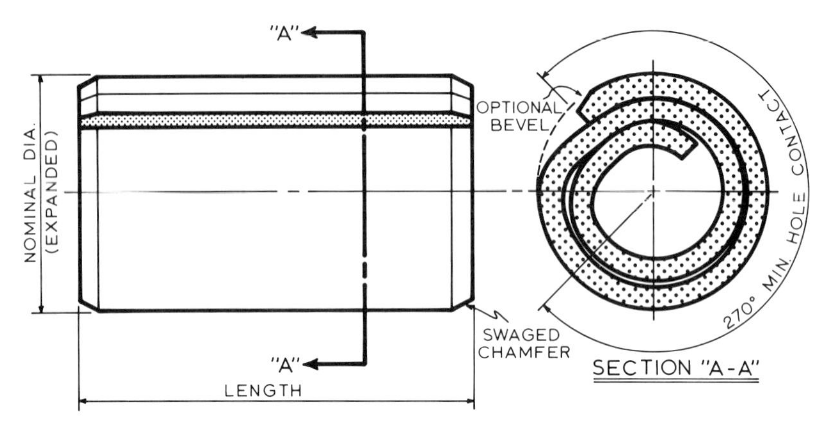

FIG. 2-5 Basic design of spirally coiled pin. Note section *A-A*, which is very close to the geometrical shape of the Spiral of Archimedes.

2-12 Shear Strength A spirally coiled pin is a true spring pin because it continues to have flexibility even after it is inserted into a hole. This concept requires a reappraisal of the engineering thinking which has been prevalent in the past. For example, there is a rule of thumb by which the static shear requirements are doubled for applications subjected to shock or dynamic situations. This rule resulted from the great difficulty in analyzing dynamic situations without actually performing simulated tests. The continued flexibility of these pins in the hole creates a closer relationship between static shear strength and dynamic loading. It has been proven that a flexible spirally coiled pin will repeatedly outlast a solid pin of equivalent or greater static shear strength.

2-13 Uniform Compression and Flexing Inherent radial spring action is best demonstrated pictorially. Figure 2-6*a* has a line scribed on a cut-off cross section. When this pin is inserted into a hole, as shown in Fig. 2-6*b*, the displacement of the lines reveals that the spring is contracting from the insertion. Note that all coils move in relation to each other.

Figure 2-7*a* depicts the pin after removal from the hole, demonstrating the recovery characteristics. This true radial-spring action results in uniform compression and flexing as shown schematically in Fig. 2-7*b*.

Uniform flexing is unique in this pin. Whatever the direction of the load, the spring pin flexes in free radial motion. This eliminates the necessity of orienting the pin either for shock-absorbing requirements or to obtain optimum shear strength.

The $2^1/_4$- to $2^1/_2$-coil spiral construction has no point of fatigue-stress concentration which could result in premature failure from working action of the pin. The multiple-coil design also permits the use of a thinner strip, which reduces the stresses imparted to the material during the forming operation in its manufacture. This not only is beneficial to the fatigue resistance of the pin but also improves its flexibility. In addition, this pin more readily assumes the shape of its hole, particularly important for tapered and out-of-round holes.

Hole tolerances for spirally coiled pins are greater than other pin designs as shown in Table 2-10.

2-14 Chamfers Spirally coiled pins have a uniform swaged chamfer with the end dimension designed for fast insertion (Fig. 2-8). Since the extreme end of the pin is basically locked, it acts as a fulcrum. As the diameter of the pin is reduced by insertion into the hole (Fig. 2-9), the chamfer shortens and the effective bearing length of the pin increases. If radial tension at the extreme end of the hole is necessary, it is recommended that the pin length be selected so the pin protrudes about a quarter-diameter from the hole end.

2-15 Circular Tolerances The expanded diameter of spirally coiled pins is within the given tolerance range for a minimum 270 degrees of its outer periphery (see Fig.

TABLE 2-8 General Data for Stock Spirol® Coiled Pins

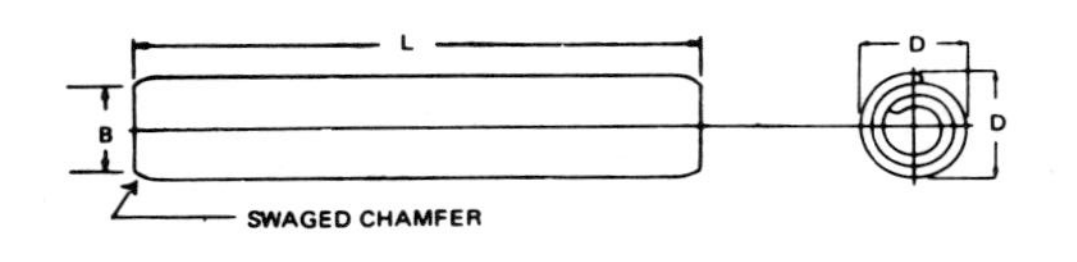

LENGTH TOLERANCES
All diameters up to and including 3/8 in.
±0.010 in. up to and including 2 in. long
±0.015 in. over 2 in. up to and including 3 in. long
±0.025 in. over 3 in. long
All diameters larger than 3/8 in.
±0.025 in. all lengths

Use ring gage to measure D and B
Use calipers to measure L

Nominal diameter	Decimal dimension	Recommended hole size		B	Standard duty				Heavy duty				Light duty			
					D		Min double shear, lb		D		Min double shear, lb		D		Min double shear, lb	
		Min	Max	(max)	Min	Max	302	1070 6150 420	Min	Max	302	1070 6150 420	Min	Max	302	1070 6150 420
1/32	0.031	0.031	0.032	0.029	0.033	0.035	65	90*								
0.039	0.039	0.039	0.040	0.037	0.041	0.044	100	135*								
3/64	0.047	0.047	0.048	0.045	0.049	0.052	145	190*								
0.052	0.052	0.051	0.053	0.050	0.054	0.057	190	250*								
1/16	0.062	0.061	0.065	0.059	0.067	0.072	265	330	0.066	0.070	360	475	0.067	0.073	160	
5/64	0.078	0.077	0.081	0.075	0.083	0.088	425	550	0.082	0.086	575	775	0.083	0.089	250	
3/32	0.094	0.093	0.097	0.091	0.099	0.105	600	775	0.098	0.103	825	1,100	0.099	0.106	360	475
7/64	0.109	0.108	0.112	0.106	0.114	0.120	825	1,050	0.113	0.118	1,150	1,500	0.114	0.121	500	650
1/8	0.125	0.124	0.129	0.121	0.131	0.138	1,100	1,400	0.130	0.136	1,700	2,100	0.131	0.139	650	826
5/32	0.156	0.155	0.160	0.152	0.163	0.171	1,700	2,200	0.161	0.168	2,400	3,100	0.163	0.172	1,000	1,300
3/16	0.187	0.185	0.192	0.182	0.196	0.205	2,400	3,150	0.194	0.202	3,500	4,500	0.196	0.207	1,450	1,900
7/32	0.219	0.217	0.224	0.214	0.228	0.238	3,300	4,200	0.226	0.235	4,600	5,900	0.228	0.240	2,000	2,600
1/4	0.250	0.247	0.256	0.243	0.260	0.271	4,300	5,500	0.258	0.268	6,200	7,800	0.260	0.273	2,600	3,300
5/16	0.312	0.308	0.319	0.304	0.324	0.337	6,700	8,700	0.322	0.334	9,300	12,000	0.324	0.339	4,000	5,200
3/8	0.375	0.370	0.383	0.366	0.388	0.403	9,600	12,600	0.386	0.400	14,000	18,000				
7/16	0.437	0.431	0.446	0.427	0.452	0.469	13,300	17,000	0.450	0.466	18,000	23,500				
1/2	0.500	0.493	0.510	0.488	0.516	0.535	17,500	22,500†	0.514	0.532	25,000	32,000†				
5/8	0.625	0.618	0.635	0.613	0.642	0.661		35,000†	0.640	0.658		48,000†				
3/4	0.750	0.743	0.760	0.738	0.768	0.787		50,000†	0.766	0.784		70,000†				

*420 only † alloy steel in place of carbon steel

MATERIALS AND HARDNESS
- Carbon steel (SAE and AISI 1070 to 1095), Rockwell C 46-53 or equivalent
- Corrosion-resistant steel (SAE 51420 – AISI 420), Rockwell C 46-55 or equivalent
- Nickel stainless steel (SAE 30302 – AISI 302), work hardened
- Alloy steel (SAE and AISI 6150), Rockwell C 43-51 or equivalent

FINISHES
Carbon steel
- Plain finish
- Phosphate-coated per Mil-P-16232, type Z, class 2
- Cadmium-plated per fed. spec. QQ-P-416a, type 1, class 3 and type 2, class 3
- Zinc-plated per fed. spec. QQ-Z-325a, type 1, class 3 and type 2, class 3

Alloy steel
- Phosphate-coated per Mil-P-16232, type Z, class 2

Corrosion-resistant and nickel stainless steel
- Passivated per fed. spec. QQ-P-35

SOURCE: CEM Company, Inc., Danielson, Conn.

TABLE 2-9 Metric Sizes of Stock Spirally Coiled Pins

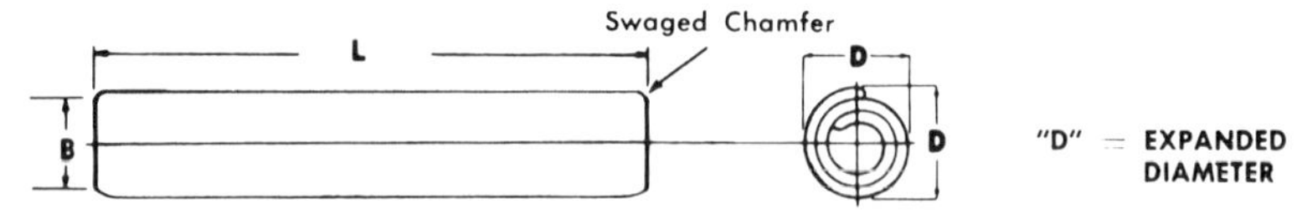

Length Tolerances

+ 0.5 up to and including 10 mm long
+ 1.0 over 10 mm up to and including 50 mm long
+ 1.5 over 50 mm long

Nominal Diameter	Recommended drilled hole tolerance		Dim. "B"	Standard duty				Heavy duty				Light duty			
				"D" dimension		Min. double shear in KG		"D" dimension		Min. double shear in KG		"D" dimension		Min. double shear in KG	
	Plus	Minus	Max.	Min.	Max.	AISI 302	SAE 1070 AISI 420	Min.	Max.	AISI 302	SAE 1070 AISI 420	Min.	Max.	AISI 302	SAE 1070 AISI 420
0.8	0.04	0.00	0.75	0.85	0.91	31	41								
1.0	0.04	0.00	0.95	1.05	1.15	46	61								
1.2	0.04	0.00	1.15	1.25	1.35	66	92								
1.5	0.10	0.00	1.40	1.62	1.73	107	148	1.61	1.71	148	194	1.62	1.75	66	—
2.0	0.10	0.01	1.90	2.13	2.25	194	255	2.11	2.21	255	357	2.13	2.28	112	—
2.5	0.10	0.01	2.40	2.65	2.78	296	398	2.62	2.73	388	561	2.65	2.82	184	235
3.0	0.10	0.01	2.90	3.15	3.30	428	561	3.12	3.25	581	775	3.15	3.35	255	337
3.5	0.12	0.02	3.40	3.67	3.84	581	765	3.64	3.79	775	1,020	3.67	3.87	347	459
4.0	0.12	0.02	3.90	4.20	4.40	775	979	4.15	4.30	1,020	1,377	4.20	4.45	449	581
5.0	0.12	0.05	4.85	5.20	5.45	1,173	1,530	5.15	5.35	1,581	2,040	5.20	5.50	714	918
6.0	0.13	0.05	5.85	6.25	6.50	1,714	2,244	6.18	6.40	2,346	3,060	6.25	6.55	1,020	1,326
8.0	0.17	0.07	7.80	8.30	8.63	3,060	3,978	8.25	8.55	4,182	5,406	8.30	8.65	1,836	2,346
10.0	0.20	0.07	9.75	10.35	10.75	4,896	6,324	10.30	10.65	6,528	8,568				
12.0	0.22	0.10	11.70	12.40	12.85	6,834	9,078	12.35	12.75	9,282	12,240				
14.0	0.25	0.15	13.60	14.45	14.95		12,240	14.40	14.85		16,830				
16.0	0.25	0.15	15.60	16.45	17.00		15,810	16.40	16.90		21,420				
20.0	0.25	0.15	19.60	20.45	21.10		25,500	20.40	21.00		34,680				

MATERIALS

B Carbon Steel (SAE 1070 - 1095)
C Corrosion Resistant Steel (Cres) (AISI 420) Magnetic
D Nickel Stainless Steel (AISI 302) Non-magnetic

FINISHES

K Plain **S** Cadmium Plated
R Phosphate Coated **T** Zinc Plated
P Passivated

SOURCE: CEM Co., Inc., Danielson, Conn.

PIN DUTY LETTERS

M Standard Duty
H Heavy Duty
L Light Duty

HARDNESS

Carbon Steel (SAE 1070 - 1095)
Rockwell C 43-52 or equivalent

Chrome Stainless Steel (AISI 420)
Rockwell C 46-55 or Equivalent

Nickel Stainless Steel (AISI 302)
Work Hardened

2-5). The diameter is not averaged out in three places as is the case with other spring pins.

2-16 Broken Edges The lip of the outer spiral is slightly broken (refer to Fig. 2-5). Although this pin design creates the optical illusion of being larger in diameter on the lip, this is not true. Since the pin, including the diameter taken across the lip, is within the diameter tolerance for a minimum of 270 degrees of its periphery, the lip will not damage the hole.

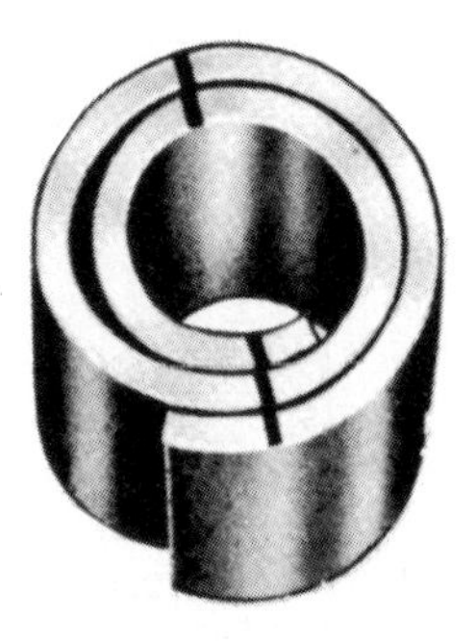
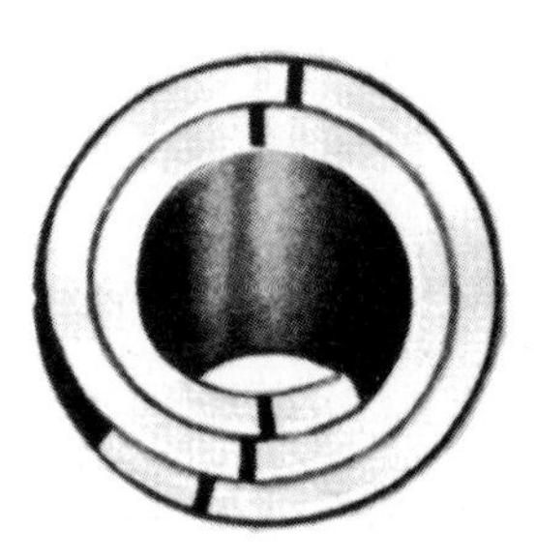

A B

FIG. 2-6 Spring action of spirally coiled pins. (*A*) Scribed line on a cut-off cross section of pin prior to insertion; (*B*) line displacement revealing spring contracting from insertion. (*CEM Co., Inc., Danielson, Conn.*)

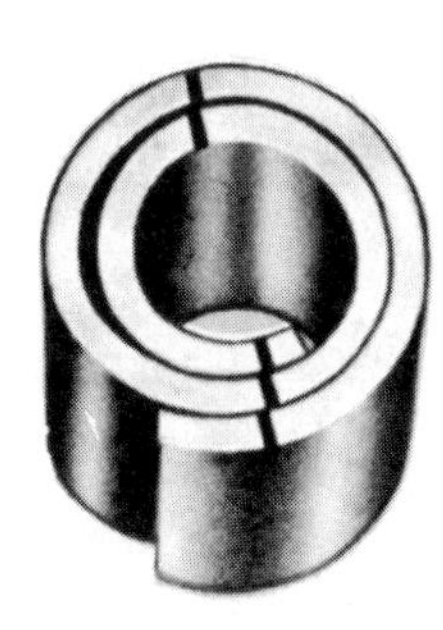
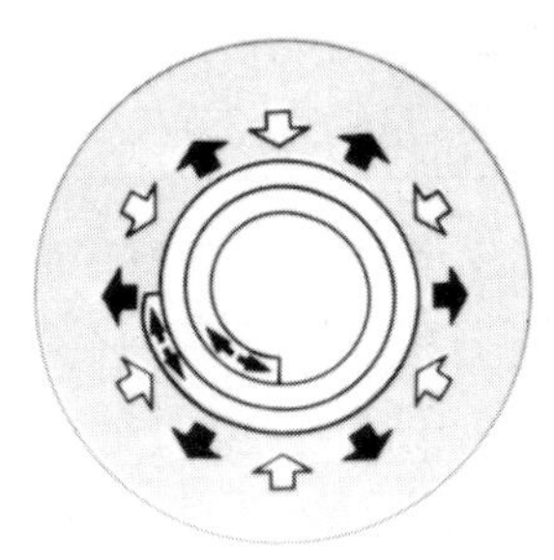

A B

FIG. 2-7 Spring action of spirally coiled pins (*continued*). (*A*) Pin after removal from hole, revealing recovery characteristics; (*B*) uniform compression and flexing, schematically shown. (*CEM Co., Inc., Danielson, Conn.*)

TABLE 2-10 Hole Tolerances for Spirally Coiled Pins

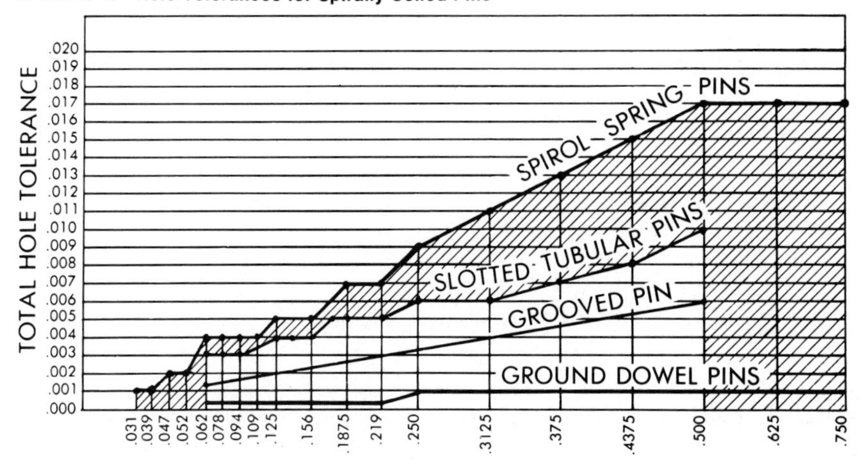

SOURCE: CEM Co., Inc., Danielson, Conn.

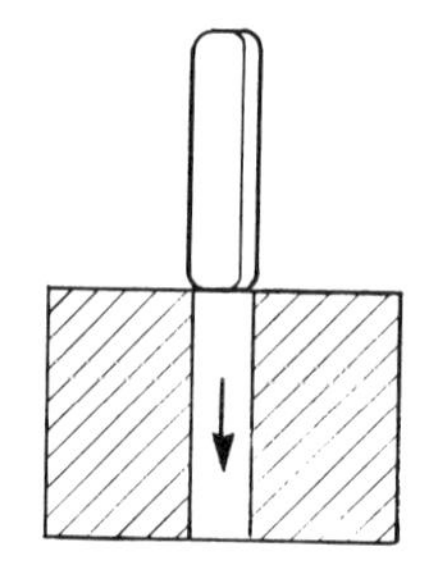

FIG. 2-8 Moment of pin insertion into hole.

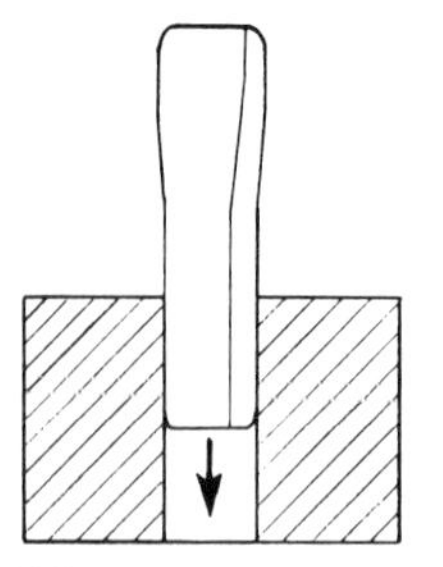

FIG. 2-9 Detail of pin insertion.

TABLE 2-11 Spirally Coiled Pin Insertion Forces (in lb/dia. length)

DUTY CODE: H = HEAVY; M = STANDARD

NOMINAL DIA.	DUTY	CARBON AND CORROSION RESISTANT STEEL			NICKEL STAINLESS STEEL		
		OPTIMUM	REC. MAX.	ABS. MAX.	OPTIMUM	REC. MAX.	ABS. MAX.
1/32 & .039	M	8	12	16	7	10	13
3/64	M	10	15	20	8	12	16
052	M	10	16	22	8	13	18
1/16	H	30	40	45	25	35	40
	M	13	20	27	10	16	22
5/64	H	40	55	65	30	45	55
	M	17	25	33	12	20	28
3/32	H	55	75	90	42	60	75
	M	25	35	45	18	29	40
7/64	H	70	100	120	55	75	90
	M	35	45	60	24	37	50
1/8	H	100	130	150	77	97	115
	M	45	60	75	30	45	60
5/32	H	150	190	220	115	140	160
	M	65	83	100	45	63	80
3/16	H	215	260	300	155	190	215
	M	90	110	130	60	80	100
7/32	H	280	340	390	200	240	275
	M	110	135	160	75	100	125
1/4	H	350	415	475	250	295	340
	M	140	170	200	100	130	160
5/16	H	490	580	650	350	405	465
	M	195	230	265	140	170	200
3/8	H	630	740	850	450	515	590
	M	250	295	340	185	220	255
7/16	H	770	900	1030	540	630	720
	M	310	360	410	230	270	310
1/2	H	850	1000	1150	640	740	840
	M	370	425	480	275	315	360

LIGHT DUTY - 50% OF THE STANDARD DUTY FORCES.

REMOVAL FORCE

INSERTED → / ← REMOVED: CARBON STEEL, CRES 420 — MINIMUM OF 70% OF INSERTION FORCE; STAINLESS 302 — MINIMUM OF 55% OF INSERTION FORCE

INSERTED → / → REMOVED: SAME AS INSERTION TO START THEN DECREASING AS REMOVAL PROGRESSES

SOURCE: CEM Co., Inc., Danielson, Conn.

2-17 Insertion Force Insertion forces are tabulated in Table 2-11 for spirally coiled pins. This table gives the insertion forces which can be expected when a pin is inserted into a nominal hardened-steel hole for its full length. This being the most unfavorable condition, it thus provides a safety factor under usual conditions. To estimate the insertion force IF in a given application, use the following formula:

$$\mathrm{IF} = \frac{\text{length of engagement}}{\text{hole diameter}} \times (\text{in lb/dia. length}) \tag{2-2}$$

2-18 Headed Spirally Coiled Pins Headed spirally coiled pins are of the same basic design as the previously detailed pin, except that they are flanged on the driven end. This provides a stop for insertion and dresses up the assembly. Removal, of course, is achieved only by a withdrawal or reversal from the entry direction. The design characteristics, however, are basically the same as the original spirally coiled pin. Table 2-12 pictures the geometry of this pin adaptation.

GROOVED PINS

2-19 Grooved Pins Grooved pins originated with a basic design—a solid pin with three parallel, equally spaced grooves impressed longitudinally on the exterior surface. Figure 2-10 illustrates the type A configuration.

When fabricating the pin grooves, the tool penetrates below the surface, displacing a carefully determined amount of metal stock. No metal is removed. The metal is displaced to each side of the groove, forming a raised portion or flute extending along each side of the groove. The crests of the flutes constitute an expanded diameter Dx which is a few thousandths of an inch larger than the nominal diameter D, as illustrated in Fig. 2-10.

When the grooved pin is forced into a drilled hole slightly larger than the nominal or specified diameter of the pin, the flutes are substantially forced back into the grooves. The resiliency of the metal forced back into the three grooves results in a powerful radial force against the hole wall, thus providing a positive self-locking fastener. The six flutes raised on the solid pin body exert a balaneed and controlled radial force.

Many variations ot the original design are now available from the manufacturer. A pictorial review of some of the more widely used variations is shown in Fig. 2-11. Refer to latest ANSI standards.

2-20 Drilling Tolerances and Pin Sizing Tolerances Drilling tolerances and pin sizing tolerances for drilled holes are shown in Table 2-13.

Recommended pin diameter for various shaft sizes and torques are listed in Table 2-14.

Table 2-15 tabulates minimum single-shear values for various materials.

2-21 Material Standard material for these pins is cold-drawn low-carbon steel. Of course, other materials such as alloy steel, brass, silicon bronze, and stainless steel are used for special purposes.

Excellent wear resistance can be obtained by surface-hardening standard low-carbon steel.

2-22 Finish The standard finish of these pins is zinc electroplated. The usual thickness of deposit is 0.00015 inch. Heavier deposits, chromate treatment, and special finishes such as brass, nickel, cadmium, and black oxide are available.

2-23 Hardness Testing Hardness testing of grooved pins is accurately achieved by preparing the pin as outlined below.

Step 1. Grind a flat surface approximately 0.0015 inch deep on the curved surface of the pin opposite a groove. This surface can be ground deeper for a check on internal hardness of pins.

Step 2. Grind a flat surface opposite and parallel to the first grinding. Be sure to grind deep enough to remove the expanded diameter over the groove.

Step 3. Place the surfaced pin with the grooved flat side down on the anvil of the Rockwell tester.

Step 4. Enter surfaced section with penetrator C Brale 150 kg for pins $\frac{1}{8}$ inch in diameter and above and read. For pins under $\frac{1}{8}$ inch in diameter, use a superficial penetrator.

TABLE 2-12 Headed Spirally Coiled Pins with Table of Stock Sizes

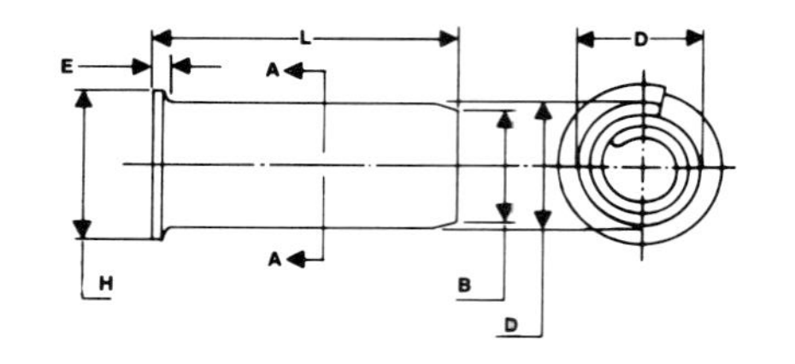

Nominal dia. D	Decimal dimension	Recommended hole tolerance +	H Min	H Max	B	D Min	D Max	Min double shear 1070	E, reference only
1/16	0.062	0.003	0.084	0.099	0.059	0.067	0.072	330	1/64
5/64	0.078	0.003	0.105	0.122	0.075	0.083	0.088	550	1/64
3/32	0.094	0.003	0.125	0.144	0.091	0.099	0.105	775	1/32
7/64	0.109	0.003	0.146	0.167	0.106	0.114	0.120	1,050	1/32
1/8	0.125	0.004	0.166	0.189	0.121	0.131	0.138	1,400	1/32
5/32	0.156	0.004	0.207	0.234	0.152	0.163	0.171	2,200	3/64
3/16	0.187	0.005	0.248	0.279	0.182	0.196	0.205	3,150	1/16

L Lengths available from stock in.
Diameters 1/16 in. through 1/8 in., L Tolerance ±0.015
Diameters 5/32 in. and 3/16 in., L Tolerance ±0.020

Nominal dia. D	3/16	1/4	5/16	3/8	7/16	1/2	9/16	5/8	11/16	3/4	13/16	7/8	15/16	1	1 1/8	1 1/4	1 3/8	1 1/2	1 5/8	1 3/4	1 7/8	2
1/16	■	■	■	■	■	■	■	■	■	■												
5/64		■	■	■	■	■	■	■	■	■	■	■										
3/32			■	■	■	■	■	■	■	■	■	■	■	■								
7/64			■	■	■	■	■	■	■	■	■	■	■	■	■	■						
1/8				■	■	■	■	■	■	■	■	■	■	■	■	■	■	■				
5/32					■	■	■	■	■	■	■	■	■	■	■	■	■	■	■	■		
3/16						■	■	■	■	■	■	■	■	■	■	■	■	■	■	■	■	■

SOURCE: CEM Co., Inc., Danielson, Conn.

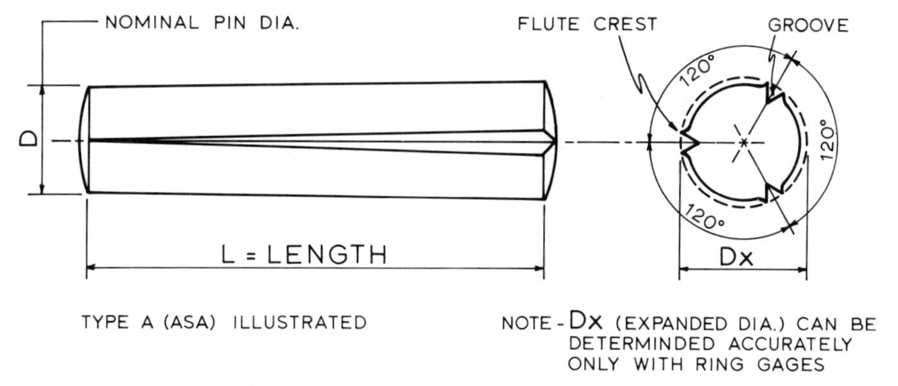

FIG. 2-10 Grooved pin. General design of typical grooved pin.

Alternatively, pins can be ground on both ends and placed on the Rockwell tester. Penetrators are to be used as in step 4.

Superficial hardness testing of spirally coiled pins (sec. 2-8) with a material thickness of more than 0.010 inch may be tested by the method noted below using the 15N scale. This includes all $^3/_{32}$ inch heavy, $^7/_{64}$ inch standard and heavy, $^1/_8$ inch standard and heavy, and larger-diameter pins in all three duties. Since the pins, however, are flexible, a hardness reading can not be taken directly against the side of the pin. The outer coil has to be removed. A hand file is recommended since the heat generated by a grinding wheel may anneal the pin. The segment should be placed under the test mandrel with the curvature up. The segment has to be narrow and flat enough so that it will not interfere with the mandrel. To avoid the longitudinal spring effect of the segment, the smallest possible specimen base should be selected.

2-24 Grooved Studs Grooved studs are widely used for fastening light metal or plastic parts to heavier members. They replace conventional screws, rivets, peened pins, and other type fasteners in applications such as attaching name plates, covers, brackets, and springs. The standard grooved stud, as shown in Fig. 2-12, has a round head. Special head types and shanks are shown in Fig. 2-13.

Three parallel grooves impressed in the shank of the pin, as noted and explained in a previous section on grooved pins and in Fig. 2-10, provide the positive lock features.

SLOTTED TUBULAR PINS

2-25 Slotted Tubular Pins Slotted tubular pins or spring pins are a well-used fastener. Tough and durable, they are a stock item. Figure 2-14 pictures the basic design of slotted tubular pins.

The slotted tubular pin is a chamfered cylindrical spring pin, heat-treated to achieve a spring with optimum toughness, resilience, and shear strength.

2-26 Fabrication When fabricated by the manufacturer, it is formed to a controlled diameter greater than the hole into which it is to be pressed. These pins are stronger than the mild-carbon-steel pins of equivalent sizes.

2-27 Assembly This self-locking reusable pin is compressed as it is driven into the assembly hole; it exerts continuous spring pressure against the sides of the hole, thus positively preventing loosening. Figure 2-15 illustrates the entry of the pin.

Holes drilled to normal production tolerances without reaming operations are satisfactory to insure a tight fit.

The slotted tubular pin not only provides an immediate replacement for many expensive types of fastener but it also can be inserted with a minimum of effort and the use of the simplest tools. Its self-locking action requires no secondary locking components or devices and can therefore be removed by a simple drift with only slightly more force than was originally needed for its insertion. Table 2-16 lists the average maximum removal forces in steel.

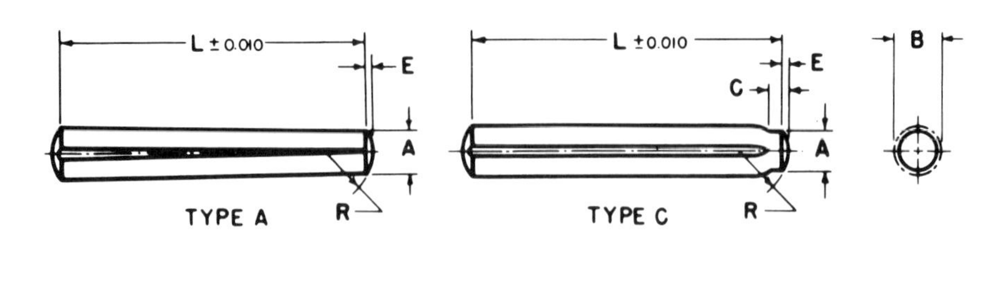

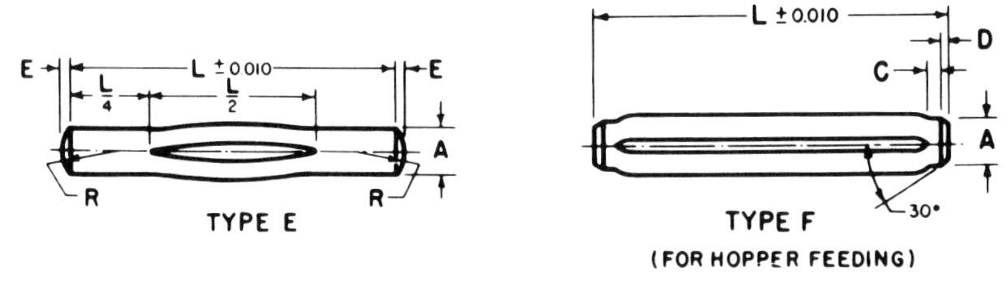

FIG. 2-11 Grooved pins. Types A, C, E, and F. (Diameters range from 1/16 in. to 1/2 in. and lengths from 1/4 in. to 4 1/2 in.)

2-28 Drill-Hole Sizing Table 2-17 graphically reports the drill-hole sizes recommended. The curve shows the relatively free hole tolerances for slotted tubular pins.

So long as the drilled holes are within the recommended tolerances, as shown in Table 2-18, the pin will self-center.

2-29 Material These pins are generally manufactured from carbon steel and type 420 corrosion-resistant steel. Beryllium copper is also used for special designs to resist corrosive attack and provide good electrical, antimagnetic, and nonsparking properties.

2-30 Finishes Carbon steel pins are customarily supplied plain, with the black oiled finish characteristic of heat-treating. Zinc- and cadmium-plated finishes are available. Where a mild degree of corrosion resistance is necessary, a phosphate coating is used.

2-31 Operating Temperatures Operating temperatures in excess of the permissible limits of the members in which the pins are anchored is not recommended. The following values may be taken as safe limits before physical properties are substantially altered.

Carbon steel	500°F
Type 420 corrosion-resistant steel	700°F
Beryllium copper	700°F

2-32 Reusability The reusability of this pin derives from its basic design. It is compressed upon insertion into the hole, and upon removal the normal elastic recovery of the pin material assures a continuing self-locking action when the pin is reinserted into holes of approximately the same diameter. As shown by Table 2-19, the insertion and removal forces fall somewhat after the first reinsertion, because of the burnishing action. They then remain substantially constant. It is interesting to note that the variations in forces with tolerance extremes are reduced with repeated use.

TABLE 2-13 Drilling Tolerances and Pin Sizing for Drilled Holes
Tolerances for drilled holes are based on depth-to-diameter ratio of approximately 5:1. Higher ratios may cause these figures to be exceeded, but in no case should they be exceeded by more than 10%. Specifications for holes having a depth-to-diameter ratio of 1:1 or less should be held extremely close; 60% of figures shown in table is recommended.

Undersized drills should never be used to produce holes for grooved pins. This malpractice results from the false assumption that the pins will "hold better." Instead, the pins bend, damage the hole wall, crack castings, and peel their expanded flutes, thus reducing or eliminating their retaining characteristics and preventing their reuse. It is recommended that holes made in hardened steel or cast iron have a slight chamfer at the entrance. This eliminates shearing of the flutes as the pins are forced in.

Drills used should be new or properly ground with the aid of an approved grinding fixture. The drilling machine spindle must be in good condition and operated at correct speeds and feeds for the metal being drilled, and suitable coolant is always recommended. Drill jigs with accurate bushings always facilitate good drilling practice.

Pin diameter, in.	Decimal equivalent	Recommended drill size, in.	Hole tolerances ADD to nominal diameter, in.
1/16	0.0625	1/16	0.002
3/32	0.0938	3/32	0.003
1/8	0.1250	1/8	0.003
5/32	0.1563	5/32	0.003
3/16	0.1875	3/16	0.004
7/32	0.2188	7/32	0.004
1/4	0.2500	1/4	0.004
5/16	0.3125	5/16	0.005
3/8	0.3750	3/8	0.005
7/16	0.4375	7/16	0.006
1/2	0.5000	1/2	0.006

SOURCE: Driv-Lok, Inc., Sycamore, Ill.

2-33 Pin Sizing Recommendations The recommended pin sizes for various shaft diameters are shown in Fig. 2-16.

KNURLED PINS

2-34 Knurled Pins Knurled pins are frequently the best fasteners to use in die-casting and plastic applications. They also provide superior fastening in the assembly of components which have thin sections, because of the many knurls in the contact area. Formed sheet-metal hinges and similar applications employ knurled pins for their excellent adaptability to nonuniform configurations. A few of the typical designs of knurled pins are represented in Fig. 2-17.

2-35 End Design All knurled pins have smooth ends to facilitate easy handling and installation. Figure 2-18 details the four most common types. The knurled section can be located anywhere along the length of the pin.

QUICK-RELEASE PINS

2-36 Quick-Release Pins Quick-release pins were conceived and thus developed to fill a basic need in the modern assembly of machinery and related applications—the need for speed. The pace of today's world requires positive quick-release functions. The aerospace, industrial, and farm machinery usage is only a part of the application of the quick-release pin. Its quick "push" insert and fast "pull" release action is a key link in many designs.

2-37 Locking A typical ring-handle pin is shown in Fig. 2-19. It employs two lockballs which are internally spring-loaded for light contact.

TABLE 2-14 Recommended Pin Diameter for Various Shaft Sizes and Torques Transmitted by Pin in Double Shear

This table is a guide in selecting the proper size grooved pin to use in keying machine members to shafts of given sizes and for specific load requirements. Torque and horsepower ratings are based on pins made of cold-finished, low-carbon steel, and a safety factor of 8 is assumed.

Shaft size, in.	Pin diameter, in.	Torque, lb-in.	Horsepower at 100 rpm
3/16	1/16	4.6	0.007
7/32	5/64	8.4	0.013
1/4	3/32	13.7	0.022
5/16	7/64	23.6	0.038
3/8	1/8	37.2	0.060
7/16	5/32	67.6	0.108
1/2	5/32	77.2	0.124
9/16	3/16	125.0	0.200
5/8	3/16	139.0	0.222
11/16	7/32	207.0	0.332
3/4	1/4	297.0	0.476
13/16	1/4	322	0.516
7/8	1/4	347	0.555
15/16	5/16	580	0.927
1	5/16	618	0.990
1 1/16	5/16	657	1.05
1 1/8	3/8	1,010	1.61
1 3/16	3/8	1,065	1.70
1 1/4	3/8	1,120	1.79
1 5/16	7/16	1,590	2.55
1 3/8	7/16	1,670	2.67
1 7/16	7/16	1,740	2.79
1 1/2	1/2	2,380	3.81

SOURCE: Driv-Lok Inc., Sycamore, Ill.

TABLE 2-15 Minimum Single-shear Values (Pounds) of Grooved Pins of Various Materials

Pin dia., in.	Cold-finished 1213-1215 steel	Shear-proof alloy steel R_c 43-48	Brass	Silicon bronze	Heat-treated stainless steel
1/16	200	363	124	186	308
5/64	312	562	192	288	478
3/32	442	798	272	408	680
7/64	605	1,091	372	558	933
1/8	800	1,443	492	738	1,230
5/32	1,240	2,236	764	1,145	1,910
3/16	1,790	3,220	1,100	1,650	2,750
7/32	2,430	4,386	1,495	2,240	3,740
1/4	3,190	5,753	1,960	2,940	4,910
5/16	4,970	8,974	3,060	4,580	7,650
3/8	5,810	12,960	4,420	6,630	11,050
7/16	7,910	17,580	6,010	9,010	15,000
1/2	10,300	23,020	7,850	11,800	19,640

SOURCE: Driv-Lok Inc., Sycamore, Ill.

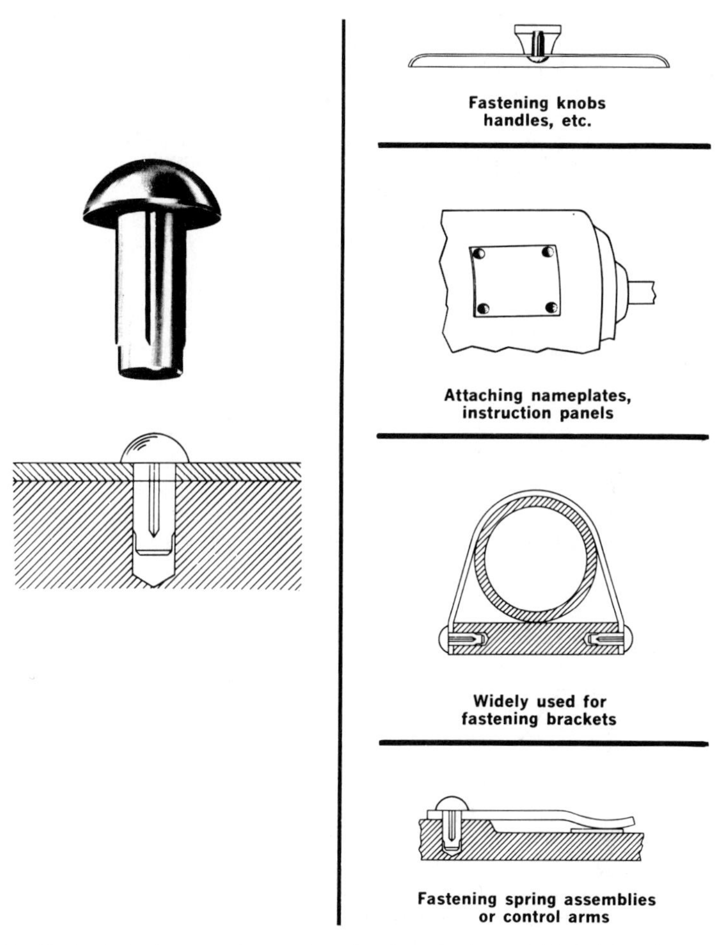

FIG. 2-12 Grooved stud. Typical grooved stud with application. (*Driv-Lok Inc., Sycamore, Ill.*)

2-38 Alternative Designs A variety of related quick-release pins have emerged upon the scene. The illustrated chart, Fig. 2-20, depicts some of the more prominent hybrid designs.

PIN INSERTION

2-39 Pin-Insertion Equipment For most fasteners, the cost of assembly and installation far exceeds the cost of the fastener itself, with most experts agreeing that assembly cost is approximately 6 times the price of the fastener. In order to reduce total in-place cost, most manufacturing engineers concentrate on assembly costs to attain substantial savings. To assist this cost-reduction effort, at least one pin manufacturer now offers pin-insertion equipment. Pin-insertion equipment is also available from builders of special assembly machines.

The main purpose and justification for pin-insertion equipment is the reduction of manual-assembly labor by mechanized or automated systems. Labor savings typically pay off the price of the machine in one year or better.

Flat head special stud with one-third length groove at lead end. Groove length can be varied.

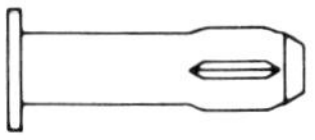

Flat head special grooved stud with shoulder. Often hardened to provide wear surface in shoulder area.

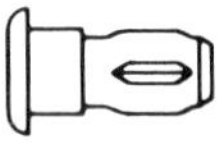

Flat head grooved stud.

Round head reverse taper groove stud.

Stud with conical head and parallel grooves.

Round head stud with parallel grooves of special length.

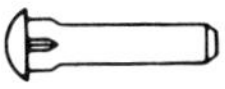

Countersunk head grooved stud.

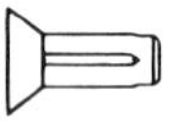

"T" head cotter used extensively in chain industry in place of cotter pins.

FIG. 2-13 Special stud pins. Eight special designs. *(Driv-Lok Inc., Sycamore, Ill.)*

Labor savings, however, are only one of the advantages of pin-insertion equipment. Some of the other benefits are: elimination of rejects, closer tolerances or more consistent quality, miniature assemblies that are impractical or impossible by hand, and the simple need for more assemblies per day.

Three typical pin inserters are shown in Figs. 2-21, 2-22, and 2-23.

2-40 Economic Considerations It is normally quite easy to justify mechanized pin assembly equipment when 300,000 or more pins per year are used. However, pin-insertion equipment can save money for users in the range of 25,000 to 50,000 assemblies per year. The key to the economic considerations is the present cost of manual versus machine assembly; this examination should be made regardless of pin volume.

Equipment prices range from a few hundred dollars for an arbor press with special fixturing all the way up to $20,000 or $30,000 for a multistation dial index table that drills holes, installs pins, inspects for proper pin assembly, and performs other related drilling, staking, machining, or inspection operations, and then often accomplishes automatic ejection of the assembly or of acceptable parts.

Straight cylindrical pins are by far the easiest item to handle, but other types such as headed pins, pins with multiple diameters, and other multidimensional objects can be handled as well. Some special pin configurations can not be oriented in conventional vibratory-bowl feeding equipment, so in these cases, the pins are fed manually. When possible, avoid nonsymmetrical pins at the design stage, especially when it appears mechanized assembly equipment would be desirable. It is best to work with a pin supplier and pin-insertion-equipment manufacturer in the early stages of design

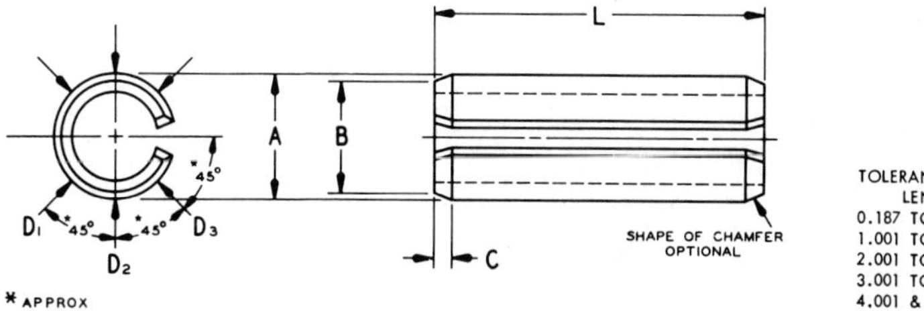

TOLERANCE ON SPECIFIED LENGTH "L"
0.187 TO 1.000 ±.015
1.001 TO 2.000 ±.020
2.001 TO 3.000 ±.025
3.001 TO 4.000 ±.030
4.001 & ABOVE ±.035

A			B	C		STOCK THICKNESS	RECOMMENDED HOLE SIZE		MINIMUM DOUBLE SHEAR STRENGTH POUNDS
NOMINAL	MAXIMUM (GO RING GAGE)	MINIMUM 1/3(D1+D2+D3)	MAX	MIN	MAX		MIN	MAX	
.062	.069	.066	.059	.007	.028	.012	.062	.065	425
.078	.086	.083	.075	.008	.032	.018	.078	.081	650
.094	.103	.099	.091	.008	.038	.022	.094	.097	1,000
.125	.135	.131	.122	.008	.044	.028	.125	.129	2,100
.140	.149	.145	.137	.008	.044	.028	.140	.144	2,200
.156	.167	.162	.151	.010	.048	.032	.156	.160	3,000
.187	.199	.194	.182	.011	.055	.040	.187	.192	4,400
.219	.232	.226	.214	.011	.065	.048	.219	.224	5,700
.250	.264	.258	.245	.012	.065	.048	.250	.256	7,700
.312	.328	.321	.306	.014	.080	.062	.312	.318	11,500
.375	.392	.385	.368	.016	.095	.077	.375	.382	17,600
.437	.456	.448	.430	.017	.095	.077	.437	.445	20,000
.500	.521	.513	.485	.025	.110	.094	.500	.510	25,800

FIG. 2-14 Design and dimensions of slotted tubular spring pins. *(ESNA Div., Amerace Corp., Union, N.J.)*

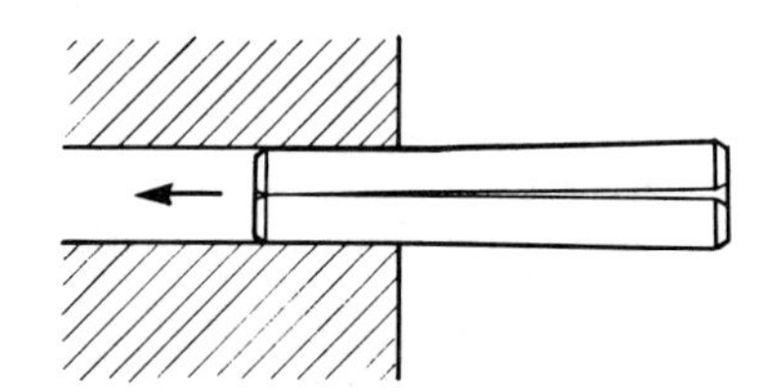

FIG. 2-15 Slotted pin entry into hole.

in order to obtain the optimum combination of part cost plus assembly cost to minimize total in-place cost.

COTTER PINS

2-41 Standard Cotter Pins Standard cotter pins are used to lock and secure clevises, nuts, and similar machine elements. The cotter pin is driven into the shaft hole and the eye will not permit passage, thus stabilizing the pin. After splitting the pin's legs, which prevents withdrawal, the pin is secure. Figure 2-24 pictures the pin designs and Table 2-20 summarizes the standard cotter-pin sizes.

TABLE 2-16 Average Maximum Removal Force in Steel (Pounds)*
Where lower insertion forces are important because pin is to be inserted in plastic or a soft metal, ask for information from manufacturer on un-heat-treated or thin-walled types.

NOMINAL DIAMETER INCHES	PIN SIZE NUMBER	ENGAGED LENGTH-INCHES							
		1/4	1/2	3/4	1	1 1/2	2	2 1/2	3
1/16	.062	125	180	230					
		45	80	105					
5/64	.078	160	310	440	540				
		100	210	300	370				
3/32	.094	250	470	620	720				
		120	230	350	460				
1/8	.125	260	500	730	950	1360			
		160	300	460	620	880			
5/32	.156	220	420	610	800	1150			
		150	300	440	580	840			
3/16	.187	650	1050	1250	1650	2300			
		450	800	1050	1250	1700			
7/32	.219	350	700	1000	1350	1900	2400		
		200	350	550	750	1100	1500		
1/4	.250	600	900	1200	1500	2100	2750		
		125	250	350	450	660	850		
5/16	.312	500	800	900	1100	1450	1800	2100	2500
		175	300	450	550	800	1050	1350	1600
3/8	.375	1000	1500	1850	2200	2800	3300	3700	4000
		200	350	500	700	1100	1500	1800	2200
7/16	.437	850	1350	1750	2200	3000	3600	4100	4400
		150	300	400	550	800	1100	1350	1600
1/2	.500	1100	2100	3000	3800	4900	6100	6800	7500
		450	850	1250	1600	2400	3000	3500	4100

* The two lines of figures to the right of each pin size number indicate the maximum hole force and minimum hole force, respectively.
SOURCE: Esna Div., Amerace Corp., Union, N.J.

TABLE 2-17 Drill-Hole Size Control
The curves illustrate the relatively free hole tolerances acceptable for slotted tubular pin fastenings plotted against the tolerance requirements of other types of pins.

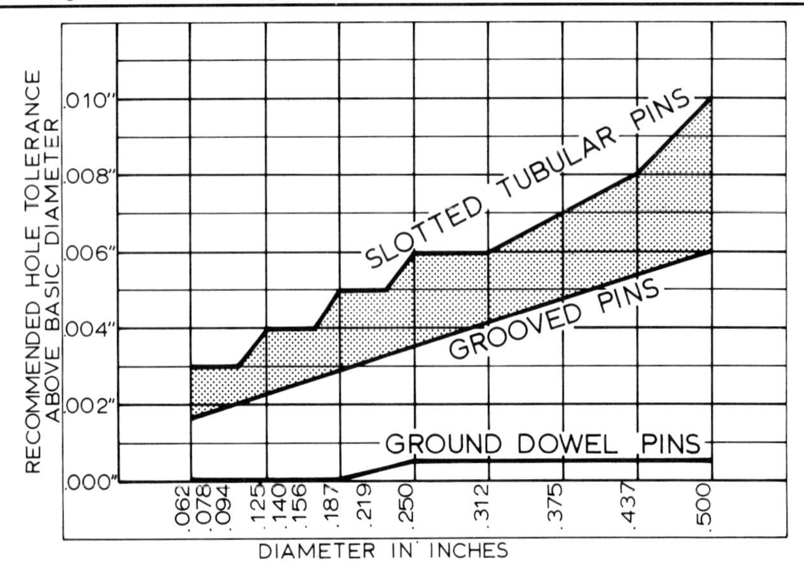

SOURCE: Esna Div., Amerace Corp., Union, N.J.

TABLE 2-18 Misalignment Table
So long as they are within recommended tolerances, holes drilled before assembly need not be redrilled or reamed to bring them to the same size. The spring-action of slotted tubular pins provides a firm grip in both.

NOMINAL PIN SIZE	MAXIMUM PERMISSIBLE MISALIGNMENT OF HOLE CENTERS
.062	.004
.078	.004
.094	.005
.125	.005
.156	.007
.187	.007
.219	.008
.250	.008
.312	.010
.375	.011
.437	.012
.500	.017

SOURCE: Esna Div., Amerace Corp., Union, N.J.

TABLE 2-19 Effects of Reuse on Insertion and Removal Forces

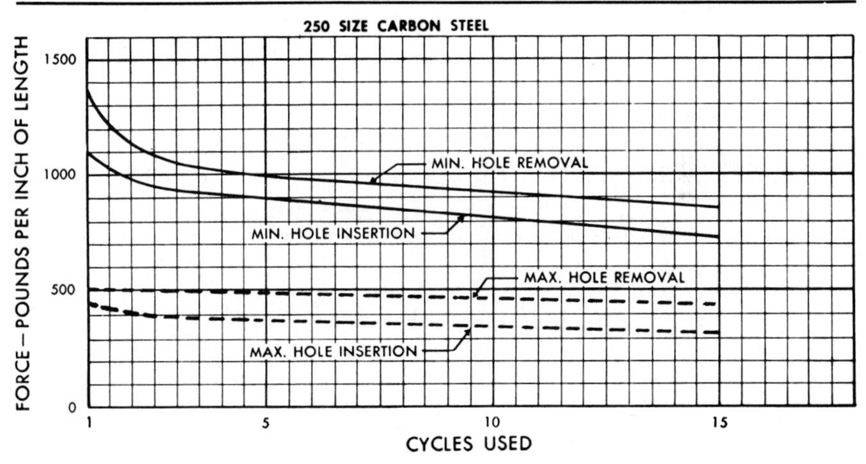

SOURCE: Esna Div., Amerace Corp., Union, N.J.

2-42 Material Unless otherwise specified, cotter pins shall be made from steel having the following chemical composition:

Carbon .... 0.15% max
Manganese .... 0.75% max
Phosphorus .... 0.04% max
Sulfur .... 0.05% max

When used as a Transverse Pin

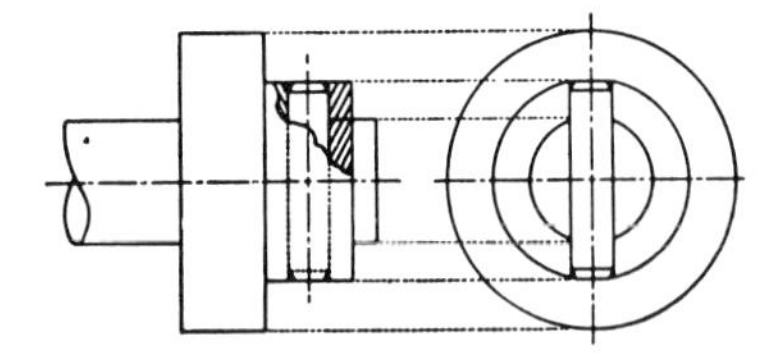

D Shaft Diameter Inches	d Pin Nominal Diameter, Inches	Pin Size Number
3/16	1/16	062
7/32	5/64	078
1/4	3/32	094
5/16-3/8	1/8	125
7/16-1/2	5/32	156
9/16 & 5/8	3/16	187
11/16 & 3/4	7/32	219
13/16	1/4	250
7/8	5/16	312
1	3/8	375
1-1/4	7/16	437
1-1/2	1/2	500

When used as a Longitudinal Key

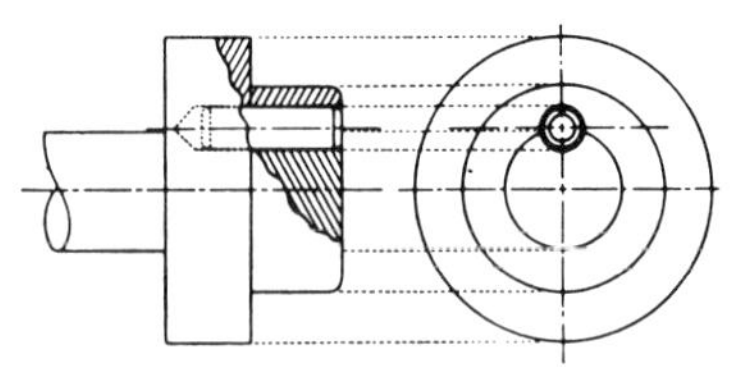

D Shaft Diameter Inches	d Pin Nominal Diameter, Inches	Pin Size Number
1/4	1/16	062
5/16	5/64	078
3/8	3/32	094
7/16-1/2	1/8	125
9/16-5/8	5/32	156
11/16-3/4	3/16	187
7/8	7/32	219
15/16-1	1/4	250
1-1/4	5/16	312
1-3/8	3/8	375
1-1/2 - 1-5/8	7/16	437
2	1/2	500

FIG. 2-16 Recommended pin sizes for shaft diameter. *(ESNA Div., Amerace Corp., Union, N.J.)*

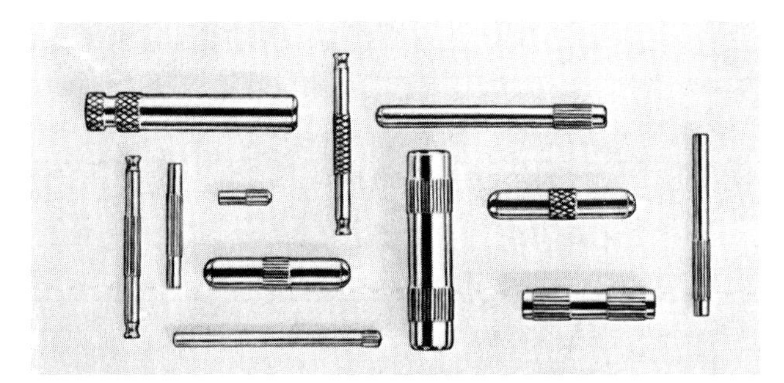

FIG. 2-17 Knurled pin designs. Variety of styles. *(Driv-Lok Inc., Sycamore, Ill.)*

Analysis determinations shall be made in accordance with methods given in ASTM Standard E30.

WIRE PINS

2-43 Wire Pins Mention should be made of the clip or formed spring-wire pin. Generally this pin is fabricated of stainless-steel material. Its use is in light applications and a basic feature is its removability and reuse.

2-44 Application This pin's general shape is shown in Fig. 2-25, which also illustrates a typical application. Note the pin's spring holding action, which stabilizes its position.

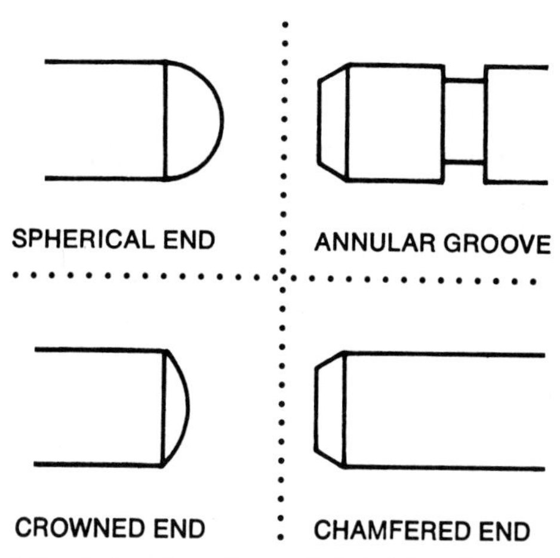

FIG. 2-18 End configurations. *(Driv-Lok Inc., Sycamore, Ill.)*

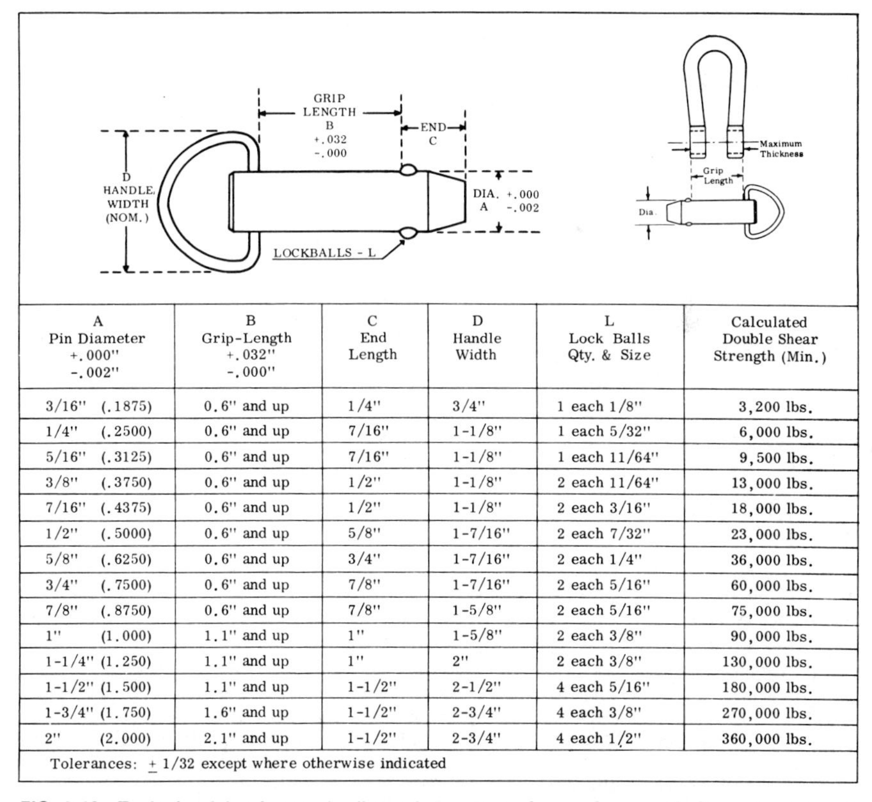

A Pin Diameter +.000" -.002"	B Grip-Length +.032" -.000"	C End Length	D Handle Width	L Lock Balls Qty. & Size	Calculated Double Shear Strength (Min.)
3/16" (.1875)	0.6" and up	1/4"	3/4"	1 each 1/8"	3,200 lbs.
1/4" (.2500)	0.6" and up	7/16"	1-1/8"	1 each 5/32"	6,000 lbs.
5/16" (.3125)	0.6" and up	7/16"	1-1/8"	1 each 11/64"	9,500 lbs.
3/8" (.3750)	0.6" and up	1/2"	1-1/8"	2 each 11/64"	13,000 lbs.
7/16" (.4375)	0.6" and up	1/2"	1-1/8"	2 each 3/16"	18,000 lbs.
1/2" (.5000)	0.6" and up	5/8"	1-7/16"	2 each 7/32"	23,000 lbs.
5/8" (.6250)	0.6" and up	3/4"	1-7/16"	2 each 1/4"	36,000 lbs.
3/4" (.7500)	0.6" and up	7/8"	1-7/16"	2 each 5/16"	60,000 lbs.
7/8" (.8750)	0.6" and up	7/8"	1-5/8"	2 each 5/16"	75,000 lbs.
1" (1.000)	1.1" and up	1"	1-5/8"	2 each 3/8"	90,000 lbs.
1-1/4" (1.250)	1.1" and up	1"	2"	2 each 3/8"	130,000 lbs.
1-1/2" (1.500)	1.1" and up	1-1/2"	2-1/2"	4 each 5/16"	180,000 lbs.
1-3/4" (1.750)	1.6" and up	1-1/2"	2-3/4"	4 each 3/8"	270,000 lbs.
2" (2.000)	2.1" and up	1-1/2"	2-3/4"	4 each 1/2"	360,000 lbs.

Tolerances: ± 1/32 except where otherwise indicated

FIG. 2-19 Typical quick-release pin dimensions. Basic design shown with dimension table of standard sizes. *(Waldick Aero-Space Devices, Inc., Long Island, N.Y.)*

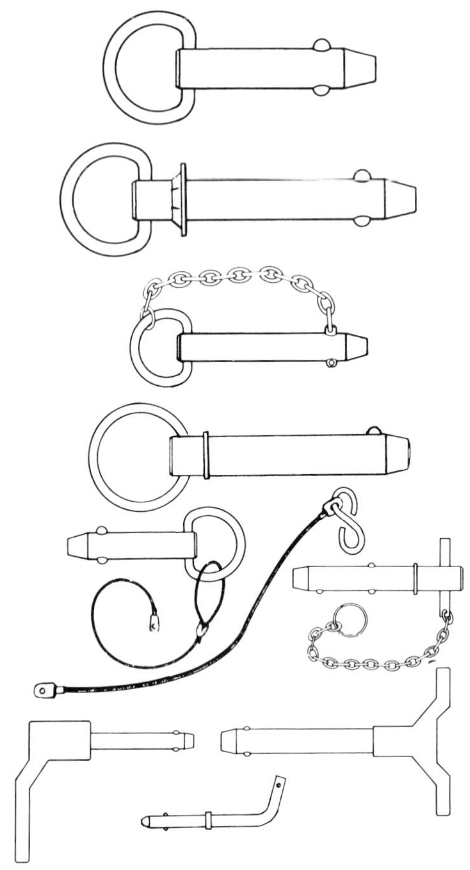

For shear loads - In standard drilled or punched holes - as found on farm and commercial vehicles, marine, aircraft, and industrial handling equipment, scaffolding, antenna masts, tow bars, hoists, shackles, etc. Detent type - Push to insert - Pull to release.

Same as basic ECO-Pins but have exclusive adjustable grip length feature. One stock pin will fit many jobs - cut inventory costs. Desirable to have on hand for prototype work.

For tension and shear loads - Double pin sets, chained together to prevent loss -- a fool proof positive lock pin design - one hand operation. A double safety pin for hoists, shackles and towing.

For shear loads - Ground shanks fit close tolerance holes. Silicone core activates lock buttons - no heat or icing problems - used on missile handling dollies - launching and ejection equipment, military vehicles. Available with ring or "T" handle.

Lanyards, chains, rings, hooks of all sizes and materials for preventing loss of pins, or for retaining miscellaneous parts to fixtures.

Special design pins - Combine features of other Waldick pins or with L and T handles - Custom made to fit your requirements. Submit sketch of shape you need, indicating diameter and grip length.

FIG. 2-20 Quick-release pin chart. Six general types of quick-release pins with descriptions. *(Waldick Aero-Space Devices, Inc., Long Island, N.Y.)*

BARBED PINS

2-45 Barbed Pins Barbed pins and studs will provide positive connections for plastics and some metals. The barbed fasteners can substitute for conventional screws, barbed nails, and other typical connectors to achieve permanent joining. Faster assembly is also a distinct advantage over standard fasteners.

2-46 Shank Design The barbs are positioned in three or four sets at 120° spacings on the pin shank. The crests of the barbs constitute the expanded shank diameter, which is a few thousandths of an inch greater than the nominal diameter.

Pins and studs have chamfered ends for easy insertion into mating holes. These fasteners are made from aluminum or low-carbon steel and are zinc-plated for resistance to corrosion. See Fig. 2-26.

Barbed pins are available in diameters from 3⁄16 in. to 3⁄8 in. in lengths from 1 in. to 2 in.

Barbed studs are made in shank diameters as small as 0.086 in. and as large as 0.250 in. in lengths 3⁄16 in. to 3⁄4 in.

JIG AND FIXTURE PINNING

2-47 Jigs and Fixtures Jigs and fixtures are used in the manufacturing industry to properly locate and hold a workpiece while a fabricating operation is performed on it. The jigs and fixtures generally have several component parts to execute their job of supporting, guiding, clamping, and gaging the workpiece. These component parts

FIG. 2-21 Rugged, precision welded, "C" frame pin inserter with automatic pin feed, foot switch actuator, and finger protection unit. Part holding fixture not shown. (*CEM Company, Inc., Danielson, Conn.*)

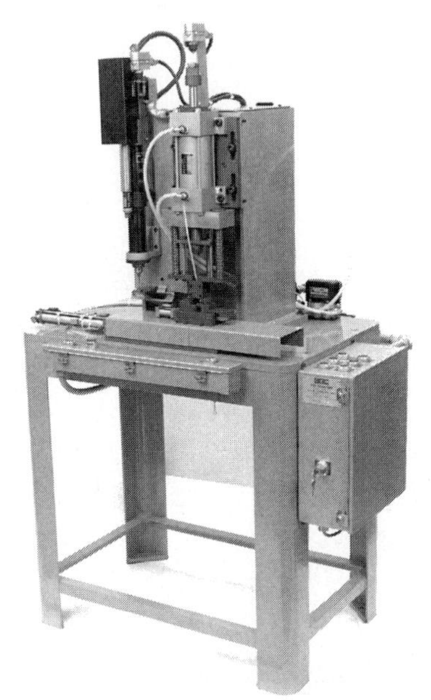

FIG. 2-22 An assembly machine that performs vertical drilling and vertical pinning. A contour fixture is mounted on a pneumatically actuated slide base. (*CEM Company, Inc., Danielson, Conn.*)

FIG. 2-23 An eight-station rotary-table assembly machine. Clockwise, the stations are: load, drill pin hole, drill hole for screw fastener, tap hole for screw fastener, insert $^{1}/_{16}$ inch × $^{5}/_{16}$ inch MBK Spirol® pin, stamp model number on part, automatically eject finished assembly; the eighth station is unused. This machine assembles a gas meter component. Four different models within the same family can be assembled by this one machine. (*CEM Company, Inc., Danielson, Conn.*)

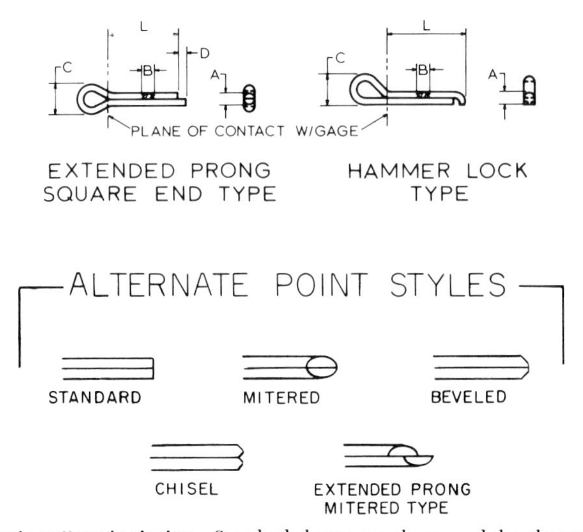

FIG. 2-24 Basic cotter pin design. Standard shapes are shown and the alternate pin points are illustrated. (Adapted from *ANSI B18.8.1-1972 and ASA B18.12-1962, as published by American Society of Mechanical Engineers.*)

TABLE 2-20 Standard Dimensions of Cotter Pins

Nominal Size[1] or Basic Pin Diameter	A Total Shank Diameter		B Wire Width		C Head Diameter	D Extended Prong Length	Recommended Hole Size
	Max	Min	Max	Min	Min	Min	
1/32 0.031	0.032	0.028	0.032	0.022	0.06	0.01	0.047
3/64 0.047	0.048	0.044	0.048	0.035	0.09	0.02	0.062
1/16 0.062	0.060	0.056	0.060	0.044	0.12	0.03	0.078
5/64 0.078	0.076	0.072	0.076	0.057	0.16	0.04	0.094
3/32 0.094	0.090	0.086	0.090	0.069	0.19	0.04	0.109
7/64 0.109	0.104	0.100	0.104	0.080	0.22	0.05	0.125
1/8 0.125	0.120	0.116	0.120	0.093	0.25	0.06	0.141
9/64 0.141	0.134	0.130	0.134	0.104	0.28	0.06	0.156
5/32 0.156	0.150	0.146	0.150	0.116	0.31	0.07	0.172
3/16 0.188	0.176	0.172	0.176	0.137	0.38	0.09	0.203
7/32 0.219	0.207	0.202	0.207	0.161	0.44	0.10	0.234
1/4 0.250	0.225	0.220	0.225	0.176	0.50	0.11	0.266
5/16 0.312	0.280	0.275	0.280	0.220	0.62	0.14	0.312
3/8 0.375	0.335	0.329	0.335	0.263	0.75	0.16	0.375
7/16 0.438	0.406	0.400	0.406	0.320	0.88	0.20	0.438
1/2 0.500	0.473	0.467	0.473	0.373	1.00	0.23	0.500
5/8 0.625	0.598	0.590	0.598	0.472	1.25	0.30	0.625
3/4 0.750	0.723	0.715	0.723	0.572	1.50	0.36	0.750

[1]Where specifying nominal size in decimals, zeros preceding decimal shall be omitted.

SOURCE: ANSI B18.8.1-1972, as published by American Society of Mechanical Engineers.

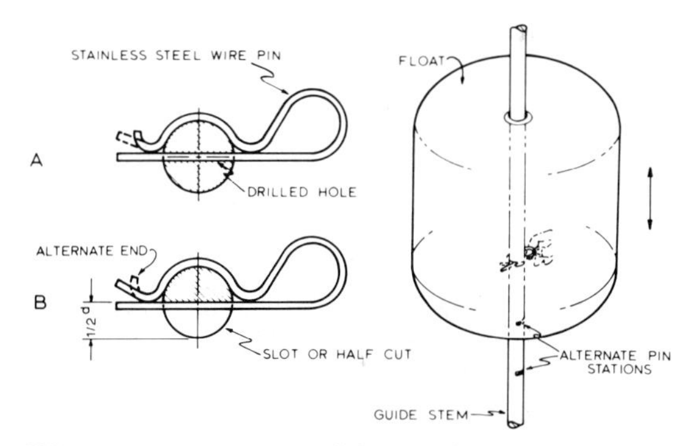

FIG. 2-25 Wire pin. General shape and hole assembly details.

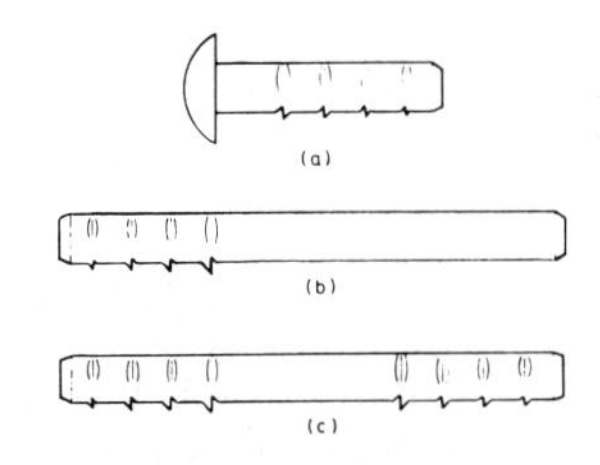

FIG. 2-26 Barbed stud and pins. *(a)* Barbed stud; *(b)* barbed pin, type 1; *(c)* barbed pin, type 2.

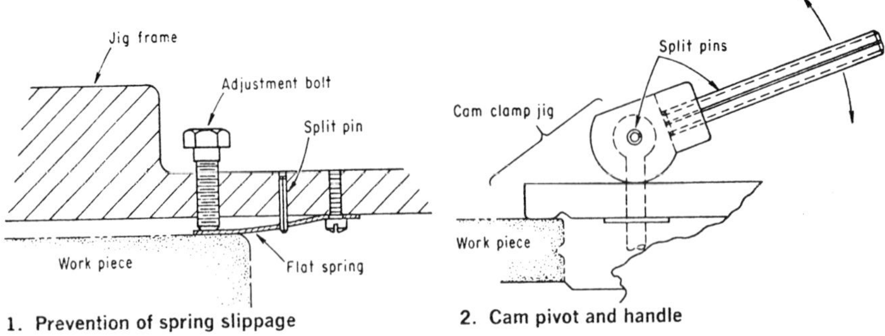

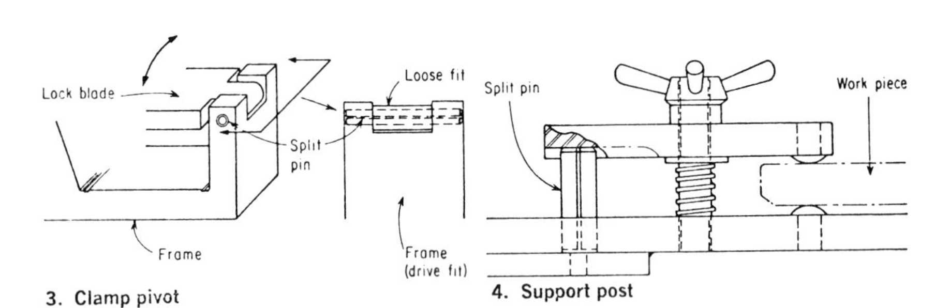

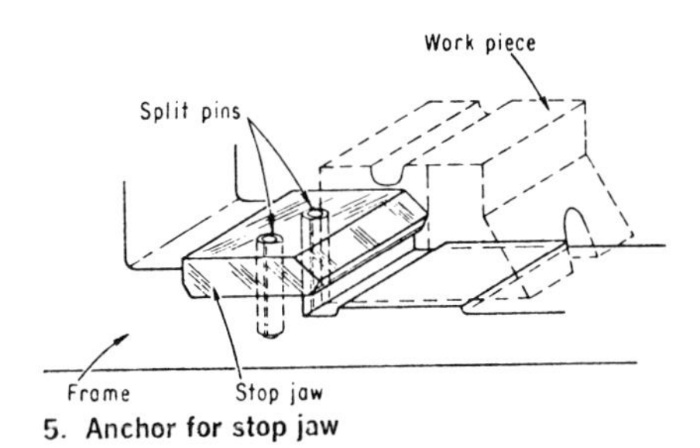

FIG. 2-27 Jig and fixture uses of slotted tubular pins (part 1). Examples show how these pins simplify assembly of jigs and fixtures. (*From R. O. Parmley, "Uses of Split Pins," in American Machinist, March 25, 1968. Reprinted by permission.*)

range from simple screws or pins to elaborate mechanisms. It is, therefore, this area where the pin and its evolved designs fill a key function.

2-48 Typical Assembly The author, motivated by the ever-increasing versatility of pins, wrote an article for *American Machinist* showing 10 examples of how tubular split pins simplify assembly of jigs and fixtures. The illustrations from this publication are reproduced here to stimulate the reader's creativity and reveal pin usage in this discipline of engineering (see Figs. 2-27 and 2-28).

2-49 Multiple Use A complex cam-operated slide clamp is pictured in Fig. 2-29. The multiple usage of a pin is elaborately revealed. The large variety and wide range of lengths and diameters of modern pins make their applications limitless.

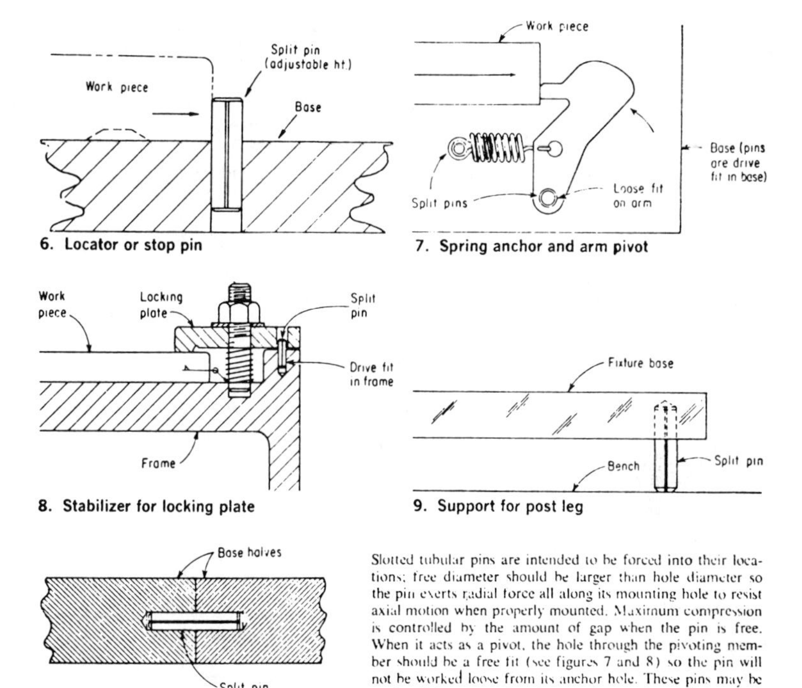

Slotted tubular pins are intended to be forced into their locations; free diameter should be larger than hole diameter so the pin exerts radial force all along its mounting hole to resist axial motion when properly mounted. Maximum compression is controlled by the amount of gap when the pin is free. When it acts as a pivot, the hole through the pivoting member should be a free fit (see figures 7 and 8) so the pin will not be worked loose from its anchor hole. These pins may be made of heat-treated carbon steel, corrosion-resistant steel, or beryllium-copper.

FIG. 2-28 Jig and fixture uses of slotted tubular pins (part 2). Five more examples of jig and fixture pinning. (*From R. O. Parmley, "Uses of Split Pins," in American Machinist, March 25, 1968. Reprinted by permission.*)

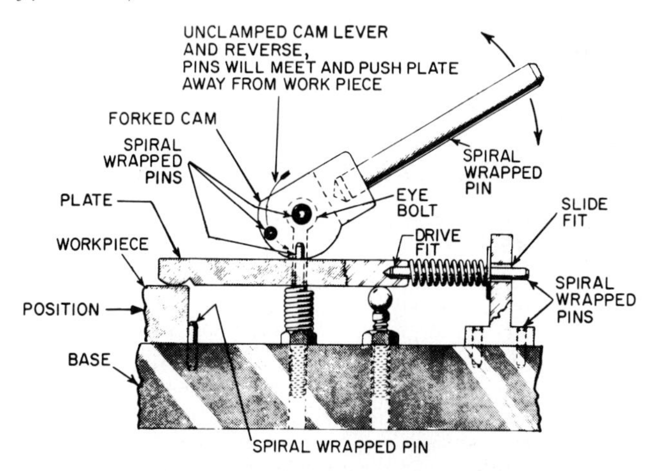

FIG. 2-29 Cam-operated slide clamp. Multiple use of spiral-wrapped pins. (*From R. O. Parmley, "Design Around Spiral-wrapped Pins," in Product Engineering, Oct. 25, 1965. Reprinted by permission. Morgan-Grampian Pub.*)

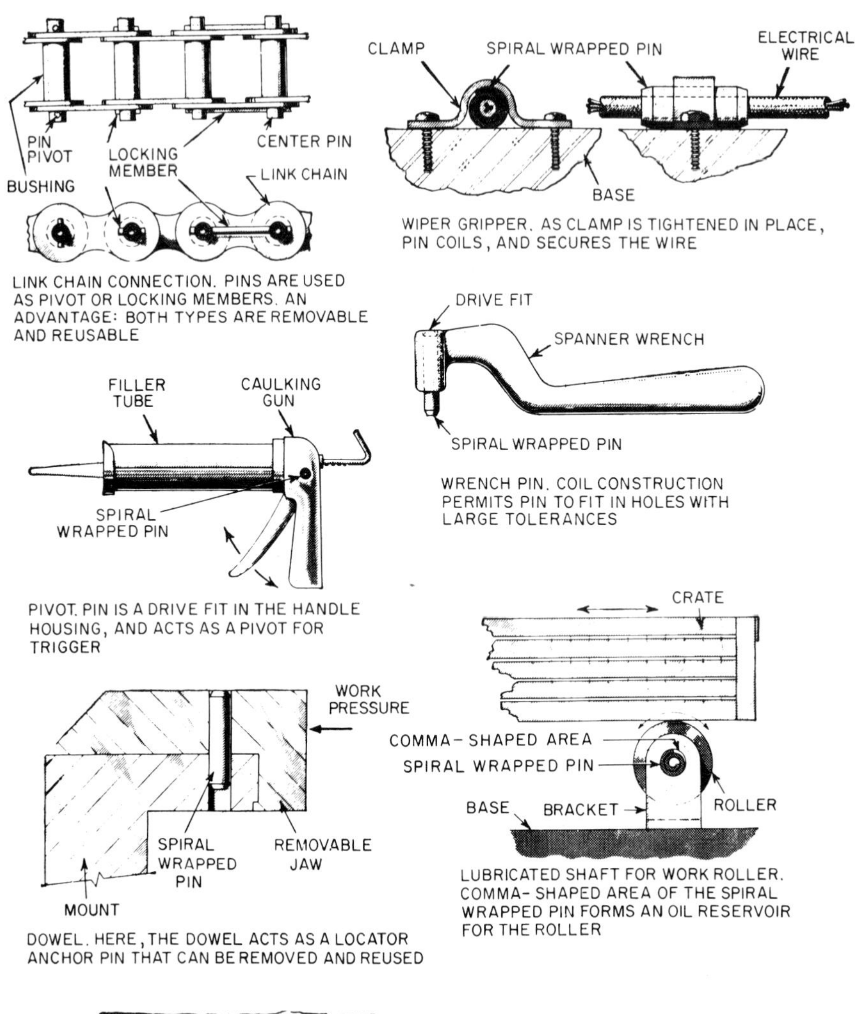
PIN PIVOT
LOCKING MEMBER
CENTER PIN
BUSHING
LINK CHAIN
LINK CHAIN CONNECTION. PINS ARE USED AS PIVOT OR LOCKING MEMBERS. AN ADVANTAGE: BOTH TYPES ARE REMOVABLE AND REUSABLE
CLAMP
SPIRAL WRAPPED PIN
ELECTRICAL WIRE
BASE
WIPER GRIPPER. AS CLAMP IS TIGHTENED IN PLACE, PIN COILS, AND SECURES THE WIRE
DRIVE FIT
SPANNER WRENCH
SPIRAL WRAPPED PIN
WRENCH PIN. COIL CONSTRUCTION PERMITS PIN TO FIT IN HOLES WITH LARGE TOLERANCES
FILLER TUBE
CAULKING GUN
SPIRAL WRAPPED PIN
PIVOT. PIN IS A DRIVE FIT IN THE HANDLE HOUSING, AND ACTS AS A PIVOT FOR TRIGGER
WORK PRESSURE
SPIRAL WRAPPED PIN
REMOVABLE JAW
MOUNT
DOWEL. HERE, THE DOWEL ACTS AS A LOCATOR ANCHOR PIN THAT CAN BE REMOVED AND REUSED
CRATE
COMMA-SHAPED AREA
SPIRAL WRAPPED PIN
BASE
BRACKET
ROLLER
LUBRICATED SHAFT FOR WORK ROLLER. COMMA-SHAPED AREA OF THE SPIRAL WRAPPED PIN FORMS AN OIL RESERVOIR FOR THE ROLLER

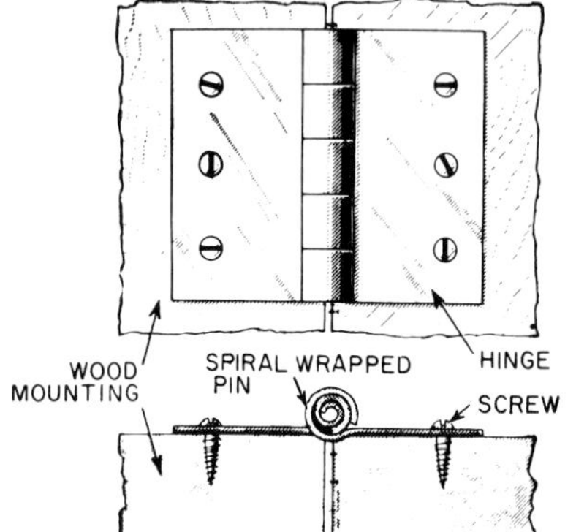
WOOD MOUNTING
SPIRAL WRAPPED PIN
HINGE
SCREW
HINGE PINS. IF HOLE SIZE IN BOTH MEMBERS IS DIFFERENT HINGE WILL BE FREE MOVING; IF THE HOLE SIZE IS THE SAME IN BOTH MEMBERS A FRICTION HINGE IS THE RESULT

FIG. 2-30 Unusual pin applications. Coil, rolled, or spiral-wrapped pins come in a wide range of lengths and diameters. Their applications are limitless; here are seven examples. (*From R. O. Parmley, "Design Around Spiral-wrapped Pins," in Product Engineering, Oct. 25, 1965. Reprinted by permission. Morgan-Grampian Pub.*)

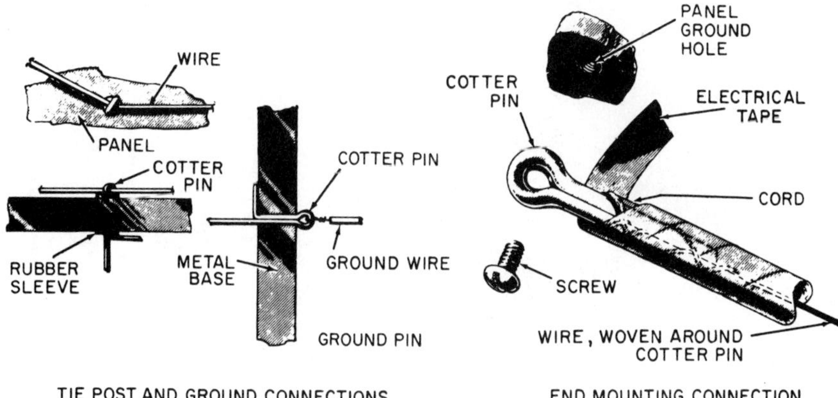

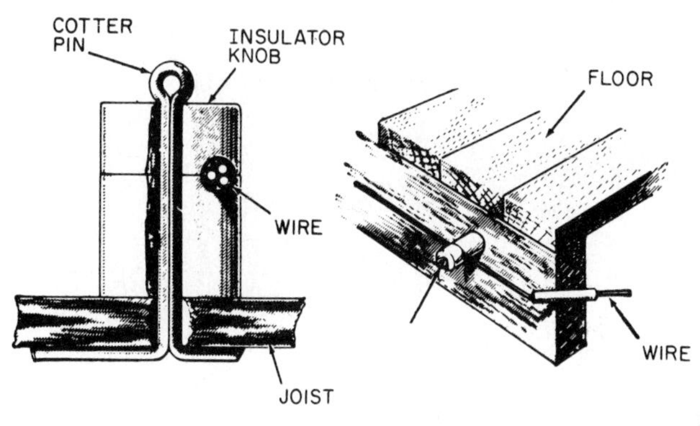

FIG. 2-31 Unusual cotter pin applications (Part 1). Inexpensive cotter pins make excellent electrical connectors. (*From R. O. Parmley, "A Penny-wise Connector—the Cotter Pin," in Product Engineering, July 19, 1965. Reprinted by permission. Morgan-Grampian Pub.*)

UNUSUAL PIN APPLICATIONS

2-50 Summary All the previous pins have many common features, but the paramount is versatility. The designer must keep this in mind when confronted with everyday fastening, joining, and locating problems. Isolate the problem and select the proper pin design and type. Major manufacturers of pins are most cooperative in assisting in solving application problems.

2-51 Examples Figures 2-30, 2-31, and 2-32 are included in this text as just a few examples of pin use. Pin applications, of course, are endless and limited space prevents complete coverage of all unusual areas. It is hoped, however, that the few examples illustrated here will shed light and expand the designer's expertise in pin usage.

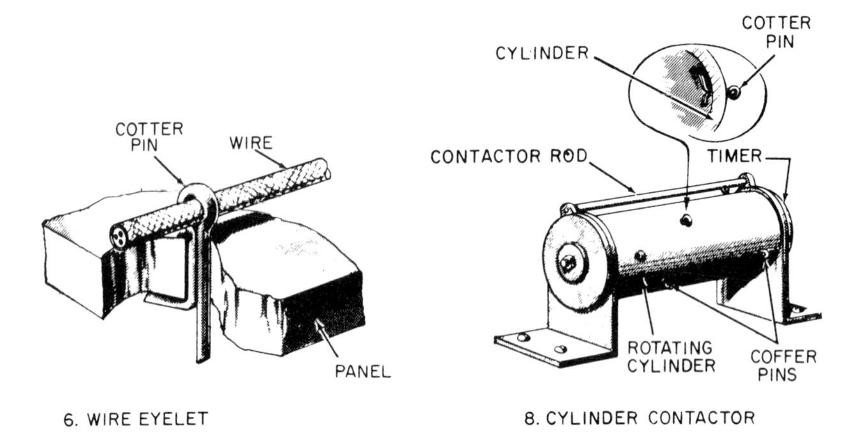

ELECTRICAL PANEL

COTTER PIN

PANEL

PRESS FIT (ALTERNATE)

BASE

STANDARD FIT

7. ELECTRICAL PANEL STABILIZER

FIG. 2-32 Unusual cotter pin applications (Part 2). Three more examples of cotter pins as electrical connectors. (*From R. O. Parmley, "A Penny-wise Connector—the Cotter Pin," in Product Engineering, July 19, 1965. Reprinted by permission. Morgan-Grampian Pub.*)

2-52 Special Pins Occasionally there are unusual design conditions that require special pins which are not standard stock items. When this situation occurs, a modified design is in order. Refer to Fig. 2-33 for a visual sampling of a wide range of special press fit pins.

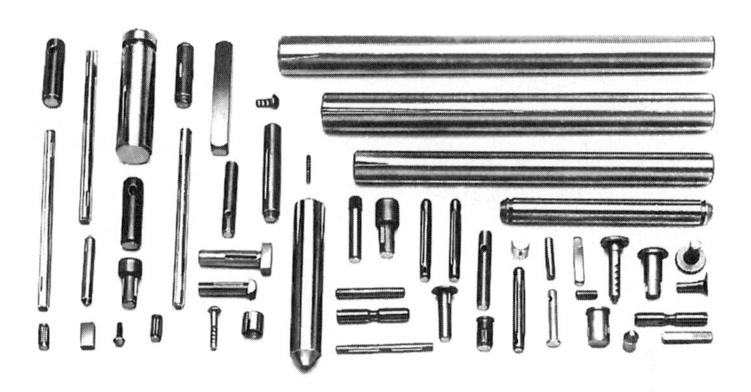

FIG. 2-33 Special Press Fit Pins. (*Courtesy: Driv-Lok, Inc.*)

ACKNOWLEDGMENTS

The author wishes to thank the following companies for their technical assistance:
CEM Co., Inc., P.O. Box 179, Danielson, CT 06239
Driv-Lok, Inc., 1140 Park Ave., Sycamore, IL 60178
Elastic Stop Nut Division of Amerace-Esna Corp., Union, NJ 07083

REFERENCES

LeGrand, Rupert, ed., *The New American Machinist's Handbook*, McGraw-Hill, New York, 1955. ANSI technical standards, available from American National Standards Institute, 1430 Broadway, New York, NY 10018, and The American Society of Mechanical Engineers.

Greenwood, D.C., ed., *Engineering Data for Product Design*, McGraw-Hill, New York, 1961.

Section 3
Retaining Rings

Part 1 Standard Retaining Rings*

EINAR A. ERIKSEN†

Manager of Engineering Services, Waldes Truarc, Inc.,
Member of the Seeger Fastening Systems Group, Long Island City, New York

signed to serve as accurately located shoulders for positioning and securing components

*All photos of product applications in sec. 3, part 1, courtesy of Waldes Truarc, Inc.

†NOTE: Original author of sec. 3, part 1, in the first edition of this handbook, was Melvin Millheiser, Chief Engineer, Truarc Retaining Ring Division, Waldes Kohinoor, Inc.

in an assembly. Most rings are made of materials having good spring properties which enable the devices to be deformed elastically to a considerable degree during installation or removal, yet spring back to their original shape. These spring characteristics permit retaining rings to be used in different ways: They may be compressed or expanded for assembly in a groove or other recess in a bore or housing or on a shaft or stud—or they may be seated on a part in a deformed condition so that they grip by means of friction. In all cases, the rings form a fixed shoulder against which other components may be abutted and prevented from moving.

The first two basic types of stamped, tapered-section rings were introduced in the United States in 1942 for use in aircraft and were manufactured to conform to National Aircraft Standards. After World War II, the rings were adapted in other industries and many different types were developed for consumer, industrial, and military product applications. Today, there are more than 50 functionally different ring types available for a variety of fastening and assembly requirements.

In most applications, rings are used to provide a *removable* means of fastening—i.e., their elastic properties permit the rings to be deformed and disassembled so that the parts they are holding in place may be dismantled without the assembly being destroyed. There are a few ring types, however, which must be destroyed for removal.

Internal-type rings are assembled in holes, bores, or housings; *external* types are installed on shafts, pins, studs, and similar parts. Both types offer a number of advantages over other fastening methods:

- Their cost is relatively low when compared with other types of fasteners.
- Their use often results in raw-material savings and simplified machining operations for other parts in the assembly.
- One retaining ring often can replace two or more parts.
- Assemblies using retaining rings generally are more compact compared with assemblies having other fastening means.
- Assembly tooling developed for retaining rings usually permits very rapid assembly of the fasteners, even by unskilled labor.

3-2 Materials and Finishes Because retaining rings depend for their function largely on their ability to be deformed elastically during assembly and disassembly, the materials of which they are made must have good spring properties. These include high tensile and yield strengths, as well as a ratio of ultimate tensile strength to modulus of elasticity which permits the required deformation without excessive permanent set. A ratio of 1:100 is satisfactory for most tapered-section ring designs.

Some of the more common materials from which retaining rings are made and which provide the necessary spring properties are carbon spring steel (SAE 1060 to 1090); precipitation-hardening stainless steel, such as PH 15-7 Mo and beryllium copper, alloy 25. Hardening of the carbon steels, stainless steels, and beryllium copper after ring fabrication is necessary to develop the spring properties needed for good ring function.

A variety of protective finishes is available for retaining rings. Among these are plain or oiled finish, zinc plating, and phosphate coating. Stainless steels always are passivated.

The selection of a particular material and finish will depend upon the conditions under which the ring is to be used: temperature, the presence of corroding media, loads to which the ring will be subjected, and other factors.

Heat Treating and Plating. Retaining rings are subjected to extremely high stresses during their installation and removal and also are under a certain degree of stress when they are seated in an assembly. Carbon-steel rings, for example, may undergo stress of as high as 250,000 lb/in.^{2} during installation. Under these conditions, improper heat-treating methods which produce brittleness–or improper plating techniques which introduce hydrogen embrittlement—can cause sudden ring failure during either ring installation or removal, or when the rings are used under even relatively light loading conditions. Special heat-treating techniques, such as austempering, and plating processes, such as mechanical plating, which avoid hydrogen embrittlement, are used to prevent this type of ring failure.

3-3 Sizes and Types Retaining rings are available in a very wide range of sizes: small enough for shafts only 0.040 inch, or 1 mm, in diameter; large enough for shafts and housings 25 inches, or 635 mm, in diameter. Specific sizes generally are available to accommodate inch and fractional-inch sizes of shafts and bores and also common millimeter sizes, corresponding to commercially available ball and roller bearing sizes.

Retaining rings may be classified as either *axially assembled* rings or *radially assembled* rings. The first are installed in a direction parallel to the axis of the shaft or bore in which the rings will be used. Radially assembled rings are installed in a radial direction to the axis of the part on which they will be used.

It is convenient to classify all rings into axial or radial groups because rings falling into either category generally have other common features. For example, most axially assembled rings have small gaps. They contact grooves in which they are seated and, when they are subjected to thrust loads, transmit the loads to the groove wall around most of the groove circumference. As a result, most axially assembled rings have relatively high thrust-load capacity.

Radially assembled rings, on the other hand, must have wide gaps for installation. Since the gap portion cannot be used for transmitting thrust loads to the groove wall, the radial types generally have lower thrust capacity than the axial rings.

3-4 Allowable Static Thrust Loads Three factors must be considered in determining the allowable static thrust capacity of a retaining ring assembly:

1. The thrust load is transmitted to the groove wall by the retaining ring and the strength of the wall must be calculated based on the type of ring being used.
2. The retaining ring is subjected to a shearing action because of the load acting upon it. Thus, the ring's shear strength also must be calculated.
3. The part abutting the retaining ring may have a corner radius or chamfer which will cause the load acting on the ring to be concentrated at a small distance away from the shaft or bore in which the ring is used. This will increase the ring's tendency to dish under load and may permit excessive movement of the retained parts or even cause the ring to "walk" out of its groove. If a corner radius or chamfer is present, therefore, the load capacity of the ring under these conditions also must be calculated.

The smallest of the three values calculated determines the static-thrust capacity of the retaining ring assembly.

Calculating Groove-Wall Strength. The strength of the groove wall may be calculated from the formula

$$P_g = \frac{C_F S d \pi s_y}{F q} \tag{3-1}$$

where P_g = allowable static thrust load on the groove wall
C_F = conversion factor (see Table 3-1)
S = shaft or housing diameter
d = groove depth
s_y = tensile yield strength of groove material, psi (see Table 3-2)
F = safety factor (a factor of 2 is satisfactory for most conditions)
q = reduction factor, taken from curve illustrated in Fig. 3-1 (dimension Z is the distance of the outer groove wall from the end of the shaft or bore, as shown in Fig. 3-2)

Calculating Shear Strength. The strength of a retaining ring in shear is calculated from the formula

$$P_r = \frac{C_F S t \pi s_s}{F} \tag{3-2}$$

where P_r = allowable thrust load of the ring, lb
C_F = conversion factor (see Table 3-1)
S = housing or shaft diameter, inches
t = ring thickness, inches
s_s = ultimate shear strength of ring material, psi (see Table 3-3)
F = safety factor

TABLE 3-1 Conversion Factor C_F for Calculating P_r and P_g

Ring type	Conversion factor C_F	
	Ring: P_r	Groove: P_g
Basic, bowed internal	1.2	1.2
Beveled internal	1.2	1.2 Use $d/2$ instead of d
Inverted internal, external	$^2/_3$	$^1/_2$
Basic, bowed external	1	1
Beveled external	1	1 Use $d/2$ instead of d
Crescent-shaped	$^1/_2$	$^1/_2$
Two-part interlocking	$^3/_4$	$^3/_4$
E ring, bowed E ring	$^1/_3$	$^1/_3$
Reinforced E ring	$^1/_4$	$^1/_4$
Locking-prong ring	See manufacturer's specifications	$^1/_2$
Heavy-duty external	1.3	2
High-strength radial	$^1/_2$	$^1/_2$
Miniature high-strength	1.33	2
Thinner-gage high-strength radial	$^1/_2$	$^1/_2$

TABLE 3-2 Tensile Yield Strength of Groove Material

Groove material	Tensile yield strength, psi
Cold-rolled steel	45,000
Hardened steel (*RC* 40)	150,000
Hardened steel (*RC* 50)	200,000
Aluminum (2024-T4)	40,000
Brass (naval)	30,000

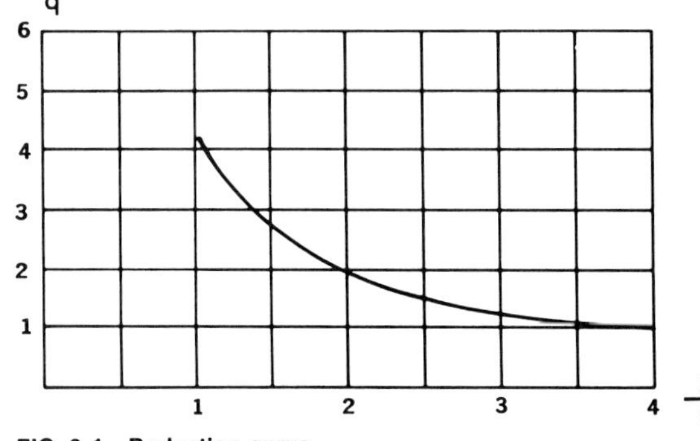

FIG. 3-1 Reduction curve.

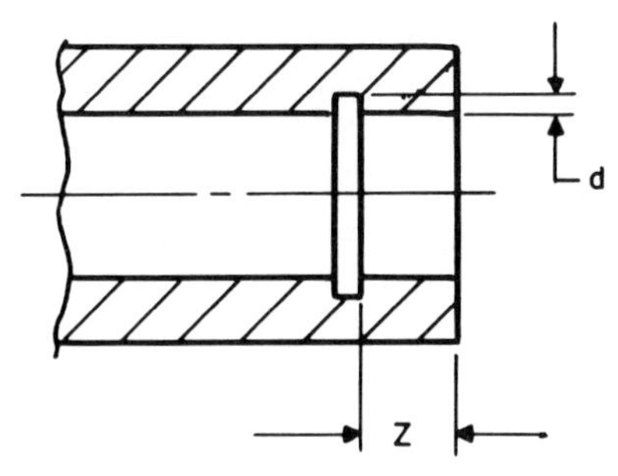

FIG. 3-2 Edge margin.

TABLE 3-3 Ultimate Shear Strength of Ring Material

Ring material	Ring type	Ring thickness, inch	Ultimate shear strength, psi
Carbon spring steel (SAE 1060–1090) and stainless steel (PH 15-7 Mo)	Basic, bowed, beveled, inverted: internal and external rings and crescent-shaped	Up to and including 0.035	120,000
		0.042 and over	150,000
	Heavy-duty external	0.035 and over	150,000
	Miniature high-strength	0.020 and 0.025	120,000
		0.035 and over	150,000
	Two-part interlocking, reinforced E ring, high-strength radial	All available	150,000
	Thinner high-strength radial	All available	150,000
	E ring, bowed E ring	0.010 and 0.015	100,000
		0.025	120,000
		0.035 and over	150,000
	Locking-prong	All available	130,000
Beryllium copper (alloy 25, CDA 172)	Basic external	0.010 and 0.015 (sizes-12 thru-23)	110,000
	Bowed external	0.015 (sizes-18 thru-23)	110,000
	E ring	0.010 (size -X4 only)	95,000

Calculating Loads with Corner Radii or Chamfers. The maximum allowable corner radius or chamfer of a retained part for various types of rings is given in Tables 3-5 through 3-8 (condensed load tables). In no case should a corner radius or chamfer exceed the value given. For the maximum values of radius and chamfer, the load capacity of the ring must be selected from the tables, listed under the symbol P'_r. If the actual corner radius or chamfer is less than the listed maximum value, the load capacity of the ring may be increased according to the formulas

$$P''_r = P'_r \frac{R_{max}}{R} \quad \text{(for radius)} \tag{3-3}$$

$$P''_r = P'_r \frac{Ch_{max}}{Ch} \quad \text{(for chamfer)} \tag{3-4}$$

where P''_r = allowable assembly load when corner radius or chamfer is *less* than listed maximum (in no case can P''_r exceed P_r)
P'_r = listed allowable assembly load with maximum corner radius or chamfer
R_{max} = listed maximum allowable corner radius
R = actual corner radius
Ch_{max} = listed maximum allowable chamfer
Ch = actual chamfer

Where static loading conditions exist, these formulas are generally sufficient to determine the allowable strength of a retaining ring assembly.

Relative Rotation. Where a part which is exerting a static load against the ring also rotates against the ring, the allowable load that may be exerted by the part on the ring may be determined by the following formula:

$$P_{rr} \leqq \frac{stE^2}{\mu 18S} \tag{3-5}$$

where P_{rr} = allowable thrust load exerted by adjacent part, lb
s = maximum working stress of ring during expansion or contraction (see Table 3-4)
t = ring thickness, inches
E = largest section of ring, inches
μ = coefficient of friction between ring and retained part or groove, whichever is higher (consult appropriate reference)
S = shaft or housing diameter, inches

TABLE 3-4 Maximum Working Stress of Ring During Expansion or Contraction

Ring material	Max. allowable working stress, psi
Carbon spring steel (SAE 1060-1090)	250,000
Stainless steel (PH 15–7 Mo)	250,000
Beryllium copper (alloy 25)	200,000

Dynamic Loading. Dynamic conditions most often encountered in retaining ring assemblies include sudden loading, impact, vibration, and relative rotation. Very often the loading pattern is cyclical in nature and may induce fatigue in the assembly. Where dynamic loads are likely to exist, it is necessary that actual tests of such applications be made by the ring user to ensure proper functioning of the assembly. The following formulas are given for calculating the ring and/or groove thrust load capacity for various conditions:

Allowable sudden load on ring, $P_{SR} \leqq 0.5\,P_r$
Allowable sudden load on groove, $P_{SG} \leqq 0.5\,P_g$
Allowable vibration loading on ring, $wa \leqq 540\,P_r$*
Allowable vibration loading on groove, $wa \leqq 400\,P_g$*

* NOTE: Actual tests should be made because of repeated or cyclic conditions.

When there is play between the ring and retained part and the part strikes the ring, the load must be calculated as impact. Formulas are:

Allowable impact loading on ring, $$I_r = \frac{P_r t}{2} \tag{3-6}$$

Allowable impact loading on groove, $$I_g = \frac{P_g d}{2} \tag{3-7}$$

Definition of terms:

P_{SR} = allowable sudden load on ring, lb
P_{SG} = allowable sudden load on groove, lb
I_g = allowable impact load on groove, in.-lb

I_r = allowable impact load on ring, in.-lb
P_g = allowable static thrust load on groove, lb
P_r = allowable static thrust load on ring, lb
w = weight of retained parts, lb
a = acceleration of retained parts, in./sec^2

For harmonic oscillation,

$$a \cong 40\,\rho f^2 \tag{3-8}$$

where ρ = amplitude of vibration, inches
f = frequency of vibration, cycles/sec

3-5 Ring Selection The selection of a particular retaining ring for an assembly will depend on such conditions as the load the ring must withstand, clearance available for ring assembly, whether the ring must provide for rigid or resilient take-up of accumulated tolerances in the assembly, whether it is possible to machine a groove in the part for the ring, the necessity of having the ring adjustable to various positions on a shaft, and other factors.

When all conditions have been considered, the choice of more than one ring may be possible. The final selection of a ring for a particular assembly should take into consideration savings which may be achieved in various parts of the assembly, the cost of installing the ring, and, of course, the ability of the ring to do the job. Ideally, the ring which will function adequately and provide for the most economical overall manufacture is the one that should be selected.

In examining the different types of rings that are available, one should consider the features of each type and the advantages and disadvantages that may result from these features.

In tapered-section retaining rings, the section height (measured in a radial direction as we proceed around the circumference of the ring from its midpoint to the open end or gap) decreases uniformly. Since the ring, when it is compressed or expanded for installation, is subjected to a varying bending moment around its circumference, the tapered section is so designed as to provide as nearly as possible a constant value of M/EI, where

M = the bending moment at the particular ring section
E = the modulus of elasticity of the ring material
I = the moment of inertia of the particular ring section

Under these conditions, the ring will keep a uniform radius during expansion or contraction and will, therefore, retain its circular shape when seated in its groove, thus providing for uniform contact all around the groove circumference. This characteristic improves the load-carrying capability of the ring and reduces the maximum bending stress in the ring for a given amount of expansion or contraction.

Uniform section rings made of circularly wound wire are subject to much higher bending stresses and to oval deformation which results in nonuniform contact with the bottom of their grooves around the groove circumference.

3-6 Axially Assembled Internal and External Rings *Basic internal rings* have a relatively small gap to permit contraction of the ring for insertion or removal in a bore or housing. The tapered section extends almost completely around the ring, providing maximum flexibility with minimum permanent set and permitting the ring to be seated in relatively deep grooves. The outside diameter of the ring is concentric with the bore and groove diameters. These features assure maximum ring contact over most of the groove circumference and provide for high thrust capacity of the assembly. Lugs with holes, on each side of the gap, permit use of assembly pliers for ring installation and removal. When the ring is seated in a groove, it forms a nonuniform shoulder against which parts may be abutted. The design allows good clearance for parts which must pass through the ring and which may be in position while the ring is being installed.

Basic external rings have only a minute gap. The tapered section extends almost completely around the ring and, as is the case with the internal type, permits maximum flexibility without excessive permanent set. Inside diameter is concentric with the diameters of the shaft and groove. When thrust loads develop, the ring contacts the groove wall around the entire groove circumference and provides high thrust resistance. Lug holes again are designed for assembly pliers. External rings also provide a nonuniform shoulder against abutting parts.

Inverted internal rings feature a relatively small gap for contraction in a bore or housing. The tapered section extends around a major portion of the circumference for maximum flexibility and seating in grooves of adequate depth. The inside diameter is concentric to the bore and groove diameters. Height of the lugs is equal to the maximum height of the tapered section at its midpoint. Because of the inverted tapering, there is less contact with the groove wall than is provided by the basic type and a somewhat reduced thrust-load capacity. But the ring provides a uniform shoulder against the retained part and greater assembly clearance. The inverted ring also provides a more uniform appearance than the basic type.

Inverted external rings have the same design features as the internal types, except that the lugs are located on the inner circumference. When the rings are installed in a groove on a shaft or similar part, they form the same type of uniform shoulder against the parts to be retained.

Heavy-duty external rings differ from the basic type in that the average radial section height and thickness are greater. The ring is designed to take a permanent set on installation and grips the bottom of the groove very tightly. It resists dishing under loads exerted by parts with large corner radii or chamfers and, because of its heavy construction, is secure against high thrust loads. The rings are useful where repeated loading and fatigue are likely to occur; they eliminate the need for support washers.

Miniature high-strength rings are thicker than standard types and have higher radial sections. They do not have lugs and are especially suitable for tamperproof assemblies. The rings provide high thrust capacity in relatively shallow grooves on small shafts.

Permanent-shoulder rings have no gap and cannot be expanded or contracted like rings made of spring material. They have notches around their circumference and are crimped into a groove by means of special crimping pliers or collet-like assembly tools. The notches deform during the crimping process so that ring diameter is reduced and the ring engages with the groove. Once installed, the ring forms a permanent shoulder on a shaft and must be destroyed to be removed. The rings are useful where a tamperproof construction is desired and have relatively high thrust capacity because they contact the groove wall completely around their circumference. Permanent-shoulder rings are made of soft materials such as unheat-treated carbon steel, brass, and aluminum.

Typical product applications of axially assembled internal and external rings are illustrated in Figs. 3-3 through 3-8. Dimensions, load capacities, and other engineering data are given in Tables 3-5 and 3-6. Figure 3-9 gives a graphic outline of rings in this range.

3-7 Radially Assembled Rings *Crescent-shaped rings* have a wide gap for radial assembly into a groove on a shaft and provide a low shoulder against a retained part. Their design is inverted—i.e., the outside diameter is concentric with the shaft and groove and the tapered ring construction is provided by an eccentric inside diameter. The tapering extends around most of the body. Small lugs on each side of the gap provide two seating points in the groove; a third point is provided by the maximum section. Assembly is accomplished by means of applicators, dispensers, and other tools illustrated in sec. 3-10, Assembly Methods (Figs. 3-30 through 3-32). These rings are particularly useful where clearance is limited and a low shoulder is required. Because of the wide gap and partial groove contact, their thrust capacity is moderate.

E rings have a large gap and large outside diameter. Three lugs inside the ring circumference seat the ring in its groove and provide contacting surfaces for the groove wall. Height of the lugs makes it possible to use somewhat deeper grooves than those for crescent-shaped rings, providing adequate groove-wall contact for moderate thrust resistance. *E* rings are used to provide relatively large-diameter shoulders on small shafts. The portions between the lugs have a uniform section so that deflection during assembly must be limited. These rings are installed with the same type of applicators, dispensers, and other tools used for the crescent type.

Reinforced E rings differ from conventional *E* rings in that the bending portions between the lugs are tapered and thus can provide a greater radial bending section than

FIG. 3-3 Tapping attachment uses *basic internal ring* (inset) to lock collet assembly and other components in housing. Smaller ring of same type (not shown) positions and secures clutch bearing and sleeve. Alternative design called for components to be locked into housing with threaded locking rings seated in internal threads in housing. Manufacturer found use of retaining rings 10 times as cost-efficient as threaded fasteners and obtained additional savings on parts, machining and assembly.

FIG. 3-4 Fire hydrant was redesigned with retaining rings to eliminate hazardous lead sealing operation used to secure nozzle assembly in cast iron body. By switching to large *basic external ring,* manufacturer sharply reduced production costs and slashed field service time for nozzle replacement from approximately 3 hours to 15 minutes.

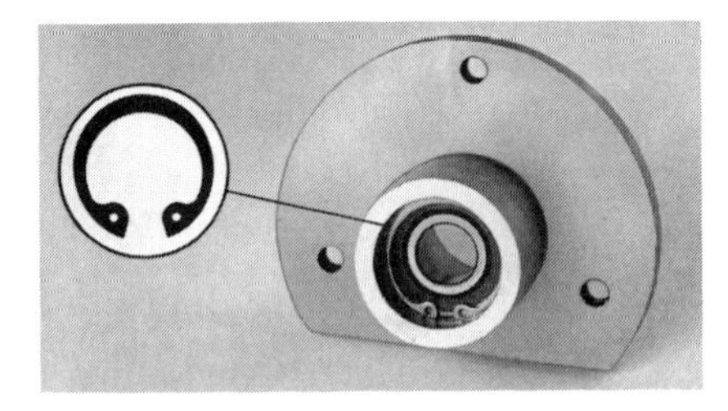

FIG. 3-5 Precision packaging machine uses star wheel shaft bearing housings, made of cast aluminum, to hold two bearings press-fit into position. Large *basic internal ring* is assembled in groove in housing, then first bearing is pressed into position until it comes to stop against ring. Housing is then turned over and second bearing pressed in until it is flush with housing. When shaft is pressed in, retaining ring keeps bearings from moving. Use of ring eliminates expensive custom-designed nest previously required and lowers assembly costs.

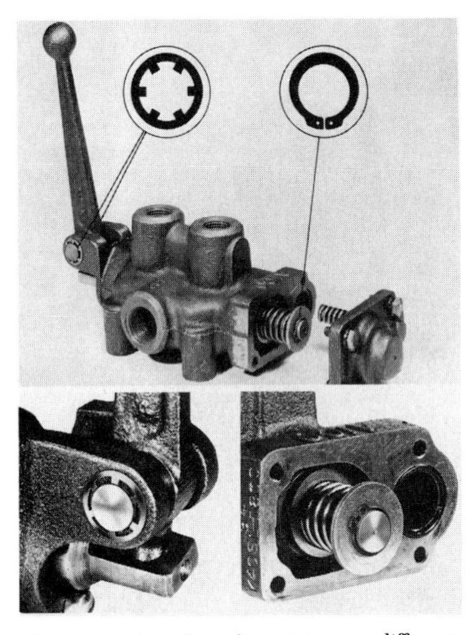

FIG. 3-6 Hydraulic valve uses two different types of retaining rings to position and secure components: *circular self-locking ring* (left-hand inset) to hold actuating lever and *heavy-duty external ring* (right-hand inset) to fasten spool end. Self-locking rings, which need no groove, replace socket-head screws and eliminate need for drilling and tapping hinge pin and casting. Heavy-duty ring replaces hex-head bolt used in original design. Use of retaining rings reduces costs for parts, machining, and assembly.

the standard type. Reinforced E rings grip their grooves very tightly and provide high resistance against radial dislodgement from the groove. They also provide a high shoulder on small shafts and have wide areas of application in many different types of assemblies.

Two-part interlocking rings are external, radially assembled rings which form a uniform shoulder completely around the shaft on which they are used. They contact the groove wall over a major portion of the groove circumference and provide high thrust capacity. The ring halves lock together by means of interengaging lugs at each end. Each half has a reduced section at the center for necessary flexibility in bending during installation or removal. When the rings are assembled in grooves, they resist high rotational speeds.

High-strength radially assembled rings have two large lugs which contact a major portion of the groove around its circumference, providing good resistance to thrust loads. The lugs are connected by a tapered bridge with two radial cutouts, one at each end, which give the bridge a longer length than would be possible otherwise. The increased bridge length makes the ring extremely flexible and permits it to be seated in grooves with large diameter tolerances. The rings are thicker than comparable-size E rings or crescent types, adding to their rigidity under acting thrust loads. The rings are seated in deep grooves and provide a large shoulder on small-diameter shafts.

Thinner-gage high-strength radial rings are the same as the high-strength rings above but their thickness is reduced. They may be used in the same-width grooves as E rings

FIG. 3-7 Diesel engine rocker arms originally were secured in heavy-duty brackets with special spiral washers which functioned as both springs and locking mechanisms. Part had to be compressed for insertion into groove on shaft, and assembly was slow and difficult. Company changed design by switching to standard wave spring secured by *basic external retaining ring,* seated in same groove used for spiral washer. Ring provides large bearing shoulder against washer, is secure against heavy thrust loads, and has reduced costs for parts and assembly.

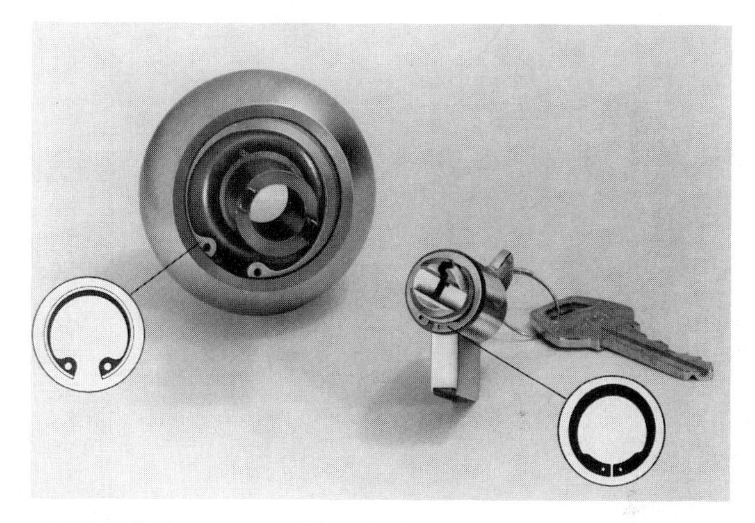

FIG. 3-8 Cylindrical lockset uses two different kinds of retaining rings to simplify design and lower costs. Knob and shank (left) are held together with *beveled internal ring* which also takes up end-play in assembly. Ring eliminates need for drilling and tapping parts and fastening with screws. Cylinder (right) uses *inverted external ring* to couple plug and housing. Ring is seated in groove on plug and, because of inverted lug and section height, forms large uniformly circular shoulder against housing. Inverted ring eliminates drilling and pinning for additional savings.

for similar shaft sizes where a higher thrust capacity is required and where the rings must accommodate large groove-diameter tolerances.

Typical product applications of radially assembled rings are illustrated in Figs. 3-10 through 3-14. Dimensions, load capacities, and other engineering data are given in Table 3-7. See Fig. 3-15 for a complete outline of this ring group.

3-8 Rings for End-play Take-up

Axially Assembled Types. *Bowed internal rings* are similar in construction to the basic internal type except that the ring body is bowed about an axis normal to the diameter bisecting the ring gap. They are assembled in grooves in bores or housings with the convex face of the ring against the abutting part. The bow serves to take up end-play caused by accumulated tolerances or wear in the assembled parts. The rings eliminate the need for spring washers, wave washers, and other tensioning devices which perform the same function but require additional parts to hold them in place.

Bowed external rings are similar in design to basic external rings and, like the bowed internal type, have a body which is bowed about an axis normal to the diameter bisect-

TABLE 3-5 Axially Assembled Tapered-Section Internal Rings

Ring type	Housing dia., dec. equiv. in.		Nom. ring thick-ness, in.	Allowable static thrust load, lb° when rings abut parts with sharp corners				Max. allow. corner radii or chamfers of retained parts, inch†				Allowable thrust load, lb, when rings abut parts with listed corner radii or chamfers
				Groove material having min tensile yield strength of 150,000 psi		Groove material having min tensile yield strength of 45,000 psi		Radii		Chamfers		
	From	Thru		From	Thru	From	Thru	From	Thru	From	Thru	P'_r
Basic internal, Bowed internal, Beveled internal in grooved housings	0.250	0.312	0.015	420	530	190	240	0.011	0.016	0.0085	0.013	190
	0.375	0.453	0.025	1,050	1,280	350	460	0.023	0.027	0.018	0.021	530
	0.500	0.750	0.035	1,980	3,000	510	1,460	0.027	0.032	0.021	0.025	1,100
	0.777	1.023	0.042	4,550	6,050	1,580	3,000	0.035	0.042	0.028	0.034	1,650
	1.062	1.500	0.050	7,450	10,550	3,050	6,000	0.044	0.048	0.035	0.038	2,400
	1.562	2.000	0.062	13,700	17,500	6,350	10,300	0.064	0.064	0.050	0.050	3,900
	2.047	2.531	0.078	22,750	27,600	10,850	15,650	0.076	0.078	0.061	0.062	6,200
	2.562	3.000	0.093	33,700	39,500	16,500	23,150	0.088	0.092	0.070	0.074	9,000
	3.062	5.000	0.109	47,100	77,000	24,100	55,000	0.097	0.158	0.078	0.126	12,000
	5.250	6.000	0.125	92,700	105,900	60,000	68,600	0.168	0.168	0.134	0.134	15,000
	6.250	7.000	0.156	137,700	154,300	74,100	93,100	0.177	0.196	0.142	0.157	23,000
	7.250	10.000	0.187	191,500	264,200	99,600	190,700	0.202	0.270	0.162	0.216	34,000
Inverted internal in grooved housings	0.750	....	0.035	1,650	....	600	....	0.050	......	0.031	......	850
	0.812	1.000	0.042	2,600	3,300	700	1,150	0.054	0.064	0.034	0.040	1,250
	1.062	1.500	0.050	4,150	5,850	1,250	2,500	0.069	0.081	0.043	0.051	1,800
	1.562	2.000	0.062	7,600	9,750	2,650	4,300	0.088	0.118	0.055	0.074	2,900
	2.062	2.500	0.078	12,650	15,300	4,500	6,500	0.125	0.144	0.078	0.090	4,600
	2.625	3.000	0.093	19,200	21,900	7,200	9,600	0.150	0.169	0.094	0.106	6,700
	3.156	4.000	0.109	27,000	34,200	10,600	16,900	0.174	0.174	0.109	0.109	9,000

Copyright © 1964, 1965, 1973, 1986, 1988 Waldes Truarc, Inc. Reprinted with permission.

° Where rings are of intermediate size—or groove materials have intermediate tensile yield strengths—loads may be obtained by interpolation.

† Approx. corner radii and chamfers limits for parts having intermediate diameters can be determined by interpolation. Corner radii and chamfers smaller than those listed will increase thrust load of rings proportionately, approaching but not exceeding allowable static thrust loads of rings abutting parts with sharp corners.

ing the ring gap. They are used to provide resilient end-play take-up of tolerances in assemblies on shafts and are positioned with the concave surface abutting the retained part.

Beveled internal rings are similar in construction to basic internal rings. They have a bevel, or taper, running around the outside circumference of the ring body. These rings are seated in grooves having a beveled wall against which the ring bevel is abutted. The retained part abuts the flat face of the ring. The beveled groove is relatively deep and—depending upon tolerances in the assembly—the ring, because of its spring properties, will seat itself at various positions in the groove to take up tolerance variations in the assembly and provide a rigid assembly without any end-play. The bevel angle of the ring and groove is generally 15 degrees.

Beveled external rings are similar in construction to basic external rings. In contrast to the internal beveled type, they have a bevel around their inside circumference and are seated in beveled grooves on shafts. They provide for rigid end-play take-up of tolerances similar to the internal type.

Radially Assembled Types. Bowed E rings resemble regular E rings in appearance, but are bowed about an axis normal to a diameter bisecting the ring gap. They are used to provide resilient take-up of tolerances in assemblies on shafts where a large shoulder is required and radial ring installation is necessary. Like other bowed rings, they can be used to eliminate noise and chatter in vibrating assemblies caused by tolerance accumulation in the retained parts.

Locking-prong rings are bowed about an axis parallel to a diameter bisecting the ring gap. They are equipped with two locking prongs on the ring body. During assembly, the rings are flattened and pushed radially into the groove. During installation the prongs bypass the groove, but when the ring is fully seated, it springs back to a bowed shape and the prongs engage the shaft. The prongs prevent the ring from backing out of the groove by providing a positive interference with the shaft. The bow expands and bridges the gap between the loaded groove wall and the abutting part, taking up end-play resiliently.

Typical product applications of rings for end-play take-up are illustrated in Figs. 3-16 through Fig. 3-20. Dimensions, load capacities, and other engineering data are given in Tables 3-5, 3-6, and 3-7. Refer to Fig. 3-21 for a graphic outline of this ring group.

3-9 Self-locking Rings (No Groove Required) *Self-locking internal rings* have a body shaped like a flat, circular rim with a series of prongs around the outer circumference. They are installed axially in housings and bores. During assembly, the prongs deflect and exert pressure against the bore. Acting thrust loads cause the prongs to increase their pressure against the bore and dig into it, preventing axial displacement of the ring in the direction in which the thrust load is acting. The rings present a flat abutting surface to the retained part and have a large inside diameter which makes them useful where clearance is a problem. Because the rings do not require any groove, they may be seated at any point in a bore or housing, automatically compensating for accumulated tolerances in the assembly.

Self-locking external rings are similar in design to the internal type, except that the prongs are located around the inside circumference of the body. When the rings are assembled axially on a shaft, stud, or similar part, the prongs deflect and engage the shaft, again providing resistance to axial displacement in the direction in which thrust loads are acting. The narrow rim provides a flat surface against which parts may be abutted and permits the ring to be used where only limited clearance is available.

Reinforced self-locking external rings have a dished, or ribbed, rim which is wider than the rim of the flat self-locking type. The prongs are in the same location around the inside circumference but are longer. When the ring is pushed over a shaft axially, the prongs deflect and grip the shaft tightly, preventing displacement in the direction in which load is applied. The longer prongs accommodate shafts with increased diameter tolerance and the dished rim reinforces the ring against buckling. The ring withstands higher thrust loads than the flat self-locking type.

Tapered-section axial clamp rings are similar in appearance to the basic external type. Their radial section height is extremely high, however, so that when the ring is spread and allowed to snap back and engage a shaft, it exerts a strong frictional hold and resists displacement by thrust loads acting in either direction. Like the basic types,

TABLE 3-6 Axially Assembled Tapered-Section External Rings

Ring Type	Shaft dia., Dec. equiv. in.		Nom. ring thickness, in.	Allowable static thrust load, lb° when rings abut parts with sharp corners				Max. allow. corner radii or chamfers of retained parts, inch†				Allowable thrust load, lb, when rings abut parts with listed corner radii or chamfers
				Groove material having min tensile yield strength of 150,000 psi		Groove material having min tensile yield strength of 45,000 psi		Radii‡		Chamfers‡		
	From	Thru		From	Thru	From	Thru	From	Thru	From	Thru	P_r'
Basic external,	§0.125	§0.156	0.010	110	130	35	55	0.010	0.015	0.006	0.009	45
Bowed external,	§0.188	§0.236	0.015	240	310	80	120	0.014	0.0165	0.0085	0.010	105
Beveled external	0.250	0.469	0.025	590	1,100	175	450	0.018	0.031	0.011	0.018	470
on grooved shafts	0.500	0.672	0.035	1,650	2,200	550	950	0.034	0.040	0.020	0.024	910
	0.688	1.023	0.042	3,400	5,050	1,000	2,250	0.042	0.058	0.025	0.035	1,340
	1.062	1.500	0.050	6,200	8,800	2,400	5,000	0.060	0.079	0.036	0.047	1,950
	1.562	2.000	0.062	11,400	14,600	5,200	8,050	0.082	0.096	0.049	0.057	3,000
	2.062	2.688	0.078	18,950	24,700	8,450	13,850	0.098	0.1115	0.059	0.067	5,000
	2.750	3.438	0.093	30,100	37,700	14,400	21,900	0.112	0.129	0.067	0.077	7,350
	3.500	5.000	0.109	44,900	64,200	22,800	37,100	0.1295	0.165	0.078	0.099	10,500
	5.250	6.000	0.125	77,300	88,300	40,800	53,800	0.169	0.184	0.101	0.110	13,500
	6.250	7.000	0.156	114,800	128,600	58,300	72,700	0.187	0.208	0.112	0.125	21,000
	7.500	10.000	0.188	165,200	220,200	84,800	149,800	0.220	0.294	0.132	0.176	30,000
Inverted external	0.500	0.672	0.035	1,100	1,450	280	470	0.051	0.065	0.032	0.041	680
on grooved shafts	0.688	1.000	0.042	2,300	3,300	500	1,050	0.066	0.091	0.042	0.057	1,000
	1.062	1.500	0.050	4,150	5,850	1,200	2,500	0.092	0.100	0.058	0.063	1,460
	1.562	2.000	0.062	7,600	9,750	2,600	4,000	0.104	0.127	0.066	0.080	2,250
	2.125	2.625	0.078	13,000	16,100	4,550	6,650	0.133	0.159	0.084	0.099	3,750
	2.750	3.346	0.093	20,100	24,500	7,200	10,500	0.165	0.194	0.103	0.121	5,500
	3.500	4.000	0.109	29,900	34,300	11,500	14,000	0.202	0.213	0.127	0.133	7,850
Heavy-duty	0.394	...	0.035	2,000	...	700	...	0.047	...	0.039	...	450
external	0.473	...	0.042	3,000	...	1,000	...	0.070	...	0.058	...	550
on grooved shafts	0.500	0.669	0.050	3,900	5,200	1,100	1,900	0.070	0.077	0.058	0.064	650–900
	0.750	1.000	0.078	9,000	11,500	2,400	4,000	0.089	0.100	0.074	0.083	2,500
	1.062	1.378	0.093	15,000	19,500	4,800	8,200	0.106	0.128	0.088	0.107	4,000
	1.500	1.772	0.109	24,500	29,000	10,000	12,400	0.128	0.128	0.107	0.107	5,000
	1.938	2.000	0.125	37,000	38,000	15,300	17,000	0.153	0.153	0.128	0.128	6,000

Miniature high-strength external on grooved shafts	0.101 0.156 0.219	0.134 0.203 0.328	0.020 0.025 0.035	250 490 1,200	330 650 1,800	60 130 220	90 200 460	0.013 0.021 0.028	0.014 0.023 0.038	0.010 0.017 0.022	0.011 0.018 0.030	200 320 ¶600
Permanent shoulder on grooved shafts	Avg. sizes shaft dia., in.			CRS SAE 1010 on soft steel shaft	Cabra 353 brass	Cabra 110 copper	Type 3003 alumi-num	Not applicable				
	0.375		0.050	900	750	600	300					
	0.500		0.062	1,200	1,200	900	450					
	0.625		0.062	1,900	1,600	1,100	650					

Copyright © 1964, 1965, 1973, 1986, 1988 Waldes Truarc, Inc. Reprinted with permission.

° Where rings are of intermediate size—or groove materials have intermediate tensile yield strengths—loads may be obtained by interpolation.

† Approx. corner radii and chamfers limits for parts having intermediate diameters can be determined by interpolation. Corner radii and chamfers smaller than those listed will increase thrust load of rings proportionately, approaching but not exceeding allowable static thrust loads of rings abutting parts with sharp corners.

‡ Exceptions: for shafts 0.551, 3.062, 3.500, 3.543, 3.625, 4.000, 4.500, 4.750, 6.000, and 6.250 dia., refer to manufacturer's specifications for data.

§ Rings for shafts 0.125 thru 0.236 dia. are made of beryllium copper only.

¶ Note: $P_r' = 700$ lb for ring used with 0.260 dia. shaft.

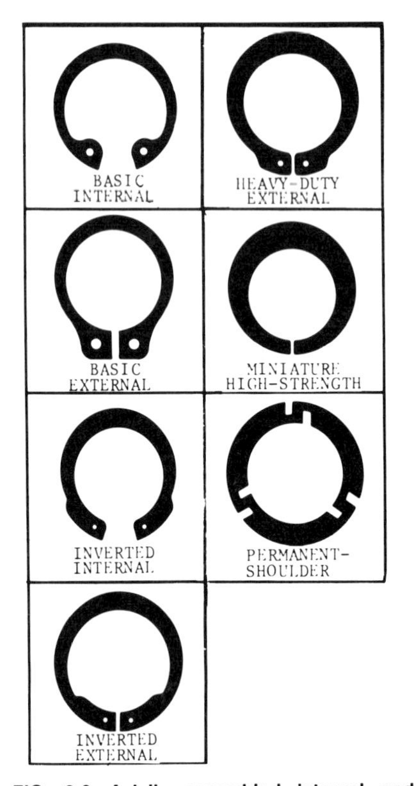

FIG. 3-9 Axially assembled internal and external rings.

the self-locking clamp rings have lugs with holes for assembly pliers. They may be adjusted to any position on a shaft to accommodate wear or tolerances in the parts.

Tapered-section radial clamp rings are designed to grip a shaft tightly without a groove or other machining and may be installed flush against the retained part. When the ring is assembled initially, it cuts shallow indentations on two sides of the shaft which increase the ring's holding power against axial displacement from either direction. It forms a high shoulder against the retained part and has notches at the free ends which permit the fastener to be spread apart with retaining ring pliers or other tools for disassembly. If the ring is disassembled and reinstalled radially, a small reduction in load capacity will result. The fasteners are designed to be assembled rapidly with the pneumatic tool shown in Fig. 3-32.

Triangular self-locking retainers have a spherically dished body and three ribbed prongs which deflect when the ring is pushed over a shaft. When the fastener is forced against a retained part, the dished body flattens. Release of the ring permits the body to snap back and causes the prongs to engage the shaft and prevent axial displacement under thrust loads. Action of the dished body when the ring is released during assembly causes the ring to be locked in place under spring tension, often eliminating the need for spring or wave washers. The triangular retainer has high thrust capacity and provides a large shoulder against which parts may be abutted.

Typical product applications of self-locking rings are illustrated in Figs. 3-22 through Fig. 3-26. Dimensions, load capacities, and other engineering data are given in Table 3-8. See Fig. 3-27 for an illustration of the self-locking rings described in this section.

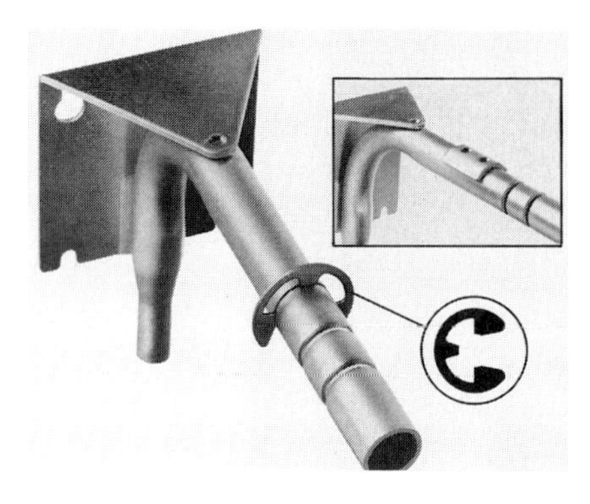

FIG. 3-10 Garden hose reel originally had welded stop (inset photo) for positioning hose connector sleeve. Welding was costly and time-consuming and often caused poorly positioned parts and sleeve breakage. To improve performance and lower costs, unit was redesigned with large *radially assembled E* ring, zinc-plated for corrosion resistance. Ring is installed in accurately located groove machined at no extra cost when two other ring grooves are formed in flow tube. Use of retaining ring assures accurate parts location, eliminates breakage.

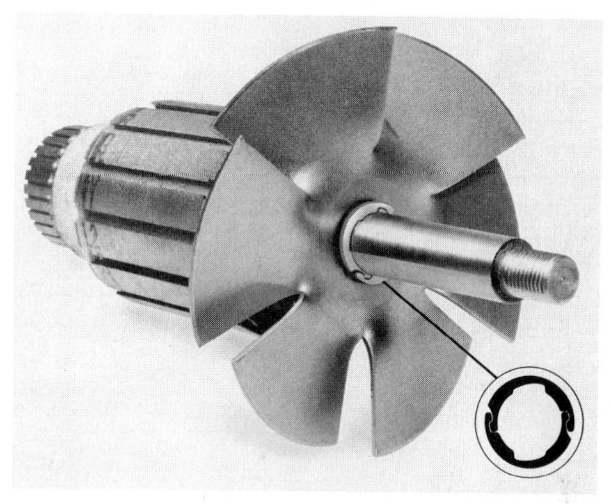

FIG. 3-11 Lawn mower motor uses *two-part interlocking ring* to position and secure fan on armature shaft. Radially installed ring is balanced to withstand high rotational speeds. Identical semicircular halves are held together by interlocking prongs at free ends. When ring is assembled on shaft, it forms high circular shoulder uniformly concentric with shaft. Fastener is secure against relative rotation between retained parts and also withstands heavy thrust loads.

3-10 Assembly Methods

Pliers. Pliers are the basic assembly tool for most axially assembled retaining rings. The tools for internal-type rings (Fig. 3-28) are used to compress rings for insertion or removal in a bore or housing; external ring pliers expand the fasteners so they may be slipped over a shaft. Tips may be straight or parallel with the plane of the pliers jaws, or may be bent at various angles to the plane of the jaws where clearance conditions in the assembly require such construction.

Pliers for assembling external rings may be equipped with an adjustable stop which can be set to limit expansion of the ring during installation, so that the ring is not overspread. Internal ring pliers have a stop which may be adjusted to set tip spacing to match the center distance of the ring lug holes, enabling an operator to insert plier tips into a ring quickly. Well-designed retaining-ring pliers feature formed handles for operator comfort and springs which return the tips to their original position after a ring has been installed.

Pliers for assembling both internal and external rings, known as convertible pliers, also are available. These have an adjustable pivot which changes the di-

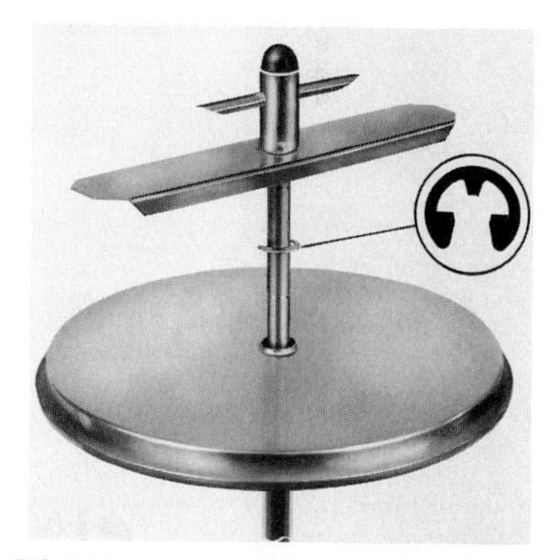

FIG. 3-12 Jelly doughnut filler has nozzles for accurately measuring and injecting two doughnuts or cakes in single operation. Large *radially assembled E* ring made of stainless steel serves as stop to limit stroke of pump piston rod, eliminating need for more expensive adjustable collars, spacer tubes, and other devices. Piston rod has three accurately located ring grooves: To change setting, operator merely pries ring out of one groove and reassembles it in another.

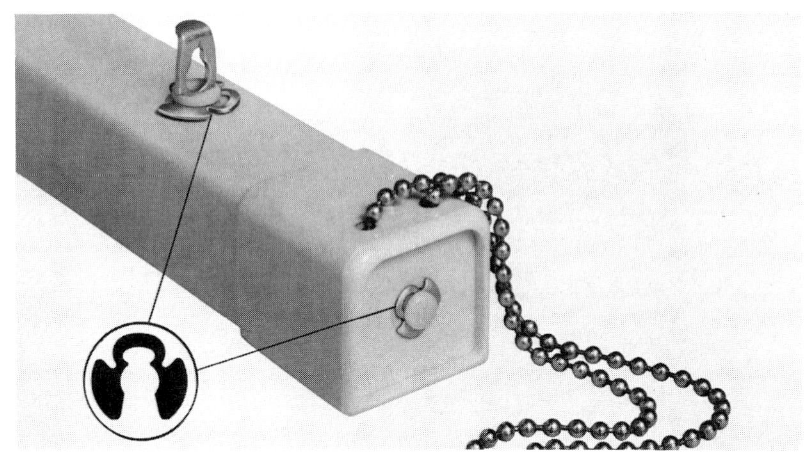

FIG. 3-13 Vertical blind track control mechanism uses *high-strength radial ring* to hold plastic drive shaft, other components. Rings provide large bearing surface against retained parts with necessary gripping strength to accommodate tolerances in premolded grooves in plastic.

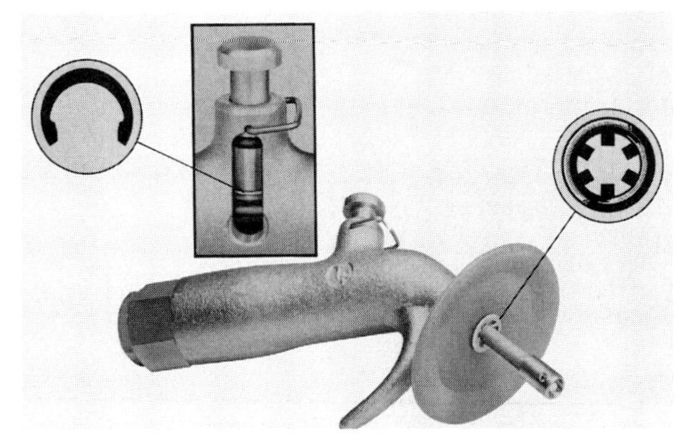

FIG. 3-14 Variable-flow air gun uses *radially assembled crescent-shaped ring* (left-hand inset) to secure push-button which activates air valve. Ring is installed in groove precut in stem and serves as stop against shoulder in housing, captivating entire assembly. Plastic guard is secured by *reinforced circular self-locking ring* (right-hand inset). Because circular ring does not require a groove or other preparatory machining, it can be installed flush against guard for tight fit, automatically compensating for tolerances in plastic part.

rection of the tip travel so the pliers may be used to expand or contract rings. The convertible pliers can handle a large variety of rings and generally are used for repair and field maintenance work. They usually are not equipped with stops or springs.

For handling larger-size retaining rings, where considerable force is necessary to compress or expand the ring for installation or removal, pliers are available equipped with a double ratchet (Fig. 3-29). The ratchet is designed so that one side or the other always is engaged by a pawl as the pliers are being manipulated to compress or expand a ring. If the operator inadvertently releases pressure on the handles, the ratchet prevents the pliers from springing open and prevents the ring from accidentally flying

TABLE 3-7 Radially Assembled External Rings

Ring type	Shaft dia., Dec. equiv. in.		Nom. ring thickness, in.	Allowable static thrust load, lb° when rings abut parts with sharp corners				Max. allow. corner radii or chamfers of retained parts, in.†				Allowable thrust load, lb, when rings abut parts with listed corner radii or chamfers P_r'
				Groove material having min tensile yield strength of 150,000 psi		Groove material having min tensile yield strength of 45,000 psi		Radii‡		Chamfers‡		
	From	Thru		From	Thru	From	Thru	From	Thru	From	Thru	
Crescent-shaped on grooved shafts	0.125	0.188	0.015	85	130	45	70	0.014	0.021	0.011	0.016	Same values as sharp corner abutment
	0.219	0.438	0.025	260	520	100	350	0.021	0.029	0.016	0.022	
	0.500	0.625	0.035	830	1,030	450	700	0.030	0.033	0.023	0.025	
	0.688	1.000	0.042	1,700	2,480	800	1,800	0.034	0.046	0.026	0.035	
	1.125	1.500	0.050	3,320	4,420	2,200	4.000	0.052	0.069	0.040	0.053	
	1.750	2.000	0.062	6,430	7,300	5,300	7,000	0.081	0.091	0.062	0.070	
Two-part interlocking on grooved shafts	0.469	0.625	0.035	2,000	2,650	620	830	0.052	0.052	0.040	0.040	610
	0.669	0.875	0.042	3,350	4,400	1,250	1,600	0.065	0.065	0.050	0.050	880
	0.984	1.500	0.050	5,850	8,950	2,900	4,450	0.086	0.086	0.066	0.066	1,250
	1.562	1.875	0.062	11,750	14,100	5,650	6,800	0.100	0.100	0.077	0.077	1,900
	1.969	2.625	0.078	18,250	24,300	9,000	12,000	0.114	0.114	0.088	0.088	3,050
	2.750	3.250	0.093	30,200	35,750	15,000	17,800	0.143	0.143	0.110	0.110	4,300
	3.375	...	0.109	43,500	...	20,600	...	0.182	0.182	0.140	...	5,950
Bowed E ring on grooved shafts	0.125	...	0.010	43	...	45	...	0.040	...	0.030	...	Same values as sharp corner abutment
	0.140	0.219	0.015	75	115	60	75	0.060	0.060	0.045	0.045	
	0.250	0.312	0.025	255	325	115	225	0.060	0.060	0.045	0.045	
	0.375	0.438	0.035	690	830	315	480	0.065	0.065	0.050	0.050	
	0.500	0.625	0.042	1,110	1,420	600	1,050	0.080	0.080	0.060	0.060	
	0.750	0.875	0.050	2,000	2,350	1,500	2,050	0.085	0.085	0.065	0.065	
E ring on grooved shafts	§0.040	0.062	0.010	13	20	6	7	0.015	0.030	0.010	0.020	Same values as sharp corner abutment
	0.094	0.140	0.015	45	75	20	45	0.040	0.040	0.030	0.030	
	0.140	0.312	0.025	170	325	60	225	0.060	0.060	0.045	0.045	
	0.375	0.438	0.035	690	830	315	480	0.065	0.065	0.050	0.050	
	0.500	0.625	0.042	1,110	1,420	600	1,050	0.080	0.080	0.060	0.060	
	0.750	1.000	0.050	2,000	2,650	1,500	1,900	0.085	0.077	0.065	0.057	
	1.188	1.375	0.062	3,450	4,100	1,500	2,350	0.090	0.090	0.070	0.070	

TABLE 3-7 Radially Assembled External Rings (*Continued*)

Ring type	Shaft dia., Dec. equiv. in.		Nom. ring thickness, in.	Allowable static thrust load, lb° when rings abut parts with sharp corners				Max. allow. corner radii or chamfers of retained parts, in.†				Allowable thrust load, lb, when rings abut parts with listed corner radii or chamfers P_r'
				Groove material having min tensile yield strength of 150,000 psi		Groove material having min tensile yield strength of 45,000 psi		Radii‡		Chamfers‡		
	From	Thru		From	Thru	From	Thru	From	Thru	From	Thru	
Reinforced E ring on grooved shafts	0.094	0.125	0.015	50	75	13	25	0.045	0.045	0.033	0.033	Same values as sharp corner abutment
	0.156	0.250	0.025	150	250	40	75	0.065	0.065	0.050	0.050	
	0.312	0.438	0.035	420	600	135	285	0.070	0.070	0.055	0.055	
	0.500	0.562	0.042	820	930	460	480	0.080	0.080	0.060	0.060	
Locking-prong ring on grooved shafts	0.092	0.156	0.010	80	120	35	100	Not applicable				
	0.188	0.312	0.015	200	350	140	300					
	0.375	0.438	0.020	550	700	450	600					
High-strength on grooved shafts	0.188	0.250	0.035	600	900	130	200	0.050	0.050	0.040	0.040	250
	0.312	0.375	0.042	1,300	1,550	250	300	0.065	0.065	0.050	0.050	350
	0.438	0.625	0.050	2,200	3,000	400	1,100	0.080	0.080	0.060	0.060	600
	0.750	...	0.062	4,600	...	1,600	...	0.085	...	0.065	...	1,000
	1.000	...	0.078	7,500	...	2,600	...	0.090	...	0.070	...	1,800
	1.250	...	0.093	11,000	...	3,500	...	0.090	...	0.065	...	2,750
	1.500	...	0.109	15,300	...	4,800	...	0.100	...	0.070	...	3,800
	1.750	2.000	0.125	20,500	23,500	8,200	9,450	0.120	0.130	0.090	0.100	5,100
Thinner-gage high-strength on grooved shafts	0.188	0.312	0.025	430	780	130	250	0.050	0.050	0.040	0.040	150
	0.375	0.438	0.035	1,300	1,850	300	400	0.065	0.065	0.050	0.050	300
	0.500	0.625	0.042	2,100	2,500	400	600	0.080	0.080	0.060	0.060	400
	0.750	1.000	0.050	3,700	4,800	1,600	2,600	0.090	0.090	0.070	0.070	1,000

Copyright © 1964, 1965, 1973, 1986, 1988 Waldes Truarc, Inc. Reprinted with permission.

° Where rings are of intermediate size—or groove materials have intermediate tensile yield strengths—loads may be obtained by interpolation.

† Approx. corner radii and chamfers limits for parts having intermediate diameters can be determined by interpolation. Corner radii and chamfers smaller than those listed will increase thrust load of rings proportionately, approaching but not exceeding allowable static thrust loads of rings abutting parts with sharp corners.

‡ Exceptions: For shafts 1.000, 1.772, 2.156 and 3.156 dia., refer to manufacturer's specifications for data. Bowed E rings for shafts 0.110, 0.140, 0.438, 0.744, and 0.750 dia. available in other sizes varying from standard design. (Refer to manufacturer's specifications for complete data.) E rings for shafts 0.062, 0.094, 0.110, 0.140, 0.438, 0.744, 0.750 and 1.000 dia. are available in one or more sizes varying from standard design. Refer to manufacturer's specifications for complete data.

§ Ring for shaft 0.040 dia. is made of beryllium copper only.

loose. The ratchet-locking action thus reduces operator fatigue and also serves as a safety device for the operator.

Applicators and Dispensers. The basic tool for installing radially assembled rings is the applicator (Fig. 3-30). It has a split, forklike blade mounted in a plastic handle. The operator uses the applicator to pick up the ring, hold it in the blade and align the ring with the groove in which the fastener is to be installed. The operator then pushes the ring into the groove and withdraws the applicator. Because the pressure between ring and groove is greater than the blade pressure on the ring, the fastener remains in the groove as the applicator is withdrawn.

Since the applicator holds only one ring at a time, it is necessary to reload it after each ring is assembled. A dispenser generally is used for this purpose. This is a device which holds a stack of rings on a flexible spring rail so that they can be picked up by the applicator blade.

To assemble a ring, the operator merely pushes the applicator forward on the dispenser base to grasp the bottom ring in the stack. The spring

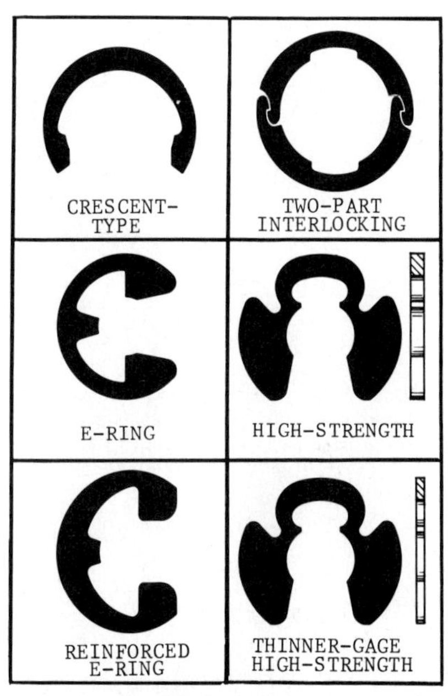

FIG. 3-15 Radially assembled rings.

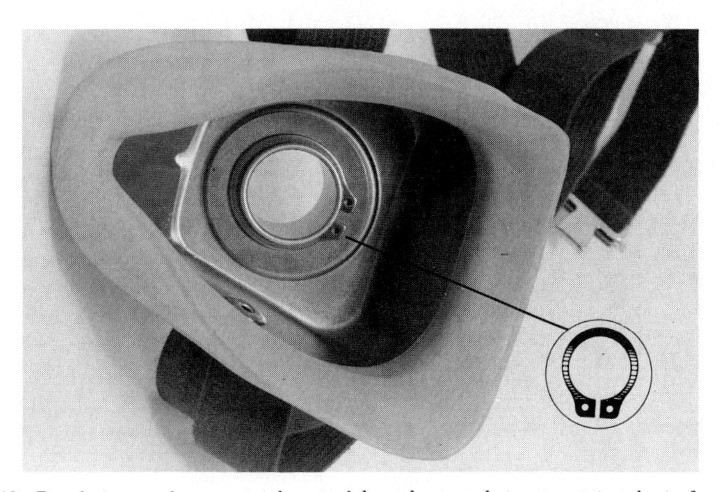

FIG. 3-16 Respirator mask uses stainless steel *bowed external ring* to retain tube in face plate. Ring must remain medically cleaner than would be possible with threaded joint, cannot vibrate loose during use. Bowed construction is necessary to assure pressure seal against O ring, prevent oxygen leakage.

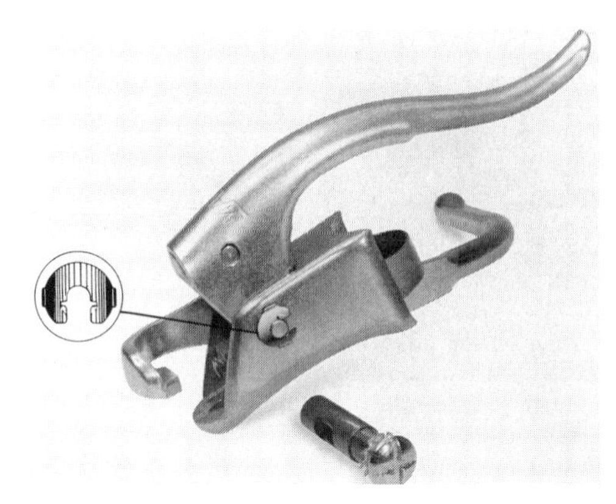

FIG. 3-17 Bicycle hand brake uses radially assembled *bowed locking-prong ring* to hold hinge pin in frame. Bowed construction assures necessary spring pressure, automatically compensates for tolerances in parts. Retaining ring replaces screw and lock washer illustrated in foreground, eliminates costly drilling and tapping operation.

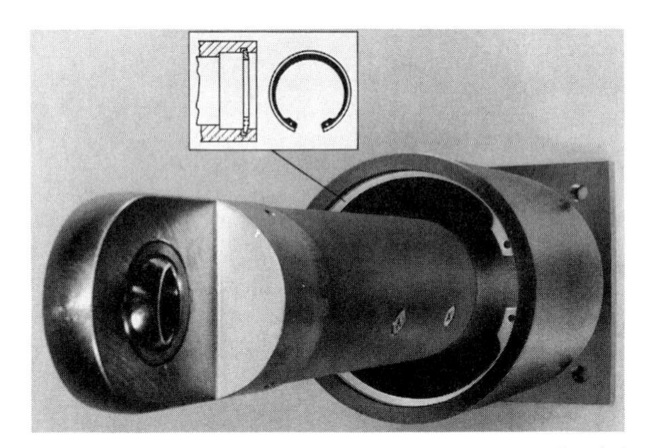

FIG. 3-18 Mechanical shock arrestor for piping and other systems uses *beveled internal ring* to secure support cylinder in housing. Ring has 15° bevel on outer circumference and is seated in groove with comparable bevel on load-bearing wall. Wedge action of ring in groove (inset drawing) provides rigid end-play take-up in assembly.

FIG. 3-19 Aerial reconnaissance camera motor uses *bowed internal ring* to position and secure aluminum cover plate and other components in housing. Ring/plate design eliminates bolted end cover, reducing size and weight of motor. Fastener is seated in accurately located groove (inset drawing) and provides resilient end-play take-up against cover plate, assuring necessary tight fit. Bowed construction automatically compensates for tolerances in parts.

FIG. 3-20 Molded plastic pump impeller is held on grooved hub with *external bowed ring.* Axially assembled fastener provides necessary spring pressure to take up end-play caused by tolerances in plastic impeller.

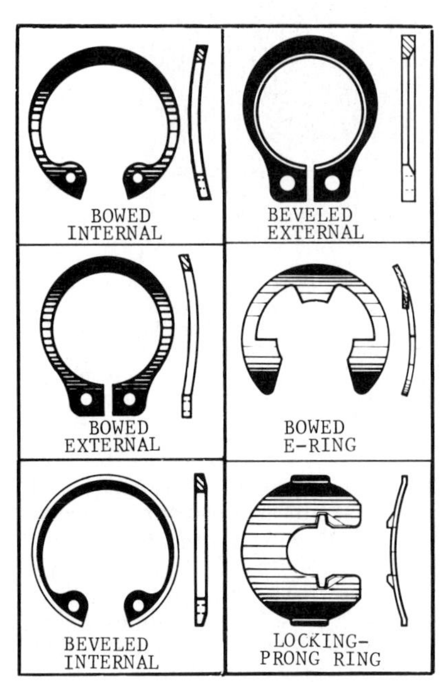

FIG. 3-21 Rings for end-play take-up.

FIG. 3-22 Folding mechanism for office tables uses *reinforced circular self-locking rings* to secure linkage systems. Rings hold stainless steel round-head rivets which serve as hinge pins. They are installed quickly with simple tubular plunger which forces ring over rivet stud so that inclined prongs spread and lock against stud. Rings' reinforced arched rim provides extra strength to resist twisting or buckling. Rivet/retaining ring system eliminates threaded studs and locknuts for substantial savings on materials and assembly time.

FIG. 3-23 Plastic strapping tool uses different types of retaining rings to achieve a compact design and reduce costs. *Four axially assembled clamp rings* hold pivot pins in seal injector assembly (inset photo, lower left). Fasteners do not require any groove: They are merely slipped over end of pin with retaining ring pliers and snapped into position. Tensioner assembly uses *heavy-duty external ring* to secure feed wheel (inset, lower right). Ring is installed in accurately located groove which assures precise seating of wheel on hub and is secure against high thrust loads. Fastener can be removed easily to permit changing wheel and is reusable following disassembly. Retaining rings replace threaded fasteners and save several drilling, tapping, and threading operations.

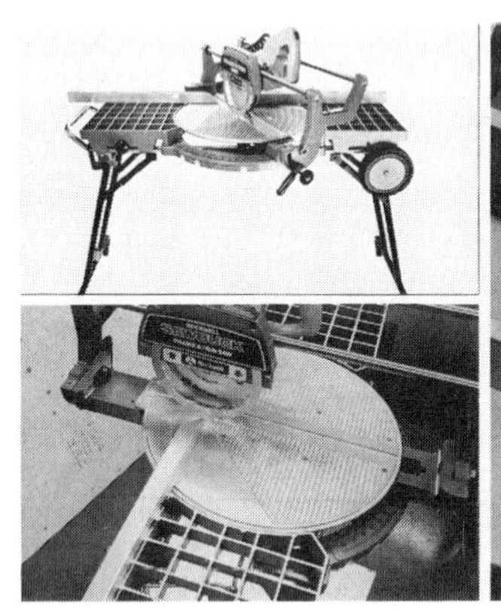

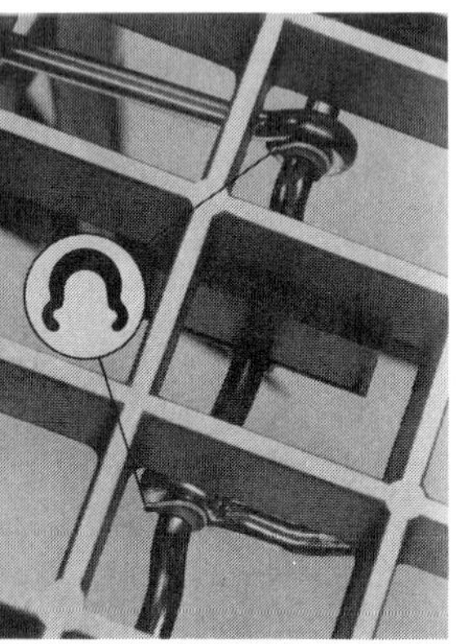

FIG. 3-24 Frame and trim saw is designed for commercial home builders, advanced "do-it-yourself" consumers, and others who require precision cutting and versatility. "Sliding fences" position work and utilize clamp mechanism activated by cammed rod linked to two eyebolts held by *radially assembled clamp rings.* In original design, engineers considered staking flat washers to rod on each side of eyebolts but this proved costly and time consuming. Clamp rings, which do not require any groove, are installed quickly and economically with pneumatic tool which forces fasteners over cammed rod so they lock into position on each side of eyebolt.

rail with the remainder of the stack swings back out of the way. As the applicator is withdrawn, the rail returns to its normal position and the next ring is ready for assembly. Tape-wrapped ring cartridges may be loaded onto the dispensers very quickly, so that ring assembly can be continued practically without interruption.

Semiautomatic Assembly Tools. Special semiautomatic tools have been developed for mass assembly of radially installed retaining rings. One such tool, illustrated in Fig. 3-31, consists of an applicator with a stack of rings held on a horizontal rail in the body of the tool. When the operator compresses the handle the applicator blade swings through an arc of some 270 degrees and picks up the first ring in the stack. Releasing the handle permits the blade to swing back to its free position. The operator then installs the ring the same way as if he were using a regular hand applicator.

Because this tool combines an applicator and dispenser in a single unit, it is especially useful for assemblies where the tool must be moved to the work.

Another type of tool, shown in Fig. 3-32, is somewhat similar to an air-driven stapler and is used for fast, economical assembly of radially installed clamp rings. The pneumatic tool is loaded with the same types of taped ring cartridges used with the hand tools. To assemble a ring, the operator positions a nose plate at the front on top of the shaft and,

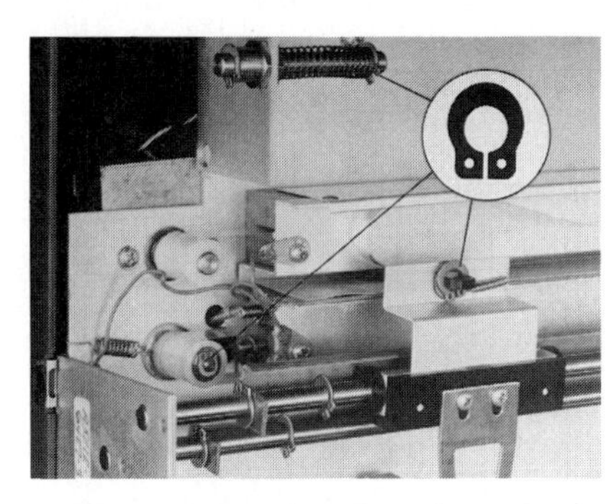

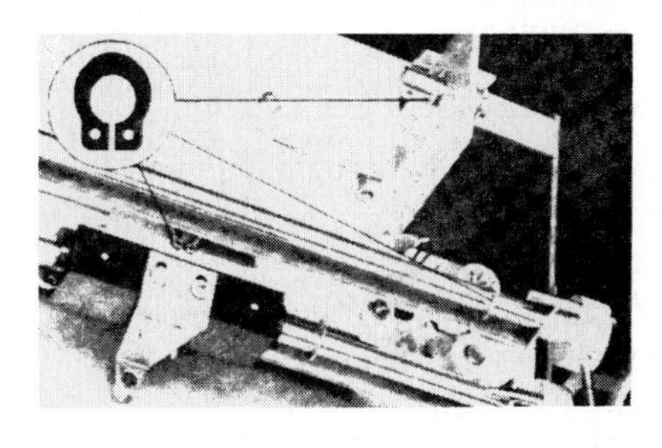

FIG. 3-25 Strip chart recorder designed for highly reliable data transcription from sensitive laboratory instruments makes extensive use of *axially assembled clamp rings.* Fasteners position and secure electrical contacts, limit stops, pulleys, spring glides, and other components. Rings need no groove: They exert frictional hold against axial displacement from either direction, automatically compensate for tolerance in parts. Because they are thin, rings permit use of shorter shafts and studs and help achieve more compact design.

holding the tool perpendicular to the shaft, squeezes the trigger. An air-driven applicator blade drives the first ring in the magazine over the shaft, abutting the retained part. The hardened steel ring cuts shallow grooves on each side of the shaft to increase the fastener's holding power against axial displacement. The tool is then removed and the assembly operation repeated with other work parts.

Mechanized Tools. Automatic tools are available for high-volume installation of both axially and radially assembled rings. One type, illustrated in Fig. 3-33, is suspended over a fixed work station or on a mechanized conveyor line. A spring-loaded or counterbalanced cable usually is used for easy handling. To install a ring, the operator places a work part in a nest, brings the tool down over the part, and squeezes a trigger. An air-driven sleeve pushes the ring down over a tapered mandrel, spreading the fastener until it rides over the work part and into its groove, as shown in the drawing. When the ring reaches the groove, it snaps into position and the operator releases the trigger. The sleeve is retracted automatically and the next ring in the stack is positioned automatically for assembly. The tool can be used for both *basic external* and *heavy-duty external* rings. Another version of the tool, for installing internal rings, uses a tapered sleeve. When the tool is activated, a plunger pushes the ring through the tapered sleeve, which has been aligned with the housing in which the ring is to be installed. The ring is compressed as it passes through the sleeve, so that it can enter the housing and snap into the groove.

Other fully automatic tools are available for radially assembled rings. These consist of air-operated applicator blades which automatically pick rings off a magazine rail and move them forward into grooves in which they are to be installed. The tools may be designed to meet a variety of assembly conditions.

TABLE 3-8 Self-Locking Rings

Ring type	Housing dia. or shaft dia., Dec. equiv. in.		Nom. ring thickness, in.	Allowable static thrust load, lb° when rings abut parts with sharp corners			
				Groove material having min tensile yield strength of 150,000 psi		Groove material having min tensile yield strength of 45,000 psi	
	From	Thru		From	Thru	From	Thru
Reinforced self-locking external on shafts, no grooves	0.094	0.375	0.010	...	...	27	65
	0.094	0.375	0.015	...	...	45	120
	0.438	1.000	0.015	...	...	120	140
Self-locking external on shafts, no grooves	0.094	0.375	0.010†	...	...	13	45
	0.438	1.000	0.015	...	...	50	65
Self-locking internal in housings, no grooves	0.312	0.625	0.010	...	...	80	45
	0.750	2.000‡	0.015	...	...	75	55
Triangular retainer on shafts, no grooves	0.062	...	0.010	...	...	25	...
	0.062	...	0.015	...	...	40	...
	0.094	0.156	0.010	...	...	60	75
	0.094	0.156	0.015	...	...	80	120
	0.188	0.312	0.015	...	...	140	200
	0.375	...	0.020	...	...	250	...
	0.437	§	0.025	...	...	270	...
Triangular nut on threaded parts	6/32	10/32	0.015	140	170	140	145
	6/32	10/32	0.020	200	220	180	190
	1/4-20	1/4-28	0.020	220	...	220	...
	1/4-20	1/4-28	0.025	220	...	220	...
Tapered-section self-locking clamp ring on shafts with or without grooves	Shaft dia., Dec. equiv. in.		Nom. ring thickness, in.	Allowable static thrust load, lb°			
				Shaft without groove		Shaft (45,000 psi) with groove¶	
	From	Thru		From	Thru	From	Thru
Inch type	0.094	0.156	0.025	8	12	...	...
	0.187	0.250	0.035	20	23	...	90
	0.312	0.375	0.042	25	30	110	180
	0.437	0.500	0.050	40	45	290	390
	0.625	0.750	0.062	60	65	570	850
Millimeter type	0.062	...	0.015	5	...	...	...
	0.079	0.118	0.024	10	15	...	...
	0.197	...	0.032	30	...	40	...
	0.236	0.276	0.039	35	40	70	100
	0.354	0.394	0.047	50	55	130	170
	0.533	0.590	0.059	75	80	340	370

Copyright © 1964, 1965, 1973, 1986, 1988 Waldes Truarc, Inc. Reprinted with permission.

° Where rings are of intermediate size—or groove materials have intermediate tensile yield strengths—loads may be obtained by interpolation.

† Ring for shaft 0.240 dia. is available only in 0.015 thickness; allowable thrust load = 40 lb.

‡ Ring for housing 1.375 dia. is available only as reinforced ring having an allowable thrust load = 150 lb.

§ Round and hex. shafts

¶ Grooved shafts are recommended *only* for rings used on shafts 0.197 in. (5.0 mm) or larger.

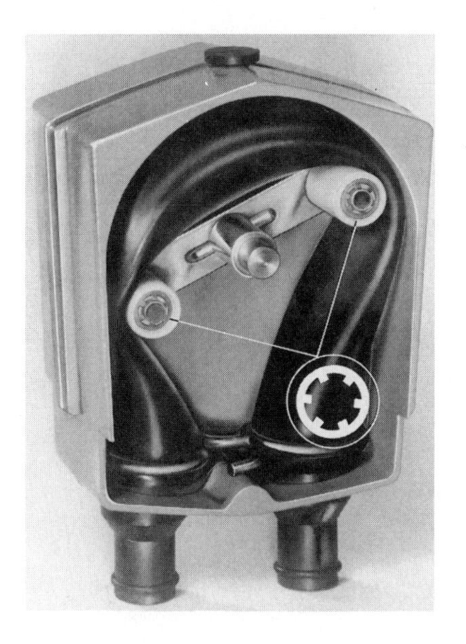

FIG. 3-26 Industrial pump uses *external circular self-locking rings* to hold plastic rollers on shafts. Fasteners are made of stainless steel for corrosion resistance and do not require any groove or other machining: They are merely pushed over ends of shafts until prongs lock into position to secure rollers. In original design, shafts were drilled and tapped and rollers secured by washers and screws. Use of retaining rings lowers costs and speeds assembly.

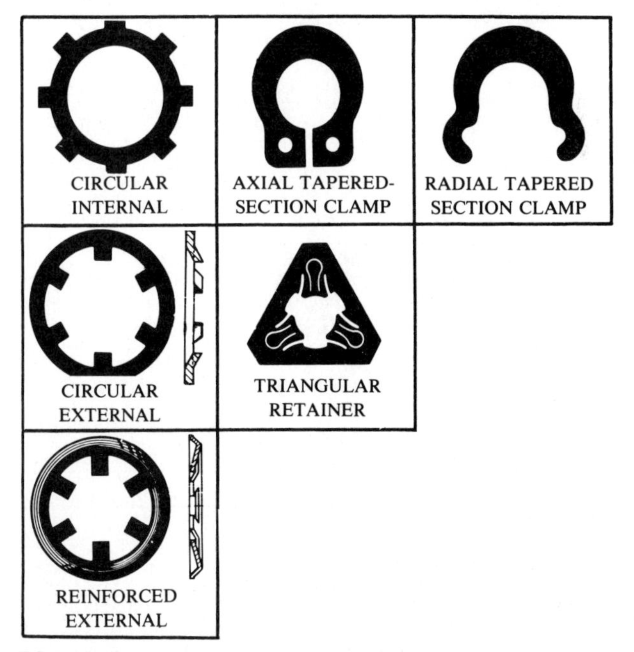

FIG. 3-27 Self-locking rings (*no grooves required*).

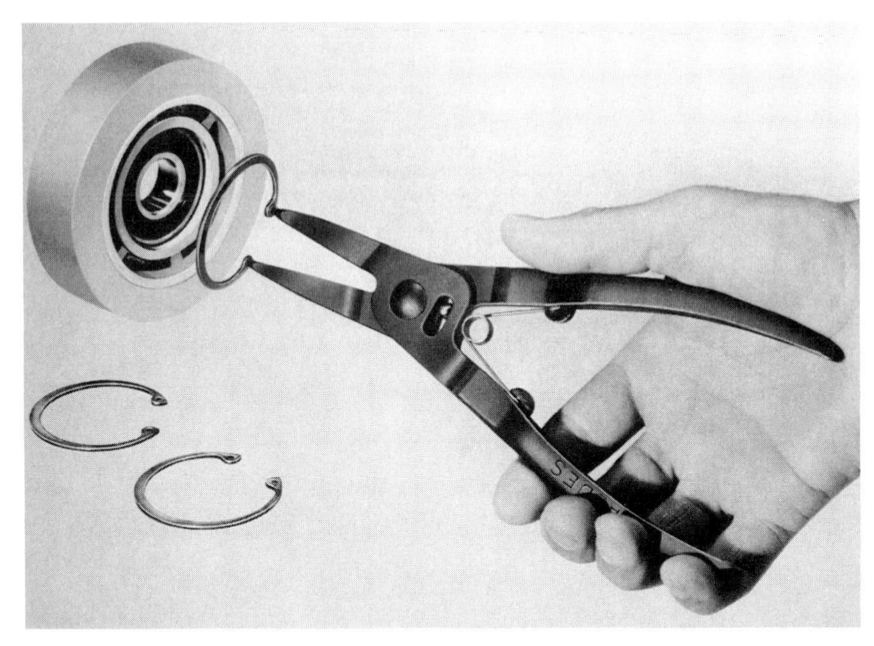

FIG. 3-28 Retaining ring pliers are among the most popular tools for assembling axially installed rings. Pliers for internal-type rings, shown in the illustration, grasp ring securely by lugs and compress fastener for insertion into bore or housing. Pliers for external rings expand fasteners so they may be assembled over shafts, studs, and similar parts. Adjustable stops limit travel of tips and are used with internal rings for automatic alignment with lug holes; on external pliers, stops prevent accidental overspreading of rings. Formed handles with plastic sleeves assure operator comfort and reduce fatigue.

FIG. 3-29 Double-ratchet pliers are used with large retaining rings which require substantial force to be compressed or expanded. By locking pliers as force is applied to handles, ratchet reduces effort to compress or expand rings and permits operator to assemble ring without maintaining pressure on handles. Ratchet also serves as safety device by preventing rings from springing loose accidentally during installation or removal.

FIG. 3-30 Applicators and dispensers are used for fast, economical assembly of various radially installed retaining rings. Fasteners are supplied in tape-wrapped cartridges which are slipped over dispenser's magazine rail. Applicator is used to grasp first ring in magazine (inset) and then push ring into groove on workpart. Because ring's gripping power on part is greater than holding power of applicator jaws, when tool is withdrawn ring remains seated on part and assembly operation can be repeated with another workpart.

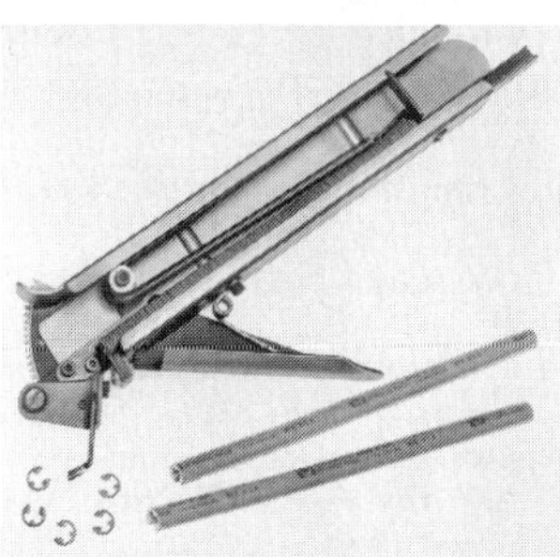

FIG. 3-31 Semiautomatic tool, also loaded with tape-wrapped ring cartridges, combines applicator and dispenser in single portable unit. To assemble ring, operator squeezes trigger to activate blade which swings upward to grasp first ring in magazine. When trigger is released, blade flips forward and operator pushes ring into groove on workpart in same manner as with hand applicator. Semiautomatic tools are ideal for assembly where parts are too large to be brought conveniently to fixed work stations or where products are assembled along mechanized conveyor line.

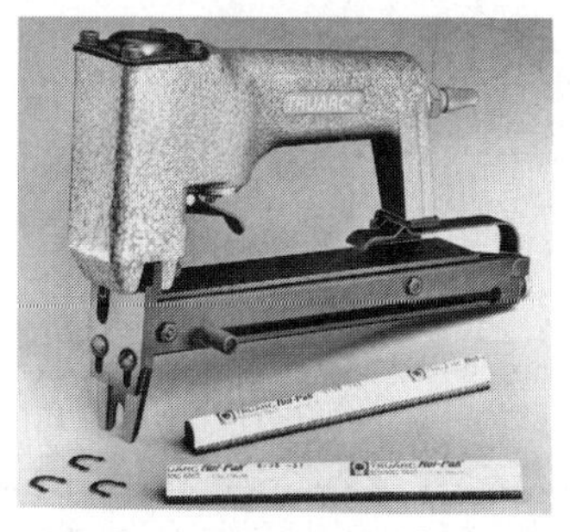

FIG. 3-32 Pneumatic tool is designed to assemble radially installed clamp rings as quickly as operator can squeeze trigger. Tool is loaded with rings furnished in tape-wrapped cartridges illustrated. To assemble ring, operator positions nose plate at front on top of shaft and, with tool held perpendicular to shaft, squeezes trigger. Air-driven applicator blade pushes first ring in magazine over shaft, abutting retained part. Hardened steel ring cuts shallow indentations on two sides of shaft, increasing ring's holding power against axial displacement from either direction. Tool is then removed and operation repeated with other assemblies.

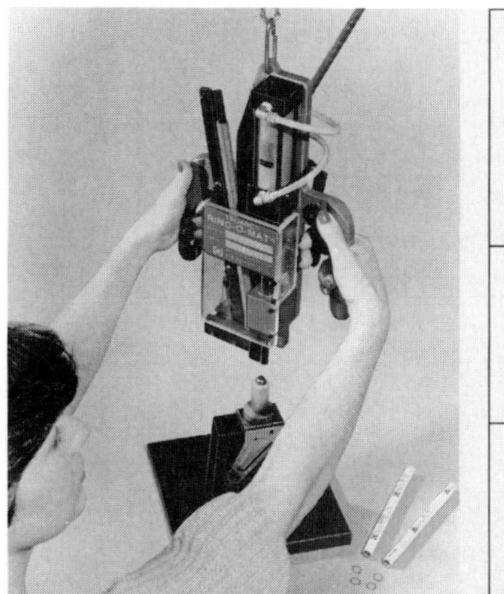

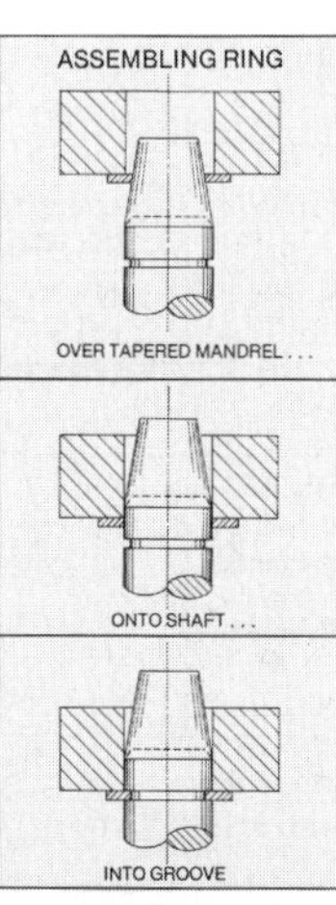

FIG. 3-33 Automatic assembly tool for high-volume installation of *basic external* and *heavy-duty external rings.* Tool can be suspended over fixed work station or on mechanized conveyor line. To install ring, operator places workpart in nest, brings tool down over part and squeezes trigger. Air-driven sleeve pushes ring down over tapered mandrel, spreading fastener until it rides over workpart and into groove, as shown in inset drawings. When ring reaches groove, it snaps into position and operator releases trigger. Sleeve is retracted automatically and next ring in stack is positioned automatically for assembly.

Other fully automatic tools are available for radially assembled rings. These consist of air-operated applicator blades which automatically pick rings off a magazine rail and move them forward into grooves in which they are to be installed. The tools may be designed to meet a variety of assembly conditions.

Part 2 Spiral-Wound Rings

edited by STEPHEN M. WARD, BSME/MSM*

Chief Engineer, Spirolox Kaydon Corp., St. Louis, Missouri

3-11 Spiral-Wound Retaining Rings These rings consist of two or more turns of rectangular, round-edge material wound on edge to provide a continuous coil. This design has several advantages:

1. The gapless ring provides 360 degrees of retention and can be easily adapted to specific applications.
2. The coil, being thin in section, is flexible. allowing ease of assembly and disassembly.
3. The ring thickness can be varied by the selection of the material thickness or by increasing the number of turns.
4. The uniform wall provides a pleasing appearance and is an asset when radial clearance is critical.
5. The laminated ring section permits a large axial deflection under thrust load without overstressing the ring.

3-12 Materials The standard materials for spiral-wound retaining rings are carbon spring steel and 302 stainless steel. Optional materials include: 316 stainles steel, NS A286 alloy, beryllium copper, and phosphor bronze (Table 3-9).

3-13 Finishes Carbon steel rings are normally given an oil-dip treatment to retard rusting. Cadmium plating (with and without supplemental dichromate dip), phosphate coating, and black oxide are readily available finishes. Platings and coatings add approximately 0.002 inch to the maximum thickness of the ring.

TABLE 3-9 Strength Characteristics of Materials for Spiral-Wound Retaining Rings

	T_1, Material thickness, inches	S, Max. work stress during installation (min. tensile strength), psi	S_S, shear strength of ring material, psi	δ, allowable working stress under load, psi	M, modulus of elasticity, psi	Maximum recommended service temp., °F
Carbon spring steel	0.0067–0.0148	269,000	153,000	242,000	30×10^6	250
	0.0150–0.0213	255,000	138,000	218,000		400
	0.0215–0.0433	221,000	126,000	199,000		400
	0.0435 plus	211,000	120,000	190,000		400
Stainless steel 302 (S)	0.0080–0.0160	210,000	119,000	178,000	28×10^6	400
	0.0162–0.0230	210,000	119,000	178,000		400
	0.0232–0.0480	200,000	114,000	170,000		400
	0.0482–0.0610	185,000	105,500	157,000		400
	0.0612 plus	175,000	100,000	157,000		400
Beryllium (M)	All sizes	175,000	100,000	149,000	17.9×10^6	225
Phosphor bronze (R)	0.0100–0.0250	106,000	60,500	90,000	14.5×10^6	225
	0.0252–0.0500	95,000	54,000	80,600		225
	0.0502 plus	90,000	51,300	76,500		225
A286 austenetic alloy (SN)	All sizes	200,000	114,000	170,000	31×10^6	900

*NOTE: Original author of sec. 3, part 2, in the first edition of this handbook, was R.L. Berkbigler, P.E., Manager, Retaining Ring Engineering, Ramsey Corporation, St. Louis, Missouri.

3-14 Sizes Standard rings are available for shafts and bores ranging from $^{15}/_{32}$- to 15-inch diameters. Special rings have been manufactured for shafts and bores up to 60-inch diameter.

3-15 Selection of Rings The following factors should be evaluated when selecting a retaining ring:

1. Assembly
2. Disassembly
3. Ring thrust capacity
4. Groove thrust capacity
5. Type of thrust load
6. Rotation between parts
7. Centrifugal force
8. Balance
9. Tolerance take-up
10. Temperature
11. Corrosion

3-16 Assembly The rings, being flexible, may be installed manually by spreading the leaves, inserting one ring end in the groove and "walking" the remainder of the ring around until the last turn snaps into the groove. Of course, this method of installation is practical only when the number of assemblies involved is small. A tapered sleeve or mandrel and a plunger is usually employed for higher-volume applications. These installation fixtures must be designed to avoid overstressing the retaining ring during its assembly. Equations for calculating the allowable expansion or contraction of the rings are

$$\textit{External rings, max. exp. ring diam.} = \left[\frac{Sd_n}{me - Sd_n}\right] \cdot \begin{matrix}\text{ring free}\\ \text{inside}\\ \text{diameter}\end{matrix} \qquad (3\text{-}9)$$

$$\textit{Internal rings, min. contr. ring diam.} = \left[1 - \frac{Sd_n}{me + Sd_n}\right] \cdot \begin{matrix}\text{ring free}\\ \text{outside}\\ \text{diameter}\end{matrix} \qquad (3\text{-}10)$$

Values for the allowable stresses are given in Table 3-9. Terms are defined in Table 3-10. Overstressed rings, unless adequate compensation has been made, will not grip the groove bottom and a decrease in thrust capacity will result.

Extremely high volume may warrant the use of completely automatic installation tools.

3-17 Disassembly Standard provisions for ring removal are shown in Fig. 3-34. In most applications, these removal devices are quite adequate and the ring can be removed by inserting a screwdriver to lift the ring end from the groove and then spiral out the remainder of the ring.

3-18 Ring Thrust Capacity A retaining ring may fail in two ways: through shear or from overstressing due to axial deflection. The type of failure occurring at the lowest thrust load limits the design.

3-19 Shear Although shear is not usually the limiting factor, it should always be evaluated. For shear failure to occur, the groove material must have a compressive yield strength greater than 45,000 psi, the load must be applied through a retained part which has a sharp corner and a yield strength greater than 45,000 psi, and the ring must be thin in section compared with the ring diameter. The equation for checking for failure due to shear is

$$P_S = \frac{AFS_S \pi}{K} \qquad (3\text{-}11)$$

Table 3-9 lists the shear strength for various ring materials. Table 3-10 defines the terms in the equation.

3-20 Axial Deflection The maximum stress on a ring which is subjected to a uniform twisting moment is a tensile stress that occurs at the inner corner of the ring. If the ring is stressed in this region to a point where the yield point is exceeded, it will tend to grow in diameter and become dished.

Generally, an excessive twisting moment is caused when the compressive yield of the groove material is exceeded, allowing the groove wall to deform and permit axial deflection of the ring (Fig. 3-35).

TABLE 3-10 Nomenclature

INTERNAL

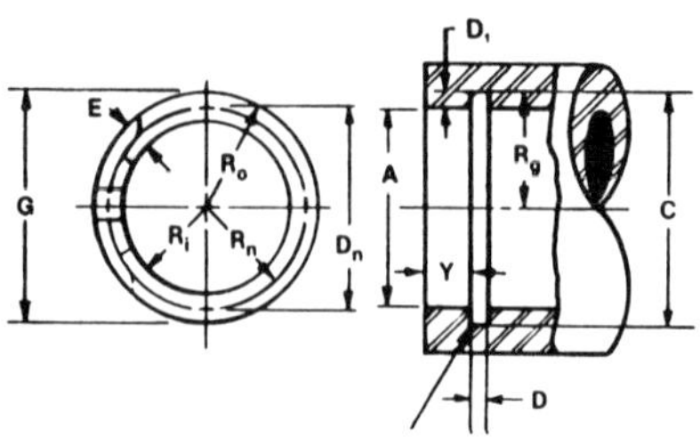

Maximum bottom groove radius is 0.005 in. for ring up to 1.063 in free diameter and 0.010 in. for larger rings.

EXTERNAL

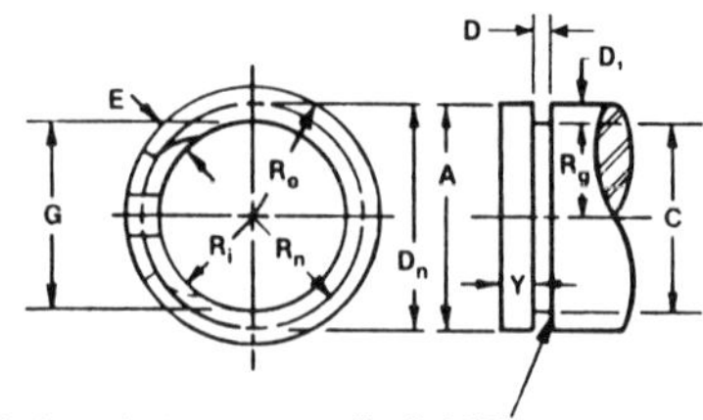

Maximum bottom groove radius is 0.005 in. for ring up to 0.946 in. free diameter and 0.010 in. for larger rings.

GROOVE DEFORMATION

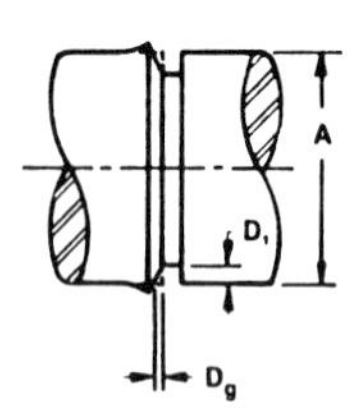

RING DEFLECTION

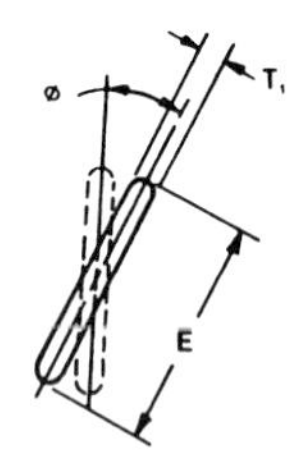

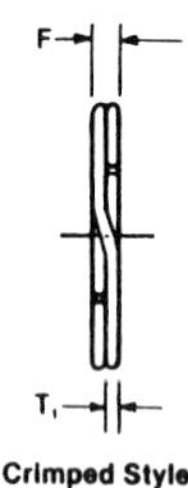

Crimped Style

F

T_1

Uncrimped Style

A = Shaft or housing diameter, in.

C = Groove diameter in.

C_h = Chamfer, in.

C_1 = Percent change of ring diameter from free state to installed state/100

D = Groove width, in.

D_g = Permanent groove deformation, in.

D_n = Ring free neutral diameter, in.

D_1 = Depth of groove, in.

E = Ring radial wall, in.

F = Ring thickness, in.

G = Ring free O.D. or I.D., in.

I = Moment of inertia $\frac{T_1 \ E^3}{12}$, in.^{4}

K = Factor of safety

L = Number of turns of ring

M = Modulus of elasticity, psi

N = Speed, rpm

P_c = Thrust load, lb. per in. of circumference

P_s = Allowable thrust load lb. based on shear strength of ring material

P_t = Thrust load, lb.

R_g = Groove radius, in.

R_i = Free inside ring radius, in.

R_n = Free neutral ring radius, in.

R_o = Free outside ring radius, in.

S = Installation stress, psi

S_c = Shear strength of groove material, psi

S_s = Shear strength of ring material, psi

S_y = Yield strength of groove material, psi

T_1 = Material thickness, in.

V = Area of material cross section, in.^{2}

X = Radius, In.

Y = Distance of groove from end of shaft or housing, in.

δ = Allowable working stress, psi

Φ = Allowable angle of deflection, deg.

τ_1 = 0.19 Frequency factors for 1st &
τ_2 = 0.32 2nd harmonics

γ = Specific weight of material in $\frac{\text{lbs.}}{\text{in.}^3}$

SF-2 = Safety factor of 2

SF-3 = Safety factor of 3

3-21 Allowable Ring Deflection The first thing to be determined in a calculation of thrust capacity based on groove deformation and ring deflection is the allowable axial deflection of the ring without overstressing. The equations for calculating the allowable angle of ring deflection are

$$\text{For external rings, } \phi = 114.6\, R_G \left[\sigma - \frac{\dfrac{C_1 EM}{(1 + C_1)\, D_N}}{MT_1} \right] \qquad \text{(3-12)}$$

$$\text{For internal rings, } \phi = 114.6\, R_G \left[\sigma + \frac{\dfrac{C_1 EM}{(1 - C_1)\, D_N}}{LMT_1} \right] \qquad \text{(3-13)}$$

3-22 Groove Deformation The next step in determining thrust capacity is to calculate the load which will deform the groove wall to the extent that the allowable angle of ring deflection is reached. The equation to determine that load is

$$P_t = \frac{AS_y D_1 \tan \phi}{0.073K} \qquad \text{(3-14)}$$

This equation is valid only if the load is applied through a retained part and groove corner which has square corners and applies the load very close to the shaft or bore diameter. Generally, a safety factor of 2 is applied to the calculated thrust load in applications where the retained part has a square corner, and a safety factor of 4 where the total allowable chamfer or radius is within the limits specified by the manufacturer.

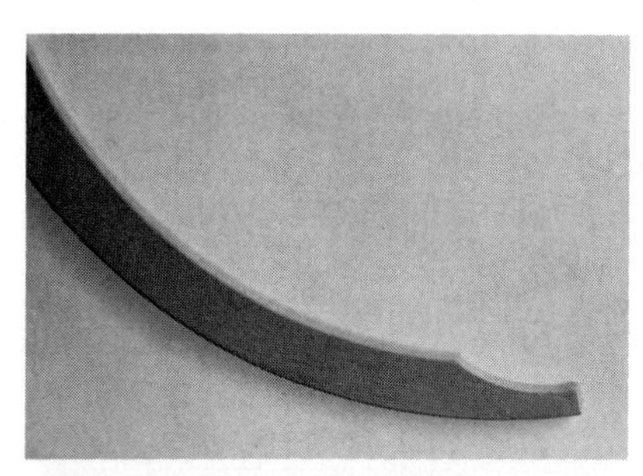

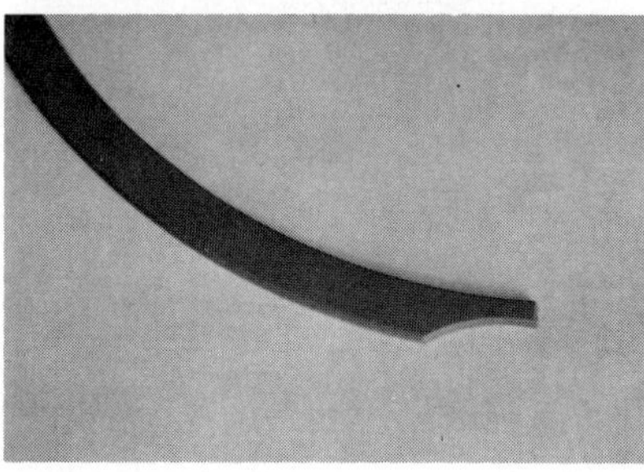

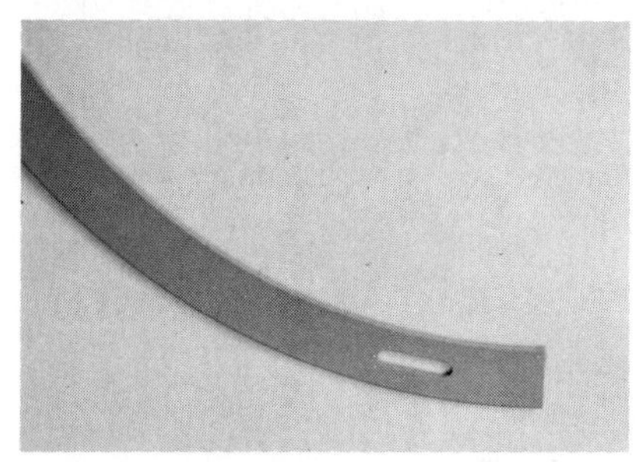

FIG. 3-34 Standard removal notches on (*a*) external and (*b*) internal rings. Slotted type (*c*) is used on external or internal rings.

3-23 Groove Thrust Capacity The groove thrust capacity is determined primarily by the following factors:

1. Groove depth
2. Amount of chamfer on retained part
3. Yield strength of groove material
4. Groove location
5. Type of thrust load (static or dynamic) The calculations above assume static load conditions.
6. Ring material used

3-24 Groove Depth The shallowest possible groove should be selected to minimize groove machining cost and ring cost and to provide ease of ring installation. The minimum permissible groove depth is governed by the thrust load to be restrained, amount of chamfer on the retained part, and the compressive yield strength of the groove material. Figure 3-36 illustrates the direct relationship of groove depth to thrust capacity. The maximum groove depth is limited by the stress on the ring during installation.

3-25 Chamfer on Retained Part Figures 3-36 and 3-37 show that excessive chamfers on retained parts cause drastic reductions in allowable thrust loads for a particular ring.

3-26 Groove Material Figure 3-38 illustrates the effect of groovc yicld strength on thrust capacity for a typical application.

It is obvious then that there are a number of ways of increasing the thrust-load capacity of a particular retaining-ring application. Summarizing briefly: increase groove depth, reduce chamfer on retained part, use a groove material with a higher yield strength.

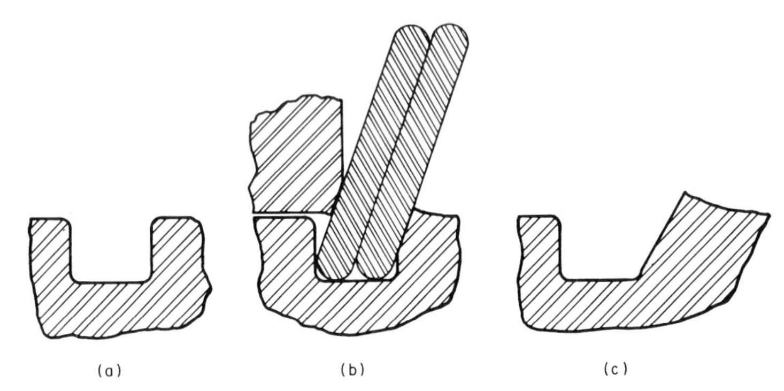

FIG. 3-35 Localized yielding under load. (*a*) Groove profile before loading; (*b*) localized yielding, under load, in both retained part and groove; (*c*) groove profile after loading beyond thrust capacity.

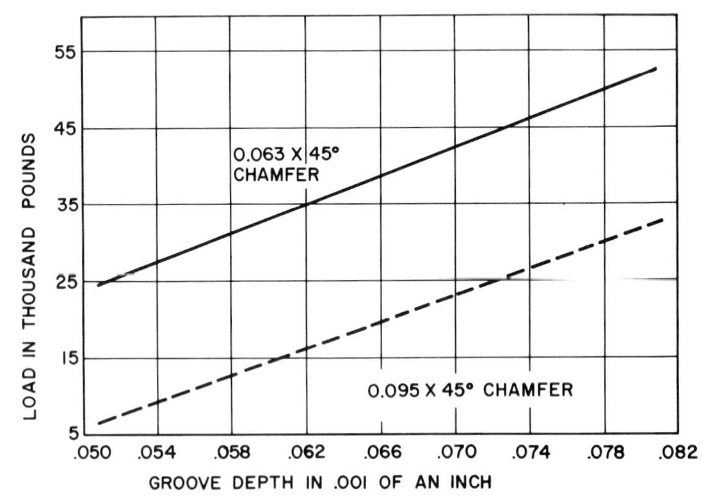

FIG. 3-36 Groove depth relationship to thrust load at which ring failed when loaded through retained parts having chamfers indicated. Yield strength of groove material, 69,000 psi. Yield strength of retained part material, 100,000 psi.

3-27 Groove Location Grooves located too close to the end of the shaft or housing may shear out. However, to facilitate removal of the retaining ring the grooves should be located as near as possible to the end of the shaft or housing bore. The following equation serves as a guide for groove location:

$$Y = \frac{KP_T}{\pi S_c A} \tag{3-15}$$

3-28 Types of Thrust Load A static, uniformly applied load is usually assumed. Dynamic and eccentric loads, however, are frequently encountered. Success of retaining rings in applications where shock or impact loads are encountered depends, in addition to the factors which affect the static-thrust capacity, upon mass and velocity of the retained part as it strikes the ring. Applications involving impact loads require testing.

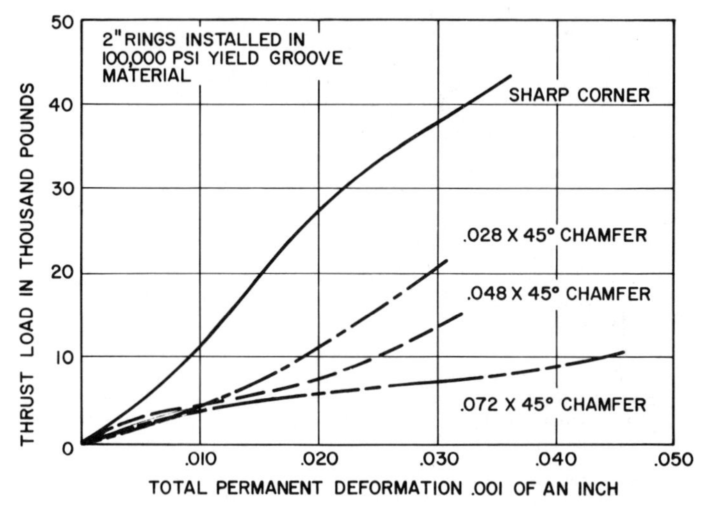

FIG. 3-37 Effect of chamfer size on thrust capacity of spiral-wound retaining rings.

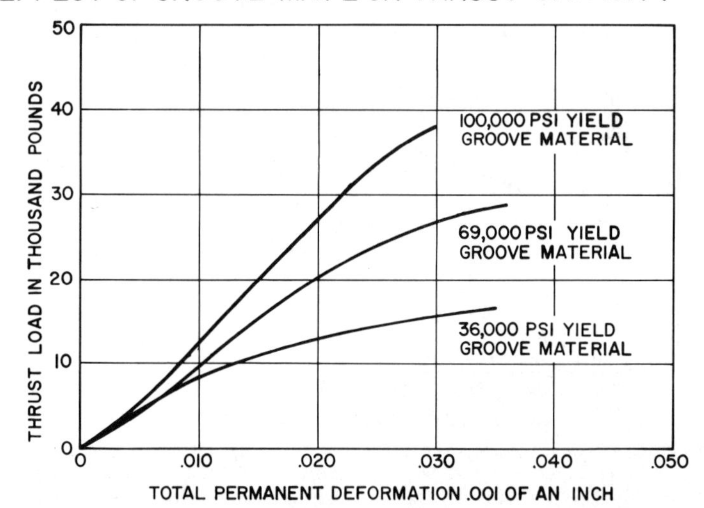

FIG. 3-38 Effect of the groove material on the thrust capacity of a retaining ring.

Eccentric loading occurs many times in practice but is not often anticipated. Eccentric loading can be caused by:

1. Face of the retained part not being machined parallel to the retaining ring.
2. Cocking of the retained part by an external force applied to it at a distance from the retaining ring. Such a case is commonly encountered when a retaining ring is used to hold a large gear or hub on a splined shaft.

3. Axial misalignment of mating parts, as in the case of a mechanical coupling. If eccentric loads are anticipated, the thickest possible ring in the deepest possible groove should be used.

3-29 Rotation Between Parts Relative rotation of a retained part against a spiral-wound retaining ring must be limited to one direction only and to applications involving light thrust loads. The direction of rotation of the retained part against the ring should tend to wind the ring into the groove. Spiral-wound rings are available in both left- and right-hand winds. Failure to observe these criteria will cause the ring to wind out of the groove.

3-30 Centrifugal Force Centrifugal force can overcome the initial cling built into an external ring and cause the ring to lose its grip on the groove bottom. Proper functioning of a retaining ring depends upon the ring remaining seated on the groove bottom. The allowable steady-state speed for a carbon-spring-steel external ring may be calculated from the equation

$$N = \left[\frac{0.466 C_I E^3 \times 10^{12}}{R_N^3 (1 + C_I)(R_O^3 - R_I^3)}\right]^{1/2} \tag{3-16}$$

Special retaining rings are available for high-speed applications. The self-locking ring (Fig. 3-39) can be used where speed is up to 5 times greater than the maximum allowable speed of standard rings.

Rings of 3-inch diameter and larger, because of their mass, are subject to spinning in the groove under conditions of sudden acceleration or deceleration. This problem may be overcome by increasing the cling of the ring within the limits of permissible installation stress or by use of a special ring having a tab which locks into a hole in the groove bottom (Fig. 3-40).

3-31 Balance A high degree of balance may be required in certain assemblies employing retaining rings. This is partially accomplished by obtaining the best balance possible in each component. A common method of balancing spiral-wound retaining rings is to remove, by slotting, an amount of material opposite the gap equal to the installed gap and the removal slots (Fig. 3-41).

3-32 Tolerance Take-up It is often desirable to maintain a continual axial force on a retained part—one of the most common reasons being to eliminate objectionable end play caused by accumulated tolerances. The spiral-wound, dished-type ring (Fig. 3-42) is one means to accomplish this purpose.

3-33 Temperature Temperature limits given in Table 3-9 are to be used as a guide. Time and stress greatly influence the allowable operating temperature. The ring stress under load should be minimized by using a square-cornered retained part,

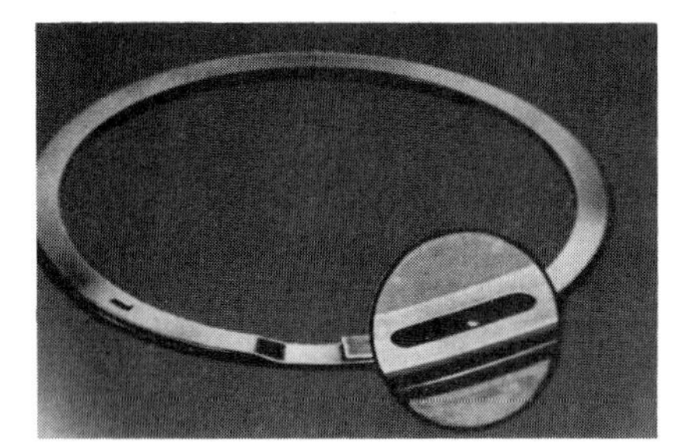

FIG. 3-39 Positive lock, provided by tab and slot arrangement, enables use of ring at rotative speeds 5 times that of standard rings.

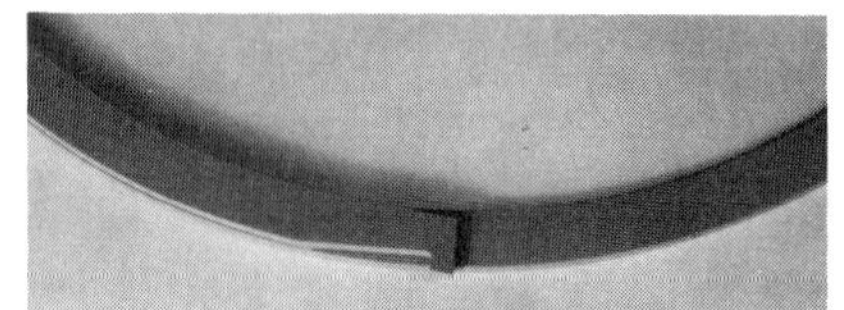

FIG. 3-40 For assemblies subject to rapid acceleration and deceleration, a tab locks into bottom of groove to prevent ring spinning.

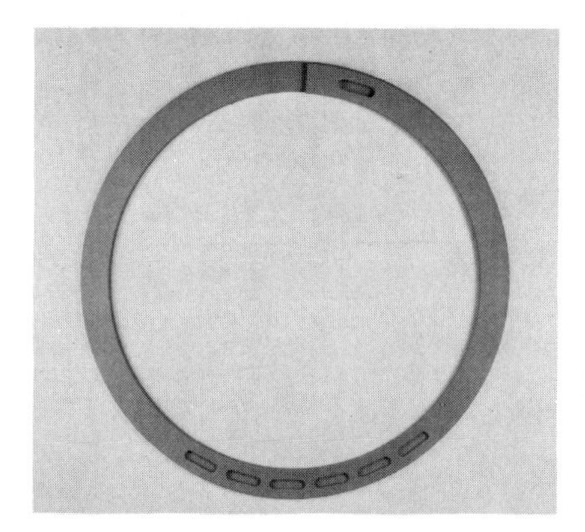

FIG. 3-41 Slots opposite the ring gap form statically balanced ring.

FIG. 3-42 Dished-type ring is a means of taking up accumulated tolerances by exerting low spring load.

a groove material with a high yield strength, and deepest possible grooves for applications operating near the maximum temperature.

3-34 Corrosion A number of materials and finishes, as previously discussed, are available for consideration where corrosion is a factor. Because of various factors affecting corrosion and corrosion rates, recommendations for retaining ring material or finish in a corrosive medium are usually based on the results of tests in the application.

Section 4
Pipe Fastening

Part 1 Construction Pipe Fastening

PATRICK J. GOLDEN, P.E.

Sales Engineer, Elk River Concrete Products,
Division of The CRETEX Companies, Inc.
Minneapolis, Minnesota

4-1 History Since long before the dawn of history, pipes have served man. A pipe is a tube used for conveying liquids, gases, slurries, or sometimes even solids. It also has structural applications in framing and support mechanisms. The earliest pipes were natural tubes such as bamboo or hollow logs. The Greeks used clay water pipes as early as 500 B.C. Romans used lead pipes to convey water from aqueducts. Copper water lines date back to the Egyptian pyramids. Julius Caesar designed a tamped-concrete-pipe water system for the city of Cologne, Germany, parts of which did service for 1,828 years. Brick sewers also date back to this era. Cast-iron pipe was first used in Versailles, France, in 1664—more than 300 years ago. Our American pioneers made water systems from logs with holes bored through their centers. To this ever-growing list of materials were added the various plastics, which date back to World War II. Many changes have taken place in the industry to meet the demand for the material which is best suited for a particular job. The selection is based on cost, availability, and suitability. Factors to consider are strength, lightness or heaviness, rigidity or flexibility, resistance to chemical attack, and ease of fabrication and installation.

4-2 Various Types of Pipe Connections The methods of joining or connecting pipes are unique for each material and type of joint. For instance, cast-iron mechanical joints are a specific bolted joint with a rubber gasket, whereas with steel pipe any coupling which is bolted into position is considered a mechanical joint. There is some cross-reference between materials and joints, but it should not be taken for granted. This section is organized in such a way that all terminology used is for a specific material.

Table 4-1 is a reference guide to selecting the specification for a material. Only the highlights of the various connections can be covered in the text. Any specific question should be obtained by referring to the specification or manufacturer.

GRAY IRON AND DUCTILE IRON

Gray iron and ductile iron are the two basic types of cast-iron pipe. Either can be used for conveying water, gas, and sewage in municipal utility and industrial piping systems. Each is basically the same in chemical nature except that gray cast iron has carbon in the form of graphite flakes whereas ductile cast iron has carbon in the form of graphite nodules. This element with an iron alloy, silicon, and appropriate amounts of manganese, sulfur, and phosphorus form cast iron. This combination of elements gives the pipe excellent corrosion resistance and strength. Ductile-cast-iron pipe has advantages over gray-cast-iron pipe in that it has the physical strength of mild steel and the ability to bend under stress. These features enable this pipe to be used in situations where greater strength is needed or a chance of bending exists.

Soil pipe is a low-pressure pipe normally used for gravity sewer services.

The following sections will describe the different types of joints available. Each has advantages which distinguish one type for a particular application. See Table 4-2.

4-3 Push-on Joints The push-on joint is a compression-type joint whose only accessory is a molded-rubber gasket. The gasket is fitted into a recessed area on the bell end of the pipe. The gasket forms a pressuretight seal with ample deflection when compressed in position by the entering beveled end of the pipe. The joint is recommended for hydraulic pressures only.

Assembly of the joint is simple, easy, and fast. Refer to Figs. 4-1 through 4-6 for assembly procedure.

Gaskets are manufactured from SBR rubber or conductive gaskets. To insure conductivity, copper straps or steel cables are attached to the outside of the pipe at the joints, or lead wedges are driven into the joints from the outside. Alternatively, conductive gaskets containing copper inserts can be used. The conductive gaskets facilitate faster installation, whereas the other methods involve an additional step in the field.

The assembly procedure for a push-on joint is as follows:

1. Thoroughly clean out the bell.
2. Brush-coat gasket-retaining groove and inner shoulder with lubricant as shown in Fig. 4-1.
3. Insert gasket with solid face toward installer. Use one hand to hold a loop in the gasket, the other to tuck the remaining portion into its groove as in Fig. 4-2.
4. Release gasket and press remaining loop firmly up into lubricated groove.
5. Check to be sure gasket is completely seated, then apply a generous coating of lubricant to the exposed gasket surface.
6. Clean the plain-end pipe and grind away all sharp edges.
7. Place plain end in companion bell and provide reasonably straight alignment. Push pipe straight home with a bar or, in case of larger pipe, one of the methods shown in Figs. 4-3, 4-4, and 4-5. These mechanical jacking devices easily pull the pipe home. The final position is shown in Fig. 4-6.

4-4 Mechanical Joints The mechanical joint is a bolted joint suitable for both hydraulic- and gas-pressure service. The joint consists of an integral bell with flange and a rubber ring gasket, follower gland, nuts, and bolts. Figure 4-7 illustrates the standard mechanical-joint dimensions. When assembled, the joint provides for normal expansion and contraction and permits adequate deflection.

Gaskets available with this joint are lead-type or SBR (plain) rubber.

Table 4-3 lists the various flange dimensions for class 125 and class 250 pipe.

The installation procedure consists of six easy steps which are clearly illustrated in Figs. 4-8 through 4-13.

4-5 Restrained Joints The various types of restrained joints available are mechanical joints with retainer glands, friction clamps, and specific-trade-name joints such

TABLE 4-1 Specifications Pertaining to Pipe Connections

MATERIAL	JOINT	ASTM	AWWA	ANSI	FEDERAL	ASSHTO	MISC.
CAST IRON & DUCTILE IRON	Push - On		C106, C111, C150	A21.6, A21.11	WW-P-421c		
	Mechanical		C106, C111, C150	A21.6, A21.11	WW-P-421c		
	Restrained			A21.52, A21.1			
	River Crossing			A21.51, A21.10			
	Calked						Procedure in text-compatible w/local codes
	Flanged		C110	A21.10, B16.1			
	Threaded			A40.5	WWP-356		
STEEL	Threaded			B2.1			
	Welded	A53, A83, A105, A106 A120, A135, A139, A181	C 201, C202	R1179, B36			
	Mechanical ●			R1179, B36, B16.5			
NON-FERROUS (COPPER, LEAD, BRASS)	Soldered			▲			▲
	Mechanical			▲			▲
	Wiped			▲			▲
NON-FERROUS (COPPER, LEAD, BRASS)	Brazed			▲			▲
	Threaded			▲			▲
	Burned Lead			▲			▲

TABLE 4-1 *(CONT.)* Specifications Pertaining to Pipe Connections

MATERIAL	JOINT	ASTM	AWWA	ANSI	FEDERAL	ASSHTO	MISC.
ALUMINUM	Welded ●						●
	Threaded			B2.1			
	Mechanical ●						●
CONCRETE	Open	C118,C412			USDA - SCS 606		
	Tongue & Groove	C14, C76, C444, C506, C478, C654, C507			SS-P-371 SS-P-375	M66,M170,M175, M176,M199,M206, M207	
	Mortar						
	Mastic				SS-S-00210		
	External Gasket ●						●
	Rubber Gasket	C443, C14, C76				M198	
CONCRETE	Bell & Spigot	C14, C76, C444, C478			SS-P-371, SS-P-375	M86, M170, M175, M199	
CONCRETE	Mortar						
	Mastic				SS-S-00210		
	Hot Poured						
	Rubber Gasket	C443, C505				M198	
	Tied ●						●
CONCRETE (PRES.)	Low Pressure w/ Steel Ring	C361 or C376	C302				
	Steel Cyl. w/Steel Ring		C300				

TABLE 4-1 *(CONT.)* Specifications Pertaining to Pipe Connections

MATERIAL	JOINT	ASTM	AWWA	ANSI	FEDERAL	ASSHTO	MISC.
	Prestressed ● w/ Steel Ring		COP Spec. SP-31			●	
	Prestressed w/Steel Ring & Steel Cyl.		C301				
	Cyl. Pretensioned Reinforced Conc.		C303				
	Tied ●						●
	Ball ●						●
CLAY	Open	C4, C498			USDA-SCS 606		
CLAY	Hot Poured ▲						▲
	Gasket	C425, C594					
ASBESTOS CEMENT	Non-Pressure	C428, C644	C400		SS-P-331b		
	Pressure	C296, C668	C400		SS-P-351c		
CORRUGATED STEEL	Bolted or Clamped				WW-P-00405	M36	
FIBERGLASS	Adhesive ▲	D2517, D2996, D2310					▲
	Mechanical ▲	D2517, D2310, D3262		ASA-B-16-1	MSS-SP-51		▲
GLASS	Armoured	▲	C599, C600, C601				
	Beaded Pressure	▲	C599, C600, C601				
	Capped	▲	C599, C600, C601				
	Conical	▲	C599, C600, C601				
	Drainage	▲	C599, C600, C601				

TABLE 4-1 *(CONT.)* Specifications Pertaining to Pipe Connections

MATERIAL	JOINT	ASTM	AWWA	ANSI	FEDERAL	ASSHTO	MISC.
PLASTIC	Sovent Welded	D313-3 ABS & PVC					
PLASTIC	P.V.C.	D2466, D2467, D3033, D3034, D2564, D2672. D2665, D2846		B72.16			
	A.B.S.	D2235, D2680. D2468, D2467, D2560, D2750 D2680		B72.23			
	C A B	D2446, D2560					
	SR	D3122, D2852					
	Gasket						
	P.V.C.	D3139, D3033, D2321, D3034					
	A.B.S.	D2321					
	Heat Fusion	D2657					
	P E † •	D2610, D2611, D2683					
	Threaded			B2.1			
	P.V.C ‡	D2464		B2.1			
	A. B. S.	D2465, D2661		B2.1			
	Mechanical	D2662, D2666, D3000					
Bituminzed Fiber	Tapered, Butt, Slip	D2818, D2316, D2311, D1861			SS-P-1540A	M158, M177	

TABLE 4-1 *(CONT.)* Specifications Pertaining to Pipe Connections

MATERIAL	JOINT	ASTM	AWWA	ANSI	FEDERAL	ASSHTO	MISC.
MISC. CONNECTORS & ADAPTORS (NON-PR)	Sleeves P.V.C. or Rubber ●						●
	Concrete Sleeves						Local Codes
	Manhole Pipe Connections ●						●
	Rubber Gasket Adaptors ●						●
MISC. CONNECTORS & ADAPTORS (PRES.)	Corporation Stop		C800				
	Service Clamp ●						●
	Expansion ●						●

NOTES:

Always refer to applicable local, state, and federal codes.

▲ Material Spec. Only

● Manufacturers Spec.

12 to 24-in. Corrugated PE, ASSHTO M294

< 12-in. Corrugated PE, ASSHTO M252 (Standard Specification)

< 12-in. Corrugated PE, ASSTM F403 (Standard Specification)

12 to 24-in. Corrugated PE, ASTM F667 (Standard Specification)

† HDPE, ASTM F894

‡ 18 to 27-in. PVC (Large Diameter), ASTM F679-86

TABLE 4-2 Outside Diameters, Inside Diameters, and Thicknesses of Various Cast-Iron Pipes

NOM. SIZE	CENTRIFUGALLY CAST PIPE ASA A21.6-1953 ASA A21.8-1953 AWWA C106-53 AWWA C108-53			PIT CAST PIPE ASA A21.2-1953 AWWA C102-53		
	O.D.	I.D.	THICK.	O.D.	I.D.	THICK.
CLASS 50 — 50 PSI — 115 FT. HD.						
3″	3.96	3.32	.32	3.80	3.06	.37
4″	4.80	4.10	.35	4.80	4.00	.40
6″	6.90	6.14	.38	6.90	6.04	.43
8″	9.05	8.23	.41	9.05	8.13	.46
10″	11.10	10.22	.44	11.10	10.10	.50
12″	13.20	12.24	.48	13.20	12.12	.54
14″	15.30	14.34	.48	15.30	14.22	.54
16″	17.40	16.32	.54	17.40	16.24	.58
18″	19.50	18.42	.54	19.50	18.24	.63
20″	21.60	20.46	.57	21.60	20.28	.66
24″	25.80	24.54	.63	25.80	24.32	.74
30″	32.00	30.42	.79	31.74	30.00	.87
36″	38.30	36.56	.87	37.96	36.02	.97
42″	44.50	42.56	.97	44.20	42.06	1.07
48″	50.80	48.68	1.06	50.50	48.14	1.18
54″				56.66	54.06	1.30
60″				62.80	60.02	1.39
CLASS 100 — 100 PSI — 231 FT. HD.						
3″	3.96	3.32	.32	3.80	3.06	.37
4″	4.80	4.10	.35	4.80	4.00	.40
6″	6.90	6.14	.38	6.90	6.04	.43
8″	9.05	8.23	.41	9.05	8.13	.46
10″	11.10	10.22	.44	11.10	10.10	.50
12″	13.20	12.24	.48	13.20	12.12	.54
14″	15.30	14.28	.51	15.30	14.14	.58
16″	17.40	16.32	.54	17.40	16.14	.63
18″	19.50	18.34	.58	19.50	18.14	.68
20″	21.60	20.36	.62	21.60	20.18	.71
24″	25.80	24.44	.68	25.80	24.20	.80
30″	32.00	30.42	.79	32.00	30.12	.94
36″	38.30	36.56	.87	38.30	36.20	1.05
42″	44.50	42.56	.97	44.50	42.00	1.25
48″	50.80	48.68	1.06	50.80	48.06	1.37
54″				57.10	54.08	1.51
60″				63.40	60.16	1.62
CLASS 150 — 150 PSI — 346 FT. HD.						
3″	3.96	3.32	.32	3.80	3.06	.37
4″	4.80	4.10	.35	4.80	4.00	.40
6″	6.90	6.14	.38	6.90	6.04	.43
8″	9.05	8.23	.41	9.05	8.13	.46
10″	11.10	10.22	.44	11.10	10.02	.54
12″	13.20	12.24	.48	13.20	12.04	.58
14″	15.65	14.63	.51	15.65	14.39	.63
16″	17.80	16.72	.54	17.80	16.44	.68
18″	19.92	18.76	.58	19.92	18.46	.73
20″	22.06	20.82	.62	22.06	20.40	.83
24″	26.32	24.86	.73	26.32	24.46	.93
30″	32.00	30.30	.85	32.40	30.20	1.10
36″	38.30	36.42	.94	38.70	36.26	1.22
42″	44.50	42.40	1.05	45.10	42.40	1.35
48″	50.80	48.52	1.14	51.40	48.44	1.48
54″				57.80	54.54	1.63
60″				64.20	60.42	1.89

as Lok-Fast Tight or Super-Lock. All have the ability to resist forces tending to pull the joints apart.

A mechanical joint with a retainer gland is a simple but effective type. It consists of a retainer gland fitted with cup-pointed steel set screws. The set screws are tightened to a specified torque, causing them to bite into the surface of the pipe. This prevents joint separation and provides electrical conductivity in the pipe. See Fig. 4-12.

Friction clamps consist of two socket clamps tightened on the pipe on both sides of the joint. The clamps cling to the pipe through friction. Tie rods are then strung between the clamps and tightened. This system is also used on bends with hooks cast into the bend to support the tie rod. See Fig. 4-13.

TABLE 4-2 Outside Diameters, Inside Diameters, and Thicknesses of Various Cast-Iron Pipes (Continued)

NOM. SIZE	CENTRIFUGALLY CAST PIPE ASA A21.6-1953 ASA A21.8-1953 AWWA C106-53 AWWA C108-53			PIT CAST PIPE ASA A21.2-1953 AWWA C102-53		
	O.D.	I.D.	THICK.	O.D.	I.D.	THICK.
CLASS 200 — 200 PSI — 462 FT. HEAD.						
3"	3.96	3.32	.32	3.80	3.06	.37
4"	4.80	4.10	.35	4.80	4.00	.40
6"	6.90	6.14	.38	6.90	6.04	.43
8"	9.05	8.23	.41	9.05	8.13	.46
10"	11.10	10.22	.44	11.10	9.94	.58
12"	13.20	12.24	.48	13.20	11.94	.63
14"	15.65	14.55	.55	15.65	14.29	.68
16"	17.80	16.64	.58	17.80	16.22	.79
18"	19.92	18.66	.63	19.92	18.22	.85
20"	22.06	20.72	.67	22.06	20.26	.90
24"	26.32	24.74	.79	26.32	24.32	1.00
30"	32.00	30.16	.92	32.74	30.36	1.19
36"	38.30	36.26	1.02	39.16	36.30	1.43
42"	44.50	42.24	1.13	45.58	42.42	1.58
48"	50.80	48.34	1.23	51.98	48.52	1.73
54"				58.40	54.60	1.90
60"				64.82	60.42	2.20
CLASS 250 — 250 PSI — 577 FT. HEAD.						
3"	3.96	3.32	.32	3.80	3.06	.37
4"	4.80	4.10	.35	4.80	4.00	.40
6"	6.90	6.14	.38	6.90	6.04	.43
8"	9.05	8.23	.41	9.05	8.05	.50
10"	11.10	10.22	.44	11.40	10.14	.63
12"	13.20	12.16	.52	13.50	12.14	.68
14"	15.65	14.47	.59	15.65	14.07	.79
16"	17.80	16.54	.63	17.80	16.10	.85
18"	19.92	18.56	.68	19.92	18.08	.92
20"	22.06	20.62	.72	22.06	20.12	.97
24"	26.32	24.74	.79	26.32	24.16	1.08
30"	32.00	30.02	.99	32.74	29.96	1.39
36"	38.30	36.10	1.10	39.16	36.08	1.54
42"	44.50	42.06	1.22	45.58	42.16	1.71
48"	50.80	48.14	1.33	51.98	47.94	2.02
54"				58.40	53.98	2.21
60"				64.82	60.06	2.38

AWWA — 1908 STANDARD

NOM. SIZE	CLASS A 43 PSI-100 FT. HD.			CLASS B 86 PSI-200 FT. HD.			CLASS C 130 PSI-300 FT. HD.			CLASS D 173 PSI-400 FT. HD.		
	O.D.	I.D.	THICK.	O.D.	I.D.	THICK.	O.D.	I.D.	THICK.	O.D.	I.D.	THICK.
3"	3.80	3.02	.39	3.96	3.14	.42	3.96	3.06	.45	3.96	3.00	.48
4"	4.80	3.96	.42	5.00	4.10	.45	5.00	4.04	.48	5.00	3.96	.52
6"	6.90	6.02	.44	7.10	6.14	.48	7.10	6.08	.51	7.10	6.00	.55
8"	9.05	8.13	.46	9.05	8.03	.51	9.30	8.18	.56	9.30	8.10	.60
10"	11.10	10.10	.50	11.10	9.96	.57	11.40	10.16	.62	11.40	10.04	.68
12"	13.20	12.12	.54	13.20	11.96	.62	13.50	12.14	.68	13.50	12.00	.75
14"	15.30	14.16	.57	15.30	13.98	.66	15.65	14.17	.74	15.65	14.01	.82
16"	17.40	16.20	.60	17.40	16.00	.70	17.80	16.20	.80	17.80	16.02	.89
18"	19.50	18.22	.64	19.50	18.00	.75	19.92	18.18	.87	19.92	18.00	.96
20"	21.60	20.26	.67	21.60	20.00	.80	22.06	20.22	.92	22.06	20.00	1.03
24"	25.80	24.28	.76	25.80	24.02	.89	26.32	24.24	1.04	26.32	24.00	1.16
30"	31.74	29.98	.88	32.00	29.94	1.03	32.40	30.00	1.20	32.74	30.00	1.37
36"	37.96	35.98	.99	38.30	36.00	1.15	38.70	35.98	1.36	39.16	36.00	1.58
42"	44.20	42.00	1.10	44.50	41.94	1.28	45.10	42.02	1.54	45.58	42.02	1.78
48"	50.50	47.98	1.26	50.80	47.96	1.42	51.40	47.98	1.71	51.98	48.06	1.96
54"	56.66	53.96	1.35	57.10	54.00	1.55	57.80	54.00	1.90	58.40	53.94	2.23
60"	62.80	60.02	1.39	63.40	60.06	1.67	64.20	60.20	2.00	64.82	60.06	2.38
72"	75.34	72.10	1.62	76.00	72.10	1.95	76.88	72.10	2.39			
84"	87.54	84.10	1.72	88.54	84.10	2.22						

SOURCE: AWWA.

Each manufacturer has its own design of a restrained push-on joint. Such joints are boltless and depend on a retainer gland surrounding lugs on the bell end of the pipe.

4-6 River-Crossing Joints Each manufacturer of ductile iron pipe has its own design of river-crossing joint. It consists of a precisely machined ball joint of the push-on type which is locked in place by bolts, pinned retainer lock, or other means. The joint design assures uniform load distribution between the restraining components even when the joint is fully deflected. The joint is assembled like a push-on joint with the additional step of locking the retainer over the lugs. Figure 4-14 illustrates a typical river-crossing joint. This joint is expensive but certain conditions make its use feasible.

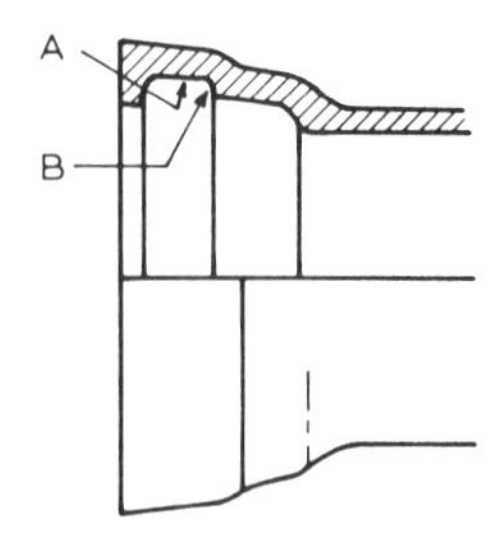

A BRUSH-COAT GASKET RETAINING GROOVE
B INNER SHOULDER

FIG. 4-1 Cross section of bell end of typical push-on pipe.

4-7 Calked Joints One method of joining cast-iron pipe is by means of a calked or percussion joint. This type of joint is made of molten lead and oakum and calked with irons to make it watertight.

FIG. 4-2 Insertion of gasket. *(American Cast Iron Pipe Co.)*

The procedure for making a vertical calk joint involves the following steps:

1. Wipe hub and spigot ends dry and clean.
2. Align and center pipe to be joined.
3. Pack oakum or similar packing material into the joint to a depth of approximately 1 inch from the top.
4. Heat lead until it appears a cherry red color then pour into the hub with a ladle.
5. The calking irons are then used to pack the oakum and lead, making the joint watertight. This is where great skill is needed because the lead must be packed carefully, but if struck

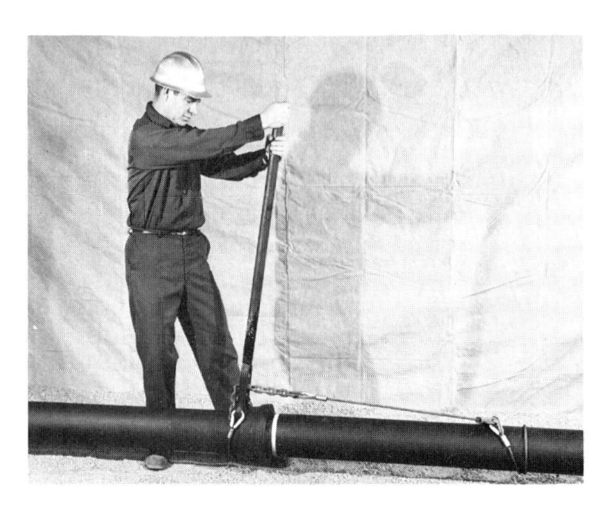

FIG. 4-3 Bar method of pushing pipe to complete connection. *(American Cast Iron Pipe Co.)*

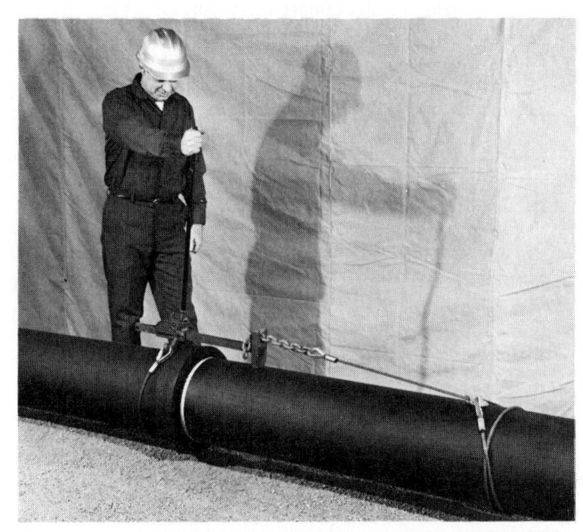

FIG. 4-4 Bar-and-rack method of jacking pipe into position. (*American Cast Iron Pipe Co.*)

FIG. 4-5 Method used to mate large-diameter pipe in place. (*American Cast Iron Pipe Co.*)

too hard it will break the bell. A compressed-air calking gun is available which speeds the operation and lessens the chance of breakage.

The process for a horizontal calked joint is similar except that a pouring rope must be used to hold the hot lead in place.

The joint may be painted, varnished, or coated after it has been tested and approved.

This joint is used quite frequently by plumbers for service connections and aboveground piping. However, most underground

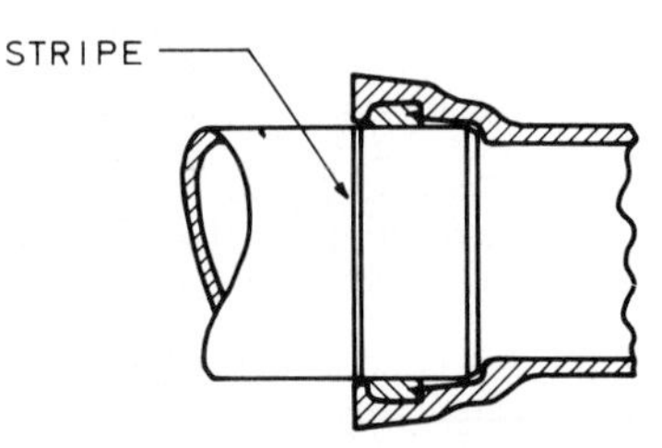

FIG. 4-6 Check the assembly. The joint is completely assembled when the "stripe" is no longer visible. Deflection should be taken after joint is assembled.

STANDARD MECHANICAL-JOINT DIMENSIONS

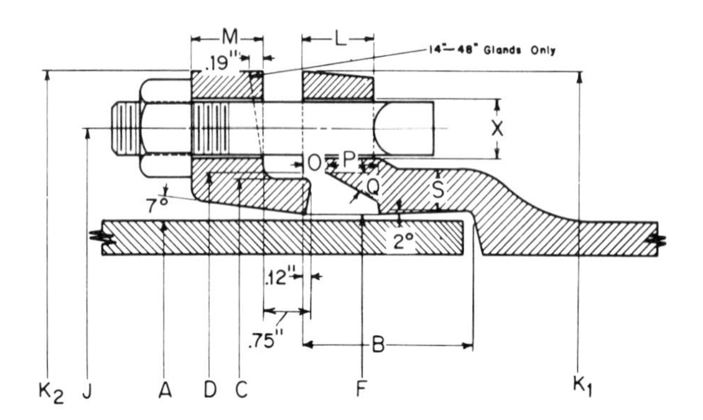

DIMENSIONS IN INCHES

Size	A	B	C	D	F	Q	X	J	K1 Centrifugal Pipe	K1 Pit Cast Pipe and Fittings	K2	L	M	O	P	S Centrifugal Pipe	S Pit Cast Pipe and Fittings	Bolts No.	Bolts Size	Bolts Lgth
3	3.96	2.50	4.84	4.94	4.06	28°	¾	6.19	7.62	7.69	7.69	.94	.62	.31	.63	.47	.52	4	⅝	3
4	4.80	2.50	5.92	6.02	4.90	28°	⅞	7.50	9.06	9.12	9.12	1.00	.75	.31	.75	.55	.65	4	¾	3½
6	6.90	2.50	8.02	8.12	7.00	28°	⅞	9.50	11.06	11.12	11.12	1.06	.88	.31	.75	.60	.70	6	¾	3½
8	9.05	2.50	10.17	10.27	9.15	28°	⅞	11.75	13.31	13.37	13.37	1.12	1.00	.31	.75	.66	.75	6	¾	4
10	11.10	2.50	12.22	12.34	11.20	28°	⅞	14.00	15.62	15.69	15.62	1.19	1.00	.31	.75	.72	.80	8	¾	4
12	13.20	2.50	14.32	14.44	13.30	28°	⅞	16.25	17.88	17.94	17.88	1.25	1.00	.31	.75	.79	.85	8	¾	4
14	15.30	3.50	16.40	16.54	15.44	28°	⅞	18.75	20.25	20.31	20.25	1.31	1.25	.31	.75	.85	.89	10	¾	4
16	17.40	3.50	18.50	18.64	17.54	28°	⅞	21.00	22.50	22.56	22.50	1.38	1.31	.31	.75	.91	.97	12	¾	4½
18	19.50	3.50	20.60	20.74	19.64	28°	⅞	23.25	24.75	24.83	24.75	1.44	1.38	.31	.75	.97	1.05	12	¾	4½
20	21.60	3.50	22.70	22.84	21.74	28°	⅞	25.50	27.00	27.08	27.00	1.50	1.44	.31	.75	1.03	1.12	14	¾	4½
24	25.80	3.50	26.90	27.04	25.94	28°	⅞	30.00	31.50	31.58	31.50	1.62	1.56	.31	.75	1.08	1.22	16	¾	5
30 *	32.00	4.00	33.29	33.46	32.17	20°	1⅛	36.88	39.12	39.12	39.12	1.81	2.00	.38	1.00	1.20	1.50	20	1	5½
36 *	38.30	4.00	39.59	39.76	38.47	20°	1⅛	43.75	46.00	46.00	46.00	2.00	2.00	.38	1.00	1.35	1.80	24	1	5½
42 *	44.50	4.00	45.79	45.96	44.67	20°	1⅜	50.62	53.12	53.12	53.12	2.00	2.00	.38	1.00	1.48	1.95	28	1¼	6
48 *	50.80	4.00	52.09	52.26	50.97	20°	1⅜	57.50	60.00	60.00	60.00	2.00	2.00	.38	1.00	1.61	2.20	32	1¼	6

*Dimensions Of 30"—48" Sizes Are Tentative

Chart taken from ASA A21.11-1953 and AWWA C111-53 Standard Specifications.

FIG. 4-7 Standard mechanical joint with dimensions. (*AWWA.*)

main line piping makes use of the faster methods of joining pipe such as the push-on joint. See Fig. 4-15.

4-8 Flanged Joints When rigid joints are required, flanged joints are used. They are manufactured by threading plain-end pipe, screwing long hub flanges in place, and refacing across both the faces of flange and the pipe ends. This operation must be done in the factory, not the field. A gasket seats itself over the pipe ends and thus pipe threading is not affected by line pressure. The adjoining flanges are bolted together and tightened to specific torques. These flanges meet ANSI A21.10 or ANSI B16.1 specifications.

4-9 Threaded Joints Threaded joints are available in cast-iron and ductile-iron pipe with the same outside diameter as steel pipe. The pipe can be cut, threaded, fitted, and installed on a job with ordinary tools. All regular-size threaded fittings are available with threading meeting ANSI B2.1.

TABLE 4-3 Cast-Iron Flange Dimensions, Class 125 and Class 250

Nominal Pipe Size	Diam of Flange	Thickness of Flange (Min)	Diam of Bolt Circle	Number of Bolts	Diam of Bolts	Diam of Bolt Holes	Length, of Bolts	Length, of Bolt-Stud With Two Nuts
1	4¼	7/16	3⅛	4	½	⅝	1¾	....
1¼	4⅝	½	3½	4	½	⅝	2	....
1½	5	9/16	3⅞	4	½	⅝	2	....
2	6	⅝	4¾	4	⅝	¾	2¼	....
2½	7	11/16	5½	4	⅝	¾	2½	....
3	7½	¾	6	4	⅝	¾	2½	....
3½	8½	13/16	7	8	⅝	¾	2¾	....
4	9	15/16	7½	8	⅝	¾	3	....
5	10	15/16	8½	8	¾	⅞	3	....
6	11	1	9½	8	¾	⅞	3¼	....
8	13½	1⅛	11¾	8	¾	⅞	3½	....
10	16	1 3/16	14¼	12	⅞	1	3¾	....
12	19	1¼	17	12	⅞	1	3¾	....
14	21	1⅜	18¾	12	1	1⅛	4¼	....
16	23½	1 7/16	21¼	16	1	1⅛	4½	....
18	25	1 9/16	22¾	16	1⅛	1¼	4¾	....
20	27½	1 11/16	25	20	1⅛	1¼	5	....
24	32	1⅞	29½	20	1¼	1⅜	5½	....
30	38¾	2⅛	36	28	1¼	1⅜	6¼	....
36	46	2⅜	42¾	32	1½	1⅝	7	....
42	53	2⅝	49½	36	1½	1⅝	7½	....
48	59½	2¾	56	44	1½	1⅝	7¾	....
*54	66¼	3	62¾	44	1¾	2	8½	10½
*60	73	3⅛	69¼	52	1¾	2	8¾	10¾
*72	86½	3½	82½	60	1¾	2	9½	11½
*84	99¾	3⅞	95½	64	2	2¼	10½	12¾
*96	113¼	4¼	108½	68	2¼	2½	11½	14

All dimensions given in inches.
*These sizes are included for convenience and do not carry a definite rating.

Extracted from American Standard Cast-Iron Pipe Flanges and Flanged Fittings (ASA B16.1—1960), with the permission of the publisher, The American Society of Mechanical Engineers, 29 West 39th Street, New York 18, N. Y.

SOURCE: ASA B16.1 and B16.2-1960, as published by ASME.

TABLE 4-3 Cast-Iron Flange Dimensions, Class 125 and Class 250 (Continued)

Nominal Pipe Size	Diam of Flange	Thickness of Flange (Min)	Diam of Raised Face	Diam of Bolt Circle	Diam of Bolt Holes	Number of Bolts	Size of Bolts	Length of Bolts	Length of Bolt Studs With Two Nuts
1	4⅞	11/16	2 11/16	3½	¾	4	⅝	2½	....
1¼	5¼	¾	3 1/16	3⅞	¾	4	⅝	2½	....
1½	6⅛	13/16	3 9/16	4½	⅞	4	¾	2¾	....
2	6½	⅞	4 3/16	5	¾	8	⅝	2¾	....
2½	7½	1	4 15/16	5⅞	⅞	8	¾	3¼	....
3	8¼	1⅛	5 11/16	6⅝	⅞	8	¾	3½	....
3½	9	1 3/16	6 5/16	7¼	⅞	8	¾	3½	....
4	10	1¼	6 15/16	7⅞	⅞	8	¾	3¾	....
5	11	1⅜	8 5/16	9¼	⅞	8	¾	4	....
6	12½	1 7/16	9 11/16	10⅝	⅞	12	¾	4	....
8	15	1⅝	11 15/16	13	1	12	⅞	4½	....
10	17½	1⅞	14 1/16	15¼	1⅛	16	1	5¼	....
12	20½	2	16 7/16	17¾	1¼	16	1⅛	5½	....
14	23	2⅛	18 15/16	20¼	1¼	20	1⅛	6	....
16	25½	2¼	21 1/16	22½	1⅜	20	1¼	6¼	....
18	28	2⅜	23 5/16	24¾	1⅜	24	1¼	6½	..
20	30½	2½	25 9/16	27	1⅜	24	1¼	6¾	....
24	36	2¾	30¼	32	1⅝	24	1½	7¾	9½
*30	43	3	37 3/16	39¼	2	28	1¾	8½	10½
*36	50	3⅜	43 11/16	46	2¼	32	2	9½	11¾
*42	57	3 11/16	50 7/16	52¾	2¼	36	2	10¼	12½
*48	65	4	58 7/16	60¾	2¼	40	2	10¾	13

All dimensions given in inches.

Extracted from American Standard Cast-Iron Pipe Flanges and Flanged Fittings (ASA B16.2—1960), with the permission of the publisher, The American Society of Mechanical Engineers, 29 West 39th Street, New York 18, N. Y.

FIG. 4-8 Preassembly position of mechanical joint. Wash socket and plain end with soapy water then slip gland and gasket over plain end. Small side of gasket and lip side of gland face the socket. (*American Cast Iron Pipe Co.*)

FIG. 4-9 Lubricate gasket of mechanical joint. Paint gasket with soapy water. *American Cast Iron Pipe Co.*)

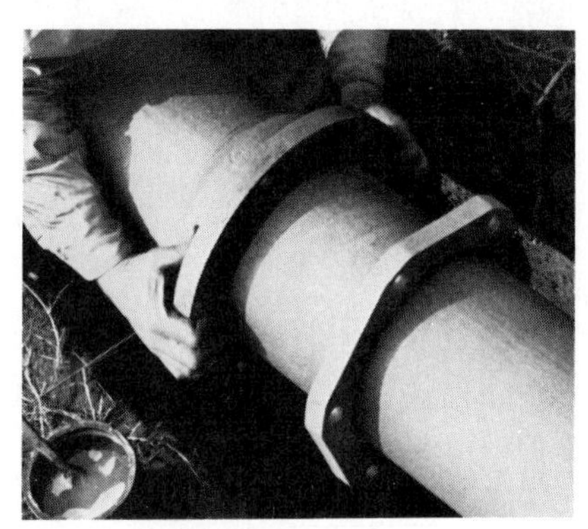

FIG. 4-10 Insert plain end into socket. Push gasket into position with fingers, making sure it is evenly seated. (*American Cast Iron Pipe Co.*)

STEEL PIPE

This material is a versatile product that has many uses. The various types of steel pipe are black, galvanized, stainless, high-strength low-alloy, and yoloy. This pipe can serve as a fluid- or gas-handling system, a structural component, or can be fabricated into a finished product. The pipe is formed into various shapes and joined by welding, threading, or clamping. It is used in many commercial, industrial, and consumer products and systems. The advantages of steel pipe are:

1. Great strength due to its higher tensile strength as compared with other materials.
2. Its ductility, which makes it easy to hot- or cold-form.
3. Ease in joining.
4. Ability to withstand high temperatures.
5. Wide choice of sizes, wall thicknesses, and weights.
6. Availability in all parts of the country.

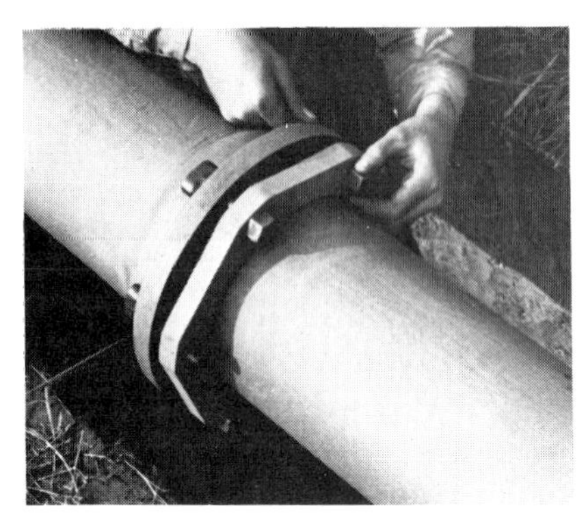

FIG. 4-11 Slide glands into position. Insert bolts and tighten nuts by hand. (*American Cast Iron Pipe Co.*)

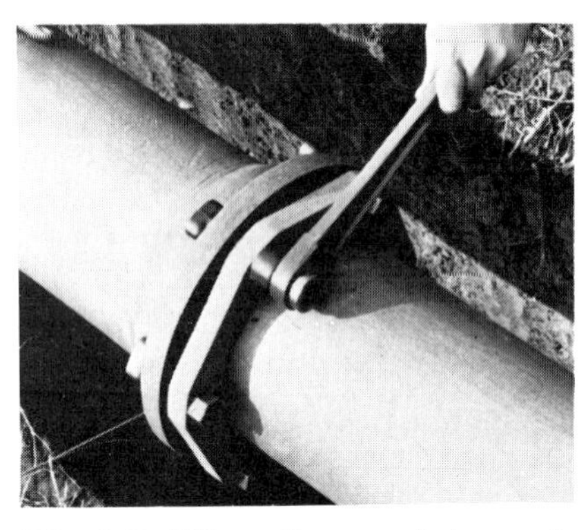

FIG. 4-12 With ordinary socket wrench, tighten bolts alternately (bottom, then top, and so on all around). Ranges of bolt torques to be applied are as follows: $^5/_8$-inch bolt, 40 to 60 lb-ft; $^3/_4$-inch bolt, 60 to 90 lb-ft; 1-inch bolt, 70 to 100 lb-ft. (*American Cast Iron Pipe Co.*)

FIG. 4-13 Completed standard mechanical joint. (*American Cast Iron Pipe Co.*)

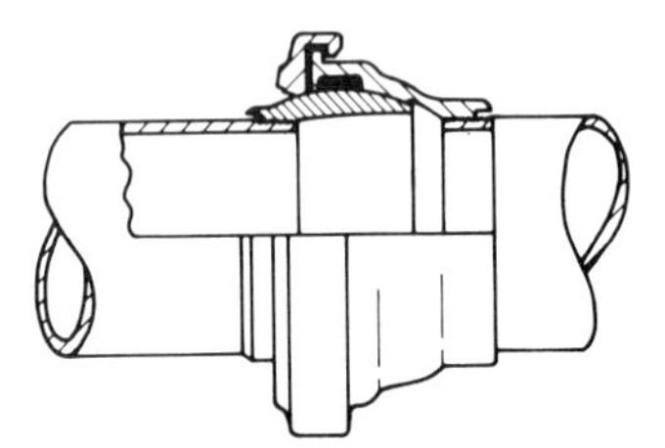

FIG. 4-14 Typical river-crossing joint. (*Clow Cast Iron Pipe.*)

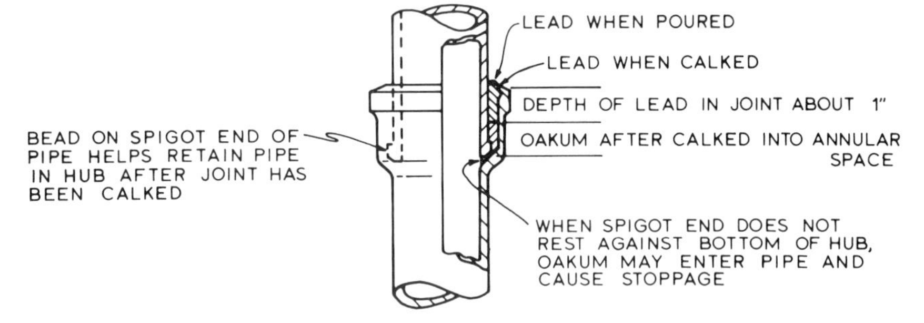
LEAD WHEN POURED
LEAD WHEN CALKED
DEPTH OF LEAD IN JOINT ABOUT 1"
OAKUM AFTER CALKED INTO ANNULAR SPACE
BEAD ON SPIGOT END OF PIPE HELPS RETAIN PIPE IN HUB AFTER JOINT HAS BEEN CALKED
WHEN SPIGOT END DOES NOT REST AGAINST BOTTOM OF HUB, OAKUM MAY ENTER PIPE AND CAUSE STOPPAGE

FIG. 4-15 Calked joint in cast-iron soil pipe.

4-10 Threaded Joints Through decades of service, joining steel pipe by threading and coupling has proved its soundness and dependability. The threading process is easily accomplished manually or mechanically and, most important, it produces sound joints. A wide variety of standard couplings and fittings are available to form any designed systems.

The threading operation for pipe up to 4 inches is done with a die that is either manually or power driven. The die is fitted with 4 to 6 thread cutters called chasers, made of tool steel. The teeth, hobbed or milled into the chasers, follow each other exactly in a spiral around the circumference of the pipe as the die is turned. Each tooth takes a tiny cut as it deepens and forms the thread to its final size and shape. To obtain a true and smooth thread, the chasers must be sharp, positioned in correct sequence, and aligned in their proper slots. The chasers must be kept sharp by sharpening or else a new set must be purchased. If the threaded pipe end is either too tight or loose when screwed hand-tight into the fitting, the chasers are not spaced the correct distance from the pipe center. This tighteness or looseness is evidenced by how the joint compares with the "hand-tight" requirements as given in the accompanying thread-specification, Table 4-4. Looseness is usually accompanied by torn threads because the chasers are set so their root diameter is less than the pipe OD.

A well-adjusted chaser with sharp cutting surface and the liberal use of cutting oil when the threads are made will produce good clean threads which will screw up well and be leakproof. The primary purposes of the cutting or threading oil are to keep the chaser cutting edges cool, serve as lubricant, and flush away the chips. When threads are cut manually, the cutting oil makes the job easier and smoother. To produce good sharp threads, clean all previously cut chips from the die before starting to thread a new section of pipe. When finished, back the die off the threads carefully to avoid cross threading or stripping, especially at the pipe end. If burrs formed by cutting the threads are present in the pipe end, delay the reaming until the threading is completed. This assures full wall strength at the end for the starting cut.

For all pipe sizes, the threads are cut at an angle of 60 degrees (the angle included between the thread flanks). In a tapered thread—the one always used in general pipe fabrication—the thread tapers 1 in 16 or 0.75 inch per foot. Straight threads are used when it is desired that the pipe ends butt together inside the coupling. The number of threads per inch varies with the pipe sizes as shown in Table 4-4.

The nominal length of thread has been cut when the front surfaces of the chasers are flush with the end of the pipe. Carrying the threading operation beyond this point produces straight-running threads from the end of the pipe to the face of the chasers. See Table 4-4, which shows proper length of thread for pipe sizes from $^1/_{16}$- through 24-inch diameters. Master gages are available also for checking pipe threads for size and length, but a well-made new coupling can be used in the field for this purpose. Standard or steam couplings always leave several threads exposed when screwed as far as they will go with a wrench. A "line" coupling generally used for underground service will screw up until the recess in either end covers the last thread on the pipe.

The advantages of threaded joints are:

1. The pipe can be cleaned inside before erection, minimizing chance of metal particles being entrapped to damage valves, etc.
2. It is the safest method to use when extending pipe runs in explosion-hazard areas, and when joining new lines to existing lines used for conveying flammable gases and liquids.
3. It is easy to dismantle and replace sections of existing threaded-and-coupled pipe systems with simple hand tools.
4. Pipe can be threaded in a shop or on the job.
5. Threading and joining with a wide variety of standard couplings is easily accomplished.

One big disadvantage of threaded couplers is the corrosive or electrolytic environment that can be created if couplings are made from dissimilar metals. A pipe "dope" or paint should be used to protect exposed threads.

4-11 Welded Joints The joining of steel pipe by welding, once limited to large pipe, now is used to join all sizes of pipe. It is possible to design a complete pipe system without threaded or bolted joints. This is done with steel fittings and valves designed for weld fabrication. Continuous-weld steel pipe and fittings may be used to weld-assemble low-pressure systems for refrigeration, air conditioning, vent and drainage, heating, and structural applications. See Fig. 4-16.

Welded pipe systems are used for transporting steam, water, oil, and air and other

TABLE 4-4 Basic Dimensions of USA (American) Standard Taper Pipe Thread, NPT

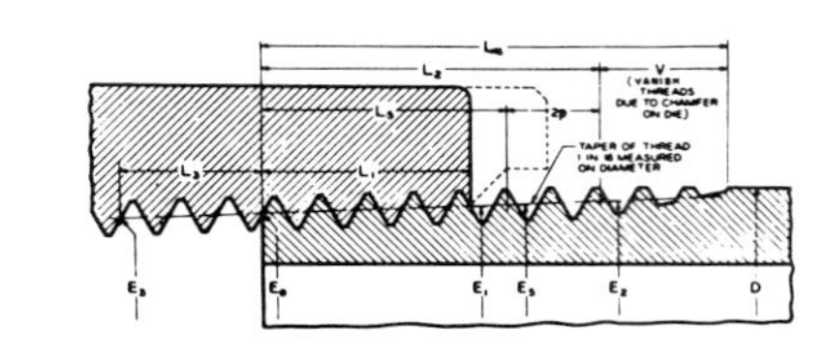

Nominal[8] Pipe Size	Outside Diameter of Pipe, D	Threads per inch, n	Pitch of Thread, p	Pitch Diameter at beginning of External Thread, E_0	Handtight Engagement: Length[2], L_1, In.	Handtight Engagement: Length[2], L_1, Thds.	Handtight Engagement: Dia[3], E_1	Effective Thread, External: Length[4], L_2, In.	Effective Thread, External: Length[4], L_2, Thds.	Effective Thread, External: Dia, E_2
1	2	3	4	5	6	7	8	9	10	11
1/16	0.3125	27	0.03704	0.27118	0.160	4.32	0.28118	0.2611	7.05	0.28750
1/8	0.405	27	0.03704	0.36351	0.1615	4.36	0.37360	0.2639	7.12	0.38000
1/4	0.540	18	0.05556	0.47739	0.2278	4.10	0.49163	0.4018	7.23	0.50250
3/8	0.675	18	0.05556	0.61201	0.240	4.32	0.62701	0.4078	7.34	0.63750
1/2	0.840	14	0.07143	0.75843	0.320	4.48	0.77843	0.5337	7.47	0.79179
3/4	1.050	14	0.07143	0.96768	0.339	4.75	0.98887	0.5457	7.64	1.00179
1	1.315	11.5	0.08696	1.21363	0.400	4.60	1.23863	0.6828	7.85	1.25630
1¼	1.660	11.5	0.08696	1.55713	0.420	4.83	1.58338	0.7068	8.13	1.60130
1½	1.900	11.5	0.08696	1.79609	0.420	4.83	1.82234	0.7235	8.32	1.84130
2	2.375	11.5	0.08696	2.26902	0.436	5.01	2.29627	0.7565	8.70	2.31630
2½	2.875	8	0.12500	2.71953	0.682	5.46	2.76216	1.1375	9.10	2.79062
3	3.500	8	0.12500	3.34062	0.766	6.13	3.38850	1.2000	9.60	3.41562
3½	4.000	8	0.12500	3.83750	0.821	6.57	3.88881	1.2500	10.00	3.91562
4	4.500	8	0.12500	4.33438	0.844	6.75	4.38712	1.3000	10.40	4.41562
5	5.563	8	0.12500	5.39073	0.937	7.50	5.44929	1.4063	11.25	5.47862
6	6.625	8	0.12500	6.44609	0.958	7.66	6.50597	1.5125	12.10	6.54062
8	8.625	8	0.12500	8.43359	1.063	8.50	8.50003	1.7125	13.70	8.54062
10	10.750	8	0.12500	10.54531	1.210	9.68	10.62094	1.9250	15.40	10.66562
12	12.750	8	0.12500	12.53281	1.360	10.88	12.61781	2.1250	17.00	12.66562
14 OD	14.000	8	0.12500	13.77500	1.562	12.50	13.87262	2.2500	18.00	13.91562
16 OD	16.000	8	0.12500	15.76250	1.812	14.50	15.87575	2.4500	19.60	15.91562
18 OD	18.000	8	0.12500	17.75000	2.000	16.00	17.87500	2.6500	21.20	17.91562
20 OD	20.000	8	0.12500	19.73750	2.125	17.00	19.87031	2.8500	22.80	19.91562
24 OD	24.000	8	0.12500	23.71250	2.375	19.00	23.86094	3.2500	26.00	23.91562

TABLE 4-4 Basic Dimensions of USA (American) Standard Taper Pipe Thread, NPT (Continued)

Nominal[8] Pipe Size	Wrench Makeup Length for External Thread, $L_2 - L_1$		Wrench Makeup Length for Internal Thread			Vanish Thread, V		Overall Length External Thread, L_4	Nominal Complete External Threads[5]		Height of Thread, h	Increase in Dia per Thread, $0.0625/n$	Basic[6] Minor Dia at Small End of Pipe, K_0
			Length, L_3		Dia[7], E_3				Length L_5	Dia, E_5			
	In.	Thds.	In.	Thds.		In.	Thds.						
1	12	13	14	15	16	17	18	19	20	21	22	23	24
1/16	0.1011	2.73	0.1111	3	0.26424	0.1285	3.47	0.3896	0.1870	0.28287	0.02963	0.00231	0.2416
1/8	0.1024	2.76	0.1111	3	0.35656	0.1285	3.47	0.3924	0.1898	0.37537	0.02963	0.00231	0.3339
1/4	0.1740	3.13	0.1667	3	0.46697	0.1928	3.47	0.5946	0.2907	0.49556	0.04444	0.00347	0.4329
3/8	0.1678	3.02	0.1667	3	0.60160	0.1928	3.47	0.6006	0.2967	0.63056	0.04444	0.00347	0.5676
1/2	0.2137	2.99	0.2143	3	0.74504	0.2478	3.47	0.7815	0.3909	0.78286	0.05714	0.00446	0.7013
3/4	0.2067	2.89	0.2143	3	0.95429	0.2478	3.47	0.7935	0.4029	0.99286	0.05714	0.00446	0.9105
1	0.2828	3.25	0.2609	3	1.19733	0.3017	3.47	0.9845	0.5089	1.24543	0.06957	0.00543	1.1441
1 1/4	0.2868	3.30	0.2609	3	1.54083	0.3017	3.47	1.0085	0.5329	1.59043	0.06957	0.00543	1.4876
1 1/2	0.3035	3.49	0.2609	3	1.77978	0.3017	3.47	1.0252	0.5496	1.83043	0.06957	0.00543	1.7265
2	0.3205	3.69	0.2609	3	2.25272	0.3017	3.47	1.0582	0.5826	2.30543	0.06957	0.00543	2.1995
2 1/2	0.4555	3.64	0.2500[7]	2	2.70391	0.4337	3.47	1.5712	0.8875	2.77500	0.100000	0.00781	2.6195
3	0.4340	3.47	0.2500[7]	2	3.32500	0.4337	3.47	1.6337	0.9500	3.40000	0.100000	0.00781	3.2406
3 1/2	0.4290	3.43	0.2500	2	3.82188	0.4337	3.47	1.6837	1.0000	3.90000	0.100000	0.00781	3.7375
4	0.4560	3.65	0.2500	2	4.31875	0.4337	3.47	1.7337	1.0500	4.40000	0.100000	0.00781	4.2344
5	0.4693	3.75	0.2500	2	5.37511	0.4337	3.47	1.8400	1.1563	5.46300	0.100000	0.00781	5.2907
6	0.5545	4.44	0.2500	2	6.43047	0.4337	3.47	1.9462	1.2625	6.52500	0.100000	0.00781	6.3461
8	0.6495	5.20	0.2500	2	8.41797	0.4337	3.47	2.1462	1.4625	8.52500	0.100000	0.00781	8.3336
10	0.7150	5.72	0.2500	2	10.52969	0.4337	3.47	2.3587	1.6750	10.65000	0.100000	0.00781	10.4453
12	0.7650	6.12	0.2500	2	12.51719	0.4337	3.47	2.5587	1.8750	12.65000	0.100000	0.00781	12.4328
14 OD	0.6880	5.50	0.2500	2	13.75938	0.4337	3.47	2.6837	2.0000	13.90000	0.100000	0.00781	13.6750
16 OD	0.6380	5.10	0.2500	2	15.74688	0.4337	3.47	2.8837	2.2000	15.90000	0.100000	0.00781	15.6625
18 OD	0.6500	5.20	0.2500	2	17.73438	0.4337	3.47	3.0837	2.4000	17.90000	0.100000	0.00781	17.6500
20 OD	0.7250	5.80	0.2500	2	19.72188	0.4337	3.47	3.2837	2.6000	19.90000	0.100000	0.00781	19.6375
24 OD	0.8750	7.00	0.2500	2	23.69688	0.4337	3.47	3.6837	3.0000	23.90000	0.100000	0.00781	23.6125

[1] The basic dimensions of the USA (American) Standard Taper Pipe Thread are given in inches to four or five decimal places. While this implies a greater degree of precision than is ordinarily attained, these dimensions are the basis of gage dimensions and are so expressed for the purpose of eliminating errors in computations.
[2] Also length of thin ring gage and length from gaging notch to small end of plug gage.
[3] Also pitch diameter at gaging notch (handtight plane.)
[4] Also length of plug gage.
[5] The length L_5 from the end of the pipe determines the plane beyond which the thread form is incomplete at the crest. The next two threads are complete at the root. At this plane the cone formed by the crests of the thread intersects the cylinder forming the external surface of the pipe. $L_5 = L_2 - 2p$.
[6] *Given as information* for use in selecting tap drills. (See Appendix E.)
[7] Military Specification MIL–P–7105 gives the wrench makeup as three threads for 3 in. and smaller. The E_3 dimensions are as follows: Size 2.5 in. 2.69609 and size 3 in. 3.31719.
[8] Designated, for example, as 3/8 NPT or 0.675 NPT.

SOURCE: B2.1-1968, as published by the American Society of Mechanical Engineers.

gases. They are also used for structures, rigging assemblies, frames, and many other applications. The structural strength of welded pipe is excellent and welding is economical. There is no loss of strength and the life expectancy of a welded system is higher than with any other method of joining.

Because all fittings and pipe are permanently fused into a single unit, the possibility of leaks due to vibration or stress caused by flow of materials is diminished.

Installation is done by cutting pipe to lengths needed, placing in position, and welding. Pipe assemblies can be conveniently fabricated in the shop or at floor level, then welded into the system.

Welded steel pipe also has the advantages of smooth flow, good appearance, and extra strength. The pipes are fused together by an equal thickness of weld metal. This eliminates obstructions in the pipe and there is no loss of metal which would curtail strength. The two methods of welding, gas and arc welding, are explained in the following paragraphs.

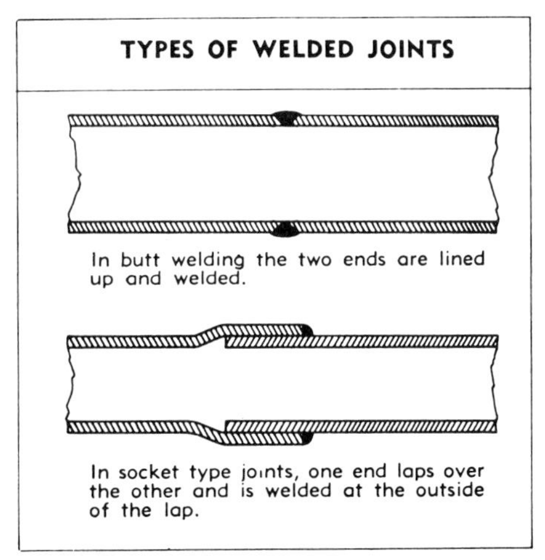

TYPES OF WELDED JOINTS
In butt welding the two ends are lined up and welded.
In socket type joints, one end laps over the other and is welded at the outside of the lap.

FIG. 4-16 Types of welded joints. (*American Iron and Steel Institute.*)

Oxyacetylene or Gas Welding Procedures. Gas and oxygen are mixed in the torch at the manufacturer's recommended pressure for the particular tip size. The welder should apply enough heat to melt the base metal. The ideal weld is one made with just enough heat to fuse the filler metal with the base metal—but no more. When pipe ends touch, or spacing is very close, the pipe ends can be "tacked" together with the torch to hold them in position. Then the filler metal may be applied as a "first pass." Most oxyacetylene welds on smaller diameter pipes can be made in one pass; $^1/_8$ to $^1/_4$ inch of weld metal can be deposited efficiently with one pass. The smallest tip that will supply enough heat for proper fusion should be used. For oxyacetylene welding use ASTM A251 GA60 filler metal.

Arc Welding Procedures. Certain skills, easily learned and perfected through practice, are required to produce a sound and good-appearing weld deposit. The development of coated, all-purpose electrodes makes it possible for welders to make ductile and sound welds with relative ease by arc welding. The weld metal is deposited in layers (or beads); each layer or bead is then cleaned of all trapped slag (coating residue) and chipped out before the next bead is applied. For electric arc welding, ASTM A233 E6010 dc reverse-polarity electrodes are universally accepted for welding pipe and pipe fittings. Other low-carbon-steel electrodes can be used under certain conditions.

- Preparation: all pipe ends, fittings, and flanges should have a beveled end for welding. Normally bevels of 30 by 37½ degrees can be used. Pipe ends should be spaced 1/16 to 1/8 inch apart.

- Position: welding may be done in all positions.
- Welding technique: manual manipulation.
- Cleaning: chip and brush free of all slag, rust, spatter.

After each pass the beads should be smooth and even, free of all flux and slag. The finished weld should be reasonably smooth and even, without excessive buildup and undercutting.

4-12 Mechanical Joints Mechanical pipe couplings can be used to join pipe assemblies quickly and easily without threaded or welded pipe ends. Various patented, leakproof couplings are available for rigid assemblies or for systems and components which must withstand vibration and angular and/or axial movement of the pipe. Some of these are Wedgelock couplings, Dresser couplings, Victaulic couplings and Naylor Shore pipe joints. Wedgelock couplings are used with a gasket to secure grooved-end or shoulder-end pipe. See Fig. 4-17. This coupling will not allow the pipe to separate, yet provides some flexibility and allowance for expansion and contraction. The wedge is simply driven home with a hammer to complete the joint.

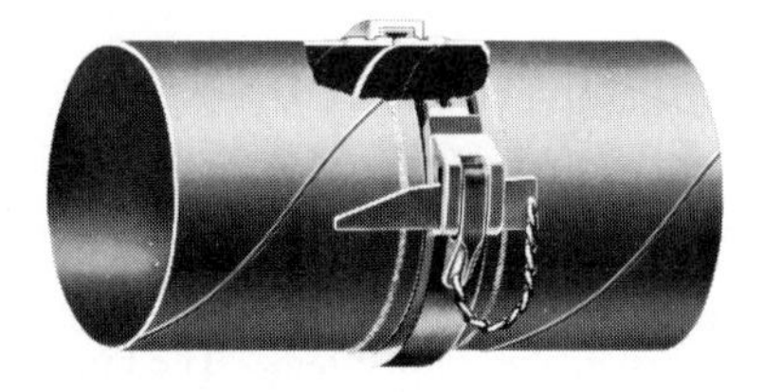

FIG. 4-17 Wedge Lock coupling. *(Naylor Pipe Co.)*

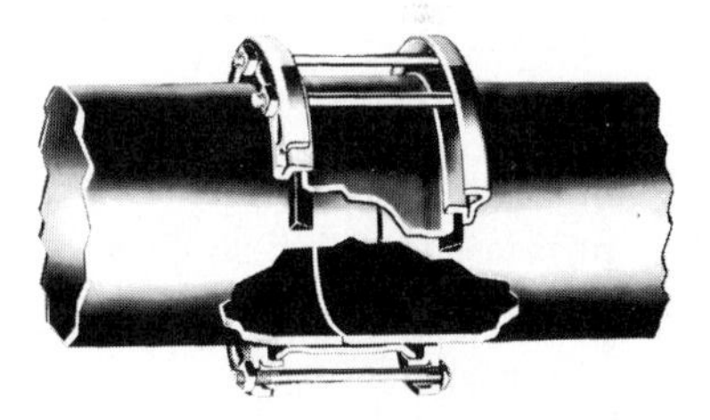

FIG. 4-18 Dresser-type coupling. *(Dresser Manufacturing.)*

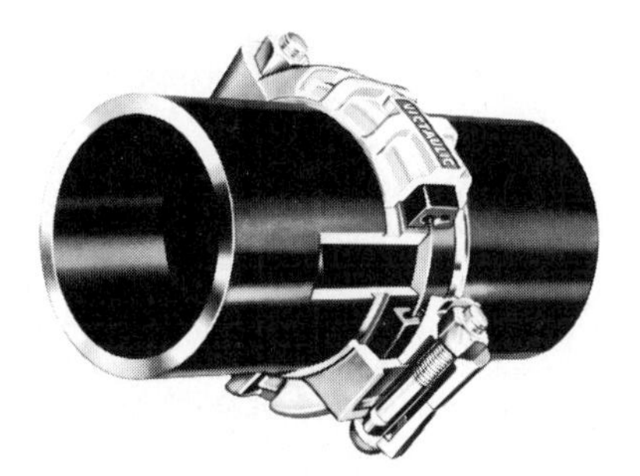

FIG. 4-19 Victaulic-type coupling. *(Victaulic Co. of America.)*

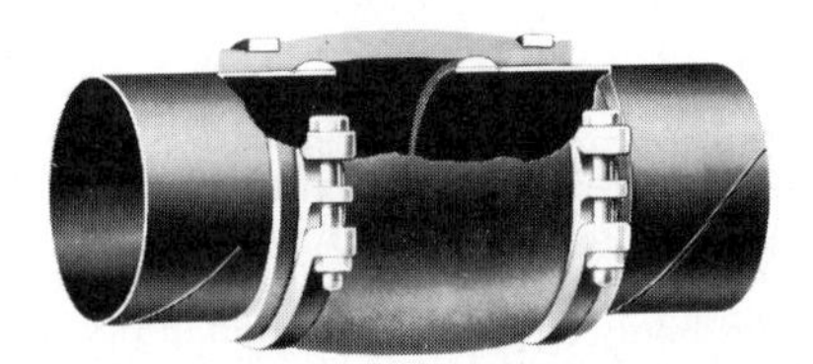

FIG. 4-20 Pontoon pipe joint for rubber sleeve connections. *(Naylor Pipe Co.)*

Victaulic and Dresser couplings are illustrated in Figs. 4-18 and 4-19. Both offer flexibility and ability to take considerable expansion and contraction.

A Pontoon pipe joint consists of a rubber sleeve held in place at the joint by two bolted clamps. This connection is quite flexible. See Fig. 4-20.

Steel flanged joints are available to meet any requirements.

Another type of coupling, generally used for structural assemblies, has either fixed-angle or adjustable-angle branches for joining two, three, four, or six lengths of pipe. Pipe ends are slipped into the coupling branches, then permanently screwed or bolted into position. See Fig. 4-21.

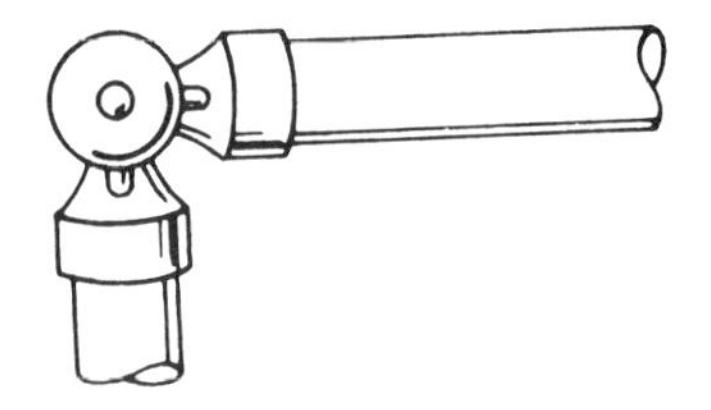

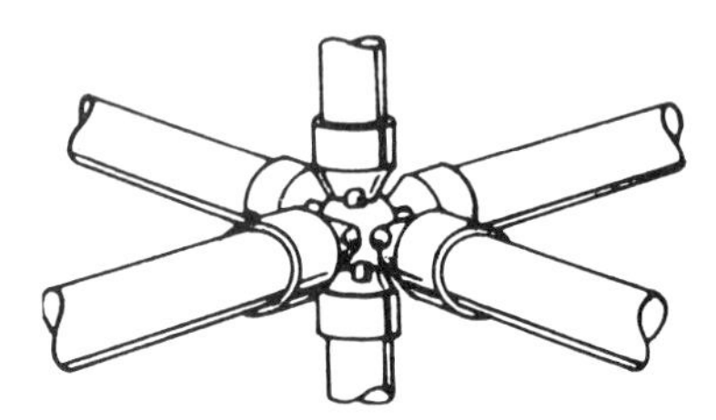

FIG. 4-21 Adjustable fittings.

NONFERROUS MATERIALS

Nonferrous materials used in sewer and water construction and drain, waste, and vent systems are brass, lead, and copper. Brass pipe is used for water installations because of its smooth interior. Brass is also used for fittings with other materials. It is an alloy of zinc and copper, materials which are expensive compared with others. The fittings used on brass pipe are of cast brass and available in all forms. The joints are normally brazed or soldered.

Lead pipe is seldom used in new construction, but may be found in existing structures or underground services. It was used for soil, waste, and vent pipe and water-supply piping. This material is easy to install because it is very ductile. It is also quite resistant to most acids. The disadvantages are its physical strength and possible breakdown if used with soft water. Most communities will not allow lead pipe in their water systems. These joints are wipe-soldered, formed by the burned-lead procedure, or are made with adapters or transition fittings.

Copper pipe and fittings may be used in plumbing installations for waste, vent, and water pipes. Type L is used extensively for water piping in above-ground installations. Type K is used for water services and types M and DWV for waste and vent lines. The joints are soldered, flared, or made with compression fittings.

4-13 Soldered or Sweat Joints Soldered joints with copper pipe are used in plumbing for both water lines and sanitary lines where the maximum temperature does not exceed 250°F. These joints depend on the solder being melted so that it flows freely and capillary action draws the molten solder into the joints. Flux acts as a wetting agent, assuring a uniform spreading of the molten solder over the surfaces to be soldered, and also prevents oxidation. Refer to Table 4-5 to select the proper solder.

A soldering procedure which produces excellent results is described in these simple steps (see Fig. 4-22):

1. Measure to assure proper depth penetration.
2. Cut tube to exact length with a square cut.
3. Remove all burrs.
4. Use sizing tool to insure true dimensions and roundness.
5. Clean surfaces to be joined using sand cloth, steel wool, or special wire brush.
6. Cover surfaces to be joined with a thin film of flux.
7. Assemble the joint by inserting the tube into the fitting and twisting slightly.
8. Apply heat with a propane gas torch or an air-acetylene torch to the fitting.
9. When tube is hot enough, remove flame. Solder should melt on contact with the tube.

TABLE 4-5 Rated Internal Working Pressure of Joints made of Copper Water Tube and Solder-type Fittings

(Pounds per Square Inch)

Solder or brazing alloy used in joints	Service temperature, °F	Water[a]					Saturated steam
		COPPER WATER TUBE NOMINAL SIZES (INCHES)					
		1/4 to 1 incl.	1 1/4 to 2 incl.	2 1/2 to 4 incl.	5 to 8 incl.	10 to 12 incl.	All
50-50 Tin-lead[b]	100	200	175	150	130	100	
	150	150	125	100	90	70	
	200	100	90	75	70	50	
	250	85	75	50	50	40	15[c]
95-5 Tin-antimony	100	500	400	300	150	150	
	150	400	350	275	150	150	
	200	300	250	200	150	140	
	250	200	175	150	140	110	15[c]
Brazing alloys (Melting at or above 1000 degrees F)	100-150-200	Note[d]	Note[d]	Note[d]	Note[d]	Note[d]	
	250[e]	300	210	170	150	150	
	350	270	190	150	150	150	120[f]

The values are based on data in the National Bureau of Standards Building Materials and Structures Reports BMS 58 and BMS 83.
[a] Including other noncorrosive liquids and gases.
[b] ASTM B32, Alloy Grade 50A.
[c] This pressure is determined by the temperature of saturated steam at 15 lb pressure or 250°F.
[d] Rated internal pressure is that of the tube being joined.
[e] For service temperatures lower than 250°F the solders as above may be used.
[f] This pressure is determined by the temperature of saturated steam at 120 lb pressure or 350°F.
SOURCE: Copper Development Assoc.

10. The solder should be drawn into the joint by the natural force of capillary action.
11. Remove residual solder and flux.
12. Allow joint to cool naturally.

This procedure should yield leakproof joints; however, all joints should be checked for leakage. Also, if copper tubing is larger than 2 inches, a more uniform heating method should be used. This can be two torches or a single torch with two nozzles.

Proper adapters or transition fittings are used to join dissimilar metals.

4-14 Mechanical Joints Mechanical joints with copper are usually flared joints and are made where the use of heat is impractical. This joint can be easily disconnected from time to time. Underground water services are generally flared joints. Soft tempered copper is required when making flared ends. The flared tube end is pulled tightly against the tapered part of the fitting by a nut which is part of the fitting, so there is a metal-to-metal contact. (Refer to Table 4-6 for rated working pressures using mechanical joints.)

Step-by-step, the procedure for impact flaring is as follows:

1. Cut the tube to the required length.
2. Remove all burrs.
3. Slip the coupling nut over the end of the tube.
4. Insert flaring tool into the tube end.
5. Drive the flaring tool by hammer strokes, expanding the end of the tube to the desired flare.
6. Assemble the joint by placing the fitting squarely against the flare. Engage

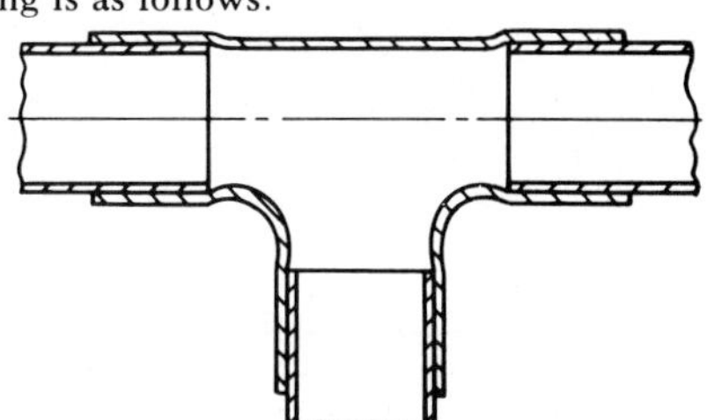

FIG. 4-22 Sweat or solder joint.

TABLE 4-6 Rated Internal Working Pressures of Copper Water Tube Using Suitable Mechanical Joints

(Pounds per Square Inch)

Standard size	Service temperature, °F									
	100 (S = 6,000 psi)			200 (S = 5,900 psi)			300 (S = 5,000 psi)		400 (S = 2,500 psi)	
	COPPER WATER-TUBE TYPE									
	K	L	M	K	L	M	K	L	K	L
1/4	1,060	900	–	1,040	880	–	880	740	440	370
3/8	1,170	800	560	1,140	780	550	960	660	480	330
1/2	920	740	510	900	730	490	760	610	380	300
5/8	760	650	–	740	640	–	630	530	310	260
3/4	880	590	420	860	570	400	730	480	360	240
1	680	510	340	660	490	330	560	420	280	210
1 1/4	550	460	340	530	440	330	450	370	220	180
1 1/2	520	430	340	500	410	330	420	350	210	170
2	450	370	300	420	370	290	350	310	170	150
2 1/2	420	350	280	400	340	270	340	280	170	140
3	410	330	260	390	320	250	330	270	160	130
3 1/2	380	320	260	370	310	250	310	260	150	130
4	370	300	260	360	290	250	300	240	150	120
5	360	280	240	350	260	230	290	220	140	110
6	370	260	230	350	250	210	300	210	150	100
8	390	280	240	380	270	220	320	230	160	110
10	390	290	240	380	270	230	320	230	160	110
12	400	270	240	380	250	230	320	210	160	100

SOURCE: Copper Development Assoc.

the coupling nut with the fitting's threads. Tighten with two wrenches, one on the nut and one on the fitting.

A screw-type flaring tool is also available. It produces joints of better quality. From steps 1 to 3, the procedure is the same as for impact flaring. Then the procedure is as follows:

4. Clamp the tube in the flaring bar so that the end of the tube is slightly above the face of the bar.

5. Place the yoke of the flaring tool on the bar so that the beveled end of the compressor cone is over the tube end.

6. Turn down firmly the compressor screw, forming the flare.

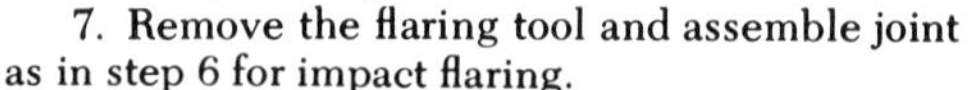

7. Remove the flaring tool and assemble joint as in step 6 for impact flaring.

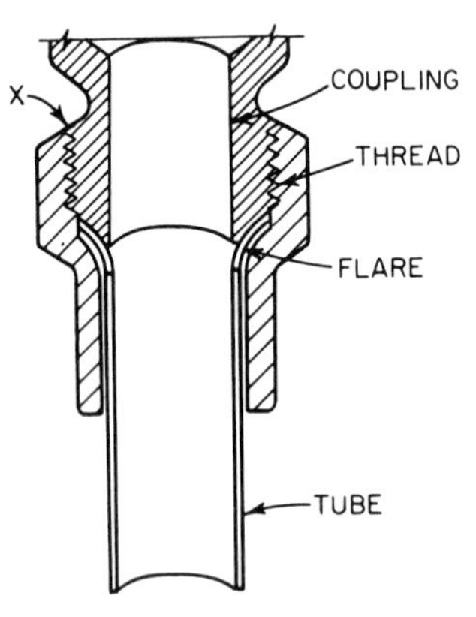

FIG. 4-23 Flared joint.

A drop of solder at point X in Fig. 4-23 will insure against possible loosening of the joint.

Compression fittings are another type of mechanical connection used for pipe. They consist of a brass, malleable, or steel body as the usage dictates, two malleable or brass end nuts, and two resilient armored gaskets, protected from damage by gasket-retainer cups. When the end nuts are tightened, the gasket is forced against the pipe, creating a watertight joint.

4-15 Wiped Joints The term *wiped joint* is normally used by plumbers to refer to a method of connecting materials made of different metals or lead. Joints in lead pipe, or between lead pipe and fittings, or between brass or copper pipe, ferrules, solder nipples, or traps can be full wiped joints. The material used is ordinary solder. Wiped joints should have an exposed surface on each side of the joint not less than 3/4 inch and at least as thick as the material being jointed. Wall or floor flange lead-wiped joints shall be made by using a lead ring or flange placed behind the joint at wall or floor. Joints between lead pipe and cast iron, steel, or wrought iron shall be made by means of a calking ferrule, soldering nipple, or bushing.

Figure 4-24 illustrates a wiped joint between lead and brass pipe.

The procedure to follow to make a wiped joint properly is as follows:

1. Square the ends of the pipe with a coarse rasp, then taper the spigot end to a feather edge for about 1/4 inch. Taper the receiving end with a rasp and then open it with a turnpin to allow the spigot end to enter about 1/8 inch.
2. Apply a thin coat of soil or paste and ground pumice to both ends of pipe.
3. When soil dries, clean or shave a distance of 1 inch on each side of the joint.
4. While supporting the pipe, apply the solder with a ladle. Holding a treated cloth under the joint, form the solder around the pipe. Allow the solder to be higher at the junction of the pipes and taper outward.

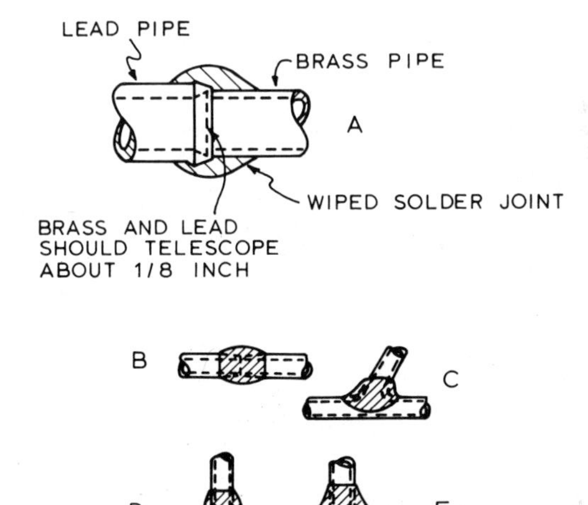

FIG. 4-24 Various types of wiped joints: (*a*) round, (*b*) horizontal, (*c*) branch, (*d*) ferrule, (*e*) floor flanged.

The process for horizontal and vertical wiped joints is similar except for the wiping process. Remember to start wiping the top of a vertical joint first because it cools first.

4-16 Brazed Joints Brazed joints are used with copper tubes where greater strength is required or where temperatures are as high as 350°F. One example is in refrigerator piping where brazing may be preferred or required. Brazing materials are sometimes called hard solders or silver solders.

The step-by-step procedure for brazing a joint is as follows:

1. Prepare joint as in the soldering process, including measuring, cutting, burr-removing, and cleaning.
2. Apply flux to the cleaned area of the tube end and fitting socket.
3. Assemble the joint by inserting the tube into the socket and butting it hard against the stop.
4. Firmly support pipe joint.
5. Apply heat to tube with oxyacetylene flame. Air-acetylene is sometimes used on smaller sizes. Heat flux until it becomes quiet and transparent.
6. Apply heat to the fitting and heat until flux appears like water.
7. Continue to keep both tube and fitting hot.
8. Feed brazing wire or rod into the opening where the tube enters the fitting.
9. Continue to heat fitting but not the wire or rod.

10. Capillary action will fill the void completely with filler metal. When making a horizontal brazed joint, start feeding the brazing wire at the top, then the sides, and finally the bottom. The sequence makes no difference with a vertical line.

11. After brazing alloy has set, clean off remaining flux with a wet brush or swab.

Brazed joints should be made in accordance with the provisions of section 6 of the Code for Pressure Piping, ASA B31.1–1951.

4-17 Threaded Joints Of the nonferrous materials used, only brass is threaded. Threads should conform to the National Taper Pipe Thread ASA B2.1. The procedure is described and a table with pertinent data, is provided in sec. 4-10.

4-18 Burned-Lead Joints This procedure requires a skilled, experienced person. It consists of lapping the lead pipes and fusing them together. The weld should be at least as thick as the pipe being joined. If done correctly, the method produces a very satisfactory job. Normally one end of the pipe is enlarged just enough to slide over the end of another pipe. With the addition of more lead, the joint is then heated and fused together.

ALUMINUM PIPE

There are numerous alloys of aluminum available, each having different physical properties. Aluminum pipe is the term used for all alloys but each application should single out one specific alloy. Aluminum alloys can be used to transport chemicals, food and drink products, gases, oil, air, and many more products. Because of its high strength, light weight, corrosion resistance, and workability it also has many structural applications. It is joined by welding, threaded couplings, mechanical couplings, and brazing. The pipe is presently manufactured in ASA sizes ⅛ inch through 20 inches.

Aluminum corrugated pipe is manufactured with band connections as shown in Fig. 4-43.

4-19 Welded Joints Refer to Sec. 6 for proper procedure and materials for welding the various aluminum alloys. Manipulation of the pipe and the various welding tools are the same as for welding steel pipe. The ends of the pipe should be beveled for sizes 2 inches and over. Backing rings may be used to insure full penetration without "icicles." The backing rings can be removable stainless steel rings or integral aluminum rings of a suitable alloy. Satisfactory welds can be made without a ring but it is difficult to obtain consistently good results. Design stresses for aluminum piping systems must be based on the strength of the welded joints. The weld strength varies with the original alloy and temper. Piping systems may be designed in accordance with the code for pressure piping, ASA B31.3. See Tables 4-7 and 4-8.

4-20 Threaded Joints Use the same procedure and threading equipment as for steel pipe. Threaded fittings are available to complete a system. Threaded joints are used in structural applications and also pressure piping. However, most pressure piping is made of welded joints.

4-21 Mechanical Joints Mechanical joints for aluminum pipe are similar to those for steel pipe. The connectors used on steel pipe for pressure applications can normally be used on aluminum, with more restrictions. Because aluminum has less tensile strength its maximum pressures will be less. All structural connectors can also be manufactured, as with steel pipe connectors.

4-22 Soldered Joints and Brazed Joints A well-designed socketed tube joint will facilitate a good soldered joint. The tapered seat at the bottom of the socket allows for the formation of a fillet on the inside of the joint. The male tube rests against this taper and provides positive alignment allowing the zinc solder to flow through to form a fillet. This assures leakproof tightness and prevents contact between flux that may be trapped in the lap and any material circulated through the tube. This joint is normally stronger than the tube itself. Use on the socket a flared end, as shown in Fig. 4-25, which is wide enough to seat the wire solder ring.

Tubes are brazed using standard practice. Normally one end of a tube is flared so the other end will fit inside. This fit should be such as to allow flux to escape. Consult the manufacturer for specific temperatures to use with a specific aluminum alloy.

4-23 Adhesive Joints Various adhesives are available which will bond aluminum to aluminum or to another metal. In structural or stressed assemblies, the ad-

TABLE 4-7 Edge Preparation—AC Gas Tungsten-Arc Welding of Pipe, Horizontal Fixed Position

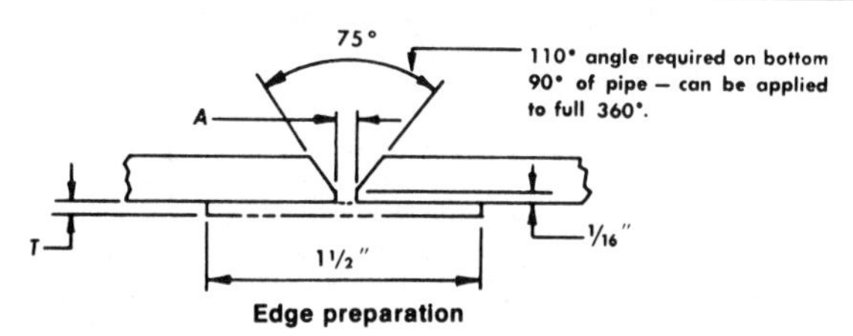

Edge preparation

$A = 0$ for no backing ring or removable backing ring

$A = 1/4$ in. maximum for integral backing ring

Nominal pipe size, in.	Wall thickness, in.	Tungsten electrode diameter, in.	Gas nozzle orifice diameter, in.	Welding rod diameter, in.	Approx. current, amp, ac	Argon flow, cu ft/hr	T, in.	No. of passes ①
1	0.133	1/8	1/2	3/32	90–110	30–80	0.072	1–2
1¼	0.140	1/8	1/2	1/8	100–120	30–80	0.072	1–2
1½	0.145	1/8	1/2	1/8	110–130	30–80	0.072	1–2
2	0.154	1/8	1/2	1/8	120–140	30–80	0.093	1–2
2½	0.203	1/8	1/2	1/8	130–150	30–80	0.093	2
3	0.216	1/8	1/2	1/8	145–165	30–80	0.093	2
3½	0.226	1/8	1/2	1/8	150–170	30–80	0.093	2
4	0.237	3/16	1/2	1/8–3/16	160–180	35–80	0.125	2
5	0.258	3/16	1/2	5/32–3/16	180–190	35–80	0.125	2
6	0.280	3/16	1/2	5/32–3/16	195–205	50–80	0.187	2
8	0.322	3/16	1/2	5/32–3/16	210–220	50–80	0.187	2–3
10	0.365	3/16	1/2	5/32–3/16	230–240	50–80	0.187	2–3
12	0.406	3/16	1/2	5/32–3/16	245–255	50–80	0.187	2–3

① Greater number of passes for bottom 90°.

SOURCE: Aluminum Co. of America.

TABLE 4-8 Edge Preparation—Vertical Fixed Position

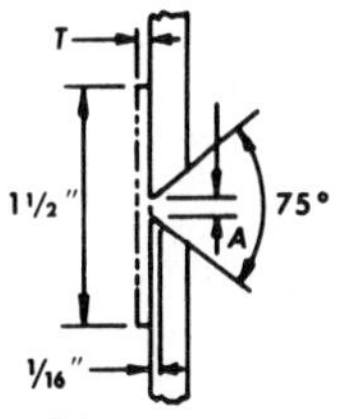

Edge preparation

$A = 0$ for no backing ring or removable backing ring

$A = 1/4$ inch maximum for integral backing ring

Nominal pipe size, in.	Wall thickness, in.	Tungsten electrode diameter, in.	Gas nozzle orifice diameter, in.	Welding rod diameter, in.	Approx. current amp, ac	Argon flow, cu ft/hr	T, in.	No. of passes
1	0.133	1/8	7/16	3/32	95–115	25–50	0.072	1–2
1¼	0.140	1/8	7/16	1/8	105–125	25–50	0.072	1–2
1½	0.143	1/8	7/16	1/8	115–135	25–50	0.072	1–2
2	0.154	1/8	7/16	1/8	125–145	30–60	0.093	2–3
2½	0.203	1/8	7/16	1/8	135–155	30–60	0.093	3–5
3	0.216	1/8	1/2	1/8	150–170	40–60	0.093	3–5
3½	0.226	1/8	1/2	1/8	155–175	40–60	0.093	3–5
4	0.237	3/16	1/2	1/8–5/32	165–185	40–60	0.125	3–5
5	0.258	3/16	1/2	5/32–3/16	185–205	50–60	0.125	3–5
6	0.280	3/16	1/2	5/32–3/16	200–220	50–60	0.187	3–5
8	0.322	3/16	1/2	5/32–3/16	215–235	60–80	0.187	5–8
10	0.365	3/16	1/2	5/32–3/16	235–255	60–80	0.187	5–8
12	0.406	3/16	1/2	5/32–3/16	250–270	70–80	0.187	6–8

SOURCE: Aluminum Co. of America.

hesive joint should have a strength equal to or stronger than that of the adherends. To obtain this strength in tube connections it is best to use a recessed tongue-and-groove, scrafed, or overlapping design rather than a plain butt design. Consult manufacturer for adhesives and various joint designs which will best serve your needs.

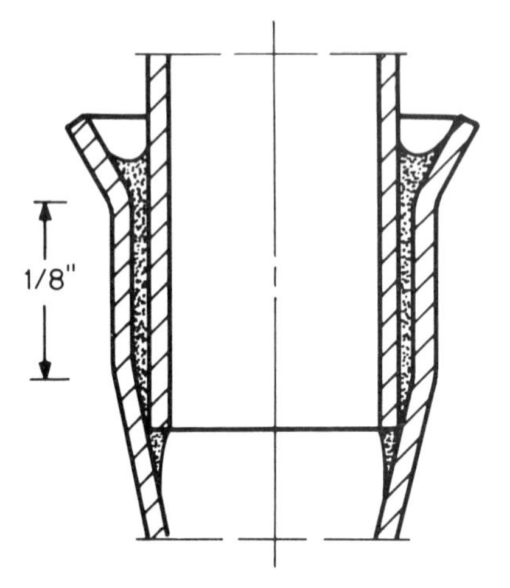

FIG. 4-25 Soldered tube joint.

CONCRETE PIPE

Over the years the joining of concrete pipe has become quite sophisticated. The joints range from the open type used with drainage tile to the bell-and-spigot joint with rubber gaskets used in joining sanitary sewer lines with low infiltration specifications.

The pipe can be designed to carry water, sewage, chemicals, or gas.

Joints are designed to provide:

1. Resistance to infiltration of ground water and/or backfill material.
2. Resistance to exfiltration of sewage or storm water.
3. Control of leakage from internal or external heads.
4. Flexibility to accommodate lateral deflection or longitudinal movement without creating leakage problems.
5. Resistance to shear stresses between adjacent pipe sections without creating leakage problems.
6. Hydraulic continuity and smooth flow.
7. Controlled infiltration of ground water for subsurface drainage.
8. Ease of installation.

The actual field performance of any pipe joint depends primarily upon the inherent performance characteristics of the joint itself, the severity of the conditions of service, and the care with which it is installed.

Since economy is important, it is usually necessary to compare the installed cost of several types of joints against pumping and treatment costs resulting from increased or decreased amounts of infiltration.

The concrete-pipe industry utilizes a number of different joints, listed in the following sections, to satisfy a broad range of performance requirements. These joints vary in cost, as well as in inherent performance characteristics. The field performance of all is dependent upon proper installation procedures.

4-24 Open Joints This type of joint is commonly used in field tile applications for draining surface and subsurface water. Its purpose is to collect water from the soil by infiltration through the joints or, in the case of perforated pipe, also through the perforations.

The recommendations for joint spacing are shown in the following table.

Joint spacing

Organic soils	1/4 to 1/2 in.
Clay soils	1/8 to 1/4 in.
Loamy soils	1/3 in.
Sandy soils	Lay to a tight fit
Mains 8 in. and over	Lay to a light fit

In sandy areas it is recommended that a suitable filter be used. The velocity of flow also enters into this determination. Filters are made of material consisting of not more than 10 percent passing the no. 60 sieve. Also, filter cloth materials such as fiber glass, spun bonded nylon fabric, and plastic filter cloth may be dropped over the pipe.

The concrete drain tile made for this purpose meets ASTM C-412.

4-25 Tongue-and-Groove Joints This joint creates a straight section of pipe through the joint with male and female ends interlocking. The packing material can be cement mortar, a preformed mastic compound, or a trowel-applied mastic compound. See Fig. 4-26. These joints have no inherent watertightness but depend exclusively upon the workmanship of the contractor. Field-poured concrete diapers or collars are sometimes used with these joints to improve performance. Joints employing mortar joint fillers are rigid and any deflection or movement after installation will cause cracks, permitting leakage. If properly applied, mastic joint fillers provide a degree of flexibility without impairing watertightness. These joints are not generally recommended for any internal or external head conditions if leakage is an important consideration.

Rubber gaskets are available for use with tongue-and-groove pipe, especially in the larger sizes. These joints would then inherit all the stricter design tolerances of bell and spigot pipe with rubber gaskets.

4-26 Mortar Joints The groove of the last pipe laid shall be buttered with cement mortar on its inner face throughout the lower half of the pipe and the tongue of pipe to be laid shall be likewise buttered on its outer face over the upper half. The mortar

Typical Cross Sections of Joints With Mortar or Mastic Packing

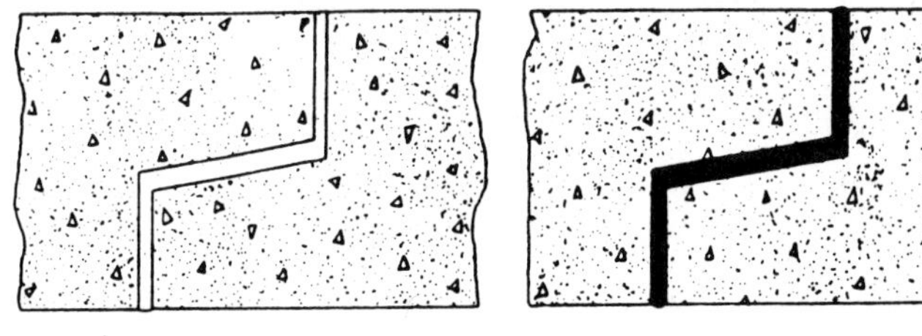

FIG. 4-26 Typical cross sections of joints with mortar or mastic packing. (*From Concrete Pipe Design Manual.*)

consists of equal parts of clean sharp sand and cement to which water is added until a low-slump mortar is obtained. The pipe shall be joined and forced together until the opening on the outside of the pipe is no less than ½ inch nor more than one-third of the length of the tongue. The outside of the joint shall be troweled with mortar and the open space filled. This troweling shall be carried around the outside lower face of the pipe as far as possible. Also, if possible, the entire circumference on the inside face of the pipe should be filled with mortar. A piece of tar paper or something similar should be placed over the outside of the pipe at the joint.

4-27 Mastic Joints A mastic joint consists of a plastic putty, rubber compound or asphaltic joint compound inserted in the joint. The procedure depends on which one is used. Each is applied differently, but the end result is the same. The mastic fills the void between the pipes at the joint. These joints are used for manhole joints and storm-sewer pipe.

4-28 External Gaskets (Tongue-and-Groove) The external wrap-around rubber-type gasket shall consist of a ribbed rubber band and a rubber-base bonding mastic coated with a water-soluble film which together form a waterproof joint seal. The length of the gasket shall be equal to the outside circumference of the pipe plus the width of the gasket to provide adequate overlap after the pipes are joined together and the gasket is on the pipe. Prewet the gasket, stretch and elongate it at least 10 percent. Clip the gasket to fasten it and then press it against the pipe to assure maximum contact with the pipe surface. If inside joint opening is greater than permissible, paint the opening with mortar.

4-29 Rubber Gaskets (Tongue-and-Groove) Concrete surfaces with or without shoulders on the tongue or groove are joined using a compression-type rubber gasket.

Although there is a wide variation in joint dimensions and gasket cross section for this type joint, most are manufactured in conformity with ASTM C443. This type joint is primarily intended for use with pipe manufactured to meet the requirements of ASTM C14 or ASTM C76. In accordance with ASTM C443, the joints covered by this specification are normally adequate for hydrostatic pressures up to 13 psi without leakage.

When a joint of this type is completed, a lubricant must be used on the gasket and also the tongue and groove. This lubricant allows the joint to slip together much easier and eliminates the chance of the gasket binding. See Fig. 4-27.

Typical Cross Sections of Basic Compression Type Rubber Gasket Joints

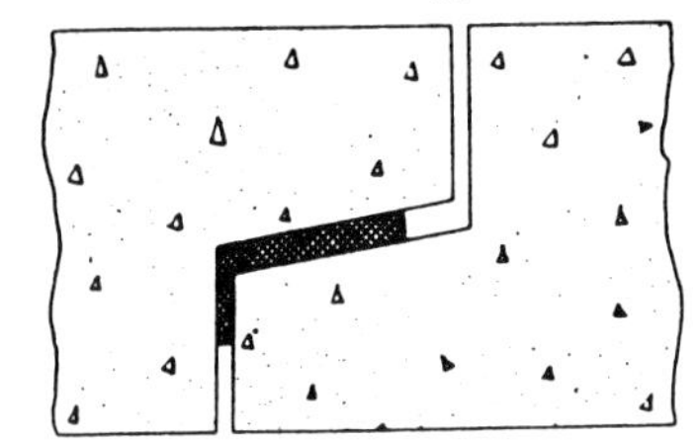

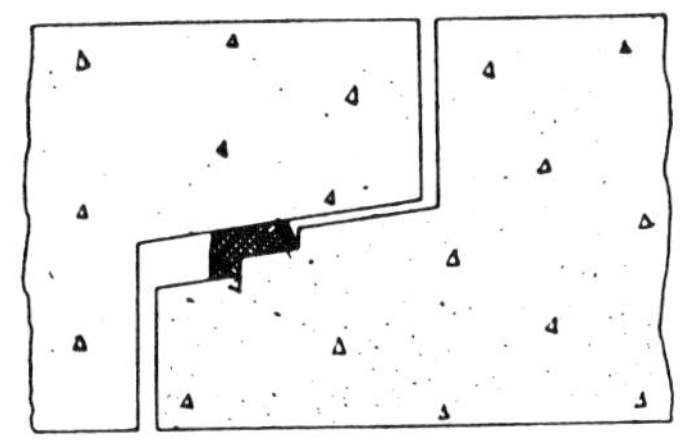

FIG. 4-27 Typical cross sections of basic compression-type rubber gasket joints. (*From Concrete Pipe Design Manual.*)

4-30 Bell-and-Spigot Joints This joint consists of a larger bell and spigot. The joint can be of the opposing-shoulder type or the spigot-groove type which can accept a rubber O ring or profile gasket. Normally this joint is used with a rubber gasket. However, it can be used with a mortar, mastic, or hot-poured compound in the joint. With rubber gaskets, these joints are frequently used for irrigation lines, water lines, sewer force mains, and gravity or low-head sewer lines where infiltration or exfiltration is a factor in design. This type joint, which provides excellent inherent watertightness in both the straight and deflected positions, may be employed to meet the joint requirements of ASTM C443, ASTM C361, and AWWA C302.

4-31 Hot-Poured Joints This type of joint can also be used with concrete pipe. Refer to sec. 4-41 on clay pipe.

4-32 Rubber Gaskets (Bell-and-Spigot) A rubber gasket is used with bell-and-spigot joints on most up-to-date jobs. The gasket can be an O ring which is circular in cross section or a type of profile gasket. The O ring is used in the joint design that has a groove in the spigot to confine the gasket. The profile gasket is used where shoulders are on the spigot and bell or just the spigot. (See Fig. 4-28.)

The joint is assembled by first of all cleaning both bell and spigot and any grooves of foreign material. The inside of the bell, outside of the spigot, and the gasket are then coated with the lubricant supplied by the manufacturer. The joint can then be pushed home with a bar or block-and-tackle arrangement, depending on the size of the pipe. If the bucket of the machine is used to support the pipe while the joint is barred home, great care must be taken to insure proper alignment and to avoid excessive forces.

4-33 Tied Joints (Gravity Pipe) If any undue forces are likely to be exerted on a culvert or storm sewer, the line may be tied together to resist these forces. For example, if movement should occur along the axis of the pipe, the system of tie rods will maintain alignment. Figure 4-29 illustrates various methods of tying joints. The added cost to the pipe is offset by the safety factor. These tied joints also have value in areas subject to many cycles of freezing and thawing. They tend to eliminate the jacking action that could occur under these conditions at the end. If proper design, drainage, and compaction are used when a pipe is installed, tied joints should not be necessary.

Typical Cross Sections of Opposing Shoulder Type Joint With O-ring Gasket

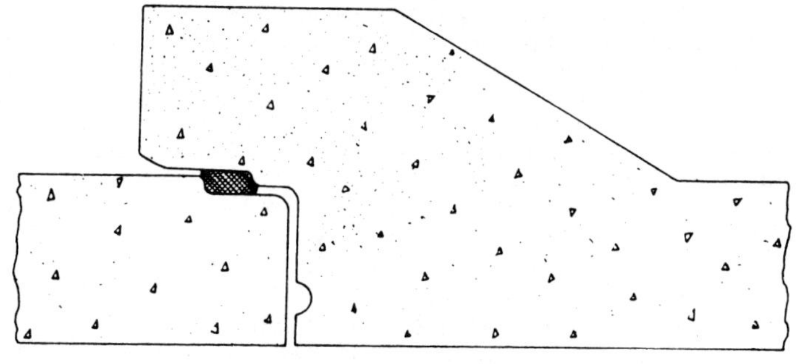

Typical Cross Section of Spigot Groove Type Joint With O-ring Gasket

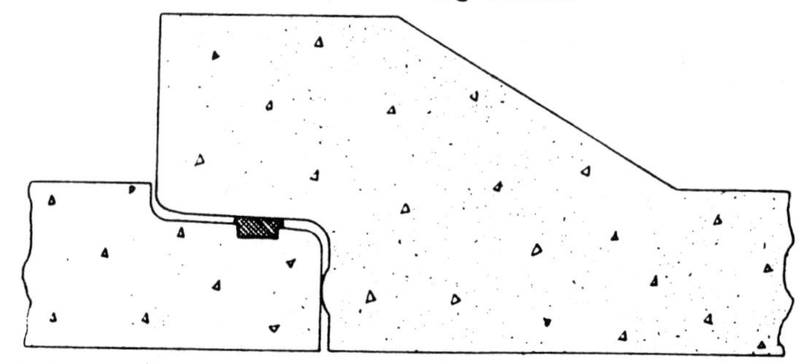

FIG. 4-28 Bell-and-spigot joints with rubber gaskets. *(From Concrete Pipe Design Manual.)*

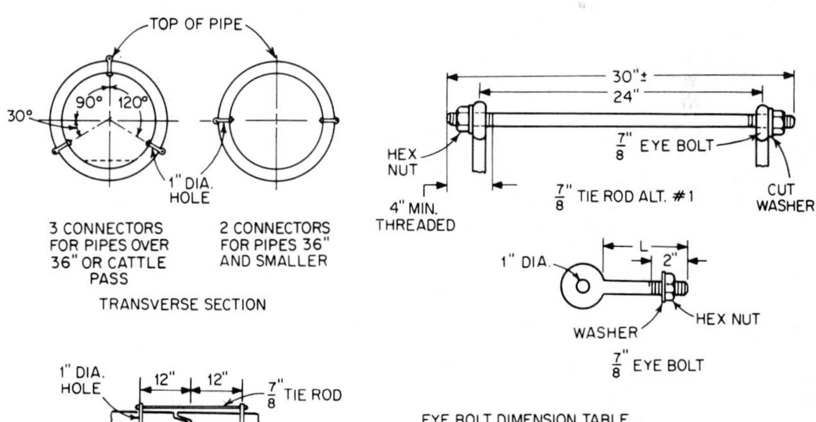

EYE BOLT DIMENSION TABLE

PIPE SIZE	LENGTH	
	TONQUE & GROOVE PIPE	MODIFIED BALL PIPE
18" TO 24"	4-1/2"	6-1/4"
30"	5"	7"
36"	5-1/2"	7"
42"	6"	
48"	8-1/2"	
60"	7-1/4"	
72"	8"	

FIG. 4-29 Tied joints.

CONCRETE PRESSURE PIPE

Concrete pressure pipe is a type of concrete pipe primarily designed for the conveyance of fluids under pressure. To classify the various joints used with each system, the types of pipe are separated into five categories: reinforced-concrete pipe with steel end rings or concrete-rubber joints, reinforced-concrete cylinder pipe, pretensioned-concrete cylinder pipe, prestressed-concrete noncylinder pipe, and prestressed-concrete cylinder pipe. In general, concrete pressure pipes have steel end rings to contain the gasket. These pressure pipes are used for transmission lines, distribution lines, irrigation lines, sewage force mains, siphons, aerial lines, industrial process lines, and subaquatic intakes, outfalls, and crossings. The typical joints are listed for the various pressure-pipe connections; however, all types of compression joints are available such as flanged joints, Dresser couplings, Victaulic couplings, and others.

4-34 Reinforced-Concrete Pipe with Steel End Rings or Concrete-and-Rubber Joints This type is reinforced-concrete pressure pipe meeting the AWWA Standard C302, *Reinforced Concrete Pressure Pipe, Noncylinder Type for Water and Other Liquids*, or ASTM C361, *Low-Head Pressure Pipe*. This pipe is similar to ASTM C76 pipe only because both are conventionally reinforced. For steel end ring pipe, the bell-and-spigot ends are formed by steel joint rings securely fastened in the wall. The rubber gasket is contained in a groove in the steel ring on the spigot and is the sole element depended upon to make the joint watertight. The gasket is manufactured from a special-composition rubber having a texture that assures a watertight and permanent seal. The concrete and rubber gasket joint is similar to that described in sec. 4-32. Pipe manufactured to ASTM C361 specifications includes 12-inch through 108-inch diameters and pressures up to 125 feet of head. Pipe manufactured to AWWA C302 standard includes 12- through 156-inch diameters and pressures up to 55 psi at working conditions and pressures up to 65 psi at transient conditions. Figure 4-30 illustrates a typical joint.

To assemble the joint, apply lubricant on the gasket and steel rings or concrete surfaces and then push the joint home. Follow the installation by pouring mortar, which is confined by an outside preapplied diaper, into the joint opening. Pointing up the inside joint opening is optional depending on local conditions.

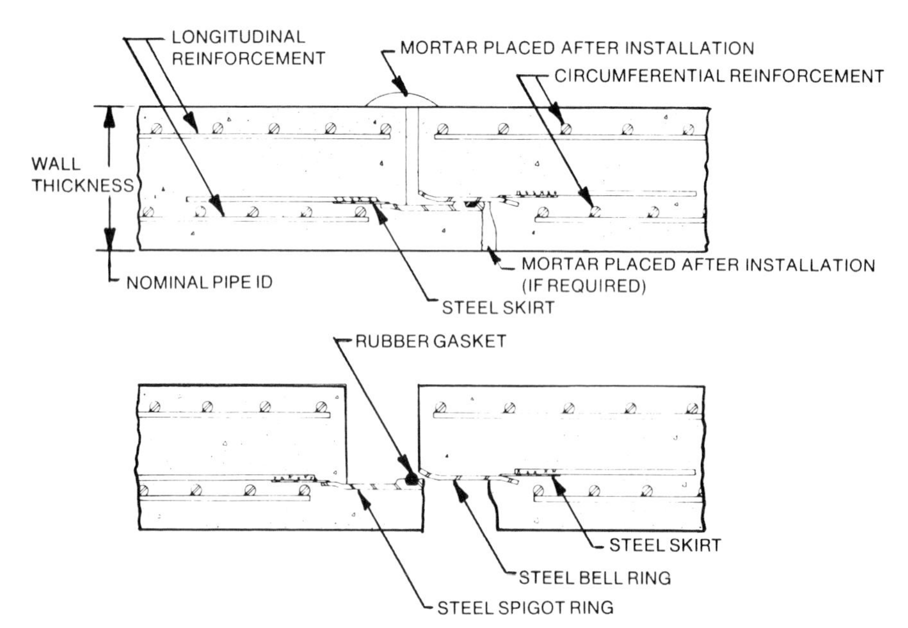

FIG. 4-30 Reinforced-concrete pipe with rubber and steel joint AWWA C302. *(CRETEX Pressure Pipe, Inc.)*

4-35 Reinforced-Concrete Cylinder Pipe with Steel End Ring Reinforced-concrete cylinder pipe consists of a steel cylinder with steel joint rings welded to it, a reinforcing cage or cages of steel rod surrounding the cylinder, and a wall of concrete encasing the cylinder and rod. This pipe is designed and manufactured in accordance with AWWA C300, *Reinforced Concrete Pressure Pipe—Steel Cylinder Type, for Water and Other Liquids.* This specification rates the pipe at an allowable pressure range from 40 psi to 250 psi in 24- to 144-inch diameters.

The joint is assembled like the previous steel-ring joint. After installation, pour the mortar into the joint as described in sec. 4-34. See Fig. 4-31.

4-36 Pretensioned-Concrete Cylinder Pipe This pipe consists of a welded steel cylinder, steel joint rings welded to a cylinder, a centrifugally cast core, a pretensioned rod wrapping wound helically around the cylinder, and an exterior mortar coating. This pipe is designed and manufactured in accordance with AWWA Standard C-303, *Reinforced Concrete Pressure Pipe—Steel Cylinder Type, Pretensioned, for Water and Other Liquids.* See Fig. 4-32.

4-37 Prestressed-Concrete Pipe with Steel End Ring Prestressed-concrete cylinder pipe is used extensively for water transmission systems, water distribution systems, water treatment plant process lines, and wastewater treatment plant process lines. This pipe consists of a welded steel cylinder, steel joint rings welded to the cylinder, a concrete core, either centrifugally lined or vertically cast, a prestressed wire wrapped helically

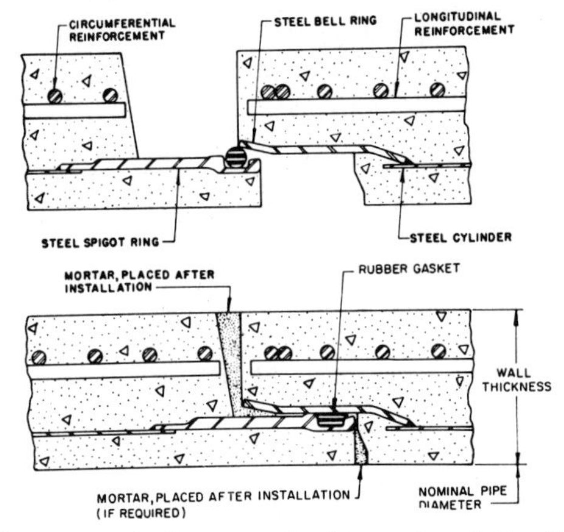

FIG. 4-31 Reinforced-concrete cylinder pipe with rubber and steel joint AWWA C300. *(CRETEX Pressure Pipe, Inc.)*

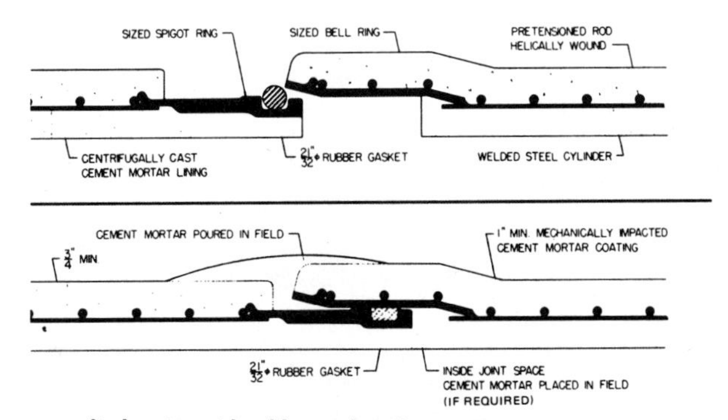

FIG. 4-32 Concrete cylinder pipe with rubber and steel joint AWWA C303. *(Gifford-Hill-American, Inc.)*

around the cylinder or core, and an exterior cement-mortar coating. The pipe is designed and manufactured to AWWA Standard C301, *Prestressed Concrete Pressure Pipe, Steel Cylinder Type, for Water and Other Liquids.* This pipe is available in 16- through 200-inch (and greater) diameters and design pressures up to 250 psi even though this type of pipe has been designed to exceed 400 psi. See Fig. 4-33, for centrifugally lined cylinder pipe.

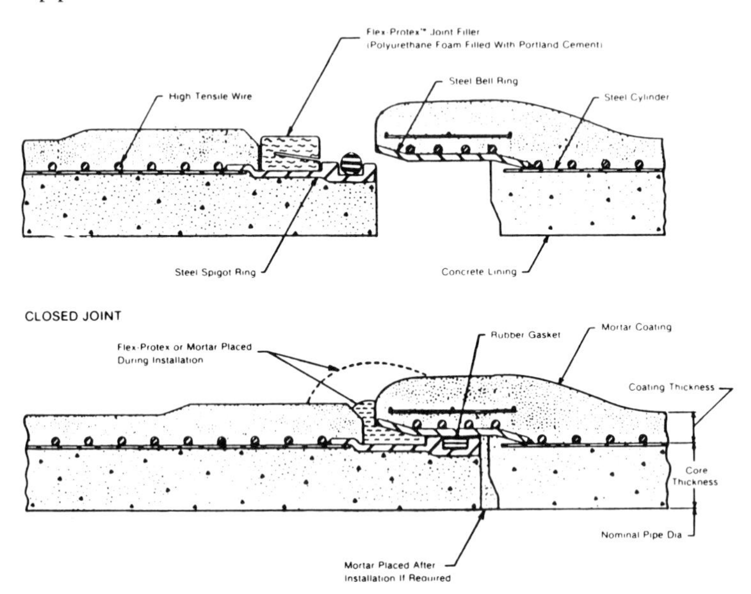

FIG. 4-33 Prestressed concrete cylinder pipe, lined cylinder type, with rubber and steel joint AWWA C301. *(CRETEX Pressure Pipe, Inc.)*

CROSS SECTION OF RUBBER AND STEEL JOINT

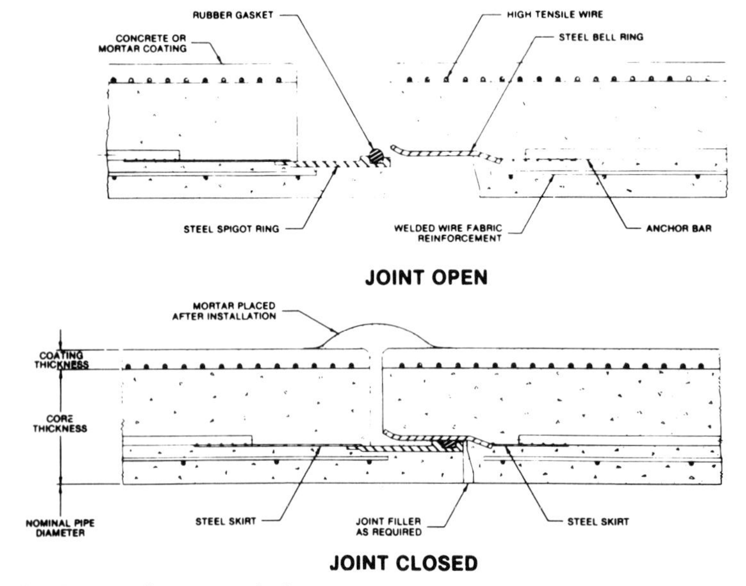

FIG. 4-34 Prestressed-concrete cylinder pipe. *(CRETEX Pressure Pipe, Inc.)*

4-38 Prestressed-Concrete Noncylinder Prestressed concrete noncylinder pipe is similar to the prestressed-concrete cylinder pipe described in sec. 4-37, except that it is manufactured without a cylinder. This pipe is manufactured in 48- through 144-inch diameters and is primarily intended for low- to moderate-pressure applications or gravity flow service such as intakes, outfalls, and the conveyance of sewage, storm, and industrial wastewaters. To date there is no AWWA standard for this pipe; however, the pipe is manufactured to AWWA Standard C301 (except for the steel cylinders) and loading conditions of ASTM C-76. See Fig. 4-34.

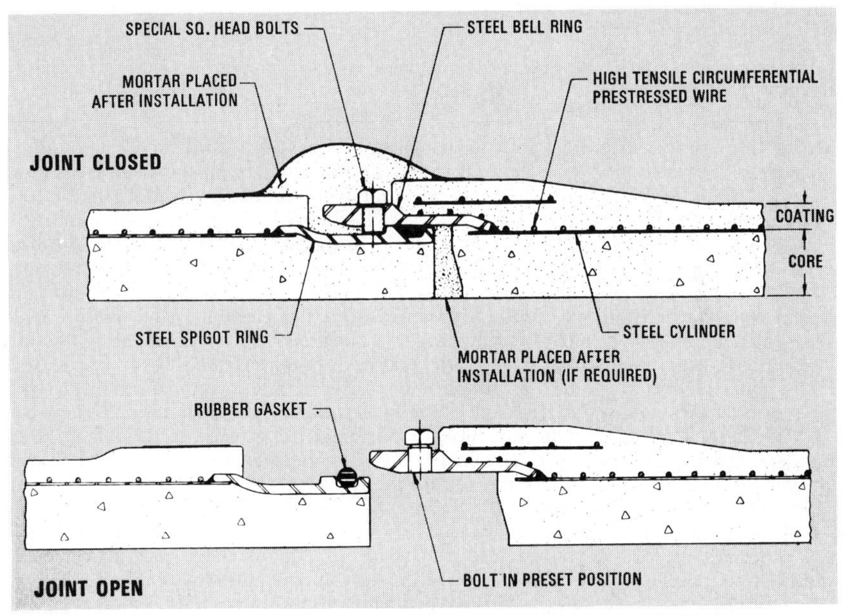

FIG. 4-35 Tied joint on prestressed concrete cylinder pipe, bell-bolt type. *(CRETEX Pressure Pipe, Inc.)*

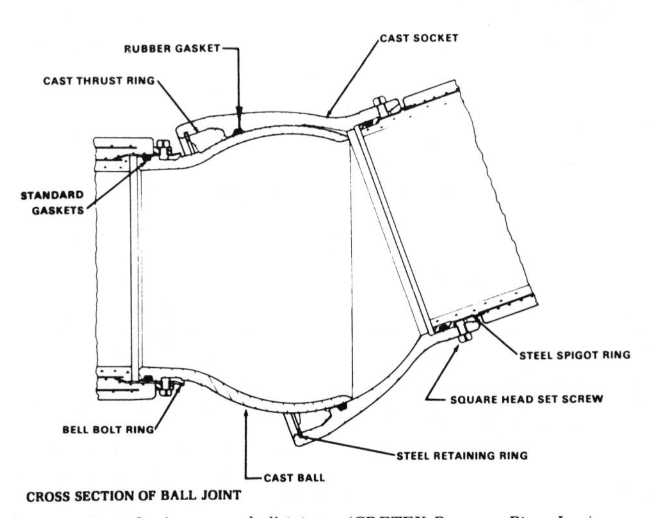

FIG. 4-36 Cross section of subaqueous ball joint. *(CRETEX Pressure Pipe, Inc.)*

4-39 Tied Joints When pressures within a pipeline cause unbalanced thrust at points of deflection in the line and thrust blocks will not resist these forces, the joints must be tied. Normally thrust blocks will not work in unstable soil areas or areas where their installation is difficult. One of the more popular types of tied joints has bolts located in the outer steel ring which are tightened down behind a ridge on the other steel ring. Figure 4-35 illustrates the joint and its assembly. Another type, used for subaqueous installations, is the flexible ball joint. This joint has the ability to roll and eliminate any undue stress in pipe or joint. If vertical deflection takes place, the joint will extend. See Fig. 4-36 for typical joint.

VITRIFIED CLAY PIPE

Vitrified clay pipe and fittings are used for the conveyance of sanitary sewage, industrial wastes, and storm water by gravity flow. Perforated clay pipe is used for underdrainage, filter fields, leaching fields, and similar subdrainage installations. Extra-strength and standard-strength clay pipe and perforated clay pipe conform to the requirements of ASTM C700. Pipe is tested according to ASTM C301 and installed according to ASTM C12. Clay pipe is one of the oldest materials known and used by humans. It is chemically inert and is highly resistant to corrosion. It requires no special coatings or linings to protect it from destructive elements such as the acids formed in sewers by the oxidation of hydrogen sulfide gas. It is a dense, hard, smooth-bore pipe offering exceptional flow characteristics. As opposed to flexible pipe, it is a rigid material with inherent pipe strength to withstand underground earth loads without support from compacted side soil. Sizes are available through 42-inch diameter.

4-40 Open Joints When clay pipe or perforated clay pipe is used for underdrainage, open joints usually are specified. This also applies to short lengths of agricultural drain tile, which meets the requirements of ASTM C4 (clay drain tile) and C498 (perforated clay drain tile).

4-41 Hot-Poured Joints Other types of jointing systems seldom are used with clay pipe, except possibly for some industrial applications where *highly* corrosive wastes are anticipated. In these instances there are many materials available that can be used to make a "field" joint. These consist of hot-poured or cold-applied acid- and alkali-resistant mortars, and other compounds utilizing sulfur, silica, tar, or resin bases, depending on the particular corrosive problem to be handled.

The hot-pour method of jointing may be used for making joints of this type:

1. Wipe bell and spigot clean and dry.
2. Pack chemically treated jute, oakum, or rope firmly into joint to center spigot and seal bottom of joint.
3. Clamp runner or "snake," or diaper, tightly against bell. Keep the "gate" or opening slightly to right or left of center so that the compound will flow around the pipe in one direction.
4. Pour the specific chemical jointing compound continuously at a steady rate until the joint is filled.
5. Remove runner or snake, or patch pouring gate of permanent diaper. See Fig. 4-37.

4-42 Joints and Couplings Compression joints and couplings are used almost exclusively today to join clay pipe. The resilient materials used for these high-compression jointing assemblies may be synthetic or natural rubber, polyester, urethane, polyvinyl chloride, or other materials. These joints utilize a sealing element that is compressed between factory-processed bearing surfaces located on both pipe ends, and meet the requirements of ASTM C425. Compression couplings for vitrified-clay plain-end pipe utilize a stop ring to properly position the pipe and coupling, and other components that, when compressed against the coupling material, form a hydrostatic seal around the entire circumference of the pipe, all meeting the requirements of ASTM C594. The joints are easily assembled in accordance with the manufacturer's recommendations. However, care should be taken to keep foreign materials from interfering

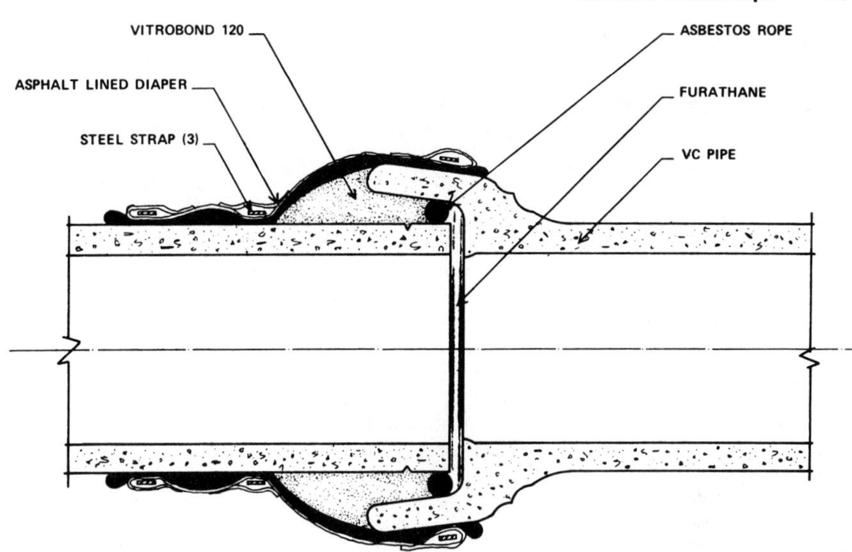

FIG. 4-37 Diaper joint. *(Atlas Minerals and Chemicals.)*

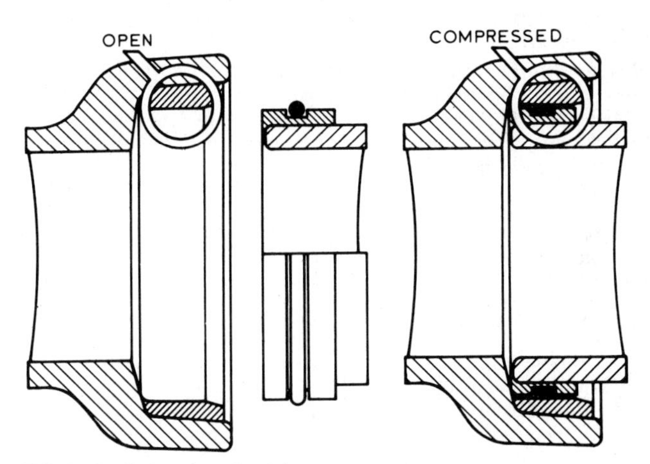

FIG. 4-38 O ring clay pipe joint.

with joint assembly. Compression joints and couplings resist shear loading, leakage, and root penetration, and are unaffected by most corrosive conditions found in sewer systems. Refer to Fig. 4-38.

ASBESTOS CEMENT PIPE

Asbestos cement pipe is manufactured by controlled blending of the basic raw materials: asbestos fibers, cement, and silica. Through autoclave curing this pipe develops strength, density, corrosion resistance, and long life expectancy. Its use varies from gravity sewer pipe, vent ducts and fittings, and air ducts and fittings to pressure water pipe. The couplings used to join water or sewer pipe are, from the design standpoint, almost the same. The distinction between nonpressure and pressure pipe is explained in the following sections.

4-43 Nonpressure Joints The joint for nonpressure transite pipe consists of a sleeve coupling sealed with rubber gaskets which slip over the ends of the pipe, which are stepped and tapered. Both ends of the pipe are lubricated and pushed home into the coupling. The rubber gaskets vary with each manufacturer; however, each is a solid cross section of chemical-resistant rubber. One type of gasket is illustrated in Fig. 4-39. The gasket sits in the groove of the coupling and is compressed when the pipe is pushed into the coupling. Normally in sizes of 10 inches and above, the coupling is factory-applied to one end of a standard length of pipe. Pipe coupling dimensions are shown in Fig. 4-40*a*. The sewer-pipe class ratings are in terms of lb/lin ft (crushing strength) test loadings based on three-edge bearing tests.

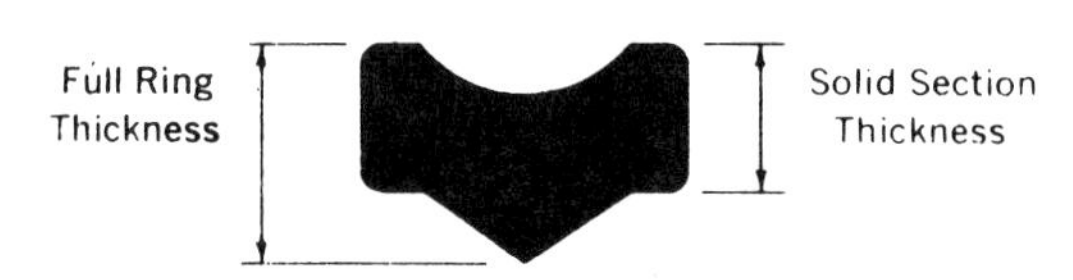

FIG. 4-39 Gasket cross section.

For vent and air-duct piping, a nongasket sleeve or ribbed coupling is used.

The pipe and joints meet the requirements of the International Association of Plumbing and Mechanical Officials' material standard UPL MS 1–58, federal specification SS-P-3316, or ASTM C644.

4-44 Pressure Joints The joint for this pipe is similar to the nonpressure pipe-coupling connection. The assembly is the same as for sewer pipe. Pressure pipe is classified according to pressure ratings. All couplings are designed to handle the requirements of their pipe classification. Again, every manufacturer has its own gasket design with its own cross-section design. Figure 4-40*b* illustrates the coupling dimensions.

PLASTIC PIPE

There are many types of plastics used for pipe production, most of which are thermoplastics. Thermoplastics are characterized by their ability to soften upon heat application and recover their original strength upon cooling or hardening. Thermoplastic piping materials in use today are:

1. Polyethylene (PE) pipe is tough, resilient, and flexible. It dominates the small-diameter market. The bulk of this material is used in diameters of 2 inches or smaller for farm, ranch, and irrigation application.

2. Polyvinyl chloride (PVC) pipe has the highest physical strength and chemical resistance. It is widely used for transmitting corrosive fluids and is the leader in sales in water, sewer, and drain-waste and vent industries.

3. Acrylonitrile butadiene styrene (ABS) exhibits good tensile strength, impact resistance, and chemical resistance. It also is widely used in irrigation, drain, waste and vent, sewer, and underground electrical conduit.

4. Cellulose acetate butyrate (CAB) is used in chemical process piping and in the oil and gas industry.

5. Rubber-modified polystyrene (SR) has had considerable usage in underground downspout drains and septic tank absorption fields. It exhibits good impact resistance and crushing strength.

6. Polypropylene is resistant to sulfur-bearing compounds and is used in the sulfur-related industry.

7. Polybutylene (PB) is a flexible, cold-flarable material used for water services.

These pipes are all referred to as *rigid plastic pipe* in the sewer and water trade.

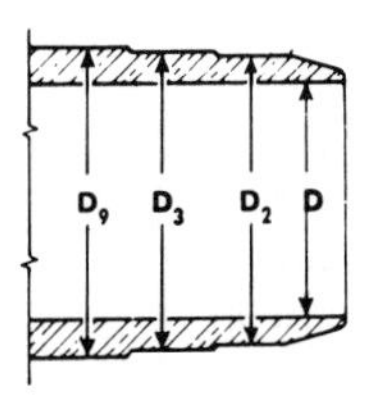

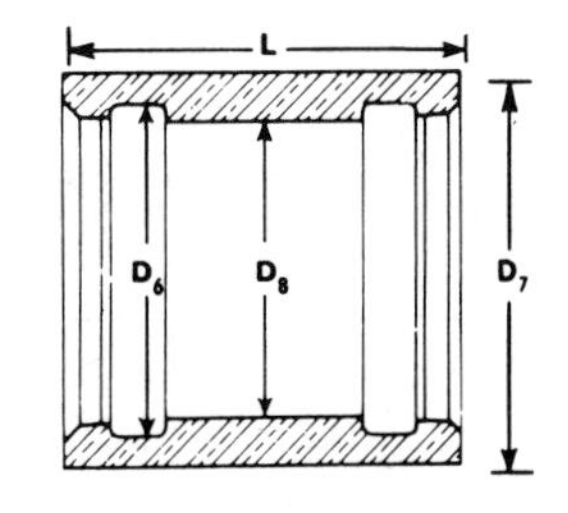

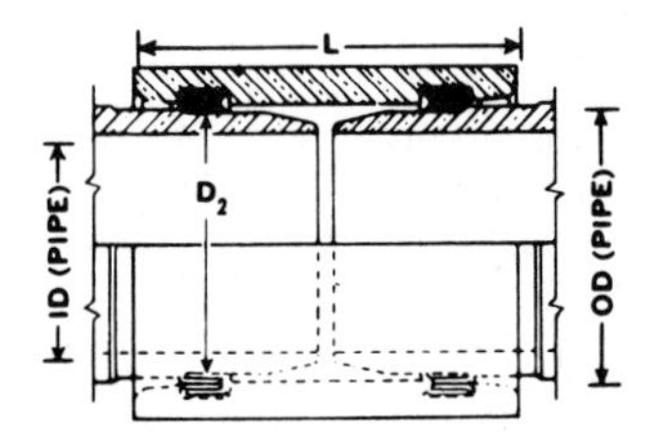

PIPE AND COUPLING DIMENSIONS (INCHES)

Class 100

Pipe Size In.	D	D_2	D_3	D_6	D_7*	D_8	D_9*	L
4	4.00	4.64	4.80	5.38	6.38	4.76	4.92	7.00
6	6.00	6.91	7.07	7.65	8.85	7.03	7.19	7.00
8	8.00	9.11	9.27	9.85	11.05	9.23	9.39	7.00
10	10.00	11.24	11.40	11.98	13.25	11.36	11.47	7.00
12	12.00	13.44	13.60	14.18	15.70	13.56	13.74	8.00
14	13.59	15.07	15.23	15.95	17.59	15.20	15.51	9.00
16	15.50	17.15	17.31	18.03	19.86	17.28	17.65	9.00
18	18.00	19.90	20.06	20.78	23.01	20.03	20.44	10.00
20	20.00	22.12	22.28	23.00	25.32	22.25	22.68	10.00
24	24.00	26.48	26.64	27.36	30.10	26.61	27.12	10.00
30	30.00	33.12	33.28	34.02	37.63	33.27	33.80	11.00
36	36.00	39.78	39.94	40.68	44.92	39.93	40.46	11.00

Class 150

Pipe Size In.	D	D_2	D_3	D_6	D_7*	D_8	D_9*	L
4	4.00	4.81	4.97	5.55	6.67	4.93	5.07	7.00
6	5.85	6.91	7.07	7.65	8.96	7.03	7.17	7.00
8	7.85	9.11	9.27	9.85	11.52	9.23	9.37	7.00
10	10.00	11.66	11.82	12.40	14.51	11.78	11.92	7.00
12	12.00	13.92	14.08	14.66	17.15	14.04	14.18	8.00
14	14.00	16.22	16.38	17.10	20.00	16.35	16.48	9.00
16	16.00	18.46	18.62	19.34	22.64	18.59	18.72	9.00
18	18.00	20.94	21.10	21.82	25.12	21.07	21.30	10.00
20	20.00	23.28	23.44	24.16	27.65	23.41	23.64	10.00
24	24.00	27.96	28.12	28.84	32.92	28.09	28.32	10.00
30	30.00	35.00	35.16	35.90	41.20	35.15	35.42	11.00
36	36.00	42.04	42.20	42.94	49.28	42.19	42.46	11.00

Class 200

Pipe Size In.	D	D_2	D_3	D_6	D_7*	D_8	D_9*	L
4	4.00	4.81	4.97	5.55	6.67	4.93	5.33	7.00
6	5.70	6.91	7.07	7.65	9.10	7.03	7.32	7.00
8	7.60	9.11	9.27	9.85	11.66	9.23	9.50	7.00
10	9.63	11.66	11.82	12.40	14.69	11.78	11.92	7.00
12	11.56	13.92	14.08	14.66	17.48	14.04	14.18	8.00
14	13.59	16.22	16.38	17.10	20.42	16.35	16.59	9.00
16	15.50	18.46	18.62	19.34	23.11	18.59	18.90	9.00
18	18.00	22.18	22.34	23.06	27.73	22.31	22.54	10.00
20	20.00	24.66	24.82	25.54	30.71	24.79	25.02	10.00
24	24.00	29.62	29.78	30.50	36.71	29.75	29.98	10.00
30	30.00	37.06	37.22	37.96	45.92	37.21	37.48	11.00
36	36.00	44.52	44.68	45.42	54.94	44.67	44.94	11.00

FIG. 4-40*a* Asbestos cement couplings with dimensions (nonpressure).

Each is, however, a flexible conduit and must rely on the side support of the surrounding soil to resist earth loads.

Because of some inherent advantages of plastic such as light weight, ease of joining, long length, ease of tapping, and bottletight joints, plastic pipe is widely accepted in the water and sewer market.

When used in higher-pressure adaptations, the pipe has thicker walls and different joints.

The basic types of joints will be explained and illustrated in the following subsections.

Pipe and Coupling Dimensions

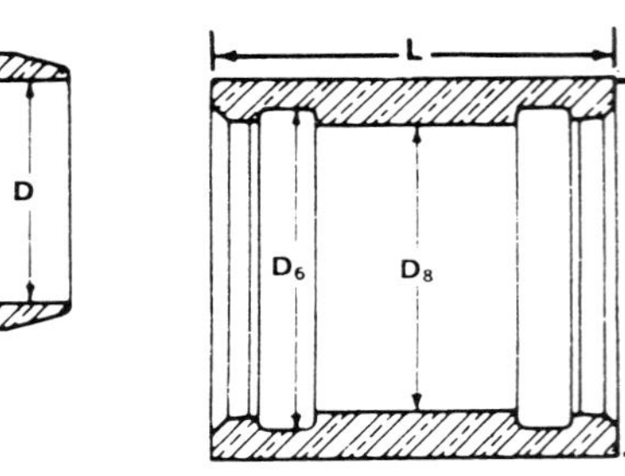

SIZE Pipe inches	ALL CLASSES D	ALL CLASSES D_2	ALL CLASSES D_3	ALL CLASSES D_6	CLASS 1500 D_7*	CLASS 2400 D_7*	CLASS 3300 D_7*	CLASS 4000 D_7*	CLASS 5000 D_7*	ALL CLASSES D_8	CLASS 1500 D_9*	CLASS 2400 D_9*	CLASS 3300 D_9*	CLASS 4000 D_9*	CLASS 5000 D_9*	L ALL CLASSES
4	4.00	4.62	4.74	5.09	5.86	5.86	6.05	—	—	4.73	4.81	4.87	5.03	—	—	4.80
5	5.00	5.66	5.78	6.13	7.00	7.00	7.19	—	—	5.77	5.90	5.97	6.15	—	—	4.80
6	6.00	6.66	6.78	7.13	7.98	7.98	8.18	—	—	6.77	6.92	6.98	7.14	—	—	4.80
8	8.00	8.80	8.96	9.38	10.31	10.31	10.31	—	—	8.91	9.02	9.04	9.22	—	—	6.80
10	10.00	10.89	11.05	11.47	12.42	12.42	12.42	12.58	12.78	11.00	11.12	11.16	11.36	11.50	11.70	6.80
12	12.00	12.99	13.15	13.57	14.66	14.66	14.66	14.82	15.04	13.10	13.22	13.26	13.50	13.64	13.86	7.80
14	14.00	15.07	15.23	15.77	16.87	16.87	16.87	17.03	17.27	15.20	15.30	15.36	15.62	15.78	16.00	8.90
16	16.00	17.15	17.31	17.85	19.05	19.05	19.05	19.23	19.49	17.28	17.38	17.46	17.72	17.90	18.14	8.90
18	18.00	19.23	19.39	19.93	—	21.23	21.23	21.69	21.69	19.36	—	19.54	19.82	20.02	20.26	9.90
20	20.00	21.29	21.45	21.99	—	23.39	23.39	23.87	23.87	21.42	—	21.62	21.92	22.12	22.38	9.90
24	24.00	25.51	25.67	26.21	—	27.79	27.79	28.31	28.31	25.64	—	25.78	26.10	26.32	26.60	9.90
27	27.00	28.74	28.90	29.44	—	—	31.16	31.70	31.70	28.87	—	—	29.22	29.46	29.76	10.90
30	30.00	31.99	32.15	32.69	—	—	34.53	35.11	35.11	32.12	—	—	32.34	32.60	32.90	10.90
33	33.00	35.21	35.37	35.91	—	—	—	38.47	38.47	35.34	—	—	—	35.72	36.04	10.90
36	36.00	38.45	38.61	39.15	—	—	—	41.85	41.85	38.58	—	—	—	38.84	39.18	10.90

FIG. 4-40*b* Asbestos cement couplings with dimensions (pressure).

4-45 Solvent-Welded Joints This kind of joint is used with PVC and ABS thermoplastic and consists of an integral bell on a length of pipe or a coupling which surrounds the spigot. The process chemically welds one pipe to another.

Square-cut ends, free of burrs, are required for proper joint fit. Proper fit between the pipe or tubing and mating socket or sleeve is essential to a good joint. Sound joints cannot be made between loose-fitting parts. The mating surfaces should be clean, dry, and free of material which might be detrimental to the joint. Cleaners and solvent cements which conform to ASTM D2564 for PVC joints and ASTM D2235 for ABS joints should be used.

After the primer has been used on both mating surfaces, apply the solvent cement evenly to the same surfaces. The pipe should be turned 90 degrees while being pushed home to distribute the solvent cement. After the joint is made, excess cement should be removed from the outside of the joint. The joint should not be disturbed until after it has properly set. In cold weather the setting time will increase. Consult manufacturer for more exact cure times. Under normal conditions the joint has sufficient strength to permit handling after 5 minutes. However, the total cure time is much more. The pressure rating of solvent-cement joints should be the same as the pipe jointed after reasonable cure time. See Fig. 4-41.

Another form of plastic pipe using a solvent joint is truss pipe. It is a composite material manufactured by extruding ABS resin into a truss shape, forming an inner and outer wall supported by webs. After extrusion the voids are filled with perlite, a lightweight concrete. A coupling is solvent-welded on the ends of the pipe to complete a joint.

4-46 Gasket Joints Each manufacturer has its own type of joint with gasket and joint variables. The basic idea, however, is the same: an O ring or profile type of gasket is compressed between the outer surface of the spigot and the inner surface of the retaining groove. Pressure applied to the system, in the range of its pressure ratings, forces the gasket into more intimate contact with the pipe and fittings so as to maintain a pressuretight joint. The joint shall be sealed by the rubber gasket so that the assembly will remain watertight under the conditions it was designed for. The rubber gasket will allow for expansion, contraction, settlement, and deformation of the pipe. Table 4-9 illustrates the SDR ratio (outside diameter divided by wall thickness) and its relation to maximum allowed pressures. The table is based on PVC piping material, which is the normal plastic used in pressure mains. Consult the manufacturer for other types of material. The advantage of a gasket joint is that proper assembly does not require as much skill as a solvent weld and can be executed under less-than-ideal field conditions.

The procedure used for a gasket joint is as follows:

1. Wipe the inside surface of the bell and outside surface of the spigot with a clean dry rag. Make sure all gasket grooves are clean.
2. Insert gasket on recessed area in bell or on spigot (depending on joint design). Some manufacturers recommend using a lubricant on gasket and spigot end to ease the installation.
3. Place the spigot into the lead end of the bell and push until contact is made with gasket. Position the two sections so that the spigot is in alignment with the bell.
4. Push pipe home by hand or else place a short length of 2 × 6-inch wood flat against open bell end of the pipe being joined.
5. Using a crow bar or pipe as a lever, push against the 2 × 6 until the joint is home. See Fig. 4-42 for gasket joint.

4-47 Heat-Fusion Joints This method is used for joining polyethylene, polybutylene, or one type of plastic to another type. PVC and ABS each require a particular solvent cement, but by using the heat-fusion method they can be joined without special solvent. Joints can be made either pipe end to pipe end or pipe end to socket fitting. The joint must be cleaned like any other plastic joint. The method involves heating the surfaces to be joined to a temperature that will permit fusion of the surfaces when brought into intimate contact. Special tools are required for this method and the procedure is outlined in the appropriate ASTM specification.

4-48 Threaded Joints Threaded joints are available in all plastic pipe with wall thickness heavier than schedule 80. The pipe can be threaded with any conventional pipe threading tool used for metal pipe, but threaded pipe should be derated to 50 percent of the pressure rating of the unthreaded pipe. A wide variety of fittings are avail-

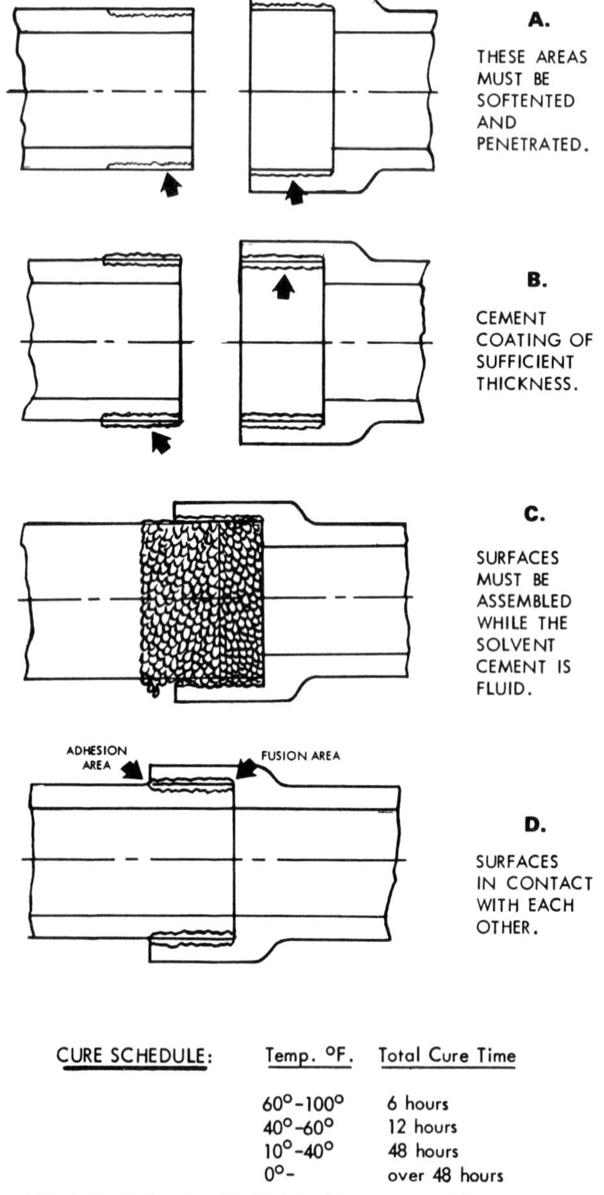
A.
THESE AREAS MUST BE SOFTENTED AND PENETRATED.
B.
CEMENT COATING OF SUFFICIENT THICKNESS.
C.
SURFACES MUST BE ASSEMBLED WHILE THE SOLVENT CEMENT IS FLUID.
ADHESION AREA
FUSION AREA
D.
SURFACES IN CONTACT WITH EACH OTHER.
CURE SCHEDULE:
Temp. °F.
Total Cure Time
60°-100° 6 hours
40°-60° 12 hours
10°-40° 48 hours
0°- over 48 hours

FIG. 4-41 Solvent-welded joint with cure schedule.

able for the various plastics. Threaded metal fittings can also be used with the plastic pipe. Thread-sealing compounds are normally not recommended; however, various thread lubricants and thread tapes are useful when used as recommended by the manufacturers. Care should be taken not to mar or distort the pipe or fitting during threading and assembly operations. Use fabric strap wrenches where possible.

4-49 Mechanical Joints A complete line of mechanical joint fittings is available for all plastics. Each joint should be properly qualified and utilized by the design

TABLE 4-9 PVC Pressure Pipe—Dimensions and Weights

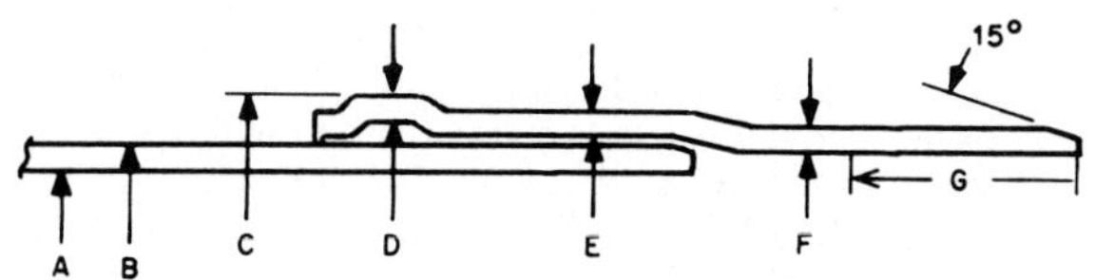

Nominal size	Pressure rating, psi at 73°F	Pipe barrel *A*, ID, in.	Pipe barrel *B*, OD, in.	OD tolerance, in. + or −	*C*, bell OD, in.	Minimum wall thickness, in. *D*, groove	Minimum wall thickness, in. *E*, socket	Minimum wall thickness, in. *F*, barrel	*G*, stop mark, in.	Weight, lb/ft, 20 ft length
SDR 26										
1½	160	1.754	1.900	0.006	2.36	0.092	0.080	0.073	3¼	0.29
2	160	2.193	2.375	0.006	2.88	0.111	0.100	0.091	3¾	0.44
2½	160	2.655	2.875	0.007	3.44	0.132	0.121	0.110	3⅞	0.63
3	160	3.230	3.500	0.008	4.14	0.159	0.148	0.135	4⅝	0.93
4	160	4.154	4.500	0.009	5.33	0.205	0.189	0.173	5⅜	1.53
6	160	6.115	6.625	0.011	7.77	0.298	0.280	0.255	5¾	3.33
8	160	7.961	8.625	0.015	10.11	0.386	0.362	0.332	5⅞	5.66
10	160	9.924	10.750	0.015	12.60	0.480	0.449	0.413	6¾	8.79
12	160	11.770	12.750	0.015	13.81	0.572	0.536	0.490	6⅞	12.44
SDR 21										
1½	200	1.720	1.900	0.006	2.40	0.114	0.101	0.090	3¼	0.34
2	200	2.149	2.375	0.006	2.94	0.140	0.127	0.113	3¾	0.53
2½	200	2.601	2.875	0.007	3.51	0.166	0.152	0.137	3⅞	0.77
3	200	3.166	3.500	0.008	4.23	0.202	0.186	0.167	4⅝	1.13
4	200	4.072	4.500	0.009	5.43	0.258	0.239	0.214	5⅜	1.87
6	200	5.993	6.625	0.011	7.92	0.377	0.352	0.316	5¾	4.09
8	200	7.805	8.625	0.015	10.29	0.496	0.451	0.410	5⅞	6.91
10	200	9.728	10.750	0.015	12.81	0.618	0.562	0.511	6¾	10.75
12	200	11.540	12.750	0.015	15.12	0.723	0.677	0.607	6⅞	15.18

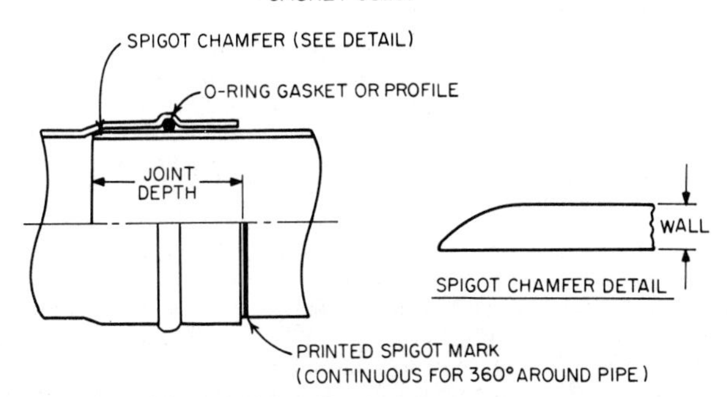

FIG. 4-42 Gasket joint with spigot chamfer detail.

engineer. When compression-type mechanical joints are used, the elastomeric gasket material in the couplings should be compatible with the plastic. The joint should be designed and installed to effectively sustain the longitudinal pullout forces caused by contraction of the piping or by external loading.

PE and PB pipe can be used for water services or other small-diameter pressure-pipe application. This pipe can be cold-flared or used with a standard compression-type mechanical fitting with an insert stiffener. Standard brass fittings would be used in conjunction with both types of joints.

Water pressure ratings at 23°C (73.4°F) for schedule 40 through 120 PVC plastic pipe are designated in the applicable ASME standards.

CORRUGATED METAL PIPE

Corrugated metal pipe is used for culverts, storm sewers, and other drainage structures. It is manufactured with galvanized steel or other material.

In selecting the types of joint to be used in a corrugated metal pipe the following factors should be analyzed. The first factor is structural considerations. All types of joint couplings will resist shear forces sufficiently. However, the stab-type connection will not offer any longitudinal tensile strength in the joints. The other types cling to the corrugations in such a way as to offset these forces.

A second factor to consider is infiltration. If this is important a flange band with a gasket, or hugger-type coupling with a gasket, should be used. These joints are designed only for gravity flow or low-head pressures up to 10 feet. The bands are drawn up tight by adjusting the bolts or using a special winch.

A third factor to consider is the beam strength required for aerial lines or where foundation conditions are less than desired. One of the positive types of connection (Fig. 4-43) should then be used.

4-50 Coupling Considerations for Steel Pipe Field joints of corrugated steel pipe and pipe-arch shall preserve the pipe alignment and prevent infiltration of the backfill. Couplers shall be made of the same zinc-coated base metal as the pipe. Unless otherwise specified they shall be of 0.064-inch thickness, but in no case shall they be less than 0.052 inch or more than 0.109 inch in thickness. Couplers may be corrugated bands, bands with projections, or sleeve type. All couplers shall lap on an equal portion of each pipe to be connected. The ends of band couplers shall overlap and be connected by galvanized angles, 2 inches by 2 inches by $^3/_{16}$ inch, fastened with $^1/_2$-inch diameter galvanized bolts, or the band couplers may have integrally or separately formed flanges or slots connected with $^1/_2$-inch diameter galvanized bolts or with a wedge lock of the same thickness as the band.

Corrugated bands for annular pipe shall mesh with the pipe corrugations and shall be not less than 7 inches wide for diameters 30 inches and under, 12 inches wide for diameters from 36 inches to 96 inches (inclusive), and 24 inches wide for diameters from 102 inches to 120 inches (inclusive).

Bands with projections for helically corrugated pipe shall have circumferential rows of projections which conform substantially in shape and size to the pipe-corrugation cross section. Spacing of projections around the circumference shall be that required to provide one projection in each corrugation of helical pipe of that diameter. Bands shall, for diameters 54 inches and under, have two circumferential rows of projections and be not less than $10^1/_2$ inches wide for $^1/_2$-inch-deep corrugations and not less than 12 inches wide for 1-inch-deep corrugations. For diameters 60 inches and greater, bands shall have four circumferential rows of corrugations and be not less than $16^1/_4$ inches wide for $^1/_2$-inch-deep corrugations and not less than 18 inches wide for 1-inch-deep corrugations.

Sleeve couplers shall have flared ends and a rectangular center stop. The center stop shall be high enough to close off the corrugations of helical pipe. Sleeves shall fit the outside pipe diameters closely and shall be not less than 6 inches wide for 6-inch through 10-inch diameters, not less than 8 inches wide for 12-inch and 15-inch diameters, not less than 10 inches wide for 18-inch and 21-inch diameters, and not less than 12 inches wide for 24-inch through 54-inch diameters.

Other equally effective types of approved couplers and/or fastening devices may be used.

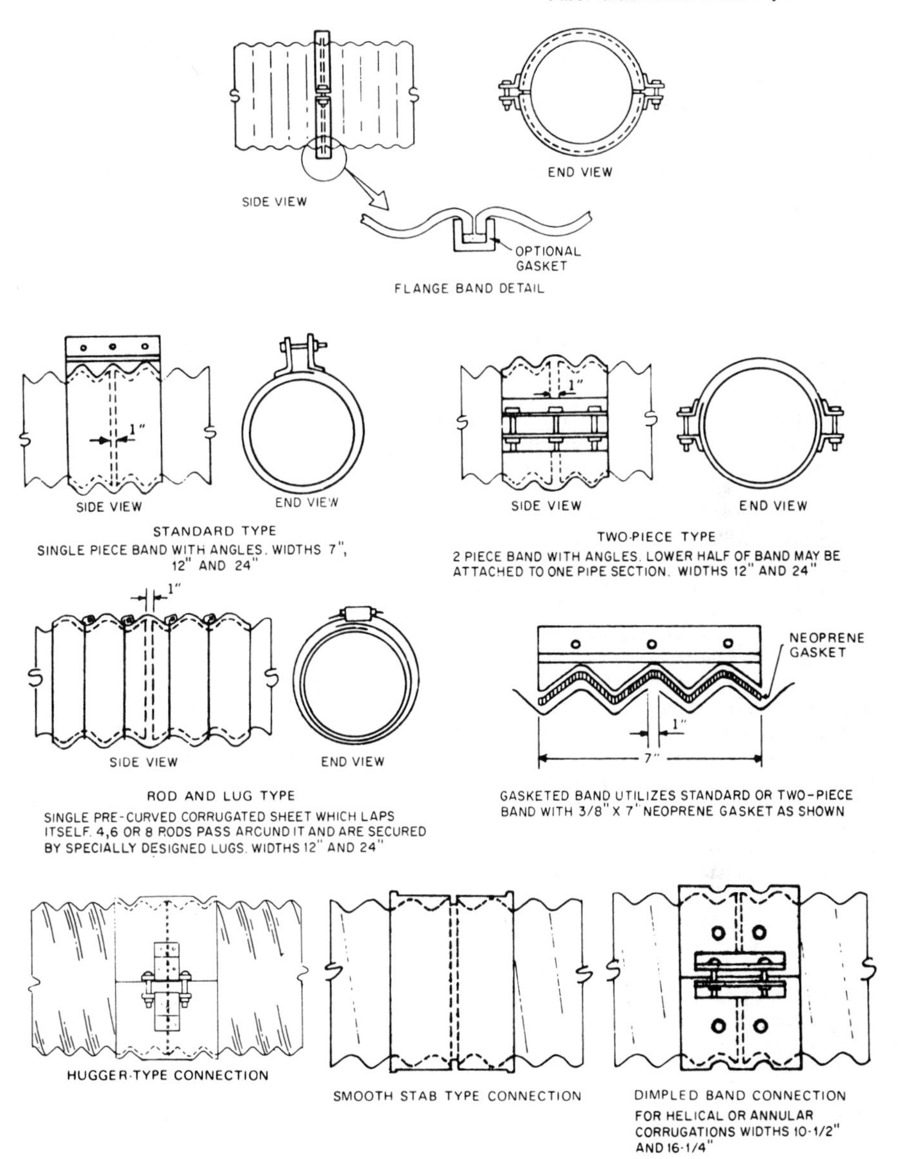

FIG. 4-43 Corrugated steel pipe connecting bands. *(Corrugated Steel Pipe Industry.)*

FIBER-GLASS-REINFORCED PIPE

Fiber-glass pipe is either filament-wound or centrifugally cast using continuous glass filaments and epoxy resin. (See Tables 4-10 and 4-11.) It is called FRP, which is short for *f*iber-glass-*r*einforced *p*ipe. The pipe is available in sizes from 2 inches to 12 ft diameter with various pressure ratings. The product has outstanding corrosion resistance, temperature capabilities, and mechanical strength. The pipe is engineered to operate at temperatures up to 300°F and pressures to 550 psi. Various ratings of pipe are available from the manufacturers. They should be contacted to recommend the appropriate pipe for the job.

TABLE 4-10 Specifications for Filament-Wound FRP Pipe—Dimensions and Weights

50 PSI FILAMENT WOUND PIPE						
Dimensions				Recommended Operating Data		
NOMINAL I. D.	NOMINAL O. D.	(NOMINAL) WALL THICKNESS INCHES	WEIGHT PER FOOT	INTERNAL PRESSURE @ 180°F	AXIAL LOAD POUNDS @ 80°F	CAPACITY GALLONS PER FOOT
14	14.34	.17	5.44	50	6250	7.997
16	16.36	.18	6.59	50	8000	10.440
18	18.38	.19	7.82	50	10250	13.220
20	20.40	.20	9.14	50	12750	16.320
24	24.42	.22	12.05	50	16750	23.500
30	30.50	.25	17.11	50	28500	36.720
36	36.56	.28	22.98	50	41000	52.880
42	42.62	.31	29.67	50	55750	71.970
48	48.68	.34	37.18	50	73000	94.000
75 PSI FILAMENT WOUND PIPE						
Dimensions				Recommended Operating Data		
NOMINAL I. D.	NOMINAL O. D.	(NOMINAL) WALL THICKNESS INCHES	WEIGHT PER FOOT	INTERNAL PRESSURE @ 180°F	AXIAL LOAD POUNDS @ 80°F	CAPACITY GALLONS PER FOOT
14	14.41	.21	6.59	75	9250	7.997
16	16.44	.22	8.07	75	12250	10.440
18	18.47	.24	9.69	75	15500	13.200
20	20.50	.25	11.45	75	19000	16.320
24	24.56	.28	15.38	75	27500	23.500
30	30.65	.33	22.29	75	43500	36.720
36	36.74	.37	30.44	75	61000	52.880
42	42.83	.42	39.82	75	84000	71.970
48	48.92	.46	50.42	75	109750	94.000
100 PSI FILAMENT WOUND PIPE						
Dimensions				Recommended Operating Data		
NOMINAL I. D.	NOMINAL O. D.	(NOMINAL) WALL THICKNESS INCHES	WEIGHT PER FOOT	INTERNAL PRESSURE @ 180°F	AXIAL LOAD POUNDS @ 80°F	CAPACITY GALLONS PER FOOT
14	14.48	.24	7.73	100	12500	7.997
16	16.52	.26	9.56	100	16250	10.440
18	18.56	.28	11.58	100	20250	13.220
20	20.60	.30	13.78	100	25500	16.320
24	24.68	.34	18.72	100	36750	23.500
30	30.80	.40	27.51	100	57250	36.720
36	36.92	.46	37.94	100	82500	52.880

SOURCE: Fibercast Co.

The resins involved are thermosetting plastics which cannot be reshaped by heating once they set or cure.

The true fiber-glass pipe should not be confused with a pipe manufactured from fiber glass polyester resin and sand called *r*einforced *p*lastic *m*ortar (RPM) pipe. This product has been redesigned and is still in the experimental stage.

4-51 Adhesive Joints One method of joining fiber-glass pipe is bonding the bell and spigot with an adhesive agent. Four major adhesives are available but the most popular is an epoxy adhesive similar to RP-34 (Bondstrand). Food and potable water applications require an epoxy adhesive RP-6A (Bondstrand). When the pipe is transporting unusual chemicals, the manufacturer should be consulted in regard to pipe and type of adhesive.

TABLE 4-11 Centrifugally Cast FRP Pipe—Basic Tables

PIPE

FIBERCAST OG 2025 PIPE — Dimensions				Recommended Operating Data				ULTIMATE INTERNAL PRESSURE
O.D.	WALL THICKNESS INCHES	REINFORCED THICKNESS INCHES	WEIGHT PER/FT.	INTERNAL PRESSURE @ 80°F.	COLLAPSE PSI @ 80°F.	AXIAL LOAD LBS. @ 80°F.	CAPACITY GALLONS PER/FT.	PSI
1.900	.175	.12	.66	400	300	2,000	.090	3,000
2.375	.20	.14	.83	375	225	3,000	.159	2,600
2.875	.25	.14	1.46	350	300	4,000	.230	3,900
3.500	.20	.14	1.40	300	70	5,000	.391	2,100
4.500	.20	.14	1.77	250	50	6,500	.685	1,800
6.625	.25	.17	3.50	250	35	9,000	1.530	1,700
8.625	.25	.17	4.50	200	20	11,000	2.690	1,350
10.750	.25	.17	5.60	160	10	15,000	4.280	1,050
12.750	.25	.17	6.70	130	6	18,000	6.120	900

FIBERCAST RB 2530 PIPE — Dimensions				Recommended Operating Data				ULTIMATE INTERNAL PRESSURE
O.D.	WALL THICKNESS INCHES	REINFORCED THICKNESS INCHES	WEIGHT PER/FT.	INTERNAL PRESSURE @ 80°F.	COLLAPSE PSI @ 80°F.	AXIAL LOAD LBS. @ 80°F.	CAPACITY GALLONS PER/FT.	PSI
2.375	.25	.17	1.25	475	550	4,000	.143	3,500
2.875	.25	.17	1.46	400	300	4,000	.230	3,900
3.500	.25	.17	1.75	325	200	5,000	.367	2,700
4.500	.25	.17	2.40	275	110	7,500	.652	2,200
6.625	.30	.22	4.15	325	65	13,000	1.480	2,800
8.625	.30	.22	5.40	250	35	17,000	2.630	1,600
10.750	.30	.22	6.70	200	16	21,000	4.200	1,250
12.750	.30	.22	8.00	160	10	25,000	6.020	1,000

FIBERCAST CL 2030 PIPE — Dimensions				Recommended Operating Data				ULTIMATE INTERNAL PRESSURE
O.D.	WALL THICKNESS INCHES	REINFORCED THICKNESS INCHES	WEIGHT PER/FT.	INTERNAL PRESSURE @ 80°F.	COLLAPSE PSI @ 80°F.	AXIAL LOAD LBS. @ 80°F.	CAPACITY GALLONS PER/FT.	PSI
1.900	.175	.12	.60	550	300	2,000	.090	4,000
2.375	.200	.14	.95	500	225	3,000	.159	3,500
2.875	.200	.14	1.27	400	150	4,000	.230	3,000
3.500	.200	.14	1.43	340	70	5,000	.391	2,550
4.500	.250	.17	2.29	360	110	7,500	.652	2,800
6.625	.300	.22	3.96	350	65	13,000	1.480	2,700
8.625	.300	.22	5.18	265	35	17,000	2.630	2,200
10.750	.300	.22	6.47	210	16	21,000	4.200	1,800
12.750	.300	.22	7.69	175	10	25,000	6.020	1,400

Operating Pressure Vs. Temperature

PIPE AND FITTINGS	
% RATED OPERATING PRESSURE	OPERATING TEMPERATURE – °F.
100%	0°F.
100%	50°F.
100%	100°F.
100%	150°F.
100%	175°F.
80%	200°F.
65%	225°F.
50%	250°F.
40%	300°F.

SOURCE: Fibercast Co.

An adhesive joint is made by assembling the fiber-glass pipe and fittings in the following manner:

1. Measure the desired length and scribe the pipe, preferably using a pipe fitter's wrap-around.
2. Shave the cut end of the pipe prior to bonding, using the pipe shaver recommended by manufacturer.
3. Sand all fittings and pipe sockets within 2 hours of assembly.
4. Thoroughly wipe the sanded socket and spigot with a clean dry cloth to remove dust particles.
5. Measure back from the spigot end of each pipe the distance recommended by manufacturer, and scribe a line using a white grease pencil or soapstone.
6. Mix adhesive for at least 1 minute and until adhesive is thoroughly mixed.
7. Apply a layer of adhesive approximately $1/32$-inch thick to the surface of the socket, including the pipe stop.
8. Apply adhesive liberally to the entire spigot surface and a thin layer of adhesive to the cut edge of the pipe.
9. Insert the pipe slowly into the spigot until the spigot end rests firmly against the pipe stop.
10. Examine the assembly for proper alignment and seating.
11. Cure the adhesive joints in accordance with manufacturer's recommendations.

This procedure may vary slightly with each manufacturer and size of pipe. See Figs. 4-44 and 4-45.

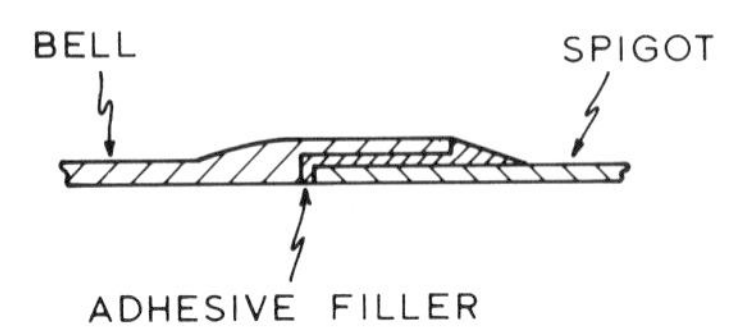

FIG. 4-44 Bell-and-spigot joint for fiber-glass-reinforced adhesive joint.

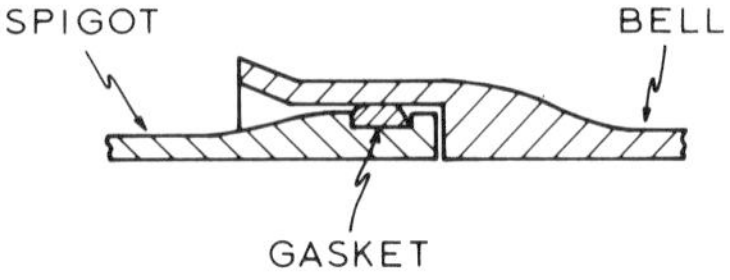

SPIGOT
BELL
GASKET

FIG. 4-45 Fiber-glass-reinforced gasket joint. *(Corban Div. of Fibreboard Corp.)*

4-52 Mechanical Joints Mechanical joints are available which conform to all ANSI standards. They consist of a flange from one pipe pressing a full-face gasket up against the flange of the joining pipe. The joint is held together by a series of bolts tightened to a specific torque.

The procedure is as follows:

1. Place full-face gasket between flanges and align.
2. Insert bolts and finger-tighten.
3. Tighten nuts in sequence as per standard practice. The idea is to build the pressure uniformly over the entire flange face with 5 lb-ft torque increments.
4. Consult manufacturer's specifications for what the maximum torque on each bolt should be when joint is completed.

Another type of mechanical joint is called the profile mechanical-gasketed joint. (See Fig. 4-45.) It is similar to a push-on joint and consists of a profile rubber gasket stretched over machined grooves between preformed stops on the spigot end of the pipe. A special lubricant is used to expel sand or other particles and ease installation.

The various fittings are available with flanges to complete this type of system.

GLASS PIPE

Glass pipe is made from a borosilicate glass with a very low alkali content. It is particularly suitable for handling acids and other active materials such as chlorinated hydrocarbons, hydrogen peroxide, bromine, and brines. The ends of all straight lengths are heat-treated to prevent warpage. The pipe is more brittle than other materials and subject to breakage. All fittings are heat-treated except for certain fittings which are annealed. Figure 4-46 illustrates the pressure and temperature difference of the five types of glass pipe. The following paragraphs explain each joint with its advantages, disadvantages, and limitations. Various adapter fittings are available to connect this

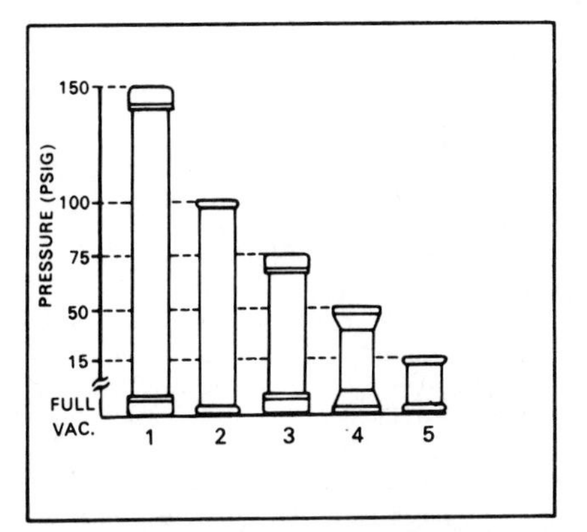

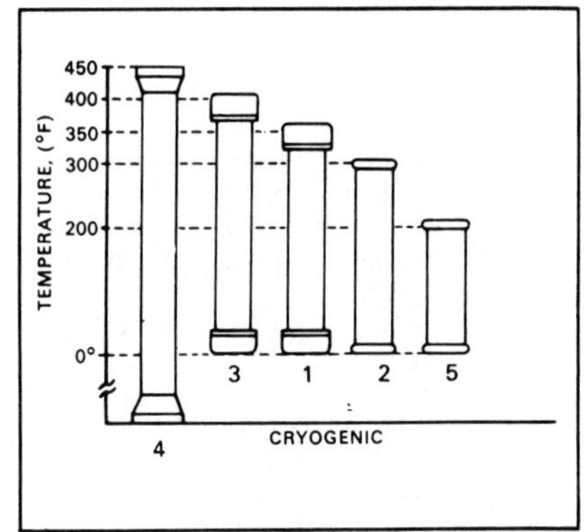

FIG. 4-46 Comparative glass piping systems: (1) armored, (2) beaded pressure, (3) capped, (4) conical, (5) drainline. (*Corning Glass Works.*)

pipe with other types of materials. This section describes only the joints connecting glass to glass.

4-53 Armored Joints The armored glass pipe system consists of a pipe and joint of the highest caliber in glass pipe. The joints are made with a ball coupling which provides flexibility, self-alignment, leak-free joints, and much higher pressure ratings. The ball coupling is quickly and easily installed and eliminates the need for flexible couplings. The pipe aligns itself while being hung and any adjustments are easily made. The ball coupling consists of three components:

1. One set of bolts making a fluorocarbon-gasket seal against one face.
2. Another set of bolts making an independent gasket seal against the other pipe face.
3 The split ball and socket, which makes use of the low coefficient of friction of the gasket, allowing angular deflections up to 3 degrees per joint. Within this 3-degree limit, the independent gasket seals remain undisturbed.

Table 4-12 illustrates qualities and limitations of this joint. Figure 4-47 illustrates a cross-section view through the joint.

TABLE 4-12 Armored Joint Table

		PROFILE ▶	pipe system	
RATINGS	maximum working pressure (all sizes can go to full vacuum)	1″-3″	150 psig	all fluids
		4″	100 psig	
		6″	60 psig	
	temperature, all sizes		0°-350° F	
	temperature shock	1″-3″	200° F	
		4″	175° F	
		6″	160° F	
MEASUREMENTS	pipe sizes		1″-6″	
	no. of standard straight lengths		13 (6″-120″)	
	deflection/joint		3°	

SOURCE: Corning Glass Works.

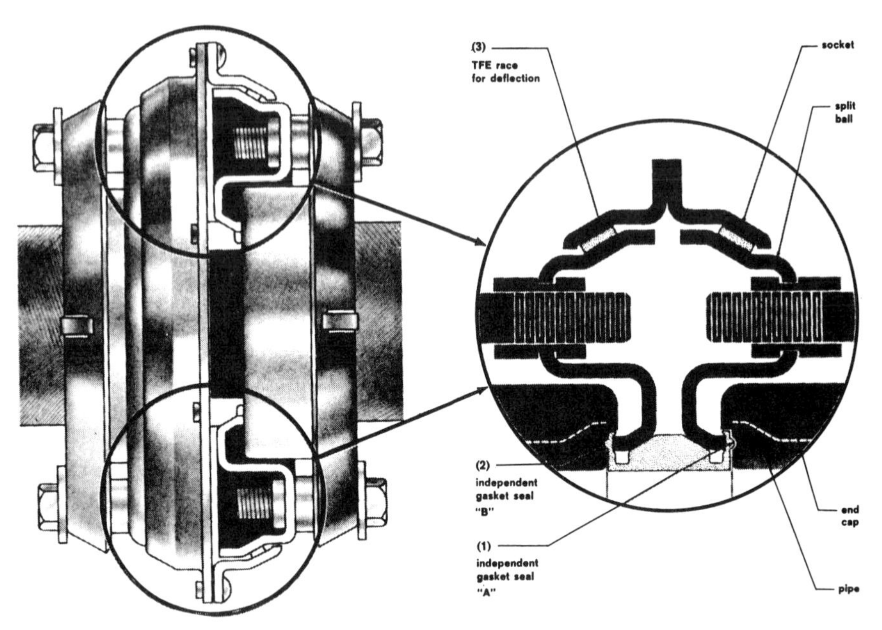

FIG. 4-47 Ball coupling for glass pipe. (*Corning Glass Works.*)

4-54 Beaded Pressure Joints The beaded pressure system is used for lower pressures and temperatures. It is priced competitively with other pipes designed to perform the same job such as plastic-lined, stainless-steel, and fiber-glass-reinforced pipe. The joint consists of a one-bolt flexible coupling which is placed over the ends of the pipe. The beads are larger than for drainline glass pipe, allowing higher pressures to be handled. These joints can be assembled in approximately 2 minutes.

4-55 Capped Joints This pipe is similar to armored but is somewhat outdated. It does not possess as high physical properties as armored; however, the joint is similar.

4-56 Conical Joints The conical joint connection between glass piping is somewhat different than for other materials. Since glass is brittle, more care should be taken to insure a good joint. Figure 4-48 is a factory-fabricated flared-end type of pipe connection. It is used on nearly all joints but must be fitted properly in regard to length and alignment. It is impossible to install a system without some adjustment. The adjustable joint is used in field application. A kit is available which allows for field fabrication. The pipe is cut and a bead is formed at the cut end. This beaded end with an adapter insert mates with the conical end and, joined by the flange, forms the field joint.

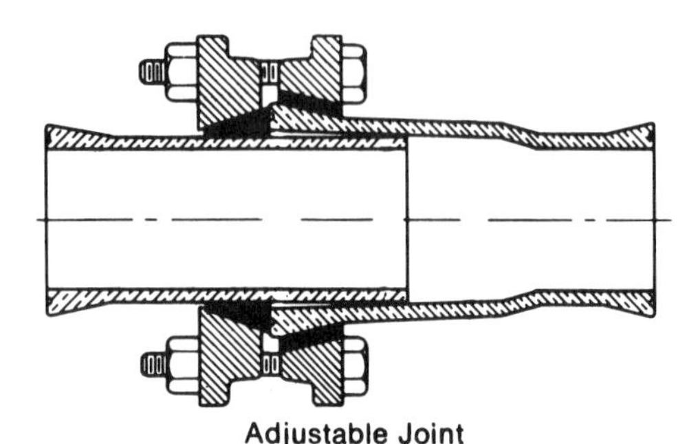

FIG. 4-48 Adjustable joint. (*Corning Glass Works.*)

However, only one half of the joint may be beaded. Two beaded ends together would give an unsatisfactory amount of gasket bearing surface.

The gasket material is available in neoprene, gum rubber, asbestos, Koroseal, and Teflon. Normally the type of chemicals carried in the pipe determines the gasket material to be used.

The manufacturers recommend a six-step approach for making a typical joint. See Fig. 4-49.

1. Metal flanges and inserts should be assembled on the glass pipe conical ends before the pipe is hung. The metal flange goes over the end first.

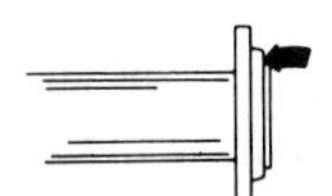

3. IMPORTANT: When the flange and insert are pulled up on the glass conical ends, be sure that the insert is flush with the metal flange face pointing toward the end of the pipe.

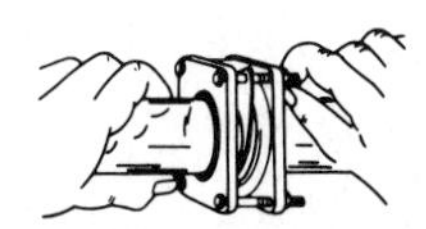

5. Line up gaskets with the pipes' inner surfaces so that no part obstructs flow. This is particularly important in milk and food installations.

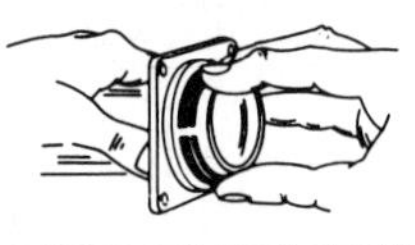

2. The flange and insert should be pulled up by hand on the glass pipe conical ends.

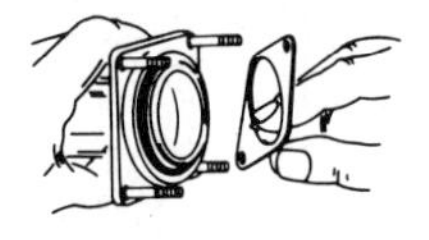

4. Keep gaskets dry during installation. Wet gaskets, especially rubber ones, may slip under line pressure.

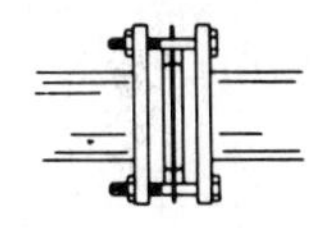

6. Be sure flange assembly is square with the line and not cocked. You should be able to see glass on each side of the gasket when the nut is tightened. When tightening nuts, turn nuts and not the bolts—if bolts rotate, they may twist the interface gasket out of alignment. Tighten bolts relatively uniformly all around. It is not necessary to "baby" the glass pipe.

FIG. 4-49 Assembly of a conical joint (six steps). *(Corning Glass Works.)*

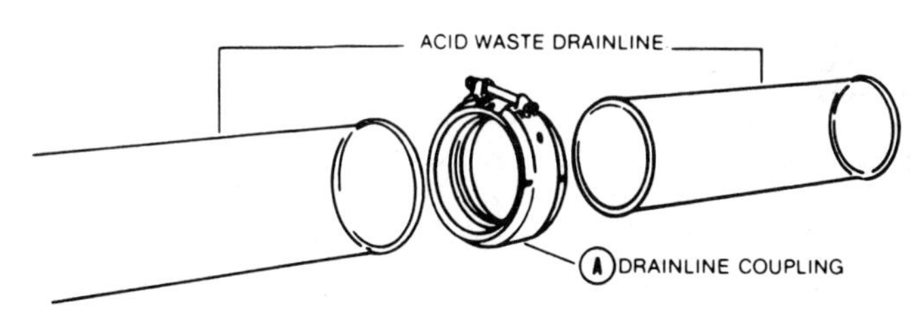

FIG. 4-50 Drainline coupling for acid waste. *(Corning Glass Works.)*

A torque wrench is recommended when beginners are making joints. Each bolt requires sufficient pound-feet of torque to obtain a good seal. If pipe is torqued considerably higher than required (in order to stop a leak) it often means misalignment.

Glass-pipe fittings are available in various bends and radii. Each has the same flared ends to connect into the pipe.

4-57 Drainline Joints Drainline pipe is used on low-pressure and also lower-temperature lines. The joint consists simply of a coupling with an adjustable bolt that fits over the beads on the ends of the pipe. This joint takes approximately 2 minutes to complete. See Fig. 4-50.

BITUMINIZED FIBER PIPE

Bituminized fiber pipe is used for gravity sewer services, sanitary drainage fields, subsurface water drainage systems, and related applications. The product is manufactured from 25 percent cellulose wood fibers and 75 percent pitch. The two combine to make a sturdy but flexible wall pipe. The pipe should be backfilled carefully because lateral support is needed on the walls of the pipe to eliminate any crushing problems. The three methods of joining this pipe are by tapered, butt, and slip joints.

4-58 Tapered Joints The tapered joint has tapered male ends on the pipe and tapered female couplings to match. The tapers shall be accurately machined or molded to insure tight joints. A coupling shall be provided for each length of pipe and for each fitting. The slope of all tapers shall be 2 degrees. The joint is made simply by inserting the coupling on one pipe and driving into position. A block of wood should be placed against the coupling to protect it against damage while driving. Never place the block against the tapered end of pipe. Wipe all joints clean but do not use any lubricant. To cut and taper odd lengths of pipe always use a taper-cutting tool supplied by the manufacturer. This joint is designed to resist 10 feet of head. Refer to Fig. 4-51.

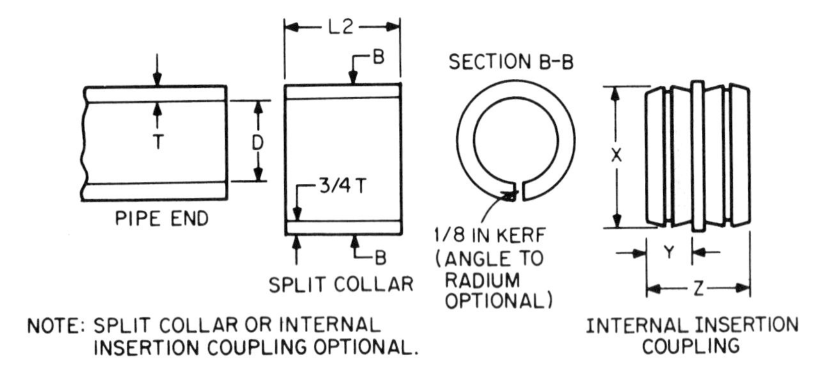

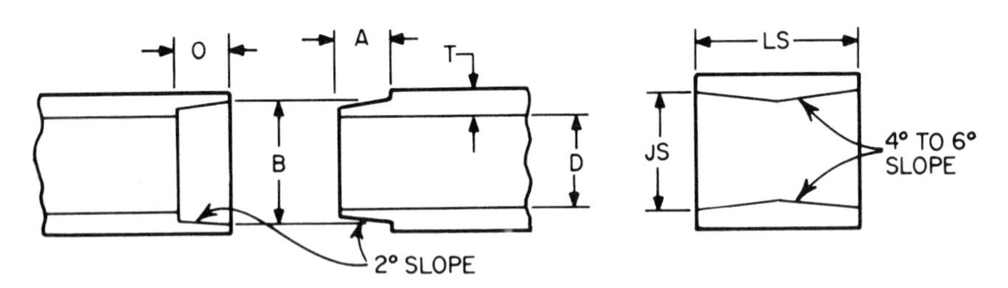

FIG. 4-51 Slip joint and 4- to 6-degree-slope couplings and typical butt joints. *(Bermico Co.)*

4-59 Butt Joints The butt joint has squarely cut ends on both pipe and fittings. A split-collar coupling or internal-insertion coupling shall be provided for each length of pipe and for each fitting. The split collar is merely slipped over the pipes and holds them together by friction. An internal insertion coupling is inserted into the square-cut end of the previously laid length and driven tight. Refer to Fig. 4-51.

4-60 Slip Joints The slip joint has a tapered male joint on one end of the pipe and a tapered female joint on the other end. This design has fittings available which have 4- to 6-degree slopes. The joint is made like a tapered joint. This joint has been nearly 100 percent replaced by internal-insertion coupling pipe on all new construction. Both the butt and slip joint are used for watertight connections.

WOOD PIPE

4-61 Wood Pipe Wood pipe is still used in some undeveloped countries to convey liquids. Wooden staves are banded together to form the pipe, as in an old-fashioned beer barrel. Sometimes joints are completed in the field by lapping wood over the joint and securing with steel straps. However, most joints are probably just butted or grooved. Very little information is available on this subject.

MISCELLANEOUS CONNECTORS AND ADAPTERS (NONPRESSURE)

Couplings and taps are used to join many types of dissimilar materials. This section is directed toward these products.

If unlike materials are being connected, the manufacturer should be consulted because adapters are available for nearly all materials. Some of the more well-known are as follows.

4-62 Connection Sleeves (PVC or Rubber) Rubber sleeve-type fittings are available to connect any materials. These sleeves are held to the pipe by stainless-steel adjustable straps. The straps can be tightened to the pipe and create an infiltration-proof joint. They should not be used on pressure piping or fittings. See Fig. 4-52.

4-63 Concrete Sleeves These are probably the oldest method of joining different types of pipe. The sleeve is made simply by butting the two pipes to be joined and pouring a concrete collar to enclose the joint. The farther the concrete can be extended along the pipes, the more rigid and tight the joint will be. This is a rigid type of connection with no flexibility.

FIG. 4-52 Connecting sleeve. *(Fernco Inc.)*

4-64 Manhole Pipe and Casting Sleeves Where sewer pipes enter manholes there is always chance for infiltration around the pipe. Normally a mortar joint is made at this point, but various types of rubber sleeves are available which provide a watertight connection which allows flexibility. One of these manhole sleeves is shown in Fig. 4-53.

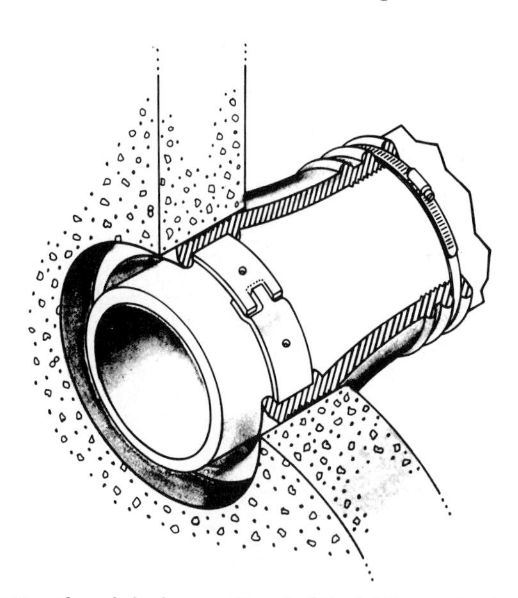

FIG. 4-53 Cross section of manhole sleeve. *(Press-Seal Gasket Corporation.)*

It is a rubber sleeve which is anchored in the manhole wall and secured to the pipe with a stainless-steel strap. Another type of manhole seal is a device called a "chimney seal" which connects the casting to a manhole cone or the adjusting ring. It is a rubber sleeve held in position by two expansion rings. These expansion rings anchor the seal in position to eliminate infiltration into a manhole. See Fig. 4-54.

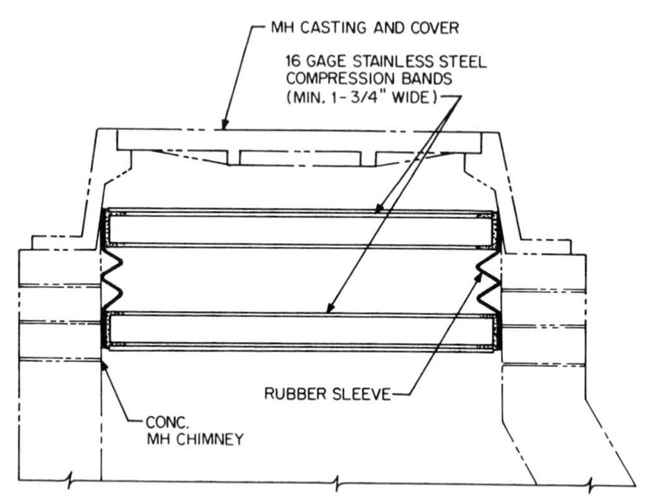

FIG. 4-54 Chimney seal. *(CRETEX Specialty Co.)*

4-65 Rubber Gasket Adapters Gasket adapters or "odapters" are used to join one pipe material to another. They consist of a preformed donut or O ring gear-grip gasket which slips over the spigot end of a pipe. This setup is then driven or pushed into a bell, hub, or straight section of another pipe. These gaskets are manufactured from specially compounded rubbers to handle all types of effluent. A specific gasket is made to fill the void between the different materials. This joint also has the advantage of flexibility. The pipes can be deflected and yet maintain a low-pressure joint. Check with the various gasket manufactures to be sure you have the proper gasket for

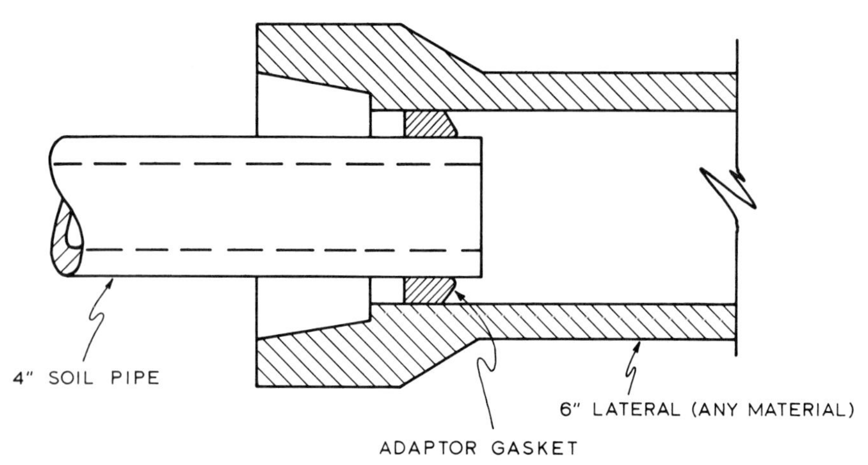

FIG. 4-55 Gear-grip gasket adapter. *(Hamilton Kent Manufacturing Co.)*

joining the given pipes. As an example, Fig. 4-55 shows a 4-inch soil pipe joined to a 6-inch lateral of concrete or clay with a gear-grip gasket.

MISCELLANEOUS CONNECTORS AND ADAPTERS (PRESSURE)

There are available connectors and adapters for changing from one material to another or reducing sizes for all pipes. The important consideration is ensuring that the transition coupling or adaptor meets the same design standards that the pipe itself is designed to meet. Included in this section are some of the most commonly used connectors.

4-66 Corporation Stops Corporation stops are installed on a water main where the house service connects with the main. They are of the same material as the service or, if not, of a compatible material which eliminates chemical action. A tapping machine is used to drill and tap the water main. Either IP threads or the manufacturer's type of threads are reamed into the wall of the main. The corporation stop also has a shutoff valve and outlet adaptor to accept the service pipe. A direct tap can be used on cast iron, ductile iron, asbestos cement, and high-pressure PVC pipe.

4-67 Service Clamps Service clamps are used when pipe wall thickness is not enough to allow at least three complete threads on the corporation stop. They are also used when working with smaller than 4-inch pipe or when the hole is larger than, say, 1 inch in a 6-inch pipe. They consist of a contour-fitting galvanized-iron fitting backed with a rubber, neoprene, or lead gasket and are clamped to the pipe with one or two steel straps. A threaded hole in the center will then accept the shutoff valve. These clamps will withstand pressures of 250 psi for one strap and 500 psi for two straps. See Fig. 4-56.

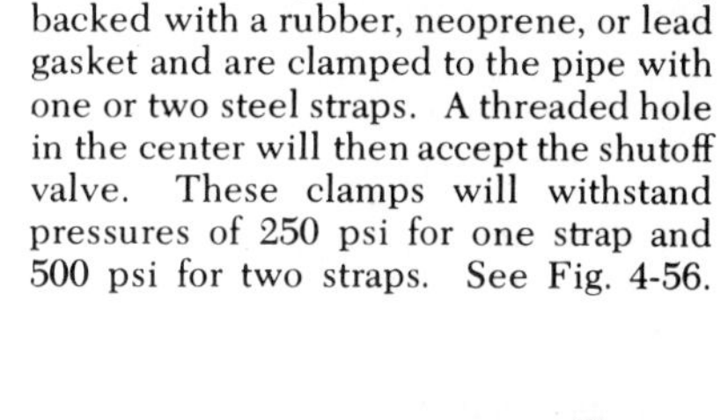

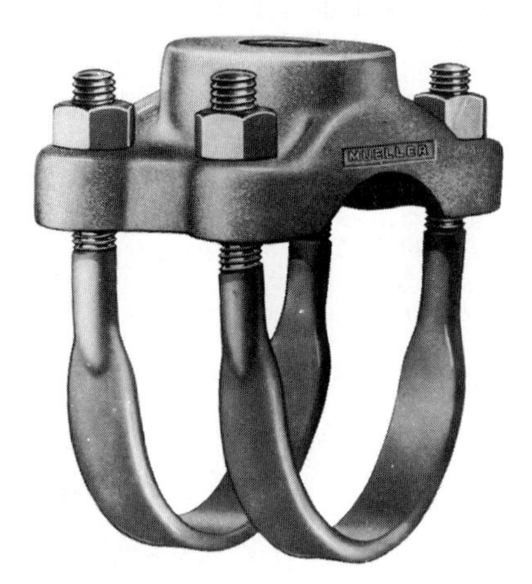

FIG. 4-56 Double-strap service clamp. (*Mueller Co.*)

FIG. 4-57 Typical expansion joint. (cast-iron or ductile-iron pipe). (*Dresser Manufacturing Division.*)

4-68 Expansion Joints Expansion joints are used when an abnormal amount of expansion or contraction will be concentrated at one point in a pipeline. The joint consists of outer core, slip pipe, follower bolts, and packing. The packing consists of alternate split rubber-compound rings (for sealing purposes) and split jute rings (for lubrication). When the bolts are tightened, the packing seals the pipe but allows it to slide. See Fig. 4-57.

Part 2 Pipe, Tube, and Hose Connections

JOSEPH F. BRIGGS
Manager, Service Engineering
Aeroquip Corporation, Jackson, Michigan

STANDARDS AND APPLICATIONS

4-69 Various Standards Various standards have been established for pipe, tube, and hose connections depending on the application in which they are being used. Since some of the terms and definitions may not be familiar to the reader, we have included the following list of standards in use today.

ASA—AMERICAN STANDARDS ASSOCIATION: This standard is still called out for pipe flanges and flanged fittings. ASA is being replaced with ANSI.

ANSI—AMERICAN NATIONAL STANDARDS INSTITUTE: The standards established by this organization are the official standards for the United States of America.

SAE—SOCIETY OF AUTOMOTIVE ENGINEERS: This organization publishes recommended standards for many tube, pipe, and hose connections used in industry today.

JIC—JOINT INDUSTRIAL COUNCIL: This group, which represented a cross section of American industry, established standards for fluid-power applications. Its standards are the same as SAE.

PTT—PARKER TRIPLE TYPE: This standard was adopted from the old AC (Air Corps) standard for aircraft. It is almost obsolete but can still be found on some diesel engines and in some mobile refrigerant systems in sizes 1 inch and 1¼ inch.

AN—AIR FORCE/NAVY AERONAUTICAL STANDARDS: These pertain to military standards. They also interchange with some of the SAE standards and/or JIC standards.

ISO—INTERNATIONAL STANDARDS ORGANIZATION: There are approximately 63 nations in this organization. Connections that carry the ISO standard are international in scope.

Although it is desirable to use the standards recommended by these groups, it is not mandatory, and there may be nonstandard connections that can be used. The most important thing to remember is that one should not intermix these standards.

4-70 Abbreviations Used in this Section

AN—Air Force/Navy Aeronautical Standards
ANSI—American National Standards Institute
ANSH—American National Standard Hose
ARP—Aerospace Recommended Practice
ASA—American Standards Association
ASME—American Society of Mechanical Engineers
ASTM—American Society for Testing and Materials
CAGI—Compressed Air and Gas Institute
CPV—Combination Pump Valve
FPTF—Fine Pipe Thread Fuel
ISO—International Standards Organization
JIC—Joint Industrial Council
Mil Spec—Military Specification
MS—Military Standard
NASA—National Aeronautics and Space Administration
NFPA—National Fluid Power Association
NPSI—National Pipe Straight Internal
NPSM—National Pipe Straight Mechanical
NPT—National Pipe Thread
NPTF—National Pipe Thread Fuel
PTT—Parker Triple Type
SAE—Society of Automotive Engineers
USAS—United States of America Standards
USASI—United States of America Standards Institute
USCG—United States Coast Guard

4-71 Application Areas The style of connection may depend on where it is to be used. A breakdown of the main areas of applications along with the recommended standards is as follows:

AUTOMOTIVE:
Systems: Coolant water, fuel, lubricating oil, air brakes, hydraulic brakes, hydraulic power steering, air conditioning, vacuum lines, gage lines, and hydraulics .... *Standards:* SAE, ANSI

GASEOUS:
Systems: Oxygen, compressed air, fuel gas, liquid propane gas, acetylene, and steam .... *Standards:* ANSI, CAGI

HYDRAULICS:
Systems: Petroleum oil, synthetic oil, and water .... *Standards:* SAE/JIC, ANSI

DOMESTIC AND INDUSTRIAL GENERAL SERVICE:
Systems: Garden hose, gasket seals .... *Standards:* SAE, ANSI, USAS

REFRIGERANT:
Systems: Air conditioning, freezers, refrigeration, industrial, mobile home .... *Standards:* SAE, ANSI

MARINE:
Systems: Sea water, fresh water, oil, fuel, hydraulics, air, etc .... *Standards:* SAE, AN, ANSI, ASA, MS, USCG

AEROSPACE:
Systems: Air, hydraulic, oil, oxygen, air conditioning, water injection, etc. .... *Standards:* SAE, ARP, AN, NASA, Mil Spec

4-72 Most Commonly Used Connections by Application

AUTOMOTIVE

Coolant water	Flexible and molded hose with band clamps.
Lubricating oil	Steel tubing and/or flexible hose with PTT, SAE 45-degree flare, SAE 37-degree flare, or SAE inverted flare to dryseal pipe threads or SAE O-ring boss.
Air brakes	Copper tubing, plastic tubing, and flexible hose with SAE inverted flared fittings to dryseal pipe threads and air-brake gladhands SAE J318.
Hydraulic power steering	Steel tubing and flexible hose with SAE inverted flared fittings or SAE 37-degree flared fittings to either dryseal pipe threads or SAE O-ring boss.
Hydraulic brakes	Steel tubing and flexible hose with SAE inverted flared fittings to dryseal pipe threads.
Air conditioning	Metal tubing and flexible hose with soldered or SAE 45-degree flared fittings to dryseal pipe threads. Self-sealing couplings tube-O.
Vacuum lines	Copper or flexible hose, with band clamps, SAE 45-degree or SAE compression to dryseal pipe threads.
Gage lines	Tubing or flexible hose with any of the above connections, depending on the system.
Hydraulics	Steel tubing or flexible hose with SAE 37-degree flared fittings, SAE flareless or SAE four-bolt split-flange fittings to dryseal pipe threads or SAE O-ring ports. Braze or weld fittings are also used.
Fuel	Tubing or flexible hose to SAE inverted, SAE 45-degree flare, SAE J520 (banjo), SAE J521a ferrule type, to NPTF pipe threads.

GASEOUS

Oxygen	American National Standard hose (ANSI B57.1), right-hand thread.
Acetylene	American National Standard hose (ANSI B57.1) left-hand thread.
General	ANSH and dryseal pipe thread (NPTF).

INDUSTRIAL HYDRAULICS	
Low-pressure return, drain, or transfer	Steel tubing or pipe and/or flexible hose with band clamps and pipe nipples, SAE 37-degree, SAE inverted, SAE four-bolt split-flange code 61, SAE adapter unions, or SAE flareless to SAE O ring, dry-seal pipe, or SAE four-bolt split-flange ports.
High pressure	Steel tubing or pipe with SAE 37-degree, SAE flareless, SAE inverted, SAE adapter unions, SAE four-bolt split-flange code 61 and code 62 to SAE O ring, NPTF dry-seal pipe threads, and both code 61 and code 62 four-bolt split-flange.
REFRIGERATION	
Home air conditioning and refrigeration	Flexible metal or annealed tubing with soldered SAE 45-degree or special self-sealing quick-disconnect couplings.
Mobile air conditioning and refrigeration	Flexible hose and/or rigid tubing with SAE 45-degree flare, PTT, NPTF, and self-sealing couplings. Soldered joints.
MARINE	
Sea water	Flexible hose or tubing with ASA-specified pipe flanges.
Fresh water	Flexible hose or tubing with ASA-specified pipe flanges.
Hydraulics	AN 37-degree flare, SAE four-bolt split-flange, NPTF dryseal pipe, Walseal, and CPV (Combination Pump Valve Co.).
Lubricating oil	SAE 45-degree, dryseal pipe and Walseal.
Fuel	SAE 45-degree, NPTF dryseal pipe thread, Aeroquip split clamp.
AEROSPACE	
All	MS flareless, Globeseal, Dynatube (Resistoflex Corp.), AN/MS flared fitting, Universal boss fitting, metal lip seal, Space Craft (Aeroquip Corp.) brazed fittings.

TYPES OF CONNECTIONS

4-73 SAE/JIC 37-Degree Flared Type (J514F) for Hydraulics, MIL-F-5509, MS33583 (Double Flare), MS33584 (Single Flare) The basic design of these fittings is derived from Air Force and Navy Standards (AN) for 37-degree flared fittings which meet a performance specification. Figure 4-58 depicts the standard three-piece flared tube fitting using 37-degree flared seat. The free-floating sleeve allows clearance between the style-B nut and the tube, permitting bends close to the fitting. Also, the sleeve aligns the flare so that it seats automatically, acts as a lock washer, supports the tube, dampens vibrations, and does not rotate during assembly so there is no twisting or wiping of the tube flare.

In preparing the tube for this type connection, the tube should be cut with a tube cutter and not a hacksaw. A hacksaw cut contaminates the tube with cuttings, does not insure a square cut, and leaves a torn edge. When flared, the tubing is more apt to

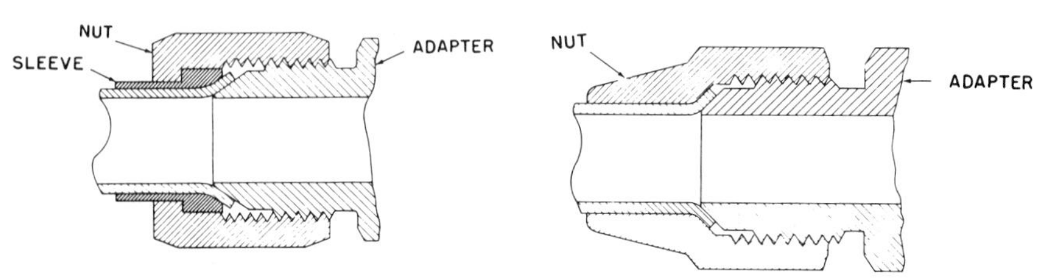

FIG. 4-58 Standard three-piece flared tube fittings.

FIG. 4-59 Standard two-piece flared tube fitting.

split and the cuttings become imbedded in the crooked flare seat. Furthermore, a hacksaw will work-harden the tube. Figure 4-59 depicts the two-piece type which consists of a style-A nut and fitting. The two-piece type is generally lower in cost. It cannot be used successfully with heavy-wall tubing. The same standard is used on flexible hose fittings for hydraulic applications.

4-74 Leak Problem Areas with SAE 37-Degree Connections *Causes:* Most of the leaks on this connection result from lack of tightening or human error. One cannot tell if the nut has been adequately tightened by just looking at the connection. Torque wrenches can be used but they are not always available. The installer must rely on memory to know if all the connections have been tightened.

Cures (for steel machined seats): Here is a foolproof method of tightening, by which anyone can tell if a joint has been tightened and how much.

1. Tighten nut finger-tight until it bottoms the seats.
2. Mark a line lengthwise on the nut and extend it onto the adapter. Use an ink pen or marker. (See Fig. 4-60.)

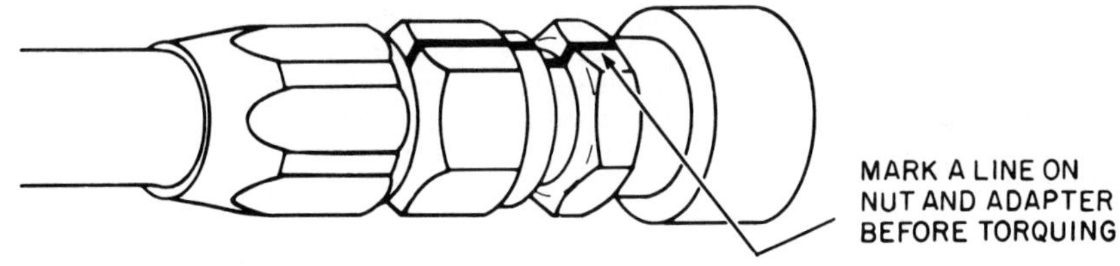
MARK A LINE ON
NUT AND ADAPTER
BEFORE TORQUING

FIG. 4-60 Pen mark lengthwise on hose connection.

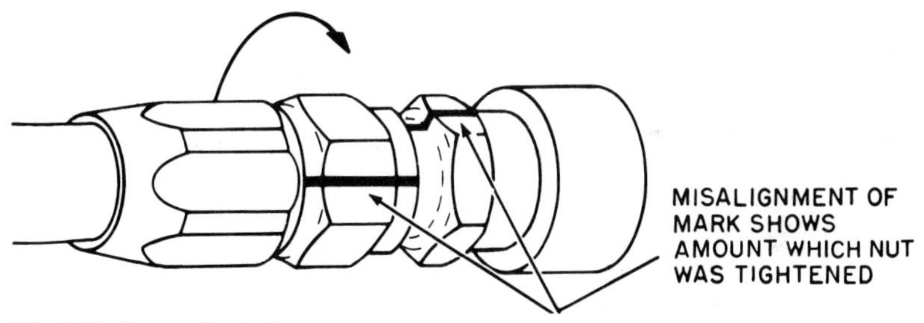
MISALIGNMENT OF
MARK SHOWS
AMOUNT WHICH NUT
WAS TIGHTENED

FIG. 4-61 Pen mark misalignment.

3. Using a wrench, rotate the nut to tighten (Fig. 4-61.) Turn the nut the amount shown in Table 4-13. The misalignment of the marks will show how much the nut has been tightened and, best of all, that it has been tightened.

TABLE 4-13 Nut Rotation

Line size		
Dash size*	Inches	Rotate number of hex flats
-4	$\frac{1}{4}$	$2\frac{1}{2}$
-5	$\frac{5}{16}$	$2\frac{1}{2}$
-6	$\frac{3}{8}$	2
-8	$\frac{1}{2}$	2
-10	$\frac{5}{8}$	$1\frac{1}{2} = 2$
-12	$\frac{3}{4}$	1
-16	1	$\frac{3}{4} = 1$
-20	$1\frac{1}{4}$	$\frac{3}{4} = 1$
-24	$1\frac{1}{2}$	$\frac{1}{2} = \frac{3}{4}$

* The size expressed in sixteenths of an inch.

What to Do if the Joint Leaks After It Has Been Tightened Properly. Disconnect the line and check for:

Condition	Remedy
1. Foreign particles in the joint	Wash them off
2. Cracked seats	Replace them
3. Seat mismatched or not concentric with threads	Replace the adapter
4. Deep nicks in the seats	Replace faulty part
5. Excessive seat impression. This indicates too soft a material for high pressures. Threads will stretch under high pressure	Replace the part
6. Phosphate treatment. This is an etching process which if overdone leaves a rough sandpaper-like surface	Replace faulty parts
7. Chatter or tool marks. High and low spots on seats	Replace faulty part
8. SAE 45-degree nuts—when connected to an SAE 37-degree male flare, fitting will leak. The SAE 45-degree nut is too long and will bottom on adapter hex in sizes 8 and 10 before the seats are tight	Use SAE 37-degree-flare parts exclusively.

4-75 SAE Flareless Hydraulic Connections (J514F), MS Flareless (ERMETO) The basic design of these fittings is derived from existing military standards MIL-F-18280, MS33514, and MS33515. In the flareless fitting, sealing is obtained by the nut forcing the sleeve into tight contact with the inside sealing diameter of the fitting body. The sleeve is retained on the tubing by a presetting operation during which the cutting edge of the sleeve is swaged onto the tube, cutting or embedding itself into the metal surface.

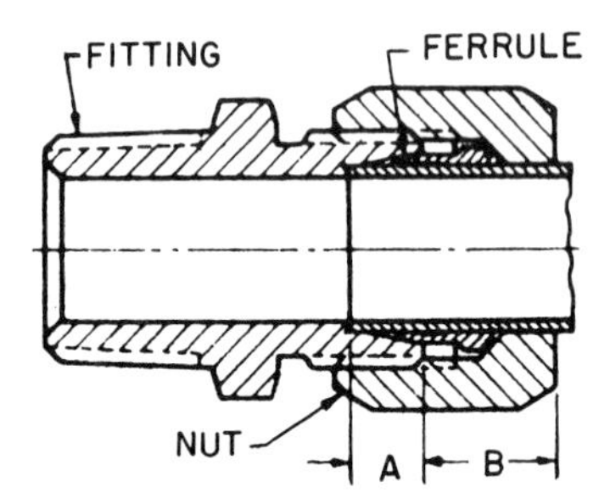

FIG. 4-62 Assembly with style A ferrule.

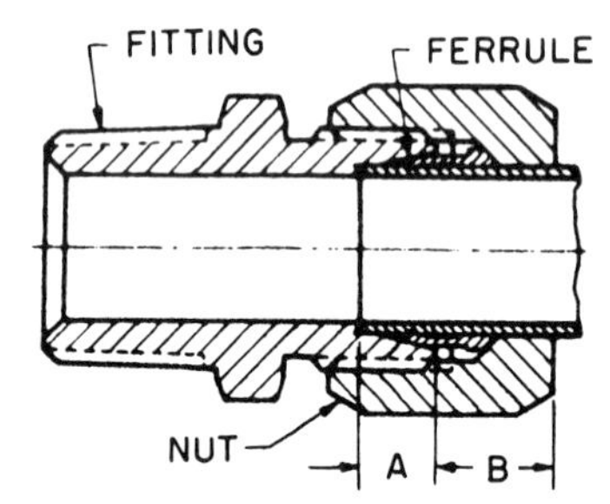

FIG. 4-63 Assembly with style B ferrule.

There are basically two standard styles in use today. The differences are in the ferrule. The assembly with the style A ferrule is shown in Fig. 4-62 and Fig. 4-63 depicts the assembly with the B-type ferrule.

4-76 Assembly Instructions for Hydraulic Flareless Tube Fittings The following instructions should be used to assure proper makeup of the fitting, which depends on the ferrule being secured to the tube by the cutting action of the ferrule into the tube.

1. Cut tube square and burr inside and outside corner (not excessively).

2. Assemble fitting by sliding nut over tubing with open end out. Slide ferrule on tubing with cutting edge out. The large head end should be inside the nut. Lubricate the ferrule and the threads on the body and nut with oil or petrolatum. Insert tube into fitting.

3. Bottom the tube in the fitting, and tighten the nut until the ferrule just grips the tube. With a little experience, the mechanic can determine this point by feel. If the fittings are bench-assembled, the gripping action can be determined by rotating the tube by hand as the nut is drawn down. When the tube can no longer be turned by hand, the ferrule has started to grip the tube.

4. After the ferrule grips the tube, tighten the nut one full turn. This may vary slightly with different tubing materials, but for general practice, a full turn is a good rule for the mechanic to follow.

5. The fittings can now be disassembled for inspection. The two styles of ferrules

differ somewhat in inspection even though the principles of makeup and application are similar. For the ferrule in Fig. 4-62, the bite or cut into the tube can be readily seen since it is on the lead edge of the ferrule. The bit into the tube should show a definite groove where the ferrule cuts into the tube and peels the metal over the lead edge of the ferrule (see Fig. 4-64 for further detail). For the ferrule in Fig. 4-63, the pilot at the end of the ferrule should be contacting or be within 0.0015 inch (0.038 mm) of the tube for hard material or not more than 0.005 inch (0.13 mm) on soft material (see Fig. 4-65 for further detail). This is an indication that the cutting edge has performed its function and has taken a secure bite in the tube. The sleeve should be slightly sprung or arched. For both styles of ferrules the rounded or lead edge should show a good seat in the fitting, and the head or shoulder end should be collapsed tight against the tube. The ferrule should have no end movement; however, the ferrule may be rotated on the tube as a result of the spring-back of the material. The performance of the fitting is not affected if the ferrule rotates.

In production, it may be preferable to use a threaded presetting tool to preform the ferrule onto the tubing. The presetting tool is a counterpart of the fitting hardened to provide good wearing properties for repeated usage. When the presetting tool is used, the assembly instructions are the same since the presetting tool takes the place of the fitting. Care should be taken to keep the cam surface of the presetting tool free of defects since they would transfer themselves to the ferrule, which would result in improper seating when the fitting is installed.

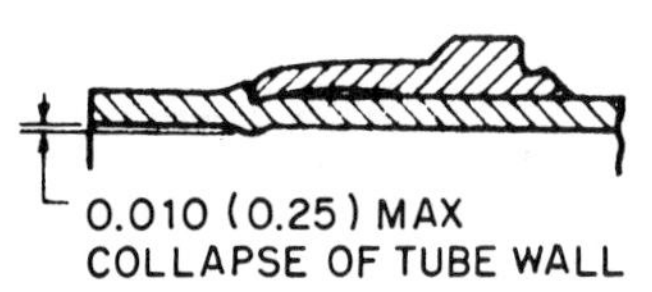

FIG. 4-64 Enlarged view of style A ferrule bite.

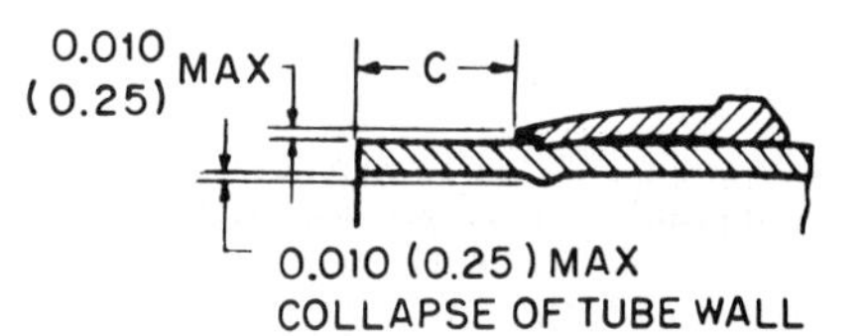

FIG. 4-65 Enlarged view of style B ferrule bite.

6. In some installations, it may be necessary to use a mandrel to support the inside of the tube when the ferrule is set. This is necessary only when the tube wall is so thin or so soft that it will not resist the biting action of the ferrule without collapsing. The mandrel in this instance supports the tube and allows the ferrule to bite into the tube without deforming or collapsing. Because the use of a mandrel allows very little give in the tubing, the setting of the ferrule may be made with slightly fewer turns than described above.

4-77 Reassembly Instructions for Flareless Fittings After disassembly of the fitting joint the flareless fitting can be reassembled by assembling the tube and ferrule into the socket of the fitting and threading the nut onto the fitting.

The operation of assembly up to the point at which the ferrule seats itself in the fitting can usually be accomplished by hand or with the use of a small wrench. If a wrench is required, only low torques are necessary to seat the ferrule.

When the ferrule is seated, an increase in the torque will be quite evident. When this point is reached, draw the nut up approximately $1/6$ of a turn minimum, but not more than $1/3$ of a turn, to complete the tightening operation.

Leak Problem Areas with SAE Flareless Connections. Cause: This fitting is sensitive to torque or tightening. It will leak if undertorqued. It will leak if overtorqued. Once it has been overtorqued, nothing can be done. It must be replaced with a new tube and sleeve. Cure: *Important*—follow the manufacturer's recommended assembly procedures and torque values very closely.

4-78 SAE Four-Bolt Split-Flange Connection (J518C) The SAE four-bolt split-flange connection (Fig. 4-66) is a face seal. The flanged head which contains the seal must fit squarely against the mating surface and be held there with even tension or all the bolts, which should be of grade 5 or better. The flanged head protrudes beyond

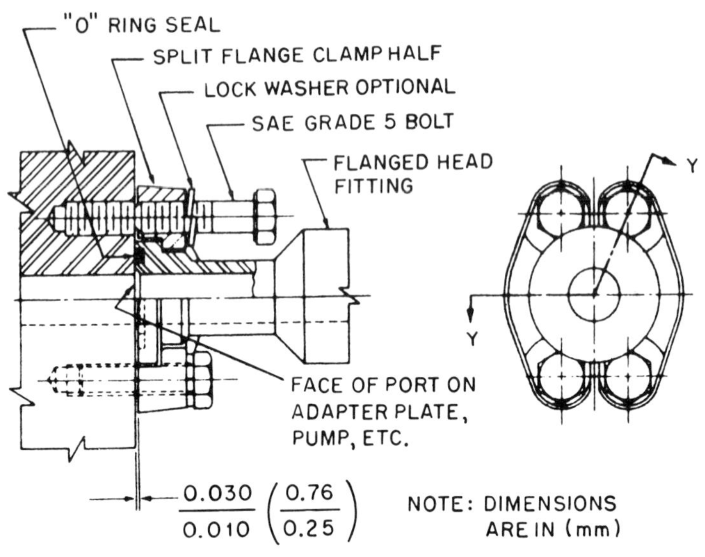

FIG. 4-66 SAE four-bolt split-flange connection.

the split-flange clamp halves from 0.010 inch (0.25 mm) to 0.030 inch (0.76 mm). This is to insure that the flanged head will be held against the face of the port under tension at all times (Fig. 4-67). It is recommended that a 90-durometer O ring be used for best results according to SAE J120 standards.

There are two series of split-flange connections used: code 61 and code 62. The pressure capabilities of the two are indicated in Table 4-14. They are *not* dimensionally interchangeable, therefore there is no possibility of mixing the two standards in the field.

The nominal size of the fitting can be determined by measuring the actual diameter of the flanged head and consulting Table 4-15.

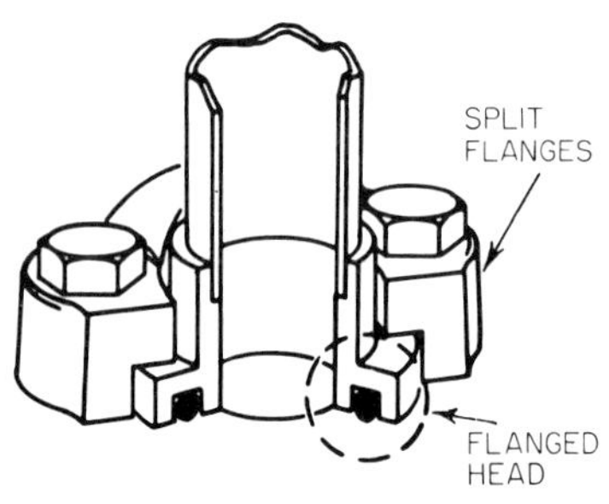

FIG. 4-67 Split flange and flanged head assembly.

4-79 Leak Problem Areas Causes: This connection is very sensitive to human error and bolt torquing. Because of the flanged head protrusion and the flange overhang, the flanges tend to tip up when the bolts are tightened on one end. This pulls the opposite end of the flange away from the flanged head and when hydraulic pressure is applied to the line, it pushes the flanged head into a cocked position (Fig. 4-68).

Cure: All bolts must be installed and torqued evenly. Finger-tightening with the use of feeler gauges will help to get the flanges and flanged head started squarely.

A second cause: When the full torque is applied to the bolts, the flanges often bend down until they bottom on the accessory. This also causes the bolts to bend outward (Fig. 4-69). Bending of the flanges and bolts tends to lift the flanges away from the flanged head in the center area between the long spacing of the bolts (Fig. 4-70). When this connection is used as a union the conditions become more severe because the spacing between mating flanges now is doubled and becomes a 0.020- to 0.060-inch gap. All conditions are now multiplied 100 percent. High torque is required on all bolts, which must be grade 5 or better, because much of the torque is lost in overcoming the bending of the flanges and bolts (Figs. 4-71 and 4-72).

TABLE 4-14 Pressure Capabilities of Split Flange Connections

Size, in.	Working pressure, psig	Size, in.	Working pressure, psig
STANDARD PRESSURE SERIES (CODE 61)			
		$2\frac{1}{2}$	2,500
$\frac{1}{2}$	5,000	3	2,000
$\frac{3}{4}$	5,000	$3\frac{1}{2}$	500
1	5,000	4	500
$1\frac{1}{4}$	4,000	5	500
$1\frac{1}{2}$	3,000		
2	3,000		
HIGH PRESSURE SERIES (CODE 62)			
$\frac{1}{2}$	6,000	$1\frac{1}{4}$	6,000
$\frac{3}{4}$	6,000	$1\frac{1}{2}$	6,000
1	6,000	2	6,000

TABLE 4-15 Flanged Head and Fitting Sizes

Actual diameter flanged head		Nominal size of fitting					
		Code 61			Code 62		
in.	mm	in.	sixteenths	mm	in.	sixteenths	mm
$1\frac{3}{16}$	30.18	$\frac{1}{2}$	-8				
$1\frac{1}{4}$	31.75				$\frac{1}{2}$	-8	13
$1\frac{1}{2}$	38.10	$\frac{3}{4}$	-12				
$1\frac{5}{8}$	41.28				$\frac{3}{4}$	-12	19
$1\frac{3}{4}$	44.45	1	-16				
$1\frac{7}{8}$	47.63				1	-16	25
2	50.80	$1\frac{1}{4}$	-20				
$2\frac{1}{8}$	53.98				$1\frac{1}{4}$	-20	32
$2\frac{3}{8}$	60.33	$1\frac{1}{2}$	-24				
$2\frac{1}{2}$	63.50				$1\frac{1}{2}$	-24	38
$2\frac{13}{16}$	71.42	2	-32	51			
$3\frac{1}{8}$	79.38				2	-32	51
$3\frac{5}{16}$	84.12	$2\frac{1}{2}$	-40	64			
4	101.60	3	-48	76			
$4\frac{1}{2}$	114.30	$3\frac{1}{2}$	-56	89			
5	127.00	4	-64	102			
6	152.40	5	-80	127			

Cure: Lubricate the O ring before assembly. All mating surfaces must be clean. All bolts must be evenly torqued. Don't tighten any one bolt fully before going to the next one (Table 4-16).

Because of the tolerance buildup in all component parts plus the bolt bending, the flange halves can move sideways in directions *A* and *B* in Fig. 4-73. This can lessen the flanged-head contact with the flanges to zero in the center area between the long bolt spacing. When flanges have a large radius on the edge *D*, the leakage problem becomes even greater with the above conditions (Fig. 4-74).

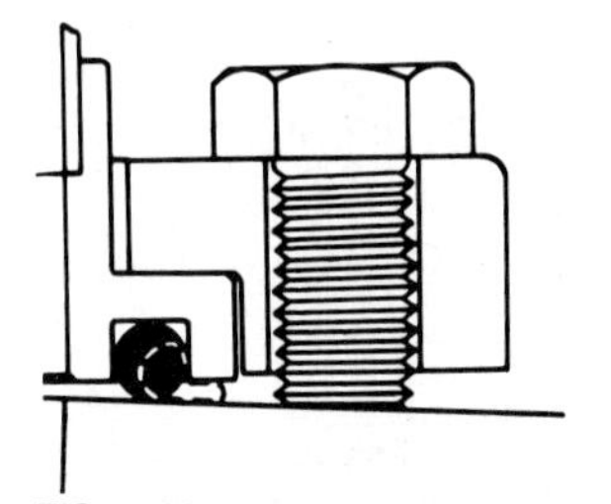

FIG. 4-68 Uneven bolt tightening produces this cocked position.

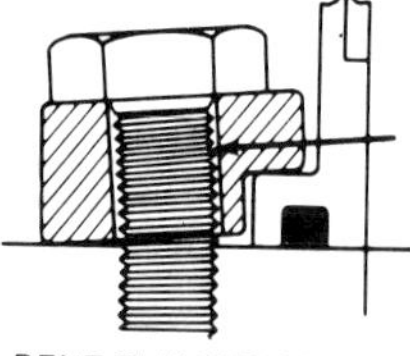

FIG. 4-69 Bolt bends outward by bent flange.

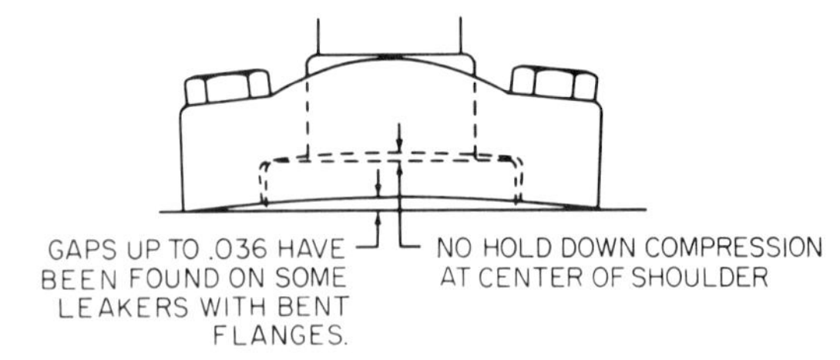

FIG. 4-70 Flange gap due to flange and bolt bending.

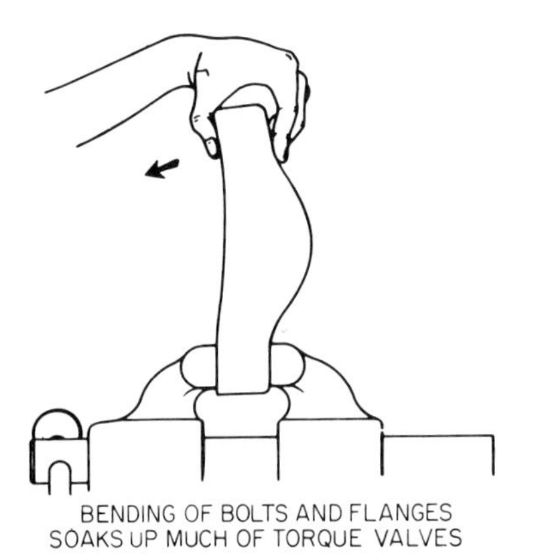

BENDING OF BOLTS AND FLANGES
SOAKS UP MUCH OF TORQUE VALVES

FIG. 4-71 Bending of bolts and flanges absorbs much of torque values.

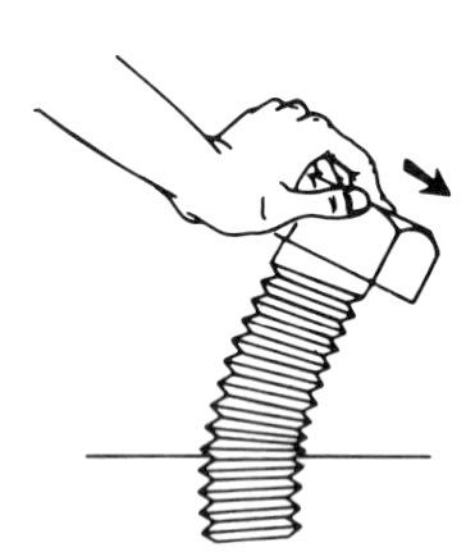

FIG. 4-72 Excessive bolt bending.

4-80 Pipe Threads All three basic types of pipe-thread fittings utilize the metal-to-metal seal in fluid-line connections:

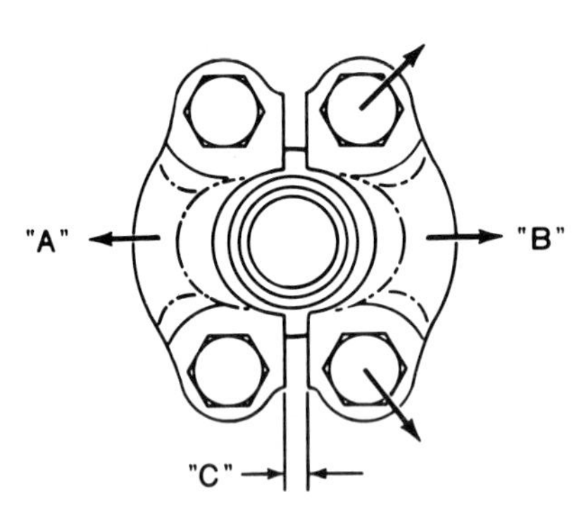

FIG. 4-73 Force direction due to tolerance buildup and bolt bending.

1. NPT (National Pipe Thread), a tapered thread connection (tapered $^1/_{16}$ inch per inch of length) which seals by thread-flank contact (Fig. 4-75), and is used for commercial low-pressure applications such as for water and air.

2. NPTF (National Pipe Thread Fuel), also known in the industry as the dryseal pipe thread. This is an SAE standard. This is also a tapered thread (tapered $^1/_{16}$ inch per inch of length), and seals by a destructive interference fit along the thread crest and thread flanks. Used for hydraulic applications, this thread has a much closer tolerance than the NPT and is less subject to leakage under higher pressures.

3. NPSM (National Pipe Straight Mechanical). This type of connection is an SAE adapter union. It does not seal on the threads. The seal takes place on a 30-degree chamfer which is machined on the inside diameter of the swivel pipe end. The function of the threads is to hold the pipe joint mechanically. The NPSM thread connection is used for both low- and high-pressure applications. It is used mainly on adapters to join male pipe hose lines to accessories (Fig. 4-75).

TABLE 4-16 Recommended Torque Values (Grade 5 Bolts)

Connection size, sixteenths of an inch	Torque, lb-ft	Connection size, sixteenths of an inch	Torque, lb-ft
-8	21	-24	90
-12	40	-32	90
-16	40	-40	90
-20	60	-48	175

NOTE: Air wrenches tend to cause flange tipping.

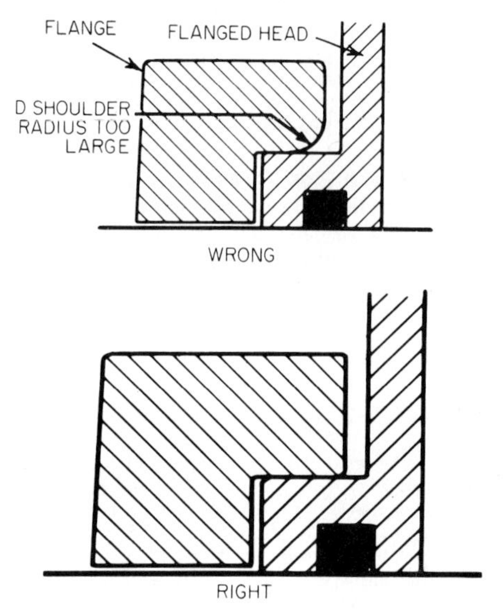

FIG. 4-74 Right and wrong flange design.

Pipe threads will perform adequately as permanent connections for many fluids, but they are not recommended for connections which must be connected and disconnected frequently. Each retightening draws the tapered threads closer together, making it more difficult to seal on the threads. Positioning and sealing become even more difficult in connections that require elbow fittings on each end, where one elbow must be adjusted to the proper angle.

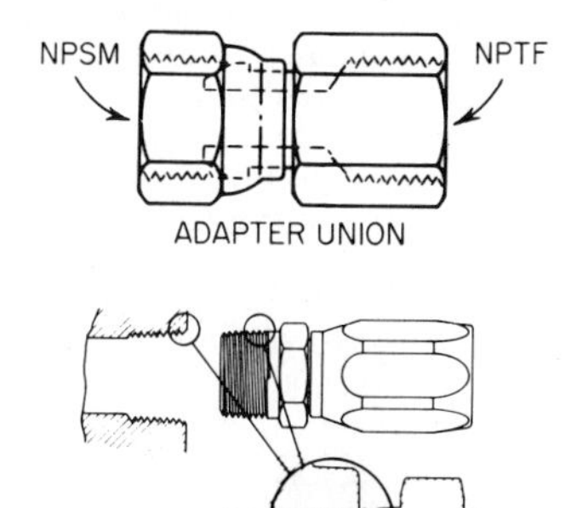

FIG. 4-75 NPSM thread connection.

Pipe threads tend to leak more at high pressures than any other style of connection. A sealing compound or pipe dope must be used on NPT threads, and should also be used on the NPTF threads, if they are reused, in order to fill voids between the male and female threads. When applying compound, be sure not to put any on the first two threads from the end of the fitting. In order to prevent the compound from being pushed into the system when the connection is made, always apply the compound to the male thread, never to the female thread.

Pipe and pipe thread size are of nominal dimensions. Whenever a number is shown following a pipe thread size as in ½-14, the 14 is the number of threads per inch. Since the number of threads per inch is standardized for each pipe size, it is not necessary to call out the number of threads. The pipe schedule number indicates the wall thickness. The pipe OD always remains the same regardless of the pipe schedule. Table 4-17 illustrates nominal pipe size (standard pipe), pipe OD, and threads per inch. Be sure to consult commercial pipe thread tables for specific dimensions based on pipe schedule numbers.

TABLE 4-17 Nominal Pipe Size, OD, and Threads per Inch

Nominal pipe size, in.	Pipe actual OD, in.	Threads per inch	Nominal pipe size, in.	Pipe actual OD, in.	Threads per inch
1/4	0.540	18	1 1/4	1.660	11 1/2
3/8	0.675	18	1 1/2	1.900	11 1/2
1/2	0.840	14	2	2.375	11 1/2
3/4	1.050	14	2 1/2	2.875	8
1	1.315	11 1/2	3	3.500	8

4-81 Leak Problem Areas with Pipe Threads Pipe threads tend to leak more at high pressure than any other style of connection. National standard pipe threads leak much more than the dryseal pipe thread. Either kind of pipe thread will leak if undertorqued or overtorqued. Use a good pipe dope on the NPT threads. When applying pipe dope do not put any on the first two threads from the end. Always put dope on the male thread—never on the female thread.

Cause of Leak	*Cure*
1. Connector not tight	Tighten
2. Cracked port or connector	Check for cracks and replace defective parts
3. Oversized threads in port	Inspect for proper thread size
4. Undersize threads on connector	Inspect for proper thread size
5. Galled threads (torn threads)	Inspect and replace if necessary
6. Damaged threads, nicks, cuts, etc.	Replace if damaged
7. Threads not dryseal standard for hydraulics	Use NPTF dryseal standard
8. Straight pipe threads instead of tapered	Use NPTF dryseal standard
9. Contaminated threads, dirt, chips, etc.	Clean and inspect
10. High vibration loosening connection	Retighten connector—check with engineering
11. Heat expansion of female threads	Retighten while hot
12. Too tight, causing thread distortion	Check and replace

4-82 SAE 45-Degree Flared Fittings SAE standards specify a flare angle for the tubing of 45 degrees from center line for nonhydraulic, automotive, and refrigeration series.

There are, however, other standards in use today (Fig. 4-76). They are (1) MS (*m*ilitary *s*tandard) 37-degree flare, which is the same as the SAE 37-degree flare but is used in military aircraft applications, and (2) PTT (*P*arker *t*riple *t*ype) 30-degree flare, which is the same as the obsolete AC811 (Air Corps) standard still used on some diesel-engine oil-line applications and mobile refrigeration systems.

Most leaks on the flared-tubing type of connection are caused by either poor flares or lack of tightening. Many of the leakage problems on this type of connection will not appear until the unit has had a few hours of service.

Several types of flare fittings are used for connecting hydraulic tubing. These are of either the single-flare or double-flare design. Standard tubing ends are formed in a

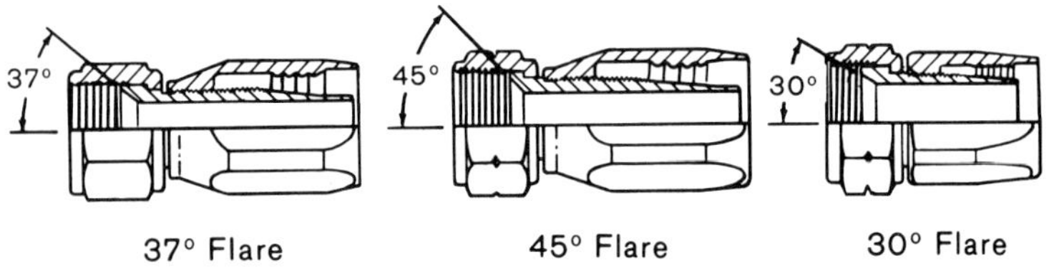

FIG. 4-76 Standard flared tubing fittings.

single flare. Thin-wall tubing, however, is doubled over to form the flare thus making a thicker flare for better retention (Fig. 4-77).

4-83 Spherical (Double-Compression) Tube Fittings, SAE J246a This style of tube fitting (Fig. 4-78) has been used for many years in automotive air-brake systems. It is designed to work with annealed copper tubing. It also may be used on heavy-wall plastic tubing along with an internal support. It is not recommended for use with steel tubing. *Caution:* Do not mix tapered-sleeve compression components with spherical-sleeve components where both standards are being used. Many manufacturers mark the spherical sleeve components with the words "air brake" for easy identification.

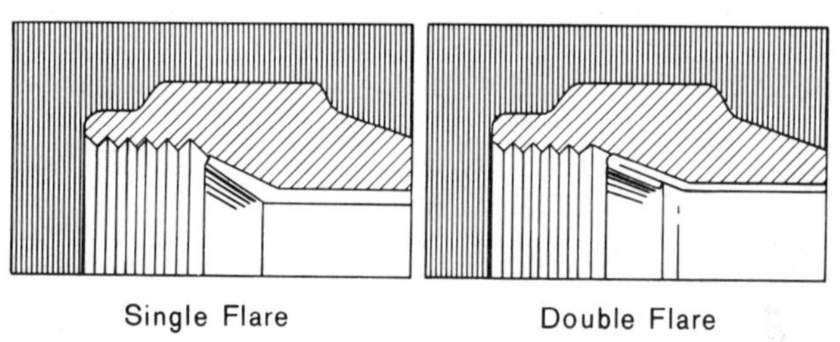

FIG. 4-77 Single- and double-flare fittings.

4-84 Tapered Sleeve (Compression-Type), SAE J512 This fitting is very similar to the spherical-sleeve compression-type fitting and there is a great risk of mixing up the two different standards when they are both being used in the same area (Figs. 4-79 and 4-80). The SAE J512 is not interchangeable with the SAE J246a standard.

This style of fitting has been listed by Underwriters Laboratories for hazardous liquids, fuel equipment, and gases. It meets the specifications of ANSI, ASME, and SAE. It may be used with copper, aluminum, and plastic tubing (soft plastic tubing requires an internal support). However, it is not recommended for use with steel tubing.

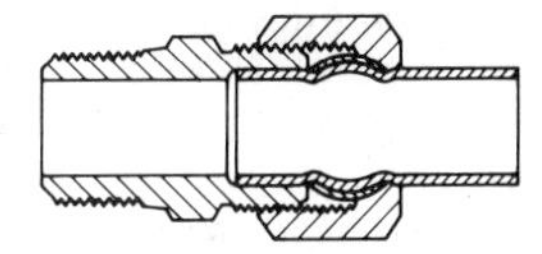

FIG. 4-78 Spherical (double-compression) tube fitting.

4-85 SAE Threaded-Sleeve Type This connection (Fig. 4-81) is used primarily in automotive applications where pressures are relatively low. It is a compression fitting, and sealing takes place when the nose of the male swivel nut is swaged on the tube as the two components are screwed together. It may be found in air systems and water systems using copper tubing or plastic tubing.

4-86 SAE Inverted Flared Fitting This style of fitting (Fig. 4-82) has been used for many years on automotive power-steering lines, fuel lines, oil lines, and hydraulic-brake lines. It is called inverted because it is an inversion or reverse order of the

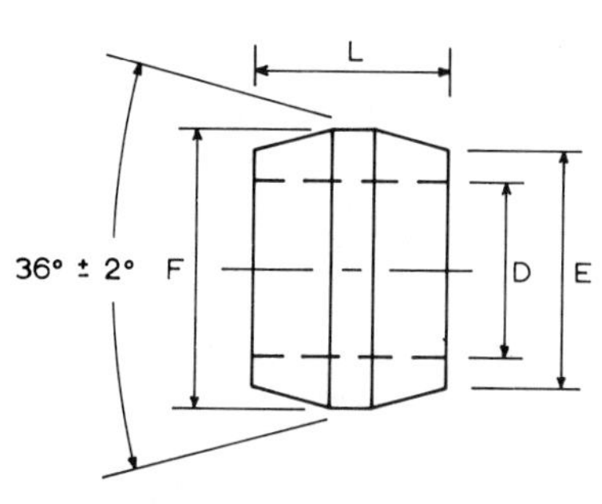

FIG. 4-79 Tapered sleeve.

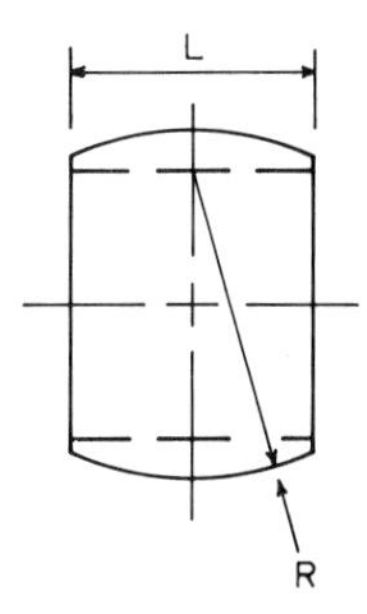

FIG. 4-80 Spherical sleeve.

standard SAE 45-degree flare. The swivel nut has male threads instead of female threads. The flared tube protrudes past the end of the swivel nut, and the adapter body is also a reverse design from the standard SAE 45-degree type. This connection is used in systems that generally do not exceed 1,500 psig operating pressures, but depending on the tubing that is used; it will withstand burst pressures up to 5,000 psi. It may be used with copper, brass, aluminum, plastic, and steel tubing that can be flared. It is listed by Underwriters Laboratories for hazardous liquids, fuel equipment, refrigerants, and gases.

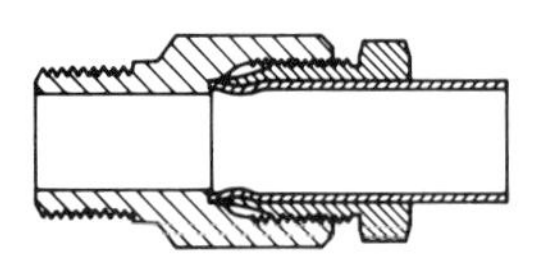

FIG. 4-81 Threaded sleeve.

4-87 Fuel-Supply Connections, SAE J520 This fitting was particularly designed for fuel connections, but a version of it may also be found in foreign hydraulic applications. In the vernacular of the trade it is often referred to as a "banjo" fitting. The design was taken from the AN standards. The fitting consists of: (1) an inlet union screw (Fig. 4-83), (2) an inlet union (Fig. 4-84), and (3) sealing gaskets (Fig. 4-85). The assembled fitting is shown in Fig. 4-86. The port or boss in the component part must be made specially to accept this style fitting.

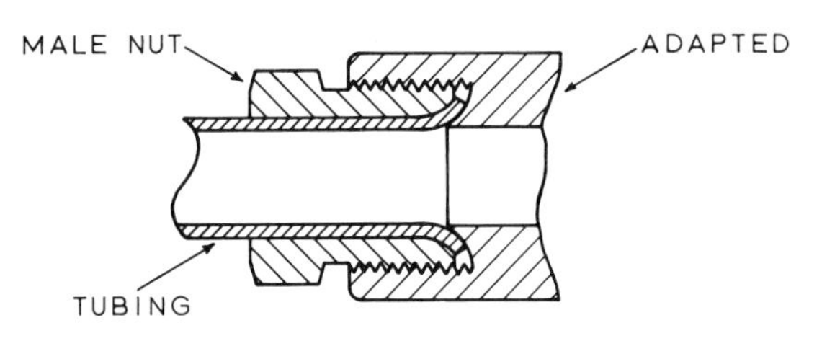

NOTE: INVERTED-FLARE FITTING HAS A 42° FLARE.

FIG. 4-82 Inverted flared fitting.

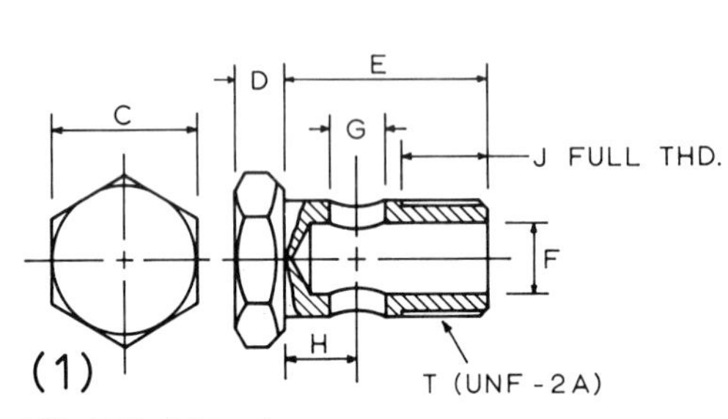

FIG. 4-83 Inlet union screw.

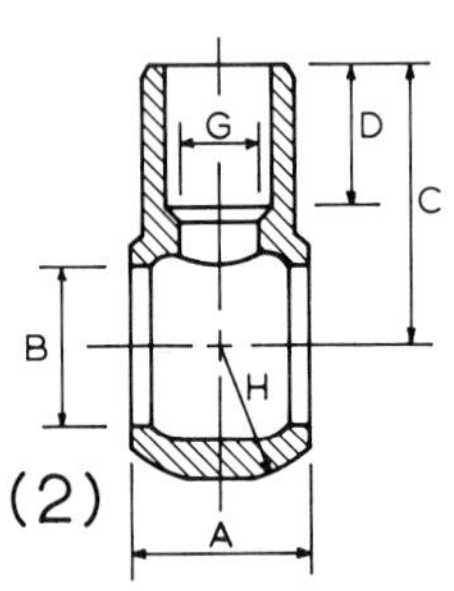

FIG. 4-84 Inlet union.

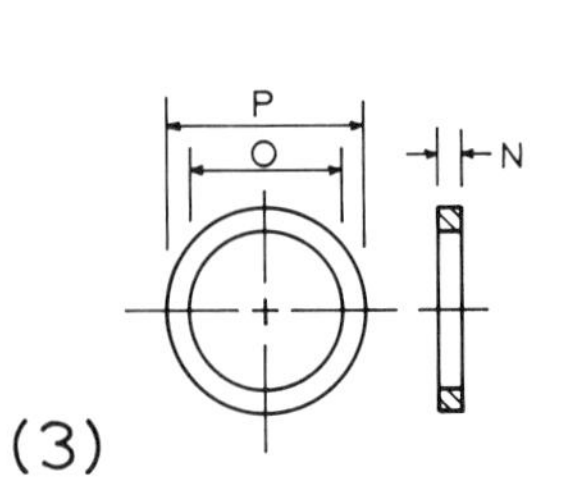

FIG. 4-85 Gasket.

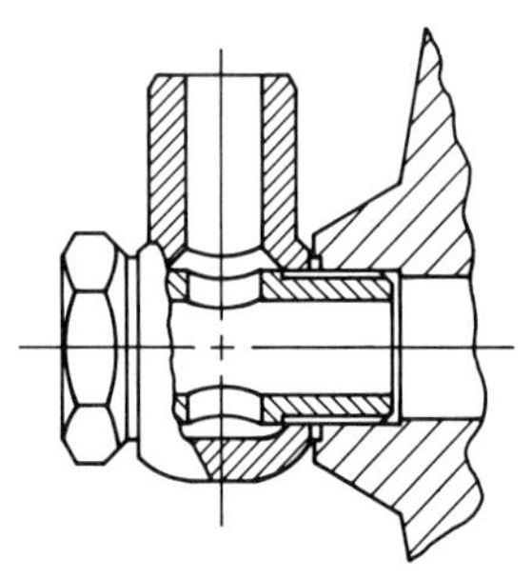

FIG. 4-86 Assembled fitting.

The fitting may be brazed onto the end of a tube, be an integral part of a hose fitting, or be made to accept a female SAE 45-degree flared fitting. One of the main advantages of this design is that because of its swivel feature it is easily positioned. Its flow characteristics are poor, however.

4-88 Types of Fuel-Injection-Tubing Connections, SAE J521a There are two basic types of connections used in fuel-injection applications. They are the ferrule type (Fig. 4-87) which consists of a loose ferrule which is slipped over the high-pressure tubing and compressed between the seat in the pump discharge or nozzle inlet fitting and the union nut, causing the sharp edge on the ferrule to bite into the tube wall. To assure a rigid, leakproof joint, the tube end must be bottomed in the counter bore of the mating seat before the nut is tightened.

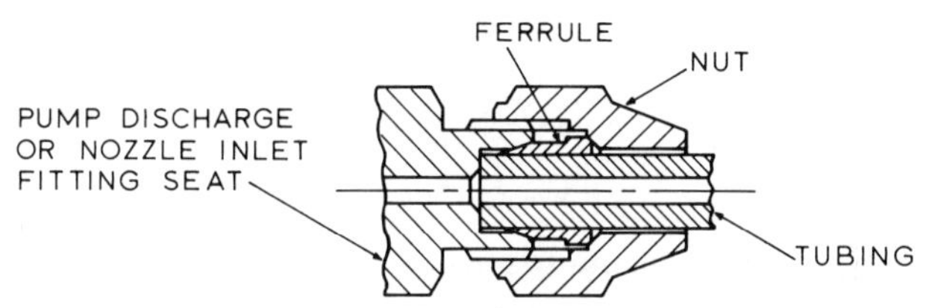

FIG. 4-87 Ferrule type (injection).

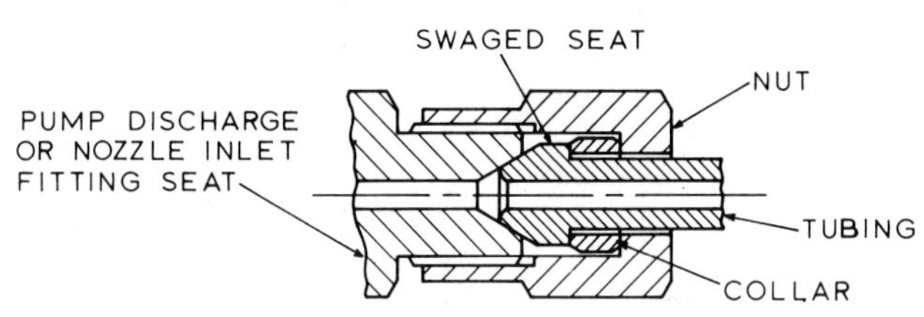

FIG. 4-88 Style A (injection).

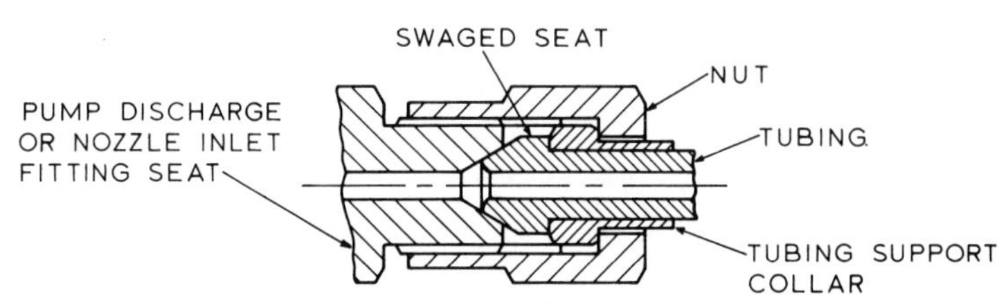

FIG. 4-89 Style B (injection).

The other style fitting requires special tooling because, after the union nut and supporting sleeve are installed on the high-pressure tube, the end of the tube is deformed or swaged to a 60-degree-included male cone seat. The end of the tube then requires reaming and deburring.

There are two different nut and sleeve combinations available for this swaged tube design. The style B (Fig. 4-89) offers more tube support than does the style A (Fig. 4-88).

4-89 Aeroquip Globeseal Connection The Aeroquip Globeseal connection is intended to mate with the SAE hydraulic flareless fittings. This connection is made up of a hollow globe-shaped nose machined on a hose nipple assembly (Fig. 4-90). It seals by the deformation of the female globe inside the female recessed-cone seat.

This design is extremely tolerant of accidental overtightening since the sealing surface is a solid, integral part of the nipple. This fitting is used in hydraulic, fuel, and oil applications, primarily in the aerospace industry where high-frequency, high- and low-amplitude vibrations place high fatigue loads on the plumbing system.

4-90 CPV (Combination Pump Valve) O-ring Seal Fitting The CPV O-ring seal fitting (a Navy standard) uses an O ring which is inserted in the packing-gland recess on the face of the union which has been silver-brazed to the end of a pipe (Fig. 4-91). The union and pipe are sometimes called a "tailpiece."

4-91 Walseal The Walseal connection is the forerunner of the CPV O-ring seal fitting. The Walseal utilizes a square compression gasket for sealing (Fig. 4-92). The

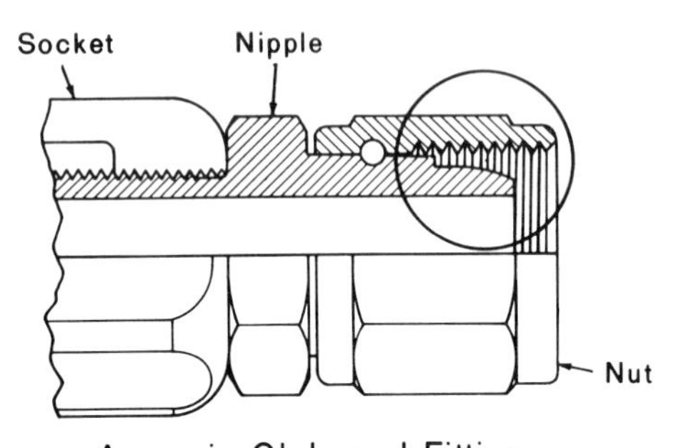

FIG. 4-90 Aeorquip Globeseal fitting.

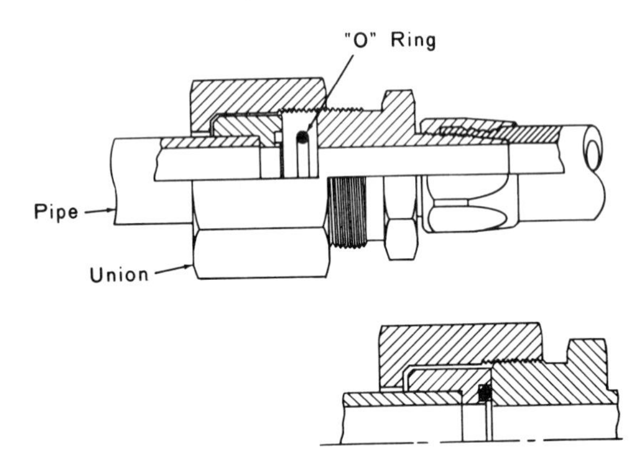

FIG. 4-91 CPV O-ring seal fitting.

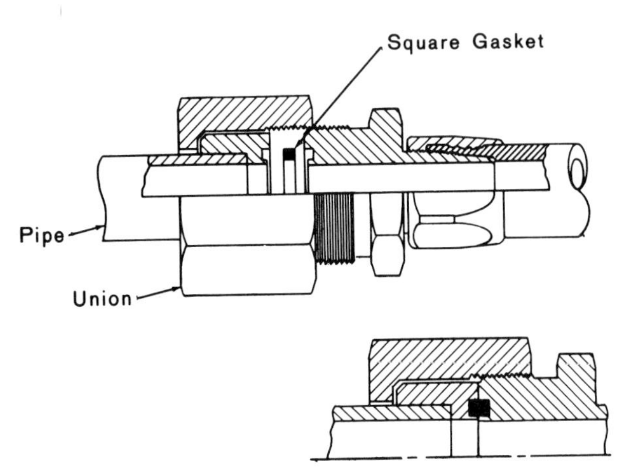

FIG. 4-92 Walseal (square-gasket type).

tailpiece is silver-brazed onto the end of the pipe after the union nut has been slid over the pipe. The square gasket is inserted in the packing gland and the union nut threaded onto the Walseal fitting until the fitting bottoms against the tailpiece. The gasket fits into both halves of the Walseal fitting to form the packing gland. A slight gap at the innermost part of the gland permits excess rubber to cold-flow out of the joint. The Walseal design is a Navy standard. It is rapidly being replaced by the CPV O-ring seal fitting.

4-92 Band Clamp The band clamp works on a compresion-sealing principle. This is a low-pressure connection. It has been in use for many years and may be the oldest type of piping-hose seal. It consists of a band clamp tightened over a rubber hose which has been inserted over a beaded nipple (Fig. 4-93). Whenever leakage

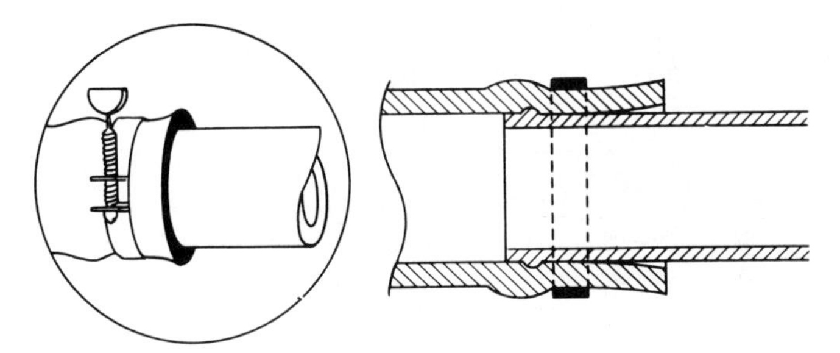

FIG. 4-93 Band clamp.

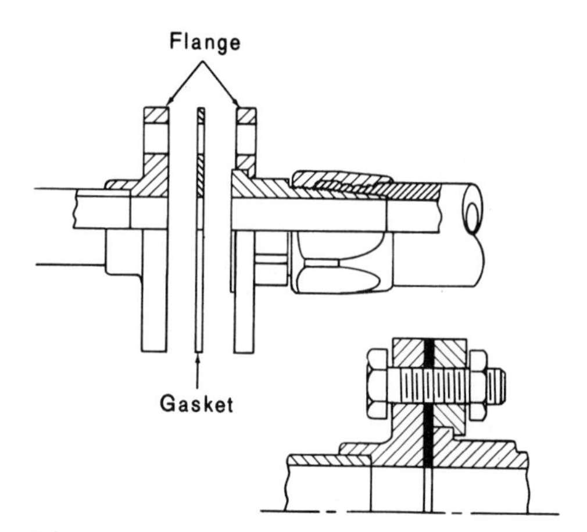

FIG. 4-94 Pipe flanges using flat gasket.

occurs, the clamp is simply tightened further. Spring-loaded clamps eliminate tightening to compensate for rubber cold-flow.

Applications range from automotive radiator lines on trucks, tractors, and construction equipment to suction and return lines in some hydraulic systems.

4-93 Pipe Flanges Many pipe flanges employ a flat gasket as a compression seal (Fig. 4-94). The are used for large sizes of pipe, particularly in medium- and low-pressure in-plant, railroad, and marine applications. The type of gasket depends on the fluid carried and the amount of pressure in the system.

Pipe flanges must be tightened evenly on all bolts just as SAE O-ring split flanges must be. If the connection is cocked, leakage will result regardless of the amount of further tightening.

There are several standards of pipe flanges for various pressures and sizes. For example, the ASA B16.5 150-lb flange is used in both medium- and high-pressure ranges and is available with either four or eight bolts, depending on application. The same is true with the B-176 (MIL-F-20024) flange, which uses anywhere from three to twelve bolts depending on size and pressure. Flat gaskets are not normally used in high hydraulic pressures with pipe flange connections, but a thin gasket is sometimes used in high hydraulic pressures between highly finished machined surfaces of accessories.

4-94 Marman Flexmaster Joint Two lengths of pipe can be joined together easily without leakage by the Flexmaster compression joint-sealing method. The pipe fits into the sleeve, while the synthetic rubber gasket creates the seal around the pipe. The gasket is compressed against the pipe and the sleeve by tightening the V-band coupling (Fig. 4-95*a*).

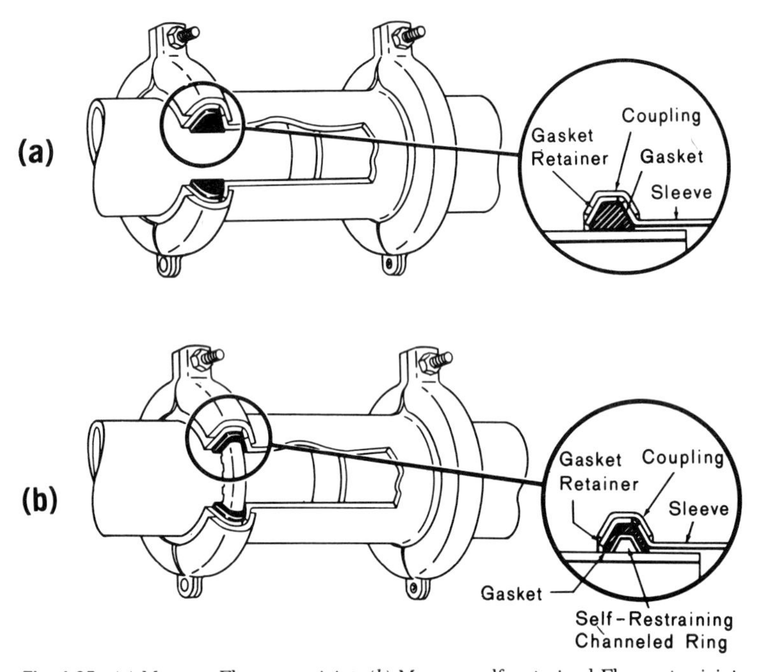

Fig. 4-95 (*a*) Marman Flexmaster joint, (*b*) Marman self-restrained Flexmaster joint.

4-95 Marman Self-Restrained Flexmaster Joint Like the Flexmaster joint, the Self-Restrained Flexmaster joint is also used to connect two pieces of tube or pipe. The Self-Restrained Flexmaster joint, however, has a notched steel ring embedded in the seal which grips the pipe or tube to form a self-restrained joint which will not slip along the pipe in case of vibration or pressure (Fig. 4-95*b*). As the V-band couplings on each end of the joint are tightened, pressure is exerted inward on the resilient gasket which in turn puts pressure on the notched stainless-steel ring beneath it. The rubber gasket seals against the pipe around the restraining ring.

4-96 Dresser Couplings The Dresser coupling provides a fast, simple way to joint two pipes or tubes together. This connection provides for up to 6-degree pipe or tube deflection while still maintaining a permanently sealed joint. It is available in a threaded union design as shown in Fig. 4-96 or a tie bolt design as depicted in Fig. 4-97.

4-97 Marman Conomaster Pipe Joint The Marman Conomaster pipe joint consists of a pair of identical welding neck flanges, a reversible metal gasket, two identical clamp halves, four bolts, and eight nuts.

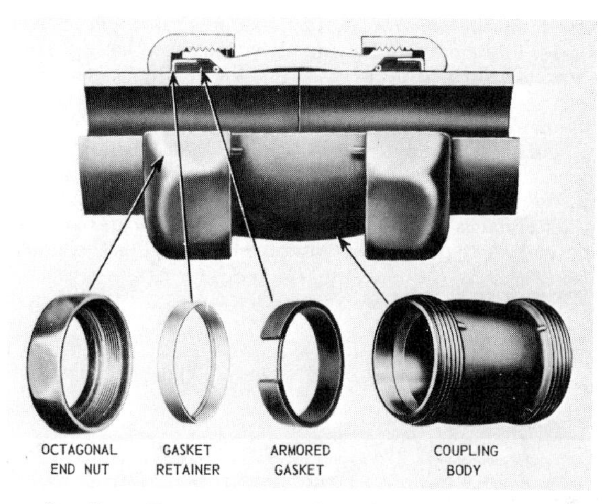

FIG. 4-96 Dresser Coupling. Cutaway section of a style 65 coupling. The threaded end nuts compress the gasket to make a flexible, permanently tight seal. *(Dresser Mfg. Co.)*

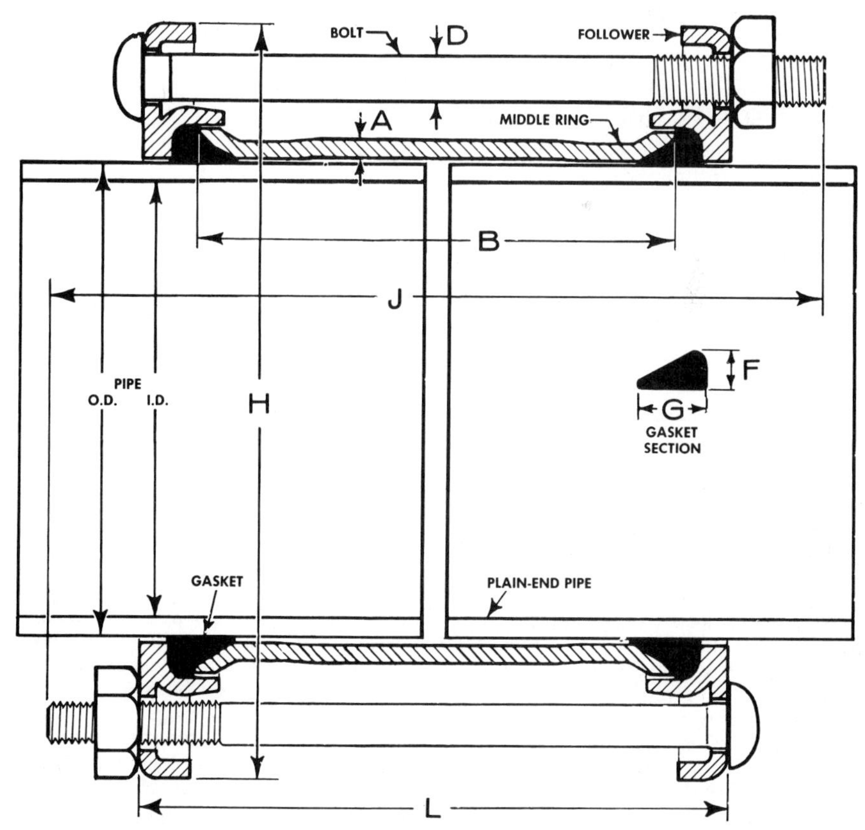

FIG. 4-97 Cross section of Dresser Coupling. Tie bolt design is available from ⅜ inch ID to 72 inch OD or larger. *(Dresser Mfg. Co.)*

Sealing is based on the frustroconically shaped gasket which rests between the mating flanges prior to clamping. During clamping these flanges are moved together axially and the gasket is loaded radially against the sealing surface of the flanges (Fig. 4-98).

Plastic flow is mechanically induced at the sealing edges of the gasket, creating a coined surface. This plastic flow insures intimate intermeshing of the surfaces of the gasket and flanges at the sealing area, providing a full circumferential metal-to-metal radial seal. This is done without the gasket brinelling the surface area of the flanges because the gasket metal is always softer than the flange material. This material condition is essential to induce plastic-flow intermeshing. When the flanges are closed the

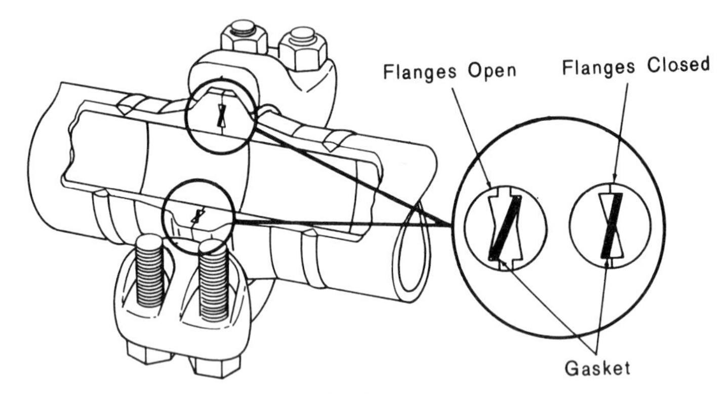

FIG. 4-98 Marman Conomaster pipe joint.

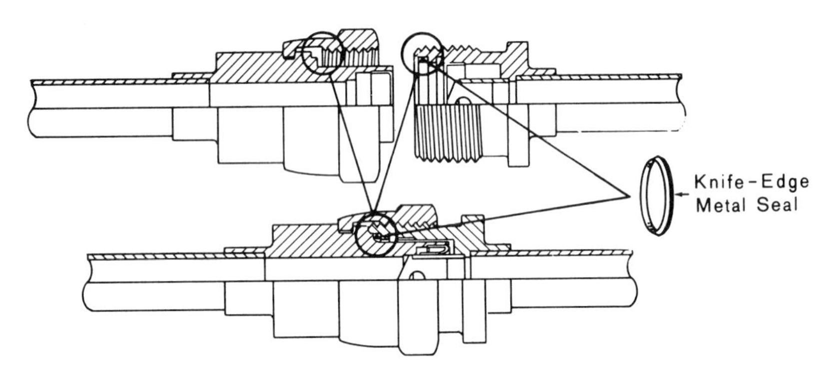

FIG. 4-99 Knife-edge metal seal.

radial column load on the gasket provides a spring effect. This spring action assures that radial contact is maintained when the flanges move because of thermal cycling.

This joint is especially suited for high-pressure steam systems, chemical processing, marine systems, petroleum and gas production and processing, and aerospace ground-support systems.

4-98 Knife-Edge Metal Seal The knife-edge metal seal is used in some varieties of refrigerant self-sealing couplings. When the two coupling halves are connected, the sharp sides of the metal ring bite into the soft-brass coupling bodies (Fig. 4-99), establishing an excellent seal to prevent the loss of refrigerant.

4-99 Welded and Brazed Joints Welded and brazed joints are intended to be permanent joints. Arc-welded joints can be butt-welded. When a brazed joint is produced using a soft material such as copper or silver, overlapping of the joined parts (Fig. 4-100) is required.

The Aeroquip Space-Craft Method of joining tubular parts by induction brazing produces permanent, lightweight, leakproof tubular systems (Fig. 4-101). In spite of the permanent nature of such joints, Space-Craft repair tooling permits modification of the systems to replace components or accessories.

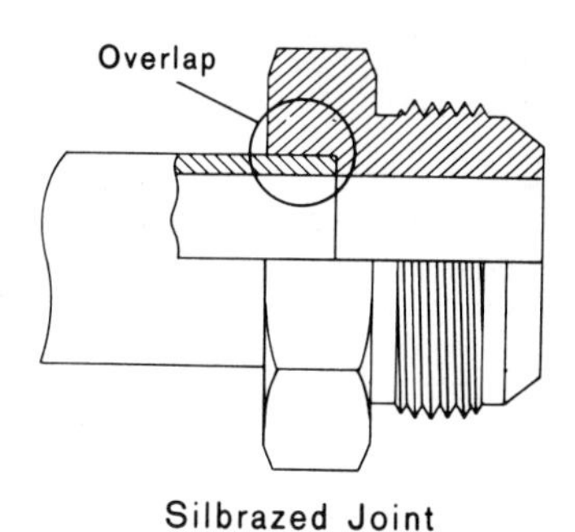

FIG. 4-100 Overlapping of joined parts.

4-100 Aeroquip Split-Clamp O-ring Seals This is a bore-seal connection (Fig. 4-102). It has an O-ring seal with a backup ring. The backup ring serves to cushion the O ring to prevent damage during pressure impulse. The backup ring is located on the low pressure side of the O-ring seal.

The O ring and backup ring are inserted in the tailpiece after it has been silver-brazed to the end of a pipe. The mating half is then slid into the tailpiece. This provides an initial squeeze on the O-ring seal. Then the clamp assembly is attached to hold the halves together. A band clamp is used around the outside of the split clamp to keep the halves together. All of the stress is taken up by the split clamp halves, not by the band clamp which holds them together.

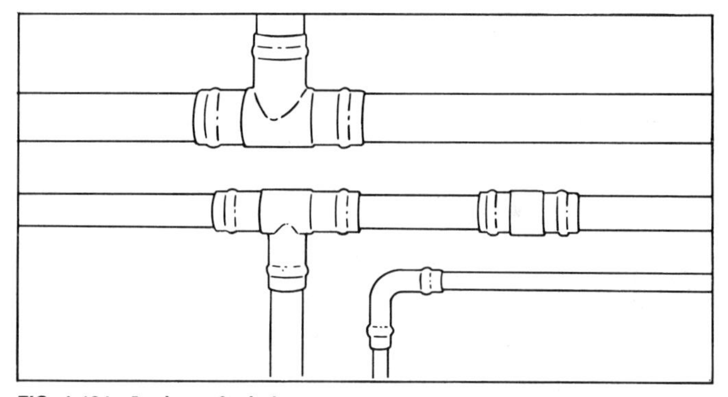

FIG. 4-101 Leakproof tubular system.

4-101 Marman Conoseal Joint The Marman Conoseal pipe and tube joint consists of a male and female flange and a frustroconical-shaped metal gasket (Fig. 4-103).

The gasket is installed between the mating flanges before the coupling has been torqued. The flanges are then compressed together by the V retainer's wedging action and the gasket is completely compressed. The mechanical leverage advantages created by both the V retainer principle and the Conoseal design induce a plastic-flow condition on the sealing edges of the gasket. This insures a leakproof metal-to-metal seal against extremely high pressures and/or temperatures and is excellent for highly corrosive conditions.

4-102 Marman Conoseal Union Fitting The sealing principle described above is also used in the Marman Cono-

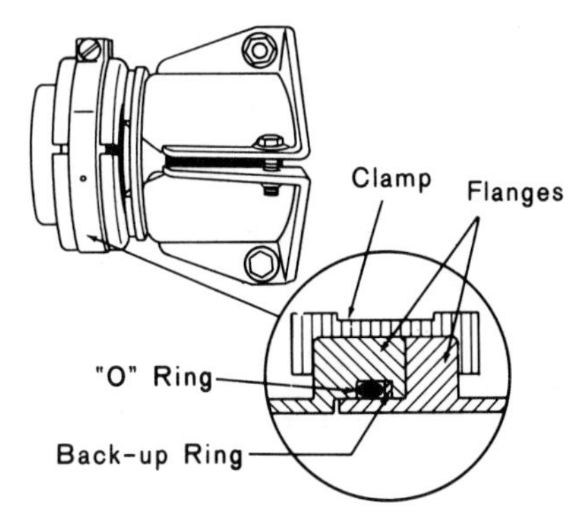

FIG. 4-102 Aeorquip spit-clamp O-ring seals.

seal union fitting. Where a V retainer coupling is used to compress the flange surfaces, a threaded union connection performs the same function here (Fig. 4-104). This fitting can be used for joining critical pneumatic, hydraulic, or fuel lines, cooling and heat transfer lines, and pressure or vacuum lines. It can be reused simply by replacing the metal gasket.

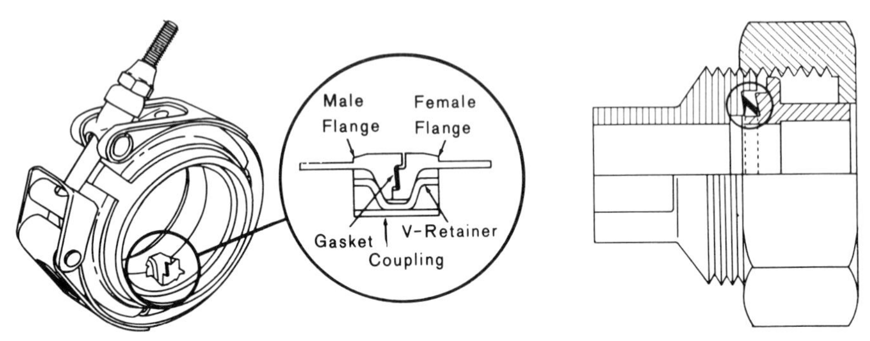

FIG. 4-103 Marman Conoseal joint.

FIG. 4-104 Marman Conoseal union fitting.

4-103 V-Band Couplings A V-band coupling is used for connecting or sealing tubing, pipe, containers, filters, regulators, or similar devices or closures where swing bolts and nuts or bolted flanges have been previously required. It is frequently used to seal high-pressure air ducting. By applying torque to the nut of the coupling, an inward radial force is created in the V retainer which wedges the joint flanges together (Fig. 4-105). Figure 4-106 shows three types of flanges used: (1) sheet metal flange with O ring; (2) formed flange with flat gasket, and (3) machined flange with O ring.

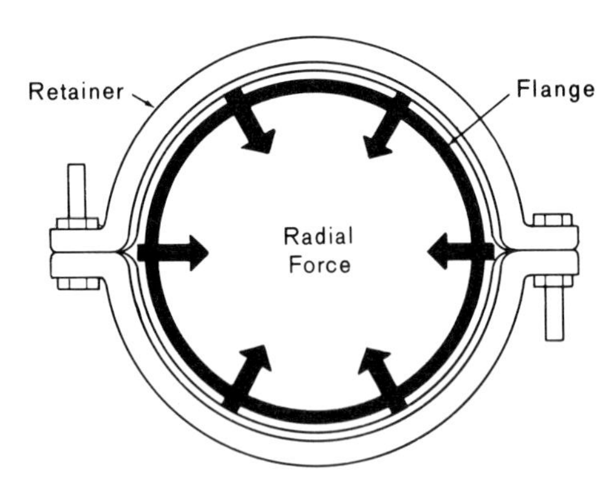

FIG. 4-105 V-band coupling radial force.

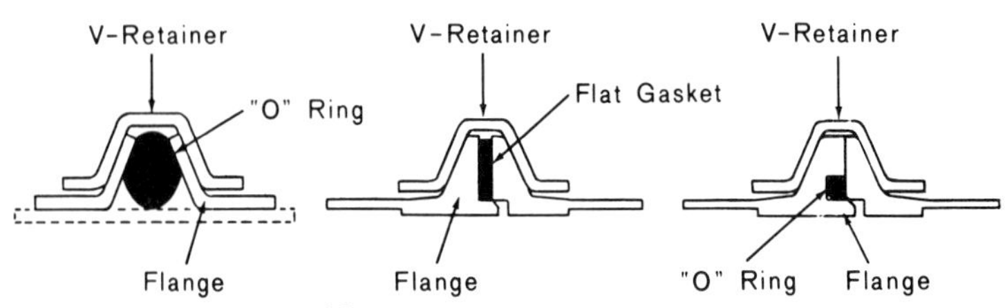

FIG. 4-106 Three types of flanges.

4-104 Self-Sealing Couplings (Quick-Disconnects) The convenience of using a self-sealing coupling in a fluid system can be compared to the convenience of an electric plug which connects into a wall socket for the operation of electrical appliances. Basically, a self-sealing coupling is a connection that seals off both halves when disconnected. It may be used on liquid or pneumatic lines for service or changing accessories, permitting quick disconnection and reconnection without loss of fluid. The ANSI symbol for a self-sealing coupling is shown in Fig. 4-107. A cross section appears in Fig. 4-108.

There are two basic methods of attachment: push-pull and screw-together. Some designs provide a breakaway feature so that a tug on the attaching line will separate the coupling. Arrangements can also be made to operate the separation of the coupling halves electrically, hydraulically, or pneumatically. These couplings are also used for connecting refrigerant lines on air-conditioning systems.

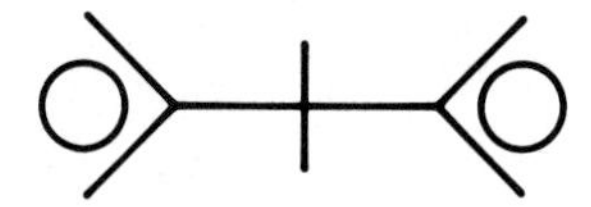

Fig. 4-107 ANSI symbol for a self-sealing coupling.

When ordering this type of connection important factors to consider are:

1. Type of fluid plus volumetric flow.
2. Operating temperatures, internal and ambient.
3. Permissible fluid loss upon disconnect, if any.
4. Breakaway method, if required.
5. Preferable method of attachment.
6. Space requirements.
7. Permissible resistance to flow through coupling in psig.
8. Service requirements.

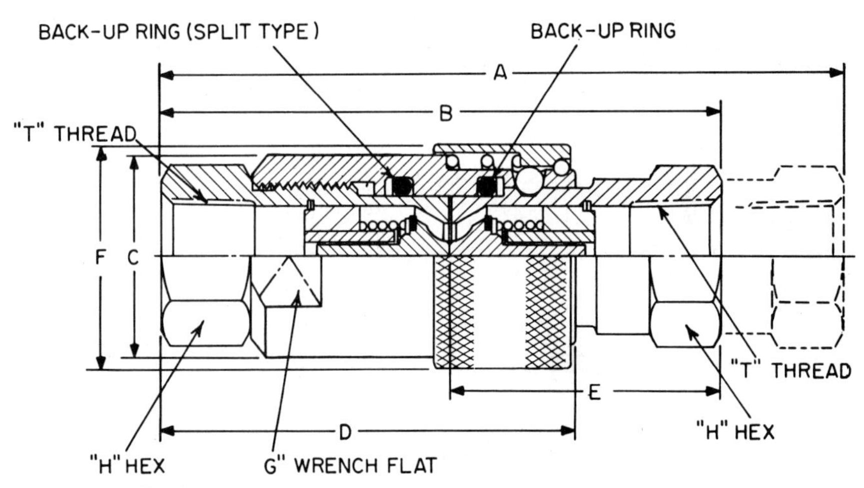

Fig. 4-108 Self-sealing coupling (quick-disconnect).

4-105 Straight-Thread O-ring Boss This connection uses an O ring to seal the high-pressure fluid. The port opening in the hydraulic component has a prominent chamfer machined into its opening. When the connection is properly made, the O ring is encapsulated between the adapter body and the chamfered area (Figs. 4-109 to 4-111). This connection is usually found on the end of the adapter that screws into the hydraulic component. The other end of the adapter would normally be SAE-JIC 37-degree flare or SAE flareless for connection to tubing or flexible hose. Many manufacturers also include an NPSM-to-straight-thread O ring (Fig. 4-112).

4-106 Assembly Instructions For best results it is improtant to carefully follow the assembly instructions for adjustable fittings as listed by SAE:

1. Lubricate the O ring by coating with a light oil or petrolatum and install it in the groove adjacent to the face of the metal backup washer, which should be assembled at the extreme end of the groove away from the port. The jam nut should be backed off also.
2. Install the fitting in the port until the metal backup washer just contacts the face of the port.
3. The fitting may be positioned by turning the fitting out of the port (counterclockwise) up to one full turn. Using two wrenches, tighten the jam nut while holding the adapter body with the other wrench.

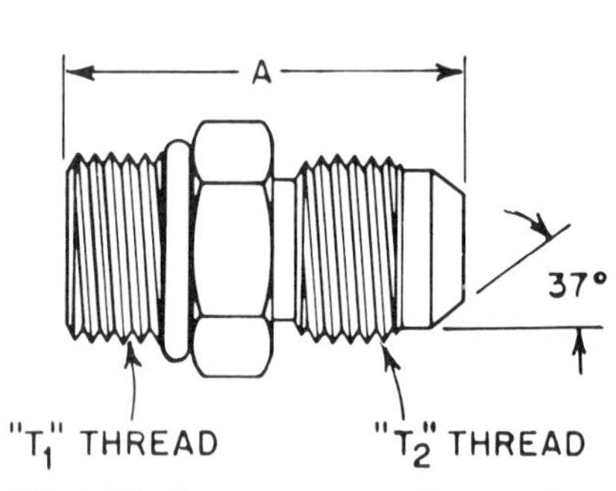

FIG. 4-109 O ring JIC-SAE 37-degree male.

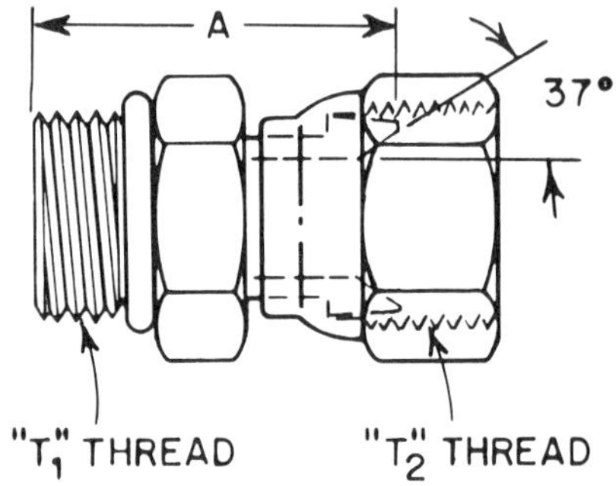

FIG. 4-110 O ring JIC-SAE 37-degree female.

Some of the problems that one may experience with this connection are as follows:

- Elbows loosen up after short service
- O-ring leakage after short service
- O-ring leakage after long service
- Instant leakage upon start up

The cause of these problems may be either human error or faulty parts.

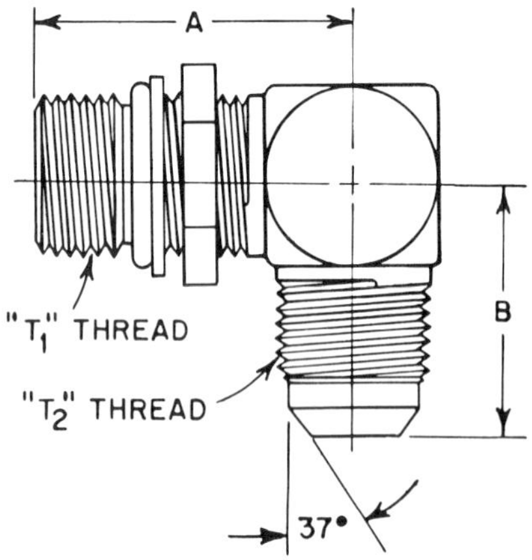

FIG. 4-111 Adjustable O-ring elbow.

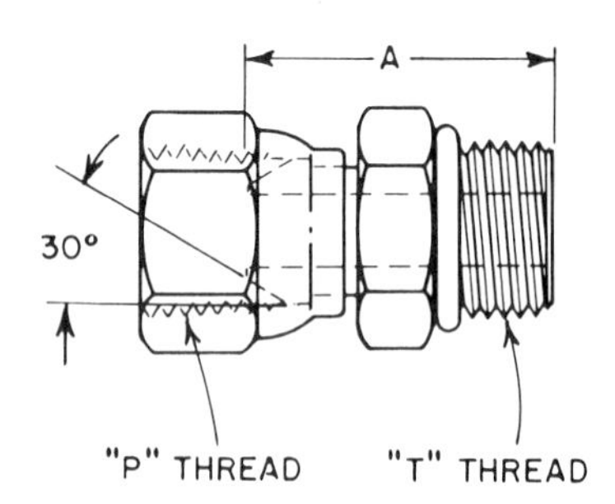

FIG. 4-112 O ring, NPSM.

Cures: Replace O-ring seal and start over. Make sure that the jam nut and metal washer are at the extreme end of the groove, away from the port. Lubricate O ring (*very important*). Follow preceding assembly procedure. If port is spot faced, make sure it is large enough to accommodate the metal washer.

Why O-ring Lubrication Is Important. When the fitting is engaged to the point where the O ring touches the face of the boss, lubrication on O ring permits it to move in direction *D* (Fig. 4-113). When the O ring and boss are dry, rotary motion of the assembly can cause friction and the O ring can move in direction *C*. The jam nut and washer can not bottom fully if the O ring is between the washer and the face of the boss.

When the jam nut and washer are not backed up prior to assembly, there is not

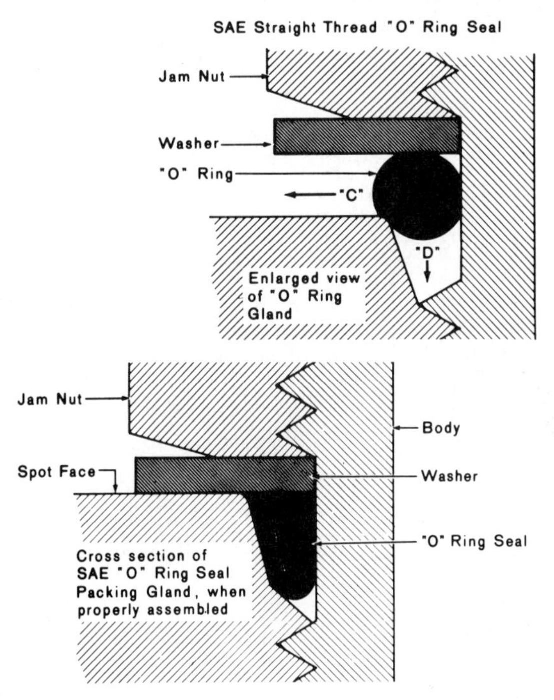

FIG. 4-113 Enlarged view of O-ring gland (before and after tightening).

enough room for the O-ring seal when the squeeze takes place. The washer can't seat properly on the face of the boss. The compressed rubber between the washer and the boss face will cold-flow out from compression and the fitting will be loose and usually leak (Fig. 4-114).

Why Some Elbows Loosen Up and Others Don't. This design depends on compression or squeeze of the jam nut, washer, face of the boss, and the body threads to hold the elbow tightly in position. If the jam nut loosens, the whole fitting becomes loose. SAE permits a chamfer on both sides of the jam nut. But when both sides of the nut are chamfered, there is much less compression area to squeeze onto the washer when it is tightened (Fig. 4-115). When the washer side of the jam nut is left flat, the contact and squeeze area is greatly increased; this makes a big difference (Fig. 4-116).

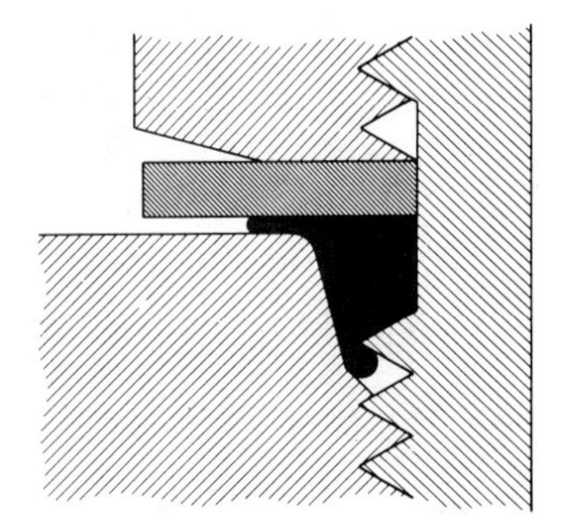

FIG. 4-114 Improper gland design.

As with all O-ring type fittings make sure the O ring is compatible with the temperature and the agent being conducted. Although this is basically a hydraulic fitting, several manufacturers are using it in die-cast lubricating filter heads in lieu of pipe threads.

4-107 Refrigerant O-ring Seal This style of fitting (Fig. 4-117) is used in mobile refrigerant systems, but can also be found in some hydraulic applications. It threads

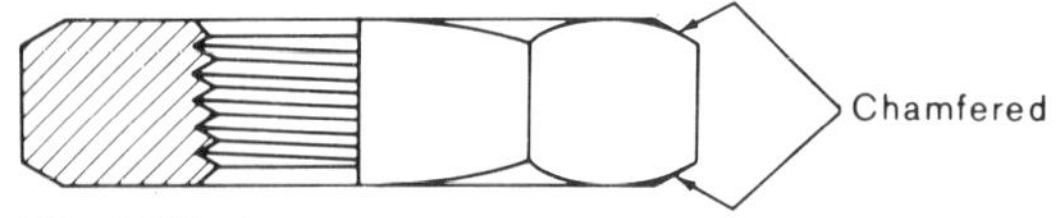

FIG. 4-115 Jam nut.

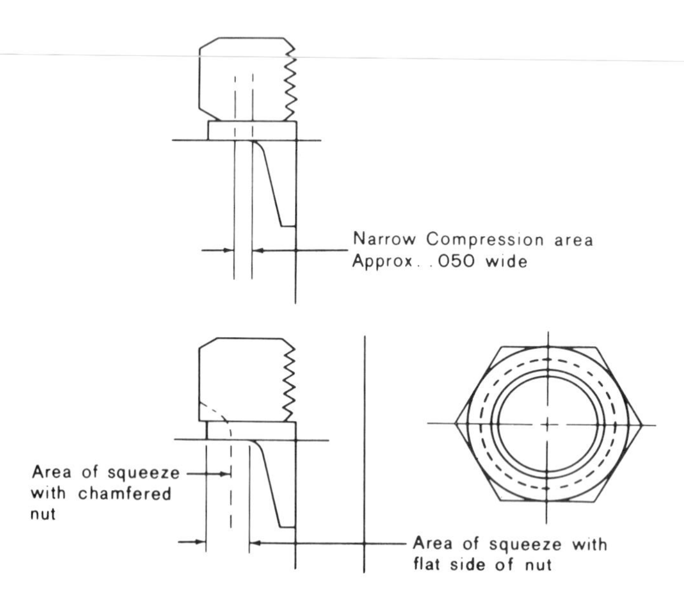

FIG. 4-116 Compression area and area of squeeze.

into an SAE O-ring boss port and seals in the bottom of the port instead of on the chamfer. This is called a "tube-O" fitting by some manufacturers.

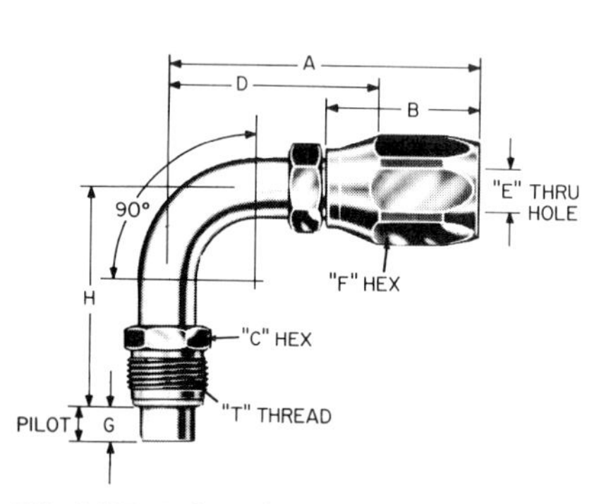

FIG. 4-117 Tube-O fitting.

4-108 Metric and BSP Threaded Fittings Metric (European) and BSP (*B*ritish *s*tandard *p*ipe) threaded fittings employ the same basic sealing principles as do other metal-to-metal type seals. Metric threaded fittings of the flareless type normally have ends with either a 24-degree or a 60-degree recessed truncated cone (Fig. 4-118). The 24-degree cone fitting may be used as a flareless tube fitting with the appropriate nut and sleeve. The female metric end is a Globeseal-type fitting which seals on the recessed cone. The 24-degree cone angle indicates a DIN3902-style fitting and the 60-degree recessed cone indicates a DIN7631 style fitting. (The DIN designation is a fitting style and thread standard, similar to SAE standards.) Metric thread sizes are designated by the thread diameter in millimeters and the pitch in millimeters. For example, if the thread diameter is 22 mm and the pitch is 1.5 mm the metric thread size is M22 × 1.5.

There are two basic British pipe thread standards: BSP taper and BSP parallel (Fig. 4-119). The BSP taper is almost identical in appearance to the American NPTF. The BSP taper seals on the thread like the NPTF dryseal. The difference between the two is in the thread pitch. There are two types of BSP parallel threaded fittings: the chamfered and the unchamfered.

The chamfered type is the more common of the two. It is similar to the BSP taper, except that it will not seal on the thread. Instead, its 30-degree chamfer seals on the recessed cone of the BSP Globeseal female. The unchamfered BSP parallel thread is

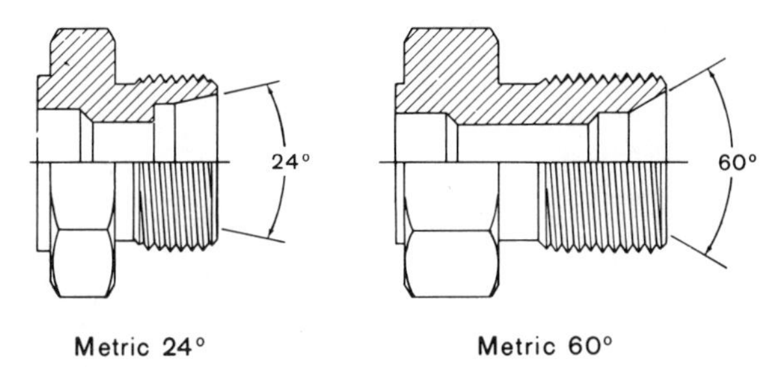

FIG. 4-118 Metric threaded fittings.

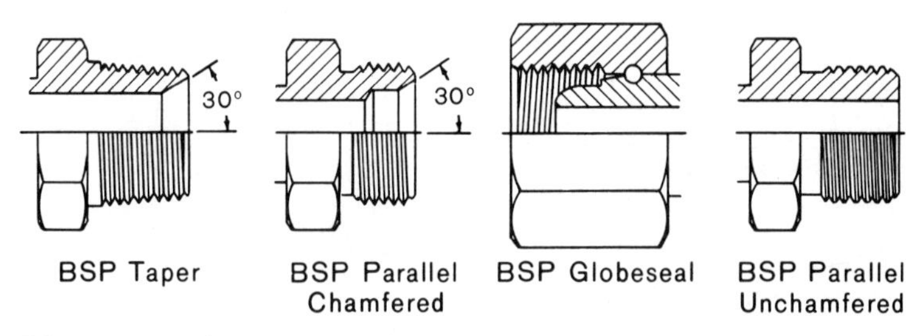

FIG. 4-119 British pipe fittings.

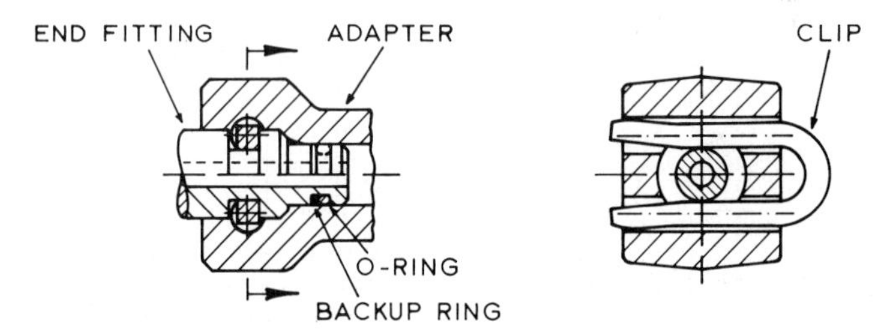

FIG. 4-120 German Steck-O connector.

used primarily for threaded ports. This type, similar to the SAE straight thread, seals with a gasket. Its threads are interchangeable with other BSP types, but it will not seal with the BSP Globeseal female since it does not have the 30-degree chamfer. BSP threads are measured and specified in the same way as the American NPTF by the nominal thread diameter and the thread pitch.

4-109 Steck-O This is a unique connection that was developed in Germany for use in hydraulic applications in mines. It is basically a radial O-ring seal with the O-ring carrier retained in the port with a U-shaped retaining clip (Fig. 4-120). The O ring is fitted with a backup washer to prevent its extrusion under high-pressure conditions. A protective cap is used to prevent dust from entering the connection.

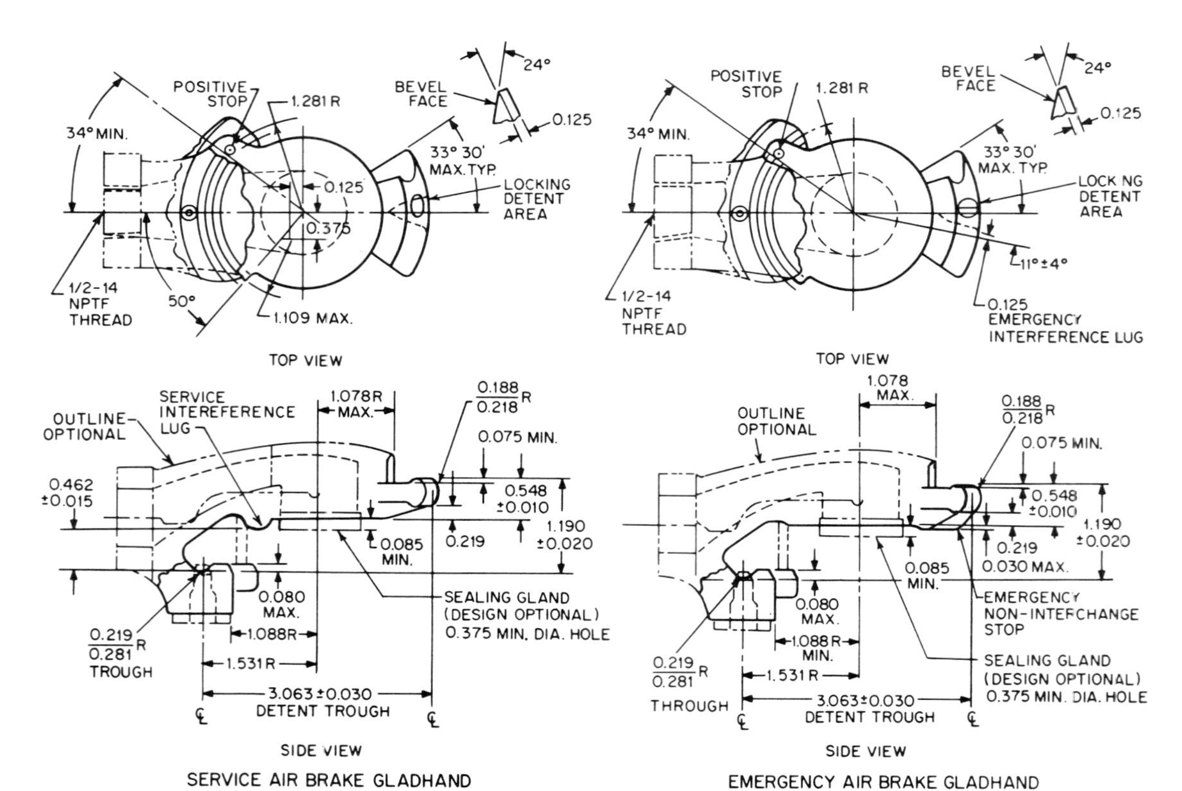

FIG. 4-121 Air-brake Gladhand (service and emergency).

It is said to be superior to American designs because there are no torque requirements to meet and it is free to rotate in the port although it is not designed for continuous rotation under pressure. Several American manufacturers are offering this connection for general hydraulic and pneumatic applications.

4-110 OHE This fitting, which also originated in Europe, is similar to the Steck-O fitting except in design details.

4-111 Air Brake Glandhand, SAE J318 This connector is primarily used for coupling air lines on trucks, truck-tractors, and trailers when these vehicles are joined to operate as a combination unit.

The Gladhand connection that has been designed for the service brake system is slightly different than the one designed for the emergency brake system so that the two will not fit together (Fig. 4-121). This eliminates the possibility of getting the two systems crossed over when joining them. The halves are usually permanently marked for identification by embossed lettering of the words *service* or *emergency* or abbreviations of these words.

The sealing gland is made of rubberlike material.

Part 3 Pipe Relining Technology

ALGER B. COLTHORP, P.E.
Insituform Technologies, Inc.,
Chesterfield, Missouri

4-112 Relining Description Use of trenchless technologies to rehabilitate — rather than to replace — deteriorated, buried piplines is now accepted and practical. Trenchless technologies make it possible to reconstruct and rehabilitate sewers and other pipelines without digging up existing pipes. Use of these technologies also eliminates joints and all the common problems inherent in joining pipes.

Trenchless technologies have been used extensively for more than 25 years to correct failing municipal sewer systems. Enhancements are now enabling these same trenchless technologies to be used for other applications, including industrial sewers, industrial process pipes, building drains, water lines, and gas mains, among others.

The most widely used trenchless technology is Insituform®, a cured-in-place pipe (CIPP) product. Other products in use today throughout the world include Paltem®, a hose lining system, and Tite Liner®, a polyethylene lining system. All three of these systems virtually eliminate joints and represent leading-edge solutions to the problems associated with faulty joints.

4-113 Cured-in-Place Pipe Insituform® is a unique process for reconstructing damaged pipeline systems in municipal and industrial applications. A new cured-in-place pipe (CIPP), or Insitupipe®, is formed inside of the existing conduit by using fluid pressure, typically from water, to install a flexible tube saturated with a liquid thermosetting resin. The water is then heated to harden the resin. This process results in a continuous, tight-fitting, pipe within a pipe. The Insituform® process is cost-effective and fast and can be used in a variety of gravity and pressure applications, such as sanitary sewers, storm sewers, process piping, drinking water systems, and ventilation systems.

The Insituform® process was invented in England in 1971. The process was brought to the United States in 1977 and has been installed worldwide in over 28 million ft of deteriorated pipeline with over 14 million ft in the United States. Several installations have been inspected after many years of service and were found to be in excellent condition. In 1991, samples were cut from the first Insituform® installation in London, England. After 20 years of exposure to domestic and industrial sewage, the Insitupipe® exceeded current industry standards for physical strength and exhibited no signs of deterioration.

4-114 CIPP Installation Although the Insituform® process may appear to be simple, it is a state-of-the-art technology involving sound engineering practices, exacting materials requirements, and experienced field installation personnel in order to manufacture quality pipe in an underground environment. A flexible, resin-absorbent fabric Insitutube®, coated on the outside with an elastomeric material, is manufactured to fit the cross section, length, and required design thickness for the damaged pipe. The Insitutube® is vacuum-impregnated (wet-out) with a liquid thermosetting resin. This work is normally done at the Insituform® installer's wet-out facility. However, for large-diameter Insitutubes® which are impractical to transport, wet-out can be accomplished at the job site with specially designed portable equipment.

The impregnated Insitutube® material is installed in the existing pipe through a manhole or other access point via a temporary inversion standpipe and inversion elbow (see Fig. 4-122, stage 1). A top inversion technique can also be used where the Insitutube® itself forms the inversion standpipe. Water from nearby hydrants, or other convenient source, is used to fill the inversion standpipe. The force of the column of water pushes the wet-out Insitutube® inside out (termed *inversion*) and into the pipe being reconstructed (see Fig. 4-122, stage 2). As the Insitutube® travels through the pipe, water is continually added to maintain a constant pressure. The inversion process results in the elastomeric coating of the Insitutube® being the new interior surface of the pipe.

Stage 1

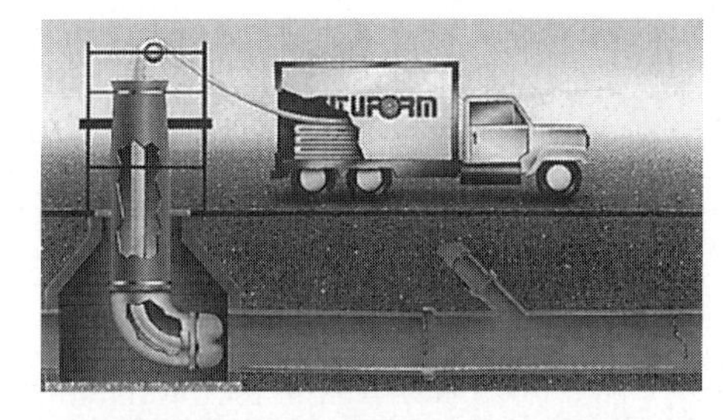

Stage 2

Stage 3

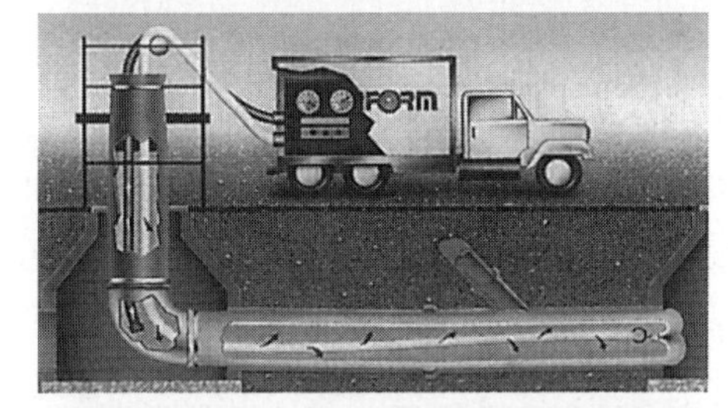

Stage 4

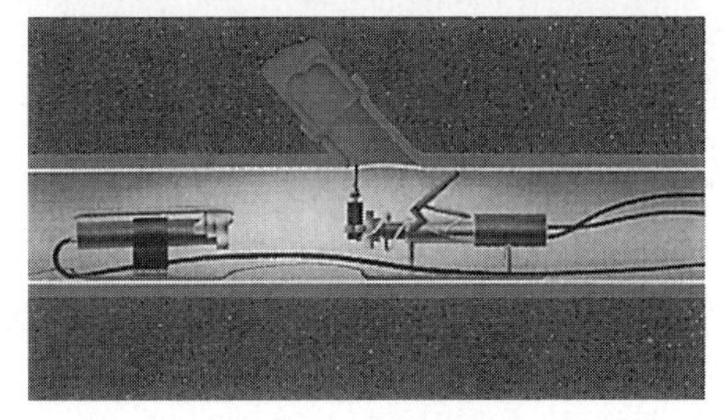

FIG. 4-122

The inversion process results in no relative movement between the Insitutube® and the deteriorated existing pipe wall, minimizing trauma and potential damage to the flexible tube material. Another benefit of the inversion technique is that incoming and standing waters are forced ahead of the inverting Insitutube® and out of the pipe reach; therefore, no water is trapped behind the Insitutube® that could inhibit proper cure of the resin or could alter the shape of the finished Insitupipe®.

After the Insitutube® reaches the termination point, the water is heated to cure the thermosetting resin (see Fig. 4-122, stage 3). Once the Insitutube® has hardened and cooled down, the water pressure is released and the ends are trimmed. Service connections are reinstated internally with a remote control cutting device or by human-entry techniques (see Fig. 4-122, stage 4). The Insituform® operation is then completed, and the newly installed pipe is ready for immediate use.

4-115 CIPP Materials Insitupipes® are custom-manufactured to meet the specific needs of each application. Insituform® installations to date have varied from 4- to 99-

in equivalent diameter with thicknesses of 0.12 to 1.65 in as required by each project. Continuous Insitupipe® lengths of up to 2700 ft have been installed with greater lengths feasible depending upon specific job conditions.

The physical characteristics of the finished Insitupipe® are largely a function of the resin system used. Resin systems approved for use in Insituform® applications are liquid thermoset materials that during the cure process change irreversibly from the liquid state to strong cross-linked solids. The resulting three-dimensional, cross-linked network is stable to heat and cannot be made to flow or melt again. Thermoset resin systems should not be confused with thermoplastic materials which can be made to soften and take new shapes by the application of heat and/or pressure.

Thermosetting systems normally used in the Insituform® process can be separated into three major groups: unsaturated polyester, vinyl ester, and epoxy. These types of resin systems have had extensive use in the reinforced plastics industry for almost 50 years in the manufacture of corrosion resistant tanks for chemical processing and fiberglass boat hulls.

The most commonly used resins in Insituform® applications are polyesters. These are mainly resin systems with high flexural moduli and low tensile elongation. These systems are particularly suitable for gravity pipe rehabilitation where a high flexural modulus and good chemical resistance are important in the structural design and service life of the Insitupipe®. Polyester resins are highly resistant to normal domestic sewage and sulfuric acid produced by hydrogen sulfide (H_2S) activity.

Vinyl ester and epoxy resins are recommended in cases where special corrosion and/or solvent resistance is required. Such cases are likely to occur in industrial sewage or industrial process lines. Vinyl ester systems demonstrate good chemical resistance to hydrolysis and halogenation. Their superior solvent resistance is a direct result of their chemical configuration. Vinyl esters offer a good balance of properties between polyesters and epoxy systems.

Epoxy resins offer enhanced chemical resistance especially against caustics and solvents. Cured properties and performance characteristics are heavily dependent on the curing agent used. For the special case of drinking water pipes, epoxies are the only resin systems certified by the National Sanitation Foundation (NSF) for drinking water applications of Insituform®.

For low- and medium-pressure pipes, such as plant water intake lines, which are suffering from internal corrosion, pinhole leaks, or faulty joints, an Insituform® pressure pipe liner can be used. Its advanced composite design incorporates a tube with two types of reinforcement and a specially engineered resin. An installed pressure pipe liner resists internal pressure and supports external loads during extended plant shutdowns.

4-116 CIPP Design In pipeline reconstruction using the conventional Insituform® process, two types of structural design considerations are normally based on the condition of the original pipe: partially deteriorated and fully deteriorated. The structural condition of existing pipelines varies widely, but some examples are shown in Fig. 4-123. In the partially deteriorated case, the condition of the original pipeline may include, but is not limited to, longitudinal cracks, some cross-sectional distortion, and displaced joints; however, the existing pipe can, and will continue to, support soil and live loads. In the fully deteriorated case, the original pipeline is not structurally sound and cannot support soil and live loads or is expected to reach this condition within the design period.

Evidence of a fully deteriorated condition includes, but is not limited to, (1) missing sections of the original pipe; (2) loss of original pipe shape; (3) severe pipeline corrosion due to the effects of the effluent, atmosphere, or soil; and/or (4) continued external corrosion of the pipe after reconstruction.

In evaluating these design conditions for Insituform® applications, design recommendations were developed on the basis of laboratory and aboveground testing, soil cell testing, extensive experience in the pipe reconstruction field, and accepted flexible pipe design analysis. This design guide presents structural analysis techniques for circular pipeline applications of Insituform®. Contact an Insituform Technologies representative for assistance on irregularly shaped pipe designs.

4-117 Pressure Pipe Relining When a pressure pipeline leaks, the gases or liquids it carries can escape, with costly and sometimes dangerous results.

Paltem® is a trenchless hose lining system used to rehabilitate all common types of

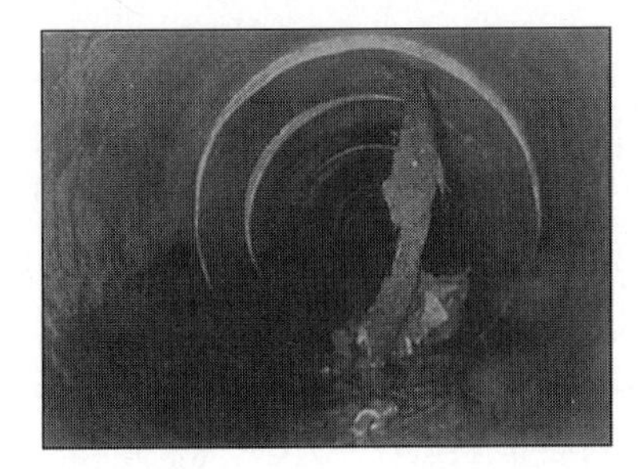

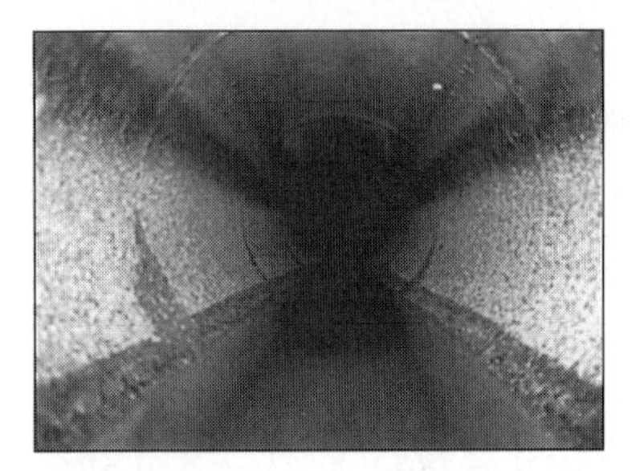

Partially Deteriorated

Partially Deteriorated

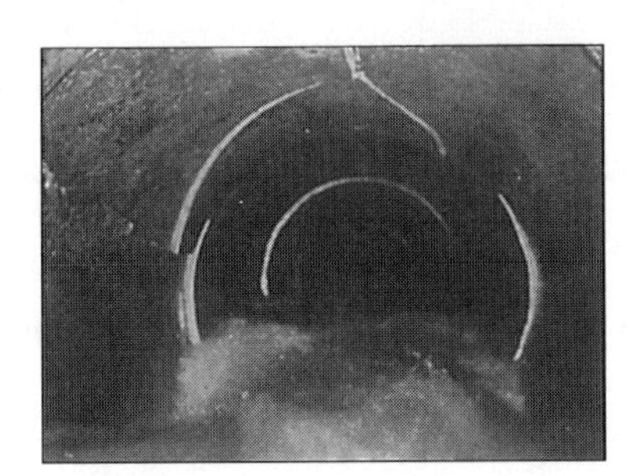

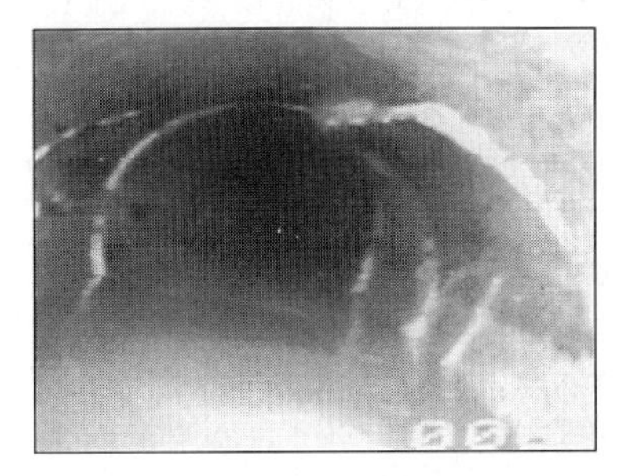

Fully Deteriorated

FIG. 4-123

pressure pipelines which have been damaged by corrosion or are experiencing leakage through joints or pinholes. Its reinforcement fibers can also provide protection against failure caused by pipe movement as a result of uneven settlement or seismic shock.

Over the past decade, Paltem® has been used to renew hundreds of miles of gas mains, industrial pressure pipes, force mains, water mains, and other pressure pipe worldwide. Because Paltem® can be installed with minimal excavation and disruption to surrounding areas, it is especially well suited for rehabilitating pressure pipes which run beneath heavily traveled streets, along bridges, and through manufacturing plants. It is environmentally safe and usually less expensive than pipe replacement.

4-118 Pressure Pipe Relining Installation Prior to installation, small access pits are dug at both ends of the pipe segment to be lined. First the host pipe must be thoroughly cleaned, to ensure a strong bond between the PAL-Liner™ and the host pipe. Mechanical scrapers are typically used to clean the pipe, although chemical methods have been used successfully. (See Fig. 4-124, stage 1.)

The primary component of the Paltem® system is the PAL-Liner™, a woven polyester hose with an elastomer coating which has been manufactured to fit the dimensions of the original pipe. The PAL-Liner™ is saturated with an epoxy resin. It

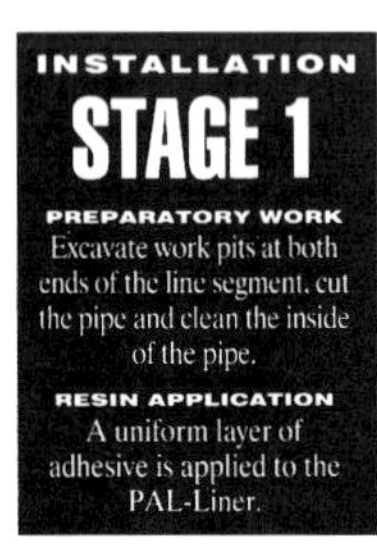

INSTALLATION
STAGE 1
PREPARATORY WORK
Excavate work pits at both ends of the line segment, cut the pipe and clean the inside of the pipe.
RESIN APPLICATION
A uniform layer of adhesive is applied to the PAL-Liner.

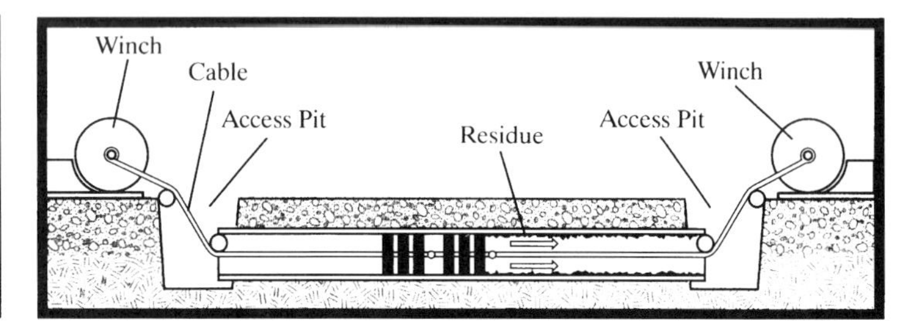

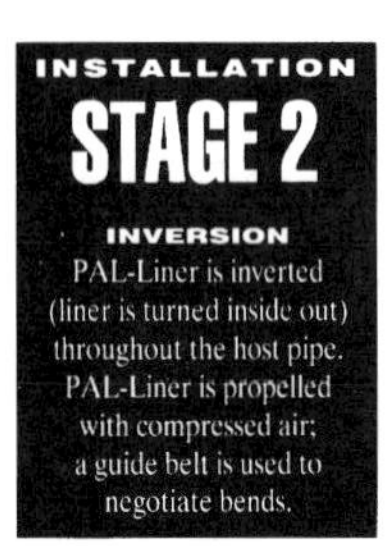

INSTALLATION
STAGE 2
INVERSION
PAL-Liner is inverted (liner is turned inside out) throughout the host pipe. PAL-Liner is propelled with compressed air; a guide belt is used to negotiate bends.

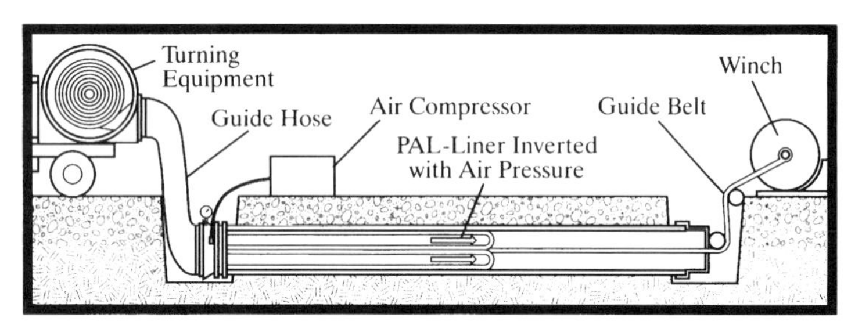

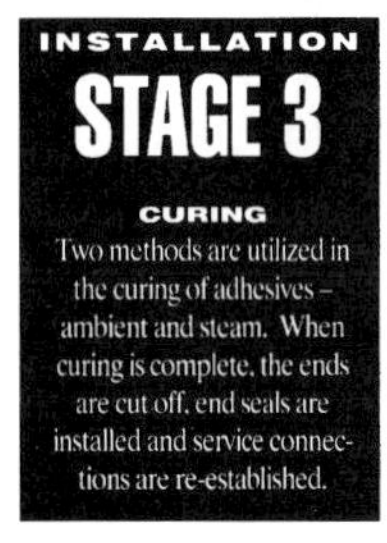

INSTALLATION
STAGE 3
CURING
Two methods are utilized in the curing of adhesives – ambient and steam. When curing is complete, the ends are cut off, end seals are installed and service connections are re-established.

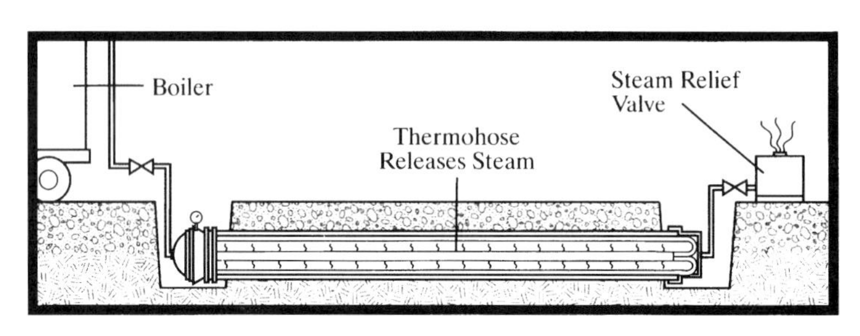

INSTALLATION
STAGE 4
REOPEN SERVICE CONNECTIONS
The remote PAL-Cutter precisely locates the dimple with the aid of a self-contained video camera, and service connections are quickly restored.

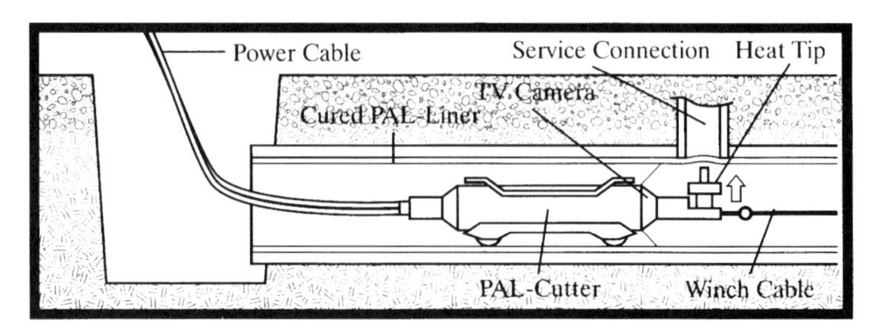

FIG. 4-124

is then stored on a reel in a pressure chamber before being inverted into the host pipe. Compressed air is used to invert and propel the PAL-Liner™ through the pipe to the receiving pit located at the opposite end of the pipe (Fig. 4-124, stage 2).

The inversion process is completed when the PAL-Liner™ reaches the receiving pit. A process to cure the epoxy resin is then initiated using either steam or ambient temperature. The PAL-Liner™ forms a smooth, pressure-resistant lining on the inside surface of the pipe (Fig. 4-124, stage 3).

The ends of the PAL-Liner™ are then cut off. Service connections, which were sealed during installation into low-pressure mains, can be opened internally using a remote-controlled cutting device aided by a small camera (Fig. 4-124, stage 4). Finally, end seals are installed to protect the transition to the existing pipe system (Fig. 4-125).

One of the advantages of Paltem® is its ability to go through forged steel elbows and offset joints in cast-iron mains. When it is necessary to navigate bends of 45° to 90°, a guide belt under tension is often used to assist the PAL-Liner™ through such pipe.

4-119 Pressure Pipe Relining Materials The PAL-Liner™ is manufactured to the requirements of the pipe to be lined. It is constructed from polyester fibers that are woven into a hoselike product. The woven hose is coated with an elastomer that is

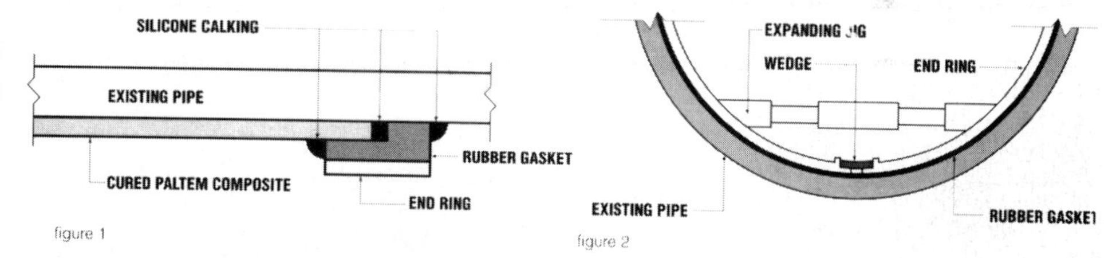

figure 1

figure 2

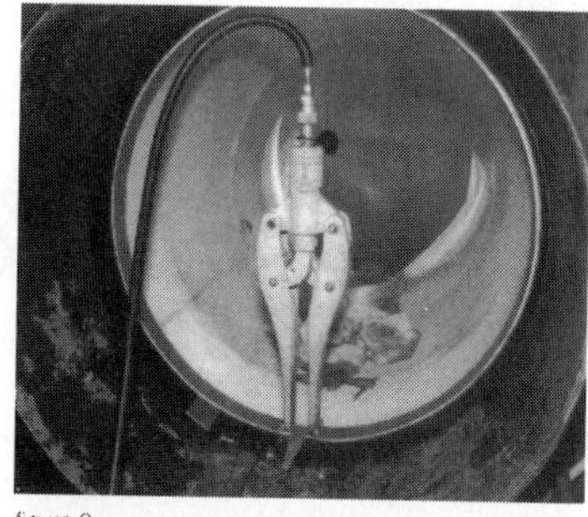

figure 3

1 The cured composite is cut evenly around the inside circumference approximately six inches from the end of the pipe. All adhesives and other materials are cleaned from the pipe end. A bead of silicone sealant is placed on the cut edge of the composite (figure 1). The rubber packing is placed over the cut edge of the composite, then the end ring is placed on the top of the gasket.

2 An expanding jig (figure 2) on pipes up to 12 inches in diameter or a hydraulic spreader (figure 3) on pipes larger than 12 inches is used to expand the end ring to fit tightly against the pipe wall.

3 After the fit of the rubber packing and end ring is checked, a wedge is tapped into the wedge guide. The wedge guide is installed at the six o'clock position in the pipe (figure 2).

4 The wedges are then cut flush with the end ring. Silicone calking is placed on both sides of the rubber packing/pipe joint as the final step in the installation (figure 4).

FIG. 4-125

FIG. 4-126

compatible with the piped products. The Paltem® system utilizes a thermoset resin, generally an epoxy.

4-120 Pressure Pipe Relining Design Testing has proved that significant holes and/or joint gaps in the existing pipe can be bridged by the seal hose while internal pressure on the system is maintained. In fact, tests have shown that Paltem® is able to sustain the line pressure with 1 m (39 in) of pipe missing for a 24-h period (a definition of earthquake resistance).

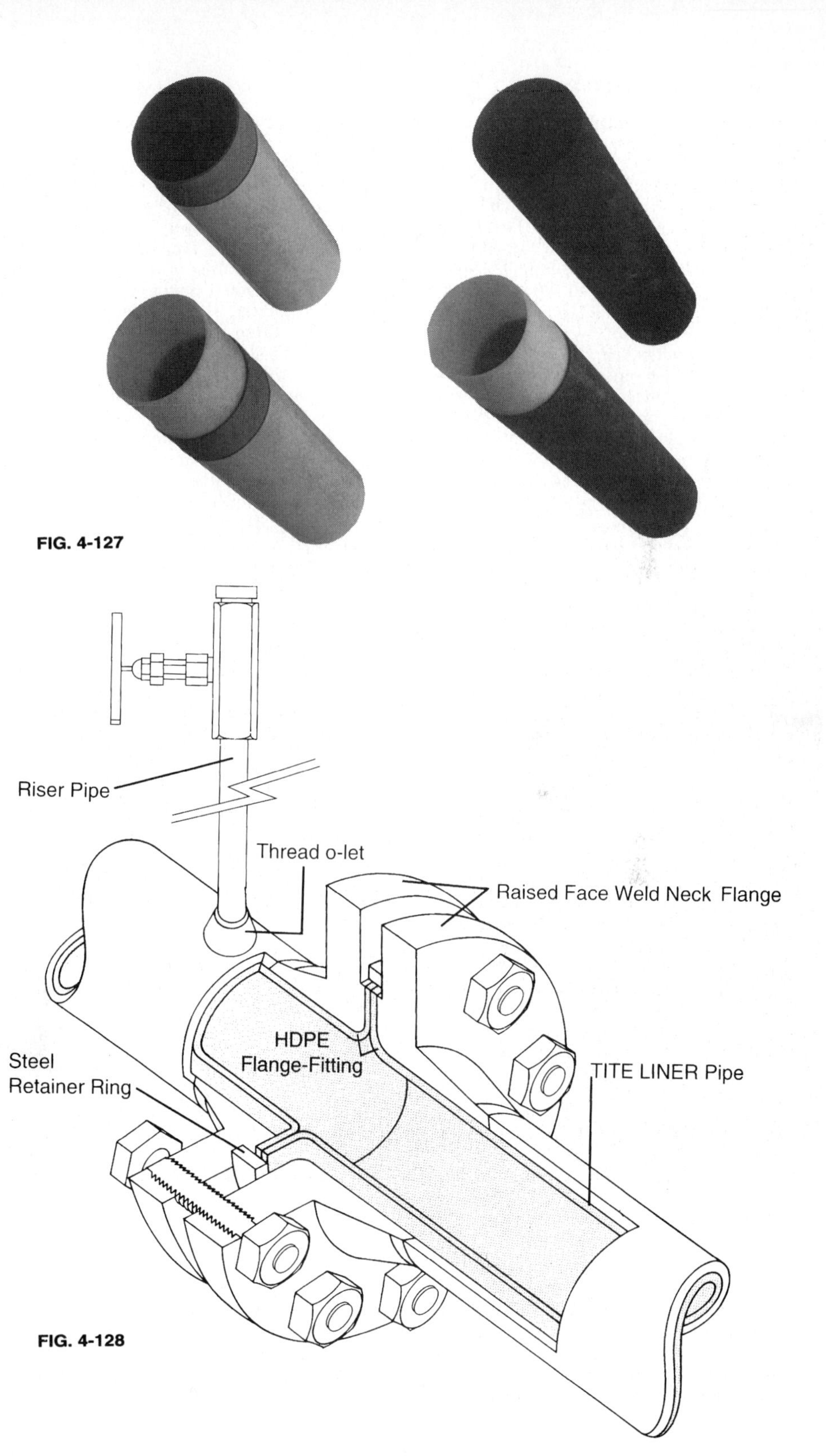

FIG. 4-127

FIG. 4-128

Other tests confirm that no "tracking" of pipe product occurs between the host pipe and the pipe lining in low-pressure applications.

In addition to the above-mentioned testing, significant testing has been done to evaluate the performance of the lining system under large strain loading events, such as earthquakes, which may pull joints apart or deflect the existing pipe joints as a result of ground movement. The Paltem® process has performed well in these tests at pressures up to 450 lb/in^2. In addition to the testing mentioned, there are hundreds of miles of Paltem® hose linings installed and successfully operating in the gas piping systems owned by gas utilities in Europe, Japan, and the United States. Both street mains and service lines to users have been successfully renewed using this technology.

For lower-pressure water and wastewater service, it is often possible for the PAL-Liner™ to withstand the full pressure (internal and external) and not be dependent on the existing pipe. In this sense, the PAL-Liner™ can be considered a pipe in a casing, even though the casing pipe may be cracked longitudinally and not capable of withstanding internal pressure.

4-121 Relining for Corrosion Protection Some steel pipelines, such as oil lines and mine slurry lines, carry products which are themselves corrosive or abrasive and which, over time, wear away at a pipe, leading to leakage and pipe failure.

The Tite Liner® system protects against such failures in two ways. First, when installed in new lines, the Tite Liner® performs as a barrier, allowing the host pipe to operate at its design pressure without being exposed to internal corrosives or abrasives. Second, Tite Liner® can be installed in existing lines to prevent further deterioration due to internal corrosion or abrasion. It can bridge pinhole leaks in these pipelines while maintaining the pressure rating of the original pipe.

4-122 Relining for Corrosion Protection — Installation The Tite Liner® system uses a high density polyethylene liner pipe having an outside diameter equal to or slightly larger than the inside diameter of the steel pipe it will protect. The liner pipe is compressed radially through a roller reduction box to allow sufficient clearance while it is pulled through the host pipe under tension (see Fig. 4-126). When the tension is released, Tite Liner® expands to create an interference fit against the host pipe's inner wall (see Fig. 4-127).

The Tite Liner® system has a simple method of connecting segments of lined pipe. (Fig. 4-128). Because Tite Liner® can be installed in segments of 3500 ft, a typical project requires only a few end connections. When it is necessary to connect line segments, a steel flange, a specially fabricated liner flange, and steel spacer ring are assembled as a unit for shipment to the field. The wraparound liner flange leaves no metal exposed within the lined section and provides the seal. The size of the spacer ring is calculated to provide the correct amount of compression of the liner flange face, giving long-term stability and a sealed connection.

The integrity of the seal can be either continuously or periodically monitored. At each end of a lined section of pipe, weld-o-lets are welded onto the steel pipe, and a small hole is drilled in the center of the weld-o-let through the steel pipe wall only. A riser pipe and valve are then connected to the weld-o-let. The valve can then be opened to verify the integrity of the seal.

4-123 Relining for Corrosion Protection — Materials The Tite Liner® system uses high-density polyethylene pipe which is specially manufactured for each project. Various SDRs (ratio between wall thickness and pipe diameter) are used depending on the project requirements, including the diameter of the host pipe. This liner pipe is typically equal to or slightly larger in diameter than the steel pipe to be lined in order to create the tight fit.

4-124 Relining for Corrosion Protection — Design The Tite Liner® system can operate at line pressures since the liner is simply compressed against the host pipe and transmits the internal pressure to the host pipe. Due to the interference fit, no radial tensile stresses exist in the liner. When the line pressure is less than 150 lb/in^2, the liner pipe is usually capable of withstanding the pressure without the existing pipe. In this case, the Tite Liner® can be considered as a stand-alone pipe or as a pipe within a casing, not dependent on the original pipe. This can be important when the problem is due to external corrosion and the internal liner is not capable of protecting the outside of the steel pipe, making it difficult to determine the life expectancy of the system.

Section 5
Expansion Joints*

WILLIAM O'KEEFE

Senior Technical Editor, *Power Magazine,*
McGraw-Hill, New York, New York

*Reprinted from William O'Keefe, "Expansion Joints for Ductwork and Piping" (Special Report), *Power Magazine,* August 1986, pp. 5·1–5·16.

5-1 Design Factors Axial expansion of fluid-handling conduits during temperature increase compels engineers to provide a way of preventing harm to the conduits and connected equipment. Large powerplant components, too, expand enough during heating to operating temperature to require means of preventing damage to connected piping or ducting. Wind, weight of contents, and fluid forces also load conduit and connected equipment variably, adding to the need for flexibility in the conduit.

For piping, at least, the design of the line itself can often meet the need to accommodate expansion. Bends, loops, and long tangents are easy to calculate today and have the advantage of a continuous wall of constant thickness.

Nevertheless, reliance on inherent flexibility has serious disadvantages, too. Added material, more conduit support, and higher heat loss, along with more space, come to mind immediately. Therefore, compact devices—expansion joints—to absorb thermal expansion will continue to have a vital place in powerplant fluid-handling systems. The concentrated flexibility or lumped softness of expansion joints will compete with inherent or distributed flexibility.

This section looks at the practical engineering side of the subject, from the viewpoint of the powerplant. The theory behind the types of expansion joints is fairly simple, but there is a large amount of practical technology in the subject.

Materials, whether metals, ceramics, elastomers, or plastomers, greatly influence life, leaktightness, and resistance to sudden unexpected failure.

Before going into details, look at an example of the loads and movements in this work. Assume a 100-ft-long steel conduit (duct or pipe) heated from 100 to 600°F. Axial expansion is about 4.4 in. If this movement were completely suppressed by end load on the conduit, the stress would be about 96,000 lb/in.^{2} compression, far above the 10,000 to 15,000-lb/in.^{2} circumferential tension allowed for pipe-wall stress.

The 96,000-lb/in.^{2} compressive stress in, say, a 12-in. Schedule 80 pipe wall corresponds to about 2.25-million-lb axial end force, a respectable load for an anchor and an impossible one for most equipment nozzles. For comparison, the end-load on a blank on the 12-in. line carrying 1000-lb/in.^{2} fluid is only 102,000 lb.

There are two main classes of expansion-joint services, each largely with its own technology:

Gas and vapor ducts (nearly always at low pressure) and low-pressure equipment connections

Piping, frequently at high pressure as well as high temperature

Changes in dimensions and positions often result in complex loading and movement of conduit and equipment. This makes the design of expansion provision an intricate task, covering lateral movement and bending at the joint.

For ducting and piping, expansion is thermally caused and is largely axial. Equipment, on the other hand, can expand both axially and radially. Wind loads can affect outdoor vessels and ducting. Equipment such as tanks and heaters can settle temporarily or permanently from the weight of water fill. Piping supports, too, are not absolutely rigid, and this can make expansion joints necessary.

The expansion joints covered in this report function according to one of three basic principles:

Slip of rigid separate concentric tubes or rotation of rigid concentric spheres past each other, with comparatively soft packing to seal the clearance

Movement of a thin-wall convoluted bellows, continuous and impervious

Folding of a fabric/elastomer/plastomer (nonmetallic) belt fastened around a uniform gap in the conduit

For the three principles, a rough analogy would be to the trombone, the accordion, and the bagpipe. Metallic joints tend to be at higher temperatures and pressures than nonmetallic, but there is overlap in application. Nonmetallic expansion joints for gas ducts in powerplants have the simplest form of any expansion joint and are the suitable starting point for this report.

DUCT JOINTS

Powerplant ducting that requires expansion joints extends from the boiler or economizer outlet to the chimney and, on branch paths, to the coal pulverizers and the furnace-air inlets (Fig. 5-1). Combustion turbines and diesel engines, too, must have expansion joints in exhaust and heat-recovery ducts.

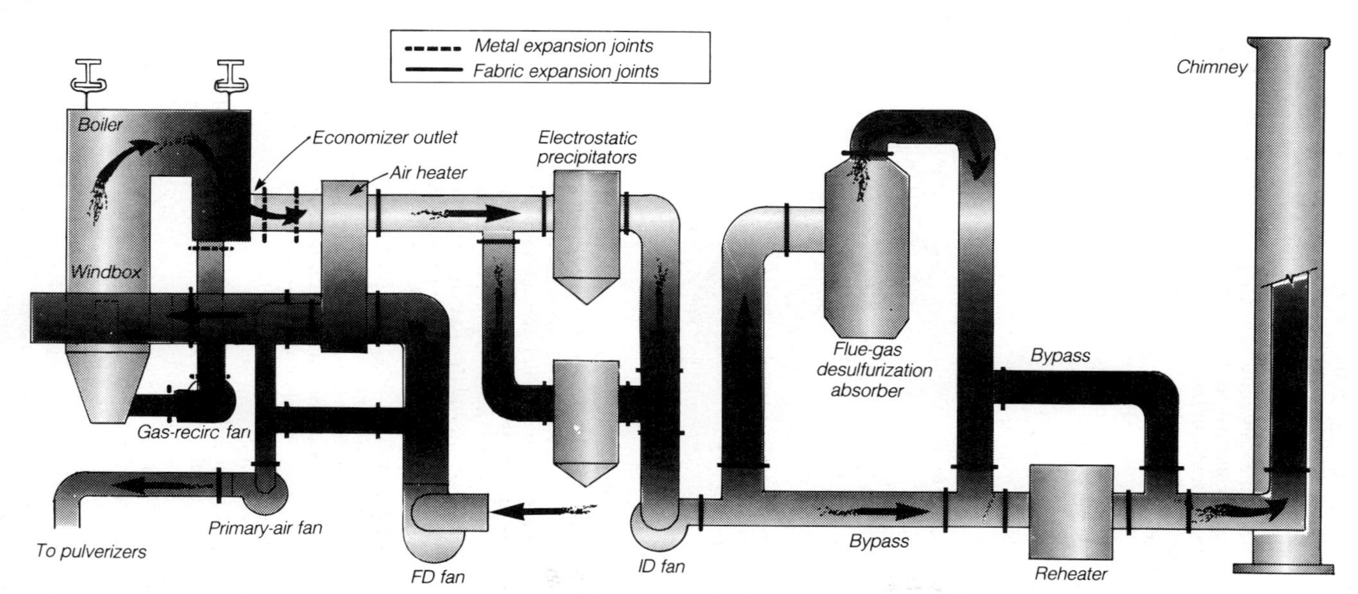

FIG. 5-1 Powerplant ducting, with long runs, multiple air or gas paths, and high temperatures, makes severe demands on expansion joints.

The ducting, of steel plate or sheet, is commonly of square or rectangular section in steam-boiler service. The duct-section size ranges from a foot or so up past 20 ft per side. Lengths between anchors or other points the designer elects to fix may be as much as 100 ft. Combustion-turbine and diesel ducting, which is more likely to be round, may be of low-alloy sheet if temperature and corrosive conditions require.

Although the primary duct movement that an expansion joint absorbs is the axial expansion and contraction resulting from temperature change, the designer may allow other relative movements of adjacent duct ends at an expansion-joint site. Lateral movement and angular movement can be in any direction and can occur in a common plane or in separate planes. Example: In a horizontal run, lateral movement can be horizontal or vertical, with one duct run rotating in horizontal or vertical plane. Torsion or twist (Fig. 5-2) is another, less common movement.

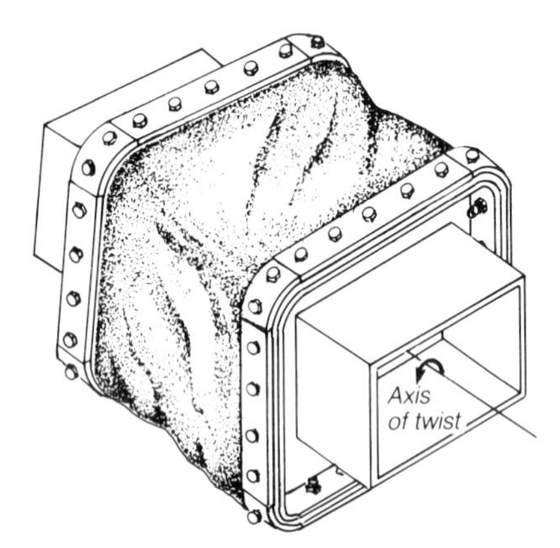

FIG. 5-2 Belt-type joint absorbs some twist, a motion very dangerous for metal joints.

5-2 Basic Gas Conditions Basic gas conditions affecting joint design and selection are temperature and pressure. Nonmetallic joints, which are much more sensitive to temperature than are metallic joints, must often be designed for a fairly accurately known operating temperature, with occasional upward excursions taken into account. If the design and installation of the entire joint are suitable, however, nonmetallic joints can withstand gas temperatures to 1200°F and higher for thousands of cycles.

Pressure in boiler-flue-gas ducting usually is below 100 in.-H_2O, so that a mild outward bow lets a flat fabric joint contain the pressure. Combustion-turbine and diesel exhausts can be at higher pressure, requiring metallic joints. Joints can experience vacuum, too. The connection between steam-turbine and condenser is a special example of an expansion or flexible joint for vacuum service.

Particle loading of the flue gas is another condition affecting joint design. High-velocity impact on the joint components is one source of damage, of course, but several other patterns of injury need consideration, too.

5-3 Chemical Attack Chemical attack on metallic and nonmetallic components of joints is frequent in flue-gas duct systems. Condensate from water formed in combustion absorbs acids which attack metal in the joint. Attack on nonmetallic materials is largely by acids and accelerates at high temperatures. Sunlight and ozone also injure nonmetallics.

Vibration can wreck expansion joints. In metallic joints, it is low-amplitude, high-frequency motion. Nonmetallic joints can flutter until the heavy fabric cracks or tears loose.

Duct-end conditions at the breach, the gap between ends which the expansion joint must seal, also are a factor in joint design. Workmanship and tolerance on the wall metal and light structural shapes at the duct end are all that can be relied upon to give uniform breach dimensions. Duct-erection tolerances, too, may result in unexpected discrepancies in alignment and breach length.

5-4 Support Support of large ductwork is often inadequate in respect to guidance at expansion joints. Unplanned twist and lateral movement, if accommodated by the joint, may reduce joint life to a fraction of the normally expected life. Both mechanical stress and thermal effects enter the pattern of destruction.

5-5 Nonmetallic Most nonmetallic joints consist of at least two materials or forms of a material. Elastomers are flexible, chemical-resistant, and can have low gas permeability. Plastomers have similar characteristics to a varying degree. Glass, in filament or fiber form, makes up abrasion-resistant sheets or thick insulating batts. Metal wire, as loose mesh or cloth, strengthens the fabric to which it is attached.

In the builtup flat fabric joint material for high temperatures (Figs. 5-3 and 5-4), the inner layer must resist dust abrasion. Outside the abrasion-resisting layer, thin layers of such compounds as TFE (tetrafluoroethylene) or other fluoroelastomers are the impermeable membranes. With insulation, this form is fairly thick, and the necessary interconnection between successive layers to hold the entire assembly together in service makes forming it into a channel shape difficult, although it can be done.

5-6 Connecting Connecting of belt ends requires splicing (Fig. 5-5). If the various layers separate easily, then they may be alternated one by one (Fig. 5-5, top). If the belt has thick layers, then a splice may require a stepped cut (Fig. 5-5, bottom).

Thinner, more homogeneous belts, which look like ordinary conveyor or power-transmission belting but are very different in construction, can be molded into various transverse profiles, such as channels, arches, and semicircles. The molding can be for corners (Fig. 5-6, upper right, lower right), too. Generally a woven fabric, in some cases with yarns running at 45° to the main axis, is impregnated with elastomer and cured into a cohesive entirety. The elastomer chosen is resistant to abrasion, gas permeation, and heat damage.

5-7 Behavior Behavior of the joint at the corners affects life and performance. Most duct-joint belts are fairly stiff, tending to hold peripheral length as the breach opens or closes. The close-spaced bolting, which prevents gas escape at the edges, also tends to restrain peripheral length change. Consequently, duct expansion forces the belt into a transverse profile in which part moves in and part out, with additional flexing and buckling at various spots.

A flat belt, builtup or of fabric-reinforced elastomer, requires a radiused corner (Fig.

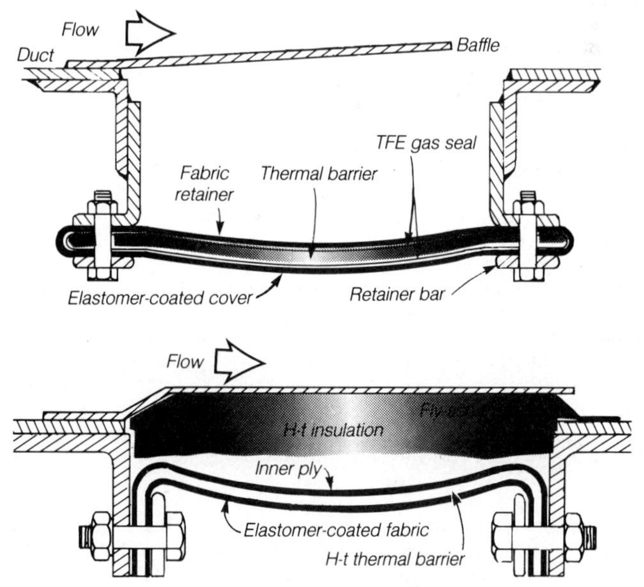

FIG. 5-3 High-temperature fabric joints may be flat (top) or have attachment flanges (bottom). Thermal barriers are vital elements.

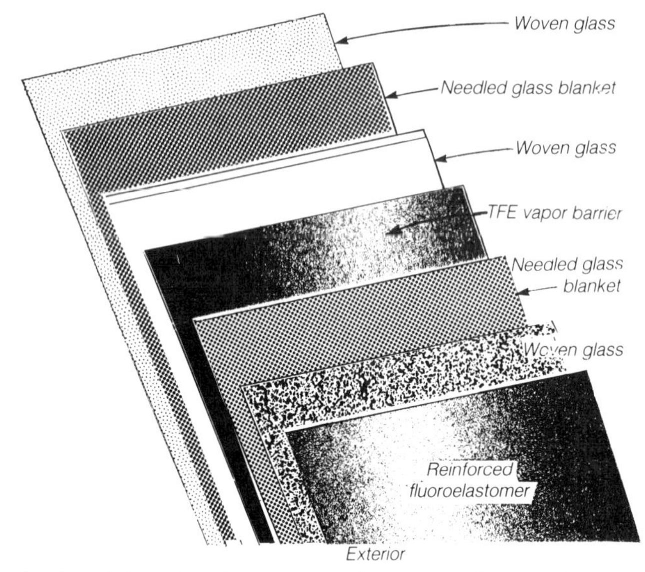

FIG. 5-4 Fiberglass cloth is often the innermost, abrasion-proof layer in a builtup joint for high-temperature fly-ash-laden flue gas.

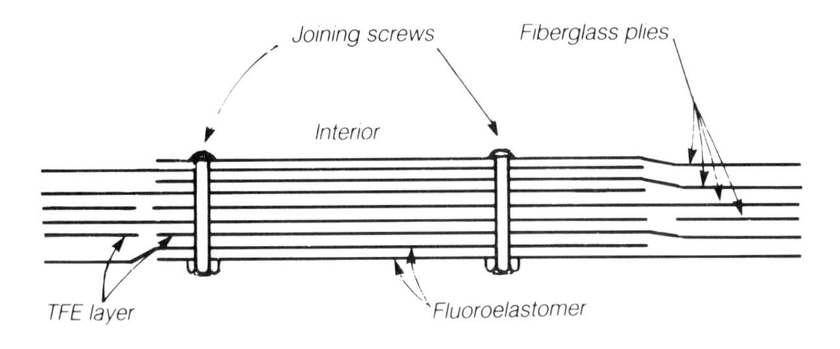

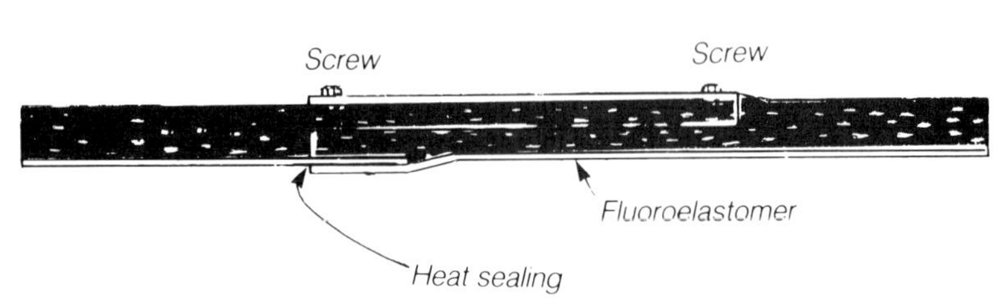

FIG. 5-5 Belt splice may involve interleaving of plies (top) or stepped cut (bottom) to retain original belt thickness and flexibility.

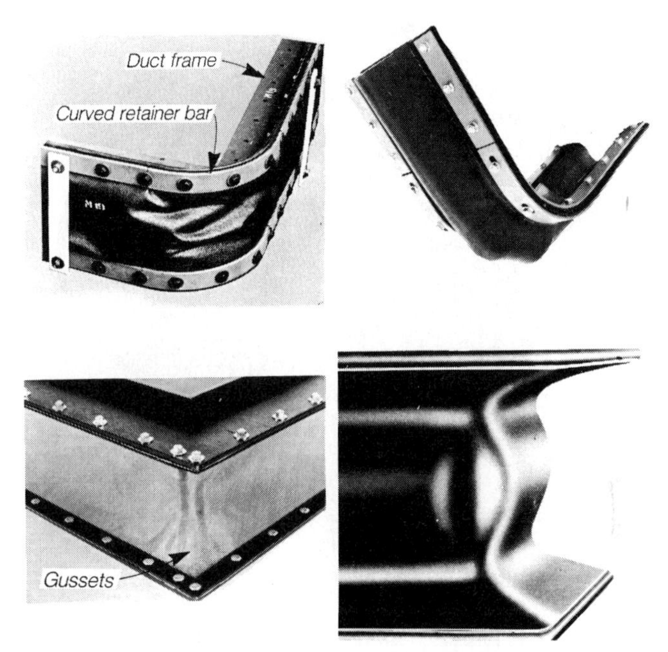

FIG. 5-6 Corners need provision for motion, with rounding of the joint framing, special molded bulges, or gussets spliced into the belt.

5-6, upper left). Gussets and bulges at corners permit more latitude, but the rule forbidding splices near the corner still holds.

The attachment of the belt to the duct ends (Fig. 5-7) may be directly to angle flanges that are welded to the duct. This serves if the belt is molded into a channel profile. For a flat belt, additional angles can be bolted or welded to the duct-end angles, permitting the bolting on of the flat belt. This design can be the choice for a flat belt with a molded V-profile, too (Fig. 5-7, lower left).

Whether the bolting is to be done from the inside or outside is a question calling for early decision. Often, both ends of the many bolts needed must be accessible to allow proper tightening, in spite of original plans to do all the work from only one side of the duct.

The details of the bolt connection need close study during evaluation of any proposed joint design. Some of the factors are illustrated in Fig. 5-8. The attachment system must hold the belt tightly to prevent flapping and scalloping which would let gas escape. Nevertheless, the connection must not pinch the belt excessively or crease it sharply at installation or in service.

The entire matter of heat sinks and heat transfer into the belt has its aspects in the connection to the duct, too. Often the belt is clamped between the mounting angle and a retainer bar, required because of belt softness and wideness of bolt spacing. The clamping blocks cooling-air flow and can shorten belt life.

5-8 Belt Deformation Deformation of the belt at the connection may cause the belt to rub against bolt heads and wear out, so some designs have carriage bolts. The tadpole of Fig. 5-8, upper left, provides a minimum radius of curvature in addition to separating the belt from the bolt head.

Increased local pressure along a narrow strip is a characteristic of several attachment designs in Fig. 5-8. Overpressuring and excessive local heat input are possible dangers in some of these.

5-9 Protection Protecting the interior belt surface from dust as well as heat is an important part of high-temperature fabric-expansion-joint design and operation. A com-

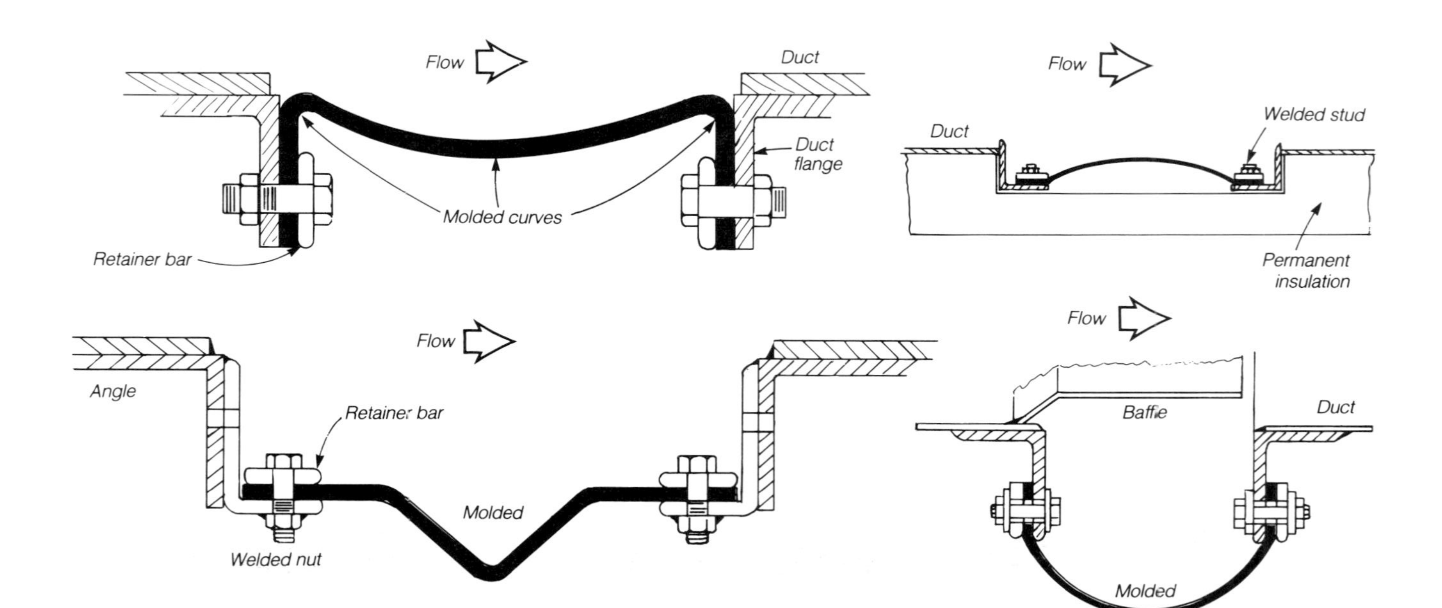

FIG. 5-7 Molded flanges, molded arches, or flat belts are profiles of joints which can be bolted down or replaced from outside or inside.

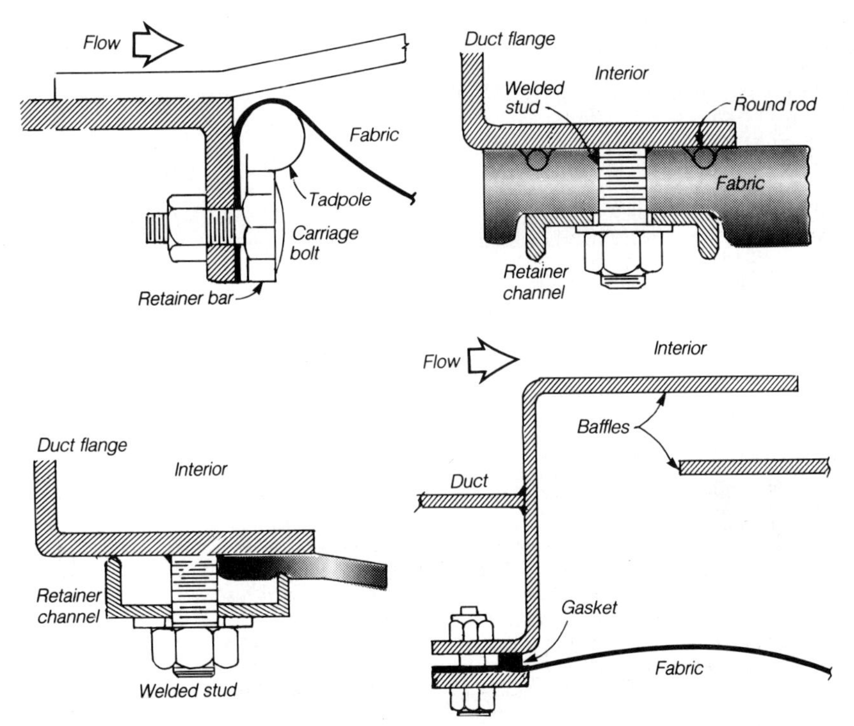

FIG. 5-8 Edge gripping and sealing have the goal of positioning belt without damage while preventing escape of flue gas. Heat sinks and air circulation are additional key factors.

mon device to shield the belt from direct impact of erosive fly-ash particles is the baffle (Fig. 5-8, lower right; Figs. 5-9 and 5-10).

Although even a short baffle prevents direct and heavy blasting against the belt, a longer baffle obviously will do a better job. If the baffle extends past the downstream side of the breach, then duct lateral movement at the joint will be restricted, to say nothing of chances for visual inspection of the belt's inner surface. Several remedies for this, such as those shown in Fig. 5-10, rely on close-fitting sliding baffles, which may have auxiliary flexible seals to close the residual gaps. As a general rule, a baffle entails closer watch over duct guiding, to assure that the movement will be that of the original design basis.

5-10 Large Ducts In large ducts, the narrow auxiliary seals are at a disadvantage because of danger of damage from impact or scraping of the heavy baffle pieces. Tolerances on baffles, in manufacture and installation, may result in too wide a variation of the gap for the lip seal to succeed.

Whatever the barrier to entry into the belt region, some fly ash and water vapor will get through. The high velocities of duct gas convert into appreciable static pressure at blockages, and the pressure can force dust through small openings and narrow gaps. Eddying at obstructions and turns is another source of the pressure differentials that can defeat optimistic predictions of the efficiency of a baffle and seal system.

Often, therefore, the objective is reduced merely to slowing the rate of filling of the joint recess. This leads to the concept of forcibly clearing the joint during operation. Figure 5-10 indicates a pipe connection for air. If the lip seals hold along most of the perimeter, purge air at an acceptable rate can keep dust at bay, although the fan-power cost over joint life must be taken into account.

Water flush is another possibility, but the water must be directed and removed. Hot water can harm some compounds in fabric-joint belts.

5-11 Void Filling Filling the void at the joint with some type of bulky flexible

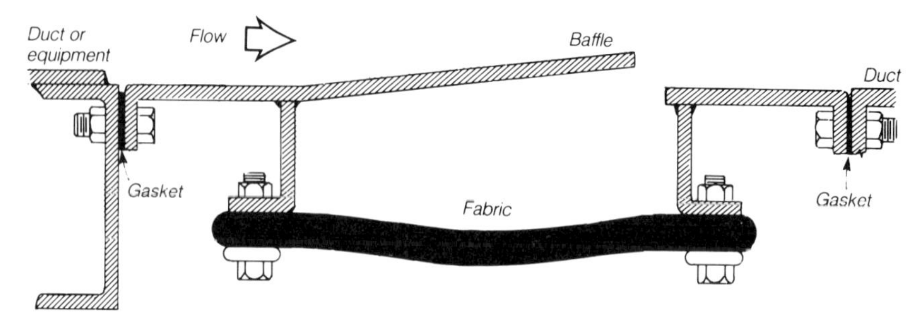

FIG. 5-9 Baffles prevent direct impingement of gas and dust on the joint. Baffle may not fully span gap.

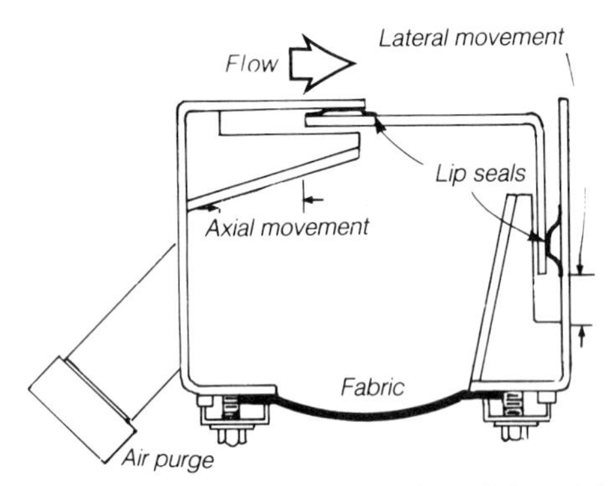

FIG. 5-10 Composite of shapes and lip seals intended to exclude all dust while still allowing axial and lateral movement of duct ends.

material is yet another way to preserve the joint in a dust-laden stream. The joint shown in Fig. 5-11 has its belt recessed well out of the gas stream, and a pad of flexible insulation fills much of the space between the baffle and belt.

The blocking pad may also be curved to follow only the inner perimeter of a boxlike joint zone (Fig. 5-12, top). This pad is retained by spiking it onto pins welded to the box interior.

The second purpose of the pads shown in Fig. 5-12 is to insulate the belt material from gas heat. The builtup type of fabric joint is a composite in which every element plays a role and is rated for a certain temperature and number of hours. Frequently, because the materials near the outer face will not withstand the duct-gas temperature, the design of the joint must base on interior insulation and circulation of outside air.

5-12 Insulation Insulation of the exterior can therefore be fatal in many high-temperature joints. The mounting flanges, which are another path of heat leakage into the belt, may have to be kept free of insulation as well. Finally, in fabric joints that absorb large compression, the folding of the fabric as the breach narrows can double over the belt and prevent outside air from removing heat at the fold.

Installation of a fabric expansion joint demands care to prevent damage to the surface or edges of the belt. While being installed in a horizontal run, a flanged or molded joint may need slings between top and bottom sides. The joint may require axial compression, too, to fit into the breach if the ambient temperature is above the expected minimum cold-weather temperature. Fabric joints in general need some arch in profile at room temperature to allow for cold-weather contraction of the duct.

A joint for a vertical duct should have a plywood platform which, carrying the compressed joint, will slide into the breach. The horizontal-duct installation, as well, profits from a plywood sheet directly under the breach.

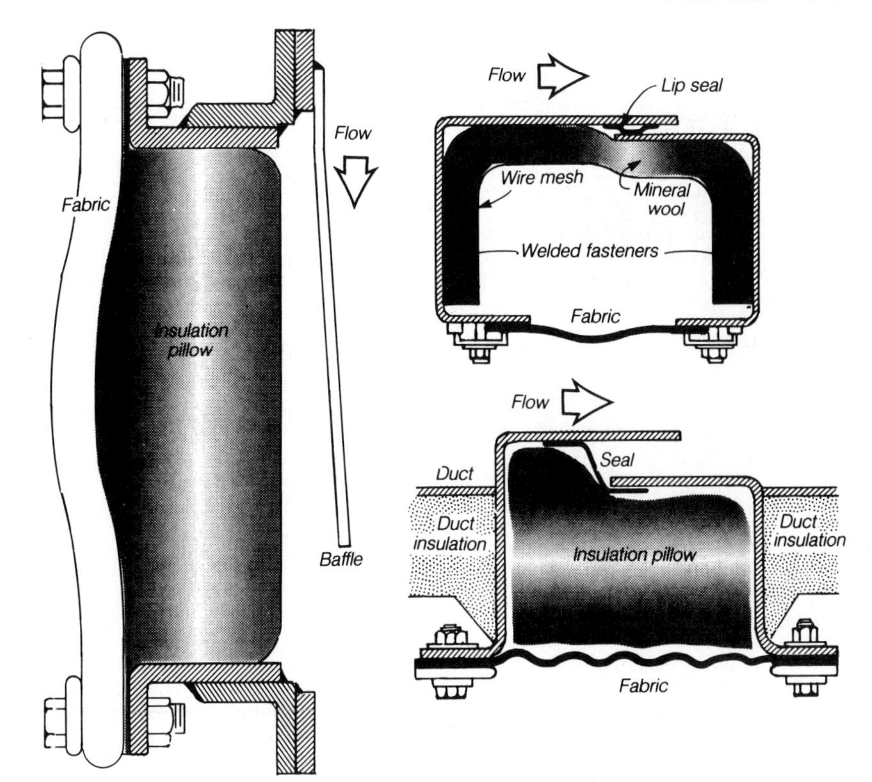

FIG. 5-11 Resilient filler has dual aim of keeping out solids and insulating builtup belt.

FIG. 5-12 Blanket liner (top) and complete filling (bottom) prevent belt overheating.

5-13 Tolerances Duct tolerances affect joint performance. One recommendation is for $\pm 1/4$ in. in all directions for breach opening and alignment. This is a fairly severe requirement on ducting as large as 20 ft in width or height, especially because the tolerance includes face flatness, section-end parallelism, and duct-support discrepancies.

Bolts and retainer bars predominate for attaching and sealing the fabric joint. Bolting-up procedure can be inside the duct or outside (Fig. 5-13). A design calling for cooperation between a worker inside and one outside has obvious drawbacks. Welded studs may be necessary to avoid this, although corrosion could be more of a problem than with through bolts.

Outside-access bolting permits baffles to stay in place during joint replacement, if the new belt can be spliced on-site. Outside bolting is easy to inspect during operation. Inside-access bolting may be easier to get at during a shutdown if structurals or piping runs close to the joint.

5-14 Fasteners The fasteners may be carriage bolts, the heads of which are less likely to cut the fabric during joint movement but which can turn during removal attempts. Stainless steel fasteners are often advisable in corrosive environments.

In operation, the voids and crevices in a fabric expansion joint in ducting that carries a heavy fly-ash concentration will eventually fill with ash. A soft filler that takes up most of the volume next to the fabric will reduce the amount of ash in the joint, but some voids will always be present, changing dimensions as the duct expands and contracts.

Packing of a joint by dry fly ash impedes joint movement and disturbs the heat-transfer patterns on which the joint designer based the work. Worse than this, however, is the potential harm from consolidation of fly ash into crusted deposits closely related to set cement. Water, as either liquid or vapor, promotes this.

FIG. 5-13 Bolting up during installation may require all work to be done from outside.

The fabric joint shown in Fig. 5-14 indicates how fly ash can settle and harden. This joint was deliberately cut open after a test period of operation. Encrusting on a lesser scale can destroy small flexible seals that were intended to keep dust out of the main joint space. Fly ash can also fill the space between a baffle and an insulating pillow. Consolidation of the ash then impairs the joint action, damaging the pillow or the joint fabric.

5-15 Water Formation Water forming in duct systems tends to pick up gases, such as sulfur dioxide, which forms sulfurous acid. The water may originate from fuel combustion or from sprays in a flue-gas desulfurization system. The latter source is the cause of considerable corrosion in ducting between the scrubbers and the stack.

Condensation forming in a duct and running along the bottom toward a joint can create a large pool at the joint (Fig. 5-15). The effect on the spot and tack welds is frequently disastrous. Periodic draining or draining at startup will help but takes time. Periodic flushing out of joints may help, but often the water will miss fly-ash deposit areas or crevices.

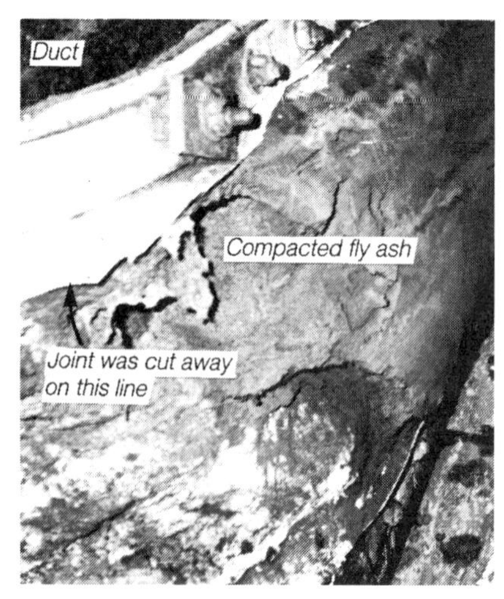

FIG. 5-14 Cut made for research revealed the extent to which fly ash can fill a joint.

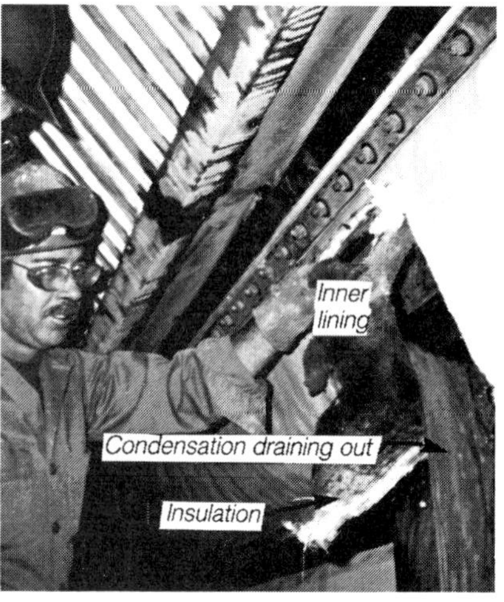

FIG. 5-15 Condensation, pouring from another test-joint opening, is a cause of rusting.

A trade association which has offered conventions and standards for nonmetallic expansion joints is the Fluid Sealing Assn., working through its ducting systems nonmetallic expansion joint division. Its specification sheet is a considerable aid to understanding the factors that influence joint selection and life.

METALLIC DUCT JOINTS

For high-temperature ducting, operating at 600–1200°F, a station may specify metallic expansion joints. Nearly all of these are of bellows type (Figs. 5-16 and 5-17). Allowable compression and extension of a single convolution are small compared with the figures for a nonmetallic joint, in which as much as 5 in. compression is possible for a 16-inch breach. The metallic joint therefore may need several convolutions of thin sheet.

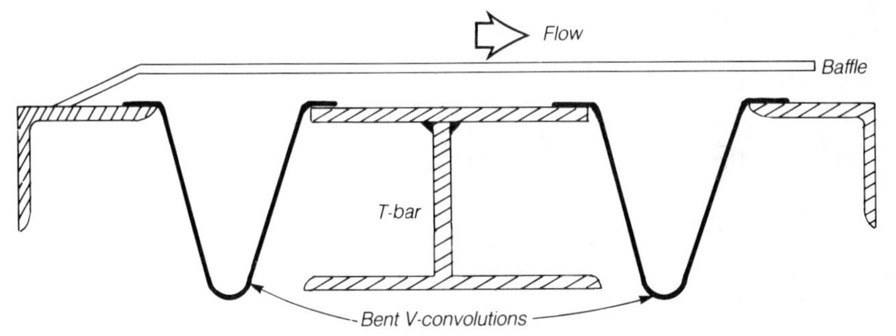

FIG. 5-16 Metal joints for rectangular ducting often have V-shaped convolutions which are simple to manufacture. Thin-gage stainless steel is most common bellows material.

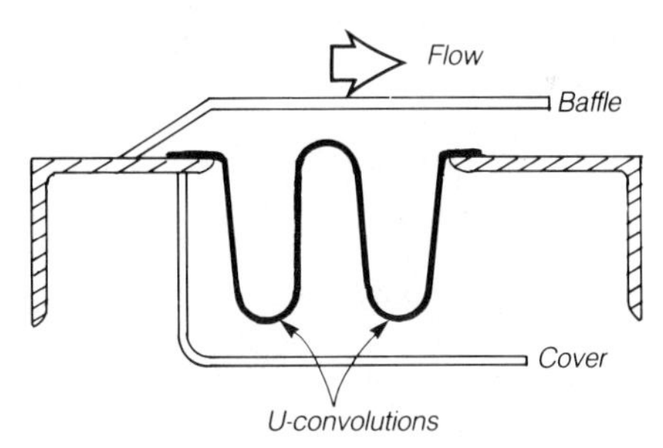

FIG. 5-17 Round ducts, like piping, have metal bellows with deep U-profile convolutions.

5-16 V-Convolution The V-convolution shown in Fig. 5-16 can be made on simple sheet-metal-forming equipment. For high corrosion resistance, Type 316 stainless steel is a frequent choice. Alloys with higher nickel content are sometimes selected. Any alloy for the purpose must be able to take the forming and welding necessary.

Metallic joints can absorb axial expansion and contraction, minor lateral motion, and some angularity. Twist is very harmful. Action at the corners of a rectangular joint throws heavy stress into the metal there. This becomes clear if the axis A-A in Fig. 5-18 is measured in neutral position and when the joint has contracted. Folding down of the convolutions shortens A-A, and the resulting stress in the corner either cracks the weld or causes local buckling along the sides.

5-17 Camera Corner The camera corner shown in Fig. 5-19 is better in respect to stress, but even that is limited in the number of cycles that it can withstand. Many metallic joints are rated at only a few thousand cycles. Although this should be enough in theory for a station's life, there are degrading factors at work, such as corrosion and vibration.

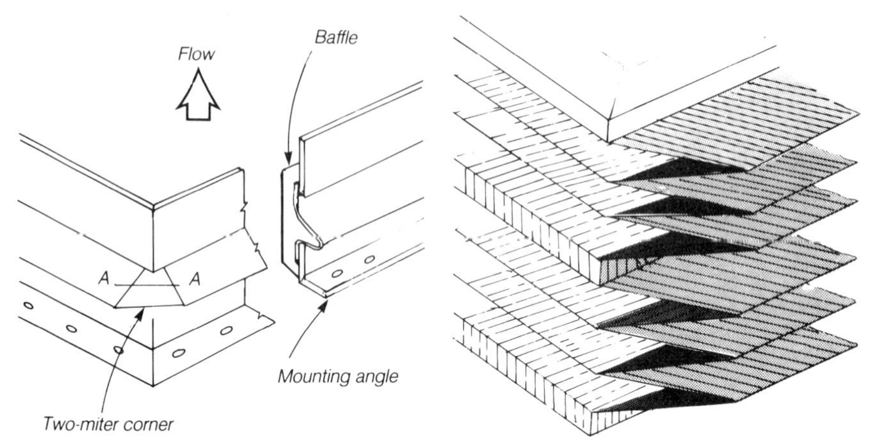

FIG. 5-18 Corner, weak spot in rectangular unit of metal, can benefit from double miter.

FIG. 5-19 Camera corner is a design that has more ability to accommodate axial motion.

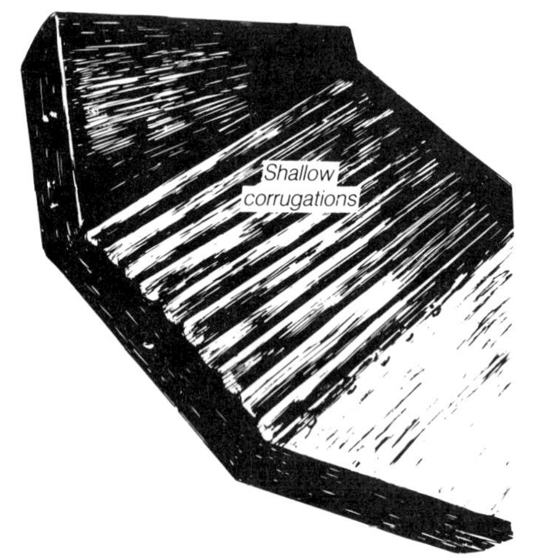

FIG. 5-20 Corrugations in mitered corner for round joint absorb length changes.

The joint shown in Fig. 5-20 has a single convolution of rectangular profile. Load on the miter welds goes in part to extend the corrugated pieces, reducing stress in other pieces of the joint.

Metallic joints often have internal baffles or external covers (Figs. 5-16 and 5-17). Lateral movement is usually small, so that the clearances can be close. If lateral movement is large, then tandem joints are commonly selected, with two convolutions or two separated groups of convolutions. Each group bends and, if the intermediate piece is long, the allowable lateral movement will be large.

5-18 Turbine/Condenser Joints A connection between turbine and condenser must be flexible enough to prevent heavy loads from transferring between the two components and also must be impermeable to air when the condenser is at nearly complete vacuum. Although the dog-bone design of Fig. 5-21 has been chosen in the past, many newer joints are flanged to permit bolting to metal flanges on condensers and turbines (Figs. 5-22, 5-23).

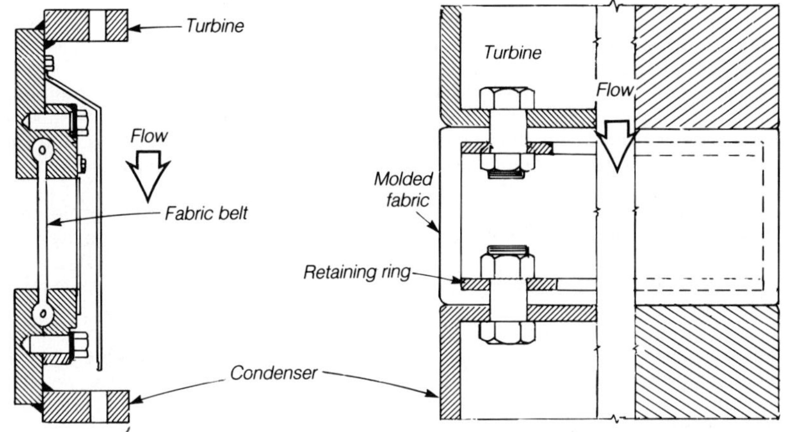

FIG. 5-21 Turbine/condenser connection takes compression and limited lateral movement.

FIG. 5-22 Internal flanges on U-type joint are accessible for bolting from the interior.

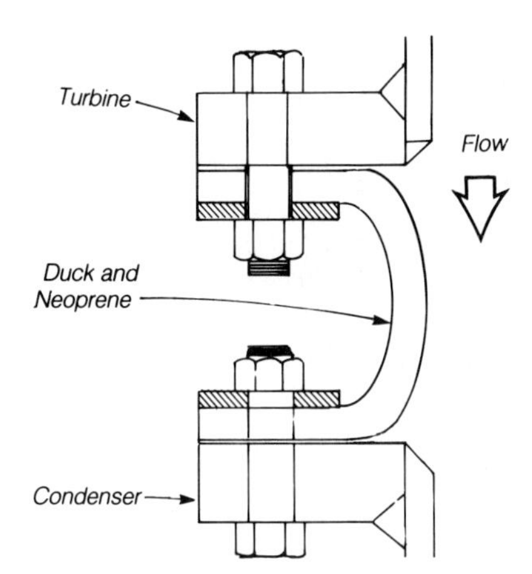

FIG. 5-23 Neoprene corner projects against oil drippings on turbine/condenser joint.

These joints can be formed to near-rectangular, oval, or round pattern, The allowable movement can be around ¾ in. axially or laterally, sufficient for a temperature range of perhaps 200°F.

Metallic bellows joints are another possibility for turbine/condenser connections. The joints are closer in design and appearance to bellows joints for piping than to ducting joints.

Several problems that are acute in piping expansion joints are not highly significant in ducting design and operation. For example, spring rate, which is the reactive force built up per inch of joint-end movement, is usually neglected with nonmetallic joints. Anchoring of ducts, however, is always necessary, to make sure that the duct holds the position specified when the joint is designed. For expansion joints in piping, the anchoring of piping and equipment, and the accurate guiding of piping expansion and contraction, are vital to the success of the joints, no matter what type or size.

PIPING EXPANSION JOINTS

Provision for thermal expansion of piping often relies on either the slip joint, with one end of a pipe length sliding inside an enlarged body attached to the opposite pipe-length end, or on the bellows joint, with its continuous convoluted wall. With any type of expansion joint for piping, the proper anchoring of the pipe and directing of expansion movement is all-important.

For this reason, look first at piping-system components that are factors in designing for thermal expansion. In Fig. 5-24, which is merely schematic, piping connects several pieces of major equipment, such as boilers, turbines, condensers, heaters, and tanks. Obviously, any piping system must be fixed in location at at least one point on the system. A fixed point is ordinarily termed an "anchor."

5-19 Load The load on an anchor may consist of the weight of nearby piping runs or spools, the fluid in the pipe, the axial thrust on the pipe from internal fluid pressure, and perhaps load from thermal expansion, if no other provision is made to absorb it. Added loads from water hammer or subsidence of piping supports may call for design estimates, too.

The equipment itself may be considered anchors, but the manufacturers of much of modern powerplant equipment, especially close-tolerance rotating machinery such as turbines and pumps, now place strict limits on forces and moments that can be applied to the equipment nozzles without voiding guarantees and causing early failures. For this reason, separate piping anchors are necessary along the system, commonly at changes in direction or near sensitive equipment.

5-20 Anchor The anchor itself ranges in form from a massive concrete block with firm embedment in the ground to shackles or welded lugs attached to structural steel far above the ground. For the purposes of expansion-joint design and selection, anchors are usually considered adequate to fix the piping. Whether they are or not could be worth looking into in a good many cases. Consider the tasks of the anchor to understand why not all anchors meet the requirements of expansion-joint design.

Although any anchor that passes even casual inspection can accept axial or lateral load in respect to either pipe run extending out from the anchor, not all anchors can prevent rotation of the pipe. In Fig. 5-24, if point A is fairly resistant to axial load from run A-B, then a high ratio of length A-B to B-C means that there is rotation about the C-C axis at anchor C. Bellows expansion joints do not handle torsion well, and a joint of that type in the C-D run would be endangered.

The design countermeasure could be an anchor at B, with expansion joint in A-B. Then the question could arise as to whether such an anchor, high in the air on structurals, would fully resist the axial thrust force.

Subsidence of anchors, or of guides, is a cause of angular movement of piping, producing a condition which can add to stresses in bellows expansion joints and destroy the effectiveness of slip joints. If the anchor is an equipment nozzle, thermal growth or loads in the equipment can cause the equivalent of subsidence of a piping anchor.

Magnification of lateral movement, because of differences in leverage arms, is another possibility calling for added anchors or guides. Guides themselves (Fig. 5-25) are the second most important element of piping restraint for expansion-joint installations.

5-21 Guide The guide is by intent a complete restriction on pipe lateral movement at the guide while allowing free sliding axially. Because the guide usually supports some pipe weight, and because of friction between spider and guide, some axial restriction develops in a guide. The exact amount can be guessed at or estimated. It may be important with some designs of joints and anchors.

Rotation of the pipe around the guide center is always possible, because of clearance of the pipe or spider in the guide sleeve and because of flexibility of the sleeve. For this reason, guides involved in expansion-joint design are installed in pairs near the expansion joint. Pairing, along with provision of other guides on a long run, assures that lateral movement at the expansion joint will be within required tolerances.

Supports for piping, although necessary, have little to do with expansion-joint installations. A support positions the piping vertically but exerts little restraint on thermal expansion and the resultant piping movement. Friction of piping sliding on plates or brackets does have some small effect in expansion-joint work, but the hanger type of support has far less effect. Once in a piping run, anchors automatically make expansion joints necessary. The introduction of expansion joints in turn often makes more anchors necessary. Figure 5-26 illustrates why.

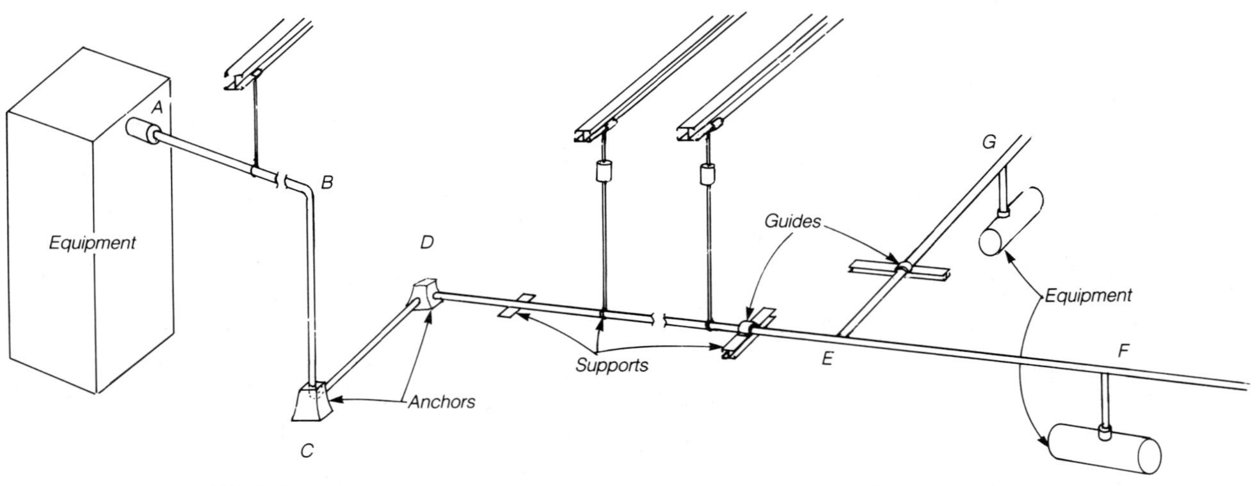

FIG. 5-24 Piping systems must have expansion joints to reduce load on anchors and equipment.

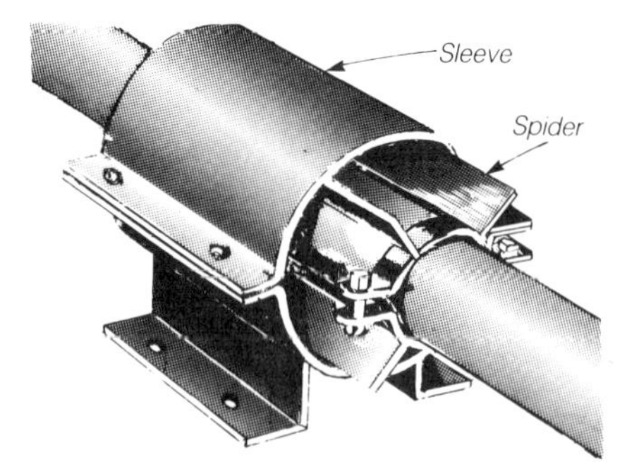

FIG. 5-25 Guides, installed in pairs near joint, confine piping movement to axial only.

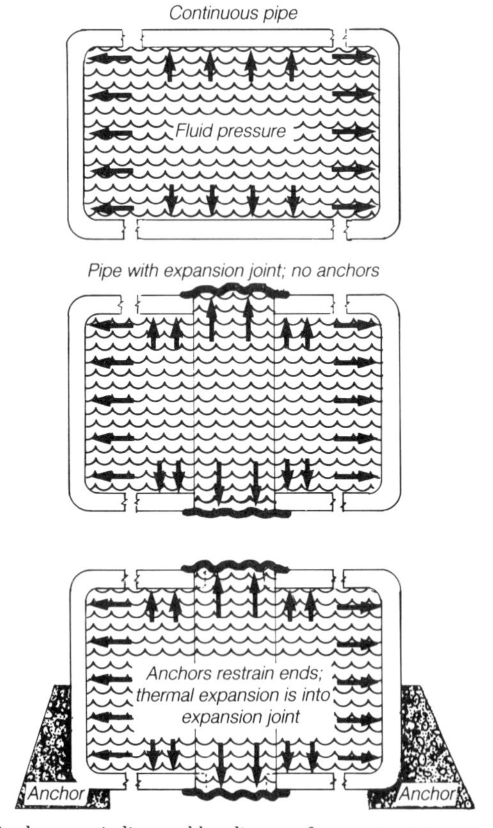

FIG. 5-26 Anchors are indispensable adjuncts of expansion joints in any pipe run.

5-22 End Thrust End thrust, always present in a pipe but resisted by axial tensile force in the pipe wall, puts a pull-apart load on an expansion joint installed in the line. Expansion joints are not intended to resist axial loads and will pull apart unless anchors are set at the pipe ends. Once the anchors are in place, the expansion joint need withstand only the circumferential wall stress from fluid pressure. Thermal expansion can easily occur into the expansion joint from either side.

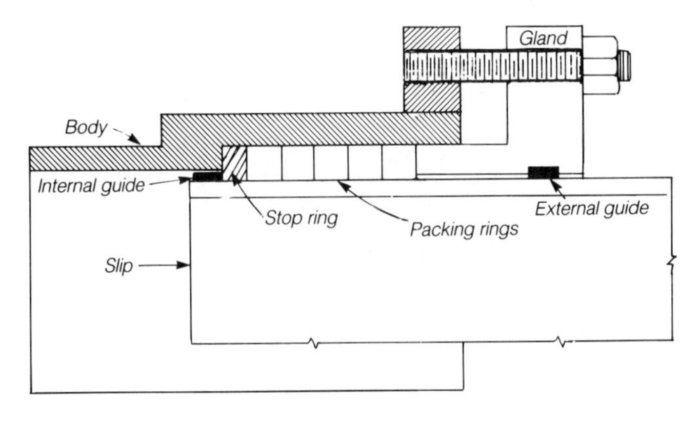

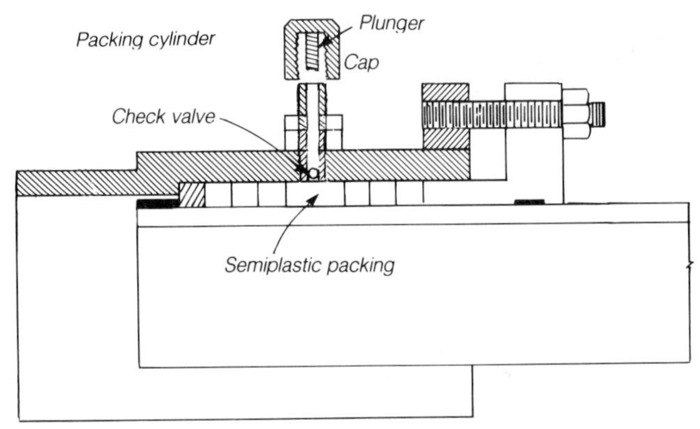

FIG. 5-27 Slip joints have in common the problem of sealing the clearance that must exist between the body, connected to one pipe-end, and the slip, attached to the adjacent pipe-end. One early method (top) called for packing rings compressed uniformly around the slip perimeter by take-up gland. This simple concept was later augmented by pressure exerted from a central zone in the packing space into which special compound is forced. Some units (bottom) have central pressurization plus the effect of gland take-up to force all the packing rings to help in sealing.

In Fig. 5-24, expansion between D and F might call for an expansion joint somewhere along the run. The equipment nozzle at F could be inadequate to restrain thrust, so an anchor would be necessary at perhaps point E, so that the E-G run could begin from a fixed point. A long E-G run would repeat the sequence, although as explained farther on, expansion joints are available to reduce load on equipment at G without making an anchor necessary.

5-23 Slip Joints The slip joints shown in Fig. 5-27 have a provision to prevent pulling apart of the joint under end thrust, but the provision is solely an emergency measure and cannot serve to replace adequate anchoring. The guides in this type of joint are no substitute for external piping guides.

Placement of guides follows rules stemming from elastic-buckling concerns (Fig. 5-28). In short runs, the expansion joint should be near one anchor. A guide at no more than 4 pipe diameters from the expansion joint is aided by another guide at no more than 14 diameters away from the first. Additional guides are set along the rest of the run as required.

Guides are rarely rigid enough to suppress rotation. Buckling at a distance from the expansion joint can therefore cause rotation and lateral movement at the expansion joint. The only cure is to prevent buckling in the run.

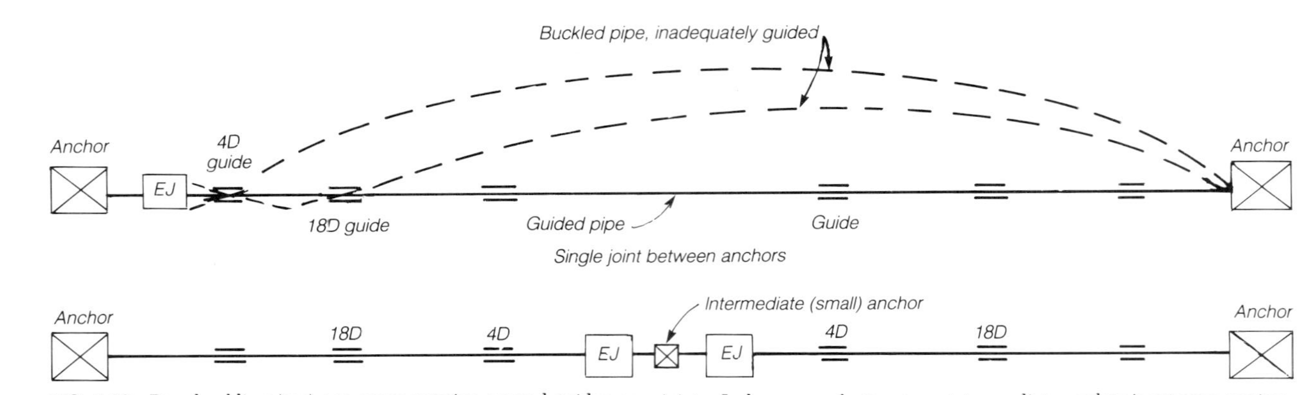

FIG. 5-28 Pipe buckling (top) can cause rotation around guides near joint. In long runs (bottom), an intermediate anchor is set near center.

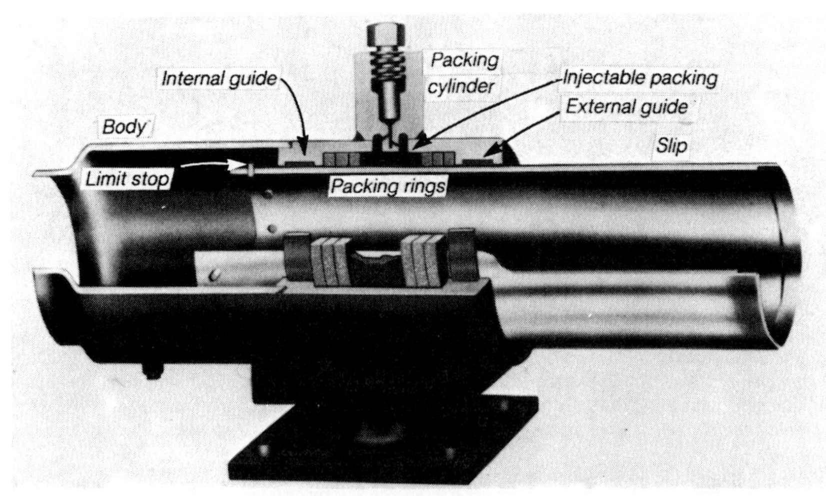

FIG. 5-29 Leakage control in this slip joint is exclusively by injection of packing into the joint through multiple cylinders around perimeter of forged-steel body mounted on anchor.

If a run is too long for the capacity of one joint, then intermediate anchors and nearby joints are placed at even spacing along the run. The intermediate anchor need resist only the force developed by the bellows spring effect and by joint and guide friction. Intermediate anchors are necessary to make sure that each of the adjacent joints takes no more than its share of expansion.

The progressive development of the slip joint has resulted in expansion-absorbing devices that are reliable and easy to inspect and maintain. One basic advantage of the slip joint is its inherent wall strength. Its body, which is an enlarged extension of one pipe end, and its slip, essentially the end of the adjacent pipe length, can be as heavy as needed to resist fluid pressure. This is very different from the bellows joint's requirement for a thin wall to assure flexibility and low spring rate.

In practice, of course, the slip, a separate piece welded or flange-connected to a pipe end, will be machined and ground, reducing its wall thickness somewhat. The body, too, may be a two-piece weldment, with a necked-down adapter welded to a sleeve that forms a packing chamber.

Two inherent problems of the slip joint are guidance of the slip's axial movement and sealing of fluid leakage. The close tolerances of slip in body mean that the 4D and 18D guides on the pipe run are not adequate. Some guidance inside the joint is necessary (Figs. 5-27 and 5-29). Special alloys for this purpose give smooth low-friction movement at temperatures up to 800°F. The slip itself is usually ground and plated to give corrosion resistance and uniform friction on the packing and interior guides.

5-24 Packing The packing or sealing of the slip joint must be done properly at installation, and the design of the packing system must allow suppression of leaks during service. Originally, packing systems for slip joints resembled valve packing boxes, with a gland that could be tightened. Changes have included devices to introduce semiplastic packing compounds to stop leaks.

The nature and action of packing in a slip joint are very complicated. Aging of the packing at high temperature and the action of steam and liquids produce changes that make the packing leak. At high pressures, a minor steam or hot-water leak can cut packing and metal quickly and force shutdown of the line. Replacement of a slip may be needed.

Because large-diameter slip joints present problems in evenness of take-up on packing, it is necessary to add semiplastic packing at several points around the perimeter if leakage is to be controlled. Special packing cylinders (Fig. 5-30) can inject semiplastic packing, which may be flake graphite, into the center of the ring packing. The semiplastic packing exerts pressure on the rings and seals the leaks, provided that the old rings or previously injected packing has not hardened or cracked. If injection fails to stop a leak, there is little except shutdown and replacement of packing which will restore the joint to service.

Injection of new semiplastic packing requires access to the entire periphery of the slip joint, but because all expansion joints require inspection access, this is not a serious comparative disadvantage. Slip joints, with thick perimeter walls, are unlikely to fail catastrophically from corrosion or fatigue and are drainable.

Total traverse available for piping expansion and contraction in slip joints can be as high as 36 in., so that even on long lines only a few slip joints are necessary. Double slip joints, with a common anchor, are another possibility to reduce the total number on a line.

For underground lines, each slip joint is usually in a vault or enlarged space to permit access. In all installations, the guiding must be accurate if the slip is to move axially. A slip joint can take pipe rotation about the longitudinal axis, but the rotation cannot be accompanied by angularity or lateral movement.

5-25 Ball Joints The ball joint shown in Fig. 5-31 resembles the slip joint in basic sealing principle, with close-fitting ductile-iron or composition rings to support the inner rotating ball which is the termination of one pipe length in the joint. The body or socket itself represents the facing pipe-end and is either flanged or has a weld end for connection to the line.

If the rings are of ductile iron, an injectable semiplastic packing, of such nature as flake graphite, is relied on for sealing against fluid escape. The iron rings position the ball in the joint, and the bolted flange shown in Fig. 5-32 is the adjustment device to set the joint for free but closely toleranced motion.

Ball joints can function on steam lines at temperatures to 800°F and at pressures to 650 psig or higher, although most of them are on considerably less demanding service. Piping expansion and contraction is fundamentally axial, a motion which the ball joint obviously cannot absorb. Consequently, two joints must be set along an offset run (Fig. 5-33) and can then handle the expansion of piping in both the connected runs.

Angularity and rotation or twist are two motions easily accommodated by the ball joint, so that guiding of the piping near the joint pair is not as critical as for other types of expansion joint. Anchoring of the lines is still important, however. The ball joint is well-protected against pulling apart, of course.

As Fig. 5-33 indicates, ball-joint pairs are often installed so that the pipe run between the joints will be at a considerable angle from 90° to the main runs at the lowest temperature of the connected piping. At the highest temperature expected, the joints will rotate to the selected angle in the other direction. This is analogous to the installation of slip

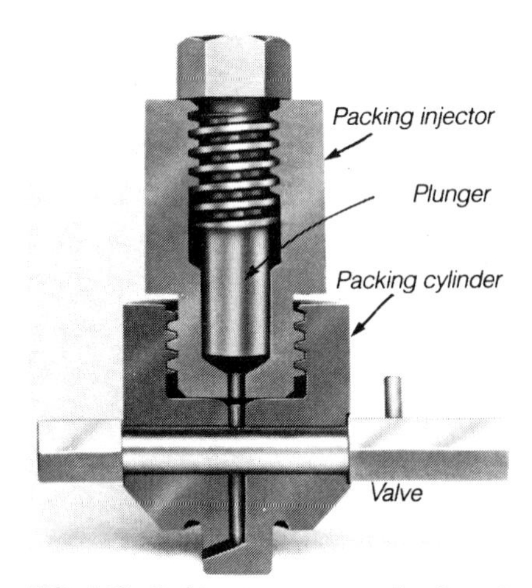

FIG. 5-30 Packing injector provides for safe drilling out of hardened packing at tip.

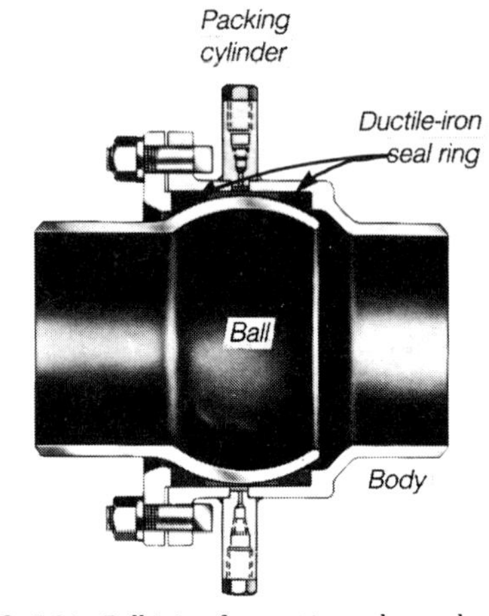

FIG. 5-31 Ball joint, for rotation only, works on a principle similar to that of slip joint.

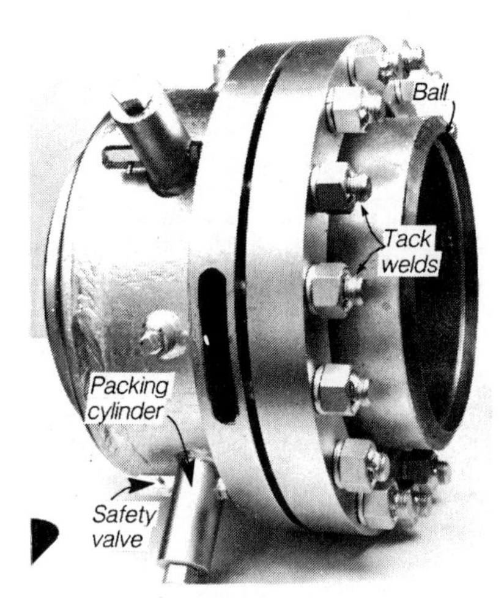

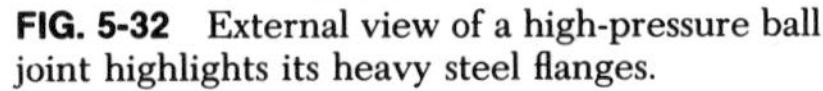
FIG. 5-32 External view of a high-pressure ball joint highlights its heavy steel flanges.

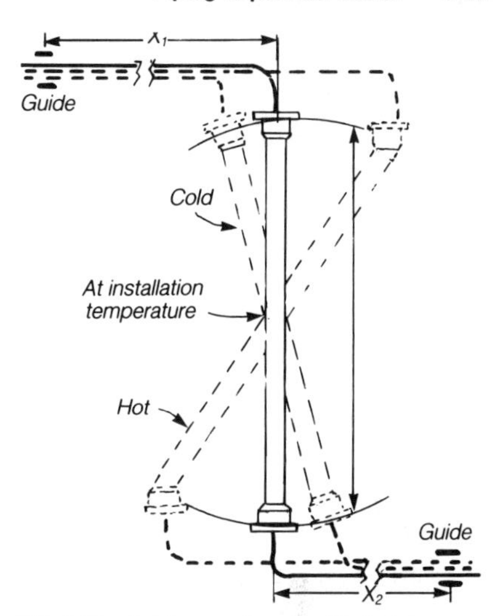

FIG. 5-33 Axial expansion of guided piping runs is taken up by paired ball joints in offset.

joints with some allowance for decrease in temperature to the minimum expected temperature in cold-weather shutdown.

At high angularity, the ball end protrudes into the flow, and for this reason the ball joint may be installed so that the flow is out of the ball rather than into it. There is no other basic reason for preferring either direction for a ball joint on clear fluids.

The distance between ball joints of a pair should be as long as practically possible, which reduces flexing torque on the joints and also the load on piping supports and guides. Although large ball joints can rotate up to 15° included angle, it is better to keep the extreme of motion considerably less.

Problems with ball joints resemble those with slip joints as far as leakage is concerned. Take-up on the flange and injection of more semiplastic packing are remedies, but continuation of high-pressure steam or water leaks means a shutdown for repacking or repair of eroded surfaces. The thick pressure-containing walls and ability to drain the joint mean that corrosion is not a serious safety consideration.

5-26 Bellows Joint A continuous metal wall is one advantage of the metal-bellows expansion joint, frequently selected for accommodation of powerplant piping expansion. Corrugations or convolutions of thin metal give axial flexibility at low spring rate and also allow the joint to take some lateral and angular displacement. The metal bellows can be part of a myriad configurations of joints and combinations of joints. The bellows itself has several forms and is manufactured in at least four different ways, each with its own advantages.

For example, convoluted diaphragms can be welded together at inner and outer perimeters (Fig. 5-34). Depending on profile of the diaphragm, internal pressures up to 1000 psig can be handled. For high pressures, the convolution of the welded diaphragm is more nearly rounded. Welding and weld inspection of this type of bellows are demanding, and most of the metal bellows now are made by cold-forming of thin-wall tubing (Fig. 5-35).

The tubing itself may have several longitudinal welds, each one mechanically worked to flatten the weld somewhat to prevent start of cracks in service. In general, no heat treating is done on the finished formed bellows.

The short cylindrical end of the bellows-joint tube is the most universal accessory, but there are many others that can be added for improved performance, protection, and safety. Some of these are indicated in Fig. 5-36. The end connection of joint to pipe calls for a ring that is much heavier than the thin bellows itself, so that the bellows must be joined, generally by welding, to the connecting rings.

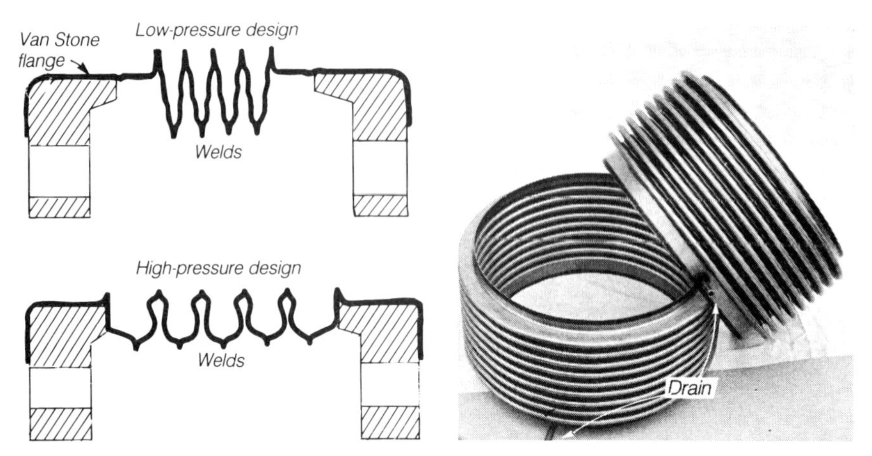

FIG. 5-34 Welded diaphragms are basic flexible elements of one type of metal-bellows joint.

FIG. 5-35 Internal pressure of force produces bellows corrugation in large welded tube.

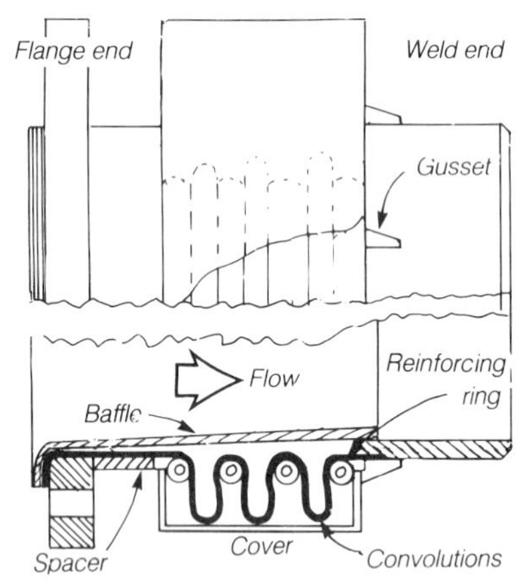

FIG. 5-36 Accessories strengthen and protect the convolutions of metal-bellows joint.

Although the metal bellows for an expansion joint is always designed by the manufacturer, who begins with the data on piping movement and fluid conditions supplied by the customer, it is often helpful to call to mind the basic ideas involved in the convoluted tube wall. Figure 5-37 simplifies a convolution to a quarter strip, on which act the fluid pressure and the adjoining metal.

Fluid pressure creates a meridional membrane stress from the reaction force at the fixed end, at the top, and a meridional bending stress because of bending moment at the fixed end. The two stresses are added together, and 0.7 of the sum is the basis of cycle-life estimates.

The force on the bellows from end loads causes deflection, too. Meridional membrane stress results from the reaction at the fixed end developed from force F. Finally, a meridional bending stress results from the bending moment at the fixed end developed from force F.

Expansion-joint fatigue life depends on the sum of the four stresses. In general, it is necessary to correlate the theoretical design for specific pressure and movement with bellows test data, however, as asserted by the Expansion Joint Manufacturers Assn. (EJMA), a trade association that has formulated standards for metal-bellows expansion joints.

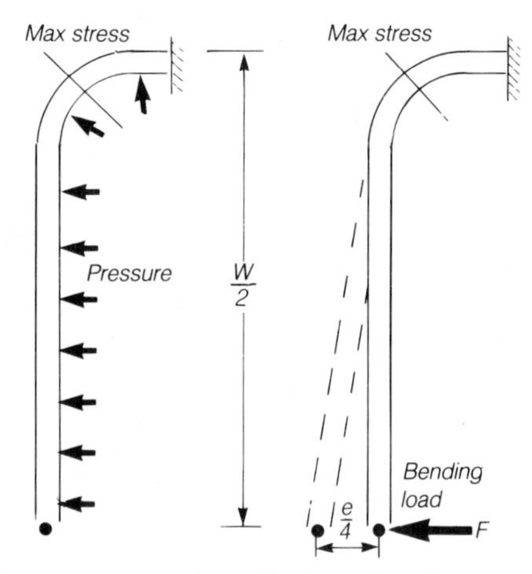

FIG. 5-37 Bellows strip must withstand loads resulting from pressure and bending.

Spring rate, the amount of axial load necessary to compress a bellows convolution a unit distance, is another important factor. It depends directly on mean diameter of the bellows, number of metal plies, and the modulus of elasticity of the metal. Spring rate depends on the cube of ply thickness, making it very sensitive to variation in metal thickness. The spring rate is inversely dependent on the cube of convolution depth, too, so that accuracy in bellows forming is essential if a spring rate must be accurately known.

Naturally, the spring rate of a group of convolutions decreases in inverse ratio to the number of convolutions, but there are limits to the number of convolutions allowed, especially for the conventional internally pressurized bellows.

Resistance to hoop stress increases if metal thickness goes up, but the addition of external reinforcing rings at convolution bottom enables thin-gage sheet to withstand higher pressure. The rings can be solid round bars, round tubes (Fig. 5-36), or high, profiled bars that support a considerable fraction of the convolution side in extreme axial compression.

Pressurization of a convolution between two reinforcing rings tends to give the originally U-shape a more nearly round form—a toroid. The toroid form is actually the design basis for some high-pressure expansion joints (Fig. 5-38). Reinforcing rings are nearly always necessary to support the thin metal in the cylindrical run, and the ring has curved edges to prevent stress concentration there.

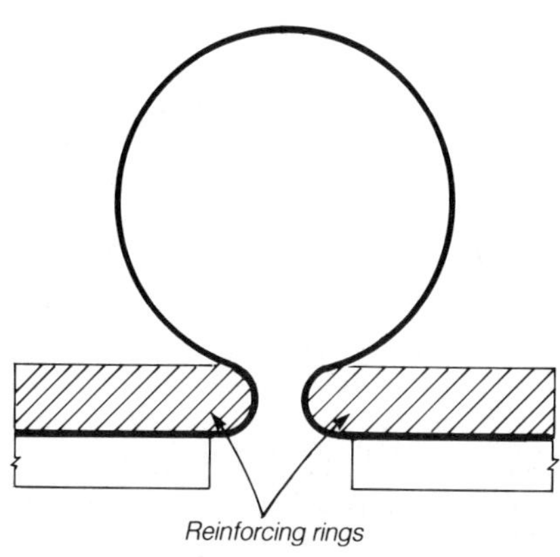

FIG. 5-38 Toroidal shape has advantage for high pressure. Rings support all convolutions.

Metal bellows with internal pressure cannot be drained when the bellows is in a horizontal pipe run, so that the concentration of corrosive agents will increase as a drained line dries, promoting pitting. Stress corrosion and fatigue corrosion are other forms of damage to metal bellows. Suitable selection of bellows metal and good manufacturing procedure help avert trouble, but periodic inspection is also recommended.

In spite of all precautions taken before operation, regular visual inspection and even advanced methods of inspection during the operating life of a bellows joint are advised. This requires access to the joint, so that direct burial is not possible. The standards of the EJMA warn that external visual inspection alone will not give warning of fatigue or internal corrosion, which can cause sudden failures with no visible or audible warning.

Design of a metal bellows depends on exact knowledge of conditions, including not only pressure, temperature, and desired axial movement, but also the lateral and angular movement, vibration, nature of the fluid, and flow velocity. Any failure to give accurate data on all these conditions can lead to inadequate performance or failure of the joint.

Reduction of joint cycle life because of increase in movement, temperature, or fluid pressure is illustrated by graphs similar to those in Fig. 5-39. A given metal has a definite temperature at which both allowable movement and allowable pressure decrease rapidly. On the other hand, if more cycles than the number designed for are wanted from a metal bellows, the reduction in movement or pressure, at a given temperature of course, is not as discontinuous.

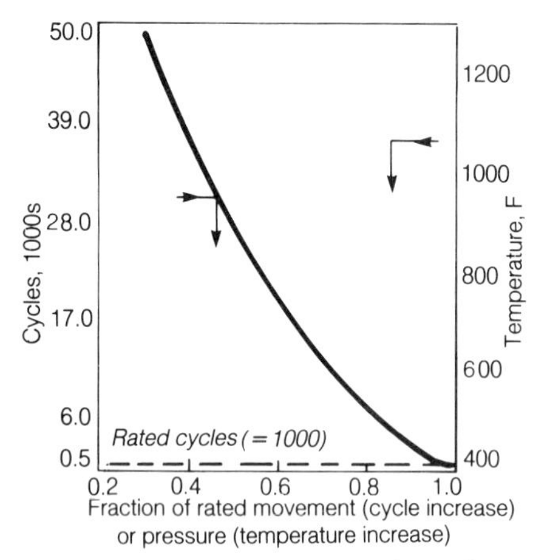

FIG. 5-39 Increase in temperature of desired cycles forces derating of metal bellows.

Lateral and angular movement added to axial movement also reduces performance. The reduction may appear as a decrease in axial movement for a given desired number of cycles. In any case, the bellows manufacturer takes the information into consideration and may decide to recommend a more complicated assembly of two bellows.

Another change can be to a bellows with multiple plies, in which the individual plies act independently with respect to spring rate. For example, replacing a given metal thickness by three plies of one-third the original thickness will cut the axial spring rate to ⅜ of the rate for the single thick ply.

Another advantage of a multiple-ply bellows is the chance for detection of leakage. A small piping connection to the external ply can draw off fluid that leaks through from a failed ply and provide warning of impending total failure (see Fig. 5-40*a*).

Lateral movement results in change in shape of the bellows as indicated in Fig. 5-40, left. One half of each convolution, except perhaps the one at the center of the joint, extends and the other half contracts. A joint with added convolutions could in theory accommodate large lateral movement and give a low spring rate in both axial and lateral directions.

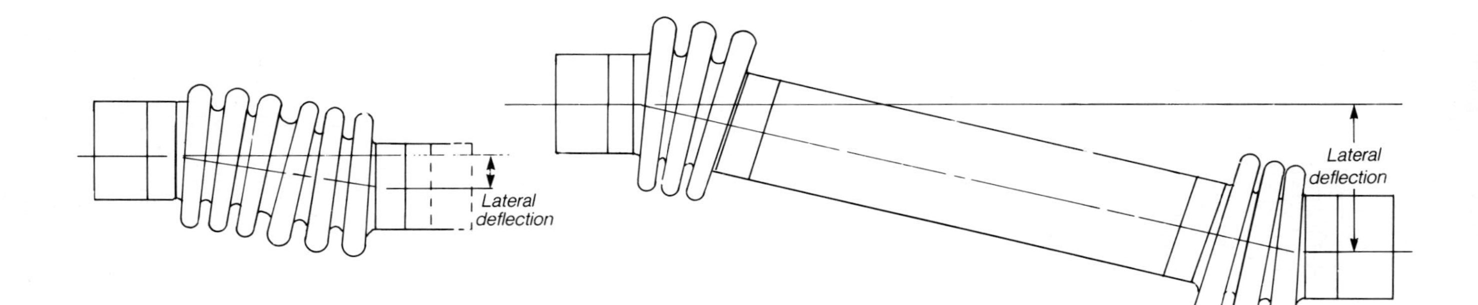

FIG. 5-40 Lateral displacement can be absorbed in a single bellows (left), but two bellows with connector (right) permit more movement.

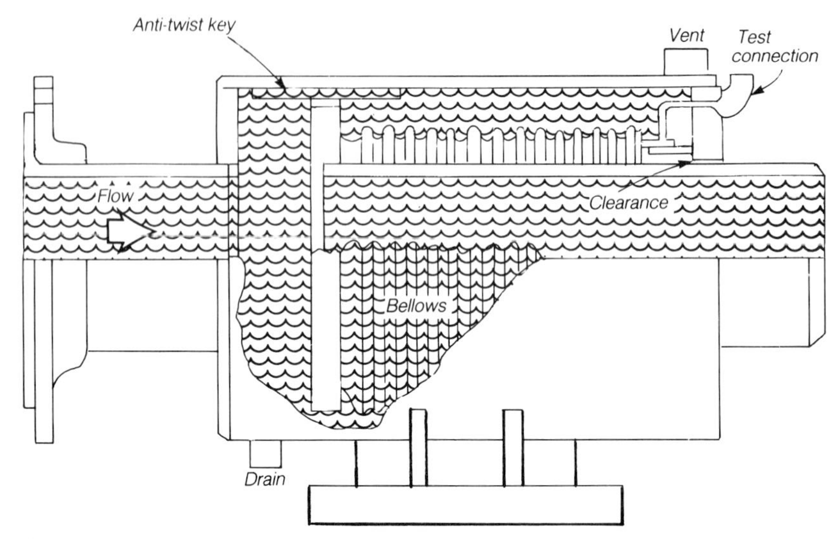

FIG. 5-40a "Squirm," a sideways buckling, is dangerous. A bellows carrying internal fluid pressure would appear at first glance to be no more likely to bend and twist itself into a coil than would a length of pipe under pressure. The high flexibility of the bellows in axial and lateral directions, however, greatly reduces the bellows' resistance to minor net sideways force produced by the fluid. The result is an unstable lateral buckling, like that of a slender column under compression, and a bellows failure. Bellows with many convolutions are endangered by the phenomenon. The fluid pressure in the bellows can be thought of as equivalent to compressive stress in an end-loaded column. In elastic buckling of the column, exceeding of a critical value for deflection sideways allows the column to continue sideways deformation without increase in load—in other words, to buckle in compression. In the bellows, a slight sideways deflection makes one side longer than the opposite. The uniform fluid pressure on the wall applies more sideways force to the longer wall, forcing it into progressively more deformation until the bellows twists and the convolutions distort out of shape permanently. Equations exist for judging whether a bellows will fail by squirm. An externally pressurized bellows (shown in this illustration) is stable and will not squirm. In its case, the slight lengthening of a side can be thought of as generating increased sideways force to return the bellows to the neutral position. Because of the stability of long externally pressurized bellows, they are comparable in total movement to slip joints.

Stability considerations for internally pressurized bellows (Fig. 5-40*a*) limit the number of convolutions, however, so that another method must be selected where lateral movement is large. Two bellows separated by a length of pipe (Fig. 5-40, right) make up the universal design. The design allows axial, lateral, and angular movement of both connected piping runs. It is still necessary to guide the connected piping and anchor it, of course, and to send complete data on desired movement to the manufacturer.

If large axial movement is inescapable and a single bellows is wanted, the externally pressurized bellows is advisable. This design is completely stable and resistant to bellows squirm. In addition, the convolutions are drainable on shutdown. Visual inspection of the bellows in the externally pressurized form of joint is very difficult, however.

Internal baffles and external covers are important features for protection of the relatively fragile bellows. A baffle prevents high-velocity fluid from striking the bellows, and this protection reduces chance for erosion and vibration.

An external cover, a sleeve with adequate clearance, protects the bellows from damage at installation and during service. For visual inspection purposes, the cover must be easily removable, a requirement which is not always observed in practice. The designer of the entire bellows joint has to take maximum joint movement into account in designing a cover or baffle. If flow is vertically upward, then a drain hole in a baffle is necessary on steam or liquid lines.

The universal joint shown in Fig. 5-40 is the simplest example of a bellows joint with

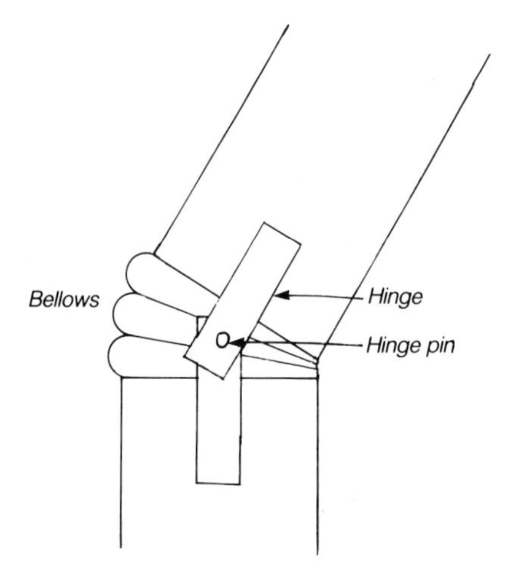

FIG. 5-41 Hinge between bellows-joint ends restricts angular motion to single plane.

more than one bellows. A large number of combinations of convolution groups with rigid or hinged restraints on movement is available, some of the combinations serving narrow applications and some meeting exceptional clearance or movement needs.

5-27 Hinge Bellows Joint The hinge bellows joint shown in Fig. 5-41 at first glance seems to destroy the capability of the bellows to take piping expansion. The hinge bars and pins, on opposite sides of the bellows, are completely resistant to axial movement. In fact, a piping run with a hinge bellows joint needs no main anchors—axial tension in the pipe wall transfers across the hinge. Torsion also bypasses the bellows.

The hinge allows only angular movement, and that must be in one plane. The joint can be installed in an offset between two piping runs, much like the ball joints shown in Figs. 5-31 and 5-33. The guides near the two hinge joints must be planar type, allowing lateral pipe movement as the offset changes length because of temperature change or angle.

Preferably, the distance between piping runs should be as long as possible, so that the lateral movement of the two runs will be small. Alternatively, a third hinge joint in one of the runs and close to the offset compensates for offset expansion and movement.

5-28 Gimbal Joint The gimbal joint (Fig. 5-42) also transfers pipe-wall tension across the bellows and in addition permits angular movement in all planes. This joint

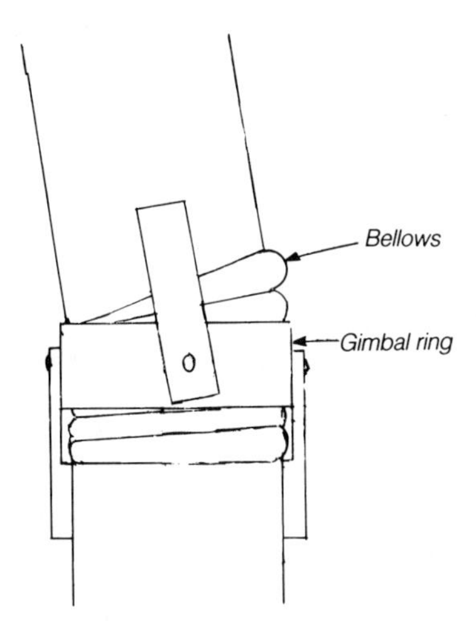

FIG. 5-42 Gimbal mount allows all-plane motion and substitutes for main piping anchors.

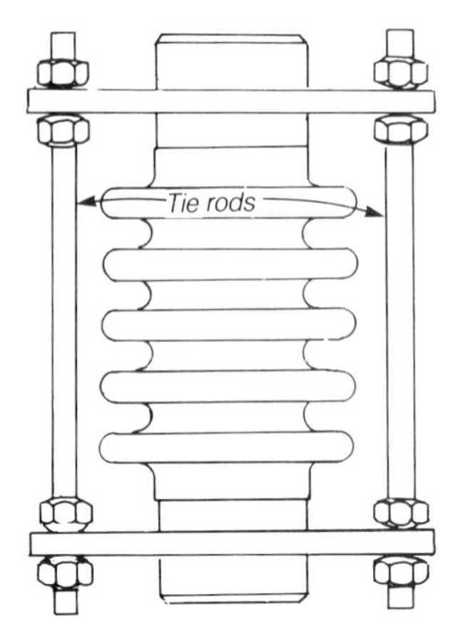

FIG. 5-43 Tie bars are another way of carrying pipe-wall tension across bellows joints.

form is equivalent to the ball joint, and it is installed in pairs, like the ball joint. A hinge joint in one run allows lateral movement as the offset changes length.

In the gimbal joint, the gimbal ring must be heavy and rigid enough to transmit load across the circumferential gap between pivot pins of opposite pairs. In a large joint, a ring to do this must be very large and heavy.

Tie bars, connecting adjacent pipe ends at a bellows (Fig. 5-43), are like a hinge in preventing pull-apart of a bellows and therefore substituting for main anchors. Tie bars, however, are more flexible than a hinge and instead of preventing angular movement in any plane but one, double tie bars allow some angular movement in all planes except one.

In universal expansion joints, with two spaced bellows, four tie bars can take the fluid-pressure thrust force while allowing angular movement. There are many possible combinations of these restraining elements. One of the most complicated is the balanced-pressure bellows joint (Figs. 5-44 and 5-45).

Where a large steam line runs to a component that cannot withstand heavy lateral load on its nozzle, the pressure-balanced joint allows steam to be supplied by a short lateral outlet from a long line without development of large sideways load on the component nozzle. In Fig. 5-44, thermal expansion of the long steam line supplying a turbine, say, might call for an expansion joint in the line. This would require an anchor separate from the turbine nozzle. By providing bellows 1 and bellows 2 of equal pitch diameter D_1 and D_2 and tying the two together by four tie bars, a central piece between the bellows can float back and forth with an ease that depends only on the spring rates of the two bellows. These spring rates can be designed to be within the limits of nozzle loading of the turbine, which is supplied with steam from a lateral off the central connector between bellows.

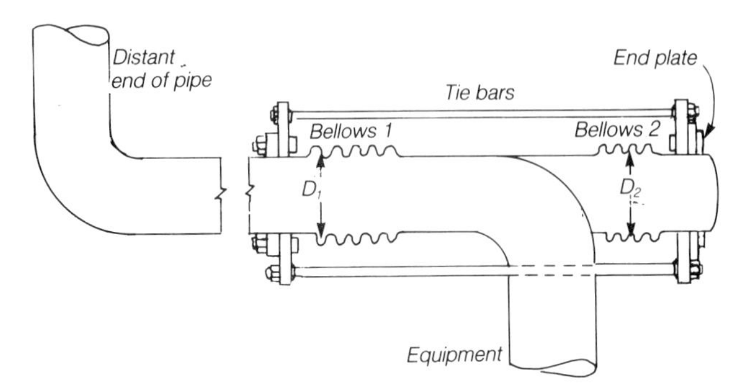

FIG. 5-44 Pressure-balance joint has lateral pipe connection that can move easily between two bellows.

The pipe connection to bellows 2 need not be large, merely enough to assure rapid equalization of steam pressure in the two bellows. The end plate and tie bars, on the other hand, must be strong enough to withstand maximum end thrust of steam in the piping.

If a pipe guide is set on the lateral connection to the turbine or other component, then the guide takes the load resulting from bellows axial stiffness and fully relieves the component nozzle, except perhaps for part of the weight of a joint assembly set above it or hung below it. The pressure-balanced joint may be useful on long vertical lines to heater tops, as well. The joint then is in a vertical attitude with the lateral in the horizontal direction.

Very long universal joints with tie bars are exemplified by connections for tanks and piping. Simple bellows here permit very large lateral movements, and the tie bars prevent fluid-pressure loads from pulling apart the joint, even though there are no main anchors. Spiders are necessary in some of these arrangements to prevent contact of tie bars with the pipe connector.

Tie bars are fixed at the ends to prevent elongation or contraction. Bars with their only firm connection one preventing elongation or contraction of the bellows past a limit are limit bars or rods, a safety measure.

Other bar or rod configurations include short bars over each bellows in a universal arrangement, with motion limits in both axial directions for each bar. This is a control-rod setup, permitting angular movement and limited elongation and compression of the individual bellows. A tie-bar spider may be on the center pipe length but float between limits on tie bars, too, to control the center piece.

Sometimes bellows expansion joints are precompressed before installation in a prepared piping gap. This allows better use of bellows movement range. Shipping bars (Fig. 5-45) are installed at the factory for this purpose. The shipping bars, which also help to protect the bellows during travel, must be removed after installation. To assure high visibility and removal, the bars are painted a bright yellow by convention.

In piping systems which get a hydrostatic test after installation, measures are necessary to protect bellows expansion joints and piping anchors and guides which have been designed chiefly for operating pressure. Applying a test pressure of perhaps 50 percent over the operating value could possibly harm the bellows joint or reduce its life, but the major peril would be indirect, coming from the failure of anchors or guides.

Bellows materials selection centers largely on type 321 stainless steel, which has a

FIG. 5-45 Pressure-balanced joints can be very large and are often chosen to protect steam-turbine connections from excessive pipe-expansion loads.

satisfactory combination of mechanical and chemical properties for most applications. Type 316 stainless has somewhat better pitting corrosion resistance, but extreme corrosion resistance is offered by alloys with even higher nickel content.

Inconel 600 and 625 and Incoloy 800 and 825 are among these, along with Hastelloy B and Hastelloy C 276. For especially high temperatures, the Incoloys and Inconels are superior to stainless steels.

5-29 Elastomer Bellows Joint The elastomer bellows joint has certain characteristics in common with the metallic bellows joint but of course cannot operate in the high pressure and temperature ranges of the metallic versions. Water, air, gas, chemical, and slurry lines are the most common applications for elastomer bellows joints in the powerplant.

One advantage of the elastomer joint over the metallic is in ability to prevent transmission of vibration and to do this without significant harm to the joint. An elastomer bellows joint may be specified for pump inlets and discharges solely to stop vibration and noise from being transmitted along the piping, either from the pump or into the pump from some other source, such as a valve.

Corrosion and erosion resistance is high for the elastomer joint, too, and this makes the joint type very suitable for slurry and acid or alkali lines, where pressures, temperatures, and fluid velocities are moderate.

To resist pressure and withstand flexing in service, the elastomer joint relies on form and reinforcement. Three examples (Figs. 5-46 through 5-48) illustrate how this principle guides design. One or more arches allow the joint to contract and extend axially and, aided by the natural flexibility of the basic cylinder, accept lateral and angular movement.

Between a tube and cover of elastomer, which protect the inner components against damage, is the carcass or body of the tube. The carcass contains fabric reinforcement to take the load of fluid pressure and help transmit piping movement to the arch.

Metal reinforcing rings may be close to the arch shoulders or be spaced partway along the tube on either side. In Fig. 5-48, instead of comparatively heavy rings, wire mesh extending along the tube is the reinforcement. All of these designs are suitable for powerplant purposes and operate satisfactorily where fluid conditions are within the limits of elastomer joints in general.

Because the water and air lines that are the major application area for elastomer joints rarely operate at over 200°F, the axial thermal expansion of the piping is low, and a single arch is usually adequate for thermal expansion. Compression in the range of ½ to 1 in.

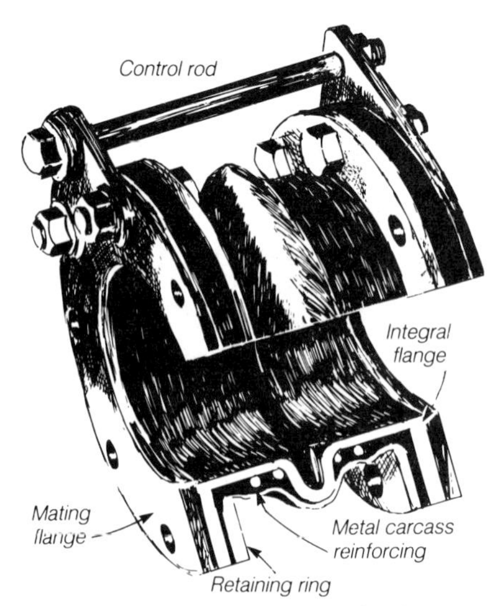

FIG. 5-46 Elastomer joint's control rods act as anchors.

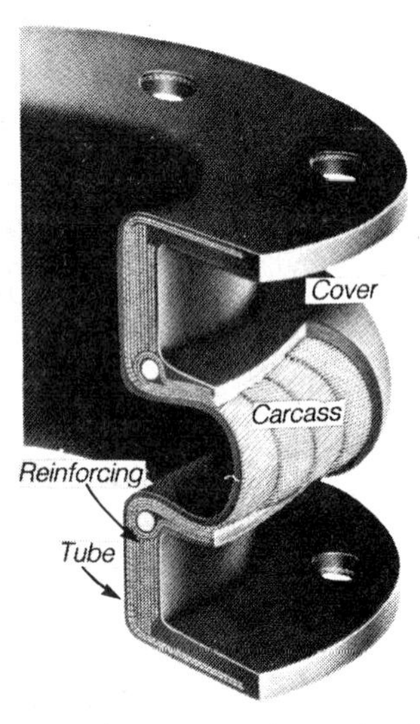

FIG. 5-47 Elastomer joint's metal reinforcement can be rings at shoulder of arch.

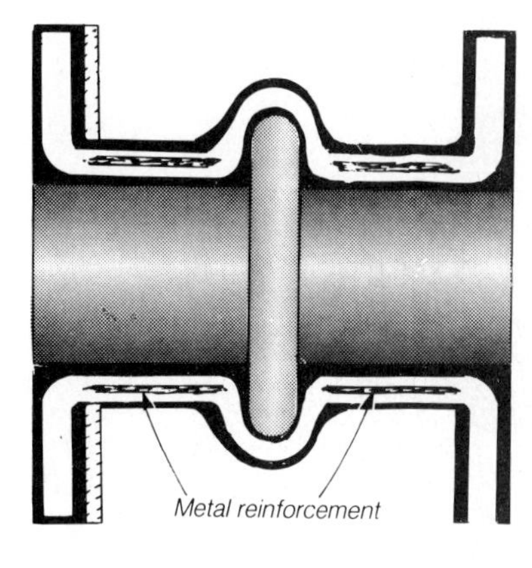

FIG. 5-48 Elastomer joint's extended bands in tube.

is common for these joints. Lateral movement of this amount or more is also accommodated, concurrently in part.

A common feature of elastomer bellows joints is the control-rod arrangement (Fig. 5-46), in which the rod end nuts limit axial elongation of the joint or sleeves on the rods prevent excessive compression of the joint. If the control-rod system is to stop angular movement in any direction, at least three rods are needed.

Although the lining or tube of the joint is resistant to chemicals and particle erosion, an advantage in slurry systems, it is possible for the arch to clog with solids, which prevents compression. An arch can be filled with flexible elastomer to counter this, but because the filled arch cannot compress as much as an open arch, the expedient is confined to strictly necessary applications.

Connection in the piping is almost always by flanges, and the elastomer flanges of the joint must be compressed against metal or plastic flanges of the piping. To assure sufficient clamping pressure without damage to the joint flange, metal retaining rings (Fig. 5-49) are necessary. The profile of these segmented rings adapts to the specific elastomer flange form and allows the joint to be bolted up without crushing the joint flange.

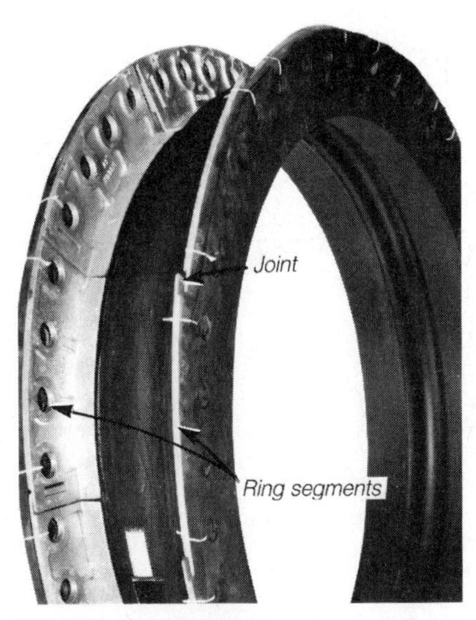

FIG. 5-49 Retaining rings.

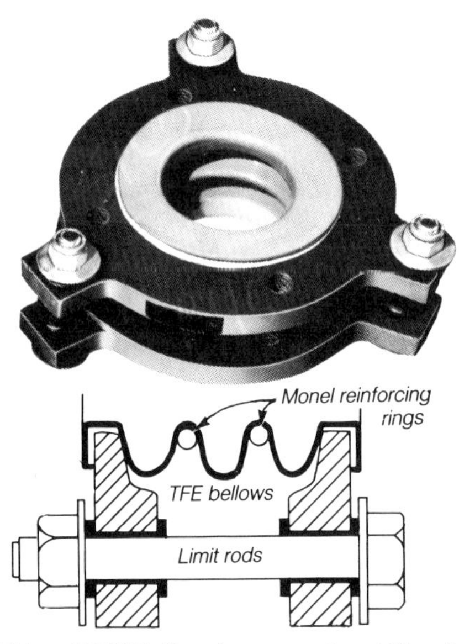

FIG. 5-50 Thin-wall TFE bellows have separate outside reinforcing rings.

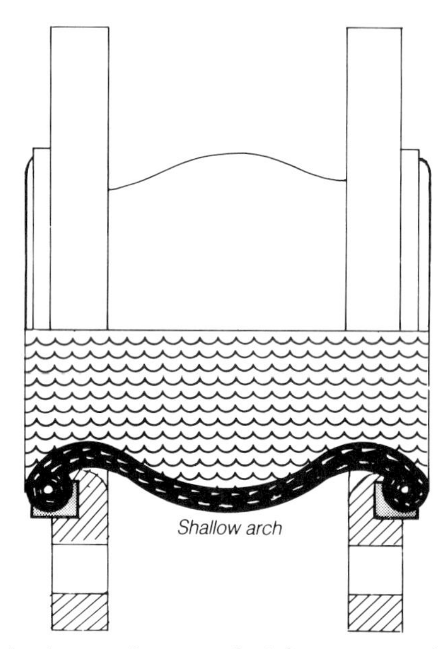

FIG. 5-51 Single low arch relies on nylon tire-cord reinforcement to resist internal liquid pressure.

Variations in elastomer joints include the TFE molded bellows, unreinforced by fabric, shown in Fig. 5-50, and the low-arch variety shown in Fig. 5-51. Pressure ratings on these are in the 100– to 225–psig span.

Concurrent movement, combining lateral or angular with axial, requires derating for the joint, and the chart shown in Fig. 5-52 indicates, in terms of maximum movement,

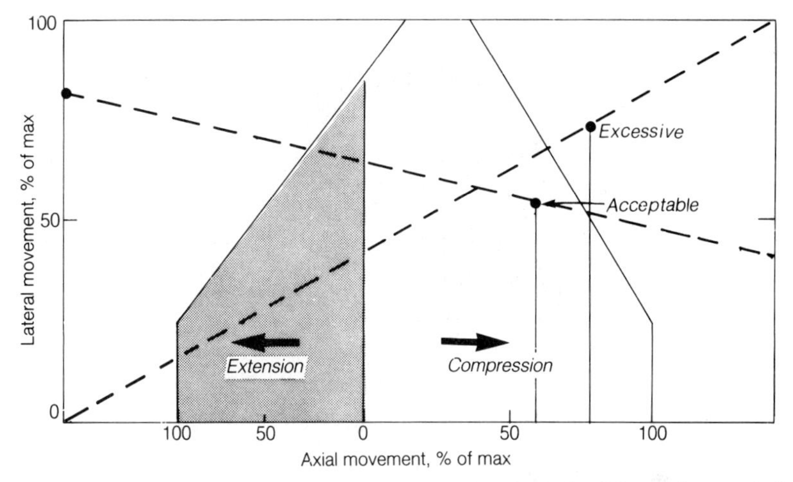

FIG. 5-52 Concurrent movement in lateral and angular form reduces the ability of elastomer bellows joints to take axial change. Chart applies to wide size range of given design.

what the amount of derating can be. Temperature limits depend on the elastomer materials principally. Most elastomer joint types are available in several different materials, sometimes with different materials for tube and cover.

Natural rubber, Neoprene, chlorobutyl, nitrile, TFE, and fluoroelastomers are prominent choices for material. One combination is natural rubber for the inner surface or tube to handle fresh or salt water at temperatures to 180°F, with a Neoprene cover for resistance to weather, oil, and ozone.

Both Neoprene and nitrile are resistant to petroleum products and therefore can be the tube material for oil and waste lines. TFE is highly resistant to chemicals, weather, and many other agencies of destruction in the expansion-joint area. It has the disadvantage, however, of being easily scratched, and this restricts its application in some cases.

Anchors and guides are as necessary with elastomer expansion joints as they are with metal joints. The joint should be near a main anchor, too. When set near the inlet and discharge of pumps and compressors, the joints protect the major components against overload from piping movement.

Section 6
Welding

Part 1 Gas Welding and Braze Welding*

MARK BOYER

Harris Calorific, Division of Emerson Electric, Cleveland, Ohio

Gas welding involves the melting of the base metal and the filler metal, if used, by means of the flame produced at the tip of the welding torch. The molten metal from the plate edges and the filler metal, if used, intermix in a common molten pool and upon cooling coalesce to form one continuous piece.

The gas flame is also used for braze welding, which differs from gas welding in that the melting temperature of the filler metal used in braze welding is below that of the base metal. This allows the braze-welding operation to be performed at a temperature below that at which the base metal would melt, but at or above the melting temperature of the filler material. Braze welding differs from brazing in that the joint design for braze welding is similar or identical to that used in gas welding, and capillary action is not a factor in the formation of a bond.

Acetylene is commonly used as the fuel gas in gas welding because the oxyacetylene flame provides the highest flame temperature of all the commonly used fuel gases. With proper adjustment, the flame can also supply the proper atmosphere to provide a protective covering over the metal in the molten pool which is maintained during welding. A 2:1 mixture of oxygen to acetylene provides flame temperatures up to 5600°F (3092°C), approximately twice the melting temperature of steel, and produces the high localized heating necessary for welding.

Hydrocarbon fuel gases such as propane, butane, natural gas, and various mixtures employing these gases as well as others, are not suitable for welding ferrous materials because of either their low flame temperature or the excessive oxidizing characteristics of the flame. These fuel gases can be used to weld some of the nonferrous metals but find greatest application in brazing and soldering. Oxyacetylene welding is by far the most widely used of the gas welding methods.

* This Section is abstracted from the *Welding Handbook* by permission of the American Welding Society.

GASES

Many processes are involved in gas welding and many fuel gases are employed. The commercial fuel gases have one common property: they all require oxygen to support combustion. A fuel gas, when burned with oxygen, must have: (1) high flame temperature, (2) high rate of flame propagation, (3) adequate heat content and (4) minimum chemical reaction of the flame with base and filler metals, in order to be used for welding operations.

Acetylene and hydrogen are the only fuel gases commercially available that possess all the desired properties. Other gases, such as propane and natural gas, have sufficiently high flame temperatures but exhibit low flame propagation rates. These other gas flames are excessively oxidizing at oxy–fuel gas ratios high enough to produce usable heat-transfer rates. Flame-holding devices, such as counterbores on the tips, are necessary for stable operation and good heat transfer even at the higher oxy–fuel gas ratios. These gases, however, can be used for brazing, soldering, and similar operations where the demands upon the flame characteristics and heat-transfer rates are not excessive.

The temperature of the flame is evaluated by either the Fahrenheit or Celsius scale. The heat quantity is measured in British thermal units (Btu) or kilocalories (kcal). These units may be converted from one to the other by use of the following relationship: 1 kcal = 3.968 Btu.

6-1 Acetylene The fuel gas most widely used in gas welding is acetylene. A compound of carbon and hydrogen (C_2H_2), the gas contains 92.3 percent carbon and 7.7 percent hydrogen by weight. It is colorless, lighter than air, and has a sweet, easily distinguishable odor similar in many respects to that of garlic. Acetylene contained in cylinders has a slightly different odor from that of pure generated acetylene. This difference is caused by the acetone vapor contained in the cylinder.

Acetylene is a highly combustible compound that liberates heat upon decomposition. Acetylene contains a higher percentage of carbon than any of the numerous other hydrocarbons. This combination gives the oxyacetylene mixture the hottest flame of any of the commercially used gases.

At temperatures above 1435°F (780°C) or at pressures above 15 psig, acetylene gas is unstable and an explosive decomposition may result even without the presence of oxygen. This characteristic has been taken into consideration in the preparation of a code of safe practices for the use of acetylene gas. It has become an accepted safe practice never to use acetylene at pressures exceeding 15 psig in generators, pipelines, and hose.

The Oxyacetylene Flame. The complete combustion of acetylene is represented by the chemical equation

$$2\,C_2H_2 + 5\,O_2 \rightarrow 4\,CO_2 + 2\,H_2O$$

This equation indicates that two volumes of acetylene (C_2H_2) and five volumes of oxygen (O_2) react to produce four volumes of carbon dioxide (CO_2) and two volumes of water vapor (H_2O). It is evident from the equation that, for complete combustion, the volumetric ratio of oxygen to acetylene is 2.5:1.

In the production of the oxyacetylene flame, the acetylene is mixed in the torch with an approximately equal volume of oxygen, and the additional 1.5 volumes of oxygen necessary to complete the combustion is obtained from the air surrounding the flame.

Since the gas mixture issuing from the torch tip does not necessarily contain enough oxygen for complete combustion, the following primary reaction takes place in the inner zone of the flame (called the inner cone) right at the welding tip:

$$2\,C_2H_2 + 2\,O_2 \rightarrow 4\,CO + 2\,H_2$$

This equation indicates that two volumes of acetylene (C_2H_2) and two volumes of oxygen (O_2) react to produce four volumes of carbon monoxide (CO) and two volumes of hydrogen (H_2). This primary reaction produces the brilliant inner cone of the flame which contributes most of the actual heating. In the outer envelope of the flame the carbon monoxide and hydrogen resulting from the primary reaction burn with oxygen from the surrounding air, forming carbon dioxide and water vapor respectively, as shown in the following two equations:

$$4\,CO + 2\,O_2 \rightarrow 4\,CO_2$$
$$2\,H_2 + O_2 \rightarrow 2\,H_2O$$

The oxyacetylene flame is easily controlled by valves on the welding torch. By a slight change in the proportions of oxygen and acetylene fed to the flame through the torch, the chemical characteristics in the inner zone of the flame—and consequently the action of the inner cone on molten metal—can be varied over a wide range. Thus, by adjusting the torch valves, it is possible to produce a neutral, oxidizing, or reducing flame.

Although acetylene is the fuel gas most widely used in welding and cutting, other fuel gases are often employed for cutting and heating. Among the more common gases used for preheating are propane, hydrogen, city gas, and natural gas. These gases, with the exception of hydrogen, are also used for soldering and brazing operations.

Other gases are available under various trade names, but these generally are commercial combinations of the basic hydrocarbon gases.

6-2 Oxygen Oxygen in the gaseous state is colorless, odorless, and tasteless. It occurs abundantly in nature. One of its chief sources is the atmosphere, which contains approximately 21 percent oxygen by volume. Although there is sufficient oxygen in the air to support combustion, the use of pure oxygen speeds up reactions and increases flame temperatures.

Oxygen may be produced from chemical sources, such as mercuric oxide, barium oxide, potassium chlorate, or other compounds containing oxygen. It may also be produced by the electrolysis of water, or it may be extracted from the atmosphere. Of these, only the last two methods are of commercial importance in the welding industry, where large quantities of oxygen are required.

Oxygen is produced from the electrolysis of water by passing an electric current through water containing an electrolyte, such as caustic soda, to increase its electrical conductivity. The water decomposes, allowing oxygen to collect at the positive pole and hydrogen at the negative pole. Limited commercial quantities of oxygen are produced by this means; the process is used principally for the production of hydrogen.

Most oxygen used in the welding industry is extracted from the atmosphere, which has the following composition by volume: nitrogen, 78.03 percent; oxygen, 20.99 percent; argon, 0.94 percent; carbon dioxide, hydrogen, neon, helium, and other rare gases, 0.04 percent.

In the extraction process, air is compressed to approximately 3000 psig (some types of equipment operate at much lower pressure), the carbon dioxide and any impurities are removed, and the air passes through coils and is allowed to expand to a rather low pressure. The air is substantially cooled during the expansion, then passes over the coils (further cooling the incoming air) until liquefaction occurs. The liquid air is sprayed on a series of evaporating trays or plates in a rectifying tower. The nitrogen and other gases boil at lower temperatures than the oxygen, and, as these gases escape from the top of the tower, high-purity liquid oxygen remains in a receiving chamber at the base. Some plants are designed to produce liquid oxygen; in other plants gaseous oxygen is withdrawn for compression into cylinders.

GAS DISTRIBUTION

Gas is distributed to various sections of an industrial plant in single cylinders, from portable manifolds, from stationary piped manifolds, from generators, and by pipelines fed from bulk-supply systems.

6-3 Single Cylinders Individual cylinders of gaseous oxygen and acetylene provide an adequate supply of gas for welding and cutting torches consuming limited quantities of gas. Cylinder trucks are used extensively to provide a convenient, safe support for a cylinder of oxygen and a cylinder of acetylene. The gas can be transported readily by such means.

Oxygen is transported to the user in individual cylinders as a compressed gas or as a liquid; there are also several bulk-distribution methods. Gaseous oxygen in cylinders is usually under a pressure of approximately 2,200 psig. Cylinders of various capacities are used, holding approximately 70, 80, 122, 244, and 300 cu ft of oxygen. Liquid oxygen cylinders contain the equivalent of approximately 3,000 cu ft of gaseous oxygen. These cylinders are used for applications which do not warrant a bulk oxygen-supply system but which are too large to be supplied conveniently by gaseous oxygen in cylinders. The liquid oxygen cylinders are equipped with liquid-to-gas converters.

A single cylinder of acetylene cannot be used if the volumetric demand is high, since acetone may be drawn with the acetylene from the cylinder. It has, therefore, become standard practice to limit the withdrawal of acetylene from a single cylinder to an hourly rate not exceeding $1/7$ the cylinder's volumetric contents.

Cylinders for liquefied fuel gas contain no filler material. These welded steel or aluminum cylinders hold the liquefied fuel gas under pressure. The pressure in the cylinder is a function of the temperature.

These liquefied-fuel-gas cylinders have relief valves which are set for 375 psig, a pressure reached at approximately 200°F (93°C). Should these temperatures be reached, the rapid discharge of fuel gas through the relief valve causes the cylinder to cool down and the relief valve to close. In a fire, the cylinder relief valve opens and shuts intermittently until all the contents of the cylinder have been discharged or the source of the extreme heating of the cylinder has been removed.

6-4 Manifolds Individual cylinders cannot supply high rates of gas flow, particularly for continuous operation over long periods of time. Manifolding of cylinders is one answer to this problem. A reasonably large volume of fuel gas is provided by this means, and can be discharged at a moderately rapid rate.

Manifolds are of two types—portable and stationary. Portable manifolds may be installed with a minimum of effort and are useful where moderate volumes of gas are required for jobs of a nonrepetitive nature, either in the shop or the field. Stationary manifolds are installed in shops where even larger volumes of gas are required. Such a manifold feeds a pipeline system distributing the gas throughout the plant. Such an arrangement enables many welding operators to work from this common pipeline system without interruption. Alternately, it may supply large automatic gas welding equipment (Figs. 6-1, 6-2, and 6-3*a*).

6-5 Safe Practices Oxygen by itself does not burn or explode, but it does support combustion. Oxygen under high pressure may react violently with oil or grease. Cylinders, fittings, and all equipment to be used with oxygen should be kept away from oil and grease at all times. Oxygen cylinders should never be stored near highly combustible materials. Oxygen should never be used in pneumatic tools, to start internal combustion engines, to blow out pipelines, to dust clothing, or to create a head pressure in tanks.

Acetylene is a fuel gas and will burn readily. It must therefore be kept away from open fire. Acetylene cylinder and manifold pressures must always be reduced through gas pressure-reducing regulators and pipeline pressures should be controlled through suitable pipeline regulators. Cylinders should always be protected against excessive temperature rises and should be stored in well-ventilated, clean, and dry locations free from combustibles. They must be stored and used with the valve end up. Loose carbide should not be scattered about or allowed to remain on floors, since it will absorb moisture from the air and generate acetylene.

Acetylene in contact with copper, mercury, or silver, especially if impurities are present, may form acetylides. These compounds are violently explosive and can be detonated by a slight shock or by the application of heat. No alloy containing copper in excess of 67 percent should be used in any acetylene system, unless such alloys have been found to be safe in the specific application by experiment or test.

Fuel-gas cylinders, other than acetylene, also contain gas under pressure and should be handled with care. These cylinders should likewise be stored in clean, dry locations.

Liquid cylinders are constructed with a vacuum between the inner and outer shell. They should be handled with extreme care to prevent damage to the internal piping and the loss of this vacuum. Such cylinders should always be transported and used in an upright position.

Cylinders can become a hazard if tipped over. The greatest care should be exercised to avoid this possibility. Acetylene cylinders, in particular, should always be used in an upright position. When gas is taken from an acetylene cylinder that is lying on its

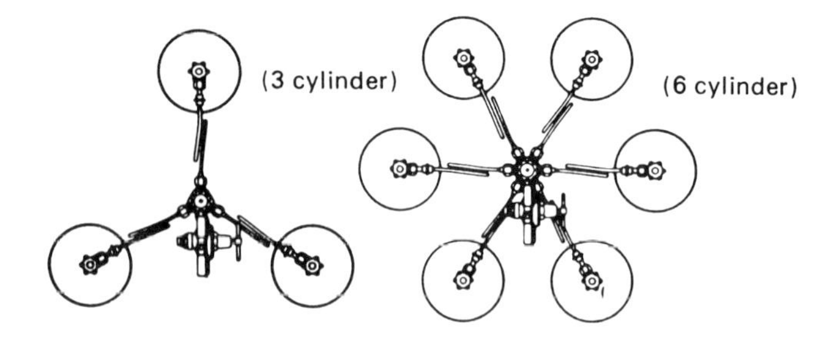

PORTABLE MANIFOLDS

Web-type manifolds have the following advantages: Master shutoff valve, circular arrangement of cylinders for maximum cylinder stability, economy of space and safety. Acetylene leader coil includes check valve, flame arrester and handle. Connection to central block is CGA-540 for non-flammable gas service and 982 for flammable gas service. Floor stand is optional for support of unit while not in use.

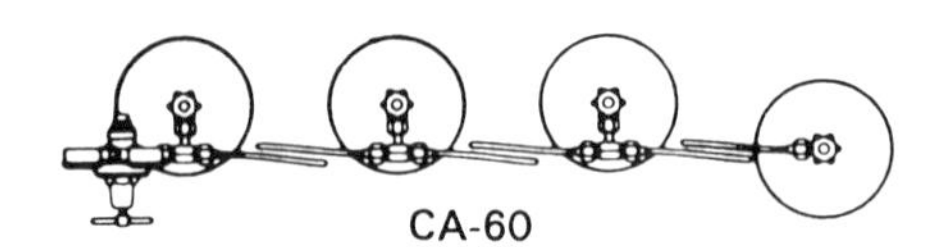

EXTENDABLE MANIFOLD (4 oxygen cylinders)

EXTENDABLE — With Control Regulator

Extendable manifolds supply a large volume of gas and eliminate frequent cylinder replacement. A manifold of any desired length can be made up by adding tee fitting and leader coil for each additional cylinder. Leader coils for flammable gases have butyl-lined fabric covered tubing with a burst strength of 12,000 psig. Non-flammable leader coils are of high strength brass tubing. Extendable manifold leader coils have the same CGA connections on both ends.

FIG. 6-1 Portable manifolds. *(Harris Calorific, division of Emerson Electric.)*

side, acetone is readily withdrawn. This contaminates the flame and will result in welds of inferior quality. It is standard practice to fasten the cylinders on a cylinder truck or to secure them against a rigid support. All hoses connecting the equipment should be kept as short as possible and free from sharp twists. Always use check valves either at the torch or regulator to prevent reverse flow of gases into the hoses that might mix there and become an explosion hazard (Fig. 6-3*b*).

EQUIPMENT

The basic tools required in gas brazing and welding are shown in Fig. 6-4, which illustrates two gas cylinders (one fuel gas, the other oxygen), gas regulators for reducing and controlling the cylinder pressures, hoses for conveying the gases to the torch, and the torch and tip combination where the gases are controlled, mixed, and burned.

Each of these units plays an essential part in the control and utilization of the heat necessary for operations such as gas welding, brazing, and heating. The heat is obtained from the combustion of the gases, but the effectiveness of the heat is governed, to a large extent by the skill of the welder. Proper training of the welder and careful selection of the equipment used are essential to the production of high-quality work.

A variety of equipment is obtainable for most welding operations. Some of this equipment is designed for general use and some is produced for specific operations. Care should be taken to select the most suitable equipment for the particular operation. It is highly important in oxyacetylene welding that the correct mixture of gases and tip size be used.

6-6 Welding Torches The welding torch is a device used for mixing and controlling the flow of gases to the tip. It also provides a means of holding and directing the tip. A simplified schematic drawing of the basic elements of a welding torch is given in Fig. 6-5. There are two throttling or control valves. The gases, after passing the valves and handle, are directed together by the gas mixer at the front end. The tip is shown as a simple tube, narrowed down at the front end to produce a suitable welding

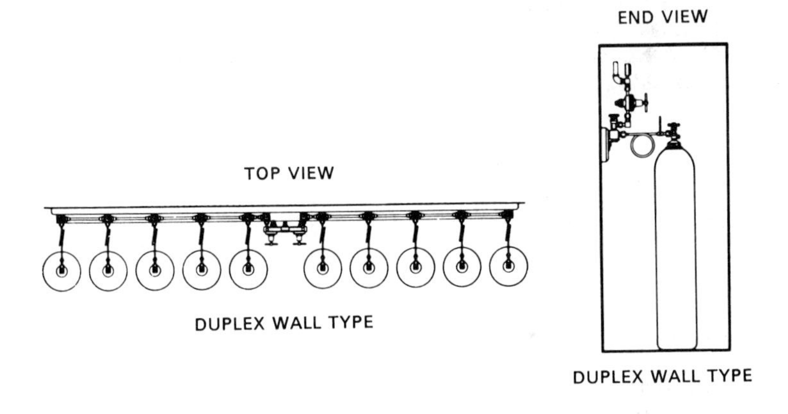

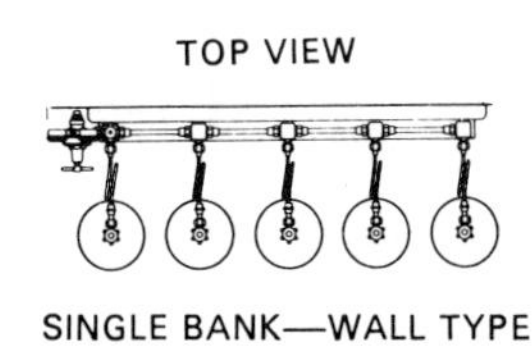

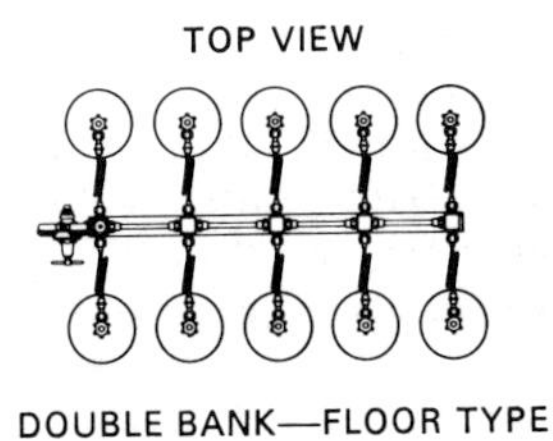

DUPLEX WALL TYPE MANIFOLDS
With Control Unit

Duplex wall type manifolds are permanent installations requiring minimum floor space. Permanent mounting permits minimum abuse to manifold and leader coils and is the safest design. It is available in double-banked design for maximum flexibility, allowing continuous operation of one bank while replacing cylinders to the empty bank.

SINGLE & DOUBLE BANK MANIFOLDS
With Control Regulator

Single and double-bank manifolds, with individual regulators, make the advantages of centralized gas supply available in many instances where the gas supply is not considered enough to warrant investment in duplex manifolds with control units.

FIG. 6-2 Duplex wall-type manifolds. *(Harris Calorific, division of Emerson Electric.)*

MANIFOLD CONTROL UNITS

Type A, a manual unit with the upper and lower bars having gas passages the entire length. The regulators between the bars are "set" to deliver the same pressures. Each bank has a master shutoff valve which can be operated independently when the demand is small.

Type B, a semiautomatic unit. The lower bar has a manually operated crossover valve to divide the lower passages. One regulator is set to a slightly higher pressure than the other, and would bring the adjoining cylinder bank into service. The other regulator automatically cuts in the opposite cylinder bank. Two gauges on the lower bar indicate contents of each bank at all times. An electric alarm gauge is available to sense diminishing cylinder pressure in either bank. Need of an attendant to stand by and change from one bank to the other is eliminated.

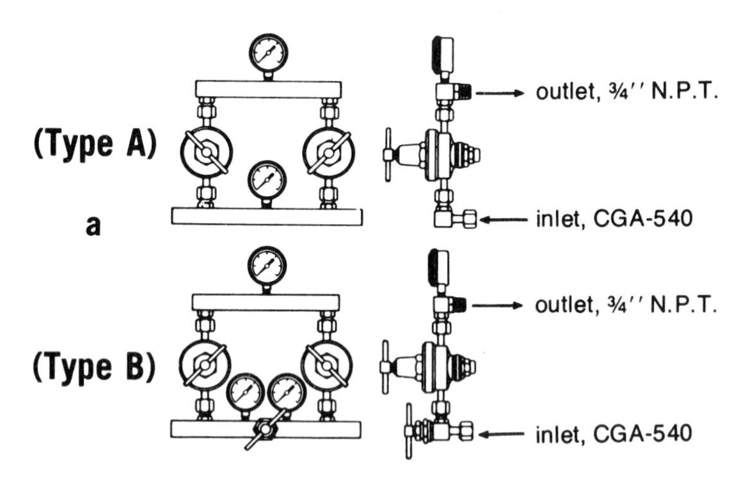

FIG. 6-3 (*a*) Manifold control units. (*b*) Check valves at torch or regulator connections help prevent explosions. (*Harris Calorific, division of Emerson Electric.*)

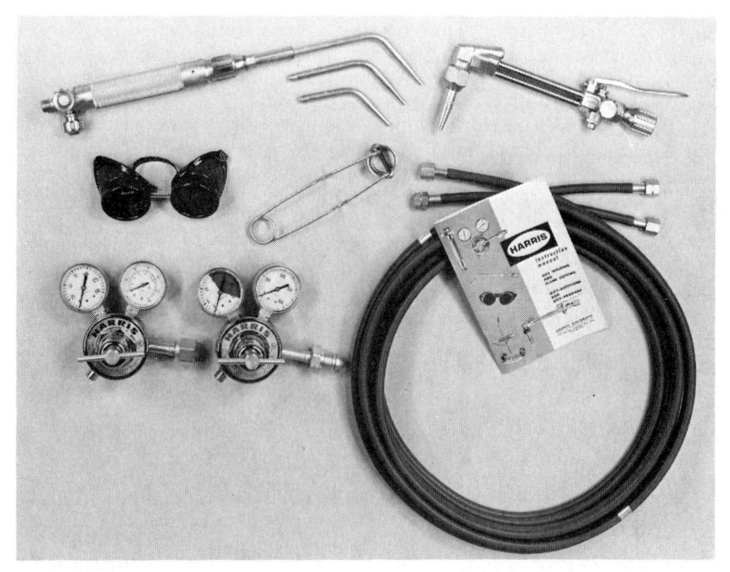

FIG. 6-4 Basic tools for gas welding and brazing. *Top:* torch, tips and cutting attachment. *Center:* goggles, lighter. *Bottom:* regulators, hoses. *(Harris Calorific, division of Emerson Electric.)*

cone. Sealing rings or surfaces have been provided in the torch head or on the mixer seats to facilitate leaktight makeup.

The fuel-gas hose fitting on all torches has a left-hand thread, making it possible to screw on only the left-hand hose nuts used on fuel-gas hose. The other fitting, used for oxygen, has a right-hand thread.

Modern torches are manufactured in a variety of sizes ranging from the small torch for extremely light (low gas flow) work to the extraheavy torches (high gas flow) used generally for localized heating operations. A typical small welding torch used for sheet metal and aircraft welding will pass acetylene at volumetric rates ranging from about ¼ to 35 cu ft/hr. The medium-sized torch is designed to provide for acetylene flows from about 1 to 100 cu ft/hr. Heavy-duty heating torches may provide for acetylene flows as high as 400 cu ft/hr.

Types of Torches. There are two general classes of torches: the positive- or equal-pressure type and the injector type.

The positive-pressure-type torch (also called medium-pressure) requires that the gases be delivered to the torch at pressures generally above 1 psig. In the case of acetylene the pressure should be between 1 and 15 psig. Oxygen generally is supplied at approximately the same pressure. There is, however, no restrictive limit on the oxygen pressure. It can, and sometimes does, range up to 25 psig on the larger sizes of tips when used in positive-pressure torches.

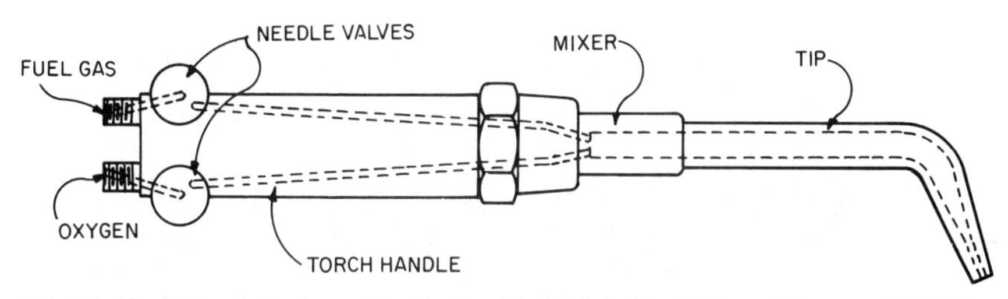

FIG. 6-5 Basic Elements of a welding torch. *(Harris Calorific, division of Emerson Electric.)*

The injector (or low-pressure) type operates at an acetylene pressure as low as 1 psig (a pressure which can easily be supplied by the low-pressure type of acetylene generator). Oxygen, on the other hand, is supplied at a pressure ranging from 10 to 40 psig, necessarily increasing as the tip size is increased. This oxygen aspirates or draws in the acetylene and mixes with it before both gases pass into the tip.

Care of Torches. Each welding torch either has a gas mixer unit as an integral part, or provision is made to attach a mixer by screwing it onto the front end of the torch (Fig. 6-6). The welding tip must be screwed into place. This means that the front portion of a welding torch should not be mishandled, or difficulty will be experienced in obtaining a perfect gas seal. When a mixer is attached to the front end of a torch, particular care should be exercised to prevent the seating surfaces from being scored. Slight damage to these seating surfaces will cause gas leaks and frequent popping sounds. Also, escaping acetylene may burn at the leak.

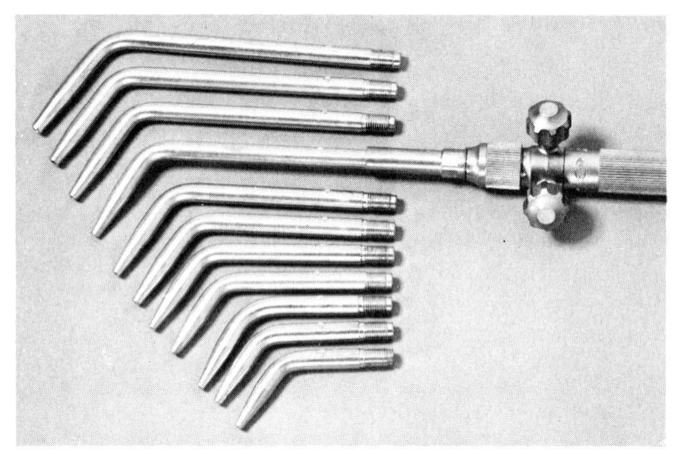

FIG. 6-6 A typical welding torch, mixer, and tips. *(Harris Calorific, division of Emerson Electric.)*

Welding torches are built to withstand rough usage, but the following rules should be observed when handling them:

1. Keep torches away from all oil and grease.
2. If the needle valve does not shut off when hand-tightened in the normal manner, do not use a wrench to tighten or seat the valve stem. If foreign matter cannot be blown off the seat, remove the stem assembly and wipe the seat clean before reassembling.
3. Never clamp the torch handle tightly in a vise, as this may collapse the handle and injure the gas tubes inside.
4. Keep the mixer seat free of dust and other foreign matter at all times.
5. Before using a torch for the first time, check the packing nuts on the needle valves to ensure that they are tight. Some manufacturers ship torches with these nuts loose.

Gas mixers. Figure 6-7 illustrates a mixing chamber wherein acetylene and oxygen come together before passing into the welding tip proper. In this respect, the action of the mixing chamber is similar to that of an automobile carburetor which mixes air and gasoline vapor. A well-designed gas-mixing chamber, or gas mixer, must be able to perform well all of the following essential functions:

1. Mix gases thoroughly for proper combustion.
2. Arrest flashbacks that might occur through improper operation.
3. Stop any flame from traveling farther back than the mixer.
4. Permit, in some designs, a range of tip sizes to be operated by the one size of mixer.

Figure 6-7 shows one type of positive-pressure mixer. The volumes of the two gases have already been controlled by the throttling or needle valves. Acetylene enters through orifice 1 while the oxygen enters at orifice 2. There are usually several holes for one of the gases, to provide for better mixing and to deter any flashbacks. The gas in the annular chamber, which feeds all holes marked 2, must be at a pressure high enough to ensure the proper volumetric gas flow into chamber 3. It can be seen that the pressure gradient of the gas along paths 2 would tend to prevent any flame from proceeding back upstream along this path of gas flow.

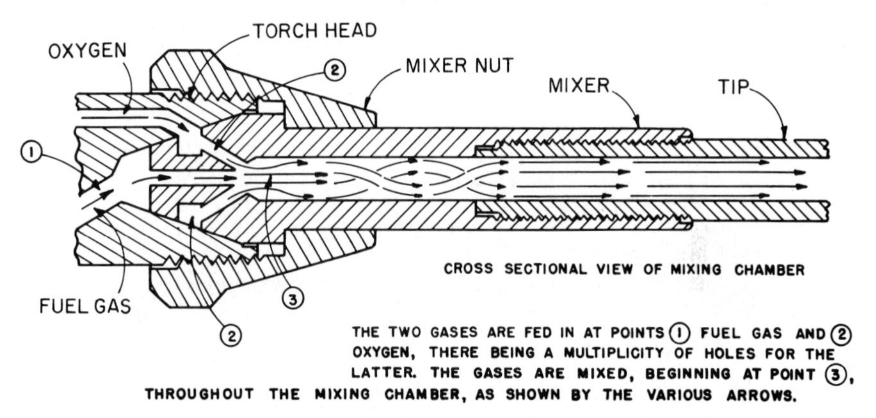
TORCH HEAD
OXYGEN
2
MIXER NUT
MIXER
TIP
1
CROSS SECTIONAL VIEW OF MIXING CHAMBER
FUEL GAS
3
2
THE TWO GASES ARE FED IN AT POINTS 1 FUEL GAS AND 2 OXYGEN, THERE BEING A MULTIPLICITY OF HOLES FOR THE LATTER. THE GASES ARE MIXED, BEGINNING AT POINT 3, THROUGHOUT THE MIXING CHAMBER, AS SHOWN BY THE VARIOUS ARROWS.

FIG. 6-7 A positive-pressure-type mixer—sometimes referred to as an equal-pressure mixer. *(Harris Calorific, division of Emerson Electric.)*

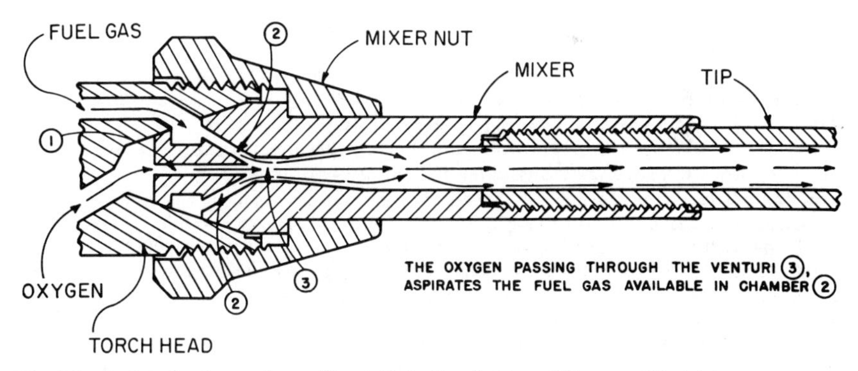
FUEL GAS
2
MIXER NUT
MIXER
TIP
1
THE OXYGEN PASSING THROUGH THE VENTURI 3, ASPIRATES THE FUEL GAS AVAILABLE IN CHAMBER 2
OXYGEN
3
2
TORCH HEAD

FIG. 6-8 An injector-type mixer. *(Harris Calorific, division of Emerson Electric.)*

The typical injector type of gas mixer shown in Fig. 6-8 is essentially similar to the positive-pressure type (Fig. 6-7). The main difference in operation is that the acetylene is at a pressure of 1 psig or lower, while the oxygen pressure may be from 10 to 40 psig. The injector is so designed that the oxygen, as it emerges from hole 1 into venturi chamber 3, creates a vacuum which aspirates or sucks the proper volume of acetylene into the gas mixture from the gas ports 2. A comparative study of Figs. 6-7, 6-8, and 6-9 will clarify the differences of construction between the two types of mixers.

6-7 Welding Tips The welding tip is that portion of the torch through which the gases pass just prior to their ignition and burning. The tip enables the welder to guide the flame and direct it to the work with maximum ease and efficiency.

Tips are generally made of a nonferrous metal such as copper or a copper alloy. Such materials have high thermal conductivity and their use reduces the danger of burning the tip at high temperatures.

Tips are manufactured mainly by drilling bar stock to the proper orifice size or by swaging tubing to the proper diameter over a mandrel. The bore in both types must be smooth in order to produce the required flame cone. The front end of the tip should likewise be shaped to permit easy use and provide a clear view of the welding operation being performed.

Welding tips are available in a great variety of sizes, shapes and constructions (Fig. 6-10). Two general methods of using tip and mixer combinations are employed. One class uses a tip and mixer unit that provides the proper mixer for each size of tip, and the other general class employs one or more mixers for the entire range of tip sizes. In the latter class the tip unscrews from its mixer, and each size of mixer has a particular thread size to prevent improper grouping of a tip and mixer.

A single mixer is used for some classes of welding. It has a so-called gooseneck into which the various sizes of tips may be screwed.

Since tips generally are made of copper or of a copper alloy (relatively soft ma-

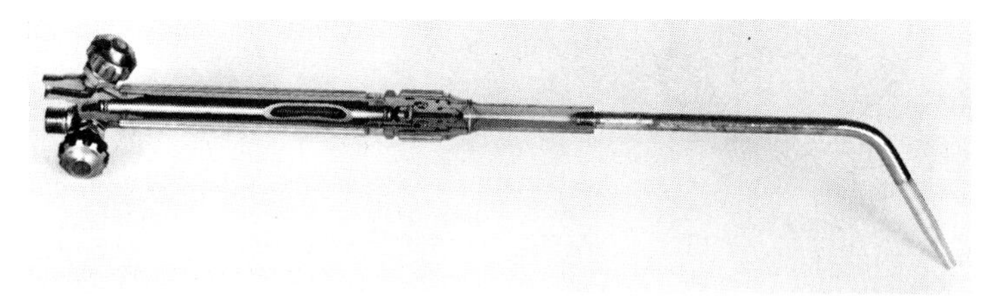

FIG. 6-9 A cutaway illustration of a welding torch. (*Harris Calorific, division of Emerson Electric.*)

terials), care must be taken to guard them against damage. The following precautions should be observed:

1. Tips should be cleaned, when necessary, with tip cleaners designed for this purpose.
2. Tips should never be used for moving or holding the work.
3. Tip seat and/or threads always should be absolutely clean and free from foreign matter in order to prevent scoring when the tip nut is tightened.

See Fig. 6-11.

When a welding operation is performed, care should also be taken to obtain the correct flame adjustment. Improper flame characteristics will nullify correct torch, mixer, and tip selection. Detailed information about the proper methods for obtaining the desired flame characteristics is given elsewhere in this chapter.

A linear relation exists between the thickness of a steel piece and the size of tip that can best be used (Fig. 6-12). When a series of welding tips is selected for a variety of thicknesses of metal, the metal thickness range covered by one tip should slightly overlap that covered by the next tip.

6-8 Volumetric Rate of Acetylene Flow The important factor in determining the usefulness of a torch tip is the action of the flame on the metal. If it is too violent it may blow the metal out of the molten pool. Under such conditions the volumetric flow rates of acetylene and oxygen have to be reduced to a point at which the metal can be welded. This point represents the maximum volumetric flow rate that can be handled by a given size of welding tip. As a general rule, the larger the volumetric rate of gas that may be handled by a specific size of tip, the greater the heat. The flame may also be too "soft" for easy welding. When the flame is too soft the gas volumetric flow rates must be increased.

Tips have a hooded or cup-shaped end designed for some of the slower-burning

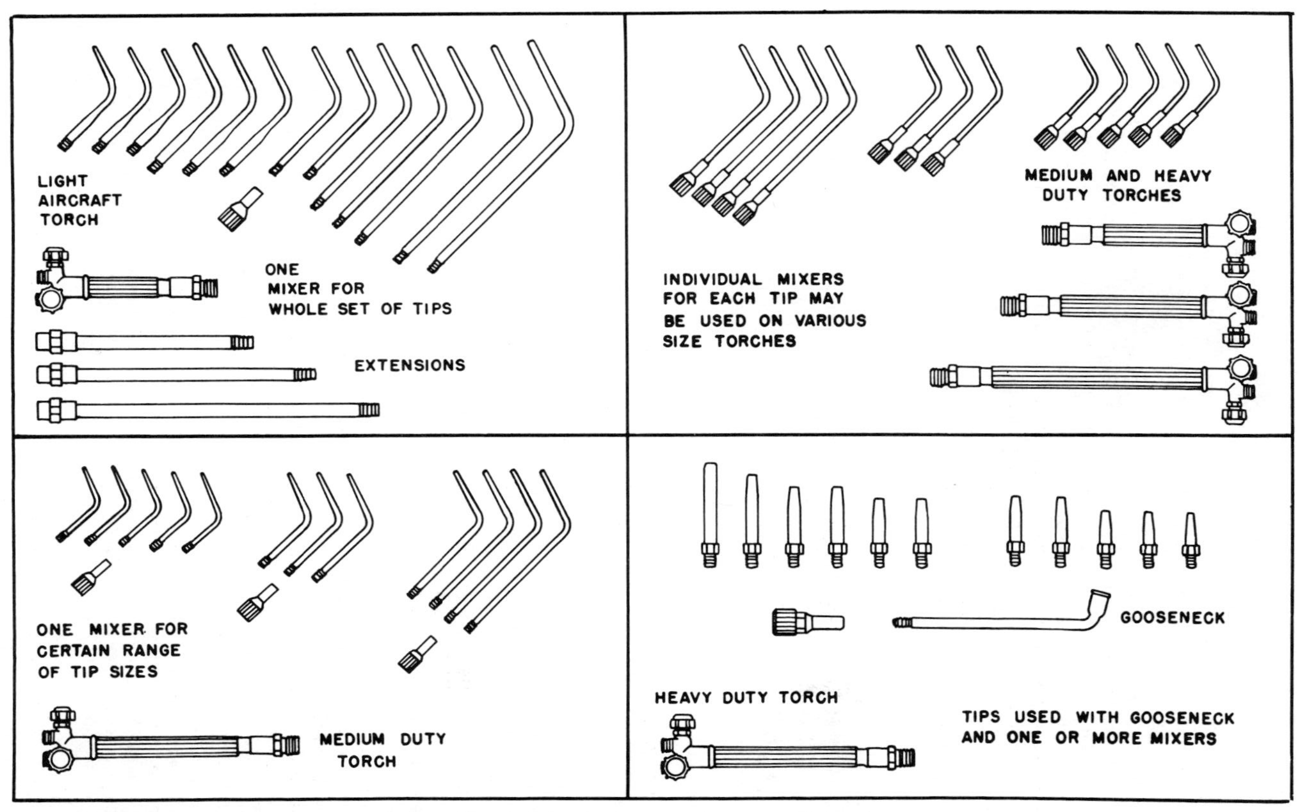

FIG. 6-10 Types of handles, mixers, and tips. *(Harris Calorific, division of Emerson Electric.)*

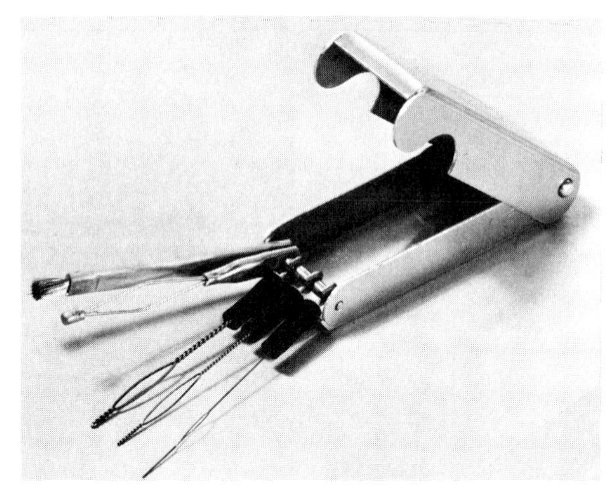

FIG. 6-11 Tip cleaners. *(Harris Calorific, division of Emerson Electric.)*

Welding Tip Chart for Equal Pressure

Metal thickness, in.	Tip size	Oxygen, PSI	Acetylene pressure
1/64	0	1	1
1/32	1	1	1
3/64	2	2	2
1/16	3	3	3
3/32	4	4	4
1/8	5	5	5
3/16	6	6	6
1/4	7	7	7
5/16	8	8	8
3/8	9	9	9
1/2	10	10	10
3/4	13	13	13
1	15	15	15
–	19	15	15
–	22	15	15

FIG. 6-12 Tip charts relate thickness of metal to size of tip and pressure. *(Harris Calorific, division of Emerson Electric.)*

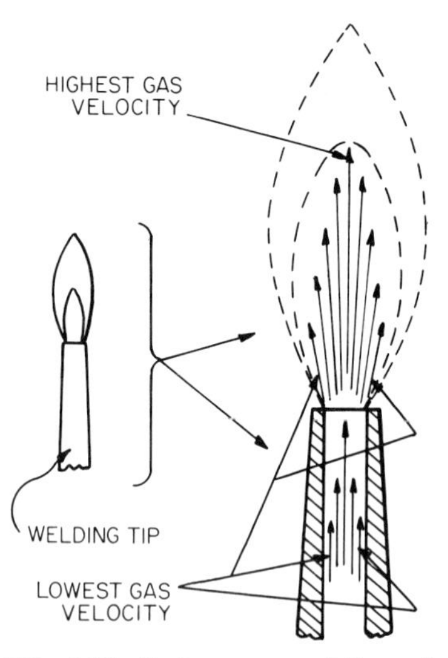

FIG. 6-13 Vector representation of laminar flow in a welding torch tip. *(Harris Calorific, division of Emerson Electric.)*

gases, such as propane, are available from most equipment manufacturers. These tips are usually used for heating, brazing, or soldering.

6-9 Flame Cones The purpose of a welding flame is to raise the temperature of metal to the point where it can be welded. This can be accomplished most easily when the welding flame (or cone) produced by the tip permits the heat to be directed easily. Consequently, the cone characteristics become important. Laminar or streamlined flow becomes of paramount importance throughout the length of the tip and especially so during the final run through the front portion of the tip. A high-velocity flame cone presents striking proof of the velocity gradient extending across a circular orifice when the existing flow is laminar (Fig. 6-13). Since the greatest velocity exists at the center of the stream, the flame length at the center is likewise the greatest. Similarly, since the velocity of the gas stream is lowest at the walls of the tip (bore), where the flowing friction is greatest, that portion of the flame bordering the wall is the shortest.

From the analysis of the principles that underlie the formation of a flame cone, it is possible to understand the flow conditions that exist along the last portion of the gas passageway in the tip. The shape of the flame cone depends upon a number of factors such as the smoothness of bore, the ratio of lead-in to final run diameter, and the sharpness of neck-down.

Generally speaking, the cone produced by a small tip will vary from a pointed to a semipointed shape. Cones from medium-sized tips will vary from a semipointed to a medium shape, and cones from a large-sized tip will vary from a semiblunt to a blunt shape.

6-10 Gas Regulators A gas regulator may be defined as a mechanical device for automatically maintaining a relatively constant reduced delivery pressure even though the inlet pressure and flow rate change. The four necessary and principal elements that constitute a pressure-reducing regulator are illustrated in Fig. 6-14 and may be listed as follows:

1. An adjusting screw that controls the thrust of the bonnet spring.
2. A bonnet spring that transmits to a diaphragm the thrust created by the adjusting screw and spring.
3. A diaphragm connected with the mating seat member.
4. A valve element consisting of a nozzle and a mating seat member.

The bonnet spring tends to hold the valve open, while the gas pressure created on the underside of the diaphragm tends to allow the valve to close. When gas is withdrawn, the pressure under the diaphragm is reduced, thus further opening the valve and admitting more gas, until the forces on either side of the diaphragm are equal.

A given set of conditions, such as constant inlet pressure, constant volumetric rate of flow, and constant outlet pressure, will produce a balanced condition whereby the nozzle and its mating seat member will remain at a fixed relationship to each other.

6-11 Basic Types of Regulators There are two basic types of pressure-reducing regulators: (1) stem type or inlet-pressure closing (sometimes referred to as inverse or negative type), illustrated in Fig. 6-14*a*, and (2) nozzle type or inlet-pressure opening (sometimes referred to as direct-acting or positive type), illustrated in Fig. 6-14*b*.

Stem-Type Regulators. In the stem-type regulator, the inlet pressure tends to close the seat member (pressure closing) against the nozzle. The outlet pressure of this type of regulator has a tendency to increase somewhat as the inlet pressure decreases. This increase is caused by a decrease of the force produced by the inlet gas pressure against the seating area, as the inlet pressure decreases.

The gas outlet pressure for any particular setting of the adjusting screw is regulated by a balance of forces between the bonnet-spring thrust and the opposing forces created by (1) the outlet pressure against the underside of the diaphragm, and (2) the force created by the inlet pressure against the seating area. When the inlet pressure decreases, its force against the seat member decreases, allowing the bonnet-spring force to move the seat member away from the nozzle. Thus, more gas pressure is allowed to build up against the diaphragm to reestablish the balanced condition.

Nozzle-Type Regulators. In the nozzle type of regulator the inlet pressure tends to move the seat member away (pressure opening) from the nozzle, thus opening the valve. The outlet pressure of this type of regulator decreases somewhat as the inlet pressure decreases, because the force tending to move the seat member away from the nozzle is reduced as the inlet pressure decreases. A small outlet pressure on the underside of the diaphragm is then required to close the seat member against the nozzle.

6-12 Two-Stage Regulators The two-stage regulator was developed for more precise regulation over a wide range of varying inlet pressures. A two-stage regulator,

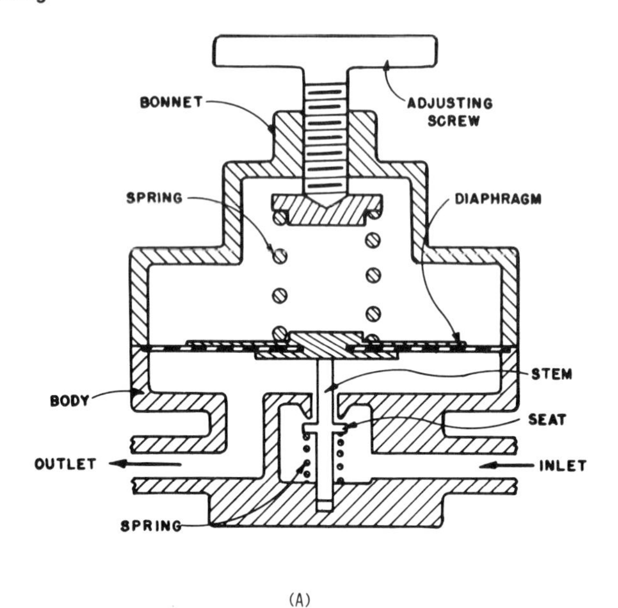

(A)

BONNET
ADJUSTING SCREW
SPRING
DIAPHRAGM
BODY
YOKE
OUTLET
INLET
NOZZLE
SEAT
SPRING

(B)

FIG. 6-14 (*a*) Single-stage stem-type regulator. (*b*) Single-stage nozzle-type regulator. (c) Cutaway illustration. *(Harris Calorific, division of Emerson Electric.)*

(C)

FIG. 6-14 (*Continued*)

as illustrated in Fig. 6-15*a* and *b* is actually two single-stage regulators in series and produced as one unit.

The outlet pressure from the first stage is usually preset to deliver a specified pressure to the inlet of the second stage. In this way, a practically constant delivery pressure may be obtained from the outlet of the regulator, even though the supply pressure should vary (as in depleting an oxygen cylinder from its initial pressure of approximately 2,200 psig to 100 psig).

The combinations employed to make a two-stage regulator are as follows:

1. Nozzle-type first stage and a stem-type second stage.
2. Stem-type first stage and a nozzle-type second stage.
3. Two stem types.
4. Two nozzle types.

Regardless of the combination used, the increase or decrease in outlet pressure is usually so slight (and apparent only at very low inlet-supply pressures) that, for all practical purposes, the variation in delivery pressure is disregarded in welding or cutting operations. Two-stage regulators are suggested for precise work, such as continuous machine cutting, in order to maintain a constant working pressure and a controlled volumetric flow rate at the welding or cutting torch. (See Figs. 6-16 and 6-17.)

6-13 Applications of Regulators in Gas Welding Many kinds of regulators are produced for various applications. When gases are utilized from single cylinders, the regulators need not have very high capacities, probably a maximum of 1,500 scfh. When gas cylinders are manifolded, or a bulk delivery system is used, high-capacity regulators to supply large-volume distribution lines may have to flow as much as 50,000 scfh or more. Distribution-line station regulators are normally low-pressure types (up to 500 psig inlet pressure) and are available in various sizes depending upon the volumetric demand at the particular station. Because acetylene distribution lines are not permitted to carry more than 15 psig, the regulators must, in many cases, have a high volumetric capacity with a relatively low inlet pressure.

6-14 Regulator Inlet and Outlet Connections The cylinder-outlet connections are of different sizes and shapes to preclude the possibility of connecting to a cylinder a regulator made for the wrong gas or pressure. Regulators must, therefore, be made with different inlet connections to fit the cylinders. The Compressed Gas Association (CGA) has formulated a complete set of noninterchangeable cylinder valve outlet connections. A few of these connections are listed in Table 6-1.

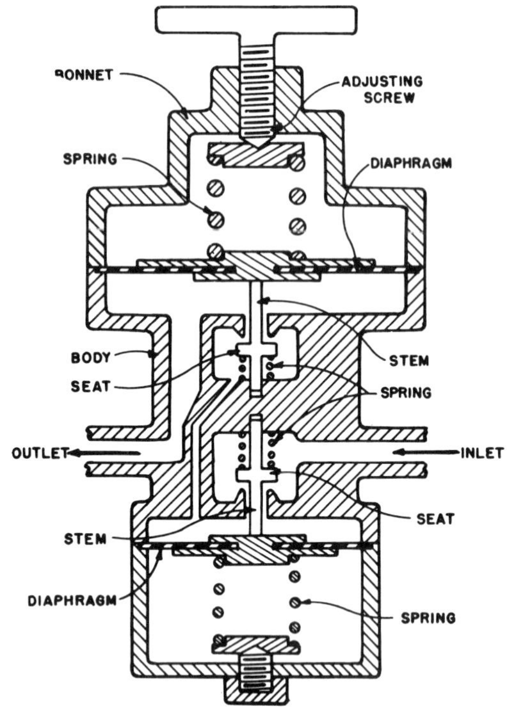

FIG. 6-15 **(*a*) Two-stage-type regulators.**

FIG. 6-15 **(*b*) Cutaway illustrations.** (*Harris Calorific, division of Emerson Electric.*)

FIG. 6-16 **Typical two-stage regulator.** (*Harris Calorific, division of Emerson Electric.*)

Regulators should be used only for the service for which they were designed. Oxygen regulators, in particular, should never be used for any other service than for oxygen. Even traces of oil or foreign matter in an oxygen regulator can cause a violent explosion.

Regulator outlet fittings also differ in size and thread, depending upon the gas and regulator capacity. Oxygen-outlet fittings have right-hand threads and fuel-outlet

FIG. 6-17 Typical gageless regulators. Gages replaced by an indicator, lower left, and pressure-setting indications on adjusting handle. Eliminates gage breakage from rough handling. (*Harris Calorific, division of Emerson Electric.*)

TABLE 6-1 Cylinder Valve Outlet Connections

Gas	CGA standard connection number	Thread size, in. and threads per in.,
Acetylene	510	0.885–14 NGO-LH-INT
Acetylene (alternate standard)	300	0.825–14 NGO-RH-EXT
Acetylene (Canada only)	410	0.850–14 NGO-LH-EXT
Acetylene (small-valve series)	200	0.625–20 NGO-RH-EXT
Acetylene (small-valve series)	520	0.895–18 NGO-RH-EXT
Argon	580	0.965–14 NGO-RH-INT
Butane	510	0.885–14 NGO-LH-INT
Butane (alternate standard)	300	0.825–14 NGO-RH-EXT
Carbon dioxide	320	0.825–14 NGO-RH-EXT
Helium	580	0.965–14 NGO-RH-INT
Helium (alternate standard)	350†	0.825–14 NGO-LH-EXT
Hydrogen	350	0.825–14 NGO-LH-EXT
Nitrogen	580	0.965–14 NGO-RH-INT
Nitrogen (alternate standard)	590	0.965–14 NGO-LH-INT
Oxygen	540	0.903–14 NGO-RH-EXT
Propane	510	0.885–14 NGO-LH-INT
Propane (alternate standard)	350†	0.825–14 NGO-LH-EXT
Propane (alternate standard)	300	0.825–14 NGO-RH-EXT

* Abbreviations: NGO—national gas outlet, RH—right hand, LH—left hand, INT—internal, EXT—external.

† Obsolete effective January 1, 1972.

From: American Standard *Compressed Gas Cylinder Valve Outlet and Inlet Connections* (B57.1–1955).

fittings have left-hand threads with grooved nuts. Table 6-2 shows pertinent hose size and thread data as revised by the Compressed Gas Association.

6-15 Recommended Practices for the Safe Use of Regulators The following safety precautions pertaining to regulators should be observed:

1. Clean cylinder valve outlets with a clean, lint-free, dry cloth and blow dust from the outlet by opening the valve momentarily before connecting the regulator to it. This is known as cracking the valve.

2. The regulator adjusting screw should be released (i.e., backed out) before the cylinder valve is opened.

3. Always open cylinder valves very slowly so that the high-pressure gas does not surge into the regulator. When doing this, stand off to one side of the regulator rather than directly in front.

4. Check gages periodically to ensure correct readings.

5. Adjusting screws should be turned in slowly to protect the regulator diaphragm from damage caused by a sudden surge of high-pressure gas.

6. Always use the correct-size wrench to connect the regulator to the cylinder valve outlet. Never force a connection.

7. Never use oil or grease on a regulator. A special lubricant sometimes is applied to equipment. The only lubricant that should be used is that specified by the manufacturer and it should be used only when specified.

8. If a leak is suspected, use a grease-free soapy water solution to detect it.

9. Regulators should be repaired by qualified, trained mechanics. Manufacturers' standard parts only should be used.

TABLE 6-2 Hose Size and Thread Data

Hose sizes (ID)	Thread size, in., and threads per in., and type
3/16, 1/8	3/8–24 NF-2
3/8, 5/16, 1/4, 3/16, 1/8	9/16–18 NF-2
1/2, 3/8, 5/16, 1/4	7/8–14 NF-2
3/4, 5/8, 1/2, 7/9	1 1/4–12 NF-2

6-16 Welding Rods The properties of the metal deposited during welding should closely match those of the base metal. Because of this requirement, welding rods of various chemical compositions have been devised for the welding of many ferrous and nonferrous materials. Obviously, it is important that the correct welding rod be selected.

The welding process itself influences the filler-metal chemistry, since certain elements tend to disappear during deposition. Filler metals that may be used to join almost all base materials in common use are available, as are filler materials for braze welding, brazing, and soldering. The standard diameters vary from 1/6 to 3/8 inch, and the standard lengths for drawn rods are 24 and 36 inches.

The chemical analysis of welding rods must be within the limits specified for that particular rod. There are many proprietary rods on the market recommended for specific applications. Filler metal should be free from blowholes, pipes, nonmetallic inclusions, and any other foreign matter. The metal should deposit smoothly.

Allowances for changes taking place during welding are made in the chemistry of good welding rods, in order that the deposited metal will be of the correct composition. These changes are to be expected, but they should occur without undue sparking or spitting. Deposits should be made with free-flowing metal that unites readily with the base metal to produce sound, clean welds (Table 6-3).

In maintenance and repair work it is not always necessary that the composition of the welding rod match that of the base metal. A steel welding rod of nominal strength can be used to repair parts made of alloy steels broken by overloading or accident. An extreme instance of this apparent side-stepping of the rules is the use of silver brazing alloys in the repair of tools such as milling cutters made of high-speed tool steel. Every effort should be made, however, to match the filler metal and base metal. Where it is necessary to heat-treat a steel part after welding, carbon can be added to a deposit of mild steel by the judicious use of the carburizing flame. It is preferable, however, to use a welding rod of low-alloy steel.

The American Welding Society Committee on Filler Metal has prepared a number of specifications. Many of the gas-welding filler metals meet these specifications.

6-17 Fluxes One of the most important factors in weld quality is the removal of oxides from the surface of the metal to be welded. Unless the oxides are removed, fusion may be difficult, the joint may lack strength, and inclusions may be present. The oxides will not flow from the weld zone but will remain to become entrapped in the solidifying metal, interfering with the addition of filler metal. These conditions may occur when the oxides have a higher melting point than the base metal, and a means must be found to remove the oxides. Fluxes are applied for this purpose.

Steel and its oxides and slags which form during welding do not fall into the above category and no flux is needed. Aluminum, however, forms an oxide with a very high

TABLE 6-3 Welding Data — Ferrous Metals

Metal	Flame adjustment	Flux	Welding rod
Steel, cast	Neutral	No	Steel
Steel pipe	Neutral	No	Steel
Steel plate	Neutral	No	Steel
Steel sheet	Neutral	No	Steel
	Slightly oxidizing	Yes	Bronze
High-carbon steel	Reducing	No	Steel
Wrought iron	Neutral	No	Steel
Galvanized iron	Neutral	No	Steel
	Slightly oxidizing	Yes	Bronze
Cast iron, gray	Neutral	Yes	Cast iron
	Slightly oxidizing	Yes	Bronze
Cast iron, malleable	Slightly oxidizing	Yes	Bronze
Cast-iron pipe, gray	Neutral	Yes	Cast iron
	Slightly oxidizing	Yes	Bronze
Cast-iron pipe	Neutral	Yes	Cast iron or base-metal composition
Chromium-nickel steel castings	Neutral	Yes	Base-metal composition 25-12 chromium-nickel steel
Chromium-nickel steel (18-8 and 25-12)	Neutral	Yes	Columbium stainless steel or base-metal composition
Chromium steel	Neutral	Yes	Columbium stainless steel or base-metal composition
Chromium iron	Neutral	Yes	Columbium stainless steel or base-metal composition

melting point, and the oxide must be removed from the welding zone before satisfactory results can be obtained. Certain substances have been found that will react chemically with the oxides of most metals, forming fusible slags at welding temperature. These substances have been developed and compounded, either singly or in combination, to make efficient fluxes.

A good flux should assist in removing the oxides occurring during welding by forming fusible slags that will float to the top of the molten puddle and not interfere with the deposition and fusion of filler metal. It should protect the molten puddle from atmospheric oxygen and prevent the absorption and reaction of other gases in the flame, without obscuring the welder's vision or hampering manipulation of the molten puddle.

During the preheating and welding periods, the flux should clean and protect the surfaces of the base metal and, in some cases, the welding rod. Flux should not be used as a substitute for base-metal cleaning during joint preparation. Flux is an excellent metal cleaner, but if its strength is used for this purpose, it will be useless for its primary functions.

Flux may be prepared as a dry powder, as a paste or thick solution, or as a coating on the welding rod. Some fluxes operate much more favorably if they are used dry. Braze-welding fluxes and fluxes for use in cast iron are usually in this class. These fluxes are applied by heating the end of the welding rod and dipping it into the powdered flux. Enough will adhere to flux adequately until the flux-coated portion of the rod is consumed. Dipping the hot rod into the flux coats another portion. Dropping some of the dry powder on the base metal ahead of the welding zone will sometimes help, especially in the repair of dirty and oil-soaked castings.

Fluxes in paste form are usually painted on the base metal with a brush, and the welding rod can either be painted or dipped. Commercial precoated rods can be used without further preparation, and when additional flux is required it can be placed on the base metal. Sometimes a precoated rod will have to be dipped in powdered flux if the flux is melted off too far from the end of the rod.

The common metals and welding rods requiring fluxes are bronze, cast iron, brass, silicon bronzes, stainless steel, and aluminum.

OXYACETYLENE WELDING

6-18 Principles of Operation The oxyacetylene welding torch serves as the medium for mixing the combustible and combustion-supporting gases and provides the means for applying the flame at the desired location. A range of tip sizes is provided for obtaining the required volume or size of welding flame, which may vary from a short, small-diameter needle flame to a flame $\frac{3}{16}$ inch or more in diameter and 2 inches or more in length.

The inner cone, or vivid blue flame, of the burning mixture of gases issuing from the tip is called the working flame. The closer the end of the flame's inner cone is to the surface of the metal being heated or welded, the more effective is the heat transfer from flame to metal. The oxyacetylene flame can be made soft or harsh by varying the gas flow. Too low a gas flow for a given tip size will result in a soft, ineffective flame sensitive to burnback. Too high a gas flow will result in a harsh, high-velocity flame that is hard to handle and will blow the molten metal from the puddle.

The chemical action of the flame on a molten pool of metal can be altered by changing the ratio of the volume of oxygen to that of acetylene issuing from the tip. Most

WELDING FLAME ADJUSTMENT

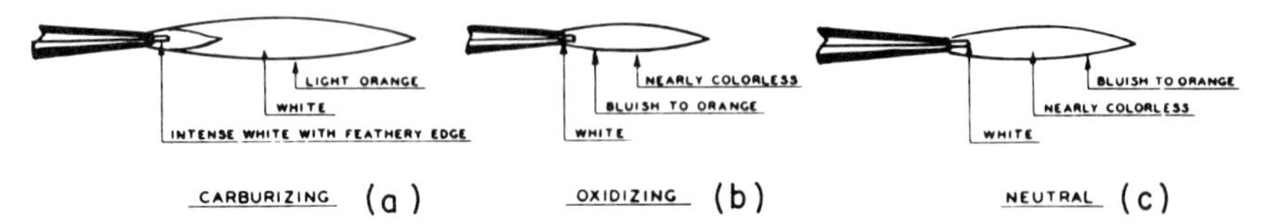

A neutral flame is used for almost all gas welding. The oxy-acetylene flame consumes all oxygen in the air around the weld area. This leaves an uncontaminated weld area resulting in maximum weld strength. An oxidizing flame is rarely used, and a carburizing flame is occasionally helpful when flame hardening, or brazing.

FIG. 6-18 Three types of flame adjustments. *(Harris Calorific, division of Emerson Electric.)*

welding is done with a neutral flame having approximately a 1:1 gas ratio. An oxidizing action can be obtained by increasing the oxygen flow, and a carburizing action will result from increasing the acetylene flow. Both adjustments are valuable aids in welding.

6-19 Flame Adjustment Torches should be lighted with a friction lighter or a pilot flame. The instructions of the equipment manufacturer should be observed when adjusting operating pressures at the regulators and torch valves before the gases issuing from the tip are ignited. Three types of flame adjustment are shown in Fig. 6-18.

The neutral flame is obtained most easily by adjustment from an excess-acetylene flame, which is recognized by the "feather" extension of the inner cone. The feather will diminish as the flow of acetylene is decreased or the flow of oxygen is increased. The neutral flame is obtained just at the point of disappearance of the feather extension.

A practical method of determining the amount of excess acetylene in a flame, when a reducing or carburizing flame is desired, is to compare the length of the feather with the length of the inner cone, measuring both from the torch tip. An excess-acetylene flame has an acetylene feather that is double the length of the inner cone. Starting with a neutral flame adjustment, the welder can produce the desired acetylene feather by increasing the acetylene flow (or by decreasing the oxygen flow).

The oxidizing-flame adjustment is sometimes given as the amount by which the length of the inner cone should be reduced—for example, $\frac{1}{10}$. Starting with the neutral flame, the welder can increase the oxygen (or decrease the acetylene) until the length of the inner cone is decreased the desired amount.

Specifications requiring a reducing flame and mentioning the length of the feather should be adhered to closely, since the carbon pickup in the weld is determined by the excess acetylene in the flame. Where oxidizing flames are necessary, the degree of adjustment is usually determined by the action of the flame on the molten metal. This is especially true in the braze welding of brass or braze welding of steel.

6-20 Technique Oxyacetylene welding may be performed with the torch tip pointed forward in the direction in which the weld progresses. This method is called the forehand technique (see Fig. 6-19). The tip can also be pointed back toward the weld that has been deposited, and this method is called the backhand technique (also shown in Fig. 6-19). Each technique has its advantages, depending upon the application.

In general, the forehand method is recommended for welding material up to $^1/_8$ inch thick, because it provides better control of the weld puddle, resulting in a smoother weld at both top and bottom. The puddle of molten metal is small and easily controlled. A great deal of pipe welding is done using the forehand technique, even in $^3/_8$ inch wall thicknesses.

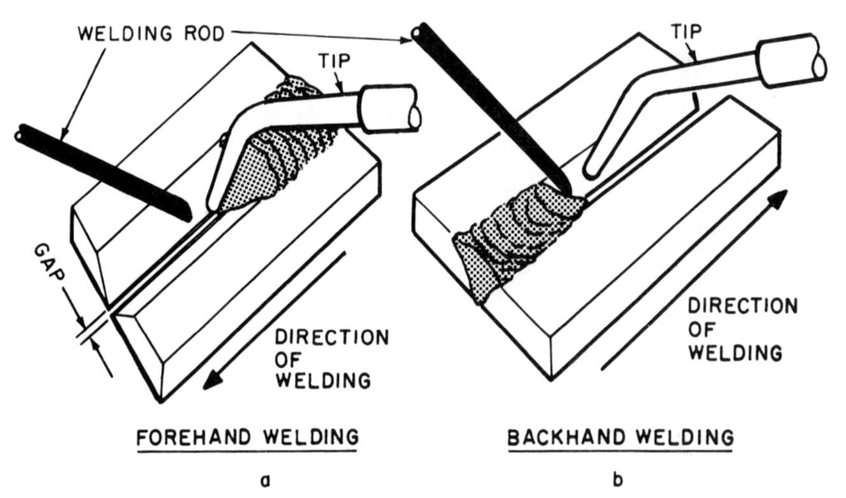

FIG. 6-19 Forehand (a) and backhand (b) welding techniques. (SOURCE: LeGrand, R., ed., *The New American Machinist's Handbook.* New York, McGraw-Hill, 1955.)

Increased speeds and better control of the puddle are possible with the backhand technique when metal $^1/_8$ inch and thicker is welded. The recommendation that material greater than $^1/_8$ inch thick be welded with the backhand technique is based on careful study of the speeds normally achieved with this technique and on the greater ease of obtaining fusion at the root of the weld. Backhand welding may be used with a reducing flame when it is desired to melt only the smallest possible amount of steel in making the joint. The increased carbon content obtained from the reducing flame lowers the melting point of a thin layer of steel and increases welding speed. This technique greatly increases the speed of making pipe joints where the wall thickness is $^1/_4$ to $^5/_{16}$ inch and the groove angle is reduced. Backhand welding is also sometimes used in surfacing operations (see Fig. 6-20).

6-21 Base Metal Preparation Cleanliness along the joint and on the sides of the base metal is of the utmost importance. Dirt, oil, and oxides can cause blowholes, incomplete fusion, slag inclusions, and porosity in the weld.

The spacing between parts to be joined should be considered carefully. The root opening for a given thickness of material should permit the gap to be bridged without difficulty, yet should be large enough to permit full penetration. Specifications for root openings should be adhered to rigidly.

The thickness of the base metal at the joint determines the amount of edge preparation for welding. Thin sheet metal is easily melted completely by the flame. Thus edges with square faces can be butted together and welded. This type of joint is limited to material approximately $^{3}/_{16}$ inch or less in thickness. For thicknesses of $^{3}/_{16}$ to $^{1}/_{4}$ inch, a slight root opening is necessary for complete penetration, but filler metal must be added to compensate for the opening.

Joint edges $^{1}/_{4}$ inch and greater in thickness should be beveled. Beveled edges at the joint provide a groove for better penetration and fusion at the sides. The angle of bevel for oxyacetylene welding varies from 35 to 45 degrees, which is equivalent to a variation in the included angle from 70 to 90 degrees, depending upon the application. A root face $^{1}/_{16}$ inch wide is provided, but feather edges are sometimes used. Plate thicknesses $^{3}/_{4}$ inch and above are double beveled when welding can be done from both sides. The root face can vary from 0 to $^{1}/_{8}$ inch. Beveling both sides reduces by approximately

OXYGEN AND ACETYLENE SETUP INSTRUCTIONS

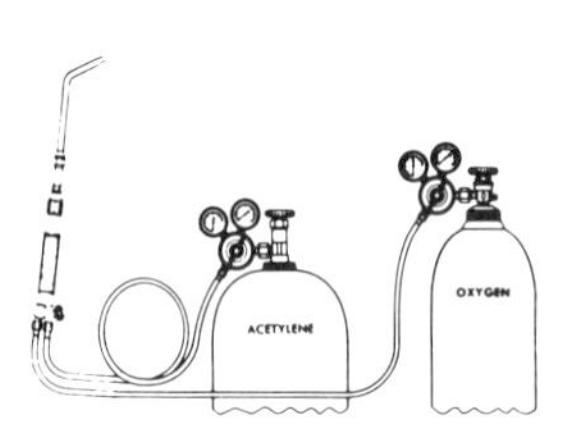

ATTACHING THE REGULATORS

Open cylinder valves slightly to blow out dirt, then close. Attach regulators and tighten the connection firmly. Attach the hoses to the regulators and tighten.

(NOTE: the acetylene hose connections are left hand threads.)

(NOTE: the oxygen hose connections are right hand threads.)

ATTACHING THE TORCH

Attach acetylene hose (red) to torch valve marked "AC", note left hand thread. Attach oxygen hose (green) to torch valve marked "OX", note right hand thread. Shut both valves on torch before opening cylinders.

FIG. 6-20 Setup instructions. *(Harris Calorific, division of Emerson Electric.)*

one-half the amount of filler metal required. Gas consumption per foot of weld is also reduced.

An edge preparation with straight sides is easiest to obtain. This edge can be machined, chipped, ground, or oxygen-cut. The thin oxide coating on an oxygen-cut surface does not have to be removed, because it is not detrimental to the welding operation or to the quality of the joint. The bevel angle can be oxygen-cut to any desired value.

6-22 Metallurgical Effects The temperature of the base metal varies during welding from melting temperature in the weld puddle to room temperature in the areas most remote from the heat. When steels are involved, the area near and in the weld is heated considerably above the transformation temperature of the steel. This results in a coarse grain structure in the weld and adjacent base metal. The coarse grain structure can be refined by a normalizing heat treatment [heating to 1500 to 1600°F (816 to 871°C) and cooling in air] after welding.

Hardening of the weld and adjacent base metal, heated to above the transformation temperature of the steel, can occur if the steel contains sufficient carbon and the cooling rate is high enough. Hardening can be avoided in most hardenable steels by using the torch to keep heat on the weld for a short time after the weld is completed. If air-hardening steels are welded, the best heat treatment is a full furnace anneal of the welded item. Hardness in the weld zone usually should be avoided because of the corresponding lack of ductility and susceptibility to cracking.

The oxyacetylene flame allows a degree of control to be maintained over the carbon content of the deposited metal and over the portion of the base metal that is heated to its melting temperature. When an oxidizing flame is used, a rapid reaction results between the oxygen and the carbon of the metal. Some of the carbon is eliminated in the form of carbon monoxide, and the steel itself and other constituents are also oxidized. When the torch is used with an excess-acetylene flame, carbon is introduced into the weld puddle.

When unstabilized austenitic stainless steel is heated to a temperature range between 800 and 1600°F (427 and 871°C), carbide precipitation occurs. These carbides gather at the grain boundaries and lower corrosion resistance in the areas adjoining the weld. If this occurs, a further heat treatment after welding is required, unless the steel has been stabilized by the addition of columbium or titanium and welded with the aid of a columbium-bearing welding rod. The columbium combines with the carbon and minimizes the formation of chromium carbide. All the chromium is left dissolved in the austenitic matrix, which is the form in which it can resist corrosion to the greatest extent.

Another factor to be considered in welding is the possible tendency toward "hot-shortness" of the material (a marked loss in strength at high temperatures). Some of the copper-base alloys have this tendency to a high degree. If the base metal has this tendency, it should be welded with care to prevent hot-cracking in the weld zone. Allowances should be made in the welding technique used with these materials, and jigging or clamping should be done with caution. Proper welding sequence and multi-layer welding with narrow string beads help to reduce hot-cracking.

6-23 Oxidation and Reduction Certain metals have such a high affinity for oxygen that oxides form on the surface almost as rapidly as they are removed. In oxy-acetylene operations these oxides are usually removed by means of fluxes. This affinity for oxygen can be a very useful characteristic in certain welding operations. Manganese and silicon, for example, are elements common to plain carbon steel. They are important in oxyacetylene welding because they react with oxygen when the metal is molten.

The reaction produces a very thin slag covering that tends to prevent any oxygen from contacting the weld metal and prevents the formation of gas pockets. When the viscosity of the slag covering is properly controlled, the molten metal may be kept in position even against the pull of gravity. The action of these elements in oxyacetylene welding is the same as in steelmaking. They are used to produce clean, deoxidized metal in open-hearth or electric furnaces. The correct manganese and silicon content in steel welding rods is therefore important.

The type of flame used in welding various materials plays an important part in securing the most desirable deposit of weld metal. The proper type of flame with the correct welding technique can be used as a shielding medium which will reduce the oxidizing and nitrogenizing effect of the atmosphere on the molten metal. Such a flame also has the effect of stabilizing the molten weld metal and preventing the burning-out of carbon, manganese, and other alloying elements.

The proper type of flame for any application is determined by the type of base and filler metal involved, the thickness of the base metal and the position of welding. For most metals a neutral flame is used. An exception is the welding of aluminum, where a slightly carburizing flame is used.

METALS WELDED

Oxyacetylene welding is used on the whole range of commercial ferrous and nonferrous metals and alloys. As in any welding process, however, physical dimensions and chemical composition may limit the weldability of certain materials and pieces.

In gas welding the temperature range through which the metal is taken is almost the same as that of the original casting procedure. The base metal in the weld area loses those properties that were given to it by heat treatment or cold working. The ability to weld such materials as high-carbon and high-alloy steels is limited by the equipment available for heat treating after welding. These metals are commonly welded with success when the size or nature of the piece permits post-heat-treating operations.

The welding procedure for wrought iron and plain carbon steels is straightforward and offers little difficulty to the welder. Sound welds are produced in other materials by appropriate variations in technique, heat treating, preheating, and fluxing.

The oxyacetylene welding process can be used for welding metal of considerable thickness and for the usual assemblies encountered in maintenance and repair. Cast-iron machinery frames that are one or more feet in thickness at the fracture point have been repaired by braze welding or by welding with a cast-iron filler rod.

6-24 Iron and Steel Low-carbon, low-alloy steels, cast steels, and wrought iron are the materials most easily welded by the oxyacetylene process. Fluxes are not required when welding these materials.

In oxyacetylene welding, straight carbon steels having more than 0.35 percent carbon are considered high-carbon steels and necessitate special care to maintain their particular properties. Alloy steels of the air-hardening type also require precautions to maintain their properties, even though the carbon content may be 0.35 percent or less. The joint area usually is preheated in order to retard the cooling of the weld caused by conduction of heat into surrounding base metal. Slow cooling prevents the hardness and brittleness associated with rapid cooling.

If hardening occurs, a post-heat-treatment is necessary. The welder should use a carburizing flame for welding and should be careful not to overheat the base metal and burn out the carbon. The preheating temperature required depends upon the composition of the steel to be welded. Temperatures of from 300 to 1000°F (149 to 538°C) have been used.

In addition to choosing proper preheat temperature, it is important that the preheat be maintained uniformly during welding. A uniform temperature is maintained by protecting the part with a covering formed by an asbestos blanket. Other means of shielding can also be used to retain the temperature in the casting. Generally the maximum temperature during welding should not exceed the preheat temperature by more than 150°F (66°C). High welding or interpass temperatures cause excessive shrinkage forces that can bring about either distortion or cracking at the weld or other sections. In the welding of circular structures made from brittle metals, such as cast iron, this type of cracking frequently occurs.

Modifications in procedures are required for stainless and similar steels. Because of their high chromium or chromium-nickel content, these steels have a relatively low heat conductivity. A smaller flame than that used for equal thicknesses of plain carbon steel is recommended. Because chromium oxidizes easily, a neutral flame is employed to minimize oxidation. A flux is used to dissolve oxides and protect the weld metal. A welding rod of high-chromium or nickel-chromium steel supplies the filler metal. Table 6-3 summarizes the basic information for welding ferrous metals.

Cast iron, malleable iron, and galvanized iron all present particular problems in welding by any method. The gray-cast-iron structure in cast iron can be maintained through the weld area by use of preheat, a flux, and an appropriate cast-iron welding rod.

Nodular iron requires materials in the welding filler metal that will assist in promoting nodularization of the free graphite to maintain ductility and shock absorption of the heat-affected area. Manufacturers of welding filler metal should be consulted to obtain information on preheat- and interpass-temperature control for the filler material being used.

Although there are instances in which cast iron materials are welded without preheat, particularly in salvage work, preheat of 400 to 600°F (204 to 316°C) with control of interpass temperature and provision for slow cooling will assure more consistent results. Care should be taken that venting or localized cooling is not allowed to occur because of exposed areas or a break in uniform covering such as a tear in the blanket.

It is also important to stress that, in the salvage of cast iron, removal of all foundry sand and slag is necessary for consistent repair results.

6-25 Nonferrous Metals The particular properties of each nonferrous alloy should be considered when selecting the most suitable welding technique. When the necessary precautions are taken, little difficulty resulting from the nature of the metal should be encountered.

Aluminum, for instance, gives no warning by changing color prior to melting, but appears to collapse suddenly at the melting point. Consequently, practice in welding is required to learn to control the rate of heat input. Aluminum and its alloys suffer from hot-shortness and should be supported adequately in all the areas heated during the welding. Finally, any exposed aluminum surface is always covered with a layer of oxide that, combining with the flux, forms a fusible slag that floats on top of the molten metal.

When copper is welded, allowances for the chilling of the weld, due to the very high thermal conductivity of copper, are necessary. Preheating is often required. Considerable distortion can be expected in copper because the coefficient of thermal expansion is higher than in other commercial metals. These characteristics obviously pose difficulties that must be surmounted for satisfactory welding.

BRAZE WELDING

The main requirement for acceptable results in braze welding is cleanliness of the joint edges. The surfaces to be joined should be clean and free of oil, dirt, and rust. In the presence of a suitable flux and the right amount of heat, the filler metal will flow and wet the joint surfaces. The joint can be completed in one pass, or successive passes can be made if the groove is large.

When malleable iron is welded, the melting of the base metal converts the heat-affected zone to white or hard cast iron. Braze welding with filler metal having a lower melting point produces a minimum change in the base-metal characteristics and effects a high-strength joint generally equivalent to the strength of the filler metal.

Part 2 Arc Welding and Its Processes

RICHARD S. SABO

Manager of Educational Services,
Lincoln Electric Company, Cleveland, Ohio

Industry depends to a great extent on arc welding as the major means for joining metals. Excellent welds can be made on a wide variety of metals, provided that certain precautions are taken to assure sound procedures. Welding is a safe occupation; however, constant attention to potential hygienic hazards is a must.

SAFETY PRACTICES IN ARC WELDING

Observing strict safety measures during welding and cutting operations is necessary to prevent injury to personnel and damage to property. Some of the recommended safe practices discussed here are mandatory and are governed by code requirements. Others are based on shop experience.

6-26 Personnel Protection Proper equipment and clothing should be used by welders, helpers, and personnel working near the welding stations to protect them from burns and spatter and from the radiant energy of the arc.

Eye and Face Protection. The welding arc should never be observed at close quarters with unprotected eyes. Failure to observe this rule can result in various degrees of eye burn. These burns do not usually cause permanent injury, but they can be very painful for several days after exposure.

A helmet-type head shield, as shown in Fig. 6-21*a* is standard equipment for protecting the welder's face and eyes from the direct rays of the arc. A hand-held face shield, as in Fig. 6-21*b*, is convenient for the use of onlookers. (Sunglasses or gas-welding goggles are not adequate protection.) These shields are generally made from a non-flammable insulating material and are black or gray in color to minimize reflection. They are shaped to protect the face, neck, and ears from direct radiant energy from the arc.

The shields are equipped with a standard-size (2 × 4¼ inch) glass window through which the welder observes the work in the area of the arc. A proper glass lens screens almost 100 percent of infrared and ultraviolet rays and most of the visible rays from the welder's eyes. The lens should be made from a tempered glass, free from bubbles, waves, or other flaws. Except for lenses that are gound for correction of vision, their flat surfaces should be smooth and parallel.

A piece of ordinary colored glass may look like a welding lens, but it does not have the necessary light-screening characteristics needed for eye protection. Special equipment is required to measure the amount of infrared and ultraviolet rays a lens will absorb. Welding lenses should be purchased only from suppliers who can be depended upon to furnish quality products.

Lenses are available in a number of shades for various types of work. Recommended shade numbers for common welding and cutting operations are listed in Table 6-4. Note that the shade number varies with the electrode size range in shielded metal-arc welding and with the thickness range of material in oxygen cutting.

FIG. 6-21 Head shields. (*a*) Helmet-type required for protecting the welder's eyes and face. (*b*) A hand-held face shield that is convenient for the use of foremen, inspectors, and other spectators. (*Lincoln Electric Co.*)

Lenses in head shields and face shields should be protected from breakage and from spatter by a cover lens or plate. This is a clear glass that is treated to resist damage from spatter. It covers the exposed surface of the lens. In addition to a helmet or shield, goggles with side shields should also be worn during arc welding or cutting operations. Goggles provide protection from spatter or rays from adjacent operations, particularly at times when the shield is removed, as is necessary when replacing electrodes, removing slag, or inspecting the weld. Goggles should be worn by welders' helpers, foremen, inspectors, and others working near the arc to protect their eyes from occasional flashes. Goggles should be lightweight, ventilated, and sterilizable, and the frames should be made of a heat-insulating material. Clear, spatter-resistant cover glasses and tinted lenses are used in the goggles.

TABLE 6-4 Guide for Shade Numbers

Welding operation	Suggested shade number*
Shielded metal-arc welding, up to 5/32 in. (4 mm) electrodes	10
Shielded metal-arc welding, 3/16 to 1/4 in. (4.8 to 6.4 mm) electrodes	12
Shielded metal-arc welding, over 1/4 in. (6.4 mm) electrodes	14
Gas metal-arc welding (nonferrous)	11
Gas metal-arc welding (ferrous)	12
Gas tungsten-arc welding	12
Plasma arc welding or cutting	12 or 14
Carbon arc welding	14
Torch soldering	2
Torch brazing	3 or 4
Light cutting, up to 1 in. (25 mm)	3 or 4
Medium cutting, 1 to 6 in. (25 to 150 mm)	4 or 5
Heavy cutting, over 6 in. (150 mm)	5 or 6
Gas welding (light) up to 1/8 in. (3.2 mm)	4 or 5
Gas welding (medium) 1/8 to 1/2 in. (3.2 to 12.7 mm)	5 or 6
Gas welding (heavy) over 1/2 in. (12.7 mm)	6 or 8

* The choice of a filter shade may be made on the basis of visual acuity and may therefore vary widely from one individual to another, particularly under different current densities, materials, and welding processes. However, the degree of protection from radiant energy afforded by the filter plate or lens when chosen to allow visual acuity will still remain in excess of the needs of eye filter protection. Filter plate shades as low as shade 8 have proved suitably radiation-absorbent for protection from the arc welding processes. NOTE: In gas welding or oxygen cutting, where the torch produces a high yellow light, it is desirable to use a filter lens that absorbs the yellow or sodium line in the visible light portion of the spectrum.

SOURCE: *Safety in Welding and Cutting,* ANSI Z49.1–1983, Table 1.

Submerged-arc welders do not need head shields but they should use goggles to protect against an accidental flash through the flux.

Protective Clothing. Ultraviolet energy from welding and cutting operations produces a skin burn, which, like a sunburn, is not immediately apparent. Thus, welders should wear clothing to protect all exposed skin areas. Woolen clothing is preferred over cotton or synthetic-fiber clothing because it is less easily ignited and it offers better protection from rapid changes in temperature. Any cotton or synthetic-fiber clothing worn for welding should be chemically treated to reduce its flammability. Dark-colored shirts are recommended to minimize reflection of rays under the helmet. Outer clothing, such as overalls, should be free from oil or grease. Low shoes with unprotected tops should not be worn; high-top safety shoes are recommended.

Shirt collars and cuffs should be kept buttoned, and pockets should be removed from the front of overalls. These measures prevent sparks or hot metal from lodging in the clothing. For the same reasons, legs of trousers or overalls should not be turned up on the outside. Minimum additional protection commonly used by welders consists of flameproof gauntlet gloves and a flameproof apron made of leather, or other suitable material.

For heavy work, fire-resistant leggings or high boots and leather sleeves or a full jacket should be used. For overhead operations, a leather cape or shoulder cover is essential. A leather skull cap or other suitable material should also be worn under the helmet to prevent head burns. Ear protection is desirable for overhead welding or for operations in confined areas. Wire-screen ear protectors are recommeneded for such operations. See Fig. 6-22.

6-27 Fire Prevention Fires connected with welding or cutting operations are usually caused by failure to keep combustible materials away from the work area. Fires rarely occur at permanent production-welding facilities; most fires involve portable equipment in areas not properly isolated or protected. If the work can be moved, it should be taken to a safe place such as a fireproof booth for cutting or welding. If the work cannot be moved from a hazardous location, welding should not be done until the area has been made safe. (See Bulletin 51B of the National Fire Protection Association.)

Welding or cutting should not be done in potentially explosive atmospheres—those containing mixtures of flammable gas, vapor, liquid, or dust with air—or near stored ignitable materials. A safe distance for welding in the general area of combustible materials is generally considered to be 35 ft.

If relocation of combustible materials is impractical, they should be protected with flameproof covers or shields with metal or asbestos screens or curtains. Edges of covers at the floor should be sealed to prevent sparks from getting underneath. Portable screens are used for isolating welding and cutting operations and for general protection of personnel from rays, spatter, and sparks.

Combustible floors in temporary welding areas should be swept clean, then protected with metal or other noncombustible material. The floor may be protected with damp sand or simply wet with water, but special care must be taken with a wet floor to protect welders and other personnel from the hazard of shock.

All openings in floors or walls should be closed so that any combustible materials on the floor below and in adjacent rooms are not exposed to sparks from the welding operation.

FIG. 6-22 Properly dressed welder. What the well dressed welder should wear: a cap or hard hat, depending on the job, safety glasses (not shown), head shield with the proper lens shade, leather jacket with long sleeves, leather gauntlet glooves, heavy trousers without cuffs, and safety shoes. (*Lincoln Electric Co.*)

Workers should be encouraged to be fire-conscious and to be on the lookout for potential fire hazards.

6-28 Precautions in Welding Containers Tanks, vessels, or other closed containers that have held combustible materials or gases should not be welded or cut unless they have been properly cleaned and marked as safe. Combustible materials include not only the common volatile petroleum products, but also:

1. Acids that react with metals to produce hydrogen.
2. Normally nonvolatile oils or solids that can release hazardous vapors when heated.
3. Fine, dustlike particles of a combustible solid that are potentially explosive.

Acceptable cleaning methods for such containers (which will not be detailed here) include water cleaning, hot-chemical-solution cleaning, and steam cleaning. The method used depends, of course, on the type of material that must be removed from the vessel. Details of approved cleaning procedures are given in AWS A6.0 and in other standards listed at the end of sec 6-29. Containers that have been cleaned and proved safe should be tagged or stenciled with the words "safe for welding and cutting," and should bear the date and the name of the person who certified the safety.

A supplementary precaution—filling the container with water or inert gas—is recommended for welding or cutting containers, even after they have been cleaned by approved methods. The container, of course, must be vented or open so the water or gas can drive out any dangerous fumes. Acceptable inert gases for this purpose are carbon dioxide (CO_2) and nitrogen. Carbon dioxide is available in pressure cylinders and in solid form ("dry ice"). Because the gas is heavier than air, it sinks to the bottom of containers that have top openings, and, as more gas is added, replaces the air or lighter-than-CO_2 fumes.

6-29 Ventilation The respiratory health hazards associated with welding operations are principally inhalation of gases, dusts, and metal fumes. The type and quantity of toxic fumes in a welding area depend on the type of welding being done, the filler and base metals used, contaminants on the base metal, solvents in the air, and the amount of air movement or ventilation in the area. Good ventilation is a primary key to avoiding or minimizing respiratory hazards.

In welding and cutting of mild steels, natural ventilation is usually considered adequate to remove fumes, provided that:

1. The room or welding area contains at least 10,000 cu ft for each welder.
2. The ceiling height is not less than 16 ft.
3. Cross ventilation is not blocked by partitions, equipment, or other structural barriers.

Spaces that do not meet these requirements should be equipped with mechanical ventilating equipment that exhausts at least 2,000 cfm of air for each welder, except where local exhaust hoods or booths or air-line respirators are used.

Welding or cutting operations that involve fluxes or other materials containing fluorine compounds, or that involve toxic metals such as zinc, lead, beryllium, cadmium, or mercury, require that hose masks, hose masks with blowers, or self-contained breathing equipment be used. Such equipment should meet U.S. Bureau of Mines standards.

Some degreasing compounds such as trichlorethylene and perchlorethylene decompose from the heat and from the ultraviolet radiation of an arc. The products of decomposition are irritating to the eyes and respiratory system. Parts that have been vapor-degreased should not be welded until all degreasing compound and vapors are completely removed.

Because of the chemical breakdown of vapor-degreasing materials under ultraviolet radiation, arc welding should not be done in the vicinity of a vapor-degreasing operation. Carbon-arc welding and gas tungsten-arc welding should be especially avoided in such areas, because they emit more ultraviolet radiation than other processes.

Exhaust Hoods and Booths. Local exhaust of welding fumes can be provided by adjustable hoods or by fixed enclosures or booths. Individual movable hoods are particularly suitable for bench welding, but can be used for any welding or cutting job provided that the hood can be moved so that it is always close to the joint being welded. These hoods are more economical to operate than a general ventilation system, particularly in cold weather, because they require less replacement air to be brought into the room and heated.

Minimum required air velocity at the zone of welding is 100 fpm when the hood is at its farthest position from the joint being welded. For a 3-inch wide flanged suction opening, this velocity requires an air volume of 150 cfm at 4 to 6 inches from the arc, and 600 cfm at 12 inches from the arc.

A ventilated booth is a second type of local exhaust arrangement. A booth is a fixed enclosure that consists of a top and at least two sides that surround the welding operation. Airflow requirements are similar to those for movable hoods—sufficient to maintain a velocity away from the welder of at least 100 fpm.

Recently exhaust equipment has become available for certain types of welding guns when semi-automatic or automatic welding is done in locations where smoke and fume accumulations are a problem for the welder and conventional exhaust systems are impractical or ineffective. The exhaust equipment consists of an exhaust blower, a system of filters, an exhaust hose, and an exhaust tube which has the open end in the immediate proximity of the welding arc. The smoke and fumes are removed at the point of origin. This protects the welder, and it also improves the general plant atmosphere.

BIBLIOGRAPHY—Welding Safety Practices

Cleaning Tanks Used for Gasoline or Similar Low-Flash Products, API Bulletin 2016, American Petroleum Institute, New York, N.Y.

Procedures for Cleaning Small Tanks and Containers, NFPA 327, National Fire Protection Association, Boston, Mass.

Purging Principles and Practices, American Gas Association, New York, N.Y.

Recommended Safe Practices for Gas-Shielded Arc Welding, AWS A6.1, American Welding Society Inc., New York, N.Y.

Safe Practices for Welding and Cutting Containers that Have Held Combustibles, AWS A6.0, American Welding Society Inc., New York, N.Y.

Safety Code for Head, Eye, and Respiratory Protection, ANSI Standard Z2.1, American National Standards Institute, New York, N.Y.

Safety in Welding and Cutting, ANSI Standard Z49.1, American Welding Society Inc., New York, N.Y.

Standard for Fire Protection in Use of Cutting and Welding Processes, NFPA 51B, National Fire Protection Association, Boston, Mass.

Standard for the Control of Gas Hazards on Vessels to be Repaired, NFPA 306, National Fire Protection Association, Boston, Mass.

ARC WELDING FUNDAMENTALS

Arc welding is one of several fusion processes for joining metals. By the application of intense heat, metal at the joint between two parts is melted and caused to intermix—directly or, more commonly, with an intermediate molten filler metal. Upon cooling and solidification, a metallurgical bond results. Since the joining is by intermixture of the substance of one part with the substance of the other part, with or without an intermediate substance, the final weldment has the potential for exhibiting at the joint the same strength properties as the metal of the parts. This is in sharp contrast to non-fusion processes of joining—such as soldering, brazing, or adhesive bonding—in which the mechanical and physical properties of the base materials cannot be duplicated at the joint.

In arc welding, the intense heat needed to melt metal is produced by an electric arc. The arc is formed between the work to be welded and an electrode that is manually or mechanically moved along the joint (or the work may be moved under a stationary electrode). The electrode may be a carbon or tungsten rod, the sole purpose of which is to carry the current and sustain the electric arc between its tip and the workpiece. Or it may be a specially prepared rod or wire that not only conducts the current and sustains the arc but also melts and supplies filler metal to the joint. If the electrode is a carbon or tungsten rod and the joint requires added metal for fill, that metal is supplied by a separately applied filler-metal rod or wire. Most welding in the manufacture of steel products where filler metal is required, however, is accomplished with the second type of electrodes—those that supply filler metal as well as provide the conductor for carrying electric current.

6-30 Basic Arc-Welding Circuit The basic arc-welding circuit is illustrated in Fig. 6-23. An ac or dc power source, fitted with whatever controls may be needed, is connected by a ground cable to the workpiece and by a "hot" cable to an electrode holder of some type, which makes electrical contact with the welding electrode. When the circuit is energized and the electrode tip touched to the grounded workpiece, and then withdrawn and held close to the spot of contact, an arc is created across the gap. The arc produces a temperature of about 6500°F at the tip of the electrode, a temperature more than adequate for melting most metals. The heat produced melts the base metal in the vicinity of the arc and any filler metal supplied by the electrode or by a separately introduced rod or wire. A common pool of molten metal is produced, called a "crater." This crater solidifies behind the electrode as it is moved along the joint being welded. The result is a fusion bond and the metallurgical unification of the workpieces.

6-31 Arc Shielding Use of the heat of an electric arc to join metals, however, requires more than the moving of the electrode in respect to the weld joint. Metals at high temperatures are reactive chemically with the main constituents of air—oxygen and nitrogen. Should the metal in the molten pool come in contact with air, oxides and nitrides would be formed, which upon solidification of the molten pool would destroy the strength properties of the weld joint. For this reason, the various arc-welding processes provide some means for covering the arc and the molten pool with a protective shield of gas, vapor, or slag. This is referred to as arc shielding, and such shielding may be accomplished by various techniques, such as the use of a vapor-generating covering on filler-metal-type electrodes, the covering of the arc and molten pool with a separately applied inert gas or a granular flux, or the use of materials within the core of tubular electrodes that generate shielding vapors.

Whatever the shielding method, the intent is to provide a blanket of gas, vapor, or slag that prevents or minimizes contact of the molten metal with air. The shielding method also affects the stability and other characteristics of the arc. When the shielding is produced by an electrode covering, by electrode core substances, or by separately applied granular flux, a fluxing or metal-improving function is usually also provided. Thus, the core materials in a flux-cored electrode may perform a deoxidizing function as well as a shielding function, and in submerged-arc welding the granular flux applied to the joint ahead of the arc may add alloying elements to the molten pool as well as shielding it and the arc.

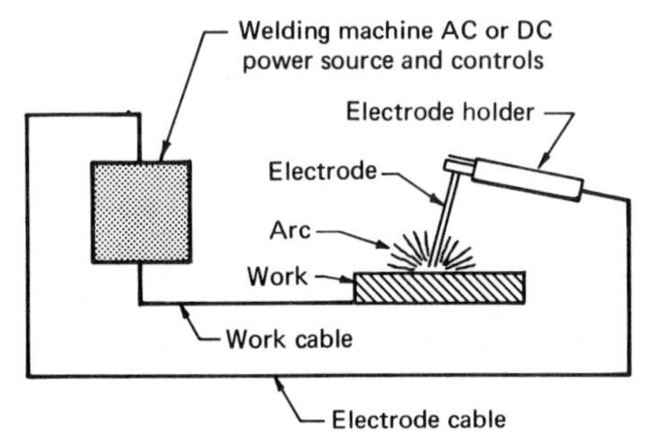

FIG. 6-23 The Basic arc-welding circuit. (*Lincoln Electric Co.*)

Figure 6-24 illustrates the shielding of the welding arc and molten pool with a covered "stick" electrode—the type of electrode used in most manual arc welding. The extruded covering on the filler metal rod, under the heat of the arc, generates a gaseous shield that prevents air from contacting the molten metal. It also supplies ingredients that react with deleterious substances on the metals, such as oxides and salts, and ties these substances up chemically in a slag that, being lighter than the weld metal, rises to the top of the pool and crusts over the newly solidified metal. This slag, even after solidification, has a protective function: it minimizes contact of the very hot solidified metal with air until the temperature lowers to a point where reaction of the metal with air is lessened.

While the main function of the arc is to supply heat, it has other functions that are important to the success of arc-welding processes. It can be adjusted or controlled

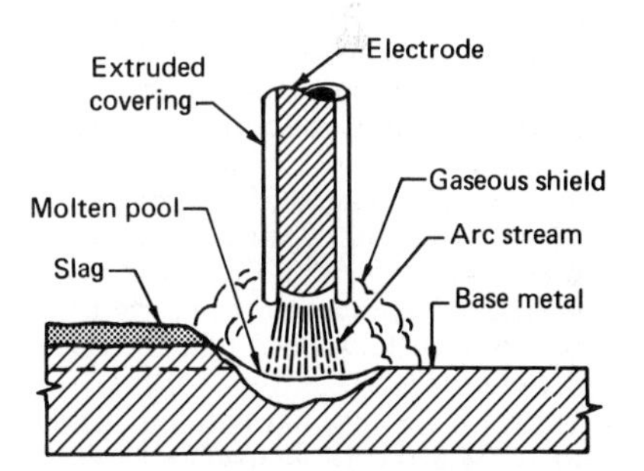

FIG. 6-24 Shielding of the weld. The arc and molten pool are shielded by a gaseous blanket developed by the vaporization and chemical breakdown of the extruded covering on the electrode in stick-electrode welding. Fluxing material in the electrode covering reacts with unwanted substances in the molten pool, tying them up chemically and forming a slag that crusts over the hot solidified metal. The slag, in turn, protects the hot metal from reaction with the air while it is cooling. (*Lincoln Electric Co.*)

to transfer molten metal from the electrode to the work, to remove surface films, and to bring about complex gas-slag-metal reactions and various metallurgical changes.

6-32 Arc Blow Arc blow is a phenomenon encountered mostly in arc welding with dc when the arc stream does not follow the shortest path between the electrode and the workpiece, but is deflected forward or backward from the direction of travel or, less frequently, to one side. Unless controlled, arc blow can be the cause of difficulties in handling the molten pool and slag, excessive spatter, incomplete fusion, reduced welding speed, porosity, and lowered weld quality.

Back blow occurs when welding toward the ground connection, end of a joint, or into a corner. Forward blow is encountered when welding away from the ground or at the start of the joint. Forward blow can be especially troublesome with iron-powder or other electrodes that produce large slag coverings, where the effect is to drag the heavy slag or the crater forward and under the arc.

There are two types of arc blow of concern to the welder. Their designations—magnetic and thermal—are indicative of their origins. Of the two, magnetic arc blow is the type causing most welding problems.

6-33 Magnetic Arc Blow Magnetic arc blow is caused by an unbalanced condition in the magnetic field surrounding the arc. Unbalanced conditions result from the fact that at most times the arc will be farther from one end of the joint than another and will be at varying distances from the ground connection. Imbalance also always exists because of the change in direction of the current as it flows from the electrode, through the arc, and into and through the workpiece.

To understand arc blow, it is helpful to visualize a magnetic field. Figure 6-25 shows a direct current passing through a conductor, which could be an electrode or the ionized gas stream between an electrode and a weld joint. Around the conductor a magnetic field, or flux, is set up, with lines of force that can be represented by concentric circles in planes at right angles to the direction of the current. These circular lines of force diminish in intensity the farther they are from the electrical conductor.

They remain circular when they can stay in one medium, say air or metal, expansive enough to contain them until they diminish to essentially nothing in intensity. But if the medium changes, say from steel plate to air, the circular lines of forces are distorted; the forces tend to concentrate in the steel where they encounter less resistance. At a boundary between the edges of a steel plate and air, there is a squeezing of the magnetic flux lines, with deformation in the circular planes of force. This squeezing can result in a heavy concentration of flux behind or ahead of a welding arc. The arc then tends to move in the direction that would relieve the squeezing (would tend to restore flux balance). It veers away from the side of flux concentration, and this veering is the observed phenomenon of arc blow.

6-34 How to Reduce Arc Blow Not all arc blow is detrimental. In fact, a small amount of arc blow can sometimes be used beneficially to help form the bead shape, control molten slag, and control penetration. See Fig. 6-26.

When arc blow is causing or contributing to such defects as undercut, inconsistent penetration, crooked beads, beads of irregular width, porosity, wavy beads, and excessive spatter, it must be controlled. Possible corrective measures have already been suggested in the preceding text. Here are some general methods that might be considered:

- If direct current is being used with the shielded metal-arc process—especially above 250 amperes—a change to alternating current may eliminate problems.
- Hold as short an arc as possible to help the arc force counteract the arc blow.
- Reduce the welding current—which may require a reduction in arc speed.
- Angle the electrode with the work opposite the direction of arc blow, as illustrated in Fig. 6-26.
- Make a heavy tack weld on both ends of the seam; apply frequent tack welds along the seam, especially if the fit-up is not tight.
- Weld toward a heavy tack or toward a weld already made.
- Use a back-step welding technique, as shown in Fig. 6-27.
- Weld away from the ground to reduce back blow; weld toward the ground to reduce forward blow.
- With processes where a heavy slag is involved, a small amount of back blow may be desirable; to get this, weld toward the ground.

■ Wrap ground cable around the workpiece and pass ground current through it in such a direction that the magnetic field set up will tend to neutralize the magnetic field causing the arc blow.

The direction of the arc blow can be observed in the open-arc process, but with the submerged-arc process must be determined by the type of weld defect.

Back blow is indicated by the following:

- Spatter
- Undercut, either continuous or intermittent
- Narrow, high bead, usually with undercut
- An increase in penetration
- Surface porosity at the finish end of welds on sheet metal

Follow blow is indicated by:

- A wide bead, irregular in width
- Wavy bead
- Undercut, usually intermittent
- A decrease in penetration

See Fig. 6-27.

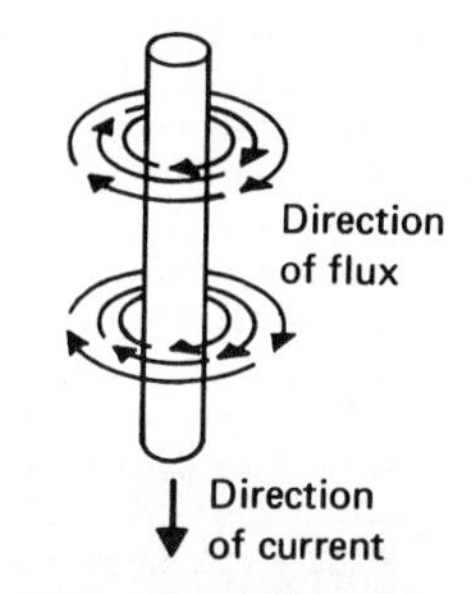

FIG. 6-25 Direct current passing through a conductor. A current through a conductor sets up a magnetic field that may be represented by planes of concentric circles—flux lines. (*Lincoln Electric Co.*)

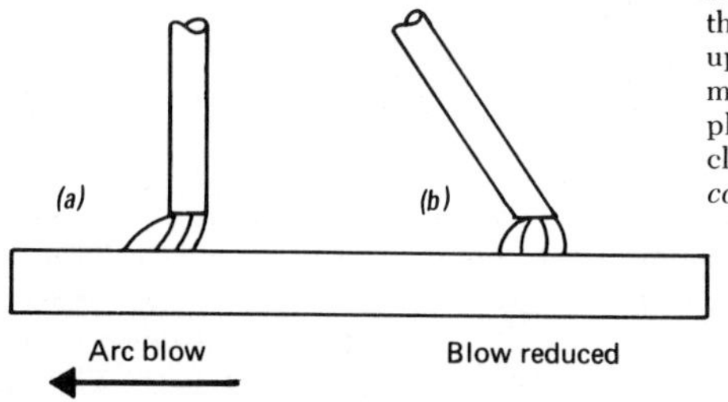

FIG. 6-26 Arc blow. Arc blow (*a*) can sometimes be corrected by angling the electrode (*b*). (*Lincoln Electric Co.*)

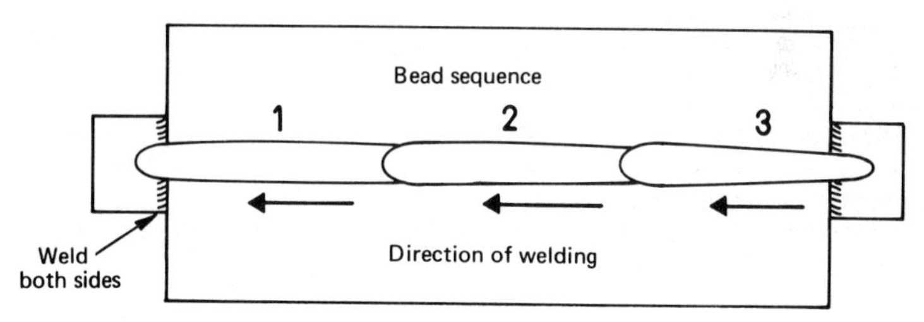

FIG. 6-27 Direction of welding. Direction of welding and the sequence of beads for the back-step technique. Note tabs on both ends of the seam. Tabs should be the same thickness as the work. (*Lincoln Electric Co.*)

WELDING PROCESSES

6-35 Shielded Metal-Arc Process The shielded metal-arc process—commonly called "stick-electrode" welding or "manual" welding—is the most widely used of the various arc-welding processes. It is characterized by application versatility and flexibility and relative simplicity in the equipment. It is the process used by the small welding shop, by the home mechanic, and by the farmer for repair of equipment. It is also a process having extensive applications in industrial fabrication, structural steel erection, weldment manufacture, and other commercial metal joining. Arc welding to persons only casually acquainted with welding usually means shielded metal-arc welding.

With this process, an electric arc is struck between the electrically grounded work and a 9- to 18-inch length of covered metal rod—the electrode. The electrode is clamped in an electrode holder, which is joined by a cable to the power source. The welder grips the insulated handle of the electrode holder and maneuvers the tip of the electrode in relation to the weld joint. When the welder touches the tip of the electrode against the work, and then withdraws it to establish the arc, the welding circuit is completed. The heat of the arc melts base metal in the immediate area, the electrode's metal core, and any metal particles that may be in the electrode's covering. It also melts, vaporizes, or breaks down chemically, nonmetallic substances incorporated in the covering for arc-shielding, metal-protection, or metal-conditioning purposes. The mixing of molten base metal and filler metal from the electrode provides the coalescence required to effect joining. See Fig. 6-28.

FIG. 6-28 Typical production scene of the shielded-metal arc process. (*Lincoln Electric Co.*)

As welding progresses, the covered rod becomes shorter and shorter. Finally, the welding must be stopped to remove the stub and replace it with a new electrode. This periodic changing of electrodes is one of the major disadvantages of the process in production welding. It decreases the operating factor, or the percent of the welder's time spent in the actual operation of laying weld beads.

Another disadvantage of shielded metal-arc welding is the limitation placed on the current that can be used. High amperages, such as those used with semi-automatic guns or automatic welding heads, are impractical because of the long (and varying) length of electrode between the arc and the point of electrical contact in the jaws of the electrode holder. The welding current is limited by the resistance heating of the electrode. The electrode temperature must not exceed the breakdown temperature of the covering. If the temperature is too high the covering chemicals react with each other or with air and therefore do not function properly at the arc. Coverings with organics break down at lower temperatures than mineral or low-hydrogen-type coverings.

However, the versatility of the process—plus the simplicity of equipment—is viewed as overriding its inherent disadvantages by many users whose work would permit some degree of mechanized welding. This point of view was formerly well taken, but now that semi-automatic self-shielded flux-cored arc welding has been developed to a similar (or even superior) degree of versatility and flexibility, there is less justification for adhering to stick-electrode welding in steel fabrication and erection wherever substantial amounts of weld metal must be placed. In fact, the replacement of shielded metal-arc welding with semi-automatic processes has been a primary means by which steel fabricators and erectors have met cost-price squeezes in their welding operations.

Principles of Operation. The shielding ingredients have various functions. One is to shield the arc—provide a dense, impenetrable envelope of vapor or gas around the arc and the molten metal to prevent the pickup of oxygen and nitrogen and the chemical formation of oxides and nitrides in the weld puddle. A second function is to provide scavengers and deoxidizers to refine the weld metal.

A third is to produce a slag coating over molten globules of metal during their transfer through the arc stream and a slag blanket over the molten puddle and the newly solidified weld. Figure 6-29 illustrates the decomposition of an electrode covering and the manner in which the arc stream and weld metal are shielded from the air.

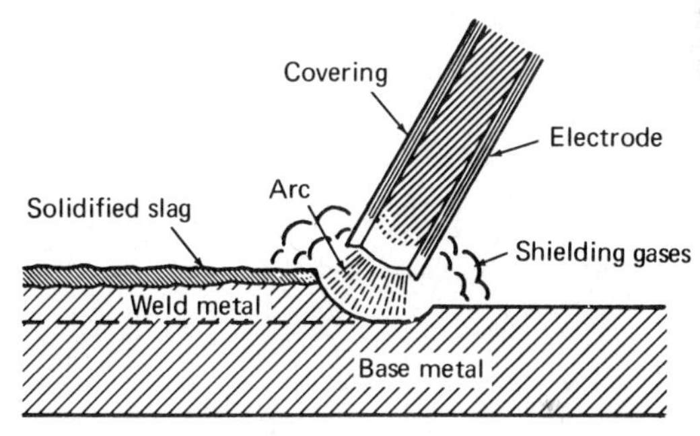

FIG. 6-29 Decomposition of an electrode covering. Schematic representation of shielded metal-arc welding. Gases generated by the decomposition and vaporization of materials in the electrode covering—including vaporized slag—provide a dense shield around the arc stream and over the molten puddle. Molten and solidified slag above the newly formed weld metal protects it from the atmosphere while it is hot enough to be chemically reactive with oxygen and nitrogen. (*Lincoln Electric Co.*)

Another function of the shield is to provide the ionization needed for ac welding. With alternating current, the arc goes out 120 times a second. For it to be reignited each time it goes out, an electrically conductive path must be maintained in the arc stream. Potassium compounds in the electrode covering provide ionized gaseous particles that remain ionized during the fraction of a second that the arc is extinguished with ac cycle reversal. An electrical path for reignition of the arc is thus maintained.

The mechanics of arc shielding varies with the electrode type. Some types of electrodes depend largely on a "disappearing" gaseous shield to protect the arc stream and the weld metal. With these electrodes, only a light covering of slag will be found on the finished weld. Other electrode types depend largely on slag for shielding. The explanation for the protective action is that the tiny globules of metal being transferred in the arc stream are entirely coated with a thin film of molten slag. Presumably, the globules become coated with slag as vaporized slag condenses on them—so the protective action still arises from gasification. In any event, the slag deposit with these types of electrodes is heavy, completely covering the finished weld. Between these extremes are electrodes that depend on various combinations of gas and slag for shielding.

The performance characteristics of the electrodes are related to their slag-forming properties. Electrodes with heavy slag formation have high deposition rates and are suitable for making large welds downhand. Electrodes that develop a gaseous shield that disappears into the atmosphere and give a light slag covering are low-deposition and best suited for making welds in the vertical or overhead positions.

Arc welding may be done with either ac or dc and with the electrode either positive or negative. The choice of current and polarity depends on the process, the type of electrode, the arc atmosphere, and the metal being welded. Whatever the current, it must be controlled to satisfy the variables—amperage and voltage—which are specified by the welding procedures.

Power Source. Shielded metal-arc welding requires relatively low currents (10 to 500 amperes) and voltages (17 to 45 volts), depending on the type and size electrode used. The current may be either ac or dc; thus, the power source may be either ac or dc or a combination ac/dc welder. For most work, a variable-voltage power source

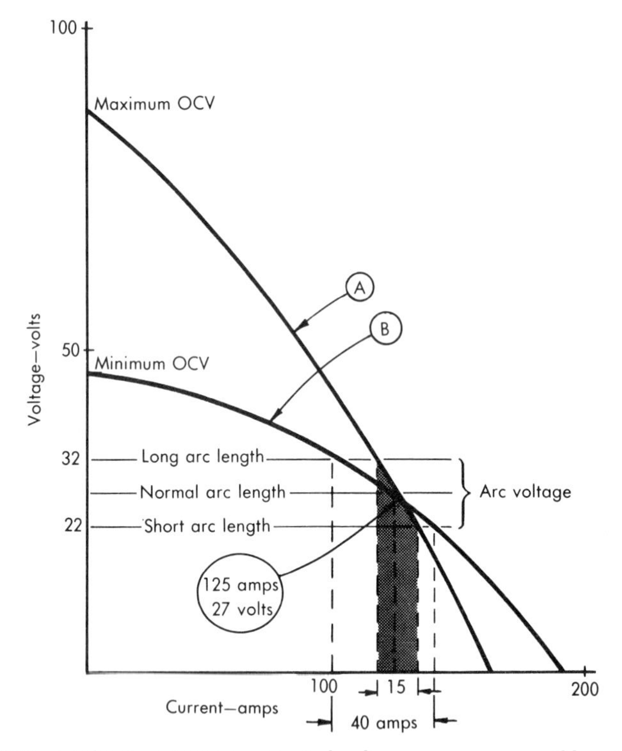

FIG. 6-30 Typical volt-ampere curves. Typical volt-ampere curves possible with a variable voltage power source. The steep curve (*a*) allows minimum current change. The flatter curve (*b*) permits the welder to control current changing the length of the arc. (*Lincoln Electric Co.*)

is preferred, since it is difficult for the welder to hold a constant arc length. With the variable voltage source and the machine set to give a steep volt-ampere curve, the voltage increases or decreases with variations in the arc length to maintain the current fairly constant. The equipment compensates for the inability of the operator to hold an exact arc length, and he is able to obtain a uniform deposition rate.

In some welding, however, it may be desirable for the welder to have control over the deposition rate—for example, when depositing root passes in joints with varying fit-up or in out-of-position work. In these cases variable-voltage performance with a flatter voltage-amperage curve is desirable, so that the welder can decrease the deposition rate by increasing the arc length or increase the rate by shortening the arc length.

Figure 6-30 illustrates typical volt-ampere curves possible with a variable-voltage power source. The change from one type of voltage-ampere curve to another is made by changing the open-circuit voltage and current settings of the machine.

The fact that the shielded metal-arc process can be used with so many electrode types and sizes, in all positions, on a great variety of materials, and with flexibility in operator control makes it the most versatile of all welding processes. These advantages are enhanced further by the low cost of equipment. The total advantages of the process, however, must be weighed against the cost per foot of weld when a process is to be selected for a particular job. Shielded metal-arc welding is a well-recognized way of getting the job done, but too-faithful adherence to it often leads to getting the job done at excessive welding costs.

6-36 Submerged-Arc Process Submerged-arc welding differs from the other arc-welding processes in that a blanket of fusible, granular material—commonly called flux—is used for shielding the arc and the molten metal. The arc is struck between the workpiece and a bare wire electrode, the tip of which is submerged in the flux. Since

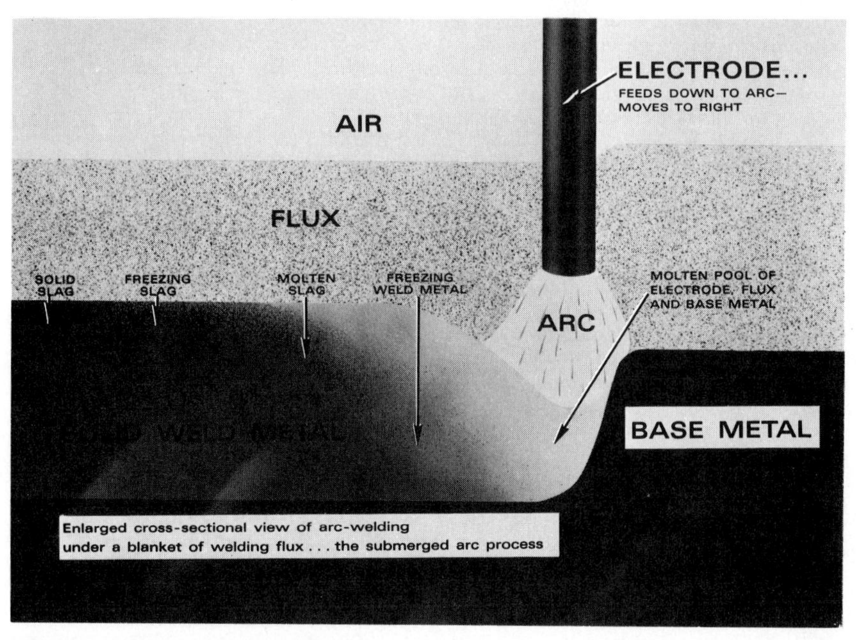

FIG. 6-31 Typical submerged-arc process. The mechanics of the submerged-arc process. The arc and the molten weld metal are buried in the layer of flux, which protects the weld metal from contamination and concentrates the heat. The molten flux arises through the pool, deoxidizing and cleansing the molten metal, and forms a protective slag over the newly deposited weld. (*Lincoln Electric Co.*)

the arc is completely covered by the flux, it is not visible and the weld is run without the flash, spatter, and sparks that characterize the open-arc processes. The nature of the flux is such that very little smoke or visible fumes are developed. See Fig. 6-31.

The Mechanics of Flux Shielding. The process is either semi-automatic or full-automatic, with electrode fed mechanically to the welding gun, head, or heads. In semi-automatic welding the welder moves the gun, usually equipped with a flux-feeding device, along the joint. Flux feed may be by gravity flow through a nozzle concentric with the electrode from a small hopper atop the gun, or it may be through a concentric nozzle tube-connected to an air-pressurized flux tank. Flux may also be applied in advance of the welding operation or ahead of the arc from a hopper run along the joint. In fully automatic submerged-arc welding, flux is fed continuously to the joint, ahead of or concentric to the arc, and full-automatic installations are commonly equipped with vacuum systems to pick up the unfused flux left by the welding head or heads for cleaning and reuse.

There are two general types of submerged-arc fluxes, bonded and fused. In the bonded fluxes, the finely ground chemicals are mixed, treated with a bonding agent and manufactured into a granular aggregate. The deoxidizers are incorporated in the flux. The fused fluxes are a form of glass made by fusing the various chemicals and then grinding the glass to a granular form. Fluxes are available that add alloying elements to the weld metal, enabling alloy weld metal or hard-facing deposits to be made with mild-steel electrodes.

Advantages of the Process. High currents can be used in submerged-arc welding and extremely high heat can be developed. Because the current is applied to the electrode a short distance above its tip, relatively high amperages can be used on small-diameter electrodes. This results in extremely high current densities on relatively small cross sections of electrode. Currents as high as 600 amperes can be carried on electrodes as small as $^5/_{64}$ inch, giving a density in the order of 100,000 amperes per square inch – 6 to 10 times that carried on stick electrodes.

Because of the high current density, the melt-off rate is much higher for a given electrode diameter than with stick-electrode welding. The melt-off rate is affected by the electrode material, the flux, type of current, polarity, and length of wire beyond the point of electrical contact in the gun or head.

The insulating blanket of flux above the arc prevents rapid escape of heat and concentrates it in the welding zone. Not only are the electrode and base metal melted rapidly, but the fusion is deep into the base metal. The deep penetration allows the use of small welding grooves, thus minimizing the amount of filler metal per foot of joint and permitting fast welding speeds. Fast welding, in turn, minimizes the total heat input into the assembly and thus tends to prevent problems of heat distortion. Even relatively thick joints can be welded in one pass by submerged arc.

Through regulation of current, voltage, and travel speed, the operator can exercise close control over penetration to provide any depth ranging from deep and narrow with high-crown reinforcement, to wide, nearly flat beads with shallow penetration. Beads with deep penetration may contain on the order of 70 percent melted base metal, while shallow beads may contain as little as 10 percent base metal. In some instances, the deep-penetration properties of the submerged-arc method can be used to eliminate or reduce the expense of edge preparation.

Multiple electrodes may be used, two side by side or two or more in tandem, to cover a large surface area or to increase welding speed. If shallow penetration is desired with multiple electrodes, one electrode can be grounded through the other (instead of through the workpiece) so that the arc does not penetrate deeply.

Deposition rates are high – up to 10 times those of stick-electrode welding. Table 6-5 shows approximate deposition rates for various submerged-arc arrangements, with comparable deposition rates for manual welding with covered electrodes.

Submerged-arc welding may be done with either dc or ac power. Direct current gives better control of bead shape, penetration, and welding speed, and arc starting is easier with it. Bead shape is usually best with dc reverse polarity (DCRP) – electrode positive – which also provides maximum penetration. Highest deposition rates and minimum penetration are obtained with dc straight polarity (DCSP). Alternating current minimizes arc blow and gives penetration between that of DCRP and DCSP.

When submerged-arc equipment is operated properly, the weld beads are smooth and uniform, so that grinding or machining are rarely required. Since the rapid heat input of the process minimizes distortion, the costs for straightening finished assemblies are reduced, especially if a carefully planned welding sequence has been followed. Submerged-arc welding, in fact, often allows the premachining of parts, further adding to fabrication cost savings.

A limitation of submerged-arc welding is that imposed by the force of gravity. In most instances, the joint must be positioned flat or horizontal to hold the granular flux.

The high quality of submerged-arc welds, the high deposition rates, the deep penetration, the adaptability of the process to full mechanization, and the comfort characteristic (no glare, sparks, spatter, smoke, or excessive heat radiation) make it a preferred process in steel fabrication. It is used extensively in ship and barge building, railroad car building, pipe manufacture, and in fabrication of structural beams, girders,

TABLE 6-5 Approximate Deposition Rate of Submerged Arc Processes on Mild Steel

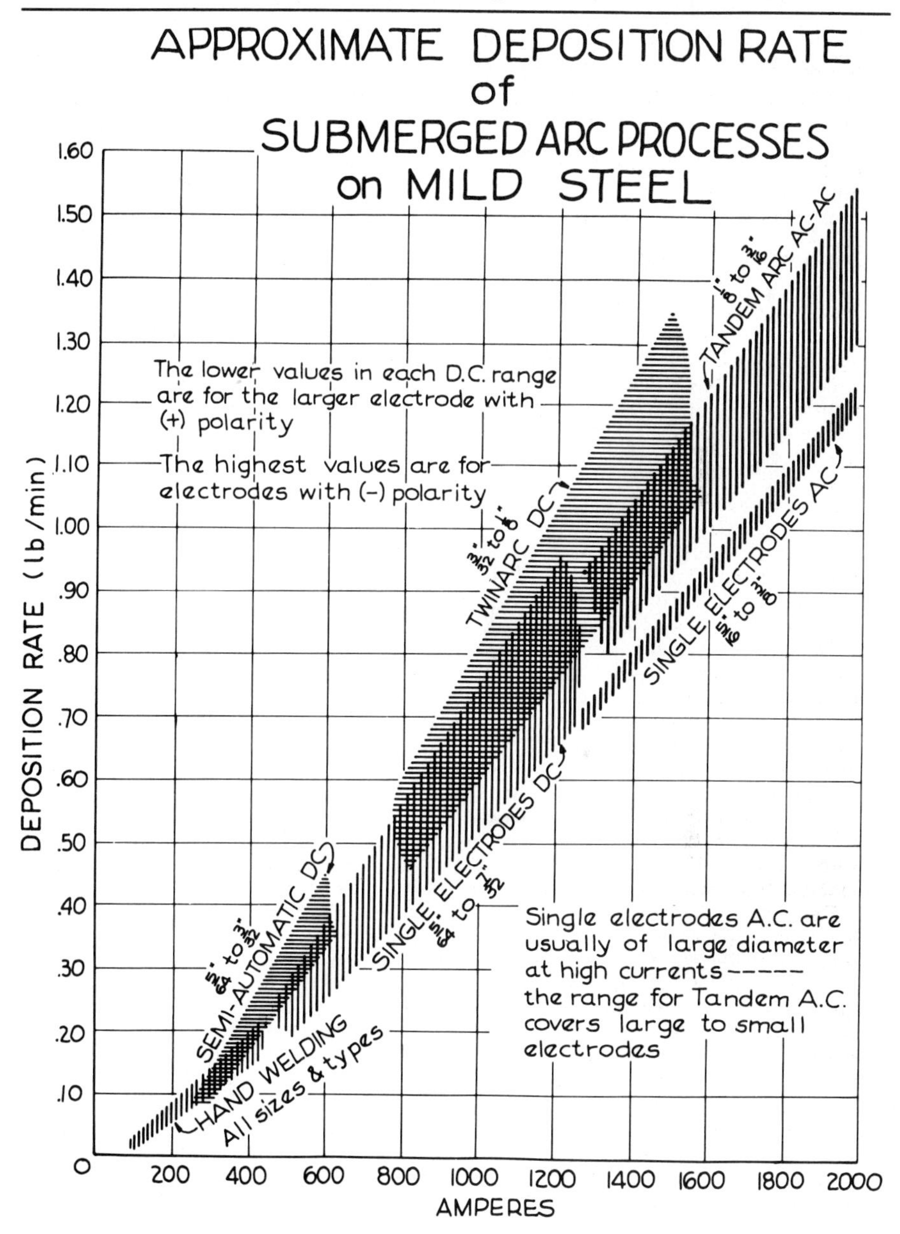

and columns where long welds are required. Automatic submerged-arc installations are also key features of the welding areas of plants turning out mass-produced assemblies joined with repetitive short welds.

The high deposition rates attained with submerged-arc are chiefly responsible for the economies achieved with the process. The cost reductions when changing from the manual shielded metal-arc process to submerged-arc are frequently dramatic.

Thus, a hand-held submerged-arc gun with mechanized travel may reduce welding costs more than 50 percent; with fully automatic multiarc equipment, it is not unusual for the costs to be but 10 percent of those attained with stick-electrode welding.

6-37 Self-Shielded Flux-Cored Process The self-shielded flux-cored arc-welding process is an outgrowth of shielded metal-arc welding. The versatility and maneuverability of stick electrodes in manual welding stimulated efforts to mechanize the shielded metal-arc process. The theory was that if some way could be found for putting an electrode with self-shielding characteristics in coil form and feeding it mechanically to the arc, welding time lost in changing electrodes and the material lost as electrode stubs would be eliminated. The result of these efforts was the development of the semi-automatic and full-automatic processes for welding with continuous flux-cored tubular electrode "wires." Such fabricated wires (Fig. 6-32) contain in their cores the ingredients for fluxing and deoxidizing molten metal and for generating shielding gases and vapors and slag coverings.

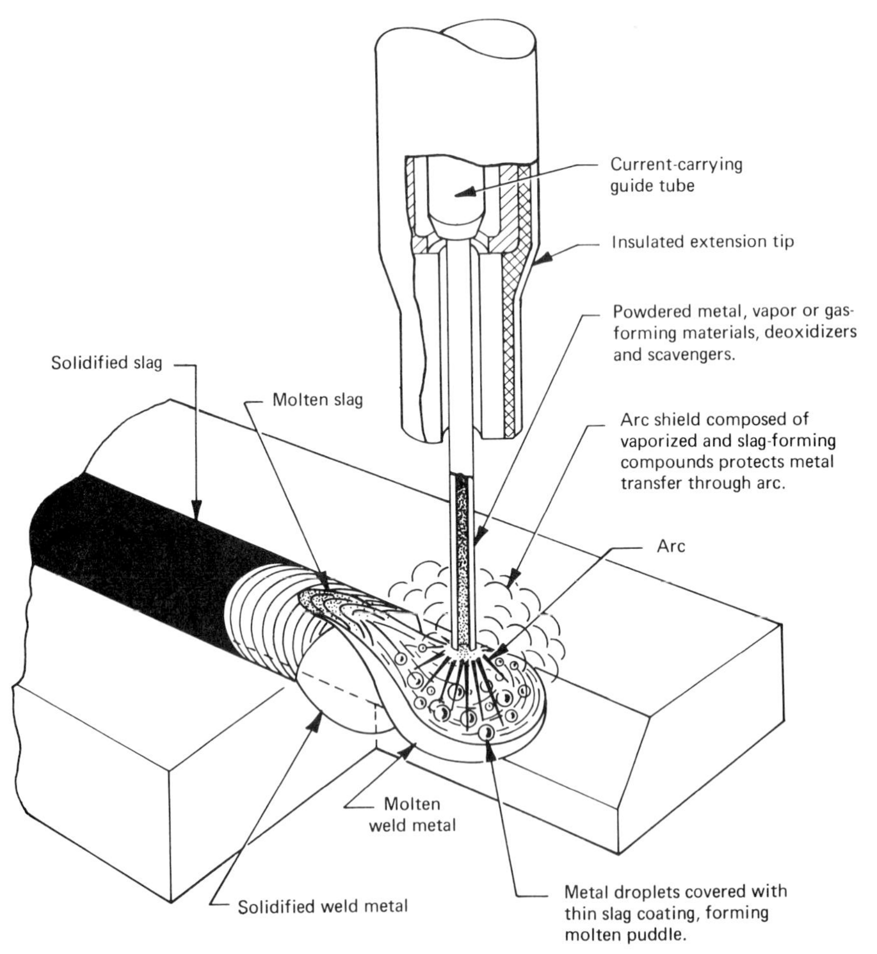

FIG. 6-32 Principles of self-shielded flux-cored arc welding. The electrode may be viewed as an "inside-out" construction of the stick electrode used in shielded metal-arc welding. Putting the shield-generating materials inside the electrode allows the coiling of long, continuous lengths of electrode and gives an outside conductive sheath for carrying the welding current from a point close to the arc. (*Lincoln Electric Co.*)

Full-automatic welding with self-shielded flux-cored electrodes is one step further in mechanization—the removal of direct manual manipulation in the utilization of the open-arc process.

Higher deposition rates, plus automatic electrode feed and elimination of lost time for changing electrodes have resulted in substantial production economies wherever the semi-automatic process has been used to replace stick-electrode welding. Decreases in welding costs as great as 50 percent have been common, and in some production welding, deposition rates have been increased as much as 400 percent.

The intent behind the development of self-shielded flux-cored electrode welding was to mechanize and increase the efficiency of manual welding. The semi-automatic use of the process does just that—it serves as a direct replacement for stick-electrode welding. The full-automatic use of the process, on the other hand, competes with other fully automatic processes and is used in production where it gives the desired performance characteristics and weld properties, while eliminating problems associated with flux or gas handling. Although the full-automatic process is important to

FIG. 6-33 Maintenance and repair welding. The semi-automatic self-shielded flux-cored arc-welding process substantially reduces cost in repair, rebuilding, and maintenance work, as well as in manufacturing, fabrication, and structural steel erection. Here, the welder is using the process to repair and rebuild shovel crawler pads. (*Lincoln Electric Co.*)

a few industries, the semi-automatic version has the wider application possibilities. In fact, semi-automatic self-shielded flux-cored welding has potential for substantially reducing costs of welding steel wherever stick-electrode welding is used to deposit other than minor volumes of weld metal. This has been proved to be true in maintenance and repair welding (Fig. 6-33) as well as in production work.

The advantages of the self-shielded flux-cored arc-welding process may be summarized as follows:

1. When compared with stick-electrode welding, gives deposition rates up to 4 times as great, often decreasing welding costs by as much as 50 to 75 percent.

2. Eliminates the need for flux-handling and recovery equipment, as in submerged-arc welding, or for gas and gas storage, piping, and metering equipment, as in gas-shielded mechanized welding. The semi-automatic process is applicable where other mechanized processes would be too unwieldy.

3. Has tolerance for elements in steel that normally cause weld cracking when stick-electrode or one of the other mechanized welding processes is used. Produces crack-free welds in medium-carbon steel with normal welding procedures.

4. Under normal conditions, eliminates the problems of moisture pickup and storage that occur with low-hydrogen electrodes.

5. Eliminates stub losses and the time that would be required for changing electrodes with the stick-electrode process.

6. Eliminates the need for wind shelters required with gas-shielded welding in field erection; permits fans and fast-air-flow ventilation systems to be used for worker comfort in the shop.

7. Enables "one-process," and even "one-process, one-electrode," operation in some shop and field applications. This, in turn, simplifies operator training, qualification, and supervision; equipment selection and maintenance; and the logistics of applying men, materials, and equipment to the job efficiently.

8. Enables the application of the long-stickout principle to enhance deposition rates, while allowing easy operator control of penetration.

9. Permits more seams to be welded in one pass, saving welding time and the time that otherwise would be consumed in between-pass cleaning.

10. Is adaptable to a variety of products; permits continuous operation at one welding station, even though a variety of assemblies with widely different joint requirements are run through it.

11. Provides the fast filling of gouged-out voids often required when making repairs to weldments or steel castings.

12. Gives the speed of mechanized welding in close quarters; reaches into spots inaccessible by other semi-automatic processes.

13. Provides mechanized welding where mechanized welding was formerly impossible, such as in the joining of a beam web to a column in building erection (Fig. 6-34).

FIG. 6-34 Typical welding for structural steel. Before all-position electrodes for self-shielded flux-cored arc welding were developed, beam web-to-column connections in building erection were made with stick electrodes. If the flange-to-column joints were made with semi-automatic equipment, the operator had to change to the stick electrode for the vertical joint, or the connection had to be made by bolting the web to an angle welded to the column. The all-position electrode wires were the development that made possible one-process mechanized welding in building erection, with all of its incidental benefits in the scheduling of erection operations. (*Lincoln Electric Co.*)

14. Enables the bridging of gaps in fit-up by operator control of the penetration without reducing quality of the weld. Minimizes repair, rework, and rejects.

6-38 Gas-Shielded Arc-Welding Processes As noted in the preceding sections, the shielded metal-arc process (stick-electrode) and self-shielded, flux-cored electrode process depend in part on gases generated by the heat of the arc to provide arc and puddle shielding. In contrast, the gas-shielded arc-welding processes use either bare or flux-cored filler metal and gas from an external source for shielding. The gas impinges upon the work from a nozzle that surrounds the electrode. It may be an inert gas—such as argon or helium—or carbon dioxide (CO_2), a much cheaper gas that is suitable for use in the welding of steels. Mixtures of the inert gases, oxygen, and carbon dioxide also are used to produce special arc characteristics.

There are three basic gas-shielded arc-welding processes that have broad application in industry. They are the gas-shielded, flux-cored process, the gas–tungsten arc (TIG) process, and the gas–metal arc (MIG) process.

6-39 Gas-Shielded, Flux-Cored Process The gas-shielded, flux-cored process may be looked upon as a hybrid between self-shielded, flux-cored arc welding and gas-metal arc welding. Tubular electrode wire is used as in the self-shielded process, but the ingredients in its core are for fluxing, deoxidizing, scavenging, and sometimes alloying additions, rather than for these functions plus the generation of protective vapors. In this respect, the process has similarities to the self-shielded, flux-cored electrode process, and the tubular electrodes used are classified by the AWS along with electrodes used in the self-shielded process. On the other hand, the process is similar to gas-metal arc welding in that a gas is separately applied to act as arc shield.

The guns and welding heads for semi-automatic and full-automatic welding with the gas-shielded process are of necessity more complex than those used in self-shielded, flux-cored welding. Passages must be included for the flow of gases. If the gun is water-cooled, additional passages are required for this purpose.

The gas-shielded, flux-cored process is used for welding mild and low-alloy steels. It gives high deposition rates, high deposition efficiencies, and high operating factors. Radiographic-quality welds are easily produced, and the weld metal with mild and low-alloy steels has good ductility and toughness. The process is adaptable to a wide variety of joints and gives the capability for all-position welding.

6-40 Gas-Metal Arc Welding Gas-metal arc welding, popularly known as MIG welding, uses a continuous electrode for filler metal and an externally supplied gas or gas mixture for shielding. The shielding gas—helium, argon, carbon dioxide, or mixtures thereof—protects the molten metal from reacting with constituents of the atmosphere. Although the gas shield is effective in shielding the molten metal from the air, deoxidizers are usually added as alloys in the electrode. Sometimes light coatings are applied to the electrode for arc stabilizing or other purposes. Lubricating films may also be applied to increase the electrode feeding efficiency in semi-automatic welding equipment. Reactive gases may be included in the gas mixture for arc-conditioning functions.

MIG welding may be used with all of the major commercial metals, including carbon, alloy, and stainless steels and aluminum, magnesium, copper, iron, titanium, and zirconium. It is a preferred process for the welding of aluminum, magnesium, copper, and many of the alloys of these reactive metals. Most of the irons and steels can be satisfactorily joined by MIG welding, including the carbon-free irons, low-carbon and low-alloy steels, high-strength quenched and tempered steels, chromium irons and steels, high-nickel steels, and some of the so-called superalloy steels. With these various materials, the welding techniques and procedures may vary widely. Thus, carbon dioxide or argon-oxygen mixtures are suitable for arc shielding when the low-carbon and low-alloy steels are welded, whereas pure inert gas may be essential when highly alloyed steels are welded. Copper and many of its alloys and the stainless steels are successfully welded by the process.

Welding is either semi-automatic, using a hand-held gun to which the electrode is fed automatically, or full-automatic.

Metal transfer with the MIG process is by one of two methods: "spray arc" or short circuiting. With spray arc, drops of molten metal detach from the electrode and move through the arc column to the work. With the short-circuiting technique—often referred to as short-arc welding—metal is transferred to the work when the molten tip of the electrode contacts the molten puddle.

Short-arc welding uses low currents, low voltages, and small-diameter wires. The molten drop short-circuits the arc an average of 100 times a second and at rates lower and much higher than this average. Metal is transferred with each short circuit, rather than across the arc as in spray-arc welding.

The technique results in low heat input, which minimizes distortion. It is useful for welding thin-gage materials in all positions and for vertical and overhead welding of heavy sections. Short-arc welding tolerates poor fit-up and permits the bridging of wide gaps.

For efficient short-arc welding, special power sources with adjustable slope, voltage, and inductance characteristics are required. These power sources produce the predictable and controllable current surges needed for successful uses of the short-arc technique.

Spray-arc transfer may be subdivided into two different types. When the shielding gas is argon or an argon-oxygen mixture, the droplets in the spray are very fine and never short-circuit the arc. When carbon dioxide or an argon-carbon dioxide mixture is used, a molten ball tends to form on the end of the electrode and may grow in size until its diameter is greater than the diameter of the electrode. These droplets, larger in size, may cause short circuits and this mode is known as globular transfer. Under conditions that cause the short circuits to occur very rapidly, the mode becomes short-circuiting transfer.

Spray-arc MIG welding produces an intensely hot, higher-voltage arc, and thus gives a higher deposition rate than short-arc welding. A high current density is required for metal transfer through the arc. The spray-arc technique is recommended for ⅛-inch and thicker sections requiring heavy single or multipass welds or for any filler-pass application where high deposition rate is advantageous.

MIG welding is a dc weld process: ac is seldom, if ever, used. Most MIG welding is done with reverse polarity (DCRP). Weld penetration is deeper with reverse polarity than it is with straight polarity. MIG welding is seldom done with straight polarity, because of arc-instability and spatter problems that make it undesirable for most applications.

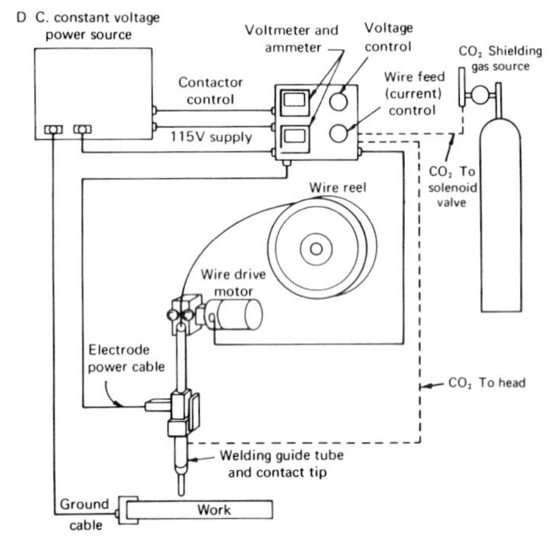

FIG. 6-35 Full-Automatic welding facility. Schematic for a full-automatic welding facility with either self-shielded or gas-shielded flux-cored electrode. The dotted line indicates the additions required with the gas-shielded version when CO_2 is the shielding gas. (*Lincoln Electric Co.*)

The gas-metal arc process can be used for spot welding to replace either riveting, electrical resistance, or TIG spot welding. It has proved applicable for spot welding where TIG spot welding is not suitable, such as in the joining of aluminum and rimmed steel. Fit-up and cleanliness requirements are not as exacting as with TIG spot welding, and the MIG process may be applied to thicker materials.

The MIG process is also adaptable to vertical electrogas welding in a manner similar to that used with the gas-shielded, flux-cored electrode process. See Fig. 6-35.

6-41 Gas-Tungsten Arc Welding The AWS definition of gas-tungsten arc (TIG) welding is "an arc welding process wherein coalescence is produced by heating with an arc between a tungsten electrode and the work." A filler metal may or may not be used. Shielding is obtained with a gas or a gas mixture.

Essentially, the nonconsumable tungsten electrode is a "torch"—a heating device. Under the protective gas shield, metals to be joined may be heated above their melting points so that material from one part coalesces with material from the other part. Upon solidification of the molten area, unification occurs. Pressure may be used when the edges to be joined are approaching the molten state to assist coalescence. Welding in this manner requires no filler metal.

If the work is too heavy for the mere fusing of abutting edges, and if groove joints or reinforcements such as fillets are required, filler metal must be added. This is supplied by a filler rod, manually or mechanically fed into the weld puddle. Both the tip of the nonconsumable tungsten electrode and the tip of the filler rod are kept under the protective gas shield as welding progresses.

Usually the arc is started by a high-frequency, high-voltage device that causes a spark to jump from the electrode to the work and initiate the welding current. Once the arc is started, the electrode is moved in small circles to develop a pool of molten metal and is positioned about 75 degrees to the surface of the puddle formed. The filler rod, held at an angle of about 15 degrees to the surface of the work, is advanced into the weld puddle. When adequate filler metal has been added to the pool, the rod is withdrawn and the torch moved forward. The cycle is then repeated. At all times, however, the filler rod is kept within the protective gas shield. For carbon steel, low-alloy steel, and copper, the touch-and-withdraw method can be used to establish the arc, but seldom, if ever, is this method satisfactory for the reactive metals. See Fig. 6-36.

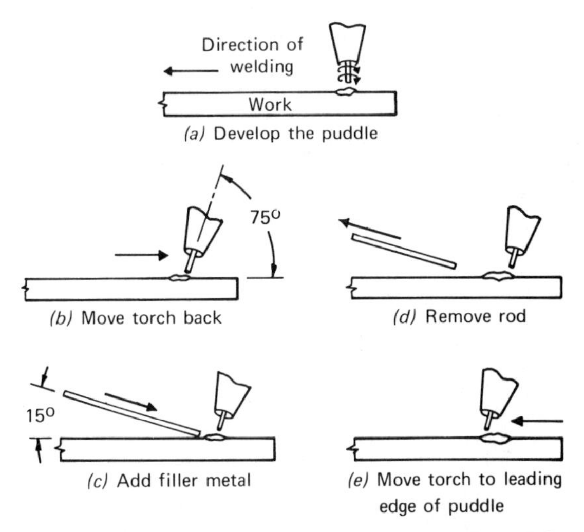

FIG. 6-36 Mode of manually feeding filler metal into the weld puddle in TIG welding. (*Lincoln Electric Co.*)

Materials weldable by the TIG process are most grades of carbon, alloy, and stainless steels; aluminum and most of its alloys; magnesium and most of its alloys; copper and various brasses and bronzes; high-temperature alloys of various types; numerous hard-surfacing alloys; and such metals as titanium, zirconium, gold, and silver. The process is especially adapted for welding thin materials where the requirements for quality and finish are exacting. It is one of the few processes that are satisfactory for welding such tiny and thin-walled objects as transistor cases, instrument diaphragms, and delicate expansion bellows.

ELECTRODE DESIGN

6-42 Electrodes The first specification for mild-steel covered electrodes, A5.1, was written in 1940. As the welding industry expanded and the number of types of electrodes for welding steel increased, it became necessary to devise a system of electrode classification to avoid confusion. The system used applies to both the mild-steel A5.1 and the low-alloy-steel A5.5 specifications. See Fig. 6-37.

Starting in 1964, AWS new and revised specifications for covered electrodes required that the classification number be imprinted on the covering, as in Fig. 6-38. However, some electrodes can be manufactured faster than the imprinting equipment can mark them and some sizes are too small to be legibly marked with an imprint. Although

AWS specifies an imprint, the color code is accepted on electrodes if imprinting is not practical.

Classifications of mild and low-allow steel electrodes are based on an E prefix and a four- or five-digit number. The first two digits (or three, in a five-digit number) indicate the minimum required tensile strength in thousands of pounds per square inch. For example, 60 = 60,000 psi, 70 = 70,000 psi, and 100 = 100,000 psi. The next-to-the-last digit indicates the welding position in which the electrode is capable of making satisfactory welds: 1 = all positions – flat, horizontal, vertical, and overhead; 2 = flat and horizontal fillet welding. The last two digits indicate the type of current to be used and the type of covering on the electrode.

6-43 Electrode Selection Choice of electrode is straightforward when high-strength or corrosion-resistant steels are welded. Here, choice is generally limited to

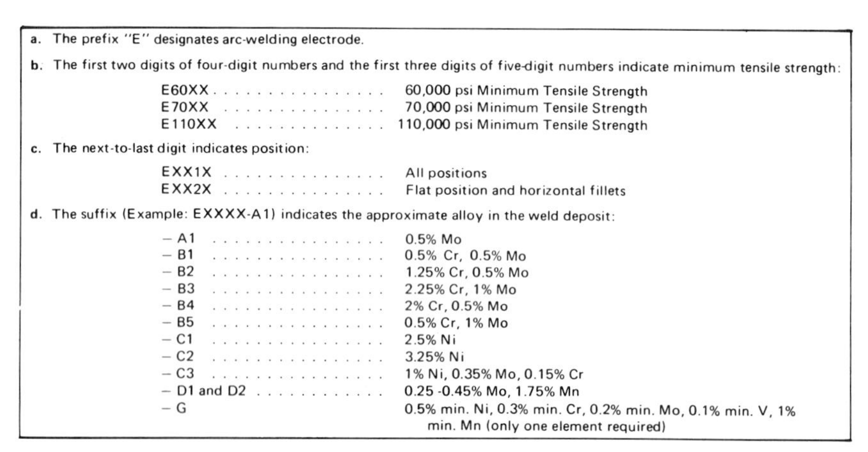

a. The prefix "E" designates arc-welding electrode.

b. The first two digits of four-digit numbers and the first three digits of five-digit numbers indicate minimum tensile strength:

E60XX	60,000 psi Minimum Tensile Strength
E70XX	70,000 psi Minimum Tensile Strength
E110XX	110,000 psi Minimum Tensile Strength

c. The next-to-last digit indicates position:

EXX1X	All positions
EXX2X	Flat position and horizontal fillets

d. The suffix (Example: EXXXX-A1) indicates the approximate alloy in the weld deposit:

– A1	0.5% Mo
– B1	0.5% Cr, 0.5% Mo
– B2	1.25% Cr, 0.5% Mo
– B3	2.25% Cr, 1% Mo
– B4	2% Cr, 0.5% Mo
– B5	0.5% Cr, 1% Mo
– C1	2.5% Ni
– C2	3.25% Ni
– C3	1% Ni, 0.35% Mo, 0.15% Cr
– D1 and D2	0.25 -0.45% Mo, 1.75% Mn
– G	0.5% min. Ni, 0.3% min. Cr, 0.2% min. Mo, 0.1% min. V, 1% min. Mn (only one element required)

FIG. 6-37 Designation for manual electrodes. *(Lincoln Electric Co.)*

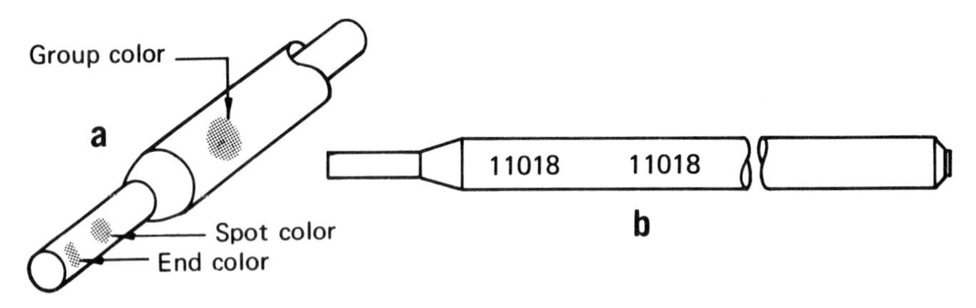

FIG. 6-38 (*a*) National Electrical Manufacturers Association color-code method to identify an electrode's classification; (*b*) American Welding Society imprint method. *(Lincoln Electric Co.)*

one or two electrodes designed specifically to give the correct chemical composition in the weld metal. But most arc welding involves the carbon and low-alloy steels for which many different types of electrodes provide satisfactory chemical compositions in the weld metal. The object is to pick, from the many possibilities, an electrode that gives the desired quality of weld at the lowest welding cost. Usually, this means the electrode that allows the highest welding speed with the particular joint. To meet this objective, electrodes are selected according to the design and positioning of the joint.

Electrodes compounded to melt rapidly are called "fast-fill" electrodes, and those compounded to solidify rapidly are called "fast-freeze" electrodes. Some joints and welding positions require a compromise between the fast-fill and fast-freeze characteristics, and electrodes compounded to meet this need are called "fill-freeze" electrodes. There are also electrodes which are classified as "fast-follow."

6-44 Fast-Freeze Electrodes Fast-freeze electrodes are compounded to deposit weld metal that solidifies rapidly after being melted by the arc, and are thus intended specifically for welding in the vertical and overhead positions. Although deposition rates are not as high as with other types of electrodes, the fast-freeze type can also be used for flat welding and is thus considered an "all-purpose" electrode that can be used for any weld in mild steel. However, welds made with fast-freeze electrodes are slow and require a high degree of operator skill. Therefore, wherever possible, work should be positioned for downhand welding, which permits the use of fast-fill electrodes.

Fast-freeze electrodes provide deep penetration and maximum admixture. The weld bead is flat with distinct ripples. Slag formation is light, and the arc is easy to control.

Applications for fast-freeze electrodes are:

- General-purpose fabrication and maintenance welding.
- Vertical-up and overhead plate welds requiring x-ray quality.
- Pipe welding, including cross-country, in-plant, and noncritical small-diameter piping.
- Welds to be made on galvanized, plated, painted, or unclean surfaces.
- Joints requiring deep penetration, such as square-edge butt welds.
- Sheet-metal welds, including edge, corner, and butt welds.

Current and Polarity. Unless otherwise specified, use DCRP and Exx10, and use ac with Exx11. Exx11 electrodes can be used on DCRP with a current about 10 percent below normal ac values. Always adjust current for proper arc action and control of the weld puddle.

Flat Welding. Hold an arc of 1/8 inch or less, or touch the work lightly with the electrode tip. Move fast enough to stay ahead of the molten pool. Use currents in the middle and high portion of the range.

Vertical Welding. Use an electrode of 3/16 inch or smaller. Vertical-down techniques are used by pipe-liners and for single-pass welds on thin steel. Vertical-up is used for most plate welding. Make the first vertical-up pass with either a whipping technique for fillet welds, or with a circular motion for V-butt joints (Fig. 6-39). Apply succeeding passes with a weave, pausing slightly at the edges to insure penetration and proper wash-in. Use currents in the low portion of the range.

Overhead and Horizontal Butt Welds. Use an electrode of 3/16 inch or smaller. These welds (Fig. 6-40) are best made with a series of stringer beads, using a technique similar to those described for first-pass vertical-up welds.

Sheet-Metal Edge and Butt Welds. Use DCSP. Hold an arc of 3/16 inch or more. Move as fast as possible while maintaining good fusion. Position the work 45 degrees downhill for fastest welding. Use currents in the middle range.

6-45 Fast-Fill Electrodes Fast-fill electrodes are compounded to deposit metal rapidly in the heat of the arc and are thus well suited to high-speed welding on horizontal surfaces. The weld metal solidifies somewhat slowly; therefore this type of electrode is not well suited for out-of-position welds. However, a slight downhill positioning is permissible. Joints normally considered fast-fill include butt, fillet, lap, and corner welds in plate 3/16 inch thick or more. These joints are capable of holding a large molten pool of weld metal as it freezes.

Arc penetration is shallow with minimum admixture. The bead is smooth, free of ripples, and flat or slightly convex. Spatter is negligible. Slag formation is heavy, and the slag peels off readily.

Applications for fast-fill electrodes are:

- Production welds on plate having a thickness of 3/16 inch or more.
- Flat and horizontal fillets, laps, and deep-groove butt welds.
- Welds on medium-carbon, crack-sensitive steel when low-hydrogen electrodes are not available. (Preheat may be required.)

The coverings of fast-fill electrodes contain approximately 50 percent iron powder. This powder increases deposition rate by helping to contain the arc heat at the electrode, by melting to add to deposited weld metal, and by permitting currents higher than those permitted by other types of coverings. The thick, iron-bearing covering also facilitates use of the drag technique in welding.

Polarity. Use ac for highest speeds and best operating characteristics. DCRP can be used, but this type of current promotes arc blow and complicates control of the molten puddle.

Flat Welding. Use a drag technique; tip the electrode 10 to 30 degrees in the direction of travel and make stringer beads. Weld with the electrode tip lightly dragging on the work so that molten metal is forced out from under the tip, thereby promoting penetration. The resulting smooth weld is similar in appearance to an automatic weld. Travel rapidly, but not too fast for good slag coverage. Stay about $^1/_4$ to $^3/_8$ inch ahead of the molten slag, as illustrated in Fig. 6-41. If travel speed is too slow, a small ball of molten slag may form and roll ahead of the arc, causing spatter, poor penetration, and erratic bead shape. Optimum current usually is 5 to 10 amperes above the center of the range for a given electrode. Do not exceed the center of the range if the weld is to be of x-ray quality.

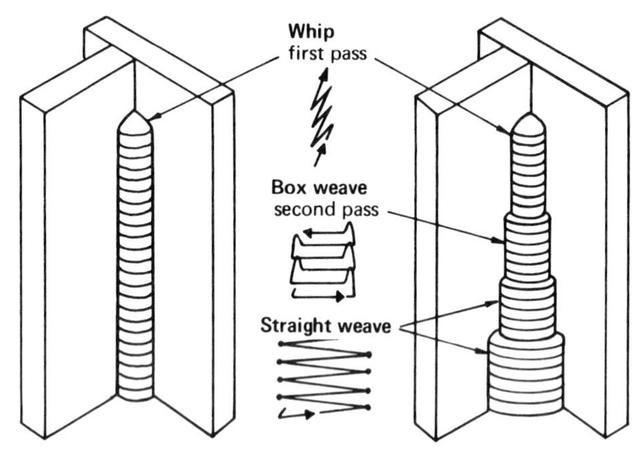

FIG. 6-39 Technique for vertical welding with fast-freeze electrodes. (*Lincoln Electric Co.*)

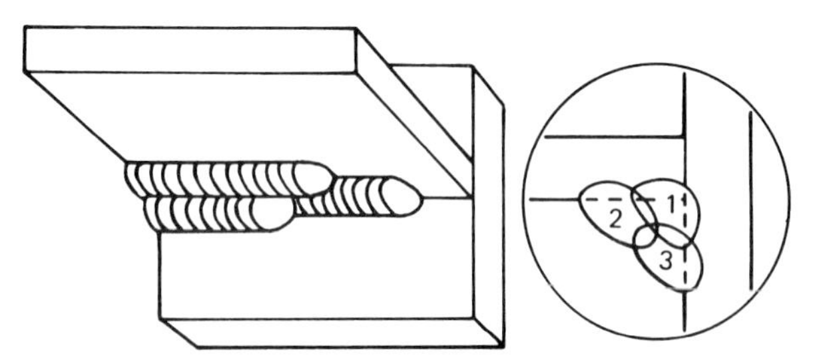

FIG. 6-40 Technique for overhead and horizontal butt welds with fast-freeze electrodes. These welds are best made with a series of stringer beads. (*Lincoln Electric Co.*)

Horizontal Fillets and Laps. Point the electrode into the joint at an angle of 45 degrees from horizontal and use the "flat" technique described above. The tip of the electrode must touch both horizontal and vertical members of the joint. If the 45-degree angle between plates is not maintained, the fillet legs will be of different sizes. When two passes are needed, deposit the first bead mostly on the bottom plate. To weld the second pass hold the electrode at about 45 degrees, fusing into the vertical plate and the first bead. Make multiple-pass horizontal fillets as shown in Fig. 6-42. Put the first bead in the corner with fairly high current, disregarding undercut. Deposit the second bead on the horizontal plate, fusing into the first bead. Hold the electrode angle needed to deposit the filler beads as shown, putting the final bead against the vertical plate.

Deep-Groove Butt Welds. To hold the large pool of molten weld metal produced by fast-fill electrodes, either a backup plate or a stringer bead made with a deeper-penetrating fast-freeze electrode is required. Deposit fast-fill beads with a stringer technique until a slight weave is required to obtain fusion of both plates. Split-weave welds are better than a wide weave near the top of deep grooves. When welding the second from the last pass, leave enough room so that the last pass will not exceed a $1/16$-inch buildup. A slight undercut on all but the last pass creates no problems, because it is melted out with each succeeding pass.

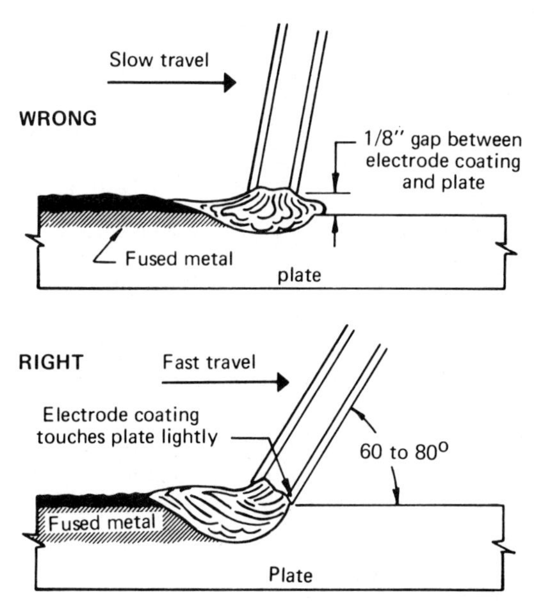

FIG. 6-41 Technique for flat welds with fast-fill electrodes. An incorrect technique is included for comparison. (*Lincoln Electric Co.*)

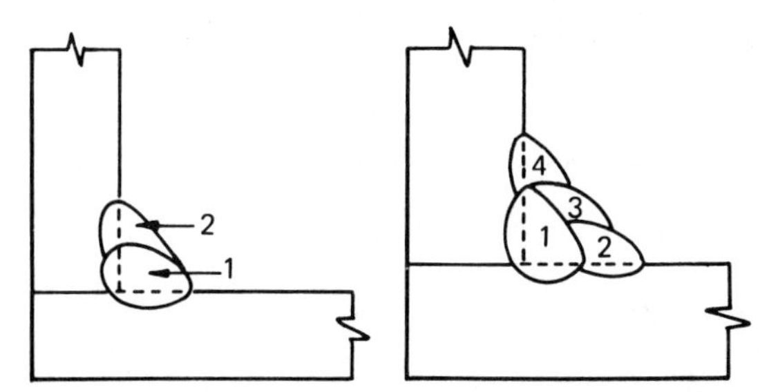

FIG. 6-42 Technique for multipass horizontal fillet welds with fast-fill electrodes. Beads should be deposited in the order indicated. (*Lincoln Electric Co.*)

6-46 Fill-Freeze Electrodes Fill-freeze electrodes are compounded to provide a compromise between fast-freeze and fast-fill characteristics, and thus provide medium deposition rates and medium penetration. Since they permit welding at relatively high speed with minimal skip, misses, undercut, and slag entrapment, fill-freeze electrodes are also referred to as fast-follow electrodes. The electrode's characteristics are particularly suited to the welding of sheet metal, and fill-freeze electrodes are thus often called sheet-metal electrodes. Bead appearance with this group of electrodes varies from smooth and ripple-free to wavy with distinct ripples. The fill-freeze electrodes can be used in all welding positions, but are most widely used in the level or downhill positions.

Applications for fill-freeze electrodes include:

- Downhill fillet and lap welds.
- Irregular or short welds that change direction or position.
- Sheet-metal lap and fillet welds.
- Fast-fill joints having poor fit-up.
- General-purpose welding in all positions.

Fast-freeze electrodes, particularly E6010 and E6011, are sometimes used for sheet-metal welding when fast-follow electrodes are not available, or when the operator prefers faster solidification. Techniques for sheet-metal welding with these electrodes are discussed in the portion of this section dealing with fast-freeze electrodes.

Welding Techniques on Steel Plate. Polarity. Use DCSP for best performance on all applications except when arc blow is a problem. To control arc blow, use ac.

Flat and Downhill. Use stringer beads for the first pass except when poor fit-up requires a slight weave. Use either stringer or weave beads for succeeding passes. Touch the tip of the electrode to the work or hold an arc length of ⅛ inch or less. Move as fast as possible, consistent with desired bead size. Use currents in the middle to higher portion of the range.

Electrode Size. Use electrodes of $^3/_{16}$-inch diameter or smaller for vertical and overhead welding.

Vertical-Down. Use stringer beads or a slight weave. A drag technique must be used with some E6012 electrodes. Make small beads. Point the electrode upward so that arc force pushes molten metal back up the joint. Move fast enough to stay ahead of the molten pool. Use currents in the higher portion of the range.

Vertical-Up. Use a triangular weave. Weld a shelf at the bottom of the joint and add layer upon layer. Do not whip or take the electrode out of the molten pool. Point the electrode slightly upward so that arc force helps control the puddle. Travel slow enough to maintain the shelf without spilling. Use currents in the lower portion of the range.

Overhead. Make stringer beads using a shipping technique with a slight circular motion in the crater. Do not weave. Travel fast enough to avoid spilling. Use currents in the lower portion of the range.

6-47 Low-Hydrogen Electrodes Conventional welding electrodes may not be suitable where x-ray quality is required, where the base metal has a tendency to crack, where thick sections are to be welded, or where the base metal has an alloy content higher than that of mild steel. In these applications, a low-hydrogen electrode may be required.

Low-hydrogen electrodes are available with either fast-fill or fill-freeze characteristics. They are compounded to produce dense welds of x-ray quality with excellent notch toughness and high ductility. Low-hydrogen electrodes reduce the danger of underbead and microcracking on thick weldments and on high-carbon and low-alloy steels. Preheat requirements are less than for other electrodes.

Low-hydrogen electrodes are shipped in hermetically sealed containers, which normally can be stored indefinitely without danger of moisture pickup. But once the container is opened, the electrodes should be used promptly or stored in a heated cabinet. Details on electrode storage and on redrying moisture-contaminated electrodes may be obtained from the manufacturer.

Applications for low-hydrogen electrodes include:

- X-ray-quality welds or welds requiring high mechanical properties.
- Crack-resistant welds in medium-carbon to high-carbon steels; welds that resist hot-short cracking in phosphorus steels; and welds that minimize porosity in sulfur-bearing steels.
- Welds in thick sections or in restrained joints in mild and alloy steels where shrinkage stresses might promote weld cracking.
- Welds in alloy steel requiring a strength of 70,000 psi or more.
- Multiple-pass, vertical, and overhead welds in mild steel.

Welding Techniques. Techniques for E7028 are the same as those described for conventional fast-fill electrodes. However, special care should be taken to clean the slag from every bead on multiple-pass welds to avoid slag inclusions that would appear on x-ray inspection. The ensuing discussion pertains to the techniques recommended for E7018 electrodes.

Polarity. Use DCRP whenever possible if the electrode size is $^5/_{32}$ inch or less. For larger electrodes, use ac for best operating characteristics (but DCRP can also be used).

Flat. Use low current on the first pass, or whenever it is desirable to reduce admixture with a base metal of poor weldability. On succeeding passes, use currents that provide best operating characteristics. Drag the electrode lightly or hold an arc of $^1/_8$ inch or less. Do not use a long arc at any time, since E7018 electrodes rely principally on molten slag for shielding. Stringer beads or small-weave passes are preferred to wide-weave passes. When starting a new electrode, strike the arc ahead of the crater, move back into the crater, and then proceed in the normal direction. On ac, use currents about 10 percent higher than those used with dc. Govern travel speed by the desired bead size.

Vertical. Weld vertical-up with electrode sizes of $^5/_{32}$ inch or less. Use a triangular weave for heavy single-pass welds. For multipass welds, first deposit a stringer bead by using a slight weave. Deposit additional layers with a side-to-side weave, hesitating at the sides long enough to fuse out any small slag pockets and to minimize undercut. Do not use a ship technique or take the electrode out of the molten pool. Travel slowly enough to maintain the shelf without causing metal to spill. Use currents in the lower portion of the range.

Overhead. Use electrodes of $^5/_{32}$ inch or smaller. Deposit stringer beads by using a slight circular motion in the crater. Maintain a short arc. Motions should be slow and deliberate. Move fast enough to avoid spilling weld metal, but do not be alarmed if some slag spills. Use currents in the lower portion of the range. See Table 6-6.

SELECTING A WELDING PROCESS

In selecting a process for production welding, a primary consideration is the ability of the process to give the required quality at the lowest cost. Here, the cost factor must include not only the operating cost, but also the amortization of the capital costs of equipment over the job and those jobs that may reasonably be expected to follow. Thus, a fabricator may have a job of running straight fillets where fully mechanized submerged-arc equipment would be the ultimate solution to low-cost, quality welding. But, unless the job involves a great amount of welding, or there is assurance that it will be followed by other jobs of a similar nature, the lowest cost to the fabricator—taking into account the feasibility of amortization—might be with a hand-held semi-automatic gun or even with the stick-electrode process. The job is thus the starting point in the selection of a welding process.

The job defines many things. It defines the metals to be joined, the number of pieces to be welded, the total length of welds, the types of joints, the preparations, the quality required, the assembly-positioning requirements—to name but a few. Process selection thus considers first the needs of the job at hand.

Since process selection is almost self-determining when the metals to be welded are aluminum, magnesium, copper, titanium, or the other nonferrous metals—and since the stick electrode is almost certainly to be the choice when steel is welded and the immediate cost is of no importance—the following text is concerned with choosing the optimum process for the production welding of steels. Here, intangibles no longer are considered, and only technology and pure economics prevail.

All of the mechanized arc-welding processes have been developed as a means of reducing welding cost. Each of these processes has certain capabilities and limitations, and selecting the best process for a particular job can be a difficult, if not confusing, task. Unless the user makes a correct selection, however, he may lose many of the benefits to be gained in moving to mechanization and thus not achieve his objective of welds that meet the application requirements at the lowest possible cost.

There are varying degrees of mechanization with all of the mechanized processes—from semi-automatic hand-held equipment to huge mill-type welding installations. Thus, one needs a "common denominator" when processes are to be compared. Semi-automatic equipment possibly provides the best common denominator. Anything added to semimechanization is extraneous to the process capability *per se* and merely amounts to putting the semi-automatic process "on wheels."

TABLE 6-6 Recommended Electrodes for ASTM Steels

ASTM Spec. No.	Description	Grades	Suggested Electrodes
A27	Carbon Steel Castings	All	E7018
A36	Structural—36,000 min. YS	All	Carbon Steel#
A131	Structural for Ships	All Ordinary Strength	Carbon Steel or E7018
		AH32, DH32 & EH32	E7018, E8018-C3
		AH36, DH36 & EH36	E7018, E8018-C3 or E8018-B2
A148	Castings for Structural Use	80-40 & -50	E8018-C3
		90-60	E8018-B2
		105-85 & 120-95	E11018-M
A184	Bar Mats for Concrete Reinforcement	See A615, A616, A617	
A202	Pressure Vessel CrMnSi	A & B	E8018-C3, E8018-B2 or E11018-M
A203	Pressure Vessel—Ni	A & B	E8018-C1
A204	Boiler & Pressure Vessel—Mo	A	E7018 or E8018-C3
		B	E7018, E8018-C3 or E8018-B2
		C	E8018-C3 or E8018-B2
A216	Carbon Steel Castings—High Temp	WCA	E7018 or E8018-C3
		WCB & WCC	E7018, E8018-C3, E8018-B2
A225	Pressure Vessel—Mn-V-Ni	C	E11018-M
A242	High Strength Structural	All	E7018
A266	Pressure Vessel Forgings	1 & 2	E7018
		2 & 3	E8018-C3
A283	Structural Plates	All	Carbon Steel#
A284	Carbon Silicon Steel Plates	All	Carbon Steel# & E7018
A285	Pressure Vessel Plate	All	Carbon Steel# & E7018
A299	Pressure Vessel Plate—MnSi		E8018-B2 or E8018-C3
A302	Pressure Vessel	A	E8018-C3 or E8018-B2
	Mn Mo and Mn Mo Ni	B, C & D	E8018-C3, E8018-B2 & E11018-M
A328	Steel Sheet Piling		E7018
A336	Pressure Vessel Forgings	F1	E7018 or E8018-C3
		F12	E8018-B2
		F30	E8018-C3
A352	Low Temp. Castings	LCA, LCB & LCC	E7018
		LC1 & LC2	E8018-C1
A356	Steam Turbine Castings	1	E7018
		2	E7018 or E8018-C3
		5 & 6	E8018-B2
A366	Carbon Steel Sheets		Carbon Steel#
A372	Pressure Vessel Forgings	I	E7018
		II	E8018-C3
		III	E8018-B2
		IV	E11018-M
A387	CrMo Pressure Vessel Plate	Gr. 2, 11 & 12	E8018-B2
A389	High Temp Castings	C23	E8018-B2
A414	Pressure Vessel Sheet	A, B, C & D	Carbon Steel#
		E, F	E7018
A424	Sheet for Porcelain Enameling		E7018
A441	High Strength Structural	All	E7018
A442	Plate with Improved Transition Properties	All	E7018
A455	C-Mn Pressure Vessel Plate	1 & 2	E8018-B2, E8018-C3 or E7018
A486	Highway Bridge Castings	70	E7018 or E8018-C3
		90	E8018-B2 or E11018-M
A487	Castings for Pressure Service	1N, 2N, 4N, 6N, 9N, 10N, 1Q, 2Q, 4Q, 7Q, 9Q	E8018-B2 or E11018-M
		A, AN, AQ, B	E7018,E8018-C3
		BN, BQ, C, CN, CQ, 11N, 12N	E8018-B2 or E11018-M

Continued

With semimechanization and the capabilities of the various processes subject to semimechanization in mind, a four-step procedure can be used in deciding which is the preferred process for the particular production welding job.

6-48 Four-Step Procedure The four steps involved in process selection are:

1. First, the joint to be welded is analyzed in terms of its requirements.

2. Next, the joint requirements are matched with the capabilities of available processes. One or more of the processes are selected for further examination.

3. A check list of variables is used to determine the capability of the surviving processes to meet the particular shop situation.

4. Finally, the proposed process or processes indicated as most efficient are reviewed with an informed representative of the equipment manufacturer for verification of suitability and for acquisition of subsidiary information bearing on production economics.

6-49 Considerations Considerations other than the joint itself have bearing on selection decisions. Many of these will be peculiar to the job or the welding shop.

TABLE 6-6 (*Continued*)

ASTM Spec. No.	Description	Grades	Suggested Electrodes
A508	Pressure Vessel Forgings Quenched & Tempered	1 & 1a	E7018, E8018-C3 or E8018-B2
		2 & 3	E8018-C3, E8018-B2 or E11018-M
		2a	E8018-B2 or E11018-M
		4 & 5	E11018-M
A514	Quenched & Tempered Plate	A-L	E11018-M
A515	High Temp. Pressure Vessel	55 & 60	E7018 or E8018-C3
		65 & 70	E7018, E8018-C3 or E8018-B2
A516	Pressure Vessel Plate	55 & 60	E7018 or E8018-C3
		65 & 70	E7018 or E8018-C3
A517	Pressure Vessel Quenched & Tempered	All	E11018-M
A521	Closed Die Forgings	AA, AB, CE, CF & CF1	E8018-C3
		AC, AD & CG	E8018-B2
		CA, CC & CC1	E7018
		AE	E11018-M
A529	Structural—42,000 Min. YS		E7018 or E8018-C3
A533	Quenched & Tempered Plate	Class 1	E8018-C3
		Class 2	E11018-M
A537	Pressure Vessel Plate	1	E7018, E8018-C3 or E8018-B2
		2	E8018-C3, E8018-B2 or E11018-M
A538	High Strength Structural	All	E7018
A541	Pressure Vessel Forgings	1	E7018, E8018-C3 or E8018-B2
		2, 3 & 4	E8018-C3, E8018-B2 or E11018-M
		5	E8018-B2 or E8018-C3
		2A	E8018-B2 or E11018-M
A543	Quenched & Tempered Plate	1 & 2	E11018-M
A570	Structural Sheet & Strip		Carbon Steel#
A572	Structural Plate, Cb-V	42, 50 & 60	Carbon Steel# & E7018
		60 & 65	E8018-C3
A573	Structural Plate	58	Carbon Steel & E7018
		65	E7018, E8018-C3
		70	E7018, E8018-C3 or E7018
A588	High Strength Structural		E8018-C3 & E7018
A592	Quenched & Tempered Fittings for Pressure Vessels	All	E11018-M
A595	Structural Tubes	A & B	Carbon Steel# & E7018
		C	E7018 or E8018-C3
A612	Pressure Vessel—Low Temp.	All	E8018-C3, E8018-C1 or E11018-M
A615	Billet Steel Bars—Concrete Reinforcement	40	Carbon Steel# & E7018
		60	E8018-B2 or E11018-M
A616	Rail Steel Bars for Concrete Reinforcement	50	E8018-C3
		60	E8018-B2 or E11018-M
A633	Normalized High Strength Low Alloy Structural	A & B	E7018 or E8018-C3
		C & D	E7018, E8018-C3 or E8018-B2
		E	E8018-C3, E8018-B2 or E11018-M
A643	Carbon & Alloy Steel Castings	A	E7018
		B	E8018-C3, E8018-B2 or E11018-M
A656	High Strength Structural	All	E8018-B2 or E11018-M
A662	Pressure Vessel—Low Temp	A & B	E7018, E8018-C1 or E8018-C3
A668	Carbon & Alloy Steel Forgings	A, B	Carbon Steel#
		C, D	E7018
		E, F	E8018-C3 or E8018-B2
		G	E8018-C3
		H, J	E8018-B2
		K, L	E11018-M

Continued

They can be of overriding importance, however, and they offer the means of eliminating alternate possibilities. When organized in a checklist form, these factors can be considered one by one, with the assurance that none is overlooked.

Some of the main items to be included on the check list are:

Volume of Production—it must be adequate to justify the cost of the process equipment. Or, if the work volume for one application is not great enough, another application may be found to help defray investment costs.

Weld Specifications—a process under consideration may be ruled out because it does not provide the weld properties specified by the code governing the work. The specified properties may be debatable as far as defining weld quality, but the code still prevails.

Operator Skill—replacement of manual welding by a semi-automatic process may require a training program. The operators may develop skill with one process more rapidly than with another. Training costs, just as equipment costs, require an adequate volume of work for their amortization.

TABLE 6-6 (*Continued*)

ASTM Spec. No.	Description	Grades	Suggested Electrodes
A678	Quenched & Tempered Plate	A	E7018, E8018-C3 or E8018-B2
		B	E8018-C3, E8018-B2 or E11018-M
		C	E11018-M
A690	H-Piles & Sheet Piling		E7018
A699	Plates, Shapes & Bars	All	E8018-B2 & E11018-M

Carbon Steel Reference: Almost any E60XX or E70XX electrodes can be used.

Auxiliary Equipment—every process will have its recommended power sources and other items of auxiliary equipment. If a process makes use of existing auxiliary equipment, the initial cost in changing to that process can be substantially reduced.

Accessory Equipment—the availability and the cost of necessary accessory equipment—chipping hammers, deslagging tools, flux laydown and pickup equipment, exhaust systems, etc.—should be taken into account.

Base-Metal Condition—rust, oil, fit-up of the joint, weldability of the steel, and other conditions must be considered. Any of these factors could limit the usefulness of a particular process—or give an alternative process a distinct edge.

Arc Visibility—with applications where there is a problem of following irregular seams, open-arc processes are advantageous. On the other hand, when there is no difficulty in correct placement of the weld bead, there are decided operator-comfort benefits with the submerged-arc process; no headshield is required and the heat radiated from the arc is substantially reduced.

Fixturing Requirements—a change to a semi-automatic process usually requires some fixturing if the ultimate economy is to be realized. The adaptability of a process to fixturing and welding positioners is a factor to take into account, and this can only be done by realistically appraising the equipment.

Production Bottlenecks—if the process reduces unit fabrication cost, but creates a production bottleneck, its value may be completely lost. For example, highly complicated equipment that requires frequent servicing by skilled technicians may lead to expensive delays and excessive overall costs.

The completed check list should contain every factor known to affect the economics of the operation. Some may be peculiar to the weld shop or the particular welding job. Other items might include:

Protection requirements	Seam length
Application flexibility	Ability to follow seams
Setup time requirements	Weld cleaning costs
Cleanliness requirements	Housekeeping cost
Range of weld sizes	Initial equipment cost

Each of these items must be evaluated realistically, with the peculiarities of the application as well as those of the process and of the equipment used with it being recognized.

Insofar as possible, human prejudices should not enter the selection process; otherwise, objectivity will be lost. At every point, when all other things are equal, the guiding criterion should be welding cost.

INTRODUCTION TO WELDING PROCEDURES

The ideal welding procedure is the one that will produce acceptable quality welds at the lowest overall cost. So many factors influence the optimum welding conditions that it is impossible to write procedures for each set of conditions. In selecting a procedure, the best approach is to study the conditions of the application and then choose the procedure that most nearly accommodates them.

For some joints, different procedures are offered to suit the weld quality—code quality, commercial quality, and strength only—that may be required.

6-50 Code-Quality Procedures Code-quality procedures are intended to provide the highest level of quality and appearance. To accomplish this, conservative currents and travel speeds are recommended.

These procedures are aimed at producing welds that will meet the requirements of the commonly used codes: AWS Structural, AISC Buildings and Bridges, ASME Pressure Vessels, AASHO Bridges, and others. Code-quality welds are intended to be defect-free to the extent that they will measure up to the nondestructive testing requirements normally imposed by these codes. This implies crack-free, pressure tight welds, with little or no porosity or undercut.

The specific requirements of these codes are so numerous and varied that code-quality procedures may not satisfy every detail of a specific code. Procedure qualification tests are recommended to confirm the acceptability of chosen procedures.

All butt welds made to code quality are full penetration; fillet welds are full size, as required by most codes. (The theoretical throat rather than the true throat is used as the basis of calculating strength.)

6-51 Commercial-Quality Procedures Commercial quality implies a level of quality and appearance that will meet the nominal requirements imposed on most of the welding done commercially. These welds will be pressure tight and crack-free. They will have good appearance, and they will meet the normal strength requirements of the joint.

Procedures for commercial-quality welds are not as conservative as code-quality procedures; speeds and currents are generally higher. Welds made according to these procedures may have minor defects that would be objectionable to the more demanding codes.

It is recommended that appropriate tests be performed to confirm the acceptability of the selected procedure for the application, prior to putting it into production.

6-52 Strength-Only Procedures The purpose of strength-only procedures is to produce the highest-speed welding at the lowest possible cost. To accomplish this, the weld appearance and quality need only be good enough to do the specific job for which it is intended. In this category, defects and imperfections may be acceptable provided that the welds perform satisfactorily under service conditions.

This quality level may apply to seal beads, partial-penetration welds, some lap welds, high-speed edge welds, and to many low-stress welded connections which have no real need for better quality at higher cost. On fillet welds the joint must be welded from both sides.

Since this is a compromise quality level, realistic tests should be made to confirm the satisfactory performance of welds made according to these procedures. Welds of strength-only quality will be free of cracks but may not be pressure tight.

6-53 The Five P's that Assure Quality After the quality standard has been established, the most important step toward its achievement is the selection of the best process and procedure. If attention is given to five "P's," weld quality will come about almost automatically, reducing subsequent inspection to a routine checking and policing activity. The five P's are:

1. Process Selection—the process must be right for the job.
2. Preparation—the joint configuration must be right and compatible with the welding process.
3. Procedures—to assure uniform results, the procedures must be spelled out in detail and followed religiously during welding.
4. Pretesting—by full-scale mock-ups or simulated specimens, the process and procedures are proved to give the desired standard of quality.
5. Personnel—qualified people must be assigned to the job.

INSPECTION AND TESTING

Whatever the standard of quality, all welds should be inspected, even if the inspection involves no more than the welder glancing back over the work after running a bead.

Testing may also be required, even in relatively rough weldments—such as leak-testing a container for liquids.

Inspection determines whether the prescribed standard of quality has been met. This function may be the responsibility of the welding supervisor or foreman, a special employee of the company doing the welding, or a representative of the purchasing organization. The formal welding inspector may have a variety of duties. These may begin with interpretation of drawings and specifications and follow each step to the analysis of test results. The inspector's operations are both productive and nonproductive, depending on where they are applied.

Inspection after the job is finished is a policing action, rather than a productive function. Important as it is to assure quality, it is a burden added to the overall production cost. No amount of after-the-job inspection will improve the weld: it merely tells what is acceptable and what must be reworked or rejected.

Inspection as the job progresses is a different matter. It detects errors in practice and defects while correction is feasible. It prevents minor defects from piling up into major defects and leading to ultimate rejection. Inspection while weld quality is in the making and can be controlled may justifiably be looked upon as a productive phase of cost, rather than an overburden.

Any program for assuring weld quality should, therefore, emphasize productive inspection and attempt to minimize the nonproductive type. This should be the guiding philosophy, even though its implementation may fall short. In most cases, such a philosophy means that visual inspection will be the main method of ascertaining quality, since it is the one method that can be applied routinely while the job is in progress.

Part 3 Thermoplastic Welding

edited by

ROBERT O. PARMLEY, P.E.

Consulting Engineer, Ladysmith, Wisconsin

Thermoplastics, particularly polyvinyl chloride, polyethylene, and polypropylene, are increasing in use throughout industry and the construction fields. In manufacturing, fabrication, construction, and plant maintenance, the possibilities for utilizing these lightweight, chemical- and moisture-resistant materials are unlimited.

6-54 Hot-Air Welding Hot-air welding and its relatively simple equipment has evolved over recent years to assure sound and uniform welds.

Welding is accomplished by means of hot air or hot gas (nitrogen) which has been compressed and then passed over an electric heating element onto the surface of the material to be bonded. A welding rod of the same kind of plastic is also heated by the stream of hot air or gas as it is forced into the joint, thus forming a permanent union.

6-55 Equipment Welding guns contain a heating unit through which the stream of compressed air or gas is passed, reaching the gun nozzle or tip at a temperature of 400°F minimum to 900°F maximum. Figures 6-43 and 6-44 show typical welding gun designs.

6-56 Typical Welding Joints Welding plastic is very similar to the welding of metal except that special protective clothing and masks are not required. Only the tools and procedures are different.

The types of joints are basically the same as those used for metal welding. Figure 6-45 illustrates the typical standard joints.

6-57 Welding Joint Analysis Good thermoplastic welds are achieved only after proper training and thus by following proven guidelines. Figure 6-46 illustrates a "no-bond" weld, a "burned" weld, and a proper weld.

Some basic points to remember are as follows:

1. Small beads should form along each side of the weld at the point where the rod meets the material to be welded.

2. The rod should hold its basic shape. (Generally rods are round; however, recently a triangular rod has proven effective for V welds or fillet joints.)
3. Neither the base material nor rod should discolor or char.
4. There should be no "stretch" in the rod over the weld.
5. Oxygen or other flammable gases should never be used.

6-58 Tack Welding Components to be welded can be joined quickly with a tack tip as shown in Fig. 6-47. Tacking allows a welder to check for proper alignment of the pieces. Tack welds may easily be broken apart without damage if necessary.

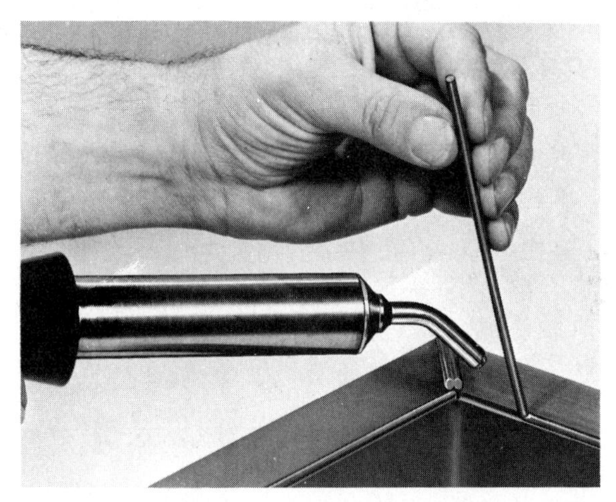

FIG. 6-43 Hand welding with round tip. Manual-operating welding gun. (*Seelye Plastics, Minneapolis, Minn.*)

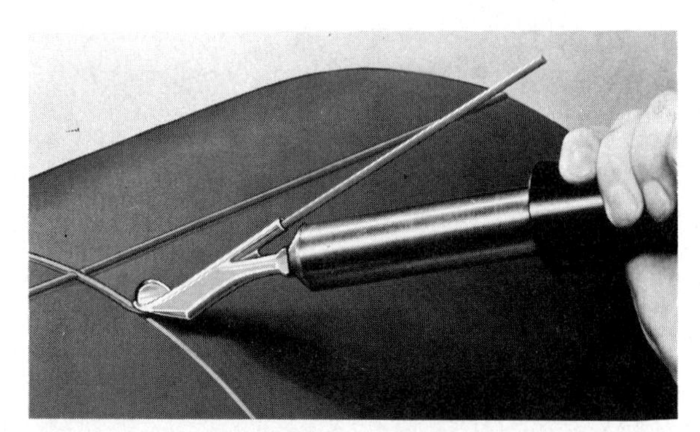

FIG. 6-44 High-speed welding gun. Shows welding gun for speed welding. Note the rod holder for automatic feeding. (*Seelye Plastics, Minneapolis, Minn.*)

6-59 Welding with Round Tip Select the proper welding rod and cut the end to a 60-degree angle with cutting pliers. Hold the rod at right angles to the work. Heat the tip of the rod and the starting point on the base material until they become tacky. Press the end of the welding rod firmly into the joint with the point of the rod away from the direction of weld. A good start is most important as this generally is the weakest and most difficult part of the weld.

After the initial bond has set, start heating the welding joint and rod together with a "fanning" motion in line with the direction of the weld. Keep a firm, downward

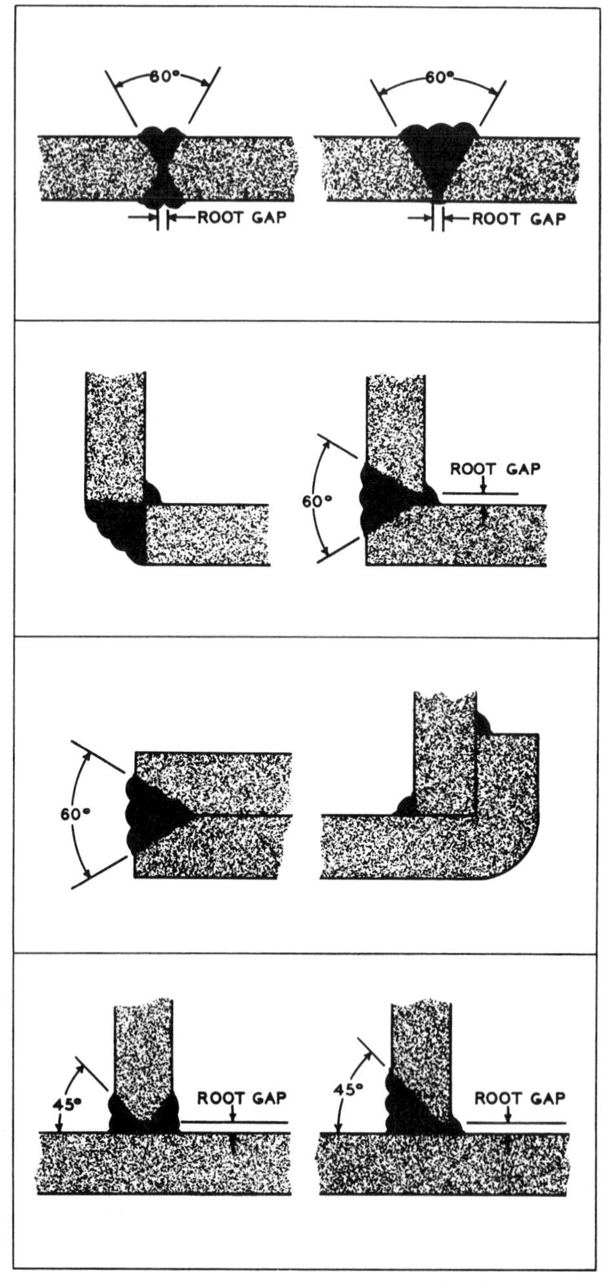

FIG. 6-45 Typical thermoplastic welding joints. Note how similar they are to standard joints for welding of metal. (*Seelye Plastics, Minneapolis, Minn.*)

pressure on the welding rod at right angles to the work. The rod will begin to move forward. The weld is fanned at a 45-degree angle between the welding rod and joint. This heats both equally. On heavy-gauge sheet with a light welding rod, direct most of the heat on the joint.

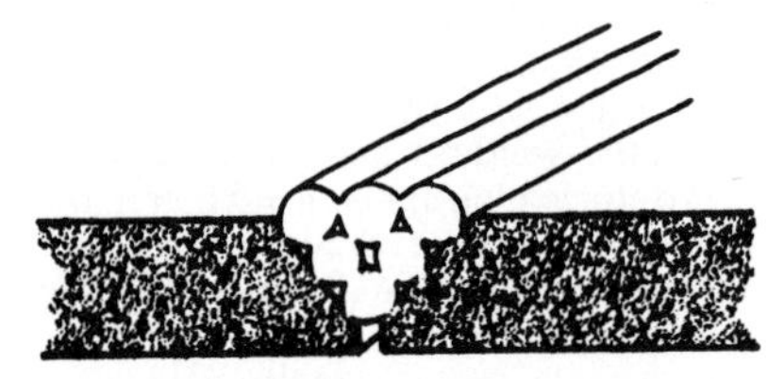

FIG. 6-46 Welding analysis. (1) No bond–weld can be pulled apart, (2) proper weld, (3) burned weld and material charred–weak weld. *(Seelye Plastics, Minneapolis, Minn.)*

FIG. 6-47 Tack tip. Nozzle used for tack welding. *(Seelye Plastics, Minneapolis, Minn.)*

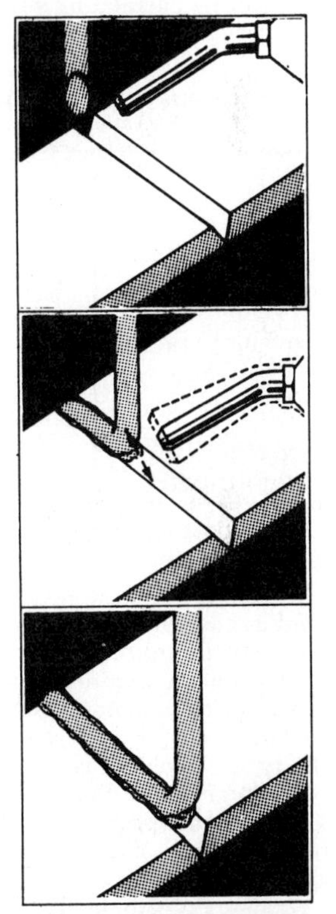

FIG. 6-48 General steps for welding with round tip. Middle illustration shows 45-degree "fanning" angle between joint and rod. *(Seelye Plastics, Minneapolis, Minn.)*

Only the outer surfaces of the rod and the materials forming the joint are heated. When pressed together they form a homogeneous bond.

PVC is a poor heat conductor. Therefore, the welding rod will retain sufficient body to exert pressure into the weld joint when proper fanning directs the correct amount of heat to the welding rod and base material.

As the welding continues, you should notice a small bead forming along both edges of the welding bead, and a small roll forming under the welding rod, both of which should continue along the weld. At the end of the weld the excess rod is cut with a knife or pliers at a 30-degree angle, which will match the start of a new welding rod to be butted to this point. See Figs. 6-43 and 6-48.

6-60 High-Speed Welding Cut the welding rod at a 6-degree angle and insert it into the feeder tube. Start the weld as usual. Hold the welder unit straight down at a 90-degree angle to the work. Press the end of the rod into the weld with the curved foot of the tip. Move the welder forward along the weld immediately, dropping the angle of the welder to the work to 45 degrees. Feed the weld rod manually into the feed tube until the weld bead is well started. The rod will then feed automatically. It is heated with the base material through the other tip opening and forced into the weld bed by the welder exerting firm downward pressure with the wrist.

You must keep moving. Watch the formation of the weld for good and bad indications, adjusting your heat or speed accordingly. To stop a high-speed weld before the rod is fully used, simply set the end of the weld with firm pressure, draw the welder off the remaining rod, and cut the weld bead with a curved knife on the top of the welding tip.

Do not allow the welding rod to remain in the feeder tube or it will melt and clog the opening. Clean the feeder foot with a soft wire brush occasionally. See Fig. 6-44.

6-61 Welding PVC Keep material clean at all times. Except on freshly beveled edges, PVC can be cleaned by wiping with MEK, or similar solvent. Welding edges

TABLE 6-7 Thermoplastic Welding Chart
Welding temperatures for various thermoplastics.

	PVC	H.D. Poly-ethylene	Poly-pro-pylene	Penton	ABS	Plexi-glass
Welding Temperature	525	550	575	600	500	575
Forming Temperature	300	300	350	350	300	350
Welding Gas	Air	WP* Nitro-gen	WP* Nitro-gen	Air	WP* Nitro-gen	Air
*W.P.—water pumped nitrogen						

SOURCE: Seelye Plastics, Minneapolis, Minn.

are beveled or offset to provide area for the welding rod and to remove polished outer surfaces, thereby permitting better adhesion. Cut bevels with a jointer, sander, router, or plane. Clamp the sections firmly in place or tack weld. Avoid backup materials that are good heat conductors. Allow a root gap in most assembly except in tack welding. Thickness, shape, size, good construction techniques, and strength required dictate the type of weld to use.

6-62 Welding Flexible PVC Welding this material is quite similar to welding regular PVC except that special strip feeders and both flat and corner tips are used. A noticeable indication of the proper welding temperature with this material will be smoking without charring at the weld point.

6-63 Welding Polyethylene and Polypropylene Procedures are the same for these materials except for the welding gas and welding temperatures. These should be varied according to Table 6-7.

To weld thermoplastics that require the use of nitrogen gas, use water-pumped nitrogen gas only, with a welding-gas flowmeter and a Y, or bypass, valve that can be obtained from local suppliers. The flowmeter should be regulated or set to 30 fpm and connected to the nitrogen tank. Couple the bypass valve to the flowmeter. This valve changes over from pure nitrogen gas to compressed air when you are not actually welding to keep the welder at the proper welding temperature and conserve nitrogen.

When you begin welding, open the nitrogen valve on the bypass and close the air valve. When you have completed your weld, or while the welder is on the stand between welds, open the air valve on the bypass and close the nitrogen valve. This will allow you to keep the welder at proper operating temperatures for immediate use.

The base material should be freshly cut or scraped and clean. With polyethylene it is important that the welding rod and base material be of the same density. Watch for stress cracking in polyethylene and polypropylene. The base material and welding rod should not be overheated, nor should the welding rod be stretched in any manner. Use only 1 ft of rod for 1 ft of weld. If the welded joint is under stress, the weld will be subject to chemical attack which would not occur under normal circumstances. This is called *environmental stress cracking*. It causes welds to fail and cracks to radiate into the sheet from the weld.

One noticeable difference in the action of polyethylene and polypropylene welding rods is that they tend to curve or loop in the direction of the weld. This is acceptable as long as you do not force the rod and add undue strain on the weld.

REFERENCE

All About Welding Plastics, Seelye Plastics, Minneapolis, Minnesota

Section 7

Concrete Fastening

ROBERT O. PARMLEY, P.E.
Consulting Engineer, Ladysmith, Wisconsin

Concrete fastening or anchoring various building components into masonry is an age-old technology which has evolved rapidly within the last two decades. Modern fasteners, devoted exclusively to this area of construction, are so numerous that this section can only briefly describe some of the more basic methods available. The Editor-in-Chief felt that at least a small section on this complex subject was warranted to further expand the original concept of the handbook.

7-1 Anchor Bolts Metal anchor bolts are a basic key element in modern structures. Generally, anchor bolts are encased in concrete walls, footings, piers, or bases with sufficient projection to permit assembly of major structural components. Figure 7-1 illustrates a typical anchor bolt detail.

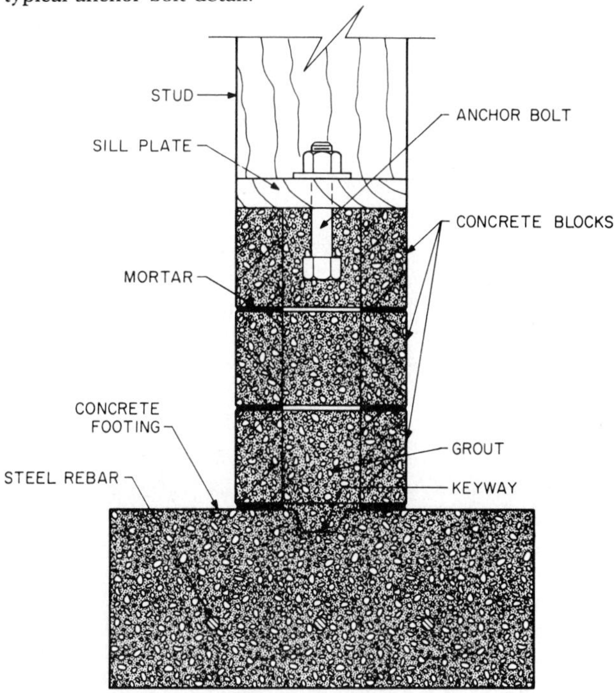

FIG. 7-1 Typical anchor bolt detail.

Proper sizing, spacing, positioning, and embedment of anchor bolts are specifically determined for each design situation. Anchor bolts that are subjected to tension forces should be designed to engage a mass of concrete which will provide resistance to uplift equal to 150 percent of the estimated force. Pull-out resistance can be achieved by the use of swedge bolts or by placement of a nut with washer or plate on the embedded end of the anchor bolt. A simpler and more cost-effective alternative is the ell or bent leg. See Fig. 7-2.

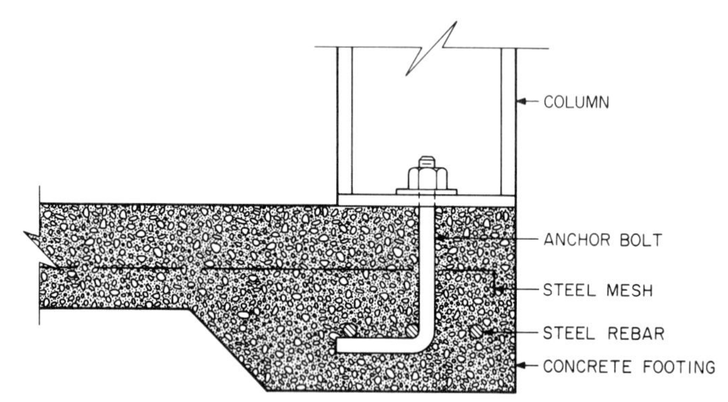

FIG. 7-2 Anchor bolt (with leg).

7-2 Anchors Securing various components to concrete can be achieved with anchors located in drilled holes. These anchors can be divided into five general categories:

1. Self-setting cone anchors secured by torque
2. Hammer-set
3. Self-drilling
4. Plastic
5. Grouted

Figure 7-3 illustrates several typical styles or commonly used anchors. It must be noted that there are hundreds of designs and variations of concrete fasteners available and space in this handbook does not provide for an expanded discussion.

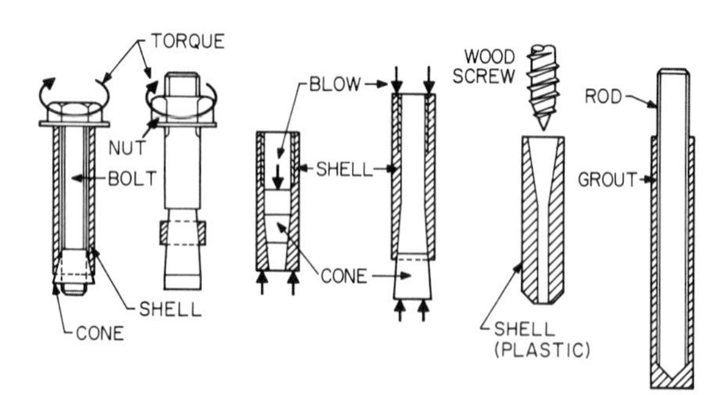

FIG. 7-3 Typical anchors. *(Courtesy: American Concrete Institute.)*

7-3 Construction Joints Construction joints in concrete are formed when freshly poured against a hardened concrete structure. These joints should be located at sections with minimum shear exposure. Construction joints usually are at the midpoint of slabs and beams, where the bending moment is at its highest. Additionally, they should be located at a place where it is convenient to stop working. Generally on large jobs, it is not practical or cost-effective to pour an entire concrete floor in one operation, so vertical joints should be positioned at the midspan. Horizontal construction joints are usually located at floor and columns. A number of typical construction joints are shown in Fig. 7-4.

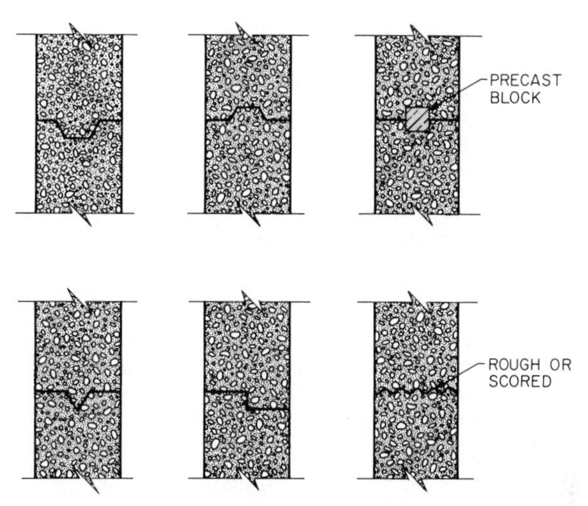

FIG. 7-4 Typical construction joints.

7-4 Contraction Joints Contraction joints, sometimes called "control joints," are placed in concrete structures to allow concrete to contract during temperature drop and to allow shrinkage during curing without producing damage from random cracking. Placement of the contraction joints should be at offsets and at changes in thickness. Maximum spacing should be 30 feet, center to center in exposed structures. See Fig. 7-5 for two types of contraction joints.

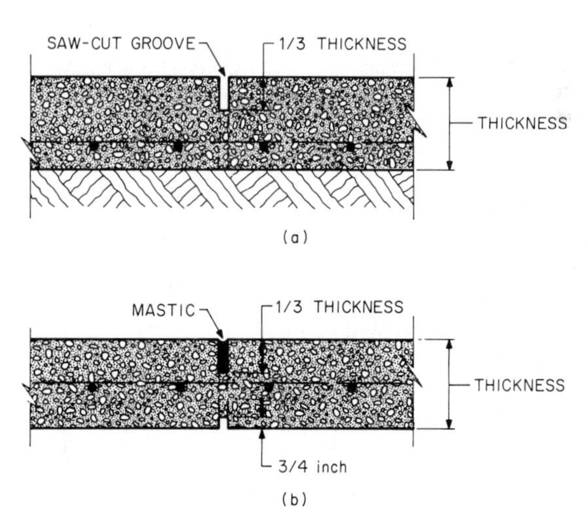

FIG. 7-5 Contraction joints. *(a)* Slab-on-grade; *(b)* wall.

7-5 Waterstop Joint Polyvinylchloride (PVC) waterstops are used at concrete joints to prevent liquid transfer. PVC waterstops are unaffected by the normal range of concrete types or additives commonly used and are highly resistant to acids, alkalis, and water-borne chemicals. These joints are suitable for above- or below-grade construction. PVC waterstops are made to be flexible and long lasting. They are manufactured in a wide range of design configurations which include serrated, splits, dumbbell, center bulb, end bulbs, tear web, and base seals. See Fig. 7-6 for a typical waterstop joint design.

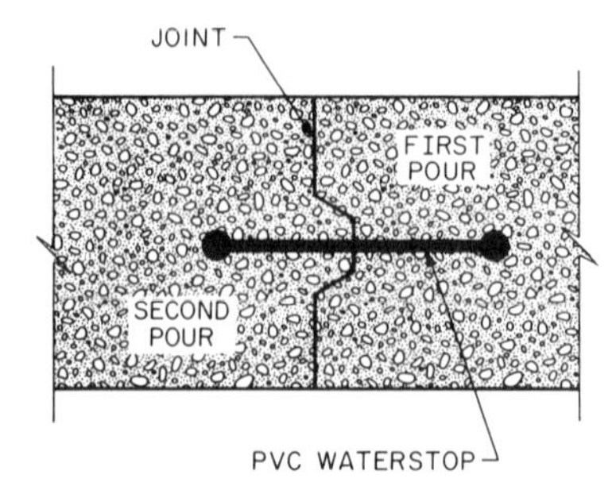

FIG. 7-6 Typical waterstop joint.

7-6 Concrete Block Connections The mortar joint is the basic connection used to join concrete blocks. Concave and V-shaped mortar joints are recommended as good engineering practice for walls of exterior concrete masonry. Raked or struck joints form small recesses which can hold storm water and thus cause weakening of the connection, especially in cold climates subject to the freeze-thaw effect. Modular-sized concrete units should have a motor joint approximately ⅜ inch. Using this thickness and employing either a concave joint or a V-joint on the wall's exterior side helps produce a neat, durable, and weathertight seal. See Fig. 7-7.

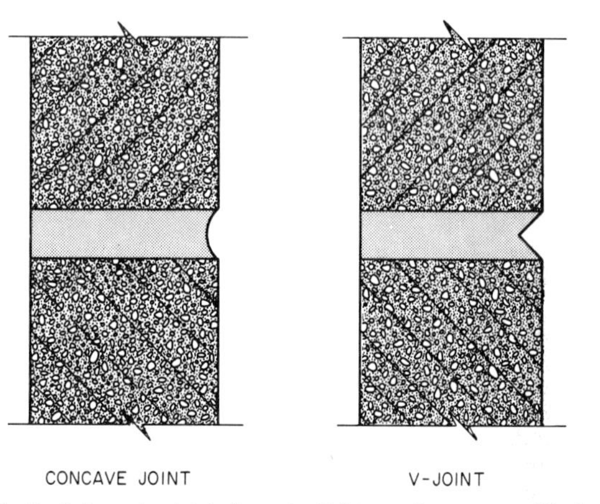

FIG. 7-7 Tooled mortar joints for watertight exterior concrete block walls.

Another critical item is the proper assembly of concrete blocks. A half-stagger has been an industry standard, and all openings should be multiples of either full- or half-size block units. See Fig. 7-8 for a typical example.

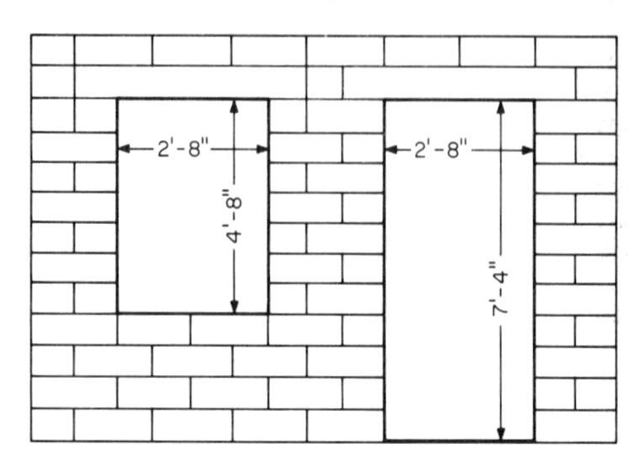

FIG. 7-8 Typical example of planning masonry wall openings.

Pilaster blocks are used to structurally reinforce a load-bearing concrete block wall. Some typical layouts are illustrated in Fig. 7-9.

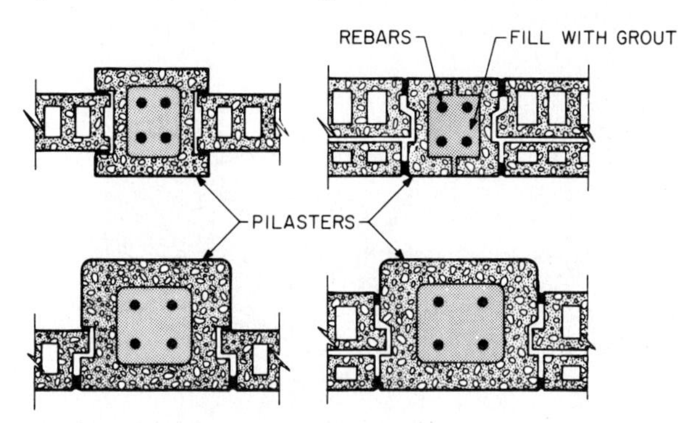

FIG. 7-9 Typical pilaster details.

Concrete block wall intersections can be structurally reinforced by using rebars grouted into the cores and placed between courses. See Fig. 7-10 for typical corner and tee intersections.

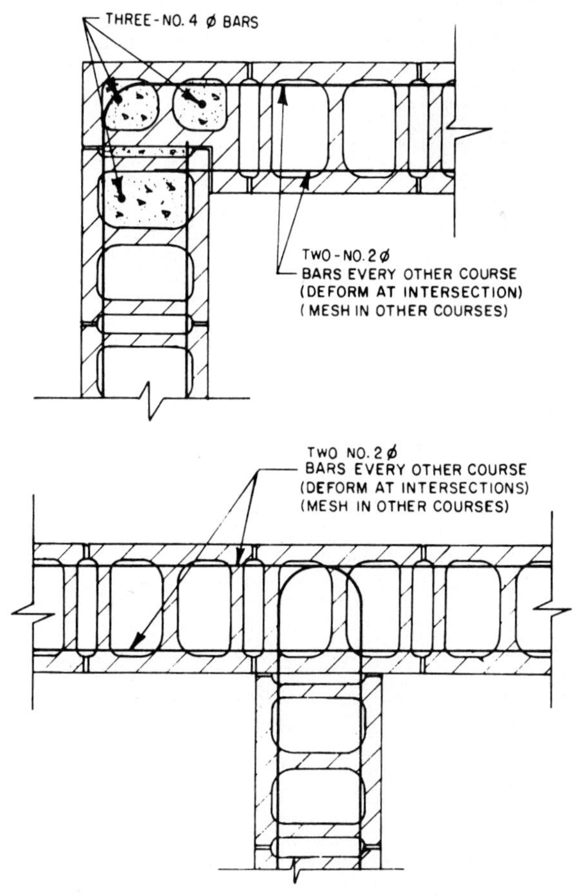

FIG. 7-10 Structurally reinforced corner and tee intersections of concrete block walls. *(Courtesy: Morgan & Parmley, Ltd.)*

Attachment of brick veneer to a concrete block wall is achieved by metal clips or extended reinforcing wires. See Fig. 7-11 for a typical method.

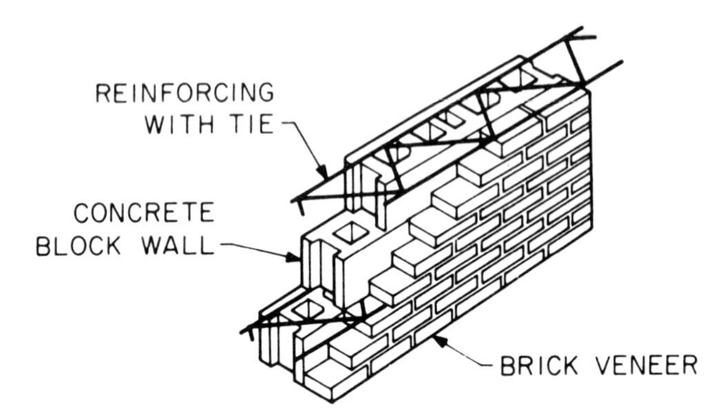

FIG. 7-11 Typical method of brick veneer attachment.

7-7 Block Bonding Patterns There are scores of bonding patterns for concrete block assembly. The main purpose of some are basically eye-pleasing, while others are designed for structural integrity. The more commonly used patterns are illustrated in Figure 7-12.

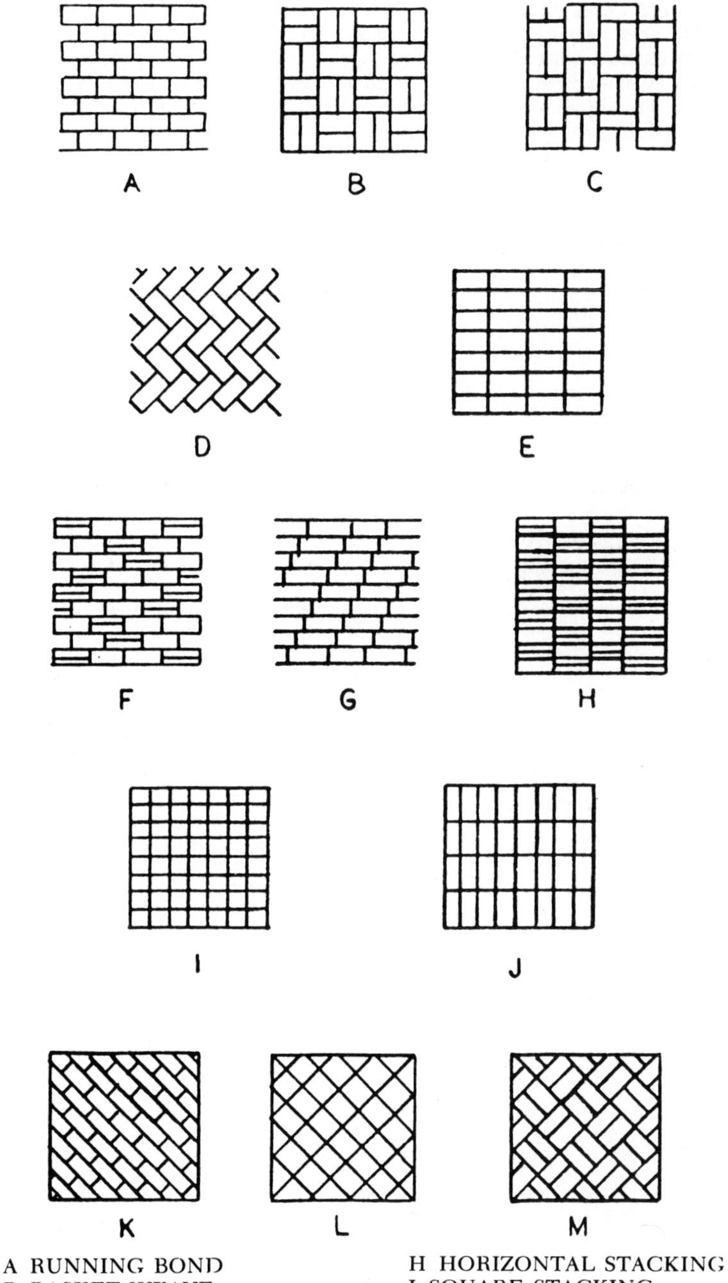

A RUNNING BOND
B BASKET WEAVE
C BASKET WEAVE VARIATION
D HERRINGBONE
E JACK-ON-JACK
F COURSED ASHLAR
G OFFSET BOND
H HORIZONTAL STACKING
I SQUARE STACKING
J VERTICAL STACKING
K DIAGONAL BOND
L DIAGONAL STACKING
M DIAGONAL BASKET WEAVE

FIG. 7-12 Concrete block assembly patterns.

7-8 Precast Concrete Joining Since its introduction into the United States in about 1950, prestressed concrete has become a major structural building component. Aside from its high structural advantage and cost-effectiveness, prestressed concrete units are easily assembled at the job site, thus reducing the construction time schedule.

There are literally scores of basic designs used to connect various prestressed units together; however, limited space in this sub-section permits mention of only a selected few. Figure 7-13 illustrates the design of attaching two double tees. Figure 7-14 shows the doweled connection of a precast column onto a pier. A single-tee roof panel can be framed into a precast concrete wall, as pictured in Fig. 7-15.

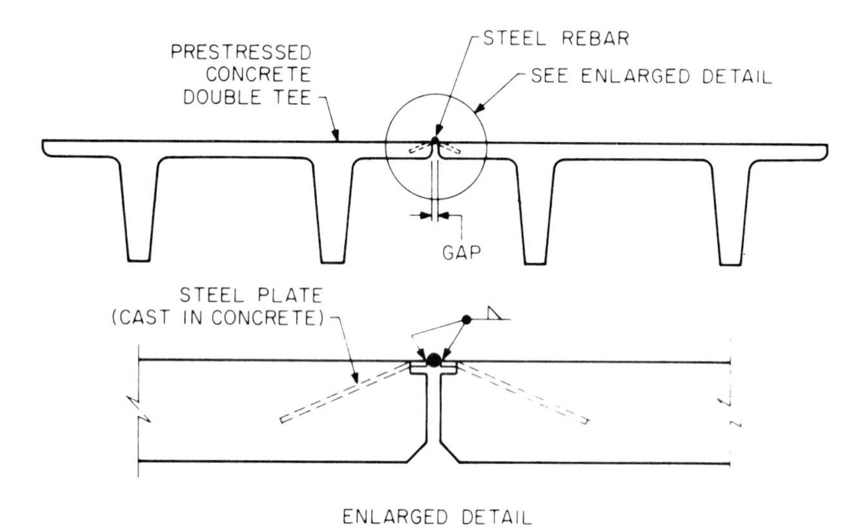

FIG. 7-13 Double-tee flange connections.

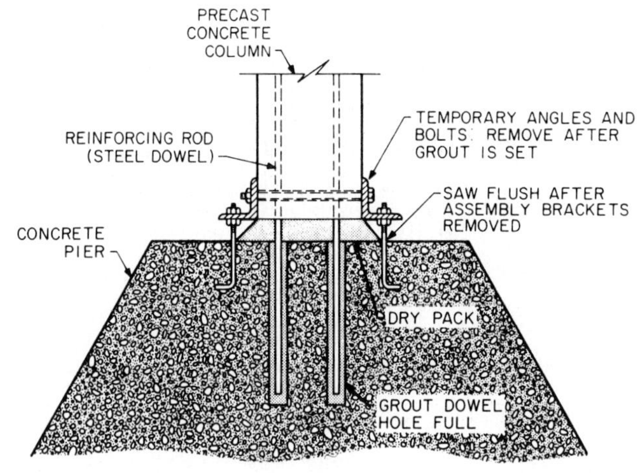

FIG. 7-14 Doweled connection.

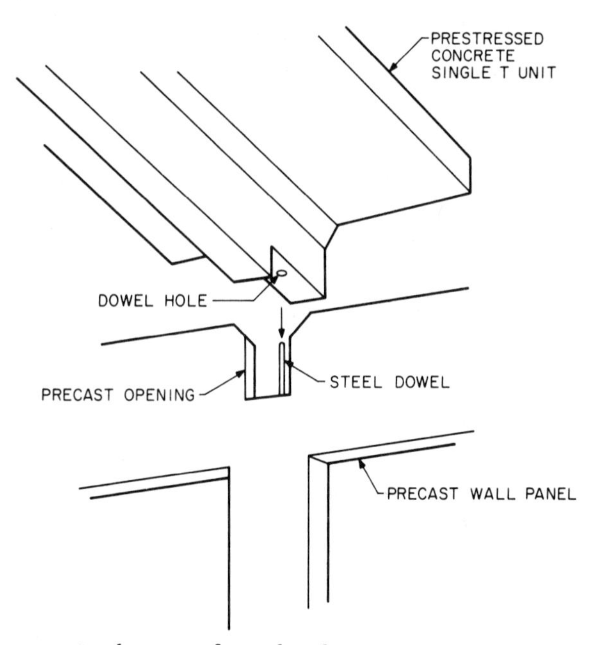

FIG. 7-15 Single-tee roof member framings into precast opening.

7-9 Pipe Connections When pipes must pass through concrete walls, either a sleeve or a wall pipe should be cast in place. This will ensure a watertight connection. See Fig. 7-16 for a standard detail.

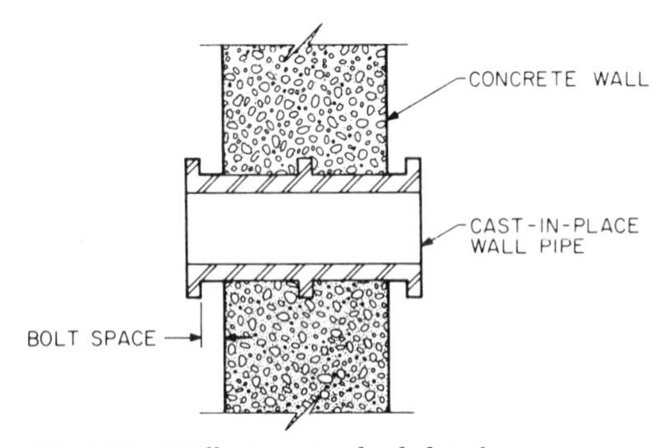

FIG. 7-16 Wall pipe standard detail.

7-10 Reinforcing Bars Steel reinforcing bars (or rods) are cast in concrete to increase strength. Standard rebars have raised ribs on their surface to provide additional contact. Figure 7-17 shows a typical layout for notations of structural reinforcement for a concrete wall.

Splicing of rebars is achieved by bending hooks, welding, wire ties, and threaded couplings.

Rebars are often bent to conform to special design conditions or to increase tying power. Figure 7-18 details the standard hook recommendations by the Concrete Reinforcing Steel Institute.

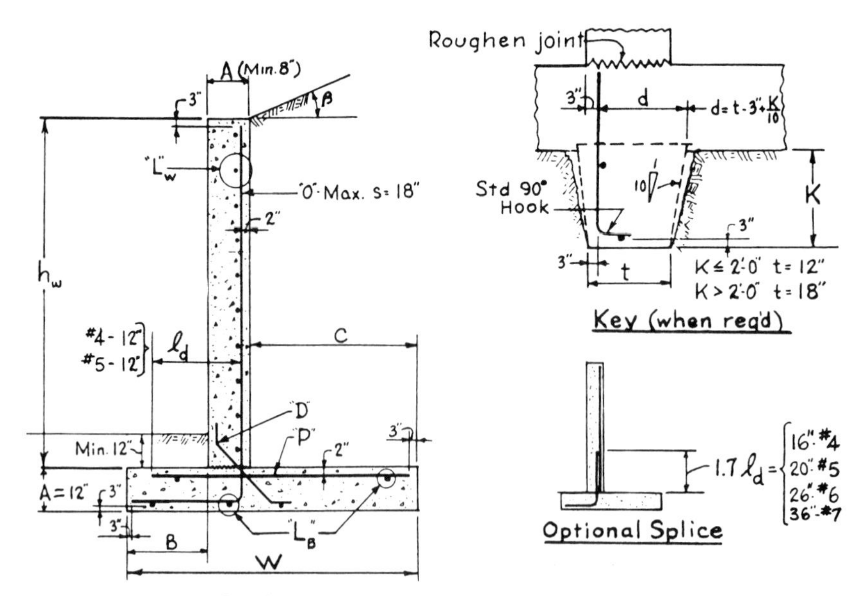

FIG. 7-17 Notation for dimensions and structural reinforcement for concrete wall. (*Courtesy: Concrete Reinforcing Steel Institute.*)

STANDARD HOOKS

All specific sizes recommended by CRSI below meet minimum requirements of ACI 318

RECOMMENDED END HOOKS
All Grades

D=Finished bend diameter

Bar Size	D	180° HOOKS		90° HOOKS
		A or G	J	A or G
# 3	2¼	5	3	6
# 4	3	6	4	8
# 5	3¾	7	5	10
# 6	4½	8	6	1-0
# 7	5¼	10	7	1-2
# 8	6	11	8	1-4
# 9	9½	1-3	11¾	1-7
#10	10¾	1-5	1-1¼	1-10
#11	12	1-7	1-2¾	2-0
#14	18¼	2-3	1-9¾	2-7
#18	24	3-0	2-4½	3-5

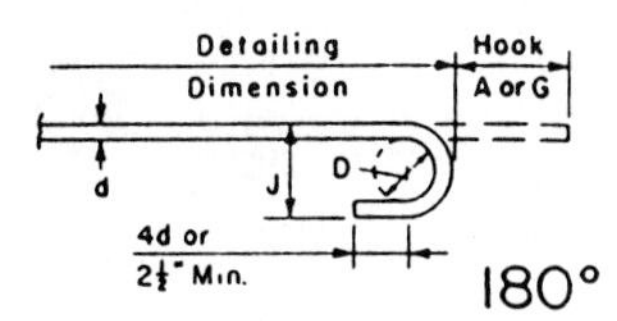

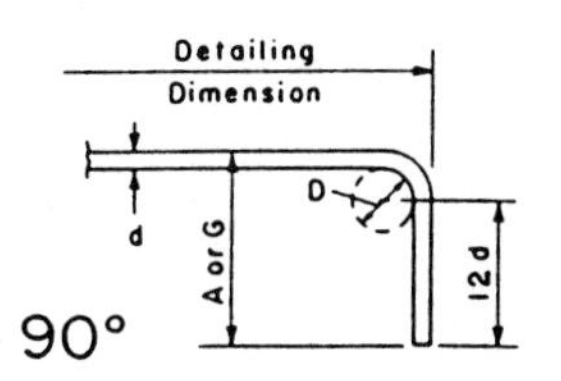

STIRRUP AND TIE HOOKS

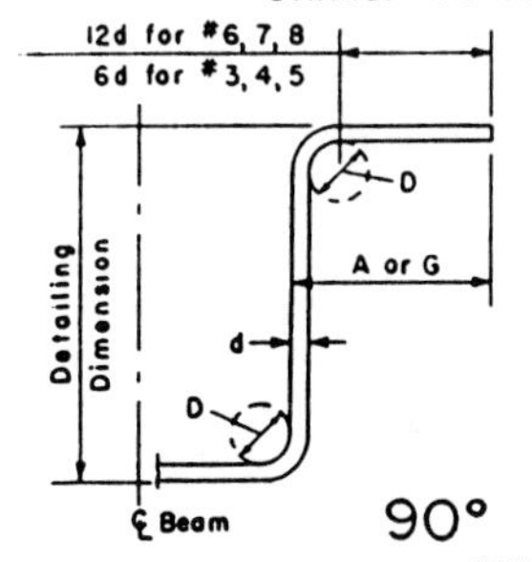

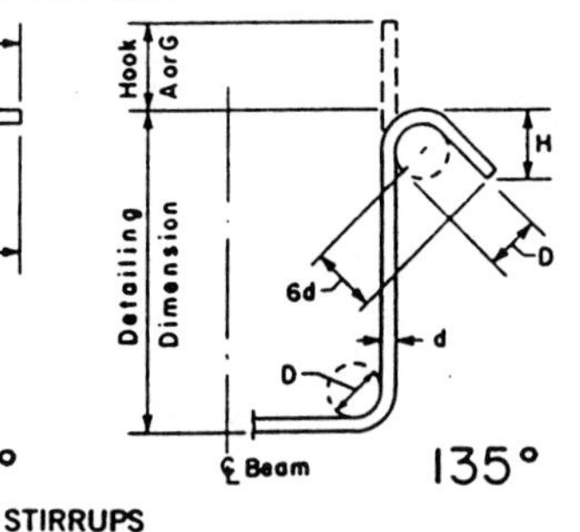

STIRRUPS
(TIES SIMILAR)

135° SEISMIC STIRRUP/TIE HOOKS

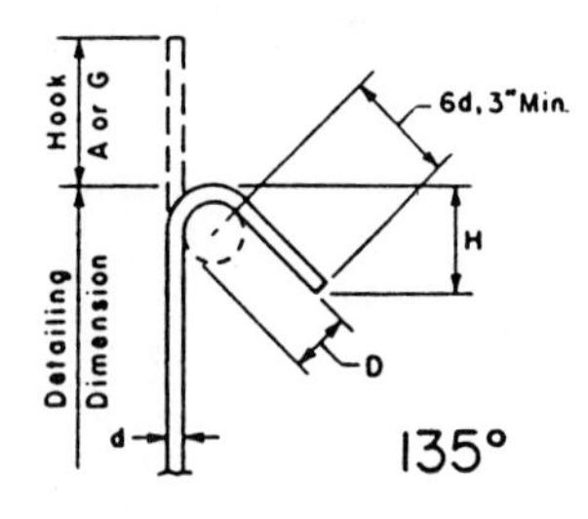

STIRRUP AND TIE HOOK DIMENSIONS
Grades 40-50-60

Bar Size	D (in.)	90° Hook	135° Hook	
		Hook A or G	Hook A or G	H Approx.
#3	1½	4	4	2½
#4	2	4½	4½	3
#5	2½	6	5½	3¾
#6	4½	1-0	8	4½
#7	5¼	1-2	9	5¼
#8	6	1-4	10½	6

135° SEISMIC STIRRUP/TIE HOOK DIMENSIONS
Grades 40-50-60

Bar Size	D (in.)	135° Hook	
		Hook A or G	H Approx.
#3	1½	4¼	3
#4	2	4½	3
#5	2½	5½	3¾
#6	4½	8	4½
#7	5¼	9	5¼
#8	6	10½	6

FIG. 7-18 Standard hooks—recommended designs. (*Courtesy: Concrete Reinforcing Steel Institute.*)

Welding of rebars generally is accomplished by lapping and then tack welding. However, when a more positive connection is required, steel stiffeners are used. See Fig. 7-19.

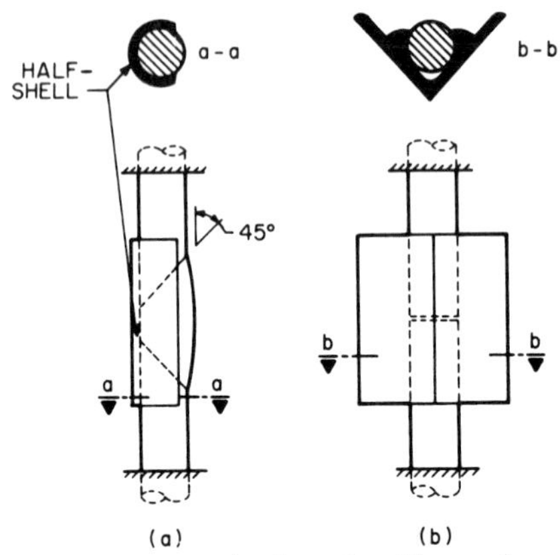

FIG. 7-19 Standard steel stiffeners for welding reinforcing rods. (*Courtesy: American Concrete Institute.*)

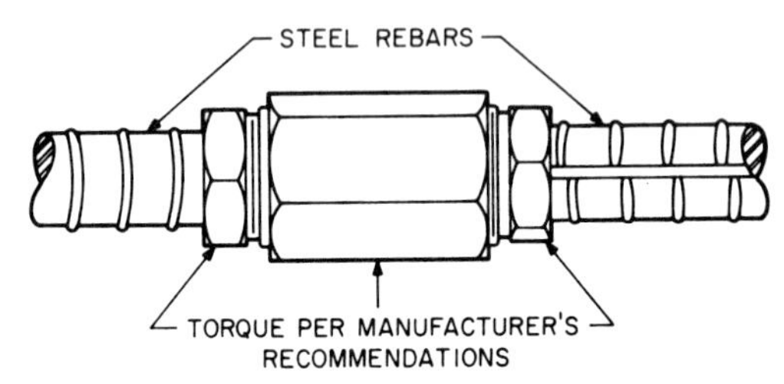

FIG. 7-20 Threaded coupling for rebar connections.

Wire ties are the most inexpensive method of connecting rebars and are used basically to hold them in position during the pouring of concrete.

Threaded couplings are employed when a stronger connection is desired. Various types have evolved over the years for a multitude of conditions. A basic type is illustrated in Fig. 7-20. It should be noted that the manufacturers supply charts for the required torque settings.

7-11 Tying Reinforcing Bars Concrete is basically weak in tension, but strong in compression. In order to overcome this deficiency, slender steel bars (or rods) are properly imbedded in the concrete to assist in safely supporting imposed loads. These steel bars may be welded wire mesh or various sizes and shapes of reinforcing bars.

Prior to positioning any reinforcing steel in forms, remove any oil, rust, grease, scale, soil, mud, slag, or any other foreign material from the reinforcing bars. This procedure will help insure a good bond between the concrete and rebar.

Tying rebars together is critical so they remain in the proper position during the pour and curing process of the concrete. The simple overlap connection of straight rebars is shown in Figure 7-21. When rebars are crossed in a typcial right angle pattern, there are several standard types of ties that are used. Refer to Figure 7-22 for six of the most common.

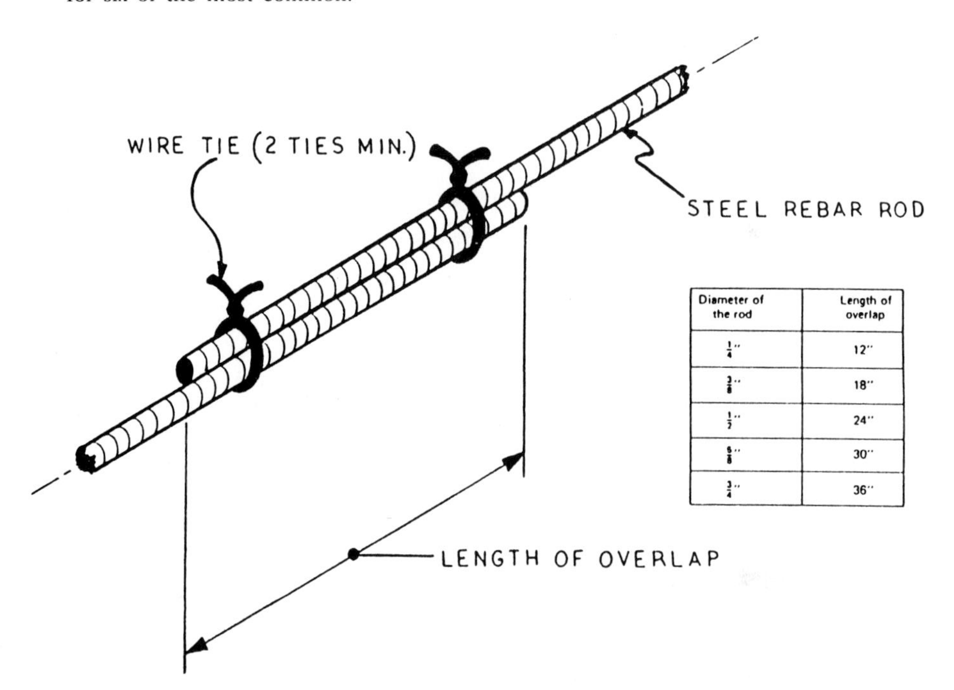

Diameter of the rod	Length of overlap
$\frac{1}{4}$"	12"
$\frac{3}{8}$"	18"
$\frac{1}{2}$"	24"
$\frac{5}{8}$"	30"
$\frac{3}{4}$"	36"

FIG. 7-21 Overlap connection.

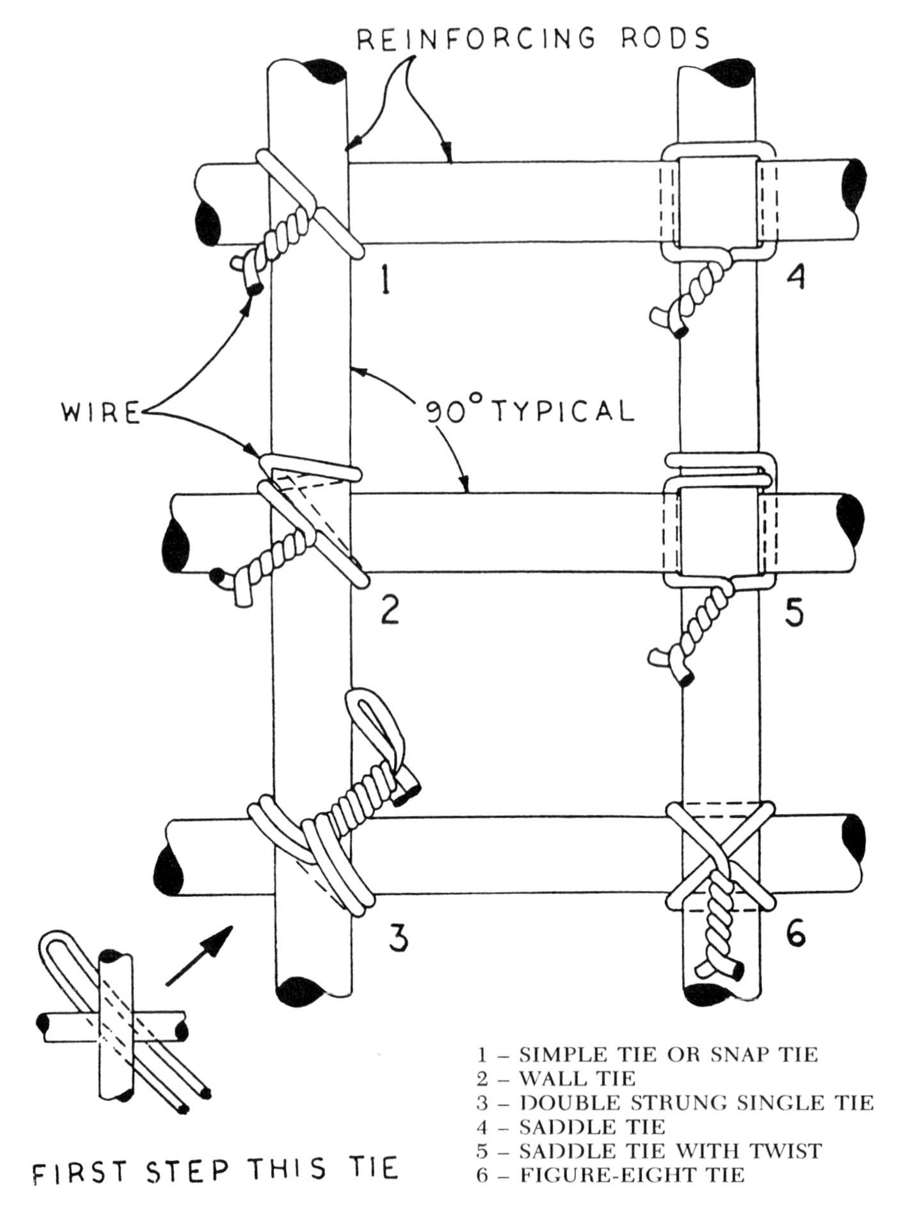

FIG. 7-22 Types of ties for reinforcing bars.

Section 8
Lumber, Timber, and Log Connections

ROBERT O. PARMLEY, P.E.

Consulting Engineer
Ladysmith, Wisconsin

A great variety of framing systems and structural fastening designs are available. Their economy, intended use, availability, configuration, structural demands, geographical location and other factors must be determined in order to select the fastening system most advantageous to a particular connection problem.

This section will be confined to the standard mechanical fastener components and log connections, plus related design factors. Glued laminated units and their adhesive technology may be found in the American Institute of Timber Construction (AITC) Standards. Section 12 of this handbook may be consulted for basic adhesive references.

Federal and Military Specifications are also available as well as technical manuals from the USDA on the gluing and laminating of lumber and timber. The *Timber Construction Manual* prepared by the American Institute of Timber Construction is an excellent source for information on glued laminated technology and related structural data.

NAILING

Nails and spikes are the oldest and, of course, most widely used fasteners today by the lumber and timber construction industry.

8-1 Types of Nails and Spikes Common nails and spikes are limited primarily by

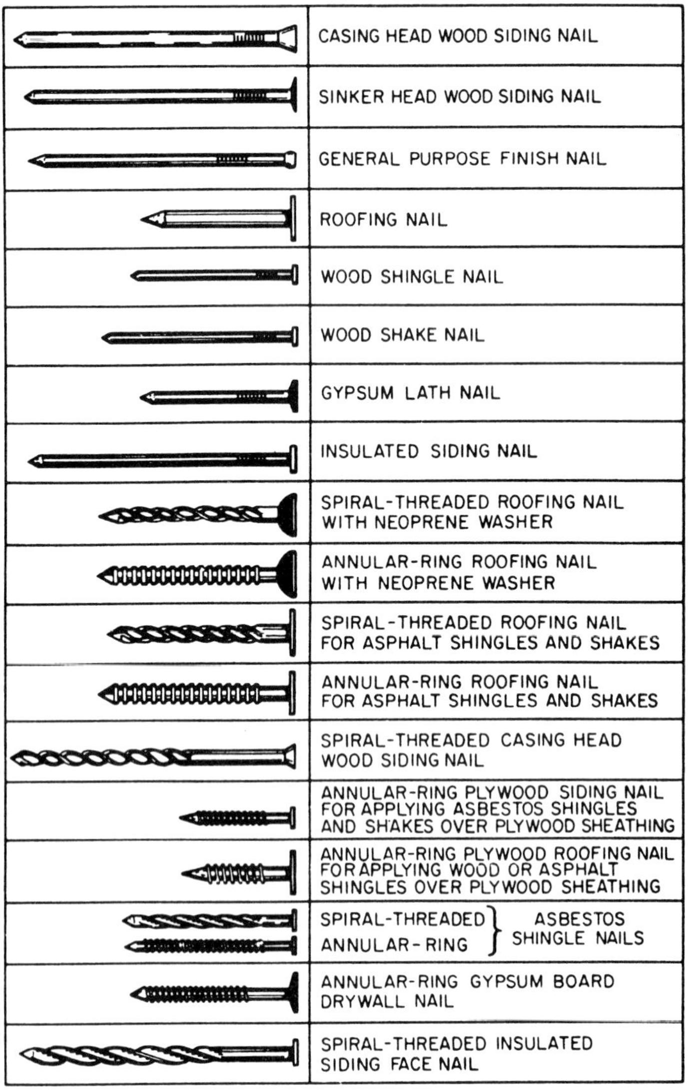

CASING HEAD WOOD SIDING NAIL
SINKER HEAD WOOD SIDING NAIL
GENERAL PURPOSE FINISH NAIL
ROOFING NAIL
WOOD SHINGLE NAIL
WOOD SHAKE NAIL
GYPSUM LATH NAIL
INSULATED SIDING NAIL
SPIRAL-THREADED ROOFING NAIL WITH NEOPRENE WASHER
ANNULAR-RING ROOFING NAIL WITH NEOPRENE WASHER
SPIRAL-THREADED ROOFING NAIL FOR ASPHALT SHINGLES AND SHAKES
ANNULAR-RING ROOFING NAIL FOR ASPHALT SHINGLES AND SHAKES
SPIRAL-THREADED CASING HEAD WOOD SIDING NAIL
ANNULAR-RING PLYWOOD SIDING NAIL FOR APPLYING ASBESTOS SHINGLES AND SHAKES OVER PLYWOOD SHEATHING
ANNULAR-RING PLYWOOD ROOFING NAIL FOR APPLYING WOOD OR ASPHALT SHINGLES OVER PLYWOOD SHEATHING
SPIRAL-THREADED } ASBESTOS
ANNULAR-RING } SHINGLE NAILS
ANNULAR-RING GYPSUM BOARD DRYWALL NAIL
SPIRAL-THREADED INSULATED SIDING FACE NAIL

FIG. 8-1 Typical nails. [*From F. S. Merritt (ed.), Building Construction Handbook, 3rd ed., McGraw-Hill, New York, © 1975.*]

their size and number which can be driven into wood without splitting it. The standard types of nails and spikes include (See Fig. 8-1 for typical nails):

Standard wire nails
Annular grooved
Cement-coated
Spirally grooved
Zinc-coated
Galvanized
Barbed
Chemically etched

Special nails have shanks which have been treated to improve withdrawal resistance.

8-2 Spacing, End, and Edge Distances For proper spacing of nails, according to the best available authority, there are no precise rules because of the variety of species of wood and their moisture content. Good judgment and experience are therefore the most reasonable methods of positioning nails to avoid unusual splitting and to ensure a structurally sound joint.

8-3 Allowable Loadings To establish the allowable lateral load (applied in any lateral direction), for one nail or spike which has been driven in the side grain it is recommended to refer to NDS® Wood Construction (Structural Lumber/Glued Laminated Timber/Timber Piles/Connections). The minimum permissible penetration is ⅓ the depth penetration required for the full allowable load.

Table 8-1 is used to determine the allowable withdrawal loads for nails or spikes. The allowable withdrawal loads for toenailed joints are 0.67 times those shown in Table 8-1. Nails clinched cross-grain have approximately 20 percent more resistance to withdrawal than those which are clinched along the grain.

For nails in double shear, where the nail fully penetrates all three members of a joint, the allowable lateral load may be increased ⅓ if each side member is not less than ⅓ the thickness of the center member. If each side member is equal to the thickness of the center member, the load value may be increased ⅔.

Toenails should be driven at an angle of about 30 degrees to the piece and enter at approximately ⅓ the nail length from the end of the member.

8-4 Standard Nailing Practices All nailed joints should be well aligned, squared, and positioned correctly. Sound stress-grade lumber and proper nails or spikes should be selected for a good structural union.

Figure 8-2 shows part of a nailing detail of a roof truss from an actual approved construction project. Note the size and location of nails, members, and gusset (plywood) plates.

THREADED CONNECTORS

Threaded connectors are detailed in Sec. 1 of this handbook. This section outlines their use in lumber and timber connections; for more detail, however, the reader is referred to Sec. 1.

8-5 Wood Screws Wood screws come in a wide range of designs. The basic head types are: flat countersunk, oval countersunk, and round. All wood screws have gimlet points. The range of material for manufacturing wood screws includes: steel, corrosion-resistant steel, brass, aluminum alloy, and others.

Wood screws are always designated by the following data:

Nominal size (number or decimal)
Screw length (fractional or decimal equivalent)
Product name (including head style and driving provision)
Protective finish

Example: 10 × 1½ slotted flat countersunk-head wood screw, steel. (See Fig. 8-2.)

The standards for wood screws are published in ANSI B18.6.1–1972, Wood Screws.

8-6 Bolts and Lag Screws Refer to Sec. 1 of this handbook for design and size of bolts and lag screws. Section 1 also relates strength factors critical to selecting the proper size and number of bolts for a good connection. Section 8-8 illustrates the allowable loads of these connectors and their joints.

8-7 Recommended Spacing Unlike nails or spikes, bolts and lag screws have a more exact method of determining the proper location and spacing at proposed joints. Bolts and lag screws must have prebored holes; therefore a more exact determination is required.

TABLE 8-1 Nail and Spike Withdrawal Design Values (W)[1]

Tabulated withdrawal design values (W) are in pounds per inch of penetration into side grain of main member.

Specific Gravity G	COMMON WIRE NAILS, BOX NAILS, and COMMON WIRE SPIKES Diameter, D															THREADED NAILS Wire Diameter, D				
	0.099″	0.113″	0.128″	0.131″	0.135″	0.148″	0.162″	0.192″	0.207″	0.225″	0.244″	0.263″	0.283″	0.312″	0.375″	0.120″	0.135″	0.148″	0.177″	0.207″
0.43	17	19	21	22	23	25	27	32	35	38	41	44	47	52	63	22	25	27	32	38
0.49	23	26	30	30	31	34	38	45	48	52	57	61	66	72	87	30	34	38	45	52
0.58	35	40	45	46	48	52	57	68	73	80	86	93	100	110	133	46	52	57	68	80
0.73	62	71	80	82	85	93	102	121	130	141	153	165	178	196	236	82	93	102	121	141

1. Tabulated withdrawal design values (W) for nail or spike connections shall be multiplied by all applicable adjustment factors (see Table 8-13).

SOURCE: American Forest & Paper Association, Washington, D.C.

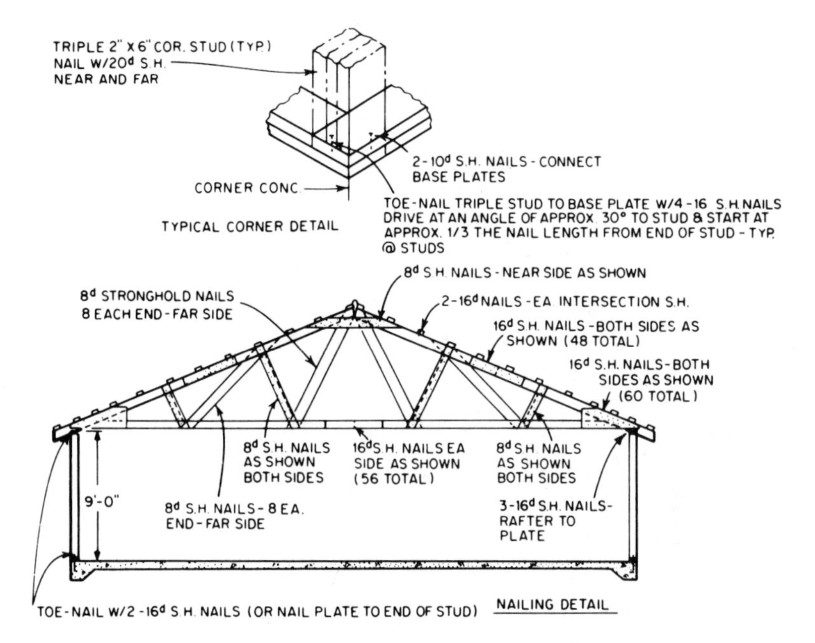

FIG. 8-2 Construction details of a wood frame building.

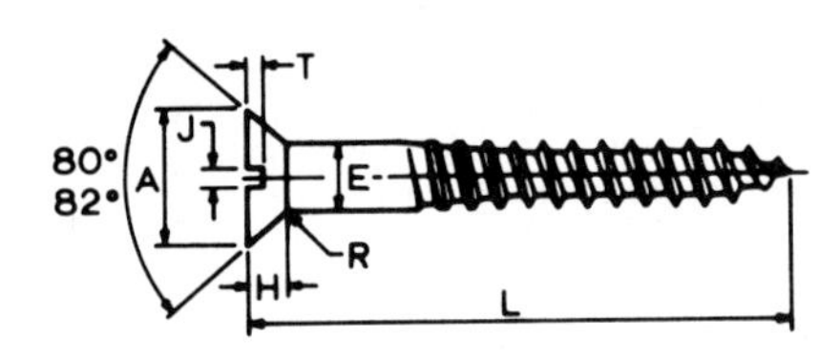

FIG. 8-3 Typical design of a slotted flat countersunk-head wood screw. A = head diameter, E = body diameter, H = head height, J = slot width, L = length, R = fillet radius, T = slot depth.

TABLE 8-2 Recommended Spacing and End Distance Values for Bolts[a]

Dimension	Parallel to Grain Loading	Perpendicular to Grain Loading
A, c–c spacing	Minimum of 4 times bolt diameter.	4 times bolt diameter unless the design load is less than the bolt bearing capacity of side members; then spacing may be reduced proportionately.
Staggered bolts	Adjacent bolts are considered to be placed at critical section unless spaced at a minimum of 8 times the bolt diameter	Staggering not permitted unless design load is less than bolt bearing capacity of side members.
B, row spacing	Minimum of 1½ times bolt diameter.	2½ times bolt diameter for l/d[b] ratio of 2; 5 times bolt diameter for l/d ratios of 6 or more; use straightline interpolation for l/d between 2 and 6.
	Spacing between rows paralleling a member may not exceed 5 in. unless separate splice plates are used for each row.	
C, end distance	In tension, 7 times bolt diameter for softwoods and 5 times bolt diameter for hardwoods. In compression, 4 times bolt diameter.	Minimum of 4 times bolt diameter; when members abut at a joint (not illustrated), the strength shall be evaluated also as a beam supported by fastenings
D, edge distance	1½ times the bolt diameter, except that for l/d[b] ratios of more than 6, use one-half the row spacing, *B*.	Minimum of 4 times bolt diameter at edge toward which load acts; minimum of 1½ times bolt diameter at opposite edge.

[a]These are minimum values for the allowable bolt loads

[b]Ratio of length of bolt in main member, *l*, to diameter of bolt, *d*.

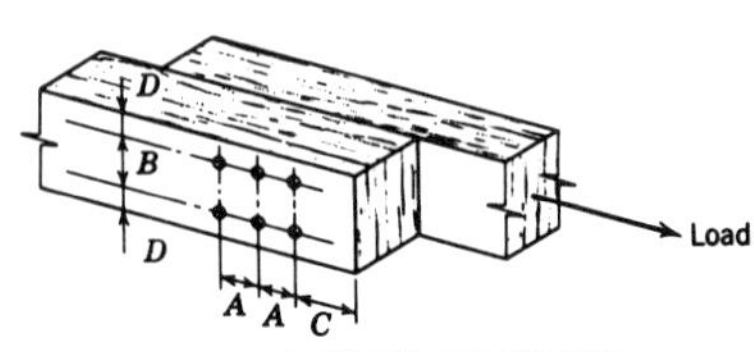

LOAD PARALLEL TO GRAIN.

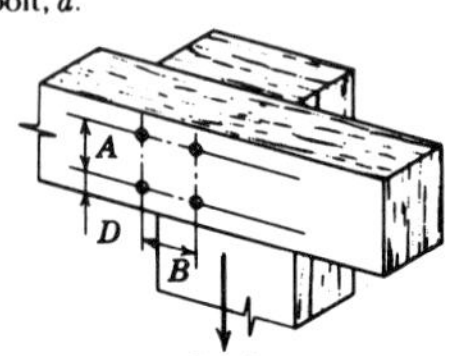

LOAD PERPENDICULAR TO GRAIN.

SOURCE: American Institute of Timber Construction, *Timber Construction Manual,* 3rd ed., John Wiley & Sons, Inc., New York, 1986.

Table 8-2 relates recommended spacing and end distance values for bolts.
Lag screws (or lag bolts) should follow the same general pattern of spacing as the bolts in Table 8-2. All lag screws require prebored lead holes as shown in Table 8-3.

TABLE 8-3 Lead Hole Diameter for Lag Bolts

	Diameter of Lead Hole (in.)			
		Threaded Portion		
Nominal Diameter of Lag Bolt (in.)	Shank (Unthreaded) Portion (in.)	Group I Species[a]	Group II Species[a]	Groups III and IV Species[a]
1/4	1/4	3/16	5/32	3/32
5/16	5/16	13/64	3/16	9/64
3/8	3/8	1/4	15/64	11/64
7/16	7/16	19/64	9/32	13/64
1/2	1/2	11/32	5/16	15/64
9/16	9/16	13/32	23/64	9/32
5/8	5/8	29/64	13/32	5/16
3/4	3/4	9/16	1/2	13/32
7/8	7/8	43/64	39/64	33/64
1	1	51/64	23/32	5/8
$1\frac{1}{8}$	$1\frac{1}{8}$	59/64	53/64	3/4
$1\frac{1}{4}$	$1\frac{1}{4}$	$1\frac{1}{16}$	15/16	7/8

[a]See Table 8-5 for species in each group.

SOURCE: American Institute of Timber Construction, *Timber Construction Manual*, 3rd ed., John Wiley & Sons, Inc., New York, 1986.

8-8 Allowable Loads *Wood Screws.* The total allowable load for a series of screws equals the sum of the loads for each screw if the spacing, wood strength, and preboring follow good engineering practice.

Lead holes for wood screws are tabulated in Table 8-4. See Table 8-5 for species load groups for fastenings.

Bolts. The allowable load for one bolt loaded in double shear in a typical timber joint under normal conditions is dependent on many factors. Refer to NDS® Wood Construction Manual. To find the allowable load for more than one bolt, multiply the given load for one bolt by the total number of bolts per joint, provided they are properly placed as to spacing and edge and end distances.

Lag Screws. Dowel bearing strength for lag screw connections is shown in Table 8-6. Lag screw design values and withdrawal values are tabulated in Tables 8-7 and 8-8 respectfully.

STEEL DOWELS

8-9 Spiral Dowels A spiral dowel is a twisted steel rod which has grooved ridges in a helix or spiral form on its entire length. The lead is sufficient to allow driving by any suitable method. Table 8-9 shows the range of diameters and length of standard spiral dowels as well as lead-hole sizes.

8-10 Allowable Loading Under normal duration and dry-use conditions, Table 8-10 may be consulted to determine allowable loads on spiral dowels. The total allowable load (for more than one dowel) equals the sum of the loads for each dowel.

TABLE 8-4 Lead Hole Diameters for Wood Screws

		Diameter of Lead Hole (in.)					
		Withdrawal Loads		Lateral Loads			
				Group I Species[a]		Groups II, III, and IV Species[a]	
Gage of Screw	Shank Diameter of Screw (in.)	Group I Species[a]	Groups II, III, and IV Species[a,b]	Shank Portion	Threaded Portion	Shank Portion	Threaded Portion
6	0.138	5/64	1/16	9/64	3/32	1/8	5/64
7	0.151	3/32	5/64	5/32	7/64	1/8	3/32
8	0.164	7/64	5/64	5/32	7/64	9/64	3/32
9	0.177	7/64	5/64	11/64	1/8	5/32	7/64
10	0.190	7/64	3/32	3/16	1/8	11/64	7/64
12	0.216	9/64	7/64	7/32	9/64	3/16	1/8
14	0.242	5/32	7/64	1/4	5/32	7/32	9/64
16	0.268	11/64	1/8	17/64	3/16	15/64	5/32
18	0.294	3/16	9/64	19/64	13/64	1/4	11/64
20	0.320	13/64	5/32	5/16	7/32	9/32	3/16
24	0.372	15/64	3/16	3/8	1/4	21/64	15/64

[a]For species in each group see Table 8-8 for sawn lumber.
[b]For group III and IV species the screw may be inserted without a lead hole.
SOURCE: American Institute of Timber Construction, *Timber Construction Manual,* 3rd ed., John Wiley & Sons, Inc., New York, 1986.

TABLE 8-5 Species Groups for Fastener Design for Sawn Lumber

			Grouping for Lag Bolts, Drift Bolts, Nails, Spikes, Wood Screws, Staples, and Metal Plate Connectors	
Species	Bolt Group[a]	Timber Connector Load Group[b] (Shear Plates and Split Rings)	Group	Specific Gravity(G)[c]
Ash, Commercial White	2	A	I	0.62
Aspen	12	D	IV	0.40
Aspen, Northern[c]	12	C	III	0.42
Beech	4	A	I	0.68
Birch, Sweet and Yellow	4	A	I	0.66
Cedar, Northern White	12	D	IV	0.31
Cedars, Western[d]	9	D	IV	0.35
Coast Species[c]	12	D	IV	0.39
Cottonwood, Black	12	D	IV	0.33
Cottonwood, Eastern	12	D	IV	0.41
Cypress, Southern	3	C	III	0.48
Douglas Fir-Larch[d]	3	B	II	0.51
Douglas Fir-Larch (dense)	1	A	II	—[f]

(continued)

TABLE 8-5 Species Groups for Fastener Design for Sawn Lumber *(Continued)*

Douglas Fir, South	6	C	III	0.48
Eastern Woods	12	D	IV	0.38
Fir, Balsam	11	D	IV	0.38
Hem-Fir[d]	8	C	III	0.42
Hemlock				
Eastern-Tamarack[e]	8	C	III	0.45
Mountain	9	C	III	0.47
Western[d]	8	C	III	0.48
Hickory and Pecan	2	A	I	0.75
Maple, Black and Sugar	4	A	I	0.66
Northern Species[e]	12	D	IV	0.35
Oak, Red and White	5	A	I	0.67
Pine				
Eastern White[d]	11	D	IV	0.38
Idaho White	11	D	IV	0.40
Lodgepole	10	C	III	0.44
Northern	9	C	III	0.46
Ponderosa[e]	11	C	III	0.49
Ponderosa-Sugar	11	C	III	0.42
Red[e]	11	C	III	0.42
Southern	3	B	II	0.55
Southern (dense)	1	A	II	—[f]
Western White	11	D	IV	0.40
Poplar, Yellow	10	C	III	0.46
Redwood				
California	3	C	III	0.42
California (open grain)	8	D	IV	0.37
Spruce				
Eastern	10	C	III	0.43
Engelmann-Alpine Fir	12	D	IV	0.36
Sitka	10	C	III	0.43
Sitka, Coast[e]	10	D	IV	0.39
Spruce-Pine-Fir[e]	10	C	III	0.42
West Coast Woods (mixed species)	12	D	IV	0.35
White Woods (Western Woods)	12	D	IV	0.35

[a]See *Timber Construction Manual* for species and density groupings applicable to bolt design.
[b]When stress graded.
[c]Based on weight and volume when oven-dry. These specific-gravity values are to be used for the determination of withdrawal design values for lag bolts, nails, spikes, and wood screws.
[d]Also applies when species name includes the designation "North."
[e]Applies when graded in accordance with National Lumber Grades Authority *Standard Grading Rules for Canadian Lumber.*
[f]The specific gravity of dense lumber is slightly higher than for medium-grain lumber. However, the design values for this group are based on the average specific gravity of the species, which is 0.51 for Douglas Fir-Larch and 0.55 for Southern Pine.
SOURCE: American Institute of Timber Construction, *Timber Construction Manual,* 3rd ed., John Wiley & Sons, Inc., New York, 1986.

TABLE 8-6 Dowel Bearing Strength for Lag Screw Connections

Species Combination	Specific[1] Gravity G	Dowel bearing strength in pounds per square inch (psi)											
		$F_{e\parallel}$	$F_{e\perp}$ D=1/4"	$F_{e\perp}$ D=5/16"	$F_{e\perp}$ D=3/8"	$F_{e\perp}$ D=7/16"	$F_{e\perp}$ D=1/2"	$F_{e\perp}$ D=5/8"	$F_{e\perp}$ D=3/4"	$F_{e\perp}$ D=7/8"	$F_{e\perp}$ D=1"	$F_{e\perp}$ D=1-1/8"	$F_{e\perp}$ D=1-1/4"
Douglas Fir-Larch	0.50	5600	4450	4000	3650	3400	3150	2800	2600	2400	2250	2100	2000
Eastern Hemlock	0.41	4600	3350	3000	2750	2550	2350	2100	1950	1800	1650	1600	1500
Mixed Oak	0.68	7600	6950	6250	5700	5250	4950	4400	4050	3750	3500	3300	3100
Southern Pine	0.55	6150	5150	4600	4200	3900	3650	3250	2950	2750	2550	2400	2300

SOURCE: American Forest & Paper Association, Washington D.C.

TABLE 8-7 Lag Screw Design Values (Z) for Single Shear (two member) Connections[1,2]

With both members of identical species

Side Member Thickness	Lag Screw Diameter	G=0.67 Red Oak			G=0.55 Mixed Maple Southern Pine			G=0.50 Douglas Fir-Larch			G=0.49 Douglas Fir-Larch (N)		
t_s inches	D inches	$Z_{\parallel}$ lbs.	$Z_{s\perp}$ lbs.	$Z_{m\perp}$ lbs.	$Z_{\parallel}$ lbs.	$Z_{s\perp}$ lbs.	$Z_{m\perp}$ lbs.	$Z_{\parallel}$ lbs.	$Z_{s\perp}$ lbs.	$Z_{m\perp}$ lbs.	$Z_{\parallel}$ lbs.	$Z_{s\perp}$ lbs.	$Z_{m\perp}$ lbs.
1/2	1/4	190	150	150	170	130	130	160	110	120	160	110	120
	5/16	270	190	210	240	140	180	220	130	170	210	120	170
	3/8	340	210	250	290	160	220	260	140	200	260	130	200
5/8	1/4	210	160	160	180	130	140	170	120	130	170	120	130
	5/16	290	210	220	250	180	190	240	160	180	240	150	170
	3/8	350	250	260	310	200	230	290	170	210	290	170	210
3/4	1/4	230	170	180	200	140	150	180	130	140	180	130	140
	5/16	310	220	240	270	190	200	250	170	190	250	170	180
	3/8	380	260	280	330	220	240	310	210	220	310	200	220
1	1/4	260	200	200	230	160	180	210	150	160	210	150	160
	5/16	360	250	270	300	210	230	280	190	210	280	190	210
	3/8	430	290	330	370	240	270	350	220	250	340	220	250
1-1/4	1/4	260	200	200	230	180	180	220	170	170	220	170	170
	5/16	370	280	280	340	230	250	320	210	240	320	210	230
	3/8	470	330	340	420	270	300	390	240	290	390	240	280
1-1/2	1/4	260	200	200	230	180	180	220	170	170	220	170	170
	5/16	370	280	280	340	250	250	320	230	240	320	230	230
	3/8	470	340	340	420	300	300	400	270	290	400	260	280
	7/16	630	430	460	570	350	400	540	320	380	540	310	380
	1/2	830	510	590	710	420	500	660	380	460	650	370	450
	5/8	1140	670	800	980	570	680	920	530	620	910	520	620
	3/4	1510	890	1040	1320	660	880	1240	590	820	1220	560	800
	7/8	1940	960	1300	1710	720	1110	1620	630	1040	1600	600	1020
	1	2450	1020	1600	2170	770	1370	2060	680	1290	2040	650	1260
	1-1/8	3030	1080	1930	2590	810	1670	2360	710	1560	2320	690	1540
	1-1/4	3520	1140	2300	2880	860	2000	2630	750	1860	2580	730	1840
2-1/2	1/4	260	200	200	230	180	180	220	170	170	220	170	170
	5/16	370	280	280	340	250	250	320	240	240	320	230	230
	3/8	470	340	340	420	300	300	400	290	290	400	280	280
	7/16	630	460	460	570	400	400	550	380	380	540	380	380
	1/2	830	590	590	750	520	520	710	480	480	710	480	480
	5/8	1290	860	880	1170	700	780	1120	630	730	1110	620	720
	3/4	1860	1060	1240	1680	870	1090	1570	800	1020	1550	780	1010
	7/8	2460	1290	1640	2100	1080	1400	1950	1000	1280	1930	970	1260
	1	2970	1550	1980	2560	1280	1660	2390	1130	1530	2360	1080	1500
	1-1/8	3550	1800	2320	3080	1350	1950	2880	1180	1800	2850	1150	1770
	1-1/4	4190	1910	2680	3650	1440	2270	3430	1250	2100	3390	1220	2070

1. Tabulated lateral design values (Z) for lag screw connections shall be multiplied by all applicable adjustment factors (Table 8-13).
2. Tabulated lateral design values (Z) are for "full diameter" lag screws (see Reference 3) inserted in side grain with lag screw axis perpendicular to wood fibers, and with the following lag screw bending yield strengths (F_{yb}):

F_{yb} = 70,000 psi for D = 1/4″
F_{yb} = 60,000 psi for D = 5/16″
F_{yb} = 45,000 psi for D ≥ 3/8″

SOURCE: American Forest & Paper Association, Washington, D.C.

TABLE 8-8 Lag Screw Withdrawal Design Values (W)[1]

Tabulated withdrawal design values (W) are in pounds per inch of thread penetration into side grain of main member. Length of thread penetration in main member shall not include the length of the tapered tip.

Specific Gravity G	Lag Screw Unthreaded Shank Diameter, D										
	1/4"	5/16"	3/8"	7/16"	1/2"	5/8"	3/4"	7/8"	1"	1-1/8"	1-1/4"
0.41	167	198	226	254	281	332	381	428	473	516	559
0.50	225	266	305	342	378	447	513	576	636	695	752
0.55	260	307	352	395	437	516	592	664	734	802	868
0.68	357	422	484	543	600	709	813	913	1009	1103	1193

SOURCE: American Forest & Paper Association, Washington, D.C.

TABLE 8-9 Dimensions and Lead Hole Diameters for Stock Sizes of Spiral Dowels in Group II Species[a]

Outside Diameter of Dowel (in.)	Minimum Length (in.)[b]	Maximum Length (in.)[b]	Maximum Driving Length (in.)[c,d]	Diameter of Lead Hole (in.)[c]
1/4	2-1/2	6		3/16
5/16	3	6-1/2		1/4
3/8	3-1/2	10		9/32
7/16	3-1/2	12	12	11/32
1/2	4	18	12	3/8
5/8	4	24	12	15/32
3/4	Lengths in these three sizes must be specially ordered		12	9/16
7/8			13	21/32
1			14	3/4

[a–d]See original source of this table for information given in these footnotes.

SOURCE: American Institute of Timber Construction, *Timber Construction Manual*, 3rd ed., John Wiley & Sons, Inc., New York, 1986.

TABLE 8-10 Design Values for Spiral Dowels for Group II Species[a,b]

Outside Diameter of Dowel (in.)	Lateral Load Design Value (lb)	Withdrawal Load Design Value[c] (lb/in. of penetration)	
		Side Grain	End Grain
1/4	111	103	59
5/16	174	123	70
3/8	249	141	81
7/16	340	158	90
1/2	444	174	99
5/8	692	206	117
3/4	998	236	135
7/8	1,360	266	151
1	1,775	293	168

[a] All applicable provisions under the heading "Spiral Dowel Design Values" must be met to develop these values.
[b] For species included in group II species see Table 8-8.
[c] For spiral dowels installed or used in unseasoned material or exposed to the weather, use 25% of these design values ($C_M = 0.25$).
SOURCE: American Institute of Timber Construction, *Timber Construction Manual,* 3rd ed., John Wiley & Sons, Inc., New York, 1986.

METAL CONNECTORS

8-11 Split Rings A popular timber fastener is the split ring. Used for wood-to-wood connections, it is manufactured in sizes ranging from 2½- to 4-inch diameters. It must be installed or seated into precut grooves conforming to the size and design of the ring. The special wedge shape of the ring and the "tongue and groove" (split) in the ring result in good bearing and tight-fitting joints.

Design and load data for these connectors are tabulated for standard sizes in Table 8-11.

Refer to Table 8-12 for geometry factors for split rings and shear plate connections.

TABLE 8-11 Split Ring Connector Unit Design Values

Tabulated design values apply to ONE split ring and bolt in single shear.

Split ring diameter	Bolt diameter	Number of faces of member with connectors on same bolt	Net thickness of member	Loaded parallel to grain (0°)				Loaded perpendicular to grain (90°)			
				Design value, P, per connector unit and bolts, lbs.				Design value, Q, per connector unit and bolt, lbs.			
inches	inches	bolt	inches	Group A species	Group B species	Group C species	Group D species	Group A species	Group B species	Group C species	Group D species
2-1/2	1/2	1	1" minimum	2630	2270	1900	1640	1900	1620	1350	1160
			1-1/2" or thicker	3160	2730	2290	1960	2280	1940	1620	1390
		2	1-1/2" minimum	2430	2100	1760	1510	1750	1500	1250	1070
			2" or thicker	3160	2730	2290	1960	2280	1940	1620	1390
4	3/4	1	1" minimum	4090	3510	2920	2520	2840	2440	2040	1760
			1-1/2"	6020	5160	4280	3710	4180	3590	2990	2580
			1-5/8" or thicker	6140	5260	4380	3790	4270	3660	3050	2630
		2	1-1/2" minimum	4110	3520	2940	2540	2980	2450	2040	1760
			2"	4950	4250	3540	3050	3440	2960	2460	2120
			2-1/2"	5830	5000	4160	3600	4050	3480	2890	2500
			3" or thicker	6140	5260	4380	3790	4270	3660	3050	2630

1. Tabulated lateral design values (P,Q) for split ring connector units shall be multiplied to all applicable adjustment factors (see Table 8-13).

SOURCE: American Forest & Paper Association, Washington, D.C.

TABLE 8-12 Geometry Factors, C_Δ, for Split Ring and Shear Plate Connectors

		2-1/2″ Split Ring Connectors & 2-5/8″ Shear Plate Connectors				4″ Split Ring Connectors & 4″ Shear Plate Connectors			
		Parallel to grain loading		Perpendicular to grain loading		Parallel to grain loading		Perpendicular to grain loading	
		Minimum for Reduced Design Value	Minimum for Full Design Value	Minimum for Reduced Design Value	Minimum for Full Design Value	Minimum for Reduced Design Value	Minimum for Full Design Value	Minimum for Reduced Design Value	Minimum for Full Design Value
Edge Distance	Unloaded Edge C_Δ	1-3/4″ 1.0	1-3/4″ 1.0	1-3/4″ 1.0	1-3/4″ 1.0	2-3/4″ 1.0	2-3/4″ 1.0	2-3/4″ 1.0	2-3/4″ 1.0
	Loaded Edge C_Δ	1-3/4″ 1.0	1-3/4″ 1.0	1-3/4″ 0.83	2-3/4″ 1.0	2-3/4″ 1.0	2-3/4″ 1.0	2-3/4″ 0.83	3-3/4″ 1.0
End Distance	Tension Member C_Δ	2-3/4″ 0.625	5-1/2″ 1.0	2-3/4″ 0.625	5-1/2″ 1.0	3-1/2″ 0.625	7″ 1.0	3-1/2″ 0.625	7″ 1.0
	Compression Member C_Δ	2-1/2″ 0.625	4″ 1.0	2-3/4″ 0.625	5-1/2″ 1.0	3-1/4″ 0.625	5-1/2″ 1.0	3-1/2″ 0.625	7″ 1.0
Spacing	Spacing parallel to grain C_Δ	3-1/2″ 0.5	6-3/4″ 1.0	3-1/2″ 1.0	3-1/2″ 1.0	5″ 0.5	9″ 1.0	5″ 1.0	5″ 1.0
	Spacing perpendicular to grain C_Δ	3-1/2″ 1.0	3-1/2″ 1.0	3-1/2″ 0.5	4-1/4″ 1.0	5″ 1.0	5″ 1.0	5″ 0.5	6″ 1.0

SOURCE: American Forest & Paper Association, Washington, D.C.

TABLE 8-13 Applicability of Adjustment Factors for Connections

		Load Duration Factor [1]	Wet Service Factor [2]	Tempera-ture Factor	Group Action Factor	Geometry Factor [3]	Penetration Depth Factor [3]	End Grain Factor [3]	Metal Side Plate Factor [3]	Diaphragm Factor [3]	Toe-Nail Factor [3]
Bolts	Z'=(Z)	(C_D)	(C_M)	(C_t)	(C_g)	($C\Delta$)	•	•	•	•	•
Lag Screws	W'=(W)	(C_D)	(C_M)	(C_t)	•	•	•	(C_{eg})	•	•	•
	Z'=(Z)	(C_D)	(C_M)	(C_t)	(C_g)	($C\Delta$)	(C_d)	(C_{eg})	•	•	•
Split Ring and Shear Plate Connectors	P'=(P)	(C_D)	(C_M)	(C_t)	(C_g)	($C\Delta$)	(C_d)	•	(C_{st})	•	•
	Q'=(Q)	(C_D)	(C_M)	(C_t)	($C\Delta$)	(C_d)	•	•	•	•	
Wood Screws	W'=(W)	(C_D)	(C_M)	(C_t)	•	•	•	•	•	•	•
	Z'=(Z)	(C_D)	(C_M)	(C_t)	•	•	(C_d)	(C_{eg})	•	•	•
Nails and Spikes	W'=(W)	(C_D)	(C_M)	(C_t)	•	•	•	•	•	•	(C_{tn})
	Z'=(Z)	(C_D)	(C_M)	(C_t)	•	•	(C_d)	(C_{eg})	•	(C_{di})	(C_{tn})
Metal Plate Connectors	Z'=(Z)	(C_D)	(C_M)	(C_t)	•	•	•	•	•	•	•
Drift Bolts and Drift Pins	W'=(W)	(C_D)	(C_M)	(C_t)	•	•	•	(C_{eg})	•	•	•
	Z'=(Z)	(C_D)	(C_M)	(C_t)	(C_g)	($C\Delta$)	(C_d)	(C_{eg})	•	•	•
Spike Grids	Z'=(Z)	(C_D)	(C_M)	(C_t)	•	($C\Delta$)	•	•	•	•	•

1. The load duration factor, C_D, shall not exceed 1.6 for connections
2. The wet service factor, C_M, shall not apply to toe-nails loaded in withdrawal
3. Specific information concerning geometry factors ($C\Delta$), penetration depth factors (C_d), end grain factors (C_{eg}), metal side plate factors (C_{st}), diaphragm factors (C_{di}) and toe-nail factors (C_{tn}) is provided in Parts VIII, IX, X, XI, XII and XIV of NDS® (Structural Lumber/Glued Laminated Timber/Timber Piles/Connections) 1991 National Design Specifications for Wood Const.

SOURCE: American Forest & Paper Association, Washington, D.C.

TABLE 8-14 Wet Service Factors, C_M, for Connections

Fastener Type	Condition of Wood[1]		Wet Service Factor C_M
	At time of fabrication	In service	
Split Ring or Shear Plate[2] Connectors	Dry	Dry	1.0
	Partially seasoned	Dry	see Footnote 3
	Wet	Dry	0.8
	Dry or wet	Partially seasoned or wet	0.67
Bolts or Lag Screws	Dry	Dry	1.0
	Partially seasoned or wet	Dry	see Footnote 4
	Dry or wet	Exposed to weather	0.75
	Dry or wet	Wet	0.67
Laterally loaded Drift Bolts or Drift Pins	Dry or wet	Dry	1.0
	Dry or wet	Partially seasoned or wet, or subject to wetting and drying	0.7
Wood Screws	Dry or wet	Dry	1.0
	Dry or wet	Exposed to weather	0.75
	Dry or wet	Wet	0.67
Common Wire Nails, Box Nails or Common Wire Spikes			
— Withdrawal Loads	Dry	Dry	1.0
	Partially seasoned or wet	Wet	1.0
	Partially seasoned or wet	Dry	0.25
	Dry	Subject to wetting and drying	0.25
— Lateral Loads	Dry	Dry	1.0
	Partially seasoned or wet	Dry or wet	0.75
	Dry	Partially seasoned or wet	0.75
Threaded, Hardened Steel Nails	Dry or wet	Dry or wet	1.0
Metal Connector Plates	Dry	Dry	1.0
	Partially seasoned or wet	Dry or wet	0.8

1. "Conditions of wood" are defined as follows for determining wet service factors for connections:
   "Dry" wood has a moisture content ≦ 19%.
   "Wet" wood has a moisture content ≥ 30% (approximate fiber saturation point).
   "Partially seasoned" wood has 19% < moisture content < 30%.
   "Exposed to weather" means that the wood will vary in moisture content from dry to partially seasoned, but is not expected to reach the fiber saturation point at times when the connection is supporting full design load.
   "Subject to wetting and drying" means that the wood will vary in moisture content from dry to partially seasoned or wet, or vice versa, with consequent effects on the tightness of the connection.
2. For split ring or shear plate connectors, moisture content limitations apply to a depth of 3/4" below the surface of the wood.
3. When split ring or shear plate connectors are installed in wood that is partially seasoned at the time of fabrication, but that will be dry before full design load is applied, proportional intermediate wet service factors shall be permitted to be used.
4. When bolts or lag screws are installed in wood that is wet at the time of fabrication, but that will be dry before full design load is applied, the following wet service factors, C_M, shall apply:

SOURCE: American Forest & Paper Association, Washington, D.C.

8-12 Toothed Rings Toothed rings are generally used only when power tools are at the job site or other connectors are not available. These rings provide a wood-to-wood connection and are standardized in 2-, 2⅝-, 3⅜-, and 4-inch diameters. Great care must be taken when installing these rings to avoid wood splitting. Slow and even penetration and good alignment are musts. See Fig. 8-4.

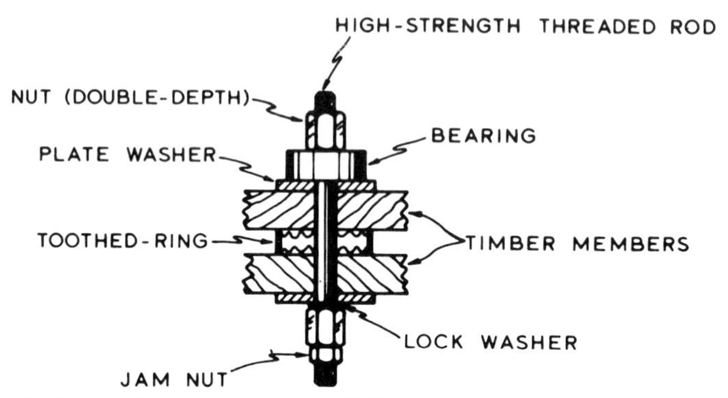

FIG. 8-4 Typical toothed-ring installation.

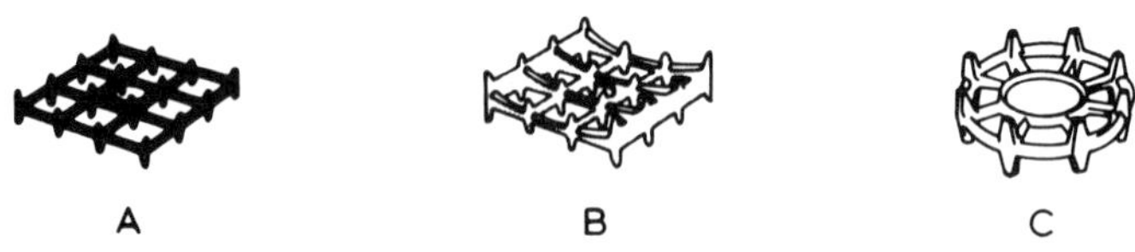

FIG. 8-5 Spike grids. *(a)* Flat, *(b)* single curve, *(c)* circular.

8-13 Spike Grids Circular, flat, or single-curve spike grids are primarily used for wood-to-wood connections on heavy timber construction, typically highway or bridge structures. They are generally installed in the same manner as toothed rings. See Fig. 8-5.

DESIGN FACTORS

8-14 Adjustment Factor Refer to table 8-13 for applicability of adjustment factors for various connections.

8-15 Wet Service Factor Table 8-14 tabulates wet service factors for the standard fasteners used in wood connections.

METAL FRAMING DEVICES

Metal framing devices or connectors have become more widely known in recent years. Firms specializing in roof-truss fabrication and contractors doing on-site fabrication are using these fastener brackets and hangers more frequently to reduce labor costs and ensure sound construction connections.

8-16 Clamping Plates and Straps Clamping plates are made from thin-gage sheet metal and are primarily for wood-to-wood connections. Generally flat clamping plates are approximately 5¼ inches square with projecting teeth on both sides to grip the timber. Flanged clamping plates are usually 8 inches wide and have teeth projecting only on the face. Installation is achieved by driving the teeth into the timber with a sledge. See Fig. 8-6.

Straps or ties are simply flat strips of metal with uniformly spaced holes. Standard sizes vary from 1¼ to 2⁵⁄₁₆ inches wide. They are fabricated from 16 to 20 gage sheet metal. Lengths vary, but units of 16 inches center to center seem to be most often observed. Heavy tie straps are ³⁄₁₆ to ¼ inch thick with a range of 2½- to 6-inch width.

Twist straps are used to secure joists to a strong back and similar structural applications. They may be purchased either with right-hand or left-hand twist.

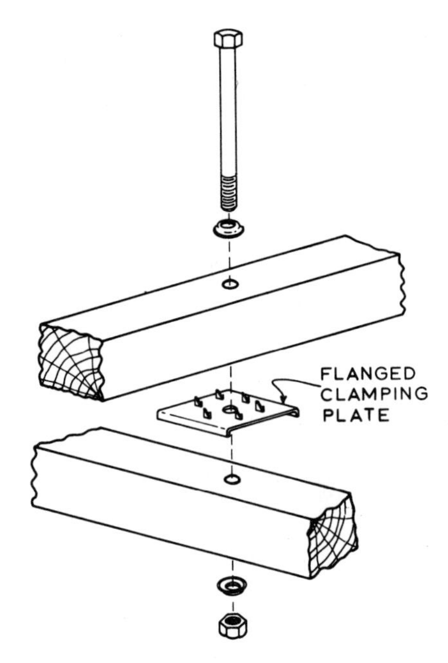

FIG. 8-6 Typical installation of flanged clamping plate.

8-17 Framing Anchors and Angles Framing anchors are generally fabricated from 18 gage galvanized steel. The specifications governing these anchors are in the Uniform Building Code of the International Conference of Building Officials.

The design of the framing anchor shown in Fig. 8-7 fills dozens of needs. It will serve as a connector for joists to beams, joists to plate, joists to rafter, beams to posts, or chimney framing.

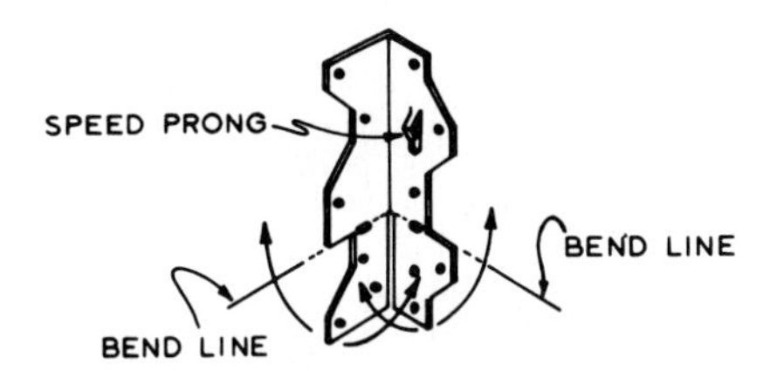

FIG. 8-7 Standard design of a framing anchor. *(Simpson Company.)*

Angle gussets (see Fig. 8-8) are exceptionally versatile. They have a three-way connection feature. Also shown in Fig. 8-8 is a heavy angle which is used in structural connections, either bolted or nailed.

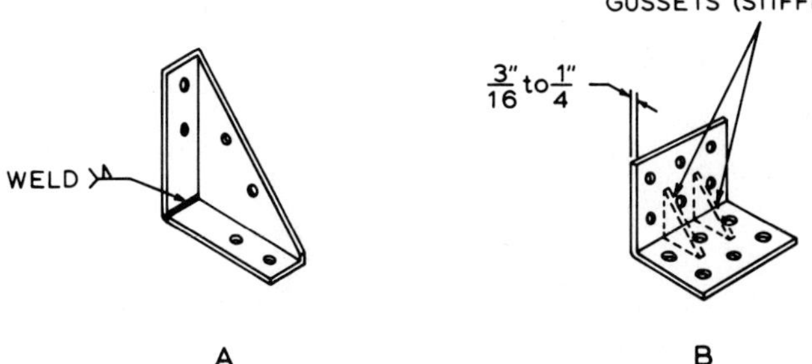

FIG. 8-8 Typical angles. *(a)* Angle gusset, *(b)* heavy angle. *(Simpson Company.)*

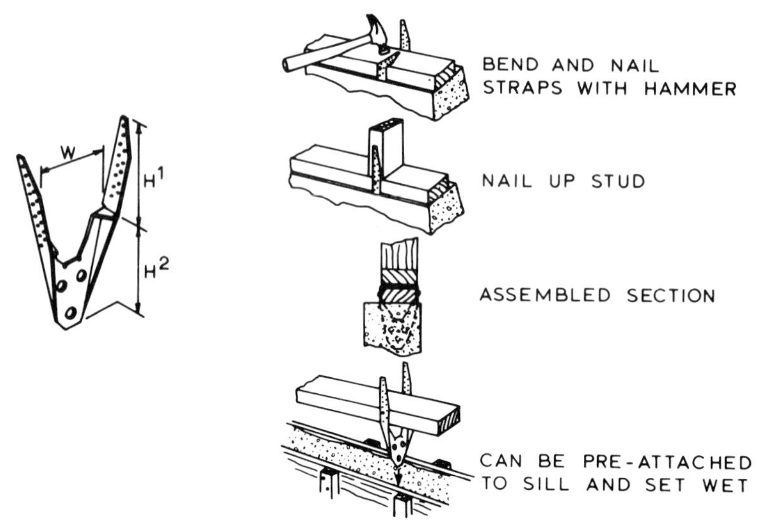

FIG. 8-9 Typical mudsill anchor and installation. *(Simpson Company.)*

8-18 Mudsill Anchors Mudsill anchors eliminate anchor bolts, drilling, special tools, and accurate settings. Figure 8-9 illustrates the installation of a mudsill.

Material of construction is 16 gage prime quality galvanized steel. Spacing of mudsills is generally on 6-ft centers, with local codes prevailing.

8-19 Base Anchors Base anchor is a general term that includes post bases, column bases, and hold-downs. See Fig. 8-10 for a pictorial review of these standard designs.

Refer to the manufacturer's technical manuals and the Uniform Building Code (Research Recommendation 1211 of the International Conference of Building Officials) for loading data.

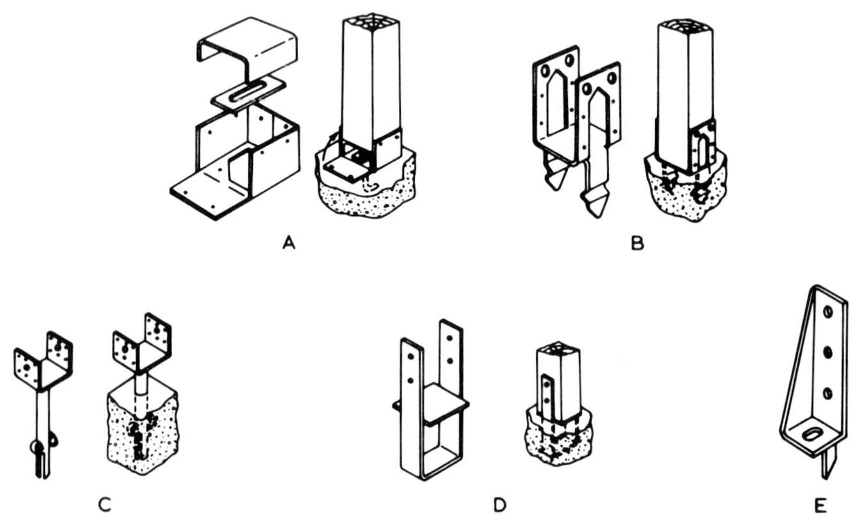

FIG. 8-10 Base anchors. *(a)* Post base (adjustable), *(b)* post base (heavy), *(c)* post base (elevated), *(d)* column base, *(e)* hold-down. *(Simpson Company.)*

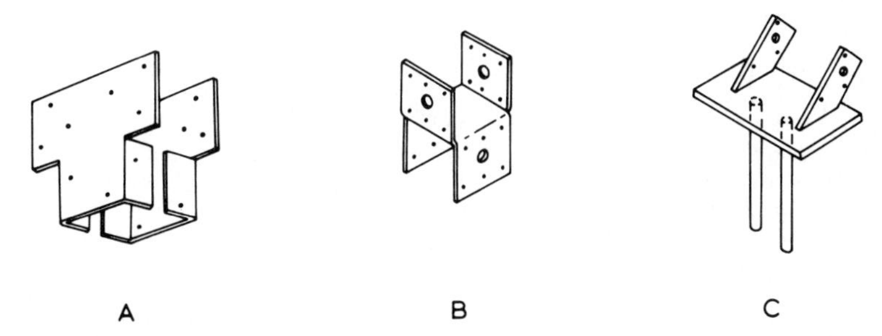

FIG. 8-11 Typical caps and seats. *(a)* Twin post cap, *(b)* post cap/base, *(c)* beam seat. *(Simpson Company.)*

8-20 Caps and Seats Another series of prefabricated fasteners includes caps and seats which are used to connect timber components.

Figure 8-11 shows a variety of designs, each suitable for a particular installation. Consult the manufacturer's catalog and technical literature for allowable loads and recommended nailing requirements.

8-21 Hangers Purlin and joist hangers are standard prefabricated connector components. Generally made from 12 gage to 16 gage sheet metal, they are galvanized to protect the base material.

Figure 8-12 illustrates some base configurations and designs. Consult manufacturers' literature for particular hanger types, sizes, and allowable loads.

8-22 Wall Bracing and Bridging Wall bracing and bridging is usually specified by architectural plans and design or by structural engineering layouts. Straps, angles, or specified weldments are employed to tie the structure together. These are not considered fasteners in the true sense. They are structural ties to reinforce critical sections.

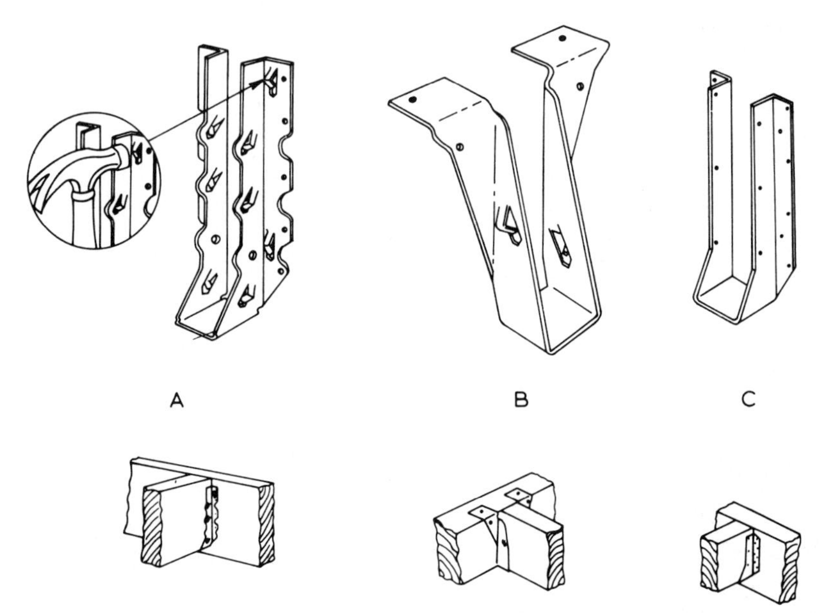

FIG. 8-12 Typical joist hangers. *(a)* Prong joist hanger, *(b)* joist and purlin hanger, *(c)* heavy-duty joist hanger. *(Simpson Company.)*

CHOICE OF FASTENING

Structural considerations, types of loads, lumber size, and assembly problems are basic factors in choosing a fastening component or connecting system.

8-23 Structural Requirements Structural requirements vary so greatly that only a general discourse can be included here. Federal, state, and local codes are the main sources for structural loading requirements. Societies and technical institutes can only test and recommend through technical publications, which must then be approved and adopted by the regulating agencies. At this point, the requirements become legal and must be followed by architects, consulting engineers, and designers. Each structural connection problem is an individual case that depends on controlling code requirements and intended application. Seismic, hurricane, frost, and wind loads, as well as general live and dead loadings, are just a few of the conditions that must be considered. Economic and construction factors are becoming more and more basic items in selecting the final design.

8-24 Type of Loading As mentioned in the preceding section, load type is a major consideration in selecting the best connection system. Dead loads, live loads, seismic forces, wind loads, overturning moments, snow loads, hurricane forces, shear resistance, bending moments, and material stresses are just a few of the loadings and forces to be dealt with when analyzing a proposed connection. Important as the various connection systems may be, the overall structure is the major unit to be considered.

8-25 Lumber Thickness Lumber thicknesses of 2 inches or greater are most generally used in structural work. All standard fasteners are suitable for these dimensions.

See Table 8-13 for standard lumber dimensions.

8-26 Assembly Considerations When joints are to be made on the site, consideration must be given to selecting fasteners which can be installed with ease and a minimum of effort.

TABLE 8-15 Standard Lumber Dimensions

TIMBER
AMERICAN STANDARD SIZES
PROPERTIES FOR DESIGNING
NATIONAL LUMBER MANUFACTURERS ASSOCIATION

Nominal Size	American Standard Dressed Size	Area of Section	Weight per Foot	Moment of Inertia	Section Modulus
In.	In.	In.^{2}	Lb.	In.^{4}	In.^{3}
2 x 4	1⅝x3⅝	5.89	1.64	6.45	3.56
6	5⅝	9.14	2.54	24.1	8.57
8	7½	12.2	3.39	57.1	15.3
10	9½	15.4	4.29	116	24.4
12	11½	18.7	5.19	206	35.8
14	13½	21.9	6.09	333	49.4
16	15½	25.2	6.99	504	65.1
18	17½	28.4	7.90	726	82.9
3 x 4	2⅝x3⅝	9.52	2.64	10.4	5.75
6	5⅝	14.8	4.10	38.9	13.8
8	7½	19.7	5.47	92.3	24.6
10	9½	24.9	6.93	188	39.5
12	11½	30.2	8.39	333	57.9
14	13½	35.4	9.84	538	79.7
16	15½	40.7	11.3	815	105
18	17½	45.9	12.8	1172	134
4 x 4	3⅝x3⅝	13.1	3.65	14.4	7.94
6	5⅝	20.4	5.66	53.8	19.1
8	7½	27.2	7.55	127	34.0
10	9½	34.4	9.57	259	54.5
12	11½	41.7	11.6	459	79.9
14	13½	48.9	13.6	743	110
16	15½	56.2	15.6	1125	145
18	17½	63.4	17.6	1619	185
6 x 6	5½x5½	30.3	8.40	76.3	27.7
8	7½	41.3	11.4	193	51.6
10	9½	52.3	14.5	393	82.7
12	11½	63.3	17.5	697	121
14	13½	74.3	20.6	1128	167
16	15½	85.3	23.6	1707	220
18	17½	96.3	26.7	2456	281
20	19½	107.3	29.8	3398	349
8 x 8	7½x7½	56.3	15.6	264	70.3
10	9½	71.3	19.8	536	113
12	11½	86.3	23.9	951	165
14	13½	101.3	28.0	1538	228
16	15½	116.3	32.0	2327	300
18	17½	131.3	36.4	3350	383
20	19½	146.3	40.6	4634	475
22	21½	161.3	44.8	6211	578

Nominal Size	American Standard Dressed Size	Area of Section	Weight per Foot	Moment of Inertia	Section Modulus
In.	In.	In.^{2}	Lb.	In.^{4}	In.^{3}
10x10	9½x9½	90.3	25.0	679	143
12	11½	109	30.3	1204	209
14	13½	128	35.6	1948	289
16	15½	147	40.9	2948	380
18	17½	166	46.1	4243	485
20	19½	185	51.4	5870	602
22	21½	204	56.7	7868	732
24	23½	223	62.0	10274	874
12x12	11½x11½	132	36.7	1458	253
14	13½	155	43.1	2358	349
16	15½	178	49.5	3569	460
18	17½	201	55.9	5136	587
20	19½	224	62.3	7106	729
22	21½	247	68.7	9524	886
24	23½	270	75.0	12437	1058
14x14	13½x13½	182	50.6	2768	410
16	15½	209	58.1	4189	541
18	17½	236	65.6	6029	689
20	19½	263	73.1	8342	856
22	21½	290	80.6	11181	1040
24	23½	317	88.1	14600	1243
16x16	15½x15½	240	66.7	4810	621
18	17½	271	75.3	6923	791
20	19½	302	83.9	9578	982
22	21½	333	92.5	12837	1194
24	23½	364	101	16763	1427
18x18	17½x17½	306	85.0	7816	893
20	19½	341	94.8	10813	1109
22	21½	376	105	14493	1348
24	23½	411	114	18926	1611
26	25½	446	124	24181	1897
20x20	19½x19½	380	106	12049	1236
22	21½	419	116	16150	1502
24	23½	458	127	21089	1795
26	25½	497	138	26945	2113
28	27½	536	149	33795	2458
24x24	23½x23½	552	153	25415	2163
26	25½	599	166	32472	2547
28	27½	646	180	40727	2962
30	29½	693	193	50275	3408

All properties and weights given are for dressed size only.
The weights given above are based on assumed average weight of 40 pounds per cubic foot.

SOURCE: American Institute of Steel Construction.

LOG CONNECTIONS

Log building construction has become very popular in recent years. Precut log home kits are fabricated by many companies and offer a wide variety of designs. Custom designs are also a key to their increased sales.

This sub-section will feature typical connections employed by conventional log construction. However, the wide variety of connections are too numerous for all of them to be included in this sub-section. The editor suggests the reader consult other publications if a more in-depth study is desired.

8-27 Basic Foundation Connections Foundations for log building construction should be made of concrete. The single-pour footing/foundation is simple and generally less expensive. This slab-on-grade floor with a perimeter footing is illustrated in Fig. 8-13 showing typical log positioning.

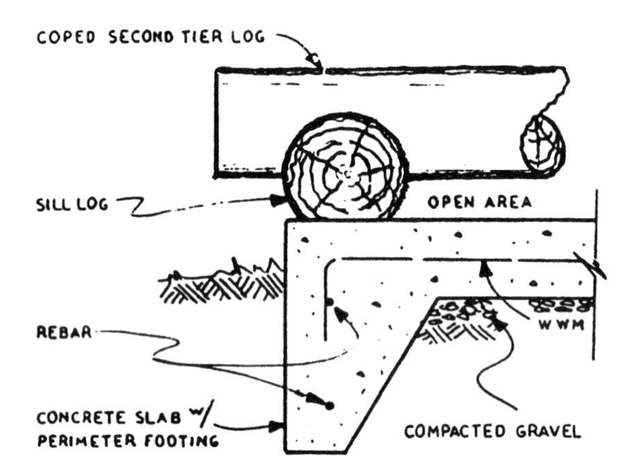

FIG. 8-13 Slab-on-grade foundation.

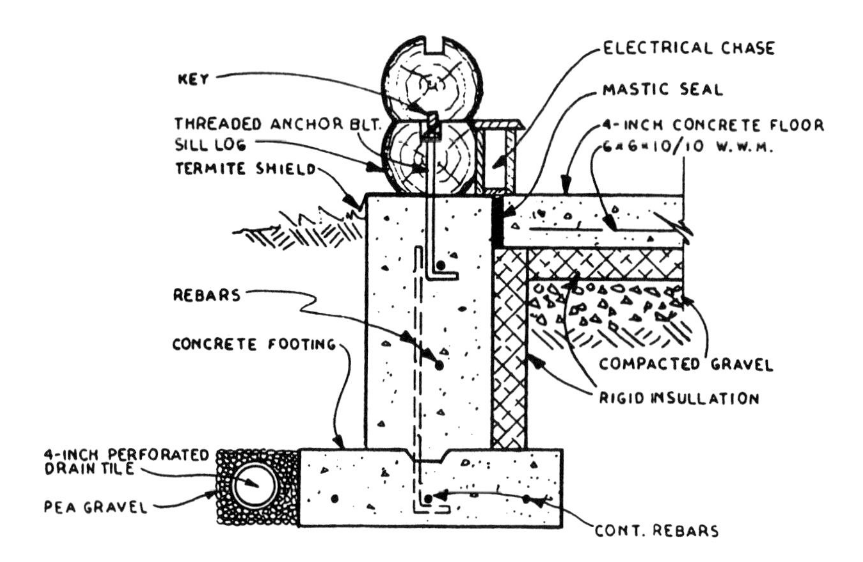

FIG. 8-14 Two-piece footing/foundation.

The two-piece footing/foundation usually requires two pours. Fig. 8-14 shows a complex design with insulation, drain tile, reinforcing bars and related components. Note that the sill log is securely fastened to the foundation with a threaded anchor bolt. The second wall log is positioned with a wood key strip. An electrical chase is conveniently located at the cove area.

Figure 8-15 details a foundation connection for a wood floor system. The sheet metal flashing and termite shield are easily installed. The sill log straddles the starter rail.

Piers can be made from a wide range of materials. However, concrete is the most lasting. Refer to Fig. 8-16 for four standard designs. Sizing and depth are site specific, but general configurations are similar to those shown.

Prior to finalizing foundation design, consult local building code and other applicable regulations.

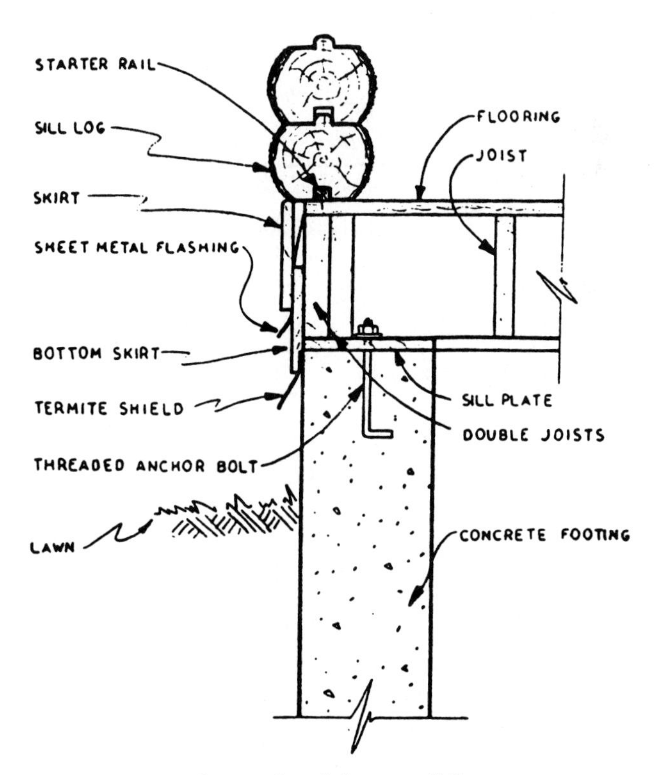

FIG. 8-15 Foundation detail for wood floor.

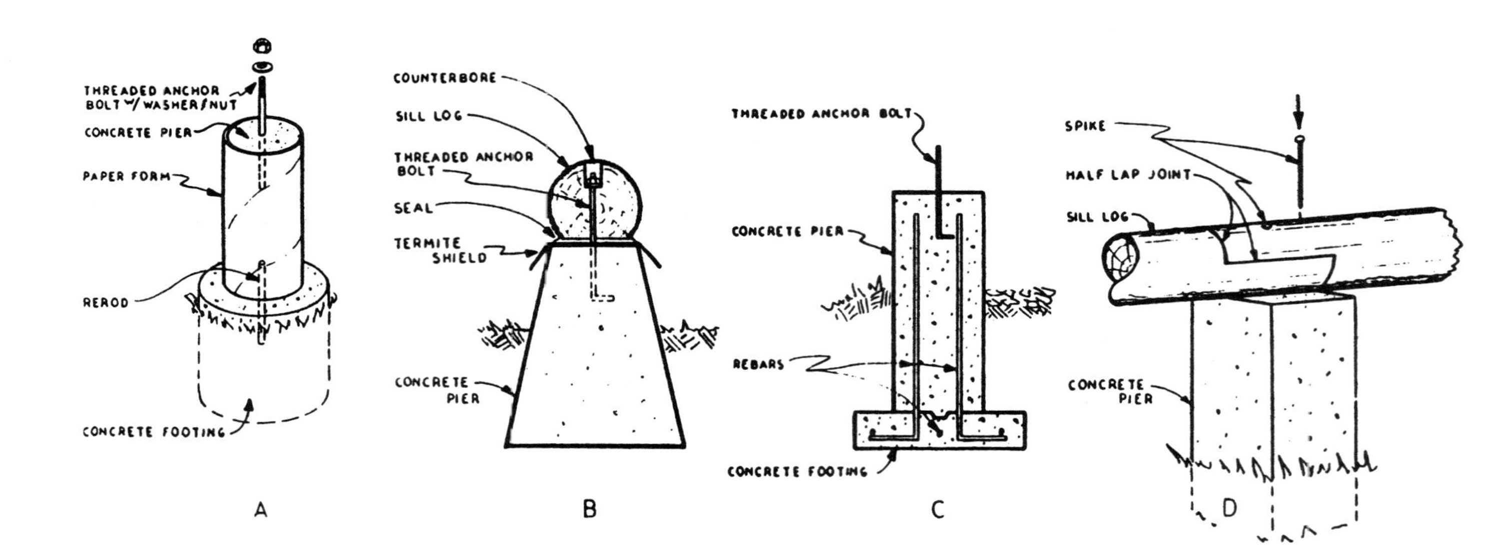

FIG. 8-16 Basic pier designs.

8-28 Corner Joints There are a wide assortment of standard corner joints used in log construction. One of the most common is the saddle notch which is illustrated in Fig. 8-17. The half-cut notch is a modified version and is pictured in Fig. 8-18.

The tenon-joint notch and the double notch corner joints, shown in Figs. 8-19 and 8-20 respectfully, require more precision and effort to obtain a good fit. However, these joints yield a positive lock.

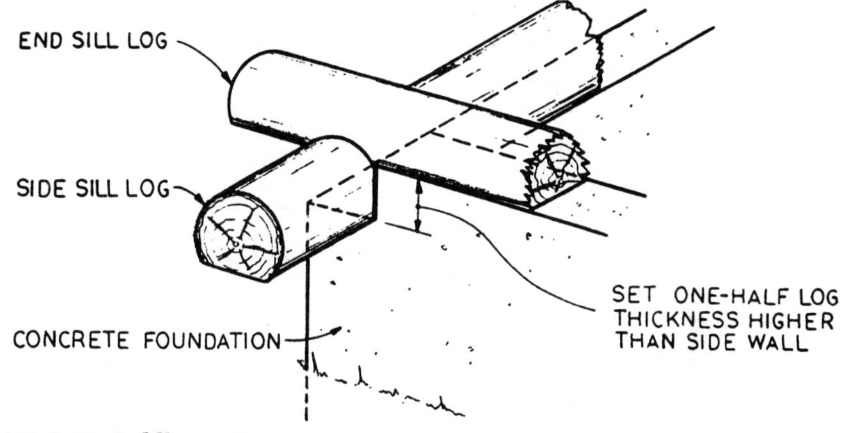

FIG. 8-17 Saddle notch or common joint.

FIG. 8-18 Half-cut notch.

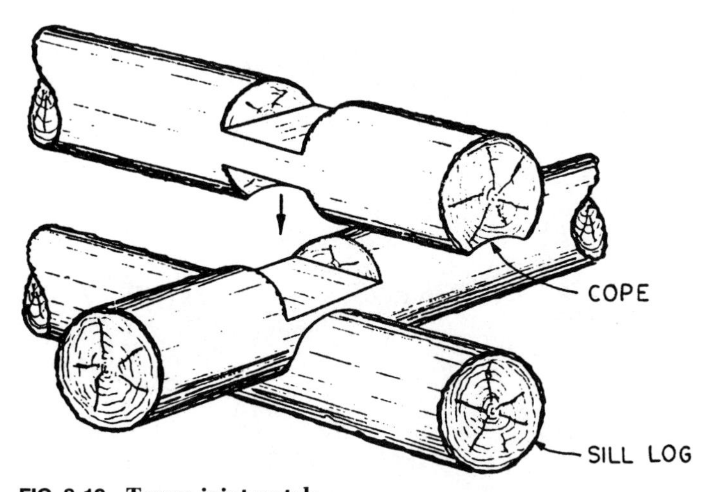

FIG. 8-19 Tenon-joint notch.

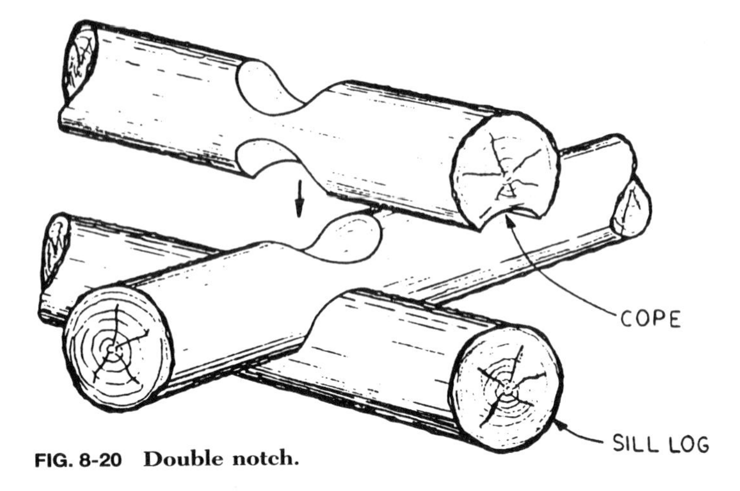

FIG. 8-20 Double notch.

Figure 8-21 pictures a dovetail notch which certainly requires expert skill to successfully fabricate.

The sharp notch is occasionally employed, but also requires precise workmanship to execute a tight fit. See Fig. 8-22.

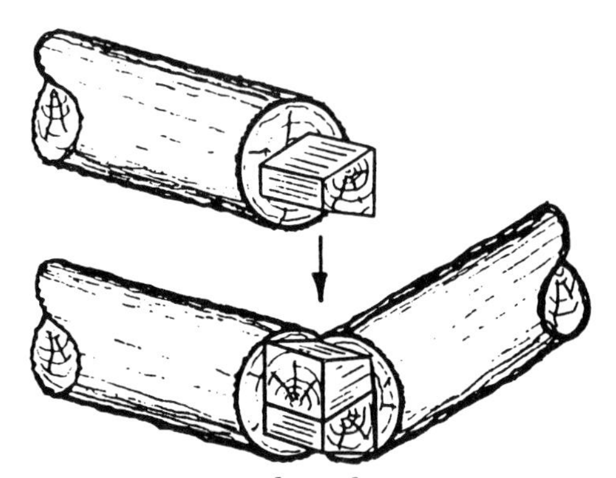

FIG. 8-21 Dovetail notch.

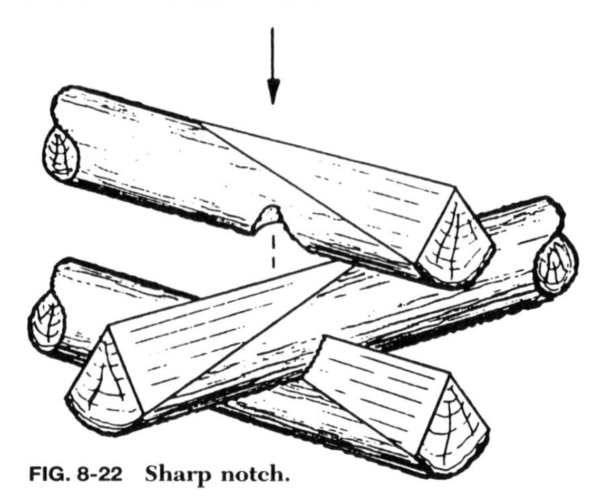

FIG. 8-22 Sharp notch.

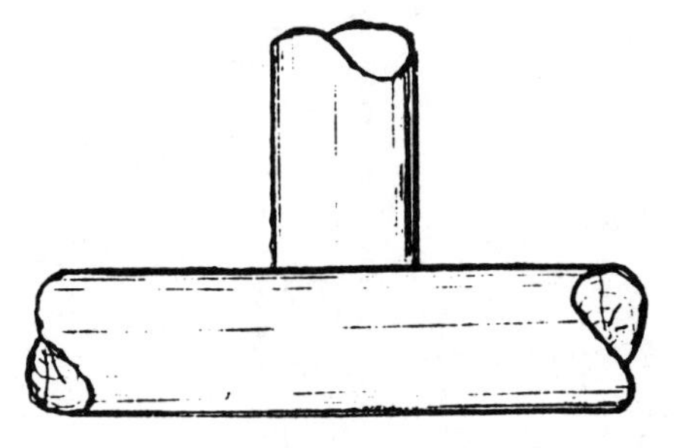

FIG. 8-23 **Butt corner notch.**

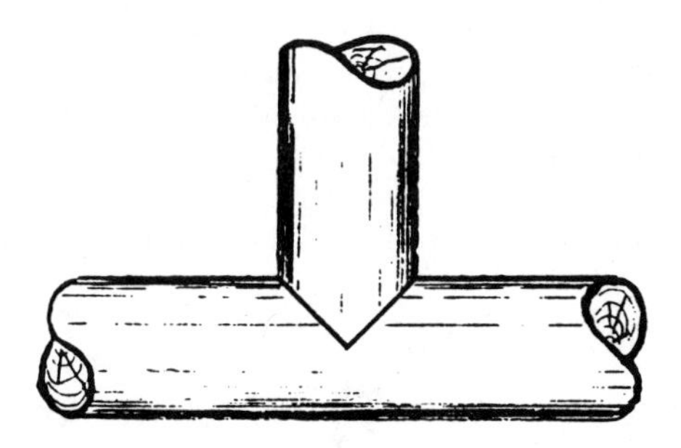

FIG. 8-24 **A and V notch.**

The butt corner notch (Fig. 8-23) and the A & V notch (Fig. 8-24) are simple connections but have some obvious drawbacks.

The plank corner post notch has some structural advantages and is relatively easy to assemble. See Fig. 8-25.

Corner post notches are used where tight fits are not as important. Refer to Fig. 8-26 for a typical detail.

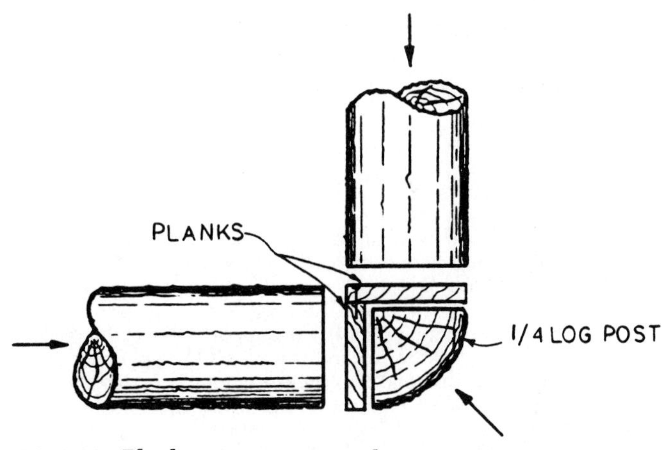

FIG. 8-25 **Plank corner post notch.**

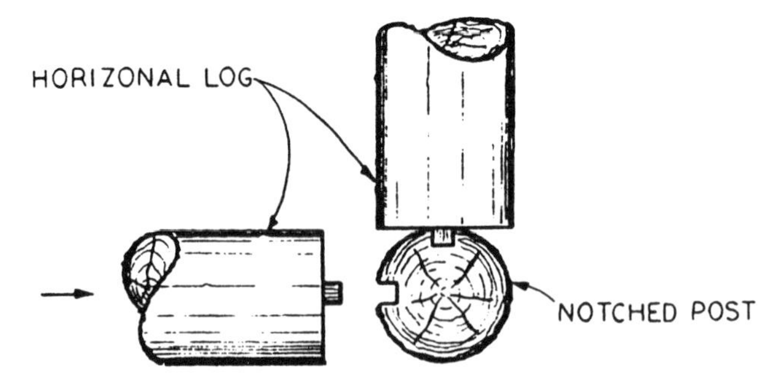

FIG. 8-26 Corner post notch.

8-29 Connecting Joists and Girders Figure 8-27 illustrates a common joist connection using dimension lumber for floor joists. Spacing of joists should be determined by structural analysis.

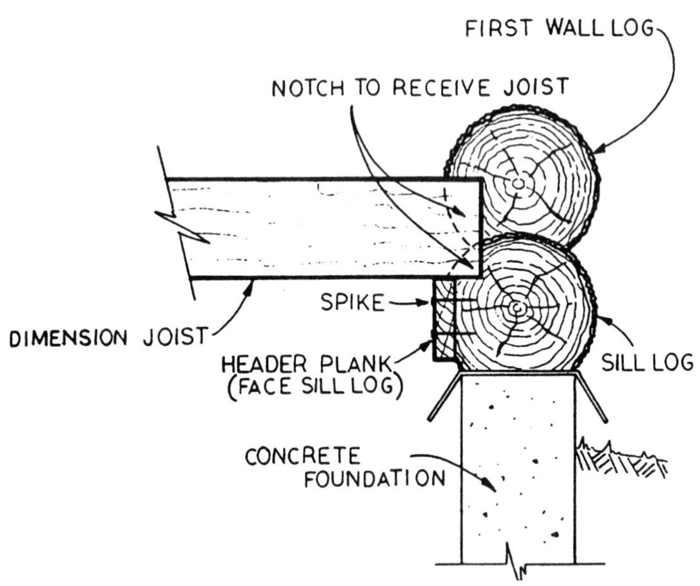

FIG. 8-27 Common joist connection.

Logs used for joists is typical in rustic log construction. The top area of the joist log should be flattened or planed smooth to receive the subflooring. Figure 8-28 depicts a tenon style used with sill logs. A typical second floor log joist can be connected as shown in Fig. 8-29.

Log girders, also, should be flattened on top to support intersecting joist and maintain a level floor. Refer to Fig. 8-30 for a combination sill log and concrete foundation connection. Utilizing girders for second floor construction is common in larger spans. Figure 8-31 pictures an often used connection when flattened wall log construction is used.

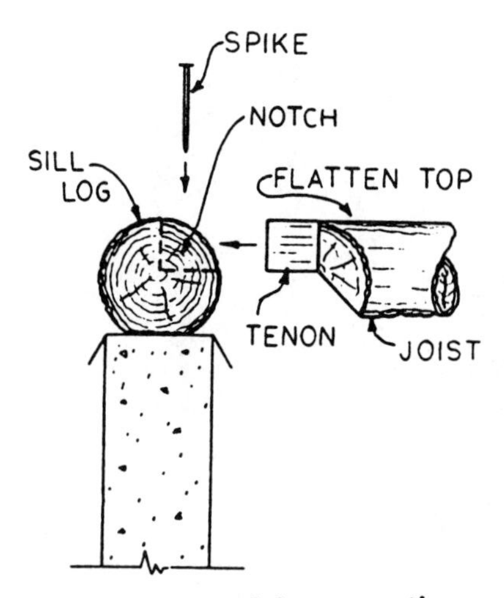

FIG. 8-28 Tenon joist connection.

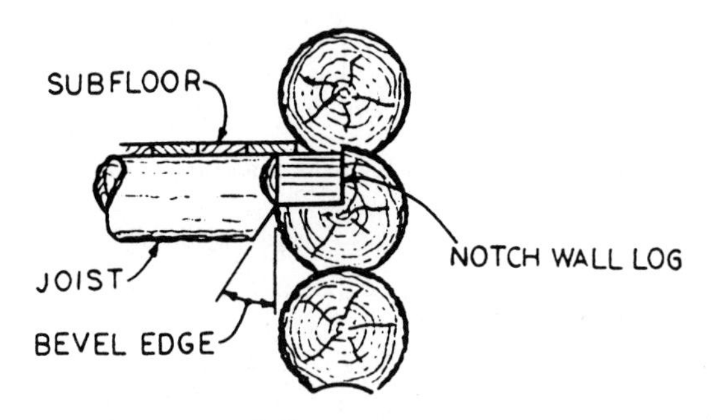

FIG. 8-29 Second floor joist connection.

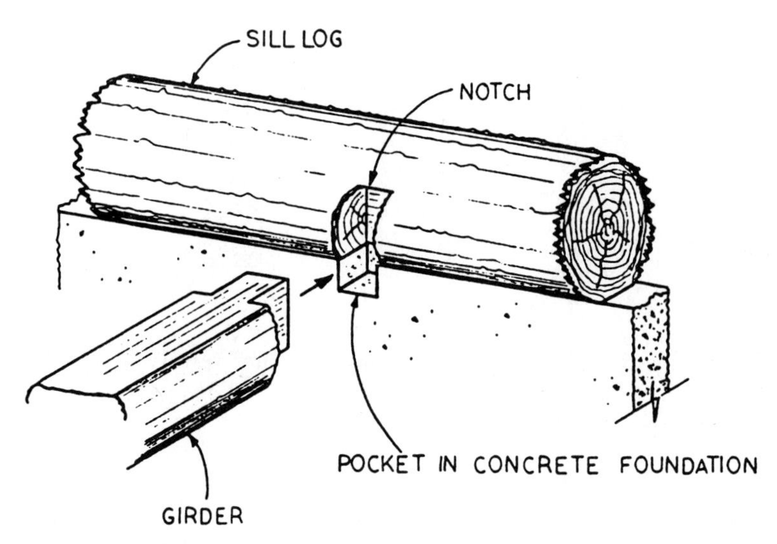

FIG. 8-30 Sill log and foundation girder connection.

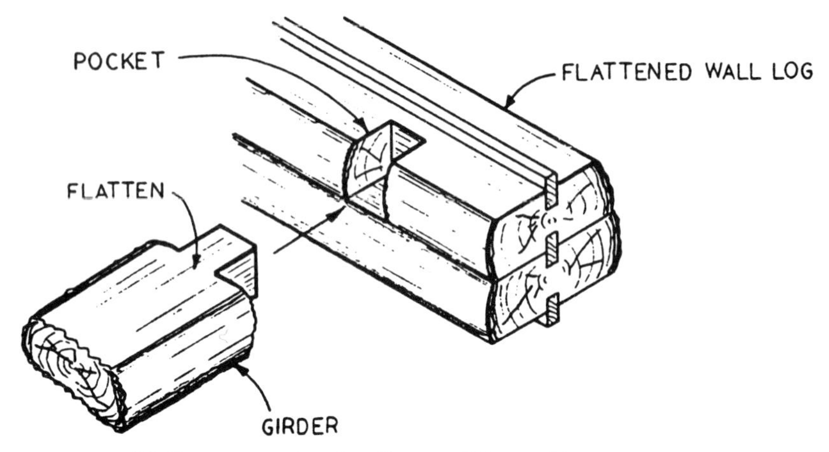

FIG. 8-31 Girder connection for flattened log wall.

8-30 Typical Wall Assembly Typical exterior log wall stacking is illustrated in Fig. 8-32. The cupped (A) vertical design requires the bottom of each log to be coped to security rest on top of previously positioned log. The next design (B) has three sides flattened or squared to stabilize the wall. The final design illustrated (C) has only the top and bottom planed.

Sline connections are liberally used in log wall construction. Refer to Fig. 8-33 for several standard styles.

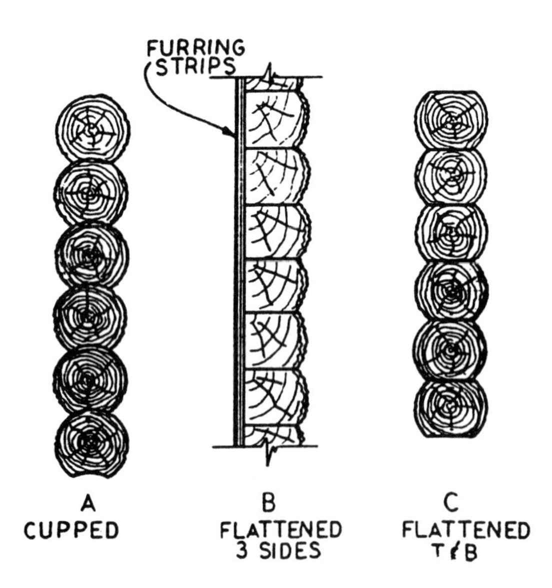

FIG. 8-32 Typical exterior log walls.

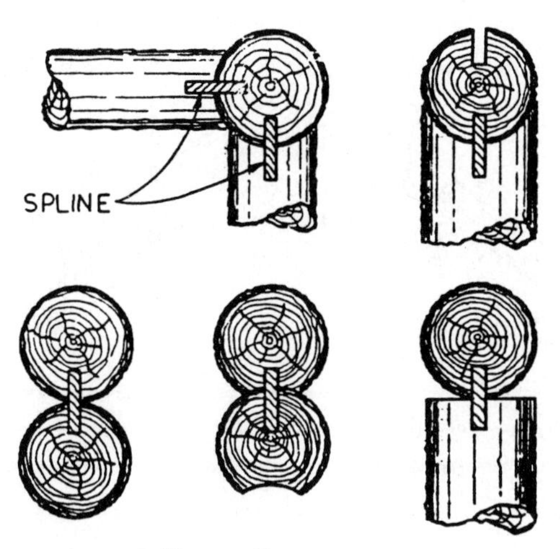

FIG. 8-33 Spline wall connections.

A more complex design is the dovetail construction. This design is much more difficult to construct. The logs must be hewed and square-milled with great care to insure proper fit. See Fig. 8-34.

The conventional mortise and tenon joint is generally used for interior walls that connect to exterior walls. Figure 8-35 pictures this type of easy-to-construct joint.

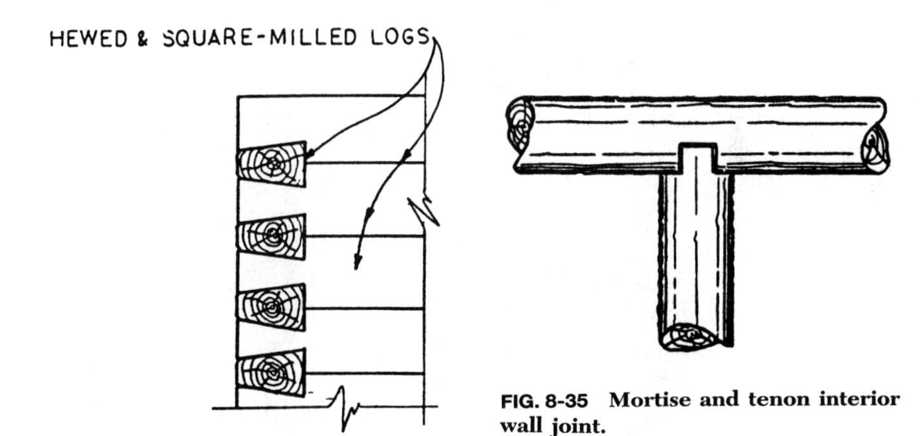

FIG. 8-35 Mortise and tenon interior wall joint.

FIG. 8-34 Dovetail connection.

8-31 Roof Members Logs used as roof framing components generally utilize the gable roof design. Experts recommend using this simple method because complex roof systems are difficult and expensive to construct. Refer to Fig. 8-36 for basic details and assembly. The sheathing and shingles are not illustrated. The purlins and center ridgepole are supported at the gable ends and at intermediate points where necessary. Sizing and spacings are dictated by the designer for specific structures. Notching styles, nailing requirements, and bolting details are also at the discretion of the designer.

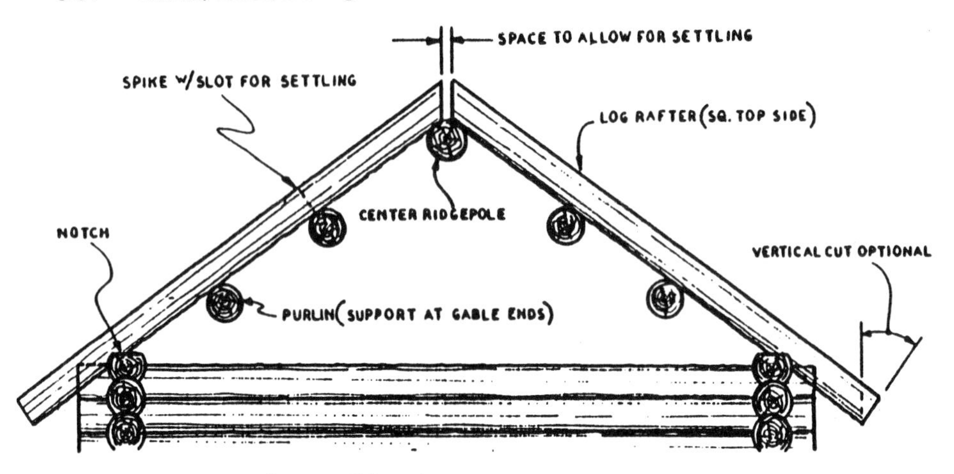

FIG. 8-36 **Common log roof framing.**

8-32 Stairs Stairs using half or hewed logs display the rustic look. See Fig. 8-37 for a typical exterior example. A staircase constructed by stacking log sections using a steel pipe support is easy to build. They can be either straight or spiral (one support pipe) if space is limited. However, it should be noted that it is difficult to maintain uniform tread and riser dimensions using logs and therefore increases the potential for accidents.

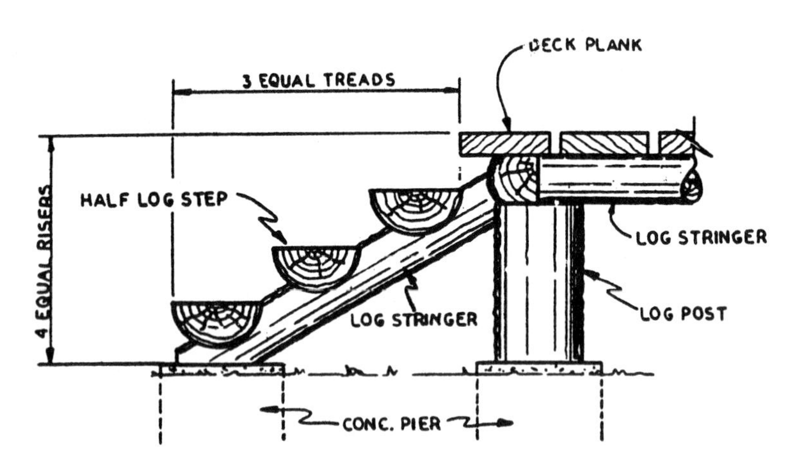

FIG. 8-37 **Rustic log stairs.**

8-33 Special Connections There are literally hundreds of special connections used for connecting logs and limited space in this section precludes their inclusion.

Section 9
Structural Steel Connections

edited by

ROBERT O. PARMLEY, P.E.

Consulting Engineer, Ladysmith, Wisconsin

This section includes all standard types of joints used to connect structural steel. Rivets, bolts, and welded connections are emphasized in other sections of this handbook. Therefore, the general topic will be an all-inclusive treatment of "connections" as standardized by the American Institute of Steel Construction.

Recommendations in this section are extracted from the AISC *Manual of Steel Construction*, 7th ed., April 1975. Typical standard connections and tabular material are shown to give the design engineer a general outline in selecting the proper type of fastening system. Additionally, detailing practices, bracing formulas, circle data, net areas, and related material are included to assist the structural steel detailer in layout work.

9-1 Bolted or Riveted Connections Structural rivets (carbon steel) are used to connect steel members (see sec. 9-11). The use of rivets is not as prevalent as in former years. They have been generally replaced by high-strength bolts.

The allowable loads for both bolted- and riveted-frame beam connections are tabulated in Tables 9-1 through 9-10.

9-2 Framed Beam Connections—Welded Framed beam connections which are welded are illustrated and the allowable loads are shown in Tables 9-11 through 9-13.

TABLE 9-1 Framed Beam Connections—Bolted or Riveted—Allowable Loads in kips (10 Rows—9 Rows)

10 ROWS

W 36

Total Shear, kips

[a]Fastener Designation	F_v ksi	Fastener Diameter ¾ Load	¾ t^b	⅞ Load	⅞ t^b	1 Load	1 t^b
A307	10.0	88.4	¼	120	¼	157	¼
A325-F A325-N A502-1	15.0	133	¼	180	¼	236	5/16
A490-F A502-2	20.0	177	¼	241	5/16	314	⅜
A325-X	22.0	194	5/16	265	5/16	346	7/16
A490-N	22.5	199	5/16	271	⅜	353	7/16
A490-X	32.0	283	7/16	385	½	503	⅝

Total Bearing,[c] kips, 10 fasteners on 1″ thick material

Fastener Diameter	F_y = 36	42	45	50	55	60	65	100
¾	365	425	456	506	557	608	658	1010
⅞	425	496	532	591	650	709	768	1180
1	486	567	608	675	743	810	878	1350

9 ROWS

W 36, 33

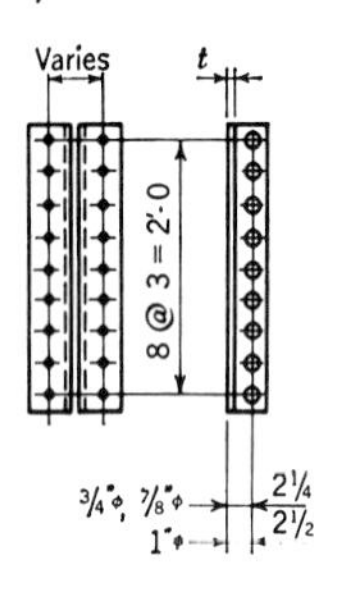

Total Shear, kips

[a]Fastener Designation	F_v ksi	Fastener Diameter ¾ Load	¾ t^b	⅞ Load	⅞ t^b	1 Load	1 t^b
A307	10.0	79.6	¼	108	¼	141	¼
A325-F A325-N A502-1	15.0	119	¼	162	¼	212	5/16
A490-F A502-2	20.0	159	¼	216	5/16	283	⅜
A325-X	22.0	175	5/16	238	5/16	311	7/16
A490-N	22.5	179	5/16	244	⅜	318	7/16
A490-X	32.0	255	7/16	346	½	452	⅝

Total Bearing,[c] kips, 9 fasteners on 1″ thick material

Fastener Diameter	F_y = 36	42	45	50	55	60	65	100
¾	328	383	410	456	501	547	592	911
⅞	383	447	478	532	585	638	691	1060
1	437	510	547	608	668	729	790	1220

[a] For description of fastener designation see AISC manual
[b] Thickness t based on connection angles of $F_y = 36$ ksi material.
[c] Use decimal thickness of enclosed web material as multiplying factor for these values.

SOURCE: American Institute of Steel Construction.

TABLE 9-2 Framed Beam Connections—Bolted or Riveted—Allowable Loads in kips (8 Rows—7 Rows)

8 ROWS

W 36, 33, 30

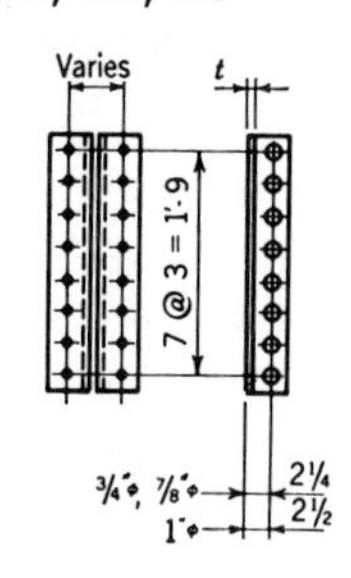

Total Shear, kips

[a]Fastener Designation	F_v ksi	¾ Load	¾ t^b	⅞ Load	⅞ t^b	1 Load	1 t^b
A307	10.0	70.7	¼	96.2	¼	126	¼
A325-F A325-N A502-1	15.0	106	¼	144	¼	188	5/16
A490-F A502-2	20.0	141	¼	192	5/16	251	⅜
A325-X	22.0	156	5/16	212	5/16	276	7/16
A490-N	22.5	159	5/16	216	⅜	283	7/16
A490-X	32.0	226	7/16	308	½	402	⅝

Total Bearing,[c] kips, 8 fasteners on 1" thick material

Fastener Diameter	F_y 36	42	45	50	55	60	65	100
¾	292	340	364	405	446	486	527	810
⅞	340	397	425	472	520	567	614	945
1	389	454	486	540	594	648	702	1080

7 ROWS

W 36, 33, 30, 27, 24
S 24*

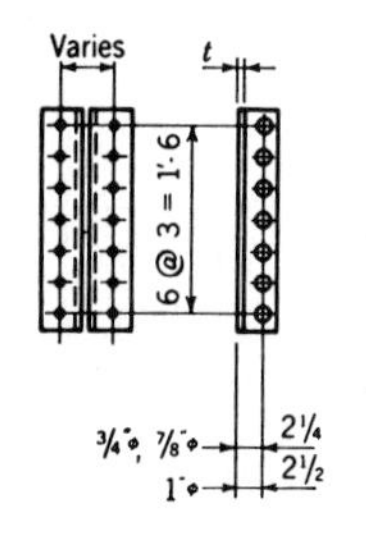

Total Shear, kips

[a]Fastener Designation	F_v ksi	¾ Load	¾ t^b	⅞ Load	⅞ t^b	1 Load	1 t^b
A307	10.0	61.9	¼	84.2	¼	110	¼
A325-F A325-N A502-1	15.0	92.8	¼	126	¼	165	5/16
A490-F A502-2	20.0	124	¼	168	5/16	220	⅜
A325-X	22.0	136	5/16	185	5/16	242	7/16
A490-N	22.5	139	5/16	189	⅜	247	7/16
A490-X	32.0	198	7/16	269	½	352	⅝

Total Bearing,[c] kips, 7 fasteners on 1" thick material

Fastener Diameter	F_y 36	42	45	50	55	60	65	100
¾	255	298	319	354	390	425	461	709
⅞	298	347	372	413	455	496	537	827
1	340	397	425	473	520	567	614	945

[a] For description of fastener designation see AISC manual
[b] Thickness t based on connection angles of $F_y = 36$ ksi material.
[c] Use decimal thickness of enclosed web material as multiplying factor for these values.
* Limited to S 24 × 79.9, 90, 100.

SOURCE: American Institute of Steel Construction.

TABLE 9-3 Framed Beam Connections—Bolted or Riveted—Allowable Loads in kips (6 Rows—5 Rows)

6 ROWS

W 36, 33, 30, 27, 24, 21
S 24

Total Shear, kips

[a]Fastener Designation	F_v ksi	¾ Load	¾ t^b	⅞ Load	⅞ t^b	1 Load	1 t^b
		Fastener Diameter					
A307	10.0	53.0	¼	72.2	¼	94.3	¼
A325-F A325-N A502-1	15.0	79.5	¼	108	¼	141	5⁄16
A490-F A502-2	20.0	106	¼	144	5⁄16	189	⅜
A325-X	22.0	117	5⁄16	159	⅜	207	7⁄16
A490-N	22.5	119	5⁄16	162	⅜	212	7⁄16
A490-X	32.0	170	7⁄16	231	½	302	⅝

Total Bearing,[c] kips, 6 fasteners on 1" thick material

Fastener Diameter	F_y = 36	42	45	50	55	60	65	100
¾	219	255	273	304	334	365	395	608
⅞	255	298	319	354	390	425	461	709
1	292	340	365	405	446	486	527	810

5 ROWS

W 30, 27, 24, 21, 18
S 24, 20, 18
MC 18

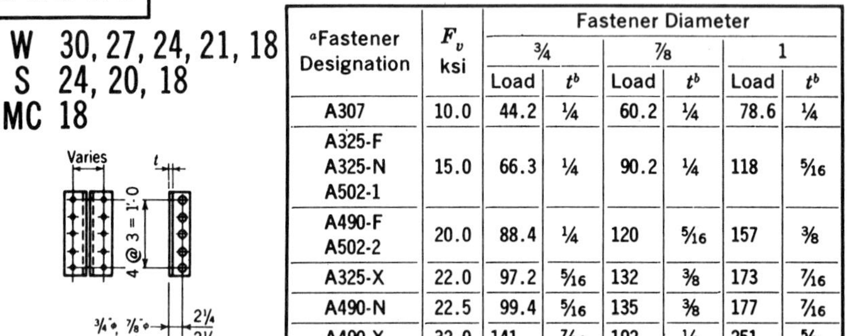

Total Shear, kips

[a]Fastener Designation	F_v ksi	¾ Load	¾ t^b	⅞ Load	⅞ t^b	1 Load	1 t^b
		Fastener Diameter					
A307	10.0	44.2	¼	60.2	¼	78.6	¼
A325-F A325-N A502-1	15.0	66.3	¼	90.2	¼	118	5⁄16
A490-F A502-2	20.0	88.4	¼	120	5⁄16	157	⅜
A325-X	22.0	97.2	5⁄16	132	⅜	173	7⁄16
A490-N	22.5	99.4	5⁄16	135	⅜	177	7⁄16
A490-X	32.0	141	7⁄16	192	½	251	⅝

Total Bearing,[c] kips, 5 fasteners on 1" thick material

Fastener Diameter	F_y = 36	42	45	50	55	60	65	100
¾	182	213	228	253	278	304	329	506
⅞	213	248	266	295	325	354	384	591
1	243	284	304	338	371	405	439	675

[a] For description of fastener designation see AISC manual
[b] Thickness t based on connection angles of F_y = 36 ksi material.
[c] Use decimal thickness of enclosed web material as multiplying factor for these values.

SOURCE: American Institute of Steel Construction.

TABLE 9-4 Framed Beam Connections—Bolted or Riveted—Allowable Loads in kips (4 Rows—3 Rows)

4 ROWS

W 24, 21, 18, 16
M 14
S 24, 20, 18, 15
C 15; MC 18

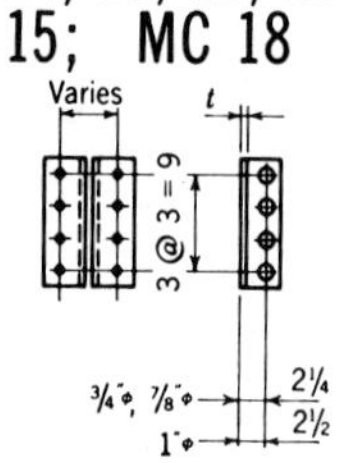

Total Shear, kips

[a]Fastener Designation	F_v ksi	Fastener Diameter 3/4 Load	t^b	7/8 Load	t^b	1 Load	t^b
A307	10.0	35.4	1/4	48.1	1/4	62.8	1/4
A325-F A325-N A502-1	15.0	53.0	1/4	72.2	1/4	94.2	5/16
A490-F A502-2	20.0	70.7	1/4	96.2	5/16	126	7/16
A325-X	22.0	77.8	5/16	106	3/8	138	7/16
A490-N	22.5	79.5	5/16	108	3/8	141	7/16
A490-X	32.0	113	7/16	154	1/2	201	5/8

Total Bearing,[c] ksi, 4 fasteners on 1" thick material

Fastener Diameter	F_y	36	42	45	50	55	60	65	100
	3/4	146	170	182	203	223	243	263	405
	7/8	170	198	213	236	260	284	307	473
	1	194	227	243	270	297	324	351	540

3 ROWS

W 18, 16, 14, 12, 10*
M 14, 12
S 18, 15, 12
C 15, 12, 10*
MC 12, 10

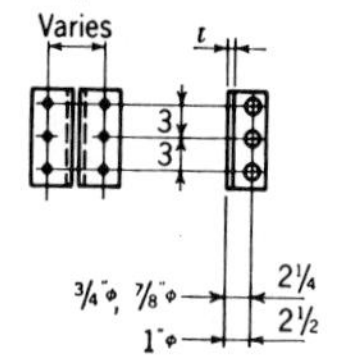

Total Shear, kips

[a]Fastener Designation	F_v ksi	Fastener Diameter 3/4 Load	t^b	7/8 Load	t^b	1 Load	t^b
A307	10.0	26.5	1/4	36.1	1/4	47.1	1/4
A325-F A325-N A502-1	15.0	39.8	1/4	54.1	1/4	70.7	5/16
A490-F A502-2	20.0	53.0	1/4	72.2	5/16	94.3	7/16
A325-X	22.0	58.3	5/16	79.4	3/8	104	7/16
A490-N	22.5	59.6	5/16	81.2	3/8	106	7/16
A490-X	32.0	84.8	7/16	115	1/2	151	5/8

Total Bearing,[c] kips, 3 fasteners on 1" thick material

Fastener Diameter	F_y	36	42	45	50	55	60	65	100
	3/4	109	128	137	152	167	182	197	304
	7/8	128	149	159	177	195	213	230	354
	1	146	170	182	203	223	243	263	405

[a] For description of fastener designation see AISC manual
[b] Thickness t based on connection angles of $F_y = 36$ ksi material.
[c] Use decimal thickness of enclosed web material as multiplying factor for these values.
* Limited to W 10 × 11.5, 15, 17, 19, 21, 25, 29; C 10 × 15.3, 20, 25, 30.

SOURCE: American Institute of Steel Construction.

TABLE 9-5 Framed Beam Connections—Bolted or Riveted—Allowable Loads in kips (2 Rows—1 Row)

2 ROWS

W 12, 10, 8
S 12, 10, 8
C 12, 10, 9, 8

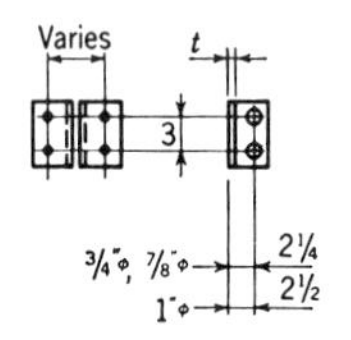

Total Shear, kips

[a]Fastener Designation	F_v ksi	3/4 Load	3/4 t^b	7/8 Load	7/8 t^b	1 Load	1 t^b
A307	10.0	17.7	1/4	24.1	1/4	31.4	1/4
A325-F A325-N A502-1	15.0	26.5	1/4	36.1	1/4	47.1	5/16
A490-F A502-2	20.0	35.3	1/4	48.1	5/16	62.8	7/16
A325-X	22.0	38.9	5/16	52.9	3/8	69.1	7/16
A490-N	22.5	39.8	5/16	54.1	3/8	70.7	1/2
A490-X	32.0	56.6	7/16	77.0	1/2	99.7*	5/8

Total Bearing,[c] kips, 2 fasteners on 1″ thick material

Fastener Diameter	F_y = 36	42	45	50	55	60	65	100
3/4	72.9	85.1	91.1	101	111	122	132	203
7/8	85.1	99.2	106	118	130	142	154	236
1	97.2	113	122	135	149	162	176	270

1 ROW

W 6, 5
M 6, 5
S 8, 7, 6, 5
C 7, 6, 5

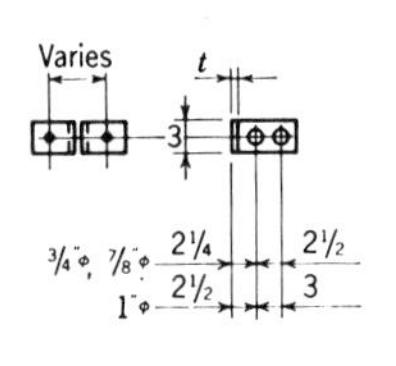

Total Shear, kips

[a]Fastener Designation	F_v ksi	3/4 Load	3/4 t^b	7/8 Load	7/8 t^b	1 Load	1 t^b
A307	10.0	8.8	1/4	12.0	1/4	15.7	1/4
A325-F A325-N A502-1	15.0	13.3	1/4	18.0	1/4	23.6	5/16
A490-F A502-2	20.0	17.7	1/4	24.1	5/16	31.4	3/8
A325-X	22.0	19.4	5/16	26.5	5/16	34.6	7/16
A490-N	22.5	19.9	5/16	27.1	3/8	35.3	7/16
A490-X	32.0	28.3	7/16	38.5	1/2	50.3	5/8

Total Bearing,[cd] kips, on 1″ thick material

Fastener Diameter	F_y = 36	42	45	50	55	60	65	100
3/4	72.9	85.1	91.1	101	111	122	132	203
7/8	85.1	99.2	106	118	130	142	154	236
1	97.2	113	122	135	149	162	176	270

[a] For description of fastener designation see AISC manual
[b] Thickness t based on connection angles of $F_y = 36$ ksi material.
[c] Use decimal thickness of enclosed web material as multiplying factor for these values.
[d] Values shown are for 1 bolt in each outstanding leg or 2 bolts in web leg.
* Indicates shear values are limited by shear capacity of 5/8″ angle of $F_y = 36$ ksi material (arbitrary limit for flexibility), and length of angle assumed to be c/c outside fasteners plus 2½″.

SOURCE: American Institute of Steel Construction.

TABLE 9-6 Heavy Framed Beam Connections—Bolted or Riveted—Allowable Loads in kips (10 Rows—9 Rows)

10 ROWS

W 36

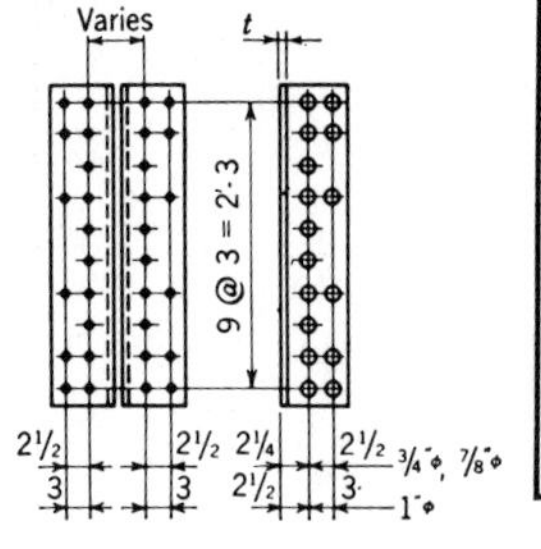

Total Shear, kips, 16 fasteners

[a]Fastener Designation	F_v ksi	Fastener diameter ¾ Load	¾ t^b	⅞ Load	⅞ t^b	1 Load	1 t^b
A325-F A325-N A502-1	15.0	212	¼	289	⅜	377	½
A490-F A502-2	20.0	283	⅜	385	½	503	⅝
A325-X	22.0	311	⅜	423	½	535*	⅝
A490-N	22.5	318	⅜	433	9/16		
A490-X	32.0	452	9/16	535*	⅝		

Total Bearing,[c] kips, 16 fasteners on 1" thick material

Fastener Diameter	F_y: 36	42	45	50	55	60	65	100
¾	583	680	729	810	891	972	1050	1620
⅞	680	794	851	945	1040	1130	1230	1890
1	778	907	972	1080	1190	1300	1400	2160

9 ROWS

W 36, 33

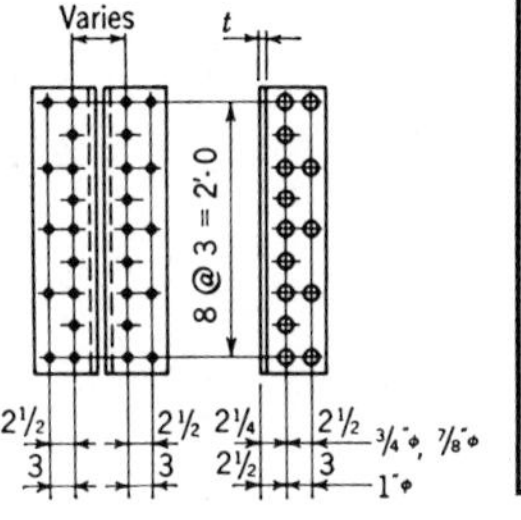

Total Shear, kips, 14 fasteners

[a]Fastener Designation	F_v ksi	Fastener diameter ¾ Load	¾ t^b	⅞ Load	⅞ t^b	1 Load	1 t^b
A325-F A325-N A502-1	15.0	186	¼	253	⅜	330	7/16
A490-F A502-2	20.0	247	⅜	337	½	440	⅝
A325-X	22.0	272	⅜	370	½	480*	⅝
A490-N	22.5	278	⅜	379	½		
A490-X	32.0	396	9/16	480*	⅝		

Total Bearing,[c] kips, 14 fasteners on 1" thick material

Fastener Diameter	F_y: 36	42	45	50	55	60	65	100
¾	510	595	638	709	780	851	921	1420
⅞	595	695	744	827	910	992	1080	1650
1	680	794	851	945	1040	1130	1230	1890

[a] For description of fastener designation see AISC manual
[b] Thickness t based on connection angles of $F_y = 36$ ksi material.
[c] Use decimal thickness of enclosed web material as multiplying factor for these values.
* Indicates shear values are limited by shear capacity of ⅝" angle of $F_y = 36$ ksi material (arbitrary limit for flexibility), and length of angle assumed to be c/c outside fasteners plus 2½".

SOURCE: American Institute of Steel Construction.

TABLE 9-7 Heavy Framed Beam Connections—Bolted or Riveted—Allowable Loads in kips (8 Rows—7 Rows)

8 ROWS

W 36, 33, 30

Total Shear, kips, 12 fasteners

[a]Fastener Designation	F_v ksi	Fastener diameter ¾ Load	¾ t^b	⅞ Load	⅞ t^b	1 Load	1 t^b
A325-F A325-N A502-1	15.0	159	¼	216	⅜	283	7/16
A490-F A502-2	20.0	212	5/16	289	7/16	377	9/16
A325-X	22.0	233	⅜	318	½	415	⅝
A490-N	22.5	239	⅜	325	½	426*	⅝
A490-X	32.0	339	½	426*	⅝		

Total Bearing,[c] kips, 12 fasteners on 1″ thick material

Fastener Diameter	F_y 36	42	45	50	55	60	65	100
¾	437	510	547	608	668	729	790	1220
⅞	510	595	638	709	780	851	921	1420
1	583	680	729	810	891	972	1050	1620

7 ROWS

W 36, 33, 30, 27, 24
S 24**

Total Shear, kips, 11 fasteners

[a]Fastener Designation	F_v ksi	Fastener diameter ¾ Load	¾ t^b	⅞ Load	⅞ t^b	1 Load	1 t^b
A325-F A325-N A502-1	15.0	146	¼	198	⅜	259	½
A490-F A502-2	20.0	194	⅜	265	½	346	⅝
A325-X	22.0	214	⅜	291	½	372*	⅝
A490-N	22.5	219	⅜	298	9/16		
A490-X	32.0	311	9/16	372*	⅝		

Total Bearing,[c] kips, 11 fasteners on 1″ thick material

Fastener Diameter	F_y 36	42	45	50	55	60	65	100
¾	401	468	501	557	613	668	724	1110
⅞	468	546	585	650	715	780	845	1300
1	535	624	668	743	817	891	965	1490

[a] For description of fastener designation see AISC manual
[b] Thickness t based on connection angles of $F_y = 36$ ksi material.
[c] Use decimal thickness of enclosed web material as multiplying factor for these values.
* Indicates shear values limited by shear capacity of ⅝″ angle of $F_y = 36$ ksi material (arbitrary limit for flexibility), and length of angle assumed to be c/c of outside fasteners plus 2½″.
** Limited to S 24 × 79.9, 90, 100.

SOURCE: American Institute of Steel Construction.

TABLE 9-8 Heavy Framed Beam Connections—Bolted or Riveted—Allowable Loads in kips (6 Rows—5 Rows)

6 ROWS

W 36, 33, 30, 27, 24, 21
S 24

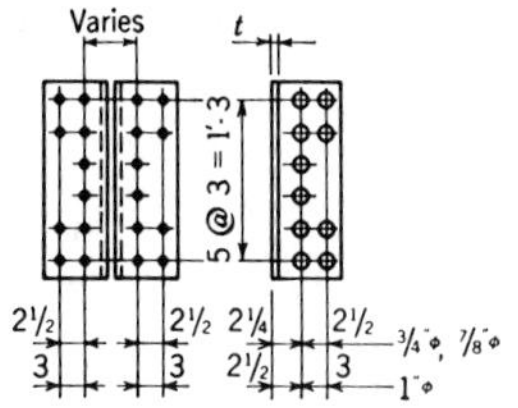

Varies; t; 5 @ 3 = 1'-3; 2½, 3, 2½, 3, 2¼, 2½, 2½, 3; ¾"ϕ, ⅞"ϕ, 1"ϕ

Total Shear, kips, 10 fasteners

[a]Fastener Designation	F_v ksi	Fastener diameter ¾ Load	¾ t^b	⅞ Load	⅞ t^b	1 Load	1 t^b
A325-F A325-N A502-1	15.0	133	5/16	180	3/8	236	1/2
A490-F A502-2	20.0	177	3/8	241	1/2	314	5/8
A325-X	22.0	194	7/16	265	9/16	317*	5/8
A490-N	22.5	199	7/16	271	9/16		
A490-X	32.0	283	9/16	317*	5/8		

Total Bearing,[c] kips, 10 fasteners on 1" thick material

Fastener Diameter	F_y 36	42	45	50	55	60	65	100
¾	365	425	456	506	557	608	658	1010
⅞	425	496	532	591	650	709	768	1180
1	486	567	608	675	743	810	878	1350

5 ROWS

W 30, 27, 24, 21, 18
S 24, 20, 18
MC 18

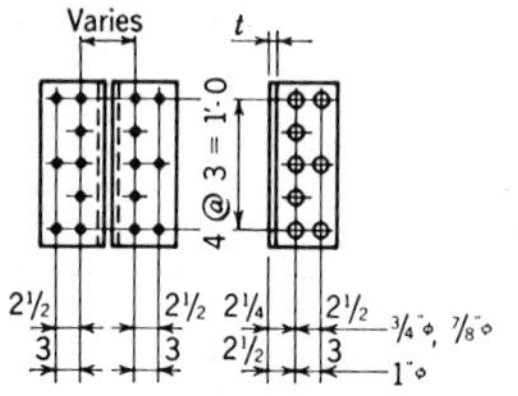

Varies; t; 4 @ 3 = 1'-0; 2½, 3, 2½, 3, 2¼, 2½, 2½, 3; ¾"ϕ, ⅞"ϕ, 1"ϕ

Total Shear, kips, 8 fasteners

[a]Fastener Designation	F_v ksi	Fastener diameter ¾ Load	¾ t^b	⅞ Load	⅞ t^b	1 Load	1 t^b
A325-F A325-N A502-1	15.0	106	5/16	144	3/8	188	1/2
A490-F A502-2	20.0	141	3/8	192	1/2	251	5/8
A325-X	22.0	156	3/8	212	9/16	263*	5/8
A490-N	22.5	159	7/16	216	9/16		
A490-X	32.0	226	9/16	263*	5/8		

Total Bearing,[c] kips, 8 fasteners on 1" thick material

Fastener Diameter	F_y 36	42	45	50	55	60	65	100
¾	292	340	365	405	446	486	527	810
⅞	340	397	425	473	520	567	614	945
1	389	454	486	540	594	648	702	1080

[a] For description of fastener designation see AISC manual
[b] Thickness t based on connection angles of $F_y = 36$ ksi material.
[c] Use decimal thickness of enclosed web material as multiplying factor for these values.
* Indicates shear values limited by shear capacity of ⅝" angle of $F_y = 36$ ksi material (arbitrary limit for flexibility), and length of angle assumed to be c/c of outside fasteners plus 2½".

SOURCE: American Institute of Steel Construction.

TABLE 9-9 Heavy Framed Beam Connections—Bolted or Riveted—Allowable Loads in kips (4 Rows—3 Rows)

4 ROWS

W 24, 21, 18, 16
M 14
S 24, 20, 18, 15
C 15; MC 18

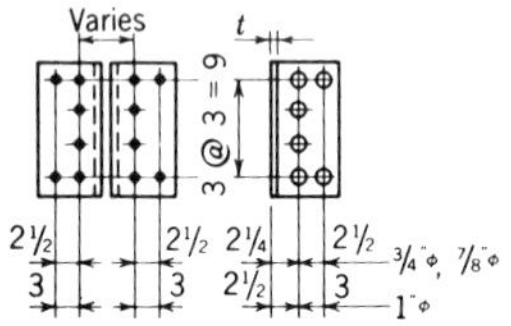

Total Shear, kips, 6 fasteners

[a]Fastener Designation	F_v ksi	Fastener diameter ¾ Load	t^b	⅞ Load	t^b	1 Load	t^b
A325-F A325-N A502-1	15.0	79.5	¼	108	⅜	141	7⁄16
A490-F A502-2	20.0	106	⅜	144	7⁄16	189	⅝
A325-X	22.0	117	⅜	159	½	207	⅝
A490-N	22.5	119	⅜	162	½	208*	⅝
A490-X	32.0	170	9⁄16	208*	⅝		

Total Bearing,[c] kips, 6 fasteners on 1″ thick material

Fastener Diameter	F_y: 36	42	45	50	55	60	65	100
¾	219	255	273	304	334	365	395	608
⅞	255	298	319	354	390	425	461	709
1	292	340	365	405	446	486	527	810

3 ROWS

W 18, 16, 14, 12, 10**
M 14, 12
S 18, 15, 12
C 15, 12; MC 12, 10

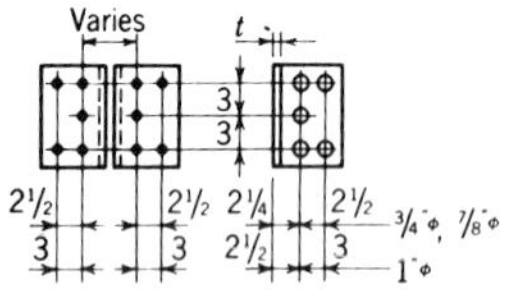

Total Shear, kips, 5 fasteners

[a]Fastener Designation	F_v ksi	Fastener diameter ¾ Load	t^b	⅞ Load	t^b	1 Load	t^b
A325-F A325-N A502-1	15.0	66.3	5⁄16	90.2	⅜	118	½
A490-F A502-2	20.0	88.4	⅜	120	½	154*	⅝
A325-X	22.0	97.2	7⁄16	132	9⁄16		
A490-N	22.5	99.4	7⁄16	135	9⁄16		
A490-X	32.0	141	⅝	154*	⅝		

Total Bearing,[c] kips, 5 fasteners on 1″ thick material

Fastener Diameter	F_y: 36	42	45	50	55	60	65	100
¾	182	213	228	253	278	304	329	506
⅞	213	248	266	295	325	354	384	591
1	243	284	304	338	371	405	439	675

[a] For description of fastener designation see AISC manual
[b] Thickness t based on connection angles of $F_y = 36$ ksi material.
[c] Use decimal thickness of enclosed web material as multiplying factor for these values.
* Indicates shear values limited by shear capacity of ⅝″ angle of $F_y = 36$ ksi material (arbitrary limit for flexibility), and length of angle assumed to be c/c of outside fasteners plus 2½″.
** Limited to W 10 × 11.5, 15, 17, 19, 21, 25, 29.

SOURCE: American Institute of Steel Construction.

TABLE 9-10 Heavy Framed Beam Connections—Bolted or Riveted—Allowable Loads in kips (2 Rows)

2 ROWS

W 12, 10, 8
S 12, 10, 8
C 12, 10, 9, 8

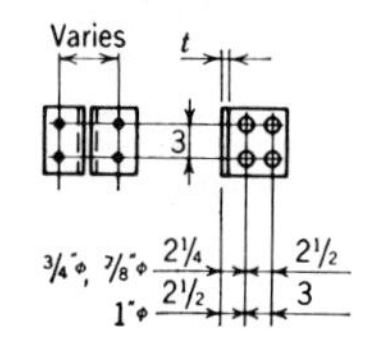

Total Shear, kips, 4 fasteners

[a]Fastener Designation	F_v ksi	Fastener Diameter					
		¾		⅞		1	
		Load	t^b	Load	t^b	Load	t^b
A325-F A325-N A502-1	15.0	26.5	¼	36.1	¼	47.1	5/16
A490-F A502-2	20.0	35.4	¼	48.1	5/16	62.8	7/16
A325-X	22.0	38.9	¼	52.9	⅜	69.2	7/16
A490-N	22.5	39.8	¼	54.1	⅜	70.7	½
A490-X	32.0	56.6	⅜	77.0	½	100*	⅝

Total Bearing,[cd] kips, on 1" thick material

	F_y	36	42	45	50	55	60	65	100
Fastener Diameter	¾	146	170	182	203	223	243	263	405
	⅞	170	198	213	236	260	284	307	473
	1	194	227	243	270	297	324	351	540

[a] For description of fastener designation see AISC manual
[b] Thickness t based on connection angles of $F_y = 36$ ksi material.
[c] Use decimal thickness of enclosed web material as multiplying factor for these values.
[d] Values shown are for 2 bolts in each outstanding leg or 4 bolts in web legs.
* Indicates shear values limited by shear capacity of ⅝" angle of $F_y = 36$ ksi material (arbitrary limit for flexibility), and length of angle assumed to be c/c of outside fasteners plus 2½".

SOURCE: American Institute of Steel Construction.

TABLE 9-11 Framed Beam Connections – Welded – E70XX Electrodes – Allowable Loads in kips

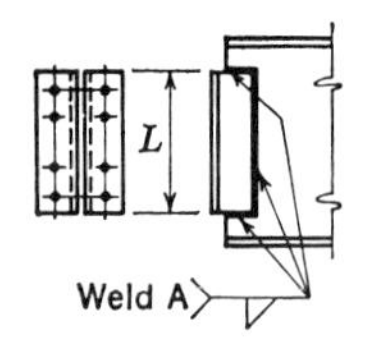

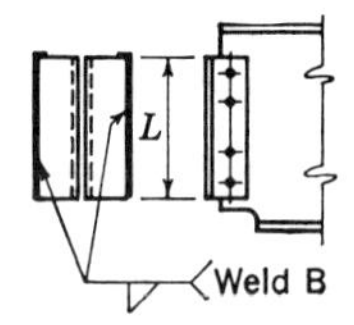

Weld A		Weld B		Angle Length L	[a]Minimum Web Thickness for Welds A		Number of Fasteners in One Vertical Row (Table I)
Capacity Kips	[b]Size In.	[c]Capacity Kips	[b]Size In.		F_y = 36 ksi F_v = 14.5 ksi	F_y = 50 ksi F_v = 20 ksi	
276	5/16	296	3/8	2′- 5½	.56	.41	10
221	¼	247	5/16	2′- 5½	.46	.33	
166	3/16	197	¼	2′- 5½	.34	.25	
246	5/16	261	3/8	2′- 2½	.55	.40	9
197	¼	217	5/16	2′- 2½	.45	.32	
148	3/16	173	¼	2′- 2½	.33	.24	
217	5/16	223	3/8	1′-11½	.54	.39	8
173	¼	186	5/16	1′-11½	.44	.32	
130	3/16	149	¼	1′-11½	.33	.24	
186	5/16	187	3/8	1′- 8½	.53	.39	7
149	¼	156	5/16	1′- 8½	.43	.31	
112	3/16	125	¼	1′- 8½	.32	.23	
157	5/16	152	3/8	1′- 5½	.52	.38	6
125	¼	126	5/16	1′- 5½	.42	.30	
94.1	3/16	101	¼	1′- 5½	.31	.23	
128	5/16	115	3/8	1′- 2½	.50	.36	5
102	¼	95.7	5/16	1′- 2½	.41	.29	
76.6	3/16	76.6	¼	1′- 2½	.30	.22	
99.0	5/16	80.1	3/8	11½	.48	.35	4
79.2	¼	66.9	5/16	11½	.39	.28	
59.4	3/16	53.4	¼	11½	.29	.21	
71.2	5/16	48.2	3/8	8½	.46	.33	3
57.0	¼	40.3	5/16	8½	.37	.27	
42.8	3/16	32.2	¼	8½	.28	.20	
44.9	5/16	21.9	3/8	5½	.44	.32	2
35.9	¼	18.3	5/16	5½	.35	.25	
27.0	3/16	14.6	¼	5½	.26	.19	

[a] When the beam web thickness is less than the minimum, multiply the connection capacity furnished by Weld **A** by the ratio of the actual web thickness to the tabulated minimum thickness. Thus, if 5/16 in. Weld **A**, with a connection capacity of 128 kips and a 1′–2½″ long angle, is considered for a beam of web thickness of 0.375″ with F_y = 36 ksi, the connection capacity must be multiplied by 0.375/0.50, giving 96.0 kips.

[b] Should the thickness of material to which connection angles are welded exceed the limits set by AISC Specifications, Sect. 1.17.5, for weld sizes specified, increase the weld size as required, but not to exceed the angle thickness.

[c] When welds are used on outstanding legs, connection capacity may be limited by the shear capacity of the supporting member as stipulated by AISC Specification Sect. 1.17.6.

Note 1: Connection Angles: Two L 4 × 3½ × Thickness × L; F_y = 36 ksi. See page **4** - 28 for limiting values of thickness and optional width of legs.

Note 2: Capacities shown in this table apply only when the material welded is F_y = 36 ksi or F_y = 50 ksi steel.

SOURCE: American Institute of Steel Construction.

TABLE 9-12 Framed Beam Connections—Welded—E70XX Electrodes

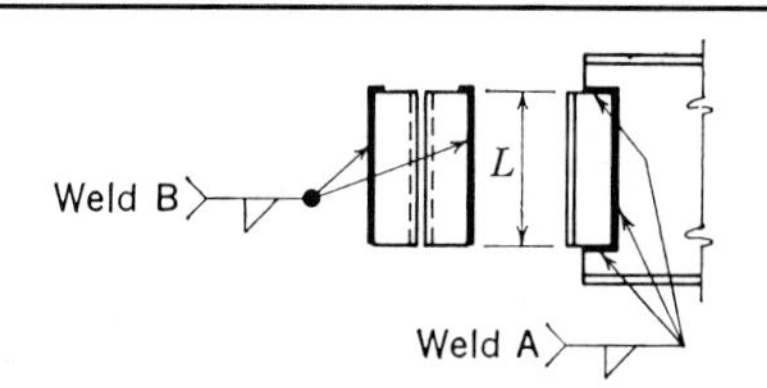

Weld A		Weld B		Angle Length L In.	Angle Size F_y = 36 ksi	[a]Minimum Web Thickness for Weld A	
Capacity Kips	[b]Size In.	[c]Capacity Kips	[b]Size In.			F_y = 36 ksi F_v = 14.5 ksi	F_y = 50 ksi F_v = 20 ksi
302	5⁄16	326	3⁄8	32	4 × 3 × 7⁄16	.57	.42
242	1⁄4	271	5⁄16	32	4 × 3 × 3⁄8	.46	.33
181	3⁄16	217	1⁄4	32	4 × 3 × 5⁄16	.34	.25
282	5⁄16	302	3⁄8	30	4 × 3 × 7⁄16	.57	.41
226	1⁄4	251	5⁄16	30	4 × 3 × 3⁄8	.46	.33
169	3⁄16	201	1⁄4	30	4 × 3 × 5⁄16	.34	.25
262	5⁄16	278	3⁄8	28	4 × 3 × 7⁄16	.56	.41
210	1⁄4	231	5⁄16	28	4 × 3 × 3⁄8	.45	.33
157	3⁄16	185	1⁄4	28	4 × 3 × 5⁄16	.34	.24
242	5⁄16	254	3⁄8	26	4 × 3 × 7⁄16	.55	.40
194	1⁄4	211	5⁄16	26	4 × 3 × 3⁄8	.45	.32
145	3⁄16	169	1⁄4	26	4 × 3 × 5⁄16	.33	.24
221	5⁄16	230	3⁄8	24	4 × 3 × 7⁄16	.55	.40
178	1⁄4	191	5⁄16	24	4 × 3 × 3⁄8	.44	.32
133	3⁄16	153	1⁄4	24	4 × 3 × 5⁄16	.33	.24
202	5⁄16	206	3⁄8	22	4 × 3 × 7⁄16	.54	.39
162	1⁄4	171	5⁄16	22	4 × 3 × 3⁄8	.44	.31
121	3⁄16	137	1⁄4	22	4 × 3 × 5⁄16	.32	.24
182	5⁄16	181	3⁄8	20	4 × 3 × 7⁄16	.53	.39
146	1⁄4	152	5⁄16	20	4 × 3 × 3⁄8	.43	.31
110	3⁄16	121	1⁄4	20	4 × 3 × 5⁄16	.32	.23
162	5⁄16	157	3⁄8	18	4 × 3 × 7⁄16	.52	.38
130	1⁄4	131	5⁄16	18	4 × 3 × 3⁄8	.42	.30
97.7	3⁄16	105	1⁄4	18	4 × 3 × 5⁄16	.31	.23
144	5⁄16	148	3⁄8	16	3 × 3 × 7⁄16	.51	.37
114	1⁄4	123	5⁄16	16	3 × 3 × 3⁄8	.41	.30
85.9	3⁄16	98.8	1⁄4	16	3 × 3 × 5⁄16	.31	.22
123	5⁄16	124	3⁄8	14	3 × 3 × 7⁄16	.50	.36
98.8	1⁄4	103	5⁄16	14	3 × 3 × 3⁄8	.40	.29
74.0	3⁄16	82.5	1⁄4	14	3 × 3 × 5⁄16	.30	.22
104	5⁄16	99.6	3⁄8	12	3 × 3 × 7⁄16	.48	.35
83.2	1⁄4	83.1	5⁄16	12	3 × 3 × 3⁄8	.39	.28
62.5	3⁄16	66.5	1⁄4	12	3 × 3 × 5⁄16	.29	.21

For footnotes, see TABLE 9-13

SOURCE: American Institute of Steel Construction.

TABLE 9-13 Framed Beam Connections—Welded—E70XX Electrodes

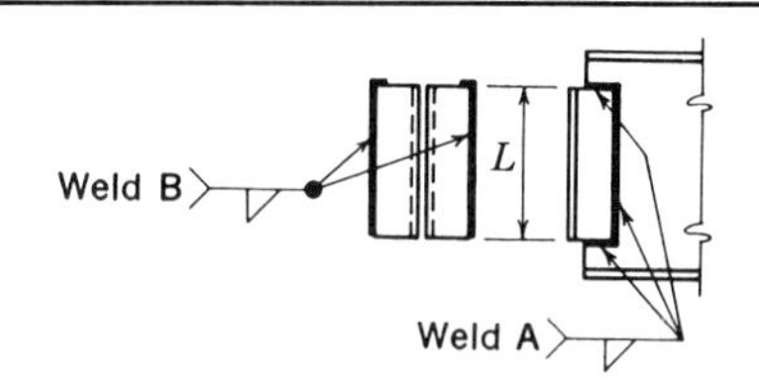

Weld **A**		Weld **B**		Angle Length L In.	Angle Size $F_y = 36$ ksi	[a]Minimum Web Thickness for Weld **A**	
Capacity Kips	[b]Size In.	[c]Capacity Kips	[b]Size In.			$F_y = 36$ ksi $F_v = 14.5$ ksi	$F_y = 50$ ksi $F_v = 20$ ksi
85.1	5/16	75.9	3/8	10	3 × 3 × 7/16	.47	.34
68.1	1/4	63.3	5/16	10	3 × 3 × 3/8	.38	.27
51.0	3/16	50.5	1/4	10	3 × 3 × 5/16	.28	.21
75.9	5/16	64.3	3/8	9	3 × 3 × 7/16	.46	.34
60.6	1/4	53.7	5/16	9	3 × 3 × 3/8	.38	.27
45.5	3/16	42.9	1/4	9	3 × 3 × 5/16	.28	.20
66.7	5/16	53.2	3/8	8	3 × 3 × 7/16	.46	.33
53.3	1/4	44.4	5/16	8	3 × 3 × 3/8	.37	.27
40.0	3/16	35.5	1/4	8	3 × 3 × 5/16	.28	.20
57.8	5/16	42.5	3/8	7	3 × 3 × 7/16	.45	.32
46.3	1/4	35.5	5/16	7	3 × 3 × 3/8	.36	.26
34.7	3/16	28.3	1/4	7	3 × 3 × 5/16	.27	.19
49.2	5/16	32.6	3/8	6	3 × 3 × 7/16	.44	.32
39.3	1/4	27.1	5/16	6	3 × 3 × 3/8	.36	.26
29.5	3/16	21.7	1/4	6	3 × 3 × 5/16	.27	.19
40.8	5/16	23.4	3/8	5	3 × 3 × 7/16	.43	.32
32.6	1/4	19.5	5/16	5	3 × 3 × 3/8	.35	.25
24.5	3/16	15.7	1/4	5	3 × 3 × 5/16	.26	.19
32.7	5/16	15.4	3/8	4	3 × 3 × 7/16	.43	.31
26.2	1/4	12.9	5/16	4	3 × 3 × 3/8	.35	.25
19.7	3/16	10.4	1/4	4	3 × 3 × 5/16	.26	.19

[a] When the beam web thickness is less than the minimum, multiply the connection capacity furnished by Welds **A** by the ratio of the actual thickness to the tabulated minimum thickness. Thus, if 5/16 in. Weld **A**, with a connection capacity of 66.7 kips and an 8″ long angle, is considered for a beam of web thickness 0.305″ and $F_y = 36$ ksi, the connection capacity must be multiplied by 0.305/0.46, giving 44.2 kips.

[b] Should the thickness of material to which connection angles are welded exceed the limits set by AISC Specification Sect. 1.17.5 for weld sizes specified, increase the weld size as required, but not to exceed the angle thickness.

[c] For welds on outstanding legs, connection capacity may be limited by the shear capacity of the supporting members, as stipulated by AISC Specification Sect. 1.17.6.

Note 1: Capacities shown in this table apply only when connection angles are $F_y = 36$ ksi steel and the material to which they are welded is either $F_y = 36$ ksi or $F_y = 50$ ksi steel.

SOURCE: American Institute of Steel Construction.

Further technical data, of course, may be obtained from the AISC *Manual of Steel Construction*, as noted in the beginning of this section.

9-3 Seated Beam Connections Seated connections should be used only when the beam is supported by a top angle. Allowable loads are given in Tables 9-14 and 9-15.

TABLE 9-14 Seated Beam Connections—Bolted or Riveted—Allowable Loads in kips

Outstanding Leg Capacity, kips (based on OSL = 4 inches)

Angle Material		F_y = 36 ksi											
Angle Length		6 inches						8 inches					
Angle Thickness, In.		3/8	1/2	5/8	3/4	7/8	1	3/8	1/2	5/8	3/4	7/8	1
Beam Web Thickness (in.) F_y = 36 ksi	3/16	7.50	10.3	13.1	15.9	18.7	18.8	8.44	11.5	14.6	17.7	18.8	18.8
	1/4	9.57	13.0	16.3	19.7	23.1	26.2	10.6	14.4	18.1	21.8	25.5	26.2
	5/16	11.3	16.3	20.3	24.2	28.1	32.0	13.1	17.9	22.2	26.5	30.8	34.7
	3/8	12.4	19.3	24.3	28.7	33.2	37.6	14.3	21.5	26.3	31.2	36.1	40.9
	7/16	13.4	21.1	28.8	33.7	38.7	43.6	15.5	23.8	30.9	36.3	41.7	47.1
	1/2	14.3	22.8	31.6	39.2	44.6	50.0	16.5	25.7	35.1	41.8	47.7	53.6
	9/16	15.2	24.4	34.0	43.8	51.0	56.9	17.5	27.5	37.8	47.8	54.1	60.5

Outstanding Leg Capacity, kips (based on OSL = 4 inches)

Angle Material		F_y = 36 ksi											
Angle Length		6 inches						8 inches					
Angle Thickness, In.		3/8	1/2	5/8	3/4	7/8	1	3/8	1/2	5/8	3/4	7/8	1
Beam Web Thickness (in.) F_y = 50 ksi	3/16	9.14	12.6	16.1	19.5	23.0	26.2	10.2	14.1	17.9	21.7	25.5	26.2
	1/4	11.9	16.1	20.3	24.5	28.7	32.9	13.1	17.7	22.4	27.0	31.6	36.2
	5/16	13.3	20.7	25.6	30.5	35.5	40.4	15.4	22.4	27.8	33.2	38.6	43.9
	3/8	14.6	23.3	31.1	36.7	42.3	47.9	16.9	26.3	33.4	39.5	45.6	51.7
	7/16	15.8	25.5	35.8	43.5	50.0	56.1	18.3	28.8	39.6	46.4	53.2	60.0
	1/2	16.9	27.7	39.0	50.6	57.9	64.8	19.5	31.1	43.2	53.9	61.4	68.8
	9/16	17.9	29.7	42.2	55.0	66.8	74.3	20.7	33.3	46.6	60.1	70.3	78.4

Fastener Capacity, kips

[a]Fastener Specification	Fastener Diameter in.	Connection Type A	B	C	D	E	F
A307	3/4	8.8	17.7	26.5	13.3	26.5	39.8
	7/8	12.0	24.0	36.1	18.0	36.1	54.1
	1	15.7	31.4	47.1	23.6	47.1	70.7
A325-F, A325-N, A502-1	3/4	13.3	26.5	39.8	19.9	39.8	59.7
	7/8	18.0	36.1	54.1	27.1	54.1	81.2
	1	23.6	47.1	70.7	35.3	70.7	—
A490-F, A502-2	3/4	17.7	35.4	53.0	26.5	53.0	79.6
	7/8	24.1	48.1	72.2	36.1	72.2	—
	1	31.4	62.8	94.3	47.1	94.3	—
A325-X	3/4	19.4	38.9	58.3	29.2	58.3	87.5
	7/8	26.5	52.9	79.4	39.7	79.4	—
	1	34.6	69.1	—	51.8	—	—
A490-N	3/4	19.9	39.8	59.6	29.8	59.6	89.5
	7/8	27.1	54.1	81.2	40.6	81.2	—
	1	35.3	70.7	—	53.0	—	—
A490-X	3/4	28.3	56.6	84.8	42.4	84.8	127
	7/8	38.5	77.0	—	57.7	—	—
	1	50.3	101	—	75.4	—	—

[a] A325-F and A490-F: Friction type connections.
A325-N and A490-N: Bearing type connections with threads included in shear plane.
A325-X and A490-X: Bearing type connections with threads excluded from shear plane.

Available Seat Angle and Thickness Range

Type	Angle Size In.	t In.
A, D	4 × 3	3/8–5/8
	4 × 3½	3/8–5/8
	4 × 4	3/8–3/4
B, E	6 × 4	3/8–7/8
	7 × 4	3/8–7/8
	8 × 4	1/2–1
C, F	[b]8 × 4	1/2–1
	9 × 4	1/2–1

[b] Suitable for use with 3/4" and 7/8" fasteners only.

SOURCE: American Institute of Steel Construction.

TABLE 9-15 Seated Beam Connections—Welded—E70XX Electrodes, Allowable Load in kips

Outstanding Leg Capacity, kips (based on OSL = 3½ or 4 inches)

Angle Material		F_y = 36 ksi											
Angle Length		6 inches						8 inches					
Angle Thickness, In.		3/8	1/2	5/8	3/4	7/8	1	3/8	1/2	5/8	3/4	7/8	1
Beam Web Thickness (in.)	3/16	7.50	10.3	13.1	15.9	18.7	18.8	8.44	11.5	14.6	17.7	18.8	18.8
	1/4	9.57	13.0	16.3	19.7	23.1	26.2	10.6	14.4	18.1	21.8	25.5	26.2
	5/16	11.3	16.3	20.3	24.2	28.1	32.0	13.1	17.9	22.2	26.5	30.8	34.7
	3/8	12.4	19.3	24.3	28.7	33.2	37.6	14.3	21.5	26.3	31.2	36.1	40.9
	7/16	13.4	21.1	28.8	33.7	38.7	43.6	15.5	23.8	30.9	36.3	41.7	47.1
F_y = 36 ksi	1/2	14.3	22.8	31.6	39.2	44.6	50.0	16.5	25.7	35.1	41.8	47.7	53.6
	9/16	15.2	24.4	34.0	43.8	51.0	56.9	17.5	27.5	37.8	47.8	54.1	60.5

Note: Values above heavy lines apply only for 4-inch outstanding legs.

Outstanding Leg Capacity, kips (based on OSL = 3½ or 4 inches)

Angle Material		F_y = 36 ksi											
Angle Length		6 inches						8 inches					
Angle Thickness, In.		3/8	1/2	5/8	3/4	7/8	1	3/8	1/2	5/8	3/4	7/8	1
Beam Web Thickness (in.)	3/16	9.14	12.6	16.1	19.5	23.0	26.2	10.2	14.1	17.9	21.7	25.5	26.2
	1/4	11.9	16.1	20.3	24.5	28.7	32.9	13.1	17.7	22.4	27.0	31.6	36.2
	5/16	13.3	20.7	25.6	30.5	35.5	40.4	15.4	22.4	27.8	33.2	38.6	43.9
	3/8	14.6	23.3	31.1	36.7	42.3	47.9	16.9	26.3	33.4	39.5	45.6	51.7
	7/16	15.8	25.5	35.8	43.5	50.0	56.1	18.2	28.8	39.6	46.4	53.2	60.0
F_y = 50 ksi	1/2	16.9	27.7	39.0	50.6	57.9	64.8	19.5	31.1	43.2	53.9	61.4	68.8
	9/16	17.9	29.7	42.2	55.0	66.8	74.3	20.7	33.3	46.6	60.1	70.3	78.4

Note: Values above heavy lines apply only for 4-inch outstanding legs.

Weld Capacity, kips

Weld Size In.	E70XX Electrodes					
	Seat Angle Size (long leg vertical)					
	4 × 3½	5 × 3½	6 × 4	7 × 4	8 × 4	9 × 4
1/4	11.5	17.2	21.8	28.5	35.6	43.0
5/16	14.3	21.5	27.3	35.6	44.5	53.8
3/8	17.2	25.8	32.7	42.7	53.4	64.6
7/16	20.1	30.1	38.2	49.8	62.3	75.3
1/2	22.9	34.4	43.6	56.9	71.2	86.1
5/8	—	43.0	54.5	71.2	89.0	—
11/16	—	47.3	60.0	78.3	—	—
3/4	—	—	65.4	85.4	—	—
Range of available seat angle thicknesses						
Minimum	3/8	3/8	3/8	3/8	1/2	1/2
Maximum	5/8	3/4	7/8	7/8	1	1

SOURCE: American Institute of Steel Construction.

9-4 Stiffened Seated Beam Connections As in the foregoing paragraph, stiffened seated connections should only be used when the beam is supported by a top angle.

Allowable capacities, Tables 9-16 through 9-18, are based on allowable bearing with steel of $F_y = 36$ ksi or $F_y = 50$ ksi in the stiffener angles.

TABLE 9-16 Stiffened Seated Beam Connections—Bolted or Riveted

Stiffener Angle Capacity, kips

Stiffener Material		$F_y = 36$ ksi ($F_p = 33$ ksi)			$F_y = 50$ ksi ($F_p = 45$ ksi)		
Stiffener Outstanding Leg, A, In.		3½	4	5	3½	4	5
Max. Length Beam Bearing, In.		3½	4¼	5¼	3½	4¼	5¼
Thickness of Stiffener Outstanding Legs In.	5/16	61.9	72.2	92.8	84.4	98.4	127
	⅜	74.3	86.6	111	101	118	152
	7/16	86.6	101	130	118	138	177
	½	99.0	116	149	135	158	203
	⅝	124	144	186	169	197	253
	¾	149	173	223	203	236	304

Use ⅜" thick seat angles with vertical legs wide enough to accommodate fastener pattern, and with outstanding legs wide enough to extend beyond outstanding legs of stiffener.

Fastener Capacity, kips

[a]Fastener Specification	Fastener Diameter In.	Number of Fasteners in One Vertical Row				
		3	4	5	6	7
A307	¾	26.5	35.4	44.2	53.0	61.9
	⅞	36.1	48.1	60.2	72.2	84.2
	1	47.1	62.8	78.6	94.3	110
A325-F A325-N A502-1	¾	39.8	53.0	66.3	79.5	92.8
	⅞	54.1	72.2	90.2	108	126
	1	70.7	94.2	118	141	165
A490-F A502-2	¾	53.0	70.7	88.4	106	124
	⅞	72.2	96.2	120	144	168
	1	94.3	126	157	189	220
A325-X	¾	58.3	77.8	97.2	117	136
	⅞	79.4	106	132	159	185
	1	104	138	173	207	242
A490-N	¾	59.6	79.8	99.4	119	139
	⅞	81.2	108	135	162	189
	1	106	141	177	212	247
A490-X	¾	84.8	113	141	170	198
	⅞	115	154	192	231	269
	1	151	201	251	302	352

[a] A325-F and A490-F: Friction type connections.
A325-N and A490-N: Bearing type connections with threads included in shear plane.
A325-X and A490-X: Bearing type connections with threads excluded from shear plane.

SOURCE: American Institute of Steel Construction.

TABLE 9-17 Stiffened Seated Beam Connections—Welded—E70XX Electrodes, Allowable Load in kips

L In.	Width of Seat, W, inches											
	4				5				6			
	Weld Size, inches				Weld Size, inches				Weld Size, inches			
	¼	5/16	⅜	7/16	5/16	⅜	7/16	½	5/16	⅜	7/16	½
6	22.7	28.4	34.0	39.7	23.5	28.1	32.8	37.5	19.9	23.9	27.9	31.9
7	29.9	37.4	44.9	52.4	31.2	37.5	43.7	50.0	26.7	32.0	37.3	42.7
8	37.8	47.2	56.7	66.1	39.8	47.8	55.8	63.7	34.3	41.1	48.0	54.8
9	46.1	57.6	69.2	80.7	49.1	59.0	68.8	78.6	42.5	51.1	59.6	68.1
10	54.9	68.6	82.3	96.0	59.0	70.8	82.6	94.4	51.4	61.7	72.0	82.3
11	63.9	79.8	95.8	112	69.4	83.3	97.1	111	60.9	73.1	85.2	97.4
12	73.1	91.4	110	128	80.2	96.2	112	128	70.8	85.0	99.2	113
13	82.5	103	124	144	91.3	110	128	146	81.1	97.4	114	130
14	92.0	115	138	161	103	123	144	164	91.9	110	129	147
15	101	127	152	178	114	137	160	183	103	123	144	165
16	111	139	167	195	126	151	176	202	115	138	160	183
17	121	151	181	212	138	165	193	221	126	151	176	201
18	131	163	196	229	150	180	210	240	137	164	192	219
19	140	175	211	246	162	194	227	259	149	179	208	238
20	150	188	225	263	174	209	243	278	161	193	225	257
21	160	200	240	280	189	223	260	298	173	207	242	276
22	169	212	254	296	198	238	277	317	185	221	258	295
23	179	224	269	313	210	252	294	336	197	236	275	315
24	189	236	283	330	222	267	311	356	209	250	292	334
25	198	248	297	347	234	281	328	375	221	265	309	353
26	208	260	312	364	247	296	345	394	233	279	326	373
27	217	272	326	380	259	310	362	414	245	294	343	392

Note: Loads shown are for E70XX electrodes. For E60XX electrodes, multiply tabular loads by 0.86, or enter table with 1.17 times the given reaction. For E80XX electrodes, multiply tabular loads by 1.14, or enter table with 87.5% of the given reaction.

SOURCE: American Institute of Steel Construction.

TABLE 9-18 Stiffened Seated Beam Connections—Welded—E70XX Electrodes, Allowable Loads in kips

L In.	Width of Seat, W, inches											
	7				8				9			
	Weld Size, inches				Weld Size, inches				Weld Size, inches			
	5/16	3/8	7/16	1/2	5/16	3/8	1/2	5/8	5/16	3/8	1/2	5/8
11	54.0	64.8	75.6	86.3	48.4	58.0	77.3	96.6	43.7	52.4	69.9	87.4
12	63.1	75.7	88.4	101	56.7	68.1	90.7	113	51.4	61.7	82.2	103
13	72.7	87.2	102	117	65.5	78.7	105	131	59.6	71.5	95.3	119
14	82.6	99.1	116	132	74.8	89.8	120	149	68.2	81.8	109	136
15	92.9	112	130	149	84.4	101	135	169	77.2	92.6	123	154
16	104	124	145	166	94.4	113	151	189	86.5	104	138	173
17	114	137	160	183	105	126	167	209	96.2	115	154	192
18	126	151	176	201	115	138	184	230	106	127	170	212
19	137	164	192	219	126	151	202	252	117	140	186	233
20	148	178	208	237	137	165	219	274	127	152	203	254
21	160	192	224	256	148	178	237	296	138	165	220	276
22	172	206	240	274	159	192	255	319	149	178	238	297
23	183	220	257	293	171	205	274	342	160	192	256	320
24	195	234	274	312	183	219	292	365	171	205	274	342
25	207	249	290	331	195	233	311	389	182	219	292	365
26	219	263	307	351	206	248	330	412	194	233	310	388
27	231	278	324	370	218	262	349	436	206	247	329	411
28	243	292	341	389	230	276	368	460	217	261	348	435
29	256	307	358	409	242	291	387	484	229	275	367	458
30	268	321	375	428	254	305	406	508	241	289	386	482
31	280	336	392	447	266	319	426	532	253	303	405	506
32	292	350	409	467	278	334	445	556	265	318	424	530

Note: Loads shown are for E70XX electrodes. For E60XX electrodes, multiply tabular loads by 0.86, or enter table with 1.17 times the given reaction. For E80XX electrodes, multiply tabular loads by 1.14, or enter table with 87.5% of the given reaction.

SOURCE: American Institute of Steel Construction.

9-5 End-Plate Shear Connections The end-plate connection consists of a plate which is less than the beam depth in length and perpendicular to the longitudinal axis of the beam. The end plate is welded to the web with fillet welds on each side.

Table 9-19 tabulates the fastener selection, allowable loadings, weld size, minimum web thickness, and the weld capacity.

TABLE 9-19 End-Plate Shear Connections

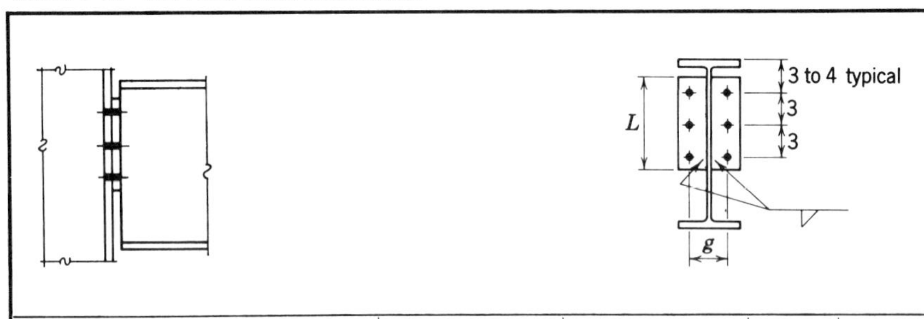

Fasteners per Vertical Line	Fastener	¾″ Diam. Total Capacity Kips	¾″ Diam. Min. Plate Thickness (t) In.	⅞″ Diam. Total Capacity Kips	⅞″ Diam. Min. Plate Thickness (t) In.	Plate Length (L) Ft-In.	Beam Depth Limits In.
1	ASTM A307 Bolts	8.8	.121	12.0	.141	3	5–8
	[a]ASTM A325 HS Bolts	13.3	.182	18.0	.212		
	[b]ASTM A325 HS Bolts	19.4	.267	26.5	.311		
2	ASTM A307 Bolts	17.7	.121	24.0	.141	5½	8–12
	[a]ASTM A325 HS Bolts	26.5	.182	36.1	.212		
	[b]ASTM A325 HS Bolts	38.9	.267	52.9	.311		
3	ASTM A307 Bolts	26.5	.121	36.1	.141	8½	12–18
	[a]ASTM A325 HS Bolts	39.8	.182	54.1	.212		
	[b]ASTM A325 HS Bolts	58.3	.267	79.4	.311		
4	ASTM A307 Bolts	35.4	.121	48.1	.141	11½	15–24
	[a]ASTM A325 HS Bolts	53.0	.182	72.2	.212		
	[b]ASTM A325 HS Bolts	77.8	.267	105.8	.311		
5	ASTM A307 Bolts	44.2	.121	60.1	.141	1′-2½	18–30
	[a]ASTM A325 HS Bolts	66.3	.182	90.2	.212		
	[b]ASTM A325 HS Bolts	97.2	.267	132.3	.311		
6	ASTM A307 Bolts	53.0	.121	72.1	.141	1′-5½	21–36
	[a]ASTM A325 HS Bolts	79.6	.182	108.2	.212		
	[b]ASTM A325 HS Bolts	116.6	.267	158.8	.311		

[a] Friction type connection, or bearing type with threads in shear planes.
[b] Bearing type connection; threads excluded from shear planes.

WELD CAPACITY

Weld Size	Minimum Web Thickness, In.		Weld Capacity, Kips (2 Fillet Welds)					
	$F_y = 36$	$F_y = 50$	3	5½	8½	11½	1′-2½	1′-5½
3/16	.389	.280	14.7	28.5	45.2	61.9	78.7	95.2
¼	.514	.370	18.6	37.1	59.4	81.6	103.9	126.1
5/16	.646	.465	22.1	45.3	73.1	101.0	128.8	156.7
⅜	.771	.555	25.1	52.9	86.3	119.8	153.4	186.6

SOURCE: American Institute of Steel Construction.

9-6 Eccentric Loads on Fastener Groups Eccentric loads on a fastener group result in unequal stresses on several fasteners within the group. Tables 9-20 through 9-23 show coefficients for several fastener groups. With the aid of standard formulas, the designer can solve the problem by plugging in known factors and the coefficient

TABLE 9-20 Eccentric Loads on Fastener Groups, Coefficients C

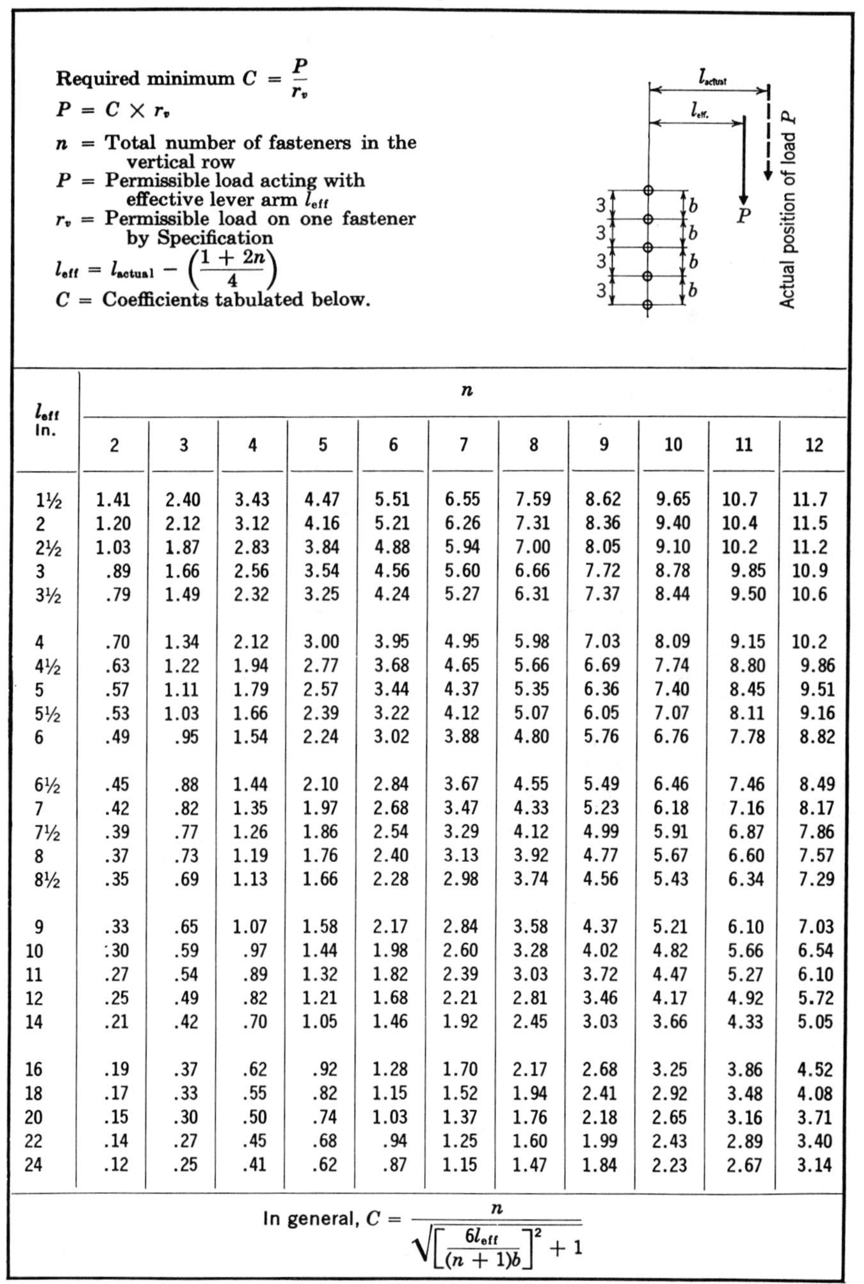

Required minimum $C = \frac{P}{r_v}$

$P = C \times r_v$

n = Total number of fasteners in the vertical row

P = Permissible load acting with effective lever arm l_{eff}

r_v = Permissible load on one fastener by Specification

$l_{eff} = l_{actual} - \left(\frac{1 + 2n}{4}\right)$

C = Coefficients tabulated below.

l_{eff} In.	n										
	2	3	4	5	6	7	8	9	10	11	12
1½	1.41	2.40	3.43	4.47	5.51	6.55	7.59	8.62	9.65	10.7	11.7
2	1.20	2.12	3.12	4.16	5.21	6.26	7.31	8.36	9.40	10.4	11.5
2½	1.03	1.87	2.83	3.84	4.88	5.94	7.00	8.05	9.10	10.2	11.2
3	.89	1.66	2.56	3.54	4.56	5.60	6.66	7.72	8.78	9.85	10.9
3½	.79	1.49	2.32	3.25	4.24	5.27	6.31	7.37	8.44	9.50	10.6
4	.70	1.34	2.12	3.00	3.95	4.95	5.98	7.03	8.09	9.15	10.2
4½	.63	1.22	1.94	2.77	3.68	4.65	5.66	6.69	7.74	8.80	9.86
5	.57	1.11	1.79	2.57	3.44	4.37	5.35	6.36	7.40	8.45	9.51
5½	.53	1.03	1.66	2.39	3.22	4.12	5.07	6.05	7.07	8.11	9.16
6	.49	.95	1.54	2.24	3.02	3.88	4.80	5.76	6.76	7.78	8.82
6½	.45	.88	1.44	2.10	2.84	3.67	4.55	5.49	6.46	7.46	8.49
7	.42	.82	1.35	1.97	2.68	3.47	4.33	5.23	6.18	7.16	8.17
7½	.39	.77	1.26	1.86	2.54	3.29	4.12	4.99	5.91	6.87	7.86
8	.37	.73	1.19	1.76	2.40	3.13	3.92	4.77	5.67	6.60	7.57
8½	.35	.69	1.13	1.66	2.28	2.98	3.74	4.56	5.43	6.34	7.29
9	.33	.65	1.07	1.58	2.17	2.84	3.58	4.37	5.21	6.10	7.03
10	.30	.59	.97	1.44	1.98	2.60	3.28	4.02	4.82	5.66	6.54
11	.27	.54	.89	1.32	1.82	2.39	3.03	3.72	4.47	5.27	6.10
12	.25	.49	.82	1.21	1.68	2.21	2.81	3.46	4.17	4.92	5.72
14	.21	.42	.70	1.05	1.46	1.92	2.45	3.03	3.66	4.33	5.05
16	.19	.37	.62	.92	1.28	1.70	2.17	2.68	3.25	3.86	4.52
18	.17	.33	.55	.82	1.15	1.52	1.94	2.41	2.92	3.48	4.08
20	.15	.30	.50	.74	1.03	1.37	1.76	2.18	2.65	3.16	3.71
22	.14	.27	.45	.68	.94	1.25	1.60	1.99	2.43	2.89	3.40
24	.12	.25	.41	.62	.87	1.15	1.47	1.84	2.23	2.67	3.14

In general, $C = \dfrac{n}{\sqrt{\left[\dfrac{6l_{eff}}{(n+1)b}\right]^2 + 1}}$

SOURCE: American Institute of Steel Construction.

TABLE 9-21 Eccentric Loads on Fastener Groups, Coefficients C

Required minimum $C = \dfrac{P}{r_v}$

$P = C \times r_v$

n = Total number of fasteners in any one vertical row

P = Permissible load acting with effective lever arm l_{eff}

r_v = Permissible load on one fastener by Specification

$l_{eff} = l_{actual} - \left(\dfrac{1+n}{2}\right)$

C = Coefficients tabulated below.

l_{eff} In.	n = 1	2	3	4	5	6	7	8	9	10	11	12
1½	1.00	2.53	4.33	6.30	8.36	10.4	12.5	14.6	16.7	18.8	20.9	23.0
2	.86	2.23	3.88	5.75	7.74	9.80	11.9	14.0	16.1	18.2	20.3	22.4
2½	.75	1.99	3.50	5.24	7.16	9.17	11.2	13.3	15.5	17.6	19.7	21.8
3	.67	1.79	3.17	4.80	6.62	8.56	10.6	12.7	14.8	16.9	19.0	21.1
3½	.60	1.63	2.89	4.41	6.13	8.00	9.97	12.0	14.1	16.2	18.3	20.5
4	.55	1.49	2.66	4.07	5.69	7.48	9.39	11.4	13.4	15.5	17.6	19.8
4½	.50	1.37	2.45	3.77	5.30	7.01	8.85	10.8	12.8	14.8	16.9	19.0
5	.46	1.27	2.28	3.51	4.96	6.58	8.34	10.2	12.2	14.2	16.3	18.4
5½	.43	1.19	2.12	3.28	4.64	6.19	7.88	9.70	11.6	13.6	15.6	17.7
6	.40	1.11	1.99	3.07	4.37	5.84	7.46	9.21	11.1	13.0	15.0	17.0
6½	.37	1.04	1.87	2.89	4.12	5.52	7.08	8.76	10.6	12.4	14.4	16.4
7	.35	.98	1.76	2.73	3.89	5.23	6.72	8.35	10.1	11.9	13.8	15.8
7½	.33	.93	1.66	2.58	3.69	4.97	6.40	7.96	9.64	11.4	13.3	15.2
8	.32	.88	1.58	2.45	3.50	4.72	6.10	7.61	9.23	11.0	12.8	14.6
8½	.30	.84	1.50	2.33	3.33	4.50	5.82	7.28	8.85	10.5	12.3	14.1
9	.29	.80	1.43	2.22	3.18	4.30	5.57	6.97	8.49	10.1	11.8	13.6
10	.26	.73	1.30	2.03	2.91	3.94	5.12	6.43	7.85	9.38	11.0	12.7
11	.24	.67	1.20	1.87	2.68	3.64	4.73	5.95	7.29	8.73	10.3	11.9
12	.22	.62	1.11	1.73	2.48	3.38	4.40	5.54	6.80	8.16	9.61	11.2
14	.19	.54	.97	1.50	2.16	2.95	3.85	4.86	5.98	7.19	8.51	9.91
16	.17	.48	.86	1.33	1.92	2.61	3.41	4.32	5.32	6.42	7.61	8.88
18	.15	.43	.77	1.19	1.72	2.34	3.07	3.88	4.79	5.79	6.88	8.04
20	.14	.39	.70	1.08	1.56	2.12	2.78	3.53	4.36	5.27	6.26	7.34
22	.13	.36	.64	.99	1.42	1.94	2.54	3.23	3.99	4.83	5.75	6.74
24	.12	.33	.59	.91	1.31	1.79	2.34	2.98	3.68	4.46	5.31	6.23

In general,
$$C = \frac{n}{\sqrt{\left[\dfrac{l_{eff}(n-1)b}{D^2 + \frac{1}{3}(n^2-1)b^2}\right]^2 + \left[\dfrac{l_{eff}\,D}{D^2 + \frac{1}{3}(n^2-1)b^2} + \frac{1}{2}\right]^2}}$$

SOURCE: American Institute of Steel Construction.

TABLE 9-22 Eccentric Loads on Fastener Groups, Coefficients C

Required minimum $C = \dfrac{P}{r_v}$

$P = C \times r_v$

n = Total number of fasteners in any one vertical row

P = Permissible load acting with effective lever arm l_{eff}

r_v = Permissible load on one fastener by Specification

$l_{eff} = l_{actual} - \left(\dfrac{1+n}{2}\right)$

C = Coefficients tabulated below.

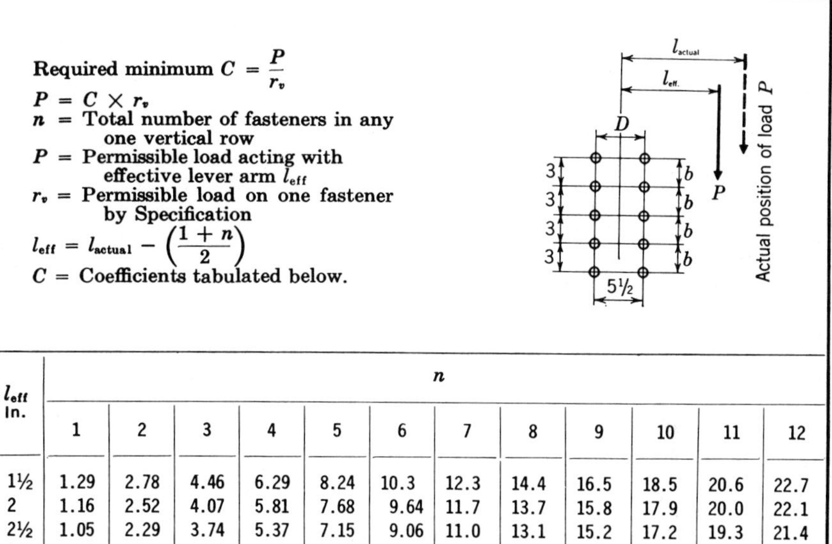

l_{eff} in.	n = 1	2	3	4	5	6	7	8	9	10	11	12
1½	1.29	2.78	4.46	6.29	8.24	10.3	12.3	14.4	16.5	18.5	20.6	22.7
2	1.16	2.52	4.07	5.81	7.68	9.64	11.7	13.7	15.8	17.9	20.0	22.1
2½	1.05	2.29	3.74	5.37	7.15	9.06	11.0	13.1	15.2	17.2	19.3	21.4
3	.96	2.11	3.45	4.98	6.67	8.51	10.4	12.4	14.5	16.6	18.7	20.8
3½	.88	1.95	3.20	4.63	6.24	7.99	9.87	11.8	13.8	15.9	18.0	20.1
4	.81	1.81	2.98	4.32	5.84	7.52	9.33	11.2	13.2	15.2	17.3	19.4
4½	.76	1.69	2.78	4.05	5.49	7.09	8.83	10.7	12.6	14.6	16.6	18.7
5	.71	1.59	2.61	3.80	5.17	6.70	8.37	10.2	12.0	14.0	16.0	18.0
5½	.67	1.49	2.46	3.58	4.88	6.34	7.94	9.67	11.5	13.4	15.4	17.4
6	.63	1.41	2.32	3.39	4.62	6.01	7.55	9.22	11.0	12.8	14.8	16.8
6½	.59	1.34	2.20	3.21	4.38	5.71	7.19	8.80	10.5	12.3	14.2	16.2
7	.56	1.27	2.09	3.05	4.16	5.44	6.86	8.40	10.1	11.8	13.7	15.6
7½	.54	1.21	1.99	2.90	3.96	5.18	6.55	8.04	9.65	11.4	13.2	15.0
8	.51	1.15	1.90	2.77	3.78	4.95	6.26	7.70	9.26	10.9	12.7	14.5
8½	.49	1.10	1.81	2.64	3.62	4.74	6.00	7.39	8.90	10.5	12.2	14.0
9	.47	1.06	1.74	2.53	3.46	4.54	5.76	7.10	8.56	10.1	11.8	13.5
10	.43	.98	1.60	2.33	3.19	4.19	5.32	6.57	7.94	9.42	11.0	12.6
11	.40	.91	1.48	2.16	2.96	3.88	4.94	6.11	7.40	8.80	10.3	11.9
12	.37	.85	1.38	2.01	2.75	3.62	4.61	5.71	6.92	8.24	9.65	11.2
14	.33	.75	1.22	1.77	2.42	3.18	4.06	5.04	6.12	7.30	8.58	9.94
16	.29	.67	1.09	1.58	2.16	2.84	3.62	4.50	5.48	6.55	7.70	8.94
18	.26	.60	.98	1.42	1.94	2.56	3.26	4.06	4.95	5.92	6.98	8.12
20	.24	.55	.89	1.29	1.77	2.33	2.97	3.70	4.51	5.41	6.38	7.42
22	.22	.51	.82	1.19	1.62	2.13	2.73	3.40	4.14	4.97	5.87	6.84
24	.21	.47	.76	1.10	1.50	1.97	2.52	3.14	3.83	4.59	5.43	6.33

In general,

$$C = \frac{n}{\sqrt{\left[\dfrac{l_{eff}\,(n-1)b}{D^2 + \frac{1}{3}(n^2-1)b^2}\right]^2 + \left[\dfrac{l_{eff}\,D}{D^2 + \frac{1}{3}(n^2-1)b^2} + \frac{1}{2}\right]^2}}$$

SOURCE: American Institute of Steel Construction.

TABLE 9-23 Eccentric Loads on Fastener Groups, Coefficients *C*

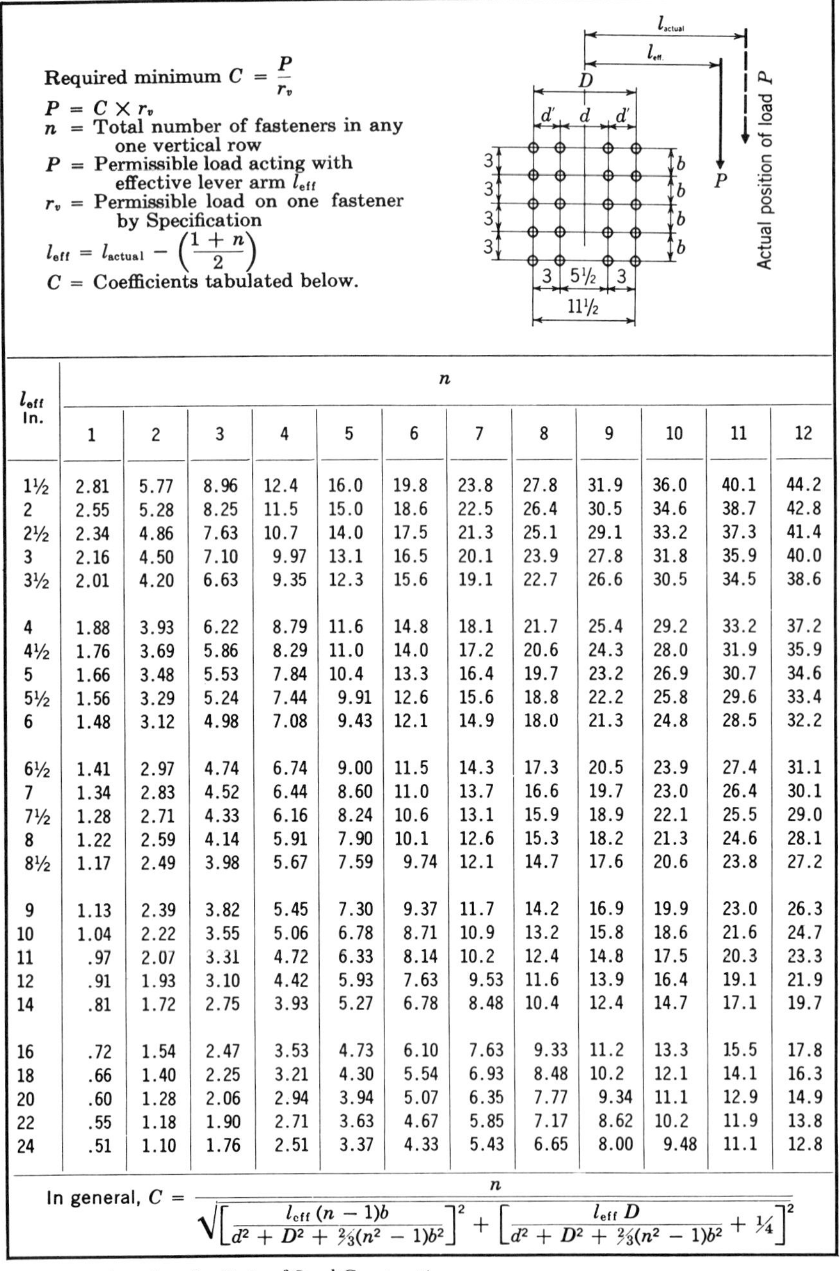

Required minimum $C = \dfrac{P}{r_v}$

$P = C \times r_v$

n = Total number of fasteners in any one vertical row

P = Permissible load acting with effective lever arm l_{eff}

r_v = Permissible load on one fastener by Specification

$l_{eff} = l_{actual} - \left(\dfrac{1+n}{2}\right)$

C = Coefficients tabulated below.

l_{eff} In.	n = 1	2	3	4	5	6	7	8	9	10	11	12
1½	2.81	5.77	8.96	12.4	16.0	19.8	23.8	27.8	31.9	36.0	40.1	44.2
2	2.55	5.28	8.25	11.5	15.0	18.6	22.5	26.4	30.5	34.6	38.7	42.8
2½	2.34	4.86	7.63	10.7	14.0	17.5	21.3	25.1	29.1	33.2	37.3	41.4
3	2.16	4.50	7.10	9.97	13.1	16.5	20.1	23.9	27.8	31.8	35.9	40.0
3½	2.01	4.20	6.63	9.35	12.3	15.6	19.1	22.7	26.6	30.5	34.5	38.6
4	1.88	3.93	6.22	8.79	11.6	14.8	18.1	21.7	25.4	29.2	33.2	37.2
4½	1.76	3.69	5.86	8.29	11.0	14.0	17.2	20.6	24.3	28.0	31.9	35.9
5	1.66	3.48	5.53	7.84	10.4	13.3	16.4	19.7	23.2	26.9	30.7	34.6
5½	1.56	3.29	5.24	7.44	9.91	12.6	15.6	18.8	22.2	25.8	29.6	33.4
6	1.48	3.12	4.98	7.08	9.43	12.1	14.9	18.0	21.3	24.8	28.5	32.2
6½	1.41	2.97	4.74	6.74	9.00	11.5	14.3	17.3	20.5	23.9	27.4	31.1
7	1.34	2.83	4.52	6.44	8.60	11.0	13.7	16.6	19.7	23.0	26.4	30.1
7½	1.28	2.71	4.33	6.16	8.24	10.6	13.1	15.9	18.9	22.1	25.5	29.0
8	1.22	2.59	4.14	5.91	7.90	10.1	12.6	15.3	18.2	21.3	24.6	28.1
8½	1.17	2.49	3.98	5.67	7.59	9.74	12.1	14.7	17.6	20.6	23.8	27.2
9	1.13	2.39	3.82	5.45	7.30	9.37	11.7	14.2	16.9	19.9	23.0	26.3
10	1.04	2.22	3.55	5.06	6.78	8.71	10.9	13.2	15.8	18.6	21.6	24.7
11	.97	2.07	3.31	4.72	6.33	8.14	10.2	12.4	14.8	17.5	20.3	23.3
12	.91	1.93	3.10	4.42	5.93	7.63	9.53	11.6	13.9	16.4	19.1	21.9
14	.81	1.72	2.75	3.93	5.27	6.78	8.48	10.4	12.4	14.7	17.1	19.7
16	.72	1.54	2.47	3.53	4.73	6.10	7.63	9.33	11.2	13.3	15.5	17.8
18	.66	1.40	2.25	3.21	4.30	5.54	6.93	8.48	10.2	12.1	14.1	16.3
20	.60	1.28	2.06	2.94	3.94	5.07	6.35	7.77	9.34	11.1	12.9	14.9
22	.55	1.18	1.90	2.71	3.63	4.67	5.85	7.17	8.62	10.2	11.9	13.8
24	.51	1.10	1.76	2.51	3.37	4.33	5.43	6.65	8.00	9.48	11.1	12.8

In general,

$$C = \frac{n}{\sqrt{\left[\dfrac{l_{eff}\,(n-1)b}{d^2 + D^2 + \frac{2}{3}(n^2-1)b^2}\right]^2 + \left[\dfrac{l_{eff}\,D}{d^2 + D^2 + \frac{2}{3}(n^2-1)b^2} + \frac{1}{4}\right]^2}}$$

SOURCE: American Institute of Steel Construction.

C from these tables (see AISC *Manual of Steel Construction*), thus obtaining the desired answer.

9-7 Bracket Plates Simple bracket plates are a basic design of connecting structural steel. Table 9-24 tabulates the net section moduli as related to number of fasteners in one vertical line (spaced 3 inches vertically), depth of plate, size of fasteners, and thickness of plate.

TABLE 9-24 Bracket Plates—Net Section Moduli

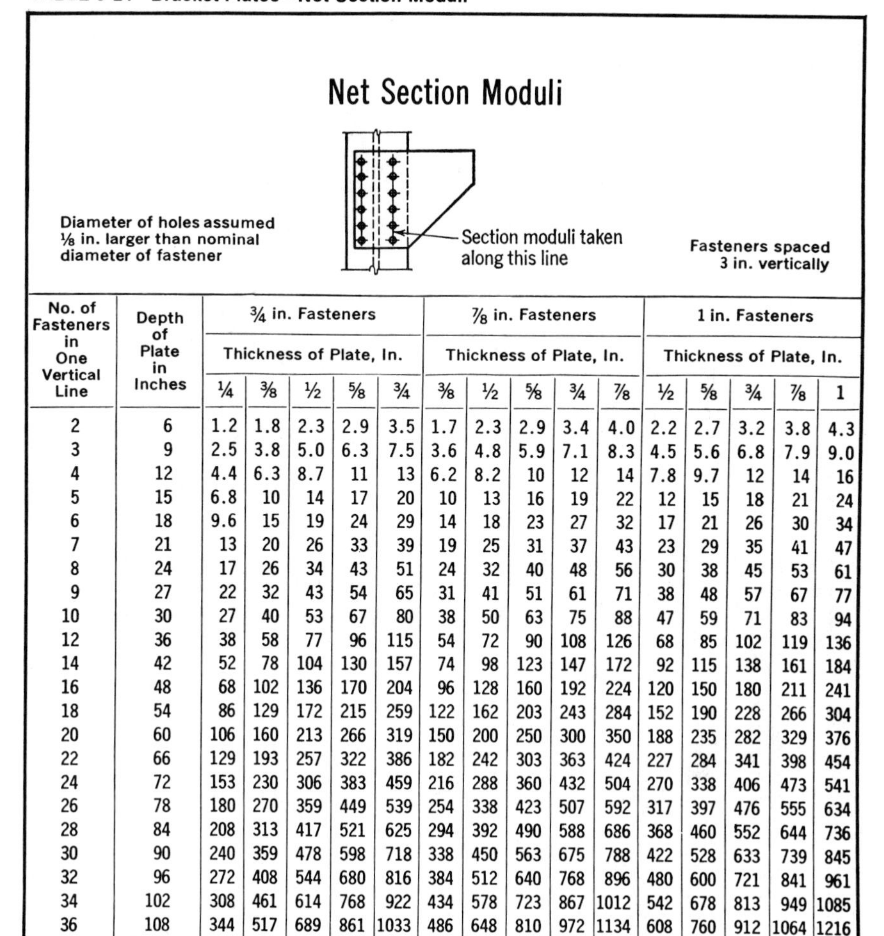

No. of Fasteners in One Vertical Line	Depth of Plate in Inches	¾ in. Fasteners					⅞ in. Fasteners					1 in. Fasteners				
		Thickness of Plate, In.					Thickness of Plate, In.					Thickness of Plate, In.				
		¼	⅜	½	⅝	¾	⅜	½	⅝	¾	⅞	½	⅝	¾	⅞	1
2	6	1.2	1.8	2.3	2.9	3.5	1.7	2.3	2.9	3.4	4.0	2.2	2.7	3.2	3.8	4.3
3	9	2.5	3.8	5.0	6.3	7.5	3.6	4.8	5.9	7.1	8.3	4.5	5.6	6.8	7.9	9.0
4	12	4.4	6.3	8.7	11	13	6.2	8.2	10	12	14	7.8	9.7	12	14	16
5	15	6.8	10	14	17	20	10	13	16	19	22	12	15	18	21	24
6	18	9.6	15	19	24	29	14	18	23	27	32	17	21	26	30	34
7	21	13	20	26	33	39	19	25	31	37	43	23	29	35	41	47
8	24	17	26	34	43	51	24	32	40	48	56	30	38	45	53	61
9	27	22	32	43	54	65	31	41	51	61	71	38	48	57	67	77
10	30	27	40	53	67	80	38	50	63	75	88	47	59	71	83	94
12	36	38	58	77	96	115	54	72	90	108	126	68	85	102	119	136
14	42	52	78	104	130	157	74	98	123	147	172	92	115	138	161	184
16	48	68	102	136	170	204	96	128	160	192	224	120	150	180	211	241
18	54	86	129	172	215	259	122	162	203	243	284	152	190	228	266	304
20	60	106	160	213	266	319	150	200	250	300	350	188	235	282	329	376
22	66	129	193	257	322	386	182	242	303	363	424	227	284	341	398	454
24	72	153	230	306	383	459	216	288	360	432	504	270	338	406	473	541
26	78	180	270	359	449	539	254	338	423	507	592	317	397	476	555	634
28	84	208	313	417	521	625	294	392	490	588	686	368	460	552	644	736
30	90	240	359	478	598	718	338	450	563	675	788	422	528	633	739	845
32	96	272	408	544	680	816	384	512	640	768	896	480	600	721	841	961
34	102	308	461	614	768	922	434	578	723	867	1012	542	678	813	949	1085
36	108	344	517	689	861	1033	486	648	810	972	1134	608	760	912	1064	1216

Interpolate for intermediate thickness of plates.
General equation for net section modulus of bracket plates:

$$S_{net} = \frac{t_P d^2}{6} - \frac{b^2 n(n^2 - 1)\,[t_P \times (\text{Bolt Diam.} + 0.125)]}{6d}$$

where

t_P = Plate thickness, inches
d = Plate depth, inches
n = Number of fasteners in one vertical row
b = Fastener spacing vertically, inches

SOURCE: American Institute of Steel Construction.

9-8 Typical Design Connections Details shown in Figs. 9-1 thru 9-11 are suggested treatments only, and are not intended to limit the use of other similar connections not illustrated. They are shown here to give the design engineer insight and stimulate creative thought processes in solving connection and fastening problems.

9-9 Bolt and Rivet Clearance Erection clearances for rivets and bolts (threaded fasteners) are shown in Tables 9-25 and 9-26. Rivet-gun and bolt-impact-wrench clearances are also illustrated to assist the designer in providing good field erection.

Details on this and succeeding pages are suggested treatments only, and are not intended to limit the use of other similar connections not illustrated.

SKEWED AND SLOPED CONNECTIONS

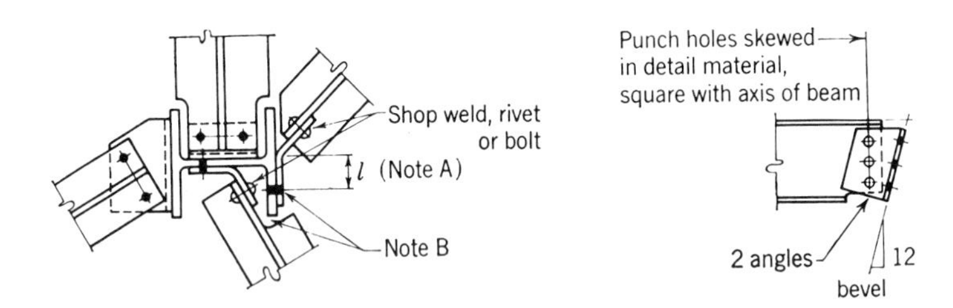

Note A: For bent plate connection, size of plate should be checked using arm l, and eccentricity in fasteners checked using tables of Eccentric Loads on Fastener Groups.

Note B: If a combination of several connections occur at one level, provide field and driving clearance.

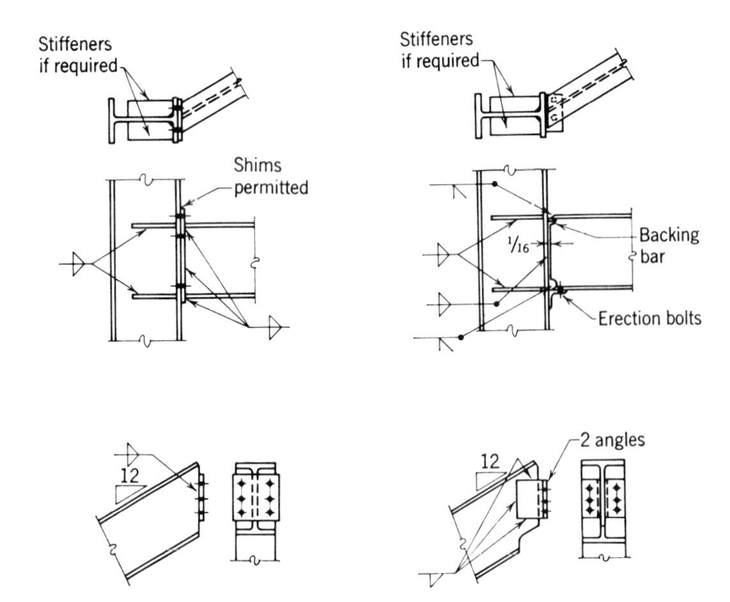

FIG. 9-1 Suggested details—beam framing (skewed and sloped connections). (*American Institute of Steel Construction.*)

MOMENT CONNECTIONS

Wind bracing connections, or connections designed to resist bending moments, are usually made with angles, structural tees or plates.

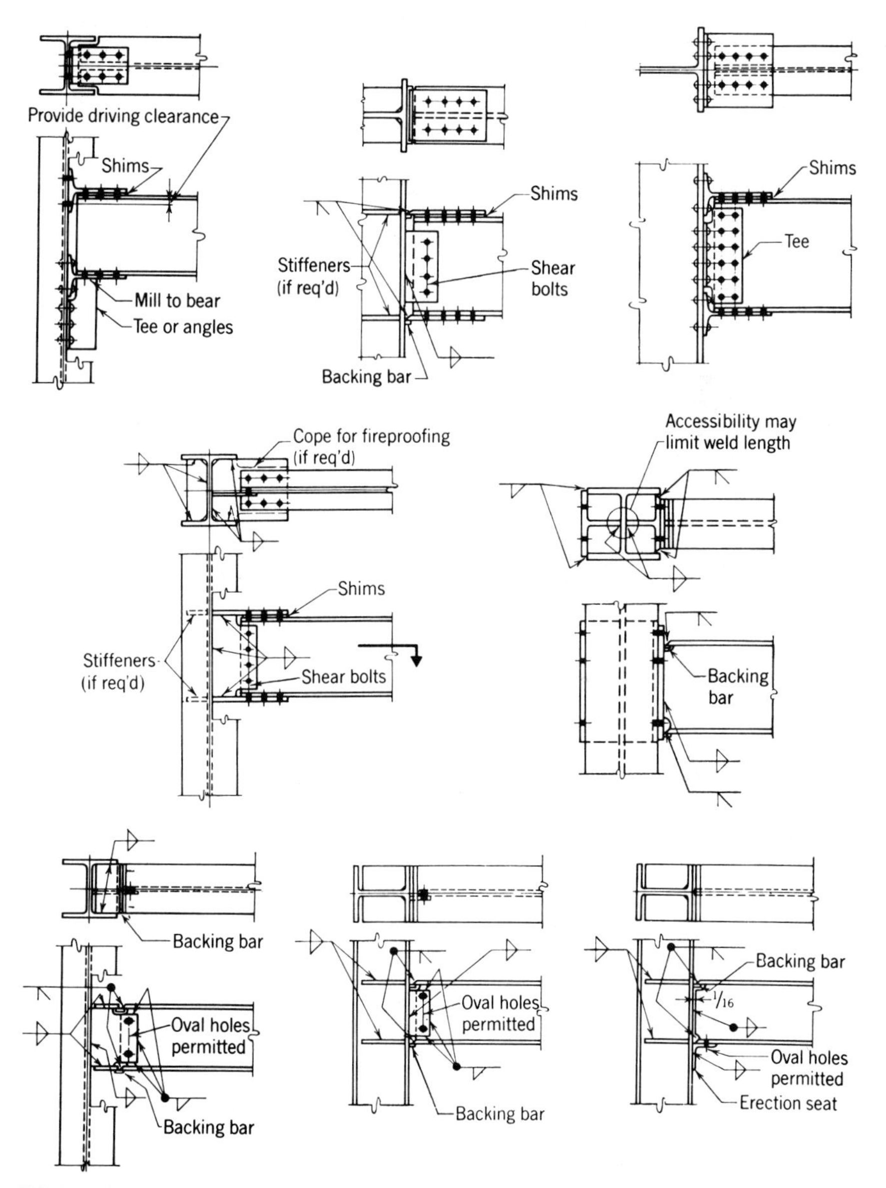

FIG. 9-2 Suggested details—beam framing (moment connections). (*American Institute of Steel Construction.*)

SHEAR CONNECTIONS

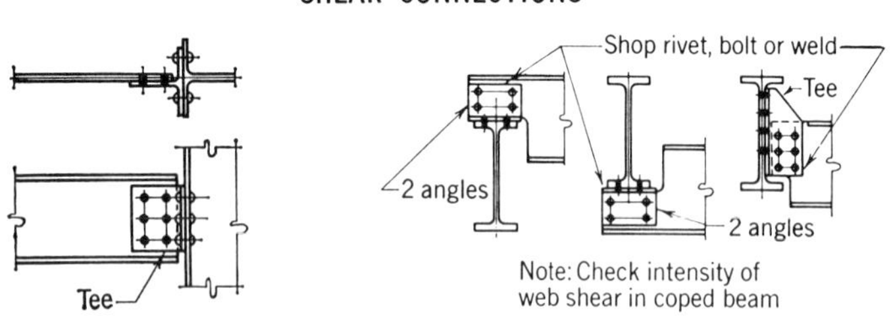

SHEAR SPLICES

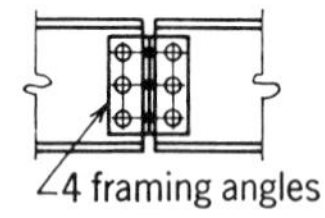

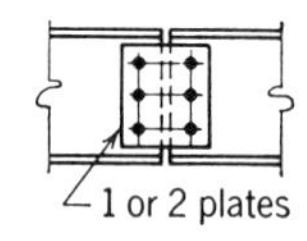

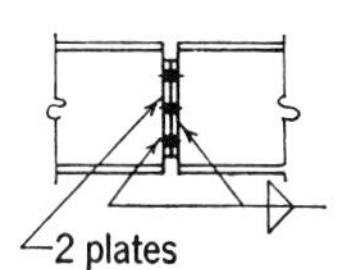

Note: Of the above types, 4 framing angles is most flexible.

BOLTED MOMENT SPLICES

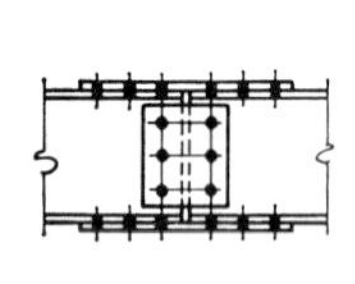

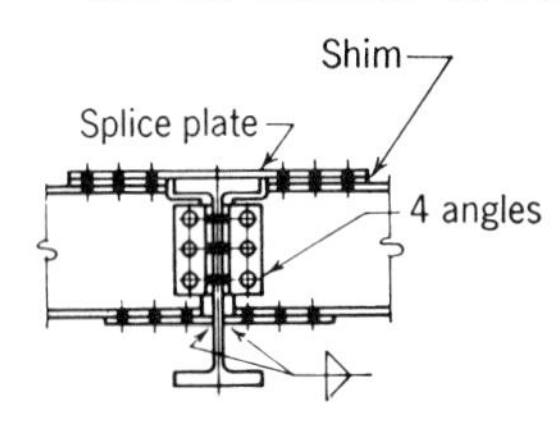

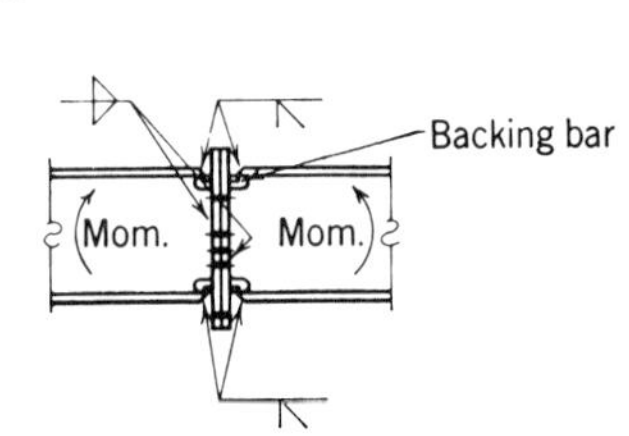

FIG. 9-3 Suggested details—beam framing (shear connections). (*American Institute of Steel Construction.*)

WELDED MOMENT SPLICES

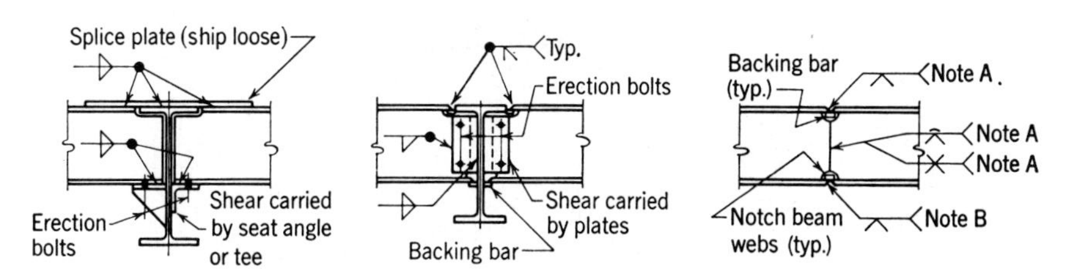

Note A: Joint preparation depends on thickness of material, and welding process.
Note B: Invert this joint preparation if beam cannot be turned over.

MOMENT SPLICE AT RIDGE (FIELD BOLTED)

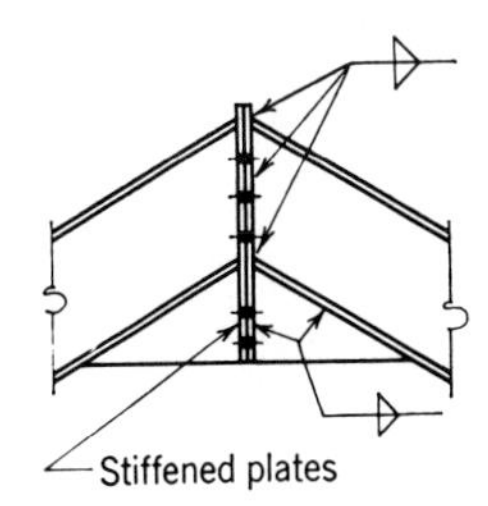

*BEAM OVER COLUMN (WITH CONTINUITY)

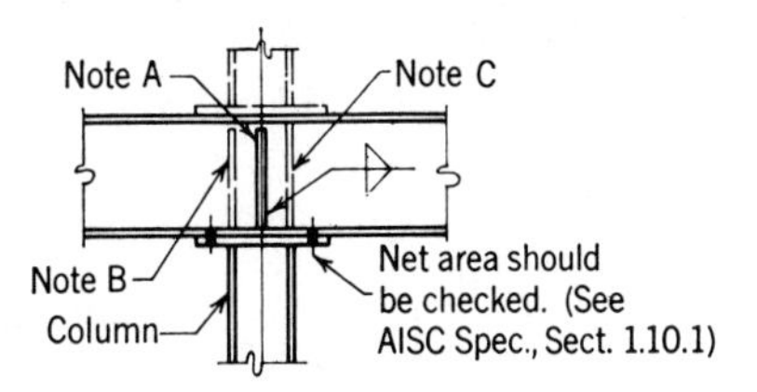

Note A: Two stiffeners, effective only if deck or slab prevents rotation of top flange.
Note B: Optional location of 2 stiffeners over supporting column flanges.
Note C: If column above, use 4 fitted stiffeners.

* For Plastic Design see Spec. Sects. 1.15.5 and 2.6.

FIG. 9-4 Suggested details—beam framing (welded moment splices). (*American Institute of Steel Construction.*)

SUGGESTED DETAILS
Column base plates

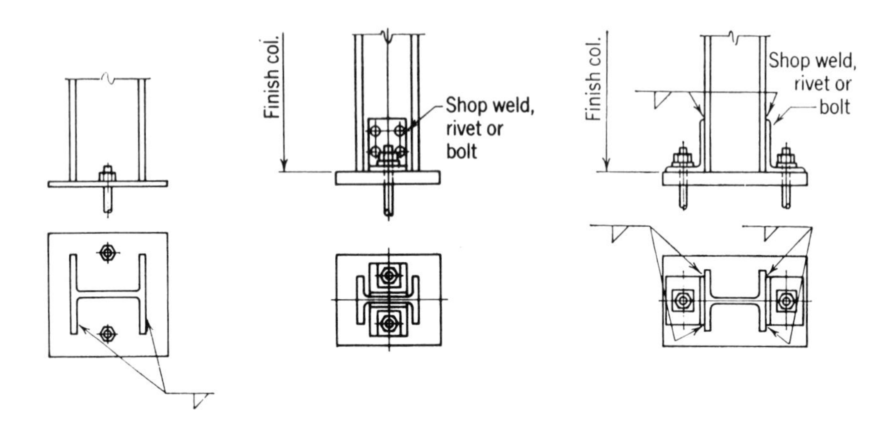

Base plate detailed and shipped separately when required.

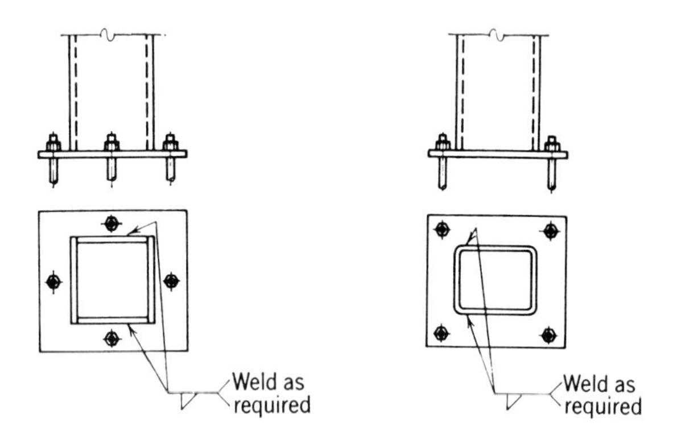

Note: Anchor bolts should be spread as far as practical for safety during erection.

FIG. 9-5 Suggested details—column base plates. *(American Institute of Steel Construction.)*

SUGGESTED DETAILS
Column base plates

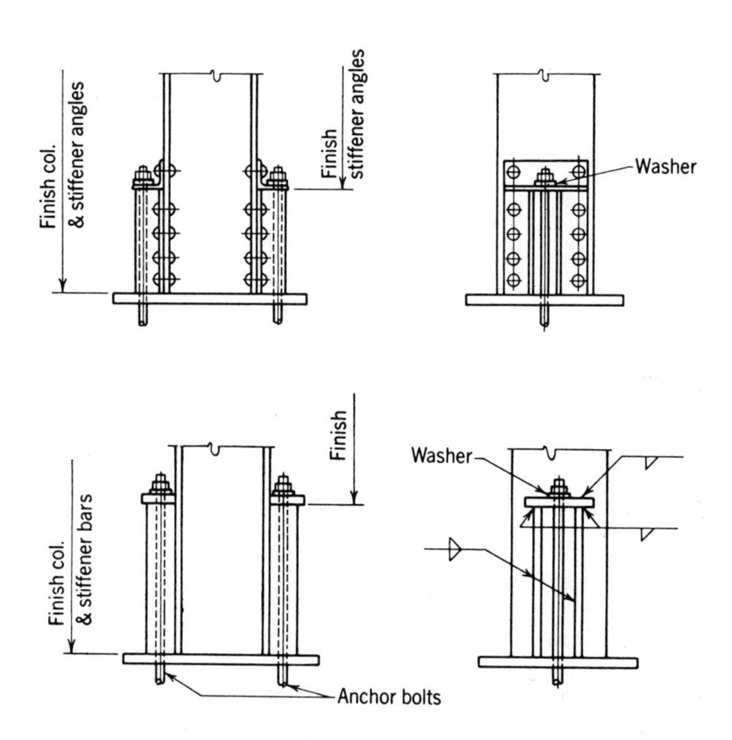

Note: Anchor bolts should be spread as far as practical for safety during erection.
Base plate detailed and shipped separately when required.

FIG. 9-6 Suggested details—column base plates. (*American Institute of Steel Construction.*)

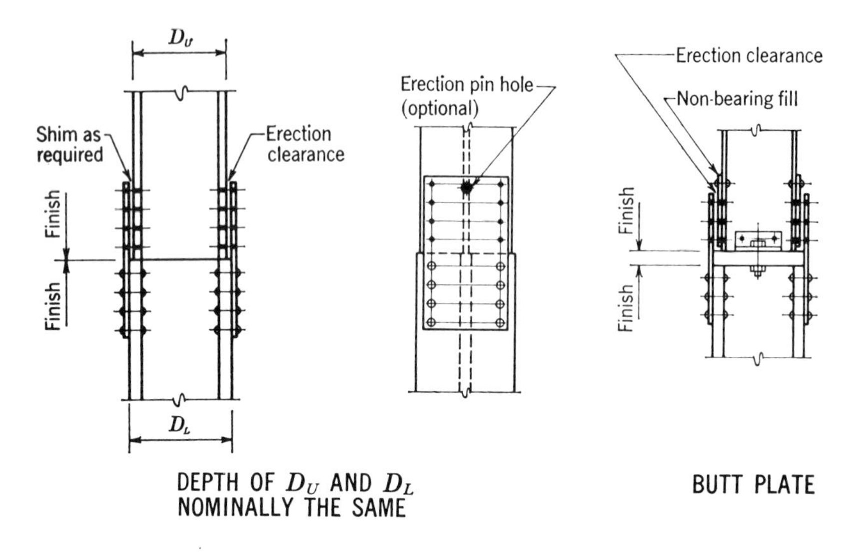

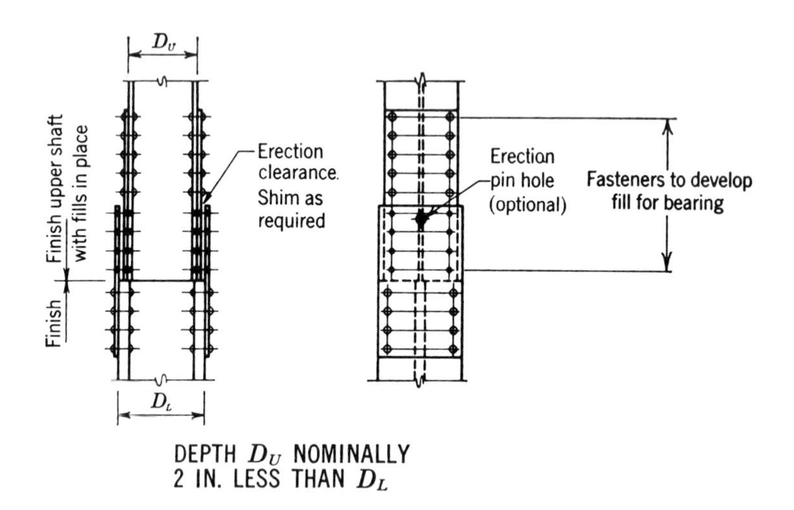

Note: Erection clearance = ⅛ in.

FIG. 9-7 Suggested details—column splices (riveted and bolted). (*American Institute of Steel Construction.*)

Welded

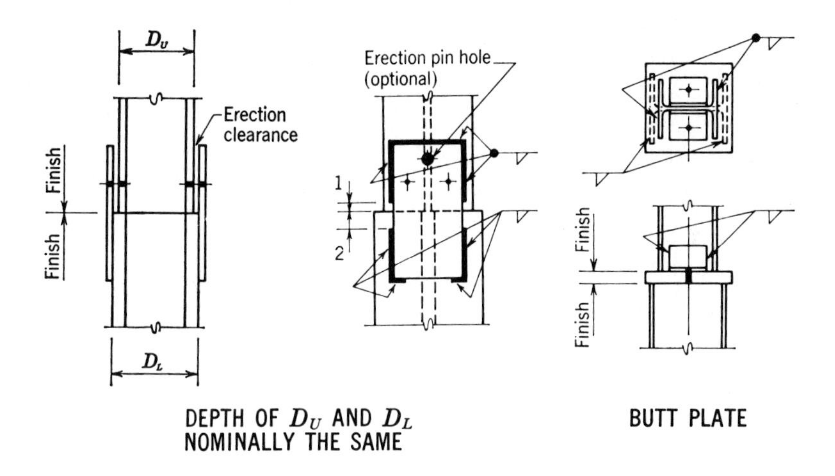

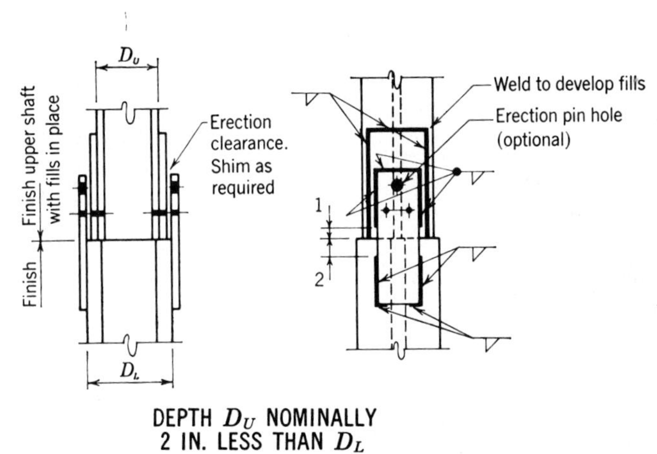

Note 1: Erection clearance = 1/16 in.
Note 2: When D_U and D_L are nominally the same and thin fills are required, shop may attach splice plate to upper section and provide field clearance over lower section.

FIG. 9-8 Suggested details—column splices (welded). (*American Institute of Steel Construction.*)

Welded

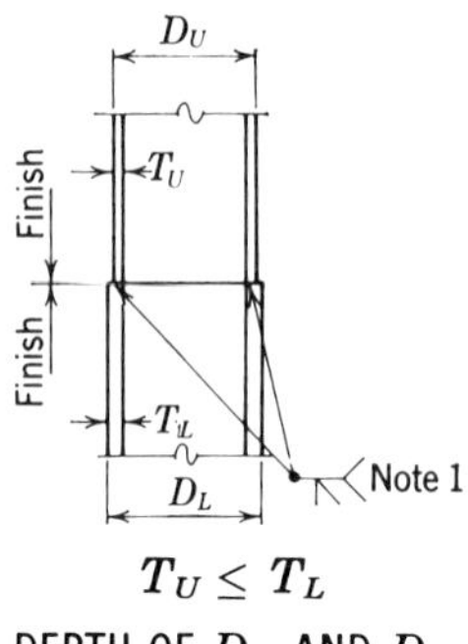

$T_U \leq T_L$

DEPTH OF D_U AND D_L
NOMINALLY THE SAME

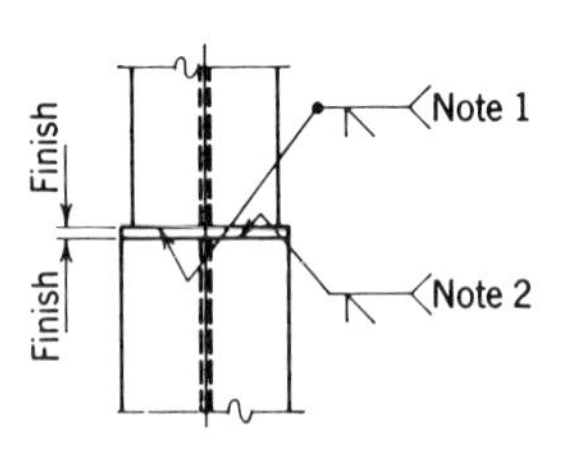

BUTT PLATE

DEPTH D_U NOMINALLY
2 IN. LESS THAN D_L

Box columns

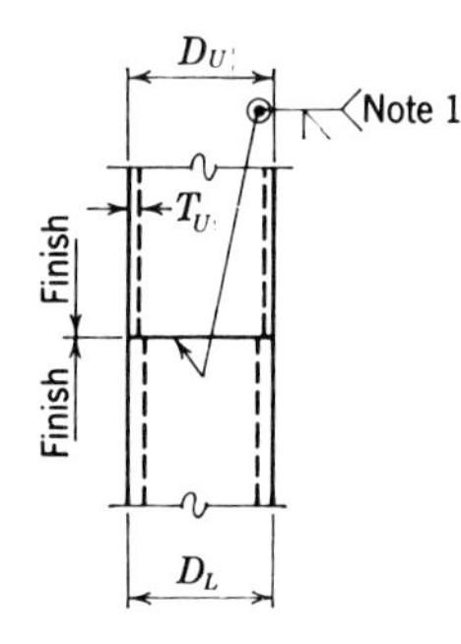

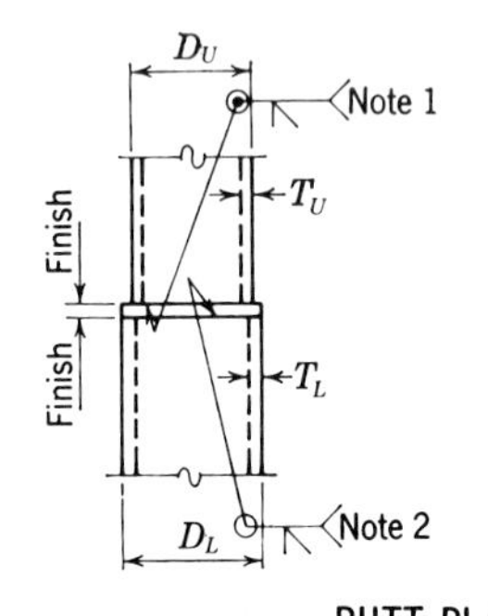

DEPTH OF D_U AND D_L
NOMINALLY THE SAME

BUTT PLATE

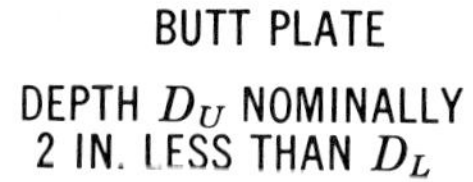
DEPTH D_U NOMINALLY
2 IN. LESS THAN D_L

Note 1: Weld size based on T_U.
Note 2: Weld size based on T_L.

FIG. 9-9 Suggested details—column splices (welded). *(American Institute of Steel Construction.)*

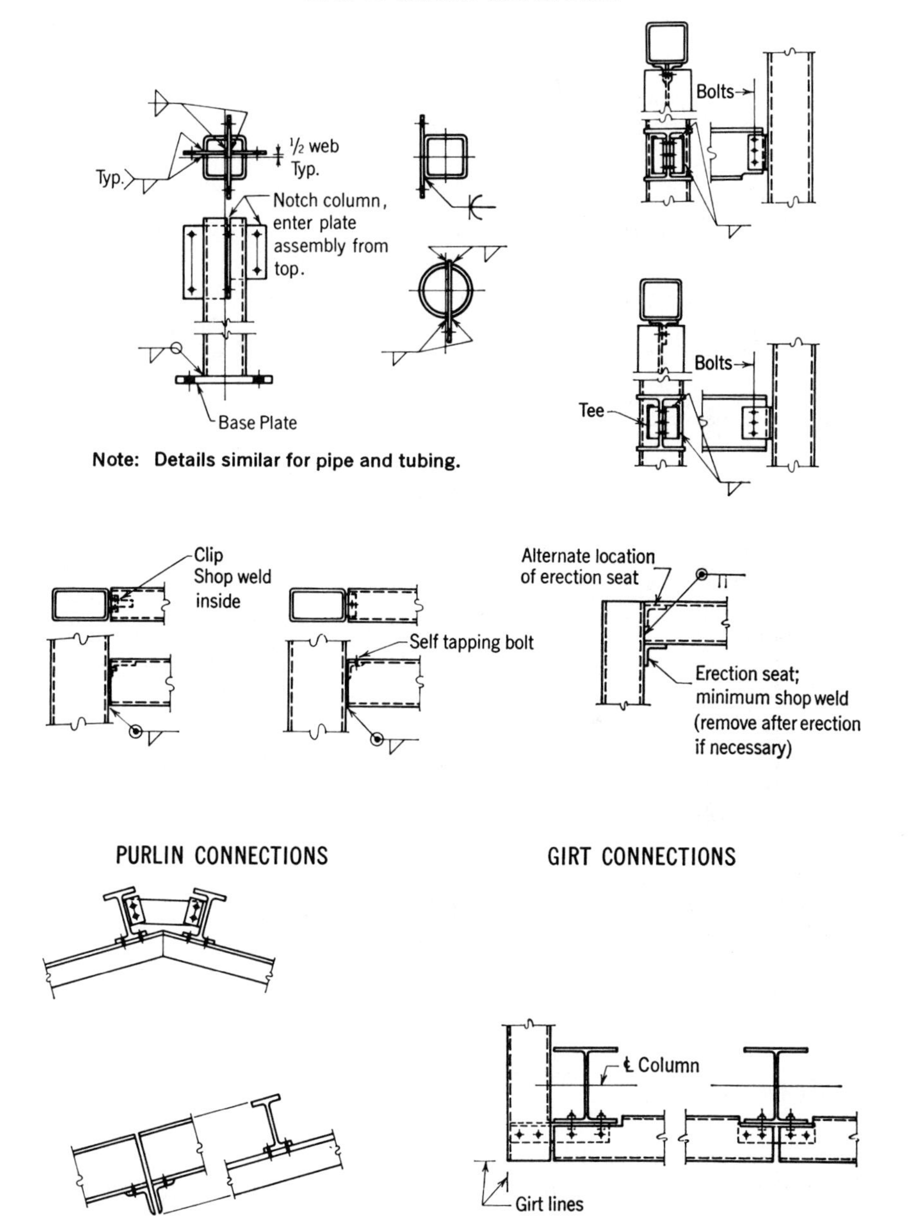

FIG. 9-10 Suggested details—miscellaneous (structural tubing and pipe beam-to-column connections). (*American Institute of Steel Construction.*)

SHELF ANGLES WITH ADJUSTMENT

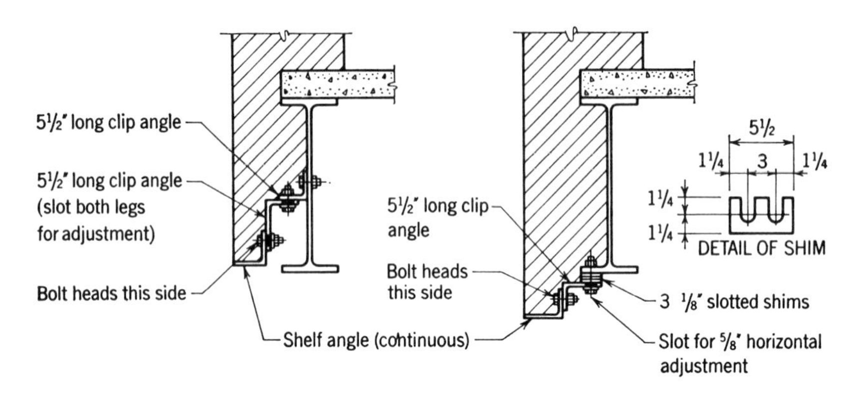

Note: Horizontal adjustment is made by slotted holes; vertical adjustment may be made by slotted holes or by shims.
For tolerance allowance in alignment, see AISC Code of Standard Practice.

TIE RODS AND ANCHORS

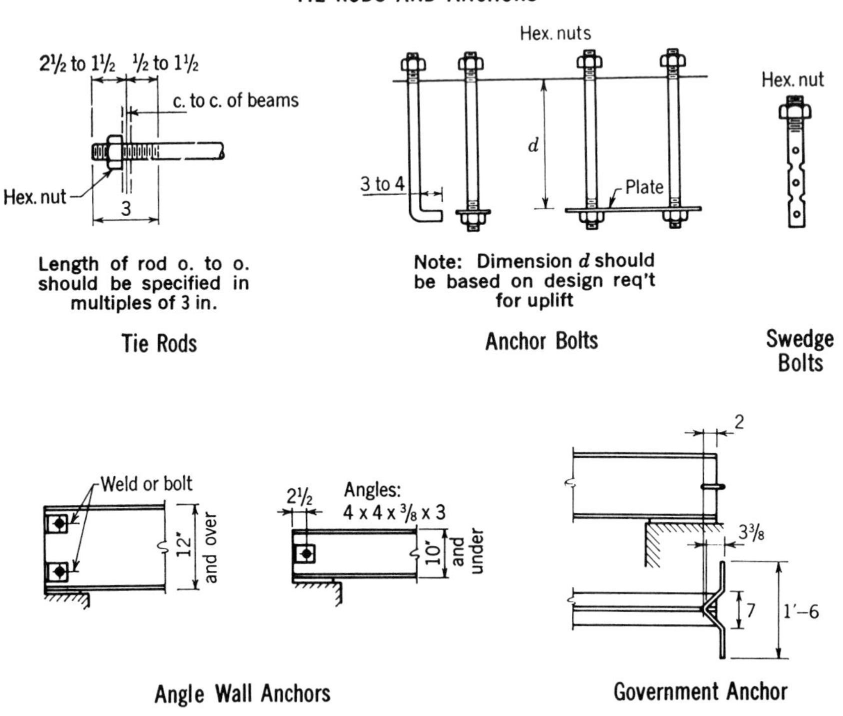

FIG. 9-11 Suggested details—miscellaneous (shelf angles with adjustment). (*American Institute of Steel Construction.*)

TABLE 9-25 Rivets and Threaded Fasteners—Erection Clearances

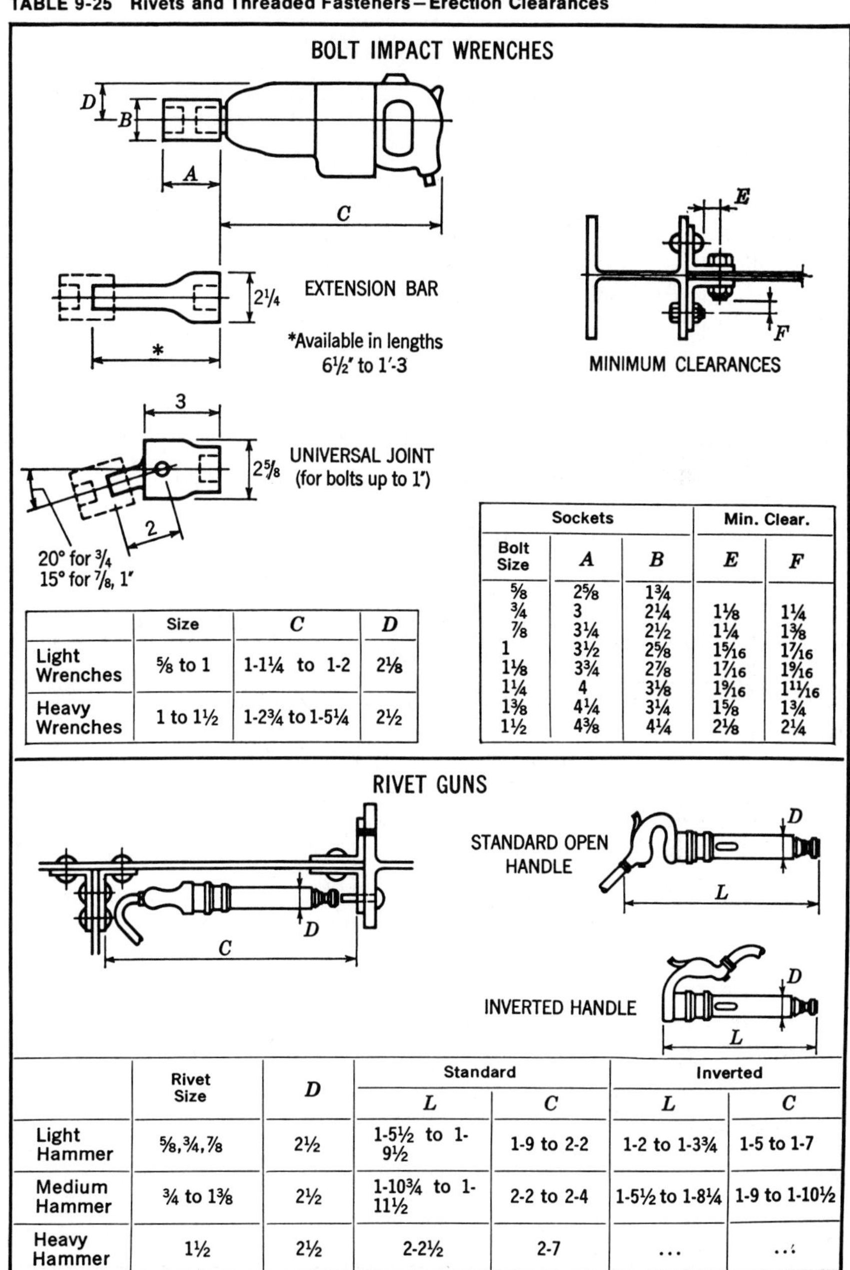

	Size	C	D
Light Wrenches	⅝ to 1	1-1¼ to 1-2	2⅛
Heavy Wrenches	1 to 1½	1-2¾ to 1-5¼	2½

	Sockets		Min. Clear.	
Bolt Size	A	B	E	F
⅝	2⅝	1¾		
¾	3	2¼	1⅛	1¼
⅞	3¼	2½	1¼	1⅜
1	3½	2⅝	1⁵⁄₁₆	1⁷⁄₁₆
1⅛	3¾	2⅞	1⁷⁄₁₆	1⁹⁄₁₆
1¼	4	3⅛	1⁹⁄₁₆	1¹¹⁄₁₆
1⅜	4¼	3¼	1⅝	1¾
1½	4⅜	4¼	2⅛	2¼

	Rivet Size	D	Standard		Inverted	
			L	C	L	C
Light Hammer	⅝, ¾, ⅞	2½	1-5½ to 1-9½	1-9 to 2-2	1-2 to 1-3¾	1-5 to 1-7
Medium Hammer	¾ to 1⅜	2½	1-10¾ to 1-11½	2-2 to 2-4	1-5½ to 1-8¼	1-9 to 1-10½
Heavy Hammer	1½	2½	2-2½	2-7	...	...

SOURCE: American Institute of Steel Construction.

TABLE 9-26 Rivets and Threaded Fasteners—Field Erection Clearances

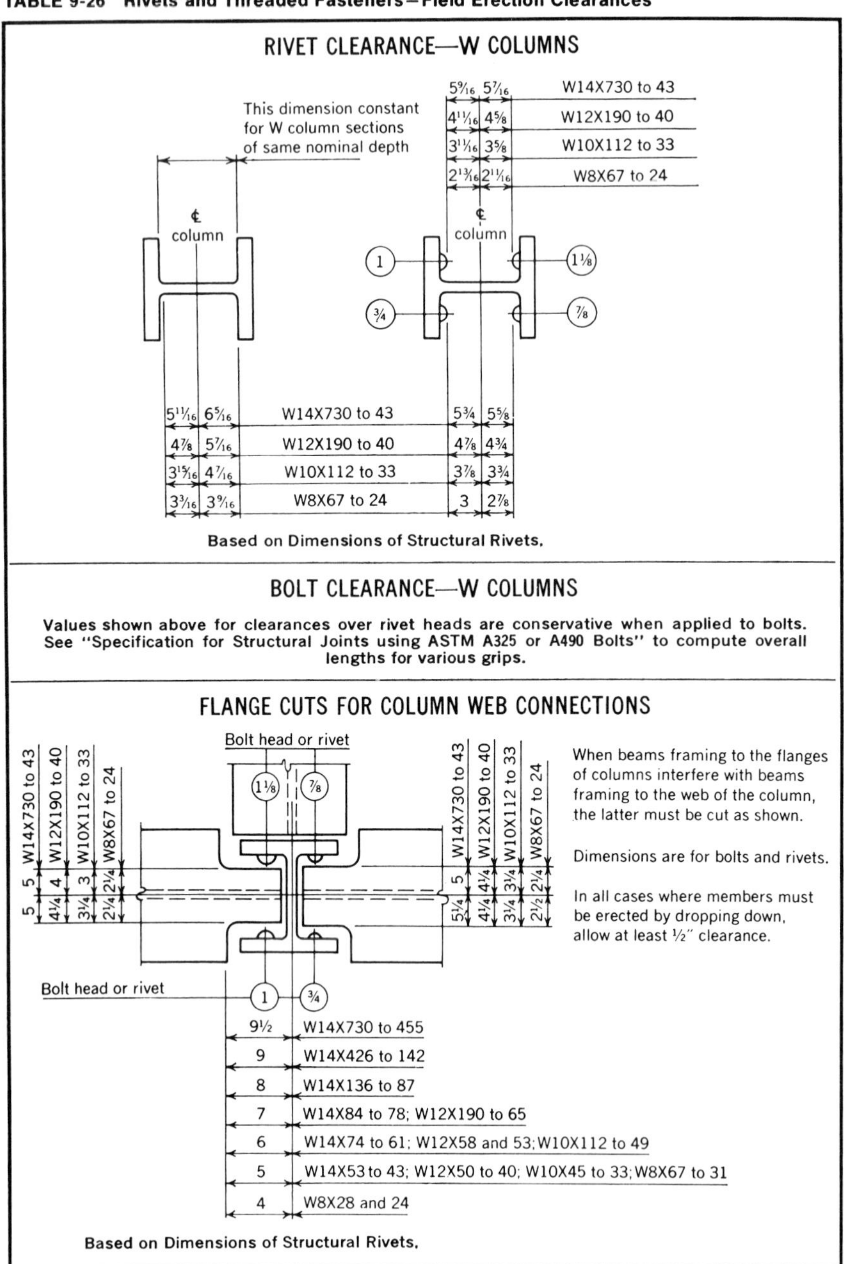
RIVET CLEARANCE—W COLUMNS
This dimension constant for W column sections of same nominal depth
℄ column
5⁹⁄₁₆ 5⁷⁄₁₆ W14X730 to 43
4¹¹⁄₁₆ 4⅝ W12X190 to 40
3¹¹⁄₁₆ 3⅝ W10X112 to 33
2¹³⁄₁₆ 2¹¹⁄₁₆ W8X67 to 24
5¹¹⁄₁₆ 6⁵⁄₁₆ W14X730 to 43 5¾ 5⅝
4⅞ 5⁷⁄₁₆ W12X190 to 40 4⅞ 4¾
3¹⁵⁄₁₆ 4⁷⁄₁₆ W10X112 to 33 3⅞ 3¾
3³⁄₁₆ 3⁹⁄₁₆ W8X67 to 24 3 2⅞
Based on Dimensions of Structural Rivets.
BOLT CLEARANCE—W COLUMNS
Values shown above for clearances over rivet heads are conservative when applied to bolts. See "Specification for Structural Joints using ASTM A325 or A490 Bolts" to compute overall lengths for various grips.
FLANGE CUTS FOR COLUMN WEB CONNECTIONS
Bolt head or rivet
W14X730 to 43
W12X190 to 40
W10X112 to 33
W8X67 to 24
When beams framing to the flanges of columns interfere with beams framing to the web of the column, the latter must be cut as shown.
Dimensions are for bolts and rivets.
In all cases where members must be erected by dropping down, allow at least ½″ clearance.
9½ W14X730 to 455
9 W14X426 to 142
8 W14X136 to 87
7 W14X84 to 78; W12X190 to 65
6 W14X74 to 61; W12X58 and 53; W10X112 to 49
5 W14X53 to 43; W12X50 to 40; W10X45 to 33; W8X67 to 31
4 W8X28 and 24
Based on Dimensions of Structural Rivets.

SOURCE: American Institute of Steel Construction.

9-10 Clevis, Turnbuckle, and Sleeve Nut Table 9-27 illustrates and tabulates the standard sizes, numbers, dimensions, and safe working loads of clevises.

TABLE 9-27 Clevises

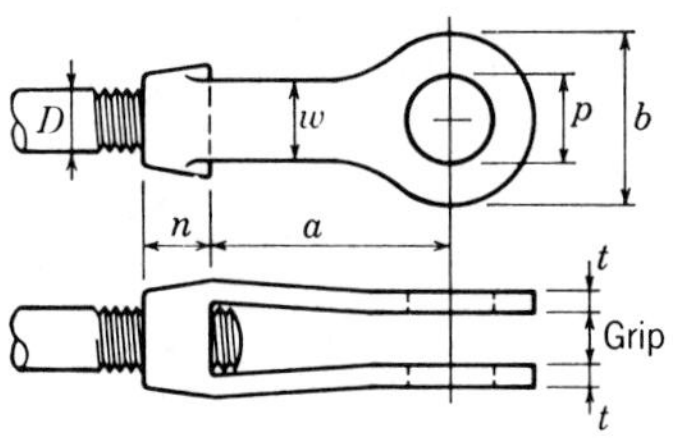

Grip = thickness plate + 1/4"

Thread: UNC Class 2B

Clevis Number	Dimensions, Inches							Weight Pounds	Safe Working Load, Kips*
	Max. D	Max. p	b	n	a	w	t		
2	5/8	3/4	1 7/16	5/8	3 7/8	1 1/16	5/16(+ 1/32 − 0)	1.0	7.0
2 1/2	7/8	1 1/2	2 1/2	1 1/8	4	1 1/4	5/16(+ 1/32 − 0)	2.0	7.5
3	1 3/8	1 3/4	3	1 5/16	5	1 1/2	1/2(+ 1/32 − 0)	4.0	15
3 1/2	1 1/2	2	3 1/2	1 5/8	6	1 3/4	1/2(+ 1/32 − 0)	6.0	18
4	1 3/4	2 1/4	4	1 3/4	6	2	1/2(+ 1/32 − 0)	8.0	21
5	2	2 1/2	5	2 1/4	7	2 1/2	5/8(+ 1/16 − 0)	16.0	37.5
6	2 1/2	3	6	2 3/4	8	3	3/4(+ 3/32 − 0)	26.0	54
7	3	3 3/4	7	3	9	3 1/2	7/8(+ 1/8 − 0)	36.0	68.5
8	4	4	8	4	10	4	1 1/2(+ 1/8 − 0)	80.0	135

* Safe working load based on 5:1 safety factor using maximum pin diameter.

CLEVIS NUMBERS FOR VARIOUS RODS AND PINS

Diameter of Tap Inches	Diameter of Pin, Inches															
	5/8	3/4	7/8	1	1 1/4	1 1/2	1 3/4	2	2 1/4	2 1/2	2 3/4	3	3 1/4	3 1/2	3 3/4	4
5/8	2	2	2 1/2	2 1/2	2 1/2	2 1/2										
3/4	...	2 1/2	2 1/2	2 1/2	2 1/2	2 1/2										
7/8	...	...	2 1/2	2 1/2	2 1/2	2 1/2										
1	...	...	...	3	3	3	3									
1 1/4	...	...	...	3	3	3	3	3 1/2								
1 3/8	...	...	...	3	3	3	3 1/2	3 1/2	4							
1 1/2	...	...	...	3 1/2	3 1/2	3 1/2	4	4	5	5						
1 3/4	...	...	...	...	4	4	5	5	5	5						
2	...	...	...	...	...	5	5	5	5	5	6	6				
2 1/4	...	...	...	...	...	...	...	6	6	6	6	6	7	7		
2 1/2	...	...	...	...	...	...	...	6	6	6	7	7	7	7	7	
2 3/4	...	...	...	...	...	...	...	...	...	7	7	7	7	8	8	
3	...	...	...	...	...	...	...	...	...	7	8	8	8	8	8	8
3 1/4	...	...	...	...	...	...	...	...	...	...	8	8	8	8	8	8
3 1/2	...	...	...	...	...	...	...	...	...	...	8	8	8	8	8	8
3 3/4	...	...	...	...	...	...	...	...	...	...	8	8	8	8	8	8
4	...	...	...	...	...	...	...	...	...	...	8	8	8	8	8	8

Above Table of Clevis Sizes is based on the Net Area of Clevis through Pin Hole being equal to or greater than 125 per cent of Net Area of Rod. Table applies to round rods without upset ends. Pins are sufficient for shear but must be investigated for bending. For other combinations of pin and rod or net area ratios, required clevis size can be calculated by reference to the tabulated dimensions.

Weights and dimensions of clevises are typical. Products of all suppliers are similar and essentially the same.

SOURCE: American Institute of Steel Construction.

Table 9-28 shows the dimensions, weights, and safe working loads of the standard turnbuckle.

Table 9-29 pictures a standard sleeve nut and lists the sizes, weights, and dimensions available.

TABLE 9-28 Turnbuckles

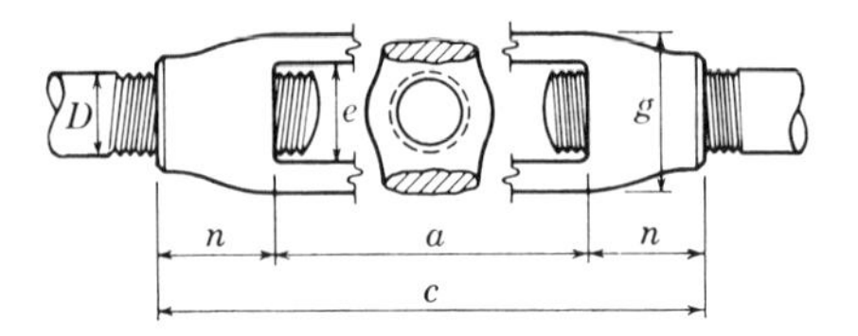

Thread: UNC and 4 UN Class 2B

Diam. D In.	Standard Turnbuckles					Weight of Turnbuckles, Pounds						Turnbuckle Safe Working Load, Kips*
	Dimensions, Inches					Length, a, Inches						
	a	n	c	e	g	6	9	12	18	24	36	
3/8	6	9/16	7 1/8	9/16	1 1/32	.41						1.2
1/2	6	3/4	7 1/2	11/16	1 5/16	.75	.80	1.00				2.2
5/8	6	29/32	7 13/16	13/16	1 1/2	1.00	1.38	1.50	2.43			3.5
3/4	6	1 1/16	8 1/8	15/16	1 23/32	1.45	1.63	2.13	3.06	4.25		5.2
7/8	6	1 7/32	8 7/16	1 3/32	1 7/8	1.85		2.83	4.20	5.43		7.2
1	6	1 3/8	8 3/4	1 9/32	2 1/32	2.60		3.20	4.40	6.85	10.0	9.3
1 1/8	6	1 9/16	9 1/8	1 13/32	2 9/32	2.72		4.70	6.10			11.6
1 1/4	6	1 3/4	9 1/2	1 9/16	2 17/32	3.58		4.70	7.13	11.30	13.1	15.2
1 3/8	6	1 15/16	9 7/8	1 11/16	2 3/4	4.50						17.4
1 1/2	6	2 1/8	10 1/4	1 27/32	3 1/32	5.50		8.00	9.13	16.80	19.4	21.0
1 5/8	6	2 1/4	10 1/2	1 31/32	3 9/32	7.50						24.5
1 3/4	6	2 1/2	11	2 1/8	3 9/16	9.50		15.25	16.00	19.50		28.3
1 7/8	6	2 3/4	11 1/2	2 3/8	4	11.50						37.2
2	6	2 3/4	11 1/2	2 3/8	4	11.50		15.25		27.50		37.2
2 1/4	6	3 3/8	12 3/4	2 11/16	4 5/8	18.00		35.25		43.50		48.0
2 1/2	6	3 3/4	13 1/2	3	5	23.25		33.60		42.38		60.0
2 3/4	6	4 1/8	14 1/4	3 1/4	5 5/8	31.50				54.00		75.0
3	6	4 1/2	15	3 5/8	6 1/8	39.50						96.7
3 1/4	6	5 1/4	16 1/2	3 7/8	6 3/4	60.50						122.2
3 1/2	6	5 1/4	16 1/2	3 7/8	6 3/4	60.50						122.2
3 3/4	6	6	18	4 5/8	8 1/2	95.00						167.8
4	6	6	18	4 5/8	8 1/2	95.00						167.8
4 1/4	9	6 3/4	22 1/2	5 1/4	9 3/4		152.0					233.8
4 1/2	9	6 3/4	22 1/2	5 1/4	9 3/4		152.0					233.8
4 3/4	9	6 3/4	22 1/2	5 1/4	9 3/4		152.0					233.8
5	9	7 1/2	24	6	10		200.0					294.7

* Safe working load based on 5:1 safety factor.

Weights and dimensions of turnbuckles are typical. Products of all suppliers are similar and essentially the same.

SOURCE: American Institute of Steel Construction.

TABLE 9-29 Sleeve Nuts

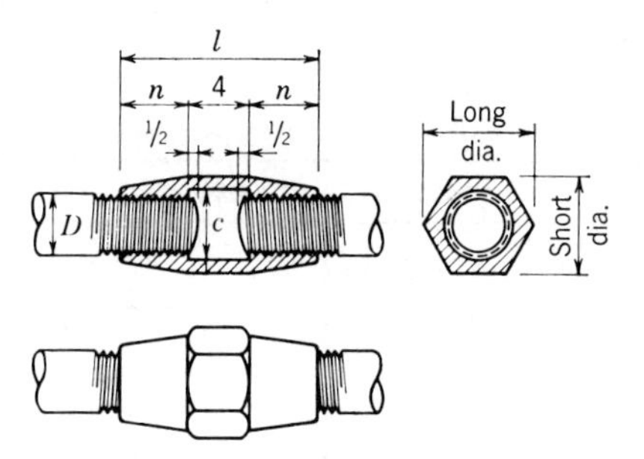

Thread: UNC and 4 UN Class 2B

Diameter of Screw D Inches	Dimensions, Inches					
	Short Diameter	Long Diameter	Length l	Nut n	Clear c	Weight Pounds
3/8	11/16	25/32	4	...	...	.27
7/16	25/32	7/8	4	...	...	.34
1/2	7/8	1	4	...	...	.43
9/16	15/16	1 1/16	5	...	...	.64
5/8	1 1/16	1 7/32	5	...	...	.93
3/4	1 1/4	1 7/16	5	...	...	1.12
7/8	1 7/16	1 5/8	7	1 7/16	1	1.75
1	1 5/8	1 13/16	7	1 7/16	1 1/8	2.46
1 1/8	1 13/16	2 1/16	7 1/2	1 5/8	1 1/4	3.10
1 1/4	2	2 1/4	7 1/2	1 5/8	1 3/8	4.04
1 3/8	2 3/16	2 1/2	8	1 7/8	1 1/2	4.97
1 1/2	2 3/8	2 11/16	8	1 7/8	1 5/8	6.16
1 5/8	2 9/16	2 15/16	8 1/2	2 1/16	1 3/4	7.36
1 3/4	2 3/4	3 1/8	8 1/2	2 1/16	1 7/8	8.87
1 7/8	2 15/16	3 5/16	9	2 5/16	2	10.42
2	3 1/8	3 1/2	9	2 5/16	2 1/8	12.24
2 1/4	3 1/2	3 15/16	9 1/2	2 1/2	2 3/8	16.23
2 1/2	3 7/8	4 3/8	10	2 3/4	2 5/8	21.12
2 3/4	4 1/4	4 13/16	10 1/2	2 15/16	2 7/8	26.71
3	4 5/8	5 1/4	11	3 3/16	3 1/8	33.22
3 1/4	5	5 5/8	11 1/2	3 3/8	3 3/8	40.62
3 1/2	5 3/8	6	12	3 5/8	3 5/8	49.07
3 3/4	5 3/4	6 3/8	12 1/2	3 13/16	3 7/8	58.57
4	6 1/8	6 7/8	13	4 1/16	4 1/8	69.22
4 1/4	6 1/2	7 1/2	13 1/2	4 3/4	4 3/8	75.00
4 1/2	6 7/8	7 15/16	14	5	4 3/4	90.00
4 3/4	7 1/4	8 3/8	14 1/2	5 1/4	5	98.00
5	7 5/8	8 7/8	15	5 1/2	5 1/4	110.0
5 1/4	8	9 1/4	15 1/2	5 3/4	5 1/2	122.0
5 1/2	8 3/8	9 3/4	16	6	5 3/4	142.0
5 3/4	8 3/4	10 1/8	16 1/2	6 1/4	6	157.0
6	9 1/8	10 5/8	17	6 1/2	6 1/4	176.0

Strengths are greater than the corresponding connecting rod when same material is used.

Weights and dimensions are typical. Products of all suppliers are similar and essentially the same.

SOURCE: American Institute of Steel Construction.

These three connecting components are a vital part of many fastening systems and must never be overlooked as a key factor in fastening tie rods, struts, and tension members.

9-11 Rivets Rivets are elaborately discussed in Sec. 13 of this handbook. How-

TABLE 9-30 Rivets—Lengths of Undriven Rivets, in Inches, for Various Grips

In inches, for various grips

Grip — 78° — Length

FULL HEAD

Grip Inches	Diameter of Rivet, Inches ½	⅝	¾	⅞	1	1⅛	1¼
½	1⅝	1⅞	1⅞	2	2⅛		
⅝	1¾	2	2	2⅛	2¼		
¾	1⅞	2⅛	2⅛	2¼	2⅜		
⅞	2	2¼	2¼	2⅜	2½		
1	2¼	2⅜	2⅜	2½	2⅝	2¾	2⅞
⅛	2⅜	2½	2½	2⅝	2¾	2⅞	3
¼	2½	2⅝	2⅝	2¾	2⅞	3	3⅛
⅜	2⅝	2¾	2¾	2⅞	3	3⅛	3¼
½	2⅞	3	3	3⅛	3¼	3⅜	3½
⅝	3	3⅛	3⅛	3¼	3⅜	3½	3½
¾	3⅛	3¼	3¼	3½	3⅝	3¾	3¾
⅞	3¼	3⅜	3⅜	3⅝	3¾	3⅞	3⅞
2	3½	3½	3⅝	3¾	3⅞	4	4
⅛	3⅝	3⅝	3¾	3⅞	4	4⅛	4⅛
¼	3¾	3⅞	3⅞	4	4⅛	4¼	4¼
⅜	4	4	4	4⅛	4¼	4⅜	4⅜
½	4⅛	4⅛	4⅛	4¼	4⅜	4½	4½
⅝	4¼	4¼	4¼	4⅜	4½	4⅝	4⅝
¾	4⅜	4⅜	4⅜	4½	4⅝	4¾	4¾
⅞	4⅝	4⅝	4⅝	4⅝	4¾	4⅞	5
3	...	4¾	4¾	4⅞	5	5	5⅛
⅛	...	4⅞	4⅞	5	5⅛	5¼	5¼
¼	...	5	5	5⅛	5¼	5⅜	5⅜
⅜	...	5⅛	5⅛	5¼	5⅜	5½	5½
½	...	5⅜	5⅜	5⅜	5½	5⅝	5⅝
⅝	...	5½	5½	5½	5⅝	5¾	5¾
¾	...	5⅝	5⅝	5⅝	5¾	5⅞	5⅞
⅞	...	5¾	5¾	5¾	5⅞	6	6
4	...	...	5⅞	6	6	6⅛	6¼
⅛	...	...	6	6⅛	6¼	6⅜	6⅜
¼	...	...	6⅛	6¼	6½	6½	6½
⅜	...	...	6⅜	6½	6½	6⅝	6⅝
½	...	...	6½	6⅝	6⅝	6¾	6¾
⅝	...	...	6⅝	6¾	6¾	6¾	6⅞
¾	...	...	6¾	6⅞	6⅞	7	7
⅞	...	...	6⅞	7	7	7⅛	7⅛
5	...	...	...	7⅛	7⅛	7¼	7¼
⅛	...	...	...	7¼	7¼	7⅜	7⅜
¼	...	...	...	7⅜	7⅜	7½	7½
⅜	...	...	...	7⅝	7⅝	7¾	7⅝
½	...	...	...	7¾	7¾	7⅞	7⅞
⅝	...	...	...	7⅞	7⅞	8	8
¾	...	...	...	8	8	8⅛	8⅛
⅞	...	...	...	8⅛	8⅛	8¼	8¼

COUNTERSUNK HEAD

Grip Inches	Diameter of Rivet, Inches ½	⅝	¾	⅞	1	1⅛	1¼
½	1	1	1⅛	1¼	1¼		
⅝	1⅛	1¼	1¼	1⅜	1⅜		
¾	1⅜	1⅜	1⅜	1½	1½		
⅞	1½	1½	1½	1⅝	1⅝		
1	1⅝	1⅝	1⅝	1¾	1¾	1⅞	1⅞
⅛	1¾	1¾	1⅞	1⅞	1⅞	2	2
¼	2	2	2	2	2	2⅛	2⅛
⅜	2⅛	2⅛	2⅛	2¼	2¼	2⅜	2⅜
½	2¼	2¼	2¼	2⅜	2⅜	2½	2½
⅝	2⅜	2⅜	2⅜	2½	2½	2⅝	2⅝
¾	2⅝	2⅝	2⅝	2⅝	2⅝	2¾	2¾
⅞	2¾	2¾	2¾	2¾	2¾	2⅞	2⅞
2	2⅞	2⅞	2⅞	2⅞	2⅞	3	3
⅛	3⅛	3	3	3	3	3⅛	3⅛
¼	3¼	3⅛	3⅛	3⅛	3¼	3¼	3¼
⅜	3⅜	3⅜	3⅜	3⅜	3⅜	3⅜	3⅜
½	3½	3½	3½	3½	3½	3⅝	3⅝
⅝	3¾	3⅝	3⅝	3⅝	3⅝	3¾	3¾
¾	3⅞	3¾	3¾	3¾	3¾	3⅞	3⅞
⅞	4	3⅞	3⅞	3⅞	3⅞	4	4
3	...	4⅛	4⅛	4⅛	4⅛	4⅛	4⅛
⅛	...	4¼	4¼	4¼	4¼	4¼	4¼
¼	...	4⅜	4⅜	4⅜	4⅜	4⅜	4⅜
⅜	...	4½	4½	4½	4½	4½	4½
½	...	4⅝	4⅝	4⅝	4⅝	4⅝	4⅝
⅝	...	4¾	4¾	4¾	4¾	4⅞	4⅞
¾	...	5	5	5	5	5	5
⅞	...	5⅛	5⅛	5⅛	5⅛	5⅛	5⅛
4	...	...	5¼	5¼	5¼	5¼	5¼
⅛	...	...	5⅜	5⅜	5⅜	5⅜	5⅜
¼	...	...	5½	5½	5½	5½	5½
⅜	...	...	5⅝	5⅝	5⅝	5⅝	5⅝
½	...	...	5¾	5¾	5¾	5¾	5¾
⅝	...	...	6	6	6	6	6
¾	...	...	6⅛	6⅛	6⅛	6⅛	6⅛
⅞	...	...	6¼	6¼	6¼	6¼	6¼
5	...	...	...	6⅜	6⅜	6⅜	6⅜
⅛	...	...	...	6½	6½	6½	6½
¼	...	...	...	6⅝	6⅝	6⅝	6⅝
⅜	...	...	...	6¾	6¾	6¾	6¾
½	...	...	...	6⅞	6⅞	6⅞	6⅞
⅝	...	...	...	7	7	7	7
¾	...	...	...	7¼	7¼	7¼	7¼
⅞	...	...	...	7⅜	7⅜	7⅜	7⅜

Above table may vary from standard practice of individual fabricators and should be checked against such standards by user.

SOURCE: American Institute of Steel Construction.

ever, we will include tables here to cover technical data for structural rivets needed for general detailing practice by the structural steel draftsman and the design engineer.

General dimensions of structural rivets, lengths of driven rivets, weights, sizes, and spacings are shown in Tables 9-30 through 9-33.

TABLE 9-31 Rivets—Weights

WEIGHT WITH ONE HIGH BUTTON (ACORN) MANUFACTURED HEAD IN POUNDS PER 100

Length Inches	Diameter of Rivet, Inches							Length Inches	Diameter of Rivet, Inches						
	½	⅝	¾	⅞	1	1⅛	1¼		½	⅝	¾	⅞	1	1⅛	1¼
								5	...	50	74	104	138	180	226
								⅛	...	51	76	106	141	183	230
1¼	11							¼	...	52	78	108	144	187	234
⅜	12							⅜	...	53	79	110	147	190	239
½	12	20	31	45	60	81	104	½	...	54	81	113	149	194	243
⅝	13	21	32	47	63	85	108	⅝	...	55	82	115	152	197	247
¾	14	22	33	49	66	88	113	¾	...	57	84	117	155	201	252
⅞	14	23	35	51	69	92	117	⅞	...	58	85	119	158	204	256
2	15	24	37	53	72	95	122	6	...	...	87	121	161	208	261
⅛	16	25	39	55	74	99	126	⅛	...	...	89	123	163	212	265
¼	17	26	40	57	77	102	130	¼	...	...	90	125	166	215	269
⅜	17	27	42	59	80	106	135	⅜	...	...	92	127	169	219	274
½	18	28	43	62	83	109	139	½	...	...	93	130	172	222	278
⅝	19	29	45	64	85	113	143	⅝	...	...	95	132	174	226	282
¾	19	31	46	66	88	116	148	¾	...	...	96	134	177	229	287
⅞	20	32	48	68	91	120	152	⅞	...	...	98	136	180	233	291
3	21	33	50	70	94	123	156	7	...	...	100	138	183	236	295
⅛	21	34	51	72	97	127	161	⅛	...	...	101	140	186	240	300
¼	22	35	53	74	99	131	165	¼	...	...	103	142	188	243	304
⅜	23	36	54	76	102	134	169	⅜	...	...	104	144	191	247	308
½	23	37	56	79	105	138	174	½	...	...	106	147	194	250	313
⅝	24	38	57	81	108	141	178	⅝	...	...	107	149	197	254	317
¾	25	39	59	83	110	145	182	¾	...	...	109	151	199	257	321
⅞	26	40	60	85	113	148	187	⅞	...	...	110	153	202	261	326
4	26	41	62	87	116	152	191	8	...	...	...	155	205	264	330
⅛	27	42	64	89	119	155	195	⅛	...	...	...	157	208	268	334
¼	28	44	65	91	122	159	200	¼	...	...	...	159	211	271	339
⅜	28	45	67	93	124	162	204	⅜	...	...	...	161	213	275	343
½	29	46	68	96	127	166	208	½	...	...	...	164	216	278	347
⅝	30	47	70	98	130	169	213	⅝	...	...	...	166	219	282	352
¾	30	48	71	100	133	173	217	¾	...	...	...	168	222	285	356
⅞	31	49	73	102	135	176	221	⅞	...	...	...	170	224	289	360

WEIGHT WITH ONE COUNTERSUNK HEAD IN POUNDS PER 100

For Countersunk Rivets, use weight given above with following deductions.	Diameter of Rivet, Inches						
	½	⅝	¾	⅞	1	1⅛	1¼
Deduction, Lb.	3	4	7	12	18	26	36

WEIGHT OF HIGH BUTTON (ACORN) HEADS AFTER DRIVING

Diameter of Rivet, Inches	½	⅝	¾	⅞	1	1⅛	1¼
Weight per 100 Heads, Lb.	4	7	12	18	26	36	48

SOURCE: American Institute of Steel Construction.

TABLE 9-32 Rivets – Spacing

MINIMUM PITCH FOR MACHINE RIVETING

Diam. of Rivet	c	k	Distance, f, inches													
			1⅛	1¼	1⅜	1½	1⅝	1¾	1⅞	2	2⅛	2¼	2⅜	2½	2¾	3
⅜	⅞	1 3/16	¼	0												
½	1	1⅜	¾	½	0											
⅝	1⅛	1 9/16	1⅛	1	¾	⅜	0									
¾	1¼	1¾	...	1¼	1⅛	1	¾	0								
⅞	1⅜	2	...	...	1½	1⅜	1⅛	⅞	⅝	0						
1	1½	2 3/16	...	...	...	1⅝	1½	1⅜	1⅛	⅞	½	0				
1⅛	1⅝	2⅜	...	...	...	...	1¾	1⅝	1½	1⅜	1⅛	⅞	0			
1¼	1¾	2⅝	...	...	...	...	...	2	1⅞	1¾	1½	1¼	1	⅝	0	
1⅜	1⅞	2 13/16	...	...	...	...	...	...	2⅛	2	1⅞	1¾	1½	1¼	½	0
1½	2	3	...	...	...	...	...	...	...	2¼	2⅛	2	1⅞	1⅝	1⅛	0

MINIMUM PITCH TO MAINTAIN 3 DIAMETERS C. TO C.

Diam. of Rivet	m	Distance, g, Inches														
		1	1¼	1½	1¾	2	2¼	2½	2¾	3	3¼	3½	3¾	4	4¼	4½
⅜	1⅛	½	0													
½	1½	1⅛	⅞	0												
⅝	1⅞	1⅝	1⅜	1⅛	⅝	0										
¾	2¼	2	1⅞	1⅝	1⅜	1	0									
⅞	2⅝	2½	2⅜	2⅛	2	1¾	1⅜	¾	0							
1	3	2⅞	2¾	2⅝	2½	2¼	2	1⅝	1⅛	0						
1⅛	3⅜	3¼	3⅛	3	2⅞	2¾	2½	2¼	2	1½	⅞	0				
1¼	3¾	3⅝	3½	3⅜	3⅜	3¼	3	2¾	2½	2¼	1⅞	1⅜	0			
1⅜	4⅛	4	4	3⅞	3¾	3⅝	3½	3¼	3⅛	2⅞	2½	2⅛	1¾	1	0	
1½	4½	4⅜	4⅜	4¼	4⅛	4	3⅞	3¾	3½	3⅜	3⅛	2⅞	2½	2	1½	0

COVER PLATE RIVETING

a In.	d In.
½	2½
1	2⅝
1½	2¾
2	2¾
2½	2⅞
3	2⅞
3½	3
4	3⅛
5	3¼
6	3⅜

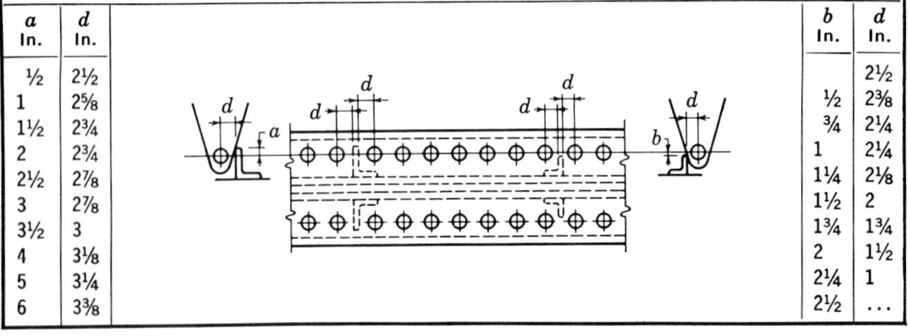

b In.	d In.
	2½
½	2⅜
¾	2¼
1	2¼
1¼	2⅛
1½	2
1¾	1¾
2	1½
2¼	1
2½	...

SOURCE: American Institute of Steel Construction.

TABLE 9-33 Rivets—Dimensions of Structural Rivets

DIMENSIONS OF STRUCTURAL RIVETS (HIGH BUTTON OR ACORN HEADS)

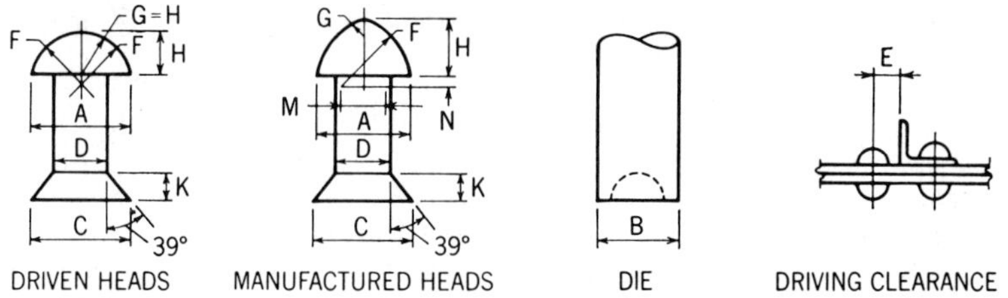

G=H F H A D K C 39° F G F H M A N D K C 39° B E

DRIVEN HEADS MANUFACTURED HEADS DIE DRIVING CLEARANCE

Diam. of Rivet, Inches				1/2	5/8	3/4	7/8	1	1 1/8	1 1/4	1 3/8	1 1/2
Driven Head Inches	Full	A	1.5 D + 1/8	7/8	1 1/16	1 1/4	1 7/16	1 5/8	1 13/16	2	2 3/16	2 3/8
		H	.425 A	3/8	7/16	17/32	5/8	11/16	3/4	27/32	15/16	1
		F	1.5 H	9/16	11/16	13/16	15/16	1 1/32	1 5/32	1 9/32	1 13/32	1 1/2
	Ctsk.	C	1.81 D	29/32	1 1/8	1 11/32	1 19/32	1 13/16	2 1/32	2 1/4	2 1/2	2 23/32
		K	.5 D	1/4	5/16	3/8	7/16	1/2	9/16	5/8	11/16	3/4
Manufactured Head Inches	Full	A	1.5 D + 1/32	25/32	31/32	1 5/32	1 11/32	1 17/32	1 23/32	1 29/32	2 3/32	2 9/32
		H	.75 D + 1/8	1/2	19/32	11/16	25/32	7/8	31/32	1 1/16	1 5/32	1 1/4
		F	.75 D + 9/32	21/32	3/4	27/32	15/16	1 1/32	1 1/8	1 7/32	1 5/16	1 13/32
		M	.50	1/2	1/2	1/2	1/2	1/2	1/2	1/2	1/2	1/2
		N	.094	3/32	3/32	3/32	3/32	3/32	3/32	3/32	3/32	3/32
		G	.75 D − 9/32	3/32	3/16	9/32	3/8	15/32	9/16	21/32	3/4	27/32
	Ctsk.	C	1.81 D	29/32	1 1/8	1 11/32	1 19/32	1 13/16	2 1/32	2 1/4	2 1/2	2 23/32
		K	.5 D	1/4	5/16	3/8	7/16	1/2	9/16	5/8	11/16	3/4
Die, In.		B		1 3/4	2	2 1/4	2 1/2	2 3/4	3	3 1/4	3 1/2	3 3/4
Driving Clearance Inches			E (min.)	3/4	7/8	1	1 1/8	1 1/4	1 3/8	1 1/2	1 5/8	1 3/4
			E (pref.)	1	1 1/8	1 1/4	1 3/8	1 1/2	1 5/8	1 3/4	1 7/8	2

CONVENTIONAL SIGNS FOR RIVETS AND BOLTS

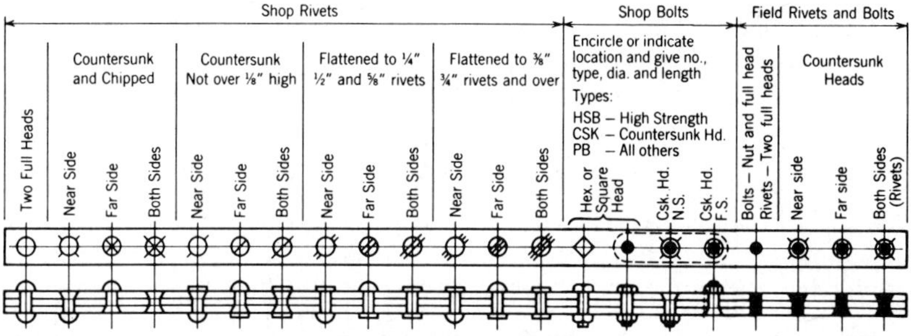

Shop Rivets — Shop Bolts — Field Rivets and Bolts

Two Full Heads

Countersunk and Chipped: Near Side, Far Side, Both Sides

Countersunk Not over 1/8" high: Near Side, Far Side, Both Sides

Flattened to 1/4" 1/2" and 5/8" rivets: Near Side, Far Side, Both Sides

Flattened to 3/8" 3/4" rivets and over: Near Side, Far Side, Both Sides

Encircle or indicate location and give no., type, dia. and length
Types:
HSB – High Strength
CSK – Countersunk Hd.
PB – All others

Hex. or Square Head, Csk. Hd. N.S., Csk. Hd. F.S.

Bolts – Nut and full head
Rivets – Two full heads

Countersunk Heads: Near side, Far side, Both Sides (Rivets)

USUAL GAGES FOR ANGLES, INCHES

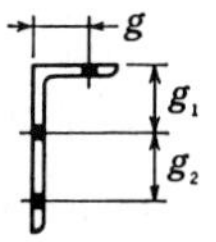

Leg	8	7	6	5	4	3 1/2	3	2 1/2	2	1 3/4	1 1/2	1 3/8	1 1/4	1
g	4 1/2	4	3 1/2	3	2 1/2	2	1 3/4	1 3/8	1 1/8	1	7/8	7/8	3/4	5/8
g_1	3	2 1/2	2 1/4	2										
g_2	3	3	2 1/2	1 3/4										

g, g_1, g_2

CRIMPS

$b = t + 1\frac{1}{2}$
Min.=2

t b

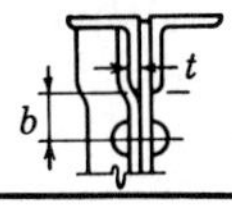

SOURCE: American Institute of Steel Construction.

9-12 Detailing Practices Design office, drafting room, and shop cooperation is a must to maintain maximum efficiency in the fabrication of modern structural steel. Drill gages, punch gages, copes, blocks, and cuts are illustrated in Fig. 9-12. Refer to other sections for additional recommended detailing practices and note that the de-

Bolted and riveted connections

Maximum efficiency in the fabrication of structural steel by modern shops is entirely dependent upon close cooperation between designing office, drafting room and shop. Designs should be favorable to, the drafting room should recognize and call for, and the shop should adapt its equipment to, the use of recurrent details which have been standardized.

Consideration should be given to duplication of details and multiple punching or drilling. Utilization of standard jigs and machine set-ups eliminates unnecessary handling of material and aids drilling or punching holes in groups.

Column gage lines should conform to the standard machine set-ups illustrated below. Once determined they should be duplicated as far as possible throughout any one job. Gages on an individual member if possible should not be varied throughout the length of that member.

DRILL GAGES

Keep gages and longitudinal spacing alike, if possible, as drilling can be done simultaneously in both flanges.

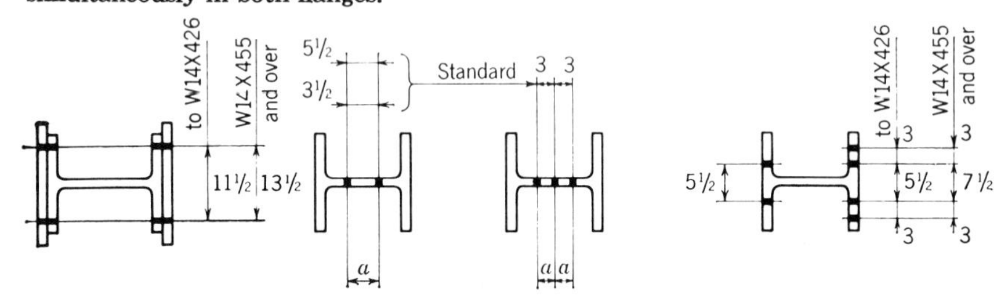

Minimum a = 3 in.; maximum a controlled by size of member. Gages other than standard should be multiples of 3 in.

PUNCH GAGES

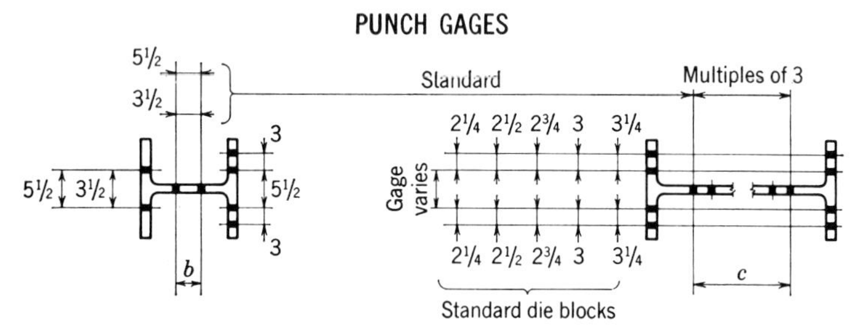

Minimum b = $2\frac{1}{4}$ in.; maximum b controlled by size of member. Gages other than standard should be multiples of 3 in. Maximum c controlled by size of member.

Longitudinal spacing of holes for both punched and drilled work should be 3 in. or multiples of 3 in. The adoption of such spacing facilitates the use of multiple drills and punches and makes possible the use of the Framed Beam Connections given on pages 4-17 to 4-26.

FIG. 9-12 Detailing practice—bolted and riveted connections. *(American Institute of Steel Construction.)*

signer, engineer, draftsman, detailer, shop foreman, and estimators should have a complete copy of the AISC *Manual of Steel Construction* to execute their total project function.

9-13 Bracing Formulas and Properties of the Circle, Parabola, and Ellipse Continuing in the general area of detail practice, Tables 9-34 through 9-37 give valuable

TABLE 9-34 Bracing Formulas

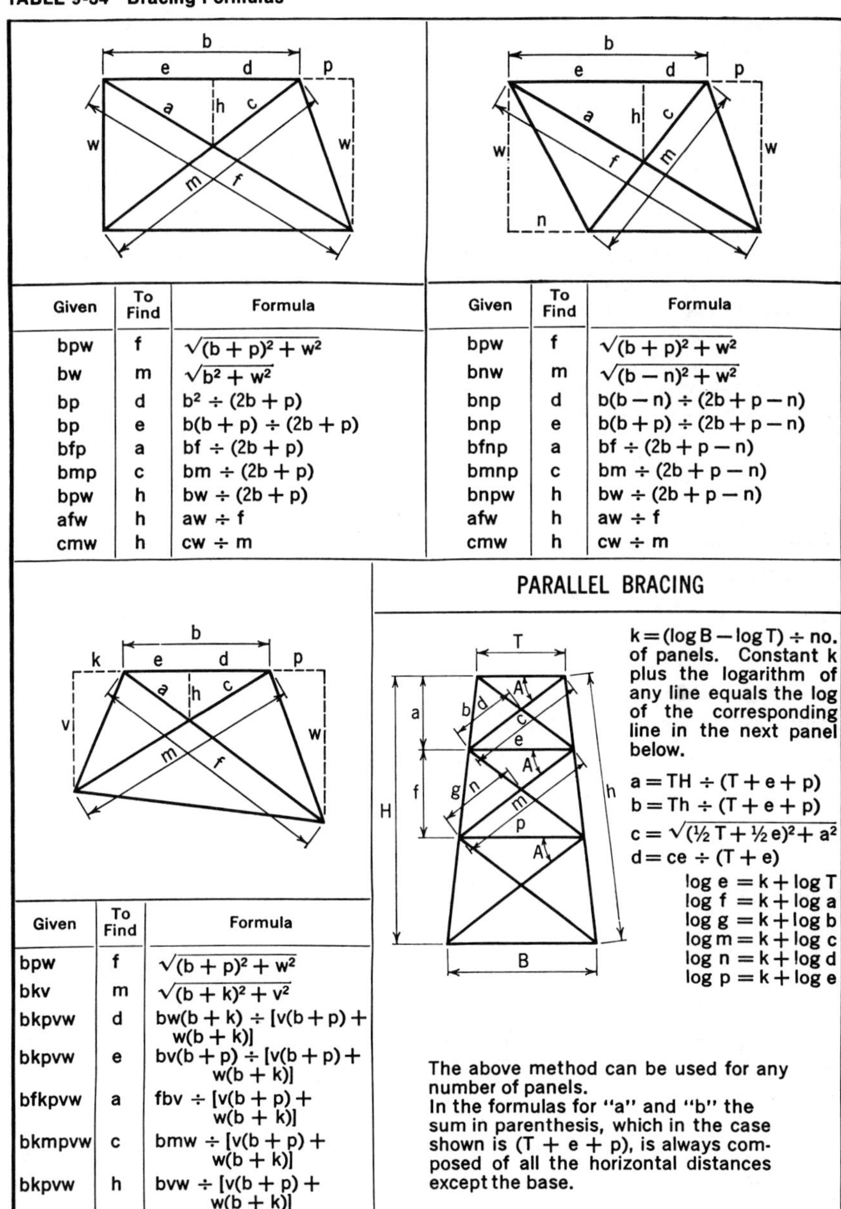

Given	To Find	Formula
bpw	f	$\sqrt{(b+p)^2+w^2}$
bw	m	$\sqrt{b^2+w^2}$
bp	d	$b^2 \div (2b+p)$
bp	e	$b(b+p) \div (2b+p)$
bfp	a	$bf \div (2b+p)$
bmp	c	$bm \div (2b+p)$
bpw	h	$bw \div (2b+p)$
afw	h	$aw \div f$
cmw	h	$cw \div m$

Given	To Find	Formula
bpw	f	$\sqrt{(b+p)^2+w^2}$
bnw	m	$\sqrt{(b-n)^2+w^2}$
bnp	d	$b(b-n) \div (2b+p-n)$
bnp	e	$b(b+p) \div (2b+p-n)$
bfnp	a	$bf \div (2b+p-n)$
bmnp	c	$bm \div (2b+p-n)$
bnpw	h	$bw \div (2b+p-n)$
afw	h	$aw \div f$
cmw	h	$cw \div m$

Given	To Find	Formula
bpw	f	$\sqrt{(b+p)^2+w^2}$
bkv	m	$\sqrt{(b+k)^2+v^2}$
bkpvw	d	$bw(b+k) \div [v(b+p)+w(b+k)]$
bkpvw	e	$bv(b+p) \div [v(b+p)+w(b+k)]$
bfkpvw	a	$fbv \div [v(b+p)+w(b+k)]$
bkmpvw	c	$bmw \div [v(b+p)+w(b+k)]$
bkpvw	h	$bvw \div [v(b+p)+w(b+k)]$
afw	h	$aw \div f$
cmv	h	$cv \div m$

PARALLEL BRACING

$k = (\log B - \log T) \div$ no. of panels. Constant k plus the logarithm of any line equals the log of the corresponding line in the next panel below.

$a = TH \div (T+e+p)$
$b = Th \div (T+e+p)$
$c = \sqrt{(\tfrac{1}{2}T+\tfrac{1}{2}e)^2+a^2}$
$d = ce \div (T+e)$
$\log e = k + \log T$
$\log f = k + \log a$
$\log g = k + \log b$
$\log m = k + \log c$
$\log n = k + \log d$
$\log p = k + \log e$

The above method can be used for any number of panels.
In the formulas for "a" and "b" the sum in parenthesis, which in the case shown is $(T+e+p)$, is always composed of all the horizontal distances except the base.

SOURCE: American Institute of Steel Construction.

TABLE 9-35 Properties of the Circle

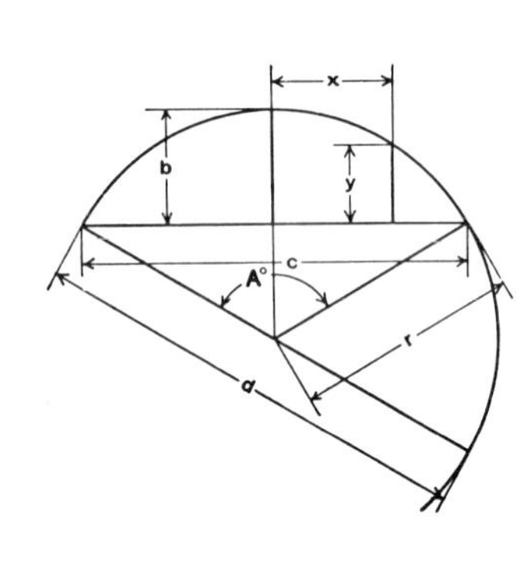

Circumference = 6.28318 r = 3.14159 d
Diameter = 0.31831 circumference
Area = 3.14159 r^2

Arc $a = \dfrac{\pi r A°}{180°} = 0.017453\ r A°$

Angle $A° = \dfrac{180°\ a}{\pi r} = 57.29578\ \dfrac{a}{r}$

Radius $r = \dfrac{4b^2 + c^2}{8b}$

Chord $c = 2\sqrt{2br - b^2} = 2r \sin\dfrac{A}{2}$

Rise $b = r - \frac{1}{2}\sqrt{4r^2 - c^2} = \dfrac{c}{2}\tan\dfrac{A}{4}$

$= 2r\sin^2\dfrac{A}{4} = r + y - \sqrt{r^2 - x^2}$

$y = b - r + \sqrt{r^2 - x^2}$

$x = \sqrt{r^2 - (r + y - b)^2}$

Diameter of circle of equal periphery as square = 1.27324 side of square
Side of square of equal periphery as circle = 0.78540 diameter of circle
Diameter of circle circumscribed about square = 1.41421 side of square
Side of square inscribed in circle = 0.70711 diameter of circle

CIRCULAR SECTOR

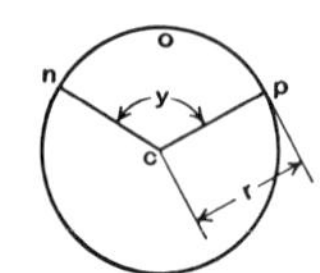

r = radius of circle y = angle ncp in degrees

Area of Sector ncpo = ½ (length of arc nop × r)

$= \text{Area of Circle} \times \dfrac{y}{360}$

$= 0.0087266 \times r^2 \times y$

CIRCULAR SEGMENT

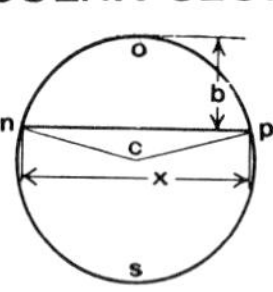

r = radius of circle x = chord b = rise

Area of Segment nop = Area of Sector ncpo — Area of triangle ncp

$= \dfrac{(\text{Length of arc nop} \times r) - x(r - b)}{2}$

Area of Segment nsp = Area of Circle — Area of Segment nop

VALUES FOR FUNCTIONS OF π

$\pi = 3.14159265359$, log = 0.4971499

$\pi^2 = 9.8696044$, log = 0.9942997 $\dfrac{1}{\pi} = 0.3183099$, log = $\bar{1}.5028501$ $\sqrt{\dfrac{1}{\pi}} = 0.5641896$, log = $\bar{1}.7514251$

$\pi^3 = 31.0062767$, log = 1.4914496 $\dfrac{1}{\pi^2} = 0.1013212$, log = $\bar{1}.0057003$ $\dfrac{\pi}{180} = 0.0174533$, log = $\bar{2}.2418774$

$\sqrt{\pi} = 1.7724539$, log = 0.2485749 $\dfrac{1}{\pi^3} = 0.0322515$, log = $\bar{2}.5085504$ $\dfrac{180}{\pi} = 57.2957795$, log = 1.7581226

Note: Logs of fractions such as $\bar{1}$:5028501 and $\bar{2}$.5085500 may also be written 9.5028501 – 10 and 8.5085500 – 10 respectively.

SOURCE: American Institute of Steel Construction.

TABLE 9-36 Properties of the Parabola and Ellipse

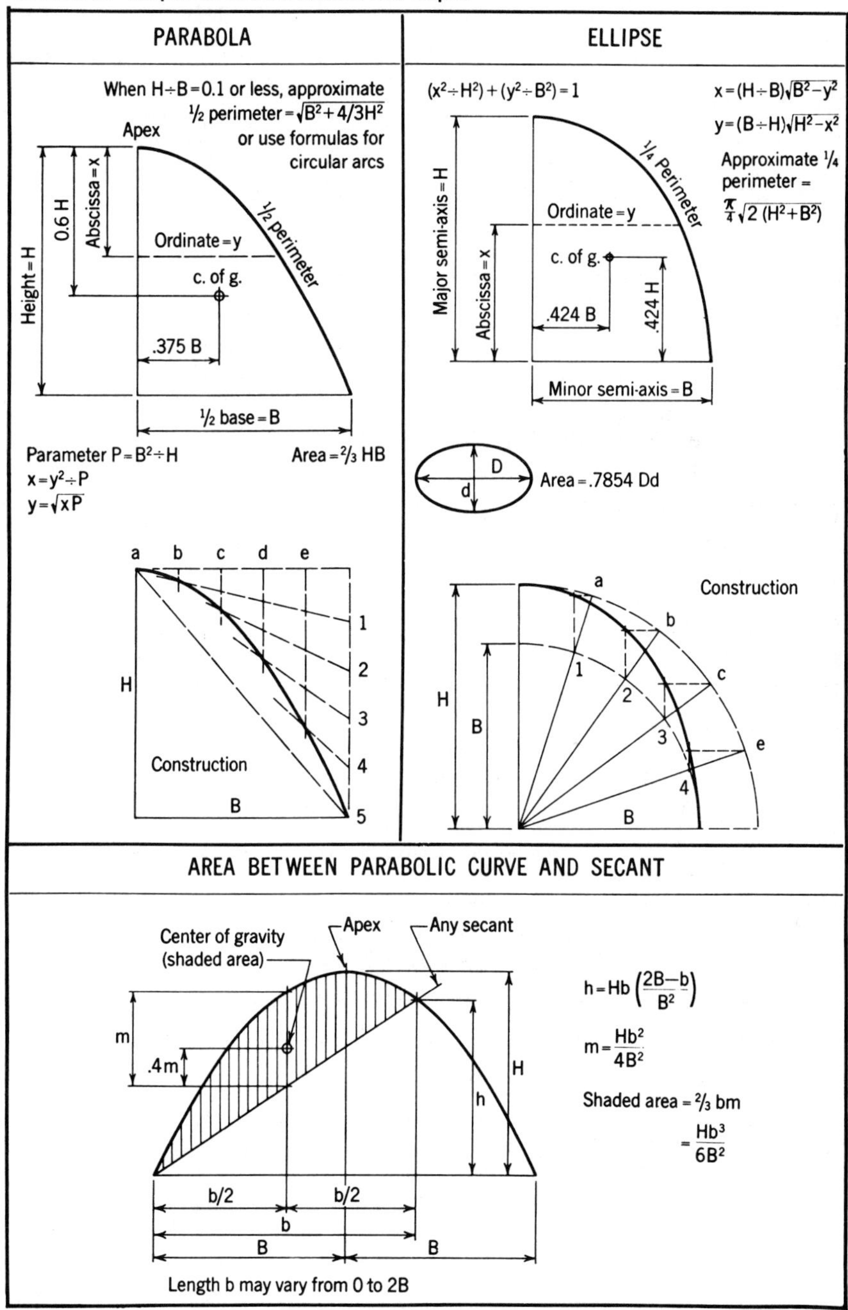

SOURCE: American Institute of Steel Construction.

data on bracing and properties of the circle, parabola, and ellipse. They are provided here as a ready reference of precalculated material and basic formulas which are not committed to memory.

TABLE 9-37 Length of Circular Arcs for Unit Radius

By the use of this table, the length of any arc may be found if the length of the radius and the angle of the segment are known.

Example: Required the length of arc of segment 32° 15′ 27″ with radius of 24 feet 3 inches.

From table: Length of arc (Radius 1) for 32° = .5585054

15′ = .0043633

27″ = .0001309

.5629996

.5629996 × 24.25 (length of radius) = 13.65 feet

DEGREES						MINUTES		SECONDS	
1	.017 4533	61	1.064 6508	121	2.111 8484	1	.000 2909	1	.000 0048
2	.034 9066	62	1.082 1041	122	2.129 3017	2	.000 5818	2	.000 0097
3	.052 3599	63	1.099 5574	123	2.146 7550	3	.000 8727	3	.000 0145
4	.069 8132	64	1.117 0107	124	2.164 2083	4	.001 1636	4	.000 0194
5	.087 2665	65	1.134 4640	125	2.181 6616	5	.001 4544	5	.000 0242
6	.104 7198	66	1.151 9173	126	2.199 1149	6	.001 7453	6	.000 0291
7	.122 1730	67	1.169 3706	127	2.216 5682	7	.002 0362	7	.000 0339
8	.139 6263	68	1.186 8239	128	2.234 0214	8	.002 3271	8	.000 0388
9	.157 0796	69	1.204 2772	129	2.251 4747	9	.002 6180	9	.000 0436
10	.174 5329	70	1 221 7305	130	2.268 9280	10	.002 9089	10	.000 0485
11	.191 9862	71	1.239 1838	131	2.286 3813	11	.003 1998	11	.000 0533
12	.209 4395	72	1.256 6371	132	2.303 8346	12	.003 4907	12	.000 0582
13	.226 8928	73	1.274 0904	133	2.321 2879	13	.003 7815	13	.000 0630
14	.244 3461	74	1.291 5436	134	2.338 7412	14	.004 0724	14	.000 0679
15	.261 7994	75	1.308 9969	135	2.356 1945	15	.004 3633	15	.000 0727
16	.279 2527	76	1.326 4502	136	2.373 6478	16	.004 6542	16	.000 0776
17	.296 7060	77	1.343 9035	137	2.391 1011	17	.004 9451	17	.000 0824
18	.314 1593	78	1.361 3568	138	2.408 5544	18	.005 2360	18	.000 0873
19	.331 6126	79	1.378 8101	139	2.426 0077	19	.005 5269	19	.000 0921
20	.349 0659	80	1.396 2634	140	2.443 4610	20	.005 8178	20	.000 0970
21	.366 5191	81	1.413 7167	141	2.460 9142	21	.006 1087	21	.000 1018
22	.383 9724	82	1.431 1700	142	2.478 3675	22	.006 3995	22	.000 1067
23	.401 4257	83	1.448 6233	143	2.495 8208	23	.006 6904	23	.000 1115
24	.418 8790	84	1.466 0766	144	2.513 2741	24	.006 9813	24	.000 1164
25	.436 3323	85	1.483 5299	145	2.530 7274	25	.007 2722	25	.000 1212
26	.453 7856	86	1.500 9832	146	2.548 1807	26	.007 5631	26	.000 1261
27	.471 2389	87	1.518 4364	147	2.565 6340	27	.007 8540	27	.000 1309
28	.488 6922	88	1.535 8897	148	2.583 0873	28	.008 1449	28	.000 1357
29	.506 1455	89	1.553 3430	149	2.600 5406	29	.008 4358	29	.000 1406
30	.523 5988	90	1.570 7963	150	2.617 9939	30	.008 7266	30	.000 1454
31	.541 0521	91	1.588 2496	151	2.635 4472	31	.009 0175	31	.000 1503
32	.558 5054	92	1.605 7029	152	2.652 9005	32	.009 3084	32	.000 1551
33	.575 9587	93	1.623 1562	153	2.670 3538	33	.009 5993	33	.000 1600
34	.593 4119	94	1.640 6095	154	2.687 8070	34	.009 8902	34	.000 1648
35	.610 8652	95	1.658 0628	155	2.705 2603	35	.010 1811	35	.000 1697
36	.628 3185	96	1.675 5161	156	2.722 7136	36	.010 4720	36	.000 1745
37	.645 7718	97	1.692 9694	157	2.740 1669	37	.010 7629	37	.000 1794
38	.663 2251	98	1.710 4227	158	2.757 6202	38	.011 0538	38	.000 1842
39	.680 6784	99	1.727 8760	159	2.775 0735	39	.011 3446	39	.000 1891
40	.698 1317	100	1.745 3293	160	2.792 5268	40	.011 6355	40	.000 1939
41	.715 5850	101	1.762 7825	161	2.809 9801	41	.011 9264	41	.000 1988
42	.733 0383	102	1.780 2358	162	2.827 4334	42	.012 2173	42	.000 2036
43	.750 4916	103	1.797 6891	163	2.844 8867	43	.012 5082	43	.000 2085
44	.767 9449	104	1.815 1424	164	2.862 3400	44	.012 7991	44	.000 2133
45	.785 3982	105	1.832 5957	165	2.879 7933	45	.013 0900	45	.000 2182
46	.802 8515	106	1.850 0490	166	2.897 2466	46	.013 3809	46	.000 2230
47	.820 3047	107	1.867 5023	167	2.914 6999	47	.013 6717	47	.000 2279
48	.837 7580	108	1.884 9556	168	2.932 1531	48	.013 9626	48	.000 2327
49	.855 2113	109	1.902 4089	169	2.949 6064	49	.014 2535	49	.000 2376
50	.872 6646	110	1.919 8622	170	2.967 0597	50	.014 5444	50	.000 2424
51	.890 1179	111	1.937 3155	171	2.984 5130	51	.014 8353	51	.000 2473
52	.907 5712	112	1.954 7688	172	3.001 9663	52	.015 1262	52	.000 2521
53	.925 0245	113	1.972 2221	173	3.019 4196	53	.015 4171	53	.000 2570
54	.942 4778	114	1.989 6753	174	3.036 8729	54	.015 7080	54	.000 2618
55	.959 9311	115	2.007 1286	175	3.054 3262	55	.015 9989	55	.000 2666
56	.977 3844	116	2.024 5819	176	3.071 7795	56	.016 2897	56	.000 2715
57	.994 8377	117	2.042 0352	177	3.089 2328	57	.016 5806	57	.000 2763
58	1.012 2910	118	2.059 4885	178	3.106 6861	58	.016 8715	58	.000 2812
59	1.029 7443	119	2.076 9418	179	3.124 1394	59	.017 1624	59	.000 2860
60	1.047 1976	120	2.094 3951	180	3.141 5927	60	.017 4533	60	.000 2909

SOURCE: American Institute of Steel Construction.

9-14 Net Areas of Tension Members Net areas for tension members in two-angle design are shown in Tables 9-38 and 9-39.

TABLE 9-38 Tension Members—Net Areas

Angle Designation	TWO ANGLES—NET AREA											
	2 Holes out				4 Holes out				6 Holes out			
	Fastener Diam., In.				Fastener Diam., In.				Fastener Diam., In.			
	¾	⅞	1	1⅛	¾	⅞	1	1⅛	¾	⅞	1	1⅛
L 6 × 4 × ⅞	14.4	14.2	14.0	13.8	12.9	12.5	12.0	11.6	11.4	—	—	—
¾	12.6	12.4	12.2	12.0	11.3	10.9	10.5	10.1	9.94	—	—	—
⅝	10.6	10.5	10.3	10.2	9.53	9.22	8.91	8.60	8.44	—	—	—
9/16	9.64	9.49	9.35	9.21	8.65	8.37	8.09	7.81	7.67	—	—	—
½	8.62	8.50	8.37	8.25	7.75	7.50	7.25	7.00	6.88	—	—	—
7/16	7.59	7.48	7.38	7.27	6.83	6.61	6.39	6.17	6.06	—	—	—
⅜	6.56	6.47	6.38	6.28	5.91	5.72	5.53	5.34	5.25	—	—	—
L 5 × 5 × ⅞	14.4	14.2	14.0	13.8	12.9	12.5	12.0	11.6	—	—	—	—
¾	12.6	12.4	12.2	12.0	11.3	10.9	10.5	10.1	—	—	—	—
⅝	10.6	10.5	10.3	10.2	9.53	9.22	8.91	8.59	—	—	—	—
½	8.62	8.50	8.37	8.25	7.75	7.50	7.25	7.00	—	—	—	—
7/16	7.59	7.48	7.38	7.27	6.83	6.61	6.39	6.17	—	—	—	—
⅜	6.56	6.47	6.38	6.28	5.91	5.72	5.53	5.34	—	—	—	—
L 6 × 3½ × ½	8.12	8.00	7.87	—	7.25	7.00	—	—	6.38	—	—	—
⅜	6.18	6.09	6.00	—	5.53	5.34	—	—	4.87	—	—	—
5/16	5.19	5.11	5.04	—	4.65	4.49	—	—	4.10	—	—	—
L 5 × 3½ × ¾	10.3	10.1	9.93	—	8.99	8.62	—	—	—	—	—	—
⅝	8.75	8.59	8.43	—	7.65	7.34	—	—	—	—	—	—
½	7.12	7.00	6.87	—	6.25	6.00	—	—	—	—	—	—
7/16	6.29	6.18	6.08	—	5.53	5.31	—	—	—	—	—	—
⅜	5.44	5.35	5.25	—	4.79	4.60	—	—	—	—	—	—
5/16	4.57	4.49	4.42	—	4.03	3.87	—	—	—	—	—	—
L 5 × 3 × ½	6.62	6.50	—	—	5.75	5.50	—	—	—	—	—	—
⅜	5.06	4.97	—	—	4.41	4.22	—	—	—	—	—	—
5/16	4.25	4.17	—	—	3.71	3.55	—	—	—	—	—	—
L 4 × 4 × ¾	9.57	9.38	9.19	9.00	8.26	7.88	—	—	—	—	—	—
⅝	8.13	7.97	7.81	7.66	7.03	6.72	—	—	—	—	—	—
½	6.62	6.50	6.37	6.25	5.75	5.50	—	—	—	—	—	—
7/16	5.85	5.74	5.64	5.53	5.09	4.87	—	—	—	—	—	—
⅜	5.06	4.97	4.88	4.78	4.41	4.22	—	—	—	—	—	—
5/16	4.25	4.17	4.10	4.02	3.71	3.55	—	—	—	—	—	—
L 4 × 3½ × ⅝	7.51	7.35	7.19	—	6.42	6.10	—	—	—	—	—	—
½	6.12	6.00	5.87	—	5.25	5.00	—	—	—	—	—	—
7/16	5.41	5.30	5.20	—	4.65	4.43	—	—	—	—	—	—
⅜	4.68	4.59	4.50	—	4.03	3.84	—	—	—	—	—	—
5/16	3.95	3.87	3.80	—	3.41	3.25	—	—	—	—	—	—
L 4 × 3 × ⅝	6.87	6.71	—	—	5.77	5.46	—	—	—	—	—	—
½	5.62	5.50	—	—	4.75	4.50	—	—	—	—	—	—
7/16	4.97	4.86	—	—	4.21	3.99	—	—	—	—	—	—
⅜	4.30	4.21	—	—	3.65	3.46	—	—	—	—	—	—
5/16	3.63	3.55	—	—	3.09	2.93	—	—	—	—	—	—
¼	2.94	2.88	—	—	2.50	2.38	—	—	—	—	—	—

Net areas are computed in accordance with AISC Specification, Section 1.14.5.

SOURCE: American Institute of Steel Construction.

TABLE 9-39 Tension Members—Net Areas

Angle Designation	TWO ANGLES—NET AREA											
	2 Holes out				4 Holes out				6 Holes out			
	Fastener Diam., In.				Fastener Diam., In.				Fastener Diam., In.			
	3/4	7/8	1	1 1/8	3/4	7/8	1	1 1/8	3/4	7/8	1	1 1/8
L 9 × 4 × 1	22.3	22.0	21.8	21.5	20.5	20.0	19.5	19.0	18.8	18.0	17.3	16.5
7/8	19.7	19.5	19.3	19.0	18.2	17.7	17.3	16.8	16.6	16.0	15.3	14.7
3/4	17.1	16.9	16.7	16.5	15.8	15.4	15.0	14.6	14.4	13.9	13.3	12.8
5/8	14.4	14.2	14.1	13.9	13.3	13.0	12.7	12.3	12.2	11.7	11.2	10.8
9/16	13.0	12.9	12.7	12.6	12.0	11.8	11.5	11.2	11.1	10.6	10.2	9.78
1/2	11.6	11.5	11.4	11.3	10.8	10.5	10.3	10.0	9.88	9.50	9.12	8.75
L 8 × 8 × 1 1/8	31.5	31.2	30.9	30.7	29.5	29.0	28.4	27.8	27.6	26.7	25.9	25.0
1	28.3	28.0	27.8	27.5	26.5	26.0	25.5	25.0	24.8	24.0	23.3	22.5
7/8	24.9	24.7	24.5	24.3	23.4	23.0	22.5	22.1	21.9	21.2	20.6	19.9
3/4	21.6	21.4	21.2	21.0	20.3	19.9	19.5	19.1	18.9	18.4	17.8	17.3
5/8	18.1	18.0	17.8	17.7	17.0	16.7	16.4	16.1	15.9	15.5	15.0	14.5
9/16	16.4	16.2	16.1	16.0	15.4	15.1	14.8	14.6	14.4	14.0	13.6	13.2
1/2	14.6	14.5	14.4	14.3	13.8	13.5	13.3	13.0	12.9	12.5	12.1	11.8
L 8 × 6 × 1	24.3	24.0	23.8	23.5	22.5	22.0	21.5	21.0	20.8	20.0	19.3	18.5
7/8	21.4	21.2	21.0	20.8	19.9	19.5	19.0	18.6	18.4	17.7	17.1	16.4
3/4	18.6	18.4	18.2	18.0	17.3	16.9	16.5	16.1	15.9	15.4	14.8	14.3
5/8	15.6	15.5	15.3	15.2	14.5	14.2	13.9	13.6	13.4	13.0	12.5	12.0
9/16	14.1	14.0	13.9	13.7	13.2	12.9	12.6	12.3	12.2	11.8	11.3	10.9
1/2	12.6	12.5	12.4	12.3	11.8	11.5	11.3	11.0	10.9	10.5	10.1	9.75
7/16	11.1	11.0	10.9	10.8	10.3	10.1	9.89	9.67	9.56	9.24	8.91	8.58
L 8 × 4 × 1	20.3	20.0	19.8	19.5	18.5	18.0	17.5	17.0	16.8	16.0	15.3	14.5
7/8	17.9	17.7	17.5	17.3	16.4	16.0	15.5	15.1	14.9	14.2	13.6	12.9
3/4	15.6	15.4	15.2	15.0	14.3	13.9	13.5	13.1	12.9	12.4	11.8	11.3
5/8	13.1	13.0	12.8	12.7	12.0	11.7	11.4	11.1	10.9	10.5	10.0	9.53
9/16	11.9	11.7	11.6	11.5	10.9	10.6	10.3	10.1	9.90	9.50	9.09	8.64
1/2	10.6	10.5	10.4	10.3	9.75	9.50	9.25	9.00	8.87	8.50	8.13	7.75
7/16	9.35	9.24	9.14	9.03	8.59	8.37	8.15	7.93	7.83	7.50	7.18	6.84
L 7 × 4 × 7/8	16.2	16.0	15.8	15.5	14.7	14.2	13.8	13.3	13.1	12.5	—	—
3/4	14.1	13.9	13.7	13.5	12.8	12.4	12.0	11.6	11.4	10.9	—	—
5/8	11.9	11.7	11.6	11.4	10.8	10.5	10.2	9.84	9.70	9.23	—	—
9/16	10.8	10.6	10.5	10.3	9.77	9.49	9.21	8.93	8.78	8.38	—	—
1/2	9.62	9.50	9.37	9.25	8.75	8.50	8.25	8.00	7.87	7.50	—	—
7/16	8.47	8.36	8.26	8.15	7.71	7.49	7.27	7.05	6.95	6.62	—	—
3/8	7.32	7.23	7.14	7.04	6.67	6.48	6.27	6.08	6.01	5.73	—	—
L 6 × 6 × 1	20.3	20.0	19.8	19.5	18.5	18.0	17.5	17.0	16.8	16.0	—	—
7/8	17.9	17.7	17.5	17.3	16.4	16.0	15.5	15.1	14.9	14.2	—	—
3/4	15.6	15.4	15.2	15.0	14.3	13.9	13.5	13.1	12.9	12.4	—	—
5/8	13.1	13.0	12.8	12.7	12.0	11.7	11.4	11.1	10.9	10.5	—	—
9/16	11.9	11.7	11.6	11.5	10.9	10.6	10.3	10.1	9.91	9.49	—	—
1/2	10.6	10.5	10.4	10.3	9.75	9.50	9.25	9.00	8.87	8.50	—	—
7/16	9.35	9.24	9.14	9.03	8.59	8.37	8.15	7.93	7.82	7.50	—	—
3/8	8.06	7.97	7.88	7.78	7.41	7.22	7.03	6.84	6.75	6.47	—	—

Net areas are computed in accordance with AISC Specification, Section 1.14.5.

SOURCE: American Institute of Steel Construction.

9-15 Net Section of Tension Members A graph or chart showing net section of tension members is pictured in Table 9-40 with an example.

TABLE 9-40 Net Section of Tension Members

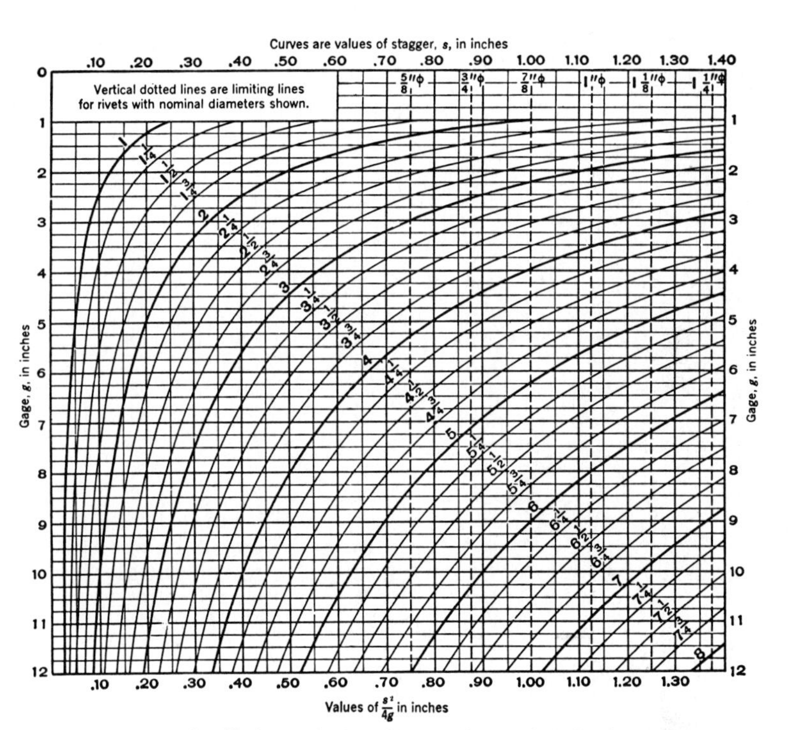

The above chart will simplify the application of the rule for net width, Sections 1.14.3 and 1.14.4 of the AISC Specification. Entering the chart at left or right with the gage g and proceeding horizontally to intersection with the curve for the pitch s, thence vertically to top or bottom, the value of $s^2/4g$ may be read directly.

Step 1 of the example below illustrates the application of the rule and the use of the chart. Step 2 illustrates the application of the 85% of gross area limitation.

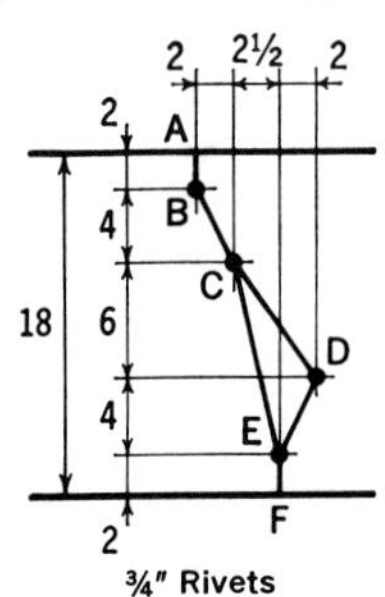

Step 1: Chain A B C E F

Deduct for 3 holes @ (¾ + ⅛)	= −2.625
BC, $g = 4$, $s = 2$; add $s^2/4g$	= +0.25
CE, $g = 10$, $s = 2\frac{1}{2}$; add $s^2/4g$	= +0.16
Total Deduction	= −2.215"

Chain A B C D E F

Deduct for 4 holes @ (¾ + ⅛)	= −3.50
BC, as above, add	= +0.25
CD, $g = 6$, $s = 4\frac{1}{2}$; add $s^2/4g$	= +0.85
DE, $g = 4$, $s = 2$; add $s^2/4g$	= +0.25
Total Deduction	= −2.15"

Net Width = 18.0 − 2.215 = 15.785".

Step 2: Net width = 18.0 × 0.85 = 15.3"
(Governs in this example)

In comparing the path CDE with the path CE, it is seen that if the sum of the two values of $s^2/4g$ for CD and DE exceed the single value of $s^2/4g$ for CE, by more than the deduction for one hole, then the path CDE is not critical as compared with CE.

Evidently if the value of $s^2/4g$ for one leg CD of the path CDE is greater than the deduction for one hole, the path CDE cannot be critical as compared with CE. The vertical dotted lines in the chart serve to indicate, for the respective rivet diameters noted at the top thereof, that any value of $s^2/4g$ to the right of such line is derived from a non-critical chain which need not be further considered.

SOURCE: American Institute of Steel Construction.

9-16 Reduction of Area for Bolt and Rivet Holes Table 9-41 shows tabular material relative to the reduction of area for bolt and rivet holes.

TABLE 9-41 Reduction of Area for Bolt and Rivet Holes

Area in square inches = assumed diameter of hole by thickness of metal. For computation purposes holes shall be taken as the nominal diameter of fastener plus ⅛ inch

Thickness of Metal Inches	Diameter of Hole, Inches					
	¾	⅞	1	1⅛	1¼	1⅜
3/16	.141	.164	.188	.211	.234	.258
¼	.188	.219	.250	.281	.313	.344
5/16	.234	.273	.313	.352	.391	.430
⅜	.281	.328	.375	.422	.469	.516
7/16	.328	.383	.438	.492	.547	.602
½	.375	.438	.500	.563	.625	.688
9/16	.422	.492	.563	.633	.703	.773
⅝	.469	.547	.625	.703	.781	.859
11/16	.516	.602	.688	.773	.859	.945
¾	.563	.656	.750	.844	.938	1.031
13/16	.609	.711	.813	.914	1.016	1.117
⅞	.656	.766	.875	.984	1.094	1.203
15/16	.703	.820	.938	1.055	1.172	1.289
1	.750	.875	1.000	1.125	1.250	1.375
1/16	.797	.930	1.063	1.195	1.328	1.461
⅛	.844	.984	1.125	1.266	1.406	1.547
3/16	.891	1.039	1.188	1.336	1.484	1.633
¼	.938	1.094	1.250	1.406	1.563	1.719
5/16	.984	1.148	1.313	1.477	1.641	1.805
⅜	1.031	1.203	1.375	1.547	1.719	1.891
7/16	1.078	1.258	1.438	1.617	1.797	1.977
½	1.125	1.313	1.500	1.688	1.875	2.063
9/16	1.172	1.367	1.563	1.758	1.953	2.148
⅝	1.219	1.422	1.625	1.828	2.031	2.234
11/16	1.266	1.477	1.688	1.898	2.109	2.320
¾	1.313	1.531	1.750	1.969	2.188	2.406
13/16	...	1.586	1.813	2.039	2.266	2.492
⅞	...	1.641	1.875	2.109	2.344	2.578
15/16	...	1.695	1.938	2.180	2.422	2.664
2	...	1.750	2.000	2.250	2.500	2.750
1/16	...	1.805	2.063	2.320	2.578	2.836
⅛	...	1.859	2.125	2.391	2.656	2.922
3/16	...	1.914	2.188	2.461	2.734	3.008
¼	...	1.969	2.250	2.531	2.813	3.094
5/16	...	2.023	2.313	2.602	2.891	3.180
⅜	...	2.078	2.375	2.672	2.969	3.266
7/16	...	2.133	2.438	2.742	3.047	3.352
½	...	2.188	2.500	2.813	3.125	3.438
⅝	...	2.297	2.625	2.953	3.281	3.609
¾	...	2.406	2.750	3.094	3.438	3.781
⅞	...	2.516	2.875	3.234	3.594	3.953
3	...	2.625	3.000	3.375	3.750	4.125

SOURCE: American Institute of Steel Construction.

TABLE 9-42 Recessed Pin Nuts and Cotter Pins

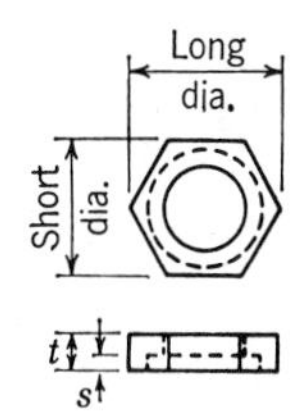

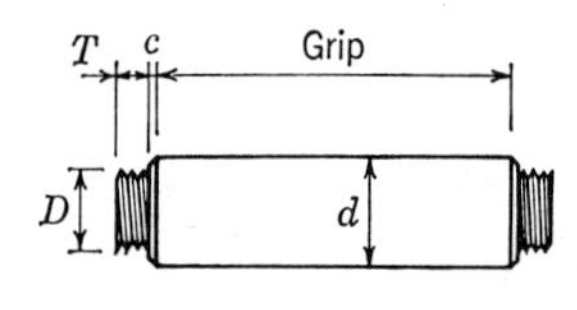

Material: Steel

Thread: 6 UN Class 2A/2B

Diameter of Pin d	PIN			NUT (Suggested Dimensions)					
	Thread		c	Thick-ness t	Diameter		Recess		Weight Pounds
	D	T			Short Diam.	Long Diam.	Rough Diam.	s	
2 2¼	1½	1	⅛	⅞	3	3⅜	2⅝	¼	1
2½ 2¾	2	1⅛	⅛	1	3⅝	4⅛	3⅛	¼	2
3 3¼ 3½	2½	1¼	⅛	1⅛	4⅜	5	3⅞	⅜	3
3¾ 4	3	1⅜	¼	1¼	4⅞	5⅝	4⅜	⅜	4
4¼ 4½ 4¾	3½	1½	¼	1⅜	5¾	6⅝	5¼	½	5
5 5¼	4	1⅝	¼	1½	6¼	7¼	5¾	½	6
5½ 5¾ 6	4½	1¾	¼	1⅝	7	8⅛	6½	⅝	8
6¼ 6½	5	1⅞	⅜	1¾	7⅝	8⅞	7	⅝	10
6¾ 7	5½	2	⅜	1⅞	8⅛	9⅜	7½	¾	12
7¼ 7½	5½	2	⅜	1⅞	8⅝	10	8	¾	14
7¾ 8 8¼	6	2¼	⅜	2⅛	9⅜	10⅞	8¾	¾	19
8½ 8¾ 9	6	2¼	⅜	2⅛	10¼	11⅞	9⅝	¾	24
9¼ 9½	6	2⅜	⅜	2¼	11¼	13	10⅝	¾	32
9¾ 10	6	2⅜	⅜	2¼	11¼	13	10⅝	¾	32

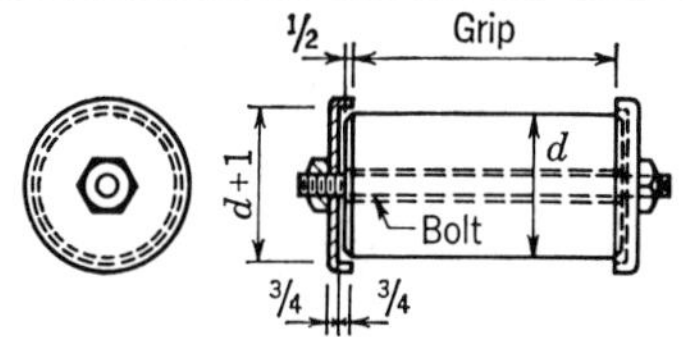

Typical Pin Cap Detail for Pins over 10 Inches in Diameter
Dimensions shown are approximate

Although nuts may be used on all sizes of pins as shown above, for pins over 10″ in diameter the preferred practice is a detail similar to that shown at the left, in which the pin is held in place by a recessed cap at each end and secured by a bolt passing completely through the caps and pin. Suitable provision must be made for attaching pilots and driving nuts.

HORIZONTAL OR VERTICAL PIN

HORIZONTAL PIN

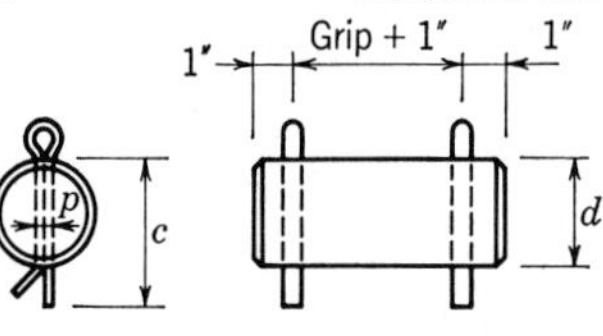

l = Length of pin, in inches.

Pin Diam. d	Pins With Heads: Head Diam. h	Pins With Heads: Weight of One (Lb.)	Cotter: Length c	Cotter: Diam. p	Cotter: Wt. per 100 (Lb.)	Pin Diam. d	Pins With Heads: Head Diam. h	Pins With Heads: Weight of One (Lb.)	Cotter: Length c	Cotter: Diam. p	Cotter: Wt. per 100 (Lb.)
1¼	1½	.19+ .35l	2	¼	2.64	2¾	3⅛	.82+1.68l	4	⅜	11.4
1½	1¾	.26+ .50l	2½	¼	3.10	3	3½	1.02+2.00l	5	½	28.5
1¾	2	.33+ .68l	2¾	¼	3.50	3¼	3¾	1.17+2.35l	5	½	28.5
2	2⅜	.47+ .89l	3	⅜	9.00	3½	4	1.34+2.73l	6	½	33.8
2¼	2⅝	.58+1.13l	3¼	⅜	9.40	3¾	4¼	1.51+3.13l	6	½	33.8
2½	2⅞	.70+1.39l	3¾	⅜	10.9						

SOURCE: American Institute of Steel Construction.

9-17 Unfinished Machine Bolts, Welded Studs, and Pins Unfinished machine bolts are sometimes used and are called *secondary connections*. These connections should be carefully defined to eliminate the possibility of wrong selection by an ironworker.

TABLE 9-43 Welded Joints—Standard Symbols

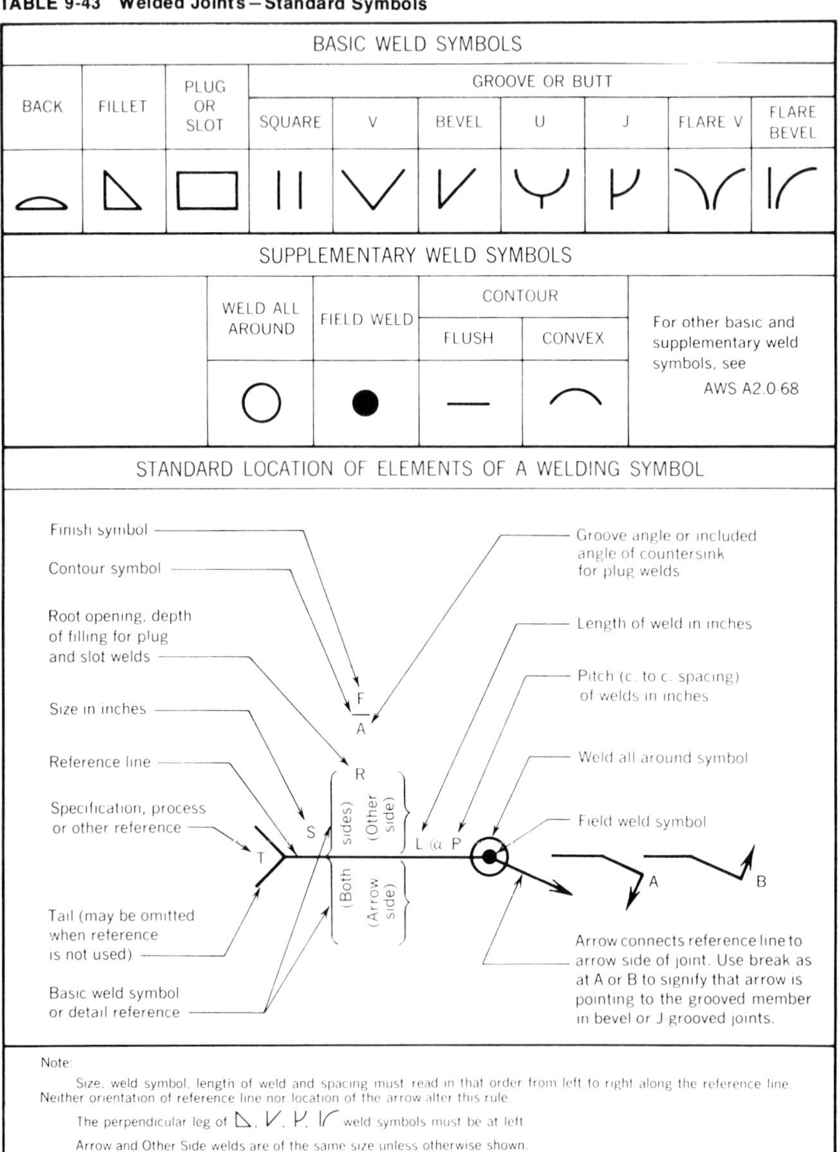

SOURCE: American Institute of Steel Construction.

Welded studs are generally threaded and used as steel-to-steel connections. Cost savings can result in the use of this type of connection. Holes may be drilled or punched to receive the stud and then plug-welded into place. Welding guns may also be employed.

Pins, used as connectors, permit rotation or a hingeing action of one or more members. Table 9-42 shows the data for recessed pin nuts and cotter pins. Refer to Sec. 2 of this handbook for additional information.

9-18 Field Welding Field welding is done on-site and sometimes under difficult conditions. The welding specifications are specified by the designer and under the regulations of all applicable codes.

Part 2 of Sec. 6 of this handbook describes the techniques of arc welding and its general processes.

The designer should avoid specifying field welding whenever possible. The chance for erection error and the difficult conditions sometimes encountered can make a significant cost increase. However, special construction problems sometimes require field cutting of structural members to fit unknown dimensions. This will require a field connection, best solved by specifying a sound field weld.

Consult the specifications prepared by the American Welding Society (AWS) for the proper procedures and recommendations. Space limits the inclusion of a more detailed welding discussion. Refer to Sec. 6 of this handbook for more details.

Table 9-43 illustrates the standard welded joint symbols used for all structural work.

Section 10
Locking Components

ROBERT O. PARMLEY, P.E.
Consulting Engineer, Ladysmith, Wisconsin

Most mechanical locking is achieved by the use of one or more standard components. This section will deal with those locking components not covered in other areas or chapters of this handbook.

Pins, retaining rings, threaded fasteners, rivets, and related data will be found elsewhere in individual sections where an elaborate technical discussion is presented. Refer to the table of contents and the index for their location. Combinations of various locking components are covered in Sec. 14, Aerospace Fastening.

Most advances in design engineering come into existence with the aid of standard components, which have long been employed as stock items and are the building blocks of the more intricate mechanisms.

This section will explore standard locking components and later illustrate some of the more complex mechanisms of which they are a vital part.

KEYS AND KEYSEATS

10-1 Keys A key is a demountable machinery component which, when positioned in an assembly, seats into the keyseat or keyway and thus provides a lock between shaft and hub for the purpose of transmitting torque.

TABLE 10-1 Key Dimensions and Tolerances
All dimensions given in inches.

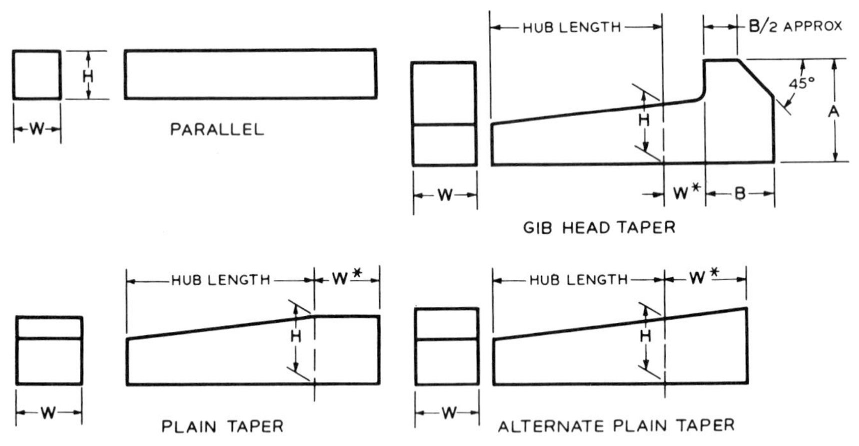

Plain and Gib Head Taper Keys Have a 1/8" Taper in 12"

KEY			Nominal key size, Width, *W*: Over	Nominal key size, Width, *W*: To (Incl)	Tolerance: Width, *W*	Tolerance: Height, *H*
Parallel	Square	Bar Stock	–	3/4	+0.000 –0.002	+0.000 –0.002
			3/4	1-1/2	+0.000 –0.003	+0.000 –0.003
			1-1/2	2-1/2	+0.000 –0.004	+0.000 –0.004
			2-1/2	3-1/2	+0.000 –0.006	+0.000 –0.006
		Keystock	–	1-1/4	+0.001 –0.000	+0.001 –0.000
			1-1/4	3	+0.002 –0.000	+0.002 –0.000
			3	3-1/2	+0.003 –0.000	+0.003 –0.000
	Rectangular	Bar Stock	–	3/4	+0.000 –0.003	+0.000 –0.003
			3/4	1-1/2	+0.000 –0.004	+0.000 –0.004
			1-1/2	3	+0.000 –0.005	+0.000 –0.005
			3	4	+0.000 –0.006	+0.000 –0.006
			4	6	+0.000 –0.008	+0.000 –0.008
			6	7	+0.000 –0.013	+0.000 –0.013
		Keystock	–	1-1/4	+0.001 –0.000	+0.005 –0.005
			1-1/4	3	+0.002 –0.000	+0.005 –0.005
			3	7	+0.003 –0.000	+0.005 –0.005
Taper	Plain or Gib Head Square or Rectangular		–	1-1/4	+0.001 –0.000	+0.005 –0.000
			1-1/4	3	+0.002 –0.000	+0.005 –0.000
			3	7	+0.003 –0.000	+0.005 –0.000

*For locating position of dimension *H*. Tolerance does not apply.

SOURCE: ANSI B17.1-1967 (Re-1973), published by the American Society of Mechanical Engineers.

10-2 Key Material Keys are made from two classes of stock presently used by industry. Bar stock is used for general purposes. It is cold-finished steel and has a negative tolerance. Key stock is cold-finished steel and has a close plus tolerance. Table 10-1 shows the standard dimensions and tolerances for parallel, plain taper, alternate plain taper, and gib-head taper keys.

10-3 Key Size Versus Shaft Diameter This is a design consideration which must not be overlooked. Table 10-2 lists the recommended nominal depths and key sizes to be used. Square keys are recommended through 6.5-inch shaft diameter and rectangular keys for any larger shafts. Sizes and dimensions in the unshaded areas of the table are preferred.

TABLE 10-2 Key Size Versus Shaft Diameter
Sizes and dimensions in unshaded areas are preferred.

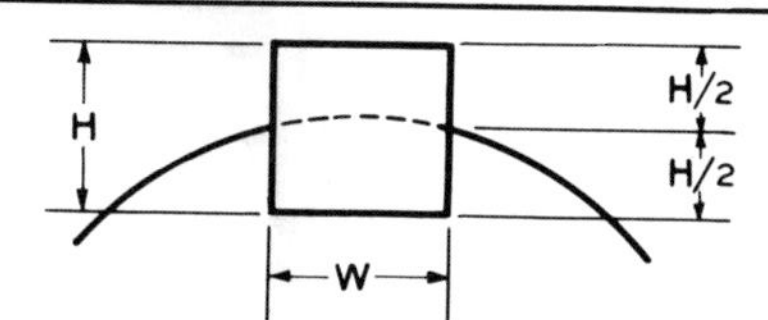

NOMINAL SHAFT DIAMETER		NOMINAL KEY SIZE			NOMINAL KEYSEAT DEPTH	
Over	To (Incl)	Width, W	Height, H		$H/2$	
			Square	Rectangular	Square	Rectangular
5/16	7/16	3/32	3/32		3/64	
7/16	9/16	1/8	1/8	3/32	1/16	3/64
9/16	7/8	3/16	3/16	1/8	3/32	1/16
7/8	1-1/4	1/4	1/4	3/16	1/8	3/32
1-1/4	1-3/8	5/16	5/16	1/4	5/32	1/8
1-3/8	1-3/4	3/8	3/8	1/4	3/16	1/8
1-3/4	2-1/4	1/2	1/2	3/8	1/4	3/16
2-1/4	2-3/4	5/8	5/8	7/16	5/16	7/32
2-3/4	3-1/4	3/4	3/4	1/2	3/8	1/4
3-1/4	3-3/4	7/8	7/8	5/8	7/16	5/16
3-3/4	4-1/2	1	1	3/4	1/2	3/8
4-1/2	5-1/2	1-1/4	1-1/4	7/8	5/8	7/16
5-1/2	6-1/2	1-1/2	1-1/2	1	3/4	1/2
6-1/2	7-1/2	1-3/4	1-3/4	1-1/2*	7/8	3/4
7-1/2	9	2	2	1-1/2	1	3/4
9	11	2-1/2	2-1/2	1-3/4	1-1/4	7/8
11	13	3	3	2	1-1/2	1
13	15	3-1/2	3-1/2	2-1/2	1-3/4	1-1/4
15	18	4		3		1-1/2
18	22	5		3-1/2		1-3/4
22	26	6		4		2
26	30	7		5		2-1/2

SOURCE: ANSI B17.1-1967 (Re-1973), published by the American Society of Mechanical Engineers.

10-4 Gib-Head Keys These keys and their nominal dimensions are tabulated in Table 10-3. This key design is valuable in special applications where disassembly may be difficult.

10-5 Depth Control Key depth is, as noted previously, a critical area and must be strictly regulated. Figure 10-1 graphically shows the chordal height, depth of shaft keyseat, and depth of hub keyseat, respectively. The depth control formulas are also represented.

Depth control values are tabulated in Tables 10-4 and 10-5. Parallel and taper (square and rectangular), parallel (square and rectangular), and taper (square and rectangular) keys are listed for the reader's reference.

10-6 Classes of Fits Fits for key-keyseat assemblies fall into three classes. Table 10-6 details them and tabulates the class 1 fit for parallel keys. Table 10-7 covers the class 2 fit for parallel and taper keys.

10-7 Alignment Tolerances Tolerances are illustrated in Fig. 10-2. Offset of center line, lead angle, and parallelism are all shown.

10-8 Chamfered Key and Filleted Keyseats These, in general practice, are not used. However, it is recognized that fillets in keyseats decrease stress concentrations

TABLE 10-3 Gib-Head Nominal Dimensions

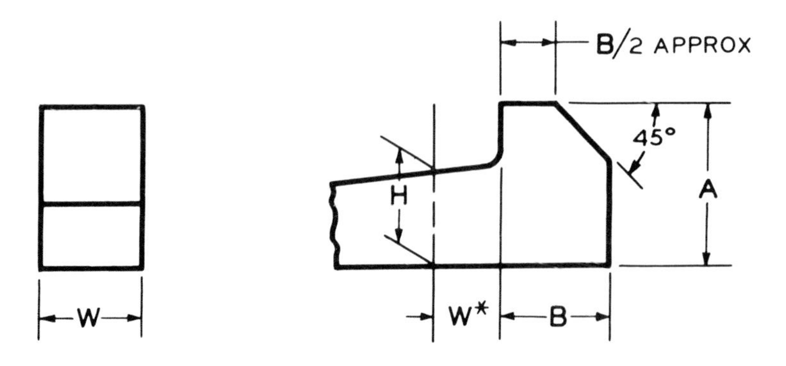

Nominal Key Size Width, W	SQUARE			RECTANGULAR		
	H	*A*	*B*	*H*	*A*	*B*
1/8	1/8	1/4	1/4	3/32	3/16	1/8
3/16	3/16	5/16	5/16	1/8	1/4	1/4
1/4	1/4	7/16	3/8	3/16	5/16	5/16
5/16	5/16	1/2	7/16	1/4	7/16	3/8
3/8	3/8	5/8	1/2	1/4	7/16	3/8
1/2	1/2	7/8	5/8	3/8	5/8	1/2
5/8	5/8	1	3/4	7/16	3/4	9/16
3/4	3/4	1-1/4	7/8	1/2	7/8	5/8
7/8	7/8	1-3/8	1	5/8	1	3/4
1	1	1-5/8	1-1/8	3/4	1-1/4	7/8
1-1/4	1-1/4	2	1-7/16	7/8	1-3/8	1
1-1/2	1-1/2	2-3/8	1-3/4	1	1-5/8	1-1/8
1-3/4	1-3/4	2-3/4	2	1-1/2	2-3/8	1-3/4
2	2	3-1/2	2-1/4	1-1/2	2-3/8	1-3/4
2-1/2	2-1/2	4	3	1-3/4	2-3/4	2
3	3	5	3-1/2	2	3-1/2	2-1/4
3-1/2	3-1/2	6	4	2-1/2	4	3

*For locating position of dimension *H*.

For larger sizes the following relationships are suggested as guides for establishing *A* and *B*.

$$A = 1.8\,H \qquad B = 1.2\,H$$

All dimensions given in inches.

SOURCE: ANSI B17.1-1967 (Re-1973), published by the American Society of Mechanical Engineers.

CHORDAL HEIGHT

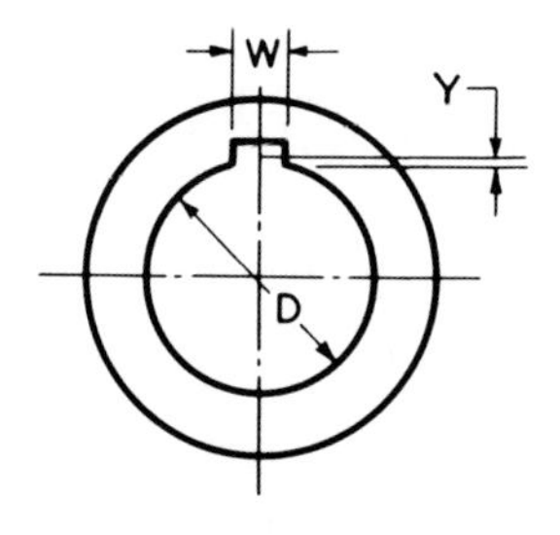

The chordal height Y is determined from the following formula:

$$Y = \frac{D - \sqrt{D^2 - W^2}}{2}$$

The distance from the bottom of the shaft keyseat to the opposite side of the shaft is specified by dimension S. The following formula may be used for calculating this dimension:

$$S = D - Y - \frac{H}{2} = \frac{D - H + \sqrt{D^2 - W^2}}{2}$$

DEPTH OF SHAFT KEYSEAT

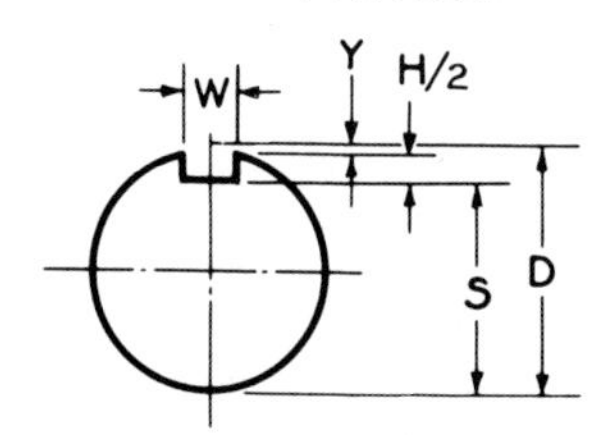

Tabulated values of S for specific shaft diameters are given in Table 3.

The distance from the bottom of the hub keyseat to the opposite side of the hub bore is specified by dimension T. For taper keyseats, T is measured at the deeper end. The following formula may be used for calculating this dimension:

$$T = D - Y + \frac{H}{2} + C = \frac{D + H + \sqrt{D^2 - W^2}}{2} + C$$

Tabulated values of T for parallel and taper keyseats for specific bores are given in Table 3.

DEPTH OF HUB KEYSEAT

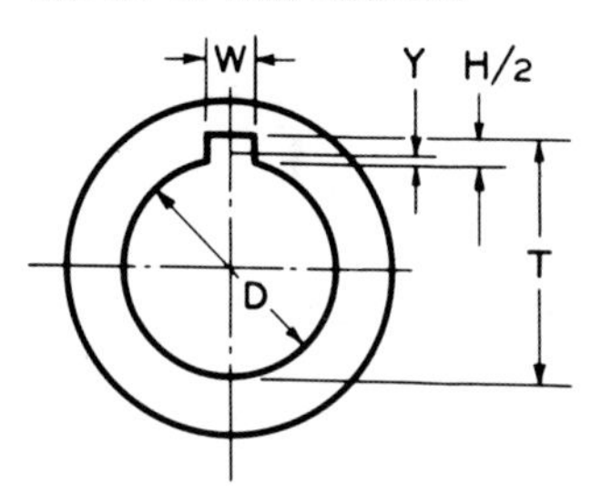

Symbols

C = Allowance
  + 0.005 inch clearance for parallel keys
  − 0.020 inch interference for taper keys
D = Nominal shaft or bore diameter, inches
H = Nominal key height, inches
W = Nominal key width, inches
Y = Chordal height, inches

FIG. 10-1 Depth-control formulas. *(From USAS B17.1–1967 (Re-1973), published by the American Society of Mechanical Engineers.)*

TABLE 10-4 Depth-Control Values for *S* and *T*

Shaft diameter ½ through 3⅛ inches

Nominal Shaft Diameter	Parallel and Taper		Parallel		Taper	
	Square *S*	Rectangular *S*	Square *T*	Rectangular *T*	Square *T*	Rectangular *T*
1/2	0.430	0.445	0.560	0.544	0.535	0.519
9/16	0.493	0.509	0.623	0.607	0.598	0.582
5/8	0.517	0.548	0.709	0.678	0.684	0.653
11/16	0.581	0.612	0.773	0.742	0.748	0.717
3/4	0.644	0.676	0.837	0.806	0.812	0.781
13/16	0.708	0.739	0.900	0.869	0.875	0.844
7/8	0.771	0.802	0.964	0.932	0.939	0.907
15/16	0.796	0.827	1.051	1.019	1.026	0.994
1	0.859	0.890	1.114	1.083	1.089	1.058
1-1/16	0.923	0.954	1.178	1.146	1.153	1.121
1-1/8	0.986	1.017	1.241	1.210	1.216	1.185
1-3/16	1.049	1.080	1.304	1.273	1.279	1.248
1-1/4	1.112	1.144	1.367	1.336	1.342	1.311
1-5/16	1.137	1.169	1.455	1.424	1.430	1.399
1-3/8	1.201	1.232	1.518	1.487	1.493	1.462
1-7/16	1.225	1.288	1.605	1.543	1.580	1.518
1-1/2	1.289	1.351	1.669	1.606	1.644	1.581
1-9/16	1.352	1.415	1.732	1.670	1.707	1.645
1-5/8	1.416	1.478	1.796	1.733	1.771	1.708
1-11/16	1.479	1.541	1.859	1.796	1.834	1.771
1-3/4	1.542	1.605	1.922	1.860	1.897	1.835
1-13/16	1.527	1.590	2.032	1.970	2.007	1.945
1-7/8	1.591	1.654	2.096	2.034	2.071	2.009
1-15/16	1.655	1.717	2.160	2.097	2.135	2.072
2	1.718	1.781	2.223	2.161	2.198	2.136
2-1/16	1.782	1.844	2.287	2.224	2.262	2.199
2-1/8	1.845	1.908	2.350	2.288	2.325	2.263
2-3/16	1.909	1.971	2.414	2.351	2.389	2.326
2-1/4	1.972	2.034	2.477	2.414	2.452	2.389
2-5/16	1.957	2.051	2.587	2.493	2.562	2.468
2-3/8	2.021	2.114	2.651	2.557	2.626	2.532
2-7/16	2.084	2.178	2.714	2.621	2.689	2.596
2-1/2	2.148	2.242	2.778	2.684	2.753	2.659
2-9/16	2.211	2.305	2.841	2.748	2.816	2.723
2-5/8	2.275	2.369	2.905	2.811	2.880	2.786
2-11/16	2.338	2.432	2.968	2.874	2.943	2.849
2-3/4	2.402	2.495	3.032	2.938	3.007	2.913
2-13/16	2.387	2.512	3.142	3.017	3.117	2.992
2-7/8	2.450	2.575	3.205	3.080	3.180	3.055
2-15/16	2.514	2.639	3.269	3.144	3.244	3.119
3	2.577	2.702	3.332	3.207	3.307	3.182
3-1/16	2.641	2.766	3.396	3.271	3.371	3.246
3-1/8	2.704	2.829	3.459	3.334	3.434	3.309

SOURCE: ANSI B17.1-1967 (Re-1973), published by the American Society of Mechanical Engineers.

TABLE 10-5 Depth-Control Values for *S* and *T*
Shaft diameter $3^3/_{16}$ through 15 inches

Nominal Shaft Diameter	Parallel and Taper		Parallel		Taper	
	Square	Rectangular	Square	Rectangular	Square	Rectangular
	S	*S*	*T*	*T*	*T*	*T*
3-3/16	2.768	2.893	3.523	3.398	3.498	3.373
3-1/4	2.831	2.956	3.586	3.461	3.561	3.436
3-5/16	2.816	2.941	3.696	3.571	3.671	3.546
3-3/8	2.880	3.005	3.760	3.635	3.735	3.610
3-7/16	2.943	3.068	3.823	3.698	3.798	3.673
3-1/2	3.007	3.132	3.887	3.762	3.862	3.737
3-9/16	3.070	3.195	3.950	3.825	3.925	3.800
3-5/8	3.134	3.259	4.014	3.889	3.989	3.864
3-11/16	3.197	3.322	4.077	3.952	4.052	3.927
3-3/4	3.261	3.386	4.141	4.016	4.116	3.991
3-13/16	3.246	3.371	4.251	4.126	4.226	4.101
3-7/8	3.309	3.434	4.314	4.189	4.289	4.164
3-15/16	3.373	3.498	4.378	4.253	4.353	4.228
4	3.436	3.561	4.441	4.316	4.416	4.291
4-3/16	3.627	3.752	4.632	4.507	4.607	4.482
4-1/4	3.690	3.815	4.695	4.570	4.670	4.545
4-3/8	3.817	3.942	4.822	4.697	4.797	4.672
4-7/16	3.880	4.005	4.885	4.760	4.860	4.735
4-1/2	3.944	4.069	4.949	4.824	4.924	4.799
4-3/4	4.041	4.229	5.296	5.109	5.271	5.084
4-7/8	4.169	4.356	5.424	5.236	5.399	5.211
4-15/16	4.232	4.422	5.487	5.300	5.462	5.275
5	4.296	4.483	5.551	5.363	5.526	5.338
5-3/16	4.486	4.674	5.741	5.554	5.716	5.529
5-1/4	4.550	4.737	5.805	5.617	5.780	5.592
5-7/16	4.740	4.927	5.995	5.807	5.970	5.782
5-1/2	4.803	4.991	6.058	5.871	6.033	5.846
5-3/4	4.900	5.150	6.405	6.155	6.380	6.130
5-15/16	5.091	5.341	6.596	6.346	6.571	6.321
6	5.155	5.405	6.660	6.410	6.635	6.385
6-1/4	5.409	5.659	6.914	6.664	6.889	6.639
6-1/2	5.662	5.912	7.167	6.917	7.142	6.892
6-3/4	5.760	•5.885	7.515	•7.390	7.490	•7.365
7	6.014	•6.139	7.769	•7.644	7.744	•7.619
7-1/4	6.268	•6.393	8.023	•7.898	7.998	•7.873
7-1/2	6.521	•6.646	8.276	•8.151	8.251	•8.126
7-3/4	6.619	6.869	8.624	8.374	8.599	8.349
8	6.873	7.123	8.878	8.628	8.853	8.603
9	7.887	8.137	9.892	9.642	9.867	9.617
10	8.591	8.966	11.096	10.721	11.071	10.696
11	9.606	9.981	12.111	11.736	12.086	11.711
12	10.309	10.809	13.314	12.814	13.289	12.789
13	11.325	11.825	14.330	13.830	14.305	13.805
14	12.028	12.528	15.533	15.033	15.508	15.008
15	13.043	13.543	16.548	16.048	16.523	16.023

SOURCE: ANSI B17.1-1967 (Re-1973), published by the American Society of Mechanical Engineers.

TABLE 10-6 Class 1—Fit for Parallel Keys

Type of Key	KEY WIDTH		SIDE FIT			TOP AND BOTTOM FIT			
			Width Tolerance		Fit Range*	Depth Tolerance			Fit Range*
	Over	To (Incl)	Key	Keyseat		Key	Shaft Keyseat	Hub Keyseat	
Square	—	1/2	+0.000 −0.002	+0.002 −0.000	0.004 CL 0.000	+0.000 −0.002	+0.000 −0.015	+0.010 −0.000	0.032 CL 0.005 CL
	1/2	3/4	+0.000 −0.002	+0.003 −0.000	0.005 CL 0.000	+0.000 −0.002	+0.000 −0.015	+0.010 −0.000	0.032 CL 0.005 CL
	3/4	1	+0.000 −0.003	+0.003 −0.000	0.006 CL 0.000	+0.000 −0.003	+0.000 −0.015	+0.010 −0.000	0.033 CL 0.005 CL
	1	1-1/2	+0.000 −0.003	+0.004 −0.000	0.007 CL 0.000	+0.000 −0.003	+0.000 −0.015	+0.010 −0.000	0.033 CL 0.005 CL
	1-1/2	2-1/2	+0.000 −0.004	+0.004 −0.000	0.008 CL 0.000	+0.000 −0.004	+0.000 −0.015	+0.010 −0.000	0.034 CL 0.005 CL
	2-1/2	3-1/2	+0.000 −0.006	+0.004 −0.000	0.010 CL 0.000	+0.000 −0.006	+0.000 −0.015	+0.010 −0.000	0.036 CL 0.005 CL
Rec-tangular	—	1/2	+0.000 −0.003	+0.002 −0.000	0.005 CL 0.000	+0.000 −0.003	+0.000 −0.015	+0.010 −0.000	0.033 CL 0.005 CL
	1/2	3/4	+0.000 −0.003	+0.003 −0.000	0.006 CL 0.000	+0.000 −0.003	+0.000 −0.015	+0.010 −0.000	0.033 CL 0.005 CL
	3/4	1	+0.000 −0.004	+0.003 −0.000	0.007 CL 0.000	+0.000 −0.004	+0.000 −0.015	+0.010 −0.000	0.034 CL 0.005 CL
	1	1-1/2	+0.000 −0.004	+0.004 −0.000	0.008 CL 0.000	+0.000 −0.004	+0.000 −0.015	+0.010 −0.000	0.034 CL 0.005 CL
	1-1/2	3	+0.000 −0.005	+0.004 −0.000	0.009 CL 0.000	+0.000 −0.005	+0.000 −0.015	+0.010 −0.000	0.035 CL 0.005 CL
	3	4	+0.000 −0.006	+0.004 −0.000	0.010 CL 0.000	+0.000 −0.006	+0.000 −0.015	+0.010 −0.000	0.036 CL 0.005 CL
	4	6	+0.000 −0.008	+0.004 −0.000	0.012 CL 0.000	+0.000 −0.008	+0.000 −0.015	+0.010 −0.000	0.038 CL 0.005 CL
	6	7	+0.000 −0.013	+0.004 −0.000	0.017 CL 0.000	+0.000 −0.013	+0.000 −0.015	+0.010 −0.000	0.043 CL 0.005 CL

*Limits of variation, CL = Clearance
All dimensions given in inches.

SOURCE: ANSI B17.1-1967 (Re-1973), published by the American Society of Mechanical Engineers.

TABLE 10-7 Class 2—Fit for Parallel and Taper Keys

Type of Key	KEY WIDTH		SIDE FIT			TOP AND BOTTOM FIT			
			Width Tolerance		Fit Range*	Depth Tolerance			Fit Range*
	Over	To (Incl)	Key	Keyset		Key	Shaft Keyseat	Hub Keyseat	
Parallel Square	—	1-1/4	+0.001 −0.000	+0.002 −0.000	0.002 CL 0.001 INT	+0.001 −0.000	+0.000 −0.015	+0.010 −0.000	0.030 CL 0.004 CL
	1-1/4	3	+0.002 −0.000	+0.002 −0.000	0.002 CL 0.002 INT	+0.002 −0.000	+0.000 −0.015	+0.010 −0.000	0.030 CL 0.003 CL
	3	3-1/2	+0.003 −0.000	+0.002 −0.000	0.002 CL 0.003 INT	+0.003 −0.000	+0.000 −0.015	+0.010 −0.000	0.030 CL 0.002 CL
Parallel Rectang-ular	—	1-1/4	+0.001 −0.000	+0.002 −0.000	0.002 CL 0.001 INT	+0.005 −0.005	+0.000 −0.015	+0.010 −0.000	0.035 CL 0.000 CL
	1-1/4	3	+0.002 −0.000	+0.002 −0.000	0.002 CL 0.002 INT	+0.005 −0.005	+0.000 −0.015	+0.010 −0.000	0.035 CL 0.000 CL
	3	7	+0.003 −0.000	+0.002 −0.000	0.002 CL 0.003 INT	+0.005 −0.005	+0.000 −0.015	+0.010 −0.000	0.035 CL 0.000 CL
Taper	—	1-1/4	+0.001 −0.000	+0.002 −0.000	0.002 CL 0.001 INT	+0.005 −0.000	+0.000 −0.015	+0.010 −0.000	0.005 CL 0.025 INT
	1-1/4	3	+0.002 −0.000	+0.002 −0.000	0.002 CL 0.002 INT	+0.005 −0.000	+0.000 −0.015	+0.010 −0.000	0.005 CL 0.025 INT
	3	Δ	+0.003 −0.000	+0.002 −0.000	0.002 CL 0.003 INT	+0.005 −0.000	+0.000 −0.015	+0.010 −0.000	0.005 CL 0.025 INT

*Limits of variation. CL = Clearance; INT = Interference
Δ To (Incl) 3-1/2 Square and 7 Rectangular key widths.
All dimensions given in inches.

SOURCE: ANSI B17.1-1967 (Re-1973), published by the American Society of Mechanical Engineers.

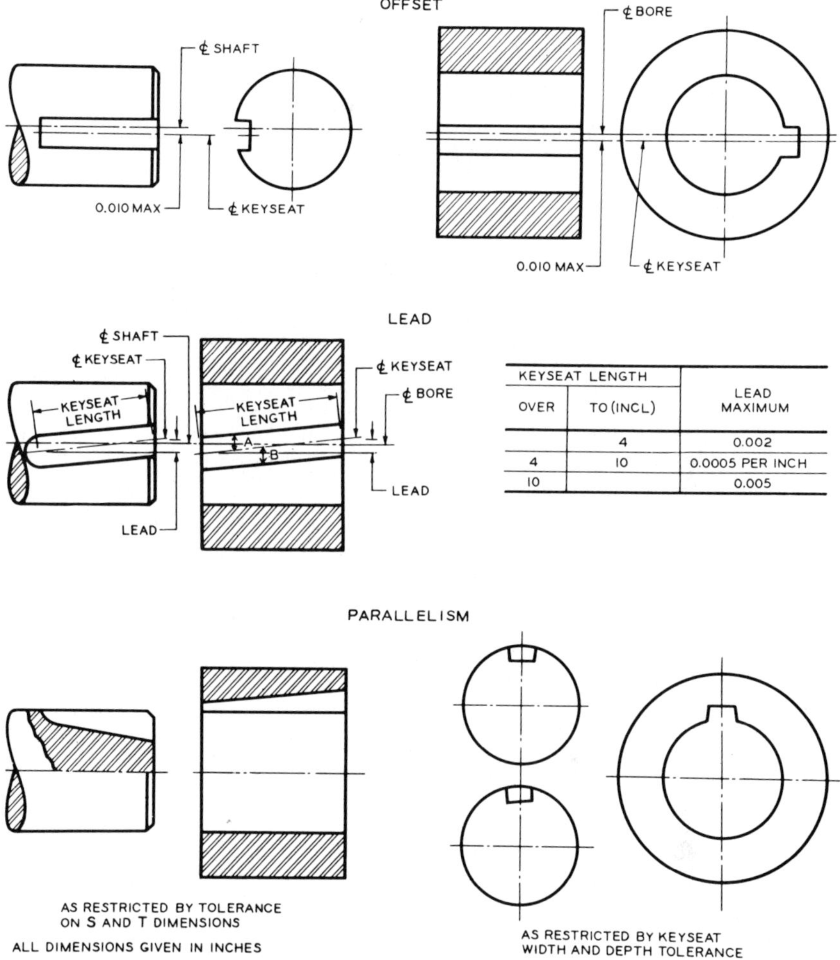

KEYSEAT LENGTH		LEAD MAXIMUM
OVER	TO (INCL)	
	4	0.002
4	10	0.0005 PER INCH
10		0.005

FIG. 10-2 Alignment tolerances. *(From USAS B17.1–1967 (Re-1973), published by the American Society of Mechanical Engineers.)*

at corners. When used, fillet radii should be as large as possible without causing excessive bearing stresses from reduced contact area between the key and its mating part (B17.1, 1967). See Table 10-8 for suggested fillet radii and key chamfer.

10-9 Setscrews Used over Keys Setscrews are to be selected from the list provided in Table 10-9, but their use must be dependent upon design considerations.

10-10 Woodruff Keys These are extensively used in the assembly of machine tools and the automotive industry. Tables 10-10 and 10-11 give the standard dimensions of Woodruff keys.

The key slot or keyseat dimensions are shown in Table 10-12 for standard Woodruff keys. Additionally, the key sizes are numbered and other basic dimensions are listed to help the designer.

10-11 Square Keys These are a valuable machine locking component. Having chamfered ends and without a head or any taper, they are finding ever-increasing use.

TABLE 10-8 Suggested Fillet Radii and Key Chamfers

CHAMFERED KEYS AND FILLETED KEYSEATS

In general practice, chamfered keys and filleted keyseats are not used. However, it is recognized that fillets in keyseats decrease stress concentrations at corners. When used, fillet radii should be as large as possible without causing excessive bearing stresses due to reduced contact area between the key and its mating parts. Keys must be chamfered or rounded to clear fillet radii. Values in Table 7 assume general conditions and should be used only as a guide when critical stresses are encountered.

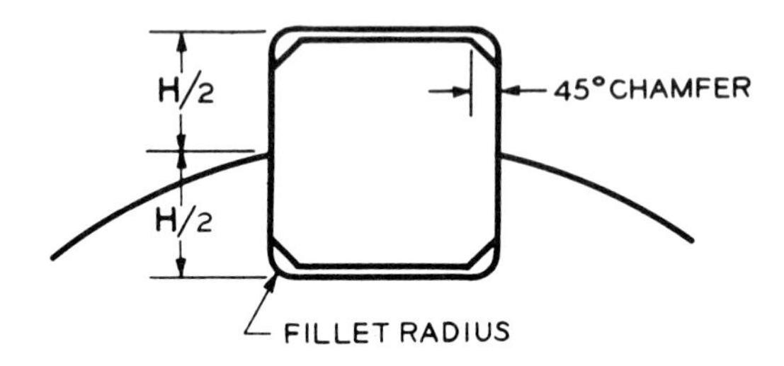

H/2 KEYSEAT DEPTH		Fillet Radius	45° Chamfer
Over	To (Incl)		
1/8	1/4	1/32	3/64
1/4	1/2	1/16	5/64
1/2	7/8	1/8	5/32
7/8	1-1/4	3/16	7/32
1-1/4	1-3/4	1/4	9/32
1-3/4	2-1/2	3/8	13/32

All dimensions given in inches.

SOURCE: ANSI B17.1-1967 (Re-1973), published by the American Society of Mechanical Engineers.

TABLE 10-9 Set Screws for Use over Keys

Set Screws for Use Over Keys

NOMINAL SHAFT DIAMETER		Nominal Key Width	Set Screw Diameter
Over	To (Incl)		
5/16	7/16	3/32	#10
7/16	9/16	1/8	#10
9/16	7/8	3/16	1/4
7/8	1-1/4	1/4	5/16
1-1/4	1-3/8	5/16	3/8
1-3/8	1-3/4	3/8	3/8
1-3/4	2-1/4	1/2	1/2
2-1/4	2-3/4	5/8	1/2
2-3/4	3-1/4	3/4	5/8
3-1/4	3-3/4	7/8	3/4
3-3/4	4-1/2	1	3/4
4-1/2	5-1/2	1-1/4	7/8
5-1/2	6-1/2	1-1/2	1

All dimensions given in inches.

SOURCE: ANSI B17.1-1967 (Re-1973), published by the American Society of Mechanical Engineers.

Figure 10-3 illustrates the basic design. Lengths are ½ through 3 inches at ¼-inch steps and 3 through 4½ inches at ½-inch steps. Width (square) sizes are ⅛, 3/16, ¼, 5/16, ⅜, 7/16, and ½ inch. They are manufactured plain or plated or fabricated from stainless steel.

10-12 Splines Splines are used when it is sound design to avoid an extra component. (Spline shafts may be used to transmit torque.) Two types of splines have become standardized by the SAE—square and involute. Figure 10-4 shows the basic splines. Further detail and tables are avoided, as splines are not strictly a component in the independent sense.

Torque capacity, in pound-inches, of a spline fitting is obtained by the formula

$$T = 1{,}000NrHL$$

where N = number of splines
r = mean radial distance from center of hole to center of spline, inches
H = depth of spline, inches
L = length of spline bearing surface, inches

The formula is based on 1,000 psi pressure on the sides of the spline.

LOCK WASHERS

10-13 Lock Washers These have a variety of designs. We will discuss the major standard types.

10-14 Helical-Spring Lock Washers These are made from many materials. They include:

- Carbon steel
- Corrosion-resistant steel (type 302 or 305)
- Aluminum-zinc alloy
- Phosphor bronze
- Silicon bronze
- K-Monel

This design is widely used in industry to provide greater bolt tension per unit of applied force (torque), protection from component loss resulting from vibration, and uniform load distribution. Table 10-13 lists the dimensions of regular helical-spring lock washers and Table 10-14 lists the dimensions for extraheavy duty. The high-collar type of helical-spring lock washers is described in Table 10-15.

TABLE 10-10 Woodruff Keys

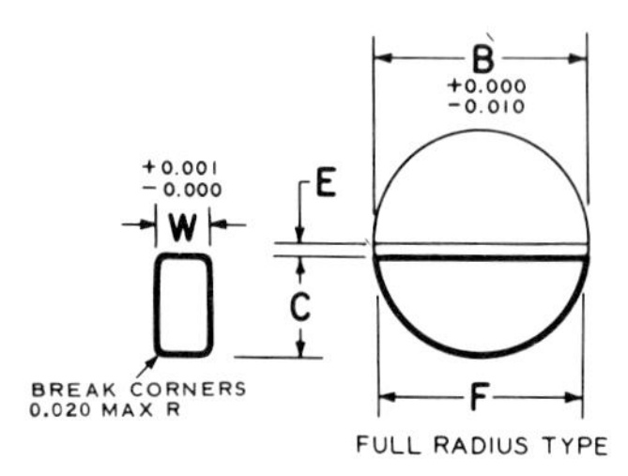

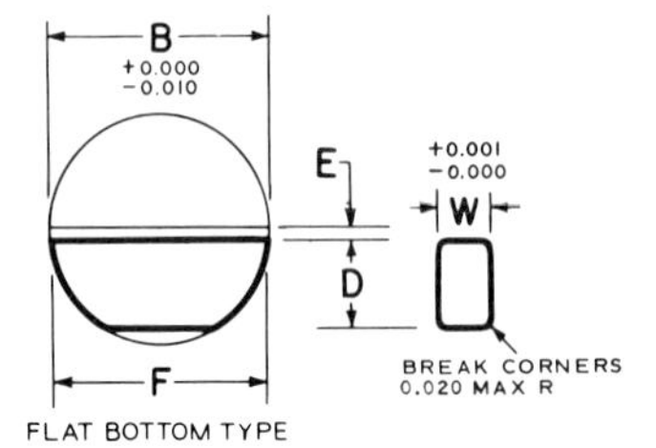

Woodruff Keys

Key No.	Nominal Key Size W × B	Actual Length F +0.000-0.010	Height of Key C Max	C Min	D Max	D Min	Distance Below Center E
202	1/16 × 1/4	0.248	0.109	0.104	0.109	0.104	1/64
202.5	1/16 × 5/16	0.311	0.140	0.135	0.140	0.135	1/64
302.5	3/32 × 5/16	0.311	0.140	0.135	0.140	0.135	1/64
203	1/16 × 3/8	0.374	0.172	0.167	0.172	0.167	1/64
303	3/32 × 3/8	0.374	0.172	0.167	0.172	0.167	1/64
403	1/8 × 3/8	0.374	0.172	0.167	0.172	0.167	1/64
204	1/16 × 1/2	0.491	0.203	0.198	0.194	0.188	3/64
304	3/32 × 1/2	0.491	0.203	0.198	0.194	0.188	3/64
404	1/8 × 1/2	0.491	0.203	0.198	0.194	0.188	3/64
305	3/32 × 5/8	0.612	0.250	0.245	0.240	0.234	1/16
405	1/8 × 5/8	0.612	0.250	0.245	0.240	0.234	1/16
505	5/32 × 5/8	0.612	0.250	0.245	0.240	0.234	1/16
605	3/16 × 5/8	0.612	0.250	0.245	0.240	0.234	1/16
406	1/8 × 3/4	0.740	0.313	0.308	0.303	0.297	1/16
506	5/32 × 3/4	0.740	0.313	0.308	0.303	0.297	1/16
606	3/16 × 3/4	0.740	0.313	0.308	0.303	0.297	1/16
806	1/4 × 3/4	0.740	0.313	0.308	0.303	0.297	1/16
507	5/32 × 7/8	0.866	0.375	0.370	0.365	0.359	1/16
607	3/16 × 7/8	0.866	0.375	0.370	0.365	0.359	1/16
707	7/32 × 7/8	0.866	0.375	0.370	0.365	0.359	1/16
807	1/4 × 7/8	0.866	0.375	0.370	0.365	0.359	1/16
608	3/16 × 1	0.992	0.438	0.433	0.428	0.422	1/16
708	7/32 × 1	0.992	0.438	0.433	0.428	0.422	1/16
808	1/4 × 1	0.992	0.438	0.433	0.428	0.422	1/16
1008	5/16 × 1	0.992	0.438	0.433	0.428	0.422	1/16
1208	3/8 × 1	0.992	0.438	0.433	0.428	0.422	1/16
609	3/16 × 1 1/8	1.114	0.484	0.479	0.475	0.469	5/64
709	7/32 × 1 1/8	1.114	0.484	0.479	0.475	0.469	5/64
809	1/4 × 1 1/8	1.114	0.484	0.479	0.475	0.469	5/64
1009	5/16 × 1 1/8	1.114	0.484	0.479	0.475	0.469	5/64

SOURCE: ANSI B17.2-1967 (Re-1972), published by the American Society of Mechanical Engineers.

TABLE 10-11 Woodruff Keys (Continued)

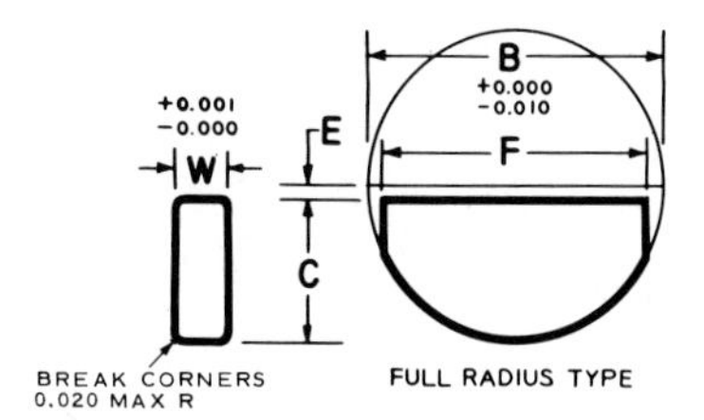

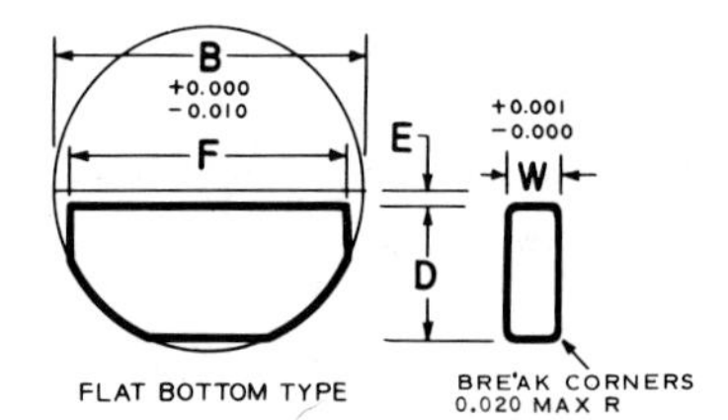

Woodruff Keys

Key No.	Nominal Key Size W × B	Actual Length F +0.000-0.010	Height of Key				Distance Below Center E
			C		D		
			Max	Min	Max	Min	
617-1	3/16 × 2 1/8	1.380	0.406	0.401	0.396	0.390	21/32
817-1	1/4 × 2 1/8	1.380	0.406	0.401	0.396	0.390	21/32
1017-1	5/16 × 2 1/8	1.380	0.406	0.401	0.396	0.390	21/32
1217-1	3/8 × 2 1/8	1.380	0.406	0.401	0.396	0.390	21/32
617	3/16 × 2 1/8	1.723	0.531	0.526	0.521	0.515	17/32
817	1/4 × 2 1/8	1.723	0.531	0.526	0.521	0.515	17/32
1017	5/16 × 2 1/8	1.723	0.531	0.526	0.521	0.515	17/32
1217	3/8 × 2 1/8	1.723	0.531	0.526	0.521	0.515	17/32
822-1	1/4 × 2 3/4	2.000	0.594	0.589	0.584	0.578	25/32
1022-1	5/16 × 2 3/4	2.000	0.594	0.589	0.584	0.578	25/32
1222-1	3/8 × 2 3/4	2.000	0.594	0.589	0.584	0.578	25/32
1422-1	7/16 × 2 3/4	2.000	0.594	0.589	0.584	0.578	25/32
1622-1	1/2 × 2 3/4	2.000	0.594	0.589	0.584	0.578	25/32
822	1/4 × 2 3/4	2.317	0.750	0.745	0.740	0.734	5/8
1022	5/16 × 2 3/4	2.317	0.750	0.745	0.740	0.734	5/8
1222	3/8 × 2 3/4	2.317	0.750	0.745	0.740	0.734	5/8
1422	7/16 × 2 3/4	2.317	0.750	0.745	0.740	0.734	5/8
1622	1/2 × 2 3/4	2.317	0.750	0.745	0.740	0.734	5/8
1228	3/8 × 3 1/2	2.880	0.938	0.933	0.928	0.922	13/16
1428	7/16 × 3 1/2	2.880	0.938	0.933	0.928	0.922	13/16
1628	1/2 × 3 1/2	2.880	0.938	0.933	0.928	0.922	13/16
1828	9/16 × 3 1/2	2.880	0.938	0.933	0.928	0.922	13/16
2028	5/8 × 3 1/2	2.880	0.938	0.933	0.928	0.922	13/16
2228	11/16 × 3 1/2	2.880	0.938	0.933	0.928	0.922	13/16
2428	3/4 × 3 1/2	2.880	0.938	0.933	0.928	0.922	13/16

All dimensions given are in inches.

The key numbers indicate nominal key dimensions. The last two digits give the nominal diameter B in eighths of an inch and the digits preceding the last two give the nominal width W in thirty-seconds of an inch.

Example:

No. 617 indicates a key 6/32 × 17/8 or 3/16 × 2 1/8
No. 822 indicates a key 8/32 × 22/8 or 1/4 × 2 3/4
No. 1228 indicates a key 12/32 × 28/8 or 3/8 × 3 1/2

The key numbers with the -1 designation, while representing the nominal key size have a shorter length F and due to a greater distance below center E are less in height than the keys of the same number without the -1 designation.

SOURCE: ANSI B17.2-1967 (Re-1972), published by the American Society of Mechanical Engineers.

TABLE 10-12 Keyseat Dimensions

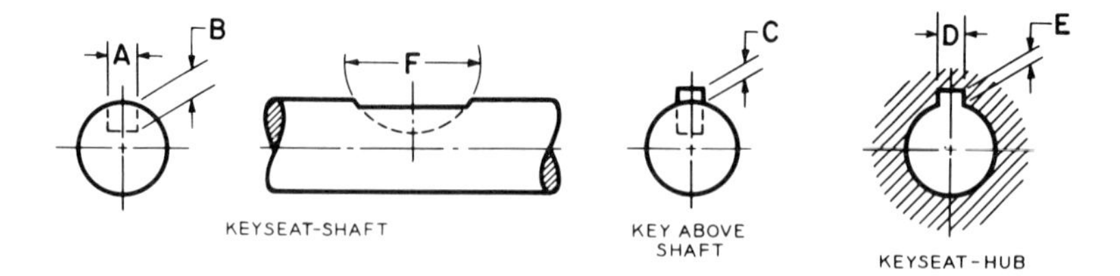

Key Number	Nominal Size Key	Keyseat – Shaft Width A* Min	Keyseat – Shaft Width A* Max	Keyseat – Shaft Depth B +0.005 -0.000	Keyseat – Shaft Diameter F Min	Keyseat – Shaft Diameter F Max	Key Above Shaft Height C +0.005 -0.005	Keyseat – Hub Width D +0.002 -0.000	Keyseat – Hub Depth E +0.005 -0.000
202	1/16 × 1/4	0.0615	0.0630	0.0728	0.250	0.268	0.0312	0.0635	0.0372
202.5	1/16 × 5/16	0.0615	0.0630	0.1038	0.312	0.330	0.0312	0.0635	0.0372
302.5	3/32 × 5/16	0.0928	0.0943	0.0882	0.312	0.330	0.0469	0.0948	0.0529
203	1/16 × 3/8	0.0615	0.0630	0.1358	0.375	0.393	0.0312	0.0635	0.0372
303	3/32 × 3/8	0.0928	0.0943	0.1202	0.375	0.393	0.0469	0.0948	0.0529
403	1/8 × 3/8	0.1240	0.1255	0.1045	0.375	0.393	0.0625	0.1260	0.0685
204	1/16 × 1/2	0.0615	0.0630	0.1668	0.500	0.518	0.0312	0.0635	0.0372
304	3/32 × 1/2	0.0928	0.0943	0.1511	0.500	0.518	0.0469	0.0948	0.0529
404	1/8 × 1/2	0.1240	0.1255	0.1355	0.500	0.518	0.0625	0.1260	0.0685
305	3/32 × 5/8	0.0928	0.0943	0.1981	0.625	0.643	0.0469	0.0948	0.0529
405	1/8 × 5/8	0.1240	0.1255	0.1825	0.625	0.643	0.0625	0.1260	0.0685
505	5/32 × 5/8	0.1553	0.1568	0.1669	0.625	0.643	0.0781	0.1573	0.0841
605	3/16 × 5/8	0.1863	0.1880	0.1513	0.625	0.643	0.0937	0.1885	0.0997
406	1/8 × 3/4	0.1240	0.1255	0.2455	0.750	0.768	0.0625	0.1260	0.0685
506	5/32 × 3/4	0.1553	0.1568	0.2299	0.750	0.768	0.0781	0.1573	0.0841
606	3/16 × 3/4	0.1863	0.1880	0.2143	0.750	0.768	0.0937	0.1885	0.0997
806	1/4 × 3/4	0.2487	0.2505	0.1830	0.750	0.768	0.1250	0.2510	0.1310
507	5/32 × 7/8	0.1553	0.1568	0.2919	0.875	0.895	0.0781	0.1573	0.0841
607	3/16 × 7/8	0.1863	0.1880	0.2763	0.875	0.895	0.0937	0.1885	0.0997
707	7/32 × 7/8	0.2175	0.2193	0.2607	0.875	0.895	0.1093	0.2198	0.1153
807	1/4 × 7/8	0.2487	0.2505	0.2450	0.875	0.895	0.1250	0.2510	0.1310
608	3/16 × 1	0.1863	0.1880	0.3393	1.000	1.020	0.0937	0.1885	0.0997
708	7/32 × 1	0.2175	0.2193	0.3237	1.000	1.020	0.1093	0.2198	0.1153
808	1/4 × 1	0.2487	0.2505	0.3080	1.000	1.020	0.1250	0.2510	0.1310
1008	5/16 × 1	0.3111	0.3130	0.2768	1.000	1.020	0.1562	0.3135	0.1622
1208	3/8 × 1	0.3735	0.3755	0.2455	1.000	1.020	0.1875	0.3760	0.1935
609	3/16 × 1 1/8	0.1863	0.1880	0.3853	1.125	1.145	0.0937	0.1885	0.0997
709	7/32 × 1 1/8	0.2175	0.2193	0.3697	1.125	1.145	0.1093	0.2198	0.1153
809	1/4 × 1 1/8	0.2487	0.2505	0.3540	1.125	1.145	0.1250	0.2510	0.1310
1009	5/16 × 1 1/8	0.3111	0.3130	0.3228	1.125	1.145	0.1562	0.3135	0.1622

TABLE 10-12 Keyseat Dimensions (Continued)

Keyseat Dimensions

Key Number	Nominal Size Key	Keyseat – Shaft					Key Above Shaft	Keyseat – Hub	
		Width A*		Depth B	Diameter F		Height C	Width D	Depth E
		Min	Max	+0.005 -0.000	Min	Max	+0.005 -0.005	+0.002 -0.000	+0.005 -0.000
610	3/16 × 1 1/4	0.1863	0.1880	0.4483	1.250	1.273	0.0937	0.1885	0.0997
710	7/32 × 1 1/4	0.2175	0.2193	0.4327	1.250	1.273	0.1093	0.2198	0.1153
810	1/4 × 1 1/4	0.2487	0.2505	0.4170	1.250	1.273	0.1250	0.2510	0.1310
1010	5/16 × 1 1/4	0.3111	0.3130	0.3858	1.250	1.273	0.1562	0.3135	0.1622
1210	3/8 × 1 1/4	0.3735	0.3755	0.3545	1.250	1.273	0.1875	0.3760	0.1935
811	1/4 × 1 3/8	0.2487	0.2505	0.4640	1.375	1.398	0.1250	0.2510	0.1310
1011	5/16 × 1 3/8	0.3111	0.3130	0.4328	1.375	1.398	0.1562	0.3135	0.1622
1211	3/8 × 1 3/8	0.3735	0.3755	0.4015	1.375	1.398	0.1875	0.3760	0.1935
812	1/4 × 1 1/2	0.2487	0.2505	0.5110	1.500	1.523	0.1250	0.2510	0.1310
1012	5/16 × 1 1/2	0.3111	0.3130	0.4798	1.500	1.523	0.1562	0.3135	0.1622
1212	3/8 × 1 1/2	0.3735	0.3755	0.4485	1.500	1.523	0.1875	0.3760	0.1935
617-1	3/16 × 2 1/8	0.1863	0.1880	0.3073	2.125	2.160	0.0937	0.1885	0.0997
817-1	1/4 ×2 1/8	0.2487	0.2505	0.2760	2.125	2.160	0.1250	0.2510	0.1310
1017-1	5/16 × 2 1/8	0.3111	0.3130	0.2448	2.125	2.160	0.1562	0.3135	0.1622
1217-1	3/8 × 2 1/8	0.3735	0.3755	0.2135	2.125	2.160	0.1875	0.3760	0.1935
617	3/16 × 2 1/8	0.1863	0.1880	0.4323	2.125	2.160	0.0937	0.1885	0.0997
817	1/4 × 2 1/8	0.2487	0.2505	0.4010	2.125	2.160	0.1250	0.2510	0.1310
1017	5/16 × 2 1/8	0.3111	0.3130	0.3698	2.125	2.160	0.1562	0.3135	0.1622
1217	3/8 × 2 1/8	0.3735	0.3755	0.3385	2.125	2.160	0.1875	0.3760	0.1935
822-1	1/4 × 2 3/4	0.2487	0.2505	0.4640	2.750	2.785	0.1250	0.2510	0.1310
1022-1	5/16 × 2 3/4	0.3111	0.3130	0.4328	2.750	2.785	0.1562	0.3135	0.1622
1222-1	3/8 × 2 3/4	0.3735	0.3755	0.4015	2.750	2.785	0.1875	0.3760	0.1935
1422-1	7/16 × 2 3/4	0.4360	0.4380	0.3703	2.750	2.785	0.2187	0.4385	0.2247
1622-1	1/2 × 2 3/4	0.4985	0.5005	0.3390	2.750	2.785	0.2500	0.5010	0.2560
822	1/4 × 2 3/4	0.2487	0.2505	0.6200	2.750	2.785	0.1250	0.2510	0.1310
1022	5/16 × 2 3/4	0.3111	0.3130	0.5888	2.750	2.785	0.1562	0.3135	0.1622
1222	3/8 × 2 3/4	0.3735	0.3755	0.5575	2.750	2.785	0.1875	0.3760	0.1935
1422	7/16 × 2 3/4	0.4360	0.4380	0.5263	2.750	2.785	0.2187	0.4385	0.2247
1622	1/2 × 2 3/4	0.4985	0.5005	0.4950	2.750	2.785	0.2500	0.5010	0.2560
1228	3/8 × 3 1/2	0.3735	0.3755	0.7455	3.500	3.535	0.1875	0.3760	0.1935
1428	7/16 × 3 1/2	0.4360	0.4380	0.7143	3.500	3.535	0.2187	0.4385	0.2247
1628	1/2 × 3 1/2	0.4985	0.5005	0.6830	3.500	3.535	0.2500	0.5010	0.2560
1828	9/16 × 3 1/2	0.5610	0.5630	0.6518	3.500	3.535	0.2812	0.5635	0.2872
2028	5/8 × 3 1/2	0.6235	0.6255	0.6205	3.500	3.535	0.3125	0.6260	0.3185
2228	11/16 × 3 1/2	0.6860	0.6880	0.5893	3.500	3.535	0.3437	0.6885	0.3497
2428	3/4 × 3 1/2	0.7485	0.7505	0.5580	3.500	3.535	0.3750	0.7510	0.3810

* Width A values were set with the maximum keyseat (shaft) width as that figure which will receive a key with the greatest amount of looseness consistent with assuring the key's sticking in the keyseat (shaft). Minimum keyseat width is that figure permitting the largest shaft distortion acceptable when assembling maximum key in minimum keyseat.

Dimensions A, B, C, D are taken at side intersection.

SOURCE: ANSI B17.2-1967 (Re-1972), published by the American Society of Mechanical Engineers.

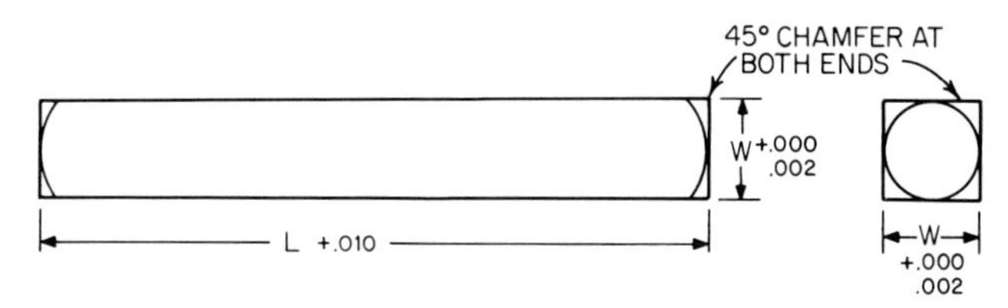

FIG. 10-3 Square machine key.

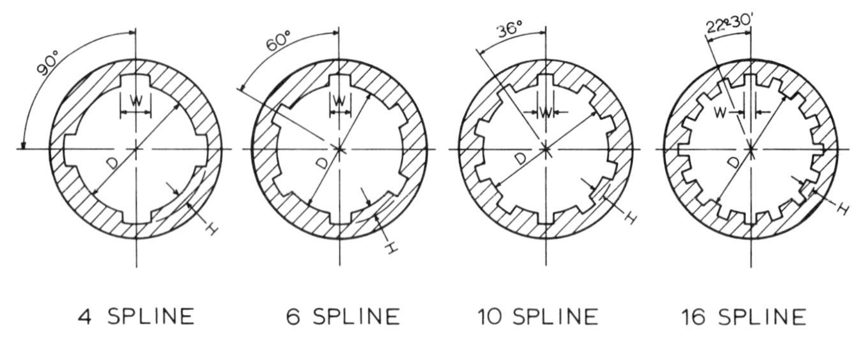

FIG. 10-4 Parallel-side splines. Basic spline designs.

10-15 Internal-Tooth Lock Washers These lock washers are illustrated and the dimensions listed in Table 10-16. Data for the "heavy" type are tabulated in the lower portion of the same table.

10-16 External-Tooth Lock Washers These washers are an excellent locking component. The teeth lock the mating elements together, thus forming a mechanical bonding. Screw and washer assemblies will be elaborated on later in the Section. Table 10-17 shows all standardized dimensions.

10-17 Countersunk External-Tooth Lock Washers These are well-explored in Table 10-18. The dished or countersunk design adds greater locking power upon compression of the teeth.

10-18 Internal–External-Tooth Lock Washers These combine two features in one unusual component that provides double gripping in an assembly. See Table 10-19.

10-19 Beveled Washers These are standardized in two types and designated as type A (malleable iron) and type B (steel). These washers are used for American Standard beams and channels in structural applications. The slope compensates for the lack of parallelism. Table 10-20 gives dimensions for both types and illustrates the washer design.

10-20 Belleville Washers Also known as conical-spring washers, these are used in many assemblies. Their basic design is shown in Fig. 10-5 with a typical example.

TABLE 10-13 Dimensions of Regular Helical-Spring Lock Washers

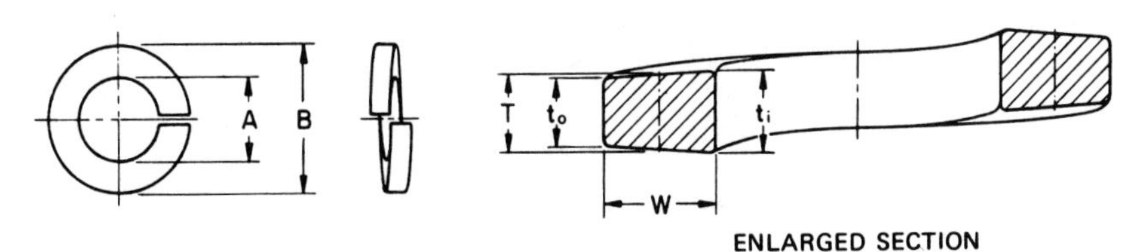

Dimensions of Regular Helical Spring Lock Washers[1]

Nominal Washer Size		A Inside Diameter		B Outside Diameter	T Mean Section Thickness $\left(\frac{t_i + t_o}{2}\right)$	W Section Width
		Max	Min	Max[2]	Min	Min
No. 2	0.086	0.094	0.088	0.172	0.020	0.035
No. 3	0.099	0.107	0.101	0.195	0.025	0.040
No. 4	0.112	0.120	0.114	0.209	0.025	0.040
No. 5	0.125	0.133	0.127	0.236	0.031	0.047
No. 6	0.138	0.148	0.141	0.250	0.031	0.047
No. 8	0.164	0.174	0.167	0.293	0.040	0.055
No. 10	0.190	0.200	0.193	0.334	0.047	0.062
No. 12	0.216	0.227	0.220	0.377	0.056	0.070
1/4	0.250	0.262	0.254	0.489	0.062	0.109
5/16	0.312	0.326	0.317	0.586	0.078	0.125
3/8	0.375	0.390	0.380	0.683	0.094	0.141
7/16	0.438	0.455	0.443	0.779	0.109	0.156
1/2	0.500	0.518	0.506	0.873	0.125	0.171
9/16	0.562	0.582	0.570	0.971	0.141	0.188
5/8	0.625	0.650	0.635	1.079	0.156	0.203
11/16	0.688	0.713	0.698	1.176	0.172	0.219
3/4	0.750	0.775	0.760	1.271	0.188	0.234
13/16	0.812	0.843	0.824	1.367	0.203	0.250
7/8	0.875	0.905	0.887	1.464	0.219	0.266
15/16	0.938	0.970	0.950	1.560	0.234	0.281
1	1.000	1.042	1.017	1.661	0.250	0.297
1 1/16	1.062	1.107	1.080	1.756	0.266	0.312
1 1/8	1.125	1.172	1.144	1.853	0.281	0.328
1 3/16	1.188	1.237	1.208	1.950	0.297	0.344
1 1/4	1.250	1.302	1.271	2.045	0.312	0.359
1 5/16	1.312	1.366	1.334	2.141	0.328	0.375
1 3/8	1.375	1.432	1.398	2.239	0.344	0.391
1 7/16	1.438	1.497	1.462	2.334	0.359	0.406
1 1/2	1.500	1.561	1.525	2.430	0.375	0.422

[1] Formerly designated Medium Helical Spring Lock Washers.
[2] The maximum outside diameters specified allow for the commercial tolerances on cold drawn wire.

SOURCE: ANSI B18.21.1-1972, published by the American Society of Mechanical Engineers.

TABLE 10-14 Dimensions of Extra-Heavy Duty Helical-Spring Lock Washers

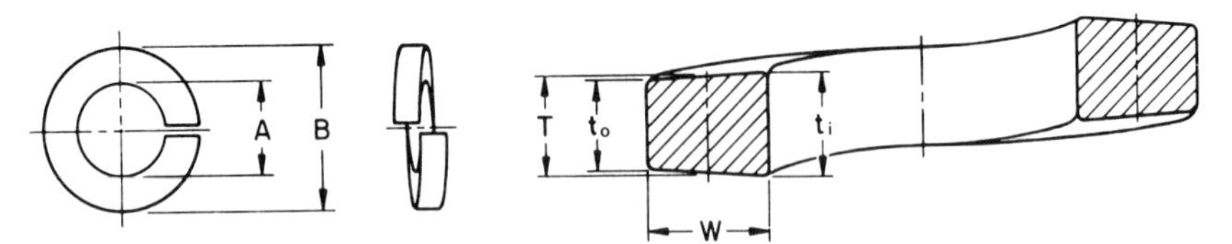

Dimensions of Extra Duty Helical Spring Lock Washers[1]

Nominal Washer Size		A Inside Diameter Max	A Inside Diameter Min	B Outside Diameter Max[2]	T Mean Section Thickness $\left(\frac{t_i + t_o}{2}\right)$ Min	W Section Width Min
No. 2	0.086	0.094	0.088	0.208	0.027	0.053
No. 3	0.099	0.107	0.101	0.239	0.034	0.062
No. 4	0.112	0.120	0.114	0.253	0.034	0.062
No. 5	0.125	0.133	0.127	0.300	0.045	0.079
No. 6	0.138	0.148	0.141	0.314	0.045	0.079
No. 8	0.164	0.174	0.167	0.375	0.057	0.096
No. 10	0.190	0.200	0.193	0.434	0.068	0.112
No. 12	0.216	0.227	0.220	0.497	0.080	0.130
1/4	0.250	0.262	0.254	0.535	0.084	0.132
5/16	0.312	0.326	0.317	0.622	0.108	0.143
3/8	0.375	0.390	0.380	0.741	0.123	0.170
7/16	0.438	0.455	0.443	0.839	0.143	0.186
1/2	0.500	0.518	0.506	0.939	0.162	0.204
9/16	0.562	0.582	0.570	1.041	0.182	0.223
5/8	0.625	0.650	0.635	1.157	0.202	0.242
11/16	0.688	0.713	0.698	1.258	0.221	0.260
3/4	0.750	0.775	0.760	1.361	0.241	0.279
13/16	0.812	0.843	0.825	1.463	0.261	0.298
7/8	0.875	0.905	0.887	1.576	0.285	0.322
15/16	0.938	0.970	0.950	1.688	0.308	0.345
1	1.000	1.042	1.017	1.799	0.330	0.366
1 1/16	1.062	1.107	1.080	1.910	0.352	0.389
1 1/8	1.125	1.172	1.144	2.019	0.375	0.411
1 3/16	1.188	1.237	1.208	2.124	0.396	0.431
1 1/4	1.250	1.302	1.271	2.231	0.417	0.452
1 5/16	1.312	1.366	1.334	2.335	0.438	0.472
1 3/8	1.375	1.432	1.398	2.439	0.458	0.491
1 7/16	1.438	1.497	1.462	2.540	0.478	0.509
1 1/2	1.500	1.561	1.525	2.638	0.496	0.526

[1] Formerly designated Extra Heavy Helical Spring Lock Washers.
[2] The maximum outside diameters specified allow for the commercial tolerances on cold drawn wire.

SOURCE: ANSI B18.21.1-1972, published by the American Society of Mechanical Engineers.

TABLE 10-15 Dimensions of High-Collar Helical-Spring Lock Washers

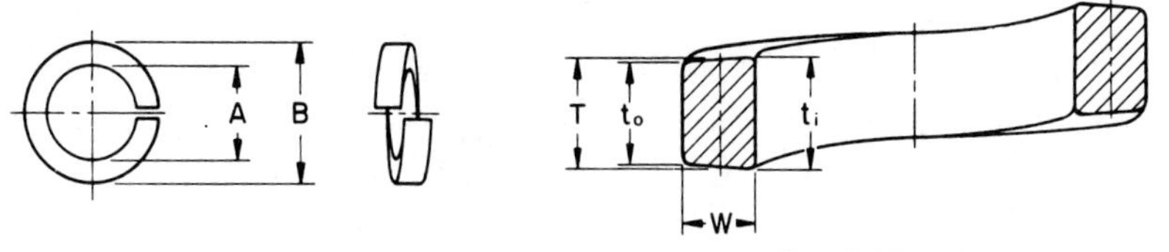

Dimensions of Hi-Collar Helical Spring Lock Washers[1]

Nominal Washer Size		A Inside Diameter		B Outside Diameter	T Mean Section Thickness $\left(\frac{t_i + t_o}{2}\right)$	W Section Width
		Max	Min	Max[2]	Min	Min
No. 4	0.112	0.120	0.114	0.173	0.022	0.022
No. 5	0.125	0.133	0.127	0.202	0.030	0.030
No. 6	0.138	0.148	0.141	0.216	0.030	0.030
No. 8	0.164	0.174	0.167	0.267	0.047	0.042
No. 10	0.190	0.200	0.193	0.294	0.047	0.042
1/4	0.250	0.262	0.254	0.365	0.078	0.047
5/16	0.312	0.326	0.317	0.460	0.093	0.062
3/8	0.375	0.390	0.380	0.553	0.125	0.076
7/16	0.438	0.455	0.443	0.647	0.140	0.090
1/2	0.500	0.518	0.506	0.737	0.172	0.103
5/8	0.625	0.650	0.635	0.923	0.203	0.125
3/4	0.750	0.775	0.760	1.111	0.218	0.154
7/8	0.875	0.905	0.887	1.296	0.234	0.182
1	1.000	1.042	1.017	1.483	0.250	0.208
1 1/8	1.125	1.172	1.144	1.669	0.313	0.236
1 1/4	1.250	1.302	1.271	1.799	0.313	0.236
1 3/8	1.375	1.432	1.398	2.041	0.375	0.292
1 1/2	1.500	1.561	1.525	2.170	0.375	0.292
1 3/4	1.750	1.811	1.775	2.602	0.469	0.383
2	2.000	2.061	2.025	2.852	0.469	0.383
2 1/4	2.250	2.311	2.275	3.352	0.508	0.508
2 1/2	2.500	2.561	2.525	3.602	0.508	0.508
2 3/4	2.750	2.811	2.775	4.102	0.633	0.633
3	3.000	3.061	3.025	4.352	0.633	0.633

[1] For use with 1960 Series Socket Head Cap Screws specified in American National Standard, ANSI B18.3.
[2] The maximum outside diameters specified allow for the commercial tolerances on cold drawn wire.

SOURCE: ANSI B18.21.1-1972, published by the American Society of Mechanical Engineers.

TABLE 10-16 Dimensions of Internal-Tooth Lock Washers

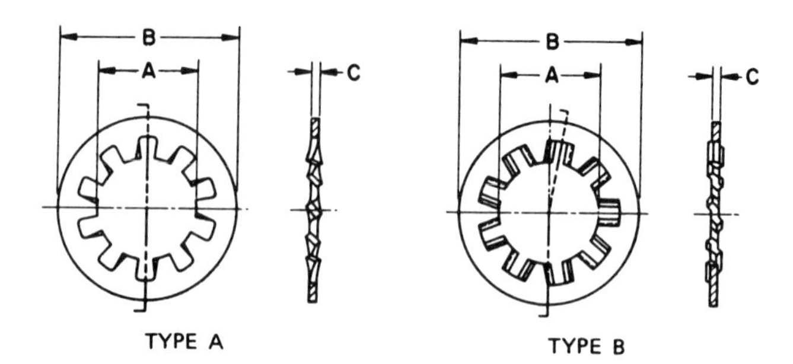

Dimensions of Internal Tooth Lock Washers

Nominal Washer Size		*A* Inside Diameter		*B* Outside Diameter		*C* Thickness	
		Max	Min	Max	Min	Max	Min
No. 2	0.086	0.095	0.089	0.200	0.175	0.015	0.010
No. 3	0.099	0.109	0.102	0.232	0.215	0.019	0.012
No. 4	0.112	0.123	0.115	0.270	0.255	0.019	0.015
No. 5	0.125	0.136	0.129	0.280	0.245	0.021	0.017
No. 6	0.138	0.150	0.141	0.295	0.275	0.021	0.017
No. 8	0.164	0.176	0.168	0.340	0.325	0.023	0.018
No. 10	0.190	0.204	0.195	0.381	0.365	0.025	0.020
No. 12	0.216	0.231	0.221	0.410	0.394	0.025	0.020
1/4	0.250	0.267	0.256	0.478	0.460	0.028	0.023
5/16	0.312	0.332	0.320	0.610	0.594	0.034	0.028
3/8	0.375	0.398	0.384	0.692	0.670	0.040	0.032
7/16	0.438	0.464	0.448	0.789	0.740	0.040	0.032
1/2	0.500	0.530	0.512	0.900	0.867	0.045	0.037
9/16	0.562	0.596	0.576	0.985	0.957	0.045	0.037
5/8	0.625	0.663	0.640	1.071	1.045	0.050	0.042
11/16	0.688	0.728	0.704	1.166	1.130	0.050	0.042
3/4	0.750	0.795	0.769	1.245	1.220	0.055	0.047
13/16	0.812	0.861	0.832	1.315	1.290	0.055	0.047
7/8	0.875	0.927	0.894	1.410	1.364	0.060	0.052
1	1.000	1.060	1.019	1.637	1.590	0.067	0.059
1 1/8	1.125	1.192	1.144	1.830	1.799	0.067	0.059
1 1/4	1.250	1.325	1.275	1.975	1.921	0.067	0.059

Dimensions of Heavy Internal Tooth Lock Washers

Nominal Washer Size		*A* Inside Diameter		*B* Outside Diameter		*C* Thickness	
		Max	Min	Max	Min	Max	Min
1/4	0.250	0.267	0.256	0.536	0.500	0.045	0.035
5/16	0.312	0.332	0.320	0.607	0.590	0.050	0.040
3/8	0.375	0.398	0.384	0.748	0.700	0.050	0.042
7/16	0.438	0.464	0.448	0.858	0.800	0.067	0.050
1/2	0.500	0.530	0.512	0.924	0.880	0.067	0.055
9/16	0.562	0.596	0.576	1.034	0.990	0.067	0.055
5/8	0.625	0.663	0.640	1.135	1.100	0.067	0.059
3/4	0.750	0.795	0.768	1.265	1.240	0.084	0.070
7/8	0.875	0.927	0.894	1.447	1.400	0.084	0.075

SOURCE: ANSI B18.21.1-1972, published by The American Society of Mechanical Engineers.

TABLE 10-17 Dimensions of External-Tooth Lock Washers

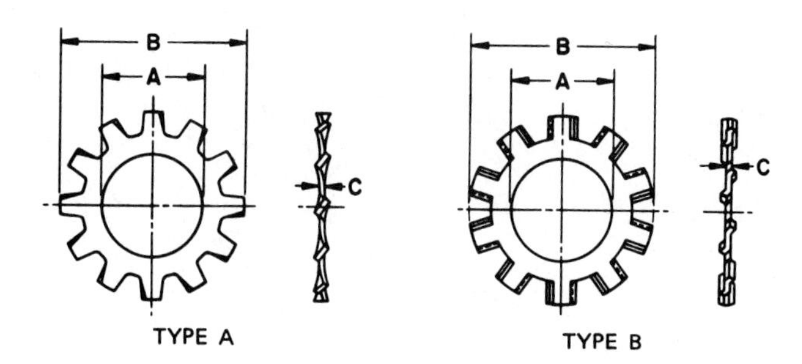

Dimensions of External Tooth Lock Washers

Nominal Washer Size		*A* Inside Diameter		*B* Outside Diameter		*C* Thickness	
		Max	Min	Max	Min	Max	Min
No. 3	0.099	0.109	0.102	0.235	0.220	0.015	0.012
No. 4	0.112	0.123	0.115	0.260	0.245	0.019	0.015
No. 5	0.125	0.136	0.129	0.285	0.270	0.019	0.014
No. 6	0.138	0.150	0.141	0.320	0.305	0.022	0.016
No. 8	0.164	0.176	0.168	0.381	0.365	0.023	0.018
No. 10	0.190	0.204	0.195	0.410	0.395	0.025	0.020
No. 12	0.216	0.231	0.221	0.475	0.460	0.028	0.023
1/4	0.250	0.267	0.256	0.510	0.494	0.028	0.023
5/16	0.312	0.332	0.320	0.610	0.588	0.034	0.028
3/8	0.375	0.398	0.384	0.694	0.670	0.040	0.032
7/16	0.438	0.464	0.448	0.760	0.740	0.040	0.032
1/2	0.500	0.530	0.513	0.900	0.880	0.045	0.037
9/16	0.562	0.596	0.576	0.985	0.960	0.045	0.037
5/8	0.625	0.663	0.641	1.070	1.045	0.050	0.042
11/16	0.688	0.728	0.704	1.155	1.130	0.050	0.042
3/4	0.750	0.795	0.768	1.260	1.220	0.055	0.047
13/16	0.812	0.861	0.833	1.315	1.290	0.055	0.047
7/8	0.875	0.927	0.897	1.410	1.380	0.060	0.052
1	1.000	1.060	1.025	1.620	1.590	0.067	0.059

SOURCE: ANSI B18.21.1-1972, published by the American Society of Mechanical Engineers.

TABLE 10-18 Dimensions of Countersunk External-Tooth Lock Washers

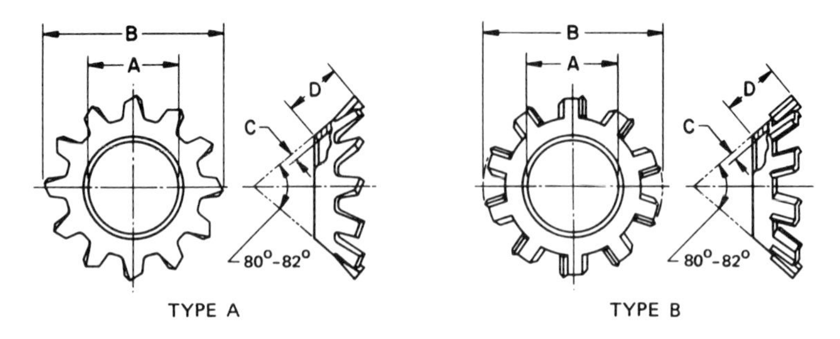

Nominal Washer Size		*A* Inside Diameter		*B*[1] Outside Diameter	*C* Thickness		*D* Length	
		Max	Min	Approx	Max	Min	Max	Min
No. 4	0.112	0.123	0.113	0.213	0.019	0.015	0.065	0.050
No. 6	0.138	0.150	0.140	0.289	0.021	0.017	0.092	0.082
No. 8	0.164	0.177	0.167	0.322	0.021	0.017	0.105	0.088
No. 10	0.190	0.205	0.195	0.354	0.025	0.020	0.099	0.083
No. 12	0.216	0.231	0.220	0.421	0.025	0.020	0.128	0.118
¼	0.250	0.267	0.255	0.454	0.025	0.020	0.128	0.113
No. 16	0.268	0.287	0.273	0.505	0.028	0.023	0.147	0.137
5/16	0.312	0.333	0.318	0.599	0.028	0.023	0.192	0.165
⅜	0.375	0.398	0.383	0.765	0.034	0.028	0.255	0.242
7/16	0.438	0.463	0.448	0.867	0.045	0.037	0.270	0.260
½	0.500	0.529	0.512	0.976	0.045	0.037	0.304	0.294

[1] For reference purposes only, not subject to inspection.

SOURCE: ANSI B18.21.1-1972, published by the American Society of Mechanical Engineers.

TABLE 10-19 Dimensions of Internal–External-Tooth Lock Washers

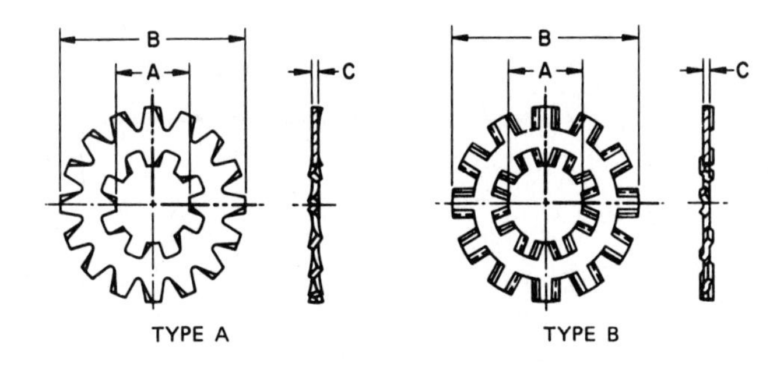

Nominal Washer Size		A		B		C	
		Inside Diameter		Outside Diameter		Thickness	
		Max	Min	Max	Min	Max	Min
No. 4	0.112	0.123	0.115	0.475	0.460	0.021	0.016
				0.510	0.495	0.021	0.017
				0.610	0.580	0.021	0.017
No. 6	0.138	0.150	0.141	0.510	0.495	0.028	0.023
				0.610	0.580	0.028	0.023
				0.690	0.670	0.028	0.023
No. 8	0.164	0.176	0.168	0.610	0.580	0.034	0.028
				0.690	0.670	0.034	0.028
				0.760	0.740	0.034	0.028
No. 10	0.190	0.204	0.195	0.610	0.580	0.034	0.028
				0.690	0.670	0.040	0.032
				0.760	0.740	0.040	0.032
				0.900	0.880	0.040	0.032
No. 12	0.216	0.231	0.221	0.690	0.670	0.040	0.032
				0.760	0.725	0.040	0.032
				0.900	0.880	0.040	0.032
				0.985	0.965	0.045	0.037
¼	0.250	0.267	0.256	0.760	0.725	0.040	0.032
				0.900	0.880	0.040	0.032
				0.985	0.965	0.045	0.037
				1.070	1.045	0.045	0.037

Nominal Washer Size		A		B		C	
		Inside Diameter		Outside Diameter		Thickness	
		Max	Min	Max	Min	Max	Min
5/16	0.312	0.332	0.320	0.900	0.865	0.040	0.032
				0.985	0.965	0.045	0.037
				1.070	1.045	0.050	0.042
				1.155	1.130	0.050	0.042
⅜	0.375	0.398	0.384	0.985	0.965	0.045	0.037
				1.070	1.045	0.050	0.042
				1.155	1.130	0.050	0.042
				1.260	1.220	0.050	0.042
7/16	0.438	0.464	0.448	1.070	1.045	0.050	0.042
				1.155	1.130	0.050	0.042
				1.260	1.220	0.055	0.047
				1.315	1.290	0.055	0.047
½	0.500	0.530	0.512	1.260	1.220	0.055	0.047
				1.315	1.290	0.055	0.047
				1.410	1.380	0.060	0.052
				1.620	1.590	0.067	0.059
9/16	0.562	0.596	0.576	1.315	1.290	0.055	0.047
				1.430	1.380	0.060	0.052
				1.620	1.590	0.067	0.059
				1.830	1.797	0.067	0.059
⅝	0.625	0.663	0.640	1.410	1.380	0.060	0.052
				1.620	1.590	0.067	0.059
				1.830	1.797	0.067	0.059
				1.975	1.935	0.067	0.059

SOURCE: ANSI B18.21.1-1972, published by the American Society of Mechanical Engineers.

TABLE 10-20 Dimensions of Beveled Washers

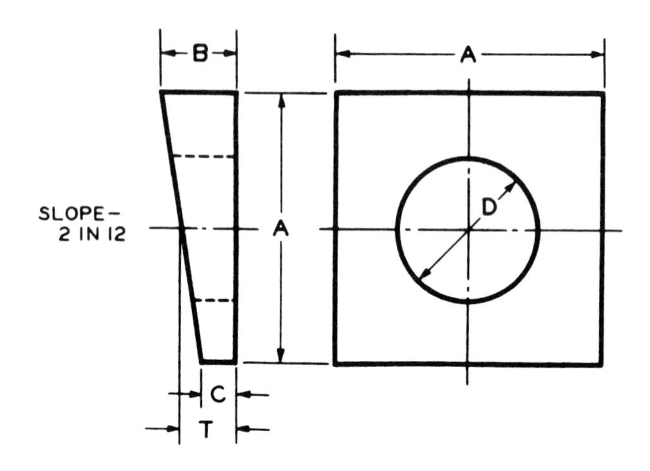

Dimensions of Beveled Washers

	Type A – Malleable Iron					Type B – Steel				
Bolt Dia	A	B	C	T	D	A	B	C	T	D
Tolerance	±0.03	±0.03	±0.03	Nominal	±0.03	±0.03	+0.02 -0.03	±0.02	Nominal	+0.03 -0.01
1/4 (0.250)	0.69	0.22	0.09	0.16	0.31	0.88	0.26	0.12	0.19	0.28
5/16(0.312)	1.00	0.31	0.16	0.23	0.38	0.88	0.26	0.12	0.19	0.34
3/8 (0.375)	1.25	0.34	0.12	0.23	0.44	0.88	0.26	0.12	0.19	0.41
1/2 (0.500)	1.25	0.34	0.12	0.23	0.56	1.75	0.45	0.16	0.31	0.53
5/8 (0.625)	1.50	0.38	0.12	0.25	0.69	1.75	0.45	0.16	0.31	0.66
3/4 (0.750)	1.50	0.44	0.19	0.31	0.81	1.75	0.45	0.16	0.31	0.81
7/8 (0.875)	2.00	0.56	0.22	0.39	0.94	1.75	0.45	0.16	0.31	0.94
1 (1.000)	2.00	0.56	0.22	0.39	1.06	1.75	0.45	0.16	0.31	1.06
1 1/8 (1.125)	2.25	0.62	0.25	0.44	1.25	2.25	0.50	0.12	0.31	1.25
1 1/4 (1.250)	2.25	0.72	0.31	0.52	1.38	2.25	0.50	0.12	0.31	1.38
1 3/8 (1.375)	2.75	0.78	0.31	0.55	1.50	2.25	0.50	0.12	0.31	1.50
1 1/2 (1.500)	3.00	0.81	0.31	0.56	1.62	2.25	0.50	0.12	0.31	1.62

SOURCE: ANSI B27.4-1967, published by the American Society of Mechanical Engineers.

SEMS

10-21 Screw and Washer Assemblies These assemblies, otherwise known as *sems*, are a combination of various standard screws and their captive washers.

Sems using helical-spring lock washers and socket-head cap screws are tabulated in Table 10-22. Table 10-23 represents their use with machine and tapping screws having UNC and UNF thread-diameter-pitch combinations.

Sems employing internal-tooth lock washers are represented in examples shown in Fig. 10-6.

External-tooth lock washer combinations are illustrated in Fig. 10-7 with various machine and tapping screws.

Table 10-24 illustrates sems employing conical-spring washers in combination with screws. Dimensions of conical-spring washers are listed for these sems.

Plain-washer sems are pictured in Table 10-25, together with dimensions for the applicable washers.

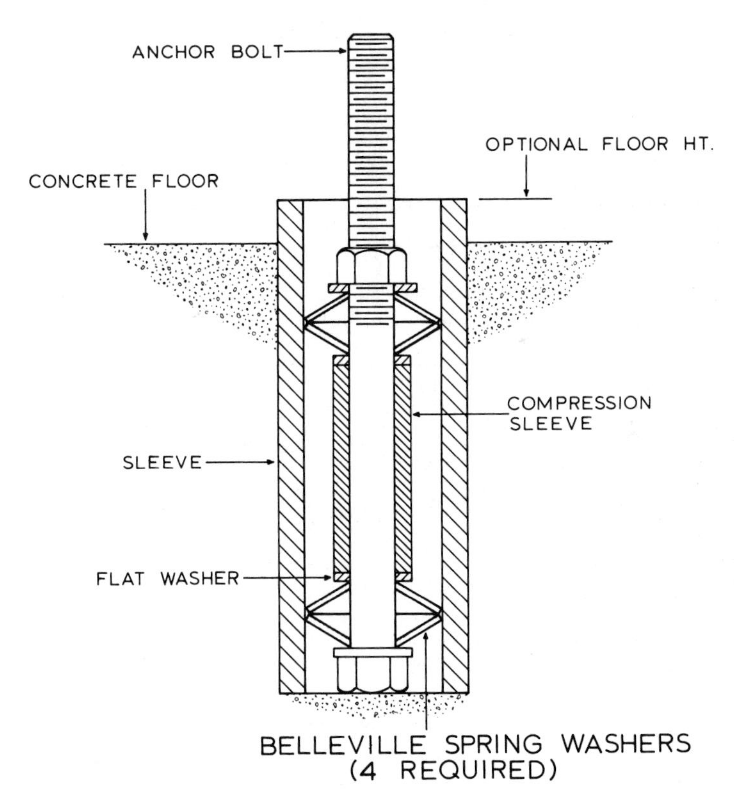

FIG. 10-5 Anchor-bolt locking assembly. Belleville washers are the key component in this locking mechanism.

TABLE 10-22 Dimensions of Helical-Spring Lock Washers for Socket-Head Cap Screw Sems

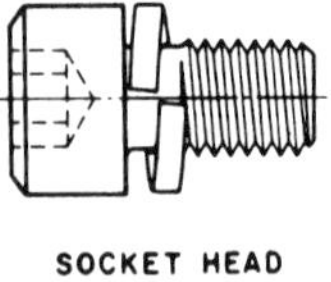

Nominal Size* or Basic Screw Diameter		Washer Inside Diameter		Washer Section		Washer Outside Diameter	
				Width	Thickness		
		Min	Max	Min	Min	Max	Min
4	0.1120	0.101	0.106	0.030	0.030	0.174	0.161
5	0.1250	0.113	0.118	0.040	0.040	0.206	0.193
6	0.1380	0.124	0.129	0.040	0.040	0.217	0.204
8	0.1640	0.149	0.155	0.047	0.047	0.257	0.243
10	0.1900	0.173	0.179	0.062	0.047	0.311	0.297
1/4	0.2500	0.230	0.238	0.062	0.062	0.370	0.354
5/16	0.3125	0.290	0.298	0.078	0.078	0.462	0.446
3/8	0.3750	0.353	0.361	0.094	0.094	0.557	0.541
7/16	0.4375	0.411	0.420	0.109	0.109	0.646	0.629
1/2	0.5000	0.473	0.482	0.125	0.125	0.740	0.723
5/8	0.6250	0.592	0.601	0.156	0.156	0.921	0.904
3/4	0.7500	0.713	0.722	0.188	0.188	1.106	1.089

*Where specifying nominal size in decimals, zeros in the fourth decimal place shall be omitted.

SOURCE: ANSI B18.13-1965, published by the American Society of Mechanical Engineers.

TABLE 10-23 Representative Examples of Helical-Spring Lock Washer Sems

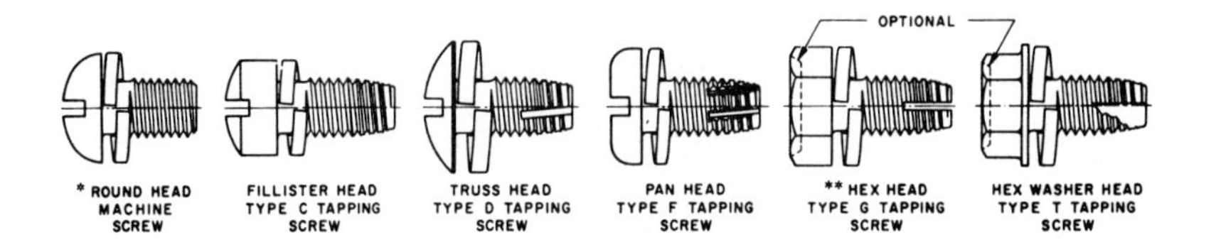

Dimensions of Helical Spring Lock Washers For Sems With Machine and Tapping Screws Having UNC and UNF Thread Diameter-Pitch Combinations

Nominal Size*** or Basic Screw Diameter		Washer Inside Diameter		Pan Head Screw**				Fillister Head Screw				Truss Head Screw			
				Washer Section		Washer Outside Diameter		Washer Section		Washer Outside Diameter		Washer Section		Washer Outside Diameter	
				Width	Thickness			Width	Thickness			Width	Thickness		
		Min	Max	Min	Min	Max	Min	Min	Min	Max	Min	Min	Min	Max	Min
4	0.1120	0.101	0.106	0.047	0.031	0.208	0.195	0.035	0.020	0.184	0.171	0.062	0.034	0.238	0.225
5	0.1250	0.113	0.118	0.047	0.031	0.220	0.207	0.035	0.020	0.196	0.183	0.093	0.031	0.312	0.299
6	0.1380	0.124	0.129	0.062	0.034	0.261	0.248	0.047	0.031	0.231	0.218	0.093	0.031	0.323	0.310
8	0.1640	0.149	0.155	0.078	0.031	0.319	0.305	0.047	0.031	0.257	0.243	0.109	0.047	0.381	0.367
10	0.1900	0.173	0.179	0.093	0.047	0.373	0.359	0.055	0.040	0.297	0.283	0.125	0.047	0.437	0.423
12	0.2160	0.196	0.203	0.109	0.062	0.429	0.414	0.062	0.047	0.335	0.320	0.136	0.070	0.483	0.468
1/4	0.2500	0.230	0.238	0.125	0.062	0.496	0.480	0.077	0.063	0.400	0.384	0.156	0.078	0.558	0.542
5/16	0.3125	0.290	0.298	0.156	0.078	0.618	0.602	0.109	0.062	0.524	0.508	0.156	0.078	0.618	0.602
3/8	0.3750	0.353	0.361	0.171	0.093	0.711	0.695	0.125	0.062	0.619	0.603	–	–	–	–
7/16	0.4375	0.411	0.420	–	–	–	–	–	–	–	–	–	–	–	–
1/2	0.5000	0.473	0.482	–	–	–	–	–	–	–	–	–	–	–	–

Nominal Size*** or Basic Screw Diameter		Round Head Screw*				Hex Head Screw**				Hex Washer Head Screw			
		Washer Section		Washer Outside Diameter		Washer Section		Washer Outside Diameter		Washer Section		Washer Outside Diameter	
		Width	Thickness			Width	Thickness			Width	Thickness		
		Min	Min	Max	Min	Min	Min	Max	Min	Min	Min	Max	Min
4	0.1120	0.047	0.031	0.208	0.195	0.035	0.020	0.184	0.171	0.062	0.034	0.238	0.225
5	0.1250	0.062	0.034	0.250	0.237	0.035	0.020	0.196	0.183	0.062	0.034	0.250	0.237
6	0.1380	0.062	0.034	0.261	0.248	0.047	0.031	0.231	0.218	0.093	0.047	0.323	0.310
8	0.1640	0.078	0.031	0.319	0.305	0.047	0.031	0.257	0.243	0.093	0.047	0.349	0.335
10	0.1900	0.093	0.047	0.373	0.359	0.062	0.047	0.311	0.297	0.093	0.047	0.373	0.359
12	0.2160	0.093	0.047	0.397	0.382	0.062	0.047	0.335	0.320	0.109	0.062	0.429	0.414
1/4	0.2500	0.109	0.062	0.464	0.448	0.109	0.062	0.464	0.448	0.125	0.062	0.496	0.480
5/16	0.3125	0.136	0.070	0.578	0.562	0.125	0.078	0.556	0.540	0.156	0.078	0.618	0.602
3/8	0.3750	0.171	0.093	0.711	0.695	0.141	0.094	0.651	0.635	0.171	0.093	0.711	0.695
7/16	0.4375	0.156	0.109	0.740	0.723	0.145	0.115	0.718	0.701	–	–	–	–
1/2	0.5000	0.160	0.133	0.812	0.793	0.160	0.133	0.812	0.793	–	–	–	–

SOURCE: ANSI B18.13-1965, published by The American Society of Mechanical Engineers.

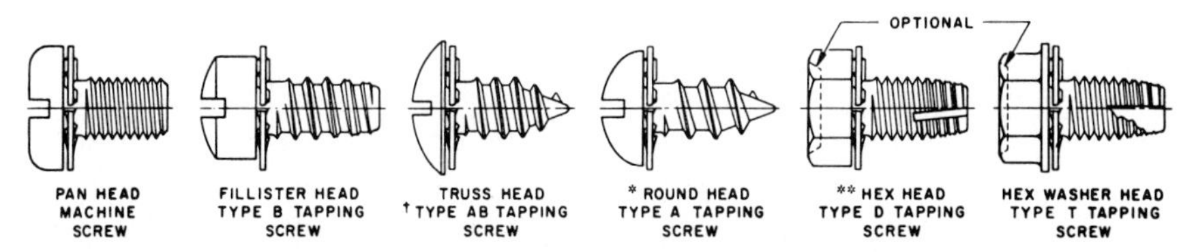

FIG. 10-6 Representative examples of internal-tooth lock washer sems. *(From ASA B18.13–1965, published by the American Society of Mechanical Engineers.)*

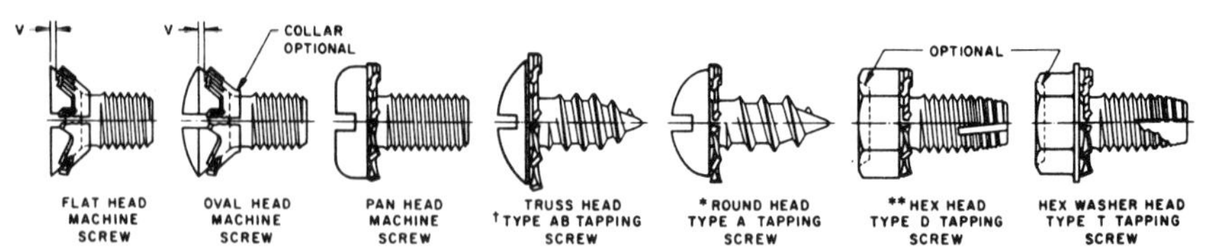

FIG. 10-7 Representative examples of external-tooth lock washer sems. *(From ASA B18.13–1965, published by the American Society of Mechanical Engineers.)*

TABLE 10-24 Representative Examples of Conical-Spring Washer Sems

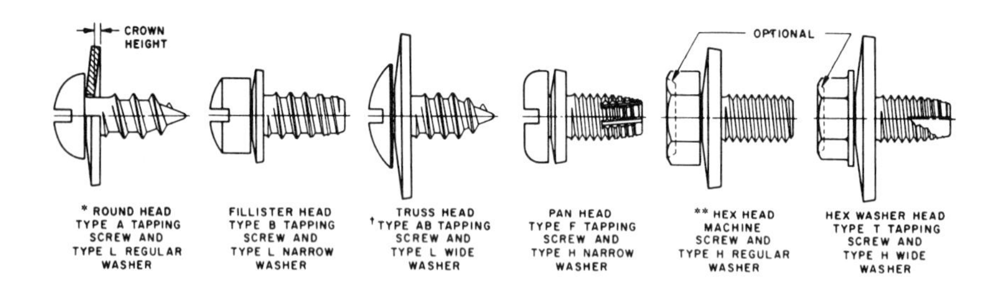

Dimensions of Conical-Spring Washers For Sems

Nominal Size*** or Basic Screw Diameter		Washer Series	Pan, Fillister, Truss, Round*, Hex** and Hex Washer Head Screws											
			Washer Outside Diameter		Type L Washer					Type H Washer				
					Thickness			Crown Height		Thickness			Crown Height	
			Max	Min	Basic	Max	Min	Min	Max	Basic	Max	Min	Min	Max
		Narrow.	0.320	0.307	0.025	0.029	0.023	0.010	0.016	0.035	0.040	0.033	0.010	0.016
6	0.1380	Regular	0.446	0.433	0.030	0.034	0.028	0.014	0.020	0.040	0.046	0.037	0.013	0.019
		Wide	0.570	0.557	0.030	0.034	0.028	0.021	0.031	0.040	0.046	0.037	0.019	0.029
		Narrow	0.383	0.370	0.035	0.040	0.033	0.010	0.016	0.040	0.046	0.037	0.010	0.016
8	0.1640	Regular	0.508	0.495	0.035	0.040	0.033	0.020	0.030	0.045	0.050	0.042	0.016	0.026
		Wide	0.640	0.620	0.035	0.040	0.033	0.027	0.037	0.045	0.050	0.042	0.030	0.040
		Narrow	0.446	0.433	0.035	0.040	0.033	0.010	0.016	0.050	0.056	0.047	0.011	0.017
10	0.1900	Regular	0.570	0.557	0.040	0.046	0.037	0.017	0.027	0.055	0.060	0.052	0.016	0.026
		Wide	0.765	0.743	0.040	0.046	0.037	0.026	0.036	0.055	0.060	0.052	0.024	0.034
		Narrow	0.446	0.433	0.040	0.046	0.037	0.011	0.017	0.055	0.060	0.052	0.010	0.016
12	0.2160	Regular	0.640	0.620	0.040	0.046	0.037	0.023	0.033	0.055	0.060	0.052	0.016	0.026
		Wide	0.890	0.868	0.045	0.050	0.042	0.034	0.044	0.064	0.071	0.059	0.023	0.033
14	0.2420	Narrow	0.515	0.495	0.045	0.050	0.042	0.014	0.024	0.064	0.071	0.059	0.011	0.021
and		Regular	0.765	0.743	0.050	0.056	0.047	0.023	0.033	0.079	0.087	0.074	0.022	0.032
¼	0.2500	Wide	1.015	0.993	0.055	0.060	0.052	0.030	0.040	0.079	0.087	0.074	0.029	0.039
5/16	0.3125	Narrow	0.640	0.620	0.055	0.060	0.052	0.016	0.026	0.079	0.087	0.074	0.016	0.026
and		Regular	0.890	0.868	0.064	0.071	0.059	0.031	0.041	0.095	0.103	0.090	0.019	0.029
20	0.3200	Wide	1.140	1.118	0.064	0.071	0.059	0.034	0.044	0.095	0.103	0.090	0.030	0.040
24	0.3720	Narrow	0.765	0.743	0.071	0.079	0.066	0.015	0.025	0.095	0.103	0.090	0.014	0.024
and		Regular	1.015	0.993	0.071	0.079	0.066	0.033	0.043	0.118	0.126	0.112	0.023	0.033
⅜	0.3750	Wide	1.265	1.243	0.079	0.087	0.074	0.037	0.047	0.118	0.126	0.112	0.035	0.045
		Narrow	0.890	0.868	0.079	0.087	0.074	0.018	0.028	0.128	0.136	0.122	0.016	0.026
7/16	0.4375	Regular	1.140	1.118	0.095	0.103	0.090	0.031	0.041	0.128	0.136	0.122	0.028	0.038
		Wide	1.530	1.493	0.095	0.103	0.090	0.049	0.059	0.132	0.140	0.126	0.039	0.049
		Narrow	1.015	0.993	0.100	0.108	0.094	0.021	0.031	0.142	0.150	0.136	0.020	0.030
½	0.5000	Regular	1.265	1.243	0.111	0.120	0.106	0.033	0.043	0.142	0.150	0.136	0.027	0.037
		Wide	1.780	1.743	0.111	0.120	0.106	0.052	0.062	0.152	0.160	0.146	0.042	0.052

SOURCE: ANSI B18.13-1965, published by the American Society of Mechanical Engineers.

TABLE 10-25 Representative Examples of Plain Washer Sems

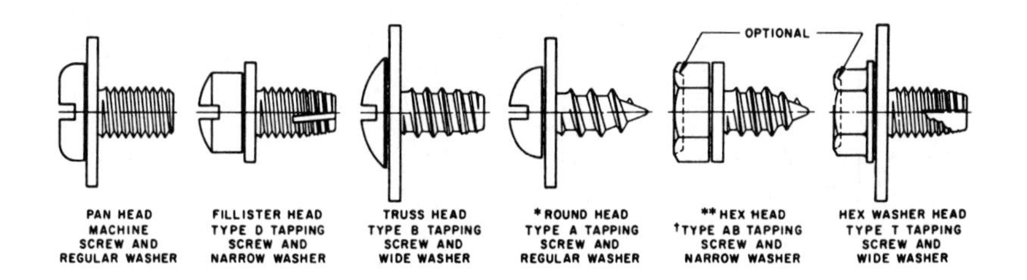

Dimensions of Plain Washers For Sems

Nominal Size*** or Basic Screw Diameter		Pan, Fillister, Truss, Round*, Hex** and Hex Washer Head Screws							Nominal Size*** or Basic Screw Diameter		Pan, Fillister, Truss, Round*, Hex** and Hex Washer Head Screws					
		Washer Series	Washer Outside Diameter		Washer Thickness						Washer Series	Washer Outside Diameter		Washer Thickness		
			Max	Min	Basic	Max	Min					Max	Min	Basic	Max	Min
		Narrow	0.188	0.183	0.025	0.028	0.022				Narrow	0.446	0.433	0.040	0.045	0.036
2	0.0860	Regular	0.250	0.245	0.032	0.036	0.028		10	0.1900	Regular	0.570	0.557	0.040	0.045	0.036
		Wide	0.312	0.307	0.032	0.036	0.028				Wide	0.749	0.727	0.063	0.071	0.056
		Narrow	0.219	0.214	0.025	0.028	0.022				Narrow	0.446	0.433	0.040	0.045	0.036
3	0.0990	Regular	0.312	0.307	0.032	0.036	0.028		12	0.2160	Regular	0.640	0.620	0.063	0.071	0.056
		Wide	0.383	0.370	0.040	0.045	0.036				Wide	0.890	0.868	0.063	0.071	0.056
		Narrow	0.250	0.245	0.032	0.036	0.028		14	0.2420	Narrow	0.515	0.495	0.063	0.071	0.056
4	0.1120	Regular	0.383	0.370	0.040	0.045	0.036		and		Regular	0.749	0.727	0.063	0.071	0.056
		Wide	0.446	0.433	0.040	0.045	0.036		¼	0.2500	Wide	1.015	0.993	0.063	0.071	0.056
		Narrow	0.281	0.276	0.032	0.036	0.028		5/16	0.3125	Narrow	0.640	0.620	0.063	0.071	0.056
5	0.1250	Regular	0.446	0.433	0.040	0.045	0.036		and		Regular	0.890	0.868	0.063	0.071	0.056
		Wide	0.508	0.495	0.040	0.045	0.036		20	0.3200	Wide	1.140	1.118	0.063	0.071	0.056
		Narrow	0.312	0.307	0.032	0.036	0.028		24	0.3720	Narrow	0.749	0.727	0.063	0.071	0.056
6	0.1380	Regular	0.446	0.433	0.040	0.045	0.036		and		Regular	1.015	0.993	0.063	0.071	0.056
		Wide	0.570	0.557	0.040	0.045	0.036		⅜	0.3750	Wide	1.280	1.243	0.100	0.112	0.090
		Narrow	0.312	0.307	0.032	0.036	0.028				Narrow	0.890	0.868	0.063	0.071	0.056
7	0.1510	Regular	0.446	0.433	0.040	0.045	0.036		7/16	0.4375	Regular	1.140	1.118	0.063	0.071	0.056
		Wide	0.570	0.557	0.040	0.045	0.036				Wide	1.499	1.462	0.100	0.112	0.090
		Narrow	0.383	0.370	0.032	0.036	0.028				Narrow	1.015	0.993	0.063	0.071	0.056
8	0.1640	Regular	0.508	0.495	0.040	0.045	0.036		½	0.5000	Regular	1.280	1.243	0.100	0.112	0.090
		Wide	0.640	0.620	0.063	0.071	0.056				Wide	1.780	1.743	0.100	0.112	0.090

SOURCE: ANSI B18.13-1965, published by the American Society of Mechanical Engineers.

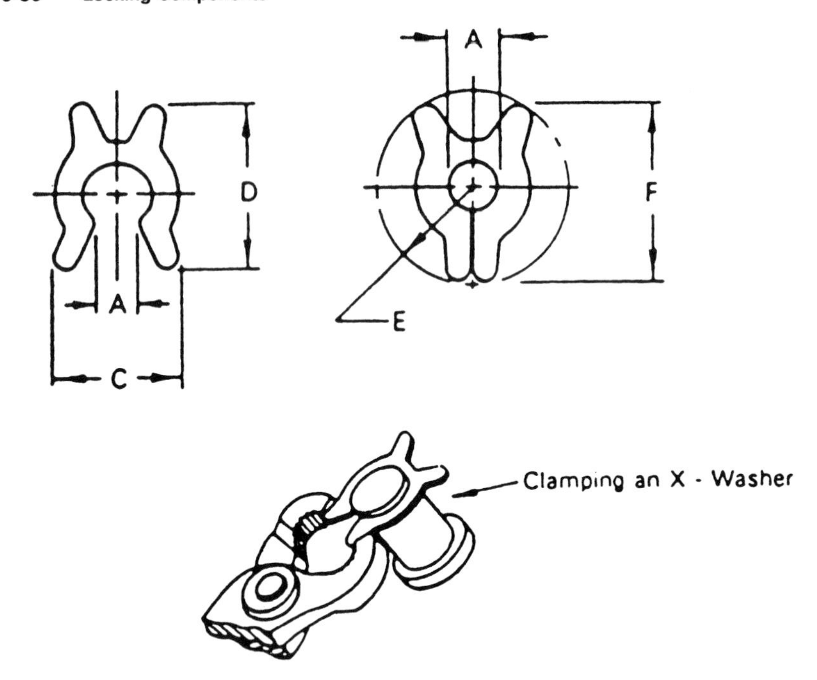

TABLE 10-21 X-Washer Dimensions (Inches)

A	B①	C	D	E	F
0.086	0.025	0.320	0.406	0.210	0.406
0.098	0.055	0.364	0.490	0.297	0.475
0.130	0.055	0.430	0.575	0.359	0.556
0.164	0.065	0.523	0.687	0.422	0.665
0.190	0.065	0.593	0.745	0.437	0.730
0.222	0.075	0.622	0.776	0.469	0.775
0.256	0.075	0.698	0.905	0.500	0.890
0.285	0.075	0.822	0.986	0.563	0.984
0.317	0.089	0.872	1.100	0.609	1.078
0.347	0.089	0.948	1.190	0.688	1.188
0.381	0.089	1.060	1.297	0.797	1.281

① = Thickness of washer

SOURCE: McGraw-Hill Manufacturing and Metalworking Handbook by R. Walsh, © 1994. Reprinted with permission of The McGraw-Hill Companies.

10-22 X-Washers The x-washer has a wide variety of uses as a locking shoulder component. Refer to Table 10-21 for a size tabulation and an illustrative display of pre-assembly versus a post-clamping configuration.

LATCHING AND LOCKING MECHANISMS

We now approach the assembly phase, where components are combined into a unit to perform the function of latching and locking as a total mechanism. The following examples are arranged into general categories.

10-23 Link-Lock Fasteners This type of fastener is a positive-locking, springless-latching device for use wherever a preloaded closure is desired. If a watertight seal is desired, a heavy pull-down pressure will insure proper sealing, if a gasket is used. Figure 10-8 pictures a typical fastener. Some link-lock designs employ springs, providing take-up for better sealing. Figure 10-9 illustrates a springless design which is recessed.

10-24 Hinge-Lock Fasteners These are pressure-tight sealing devices used primarily on hinged-cover transit and equipment cases (see Fig. 10-10).

10-25 Hook-Lock Fasteners These are springless positive-locking latching devices which are ideally suited for use on rigidly specified military transit cases (Fig. 10-11).

10-26 Rotary-Lock Mechanism This mechanism is shown in Fig. 10-12 in open and closed positions. Note that an allen wrench is the key. This mechanism will draw panels together at sufficient pressure to establish an airtight and watertight seal. It will carry high-tension loads too. This locking device is positve and the tapered cam design will also carry heavy shear loads.

10-27 Spring-Lock Fasteners These are a one-piece blind-river design with a variety of head styles. The most common are round, hex, and wing heads as shown in Fig. 10-13. Installation of the slotted roundhead design is illustrated in Fig. 10-14. The fasteners are self-adjusting to compensate for various material thicknesses within their range.

10-28 Trigger-Action Latches These are fasteners of the keeper variety which are released by depressing a trigger (Fig. 10-15). The main feature of the typical trigger-action latch is, of course, rapid access. When the spring-loaded trigger is pushed, the bolt flies open and the panel or lid is released.

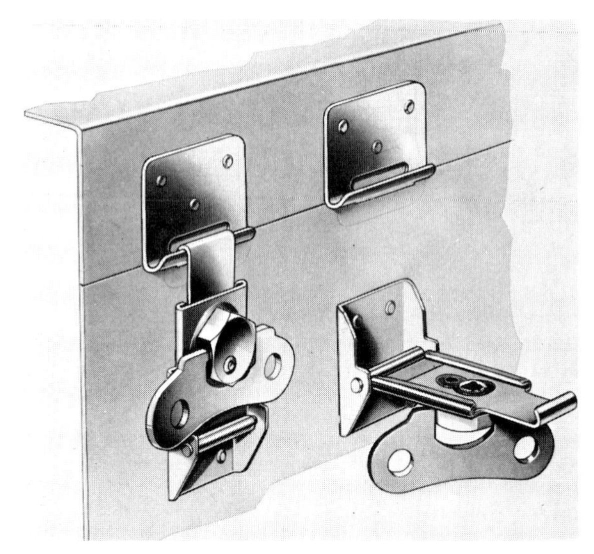

FIG. 10-8 Link lock is the ideal latching device where heavy locking pressure is necessary. (*Simmons Fastener Corp., Albany, N.Y.*)

FIG. 10-9 Recessed link lock minimizes protrusion, ensures pressuretight seal. May be bonded with adhesive or riveted to any sheet, sandwich material, or plywood. (*Simmons Fastener Corp., Albany, N.Y.*)

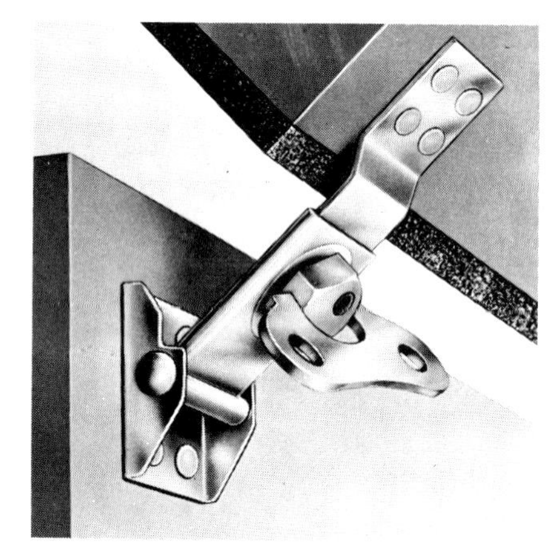

FIG. 10-10 Hinge lock is a springless pressure hinge which provides pressuretight sealing of gaskets on the hinge line of transit and equipment cases. Has high preload, excellent impact and drop-test resistance, and withstands arctic temperatures. Fastener is actuated by finger-tip pressure on wing nut. A half turn opens or closes. (*Simmons Fastener Corp., Albany, N.Y.*)

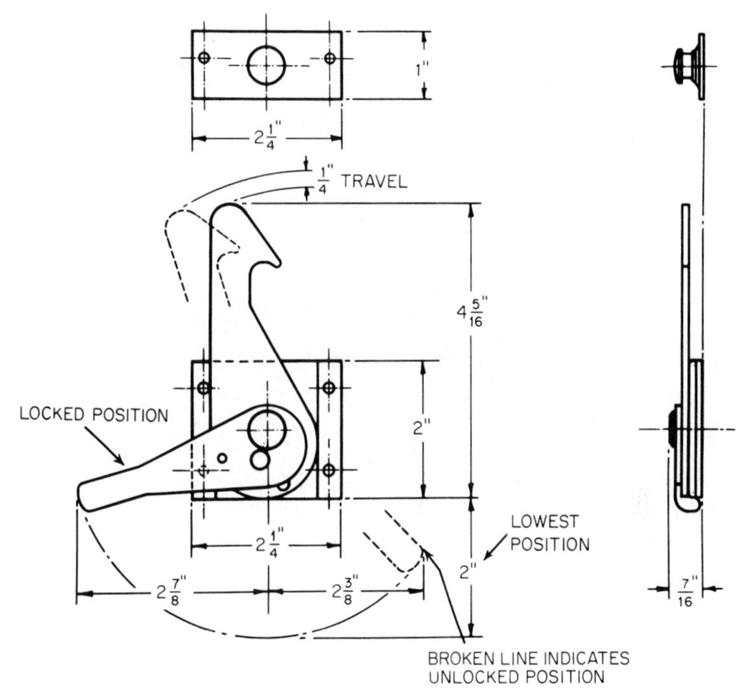

FIG. 10-11 Hook-lock fastener. This is a manufacturer's certified mounting drawing detailing the exact dimensions used for proper installation. (*Simmons Fastener Corp., Albany, N.Y.*)

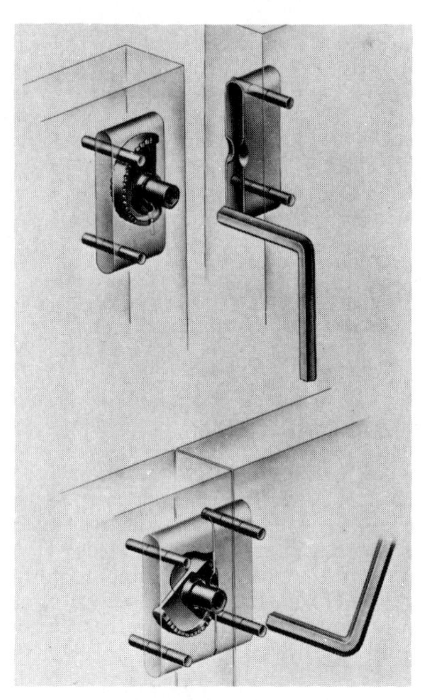

FIG. 10-12 Roto-lock is shown in the open position above and in the closed position below. (*Simmons Fastener Corp., Albany, N.Y.*)

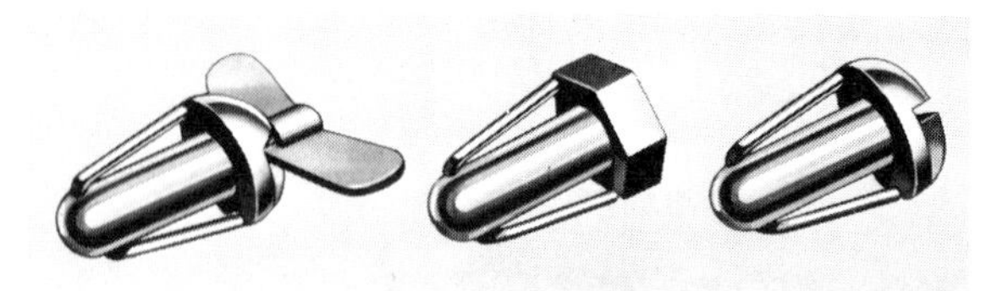

FIG. 10-13 Spring-lock fastener head styles: left—wing; middle—hex; right—round. (*Simmons Fastener Corp., Albany, N.Y.*)

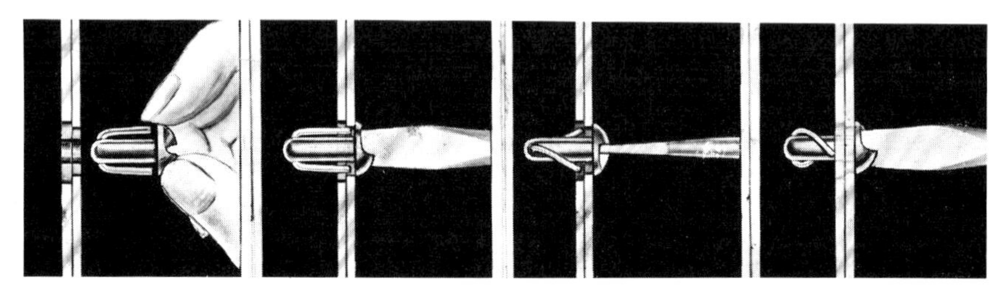

FIG. 10-14 Spring-lock fastener. The four-step installation is shown here: insert, apply screw driver, quarter turn to twist spring, another quarter turn to lock. (*Simmons Fastener Corp., Albany, N.Y.*)

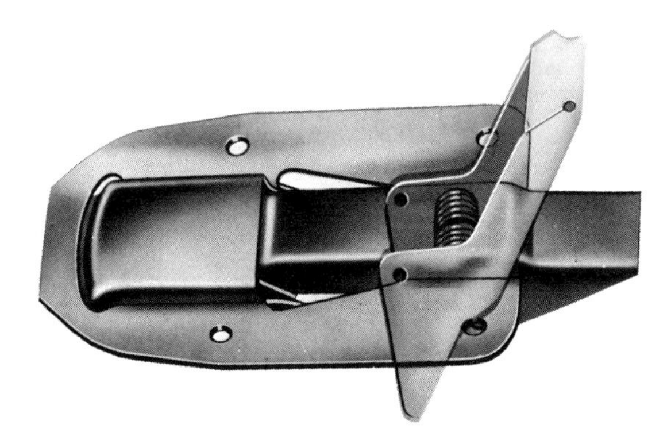

FIG. 10-15 Trigger-lock fastener. Flush mounting, positive action, ruggedness, quick opening, and sure closing are the main features. (*Hartwell Corp., Los Angeles, Calif.*)

AEROSPACE FASTENERS AND MECHANISMS

10-29 Missile Air-Sealed Latches These fasteners are environmentally sealed keeper-type latches. Figure 10-16 illustrates a typical latch. The over-center toggle engagement affords excellent shock and vibration characteristics. The gasketed buttons seal against internal and external pressure differentials. The design protects against environmental hazards. Fingertip actuation and latching permit access without tools.

10-30 Flush Handles These handles are employed in aircraft design. Safety, in being flush with the unit contour, and the light weight are in the main advantages of the modern locking handle. (See Fig. 10-17).

CLAMPING DEVICES

Clamping devices are manufactured by a host of companies for a multitude of applications. A few of them are shown here to acquaint the reader with their basic design.

10-31 Vertical Handle Hold-down Clamp This U-shaped hold-down bar clamp is furnished with a vinyl-tipped adjusting spindle, flanged washers, and lock nuts (see Fig. 10-18).

10-32 Heavy-Duty Work-Holding Clamp This clamp has a square cold-drawn-steel hold-down bar. Pivot points have serrated hardened bushings and pivot pins. Its holding capacity is 1,000 lb max (see Fig. 10-19).

10-33 Jaw Clamp This clamp is furnished with two adjustment spindles. The pressure-spring-loaded spindle plus the hex-head spindle are the main features of this two-way trigger-release manual clamp. Figure 10-20 pictures its basic geometry.

FIG. 10-16 Missile air-sealed latches. Protection against environmental hazards. (*Hartwell Corp., Los Angeles, California.*)

FIG. 10-17 Flush handles. These mechanisms are generally used in the aerospace industry. Their light weight is one of the main advantages. (*Hartwell Corp., Los Angeles, Calif.*)

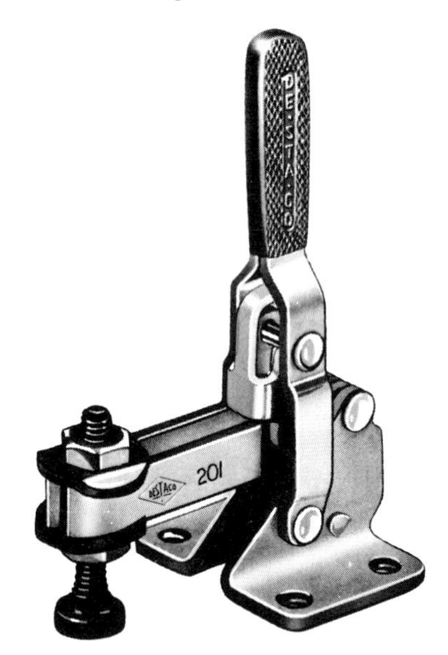

FIG. 10-18 Vertical handle hold-down clamp.
(De-Sta-Co Div., Dover Corp., Detroit, Mich.)

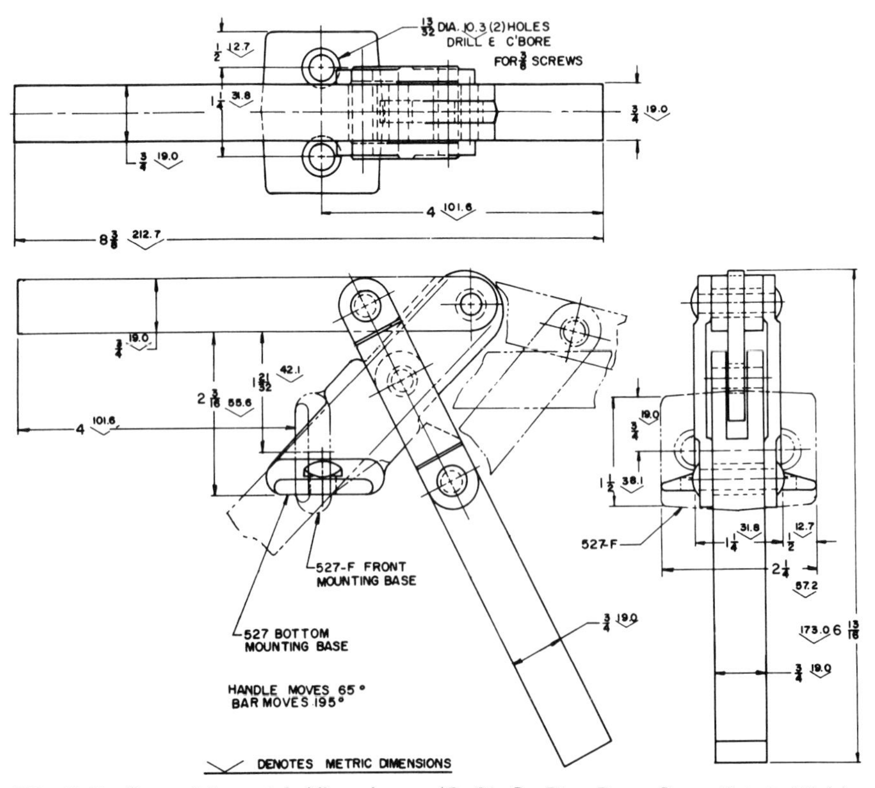

FIG. 10-19 Heavy-duty work-holding clamp. *(De-Sta-Co Div., Dover Corp., Detroit, Mich.)*

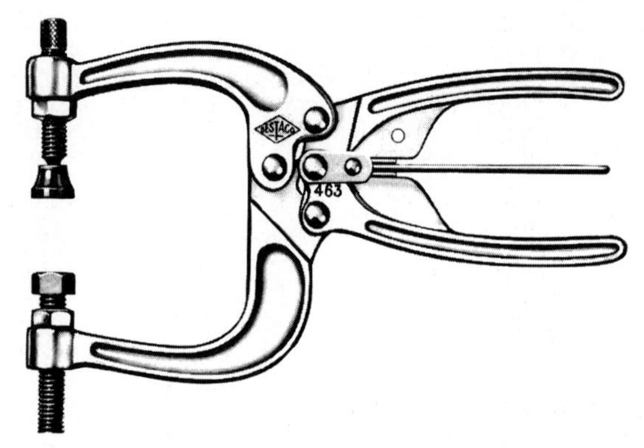

FIG. 10-20 Jaw clamp. *(De-Sta-Co Div., Dover Corp., Detroit, Mich.)*

10-34 Palletized Clamping These clamping or fixturing assemblies function with a ride along hydraulic pressure storage system for fixtures that must be disconnected from the power source during portions of the machining process. This design completely eliminates the umbilical from the power supply after loading. This system is suitable for fixtures being used on pallet changers, rotary tables or indexing. Figure 10-21 illustrates a typical example.

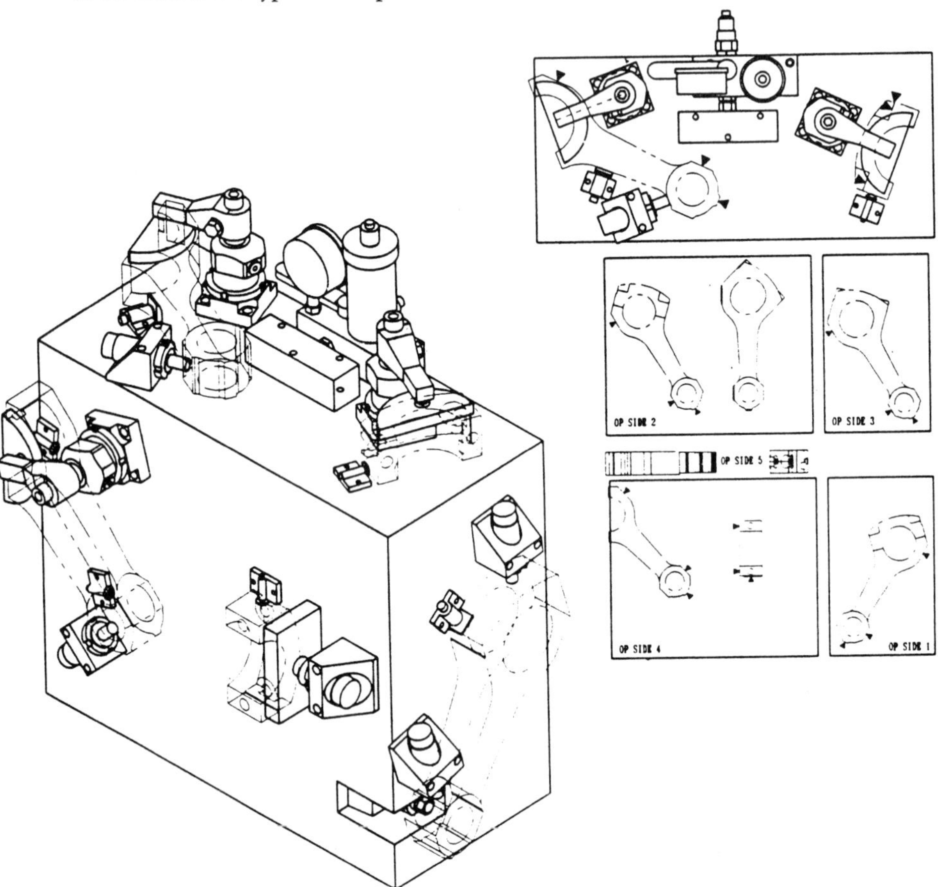

FIG. 10-21 Palletized clamping fixture. (*Courtesy Vektek, Inc., Emporia, Kansas.*)

PLASTIC INSERTS

10-35 Plastic Insert Fastener This basic plastic insert fastener (Fig. 10-22) allows the insertion of a grommet into a hole, followed by the partial pressing of a plunger into the grommet, thus positioning the assembly to be installed in the fixed panel.

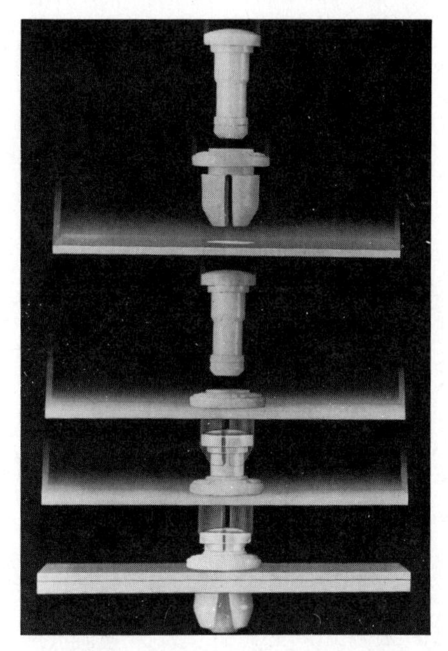

FIG. 10-22 Plastic insert fastener. The sequence of installation is shown here in four steps. (*Hartwell Corp., Los Angeles, Calif.*)

The grommet is available molded from a newly developed low-viscosity polycarbonate resin. This high-impact material yields 4 to 5 times the shear-impact resistance available with former material.

This combination of grommet and plunger is nonconductive and easy to install. It needs no tools and compensates for any minor hole misalignment. It is ideal for electronic-equipment, office-machinery, and computer assembly.

PIPE SEALING

10-36 Permanent Mechanical Seal When piping is required to penetrate through concrete walls, permanent sealing is generally specified. Grouting or mastic sealing often are not positive enough or long lasting.

Modular mechanical seals offer the best solution because they can seal any pipe whether cast iron, concrete, copper, steel, plastic, electrical or telecommunications cable. The seal is permanent, and when properly installed has a positive hydrostatic resistance up to 40 feet of head. Refer to Fig. 10-23 for a cut-away view and Fig. 10-24 for a cross-section detail.

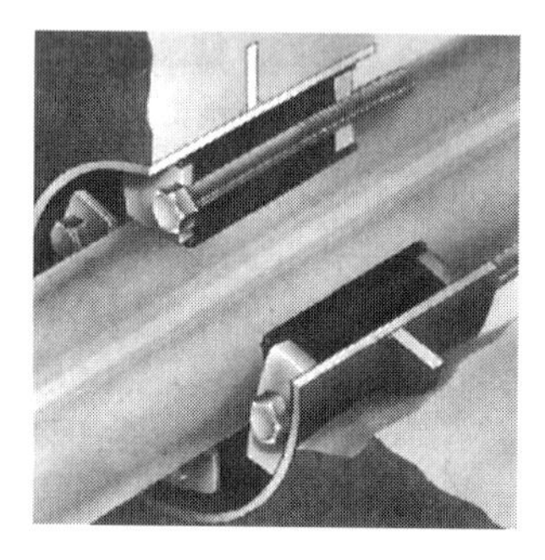

FIG. 10-23 Cut-away view of link-seal.® *(Courtesy Thunderline/Link-Seak, A Division of PSI.)*

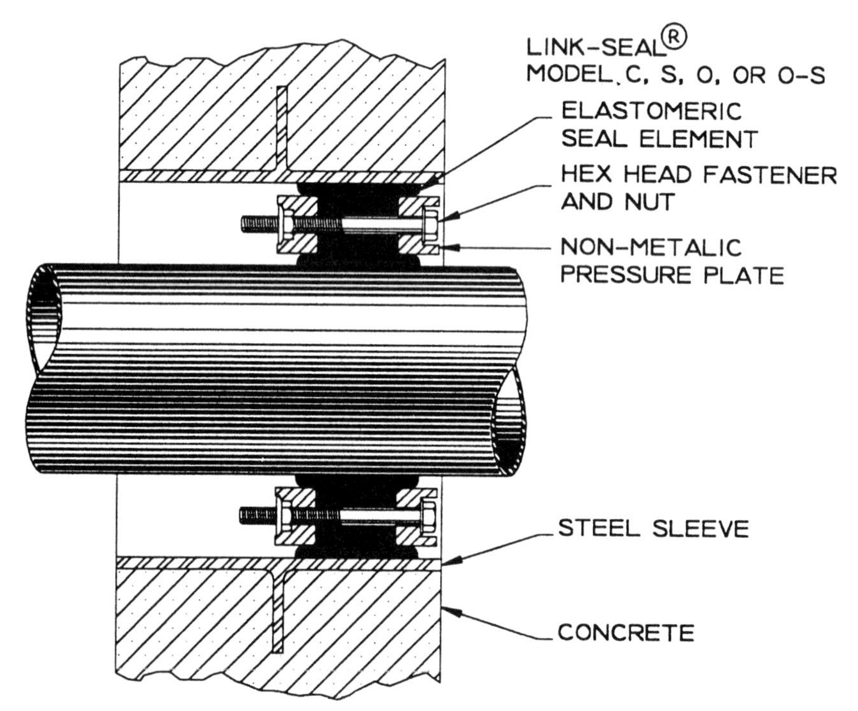

FIG. 10-24 Cross-sectional view of link-seal.® *(Courtesy Thunderline/Link-Seak, A Division of PSI.)*

1 Link-Seal is shipped as a belt of interconnected rubber links.

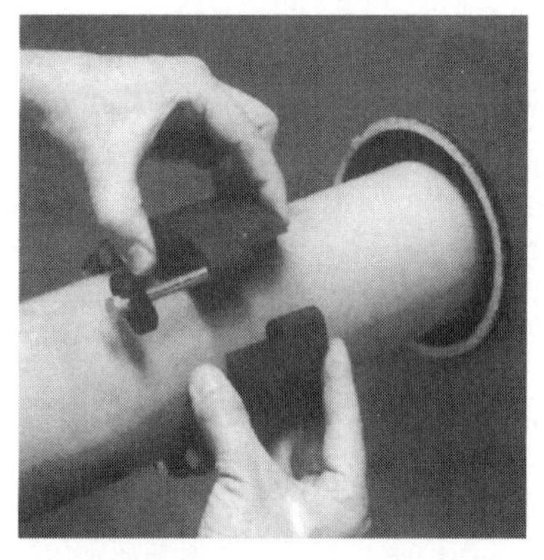

2 Wrap the belt around the pipe. Then connect the first and last links.

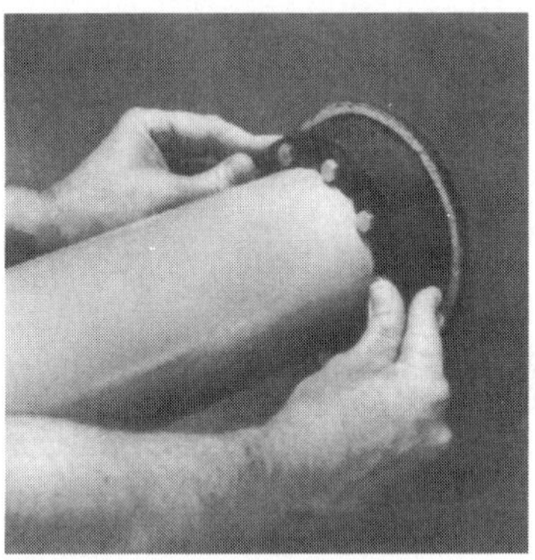

3 Slide the assembly into the space between the pipe and wall.

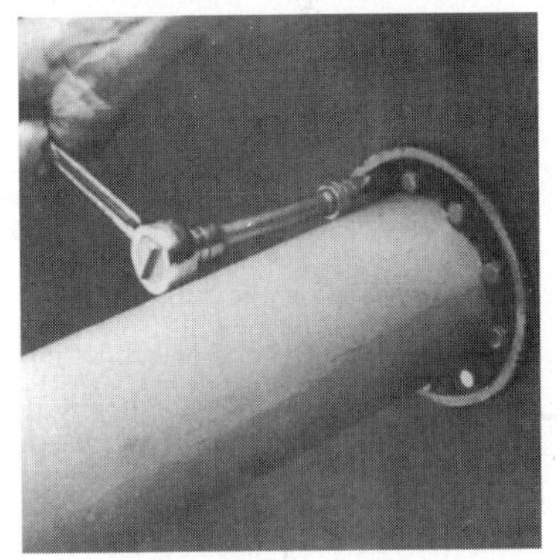

4 When the bolts are tightened, Link-Seal expands to create a gas and water tight seal.

FIG. 10-25 Assembly method for link-seal.® (*Courtesy Thunderline/Link-Seak, A Division of PSI.*)

A wide range of models and sizes are provided for a broad range of applications. Installation is simple and no special tools are required. See Fig. 10-25 for a step-by-step typical installation.

Typical highway and railroad crossings that require carrier pipes to be encased in a casing pipe generally specify that each end of the casing pipe be sealed. The modular mechanical seal is ideal for this application. This installation provides a positive, hydrostatic protection from water entry, soil or backfill material. The annular space between the carrier pipe and casing pipe will not be violated.

Pipes that pierce curved walls, which are found in manholes, can be successfully sealed with modular mechanical seals. Refer to Fig. 10-26 for a sketch showing two applications. Note that a relatively thick wall allows the standard installation, while the thin wall requires the seal to be adjusted to the curvature. However, the latter installation requires special techniques and must be coordinated with the manufacturer for proper sealing.

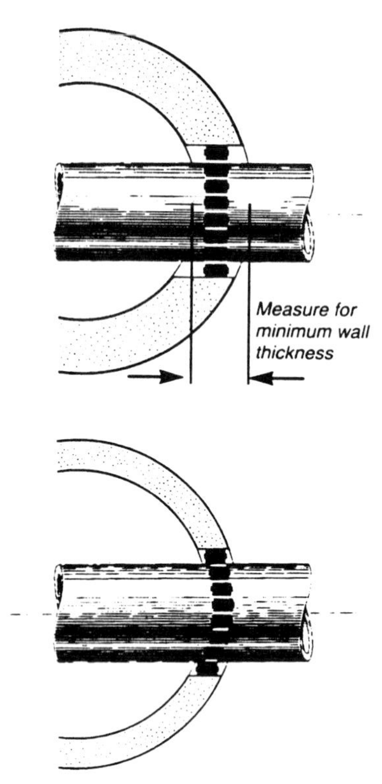

FIG. 10-26 Installation in curved wall. (*Courtesy Thunderline Link-Seak, A Division of PSI.*)

Section 11
Electrical Connections

MERLIN T. LEBAKKEN, P.E.*

President, Power System Engineering, Inc.,
Madison, Wisconsin

11-1 Scope The making of electrical connections or the fastening of electrical conductors involves concern for the mechanical and chemical properties of the conductors as well as the electrical properties. Almost all conductors used in the transportation of electrical energy and electric impulses are made of copper or aluminum or their alloys. This chapter is concerned only with the joining together of these two types of conductors.

The actual fastening of two metal conductors is a mechanical process and is usually accomplished by pressure with the aid of a screwdriver, wrench, or compression tool. Fastening by thermal means such as welding and soldering is sometimes used to minimize contact resistance for long periods of time. In the fastening together of any two conductors by any means, it is essential that the contact surfaces be clean to minimize contact resistance.

Use of copper as an electrical conductor is longstanding, partly because of the metal's properties, which allow simple and problem-free connections. Aluminum has been widely used as an electrical conductor and, by reason of its low cost, its use is increasing rapidly. Making electrical connections with aluminum is somewhat more difficult because of "cold flow" (movement or creep under a constant force) and its high solution

* Author of original section of first and second edition.
Editor-in-Chief added material for third edition.

potential (because of which, contact with other conductors such as copper in the presence of a corrosive medium will cause a deterioration of the aluminum). The latter problem can be solved by avoiding copper-to-aluminum connections or by making such connections with the aid of a bimetallic connector, with a proper listed connector in a dry location without any other corrosive medium present, or with a proper listed connector and enclosing with an antioxident compound.

Also of concern when electrical connections are made is the voltage level under which the conductor must operate and the integrity of the insulation used as a covering for the electrical conductor. These considerations will be discussed in more detail in the following subsections.

This section is divided into subsections based on the use of the conductor. The first subsection is a discussion of the fastening of electrical conductors for the distribution of electricity below 600 volts in homes and commercial establishments. It is concerned with circuits for home lighting and appliances up to 250 volts. The second subsection is concerned with the fastening of both overhead and underground conductors for the distribution of electric power for commercial, industrial, and public utility applications. The third subsection is concerned with electrical connections for electronic applications.

HOME AND COMMERCIAL LIGHTING

11-2 Methods of Connection There are three basic methods of fastening two or more electrical conductors together for the distribution of electrical energy in homes and other buildings. They are: (1) placing two or more conductors on a flat surface under a flathead screw, (2) using solder and tape, and (3) using mechanical wire connectors.

11-3 Screw Terminations Placing an electrical conductor between a flat surface and a flathead screw is usually reserved for terminating the electrical conductor on an appliance or an electrical outlet or switch. Its use for any other purpose should be avoided. This means of fastening or termination involves stripping the proper amount of insulation and applying the proper mechanical torque with the aid of a screwdriver. Special care in workmanship, especially when aluminum conductor is used, should be taken to cause satisfactory termination performance and avoid dangerous overheating. Figure 11-1 and the following steps illustrate the correct method of termination at wire-binding screw terminals of receptacles and snap switches.

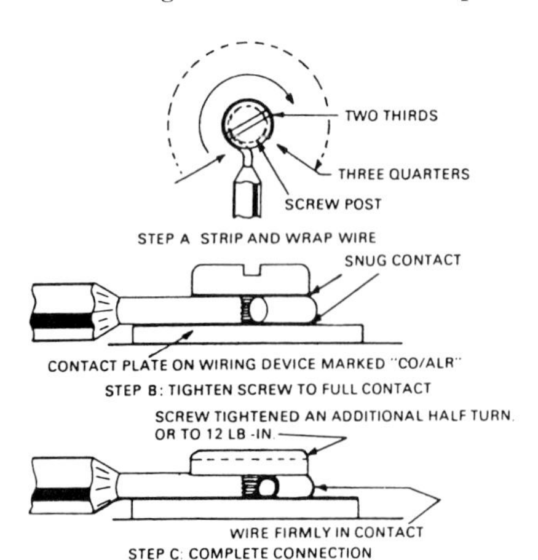

FIG. 11-1 Aluminum conductor on terminal post. Correct method of terminating aluminum wire at wire-binding screw terminals of receptacles and snap switches. (*Underwriters Laboratories, Inc.*)

1. Wrap the freshly stripped end of the wire $^2/_3$ to $^3/_4$ of the distance around the wire-binding screw post as shown in step *A*. The loop is made so that the rotation of the screw as it is tightened will tend to wrap the wire around the post rather than unwrap it.
2. Tighten the screw until the wire is snugly in contact with the underside of the screw head and with the contact plate on the wiring device, as shown in step *B*.
3. Tighten the screw an additional $^1/_2$ turn thereby providing a firm connection. Where torque screwdrivers are used, tighten to 12 lb-in. See step *C*.
4. Position the wires behind the wiring device so as to decrease the likelihood of the terminal screws loosening when the device is positioned in the outlet box.

There are bare and vinyl- or nylon-insulated crimp-type terminals that can make a more secure termination. Lock-spade terminals that snap around studs add security to an open-end terminal. A variety of terminals available is shown in Fig. 11-2.

11-4 Solder and Tape Use of solder and tape provided an excellent means for joining two conductors for more than 60 years; however, its acceptance has decreased rapidly, partly because of improved alternative methods, equipment, and materials. The twisting of the electrical conductors makes the joint mechanically secure whereas the addition of solder helps ensure good electrical conductivity. The steps to follow in making a safe soldered tape joint are:

1. Strip the wires a minimum of 1 inch. If more than 3 wires are joined, increase the stripping length to a minimum of $1^1/_4$ inch.
2. Twist the conductors. The conductors must be twisted to get a good, tight, mechanically secure joint.
3. Solder the twisted splice. This operation requires both care and skill to avoid cold joints. In recent years the propane torch has all but replaced the dip or solder pot. The solder should be preferably a 60-40 analysis with a resin core or paste for flux. Acid-core solder should never be used for this application.
4. Tape the joint. The generally accepted method of taping is first to slide the end of the tape in between the two or more wires being joined, just as a convenience to hold it in position when you start your wrap. Wrap the tape down toward and over the

FIG. 11-2 Crimp terminals. Bare, vinyl-insulated, or nylon-insulated spade or ring terminals and bare or vinyl-insulated splices. (*Ideal Industries, Inc., Sycamore, Ill.*)

end of the conductors. Proceed in a circular motion until tape extends beyond the cut ends of the wires. Fold plastic tape over, then tape back up the splice $^1/_2$ inch minimum over the wire insulation, and the joint is complete.

11-5 Wire Connectors Mechanical wire connectors are available in three basic designs: crimp type, setscrew type and screw-on type, as shown in Fig. 11-3. The insulated connectors are tested and listed by Underwriters Laboratories, Inc., (UL) for use at 300 and 600 volts maximum, and for use at temperature maximums from 75 to 150°C.

Crimp-type connectors are small tubular copper or steel sleeves which are placed over the conductor twisted and crimped with a special tool. An insulating cap placed over the connection completes the splice. This style is not widely used where time is an essential factor. The hand-crimping operation and additional step of installing the cap is slower than other connecting methods.

The setscrew type, as shown in Fig. 11-4, is made of a small tubular sleeve (usually brass or bronze) with a built-in setscrew. To form the connection, this screw is tightened on the wire and an insulating cap placed over the splice. This connector is used on many small applications but, like the crimping type, requires several steps which use up valuable time. It does, however, offer opportunity for easy disconnection and reuse of devices.

The screw-on type connectors are the most popular in use today by reason of their low cost and fast, simple application. They consist of an inner coil spring surrounded by a plastic insulating shell, as shown in Fig. 11-5. As the connector is applied, the screw action of the spring's spiral threads forces the wires into the insert. The insert's tapered shape acts as a wedge, gripping the wires tightly. Under the leverage of the shell this "cold forging" action on the wires makes a joint as strong and as permanent

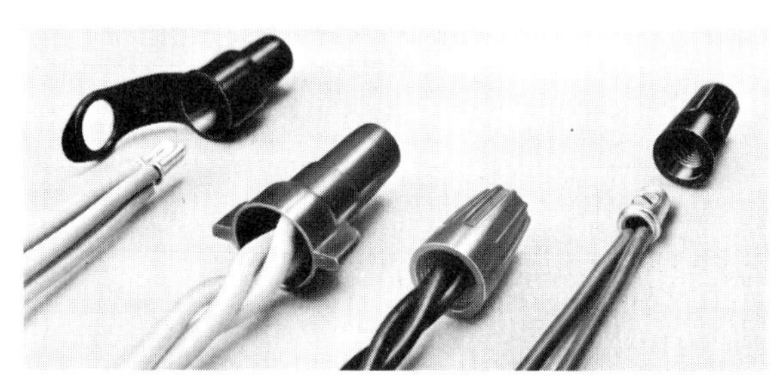

FIG. 11-3 Wire connectors. Ideal's line of solderless pigtail splicing connectors including crimp, Wing-Nut,® Wire-Nut,® and setscrew connectors. *(Ideal Industries, Inc., Sycamore, Ill.)*

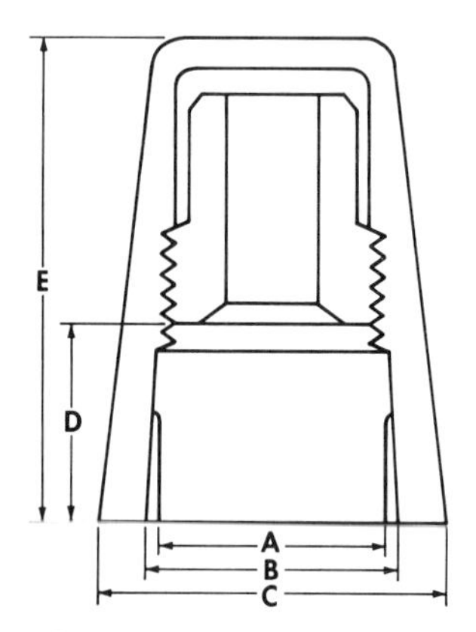

FIG. 11-4 Setscrew connector. Setscrew connector allows visual inspection of the splice. *(Ideal Industries, Inc., Sycamore, Ill.)*

as the wire itself. The basic connection is wire-to-wire and is not dependent on any metal part of the connector for the current-carrying function.

In using a manually torqued connector such as a Wire-Nut® or a Wing-Nut® wire connector, one should adhere to the specifications and instructions published by the manufacturer.

To assure that the connector is properly applied, five simple steps should be followed:

1. Strip wire to length specified by the manufacturer of the connector.
2. Keep conductor ends evenly aligned when preparing to apply the connector.
3. Place the connector over the ends and give a small push to start it on the wires. Don't ram the connector on—let the connector do the work of pulling the wires tightly into the spring.

4. Turn the connector until it is fastened snugly on the wires. A good connection has been made when the wires outside the connector go into a two-turn pigtail as shown in Fig. 11-6. As long as the pigtail twist has been made, the connection is sound.

5. After the connection has been completed, a simple way of checking its soundness is to give each individual conductor a quick pull. Once the connector has been turned tight a safe, trouble-free joint is assured.

In screw-on connectors it is not necessary to pretwist the wire. With some of the larger combinations, however, a partial twist may help hold the wires together and get the connector started for the splice. In joining stranded and single-conductor wire, the stranded conductor should lead the solid into the connector. The stranded should be stripped $^1/_{16}$ inch longer than the solid.

FIG. 11-5 Screw-on connector. Wire-Nut® connectors for branch-circuit and fixture-splicing applications screw on like a nut on a bolt. (*Ideal Industries, Inc., Sycamore, Ill.*)

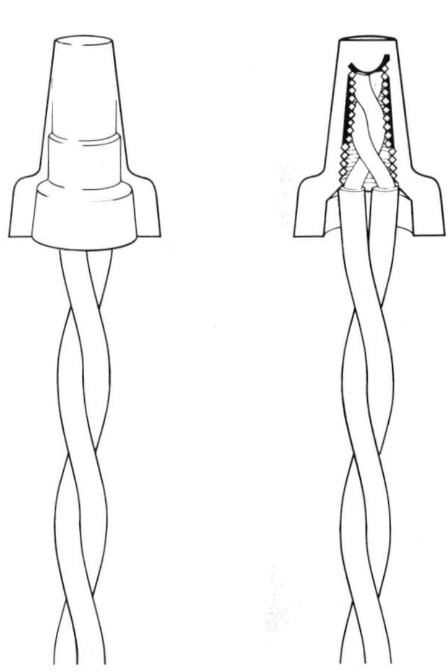

FIG. 11-6 Screw-on splice. Built-in wrench makes turning easy. Live-action, square-edged spring grips wires tight for a strong, low-resistance joint. (*Ideal Industries, Inc., Sycamore, Ill.*)

COMMERCIAL, INDUSTRIAL, AND UTILITY

11-6 Methods of Connection The making of adequate and reliable electrical connections for the distribution of electrical energy in commercial, industrial, and public-utility applications is important to satisfactory business operation. These connections employ a variety of connector and terminal-connector devices and, for the most part, are made by pressure applied by any one of several tools or devices (with the exception of hand-torqued pigtail pressure wire connectors).

The electrical connection should provide current-carrying capacity equal to or greater than the conductors being joined, provide a low-resistance connection, be able to withstand continually changing environmental conditions, and withstand surge currents resulting from system faults. To make such an adequate and reliable connection, the installer must: (1) properly prepare the conductors to be joined, (2) use the proper size and type of connector or terminal device, (3) use the proper tool for the connector device to accomplish necessary mechanical pressure, and (4) use proper workmanship.

11-7 Conductor Preparation There are several steps to follow in preparing and maintaining the conductors to be joined in a usable connection. They are:

1. Strip the proper amount of insulation, if any, from the conductor without nicking the conductor.
2. Clean the contact surfaces with a wire brush for trouble-free connections. This is especially important where aluminum is one or both of the conductors. The aluminum oxide must be removed to ensure a low-resistance connection. Once the conductor is cleaned, the connection should be made as soon as possible.
3. If aluminum, protect the clean surface with an oxide-penetrating inhibitor. A good inhibitor should seal out moisture and other atmospheric conditions and transfer the current.
4. Seal the finished connection against the entrance of air and moisture with one of the many types of sealers available.
5. Replace or add required insulation.

Two problems associated with metals that must be considered when making connections are galvanic corrosion and oxidation. Galvanic corrosion is a phenomenon of electrochemical action which converts chemical energy into electrical energy. This is best illustrated by drawing an analogy between an ordinary storage cell and a bimetallic bolted-type connector. When aluminum or copper are placed in the presence of an electrolyte, a cell is formed where the aluminum becomes the anode and the copper the cathode. In a bimetallic connection the same holds true, with moisture being the electrolyte. The aluminum displaces itself on the copper, thereby considerably weakening the aluminum. This problem is virtually eliminated by maintaining a space between the two conductors to be joined and by applying an inhibitor.

Oxidation has a different effect on copper than on aluminum. Oxygen unites with copper to form copper oxide, which has a relatively high conductivity. When oxygen and aluminum unite, aluminum oxide is formed, which is not only extremely hard but also forms a semi-insulator with high electrical resistance. Aluminum oxide is formed almost instantaneously upon exposure, with atmospheric conditions and time governing the rate of growth. To ensure continuity of service in a connection the oxide must be brushed off and the remaining molecular thickness penetrated by an oxide-penetrating inhibitor.

11-8 Connector Selection Current-carrying connectors and terminal connectors can be generally categorized as either compression or noncompression types. The noncompression type makes use of either a threaded fastener or bolted connection. Mechanical pressure in the noncompression type is provided by a wrench, screwdriver or other tool giving a mechanical advantage. The compression connector requires use of a special mechanical-advantage tool or a hydraulic tool.

Connectors for use in commercial, industrial, and housing systems are usually subject to the National Electrical Code requirements and therefore are listed by Underwriters Laboratories (UL). The National Electrical Manufacturers Association (NEMA) is concerned only with specifications and standards. Developing public utility systems connector specifications is one of the functions of the Edison Electric Institute (EEI).

It is most important that the proper size and type of connecting device is used. A "make do" installation is almost certain to result in system failure. The manufacturers of these connector devices have done a remarkable job in marking or color-coding the devices for size and application and detail installation procedures are usually provided with the devices.

In general, a connector device should:

1. Have sufficient contact surface and low contact resistance for efficient transfer of current.
2. Have proper contour for cold-flow and creepage protection.
3. Distribute current for surge protection.
4. Be compatible with the conductors for corrosion protection, or provide space separation.
5. Have resiliency in design to handle the thermal expansion and contraction of the conductor.
6. Be sufficiently massive that the connection will run cooler than the conductor.

There are many different types of both compression and noncompression connector devices and therefore one is available for almost any situation. A variety of connector materials and plating is used to prevent electrolytic action. The connection used should be the same material as the conductor, be plated with the same material as the conductor or be plated with a tin or tin alloy material to prevent galvanic action.

It is not possible to determine visually whether a connector is suitable for aluminum by appearance alone. The fact that a connector is plated does not necessarily mean that it is usable with aluminum conductors. The best guide is the UL listing mark. A connector which has been found suitable for aluminum by UL will be marked AL or AL/CU. The AL designation by itself means the connector is suitable for use only with aluminum and the AL/CU designation indicates suitability for both aluminum and copper conductors.

The connector should be designed so that with application of the proper pressure, the initial contact resistance is overcome and stability is ensured even though there is force relaxation due to thermal expansion and contraction and cold flow of the metal. Cold flow or creep is caused by different coefficients of expansion plus pressure.

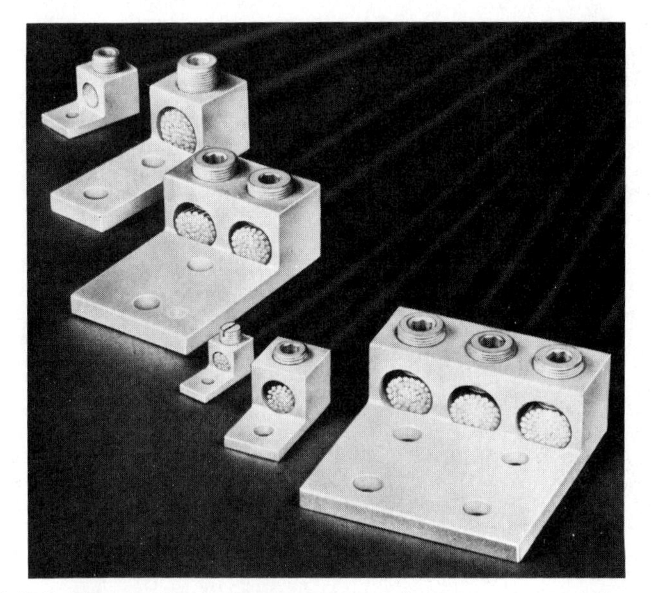

FIG. 11-7 Threaded terminal connector. *(Burndy Corporation, Norwalk, Conn.)*

A properly designed connector has built in it the ability to compensate for some relaxation of force on the conductor. Therefore, even if it is apparent that there has been a reduction in torque after the initial tightening, the connector should operate properly. The recommendation normally given on retightening is to specify one retightening within a few days of the initial installation. This also serves to assure that connections which had not been originally tightened properly are found. Frequent retightening of connections on electrical conductors is not recommended since this will cause eventual damage to the conductor and jeopardize the integrity of the connection.

11-9 Low-Voltage Terminal Connections Two types of pressure-applied terminal connectors are available. One uses a threaded fastener as shown in Fig. 11-7 to apply contact pressure either directly on the conductor or on the pressure member which bears on the conductor. Some types require the use of an allen wrench for applying the pressure, others require a screwdriver or wrench to turn a bolt-type fastener. A bolted connection is then made to a bus or piece of equipment.

The other type of terminal connector is the compression type (shown in Fig. 11-8), where pressure is applied by an external tool which causes a change in the shape and size of the connector body around the conductor to form one homogenous mass. A bolted connection is then made to a bus or piece of equipment.

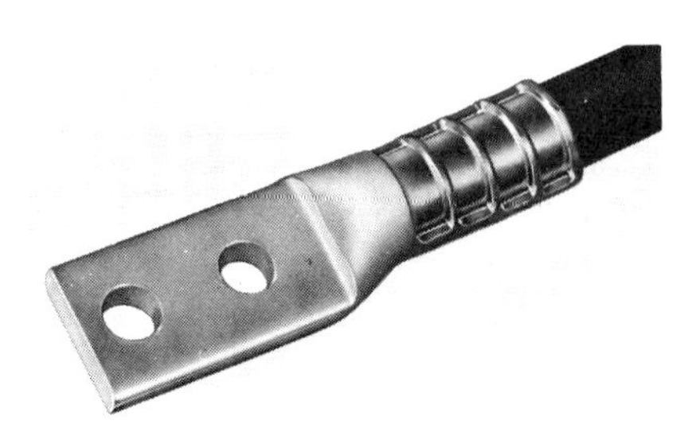

FIG. 11-8 Compression terminal connector. *(Burndy Corporation, Norwalk, Conn.)*

There are many varieties of these terminal connectors on the market to fit almost any situation. Connectors come in various configurations to accommodate various sizes and numbers of conductors, various bolt-hole configurations and various ways to apply the pressure connection. The same conductor preparation and procedure as discussed in sec. 11-7 is required for the installation of any of these terminal connectors. Use of the proper size and type of connector device is essential.

Recommended torque values for screwdriver- or wrench-applied connections are usually provided by the manufacturer with the devices or can be obtained from the manufacturer.

There may be situations where aluminum conductors are to be connected to equipment which do not have terminals for aluminum. Methods for resolving this situation are:

1. Run a copper conductor to the equipment terminal and connect the copper conductor to the aluminum wiring system in an improved location and with an AL/CU connector as shown in Fig. 11-9.

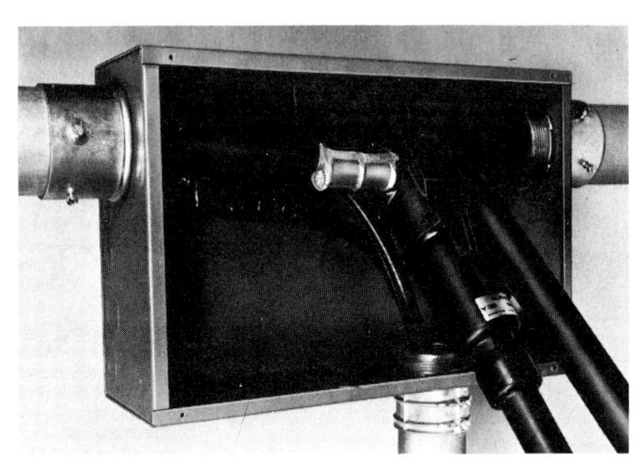

FIG. 11-9 Aluminum/copper connector. *(Burndy Corporation, Norwalk, Conn.)*

2. Replace the equipment terminal with a terminal listed for aluminum. This sometimes is difficult since the terminal may not be removable or since there may not be a suitable terminal for use as a replacement.

3. Attach to the aluminum conductor an adapter which will permit a suitable transition to the questionable terminal, as shown in Fig. 11-10. Such adapters or similar adapters are sometimes required to accomplish the physical transition from a conductor to an equipment terminal, even though it is U.L. listed for aluminum.

11-10 Low-Voltage Splices and Taps There are a variety of connectors available for the joining of two conductors and for the taking of branch circuits off from a main circuit. Fig. 11-11 shows a compression splice which joints two conductors of equal size and material. Similar connectors are available to join conductors of different sizes, insulation materials, or conductor materials. Figure 11-12 shows the joining of two or more conductors with a bolted or threaded connector. Several taps can be taken from one source for the distribution of electrical power by use of a multitap terminal, shown in Fig. 11-13.

Branch circuits can be taken from a main circuit by use of special connectors as shown in Fig. 11-14. These connectors can be either the bolted type where the body of the connector makes a contact with the conductor or threaded screw type where the screw makes the contact. There are also compression-type connectors for these installations.

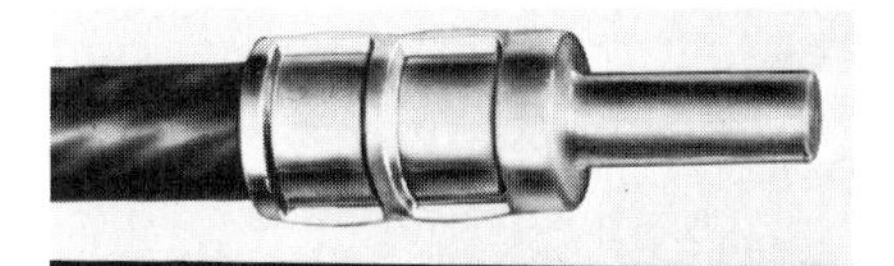

FIG. 11-10 Compression equipment adapter. *(Burndy Corporation, Norwalk, Conn.)*

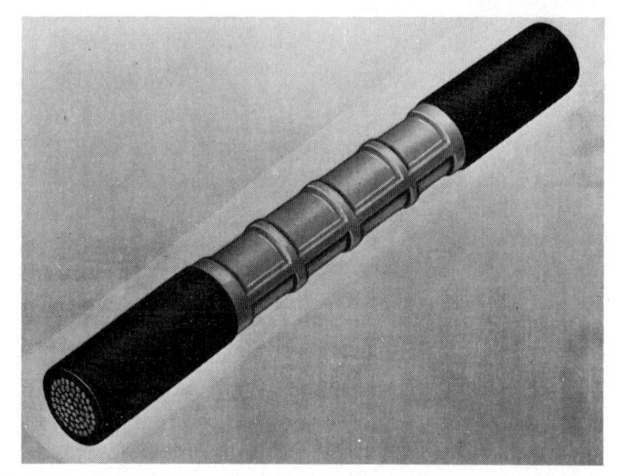

FIG. 11-11 Compression splice. *(Burndy Corporation, Norwalk, Conn.)*

FIG. 11-12 Threaded multitap connector. *(Burndy Corporation, Norwalk, Conn.)*

In all cases the connector can be covered either with insulating tape or insulated boots made for the purpose. The same conductor preparation and installation procedures discussed in sec. 11-7 apply to splice and tap connections.

11-11 Bus-Bar Connections A bus-bar connection refers to the joining of two or more flat or circular bus members for electrical conductivity purposes. Connection of a connector tang to a bus bar or equipment pad is included. The rule concerning clean contact surfaces and adequate pressure still applies for bus-bar connections. If the contact surfaces are plated, abrasion should not be used to clean the surfaces; however, solvents may be applied. Unplated contact surfaces should be scratch-brushed to ensure cleanliness.

FIG. 11-13 Multitap terminal. *(Fargo Manufacturing Company, Poughkeepsie, N.Y.)*

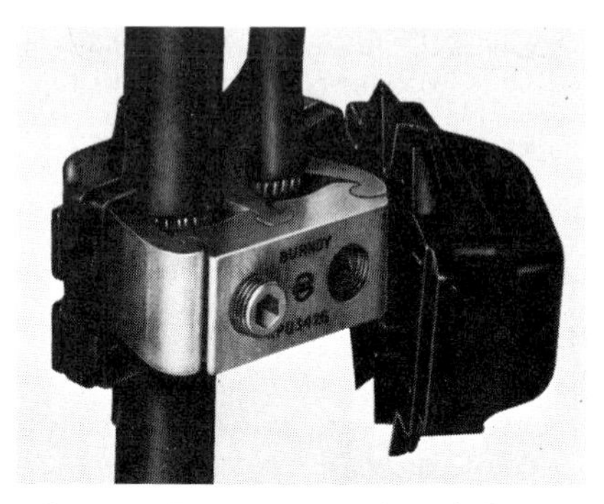

FIG. 11-14 Tap connector. *(Burndy Corporation, Norwalk, Conn.)*

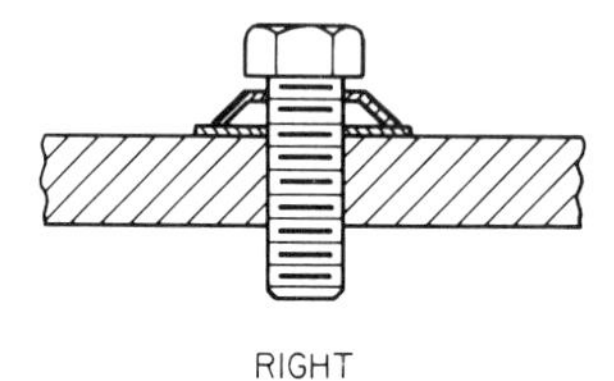

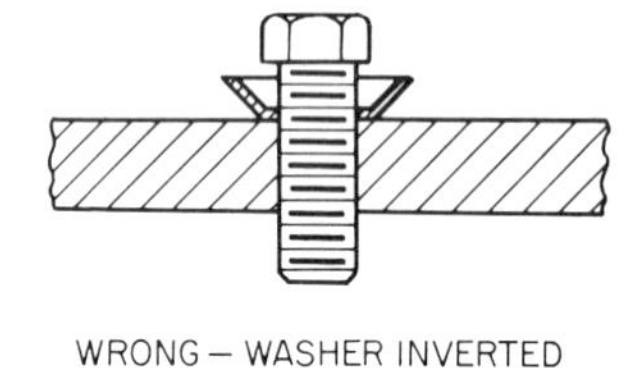

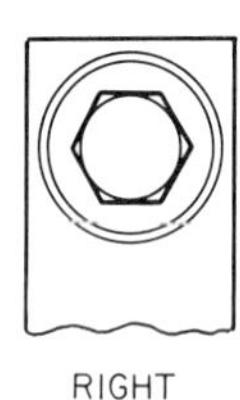

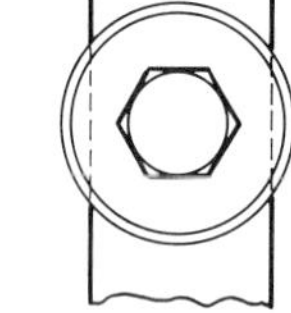

FIG. 11-15 Belleville washer installation. To obtain proper pressure, tighten fastener until belleville is flat. *(Burndy Corporation, Norwalk, Conn.)*

Bus-bar connections can be made either mechanically with connector devices or by welding (metal fusion). Although welding results in no contact surface or resistance, along with other advantages, most joints are mechanically made with a variety of devices available for the specific situation. The following discussion is concerned with these connections.

It is always necessary to apply joint compound to a bare aluminum connection and there is no harm and probably an advantage to the use of a compound on bare copper surfaces. Compounds are also excellent lubricants for aluminum conduit threads and help to enhance the electrical continuity of the conduit system. Excessive compound should be removed after the connection is made.

Generally, aluminum hardware should be used with aluminum connectors. In the early usage of aluminum bus bars the aluminum bar material was very soft and flowed easily under pressure; therefore belleville washers were recommended when steel hardware was used. Presently available aluminum alloys are much more resistant to this flow and use of belleville washers is not always required. The use of belleville washers as shown in Fig. 11-15 has both advantages and disadvantages. If the proper belleville washer is used and applied in a correct manner, it can compensate for temperature variation and metal flow. The washer must flatten at a sufficient force to be meaningful, not be larger in diameter than the bus-bar width, and must have a proper finish for the particular environment in which it will be exposed. The washer must be right side up to avoid digging into the aluminum material.

Figure 11-16 shows various methods and devices for making bus-bar connections.

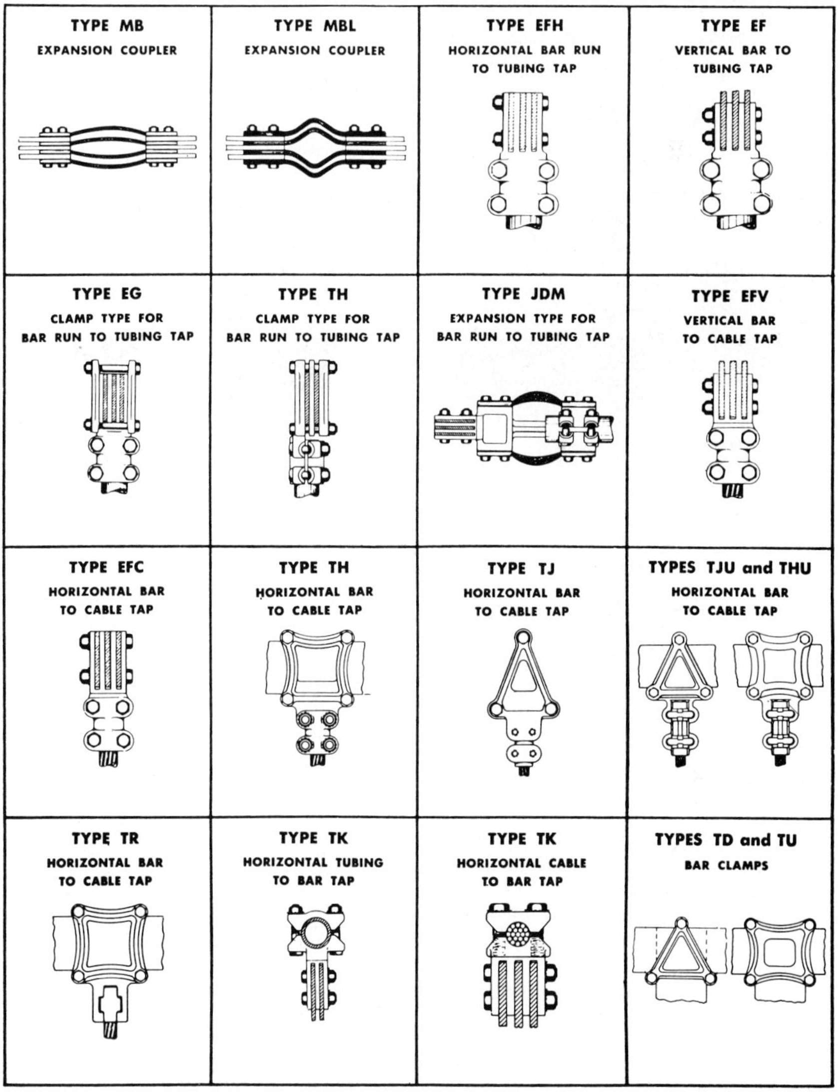

TYPE MB EXPANSION COUPLER	TYPE MBL EXPANSION COUPLER	TYPE EFH HORIZONTAL BAR RUN TO TUBING TAP	TYPE EF VERTICAL BAR TO TUBING TAP
TYPE EG CLAMP TYPE FOR BAR RUN TO TUBING TAP	TYPE TH CLAMP TYPE FOR BAR RUN TO TUBING TAP	TYPE JDM EXPANSION TYPE FOR BAR RUN TO TUBING TAP	TYPE EFV VERTICAL BAR TO CABLE TAP
TYPE EFC HORIZONTAL BAR TO CABLE TAP	TYPE TH HORIZONTAL BAR TO CABLE TAP	TYPE TJ HORIZONTAL BAR TO CABLE TAP	TYPES TJU and THU HORIZONTAL BAR TO CABLE TAP
TYPE TR HORIZONTAL BAR TO CABLE TAP	TYPE TK HORIZONTAL TUBING TO BAR TAP	TYPE TK HORIZONTAL CABLE TO BAR TAP	TYPES TD and TU BAR CLAMPS

FIG. 11-16 Bus-bar connections. *(Southern States, Inc., Hampton, Georgia.)*

11-12 Insulated Low-Voltage Connections There are occasions where it is desired to insulate the connections discussed in the previous three sections. This is almost always the case with connections of insulated conductors in junction boxes or other locations where a splice or tap is made. There are several types of insulating systems available.

The first and most common type of insulating system is one using insulated tape. Taping of a joint can provide electrical insulation and protection against the environment and safety to the public. The choice of tape depends on the voltage of the system, the ambient temperature during applications, and the mechanical and corrosive environment involved. If good sealing and smoothing of irregularities are desired, a mastic or puttylike material is used with or without a wrap of standard electrical tape. In taping splices above 600 volts and especially above 5,000 volts it is extremely important that the correct material be properly applied. This will be discussed in a subsequent subsection.

A second means of insulation is by use of snap-on covers which are available for specific designs. They can be snapped on in tight quarters where taping would be difficult, allow for inspection of a joint without destruction of the insulation, and are reusable. One type of such a snap-on cover is shown in Fig. 11-17.

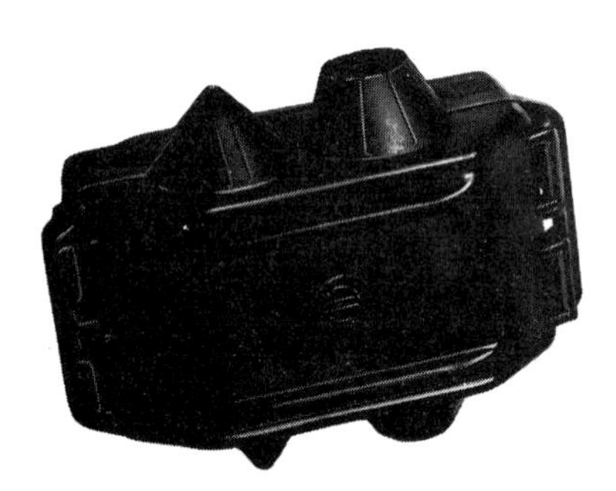

FIG. 11-17 Snap-on cover. *(Burndy Corporation, Norwalk, Conn.)*

FIG. 11-18 Heat-shrinkable pads. *(3M Company, St. Paul, Minn.)*

There are covers which are heat-shrinkable and also elastomeric units or pads which slip over or wrap around the connection and the conductor. Figure 11-18 shows an example of this type of insulation.

Utility companies make extensive use of a variety of low-voltage insulated-splice kits, shown in Fig. 11-19. A compression connection is made with the tight-fitting elastomeric cover, giving a watertight splice.

11-13 Outdoor Aerial Connections Most of the outdoor aerial connections are made by public utilities or electrical contractors in commercial and industrial situations. As in other areas, there are both compression and threaded- or bolted-type connector devices. Again, the same type of preparation and procedures as outlined in sec. 11-7 apply to these installations. A variety of noncompression-type fasteners are available for different situations. A few often-used types are discussed in the following paragraphs.

A *hot-line clamp*, shown in Fig. 11-20, is recommended for installations where the connections are likely to be made or removed while the electrical system is energized. Primary uses are for apparatus taps, branch-line taps, deadend corner jumpers where a power line turns a corner, and as a sectionalizing device where the installation is simply a place to break the circuit.

A *parallel-groove clamp*, shown in Fig. 11-21, is an easy and inexpensive way to make light-duty connections such as service taps, branch taps, taps to lightning arresters or other equipment, secondary connections of dissimilar conductors or conductors of different sizes, ground connections, and neutral connections. These connectors can be used with a variety of conductor sizes over a wide range.

Two- and three-bolt parallel-groove clamps, shown in Fig. 11-22, are used for heavier conductors where more stability and greater contact and clamping force are required.

U-bolt connectors, shown in Fig. 11-23, are used for major connections where maximum joint efficiency, exceptional thermal stability, and substantial surge capacity are required. Typical installations are substation connections, major power-transformer

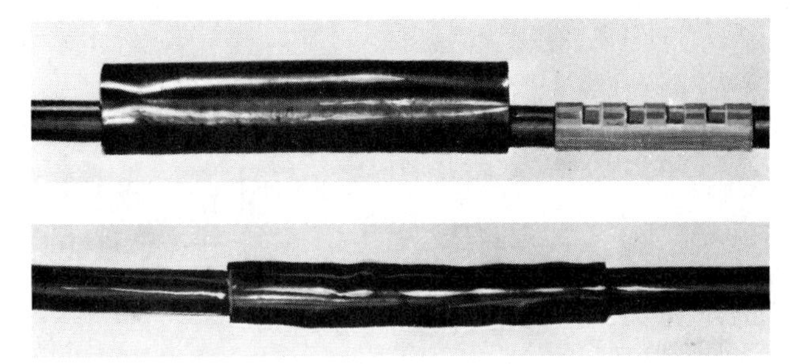

FIG. 11-19 Heat-shrinkable covers. (*Sangamo, Springfield, Ill.*)

FIG. 11-20 Hot-line clamp. (*A. B. Chance Company, Centralia, Mo.*)

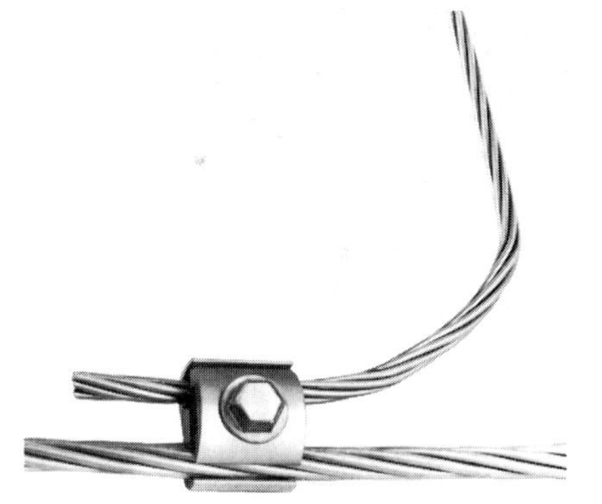

FIG. 11-21 Parallel-groove clamp. (*IIT Blackburn, St. Louis, Mo.*)

FIG. 11-22 Bolted clamp. (*IIT Blackburn, St. Louis, Mo.*)

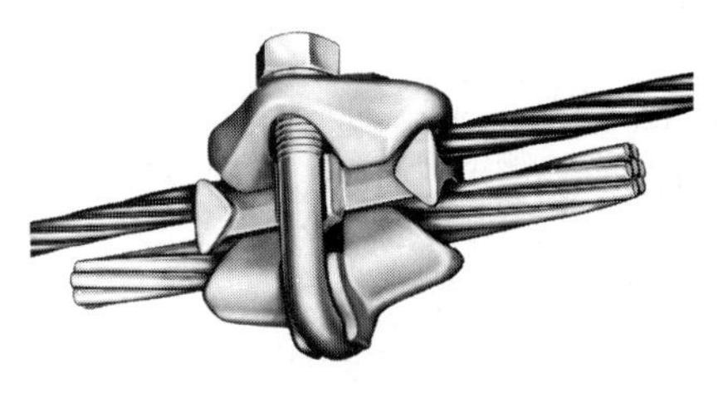

FIG. 11-23 U-bolt clamp. (*IIT Blackburn, St. Louis, Mo.*)

banks, transmission-line connections, and heavy-duty distribution-line connections. These connectors make a tight, low-resistance, and heavy-duty connection.

The *tee* or *cross clamp*, Fig. 11-24, is a multipurpose connector for low-voltage and primary-distribution connections. The design of the clamp makes it easy to tape where location of the connection so requires.

Split-bolt connectors such as shown in Fig. 11-25 have many uses where light-duty connections are required. Some uses are light-duty, high-voltage connections, secondary connections, service connections, and internal wiring connections to lightning arresters and other equipment.

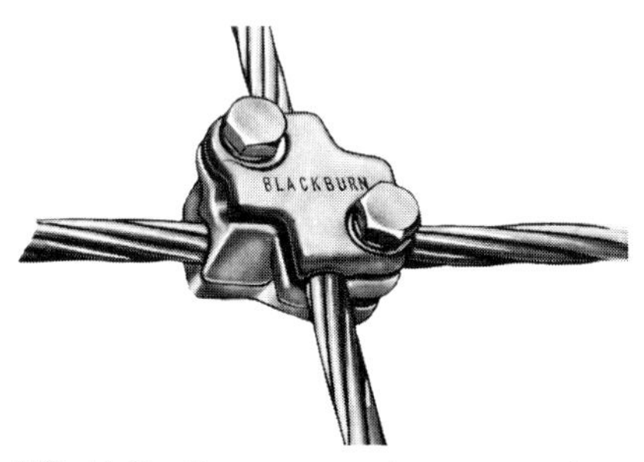

FIG. 11-24 Tee or cross clamp. *(IIT Blackburn, St. Louis, Mo.)*

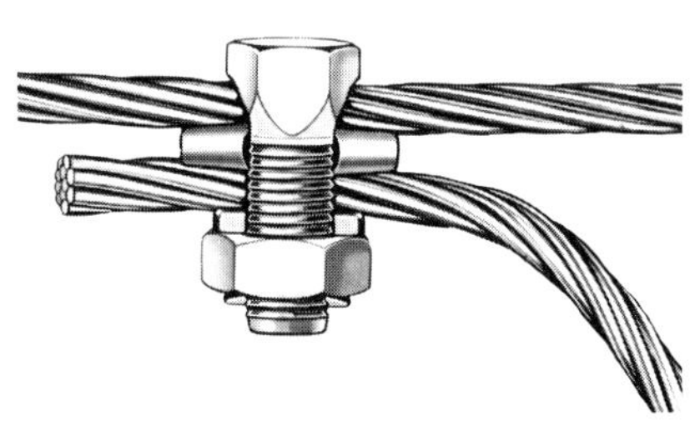

FIG. 11-25 Split-bolt connector. *(IIT Blackburn, St. Louis, Mo.)*

FIG. 11-26 Parallel-groove connector. *(IIT Blackburn, St. Louis, Mo.)*

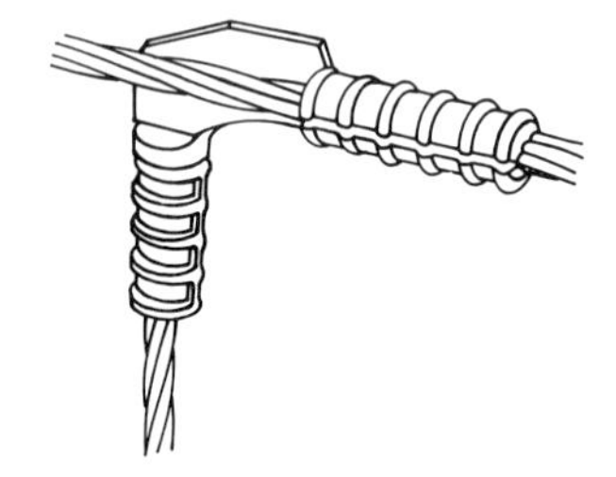

FIG. 11-27 L-tap connector. *(Kearney Div. of Kearney National, Inc., Atlanta, Ga.)*

There is also a variety of compression-type fasteners available for different situations. A few often used are discussed in the following paragraphs.

A *parallel-groove* or H-shaped compression connector which meets all mechanical and electrical requirements for aluminum-to-aluminum and aluminum-to-copper connections is shown in Fig. 11-26. With dissimilar conductor materials, such as copper and aluminum, the copper conductor must be located below the aluminum conductor. If a copper conductor is located above an aluminum conductor, the copper salts will drain down on the aluminum conductor and cause excessive corrosion due to electrolytic action. This compression sleeve has built-in spacers and individual form-fitting grooves for distortionless seating of the conductors and has wraparound-groove jaws for full-circumference contact of each conductor. It has many uses such as in service, branch, and equipment taps.

Compression L *taps*, shown in Fig. 11-27, are primarily for connecting dissimilar metals where a highly contaminated atmospheric condition exists. Long heavy-walled run sections securely grip the line wire. The socket for the tap wire is filled with an inhibitor.

Compression T *taps*, shown in Fig. 11-28, are to accommodate large-size conductors and for use in contaminated atmospheric areas. The tap socket is usually filled with an inhibitor.

T-*tap stirrups,* shown in Fig. 11-29, provide complete protection to aluminum conductors from arc damage when hotline clamps are installed or removed, as discussed earlier in this section. They help reduce costs on large conductors because smaller and less expensive clamps may be used.

Service-entrance sleeves as shown in Fig. 11-30 are used for the splicing of service connections and are not meant to be full-tension devices. The sleeve is made of drilled rod stock and incorporates an integral solid-metal barrier, completely eliminating galvanic corrosion between unlike conductors. The barrier provides an individual

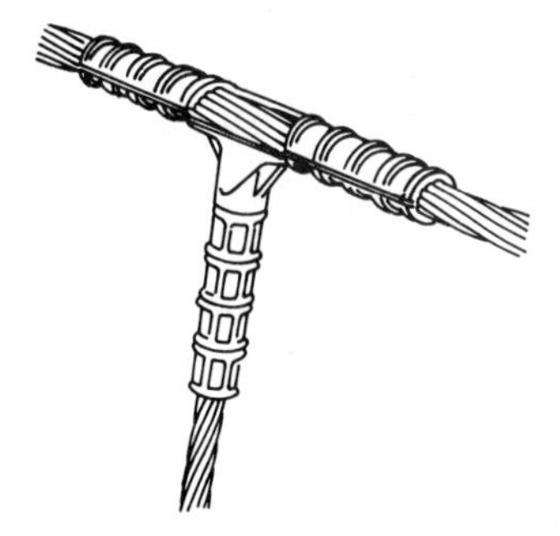

FIG. 11-28 T-tap connector. (*Kearney Dic. of Kearney National, Inc., Atlanta, Ga.*)

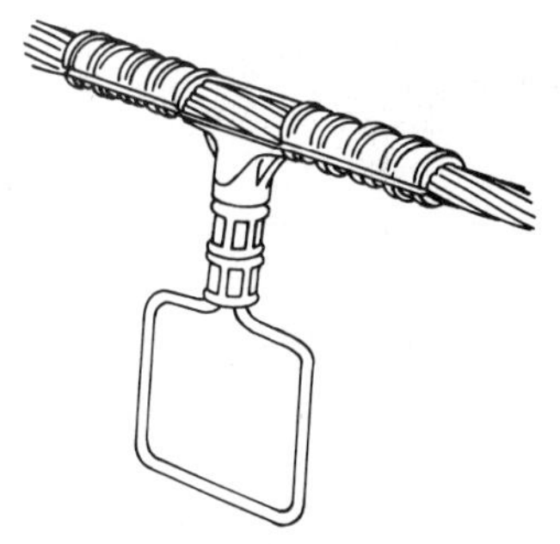

FIG. 11-29 T-tap stirrup. (*Kearney Div. of Kearney National, Inc., Atlanta, Ga.*)

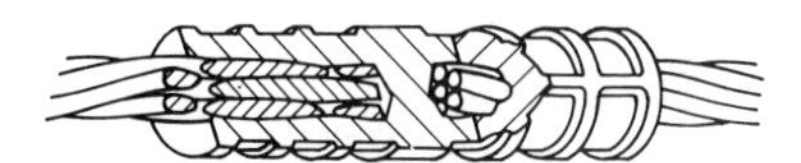

FIG. 11-30 Service-entrance sleeve. (*Kearney Div. of Kearney National, Inc., Atlanta, Ga.*)

socket for each conductor and prevents overinsertion of either one. Preinsulated service sleeves can be used on insulated services to reduce installation time.

Serve-En adapters, shown in Fig. 11-31, are used when an aluminum tap is connected to the bronze or copper terminals of transformers, cutouts, and capacitors and incorporate a 7-inch piece of soft-drawn copper wire which is factory-installed and covered with a highly weather-resistant plastic to prevent galvanic corrosion.

Compression *ground-rod fittings* such as that shown in Fig. 11-32 have permanent, low-resistance, nonloosening, tamperproof connections to ensure good grounding.

Compression *sleeves* such as shown in Fig. 11-33 are available for splicing ACSR

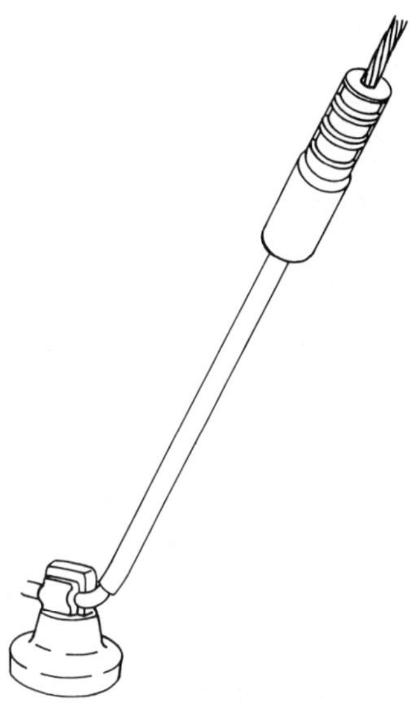

FIG. 11-31 Serve-En adapter. (*Kearney Div. of Kearney National, Inc., Atlanta, Ga.*)

conductors 1/0 and larger, the aluminum sleeves have two holes which are used as sighting holes for centering the steel sleeve and for the application of an inhibitor with a calking gun. One-piece sleeves may be used as full-tension or jumper connectors on small sizes of ACSR, aluminum, copper, copperweld, and steel conductors to save installation time.

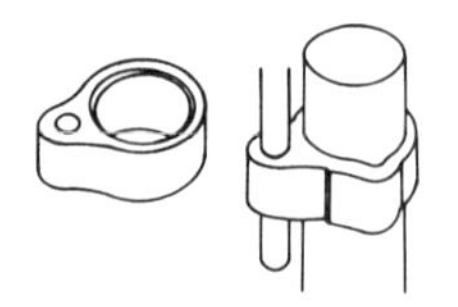

FIG. 11-32 Ground-rod fitting. *(Kearney Div. of Kearney National, Inc., Atlanta, Ga.)*

11-14 Splicing and Terminating High-Voltage Cables

The jointing of high-voltage cable could be defined as the manufacture of a short piece of high-voltage power cable under conditions that are far from perfect by technicians who are, one hopes, skilled and highly trained. Until recently, cable splices and/or cable terminations were fabricated in the field by skilled cable splicers using semiconduction and insulating tapes. Splices and terminations thus made are both expensive and time-consuming. In addition they could not be adequately tested for electrical integrity and corona performance.

With the required increase in use of underground high-voltage cables in the 1960s, a new breed of preformed plug-in type terminators, connectors, and splices appeared. These devices were manufactured and tested under controlled conditions and ensure a degree of installed reliability which was not provided by previously available methods. As a result they have gained outstanding universal acceptance for use with concentric neutral power cables and tape-shielded jacketed power cables. Because of this universal acceptance this section will concentrate on the installation of these devices.

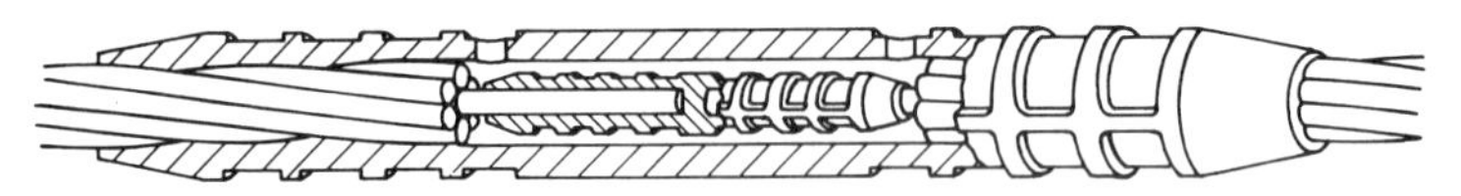

FIG. 11-33 Splicing sleeve. *(Kearney Div. of Kearney National, Inc., Atlanta, Ga.)*

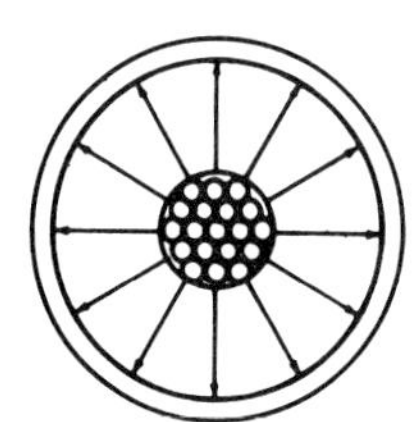

FIG. 11-34 Flux lines. Radial stress on a cross-sectional plane. *(Elastimold® Div., Amerace Corp., Hackettstown, N.J.)*

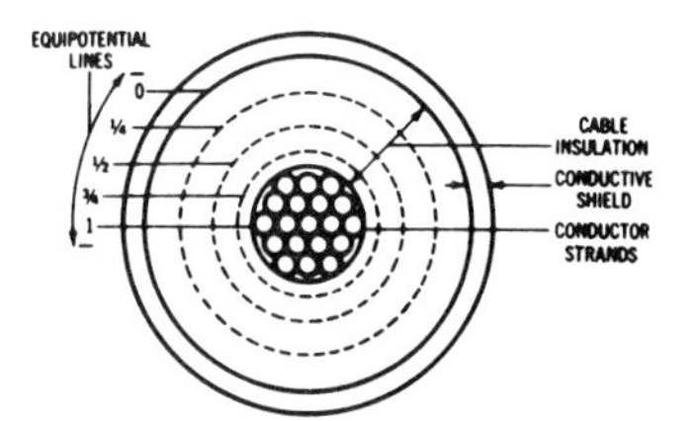

FIG. 11-35 Equipotential lines. Points of equal magnitude at a constant distance from the conductor. *(Elastimold® Div., Amerace Corp., Hackettstown, N.J.)*

These plug-in devices are generally used for high-voltage systems between 2.4 and 34.5 kV. Special training is required for terminating and splicing of high-voltage cables above 34.5 kV and hence they will not be discussed in this section. Consultation with the manufacturer will certainly be required in these cases.

The splicing or terminating of a high-voltage cable requires the fastening of the two metal conductors with the same care and procedures as outlined in sec. 11-7 and/or as depicted in this section.

Because high-voltage stresses are contained in the cables suitable for use with preformed connectors we must be concerned with the proper maintenance of insulation integrity to ensure satisfactory control of these stresses. Therefore, those involved with engineering, installing, operating, and maintaining high-voltage cable installations should be aware of the characteristics of both the cable and the connectors.

Each high-voltage cable consists of a conductor, strand shield, insulating medium, insulating shield, and shield-drain assembly. The design of the cable using these elements is primarily based on the control of the two types of electrical stresses which are radial within the insulation wall and on its surface. These dielectric stresses are dependent on the voltage gradients between the conductors and can be represented as flux lines and equipotential lines. Figure 11-34 shows the flux lines and radial stress on a cross-sectional plane. These stresses diminish in intensity as the lines diverge and approach the ground-electrode outer shield. Maximum stress occurs at the conductor shield.

Figure 11-35 illustrates equipotential lines which are points of equal magnitude at a constant distance from the conductor. The highest stress level is closest to the conductor and zero at the outer surface of the insulation. Stated differently, the basic design of a cable results in a uniform stress on the cable insulation. There is no point of stress concentration on the inner-strand shield.

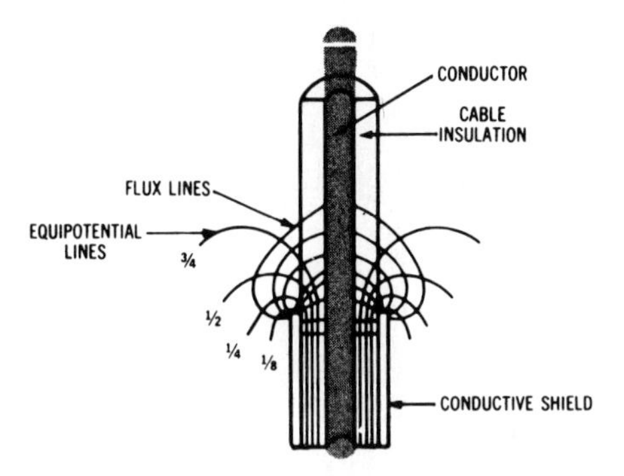

FIG. 11-36 Field stresses. With conductive shield removed. (*Elastimold® Div., Amerace Corp., Hackettstown, N.J.*)

When a cable is terminated, it is necessary to remove the shield from the proximity of the conductor, causing an abrupt change in the dielectric field. The radial and longitudinal stresses are no longer completely controlled by the shield. The result is a concentration of these stresses at the termination of the cable shield. This concentration of stress results in a minimum dielectric strength of the cable system at this point, a situation which could result in failure unless measures are taken to control the stress and reduce it to tolerable limits. Figure 11-36 shows the stresses with the cable shield removed.

To prevent failure, the cable must be terminated in such a manner that uniform stresses on the cable insulation are continued. The most common method of accomplishing this is to gradually increase the wall thickness of the insulation beyond the termination of the cable shield. This insulation buildup is shaped in the form of a cone. The cable shielding is then extended up the surface of the cone and terminated at a point of increased insulation thickness over that of the original cable. Stress concentration is thereby reduced to an acceptable level. This construction is called a stress-relief cone. A stress-relief cone can be constructed with insulating and conducting shielding tapes, with partially fabricated tape pennants, or with factory-molded devices.

The simplest type of molded stress cone is shown in Fig. 11-37. With the shielding firmly bonded to the rubber insulation, the preformed stress-relief cone provides for a corona-free termination. The lower end is the conducting portion which connects the cable shield and flares out to relieve the stress. The insulation is attached to the conductor part to eliminate all air from the high-stress region. To assemble, it is slightly stretched and slipped over the cable. This interference fit provides a watertight seal between the stress cone and cable insulation. These devices are low in cost and easy to install. Most can be used for indoor terminations or in conjunction with a pothead or tape-covered or molded terminator outdoors.

When a cable is terminated in air on a pole or switchgear, it is exposed to contaminants, many of which are corrosive, erosive, and/or conductive. When deposited on the surface of the cable insulation they cause flashover or arcing from contaminants to the closest conductor, resulting in degradation of the cable shield and ultimately a failure of the termination. Therefore, when a cable is terminated in air, a stress relief and extended creep path dependent on the applied voltage are required. Preformed terminators range from stress cones molded of conducting and insulating materials to dielectric-filled porcelain and aluminum housings. The simplest and most easily installed device is the molded stress-cone type shown in Fig. 11-38. Figure 11-39 shows an installation of a porcelain pothead giving the added creep distance required.

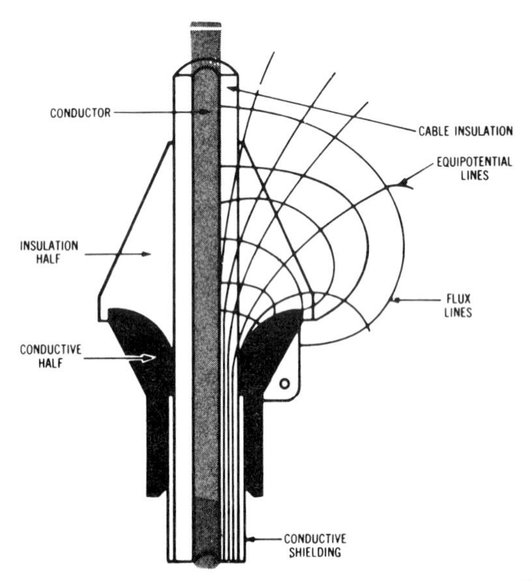

FIG. 11-37 Molded indoor stress cone. *(Elastimold® Div., Amerace Corp., Hackettstown, N.J.)*

Preformed splices, apparatus connectors, and other cable accessories are in essence an extension of the stress-cone design. They in general consist of separable products that are joined by mating surfaces with an interference fit. Each contains stress relief to which external shielding, insulation, and connector shielding have been added.

These products and their design can best be illustrated by the straight splice shown in Fig. 11-40. In a conventional tape splice, abrupt changes in the conductor and insulation configuration are avoided to maintain symmetrical stress distribution and preclude high voltage gradients within the splice. In the preformed splice the design is predetermined and stress control is achieved by the principle referred to as connector shielding. The connector shield is a conductive rubber insert that extends over the connector area and onto the cable insulation. The insert is energized at line potential. The conducting parts are completely surrounded by an internal conductive rubber sleeve which is smooth and symmetrical and eliminates corona caused by trapped air. In effect, it removes electrical stress from all air in the system and confines this electric field to the insulating material between the two conductive portions. Thus, excellent electrical strength and corona characteristics can be achieved. The inside diameter of the connector housing is made slightly less than the diameter of the cable which is applied to effect an interference fit. When the connector is pushed into place, the elas-

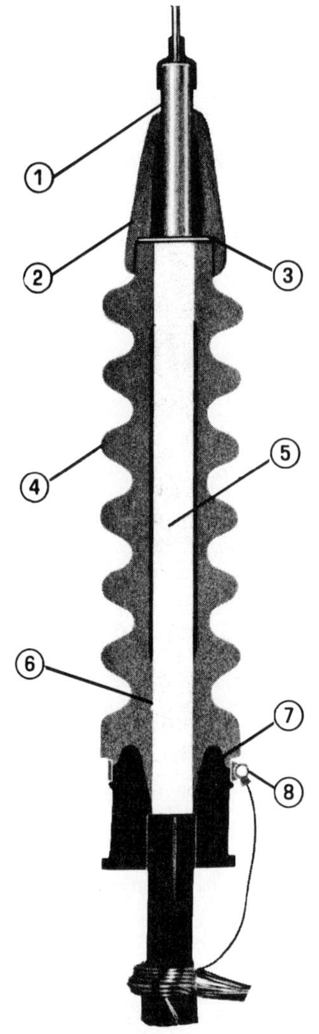

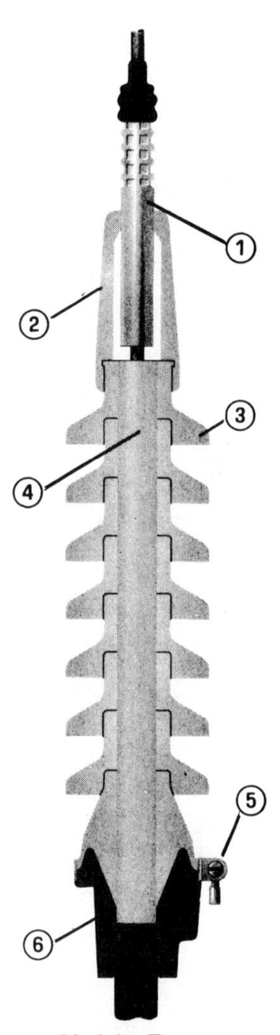

Single Piece Type

1. Terminal Assemblies
2. Molded Rubber Cap
3. Retaining Washer
4. Non-Tracking Rubber
5. Cable Insulation
6. Designed Interference Fits
7. Stress Relief
8. Ground Clamp

Modular Type

1. Two Metal Contacts
2. Rubber Cap
3. Non-Tracking Rubber Modules
4. Cable Insulation
5. Ground Clamp
6. Stress Relief

FIG. 11-38 Molded outdoor stress cone. *(Elastimold® Div., Amerace Corp., Hackettstown, N.J.)*

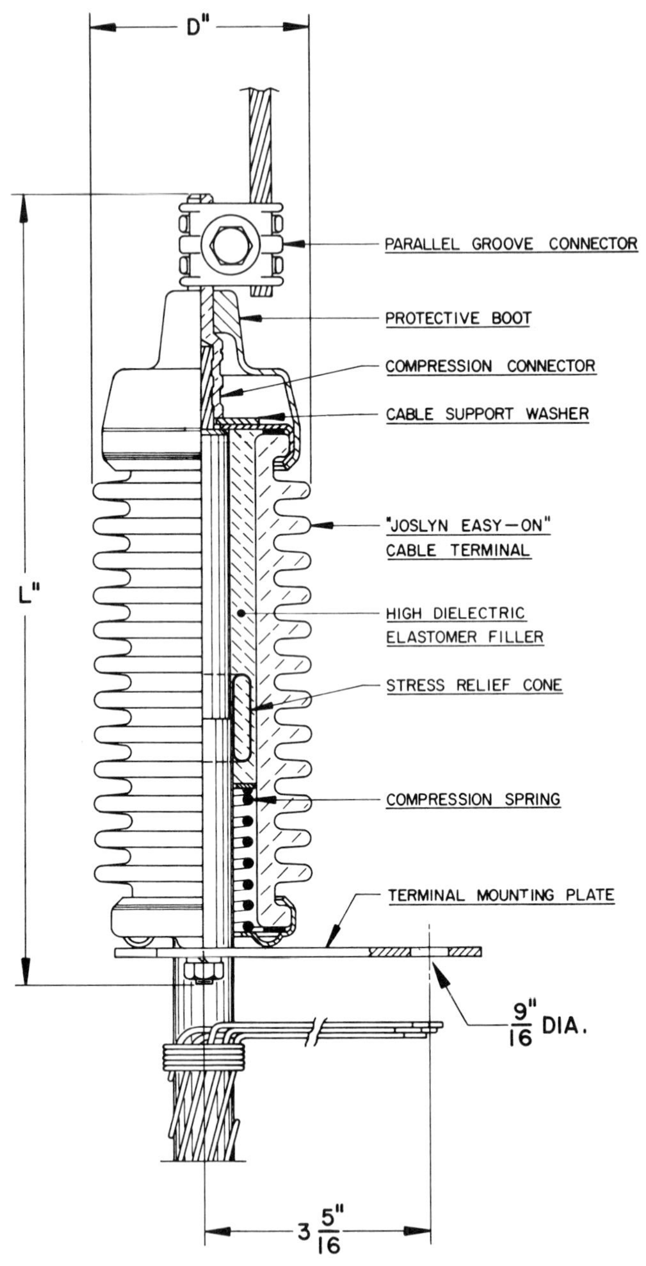

FIG. 11-39 Porcelain outdoor stress cone. (*Joslyn Manufacturing & Supply Co., Chicago, Ill.*)

tomer is expanded and exerts a uniform 360-degree pressure on the cable and housing and significantly increases the electrical strength of the creep path between the conductor and the external shield. This pressure is maintained even through thermal expansion and contraction cycles of the cable. This type of design reduces the required creep-path length considerably as compared with taping. In some designs taping or epoxy-filling of the stress-relief area is required.

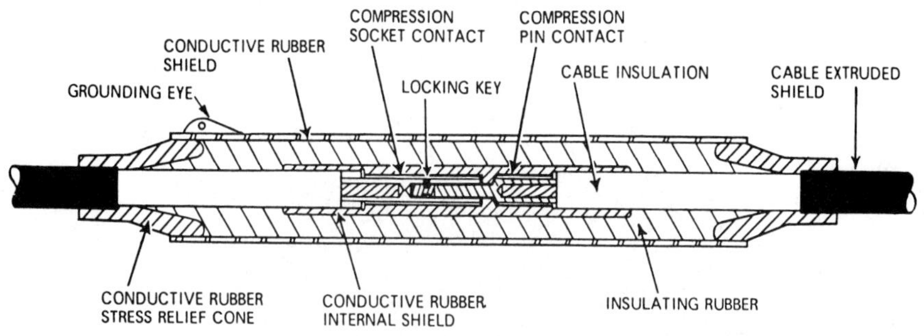

FIG. 11-40 Molded splice. (*Elastimold® Div., Amerace Corp., Hackettstown, N.J.*)

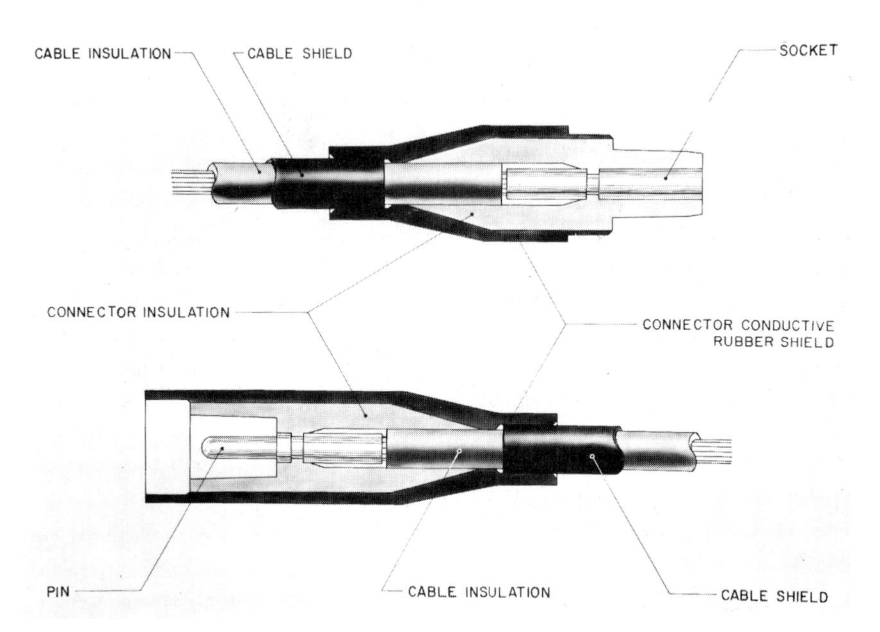

FIG. 11-41 Plug-in receptacle. Disconnected line splice. (*Elastimold® Div., Amerace Corp., Hackettstown, N.J.*)

A disconnectable plug-in receptable, shown in Fig. 11-41, is very similar to the straight splice. The mating pin and socket connectors are first crimped on the end of the cables to be joined. The conducting jacket is then removed from the end of each cable and the rubber plug or receptacle is replaced over the cable and slid back until a locking ring at the connector falls into a slot in the insulated fitting. The conducting-rubber portion of the device is sealed against the cable jacket and the insulating material sealed to the cable insulation. The insulated and shielded plug-in receptable parts are then joined using the before-mentioned interference-fit principle. Note the conducting rubber molded to the inner surface of the insulation which surrounds the conductor and eliminates any potential gradient across the air spaces within the assembly. This principle can also be used for T or Y receptacles.

The elbow connector and mated apparatus shown in Figs. 11-42 and 11-43 is in essence an extension of the basic plug and receptacle splice concept. The bushing-plug

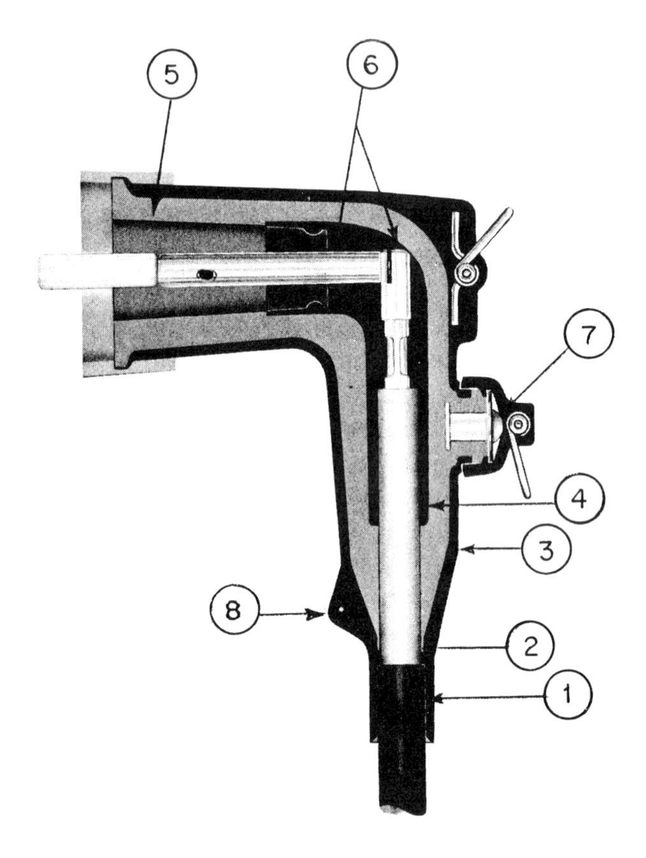

1. Creep path integrity at the cable-elbow interface.
2. Stress relief.
3. External shield.
4. Internal shield.
5. Creep path integrity at the bushing elbow interface.
6. Electrical circuit.
7. Voltage test point.
8. Grounding point.

FIG. 11-42 Elbow connector. *(Elastimold® Div., Amerace Corp., Hackettstown, N.J.)*

conductive shield has been both terminated for stress relief and flanged for assembly in the transformer tank. This combination provides low-cost, completely shielded, disconnectable, and submersible transformer connections and/or deadfront, pad-mount cable terminations.

This combination of elbow connector and bushing can be used for sectionalizing high-voltage systems and, in addition, has deadfront-transformer applications. Figure 11-44 shows a multiple bushing with a common bus that provides a line, load, and tap position at a sectionalizing point in the system. These devices are available with up to four points of connection.

With the increasing concern for safety of both the public and operators of high-voltage systems, the elbow connector and bushings are becoming popular for connecting all kinds of equipment and devices to underground power systems. Numerous

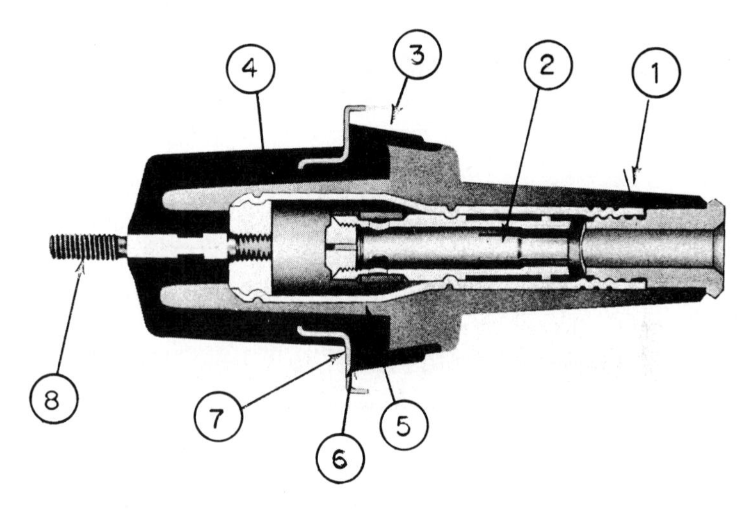

BUSHING
1. Creep path integrity.
2. Electrical circuit.
3. Creep path integrity.
4. External shield.

BUSHING WELL "FLOWER POT"
5 Creep path integrity.
6 Conductive rubber ring.
7 Stainless steel flange and hold-down brackets.
8 Electrical circuit.

FIG. 11-43 Bushing and bushing well. (*Elastimold® Div., Amerace Corp., Hackettstown, N.J.*)

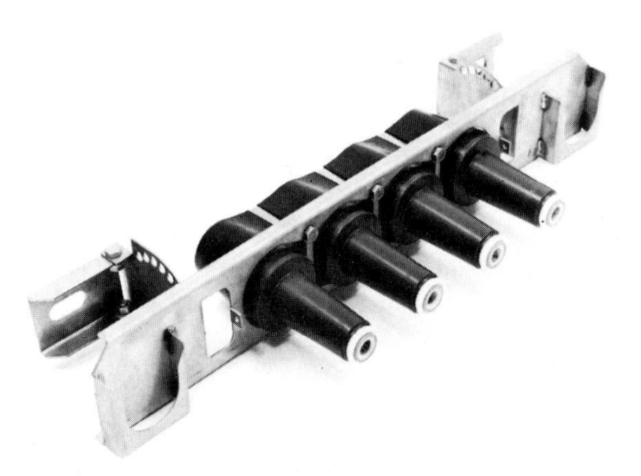

FIG. 11-44 Molded junction. (*Elastimold® Div., Amerace Corp., Hackettstown, N.J.*)

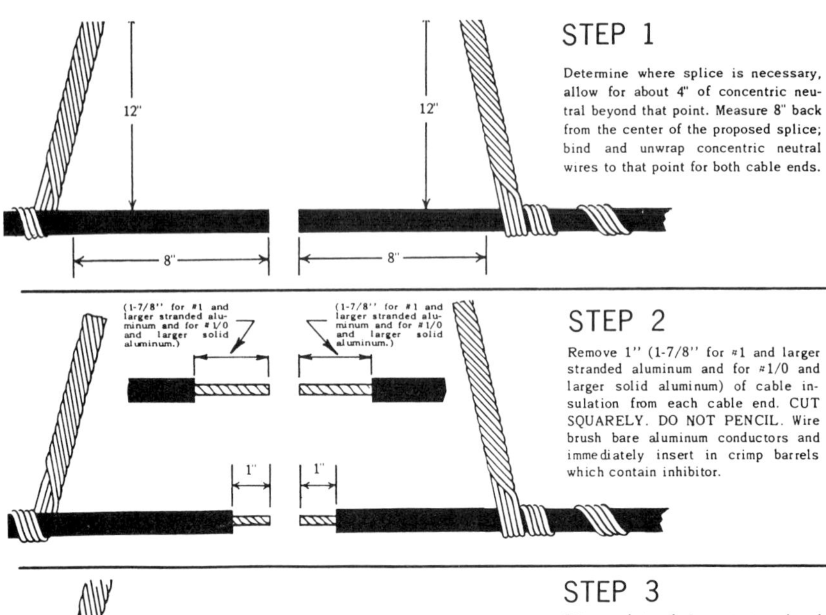

STEP 1

Determine where splice is necessary, allow for about 4" of concentric neutral beyond that point. Measure 8" back from the center of the proposed splice; bind and unwrap concentric neutral wires to that point for both cable ends.

STEP 2

Remove 1" (1-7/8" for #1 and larger stranded aluminum and for #1/0 and larger solid aluminum) of cable insulation from each cable end. CUT SQUARELY. DO NOT PENCIL. Wire brush bare aluminum conductors and immediately insert in crimp barrels which contain inhibitor.

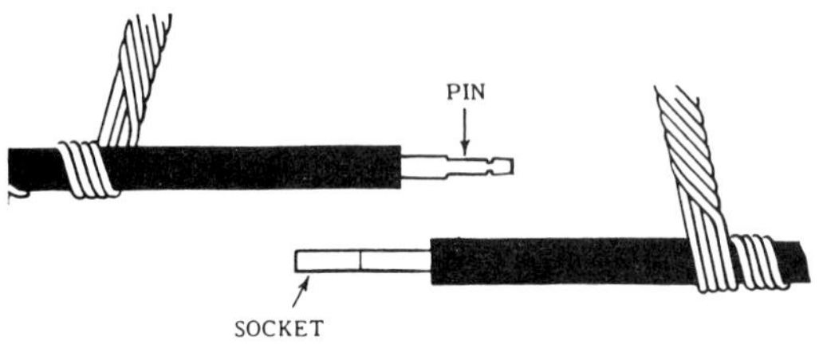

STEP 3

Crimp socket and pin contacts on bared conductors, being careful to hold the base of the crimp barrel against the cable insulation. See crimp chart packed with contact. Start the crimps 1" from the end of the crimp barrels (1-7/8" for #1 stranded or #1/0 solid or larger aluminum cable) and rotate each successive crimp or indent 90°. When using aluminum cable, carefully wipe excess inhibitor from contacts and cable.

SOCKET SIDE

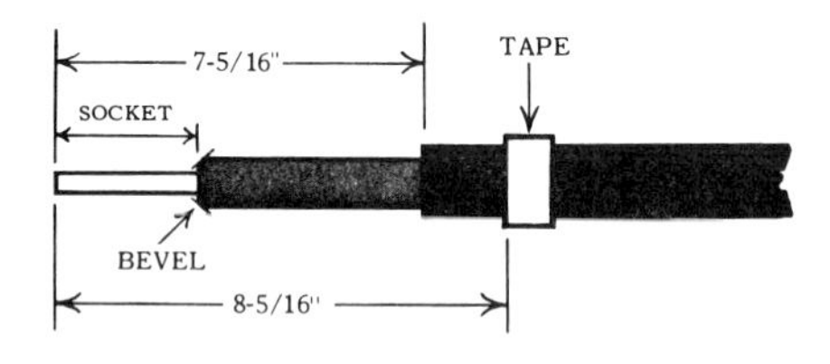

STEP 4

SOCKET SIDE Of Cable End – cut semi-con shield 7-5/16" back from the front end of the socket with a smooth, straight, squared cut. DO NOT CUT OR NICK THE INSULATION. Remove the semi-con shield to this point. Bevel the insulation no more than 1/8" back at 45°. At a point 8-5/16" back from the far end of the socket, put on a turn of electrical tape as an indicator on the cable shield.

PIN SIDE

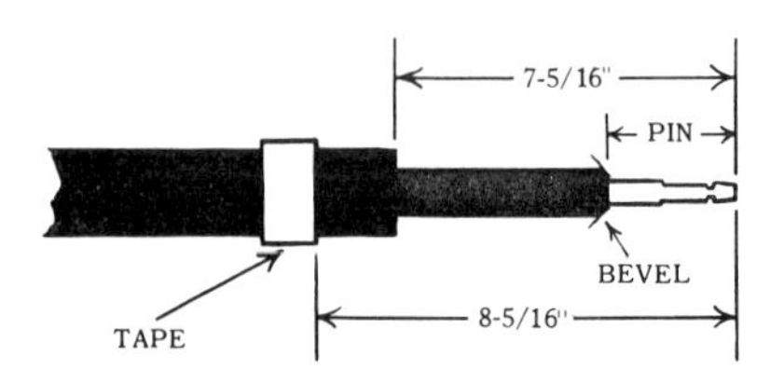

STEP 5

PIN SIDE Of Cable End – cut semi-con shield 7-5/16" back from the front end of the pin with a straight, smooth, "squared" cut. DO NOT CUT OR NICK THE INSULATION. CAUTION: KEEP INSULATION CLEAN. Remove the semi-con shield to this point. Bevel the insulation no more than 1/8" back at 45°. At a point 8-5/16" back from the front end of the pin, put on a turn of electrical tape as an indicator on the cable shield.

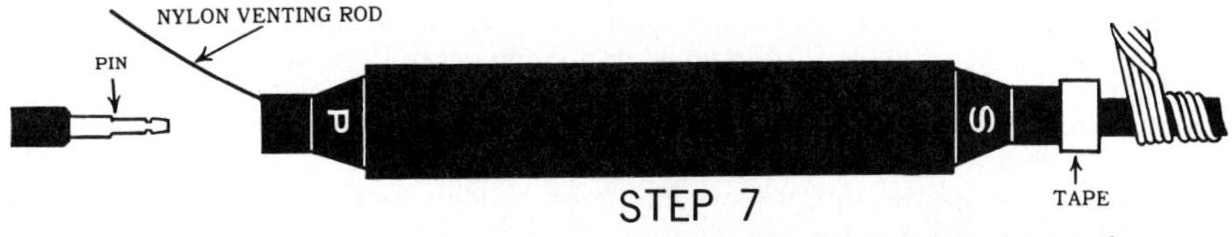

STEP 6

Lubricate the inside of the housing with lubricating grease supplied. (Use silicone grease. Do Not Substitute.)

STEP 7

Examine the splice housing and note that one side is engraved with a "P" (for pin contact), and the other side with an "S" (for socket contact). One man holds the socket end of the cable while the other man slips the "S" side of the housing over the socket end. Twist the housing and push socket end into housing until the tape indicator lines up with the end of the housing. Next insert the nylon venting rod apptoximately half way into "P" side of splice housing.

STEP 8

Slip the pin end of the cable into the "P" side of the housing. While both men are pushing the cable ends together, twist or rotate the housing until the tape indicator on the cable shield is approximately 1" away from the splice housing. Remove the venting rod from the housing. Continue to push the pin end into the housing until the tape indicator lines up with the end of the splice housing. (This indicator may be as far as 1/8" from the splice housing depending upon cable insulation size. The housing shrinks slightly as cables are installed.) Check splice by pulling both ends of cable. Key located in socket will prevent splice from pulling apart, if properly connected. Remove tape marker.

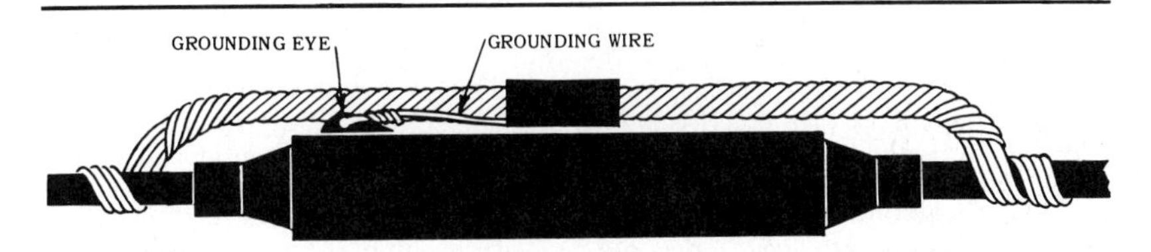

STEP 9

A. Twist the concentric neutral wires together and train them over the splice housing.

B. Insert one end of a short piece of wire equivalent to #14 copper through the grounding eye and twist to make a small loop taking care not to damage the eye.

C. Crimp the neutral wires and the wire used for the grounding eye connection together with an appropriate connector.

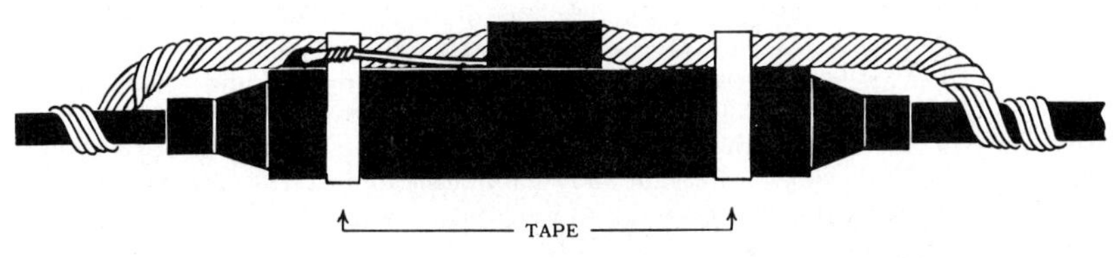

STEP 10

Tape concentric neutral wires to housing. Splice is complete.

FIG. 11-45 Splice installation. (*Elastimold® Div., Amerace Corp., Hackettstown, N.J.*)

manufacturers now have splice and termination kits available with detailed installation instructions. Figure 11-45 shows one manufacturer's instructions for making a straight splice. Included in the kit are the splice housing, pin contact, socket contact, lubricant-wiping cloth, and a nylon venting rod. The installer should use the instructions provided with the particular splice or termination kit.

Because tape splices are presently the only available method for splicing very high-voltage cables, we will discuss them briefly. The premolded type terminators and splices are available up to 34.5 kV and work is in progress to extend their use to 138 kV when shielded cable becomes economically available for higher voltages.

The design and construction of tape joints approximates as closely as possible the design and construction of shielded power cable. The insulation and shielding of a tape joint is designed to avoid high voltage gradients within the joint. High radial stresses are found at the connector and conductor shield diameters. Longitudinal stresses occur between the conductor shield in the area where the cable insulation has been removed, across the exposed cable insulation where the insulation shield has been removed, and the termination of the insulation shield. Tangential stresses occur at the shoulder of the connector shield, the base of the cable-insulation pencil, and at the base of the insulation-belt taper. These stresses are controlled by proper design and shaping of the joint.

Shielding offers an effective means of distributing stresses through the insulation wall between the line and ground electrodes in the joint. Abrupt changes in electrode configuration result in high voltage gradients. To avoid this, a connector with tapered rather than square shoulders is used to joint the conductors. Conducting shielding tape is used to replace the factory-applied conductor shield and to provide as smooth as possible a transition between the conductor and connector profiles. Shielding is also applied evenly and gradually from the termination of the cable insulation shield, up the belt taper, and across the splice. The configuration of this joint-insulation shield is determined by the shape of the joint insulation.

In order to minimize voltage gradients in the joints, abrupt changes in insulation sections are avoided. To accomplish this the cable insulation is gradually tapered or penciled where the insulation joins the exposed conductor. The hand-applied tape insulation is formed to provide stress-relief cones at both cable-insulation-shield terminals. Mathematical calculations determine that these stress-relief cones should have curved slopes of the log-log type. This degree of accuracy is not essential at voltages below 69 kV. In practice these slopes are straight lines gradually rising to a point that is mathematically determined to reduce average voltage gradients to the desired level. Although the straight-line slope is expedient, it does not provide the high degree of stress control that the logarithmic slope provides at this critical area.

Longitudinal stresses exist along the axis of the joint. The magnitude of these stresses may be determined by successive calculations through the insulation at various sections. The difference in potential along the surface between two points divided by the distance between the points will give the average gradient between the points. Determination and selection of desired longitudinal gradient levels determines the overall length of the joint. Laminated insulations have a higher ratio of radial-to-longitudinal dielectric strength as compared with fusing materials, and thus the type of material being used must also be considered. Stresses along the boundary layers of insulation materials are also affected by the dielectric constants of these materials.

Maximum radial stress occurs at the diameter of the conductor and connector shield. Minimum radial stress occurs at the insulation shield. The average stress is the mathematical average between the two points. Determination and selection of desired levels of radial stress determine the outside diameter of the splice. Longitudinal stress determines the length of cable-insulation pencil and the connector shoulder. Proper stress distribution in the joint should preclude ionization and corona discharge within the joint over the operating life of the joint.

The properties of the insulating materials should match as closely as possible the properties of the cable insulation. Shielding materials should have conductance that is sufficient to handle charging and leakage currents. Provision should also be made to prevent damage to the joint by fault current.

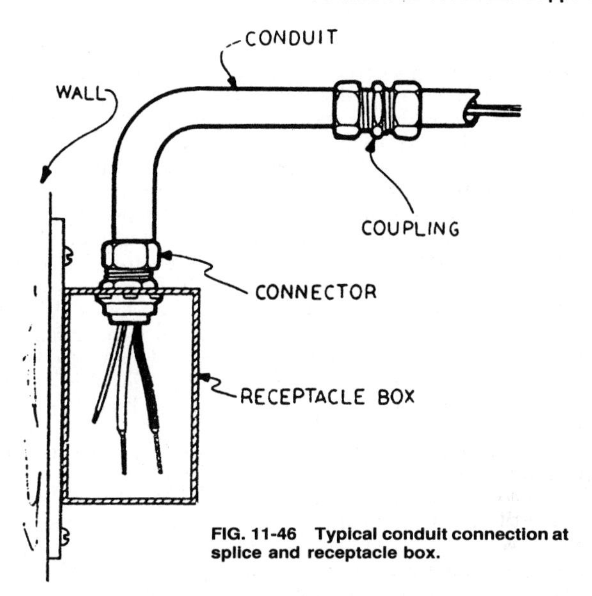

FIG. 11-46 Typical conduit connection at splice and receptacle box.

11-15 Conduit Connection and Support Sizes, uses, and types of material vary considerably for electrical conduit. Therefore, conduit connections are extensive; many of which are described elsewhere in this handbook under different headings. Heavy duty piping often utilize standard plumbing fittings, while PVC (plastic) conduits generally employ solvent cement to attach their fittings at the joints. Thin wall metal conduit use compression type fittings and heavy wall conduit thread their connections. Refer to Fig. 11-46 for an illustration of a typical conduit connection at a splice and a mounting at a standard receptacle box.

Support of electrical conduit is extremely important and strictly mandated by various regulations and the NEC. Conduit in all sizes is required by NEC to be properly supported within 3 feet of each box, fitting, or cabinet as well as at a minimum of 10 feet intervals on horizontal runs. The reader is advised to consult proper codes for site specific requirements.

Some of the most common fasteners used to support conduit are shown in Fig. 11-47 for the reader's convenience.

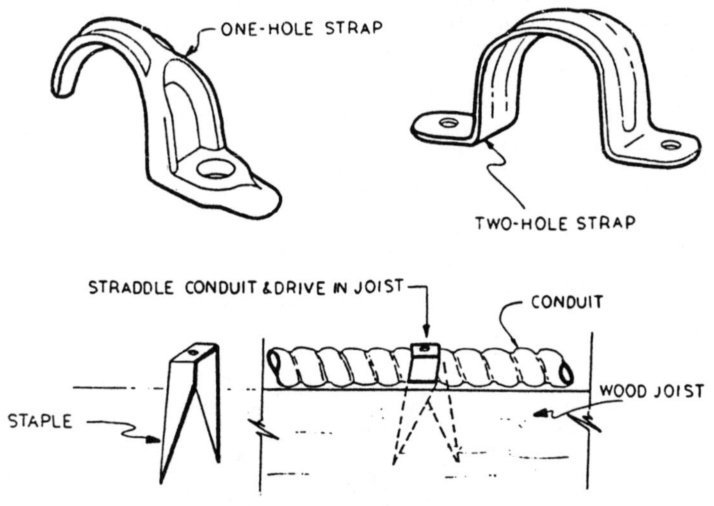

FIG. 11-47 Supporting fasteners for conduit.

ELECTRONIC

11-16 Electronic Connections Because of the high frequencies and low voltages and currents associated with electronic circuits, the joining or terminating of electrical conductors becomes very important with respect to minimizing contact resistance. The means and considerations for joining and fastening conductors for electronic uses will be only briefly mentioned here. It will be left to the reader interested in this area to further search the literature.

Two basic methods for joining and terminating conductors for electronic use are soldering and wrapping joints. Solder will, of course, minimize contact resistance, provided that a cold joint is avoided. There has been a long history of solder joints and terminations and they now are made by machine and other mass-production techniques.

The wrapped joint is produced by wrapping wire under tension around a sharp-cornered terminal. The notches created in the wire by the terminal corners effectively prevent relaxation from the as-wrapped geometry. This kind of connection was first used in the telecommunications field but has gained acceptance in many other fields of the electronic industry.

There are separable connectors using pin-and-socket-type or clip connections for electronic circuits and research is substantial in this area.

The making and breaking of electrical contacts in electronic applications is an extensive field in itself involving relays, switches, and connectors. Most of these make and break operations use varying degrees of mechanical sliding or wiping of electric contacts to ensure reliable electrical performance. Such wiping action not only breaks up any oxide or contaminant films that may have formed on a contact but also effectively dislodges foreign particles that may have adhered to the contact surface. This wiping and resulting mechanical friction and wear accompanying the sliding action will reduce the useful life of the contact. The useful life of the contact is sometimes extended through the application of suitable lubricants.

REFERENCES

Skroski, J. C., "The New Breed Preformed Slip-On Cable Accessories," Elastimold Div., Amerace Corp.

Fay, Edgar, "Basic Design, Theory and Construction of High Voltage Cable Joints," Elastimold Div., Amerace Corp.

ACKNOWLEDGMENTS

The author wishes to thank the following persons and their companies for their contributions:
L. E. Peterson and William Scott, Ideal Industries, Inc., Sycamore, Illinois
J. Fitzhugh, Elastimold Div., Amerace Corp., Hackettstown, New Jersey
R. Shackman and M. R. Monaskkin, Burndy Corp., Norwalk, Connecticut
John Thornton, Kearney Div. of Kearney National, Inc., Atlanta, Georgia

The author wishes to thank the following companies for use of materials and pictures:
Underwriters Laboratories, Inc., Chicago, Illinois
A. B. Chance Co., Centralia, Missouri
IIT Blackburn, St. Louis, Missouri
Fargo Manufacturing Co., Poughkeepsie, New York

Section 12
Adhesive Bonding

GERALD L. SCHNEBERGER, Ph.D., P.E.
President, Training Resources, Inc.
Flint, Michigan

12-1 Introduction Adhesives are defined as substances capable of holding materials together by surface attachment. The earliest adhesives predate history. Then as now adhesives were problem-solving materials. The number of specific adhesive formulations has increased rapidly in recent years, because of the expanding demands of the joining industry and the increasing sophistication of synthetic polymer chemistry. Figure 12-1 illustrates several new uses for adhesives.

A number of theories have been proposed over the years to explain the nature of adhesive attraction. These include theories based on diffusion, chemical reaction, mechanical interlocking, and electrical attraction. The most widely accepted explanation for adhesion is based upon the attraction of unlike electrical charges. This concept holds that the electrons of any material are unevenly distributed across its surface. Thus electron-rich (electrically negative) surface areas exist intermixed with electron-poor (electrically positive) areas. When any two surfaces are brought together, the electrically positive and negative regions attract one another and adhesion results. This explanation is supported by the fact that adhesion occurs if virtually any two surfaces are brought into intimate contact.

The practical significance of this theory of adhesion is simply that intimate surface contact is essential for a strong bond. Many attractive forces decrease as the seventh power of the distance between their charge centers. Thus many of the surface preparation techniques used with adhesives are designed to promote intimate contact of the adhesive with the surfaces.

12-2 The Pros and Cons The primary advantages of adhesives are:

1. The entire bond surface area contributes to the strength of the joint. Often an adhesive of lower bond strength per unit area can be used because there is sufficient area involved to bear the entire applied load.

Carpet Seaming

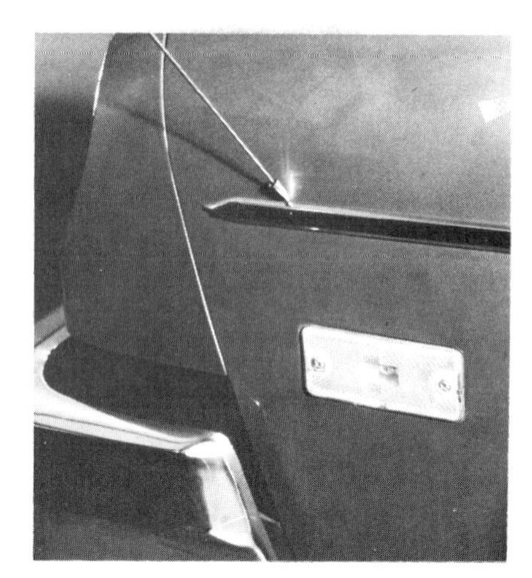

Automotive Trim Attachment

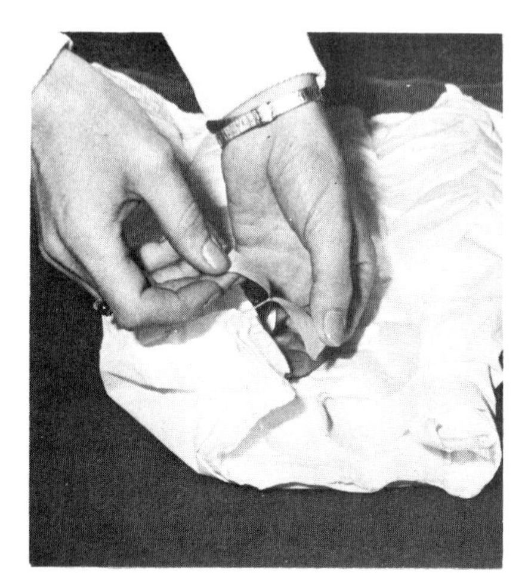

Adhesively Secured Diapers

Bonded Beverage Can Seams

FIG. 12-1 Recent innovations in the use of adhesives.

2. Stress concentrations are minimized. Since the entire bond area in a well-designed joint is utilized, localized stress concentrations are avoided around screws, rivets, welds, and bolts.

3. Vibration damping is possible. Most adhesives are inherently less rigid than metals. Thus vibration is damped out rather than transmitted across bond lines. The excellent vibration-fatigue resistance of adhesively assembled helicopter rotor blades illustrates this behavior.

4. With very few exceptions adhesives excel at joining dissimilar materials. Glass-to-rubber bonds are probably more easily achieved with adhesives than with any other assembly technique.

5. Smooth contours are possible. Bolts, rivet heads, or spot welds may increase wind resistance or detract from appearance. Adhesives have reduced this problem in the airframe and automotive industries.

6. Adhesives may be used to electrically insulate joined materials from one another or to seal the joint against air or liquid leaks.

7. Adhesives often result in weight savings. Adhesively bonded truck tractors and trailers illustrate this point.

8. Ease of assembly is a frequent advantage in adhesive bonding. The construction industry finds that gluing down subfloors is much more rapid than nailing. Figure 12-2 illustrates the use of adhesively attached telephone equipment.

9. Cost reductions may be possible when adhesives are substituted for other joining methods. This is often because inexpensive adhesives may be used, drilling or forming may be eliminated, or assembly operations may be automated.

FIG. 12-2 Adhesive attachment of telephone equipment speeds installation.

Since there is no perfect joining method, it is well to be aware of some of the limitations of adhesive bonding. These may be summarized as follows:

1. There is no universal adhesive. If a variety of materials are being joined for a number of different service conditions, several adhesives may have to be purchased, stored, tested, mixed, applied, and cured.

2. A trade-off between high tensile strength and good impact resistance may be necessary. Adhesives rarely excel at both; the strong ones may be brittle while the resilient ones may creep.

3. Fixturing of substrates may be required during cure. The time involved may be from seconds to hours.

4. Service temperature limits of adhesives may preclude their use. Most adhesives lose strength above 350°F (177°C). Adhesives useful above 400°F (204°C) are available but their cost is high and their versatility may be limited.

5. Allergenic reactions may be a problem with some of the curing agents used for high-strength adhesives.

6. Shelf life may make close control of inventory and usage important. Some adhesives must be used within six months unless refrigerated.

7. Surface preparation of the parts to be bonded is often more critical than for other joining methods.

8. Some adhesives use solvents which are toxic or flammable. Thus ventilation and fire-extinguishing systems may be required.

12-3 Types of Adhesives Adhesives may be classified according to their function, chemical structure, or method of application or on the basis of the substrates joined. Functional categories include structural, holding, and sealing adhesives. Structural adhesives are those which, by design, endure mechanical loading. Thiokol automobile-rear-window adhesives exemplify this group. Holding adhesives are those which bear little structural load but do form unstressed joints between parts. Household appliance trim and many types of labels typically use holding adhesives. Sealing adhesives are those used to exclude gases or liquids from a joint. Calking compounds are typical examples. In many applications an adhesive may perform two or even all three of the structural, holding, and sealing functions.

The chemical structure of an adhesive resin is often used as a basis for classification. Thermosetting adhesives are those which cure to form an insoluble, infusible film. This process is known as crosslinking. Thermoplastic adhesives are those whose molecules remain chemically separate; i.e., they do not form chemical bonds between molecules. These adhesives yield films which are generally heat-softenable and soluble in the proper solvent. A more specific type of chemical structure classification is based upon the repetitive structure of the adhesive polymer. Thus we have epoxies, polyamides, polyurethanes, and polyacrylates, to name only a few.

The method of application or cure may also be used to classify adhesives. Solvent cements are those adhesives in which the resin is applied in solution and allowed to dry to a tacky state before the bond is formed. Hot melts are applied in their molten state. They cure, i.e., form a bond, by cooling to a temperature below their softening point. Two-part adhesives have two separate components which must be mixed prior to use; one-part systems are just that—they are used as supplied. Room-temperature-curing adhesives do not require heating, while hot-curing systems do. Tape and film adhesives are supplied in sheet form and are simply laid in place between adherends. Pressure-sensitive adhesives are supplied as tapes or sheets and form bonds immediately upon contact with the substrate. Table 12-1 lists the commonly available forms of a number of important adhesives.

TABLE 12-1 Commonly Used Forms of Various Adhesives*

	Film	Hot melt	Liquid	Paste	Solid
Epoxies	X		X	X	
Nitrile–phenolic	X				X
Urea–formaldehyde	X				X
Epoxy–phenolic	X		X		
Polyester			X		
Silicone				X	X
Polyvinyl acetate		X			X
Polyamide	X	X			X
Polyurethane			X	X	
Polyethylene	X	X			
Polyacrylate			X		X
Alpha-cyanoacrylate			X		
Polysulfide			X		
Neoprene			X		X
Butadiene–styrene		X	X		X

* Suppliers should be consulted for more detailed information.

Still another method of classifying adhesives is based on the nature or properties of the substrates joined. Thus we have metal-to-metal, metal-to-plastic, plastic-to-glass, and similar adhesive types. Similarly, adhesives may be described as heat stable or as rigid, flexible, or elastomeric.

Often the classification systems just described are combined. This leads to adhesive categories such as structural two-part, flexible plastic-to-plastic, or elastomeric pressure-sensitive. Section 12-4 gives a summary of typical properties associated with many common types of adhesives.

12-4 Choosing an Adhesive The choice of an adhesive for a particular application is often complicated by the fact that the people who buy and use adhesives are not those who formulate and sell them. This division of expertise requires clear and complete communication between adhesive supplier and user. It also requires an understanding of how adhesives function.

Service requirements, production limitations and cost considerations are the three major issues which must be resolved before an adhesive can be selected for mass production use.

Experience has shown that the following recommendations are helpful in avoiding problems and minimizing costs in choosing and using adhesives.

1. If you are working through a vendor, give as much information as possible about how your product is made, what materials are used, and how the product is expected to perform. Avoid general statements like "metal-to-plastic adhesive"; it is much more useful to say "1010 steel to high-density polyethylene." The thermal history of your materials prior to bonding may be especially important in helping your supplier recommend the proper surface preparation.

2. Do not overdesign. Considerable money is wasted if joints are specified which are stronger or longer-lasting than necessary.

3. Remember that heat-curing adhesives can be used only if the temperature tolerance of the parts is above the cure temperature.

4. Interpret bond-strength values found in sales literature with the knowledge that they pertain to specific adherends prepared in a specific manner. If your substrates, surface preparation, or cure conditions are different, the bond strengths may be markedly affected. Literature bond strengths are often reported on aluminum specimens because the values are usually high and aluminum is easy to work with. The same adhesive used on steel or plastic may give quite different results. You must check this out in your plant with your parts.

5. Strength-retention values with aging, obtained from commercial literature, are often for unstressed specimens. If joints are aged under load, particularly cyclic load, their strength-retention behavior may be grossly affected. You should determine strength retention behavior on joints made in your plant under conditions which simulate the service environment of your assemblies.

6. Remember that production-line bonds usually show less consistent strength than bonds made in your process-engineering lab. Process-development people usually have the time to be more careful about surface preparation, pressure, cure time, and bond-line thickness than is possible in production. Do not choose an adhesive for production which is only marginally acceptable in a pilot operation.

7. Be aware that some adhesives contain chemicals which may affect some people physically. Good ventilation and minimum skin contact are often essential.

8. One-part adhesives are preferred over two-part formulations. Mixing, metering, and dispensing operations allow operator and equipment errors to occur.

9. Tape and film adhesives are preferred over liquid and paste systems because mixing, handling, outgassing, and shrinkage problems are reduced. The result is higher, more uniform bond strengths.

10. Be aware that differences in thermal expansion properties of the adherends can produce severe stresses at the bond line when the joint undergoes temperature changes.

11. It is unrealistic to expect an adhesive bond to have high tensile strength and high peel strength. One or the other, or a compromise, is the rule. Problems will be minimized if the joint is designed with this trade-off in mind.

12. Remember that the most common cause of poor joint performance is failure to properly clean the surface. The second most common cause is similar—failure to keep the surface clean.

13. An adhesive which is more expensive per pound may be cheaper to use because less material is needed for each bond or because curing requires less time, space, or energy.

14. Time spent on training your work force to properly prepare, assemble, and cure bonds is well worth the cost and effort. This is especially true if you are converting from some other assembly method to adhesive bonding. The natural human tendency to resist change can make such a conversion extremely difficult. A good training program can minimize these problems.

15. Adhesives with the least critical surface cleanliness requirements are preferred. If possible avoid the use of dip- or spray-cleaning operations where solutions are recycled. If economics dictates the use of these processes, be aware that they are subject to gradual or sudden contamination.

Table 12-2 is a partial list of adhesive types commonly used to bond various substrates. Table 12-3 gives typical cure conditions and descriptive comments about each of the adhesives listed in Table 12-2. These tables are not definitive. Adhesive suppliers should be consulted for additional information about specific products or bonding techniques. Table 12-4 gives additional information about the cure of common epoxy adhesives.

12-5 Joint Design Adhesive bonds undergo a variety of stresses. These are usually a combination of shear, tension, compression, cleavage, or peel, as shown in Fig. 12-3.

Of these shear, tension, and compression stresses are the most easily endured. This is because the entire adhesive-bond area can aid in withstanding the stress. Cleavage and peel forces are much less desirable because the load is concentrated at the end of the bond and the adhesive fails gradually from that point.

The origin of the stresses discussed above is usually mechanical but it may also be thermal. When bonded materials of different thermal expansion coefficients undergo a

TABLE 12-2 Adhesives Commonly Used for Joining Various Materials*

Material	Adhesive	Table 12-3 reference
ABS	Polyester	*a*
	Epoxy	*e*
	Alpha cyanoacrylate	*c*
	Nitrile-phenolic	*b*
	Acrylics	*w*
Aluminum and its alloys	Epoxy	*e*
	Epoxy-phenolic	*d*
	Nylon-epoxies	*f*
	Polyurethane rubber	*g*
	Polyesters	*a*
	Alpha-cyanoacrylate	*c*
	Polyamides	*h*
	Polyvinyl-phenolic	*i*
	Neoprene-phenolic	*b*
	Acrylics	*w*
Brick	Epoxy	*e*
	Epoxy phenolic	*d*
	Polyesters	*a*
Ceramics	Epoxy	*e*
	Cellulose esters	*j*
	Vinyl chloride-vinyl acetate	*k*
	Polyvinyl butyral	*l*
	Acrylics	*w*
Chromium	Epoxy	*e*
Concrete	Polyester	*a*
	Epoxy	*e*
Copper and its alloys	Polyesters	*a*
	Epoxy	*e*
	Alpha-cyanoacrylate	*c*
	Polyamide	*h*
	Polyvinyl-phenolic	*i*
	Polyhydroxyether	*m*
	Acrylics	*w*
Fluorocarbons	Epoxy	*e*
	Nitrile-phenolic	*b*
	Silicone	*t*
Glass	Epoxy	*e*
	Epoxy-phenolic	*d*
	Alpha-cyanoacrylate	*c*
	Cellulose esters	*j*
	Vinyl chloride-vinyl acetate	*k*
	Polyvinyl butyral	*l*
	Acrylics	*w*
Lead	Epoxy	*e*
	Vinyl chloride-vinyl acetate	*k*
	Polyesters	*a*
Leather	Vinyl chloride-vinyl acetate	*k*
	Polyvinyl butyral	*l*
	Polyhydroxyether	*m*
	Polyvinyl acetate	*n*
	Flexible adhesives	*g*
Magnesium	Polyesters	*a*
	Epoxy	*e*
	Polyamide	*h*
	Polyvinyl-phenolic	*i*
	Neoprene-phenolic	*b*
	Nylon-epoxy	*f*
	Acrylics	*w*
Nickel	Epoxy	*e*
	Neoprene	*g*
	Polyhydroxyether	*m*

TABLE 12-2 Adhesives Commonly Used for Joining Various Materials* (*Continued*)

Material	Adhesive	Table 12-3 reference
Paper	Animal glue	*o*
	Starch glue	*p*
	Urea-, melamine-, resorcinol-, and phenol-formaldehyde	*q*
	Epoxy	
	Polyesters	*a*
	Cellulose esters	*j*
	Vinyl chloride-vinyl acetate	*k*
	Polyvinyl butyral	*l*
	Polyvinyl acetate	*n*
	Polyamide	*h*
	Flexible adhesives	*g*
Phenolic and melamine	Epoxy	*e*
	Alpha-cyanoacrylate	*c*
	Flexible adhesives	*g*
Polyamide	Epoxy	*e*
	Flexible adhesives	*g*
	Phenol- and resorcinol-formaldehyde	*q*
	Polyester	*a*
Polycarbonate	Polyesters	*a*
	Epoxy	*e*
	Alpha-cyanoacrylate	*c*
	Polyurethane rubber	*g*
	Acrylics	*w*
Polyester, glass-reinforced	Polyester	*a*
	Epoxy	*e*
	Polyacrylates	*r*
	Nitrile-phenolic	*b*
Polyethylene	Polyester, isocyanate-modified	*a*
	Butadiene-acrylonitrile	*g*
	Nitrile-phenolic	*b*
Polyformaldehyde	Polyester, isocyanate-modified	*a*
	Butadiene-acrylonitrile	*g*
	Nitrile-phenolic	*b*
Polymethylmethacrylate	Epoxy	*e*
	Alpha-cyanoacrylate	*c*
	Polyester	*a*
	Nitrile-phenolic	*b*
	Acrylics	*w*
Polypropylene	Polyester, isocyanate-modified	*a*
	Nitrile phenolic	*b*
	Butadiene-acrylonitrile	*g*
Polystyrene	Vinyl chloride-vinyl acetate	*k*
	Polyesters	*a*
Polyvinyl chloride, flexible	Butadiene-acrylonitrile	*g*
	Polyurethane rubber	*g*
Polyvinyl chloride, rigid	Polyesters	*a*
	Epoxy	*e*
	Polyurethane	*g*
Rubber, butadiene-styrene	Epoxy	*e*
	Butadiene-acrylonitrile	*g*
	Urethane rubber	*g*
Rubber, natural	Epoxy	*e*
	Flexible adhesives	*h*
Rubber, neoprene	Epoxy	*e*
	Flexible adhesives	*h*
Rubber, silicone	Silicone	*t*

TABLE 12-2 Adhesives Commonly Used for Joining Various Materials* (*Continued*)

Material	Adhesive	Table 12-3 reference
Rubber, urethane	Flexible adhesives	*h*
	Silicone	*t*
	Alpha-cyanoacrylate	*c*
Silver	Epoxy	*e*
	Neoprene	*g*
	Polyhydroxyether	*m*
Steel	Epoxy	*e*
	Polyesters	*a*
	Polyvinyl butyral	*l*
	Alpha-cyanoacrylate	*c*
	Polyamides	*h*
	Polyvinyl-phenolic	*i*
	Nitrile-phenolic	*b*
	Neoprene-phenolic	*b*
	Nylon-epoxy	*d*
	Acrylics	*w*
Stone	See brick	
Tin	Epoxy	*e*
Wood	Animal glue	*o*
	Polyvinyl acetate	*n*
	Ethylene-vinyl acetate	*u*
	Urea-, melamine-, resorcinol-, and phenol-formaldehyde	*q*

*Compiled from various sources. Adhesive suppliers should be consulted for additional information.

temperature change, the more rapidly expanding (or contracting) material tries to slide over the other material. Thus a shear stress is generated.

It is also possible for tensile loads to generate peel forces. When the lap joint shown in Fig. 12-4 is stressed in tension, there is a tendency for the stress axes to become colinear. This requires that the bonded members deform as shown. The result is the appearance of peel forces at the bond ends.

In order for a joint to be properly designed, especially a load-bearing one, the magnitude and direction of the stresses which the joint will experience must be known. Often a welded or mechanically fastened joint will be completely unsuitable for adhesive bonding and must be redesigned. Simple butt joints, for example, are rarely suitable for bonding, although they may be completely appropriate for welding.

A number of commonly used joint designs are shown in Fig. 12-5. Some are less practical than others because of machining requirements or because they may allow excessive cleavage or peel forces.

Corner joints may be subject to peel if the members are thin, or to cleavage if the members are thick, as shown in Fig. 12-6. Designs which minimize these stresses are shown in Fig. 12-7. Standard woodworking joints, such as dovetails or mortise-and-tendon designs, may also be used if preparation costs are not prohibitive. When a thin and thick member form a corner, the designs shown in Fig. 12-8 are possible.

Another approach to overcoming peel and cleavage stresses is to combine adhesive bonding with more traditional fastening methods. Rivets may be used at the ends of a lap joint, for example. The adhesive greatly increases the fatigue life of the joint compared with that for rivets alone. The rivets help secure the ends of the bond against cleavage or peel failure and may replace clamps and fixtures while the adhesive cures.

Weld bonding uses welds to hold metal parts together while a previously applied adhesive is cured. The technique protects bond edges against cleavage and peel failure and can often eliminate clamps and fixtures during the curing process.

In addition to joint geometry, the designer may have some control over adherend

TABLE 12-3 Adhesive Commentary

Adhesive type	Table ref.	Comments	Typical cure conditions
Polyesters and their variations	*a*	Used primarily for repairing fiber-glass-reinforced polyester resins, ABS, and concrete. Generally unsaturated esters are polymerized with a catalyst such as methyl ethyl ketone (MEK) peroxide and an accelerator such as cobalt naphthenate. A coreactant-solvent such as styrene may be present. Bonds are strong. Sometimes combined with polyisocyanates to control shrinkage stresses and reduce brittleness. Unreacted monomer, if present, keeps viscosity low for application, provides good wetting, enhances crosslinking. Occasionally used on metals.	Minutes to hours at room temperature
Nitrile-phenolic, neoprene-phenolic	*b*	These adhesives are a blend of flexible nitrile or neoprene rubber with phenolic novolac resin. They combine the impact resistance of the rubber with the strength of the crosslinked phenolic. They are inexpensive and produce strong, durable bonds which resist water, salt spray, and other corrosive media well. They are the workhorses of the adhesive-tape industry although they do require high-pressure, relatively long high-temperature cures. They are used for metals and some plastics including ABS, polyethylene, and polypropylene. Airframe components and automotive brakes are typical examples.	Up to 12 hours at 250–300°F (120–150°C)
Alpha-cyanoacrylate	*c*	These low-viscosity liquids polymerize or cure rapidly in the presence of moisture or many metal oxides. Thus most surfaces can be bonded. The bonds are fairly strong but somewhat brittle. Used widely for the assembly of jewelry and electronic components.	0.5-5 minutes at room temperature
Epoxy-phenolic	*d*	A combination of epoxy resin with a resol phenolic. Noted for strength retention at 300–500°F (150–250°C), strong bonds, and good moisture resistance. Normally stored refrigerated. Used for some metals, glass, and phenolic resins.	1 hour, 350°F (175°C)
Epoxy; amine, amide, and anhydride-cured	*e*	As a class epoxies are noted for high tensile and low peel strengths. They are crosslinked and in general have good high-temperature strength, resistance to moisture, and little tendency to react with acids, bases, salts, or solvents. There are important exceptions to these generalizations, however, which are often the result of the curing agent used. Primary amines give faster-setting adhesives which are less flexible and less moisture resistant than is the case when polyamide curing agents are used. Anhydride cured epoxies generally have good high-temperature strength but are subject to hydrolysis, especially in the presence of acids or bases. Other important features of epoxies are their low shrinkage upon cure, their compatibility with a variety of fillers, their long life when properly applied, and their easy modification with other resins. Crosslink density is easily varied with epoxies, thus some control over brittleness, vapor permeation, and heat deflection is possible. These resins are widely used to bond metal, ceramics, and rigid plastics (not polyolefins).	See Table 12-4

TABLE 12-3 Adhesive Commentary (*Continued*)

Adhesive type	Table ref.	Comments	Typical cure conditions
Nylon-epoxy	*f*	Tensile shear strengths above 6000 psi (41.4 MPa) and peel strength above 100 lb/in (18 kg/cm) are possible when epoxy resins are modified with special low-melting nylons. These gains, however, are accompanied by loss of strength upon exposure to moist air, a tendency to creep under load, and poor low-temperature impact behavior. A phenolic primer may increase bond life and moisture resistance. Used primarily for aluminum, magnesium, and steel.	1 hour, 300–350°F (150–175°C)
Flexible adhesives: natural rubber, butadiene-acrylonitrile, neoprene, polyurethane, polyacrylates, silicones	*g*	These adhesives are flexible. Thus their load-bearing ability is limited. They have excellent impact and moisture resistance. They are easily tackified and are used as pressure-sensitive tapes or as contact cements. Urethane and silicone adhesives are lightly crosslinked, which gives them reasonable hot strength. They are also compatible with many surfaces but are somewhat costly and must be protected against moisture before use. They have good low-temperature tensile, shear, and impact strength. The urethanes are two-part products which require mixing before use. Silicones cure in the presence of atmospheric moisture. (See entries *r* and *t*.)	Pressure-sensitive tape or solvent cements. Low-temperature bake for urethane. Ambient cure for silicones.
Polyamides	*h*	These adhesives, which are chemically similar to nylon resins, have good strength at ambient temperatures and are fairly tough. They are available in a variety of molecular weights, softening ranges, and melt viscosities. Often applied as hot melts, they have good adhesion to a variety of surfaces. The higher-molecular-weight varieties often have the best tensile properties. Lower-molecular-weight polyamides may be applied in solution.	Hot melt—cures by cooling
Polyvinyl-phenolic	*i*	These resins, which combine a resol phenolic resin with polyvinyl formal or polyvinyl butyral, were the first important synthetic structural adhesives. A considerable range of compositions is available with hot strength and tensile properties increasing at the expense of impact and peel strength as the phenolic content rises. The durability of vinyl–phenolics is generally excellent. They are often selected for low-cost applications where heat and pressure curing can be used.	1 hour, 300°F (150°C)
Cellulose esters	*j*	Cellulose ester adhesives are usually high-viscosity, inexpensive, rigid materials. They do not have high strength and are sensitive to heat and many solvents. Normally used for holding small parts or repairing wood, cardboard, or plastic items. Model airplane cement is a common example.	Air dry

Vinyl chloride–vinyl acetate	*k*	This is a combination of two resins which are sometimes used alone. They may be used as hot melts or as solution adhesives. Since thin films of vinyl chloride–vinyl acetate are somewhat flexible, they are often used for bonding metal foil, paper, and leather. A range of compositions is available with a corresponding variety of properties.	Cooling (hot melt) or solvent loss
Polyvinyl butyral	*l*	A tough, transparent resin which is used as a hot-melt or heat-cured solution adhesive. It has good adhesion to glass, wood, metal, and textiles. It is flexible and can be modified with other resins or additives to give a range of properties. Not generally used as a structural adhesive, although structural phenolics sometimes incorporate polyvinyl butyral to give better impact resistance.	Cooling (hot melt), heating under pressure
Polyhydroxyether	*m*	These are resins based on hydroxylated polyethylene oxide polymers. Generally used as hot melts, they have only moderate strength, but are flexible and have fairly good adhesion.	Hot melt—cures by cooling
Polyvinyl acetate	*n*	This adhesive is generally supplied as a water emulsion (white glue) or used as a hot melt. It dries quickly and forms a strong bond. It is flexible and has low resistance to heat and moisture. Porous substrates are required when the resin is used as an emulsion.	Hot melt—cure by cooling; emulsion—air dry
Animal glue	*o*	Chemically, animal glues are proteins; they are polar water-soluble polymers with high affinity for paper, wood, and leather surfaces. They easily form strong bonds but have poor resistance to moisture. They are being replaced in many areas by synthetic resin adhesives but their low cost is often an important advantage. They are usually applied as highly viscous liquids.	Air dry under pressure
Starch glue	*p*	These products, based on corn starch, have high affinity for paper but are used for little else. They are moisture sensitive and are applied as water dispersions.	Low-temperature dry
Urea-formaldehyde Melamine-formaldehyde Resorcinol-formaldehyde, Phenol-formaldehyde	*q*	These thermosetting resins are widely used for wood bonding. Urea-formaldehyde is inexpensive but has low moisture resistance. It can be cured at room temperature if a catalyst is used. Melamine-formaldehyde resins have better moisture resistance but must be heat-cured. Phenol-formaldehyde adhesives form strong, waterproof wood-to-wood bonds. The resorcinol-formaldehyde resin will cure at room temperature while phenol-formaldehyde requires heating. These resins are often combined resulting in an adhesive with intermediate processing or performance characteristics.	Up to 300°F (149°C) and 200 psi (1.38 MPa)
Polyacrylate esters	*r*	These resins are n-alkyl esters of acrylic acid. They have good flexibility and find frequent use for high-quality pressure-sensitive tapes and foams. They are not suitable for structural applications because of their poor heat resistance and their cold-flow behavior. Frequently used on flexible substrates.	Pressure sensitive

TABLE 12-3 Adhesive Commentary (*Continued*)

Adhesive type	Table ref.	Comments	Typical cure conditions
Polysulfides	*s*	These resins have good moisture resistance and can range from thermoplastic to thermosetting, depending on the degree of crosslinking which is developed during cure. They are two- or three-part systems, the third part being a catalyst. Ventilation is generally required. They make excellent adhesive sealants for wood, metal, concrete, and glass. Polysulfide resins may be combined with epoxies to flexibilize the latter.	Low pressures, moderate temperature
Silicones (see also flexible adhesives)	*t*	These expensive adhesives have high peel strength and excellent property retention at high and low temperatures. They resist all except the most corrosive environments and will adhere to nearly everything. They are usually formulated to react with atmospheric moisture and form lightly crosslinked films.	Low pressure, room temperature
Ethylene–vinyl acetate	*u*	This copolymer is widely used as a hot-melt adhesive because it is inexpensive, adheres to most surfaces, and is available in a range of melting points. It is widely used for bookbinding and packaging.	Hot melt—cures by cooling
Urethanes, rigid	*v*	Rigid urethanes are highly crosslinked. While somewhat expensive they adhere well to most materials, especially plastics, and have good impact strength. Structural urethanes are two-part systems and have good low-temperature strength retention.	Low pressures, up to 300°F (149°C)
Acrylics	*w*	Acrylics are versatile structural adhesives which are becoming increasingly popular. They cure at room temperature and are applied as a conventional two-part system or by coating one substrate with the resin and the other with the catalyst. Impact resistance is controllable since acrylics may vary from rigid to flexible. Floor ventilation is often recommended because they may release heavier-than-air monomer vapors which are odorous.	Room temperature or up to 130°F (54°C); 10–20 minutes with only fixturing pressure

TABLE 12-4 Epoxy-Adhesive Curing Conditions

Adhesive type	Time*	Temperature*	Pot life*
Two-part, room-temp. curable			
Slow	8–24 hr	Room	2–4 hours
Fast	2–4 hr	Room	0.5 hour
Very fast	2–5 min	Room	1–2 minutes
Two-part, heat-cured			
Slow	3–5 hr	300°F (146°C) days	
Fast	0.5–2 hr	250°F (224°C) hr	
One-part, heat-cured			
Slow	0.5–1 hr	350°F (175°C)	
Fast	5 min	300°F (150°C)	

* Time and temperature values given are approximate. They may vary considerably depending on the curing agent and accelerator, if any.

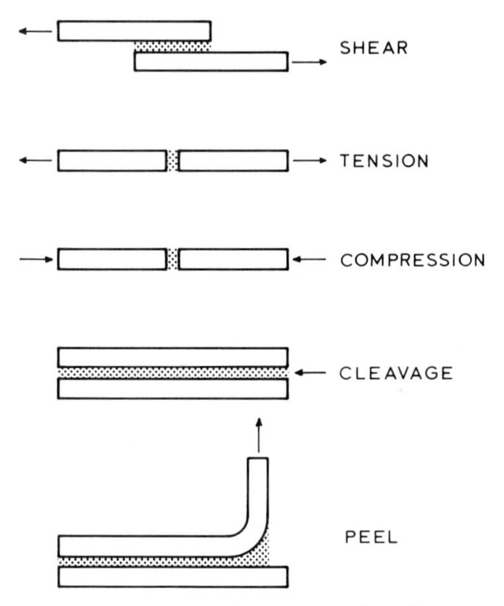

FIG. 12-3 Joints undergoing shear, tension, compression, cleavage, and peel forces.

width, thickness, or overlap. The influence of these factors on lap shear strength is generally as shown in Fig. 12-9. The practical significance of these relationships is simply that a stronger lap bond results when width, overlap, or adherend thickness is increased. With thin or low-modulus adherends, it is easily possible for the bond strength to exceed the yield strength of the material. Joints having such strength are undesirable since they use more adhesive than is necessary.

Large internal stresses at the adhesive-adherend interface may result when materials of differing elastic moduli are bonded. This is because of their differing responses to an applied load. These stresses can usually be equalized by tapering the adherends in a double scarf joint in proportion to their moduli.

12-6 Surface Preparation A clean, dry surface which resists physical disintegration is essential for routine high-quality bonding operations. Most surfaces are far from smooth on an atomic or molecular level. Their "hills and valleys" are usually contaminated with water, metal oxides, adsorbed gases, processing lubricants, and perhaps loosely held products of a reaction between the surface and its surroundings. These materials must be removed or at least reduced in quantity if strong, durable bonds are to be achieved.

Surfaces may be prepared for bonding by abrasion, solution-cleaning, or conversion techniques. Often a combination of methods is used. Abrasive techniques include wet and dry blasting, sanding, and brushing. They are often used when objects are large or when heavy layers of rust or other coatings must be removed. Solution cleaning methods include hot alkaline washing, solvent wiping, and vapor degreasing. They

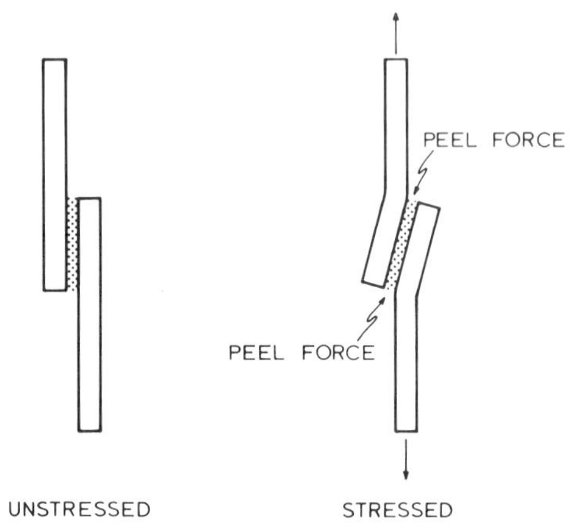

FIG. 12-4 Peel forces may be generated at the bond ends when lap joints are stressed in tension.

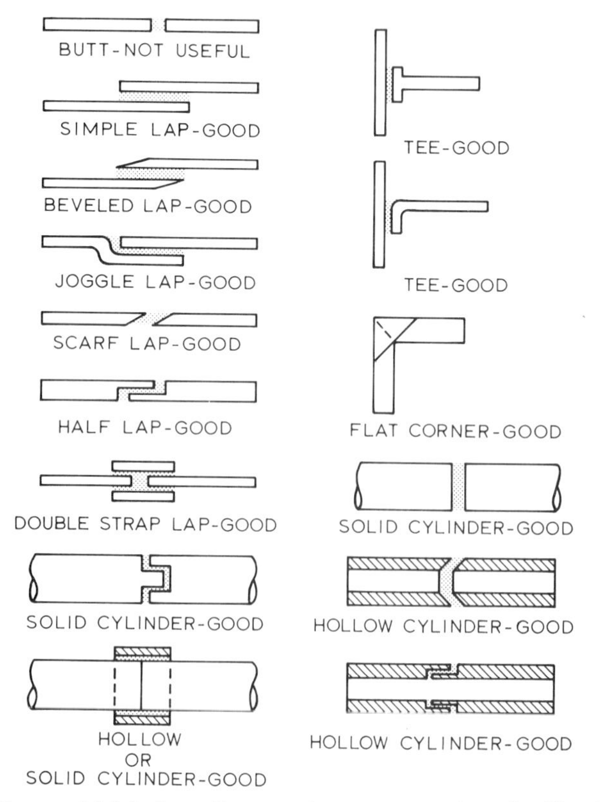

FIG. 12-5 Common joint designs. Some require more preparation or handling than others.

are economical for high production and are most effective for loosely held contaminants. Vigilance is required to prevent contamination of wipe rags or solution tanks. Conversion techniques are generally controlled, severe chemical treatments in which the contaminant is removed and the base material is converted to a chemically different structure. Conversion processes include anodizing, phosphating, and chemical cross-

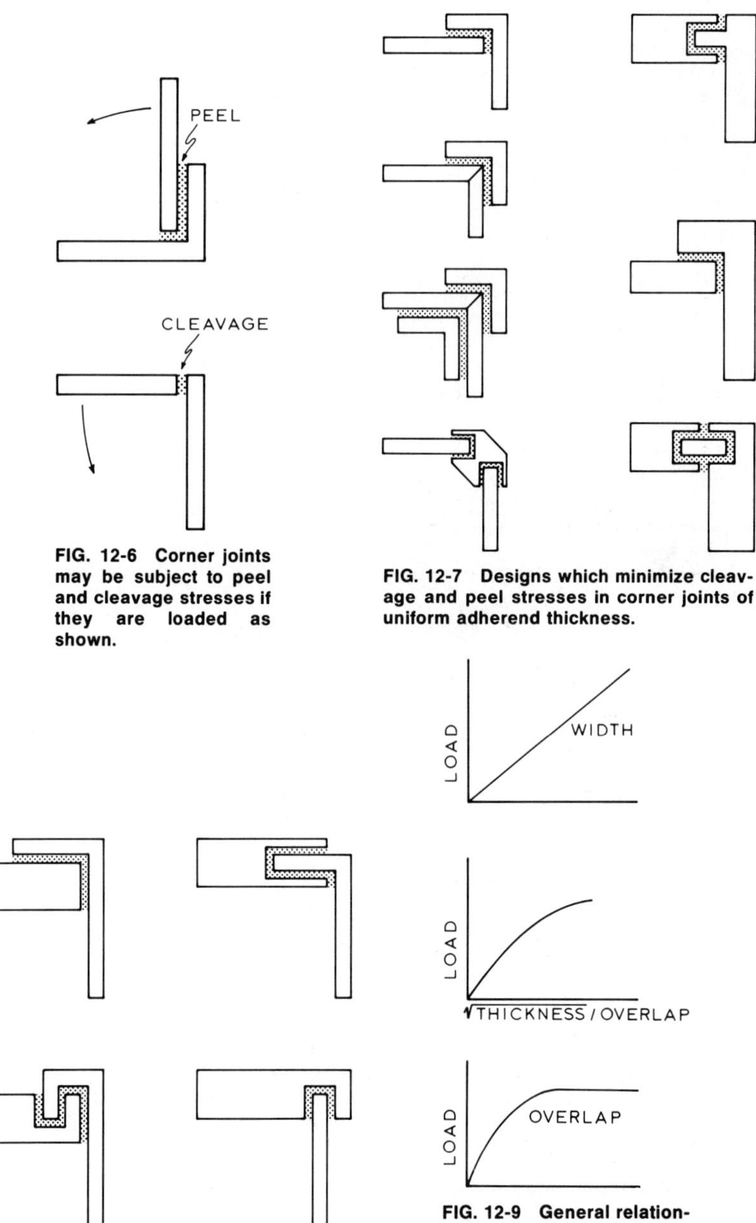

FIG. 12-6 Corner joints may be subject to peel and cleavage stresses if they are loaded as shown.

FIG. 12-7 Designs which minimize cleavage and peel stresses in corner joints of uniform adherend thickness.

FIG. 12-8 Designs which minimize cleavage and peel stresses in corner joints involving adherends of different thickness.

FIG. 12-9 General relationship between lap-shear strength and adherend width, thickness, and overlap.

linking for certain plastics. Technically speaking, pickling (mild acid treatment) and etching (severe acid or alkali treatment) are conversion methods but they are generally classified as solution treatments.

Surface cleanliness is difficult to measure with certainty. The water break-free test is often employed because it is rapid and easy to interpret. In this test a small quantity of water is poured over the surface and observed to see if it breaks up into droplets as shown in Fig. 12-10. If individual droplets form, it is usually because oil or grease contamination is present. The test will not detect salt or metal oxide contamination and some surfaces such as polyolefins and Teflon* break up a water layer even when they are clean.

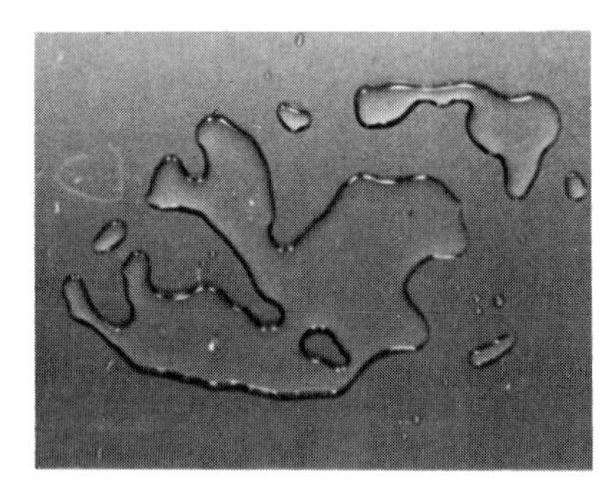

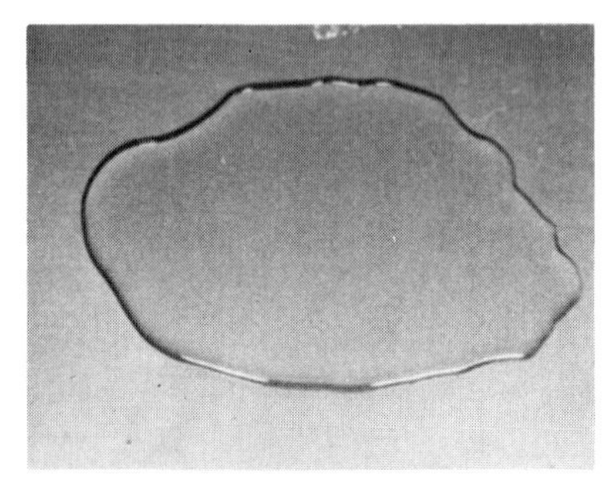

FIG. 12-10 The water break-free test to check for surface cleanliness. A dirty surface breaks the water into droplets.

Table 12-5 lists typical surface treatments for common metals and plastics. Other methods may be used. Adhesive suppliers and texts on surface treatment should be consulted for additional information.

12-7 Applying Adhesives The method chosen to apply an adhesive often makes the difference between a routine, economical, trouble-free operation and one which is plagued by breakdown, wastefulness, and worker dissatisfaction. The best application method for a particular job depends primarily on the size and shape of the parts, the production rate required, and the nature of the adhesive. Commonly used methods may be classified as manual, machine, or spray.

Manual application methods for liquid adhesives are the simplest to use. Nearly all types of adhesives can be applied manually if provision is made for properly handling toxic or flammable products. Brushes, flow brushes, rollers, squeeze bottles, spatulas, or flow guns may be used. These devices have low capital cost and are easily adapted to varying part shapes. They depend upon operator skill for control of adhesive thickness and they may be somewhat wasteful, especially if production is on an intermittent basis or if operations are unskilled. They are ideal for low production rates with large or complex parts, and can often be made portable for use in various plant or on-site locations.

Film and tape adhesives may be applied by hand or with hand-held dispensers. These methods are often used when one of the adherends is supplied with a pressure-sensitive adhesive backing.

Machine application methods are generally used for high production rates involving flat surfaces. They tend to waste very little material. Roll coaters, curtain coaters, and extrusion devices are the most common types of application machines. They may be bench- or floor-mounted. Roll and curtain coaters are ideal for use with large areas of flat, rigid, or flexible substrates. For maximum efficiency they should be used on a early continuous basis.

Roll coaters are produced in a variety of styles. They all employ one or more rollers to transport a liquid adhesive from a reservoir to the surface being coated. Film

* Teflon® is a registered trademark of the E. I. DuPont Co., Wilmington, Delaware.

TABLE 12-5 Commonly Used Methods for Preparing Surfaces for Adhesive Bonding

Surface	Treatment
METALS:	
Aluminum and its alloys	Blast or solvent wipe.
	Immerse for 10 minutes at 150–160°F (66–71°C) in a solution of (by weight) 30 parts deionized water, 10 parts concentrated sulfuric acid, and 1 part sodium dichromate.
Copper and its alloys	Vapor degrease, abrade, degrease.
	Immerse for 2 minutes at 77°F (25°C) in a solution of (by weight) 15 parts of 42% aqueous ferric chloride, 30 parts concentrated nitric acid, and 197 parts distilled water. Rinse in cold running distilled water; dry immediately with a room temperature dry air stream.
Stainless steel	Degrease or sandblast or both; then immerse in 150°F (66°C) solution composed of (by volume) 35 parts saturated sodium dichromate and 100 parts concentrated sulfuric acid. Rinse well with water and fan dry. *Caution:* Add the acid slowly and while stirring to the sodium dichromate solution.
	Pickle in a solution of (by volume) 10 parts sulfuric acid, 10 parts nitric acid, and 80 parts tap water. Rinse with tap water, then dip for 45 seconds in a solution of (by volume) 55 parts concentrated hydrochloric acid, 2 parts 30% hydrogen peroxide, and 43 parts tap water. Hot-water-rinse and fan dry. Bond or prime as soon as possible.
Zinc (and galvanized)	Vapor degrease, abrade, degrease again for stronger bonds. Follow with a 3-minute dip at 77°F (35°C) in a 20% (by weight) concentrated hydrochloric acid solution. Rinse with running water. Oven dry at 150°F (66°C). Bond as soon as possible.
POLYMERS:	
Acrylic Polycarbonate Polystyrene	Degrease with methyl alcohol, abrade, degrease again with methyl alcohol.
Acetal Chlorinated polyether	After acetone or methyl ethyl ketone wiping, etch in a chromic acid solution made of (by weight) 75 parts potassium dichromate, 120 parts distilled water and 1,500 parts concentrated sulfuric acid. *Caution:* Dissolve the potassium dichromate in the distilled water and add the acid slowly while stirring.
Polyethylene Polypropylene	Etch times are polyethylene, 1 minute at 160°F (71°C); acetals, 10 seconds at 77°F (25°C), and chlorinated polyethers, 5 minutes at 160°F (71°C).
Rubber, natural and synthetic, but not silicone materials	Degrease with methyl alcohol, etch with room-temperature sulfuric acid until numerous small cracks appear when the surface is flexed (5 to 15 minutes is normally required). Rinse with distilled water, neutralize with 0.2% caustic for 5 to 10 minutes. Rinse with distilled water, dry thoroughly.
Nylon, epoxies, fiber-glass composites	Solvent-wipe with acetone or methyl ethyl ketone, abrade with emery paper, repeat the solvent wipe.
CERAMICS	
Glass	Degrease, sandblast, or abrade with wet carborundum, degrease again, oven dry for 30 minutes at 212°F (100°C).
Jewels	Degrease only.
Porcelain and glazed china	Degrease, sandblast, or abrade; degrease with methyl ethyl ketone.
Concrete	Detergent wash, rinse with water, sandblast away all loose surface material. If blasting is impractical, etch with 15% (by weight) hydrochloric acid until bubbling subsides. Rinse with water until neutral (litmus paper), allow to dry, then bond.

thickness is controlled by roller spacing, doctor blades, and substrate feed rates. Most low- and medium-viscosity adhesives can be applied to flexible or rigid materials up to several feet in width. Viscosity control is important for consistent film thickness.

Curtain coaters work by moving the surface to be coated through a falling sheet of liquid adhesive. A typical curtain coater, which is also known as a flow coater, is shown in Fig. 12-11. Film thickness is controlled by the rate of adhesive flow and substrate feed. Excess adhesive flowing over the edges of the substrate is collected in a drain pan and returned to the feed reservoir.

Spray application of adhesives is often the best method for use on large, complex, rigid parts. Spray techniques are suitable for intermittent operation but they generally waste more material than do manual or machine methods. Also, greater operator skill is required, more ventilation is needed, and general housekeeping requirements are more stringent.

Extrusion equipment pumps adhesive through one or more nozzles in the form of a liquid bead. Typically the nozzle moves over the substrate, although in some cases, the surface to be bonded can be moved beneath the nozzle. Extrusion nozzles are normally

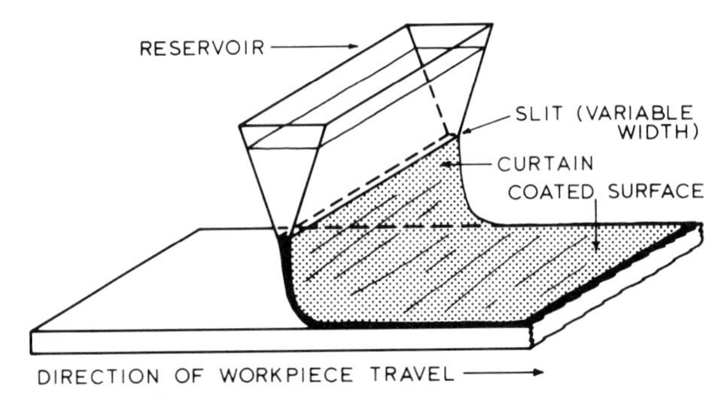

FIG. 12-11 Operation of a curtain coater.

guided by some combinations of gears, belts, cams, etc. In recent years, a number of extrusion heads have been fitted to robots, which allows them to be moved in a variety of preprogrammed paths.

Adhesives may be applied via air or airless spray of heated or ambient-temperature material with automatic or manual guns. Air spray uses compressed air to atomize the adhesive and propel it toward the desired surface as shown in Fig. 12-12. Airless or hydraulic spray uses pressure in excess of 6.894 MPa. (1,000 psi) applied directly to the adhesive. Atomization occurs as a result of the pressure release when the adhesive enters the atmosphere at the gun tip. The safety, maintainance, and capital cost factors associated with high pressures may be compensated for by the less turbulent—and thus less wasteful—spray cloud common to airless guns. Automatic stationary or movable spray guns are ideal for high production rates and simple shapes. Manual guns are better for complex shapes and intermittent operation.

Adhesives are sometimes heated to over 50°C (122°F) for spray application. These heated materials can be sprayed at a higher-percent solids content and with a lower energy requirement, per unit weight, especially for airless (hydraulic) spray systems, than is possible at ambient temperatures.

12-8 Curing Adhesives *Cure* means the process by which an adhesive is converted from its applied condition to its final solid state. Adhesives may cure at ambient or elevated temperatures. The process may involve simple solvent evaporation, as with a contact cement, or a complex chemical reaction between two or more components, as with epoxies and urethanes.

Heat and pressure are often used to obtain rapid production-line curing. Heating speeds the cure while the pressure controls bond-line thickness and immobilizes the parts while the adhesive solidifies. There is an optimum curing pressure for most adhesives. If too much pressure is used, the adhesive will be squeezed out of the bond line and a "starved" joint will result. Too little pressure permits an excessively thick bond line which increases the probability that a major flaw within the solid adhesive will lead to failure by cracking.

Figure 12-13 shows that the time required for curing is very much a function of temperature. Higher temperatures invariably mean shorter cure times. Temperature must be selected, however, with a view to the thermal stability of the adherends.

Pressure-applying equipment may range from simple weights and clamps to sophisticated springs, fixtures, and autoclaves. Often the exact pressure applied is less important than maintaining a constant pressure during cure.

Adhesive bonds are usually heated in gas or steam hot-air ovens. Occasionally infrared, dielectric, or induction heating may be employed. Autoclaves, mentioned earlier, can simultaneously apply heat and pressure. Heated-platen presses can also be used to apply heat and pressure if the shape of the adherend permits contact with the platen.

A newer concept which seems to be gaining acceptance is that of weld-bonding. In this process metal is spot-welded through an adhesive layer to immobilize the parts. The adhesive is then cured in a subsequent processing stage, perhaps in a paint bake oven. An obvious advantage is that clamps or other pressure devices need not be continuously recycled.

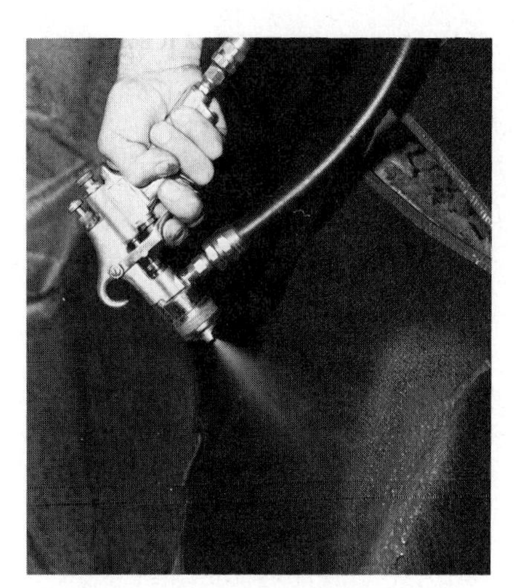

FIG. 12-12 Adhesive being applied by compressed air spray. (*DeVilbiss Company.*)

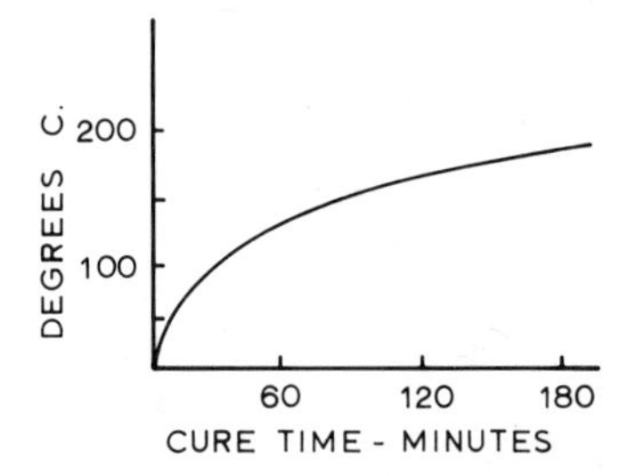

FIG. 12-13 Effect of temperature on the cure time of a typical epoxy adhesive.

From a production point of view the best adhesives are those requiring the simplest cure. This explains the rapid increase in the use of pressure-sensitive and hot-melt adhesives. At the present time, however, these adhesives are limited to nonstructural applications, because of their tendency to creep under load.

12-9 Testing Adhesives Adhesives may be tested to ensure that the adhesive material or the bonding process meets desired standards. Testing is also done to qualify a new product or process for production use. Test procedures are frequently those developed by the American Society for Testing and Materials (ASTM). Such tests should be used without modification whenever possible because they represent standards agreed upon by adhesive suppliers and consumers. If they are modified in any way, their validity may be jeopardized.

Routine tests on incoming adhesive materials often include tests for percent solids, viscosity, density, color, flow, tack, pot life, elongation, and bond strength.

These tests, although routine, are critically important because they prevent the use of misformulated or over-age products. The importance of careful, consistent attention to the details of the particular test being used can not be overemphasized.

Tests to destruction of actual bonded assemblies are often carried out to ensure that the proper production procedures have been followed, or to ascertain the effectiveness of a particular joint design (sec. 12-5).

The following paragraphs will examine the scope and limitations of a few widely used adhesive tests.

The tensile lap-shear test (ASTM-1002-72) is probably the most common adhesive bond test in use today. This test uses a flat overlap of 0.0127 meter (0.50 inch) for parallel specimens 0.0254 meter (1.00 inch) wide and 0.0016 meter (0.064 inch) thick. The specimen is stressed to failure in tension as shown in Fig. 12-14. This test is widely used because it is simple and economical. There is serious question, however, as to whether the lap-shear test effectively simulates service conditions. One must never

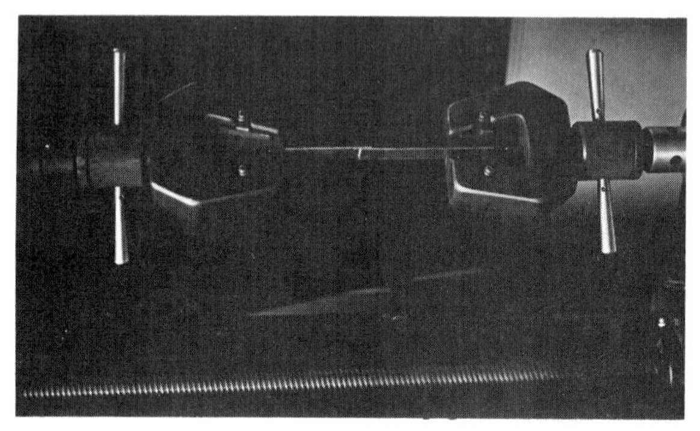

FIG. 12-14 Typical lap-shear test.

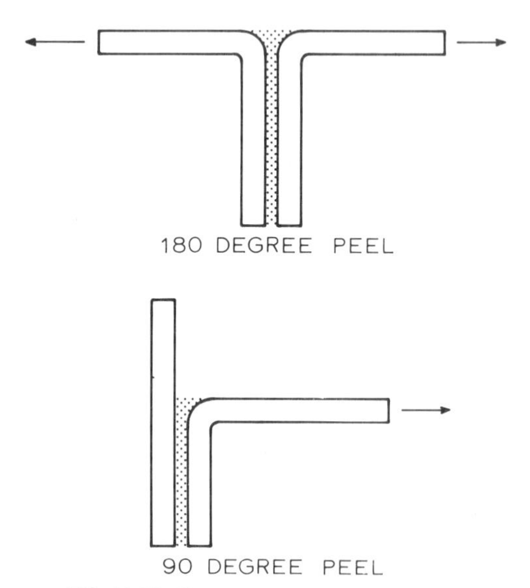

FIG. 12-15 Common tests of peel strength.

try to extrapolate lap-shear test results to product performance. Nonetheless, the test is widely used to qualify industrial adhesives. It is most valuable in determining the relative performance of a series of adhesives or process variations.

The direct tensile strength of a bond is often determined according to ASTM D-897. In this test, cylindrical specimens of 0.000645 sq m (1.00 sq in.) for metals or 0.000322 sq m (0.5 sq in.) for wood are subjected to direct tension until the adhesive or adherend fails. The test is useful but it requires that the glued surfaces be as parallel as possible to avoid introducing cleavage stresses.

Peel tests of bonds involving one or two flexible adherends are typified by the 180-degree peel test (ASTM D903) and the T-peel test (ASTM D-1876). These tests subject the bond to peeling stresses as indicated in Fig. 12-15. Failing loads are reported in force per unit of bond width for peeling at a given rate.

Cleavage tests involve pulling a bond specimen apart with tensile force applied at one edge. ASTM D-1062 uses an assembly of the shape shown in Fig. 12-16. The failing load is reported as force per unit of bond width. Cleavage strengths are often a fraction of tensile or lap-shear values, especially for structural adhesives.

Creep testing of adhesive bonds is carried out to determine their service behavior under load. Creep-testing methods such as ASTM D-2293 and D-2294 use spring-loaded devices to stress lap joints, in compression or tension. A fine line is scribed across the edge of the glue line and examined with a measuring microscope after various test times. Creep is expressed by the amount of movement of the scribed lines. This

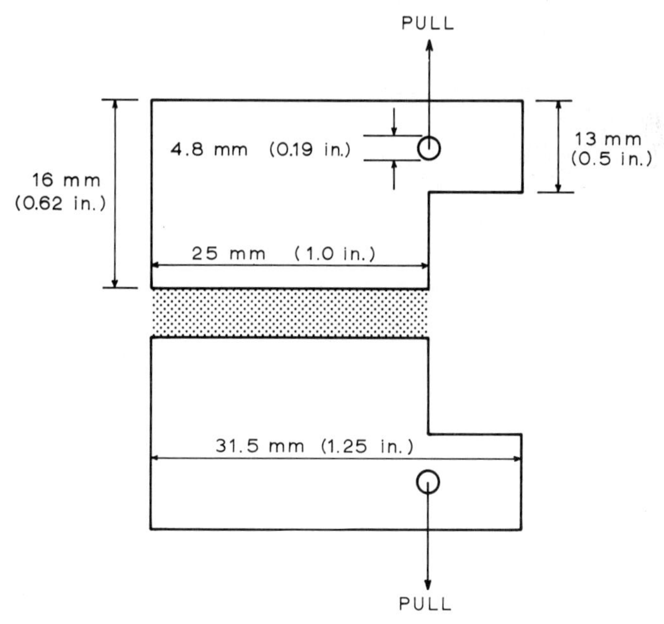

FIG. 12-16 Geometry and method of loading a specimen for cleavage testing according to ASTM D-1062.

creep distance is compared with the bond line thickness in the case of compression testing.

Environmental and age testing of bonds is performed by holding the specimen under the desired conditions for various times and then using the appropriate test method to determine the effect of the environment-age combination. Temperature and chemical exposure tests are particularly common. Unfortunately many tests are carried out using unstressed specimens, i.e., specimens which are placed in the desired environment but not cyclicly loaded and unloaded to simulate service conditions. This omission may sometimes be serious and should be avoided if possible.

Nondestructive testing of bonded assemblies frequently employs ultrasonic techniques. An ultrasonic vibration is passed through one adherend and is reflected by the glue line and the other adherend. The reflected vibrations are detected and monitored for consistency. When a debond or no-bond area is encountered, the reflected wave will differ significantly from the expected and the part can then be removed for repair.

Section 13
Industrial Riveting

Part 1
Semitubular Riveting and Cold-Headed Parts

TECHNICAL STAFF

The Milford Rivet & Machine Co.
Milford, Connecticut

RIVETING AS A FASTENING METHOD

Riveting is one of the oldest fastening and joining methods and yet today its importance as an assembly technique is greater than ever before. The use of rivets and machine riveting has been constantly expanding for both job-shop and production-line operations. The automotive, appliance, electronic, furniture, hardware, military, sheet metal, wiring device, and business machine fields are among the many in which riveting is a popular fastening method. This growth in riveting has occurred because of further refinement of high-speed riveting techniques and because, over the years, its inherent advantages have become more and more valuable with changes in manufacturing methods and economics.

13-1 Advantages of Rivets Because they are produced in tremendous quantities in high-speed heading machines, rivets (Fig. 13-1) are themselves much lower in cost than comparable threaded or proprietary fasteners. Unit labor costs for setting rivets are low because riveters are automatic-feeding, high-speed machines, capable of fastening more than 1,000 assemblies per hour. At the same time, machine operation does not require great skill and can be taught quickly.

Rivets can be used to join dissimilar materials in various thicknesses. As long as its shank length is sufficient, a rivet can fasten as many parts at one point as necessary. Any material that can be cold-worked can be made into rivets. And a wide variety of surface finishes are available.

Special considerations in the overall product design are seldom required if rivets are specified. Changes in the product design or new models seldom make riveting obsolete as long as required clearances are retained in the new designs. In addition to their basic function as fasteners, rivets can also serve as pivot shafts, spacers, electric contacts, stops, and inserts. They can be clinched in assemblies before the components are cleaned or they can be used to fasten parts that have already received final painting or other finishing.

The appearance of the rivet head, in comparison with other fasteners, may be an advantage. In addition, the rivet head—either a standard type or one with a special shape or embossing—may be part of a functional or decorative design or made integral with the configuration of a part.

A riveted joint is a positive joint, and thus quality is determined by visual inspection. Quality is immediately apparent on completion of the operation. If the rivet cracks or splits, if the assembly is loose, or if some other undesirable condition exists, it can be seen by the operator and corrective action taken.

FIG. 13-1 Assortment of standard rivets.

13-2 Limitations of Rivets Rivets have some limitations, too. Their tensile and fatigue strengths are lower than those of bolts: high enough tensile loads can pull out the clinch, and severe vibrations can loosen the joint. Because of the forged grain structure of rivets, they have higher compression and shear strengths than many other solid fasteners. The strength of a riveted joint is usually many times more than required by most applications.

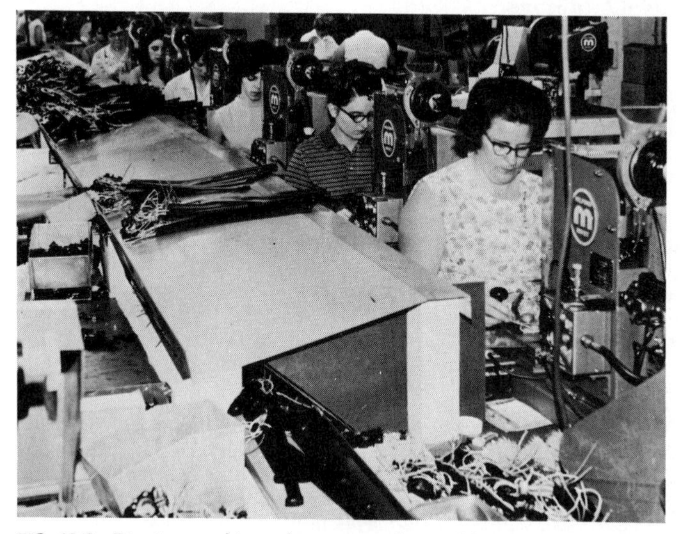

FIG. 13-2 Riveting machines along a typical assembly line.

Riveted joints may not be either watertight or airtight, although they can be made so by the application of a sealing compound or through the use of rivets of special design. Nor can the part, once riveted, be easily disassembled for maintenance, and this limits the application of rivets in some subassemblies and products.

13-3 Riveting Machines Riveting machines (Fig. 13-2) consist essentially of a base or pedestal on which is mounted the head which contains the drive mechanism and the hopper feed. Rivets are fed one at a time to the jaws which are positioned above the roll set (or clinching tool) mounted in a bracket projecting from the base or pedestal of the machine. Machines are activated by electrical-clutch trip mechanisms that are operated by foot pedals, palm switches, or automatic controls. They are basically mechanical in nature (although hydraulically and pneumatically actuated machines are available), relatively inexpensive, rugged, and easily maintained.

13-4 Objectives in Riveting Once the number and location of rivets in an assembly are known, the goal is to set these rivets properly at the lowest in-place cost and with a satisfactory rate of return on machine investment. Riveting procedures, types of rivet-setting machines, riveting fixtures, machine layouts, and maintenance procedures must all be established. These factors must be considered with respect not only to the particular riveting job but also to the overall riveting capabilities and volume requirements of the assembly plant.

Before any decisions can be made on rivet-setting procedures or equipment, certain facts about the assembly are needed. Specifically, these questions should be answered:

1. How many assemblies must be riveted over what period of time?
2. What is the maximum production rate that may be required?
3. How many different rivet sizes—diameter and length—are involved in the assembly?
4. Will the parts to be riveted require special handling or fixturing because of their size, weight, or shape?
5. Are there any possible problems of access to the rivet locations in the parts?
6. Are any of the parts to be riveted made of fragile materials?
7. Are any unusually large material-thickness variations likely?
8. If the riveting task cannot be done in one sequence of operations at one station, how should it be divided?

13-5 Productivity and Safety Simplicity of the rivet and the versatility of the riveting machine offer unusual freedom in planning assembly operations for minimum in-place costs, whatever the fastening task and volume of production. A standard riveter generally can do many different fastening jobs with the same rivet—or at least with different lengths of the same rivet diameter—with minimum adjustment. The riveting machine can also be custom-built to match the special needs of a particular fastening task, perhaps being combined with additional riveters and one or more other types of assembly machines to provide increased productivity.

The continuing search for ways to increase productivity in riveting has had an associated benefit in the vital area of safety. Many developments in the design of riveting equipment and methods—directed primarily at increasing the productivity of machine and operator—have also increased the safety of operation. Conversely, the additional cost of making a riveting station safe can often be paid for in a short time through lower unit cost of the assembly task that the station is performing.

THE RIVET

13-6 Small Rivets Small rivets are divided into several categories, the most important of which are solid and tubular. Tubular rivets may themselves be classified as *full tubular, semitubular, compression,* and sometimes *self-piercing.* The shapes in a third category of fasteners vary so widely that they are best referred to generally as *cold-headed parts* and will be treated separately in this section.

In both solid and tubular rivets, two or more parts are fastened together by the rivet head on one side and a formed shape at the end of the rivet shank on the other side. In solid rivets (Fig. 13-3), the formed shape is produced by heading over, hammering, or spinning the solid end of the shank. In tubular rivets (Fig. 13-3), the formed shape—commonly called a clinch—is produced as the material around the edge of a hole in the end of the shank is rolled over against the surface of one of the pieces being joined. Solid rivets, which are usually installed with relatively slow manual or semimanual methods, are widely used in aircraft, transportation equipment, and other products where an extremely high joint strength is required. Tubular rivets, which are used in nearly every conceivable type of appliance, electromechanical device, and assembled product, are large-volume fasteners which are set on high-production semi-automatic or automated riveting machines.

The publication "Dimensional Standards for General-Purpose Semitubular Rivets, Full Tubular Rivets, Split Rivets and Rivet Caps," published by the Tubular Rivet &

Machine Institute, tabulates standard nominal rivet shank diameters up to 0.310 inch and rivet head diameters up to 0.570 inch. These dimensions will be assumed in Section 13 as defining the maximum limits of what will be considered small rivets.

The semitubular rivet (Fig. 13-4) is the most common tubular type used in riveted assembly. The depth of the hole at the end of the shank in semitubular rivets, by definition, does not exceed 112 percent of mean shank diameter. Full tubular rivets, which are used mostly in self-piercing applications (see below), have a hole depth that exceeds 112 percent of mean shank diameter and often extends the full length of the rivet shank. The walls of the holes in most semitubular rivets are either straight or very slightly tapered. Semitubular rivets are automatically inserted and then clinched by the riveting machine into prepunched and prealigned holes in the parts to be fastened. When a semitubular rivet is clinched, it is essentially a solid member, with shear and compressive strengths equivalent to those of a solid rivet.

13-7 Self-Piercing Types Self-piercing tubular rivets, in contrast, pierce their own holes through the aligned parts as they are driven down in the jaws of the riveting machine. While semitubular rivets can fasten any prepunched parts if they are long enough (and have an acceptable length/diameter ratio), self-piercing rivets are limited with respect to the toughness and total thickness of the parts they are able to pierce.

There are three basic types of self-piercing rivets (Fig. 13-4): *full tubular*, *bifurcated* (or split), and *metal-piercing*. The full tubular rivet can pierce its own hole and the slug of material can be retained inside the larger tubular cavity. The full tubular type is generally used for assembling leather, plastics, wood, fabric, or similar soft materials.

Bifurcated rivets, often called split rivets, can be recognized by their "pronged" appearance. These fasteners are sawed or punched in a secondary broaching or press operation. Since the legs are not distorted as much as when they are punched, bifurcated rivets that are sawed can be used for joining light-gage metals or heavier sections of fiber, wood, or plastic. Punched types, on the other hand, are only suitable for lighter piercing operations.

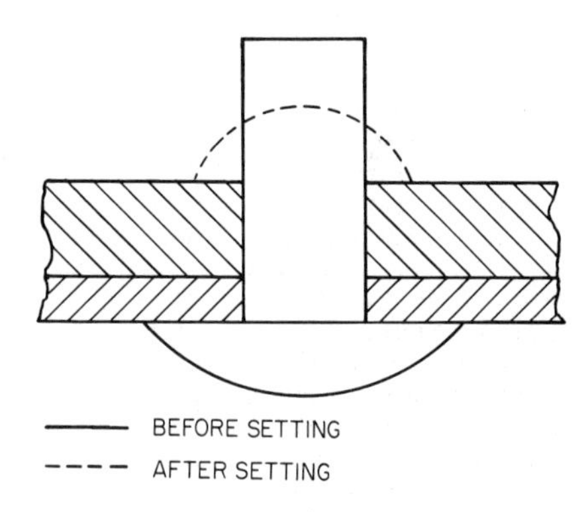

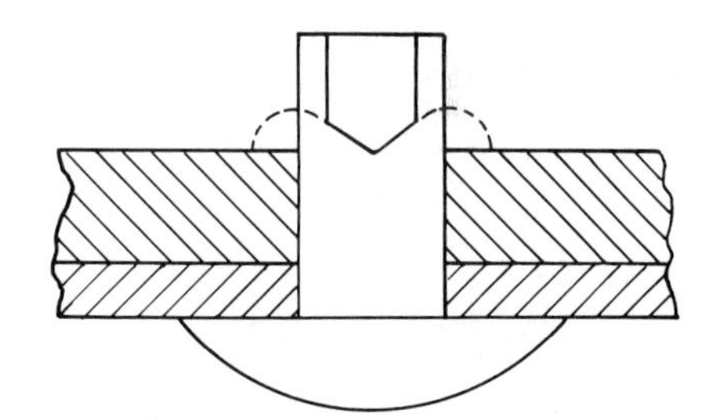

FIG. 13-3 Comparison of cross-sectional shapes of solid and tubular rivets before and after setting.

Metal-piercing rivets can fasten any of the standard low-carbon steels, aluminum alloys, and stainless steels—as long as the material is not harder than about R_B 50.* Although they look very much like standard semitubular rivets, their shape and material are designed to enhance their column strength. Depending on their size and hardness relative to the material, metal-piercing rivets are capable of piercing total material thicknesses of 0.150 inch and more. When split rivets or standard semitubular rivets are used for piercing metals, the riveted sections must be much thinner. In addition, split rivets are seldom used in fastening metals because they usually cost more than semitubular metal-piercing rivets for the same tasks.

* Rockwell hardness number determined by ball indent.

13-8 Compression Rivets Compression (or cutlery) rivets (Fig. 13-4) are formed of two members: a solid or blank rivet and a deep-drilled tubular member. The diameters of the solid shank and the drilled hole are selected so as to produce a compression or press fit when the parts are mated. Compression rivets are used in the cutlery field because the heads fit evenly into counterbored holes and do not allow food or dirt particles to collect in the crevices. Since the pressure required to mate the two members can be closely controlled, they can be specified to minimize splitting of wood- or plastic-composition parts.

Full tubular and bifurcated rivets are often used in combination with a cap (Fig.

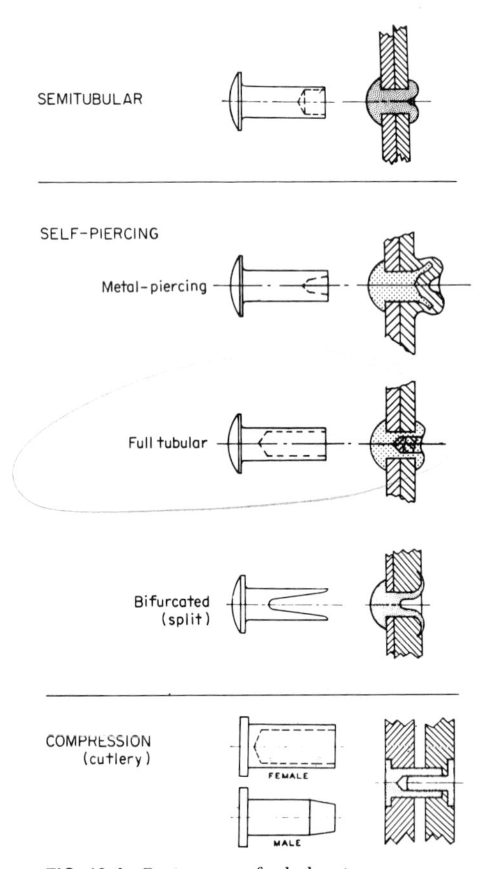

FIG. 13-4 Basic types of tubular rivets.

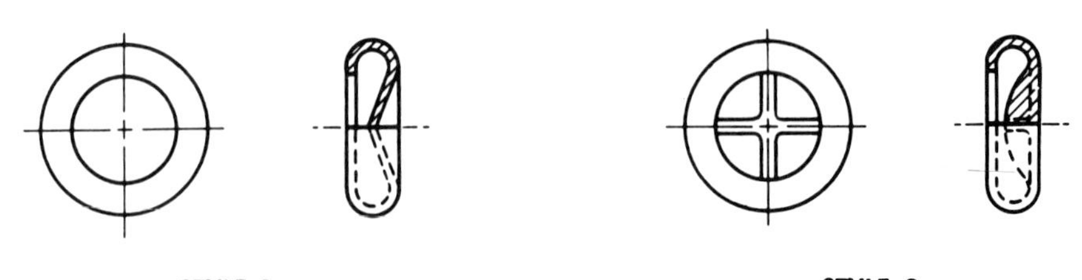

FIG. 13-5 Two styles of caps used with full tubular and bifurcated rivets.

13-5). This acts as a washer to prevent the clinch from snagging or tearing soft materials. The assembly is also given a more finished appearance.

13-9 Cold-Headed Parts Cold-headed parts (Fig. 13-6) may be tubular or solid and made with any of a variety of special features such as: tenons or shoulders to simplify location and assembly; fluted or knurled shanks, which are common when the part is to be used as an insert in a molding or casting; ornamental heads for cosmetic hardware or such products as fountain pens or pocket lighters; or special heads such as balls, cams, or hexagonal configurations to permit the rivet to perform a secondary function in addition to acting as a fastener. Gage pointers for television channel selectors, electrical contacts, bearing pivots for gages and instruments, and similar devices are typical of these applications.

FIG. 13-6 Assortment of cold-headed parts.

13-10 Rivet Heads Rivets in the full-tubular or semitubular classifications are made in oval, truss, or countersunk types of heads (Fig. 13-7). Bifurcated rivets also are available in these three general head types, although oval head designs are far more common. All types of rivets can also be specified in an almost limitless variety of special or custom decorative heads (Fig. 13-8).

13-11 Materials Small rivets are most frequently made of steel, aluminum, or brass. Most fastener manufacturers consider these as standard materials.

Of the ferrous materials, low-carbon grades from about A1S1 1006 to 1019, medium-carbon grades from about 1013 to 1023, and high-carbon grades from 1023 to 1040 provide both strength and economy. The lower the carbon content, the easier the forming. Also, since the holes in rivets made of higher-carbon-content materials cannot be extruded but must be drilled, low A1S1 grades are generally recommended.

Small rivets are also produced in all grades of brass, from rich low alloys to 35–65 compositions. They are also made of copper, electrolytic or oxygen free, which is often specified by the electronics industry.

Almost any soft grade of aluminum – 1100-S, 3003-S, 2017-S, 2024-S, 5052-S, or even higher strength alloys – can be cold-formed into rivets.

The major rivet materials, their reasons for use, advantages and typical applications are summarized in Table 13-1.

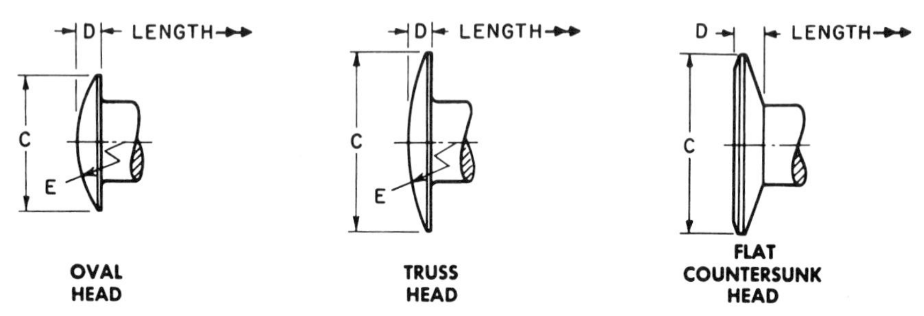

FIG. 13-7 Basic types of full tubular or semitubular rivet heads: oval, truss, and countersunk.

FIG. 13-8 Assortment of decorative heads.

13-12 Standard Sizes Standard rivets are often available from stock – or can be delivered on relatively short lead times – and cost less than special types. The user who specifies a standard type, therefore, saves time and money. Slight changes in tolerances, dimensions, or such a seemingly insignificant characteristic as fillet radius often require the development of completely new dies. This almost always results in higher rivet costs unless the production run is high enough to absorb tooling charges.

For these reasons, serious consideration should be given to standard rivets before deciding on special sizes. Shank diameters can be maintained within ±0.001 inch on rivets up to $\frac{1}{16}$ inch diameter and ±0.003 inch on diameters $\frac{1}{4}$ inch or over. In Fig. 13-9, standard dimensional tolerances are given for the most commonly specified rivet diameter (nominally 0.123 inch but customarily called $\frac{1}{8}$ inch). Tolerances of ±0.005 inch

TABLE 13-1 The Major Rivet Materials

	Reasons for use	Advantages	Typical applications
Low- and medium-carbon steel	Account for perhaps 95 percent of all applications of ferrous materials, and, because of wide usage of low-carbon sheet for metal fabricated products, for the majority of all rivet applications.	Low cost, high strength, readily formed, can be used as a pivot or other functional wear part without danger of seizing, can be electroplated or treated with a variety of chemical coatings.	Automotive assemblies, business equipment, toys, hardware, photographic equipment, sheet-metal assemblies of all types, appliances, motor generators, transportation equipment, etc.
Brass and copper alloys	Generally used for fastening of copper-alloy assemblies, but also when appearance, good electrical properties, and corrosion resistance are important.	Good cold-heading characteristics. Also lend themselves well to secondary operations such as thread rolling, beveling, etc.	Lighting fixtures, electrical and electronic assemblies, luggage, cosmetic cases, ornamental jewelry, screw fasteners, cutlery, etc.
Aluminum	Increasing use of aluminum for fabricated metal products has expanded use of aluminum fasteners.	Lowest cost per 1,000 of any of the metals. Readily formed. Bright appearance. Resists oxidation in corrosive atmospheres.	Die castings, tubular furniture, jalousies, storm windows and doors, electrical assemblies, automotive parts, kitchen utensils, transportation equipment, business equipment, toys, lighting fixtures, electrical assemblies, etc.
High-alloy and stainless steel	Used primarily when high strength and/or good wear characteristics are required, and, in the case of stainless, when good corrosion resistance is necessary. Stainless steel rivets also are used on occasion instead of plated fasteners.	High strength. In the stainless grades, good corrosion resistance, excellent appearance.	Boat hardware, kitchen utensils, lawn furniture, automotive assemblies, cutlery, medical and surgical equipment, food chemical equipment, ornamental applications.
Precious metals	Primarily for jewelry applications or electronic-electrical assemblies when maximum conductivity is required. Silver alloys are most commonly employed.	Good electric conductivity, excellent appearance, ductile and easy to form. However, are considered "specials" since cost is high.	Electrical contacts, jewelry.
Other materials	Nickel, Monel, lead, zinc, and many other alloys can be used for rivets, but applications are generally specialized.	Lead, zinc, and other low-melting-point alloys are used for fusible plugs. Lead often used as inserts for balancing rotating equipment, such as fan blades. Zinc sometimes used for die-cast products.	Nickel and Monel applied when good corrosion resistance is required—as, for example, when electrolytic corrosion is a factor.

on head diameter and shank length generally are recommended on most rivet sizes.

Standard sizes of semitubular, bifurcated, and compression rivets—all available from the major rivet manufacturers—are given in Tables 13-2, 13-3 and 13-4, which are abstracted from standards published by Milford Rivet & Machine Company. The sizes of metal-piercing rivets given in Table 13-5 are typical, rather than standard, because this type of rivet is relatively new and no industry standards have yet been established.

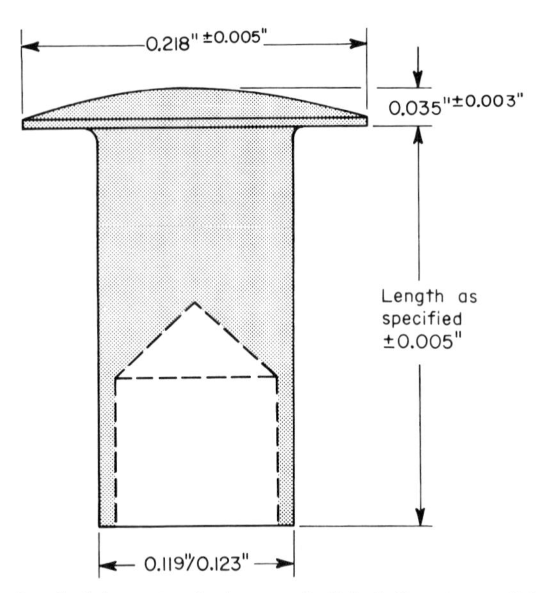

FIG. 13-9 Standard dimensional tolerances for ⅛-inch-diameter semitubular rivet.

TABLE 13-2 Standard Sizes of Semitubular Rivets (Oval Head)

SHANK DIAMETER A	HEAD DIAMETER B	HEAD HEIGHT C	STRAIGHT HOLE		MAX. CLINCH ALLOW-ANCE	MIN. REC. HOLE SIZE
			D	E		
.058-.061	.104-.114	.015-.019	.043	.041-.049	.038	.063
.058-.061	.120-.130	.015-.019	.043	.041-.049	.038	.063
.085-.089	.142-.152	.020-.026	.063	.061-.071	.058	.093
.085-.089	.157-.163	.022-.026	.063	.061-.071	.058	.093
.095-.099	.182-.192	.026-.032	.071	.069-.079	.065	.103
.119-.123	.213-.223	.032-.038	.088	.088-.098	.084	.129
.119-.123	.276-.286	.032-.038	.088	.088-.098	.084	.129
.141-.146	.229-.239	.039-.045	.105	.105-.115	.101	.154
.141-.146	.276-.286	.032-.038	.105	.105-.115	.101	.154
.141-.146	.306-.318	.039-.045	.105	.105-.115	.101	.154
.141-.146	.369-.381	.059-.065	.105	.105-.115	.101	.154
.182-.188	.306-.318	.059-.065	.136	.137-.147	.129	.196
.182-.188	.369-.381	.059-.065	.136	.137-.147	.129	.196
.246-.252	.430-.444	.068-.074	.182	.188-.198	.156	.265
.246-.252	.493-.507	.078-.084	.182	.188-.198	.156	.265
.302-.310	.554-.570	.093-.099	.224	.229-.243	.156	.325

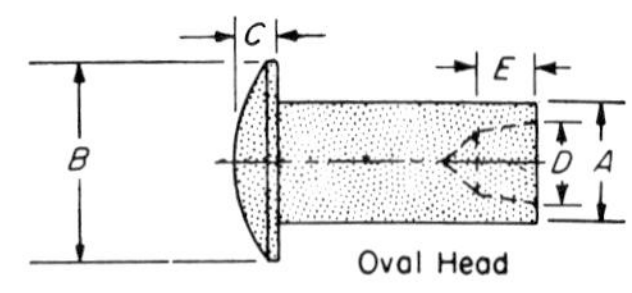

TABLE 13-3 Standard Sizes of Bifurcated (Split) Rivets

SHANK DIAMETER A	HEAD DIAMETER B	HEAD HEIGHT C	CLINCH ALLOW-ANCE
.090-.092	.145-.151	.019-.025	.091
.078-.082	.153-.159	.019-.025	.080
.120-.123	.215-.221	.031-.035	.119
.117-.120	.307-.317	.052-.058	.119
.144-.148	.309-.315	.043-.049	.146
.144-.148	.370-.380	.060-.065	.146

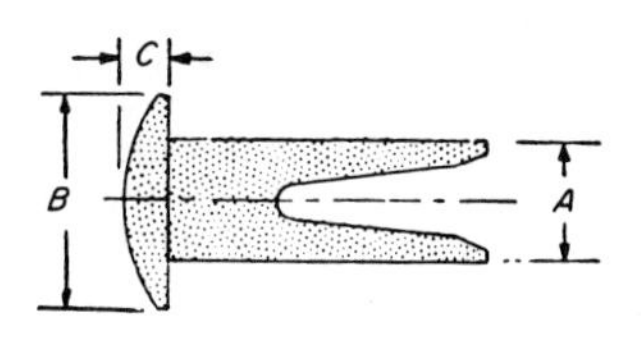

TABLE 13-4 Standard Sizes of Compression Rivets

MALE SHANK DIAMETER A	FEMALE SHANK DIAMETER A1	HEAD DIAMETER B	HEAD HEIGHT C	RECOMMENDED HOLE SIZE Blade Handle
.123-.127 —	— .149-.153	.244-.250	.042-.046	.162-.172
.123-.127 —	— .149-.153	.306-.312	.042-.046	.162-.172
.123-.127 —	— .149-.153	.338-.344	.042-.046	.162-.172
.123-.127 —	— .149-.153	.369-.375	.042-.046	.162-.172
.144-.148 —	— .193-.197	.306-.312	.048-.052	.206-.216
.144-.148 —	— .193-.197	.338-.344	.048-.052	.206-.216
.144-.148 —	— .193-.197	.369-.375	.048-.052	.206-.216
.144-.148 —	-- .193-.197	.432-.438	.048-.052	.206-.216

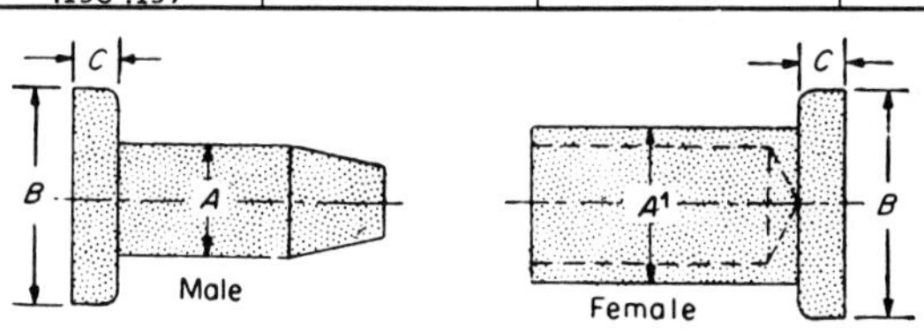

TABLE 13-5 Typical Sizes of Semitubular Metal-Piercing Rivets*

Nominal shank diameter† in.	Shank length in.	Total sheet-steel thickness range§ in.
1/8	5/32	Up to 0.055
1/8‡	5/32	0.055 to 0.083
1/8‡	3/16	0.075 to 0.100
1/8‡	7/32	0.090 to 0.125
1/8‡	1/4	0.120 to 0.150
9/64	5/32	Up to 0.055
9/64‡	5/32	0.055 to 0.083
9/64‡	3/16	0.075 to 0.100

* A 3/16-in. diameter metal-piercing rivet has been added since this writing.

† Actual diameters are 0.122 in. for 1/8 in. nominal and 0.143 in. for 9/64 in. nominal.

§ For material hardness of R_B 50 or lower.

‡ Hardened after cold forming.

PLANNING FOR SEMITUBULAR RIVETING

There are four basic design steps in planning for semitubular riveting:

- Analyze required joint strength
- Select rivet characteristics
- Locate the rivets
- Specify the holes for rivets

13-13 Analyze Required Joint Strength In small assemblies rivets are rarely required to sustain loads that are any more than a small fraction of their actual capacity. However, it is good practice always to confirm this fact, and it may be necessary to determine if the joint strength required will affect rivet selection.

The bearing strength at the rivet holes in the materials being joined may be more important than the strength of the rivet itself, particularly in plastics and thin metal sections. Thus, analysis of joint strength in a load-carrying application also requires study of the strength of the wall of the rivet holes and the surfaces of the material on the clinch and head sides.

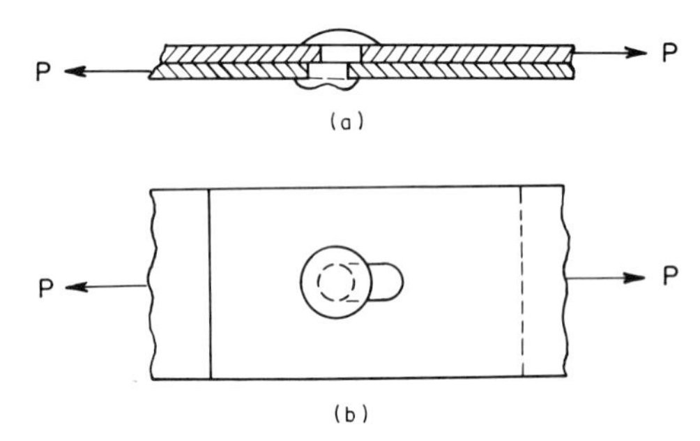

FIG. 13-10 (*a*) Failure of riveted joint because of shearing of rivet shank. (*b*) Failure of riveted joint because of tearing of the riveted material, which elongates the rivet hole.

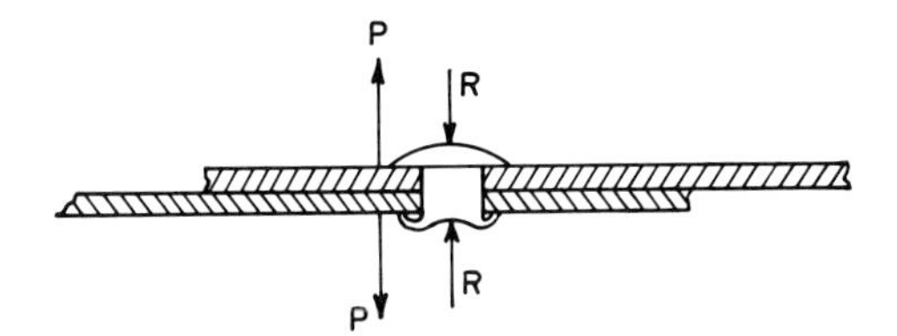

Fig. 13-11 The pull-apart force (P) will separate the riveted parts if it exceeds the retention strength (R) of the joint.

The three loads on a riveted joint are *shear*, *pull-apart*, and *torque*.

Shear failure of a single riveted joint can occur in one of two ways. The rivet shank may shear (Fig. 13-10*a*) or the material itself may give, elongating the rivet hole (Fig. 13-10*b*). Both can be prevented by selecting a sufficiently large rivet (and hole) diameter. Pull-apart forces are parallel to the rivet center line and tend to separate the riveted parts (Fig. 13-11). The retention strength of a riveted joint is defined as the maximum pull-apart force that can be sustained without failure of the rivet or the riveted material.

In fastening plastic or other nonmetallic materials, the retention strength of the material around the rivet hole is usually lower, and therefore more significant, than that of the rivet. In these cases, the rivet head or clinch can actually be drawn through the material by a sufficiently high pull-apart force. The resistance of the material to pull-apart forces is proportional to material thickness, circumference of the clinch and rivet heads, and the shear strength of the material perpendicular to its surface plane (in materials such as plastic laminates, this shear strength may be quite different from shear strength parallel to the surface plane).

The best procedure is to test actual prototype assemblies to determine if the retention strength (with safety factor) is higher than the maximum anticipated pull-apart force.

The most common form of rivet failure is rolling back or stripping of the clinched end. It is good practice to test the joint to determine whether the riveted metal or the rivet is the weaker under pull-apart forces and to ensure that ultimately both the metal parts and rivet can safely sustain in-use forces.

The retention strength of a rivet depends primarily on the compressive strength of the rivet head and clinched end against the surfaces of the components. Tearing of the rivet head is rarely the reason for pull-apart failure. The retention strength is instead determined by the strength of the clinched end of the rivet.

The amount of material that is rolled over to form the clinch is known as the clinch allowance (Fig. 13-12). It is the difference between the rivet shank length and the total material thickness. Clinch allowance is often expressed as a percentage of the rivet shank diameter.

The curve in Fig. 13-13 for low-carbon-steel semitubular rivets give ultimate retention strength for various shank diameters as a function of clinch allowance. The data are given only for clinch allowances between 50 and 70 percent, the range suggested by most rivet manufacturers (Fig. 13-14). The basic relationship between retention strength and clinch allowance also applies outside this range. Note that the data in Fig. 13-13 are given in terms of ultimate strength, which is similar to ultimate shear because of the way in which pull-

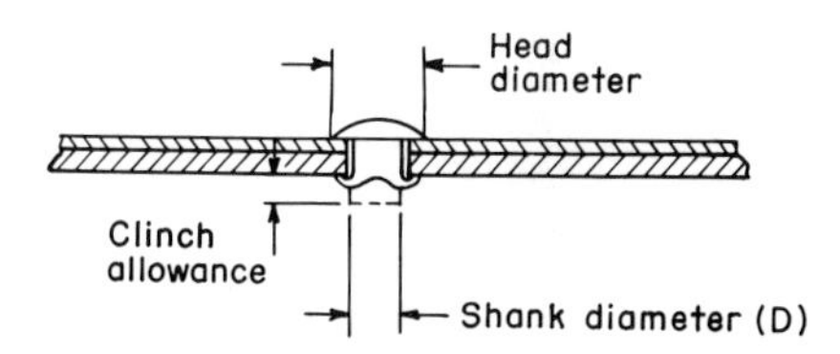

FIG. 13-12 Clinch allowance is the difference between the rivet shank length and total material thickness.

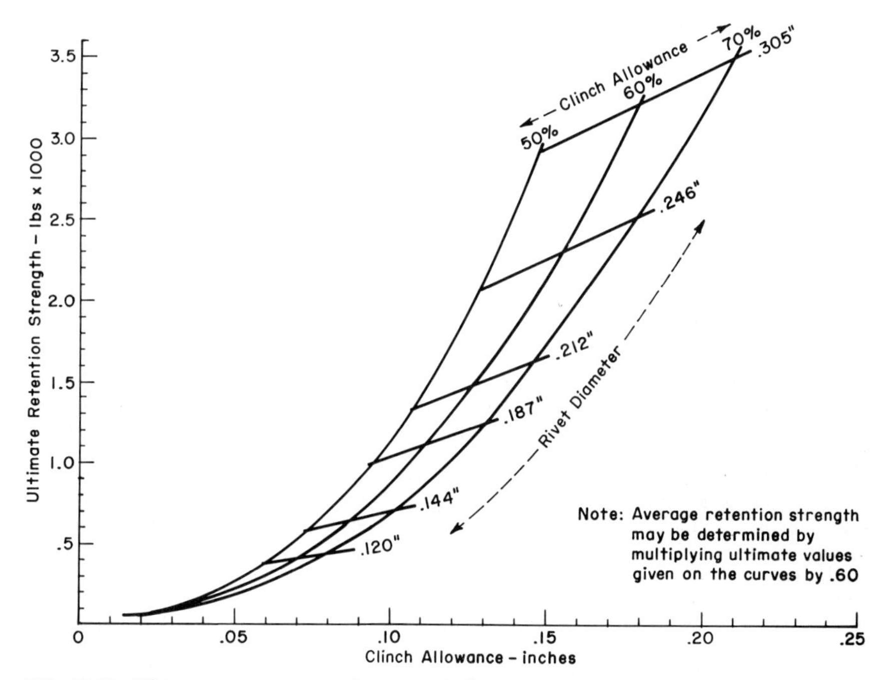

FIG. 13-13 Ultimate retention strength vs. clinch allowance.

apart failure occurs. Therefore, average retention strength is found by multiplying the ultimate value by 0.60. The appropriate safety factor may then be used.

A common method for determining the relative retention strength of riveted joints is the push-out test. A push-out rod is placed against the clinched end of the rivet and an increasing load applied until the clinch fails and the rod pushes the rivet out. Tests indicate that push-out force and retention strength may usually be assumed equal for semitubular rivets.

Torsional failure of single riveted joints may be caused by insufficient clinch allowance. Maximum torque resistance is attained with clinch allowances around 70 percent. If torque strength is critical, it is good practice to design mating parts at the rivet location to form a locked joint or to specify two rivets in the joint. As an alternative, a riveted joint with square or hexagonal holes may be tried. If a relatively soft

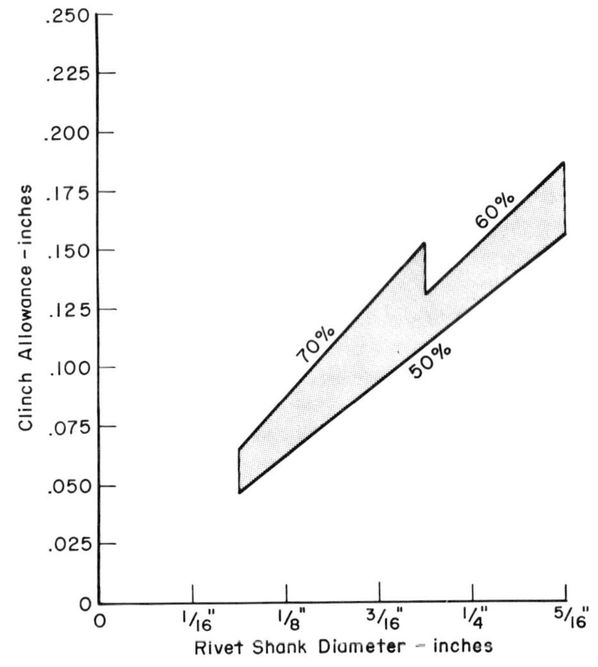

FIG. 13-14 Clinch allowance vs. rivet shank diameter.

rivet is used, the material will flow radially in the joint during clinching to produce a locked condition.

13-14 Select Rivet Characteristics The optimum rivet diameter in non-load-carrying joints is determined by economics—the total cost of the rivet itself and the labor in clinching it.

Initial costs can be minimized by specifying a rivet of standard diameter, length increment, and tolerance (Tables 13-2, 13-3, and 13-4). Not only is the actual cost of standard rivets lower than that of special sizes, but delivery may be expected to be faster. In addition, standard riveting tools and fixtures are more likely to be readily available.

Proper selection of rivet size and material can lead to substantial savings in production costs. For example, large rivet diameters mean higher rivet-feeding efficiency, and carbon steel and brass are materials most easily handled in riveters. In addition, it is helpful to limit the rivet shank length–diameter ratio to a maximum of 6:1 (such as 3/4 inch maximum length for the standard 1/8 inch rivet in Fig. 13-9) so that standard barrel hoppers can be used. At higher ratios, it may be necessary to use special feeding devices. For rivets above the 6:1 ratio, the head diameter should be at least twice the shank diameter to facilitate efficient automatic handling.

The *D/T* relationship, which is the ratio of rivet shank diameter to material thickness, may affect the selection of a particular standard rivet diameter. A large-diameter rivet fastening one or more thin parts (large *D/T*) could cause buckling of the material during the clinching process. On the other hand, a small-diameter rivet fastening two very thick sections (small *D/T* ratio) would cause buckling of the long, thin rivet. As a rule of thumb, the ratio *D/T* should be at least 1 and not over 3 when fastening metal sections. A *D/T* of 1 is recommended for riveted joints in plastic materials, although the ratio may be larger for very thin sections. There is no loss of shear strength in double-shear joints (Fig. 13-15) with small *D/T* ratios. However, there may be a shear strength loss up to about 20 percent when *D/T* is as high as 3 in a double-shear joint.

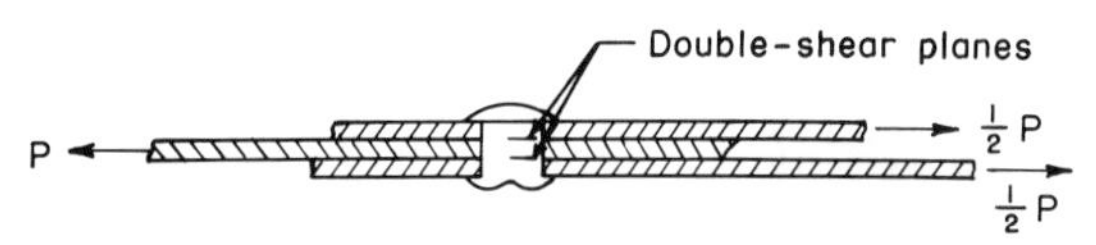

FIG. 13-15 A double-shear joint loses a significant amount of shear strength at high ratios of shank diameter to total material thickness *T*.

13-15 Edge Distance *The edge distance* (Fig. 13-16) is the interval between the edge of the part and the center line of the rivet. If the edge distance is too small, the bearing strength of the material will be significantly reduced and the joint may fail.

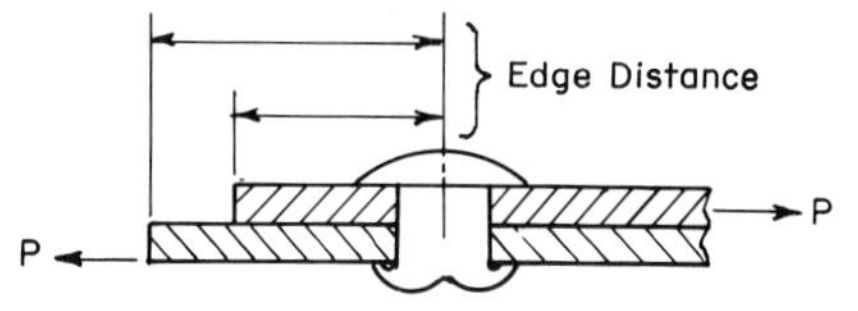

FIG. 13-16 Edge distance is the interval between the edge of the part and center line of the rivet.

Edge distance should be at least twice the rivet shank diameter. Below this distance there is a roughly proportional decrease in the bearing strength of the joint when shear forces are acting. It is good practice to retain an edge distance of twice the rivet shank diameter, even in the absence of shear load, in order to avoid possible buckling of the material under clinching pressures. The recommended edge distance for plastic materials, either solids or laminates, is as high as 3 rivet diameters, depending on the thickness and strength of the material.

It is also important not to locate rivets too far away from the edges of the parts. The larger the distance between the edge and the rivet hole, the longer it is apt to take the machine operator to locate the hole on the riveting fixture.

13-16 Pitch Distance *The pitch distance* (Fig. 13-17), defined as the interval between the center lines of adjacent rivets, should be large enough to avoid unnecessarily high stress concentrations in the riveted material and to prevent buckling at adjacent unriveted holes. The pitch distance in metals should be at least 3 times the shank diameter of the largest rivet in the assembly. For plastics it should be greater than 5 times the rivet shank diameter.

Rivet location is affected by clinching requirements primarily because the rivet holes must be accessible to the rivet-setting machine. If a tool-clearance problem is anticipated in setting a rivet adjacent to a flange, a distance of 2½ shank diameters between the rivet center line and the vertical portion of the flange will usually avoid interference on the clinch side of the rivet. Clearances may have to be much greater where flanges are adjacent to the rivet-head side of the joint.

Joint strength, ease of access in production, or appearance may determine on which sides of the joint the rivet head and clinch are to be located. For example, the rivet head should be against the thinner of two sections in a joint to minimize stress concentration in the material. As an exception, however, the head should be placed against the more crushable or fragile of the two materials, whatever the relative thickness, because of possible damage due to the setting pressure on the clinch side. If one of the surfaces being riveted is located in a recess or cavity, the clinch side should usually be placed in the cavity.

13-17 Specify the Holes for Rivets Proper size of holes for rivets contributes greatly to the quality of the joint. If the hole for the rivet is too small for the particular shank diameter selected (Fig. 13-18*a*), the rivet end may lock in the hole before the rivet head is fully seated. As a result, the clinch is incomplete and the riveted joint must be rejected. If the hole is too large (Fig. 13-18*b*), the clinch may be off-center and the joint will move too easily. The rivet does not swell to contact the wall fully if the hole diameter is too large, thus reducing the retention and torque strengths of the joint.

Recommended rivet-hole diameters (Tables 13-2, 13-3, and 13-4) are normally minimum values. In general, the diameter of the rivet hole should be between 105 and 110 percent of the rivet shank diameter. A good procedure is to make the minimum hole diameter equal to 105 percent of the high diameter limit for the rivet size selected (such as a hole diameter of 0.129 inch for the rivet in Fig. 13-9).

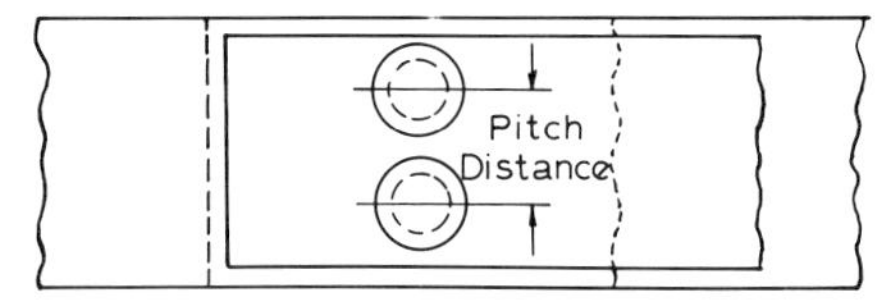

FIG. 13-17 Pitch distance is the interval between the center lines of adjacent rivets.

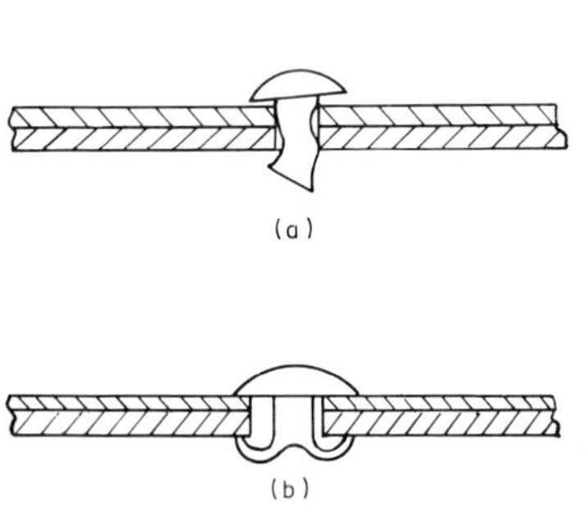

FIG. 13-18 (*a*) If the hole for the rivet is too small, the rivet end may lock in the hole before the rivet head is fully seated. (*b*) If the hole for the rivet is too large, the clinch may be off-center and the joint will move too easily.

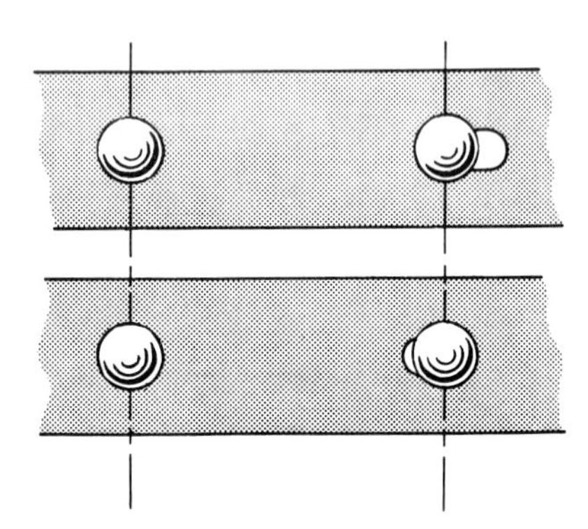

FIG. 13-19 Slotted holes in the parts may eliminate the possibility of interference between misaligned rivet holes.

Counterbores or countersinks may be specified to provide a flush surface either for better appearance or for functional reasons, such as in circuit breakers which must be closely stacked. A countersink is often used when there is insufficient material thickness to accept a counterbore. In such cases, a flat countersunk rivet head (Fig. 13-7) should be used.

A counterbore may be specified to decrease the total material thickness and thus call for a shorter, and possibly less expensive, rivet. This may be done either to facilitate multiple-head riveting by using rivets of the same diameter and length in each head, or simply to standardize rivet sizes to reduce initial cost or minimize inventory. Counterbores, countersinks, or raised bosses are easily designed into molded plastic parts and castings. However, unless there are very good functional reasons for special hole treatment, the cost of additional machining or forming of metal parts is usually not justified.

Slight shifts in rivet-hole location caused by normal manufacturing variations, particularly in sheet metal components, may cause misalignment of holes in the part. This may be prevented by specifying either larger rivet holes or requiring that one set of holes be slotted. Nominal hole sizes from 110 to 115 percent of rivet shank diameter are often enough to eliminate the possibility of interference. Rivet holes in this size range also help to avoid production problems because of hole clogging in parts that have been painted or otherwise finished before being assembled. Slotted holes (Fig. 13-19) in one of the pieces will usually allow sufficient leeway and yet provide adequate surface contact on the clinch side.

PLANNING FOR METAL-PIERCING RIVETING

A metal-piercing rivet (Fig. 13-20) is different in both shape and material than a conventional semitubular rivet. Both differences are designed to increase the column strength of the shank wall and so reduce the possibility of buckling when the rivet is driven down through the metal sandwich. The shank walls of metal-piercing rivets are as much as 50 percent thicker than those of conventional rivets. The rivet material—carbon steel, stainless steel, or aluminum—is harder than standard rivets and yet not so hard that it cannot be properly headed or lacks adequate ductility to form a satisfactory clinch. Metal-piercing rivets may be used as cold-formed or later hardened for additional piercing capacity (Table 13-5).

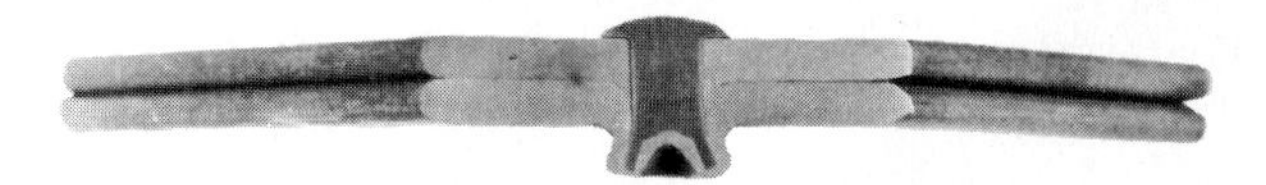

FIG. 13-20 Cross section of a metal joint that has been fastened by a semitubular metal-piercing rivet.

FIG. 13-21 Progressive cross sections in a metal-pierced joint show how the rivet pierces the metal parts and a clinch is formed.

13-18 Advantages of Metal-Piercing Rivets The advantages of metal-piercing rivets with respect to standard semitubular rivets are:

1. *Reduced component cost*—no need for separate prepunching operations in components that are roll-formed or extruded; simpler tooling for blanking operations.
2. *Reduced assembly cost*—cut riveting cycle time where prepunched holes that are misaligned, paint-filled, or burred would otherwise slow up placing components over anvil pins.
3. *Improved appearance*—minimize possible surface damage to precoated stock while fishing for holes.
4. *Extended tool life*—with no moving anvil pin, as used in conventional semitubular riveting, the riveting tool lasts longer, reducing replacement tooling cost.

The initial cost of metal-piercing rivets is generally equal to or slighly more than standard semitubular rivets. The benefits of metal-piercing therefore can often be obtained without additional fastener cost. Metal-piercing rivets can usually be clinched on the same riveting machines as standard semitubular rivets of the same size, at least for applications involving thinner gauges of stock.

13-19 Piercing and Clinching Action The piercing and clinching action of a semitubular metal-piercing rivet is, as might be expected, more complicated than clinching a standard rivet through aligned holes in the parts. As shown in the progressive cross sections in Fig. 13-21, the material in both the upper and lower parts (assuming there are only two) is stretched downward before being pierced in a circular pattern as the cylindrical end of the shank continues to advance. The shank end spreads as it nears the end of its downward motion, both because of material buildup in the hole at the end of the rivet and the upward force from the riveting anvil.

In joining the ductile materials best-suited for metal-piercing rivets, a small slug of material from the upper part is forced into the hole in the rivet as it draws the material downward into the cavity of the anvil. This slug normally presents no problem because it is retained in the hole by the drawn material of the lower part that has been crimped over the end of the clinch.

13-20 The Good Pierced Clinch A well-designed pierced clinch like that in Fig. 13-20 is usually stronger than a clinch made by a semitubular rivet set through a prepunched hole. Its shear strength depends on the bearing strength of the parts being riveted. Because of the extra material drawn into the clinch, the total bearing area in the part—the bearing strength of the joint—is increased considerably. In addition, some of the material of the two parts is drawn across the plane of the shear force (the touching surfaces of the parts) and so adds to the strength of the rivet itself. Its tensile strength is higher because a great deal more force is needed to pull the flared end of the heavy-walled metal-piercing rivet (reinforced by the metal slug in the hole) back through the sandwich than to unroll the clinch on a standard semitubular rivet.

FIG. 13-22 The inherent strength of a metal-pierced joint is demonstrated by bending the two 0.075-inch-thick steel pieces back over the rivet head.

The inherent strength of a metal-pierced joint is demonstrated in Fig. 13-22, in which two 0.075-inch-thick sections of low-carbon steel have been joined with a metal-piercing rivet and then bent back around the rivet head. In spite of the extreme tensile load on the rivet, as well as a lesser shear load, the joint in this case did not even loosen.

One of the production advantages of metal-piercing rivets is that, as in conventional semitubular rivets, the quality of the joint can be established very quickly with only a glance at the clinch. The characteristic swelling on the circumference of the clinch indicates that the rivet wall has properly spread and the part material has fully crimped over.

The prominent clinch in a metal-pierced joint may be undesirable either because of its appearance or the possibility of snagging on the clinch. However, neither appearance or snagging is a problem at all when the joint is designed so that the clinch side is hidden, as is normally done even with semitubular rivets. On the head side, of course, there is little difference between the two types of rivets.

13-21 Selecting Metal-Piercing Rivet Size Selection of a metal-piercing rivet for a particular fastening task is very much the same as selecting a semitubular rivet. In Table 13-5, total thickness of sheet steel determines shank length. Shank diameter is not critical; a $^1/_8$-inch metal-piercing rivet, for example, is capable of piercing the same thicknesses as a $^9/_{64}$-inch rivet of the same shank length. The only reason for selecting a larger rivet diameter at a higher cost is to provide more bearing surface under the head and in the rivet hole for higher tensile and shear strengths.

At R_B 50 and below and about 0.050-inch sandwich thickness and below, unhardened metal-piercing rivets are as effective as hardened rivets. Over R_B 50 and/or over 0.050 inch, hardened rivets are worthwhile for superior consistency. In any case, it's always wise to consult a specialist in metal-piercing rivets for recommendations on the proper rivet, tools, and machinery.

PLANNING FOR COLD-HEADED PARTS

Cold-headed parts (Fig. 13-23) are formed from short lengths or blanks of round wire in heading machines and, while they can be made in an infinite variety of shapes, are usually symmetrical about the wire axis. Many of these parts have design features such as heads, shanks, collars, and shoulders that are usually associated with standard rivets.

13-22 The Cold-Heading Process In a heading machine, a blank of required length is cut from a coil of wire, aligned with a die cavity, and formed by one or more blows at room temperature. The three sketches in Fig. 13-24 show how a rivet or cold-headed part is formed in a two-blow cold header. In the first blow, the punch at right forces the blank into the die at left to form the end of the shank; at the same time, the

FIG. 13-23 Assortment of cold-headed parts.

head of the part is partially formed by the punch. In the second blow, a different punch tool is brought into position and completes the head shape.

The five basic operations performed by cold-heading machines are shown in Fig. 13-25. By impact, *upsetting* flows the material into a shape with diameters larger than the wire (for example, the head of a fastener or special rivet). *Forward extrusion* reduces the wire to a smaller diameter in the direction of the impact. *Backward extrusion* squirts the metal back between the punch and die opposite to the direction of the blow. It is useful in forming holes in tubular parts and is often done with forward extrusion in the same blow. Extrusion is beneficial in that it cold-works the metal and improves its physical properties. *Trimming* usually shapes the head of a part, while *piercing* produces through-holes automatically on the header.

While modern multiblow headers can be tooled to produce almost any cylindrical shape, the primary limitation is in the maximum wire diameter that a particular cold-heading machine can accept. Cold-headed parts may be made from wire as small as $\frac{1}{32}$ inch in diameter and up to $1\frac{5}{16}$ inch.

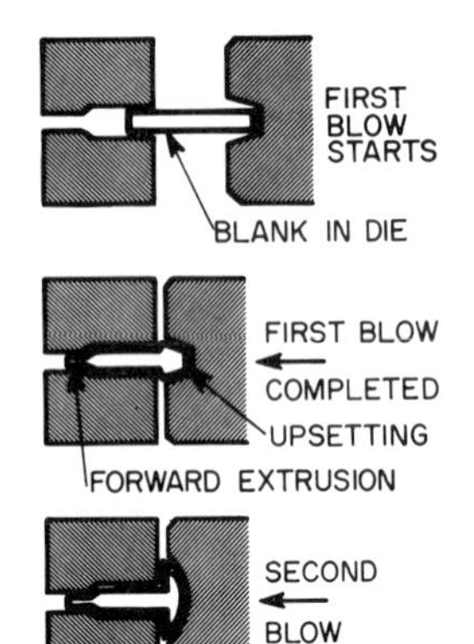

FIG. 13-24 How a rivet or cold-headed part is formed in a two-blow cold header.

13-23 Advantages of Cold Heading The major advantages of cold heading with respect to machining as a method of producing small metal parts are:

1. *No scrap loss*—upsetting and extruding forms metal, working with all the material in the blank. In contrast, machining shapes a part by removing metal and so produces scrap, which may comprise a large proportion of the material in the blank.

2. *High production speed*—cold heading shapes instantly by impact, in contrast to gradual removal of metal. Depending on the part design and the material being used, the production rate of a header ranges from about 50 to 450 pieces per minute.

3. *High material strength*—a continuous, unbroken grain structure (Fig. 13-26) produces high fatigue strength and impact resistance, and cold working improves the tensile strength of the material. On the other hand, the material is weakened wherever machining cuts across the grain structure.

Any savings in unit-part cost of cold heading with respect to machining is primarily in eliminating scrap loss and reducing unit labor and machine cost by operating at a higher production speed. The relative importance of material savings and high output rate, of course, depends on the particular material and part design.

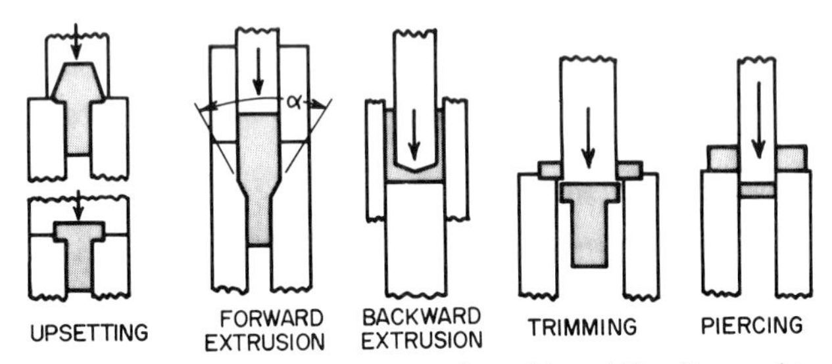

FIG. 13-25 The five basic operations performed by cold-heading machines.

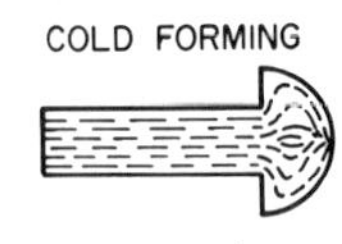

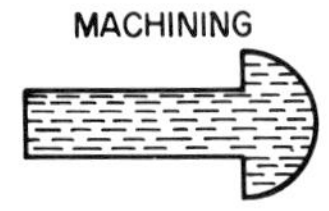

FIG. 13-26 Comparison of the grain structure in identically shaped cold-formed and machined parts.

13-24 Dimensional Tolerances Dimensional tolerances in cold heading are a matter of economics rather than inherent process capacity. As allowable tolerances become extremely tight, the additional cost of frequently replacing worn punches and dies can substantially increase unit-part cost. The relative difficulty of maintaining dimensional control and the rate of tool wear depend very much on the forming characteristics of the material and the shape of the part.

Typical tolerances on major dimensions of cold-headed parts are:

1. *Shank diameter*—tolerance ranges for shank diameter are the same as for rivets: ±0.001 inch for parts up to $^1/_{16}$ inch diameter, and ±0.003 inch on diameters $^1/_4$ inch or over.

2. *Head diameter and thickness*—head diameter tolerances depend on the type of head. Head diameter tolerances increase from ±0.004 to ±0.020 inch for standard oval heads of parts with shank diameters between $^1/_{32}$ and $^7/_{16}$ inch. Head thickness tolerances vary from roughly ±0.004 to ±0.010 inch for shank diameters between $^1/_{32}$ and $^7/_{16}$ inch.

3. *Shank length*—measured as the distance from the extreme end of the part to the largest diameter of the upset portion. Length tolerance varies with the ratio of the shank length L to shank diameter D. Shanks have a length tolerance of ±0.011 inch through a ratio of $L/D = 12$, ±0.016 inch through $L/D = 18$, and ±0.022 inch for L/D over 18.

The publication "Dimensional Standards for General Purpose Semitubular Rivets, Full Tubular Rivets, Split Rivets and Rivet Caps," published by the Tubular Rivet & Machine Institute, gives standard tolerances that can be expected to be achieved with normal production procedures on cold-headed parts. Tighter tolerances that may be necessary for proper function in a specific application are obtainable, of course, at higher unit cost.

13-25 Shapes, Surfaces, and Secondaries Some of the most common shapes and surfaces that can be produced automatically in a header are:

1. Cupping (or dimpling)
2. Ribbing
3. Lugs and fins
4. Tapers
5. Struck slots
6. Phillips punches
7. Cross slots

Ribs and serrations are often specified to prevent the cylindrical shank from turning, while ornamental head designs are specified for appearance and identification of parts.

Many types of machining can, of course, be done as a secondary operation. The most common secondaries that are specified for cold-headed parts are:

1. Slotting
2. Turning
3. Counterboring
4. Roll threading
5. Tapping

However, as might be expected, cold-headed parts are designed to be shaped as much as possible on the heading machine itself. The cost of secondary machining in machine time, labor, and material handling is high enough that it often is economical to use special forming tools, sometimes quite intricate, in order to eliminate altogether the need for a secondary.

The assortment of cold-headed parts in Fig. 13-23 includes many that have been made with a variety of secondary operations. The most common shapes to be seen are shoulders and collars, knurled and threaded shanks or shoulders, and drilled and tapped heads and ends.

13-26 Materials and Finishes The range of metals which can be cold-headed (Table 13-6) offers the same broad variety of material application characteristics as machined stock. Formability depends on the material. For a given material, the brittleness of the finished part depends largely on the degree of upset. Stresses can be

TABLE 13-6 The Major Heading Materials: Types and Main Characteristics

Heading material	Types	Main characteristics
Carbon steels	C1006–C1040	Low cost, wide tensile range, high strength
Stainless steels	300, 400 series	High strength, excellent appearance, corrosion resistance, resistance to scaling at high temperature
Copper alloys	65/35, 70/30, 80/20 Cu/Zn	Electrical conductivity, nonmagnetic, excellent corrosion resistance
Aluminum alloys	1100, 2000, 3000, 6000	Electrical conductivity, corrosion resistance
Precious metals	Fine silver, coin silver	Electrical conductivity, excellent appearance, ductile, easy to form
Other materials	Nickel, Monel, lead, zinc, etc.	Corrosion resistance

relieved by annealing at very little extra cost. Since cold heading shapes by impact, finishes are normally excellent without secondary treatment. Any type of finish or plating appropriate for the formed material may be specified.

13-27 Practical Shape Limitations The range of proportions of cold-headed parts is determined by the type of cold header (all ratios given here are for two-blow headers, the most common type). The ratio of head diameter to shank diameter (HD/D in Fig. 13-27) indicates the degree of spreading that occurs during forming. A ratio HD/D = 3 is considered severe. Above that ratio, the metal is being spread over too great an area, so that there may be a loss in concentricity and the grain structure is extended to the point where radial cracking is liable to occur. The ratio of head thickness HT to shank diameter D represents the relative impact force between the header and die faces; the smaller the ratio, the more severe the impact. The ratio HT/D should not be less than $^1/_4$ in a cold-formed part.

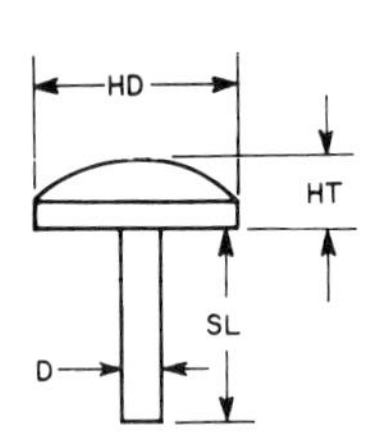

FIG.13-27 The recommended limits on the proportions of cold-headed parts depend on the type of cold header.

The allowable ratio of shank length SL to diameter D is very closely related to the production rate of the header. Maximum shank length (including shoulder) is 8D for normal production rates but may be up to 12D when the header is operated at a lower speed. Special tooling is required for shank lengths exceeding 12D; it may be possible to go as high as 25D with special forming innovations.

The ratio SL/D is an important element in determining the relative cost of cold heading and machining. While impact forming takes roughly the same time for any length of shank, machining time usually increases proportionately to shank length. For shank lengths under 3D, particularly with tight dimensional tolerances, machining is likely to be the lower-cost method. When shank length exceeds 3D, the cost of cold heading is likely to be lower. Of course, the 3D rule of thumb considers only the basic head/shank shape; it does not account

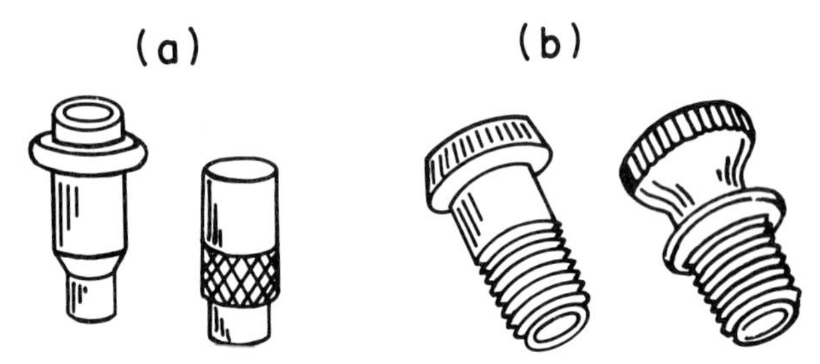

FIG. 13-28 (*a*) The cold-headed part (left) and machined part (right) are functionally identical inserts for molded distributor caps. (*b*) The cold-headed part (left) and machined part (right) are functionally identical file cabinet guide rod knobs.

for the relative cost of other part design features such as ribbing, knurling, and slots.

Sharp corners should be avoided. Cold-headed parts should have fillets wherever the upset portion joins the shank. The radii on all internal and external corners should be as large as possible and no less than 10 percent of shank diameter without stress relief. Tapers and swell necks from head to shank must be shallow and not much longer than 1 shank diameter under the head.

If a part cannot be cold-headed as is, a change in shape or size (or both) may make cold heading possible. In any case, even if the part can be cold-headed as is, it is good practice to make any design changes that will improve its formability and reduce secondary operations without affecting its function. Here are a few of the design questions that might be considered as possibilities in improving the formability of the part:

1. Can the whole part be made smaller?
2. Can a particular portion of the part, such as the head or shoulder, be made smaller?

3. Is a small radius or fillet permissible on sharp corners?
4. Can a square step between two diameters be replaced by a taper?

The parts in Fig. 13-28 demonstrate how design changes can both improve formability and eliminate secondary operations without changing the function of the part. The cold-headed (left) and machined (right) inserts in Fig. 13-28*a* are both embedded in molded distributor caps. Each is the same overall length and has an identical cavity length and diameter. Rather than being knurled like the machined insert, however, the cold-headed part has an annular bead which can be produced in the header and which serves equally well for securing the part.

In Fig. 13-28*b*, the cold-headed file-cabinet guide-rod knob at left performs the same function as the machined knob at right. However, the inwardly curved section in the machined knob was eliminated so that the cold-headed part could be removed from the forming die and a secondary operation could be avoided. In addition, while it was possible to form the knurl for the thumb grip, it had to be tapered slightly so that the part could be readily removed from the header. The shoulder diameter was also reduced in order to facilitate feeding during a secondary drilling operation.

13-28 Order Quantity and Cost Because cold heading is inherently a high-speed parts-production method, order quantity is an important factor in establishing relative unit cost with respect to machining. Simple cold-headed parts, those in which the tooling cost may be considered relatively small, can be economically produced in quantities as low as 10,000 pieces. Orders from 25,000 to 50,000 pieces may be needed to absorb tooling and setup costs for more complex cold-headed parts, particularly those with secondaries. However, the importance of quantity should not be prejudged because the part design may be such that cold heading is more economical than machining even in orders below 10,000 pieces.

RIVETING MACHINES

All riveting machines operate in essentially the same way. Rivets are stored in a hopper located behind a rivet driver in the head. The rivets are delivered to the required clinching position by a feeding system synchronized with the driver action. The driver propels each rivet down through aligned holes in the parts (metal-piercing rivets form the holes themselves) for clinching of the tubular shank end in a roll-set tool below.

Riveting machines may be classified under four basic types:

1. Single-spindle automatic riveters
2. Double-spindle riveters
3. Multiple-machine groups
4. Automated riveting systems (automatic fixtures and rotary tables).

These types of riveters are listed in general order of increasing output rate, complexity, and initial cost.

13-29 Single-Spindle Automatic Riveters The single-spindle automatic machine is the simplest, most flexible, and most widely used type of riveter. The machine may be either mechanically, pneumatically, or hydraulically operated. The most common mechanical machines (pedestal type in Fig. 13-29*a*, bench type in Fig. 13-29*b*) may have either a mechanical trip or a solenoid-operated clutch mechanism, although the latter is becoming predominant because it permits superior safety features. The pneumatic machines (pedestal type in Fig. 13-30*a*, bench type in Fig. 13-30*b*) are particularly suited to multiple-head riveting because they have very narrow heads and can be tightly grouped.

Most pneumatic machines are used for multiple-machine riveting or for fastening parts that are fragile or vary widely in thickness. A pneumatic riveter has a squeezing action in clinching the rivet, rather than an impact, and therefore is more gentle with easily fractured materials because stroke speed and length are accurately controlled. However, mechanical riveters are specified on most other assemblies because they can do exactly the same job as the pneumatic machine, with lower machine investment for equivalent capacity.

All the machines in Figs. 13-29 and 13-30 are vertical in that the drivers move the rivets down vertically for clinching. The drivers of horizontal riveters (Fig. 13-31) propel the rivets horizontally, which can be useful in eliminating reorientation of parts

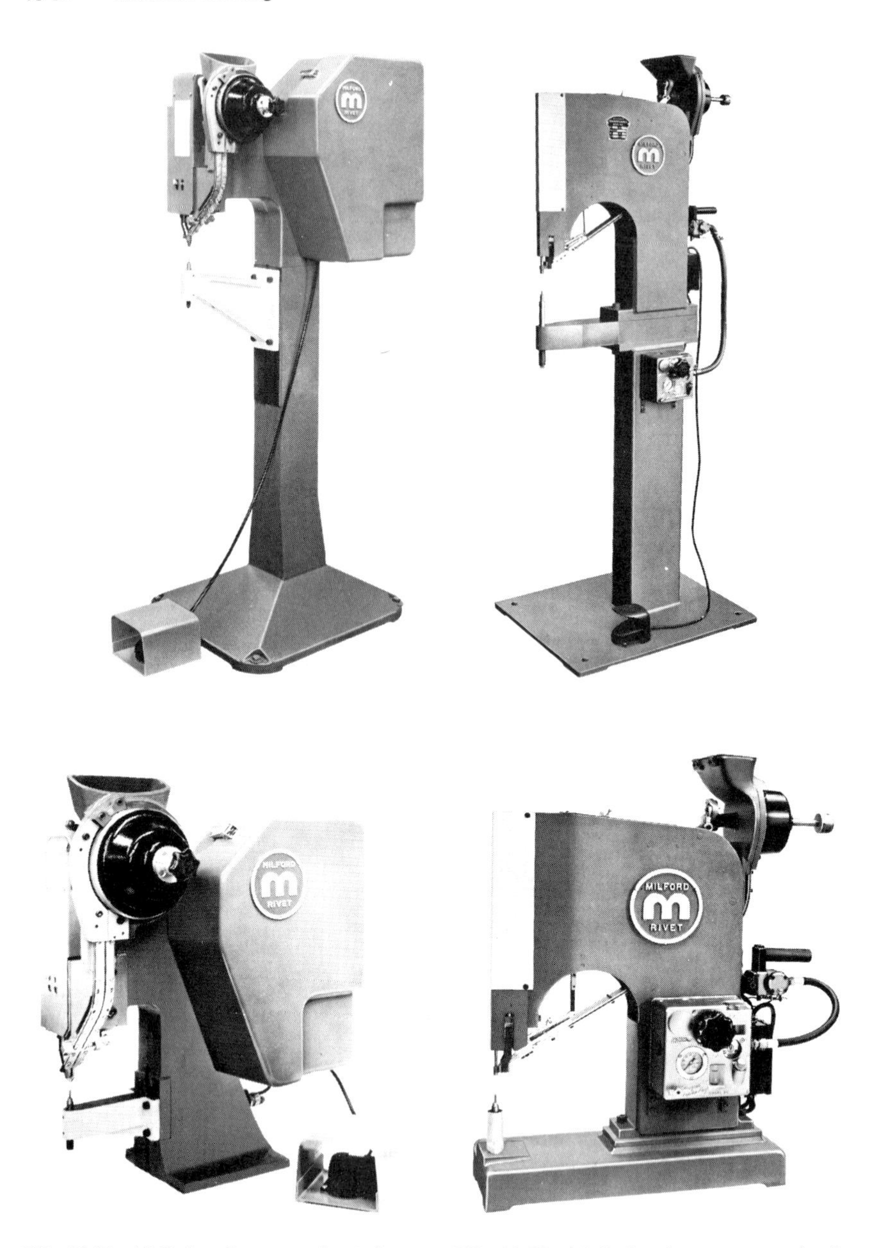

FIG. 13-29 (*a*) Pedestal-type mechanical riveter. (*b*) Bench-type mechanical riveter.

FIG. 13-30 (*a*) Pedestal-type pneumatic riveter. (*b*) Bench-type pneumatic riveter.

in automated riveting stations and in permitting the setting of rivets from opposite sides at the same time.

13-30 Double-Spindle Riveters Double-spindle riveting machines (Fig. 13-32) set two rivets simultaneously; the rivets need not be the same shank diameter and length or pass through the same parts sandwich. The center spacing between the rivets

FIG. 13-31 Two horizontal riveters are used in an automated assembly station.

FIG. 13-32 Double-spindle riveting machine.

may either be fixed or adjustable. Because of its lower cost, a fixed-center riveter is normally specified for long-run riveting jobs where no change in spacing is expected. Standard adjustable-center machines are available which clinch two rivets simultaneously on centers from $^3/_8$ inch to 13 inches or more apart (special machine design may reduce the minimum spacing to the head diameters of the rivets). A simple hand adjustment is sufficient when center distance only is changed between production runs. Double-spindle machines are useful because of the large variety of assemblies in which adjacent rivets can be set in pairs in considerably less time than required to set the rivets sequentially on a single-spindle machine.

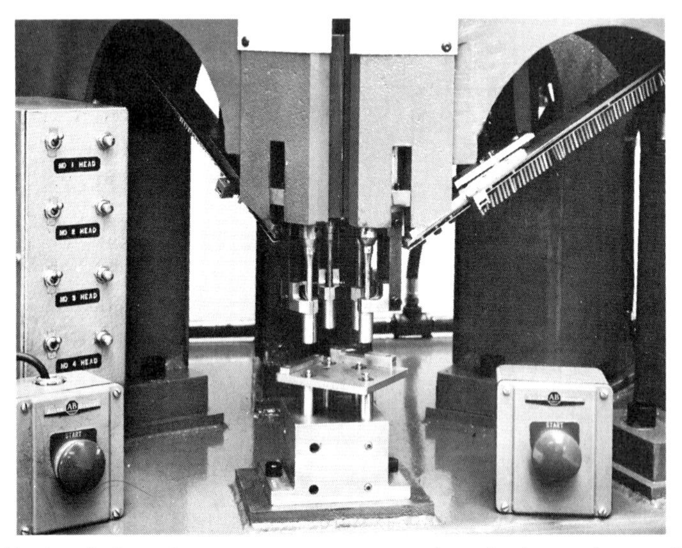

FIG. 13-33 A multiple-machine group sets two or more rivets simultaneously or, with powered fixtures, in rapid automatic sequence.

TABLE 13-7 Factors Involved in Evaluating Multiple-Machine Riveting and Single-Spindle Riveting

Production factor	Multiple machine riveting	Single-spindle riveting
Output volume	Larger runs	Smaller runs
Labor cost	Reduced substantially—one operator for each group of heads	One operator for each head
Product mix	One assembly or several similar assemblies	Variety of assemblies
Capital investment (buy or rent)	Larger—more machines, engineering and special construction needed	Smaller—nearly always standard machine
Scheduling flexibility	Should plan on long runs, few changes	For given rivet diameter, can change immediately
Location of rivets in product	All rivets in same (or nearly same) direction or plane	No restriction in direction as long as clearance OK
Complexity of product	May require higher-cost fixture to position intricate parts	Doesn't matter, as long as clearance OK
Operator skill	Same	
Space occupied	Negligible difference	

13-31 Multiple-Machine Groups Multiple-machine groups (Fig. 13-33) are designed to set two or more rivets in an assembly simultaneously or, with powered fixtures, in rapid automatic sequence. The initial cost of such arrangements is usually higher than the total cost for the same number of separate machines because more elaborate controls, special machine design, and fixtures are involved. Multiple-machine riveters are rarely used for more than one unrelated riveting task (although the group may be disassembled and the machines regrouped or operated individually).

Extremely high production rates can be obtained with multiple-machine groups at much-reduced direct labor and overhead costs, often paying for the higher initial investment in a matter of a few months. Since the basic cycling rate of the machines is so much faster than the parts can be handled, the production rate is limited only by the speed with which the operator can set the assembly parts in the riveting fixture and move them into position for riveting. The factors involved in evaluating multiple-machine riveting and single-spindle riveting as alternatives are summarized in Table 13-7.

Bench-type pneumatic riveters (Fig. 13-30*b*) are particularly well suited to multiple-machine riveting in several ways:

1. Heads are narrow and shaped for grouping.
2. There are no flywheels and clutches, as in mechanical riveters (Fig. 13-29*b*) to interfere with close mounting of riveting heads.
3. Rear feed, as contrasted with side feed in mechanical machines, permits a thin head configuration.
4. Pneumatically powered heads offer greater flexibility in machine cycling control than mechanical machines.

Most standard pneumatic machines are able to set rivets on centers as small as $^3/_4$ inch; machines with special offset drivers can set rivets on centers as small as $^3/_8$ inch. In multiple-machine riveting, all the rivets may be of different sizes and may be set at different levels of the assembly. It is even possible for one or more rivets to be set at a different angle than the others.

13-32 Automated Riveting Systems Automated riveting systems are based primarily on multiple-fixture indexing mechanisms (rotary in Fig. 13-34*a*, linear in Fig. 13-34*b*), in which fixtures containing the roll sets and parts to be riveted are automatically cycled under the rivet drivers. These systems may also require very elaborate and expensive controls. However, they may be justified by either the high production rate attained or the substantially improved operator safety, or both.

13-33 Riveter Characteristics The major components of almost all rivet-setting machines are shown on the drawing of a pedestal-type mechanical riveter in Fig. 13-35.

Throat depth (the horizontal distance back from the roll set pin to the front of the machine trunk) and loading room (the vertical distance between the top of the roll set pin and the riveter jaws in a retracted position) are two machine dimensions which can limit the sizes of the parts being riveted. While standard riveters are available with a wide variety of dimensions and operating features, it is often necessary or advantageous to seek the help of a machine manufacturer to determine if a standard or a special machine for riveting a particular assembly is required.

There are two methods by which a machine can be actuated to begin the riveting cycle. On either pedestal or bench-type machines, the clutch may be mechanically engaged with the flywheel assembly by pressing a foot treadle. The foot treadle may be fixed or on a flexible cable so that it can be located at a position convenient to the operator. Solenoid-clutch trip mechanisms are specified in systems involving automatically indexing fixtures (Fig. 13-34*a*). They are also being offered as standard equipment by many riveter manufacturers because they are compatible with many safety devices, such as non-tiedown palm switches (Fig. 13-36) and the safety ring guard (Fig. 13-54).

The choice of a pedestal-type or bench-type riveter depends mainly on the size of the parts being assembled. Large or heavy parts are more conveniently handled and placed in position for riveting at a pedestal-type machine. Most luggage and larger sheet-metal parts could not be placed in position for riveting without the extra space between the anvil bracket and floor. On the other hand, most electromechanical devices are small enough for assembly at bench-type machines. Cost is not particularly

FIG. 13-34 (*a*) Rotary multiple-fixture indexing mechanism. (*b*) Linear multiple-fixture indexing mechanism.

important, because the difference in the prices of pedestal-type and bench-type machines of the same capacity is quite small.

The type of hopper depends on the shape and size of the rivets. The rotary hopper (Figs. 13-29 and 13-30) costs least and is standard equipment on most riveting machines. It is capable of handling all but unusually long or odd-shaped rivets, but it becomes quite massive when designed to handle larger rivets. The vibratory hopper (Fig. 13-37) has electric-eye control and is used to feed very small, short, long, or odd-shaped rivets that cannot be properly oriented in rotary hoppers.

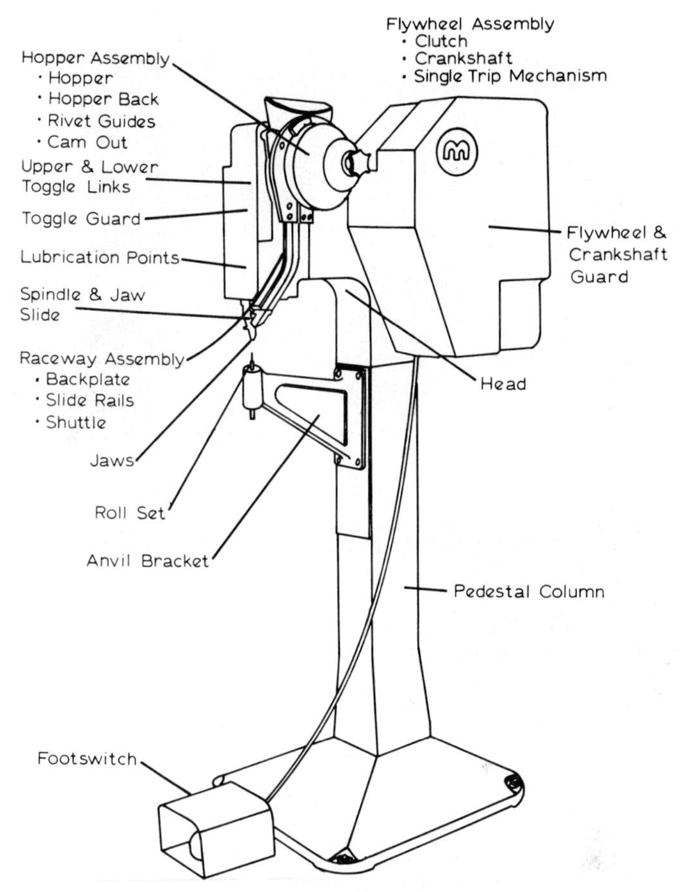

FIG. 13-35 Major components of all rivet-setting machines identified on a pedestal-type mechanical riveter.

RIVETING FIXTURES

Continuing development of improved rivet-setting machines has made it possible to produce superior fastened joints at increasingly higher production speeds. The capacity to eliminate many or all manual steps before and after riveting and to adjust the riveting cycle to match the flow requirements of associated equipment is particularly important in planning automated production. The performance of rivet-setting machines depends very much on how well the riveting fixture has been designed to prepare the parts for the riveting action itself.

Production volume, the nature of the parts to be riveted, and the rivet size usually determine the optimum type of fixturing. The higher the expected production volume over the life of the product, the more costly the riveting station is likely to be. This is not necessarily because of its greater complexity but because more rugged, heavier construction is required in order to minimize maintenance. In addition, requirements for operator safety, as represented by both government regulations and established company practice, can very much affect the ultimate fixture design.

13-34 Nesting Fixtures Nesting fixtures (Fig. 13-38) improve riveted assembly by precisely orienting parts for riveting or simply by making it easier to hold two or

FIG. 13-36 Double palm switches, which must be simultaneously pressed to activate the machine, keep the operator's hands clear of the driver area during the cycling time.

FIG. 13-37 A vibratory hopper feeds very small, short, long, or odd-shaped rivets that cannot be properly oriented in rotary hoppers.

FIG. 13-38 The shapes of the parts that are oriented for riveting in nesting fixtures may be simple (*a*) or complex (*b*).

more parts in the machine. They may be stationary devices located under the driver or may be mounted on a slide or rotary table. The shape of the nested parts may be simple (Fig. 13-38*a*) or complex (Fig. 13-38*b*).

Nesting fixtures may be relatively complex because of the many machining operations that may be involved in matching the nest to the part shape. The fixture may have to be chamfered and tapered to allow the part to be easily placed in and removed from the nest. In addition, surface finishes in nests must be especially smooth to help minimize friction drag and the part's catching on rough or sharp edges. Machining tolerances must be close in order to locate the part accurately and reduce play.

Regardless of the complexity of the nesting fixture, how it is combined with other fixturing components determines the extent to which available safety accessories may be applied. For example, if the nesting fixture is located directly under the machine driver (the point of operation), double palm switches or a point-of-operation sensing device may be required to protect the operator. On the other hand, if the nesting fixture is mounted on a hand-operated slide, so that all parts are loaded and unloaded away from the point of operation, then a single palm switch may suffice. If the fixture is electrically interlocked with the single palm switch, the riveter is cycled only when the fixture closes a limit switch as it reaches its riveting position at the same time that the operator strikes the palm switch with his free hand.

13-35 Slide Fixtures Slide fixtures (Fig. 13-39) are mainly used for improved operator safety. Other reasons include lower assembly cost, which is achieved by reducing parts handling time and thereby increasing the riveting rate. There may also be inadequate clearance over a stationary roll set so that parts must be loaded on a slide fixture in an "out" position, away from the machine driver. After loading, slide fixtures are then moved into a riveting position under the driver manually or with air cylinders or electric motors.

Manually positioned slide fixtures (Fig. 13-39*a*), as might be expected, are slower but less expensive than automated fixtures (Fig. 13-39*b*). However, safety interlocks can be designed into both manual and automated types.

A lock-and-trip slide fixture permits two or more rivets to be clinched in two or more exact positions, at each of which the fixture locks in place and trips a limit switch. The lock-and-trip arrangement is similar to the hand-operated slide fixture, except that by shifting the handle to one side the operator moves a sliding key into a slot to lock the fixture in riveting position. If the lower production rate is acceptable, it may be possible to use one riveting head in conjunction with a multiple-station lock-and-trip fixture instead of a more expensive multiple-head riveting station.

13-36 Hold-down Mechanisms A hold-down mechanism may be specified to secure the parts firmly in their riveting positions. The reason for this is that the conventional riveting roll set (Fig. 13-40) contains a spring-loaded plunger pin. As the parts are located on the fixture, the plunger pin passes through the riveting holes and protrudes above the parts to pick up and align the rivet. Without a hold-down mechanism, the spring-loaded plunger pin is depressed by the rivet as it is being clinched and, when released, pushes the riveted assembly up.

A stationary hold-down bar may be used in conjunction with either a stationary fixture, pivoting over the parts, or with a slide fixture (Fig. 13-41). The bar is mounted on the base of the fixture and set to the same height, with some clearance, as the top of the workpiece on the fixture. There is usually no downward pressure by the bar on the parts at point *a* in the riveting cycle. At point *b* in the cycle, the spring-loaded roll set plunger pin moves partially upward to hold the riveted parts against the hold-down bar. When the slide is pulled further out, the upward pressure on the riveted parts is gradually released as the assembly rides under the raised portion of the bar.

A manually operated clamp (Fig. 13-42) holds the work down more positively than a hold-down bar and interferes less with loading and unloading the fixture. Automatic hold-down mechanisms may be designed with cams, air cylinders, or solenoids to give the riveted assembly a smooth release from the fixture. Nylon inserts can be located at contact points to prevent marring of parts.

13-37 Friction-type Roll Set A friction-type roll set (Fig. 13-43) is an alternative to a hold-down mechanism. This roll-set design eliminates the spring that returns the plunger pin to the extended position. The pin extends and retracts but does not itself

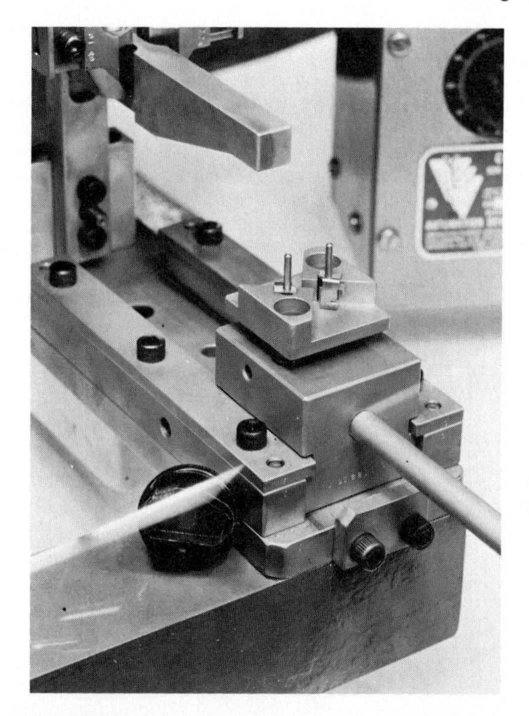

FIG. 13-39 (*a*) A manually positioned slide fixture for riveting. (*b*) An automatically positioned slide fixture for riveting.

return to the extended position. It is usually a two-piece pin in construction—a standard plunger pin supported by another pin that projects out of the bottom of the anvil. Once the plunger pin has been depressed by riveting, a cam, air cylinder, or other device must return the pin to the extended position before riveting occurs in the next cycle.

Since the pin in a friction-type roll set can be rigidly supported at the loading position, it may thus make it easier for the operator to load parts. The friction-type roll set may also be preferable in automated riveting stations since hold-down bars may interfere with loading or ejection mechanisms.

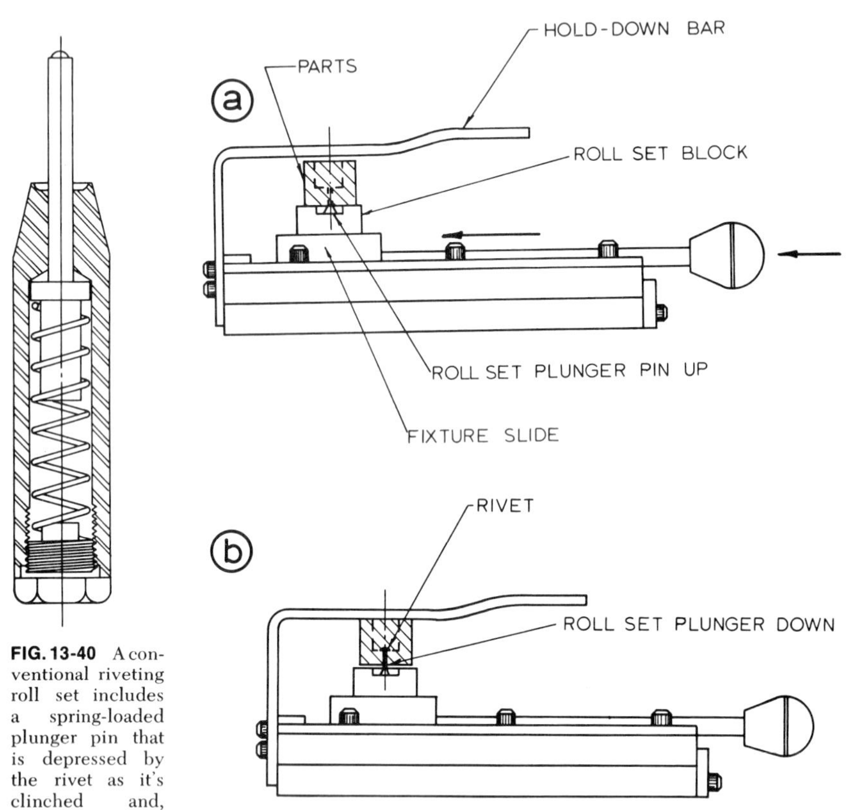

FIG. 13-40 A conventional riveting roll set includes a spring-loaded plunger pin that is depressed by the rivet as it's clinched and, when released, pushes the riveted assembly up.

FIG. 13-41 A stationary hold-down bar is usually designed to clear the parts before riveting (*a*) and, after riveting (*b*), to hold down the assembly against the upward pressure of the spring-loaded roll set plunger pin.

13-38 Rotary Tables Rotary tables (Fig. 13-34*a*) contain two or more identical riveting fixtures that are indexed into clinching position by an automatic drive unit. They are very popular fixtures in high-speed assembly because they enhance safety by isolating the operators from where the actual riveting is done and they increase production rate substantially by absorbing much of the parts handling time as part of the indexing time of the table. And the dwell time (for loading fixtures) absorbs the cycle time of the riveting machine and the time for ejecting the riveted assembly.

The same improvements in safety and productivity can be attained by mounting the fixtures on a linear conveyor (Fig. 13-34*b*) in an automated riveting station. However, a rotary table occupies less floor space—and usually involves a smaller investment—for a given number of operator stations.

A wide variety of commercial drives for rotary or linear indexing stations is available in numerous sizes and types, including geneva motions, ratchet and pawl devices, and cam-operated mechanisms. The best drive depends on the types of loads and indexing speeds. It must have the proper dynamic characteristics to function reliably at a given production rate.

When there are several rivets clinched into the parts by more than one riveting head at separate rotary-table stations, a hold-down mechanism is needed to prevent the partly riveted parts from lifting up after the first rivet has been clinched. Individual hold-down mechanisms may be incorporated into each fixture on the rotary table, actuated by a cam utilizing the rotary motion of the table. Friction-type roll sets (Fig. 13-43) can also be used in rotary tables instead of hold-downs.

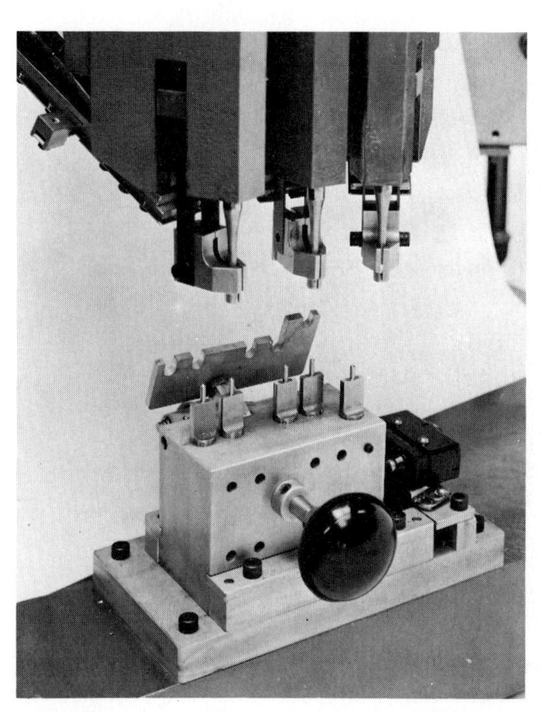

FIG. 13-42 A manually operated clamp holds the work down more positively than a hold-down bar and interferes less with loading and unloading the riveting fixture.

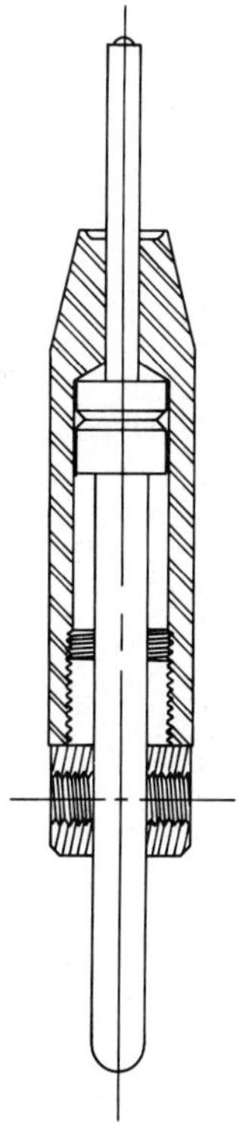

FIG. 13-43 A friction-type roll set, without a plunger pin spring, has a plunger pin that stays down after clinching until extended again by a cam or other device for the next riveting cycle.

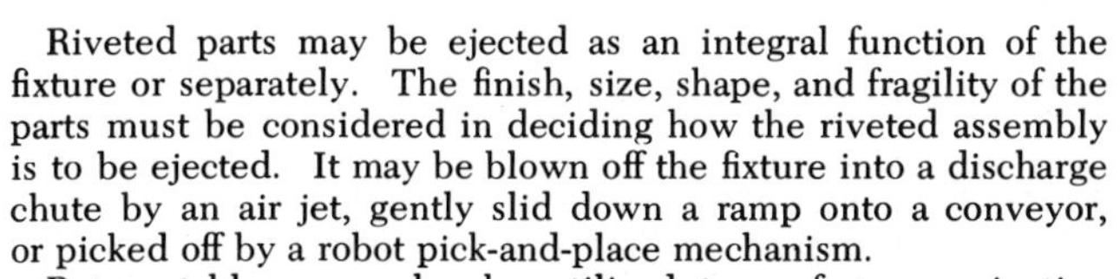

Riveted parts may be ejected as an integral function of the fixture or separately. The finish, size, shape, and fragility of the parts must be considered in deciding how the riveted assembly is to be ejected. It may be blown off the fixture into a discharge chute by an air jet, gently slid down a ramp onto a conveyor, or picked off by a robot pick-and-place mechanism.

Rotary tables may also be utilized to perform nonriveting operations on the parts to be fastened. Qualifying stations can ensure that the parts are positioned properly prior to riveting and automatically shut off the machine if they are not loaded correctly. Inspection stations can automatically determine if the parts have been fastened properly and sort assemblies into "accept" and "reject" bins.

The controls for operating rotary tables are usually mounted in one master control panel (Fig. 13-44) with all run- and jog-cycle features accessible to the operator. Hand-operated selectors and pushbutton switches are provided for various modes of operation. If there is more than one operator, an emergency stop switch should be installed at each operator's station. Rotary tables may be guarded by a shroud-type barrier (Fig. 13-44) to protect operators from the indexing fixtures.

Besides the use of nesting fixtures, slide fixtures, and rotary tables to improve safety and productivity, special riveting fixtures may be required by the shape, size, or fragility of the part itself. In addition to hold-down mechanisms and friction-type roll sets, the most common special riveting devices are lever-action roll sets, retractable plunger pins, pressure pads, and material thickness compensators.

FIG. 13-44 The controls for operating rotary tables are usually mounted in one master control panel.

EVALUATION OF CLINCH QUALITY

13-39 Acceptability of a Rivet Clinch For most rivet applications, acceptability can be confirmed quickly and positively by immediate inspection after the clinching operation. Does the clinch look all right? Does the fastened joint seem tight? However, in order to troubleshoot an operating condition that is producing unacceptable clinched rivets, it is necessary to know a good deal more than that about what makes a good clinched rivet good. Assuming that the manufactured rivet is right—proper design, material, and finish—for the application, here are the major characteristics of a well-clinched rivet (Fig. 13-45):

1. A full uniform roll—the correct rivet shank length to achieve a full clinch is the total material thickness plus a clinch allowance of 50 to 70 percent of the shank diameter.

2. Roll evenly against top surface—ideally, there should be no gap at all in fastening nonrotating components.

3. Roll free of cracks—an ideal objective which no rivet manufacturer can guaran-

tee. Minor cracks (those that don't reach the rivet shank) seldom affect the function of the riveted joint.

4. Correct hole depth/diameter in shank end of rivet.
5. Correct anvil form in the roll set.
6. Rivets properly lubricated.

A clinched rivet that looks and feels right can be expected to have these characteristics, although it is possible under some circumstances to have an acceptable clinched rivet with a clinch allowance outside the recommended range, with some cracking and without proper lubrication. In general, well-clinched semitubular rivets all have roughly the same dimensional relationships to the shank diameter:

Clinch diameter *F* about 150 percent of shank diameter *A*
Clinch height *G* about 20 percent of shank diameter *A*

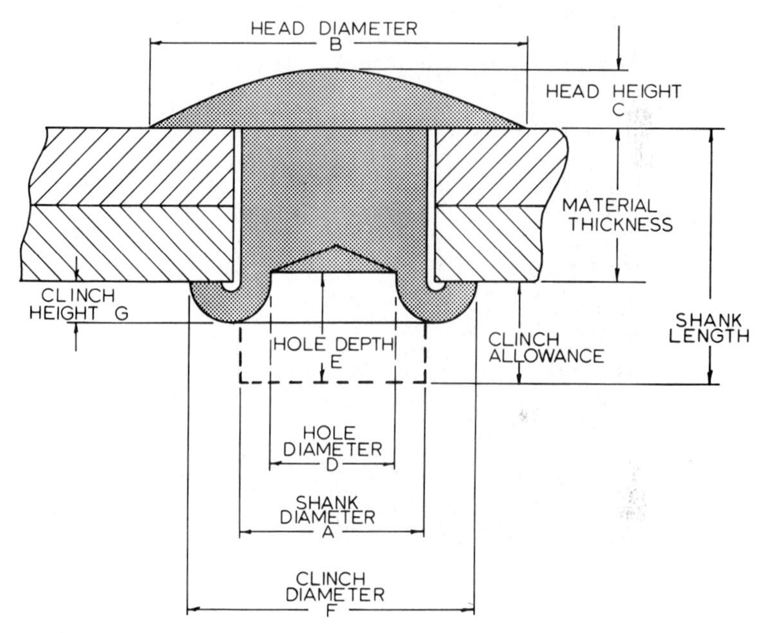

FIG. 13-45 All well-clinched semitubular rivets have roughly the same dimensional relationships.

Table 13-8 gives the standard hole diameter *D* and depth *E* for straight-hole semitubular rivets as published by the Milford Rivet & Machine Company. These standard dimensions provide the proper amount of shank material for rolling over to produce a good clinch in most properly designed riveting operations. Fastening specialists at a rivet manufacturer should be consulted before deviating from these standard dimensions.

The troubleshooting chart in Table 13-9 presents the most common types of poor clinched rivets, all of them easily detectable by eye. These riveting conditions may either be encountered immediately in testing a riveting setup before production or at any time later in an ongoing riveting operation. Many of the possible causes can be pinpointed by measuring dimensions of the rivets (the source of most troubles), holes in the parts, or the roll set. The solution, which may require consultation with the rivet or machine manufacturer, usually involves the rivet rather than the riveter.

TABLE 13-8 Standard Hole Diameter and Depth for Semitubular Rivets

As published by the Milford Rivet & Machine Company.

STRAIGHT HOLE	
D	E
.043	.041-.049
.043	.041-.049
.063	.061-.071
.063	.061-.071
.071	.069-.079
.088	.088-.098
.088	.088-.098
.105	.105-.115
.105	.105-.115
.105	.105-.115
.105	.105-.115
.136	.137-.147
.136	.137-.147
.182	.188-.198
.182	.188-.198
.224	.229-.243

MACHINE LAYOUT

13-40 Production Capacity The capacity of a rivet-setting machine is limited by the parts-handling time rather than by the cycling speed of the machine. The speed of a mechanical riveter may be as high as 280 cycles/min or more (the rotational speed of its flywheel) and pneumatic rivets can operate at up to 120 cycles/min. Even when only one or two simple parts need to be handled and powered slide fixtures are being used, time still is needed to place parts on the fixture or roll set.

Because parts handling time represents such a large portion of the riveting cycle, an important goal in production is to reduce this time through careful arrangement of machines, parts, and materials.

The simplest riveting arrangement is one operator at a single-spindle machine (Fig.

TABLE 13-9 Poor Rivet Clinches—Causes and Solutions

Types of poor clinched rivet	Possible causes	Solutions
A. Rivet clinch end only flared, not flush against part surface	1. Parts tight—rivet length too short. 2. Parts loose—roll set adjusted too low. 3. Rivet buckling in work or under head—hole diameter in part too small.	1. Correct rivet length: total material thickness plus 50%–70% of shank diameter. 2. Correct clinch adjustment. 3. Correct hole diameter: 107% of shank diameter.
B. Rivet head not seated against top surface	See possible causes, parts 2 and 3 of condition A.	See solutions, parts 2 and 3 of condition A.
C. Rivet shank buckling	1. Rivet too long. 2. Incorrect hole dimensions in semitubular end of rivet. 3. Rivet inadequately supported by parts. 4. End of rivet locking in holes in parts. 5. Worn roll set. 6. Rivets need lubrication. 7. Rivet material too soft. 8. Misalignment in riveter tools.	1. Correct rivet length: total material thickness plus 50%–70% of shank diameter. 2. Use correct hole dimension (see standard). 3. Use larger hole in semitubular end so that rivet rolls back more easily. 4. Correct hole diameter or use pressure pads. 5. Replace roll set. 6. Light waxing of rivets. 7. Check rivet supplier. 8. Check alignment by bringing driver down to plunger pin. If necessary, realign as instructed by riveter manufacturer.
D. Excessive cracking in clinch roll	1. Flaws in rivet shank. 2. Poor cutoff on rivet end or hole off center with rivet shank. 3. Form in roll-set tool too broad. 4. Rivet material too hard. 5. Hydrogen embrittlement of plated rivets. 6. Plunger pin setting of roll set too high when in retracted position.	1. Check rivet supplier. 2. Manufacture of better quality rivet by rivet supplier. 3. Use a tighter form in roll-set tool. Cracks may still exist, but will be closed up. 4. Check rivet supplier. 5. Check rivet supplier. 6. Check tool supplier for chart of proper plunger pin settings for roll sets.
E. Rivet shank buckling behind the clinch or jammed back	1. Incorrect hole dimension in semitubular end of rivet. 2. Rivets not lubricated. 3. Rivet locking in material before fully through. 4. Plunger pin set too low in relation to the form in the anvil when the pin is in the retracted position. 5. Worn roll set. 6. Loose roll-set nut. 7. Incorrect roll-set form.	1. Use correct hole dimension (see standard). 2. Light waxing of rivets. 3. Pressure pad required. Consult riveter manufacturer. 4. Adjust roll set as instructed by manufacturer. Check rivet supplier for chart of proper plunger-pin setting for roll sets. 5. Replace roll set. 6. Tighten roll-set nut. 7. Replace with correct roll set. Consult riveter manufacturer.

13-46). Stock boxes are located in front of the operator on either side of the machine head. The riveted assembly is either placed by the operator in a tote box or dropped in a chute to slide into the box. Materials handling time is minimized by reducing the amount of reaching necessary to pick up the parts and place them on the roll set. For example, gravity feed can be used to bring parts to the roll set without elaborate loading devices.

13-41 Multiple Layout A multiple-machine layout (Fig. 13-47) is simply the grouping of two or more separate machines, either pedestal or bench type (or mixed). The parts handling at each machine is the same as for the single-station arrangement in Fig. 13-46. While riveting machines are usually arranged in job-lot groups in this manner, many companies have reduced their assembly costs by grouping the necessary riveters, welders, stakers, presses, and other production machinery together for in-line flow in assembling a specific product.

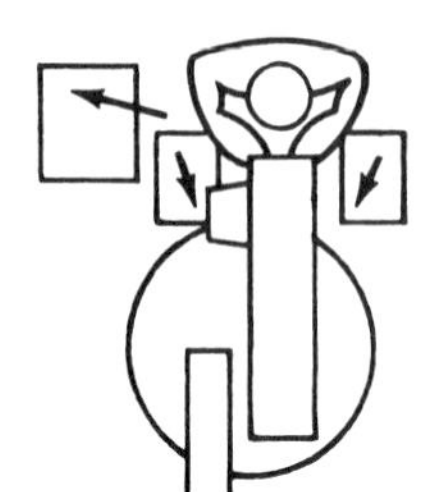

FIG. 13-46 The simplest riveting arrangement is one operator at a single-spindle machine.

13-42 Conveyorized Riveting Conveyorized riveting (Fig. 13-48) eliminates the time needed to transfer and replace tote boxes full of finished assemblies. Stocks of parts are supplied in boxes to each riveter, but the riveted assemblies are placed on a conveyor for distribution to storage or to subsequent assembly stations. If there is some arrangement for separating the riveted assemblies down the line, the machines arranged along a conveyor need not perform identical riveting operations.

Subassemblies which are too large to be handled by stock boxes are placed on the conveyor by the operators for flow between riveters (Fig. 13-49). The stock boxes contain smaller components which are riveted to the subassemblies. In contrast to the layout in Fig. 13-48, the production rate of the conveyor line in Fig. 13-49 is determined by the linear speed of the conveyor. The conveyor moves the subassemblies at a rate that is within the capacity of the four riveters. The subassemblies are picked up by one of the several operators, riveted, and then placed back on the conveyor for delivery to final assembly.

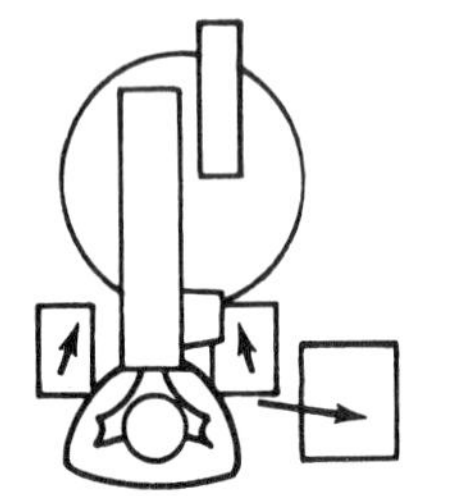

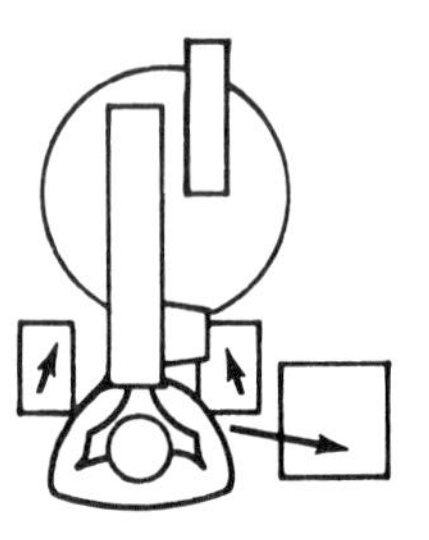

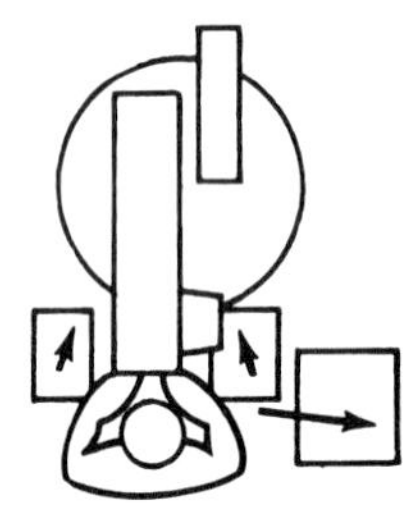

FIG. 13-47 A multiple-machine riveter layout.

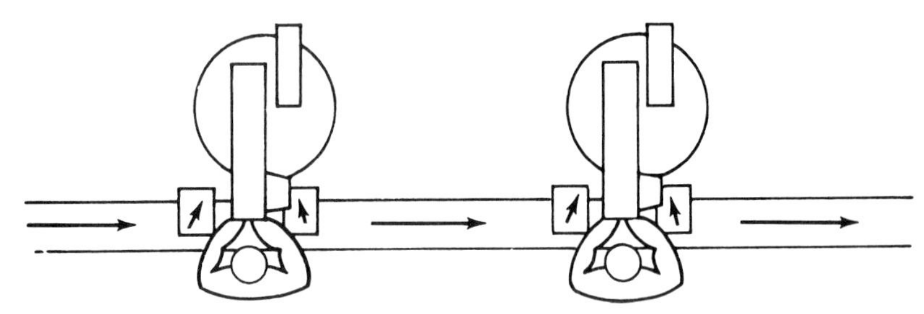

FIG. 13-48 In this form of conveyorized riveting, parts are received at each station in boxes and finished assemblies are placed on the conveyor.

13-43 Mixed Conveyorized Assembly Riveting and manual operations are combined along a conveyor in Fig. 13-50. Subassemblies are produced by an initial riveting operation (at left), move on to other stations for one or more manual operations, and are then conveyed to another riveter for a final assembly operation. This type of line is frequently used when several variations of the same type of assembly must be produced in high volume.

The production rate of the combination assembly line in Fig. 13-51 is established by the speed of the conveyor belt. Subassemblies are received by the first operator (left), and the remaining two operators each use the previous operator's assemblies as stock.

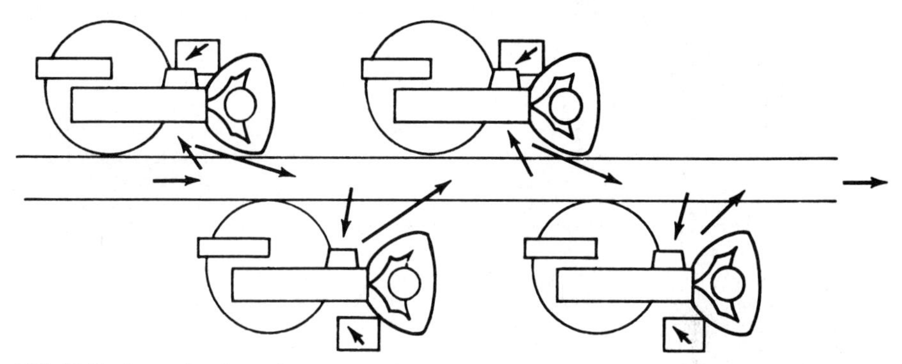

FIG. 13-49 In another form of conveyorized riveting, subassemblies are picked up by any one of the operators, riveted, and then placed back on the conveyor.

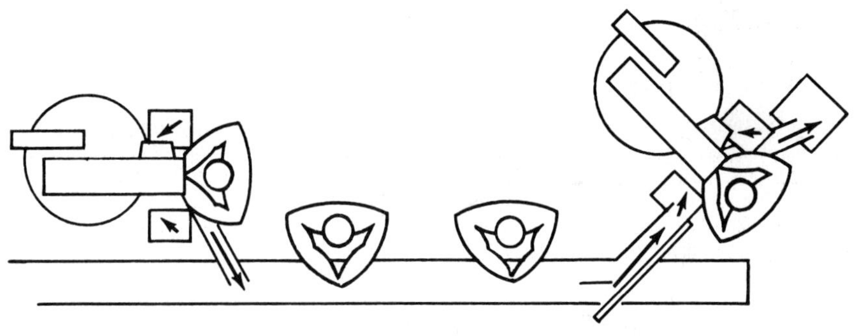

FIG. 13-50 Riveting and manual operations may be combined along a conveyor.

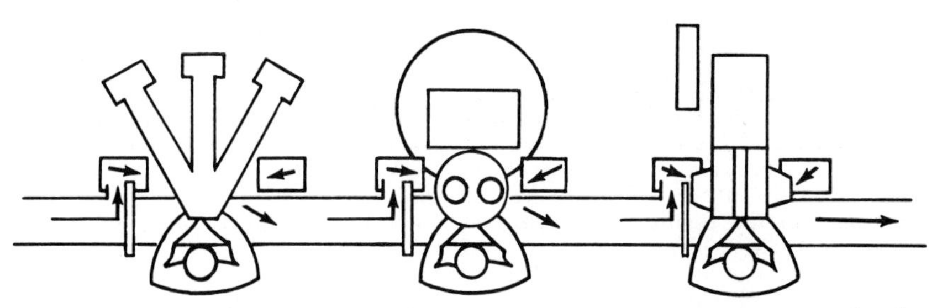

FIG. 13-51 In a combination assembly line, each operator uses the previous operator's subassembly as stock in his assembly task and then places it on the conveyor for the next operator.

A controlled line like this will not run efficiently unless there is a balanced work pattern in which each operation takes almost exactly the same amount of time. To avoid piling up of subassemblies, the belt speed must be timed to the slowest operation. If these operations are not balanced, a significant portion of the capacity of the other operators and machines will be wasted.

Two typical machine layouts using rotary indexing tables are shown in Fig. 13-52. In Fig. 13-52*a*, the rotary table indexes automatically at a predetermined rate and controls the cycling of the single-spindle riveter. The three operators at loading stations around the table place parts on eight identical fixtures at the table's indexing rate.

The more elaborate rotary-table arrangement in Fig. 13-52*b* requires only one operator. It includes an automatic parts-feed station and a secondary operation such as welding or brazing. Four rivets are set simultaneously at one station by a multiple-head riveting group.

FIG. 13-52 Two typical machine layouts using rotary indexing tables.

SAFETY IN RIVETING

13-44 Sources of Injury Safety in riveting is primarily concerned with two sources of possible injury to riveter operations: (1) moving components of the drive train while the machine power is switched on and (2) action of the riveter's driver at the point of operation during the instant that the clinching takes place. The American National Standard ANSI B154.1, "Safety Requirements for the Construction, Care, and Use of Rivet Setting Equipment," was initiated by the Tubular Rivet & Machine Institute, 331 Madison Ave., New York, N.Y. and its member firms. "Pinch point" guarding of moving components of the drive train is covered in Sec. 3 of Standard B154.1. Protection of the operator at the point of operation is covered in Sec. 5 of B154.1.

13-45 Guarding the Drive Train The two mechanical riveters in Fig. 13-53 (bench-type in Fig. 13-53*a*, pedestal-type in Fig. 13-53*b*) provide full guarding of the drive train, which consists of the flywheel, crankshaft, connecting rod, solenoid-actuated clutch trip mechanism, V belt, motor, and pulley. In Fig. 13-53*a* (insert), the foot pedal which actuates the solenoid of the clutch trip mechanism is fully shrouded to prevent accidental tripping.

13-46 Protection at Point of Operation The point of operation (dashed circles in Fig. 13-53) of a riveter must necessarily be exposed to permit the operator to position parts on the roll set or fixture for riveting. There are four ways to prevent possible injury to the operator's fingers or hands at the point of operation:

1. Prevent the machine from cycling, even if it is actuated by the operator, when a hand is under the driver.
2. Provide that hands be clear of the driver area during the cycle time.
3. Keep hands out of the driver area entirely, in or outside the cycle time.
4. Load and cycle the riveter automatically.

The riveter can be prevented from cycling if there is any obstruction under the machine driver by use of a electromechanical safety device such as the safety ring guard (Fig. 13-54). The safety ring guard pivots down under the force of gravity when the machine foot pedal is depressed. When the ring reaches a preselected point—less than a finger's width above the work—the solenoid activates the clutch trip mechanism. If an obstruction prevents this ring from reaching the preselected "down" position, the clutch trip mechanism cannot operate.

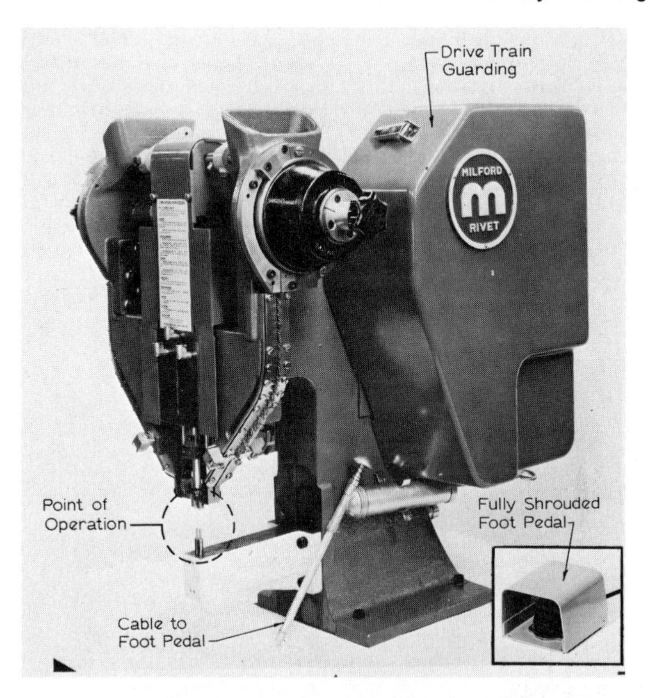

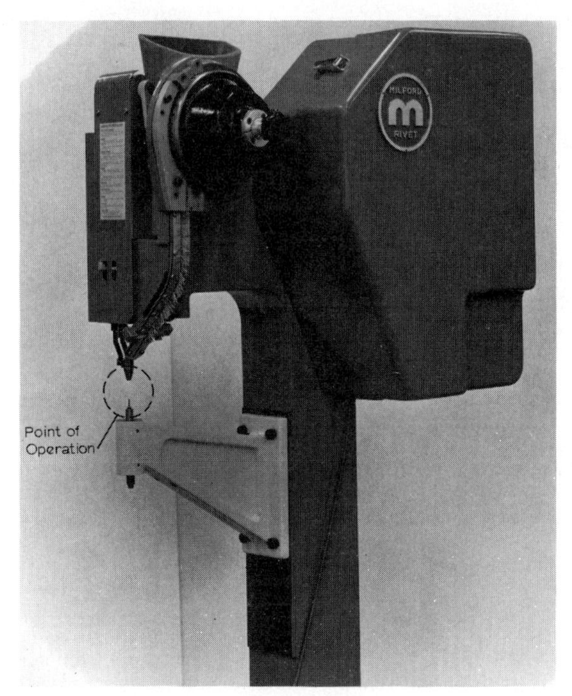

FIG. 13-53 Full guarding of the drive train is provided in safety-oriented bench-type (*a*) and pedestal-type (*b*) mechanical riveters.

Another way to accomplish the same objective is to have an electric eye set so that, even though the machine is activated, the driver of the riveter will not move down while there is an obstruction above the work. An all-electronic personnel safety guard, which is somewhat more expensive, operates by using low-frequency radio waves to generate a presence-sensing field around the point of operation.

The most positive way of ensuring that the operator's hands are clear of the driver area during the cycling time is to make it necessary to have the hands elsewhere in order to cycle the machine at all. Double palm switches (Fig. 13-36), which have been available as riveter accessories for many years, require that each switch be separately and simultaneously pressed in order to cycle the machine. Palm switches may be placed below the table surface, above the riveting head, or to the left and right of the drivers.

FIG. 13-54 The safety ring guard prevents the riveter's clutch trip mechanism from operating if there is any obstruction under the machine driver.

As might be expected, the time required for the operator to move both hands to the palm switches adds to the riveting cycle time. In this respect, the increase in safety over foot pedal actuation tends to reduce the productivity of the riveting station.

There are two methods of keeping the operator's hands out of the driver area entirely. The first method is the slide fixture (Fig. 13-39), as discussed previously. Parts are assembled on the fixture in an "out" position, and the slide is moved into "in" position for riveting either manually (Fig. 13-39*a*) or automatically (Fig. 13-39*b*).

The second method of keeping the operator's hands out of the driver area entirely—a method which substantially increases productivity—is the rotary work transfer table (Fig. 13-55). The operator is using the indexing time of the table (fixtures) and cycling time of the riveter to reach for parts, thus increasing the productivity of the station. The operator is able to accomplish this while entirely isolated from the point of operation of the riveter.

Most slide fixtures and rotary tables are loaded manually by the operator. One way to improve safety even further is to provide automatic devices to load parts on fixtures. Most special assembly machines having automatic feeding and parts-positioning capabilities are designed and constructed by the user or a special machinery builder.

ECONOMICS OF RIVETING

The best way of evaluating different fasteners and fastening methods is to compare their respective in-place fastening costs. Because this cost can be a significant part of an assembly's total manufacturing cost, it deserves careful analysis in both existing and new products.

When additional production equipment is needed to fasten a new or modified assembly, its purchase usually must be justified by factors such as a specified maximum payoff period or minimum first-year return on investment. This information can be determined readily from the in-place costs of the existing fastening method and the proposed new method.

FIG. 13-55 A rotary work-transfer table keeps the operator's hands entirely out of the driver area of the riveter.

13-47 In-Place Fastener Cost In-place fastener cost per assembly covers more than just the initial cost of the fasteners. Direct labor and factory burden also must be included. As might be expected, direct labor is the largest and most critical component of in-place cost. Factory burden varies widely with the cost-accounting practices of each manufacturer and with the type of production equipment involved. Many companies assign an overall corporate factory burden to all departments as a percent of direct labor. Other companies assign a different burden figure for each type of production equipment, a method that involves more bookkeeping but ensures more accurate cost estimates. An overall figure for factory burden as a percent of direct labor is usually adequate. A common rule of thumb in assembly operations is to assume factory burden equal to direct labor cost.

In-place cost per assembly can be computed in

$$C_A = \frac{C_L(1 + K_B/100)}{R_P} + C_F \tag{1}$$

where

C_A = in-place cost per assembly, dollars
C_L = cost of all direct labor in the production group assigned to a given assembly, dollars per hour
K_B = factory burden, percent of the hourly direct labor cost
R_P = production rate, number of assemblies produced per hour
C_F = total fastener cost per assembly, dollars

Total fastener cost per assembly may range from less than ⅓ cent for, say, a small semitubular rivet bought in large quantities to 50 cents or more for a number of patented special fasteners. Fastener cost of semitubular rivets, incidentally, is rarely more than a small fraction of the in-place cost.

The nomogram in Fig. 13-56 can be used to determine an approximate in-place cost

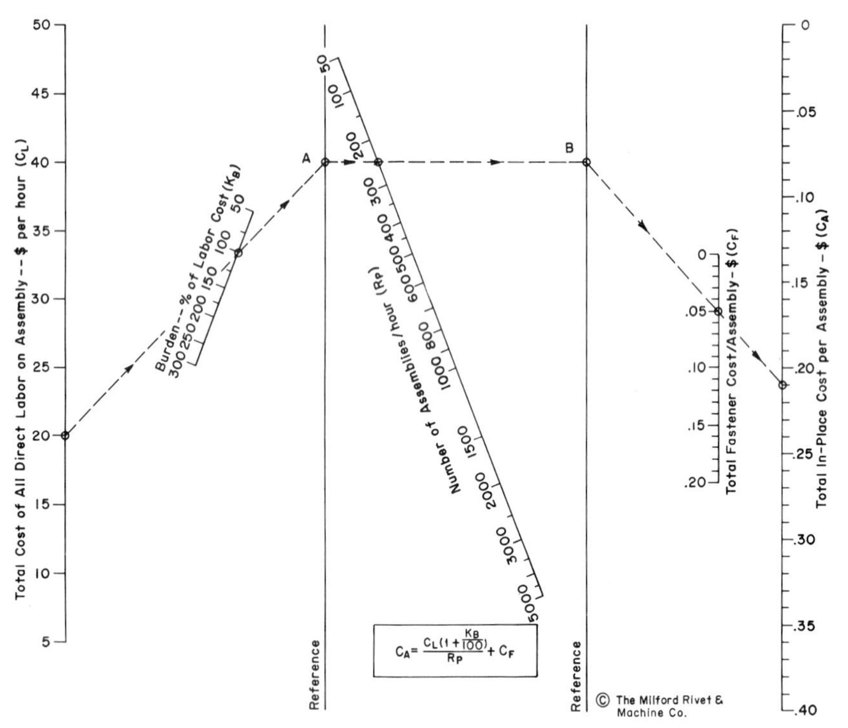

FIG. 13-56 Nomogram for determining the approximate in-place cost of any fastening operation.

more quickly, although less accurately, than by using Eq. (1) for C_A. In an example (dotted lines in Fig. 13-56), assume that several electromechanical parts are being fastened with 10 semitubular rivets purchased at a price of $5 per thousand. The applicable wage scale is $4 per hr, and five operators are needed for the assembly task. The planned production rate is $R_P = 250$ per hr.

The total direct labor cost is $C_L = 5 \times \$4 = \$20$ per hr, and the total fastener cost per assembly $C_F = \$5/1{,}000 \times 10 = \$0.05$. The factory burden is $K_B = 100$ percent (factory burden equals direct labor). Followed from left to right in Fig. 13-56, the dotted lines show the total in-place cost per assembly $C_A = \$0.21$.

The nomogram is also useful in evaluating the effects of changes in assembly arrangement. Continuing with the same example, two new single-spindle rivet-setting machines are added to two other machines already on the assembly line to make two multiple-machine groups, each pair of riveters run by one operator inserting two rivets simultaneously (rather than one after another as was done previously on the single-spindle machines). While the number of operators and C_L would remain the same, R_P might then be increased from 250 to 300 per hour. The nomogram shows that, as a result, C_A would be decreased from 21 to 18.3 cents per assembly.

13-48 Machine Investment First-year return on investment in riveting machinery is

$$P_a = \frac{100Q_a(C_{A2} - C_{A1})}{C_M + C_F} \tag{2}$$

where

Q_a = number of assemblies produced in the first year
C_{A2} and C_{A1} = higher and lower in-place costs, respectively, for the previous and new methods of doing the same fastening job
C_M and C_F = additional investments required in machines (C_M) and fixtures (C_F)

C_M represents either additional annual rent or depreciation during the assigned machine life, exceeding the amount which would otherwise be included in factory burden. C_F is almost always absorbed as expense in the first year. The payoff period in years is given as $100/P_a$.

Consider again the previous example of in-place cost calculations. Assume that the two new rivet-setting machines and the additional slide fixtures required for the two multiple-machine pairs are purchased outright. The expenses incurred are $C_M = \$4{,}800$ and $C_F = \$1{,}800$. Assume also that the purchase of the machines and fixtures must be justified by an expected volume (Q_a) of 400,000 of one type of assembly, even though the machines may be assigned to other jobs in the free time available. Then, the reduction in in-place cost previously calculated, from 21 to 18.3 cents, is used in Eq. (2) to calculate the gross first-year return on investment:

$$P_a = \frac{100(400{,}000)(0.210 - 0.183)}{4{,}800 + 1{,}800} = 164\%$$

The payoff period, therefore, is 100/164 = 0.61 years or slightly over 7 months.

13-49 Reducing Handling Time The fundamental approach for reducing in-place cost of riveting is to determine the most economical combination of factory burden and direct labor for the assembly. How much machine investment, which adds to burden, is justified to reduce how much parts-handling time, which is mostly direct labor? There are three basic methods for reducing parts-handling time:

1. Reduce materials-handling time by upgrading fixturing and machine layout.
2. Increase operator productivity through multiple riveting.
3. Overlap machine cycle and parts-handling time through the use of automatic work-transfer devices.

While methods 2 and 3 can involve substantial machine investment, method 1 usually will add very little, if anything, to the original investment. Fixturing and machine layout therefore receive special attention in assembly-cost-reduction efforts.

Evaluation of the relative in-place costs of two methods for clinching the same rivets at the same point usually does not involve quality of the fastening at all. For example, two rivets can usually be expected to be equally effective whether they are set one after another on the same machine or simultaneously on two identical machines. However, when the in-place cost approach is applied to two different types of fastening or joining methods (rivets and spot welding, for example), relative quality of the joints may be extremely important. Except in those rare cases where some compromise in quality is permitted, no advantage will be gained unless the joint quality of the new method is equal to or better than it is with the existing method under review.

Part 2 Blind Fastening

HARRY S. BRENNER, P.E.

President
Almay Research and Testing Corporation
Los Angeles, California

AN INTRODUCTION TO BLIND FASTENING

The special class of fasteners identified as *blind fasteners* constitutes one of the more innovative and invaluable systems used for modern mechanical joining. Put another way, many of today's advanced design structures and equipment simply could not be assembled without the aid of blind fasteners. While blind rivets are perhaps the most commonly recognized fastener in this class, this fastener category has been broadened to include other specialized systems specifically designed to fill a need where blind structural joining is a necessity. Actually, the blind rivet was originally developed as a replacement fastener for solid rivets where service repair was required. Since the first introduction of this unique fastening system though, blind rivets and other blind fasteners have been established as versatile and important types of fastening devices in their own right, and are now widely called for in original design applications, in addition to supporting their initial role as a vital repair fastener.

13-50 Solid Rivet System As emphasized earlier, solid rivets and specials, such as split and tubular rivets, represent a significant and valuable technique for joining materials and light structures. They are low-cost fasteners and are relatively economical to install. For successful use, the key consideration is that the rivets be fabricated from a ductile material, to permit necessary formation of the second or upset head. As a corollary to this requirement, the solid rivet system is essentially intended as a shear-type fastener and offers certain advantages in minimizing joint slip by virtue of hole-filling capability during installation. When installed, the solid rivet provides a basic permanent fastener. Critical to the installation is the requirement that there be access and clearance on both sides of the structure, to allow the driving tools and/or installation equipment to upset the rivet, as shown in Fig. 13-57.

Even though rivets are designed to be ductile, pressure is needed to fill the hole in the joint and to form the upset head, to effect a good design joint. Pneumatic-hammer riveting involving two-man teams (one on each side of the structure) has been common where low-cost tooling or on-site assembly have been necessary. It should be noted that some government concern has been expressed, particularly by OSHA, at the noise levels associated with hammer or impact riveting. To lower noise levels, new concepts of single-blow riveting are being developed and introduced to comply with government regulations.

Where access on larger-size assemblies is available, squeeze or pressure riveting can be employed to advantage by using equipment of the type illustrated in Fig. 13-58. However, many major production organizations concerned with the design and fabrication of extensive riveted structures (such as in the aerospace industry) have successfully used automatic riveting machines which are capable of being programmed and which automatically drill the structure, insert the rivet, drive the rivet, and shave the rivet where required, as in the case of a flush-head type. Figure 13-59 illustrates such a riveting installation.

The high-speed, automatic solid riveting system has been instrumental in assuring a high level of uniformity and in substantially lowering the cost of complex and large assembled structures. The benefit of such riveting systems is limited to original fabricated assemblies which can support the high equipment investment. There is, of course, a sacrifice in mobility and ability to bring the riveting machine to the structure as well as to utilize the system for repair work. In any event, whether the mobile pneumatic riveting tool or the stationary automatic riveting machine is used, the ultimate value of the solid rivet system directly depends on access to both sides of the structure to properly drive the rivet.

13-51 Blind Rivet System As distinguished from a solid rivet, the blind rivet can be inserted and fully installed in a joint from only one side of a structure. The back, or "blind" side, of this type of rivet is mechanically expanded to form a bulb or upset head, resulting in a permanently installed fastener which duplicates and sometimes exceeds the performance criteria for a comparable solid rivet installation.

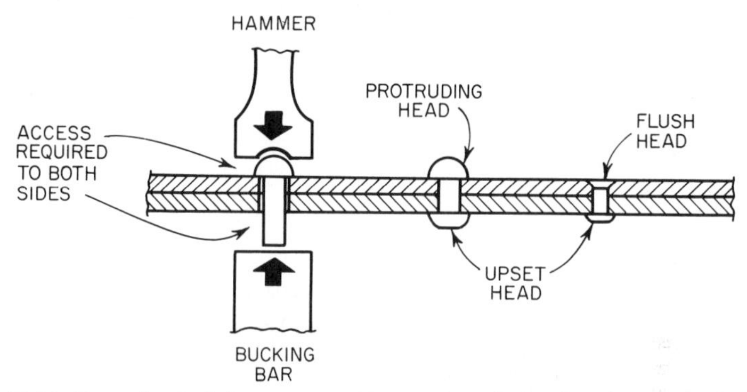

FIG. 13-57 Principle of solid rivet upset. Note access to both sides of structure required.

FIG. 13-58 Production-type squeeze riveter for driving solid rivets. (*Northrop Corp./Aircraft Div.*)

Because of the ability to install only from the accessible side of the structure, the blind fastener has lent itself to many unique joining applications. Some facets of modern design have stressed miniaturization and streamlining. Another factor in today's technology is the availability of materials in a range of structural shapes (i.e., channel, tubing, stiffeners, forms, etc.) which are widely used and which have contributed to the integrity of lightweight, built-up structural assemblies. The combination of these conditions can practically limit available accessibility to the structure and working space for assembly, as illustrated in typical examples in Fig. 13-60. To contend with the major problem of accessibility, blind fastening has found application as a prime design technique for joining materials and components which otherwise would not be capable of economical mechanical assembly. Figure 13-61 illustrates a current aircraft structure emphasizing complexities of modern design and the need for blind fastening to facilitate assembly. Similarly, reliance on the use of blind fasteners has gained general design acceptance in commercial and industrial applications as well.

Also, there has been a trend toward use of thinner structural sheet gages for many industrial applications. For very thin sheet gages, this has posed the problem of sheet damage and coining when attempting to join by means of solid riveting. The function of the blind rivet and the method of its installation often minimize this condition.

A different problem exists with completed structures or equipment already in service which may require structural repairs or modifications. Aside from the handicap of providing the proper field installation tooling to drive solid rivets, backside access of existing built-up structure may not be available, or at best may be limited to permit satisfactory installation of replacement solid rivets. In such cases, blind rivets can be used to advantage as a replacement fastener.

FIG. 13-59 Automatic-type riveting installation for riveting large-size structures and assemblies. (*Northrop Corp./Aircraft Div.*)

Blind rivets are commercially available in a comprehensive range of structural materials in both flush- and protruding-head configurations. But it should be understood that blind fasteners represent a special category and their use requires compatible installation tooling, in order to upset the blind rivet or blind fastener. Essentially, consideration of blind rivets requires a systems approach to support proper hole preparation, grip range availability, and optimum installation tooling for the specific fastening job.

13-52 Types of Blind Rivets Mechanically expanded blind rivets are the most common type employed for general industrial riveting. Included in this category are the pull-type blind rivet and the drive-pin-type blind rivet.

For the pull-type blind rivet design, the rivet body normally incorporates a through hole. The mating mandrel is positioned in the rivet body, which includes a preformed head with the mandrel extending above the rivet head. A mating installation tool is required to effect the driving of the blind rivet. The tool grips and pulls the mandrel while reacting, to hold the rivet in the structure. The action of drawing the mandrel through the rivet body expands the rivet to fill the joint hole and also expands the portion of the rivet sleeve extending on the backside of the structure to form an upset or blind head.

The pull-type blind rivet is available in two configurations, depending on the intended use of the fastener. In the self-plugging design, a portion of the mandrel is permanently retained in the rivet body, contributing additional shear-strength properties to the installed fastener. The self-plugging blind rivet is generally intended for structural applications, or where higher fastener shear strength is necessary by virtue of joint design loadings. The second category includes the pull-through blind rivet, where the mandrel is completely drawn through the rivet body after expanding the rivet. This type of blind rivet is generally used for lightly loaded or nonstructural joining applications. Representative types of pull-type blind rivets are graphically illustrated in Fig. 13-62.

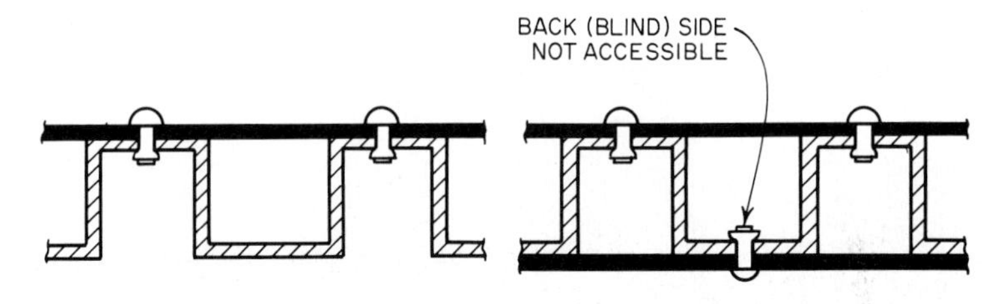

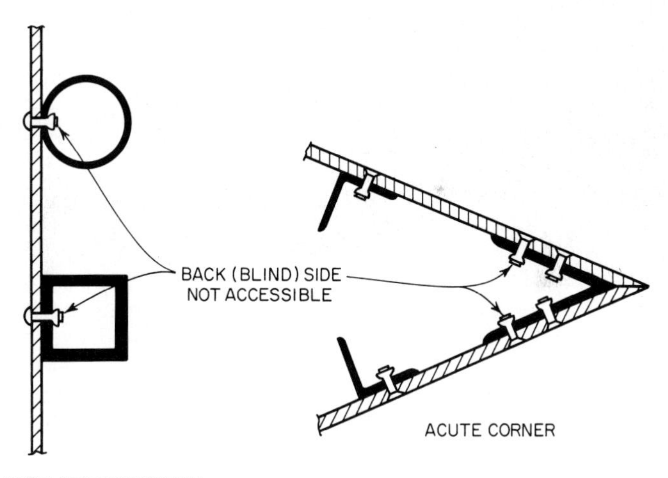

FIG. 13-60 Typical joining applications where backside access is not available.

The drive-pin rivet design includes a partial hole in the rivet body and a mating protruding pin which is positioned in the hole. In this case, the rivet body is slotted, and the pin is driven into the rivet, which results in flaring of the rivet and the resultant blind side upset. This principle of operation is also shown in Fig. 13-62.

For aerospace applications, special versions of the pull-type blind rivet have been developed for high vibration resistance and for higher shear- and fatigue-strength properties than would normally be associated with general commercial and industrial applications. The method of installation remains the same for this class of pull-type fastener. However, special design features include the ability to mechanically lock the mandrel during installation, to prevent possible loss under severe vibration environments. Typical aerospace mechanical lock-type blind rivets are shown in Fig. 13-63. The principle of operation of a lock-type blind fastener is illustrated in Fig. 13-64.

13-53 Blind Fastener Standards The prevailing use of standard blind rivets is in the $^1/_8$-, $^5/_{32}$-, $^3/_{16}$-, and $^1/_4$-inch-diameter sizes. Some rivet configurations are available in $^3/_{32}$-inch diameter, and specials are occasionally used with diameters larger than $^1/_4$ inch. Distinction is made for standard blind rivets intended for use in standard nominal hole sizes. Some series of blind rivets are designed as oversize shank diameter fasteners to replace standard-size rivets during repair operations, in lieu of going to the next full, larger rivet size. Such fasteners are valuable for repair functions but are not normally considered as representing prime standards. Structural blind fasteners, including blind bolts, are available in various configurations up to $^1/_2$-inch diameter.

FIG. 13-61 Modern airplane structure illustrating complex design and access problems requiring extensive use of blind fasteners. (*Cherry Fasteners.*)

Significant National Standards for blind rivet and blind fastener systems are reflected in specifications and standards released by the Department of Defense, the National Aerospace Standards Committee, and the Industrial Fasteners Institute. A summary of the specifications of major interest is noted in Table 13-10. Prominent industrial standards for blind rivets are illustrated for information in Tables 13-11 through 13-22.

As with other classifications of fasteners, many "specials" and proprietary designs have been adapted to specific industry applications without National Standards being issued. In some cases, limited industry standards have been released. The listing of the National Standards for blind rivets is intended as a guide to the major types and styles commercially available and generally recognized as standards for use by industry.

13-54 Blind Rivet Materials For the blind rivet system to work, the pull mandrel or drive pin is designed to be stronger than the rivet body, which must be relatively ductile to permit blind end expansion without cracking. This requires the use of multi-materials in a single fastener system, which has provided some versatility in tailoring the blind rivet to the intended application.

Aluminum alloys, low-carbon steels, nickel-copper alloys, and stainless steels have been the prime materials used for fabrication of the rivet bodies. Selection of the higher-strength mandrel materials depends in part on the type of blind rivet, i.e., self-plugging or pull-through. For the self-plugging type of blind rivet, the mandrel material should be as compatible as possible with the rivet body, to minimize potential

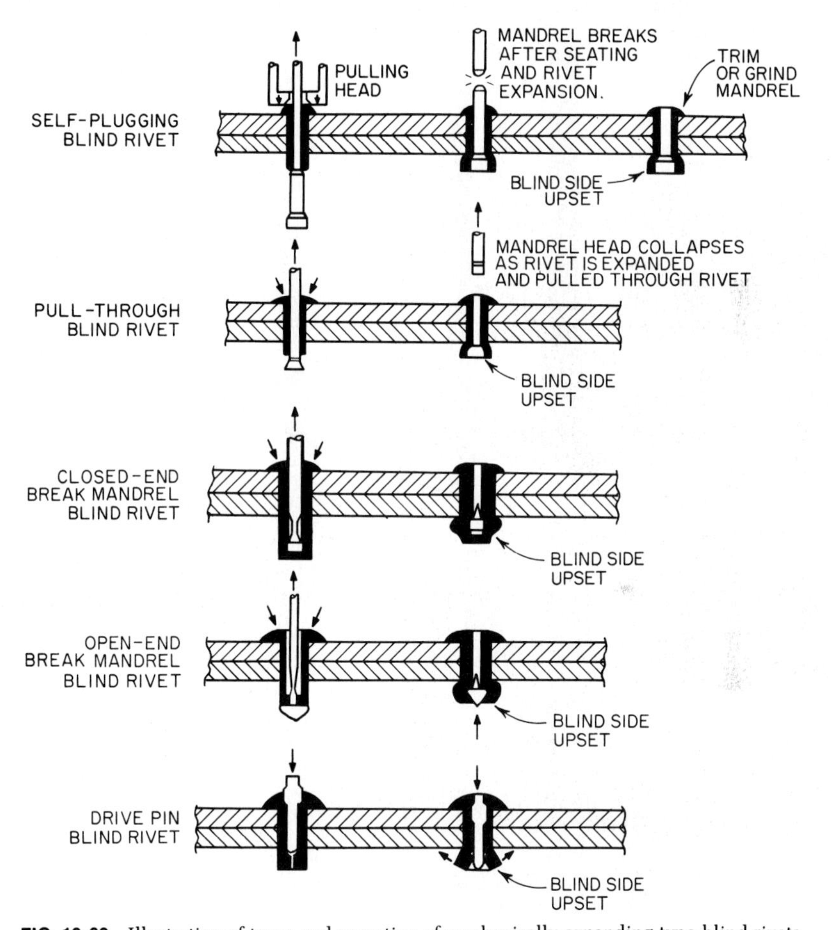

FIG. 13-62 Illustration of types and operation of mechanically expanding type blind rivets.

effects of dissimilar-metal corrosion, and also to provide balanced strength properties for the installed blind fastener. Standard blind rivets covered by Industrial Fasteners Institute Standards have been defined by grade designation for the various rivet types and rivet body and mandrel material call-outs. The grade designations are summarized in Table 13-23.

Many of the rivet materials used for commercial standards are similarly employed in the fabrication of blind rivets intended for aerospace applications. In addition, for places where higher strength is needed, or where high temperatures may be encountered, aerospace blind rivets have been fabricated from A286 corrosion- and heat-resistant alloy. Recent blind rivet systems have also been introduced which are fabricated from titanium for light weight and high-strength performance.

The great selection of blind rivets available in different structural materials requires that some care be exercised in selecting the right rivet for the right application. Earlier sections of this handbook have reviewed the problems of corrosion which may be encountered when dissimilar metals are subjected to various types of environmental exposure. The same considerations should be exercised when specifying blind rivets for installation in parent sheet materials.

FIG. 13-63 Typical high-performance mechanically locked blind rivets intended for use in aerospace applications. (*Voi-Shan/Div. of VSI Corp.*)

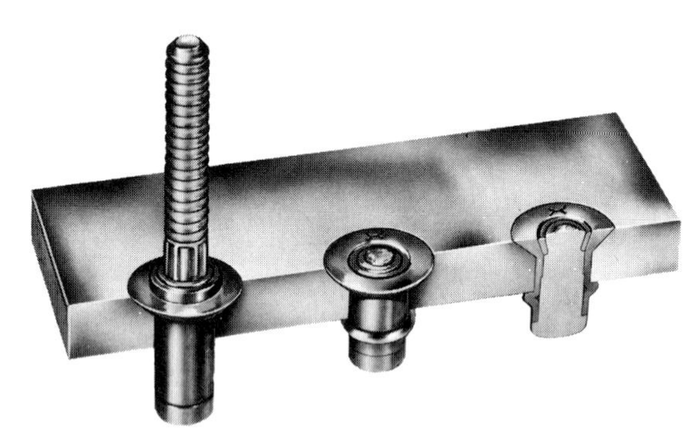

FIG. 13-64 Blind fastener illustrating mandrel mechanical lock feature. (*Huck Manufacturing Company.*)

13-55 Structural-Strength Properties Blind rivets, like solid rivets, are designed to carry primarily shear-type loadings. While structural-type blind rivets are capable of sustaining modest tensile strength, it is recognized that rivets per se are not intended as tensile-type fasteners.

Blind rivet strength, of course, will vary depending on the materials used and on the specific type or design of fastener. Ultimate shear- and tensile-strength ratings for

TABLE 13-10 Summary of Major Specifications for Blind Rivets and Blind Fastener Systems

SPECIFICATION NUMBER	TITLE	ISSUING AGENCY
MIL-R-7885	Pull Stem and Chemically Expanded Structural Blind Rivet	Dept. of Defense
MIL-R-8814	Non-Structural Type Blind Rivet	Dept. of Defense
MIL-F-8975	Corrosion and Heat Resistant Steel Positive Mechanical Lock Pull Type High Strength Blind Fasteners	Dept. of Defense
MIL-R-24243	General Specification for Retained Mandrel Non-Structural Blind Rivet	Dept. of Defense
MIL-R-27384	Drive Type Blind Rivet	Dept. of Defense
MIL-F-81177	Positive Mechanical Lock Pull Type High Strength Blind Fastener	Dept. of Defense
MIL-F-81942	Self-Locking Threaded High Strength Blind Fastener	Dept. of Defense
NAS 1400	Mechanically Locked Spindle Self-Plugging Blind Rivet	National Aerospace Standards Committee
NAS 1675	Self-Locking External Sleeve, Internally Threaded Blind Fastener	National Aerospace Standards Committee
NAS 1740	Bulbed Mechanically Locked Spindle Self-Plugging Blind Rivet	National Aerospace Standards Committee
IFI-114	Standard for Break Mandrel Blind Rivets	Industrial Fasteners Institute
IFI-116	Standard for Structural Self Plugging Pull Mandrel Blind Rivets - Type 2B	Industrial Fasteners Institute
IFI-117	Standard for Pull Through Mandrel Blind Rivets	Industrial Fasteners Institute
IFI-119	Standard for Structural Flush Break Pull Mandrel Self Plugging Blind Rivets, Type 2A	Industrial Fasteners Institute
IFI-123	Standard for Drive Pin Blind Rivets	Industrial Fasteners Institute
IFI-126	Standard for Break Mandrel Closed End Blind Rivets	Industrial Fasteners Institute

prominent grades of blind rivets covered by Industrial Fastener Institute Standards are noted for information in Table 13-24.

For many structural blind rivets falling under aerospace standards, design-load allowable shear-strength values are included in MIL-HDBK-5, "Metallic Materials and Elements for Aerospace Vehicle Structures." Attention has been focused on the strength of the total joint, which encompasses not only the blind rivet but also the type and gage of sheet material in which the rivet is installed. In heavier sheet gages, design may be predicated on developing the full rated shear strength of the blind rivet. In thinner sheet gages, though, sheet failure may be encountered at load levels well below the full shear-strength capacity of the fastener. The joint design strength values reflected in the Military Handbook are based on extensive support testing and recognition of factors of safety appropriate to aerospace structural design.

13-56 Tooling Requirements The principle of operation of the blind rivet requires the use of a mating installation tool to hold the rivet body in place while the mandrel is pulled into, or through, the rivet. The installation tooling should be considered as part of the fastening system and must be compatible with the type of blind rivet being used.

TABLE 13-11 IFI-114 Standard for Protruding-Head Break Mandrel Blind Rivets

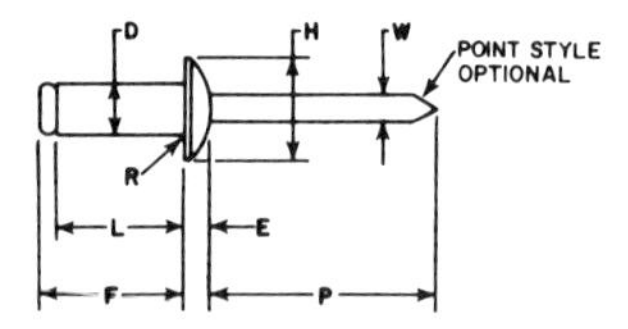

DIMENSIONS OF REGULAR AND LARGE PROTRUDING HEAD STYLE BREAK MANDREL BLIND RIVETS

RIVET SERIES NO.	NOM RIVET SIZE	D		H		E	H		E	R	W	P	F
		BODY DIA		STYLE 1 – REGULAR HEAD			STYLE 2 – LARGE HEAD			RADIUS OF FILLET	MAN-DREL DIA	MAN-DREL PROTRU-SION	BLIND SIDE PROTRU-SION
				HEAD DIA		HEAD HEIGHT	HEAD DIA		HEAD HEIGHT				
		Max	Min	Max	Min	Max	Max	Min	Max	Max	Nom	Min	Max
3	3/32 0.0938	0.096	0.090	0.198	0.178	0.032	0.293	0.269	0.040	0.015	0.057	1.00	L + .100
4	1/8 0.1250	0.128	0.122	0.262	0.238	0.040	0.390	0.360	0.065	0.020	0.076	1.00	L + .120
5	5/32 0.1562	0.159	0.153	0.328	0.296	0.050	0.488	0.448	0.075	0.020	0.095	1.06	L + .140
6	3/16 0.1875	0.191	0.183	0.394	0.356	0.060	0.650	0.600	0.092	0.025	0.114	1.06	L + .160
8	1/4 0.2500	0.255	0.246	0.525	0.475	0.080	0.780	0.720	0.107	0.030	0.151	1.25	L + .180
See Notes 3										4			5

NOTES: 1. All dimensions are in inches.

2. For application data see below.

3. Rivet series numbers represent the nominal sizes of rivets in 1/32 in.

4. The junction of head and shank shall have a fillet with a max radius as shown. For Grades 40, 50, 51 and 52 rivets, the max fillet radius for 3/16 in. rivets shall be 0.035 in., and for 1/4 in. rivets shall be 0.060 in.

5. When computing the blind side protrusion (F), the max length of rivet (L) as given below for the applicable grip shall be used. Minimum blind side clearance may be calculated by subtracting the actual grip (G), (i.e. total thickness of the material to be joined), from the specified blind side protrusion (F). (Example: To join two plates, each .100 in. thick, with a 5/32 in. rivet, a No. 54 rivet would be used. Minimum blind side clearance necessary to permit proper rivet setting would be L + .140 – G, which is .425 + .140 – .200, and equals .365 in.).

APPLICATION DATA FOR PROTRUDING HEAD STYLE BREAK MANDREL BLIND RIVETS

RIVET SERIES NO.	NOM RIVET SIZE	RECOM-MENDED DRILL SIZE	RECOMMENDED HOLE SIZE		RIVET NO.	GRIP RANGE	RIVET LENGTH L
			Max	Min			Max
3	3/32 0.0938	#41	0.100	0.097	32	.020- .125	.250
					34	.126- .250	.375
					36	.251- .375	.500
4	1/8 0.1250	#30	0.133	0.129	41	.020- .062	.212
					42	.063- .125	.275
					43	.126- .187	.337
					44	.188- .250	.400
					,5	.251- .312	.462
					46	.313- .375	.525
					48	.376- .500	.650
					410	.501- .625	.775
5	5/32 0.1562	#20	0.164	0.160	52	.020- .125	.300
					53	.126- .187	.362
					54	.188- .250	.425
					56	.251- .375	.550
					58	.376- .500	.675
					510	.501- .625	.800
6	3/16 0.1875	#11	0.196	0.192	62	.020- .125	.325
					63	.126- .187	387
					64	.188- .250	.450
					66	.251- .375	.575
					68	.376- .500	.700
					610	.501- .625	.825
					612	.626- .750	.950
					614	.751- .875	1.075
					616	.876-1.000	1.200
					618	1.001-1.125	1.325
8	1/4 0.2500	F	0.261	0.257	82	.020- .125	.375
					84	.126- .250	.500
					86	.251- .375	.625
					88	.376- .500	.750
					810	.501- .625	.875
					812	.626- .750	1.000
					814	.751- .875	1.125
					816	.876-1.000	1.250
					818	1.001-1.125	1.375
					820	1.126-1.250	1.500
See Notes		3			2		

NOTES: 1. All dimensions are in inches.

2. The first numeral in the rivet number designates the rivet series number, the last one or two numerals give the maximum grip in 1/16 in. which the rivet is capable of joining.

3. Recommended drill sizes are those which normally produce holes within the specified hole size limits.

SOURCE: Industrial Fasteners Institute.

TABLE 13-12 IFI-114 Standard for Flush-Head Break Mandrel Blind Rivets

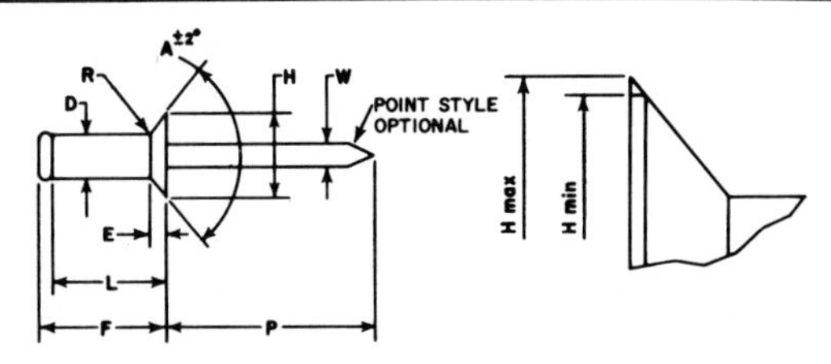

DIMENSIONS OF 100 DEG AND 120 DEG FLUSH HEAD STYLE BREAK MANDREL BLIND RIVETS

		D		A	H		E	A	H		E	R	W	P	F
				STYLE 3 – 100 DEG HEAD				STYLE 4 – 120 DEG HEAD							
RIVET SERIES NO.	NOM RIVET SIZE	BODY DIA		HEAD ANGLE	HEAD DIA		HEAD HEIGHT	HEAD ANGLE	HEAD DIA		HEAD HEIGHT	RADIUS OF FILLET	MANDREL DIA	MANDREL PROTRUSION	BLIND SIDE PROTRUSION
		Max	Min	Deg Nom	Max	Min	Ref	Deg Nom	Max	Min	Ref	Max	Nom	Min	Max
3	3/32 0.0938	0.096	0.090	100	0.187	0.161	0.039	120	0.187	0.161	0.027	0.020	0.057	1.00	L+.100
4	1/8 0.1250	0.128	0.122	100	0.233	0.207	0.045	120	0.233	0.207	0.031	0.025	0.076	1.00	L+.120
5	5/32 0.1562	0.159	0.153	100	0.294	0.268	0.058	120	0.294	0.268	0.040	0.030	0.095	1.06	L+.140
6	3/16 0.1875	0.191	0.183	100	0.361	0.335	0.073	120	0.361	0.335	0.050	0.035	0.114	1.06	L+.160
See Notes 3					4		5		4		5				6

NOTES: 1. All dimensions are in inches.
2. For application data see below.
3. Rivet series numbers represent the nominal sizes of rivets in 1/32 in.
4. Max head diameter is calculated on nominal rivet diameter and nominal head angle extended to sharp corner. Min head diameter is absolute.
5. Head height is given for reference purposes only. Variations in this dimension are controlled by the diameters (H) and (D) and the included angle of the head.
6. When computing the blind side protrusion (F), the max length of rivet (L) as given below for the applicable grip shall be used. Minimum blind side clearance may be calculated by subtracting the actual grip (G), (i.e. total thickness of the material to be joined), from the specified blind side protrusion (F). (Example: To join two plates, each .187 in. thick, with a 3/16 in. rivet, a No. 66 rivet would be used. Minimum blind side clearance necessary to permit proper rivet setting would be L+.160−G, which is .575+.160−.374 which equals .361 in.).

APPLICATION DATA FOR FLUSH HEAD STYLE BREAK MANDREL BLIND RIVETS

RIVET SERIES NO.	NOM RIVET SIZE	RECOMMENDED DRILL SIZE	RECOMMENDED HOLE SIZE		RIVET NO.	GRIP RANGE	RIVET LENGTH L
			Max	Min			Max
3	3/32 0.0938	#41	0.100	0.097	32	.079-.125	.250
					34	.126-.250	.375
4	1/8 0.1250	#30	0.133	0.129	42	.092-.125	.275
					43	.126-.187	.337
					44	.188-.250	.400
					45	.251-.312	.462
					46	.313-.375	.525
					48	.376-.500	.650
5	5/32 0.1562	#20	0.164	0.160	53	.120-.187	.362
					54	.188-.250	.425
					56	.251-.375	.550
					58	.376-.500	.675
6	3/16 0.1875	#11	0.196	0.192	63	.151-.187	.387
					64	.188-.250	.450
					66	.251-.375	.575
					68	.376-.500	.700
					610	.501-.625	.825
See Notes		3			2		

NOTES: 1. All dimensions are in inches.
2. The first numeral in the rivet number designates the rivet series number, the last one or two numerals give the maximum grip in 1/16 in. which the rivet is capable of joining.
3. Recommended drill sizes are those which normally produce holes within the specified hole size limits.

SOURCE: Industrial Fasteners Institute.

TABLE 13-13 IFI-116 Standard for Protruding-Head Self-Plugging Blind Rivets

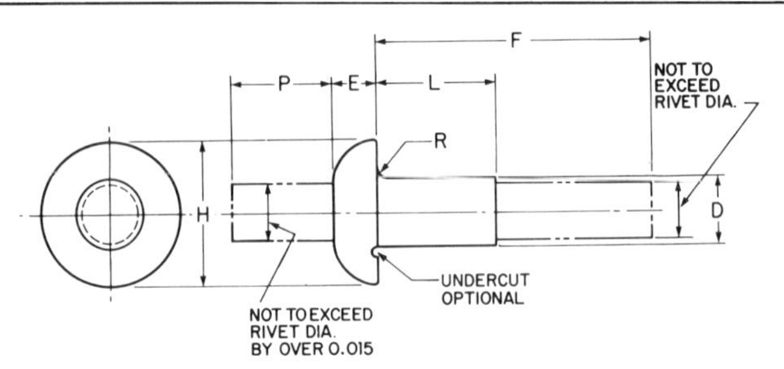

DIMENSIONS OF TYPE 2B PROTRUDING HEAD STRUCTURAL SELF PLUGGING PULL MANDREL BLIND RIVETS

RIVET SERIES NO.	NOM RIVET SIZE	D BODY DIAMETER		H HEAD DIAMETER		E HEAD HEIGHT		R RADIUS OF FILLET MAX		P MANDREL PROTRUSION	F BLIND SIDE PROTRUSION
		Max	Min	Max	Min	Max	Min	For Grades 13, 14	For Grades 30, 41	Min	Max
4	1/8 0.1250	0.128	0.124	0.262	0.238	0.064	0.054	0.01	0.02	0.75	2L + .170
5	5/32 0.1562	.159	.155	.328	.296	.077	.067	.01	.02	0.75	2L + .205
6	3/16 0.1875	.190	.186	.394	.356	.090	.080	.01	.02	0.75	2L + .239
8	1/4 0.2500	.253	.249	.525	.475	.117	.107	.01	.02	0.75	2L + .200
See Note 3		4, 5									6

NOTES:
1. All dimensions are in inches.
2. For application data see below.
3. Rivet series numbers represent the nominal size in 1/32 inch.
4. Maximum body diameter may be increased by 0.001 inch within 0.100 inch of underside of head.
5. Maximum body diameter of Grade 30 rivets may be increased by 0.001 inch when rivets are plated or coated.
6. When computing the blind side protrusion (F), the maximum length of rivet (L) as given below. for the applicable grip shall be used. Minimum blind side clearance may be calculated by subtracting the actual grip (G), (i.e., total thickness of the material to be joined), from the specified blind side protrusion (F). (Example: To join two plates, each .100 inch thick, with a 5/32 inch rivet, a No. 54 rivet would be used. Minimum blind side clearance necessary to permit proper rivet setting would be 2L + .205 - G, which is 2 x .379 + .205 - .200, and equals .763 inch.)

APPLICATION DATA FOR TYPE 2B PROTRUDING HEAD STRUCTURAL SELF PLUGGING PULL MANDREL BLIND RIVETS

RIVET SERIES NO.	NOM RIVET SIZE	RECOMMENDED DRILL SIZE	RECOMMENDED HOLE SIZE		RIVET NO.	GRIP RANGE	RIVET LENGTH L
			Max	Min			Max
4	1/8 0.1250	#30	0.133	0.129	41	To .062	.170
					42	.063 – .125	.232
					43	.126 – .187	.295
					44	.188 – .250	.357
					45	.251 – .312	.420
					46	.313 – .375	.482
					47	.376 – .437	.545
					48	.438 – .500	.607
5	5/32 0.1562	#20	0.164	0.160	51	To .062	.192
					52	.063 – .125	.254
					53	.126 – .187	.317
					54	.188 – .250	.379
					55	.251 – .312	.441
					56	.313 – .375	.503
					57	.376 – .437	.567
					58	.438 – .500	.629
6	3/16 0.1875	#11	0.196	0.192	61	To .062	.215
					62	.063 – .125	.277
					63	.126 – .187	.340
					64	.188 – .250	.402
					65	.251 – .312	.465
					66	.313 – .375	.527
					67	.376 – .437	.590
					68	.438 – .500	.652
					69	.501 – .562	.715
					610	.563 – .625	.777
					611	.626 – .687	.840
					612	.688 – .750	.902
8	1/4 0.2500	F	0.261	0.257	83	.125 – .187	.385
					84	.188 – .250	.447
					85	.251 – .312	.510
					86	.313 – .375	.572
					87	.376 – .437	.635
					88	.438 – .500	.697
					89	.501 – .562	.760
					810	.563 – .625	.832
					811	.626 – .687	.885
					812	.688 – .750	.947
					813	.751 – .812	1.010
					814	.813 – .875	1.072
See Note		3			2		

NOTES:
1. All dimensions are in inches.
2. The first numeral in the rivet number designates the rivet series number, the last one or two numerals give the maximum grip in 1/16 inch which the rivet is capable of joining.
3. Recommended drill sizes are those which normally produce holes within the specified hole size limits.

SOURCE: Industrial Fasteners Institute.

TABLE 13-14 IFI-116 Standard for Flush-Head Self-Plugging Blind Rivets

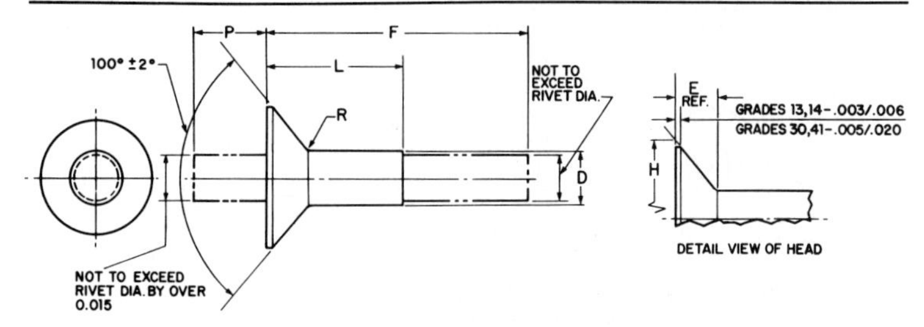

DIMENSIONS OF TYPE 2B 100 DEG FLUSH HEAD STRUCTURAL SELF PLUGGING PULL MANDREL BLIND RIVETS

RIVET SERIES NO.	NOM RIVET SIZE	D BODY DIAMETER Max	D BODY DIAMETER Min	H HEAD DIAMETER Max	H HEAD DIAMETER Min	E HEAD HEIGHT Ref	R RADIUS OF FILLET MAX For Grades 13, 14	R RADIUS OF FILLET MAX For Grades 30, 41	P MANDREL PROTRUSION Min	F BLIND SIDE PROTRUSION Max
4	1/8 0.1250	0.128	0.124	0.229	0.206	0.042	0.01	0.02	0.75	2L + .170
5	5/32 0.1562	.159	.155	.290	.265	.055	.01	.02	0.75	2L + .205
6	3/16 0.1875	.190	.186	.357	.329	.070	.01	.02	0.75	2L + .239
8	1/4 0.2500	.253	.249	.480	.447	.095	.01	.02	0.75	2L + .200
See Note 3		6		4		5				7

NOTES:

1. All dimensions are in inches.
2. For application data see below.
3. Rivet series numbers represent the nominal size of rivets in 1/32 inch.
4. Maximum head diameter is calculated on nominal rivet diameter and nominal head angle extended to sharp corner. Minimum head diameter is absolute.
5. Head height is given for reference purposes only. Variations in this dimension are controlled by the diameters (H) and (D) and the included angle of the head.
6. Maximum body diameter of Grade 30 rivets may be increased by 0.001 inch when rivets are plated or coated.
7. When computing the blind side protrusion (F), the maximum length of rivet (L) as given below for the applicable grip shall be used. Minimum blind side clearance may be calculated by subtracting the actual grip (G), (i.e., total thickness of the material to be joined), from the specified blind side protrusion (F). (Example: To join two plates, each .187 inch thick with a 3/16 inch rivet, a No. 66 rivet would be used. Minimum blind side clearance necessary to permit proper rivet setting would be 2L + .239 - G, which is 2 x .527 + .239 - .374 which equals .919 inch.)

APPLICATION DATA FOR TYPE 2B 100 DEG FLUSH HEAD STRUCTURAL SELF PLUGGING PULL MANDREL BLIND RIVETS

RIVET SERIES NO.	NOM RIVET SIZE	RECOMMENDED DRILL SIZE	RECOMMENDED HOLE SIZE Max	RECOMMENDED HOLE SIZE Min	RIVET NO.	GRIP RANGE	RIVET LENGTH L Max
4	1/8 0.1250	#30	0.133	0.129	41	To .062	.170
					42	.063 – .125	.232
					43	.126 – .187	.295
					44	.188 – .250	.357
					45	.251 – .312	.420
					46	.313 – .375	.482
					47	.376 – .437	.545
					48	.438 – .500	.607
5	5/32 0.1562	#20	0.164	0.160	52	.065 – .125	.254
					53	.126 – .187	.317
					54	.188 – .250	.379
					55	.251 – .312	.441
					56	.313 – .375	.503
					57	.376 – .437	.567
					58	.438 – .500	.629
6	3/16 0.1875	#11	0.196	0.192	62	.080 – .125	.277
					63	.126 – .187	.340
					64	.188 – .250	.402
					65	.251 – .312	.465
					66	.313 – .375	.527
					67	.376 – .437	.590
					68	.438 – .500	.652
					69	.501 – .562	.715
					610	.563 – .625	.777
					611	.626 – .687	.840
					612	.688 – .750	.902
8	1/4 0.2500	F	0.261	0.257	83	.125 – .187	.385
					84	.188 – .250	.447
					85	.251 – .312	.510
					86	.313 – .375	.572
					87	.376 – .437	.635
					88	.438 – .500	.697
					89	.501 – .562	.760
					810	.563 – .625	.822
					811	.626 – .687	.885
					812	.688 – .750	.947
					813	.751 – .812	1.010
					814	.813 – .875	1.072
See Note		3			2		

NOTES:

1. All dimensions are in inches.
2. The first numeral in the rivet number designates the rivet series number, the last one or two numerals give the maximum grip in 1/16 inch which the rivet is capable of joining.
3. Recommended drill sizes are those which normally produce holes within the specified hole size limits.

SOURCE: Industrial Fasteners Institute.

TABLE 13-15 IFI-117 Standard for Protruding-Head Pull-Through Blind Rivets

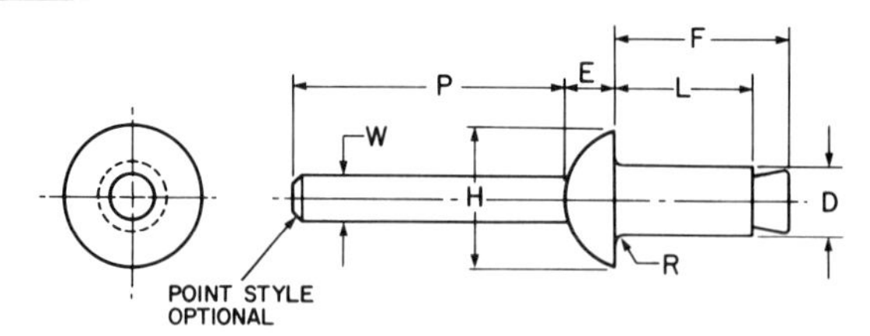

DIMENSIONS OF REGULAR AND LARGE PROTRUDING HEAD PULL THROUGH MANDREL BLIND RIVETS

RIVET SERIES NO.	NOM RIVET SIZE	D BODY DIAMETER		H STYLE 1 REGULAR HEAD HEAD DIAMETER		E STYLE 1 HEAD HEIGHT	H STYLE 2 LARGE HEAD HEAD DIAMETER		E STYLE 2 HEAD HEIGHT	R RADIUS OF FILLET	W MANDREL DIA	P MANDREL PROTRUSION	F BLIND SIDE PROTRUSION
		Max	Min	Max	Min	Max	Max	Min	Max	Max	Nom	Min	Max
4	1/8 0.1250	0.128	0.122	0.262	0.238	0.064	0.390	0.360	0.065	0.025	0.085	0.75	L + .093
5	5/32 0.1562	0.159	0.153	0.328	0.296	0.077	0.488	0.448	0.075	0.025	0.107	0.75	L + .125
6	3/16 0.1875	0.190	0.184	0.394	0.356	0.090	0.650	0.600	0.092	0.030	0.126	0.75	L + .141
8	1/4 0.2500	0.253	0.247	0.525	0.475	0.117	0.780	0.720	0.107	0.035	0.169	0.75	L + .156
See Note 3		4, 5								6			7

NOTES:
1. All dimensions are in inches.
2. For application data see below.
3. Rivet series numbers represent the nominal size in 1/32 inch.
4. Maximum body diameter may be increased by 0.001 inch within 0.100 inch of underside of head.
5. Maximum body diameter of Grade 30 rivets may be increased by 0.001 inch when rivets are plated or coated.
6. For Grade 40 rivets, the maximum fillet radius for 3/16 inch rivets may be 0.035 inch, and for 1/4 inch rivets may be 0.060 inch.
7. When computing the blind side protrusion (F), the maximum length of rivet (L) as given below for the applicable grip shall be used. Minimum blind side clearance may be calculated by subtracting the actual grip (G), (i.e., total thickness of the material to be joined), from the specified blind side protrusion (F). (Example: To join two plates, each .100 inch thick, with a 5/32 inch rivet, a No. 54 rivet would be used. Minimum blind side clearance necessary to permit proper rivet setting would be L + .125 - G, which is .380 + .125 - .200, and equals .305 inch.)

APPLICATION DATA FOR PROTRUDING HEAD PULL THROUGH MANDREL BLIND RIVETS

RIVET SERIES NO.	NOM RIVET SIZE	RECOMMENDED DRILL SIZE	RECOMMENDED HOLE SIZE Max	RECOMMENDED HOLE SIZE Min	RIVET NO.	RECOMMENDED GRIP RANGE	RIVET LENGTH L Max
4	1/8 0.1250	#30	0.133	0.129	41	To .062	.170
					42	.063 – .125	.232
					43	.126 – .187	.295
					44	.188 – .250	.357
					45	.251 – .312	.420
					46	.313 – .375	.482
					47	.376 – .437	.545
					48	.438 – .500	.607
					49	.501 – .562	.670
					410	.563 – .625	.732
5	5/32 0.1562	#20	0.164	0.160	51	To .062	.192
					52	.063 – .125	.255
					53	.126 – .187	.317
					54	.188 – .250	.380
					55	.251 – .312	.442
					56	.313 – .375	.505
					57	.376 – .437	.567
					58	.438 – .500	.630
					59	.501 – .562	.692
					510	.563 – .625	.755
6	3/16 0.1875	#11	0.196	0.192	61	To .062	.215
					62	.063 – .125	.277
					63	.126 – .187	.340
					64	.188 – .250	.402
					65	.251 – .312	.465
					66	.313 – .375	.527
					67	.376 – .437	.590
					68	.438 – .500	.652
					69	.501 – .562	.715
					610	.563 – .625	.777
					611	.626 – .687	.840
					612	.688 – .750	.902
					613	.751 – .812	.965
					614	.813 – .875	1.027
					615	.876 – .937	1.090
					616	.938 – 1.000	1.152
8	1/4 0.2500	F	0.261	0.257	82	.063 – .125	.322
					83	.126 – .187	.385
					84	.188 – .250	.447
					85	.251 – .312	.510
					86	.313 – .375	.572
					87	.376 – .437	.635
					88	.438 – .500	.697
					89	.501 – .562	.760
					810	.563 – .625	.822
					811	.626 – .687	.885
					812	.688 – .750	.947
					813	.751 – .812	1.010
					814	.813 – .875	1.072
See Note		3			2	4	

NOTES:
1. All dimensions are in inches.
2. The first numeral in the rivet number designates the rivet series number, the last one or two numerals give the maximum grip in 1/16 inch which the rivet is capable of joining.
3. Recommended drill sizes are those which normally produce holes within the specified hole size limits.
4. If economically feasible, and if blind side clearances permit, rivets with lengths longer than those recommended for a given grip may be substituted. In this way, the number of different inventory items may be reduced.

SOURCE: Industrial Fasteners Institute.

TABLE 13-16 IFI-117 Standard for Flush-Head Pull-Through Blind Rivets

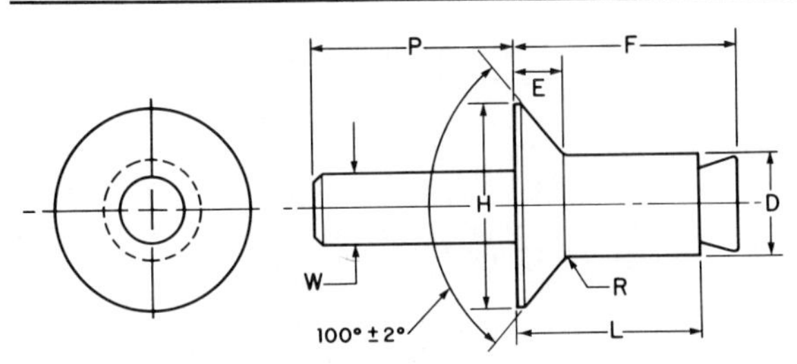

DIMENSIONS OF 100 DEG FLUSH HEAD PULL THROUGH MANDREL BLIND RIVETS

		D		H		E	R	W	P	F
RIVET SERIES NO.	NOM RIVET SIZE	BODY DIAMETER		HEAD DIAMETER		HEAD HEIGHT	RADIUS OF FILLET	MANDREL DIA	MANDREL PROTRU-SION	BLIND SIDE PROTRU-SION
		Max	Min	Max	Min	Ref	Max	Nom	Min	Max
4	1/8 0.1250	0.128	0.122	0.233	0.206	0.045	0.025	0.085	0.75	L + .093
5	5/32 0.1562	0.159	0.153	0.294	0.265	0.058	0.030	0.107	0.75	L + .125
6	3/16 0.1875	0.190	0.184	0.361	0.329	0.073	0.035	0.126	0.75	L + .141
8	1/4 0.2500	0.253	0.247	0.484	0.447	0.098	0.040	0.169	0.75	L + .156
See Note 3		6		4		5				7

NOTES:

1. All dimensions are in inches.
2. For application data **see below**.
3. Rivet series numbers represent the nominal size of rivets in 1/32 inch.
4. Maximum head diameter is calculated on nominal rivet diameter and nominal head angle extended to sharp corner. Minimum head diameter is absolute.
5. Head height is given for reference purposes only. Variations in this dimension are controlled by the diameters (H) and (D) and the included angle of the head.
6. Maximum body diameter of Grade 30 rivets may be increased by 0.001 inch when rivets are plated or coated.
7. When computing the blind side protrusion (F), the maximum length of rivet (L) as given **below** for the applicable grip shall be used. Minimum blind side clearance may be calculated by subtracting the actual grip (G), (i.e., total thickness of the material to be joined), from the specified blind side protrusion (F). (Example: To join two plates, each .187 inch thick with a 3/16 inch rivet, a No. 66 rivet would be used. Minimum blind side clearance necessary to permit proper rivet setting would be L + .141 - G, which is .527 + .141 - .374 which equals .294 inch.)

APPLICATION DATA FOR 100 DEG FLUSH HEAD PULL THROUGH MANDREL BLIND RIVETS

RIVET SERIES NO.	NOM RIVET SIZE	RECOM-MENDED DRILL SIZE	RECOMMENDED HOLE SIZE		RIVET NO.	RECOMMENDED GRIP RANGE	RIVET LENGTH L
			Max	Min			Max
4	1/8 0.1250	#30	0.133	0.129	41	To .062	.170
					42	.063 – .125	.232
					43	.126 – .187	.295
					44	.188 – .250	.357
					45	.251 – .312	.420
					46	.313 – .375	.482
					47	.376 – .437	.545
					48	.438 – .500	.607
					–	.501 – .562	–
					410	.563 – .625	.732
5	5/32 0.1562	#20	0.164	0.160	52	.063 – .125	.255
					53	.126 – .187	.317
					54	.188 – .250	.380
					55	.251 – .312	.442
					56	.313 – .375	.505
					57	.376 – .437	.567
					58	.438 – .500	.630
					59	.501 – .562	.692
					510	.563 – .625	.755
6	3/16 0.1875	#11	0.196	0.192	62	.063 – .125	.277
					63	.126 – .187	.340
					64	.188 – .250	.402
					65	.251 – .312	.465
					66	.313 – .375	.527
					67	.376 – .437	.590
					68	.438 – .500	.652
					69	.501 – .562	.715
					610	.563 – .625	.777
					611	.626 – .687	.840
					612	.688 – .750	.902
					613	.751 – .812	.965
					–	.813 – .875	1.027
					–	.876 – .937	1.090
					–	.938 – 1.000	1.152
8	1/4 0.2500	F	0.261	0.257	82	.063 – .125	.322
					83	.126 – .187	.385
					84	.188 – .250	.447
					85	.251 – .312	.510
					86	.313 – .375	.572
					87	.376 – .437	.635
					88	.438 – .500	.697
					89	.501 – .562	.760
					810	.563 – .625	.822
					811	.626 – .687	.885
					812	.688 – .750	.947
					813	.751 – .812	1.010
					814	.813 – .875	1.072
See Note		3			2	4	

NOTES:

1. All dimensions are in inches.
2. The first numeral in the rivet number designates the rivet series number, the last one or two numerals give the maximum grip in 1/16 inch which the rivet is capable of joining.
3. Recommended drill sizes are those which normally produce holes within the specified hole size limits.
4. If economically feasible, and if blind side clearances permit, rivets with lengths longer than those recommended for a given grip may be substituted. In this way, the number of different inventory items may be reduced.

SOURCE: Industrial Fasteners Institute.

TABLE 13-17 IFI-119 Standard for Protruding-Head Self-Plugging Blind Rivets

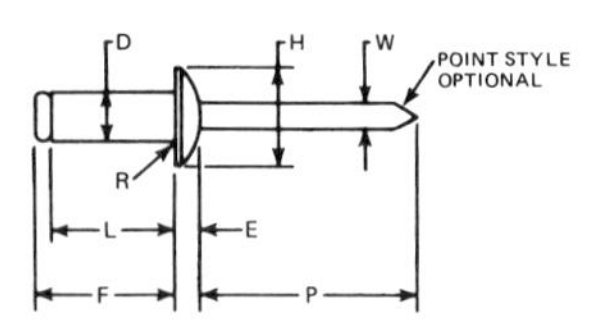

DIMENSIONS OF REGULAR AND LARGE PROTRUDING HEAD STYLE STRUCTURAL FLUSH BREAK PULL MANDREL SELF PLUGGING BLIND RIVETS, TYPE 2A

RIVET SERIES NO.	NOM RIVET SIZE	D		H		E	H		E	R	W	P	F
		BODY DIA		STYLE 1 - REGULAR HEAD HEAD DIA		STYLE 1 - REGULAR HEAD HEAD HEIGHT	STYLE 2 - LARGE HEAD HEAD DIA		STYLE 2 - LARGE HEAD HEAD HEIGHT	RADIUS OF FILLET	MAN-DREL DIA	MAN-DREL PROTRU-SION	BLIND SIDE PROTRU-SION
		Max	Min	Max	Min	Max	Max	Min	Max	Max	Nom	Min	Max
4	1/8 0.1250	0.128	0.122	0.262	0.238	0.040	0.390	0.360	0.065	0.020	0.076	1.00	L + .187
5	5/32 0.1562	0.159	0.153	0.328	0.296	0.050	0.488	0.448	0.075	0.020	0.095	1.06	L + .203
6	3/16 0.1875	0.191	0.183	0.394	0.356	0.060	0.650	0.600	0.092	0.025	0.114	1.06	L + .218
8	1/4 0.2500	0.255	0.246	0.525	0.475	0.080	0.780	0.720	0.107	0.030	0.151	1.25	L + .250
See Notes 3										4			5

NOTES:

1. All dimensions are in inches.
2. For application data see below.
3. Rivet series numbers represent the nominal sizes of rivets in 1/32 in.
4. The junction of head and shank shall have a fillet with a max radius as shown. For Grades 40, 50, 51 and 52 rivets, the max fillet radius for 3/16 in. rivets shall be 0.035 in., and for 1/4 in. rivets shall be 0.060 in.
5. When computing the blind side protrusion (F), the max length of rivet (L) as given below for the applicable grip shall be used. Minimum blind side clearance may be calculated by subtracting the actual grip (G), (i.e. total thickness of the material to be joined), from the specified blind side protrusion (F). (Example: To join two plates, each .100 in. thick, with a 5/32 in. rivet, a No. 53 rivet would be used. Minimum blind side clearance necessary to permit proper rivet setting would be L + .203 – G, which is .362 + .203 – .200, and equals .365 in.).

APPLICATION DATA FOR PROTRUDING HEAD STYLE STRUCTURAL FLUSH BREAK PULL MANDREL SELF PLUGGING BLIND RIVETS, TYPE 2A

RIVET SERIES NO.	NOM RIVET SIZE	RECOM-MENDED DRILL SIZE	RECOMMENDED HOLE SIZE		RIVET NO.	GRIP RANGE		RIVET LENGTH L
			Max	Min		Min	Max	Max
4	1/8 0.1250	#30	0.133	0.129	41	0.047	0.078	0.212
					41.5	0.078	0.109	0.275
					42	0.109	0.141	0.275
					42.5	0.141	0.172	0.337
					43	0.172	0.203	0.337
					43.5	0.203	0.234	0.400
					44	0.234	0.266	0.400
					44.5	0.266	0.297	0.462
					45	0.297	0.328	0.462
					45.5	0.328	0.359	0.525
					46	0.359	0.391	0.525
					46.5	0.391	0.422	0.587
					47	0.422	0.453	0.587
					47.5	0.453	0.484	0.650
					48	0.484	0.516	0.650
					48.5	0.515	0.546	0.712
					49	0.546	0.578	0.712
					49.5	0.578	0.609	0.775
					410	0.609	0.640	0.775
5	5/32 0.1562	#20	0.164	0.160	52	0.109	0.141	0.300
					52.5	0.141	0.172	0.362
					53	0.172	0.203	0.362
					53.5	0.203	0.234	0.425
					54	0.234	0.266	0.425
					54.5	0.266	0.297	0.487
					55	0.297	0.328	0.487
					55.5	0.328	0.359	0.550
					56	0.359	0.391	0.550
					56.5	0.391	0.422	0.613
					57	0.422	0.453	0.613
					57.5	0.453	0.484	0.675
					58	0.484	0.516	0.675
					58.5	0.515	0.546	0.737
					59	0.546	0.578	0.737
					59.5	0.578	0.609	0.800
					510	0.609	0.640	0.800
6	3/16 0.187	#11	0.196	0.192	61	0.046	0.078	0.325
					62	0.093	0.156	0.325
					63	0.156	0.218	0.450
					64	0.218	0.281	0.450
					65	0.281	0.343	0.575
					66	0.343	0.406	0.575
					67	0.406	0.468	0.700
					68	0.468	0.531	0.700
					69	0.531	0.593	0.825
					610	0.593	0.656	0.825
					611	0.656	0.719	0.950
					612	0.719	0.781	0.950
					613	0.781	0.843	1.075
					614	0.843	0.906	1.075
					615	0.906	0.968	1.200
					616	0.968	1.032	1.200
8	1/4 0.2500	F	0.261	0.257	82	0.093	0.156	0.375
					83	0.156	0.218	0.500
					84	0.218	0.281	0.500
					85	0.281	0.343	0.625
					86	0.343	0.406	0.625
					87	0.406	0.468	0.750
					88	0.468	0.531	0.750
					89	0.531	0.593	0.875
					810	0.593	0.656	0.875
					811	0.656	0.719	1.000
					812	0.719	0.781	1.000
					813	0.781	0.843	1.125
					814	0.843	0.906	1.125
					815	0.906	0.968	1.250
					816	0.968	1.032	1.250
See Notes		3			2	4		

NOTES:

1. All dimensions are in inches.
2. The first numeral in the rivet number designates the rivet series number, the last one or two numerals give the nominal grip in 1/16 in. which the rivet is capable of joining.
3. Recommended drill sizes are those which normally produce holes within the specified hole size limits.
4. The mean grip between specified min and max grips will give a mandrel break plane closest to being flush with the top surface of rivet head.

SOURCE: Industrial Fasteners Institute.

TABLE 13-18 IFI-119 Standard for Flush-Head Self-Plugging Blind Rivets

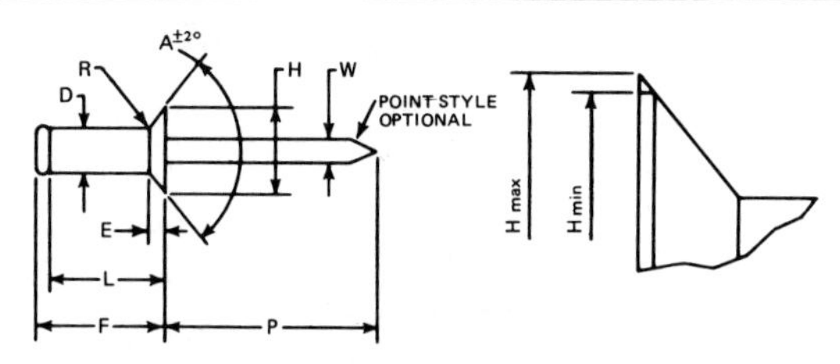

DIMENSIONS OF 100 DEG AND 120 DEG FLUSH HEAD STYLE STRUCTURAL FLUSH BREAK PULL MANDREL SELF PLUGGING BLIND RIVETS, TYPE 2A

RIVET SERIES NO.	NOM RIVET SIZE	D		A	H		E	A	H		E	R	W	P	F
		BODY DIA		STYLE 3 - 100 DEG HEAD				STYLE 4 - 120 DEG HEAD				RADIUS OF FILLET	MAN-DREL DIA	MAN-DREL PROTRU-SION	BLIND SIDE PROTRU-SION
				HEAD ANGLE	HEAD DIA		HEAD HEIGHT	HEAD ANGLE	HEAD DIA		HEAD HEIGHT				
		Max	Min	Deg Nom	Max	Min	Ref	Deg Nom	Max	Min	Ref	Max	Nom	Min	Max
4	1/8 0.1250	0.128	0.122	100	0.233	0.207	0.045	120	0.233	0.207	0.031	0.025	0.076	1.00	L + .187
5	5/32 0.1562	0.159	0.153	100	0.294	0.268	0.058	120	0.294	0.268	0.040	0.030	0.095	1.06	L + .203
6	3/16 0.1875	0.191	0.183	100	0.361	0.335	0.073	120	0.361	0.335	0.050	0.035	0.114	1.06	L + .218
See Notes 3					4		5		4		5				6

NOTES:
1. All dimensions are in inches.
2. For application data see below.
3. Rivet series numbers represent the nominal sizes of rivets in 1/32 in.
4. Max head diameter is calculated on nominal rivet diameter and nominal head angle extended to sharp corner. Min head diameter is absolute.
5. Head height is given for reference purposes only. Variations in this dimension are controlled by the diameters (H) and (D) and the included angle of the head.
6. When computing the blind side protrusion (F), the max length of rivet (L) as given below for the applicable grip shall be used. Minimum blind side clearance may be calculated by subtracting the actual grip (G), (i.e. total thickness of the material to be joined), from the specified blind side protrusion (F). (Example: To join two plates, each .187 in. thick, with a 3/16 in. rivet, a No. 66 rivet would be used. Minimum blind side clearance necessary to permit proper rivet setting would be L + .218 – G, which is .575 + .218 – .374 which equals .419 in.).

APPLICATION DATA FOR FLUSH HEAD STYLE STRUCTURAL FLUSH BREAK PULL MANDREL SELF PLUGGING BLIND RIVETS, TYPE 2A

RIVET SERIES NO.	NOM RIVET SIZE	RECOM-MENDED DRILL SIZE	RECOMMENDED HOLE SIZE		RIVET NO.	GRIP RANGE		RIVET LENGTH L
			Max	Min		Min	Max	Max
4	1/8 0.1250	#30	0.133	0.129	42	0.109	0.141	0.275
					42.5	0.141	0.172	0.337
					43	0.172	0.203	0.337
					43.5	0.203	0.234	0.400
					44	0.234	0.266	0.400
					44.5	0.266	0.297	0.462
					45	0.297	0.328	0.462
					45.5	0.328	0.359	0.525
					46	0.359	0.391	0.525
					46.5	0.391	0.422	0.587
					47	0.422	0.453	0.587
					47.5	0.453	0.484	0.650
					48	0.484	0.516	0.650
5	5/32 0.1562	#20	0.164	0.160	53	0.172	0.203	0.362
					53.5	0.203	0.234	0.425
					54	0.234	0.266	0.425
					54.5	0.266	0.297	0.487
					55	0.297	0.328	0.487
					55.5	0.328	0.359	0.550
					56	0.359	0.391	0.550
					56.5	0.391	0.422	0.613
					57	0.422	0.453	0.613
					57.5	0.453	0.484	0.675
					58	0.484	0.516	0.675
6	3/16 0.187	#11	0.196	0.192	63	0.156	0.218	0.450
					64	0.218	0.281	0.450
					65	0.281	0.343	0.575
					66	0.343	0.406	0.575
					67	0.406	0.468	0.700
					68	0.468	0.531	0.700
					69	0.531	0.593	0.825
					610	0.593	0.656	0.825
See Notes		3			2	4		

NOTES:
1. All dimensions are in inches.
2. The first numeral in the rivet number designates the rivet series number, the last one or two numerals give the nominal grip in 1/16 in. which the rivet is capable of joining.
3. Recommended drill sizes are those which normally produce holes within the specified hole size limits.
4. The mean grip between specified min and max grips will give a mandrel break plane closest to being flush with the top surface of rivet head.

SOURCE: Industrial Fasteners Institute.

TABLE 13-19 IFI-123 Standard for Protruding-Head Drive Pin Blind Rivets

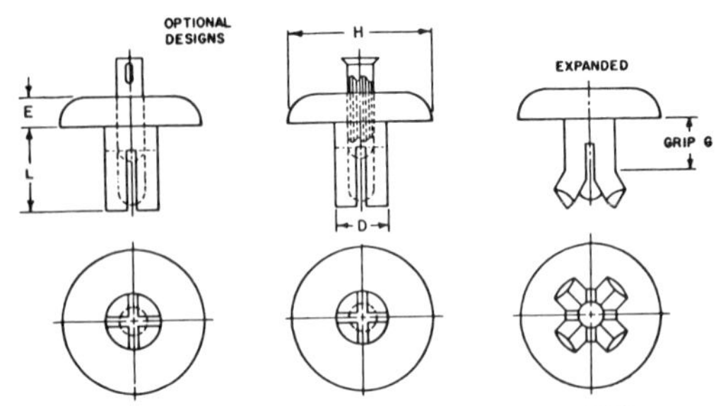

DIMENSIONS OF PROTRUDING HEAD DRIVE PIN BLIND RIVETS

RIVET SERIES NO.	NOM RIVET DIA	D BODY DIA		H HEAD DIA		E HEAD HEIGHT	
		Max	Min	Max	Min	Max	Min
4	1/8 0.1250	0.127	0.121	0.262	0.238	0.064	0.054
5	5/32 0.1562	0.158	0.152	0.328	0.296	0.077	0.067
6	3/16 0.1875	0.190	0.184	0.394	0.356	0.090	0.080
8	1/4 0.2500	0.252	0.246	0.525	0.475	0.117	0.107
See Note 3							

NOTES: 1. All dimensions are in inches.
2. For application data, see below.
3. Rivet series numbers represent the nominal size in 1/32 in.
4. The slotted shanks of No. 4 rivets may produce 2, 3, or 4 segments. Slotted rivet shanks of larger sizes may produce 3 or 4 segments.
5. Illustrations are for basic dimensioning purposes only and are not intended to restrict designs and shapes of rivets otherwise conforming to the dimensional requirements.

APPLICATION DATA FOR PROTRUDING HEAD DRIVE PIN BLIND RIVETS

RIVET SERIES NO.	NOM RIVET SIZE	RECOMMENDED DRILL SIZE	RECOMMENDED HOLE SIZE Max	RECOMMENDED HOLE SIZE Min	RIVET NO.	GRIP RANGE	RIVET LENGTH +.030 -.010 L
4	1/8 0.1250	#30	0.133	0.129	42	.046 – .078	0.156
					43	.079 – .109	.188
					44	.110 – .140	.219
					45	.141 – .171	.250
					46	.172 – .203	.281
					47	.204 – .234	.312
					48	.235 – .265	.344
					49	.266 – .296	.375
					410	.297 – .328	.406
					411	.329 – .359	.438
					412	.360 – .390	.469
					413	.391 – .421	.500
5	5/32 0.1562	#20	0.164	0.160	52	.046 – .078	.188
					53	.079 – .109	.219
					54	.110 – .140	.250
					55	.141 – .171	.281
					56	.172 – .203	.312
					57	.204 – .234	.344
					58	.235 – .265	.375
					59	.266 – .296	.406
					510	.297 – .328	.438
					511	.329 – .359	.469
					512	.360 – .390	.500
					513	.391 – .421	.531
					514	.422 – .453	.562
					515	.454 – .484	.593
					516	.485 – .515	.625
					517	.516 – .546	.656
					518	.547 – .578	.688
					519	.579 – .609	.718
					520	.610 – .640	.750
6	3/16 0.1875	#11	0.196	0.192	62	.046 – .109	.250
					64	.110 – .171	.281
					66	.172 – .234	.344
					68	.235 – .296	.406
					610	.297 – .359	.469
					612	.360 – .421	.531
					614	.422 – .484	.594
					616	.485 – .546	.656
					618	.547 – .609	.719
					620	.610 – .671	.781
8	1/4 0.2500	F	0.261	0.257	84	.110 – .171	.281
					86	.172 – .234	.344
					88	.235 – .296	.406
					810	.297 – .359	.469
					712	.360 – .421	.531
					814	.422 – .484	.594
					816	.485 – .546	.656
					818	.547 – .609	.719
					820	.610 – .671	.781
See Note		3			2		4

NOTES: 1. All dimensions are in inches.
2. The first numeral in the rivet number designates the rivet diameter in 1/32 inches. (See table for specified grip range.)
3. Recommended drill sizes are those which normally produce holes within the specified hole size limits.
4. Minimum blind side clearance necessary to permit proper rivet setting equals max rivet length minus total thickness of material to be joined (grip).

SOURCE: Industrial Fasteners Institute.

TABLE 13-20 IFI-123 Standard for Flush-Head Drive Pin Blind Rivets

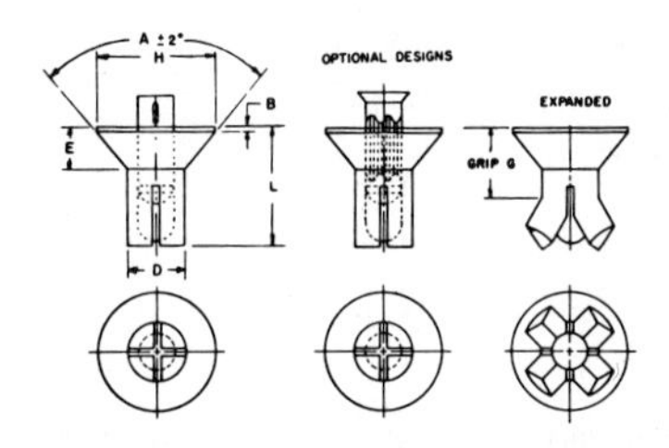

DIMENSIONS OF FLUSH HEAD DRIVE PIN BLIND RIVETS

RIVET SERIES NO.	NOM RIVET SIZE	D		A	H		E	A	H		E	B		
		BODY DIA		HEAD ANGLE	HEAD DIA		HEAD HEIGHT	HEAD ANGLE	HEAD DIA		HEAD HEIGHT	FLAT ON EDGE OF HEAD		
												Grades 12, 60, 61		Grades 30, 31
		Max	Min	Deg Nom	Max	Min	Ref	Deg Nom	Max	Min	Ref	Max	Min	Max
4	1/8 0.1250	0.127	0.121	100	0.229	0.204	.042	78	0.229	0.210	0.062	0.006	0.002	0.008
5	5/32 0.1562	0.158	0.152	100	0.291	0.262	.055	78	0.290	0.264	0.078	0.007	0.003	0.010
6	3/16 0.1875	0.190	0.184	100	0.359	0.324	.070	78	0.343	0.316	0.094	0.009	0.003	0.010
8	1/4 0.2500	0.252	0.246	100	0.484	0.439	.095	78	0.456	0.426	0.125	0.011	0.005	0.012
See Note 3					6		7		6		7			

NOTES: 1. All dimensions are in inches.
2. For application data see **below**.
3. Rivet series numbers represent the nominal size in 1/32 inches.
4. The slotted shanks of No. 4 rivets may produce 2, 3, or 4 segments. Slotted rivet shanks of larger rivets may produce 3 or 4 segments.
5. Illustrations are for basic dimensioning purposes only and are not intended to restrict designs and shapes of rivets otherwise conforming to the dimensional requirements.
6. Max head diameter is calculated on nominal body diameter and nominal head angle extended to a theoretical sharp corner. Min head diameter is absolute.
7. Head height is given for reference purposes only. Variations in this dimension are controlled by the diameters (H) and (D) and the included angle of the head.

APPLICATION DATA FOR 100 DEG FLUSH HEAD DRIVE PIN BLIND RIVETS

RIVET SERIES NO.	NOM RIVET SIZE	RECOMMENDED DRILL SIZE	RECOMMENDED HOLE SIZE		RIVET NO.	GRIP RANGE	RIVET LENGTH +.030 -.010 L
			Max	Min			
4	1/8 0.1250	#30	0.133	0.129	44 45 46 47 48 49 410 411 412 413	.110 – .140 .141 – .171 .172 – .203 .204 – .234 .235 – .265 .266 – .296 .297 – .328 .329 – .359 .360 – .390 .391 – .421	0.219 .250 .281 .312 .344 .375 .406 .438 .469 .500
5	5/32 0.1562	#20	0.164	0.160	54 55 56 57 58 59 510 511 512 513 514 515 516 517 518 519 520	.109 – .140 .141 – .171 .172 – .203 .204 – .234 .235 – .265 .266 – .296 .297 – .328 .329 – .359 .360 – .390 .391 – .421 .422 – .453 .454 – .484 .485 – .515 .516 – .546 .547 – .578 .579 – .609 .610 – .640	.250 .281 .312 .344 .375 .406 .438 .469 .500 .531 .562 .594 .625 .656 .688 .719 .750
6	3/16 0.1875	#11	0.196	0.192	66 68 610 612 614 616 618 620	.172 – .234 .235 – .296 .297 – .359 .360 – .421 .422 – .484 .485 – .546 .547 – .609 .610 – .671	.344 .406 .469 .531 .594 .656 .719 .781
8	1/4 0.2500	F	0.261	0.257	86 88 810 812 814 816 818 820	.172 – .234 .235 – .296 .297 – .359 .360 – .421 .422 – .484 .485 – .546 .547 – .609 .610 – .671	.344 .406 .469 .531 .594 .656 .719 .781
See Note		3			2		4

NOTES: 1. All dimensions are in inches.
2. The first numeral in the rivet number designates the rivet diameter in 1/32 inches. (See table for specified grip range.)
3. Recommended drill sizes are those which normally produce holes within the specified hole size limits.
4. Minimum blind side clearance necessary to permit proper rivet setting equals max rivet length minus total thickness of material to be joined (grip).

SOURCE: Industrial Fasteners Institute.

TABLE 13-21 IFI-126 Standard for Protruding-Head Closed End Blind Rivets

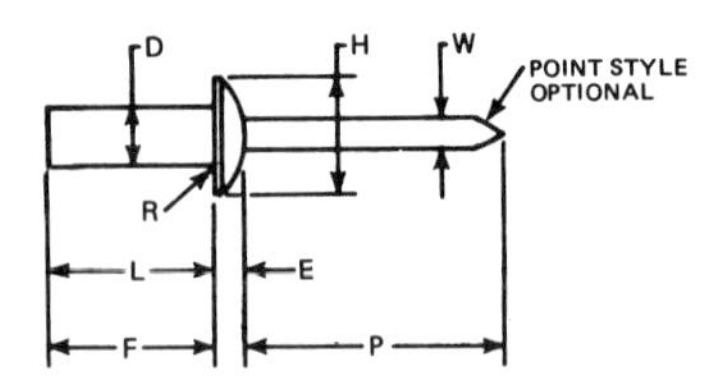

DIMENSIONS OF PROTRUDING HEAD STYLE BREAK MANDREL CLOSED END BLIND RIVETS

RIVET SERIES NO.	NOM RIVET SIZE	D BODY DIA		H STYLE 1 - REGULAR HEAD HEAD DIA		E STYLE 1 - REGULAR HEAD HEAD HEIGHT	R RADIUS OF FILLET	W MAN-DREL DIA	P MAN-DREL PROTRU-SION	F BLIND SIDE PROTRU-SION
		Max	Min	Max	Min	Max	Max	Max	Min	Max
4	1/8 0.1250	0.128	0.122	0.252	0.224	0.050	0.025	0.073	1.00	EQUAL TO
5	5/32 0.1562	0.159	0.153	0.328	0.296	0.065	0.025	0.091	1.06	"L"
6	3/16 0.1875	0.191	0.183	0.394	0.356	0.080	0.025	0.109	1.06	RIVET
8	1/4 0.2500	0.255	0.246	0.525	0.475	0.100	0.025	0.146	1.06	LENGTH
See Notes 3							4			5

NOTES:
1. All dimensions are in inches.
2. For application data see below.
3. Rivet series numbers represent the nominal sizes of rivets in 1/32 in.
4. The junction of head and shank shall have a fillet with a max. radius as shown.
5. The blind side protrusion (F) equals the max length of rivet (L) as given below for the applicable grip. Minimum blind side clearance may be calculated by subtracting the actual grip (G) (i.e. the total thickness of the material to be joined) from the blind side protrusion (F). (Example: To join two plates each .100 in. thick with a 5/32 in. rivet, a No. 54 rivet would be used. Minimum blind side clearance necessary to permit proper rivet setting would be L-G, which is .500 - .200 and equals .300 in.).

APPLICATION DATA FOR PROTRUDING HEAD STYLE BREAK MANDREL CLOSED END BLIND RIVETS

RIVET SERIES NO.	NOM RIVET SIZE	RECOM-MENDED DRILL SIZE	RECOMMENDED HOLE SIZE		RIVET NO.	GRIP RANGE	RIVET LENGTH L
			Max	Min			Max
4	1/8 0.1250	#30	0.133	0.129	41 42 43 44 45 46 48	.020 – .062 .063 – .125 .126 – .187 .188 – .250 .251 – .312 .313 – .375 .376 – .500	0.297 0.360 0.422 0.485 0.547 0.610 0.735
5	5/32 0.1562	#20	0.164	0.160	52 53 54 55 56 58	.020 – .125 .126 – .187 .188 – .250 .251 – .312 .313 – .375 .376 – .500	0.375 0.437 0.500 0.562 0.625 0.750
6	3/16 0.1875	#11	0.196	0.192	62 63 64 66 68 610 612	.020 – .125 .126 – .187 .188 – .250 .251 – .375 .376 – .500 .501 – .625 .626 – .750	0.406 0.468 0.531 0.656 0.781 0.906 1.026
8	1/4 0.2500	F	0.261	0.257	82 84 86 88 810 812 814 816	.020 – .125 .126 – .250 .251 – .375 .376 – .500 .501 – .625 .626 – .750 .751 – .875 .876–1.000	0.445 0.570 0.695 0.820 0.945 1.070 1.195 1.320
See Notes		3			2		

NOTES:
1. All dimensions are in inches.
2. The first number in the rivet number designates the rivet series number, the last one or two numbers give the maximum grip in 1/16 in. which the rivet is capable of joining.
3. Recommended drill sizes are those which normally produce holes within the specified hole size limits.

SOURCE: Industrial Fasteners Institute.

TABLE 13-22 IFI-126 Standard for Flush-Head Closed End Blind Rivets

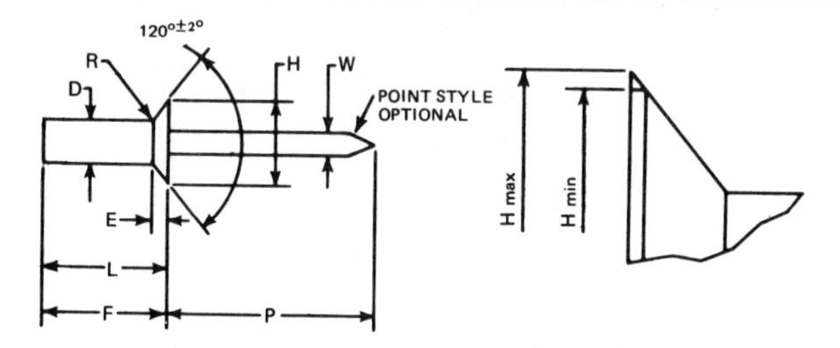

DIMENSIONS OF 120 DEGREE FLUSH HEAD STYLE BREAK MANDREL CLOSED END BLIND RIVETS

RIVET SERIES NO.	NOM RIVET SIZE	D BODY DIA		H STYLE 4 - 120 DEG HEAD HEAD DIA		E HEAD HEIGHT	R RADIUS OF FILLET	W MANDREL DIA	P MANDREL PROTRUSION	F BLIND SIDE PROTRUSION
		Max	Min	Max	Min	Ref	Max	Max	Min	Max
4	1/8 0.1250	0.128	0.122	0.245	0.221	0.042	0.025	0.073	1.00	EQUAL TO
5	5/32 0.1562	0.159	0.153	0.328	0.296	0.051	0.025	0.091	1.06	"L"
6	3/16 0.1875	0.191	0.183	0.394	0.356	0.060	0.025	0.109	1.06	RIVET
8	1/4 0.2500	0.255	0.246	0.525	0.475	0.080	0.025	0.146	1.06	LENGTH
See Notes 3				4		5				6

NOTES:
1. All dimensions are in inches.
2. For application data see below.
3. Rivet series numbers represent the nominal sizes of rivets in 1/32 in.
4. Max. head diameter is calculated on nominal rivet diameter and nominal head angle extended to sharp corner. Min. head diameter is absolute.
5. Head height is given for reference purposes only. Variations in this dimension are controlled by the diameters (H) and (D) and the included angle of the head.
6. The blind side protrusion (F) equals the max length of rivet (L) as given below for the applicable grip. Minimum blind side clearance may be calculated by subtracting the actual grip (G) (i.e. the total thickness of the material to be joined) from the blind side protrusion (F). (Example: To join two plates each .100 in. thick with a 5/32 in. rivet, a No. 54 rivet would be used. Minimum blind side clearance necessary to permit proper rivet setting would be L-G, which is .550 - .200 and equals .350 in.).

APPLICATION DATA FOR FLUSH HEAD STYLE BREAK MANDREL CLOSED END BLIND RIVETS

RIVET SERIES NO.	NOM RIVET SIZE	RECOMMENDED DRILL SIZE	RECOMMENDED HOLE SIZE Max	RECOMMENDED HOLE SIZE Min	RIVET NO.	GRIP RANGE	RIVET LENGTH L Max
4	1/8 0.1250	#30	0.133	0.129	41 42 43 44 45 46 48	.031 – .062 .063 – .125 .126 – .187 .188 – .250 .251 – .312 .313 – .375 .376 – .500	0.332 0.395 0.457 0.520 0.582 0.645 0.770
5	5/32 0.1562	#20	0.164	0.160	52 53 54 55 56 58	.063 – .125 .126 – .187 .188 – .250 .251 – .312 .313 – .375 .376 – .500	0.425 0.487 0.550 0.612 0.675 0.800
6	3/16 0.1875	#11	0.196	0.192	62 63 64 66 68 610 612	.063 – .125 .126 – .187 .188 – .250 .251 – .375 .376 – .500 .501 – .625 .626 – .750	0.471 0.538 0.601 0.736 0.851 1.026 1.101
See Notes		3			2		

NOTES:
1. All dimensions are in inches.
2. The first numeral in the rivet number designates the rivet series number, the last one or two numerals give the maximum grip in 1/16 in. which the rivet is capable of joining.
3. Recommended drill sizes are those which normally produce holes within the specified hole size limits.

SOURCE: Industrial Fasteners Institute.

TABLE 13-23 Grade Designations—IFI Standards

IFI Standard Number	Title	Grade Designation	Rivet Body Material	Mandrel Material
IFI-119	Standard for Structural Flush Break Pull Mandrel Self Plugging Blind Rivets, Type 2A	11	Aluminum Alloy 5052	Aluminum Alloy 7178 or 2024
		12	Aluminum Alloy 5056	Aluminum Alloy 7178 or 2024
		19	Aluminum Alloy 5056	Carbon Steel
		30	Low Carbon Steel	Carbon Steel
		40	Nickel-Copper Alloy	Carbon Steel
		42	Nickel-Copper Alloy	Stainless Steel (300 Series)
		50	Stainless Steel (300 Series)	Carbon Steel
		51	Stainless Steel (300 Series)	Stainless Steel (300 Series: A286 or equivalent)
		52	Stainless Steel (300 Series)	Stainless Steel (400 Series)
IFI-116	Standard for Structural Self Plugging Pull Mandrel Blind Rivets – Type 2B	13	Aluminum Alloy 5056	Aluminum Alloy 2017
		14	Aluminum Alloy 2117	Aluminum Alloy 2017
		30	Low Carbon Steel	Carbon Steel
		41	Nickel-Copper Alloy	Nickel-Copper Alloy
IFI-114	Standard for Break Mandrel Blind Rivets	10	Aluminum Alloy 5050	Aluminum Alloy 7178 or 2024
		11	Aluminum Alloy 5052	Aluminum Alloy 7178 or 2024
		18	Aluminum Alloy 5052	Carbon Steel
		19	Aluminum Alloy 5056	Carbon Steel
		20	Copper Alloy No. 110	Carbon Steel
		21	Copper Alloy No. 102	Carbon Steel
		30	Low Carbon Steel	Carbon Steel
		40	Nickel-Copper Alloy (Monel)	Carbon Steel
		50	Stainless Steel (300 Series)	Carbon Steel
		51	Stainless Steel (300 Series)	Stainless Steel (300 Series: A286 or equivalent)
		52	Stainless Steel (300 Series)	Stainless Steel (400 Series)
IFI-126	Standard for Break Mandrel Closed End Blind Rivets	15	Aluminum Alloy 1100	Aluminum Alloy 7178 or 2024
		19	Aluminum Alloy 5056	Carbon Steel
		20	Copper Alloy No. 110	Carbon Steel
IFI-117	Standard for Pull Through Mandrel Blind Rivets	17	Aluminum Alloy 2117	Carbon Steel
		19	Aluminum Alloy 5056	Carbon Steel
		30	Low Carbon Steel	Carbon Steel
		40	Nickel-Copper Alloy (Monel)	Carbon Steel
IFI-123	Standard for Drive Pin Blind Rivets	12	Aluminum Alloy 5056	7178, 7075, or 2024 Aluminum Alloy
		30	Low Carbon Steel	Carbon Steel
		31	Low Carbon Steel	300 Series Stainless Steel
		60	Aluminum Alloy 2117	300 Series Stainless Steel
		61	Aluminum Alloy 5056	300 Series Stainless Steel

SOURCE: Industrial Fasteners Institute.

TABLE 13-24 Ultimate Shear and Tensile Strength Ratings for Prominent Blind Rivet Grades

IFI-119—ULTIMATE SHEAR AND TENSILE STRENGTHS OF STRUCTURAL FLUSH BREAK PULL MANDREL SELF-PLUGGING BLIND RIVETS, TYPE 2A

No.	Ultimate shear strength Min lb							Ultimate tensile strength Min lb						
	Grade 11	Grade 12	Grade 19	Grade 30	Grade 40	Grade 42	Grade 50,51,52	Grade 11	Grade 12	Grade 19	Grade 30	Grade 40	Grade 42	Grade 50,51,52
4	220	220	340	450	525	475	550	180	180	275	350	450	400	630
5	340	340	520	525	700	630	850	300	300	425	500	700	630	820
6	480	480	650	900	1200	1080	1325	400	400	600	700	1000	900	1200
8	900	900	1200	1750	2000	1800	2250	650	650	1000	1400	2000	1800	2100

IFI-116–ULTIMATE SHEAR AND TENSILE STRENGTHS OF TYPE 2B STRUCTURAL SELF-PLUGGING BREAK MANDREL BLIND RIVETS

Nom rivet size, in.	Ultimate shear strength Min lb				Ultimate tensile strength Min lb			
	Grade 13	Grade 14	Grade 30	Grade 41	Grade 13	Grade 14	Grade 30	Grade 41
1/8 0.1250	340	350	490	630	230	240	280	540
5/32 0.1562	550	570	750	970	375	390	490	860
3/16 0.1875	760	790	1090	1400	540	570	700	1240
1/4 0.2500	1380	1440	1970	2500	980	1030	1250	2300

IFI-114—ULTIMATE SHEAR AND TENSILE STRENGTHS OF BREAK MANDREL BLIND RIVETS

Nom Rivet size, in.	Ultimate shear strength Min lb					Ultimate tensile strength Min lb				
	Grades 10, 11, 18	Grade 19	Grade 30	Grade 40	Grades 50, 51, 52	Grades 10, 11, 18	Grade 19	Grade 30	Grade 40	Grades 50, 51, 52
3/32 0.0938	70	90	130	200	230	80	120	170	250	280
1/8 0.1250	120	170	260	350	420	150	220	310	450	530
5/32 0.1562	190	260	370	550	650	230	350	470	700	820
3/16 0.1875	260	380	540	800	950	320	500	680	1,000	1,200
1/4 0.2500	460	700	1,000	1,400	1,700	560	920	1,240	1,850	2,100

NOTE: Grades 20 and 21 rivets are not subject to shear and tensile testing.

TABLE 13-24 Continued

IFI-126—ULTIMATE SHEAR & TENSILE STRENGTH OF BREAK MANDREL CLOSED END BLIND RIVETS

Nom Rivet size, in.	Ultimate shear strength Min lb			Ultimate tensile strength Min lb		
	Grade 15	Grade 19	Grade 20	Grade 15	Grade 19	Grade 20
1/8 0.1250	100	240	220	110	280	300
5/32 0.1562	130	350	–	160	480	–
3/16 0.1875	210	500	–	250	690	–
1/4 0.2500	–	900	–	–	1100	–

IFI-117—ULTIMATE SHEAR AND TENSILE STRENGTH OF PULL-THROUGH MANDREL BLIND RIVETS

Nom rivet size, in.	Ultimate shear strength Min lb				Ultimate tensile strength Min lb			
	Grade 17	Grade 19	Grade 30	Grade 40	Grade 17	Grade 19	Grade 30	Grade 40
1/8 0.1250	140	120	200	250	220	180	300	400
5/32 0.1562	220	190	340	400	350	290	510	650
3/16 0.1875	330	270	500	620	530	420	670	1050
1/4 0.2500	600	490	900	1150	960	780	1300	1900

SOURCE: Industrial Fasteners Institute.

As part of the system, installation tools are generally available from the blind rivet manufacturer. While installed blind rivets are designed to perform the same structural function, there are different designs of pulling mandrels which specifically require a matching pulling head to effect proper upset and installation. It should be noted that mandrel configurations are not standardized, and care should be taken to assure that the right pulling head is being used at all times.

The installation system usually consists of a reaction or pull-type tool, and a pulling head which grips and holds the mandrel during the pulling or expansion cycle. Common tooling systems include pneumatic, manual, and electric-powered tools. They have the advantage of being portable and capable of operation by a single person, making it possible to bring the required tooling to the assembly as needed. A full range of pulling heads and nose pieces have been developed, to further permit rivet installation in otherwise inaccessible locations. Various types of power and manual installation tools and pulling heads are shown in Fig. 13-65, and illustrate some of the options in adapting tooling to drive the blind rivet system.

Pneumatic tools are perhaps the most common method for installing blind rivets. They permit quick interchange of pulling heads, to adapt to different rivet sizes and configurations, and are fast operating. Figure 13-66 illustrates a pneumatic tool in operation. Hand, or manual-type, tools are generally limited to installation of smaller-

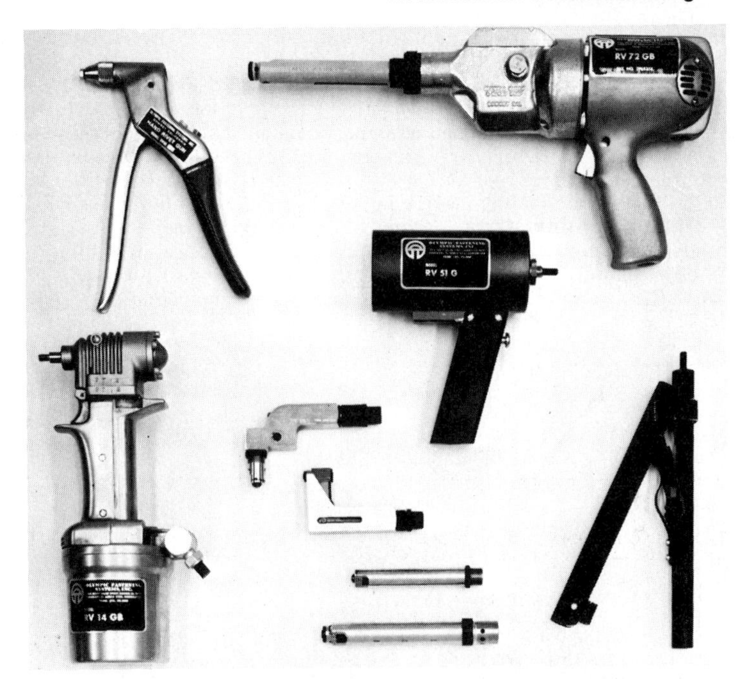

FIG. 13-65 Typical power and manual tools for installation of blind rivet systems. *(Olympic Fastening Systems.)*

FIG. 13-66 Pneumatic-type installation gun for driving blind rivets. *(Cherry Fasteners.)*

diameter sizes, or to blind rivets fabricated from softer materials, such as aluminum alloy. Hand tools provide versatility for on-site application, as illustrated in Fig. 13-67.

As with any specialty fastener, the expense of providing a support tooling system for installation must be taken into account. The advantage of the blind rivet is that it can be installed by only one operator, and with the aid of fast-operating tooling, production installation rates can be significant. The combination of these advantages often contributes to a lower installed cost for the fastener, even though initial cost of the blind rivet and the installation tooling may be higher than for conventional fasteners.

13-57 Design Considerations Many of the design principles and structural-strength concepts employed for solid rivets are equally applicable to blind rivets as well. There are, however, certain additional requirements which must be observed for blind rivet application because of their unique function and operation.

FIG. 13-67 Installation of blind rivet using manual tool. (*Cherry Fasteners.*)

Backside Clearance. Since the blind rivet is inserted from one side of the structure, a portion of the rivet body and mandrel will extend beyond the blind side of the work being assembled. As illustrated in Fig. 13-68, the backside clearance is the distance the blind fastener protrudes beyond the structure. The backside clearance will vary, depending on rivet style and diameter. When considering original design, adequate room should be provided to include the backside clearance for the blind rivet. As discussed earlier, these clearance dimensions are substantially less than is required for solid rivet upset, but are still an important feature which must be accommodated.

Grip Range. Blind rivets are designed to accommodate various sheet thicknesses. Any one rivet, though, is limited to anywhere from $^1/_{16}$- to $^1/_8$-inch variation in structural thickness. Installation of a blind rivet in a grip condition below minimum may result in the mandrel pulling through or in an unsatisfactory upset. A blind rivet installed beyond the maximum grip may not upset satisfactorily or develop desirable hole-filling properties. Representative upsets in the specified grip range are shown in Fig. 13-69. For a structural application, it is important to know the sheet thickness that is being joined or the grip condition before specifying the fastener call-out. Grip range capability is normally referenced on applicable standards.

Hole Filling. The expansion of the blind rivet during installation is designed to fill the hole for maximum rivet performance. Hole-size tolerances are critical for this type of fastener and should be observed in accordance with applicable standards or the manufacturer's recommendations. Too large a hole for the fastener may not permit satisfactory upset and hole filling, and may contribute to loose rivets in the installation.

Clinch Action. Many blind rivet designs are able to draw sheets together during the installation cycle. When assembling a large structure, it is advisable to assist the blind fasteners by using blind clamps or other similar technique, to prevent undue gaps in the structure. The clinch action is indicative of favorable clamp up and tensile properties in the joint.

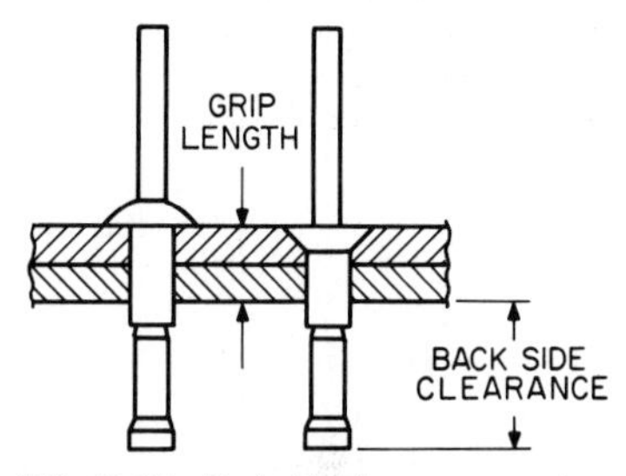

FIG. 13-68 Backside clearance requirement for proper use of blind rivet.

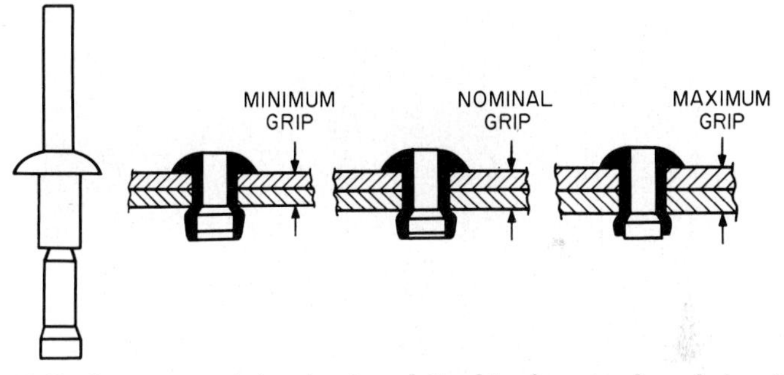

FIG. 13-69 Grip range variation showing relationship of upset and mandrel position.

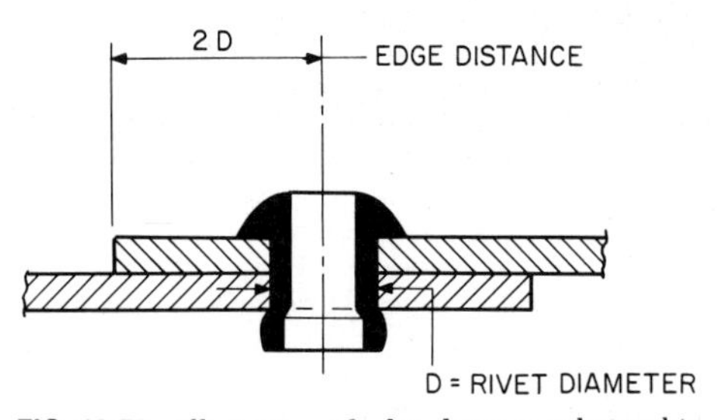

FIG. 13-70 Illustration of edge distance relationship.

Edge Distance. As with solid rivets, care should be exercised in positioning blind rivets too close to the edge of a joint subjected to structural loading. A general rule of thumb has been to locate the centerline of the rivet hole at an edge distance of $2D$, where D is equal to the blind rivet diameter. This condition is illustrated in Fig. 13-70. In some cases where solid rivets are removed and holes enlarged to accommodate a replacement blind rivet, it is sometimes permissible to alter the relationship to $1.5D$.

It is not recommended that edge distance be permitted to fall below 1.5*D* for structural assembly.

Knife Edge. Flush-head rivets are often used where streamlining or flush surface is required in joining applications. In very thin sheet applications, care should be taken to avoid the combination where the height of the countersunk head is equal to the thickness of the sheet being joined, as illustrated in Fig. 13-71. The machine counter-

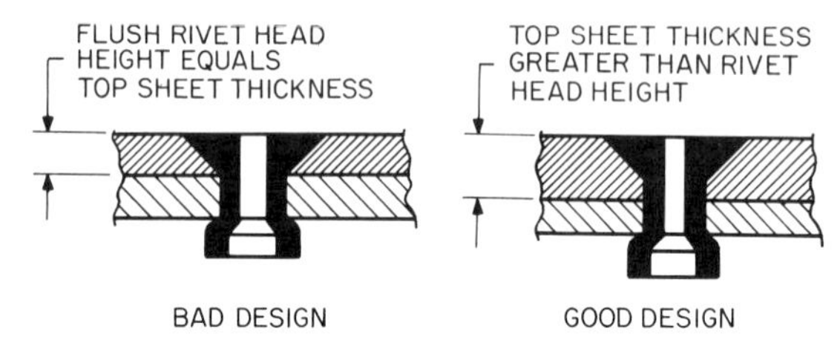

FIG. 13-71 Illustration of knife-edge condition for flush-head-type rivets.

FIG. 13-72 Pull-type blind fastener. (*Huck Manufacturing Company.*)

FIG. 13-73 Screw-type blind fastening system. (*Monogram/Aerospace Fasteners.*)

sunk sheet forms a knife edge, reducing strength properties of the joint. Preferential design suggests sheet thicknesses greater than the height of the countersunk-head blind rivet where machine countersinking is employed.

Corrosion. As discussed earlier, the selection of rivet materials is based on a combination of developing usable strength properties with materials which are ductile and capable of proper upset and expansion. Rivet standards specify protective finish combinations to minimize galvanic corrosion. It is important to also consider the rivet material in conjunction with the parent sheet material for maximum galvanic compatibility. In some cases with self-plugging-type rivets where the mandrel was fractured after rivet installation, additional painting or corrosion protection may be advisable if the environmental exposure is considered severe.

13-58 Other Blind Fasteners Primary discussion has covered blind rivets and their application. Blind rivets are relatively small diameter fasteners. Since the advent of the blind fastening concept, new families of blind structural fasteners have been introduced which cover larger size capability and higher strength properties by virtue of material selection. Such blind fastening systems include both the pull-type configuration, as illustrated in Fig. 13-72, and the screw-type system, as shown in Fig. 13-73. Both types are similar to the blind rivet in that they can be installed from only one side of the structure. For fasteners fabricated from alloy steels, higher-

FIG. 13-74 Illustration of sheet pull-up (clinch) capability of blind fastener. (*Huck Manufacturing Company.*)

capacity installation tooling is required. Figure 13-74 is indicative of the clinch action observed with some blind fastening systems.

There are current efforts to define standards for blind fasteners as distinguished from blind rivets, primarily within the Department of Defense. Many products are still proprietary and are available as manufacturer's standards. There is an increasing application for blind fasteners approximating standard bolt sizes, and available systems may be of particular help where blind or special design consideration warrant their use.

Section 14
Aerospace Fastening

edited by
ROBERT O. PARMLEY, P.E.
Consulting Engineer, Ladysmith, Wisconsin

The National Aeronautics and Space Administration and the Atomic Energy Commission have established a Technology Utilization Program for the dissemination of information on technological developments which have potential utility outside the aerospace community.

The following devices, methods, and techniques of aerospace fastening have been extracted from a document entitled "Fasteners and Fastening Techniques," NASA SP-5906 (03), published by NASA. Special permission has been granted for its reprint in this handbook.

Fasteners and fastening techniques as employed and developed in the space program are, of course, an outgrowth of previously established designs and standard hardware.

Unusual requirements in aerospace technology have resulted in many mating or composite applications. Some of the best examples are presented here to give the reader a general coverage and background.

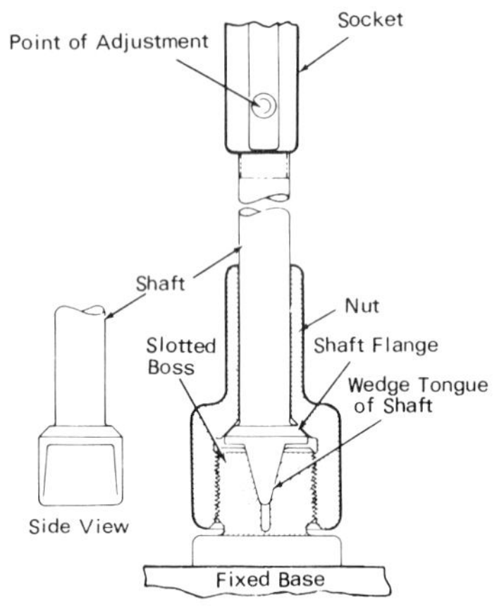

FIG. 14-1 Connect-disconnect coupling. A new coupling enables a rigid shaft to be connected to or disconnected from a fixed base without disturbing the adjustment point of the shaft in a socket or causing the shaft to rotate. The coupling consists of an externally threaded, internally slotted boss extending from the fixed base, and a nut that mates with the boss.

The slot in the boss is wedge-shaped to engage a wedge-shaped, flanged tongue at the end of the shaft. When the nut is tightened, the shaft is secured to the coupling in a rigid, nonrotating assembly. After the shaft is initially locked in the socket at the point of adjustment, the shaft can be disconnected from or reconnected to the boss in the fixed base. This is effected by loosening or tightening the nut as required, without causing the shaft to rotate or to change its initial adjustment point in the socket. [A. Holmberg and F. W. Bajkowski of North American Rockwell Corp., under contract to Manned Spacecraft Center, (MSC-15470).]

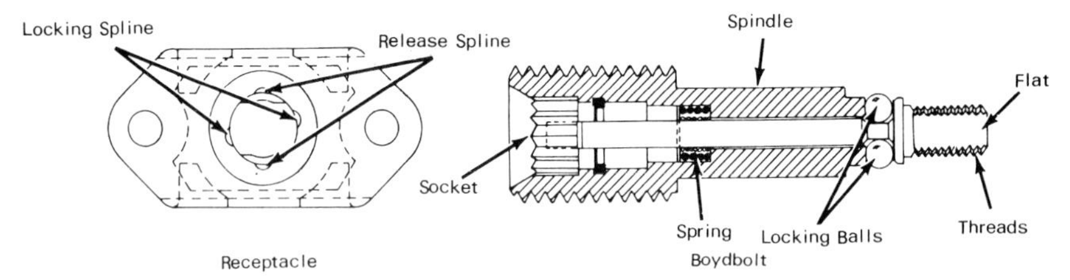

FIG. 14-2 Boydbolt positive-lock fastener. This fastener remains in a locked position under high dynamic preload but is easily removed by the application of only a small force. It has positive lock and release features to prevent accidental operation, and can be fabricated in a variety of sizes for a wide range of applications.

A floating receptacle, designed to receive the Boydbolt, is riveted or screwed to a support structure. This receptacle contains two locking and two release splines in its outer section. The inside of the receptacle is machined to form a four-lead thread that has opposing $\pi/2$ rad (90-degree) slots to provide a breach-type engagement for the bolt. The forward portion of the bolt is threaded in a mating pattern; i.e., flats are machined on approximately $\pi/2$ rad opposing slots so that the bolt may be freely inserted into the receptacle. Two locking balls whose action is controlled by the position of a central spindle are positioned as shown in the figure. In the raised position, the spindle prevents the balls from moving toward the center of the bolt, thus providing for positive engagement of the balls with either the locking or release splines. The spindle is maintained in its free position by a spring. The top of the bolt is machined to a double hexagonal socket in order to receive a lock or to release a hexagonal headed hand tool.

To assemble the Boydbolt fastener, the hand tool is inserted in the bolt socket, depressing the spindle and releasing the locking balls. The bolt assembly is then inserted in the receptacle with the locking balls oriented colinearly with the release spline. The bolt is rotated clockwise until the balls engage the locking spline and stop the bolt from turning further. The threads of the bolt and receptable are now fully engaged. Upon withdrawal of the hand tool from the bolt, the spindle moves in to force the balls into the locking splines, thus preventing rotation of the bolt assembly under vibration. [W. Hamill, J. Brueger, M. Katz, and T. Fenske of the Bendix Corp., under contract to Manned Spacecraft Center (MSC-13061).]

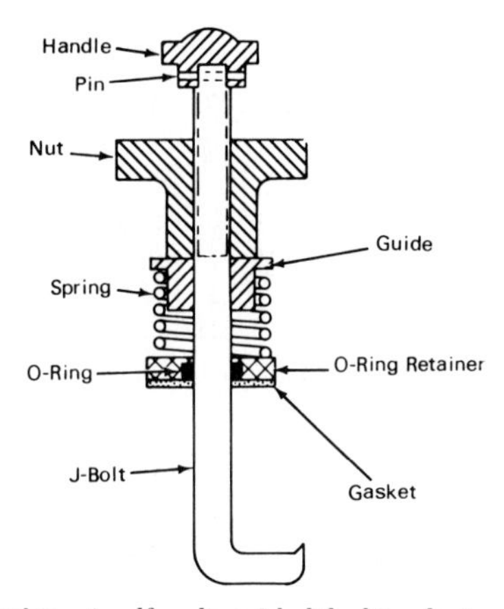

FIG. 14-3 J-Bolt locking device. A self-sealing, J-bolt locking device can be inserted in apertures where no threaded holes exist. Because of its hook-shaped end, the device can be anchored to an internal reinforcement to fasten a sealing plate or plug. This innovation can be used to seal openings in engine blocks and various castings.

The device consists of a J-shaped bolt threaded only at the straight end, a gasket to seal the attaching surface, an O ring with a retainer to seal around the bolt, a compression spring for providing adequate load on the gasket, and a guide bushing. A handle on the threaded end makes adjustments easier. To lock a plug or plate, the device is inserted into the aperture of the part being sealed, rotated to grip a reinforcement, and adjusted for adequate sealing. [D. L. Dickinson, North American Rockwell Corp., under contract to Marshall Space Flight Center (MFS-14275).]

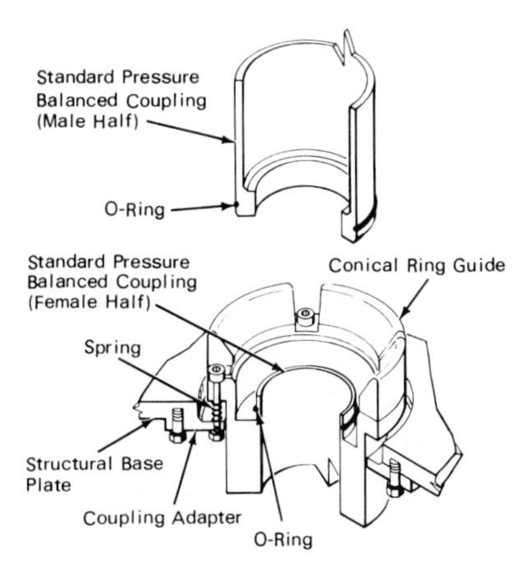

FIG. 14-4 Modified pneumatic umbilical coupling. The coupling and uncoupling of large pneumatic umbilical lines can be very difficult for remote manipulators, especially when the male and female connectors are out of alignment. To overcome this, the female half of the umbilical connector was modified to include a floating conical tapered guide that prealigns the male section, and four springs that restrain the axial and lateral movements of the female section.

The female flange, normally bolted to a structural baseplate, is bolted to an adapter ring in the modified coupling. The ring arrangement provides the clearance necessary for free movement of the female section, which is restrained by four spring-loaded bolts. These bolts also hold a conical ring guide whose inside diameter is considerably greater than the outside diameter of the male section. The use of the ring guide eliminates the problem of center-line mismatch, and the spring loading of the entire female section compensates for angular mismatch. [P. A. Griffin of Aerojet-General Corp., under contract to AEC-NASA Space Nuclear Systems Office (NUC-0037).]

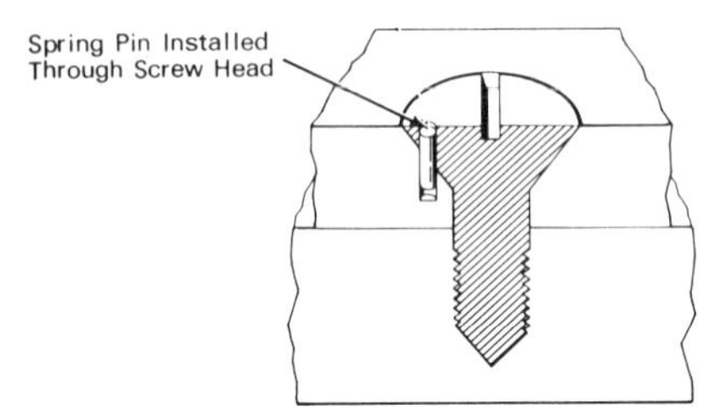

FIG. 14-5 Countersunk head screw retainer. A technique has been proposed for retaining a countersunk screw (under dynamic conditions) when the use of a self-locking device is not feasible and a flat surface is desired. This innovation should interest the manufacturers of fasteners, as well as personnel working in the machinery, aircraft, automotive, and aerospace industries.

In an environment where countersunk-head screws cannot be held by either self-locking inserts or lock wiring, a spring pin can be used. A hole is drilled through one side of the screw head and into the component. The spring pin is then inserted to form a flat surface. A pin installed in this fashion performs adequately under dynamic conditions. [R. S. Totah of North American Rockwell Corp., under contract to Marshall Space Flight Center (MFS-16481).]

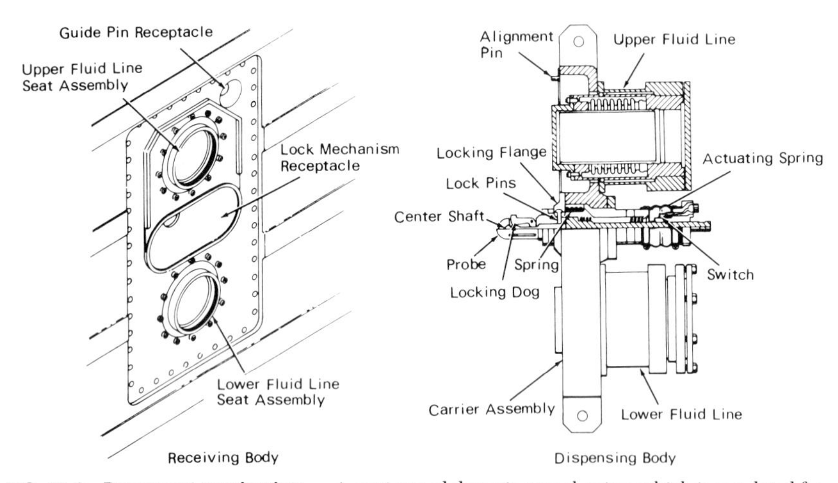

FIG. 14-6 Reconnect mechanism. A mating and demating mechanism which is regulated from a central control system provides a means for locating, mating, and locking two bodies, and for remotely demating the bodies by unlocking and separating them. The mechanism is designed for use in the transfer of fluids from a dispensing body to a receiving body.

The locking sequence begins when the dispensing body lock-mechanism probe contacts the lock-mechanism receptacle on the receiving body. The probe is guided, by the configuration of the receptacle, to a hole in the receptacle bottom and is aligned with the assistance of a guide pin receptacle in the receiving body plate. Three locking dogs on the probe are held in retracted position by a system of springs and lock pins. As the probe enters the hole in the receptacle, the locking flange is pushed back, compressing the actuating spring. When the probe reaches maximum depth in the receptacle, the locking flange releases the lock pins, permitting the center shaft to move forward, and forcing the locking dogs to pivot and lock on the inside surface of the receiving body plate.

To demate the two bodies, a pneumatic system (not shown) retracts the center shaft, withdrawing the locking dogs and releasing the two bodies. When the bodies are free of one another, a relatively high pneumatic pressure is applied to separate the two bodies at a controlled rate. [D. L. Moore of the Boeing Co., under contract to Marshall Space Flight Center (MFS-12968).]

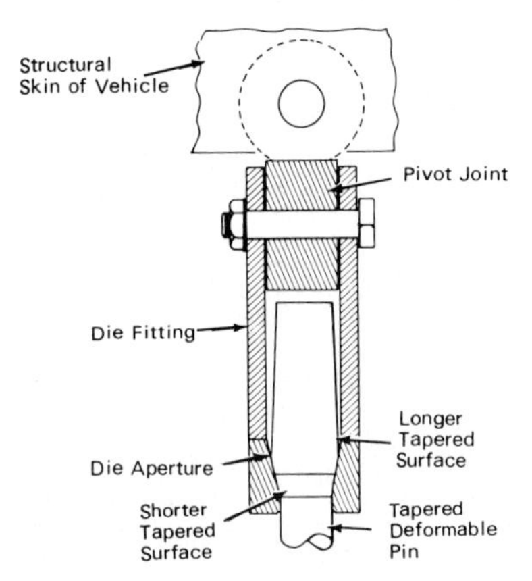

FIG. 14-7 Controlled-release device. A controlled-release device can be used to retard motion by extruding or drawing a tapered ductile pin through a die. The device prevents damage from dynamic stresses imposed on a high-thrust vehicle which is instantaneously released at full thrust. This innovation could be used as a fail-safe system for tension loads, a deceleration mechanism for elevators, and other applications where loads must be limited to specific values or where given amounts of energy must be absorbed.

The device consists of a pivot joint, die fitting, and tapered deformable pin. The pivot joint is linked to the die fitting and is bolted to the structural skin at the base of the vehicle. The double tapered segment of the ductile pin fits within the die fitting, and the straight shanked segment passes through the die aperture and extends outside the fitting.

The aperture of the die is tapered with the small end facing outward, and the shorter tapered surface of the pin seats against this aperture. Since the minor diameter of the aperture is smaller than the major diameter of the pin, the ductile pin deforms as it is pulled through the aperture. Resistance to movement diminishes as the pin is pulled through the die because the longer tapered surface of the pin has a decreasing diameter. [T. W. Burcham, Kennedy Space Center, (KSC-66-14).]

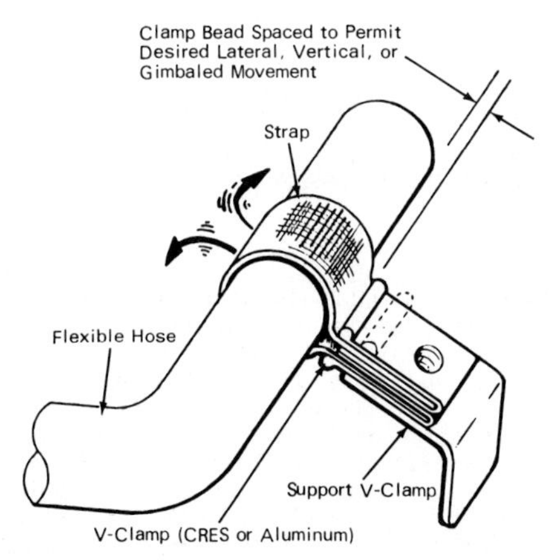

FIG. 14-8 Universal flexible clamp. A concept has been advanced for a flexible support clamp that can be adapted to hold hoses (ducts) of various sizes. The clamp would firmly support the hose while allowing lateral, vertical, or gimballed movement.

The clamp can be made of two metallic V clamps (CRES or aluminum) and a metallic or nonmetallic flexible strap, and can be quickly assembled at the time of installation. The use of a sizing table during assembly allows for a full range of sizes. This conceptual device provides adequate support without hampering the flexed or gimballed movement of the hose, and without interfering with adjacent components. [R. A. Kotler and W. L. Owens of North American Rockwell Corp., under contract to Manned Spacecraft Center, (MSC-17204).]

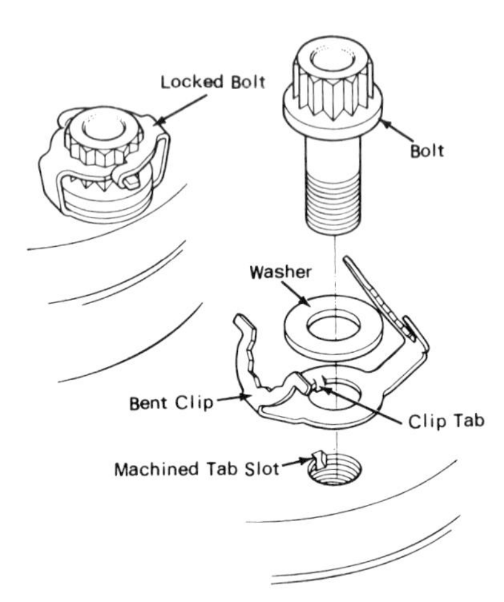

FIG. 14-9 Single-unit locking device. A single-unit "bent-clip" locking device concept provides a positive lock for wrench-type bolts. The device consists of a single stamped part with two open-end wrench arms and a washer. The device would eliminate safety wires, which present a significant disadvantage during maintenance and cannot be installed in confined areas because of the inaccessibility.

The stamped part and washer are installed and the bolt is torqued. The two locking arms are bent in such a way that they meet and interlock around the bolt-head points, providing a positive lock. The locking device is prevented from rotating by the clip tap, which is bent into a previously machined slot. [J. P. Jenson of North American Rockwell Corp., under contract to Marshall Space Flight Center (MFS-13545).]

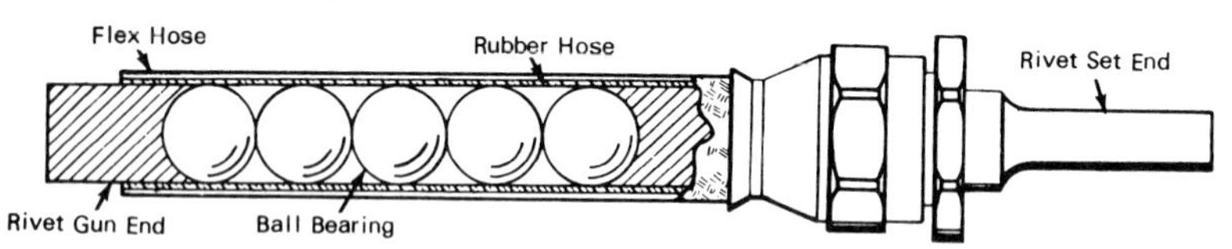

FIG. 14-10 Flexible rivet-set device. A very simple device can be used to set rivets in confined places where the head of a conventional gun cannot be laid on the rivet. The rivet-set device may interest riveting gun users such as the builders and repairers of aircraft, ships, radios, and tanks. Previously, a special set had to be fabricated for each different situation, whereas this device suffices for all.

A typical device consists of a 10 cm (4 inch) length of rubber hose, with a 1.58 cm ($^5/_8$ inch) inner diameter, encased in a similar length of braided metal hose. An anvil for the riveting gun is set in the driven end of the rubber hose which is loaded with five steel ball bearings of 1.58 cm ($^5/_8$ inch) diameter; a rivet set is mounted in the other end of the device.

When the rivet-set tool is flexed to any degree between the head of a riveting gun (or any impact tool) and a rivet, the loss of impact is negligible. The tool may be made in any of many diameters and lengths, and its principle and use are not restricted to riveting. The ball-to-ball line of contact might be improved by the insertion of spacers. [W. H. Hespenhide of McDonnell Douglas Corp., under contract to Marshall Space Flight Center (MFS-20317).]

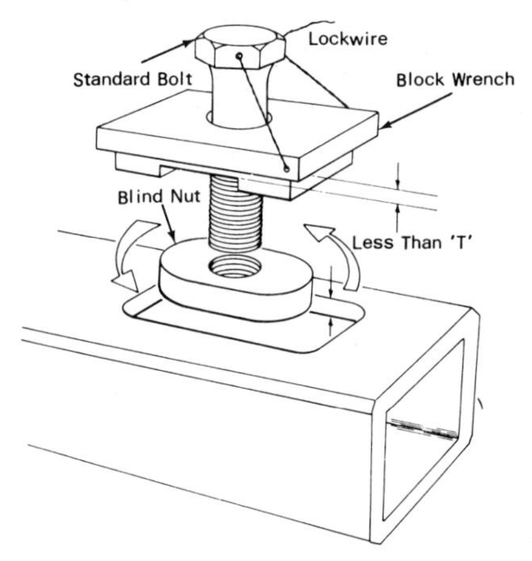

FIG. 14-11 Simplified blind nut. A universal blind nut assembly provides a simple, inexpensive means for fastening plates, brackets, clamps, and other parts to a surface accessible only from one side. Neither riveting nor welding is necessary, and only minimum machining is required. The assembly consists of a standard bolt, an elongated nut, and a special cast or machined block designed for use in blind areas. This innovation can be used to fasten accessories and parts to tubing, box channels, ceilings, walls, floors, ducts, and any surface that cannot be tapped.

To obtain a positive lock, the bolt is inserted in the assembly, rotated $\pi/2$ rad (90 degrees), pulled tight, and torqued as needed. A lockwire is then attached as shown. The assembly can be removed at any time without destroying its components. [C. H. Held of North American Rockwell Corp., under contract to Marshall Space Flight Center (MFS-14319).]

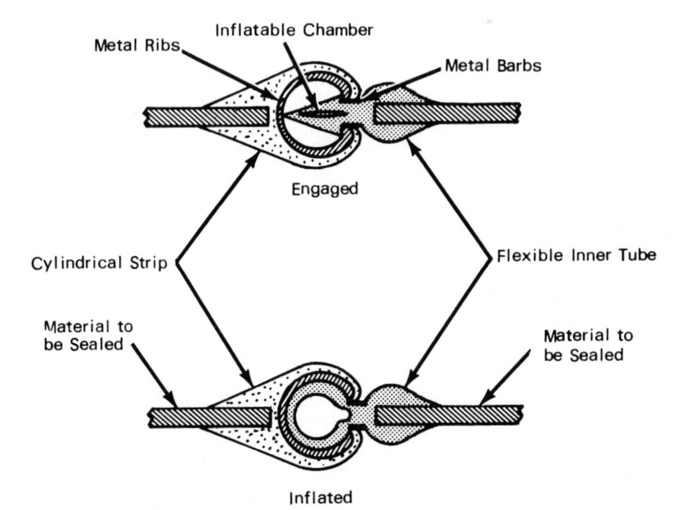

FIG. 14-12 Flexible fluidtight fastener. A conceptual flexible fastener can be quickly joined or separated and will prevent the passage of fluids. A flexible, flat, preformed inflatable tube is inserted into a flexible, $^3/_4$-round strip receptacle. The tube is attached to one side of the material to be sealed and the receptacle is attached to the other side to form the closure. This innovation might be used with such equipment as underwater suits, contamination protection outfits, masks, tents, and doors.

The open side of the $^3/_4$-round strip receptacle (made of plastic or rubber), which is reinforced with metal ribs, faces directly away from the material to which it is attached. The inflatable tube portion, opposite the material to which the receptacle is attached, is arrow-shaped for easy insertion into the receptacle. The metal plates along the tube base adequately secure it in place.

The inflatable tube is inserted into the receptacle by pressing in either end and then gradually working the tube in lengthwise. After insertion, the tube is inflated through a releasable check valve. To break the closure, the check valve is opened to release the pressure, the metal plates at either end of the inflatable tube are gripped and compressed, and the tube is easily stripped from the receptacle. [D. L. Nay, Jet Propulsion Laboratory (JPL-684).]

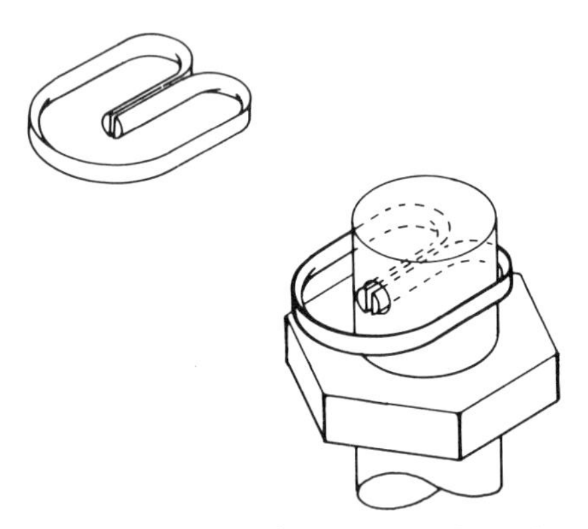

FIG. 14-13 Improved cotter pin. An improved cotter pin can be easily installed or removed without the use of hand tools. The pin, made of high-strength materials, does not easily shear and provides a stronger joint than those with malleable standard pins.

In the installation or removal of conventional cotter pins, considerable time is spent forming the ends with a hand tool. When cotter pins must be installed or removed in confined areas where the free use of hand tools is restricted, the operation is even more time-consuming. The improved cotter pin can be easily installed or removed by hand.

The pin is installed by placing the loop edge on the stud, with the two cotter legs lined up with the drilled hole. The legs are inserted into the hole by pushing the pin downward, and the entire pin is then pushed toward the stud. The loop of the pin springs down over the stud to retain the pin in place.

To remove the pin, the looped end is raised to clear the end of the stud. The pin is then pushed away until the legs clear the hole. [R. M. Cobiella of the Boeing Co., under contract to Kennedy Space Center (KSC-10170).]

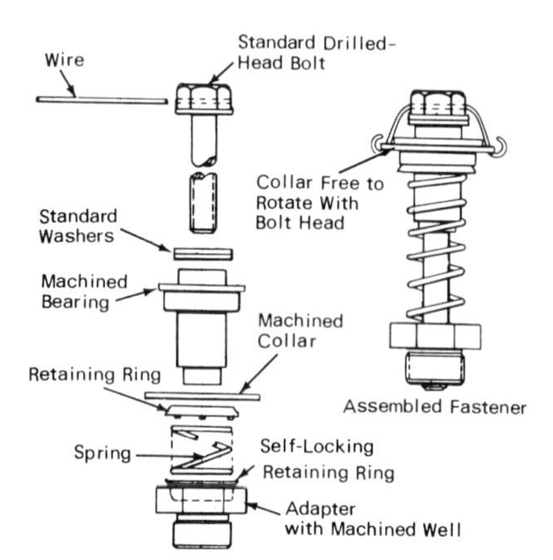

FIG. 14-14 Retractable captive fastener. A retractable captive fastener, consisting of a standard drilled-heat bolt and a spring to retract the bolt, held together as shown, provides a positive means

for attaching bolts without modifying test plates or panels. The fastener could be used in steam-engine caps, hydraulic lines, oil-well caps, and various system test plates. It is particularly adaptable to heavy-duty test plates used for system pressurization in either through- or blind-hold installations. This means of fastening is better than conventional ones because it allows the use of bolts of various lengths and diameters, eliminates the hazard of loose or worn hardware on the far side of a hole in a test plate, and allows work to be performed on test plates without structural damage to the plate.

The linear movement of the spring must be coordinated with the rotational-linear movement of the bolt. To accomplish this, the hole through the bolt head is used to attach the bolt to a free-moving machined collar that carries part of the load exerted on the bolt.

To retract the bolt, a spring is attached. One end of the spring is connected to the bearing shank and the other end is retained in a machined well. The well may be in an adapter, making the assembly a self-contained unit, or it may be machined directly into a plate. As the spring is compressed, the bearing comes in contact with the test plate, transmitting the load to the plate. [D. L. Dickinson of North American Rockwell Corp., under contract to Marshall Space Flight Center (MFS-14261).]

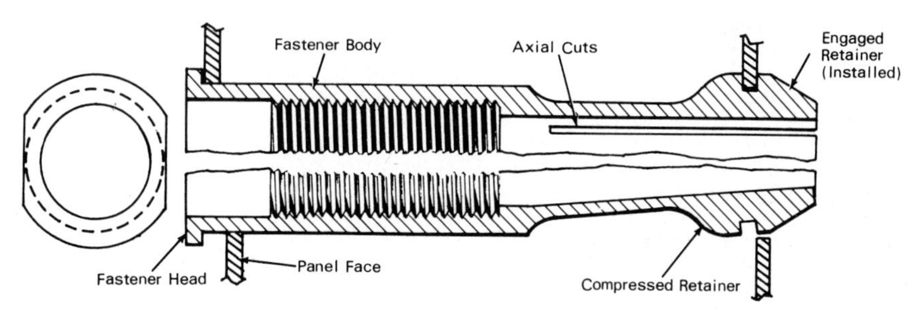

FIG. 14-15 Honeycomb panel fastener. A conceptual fastener may be used in mounting lightweight electronic components on fiberglass or aluminum honeycomb panels. The one-piece fastener is retained by the spring action of tangs formed by four axial cuts through the fastener's blind end. Presently, panels must be removed prior to mounting electronic equipment. This procedure results in costly electric wiring and checkout.

The new fastener, easily inserted without removing the panel, would support one-half of a 0.9 kb (2 lb) component. The blind end of the fastener is forced through a prepared hole in the panel until the fastener snaps in place. This innovation should be of interest to the manufacturers of air frames and general-purpose hardware. [R. M. Johnson and R. K. Shogren of McDonnell Douglas Corp., under contract to Marshall Space Flight Center (MFS-20570).]

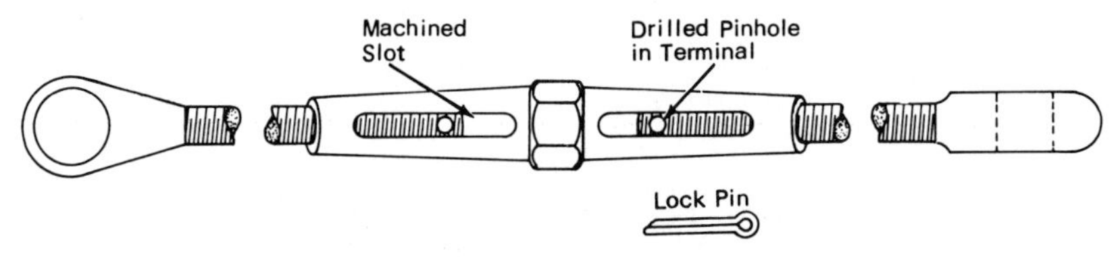

FIG. 14-16 Safety-locking turnbuckle. A new safety-locking turnbuckle which does not need a lockwire provides a safer, more effective locking capability at reduced cost. The turnbuckle can be installed more easily and in less time than the lockwire type. The device should be particularly useful in the marine and construction industries.

The lock pin is inserted through drilled holes or slots in the barrel and then through a hole drilled in the end of the threaded terminal, locking the terminal and barrel together. [F. Broadwick of Chrysler Corp., under contract to Marshall Space Flight Center (MFS-14645).]

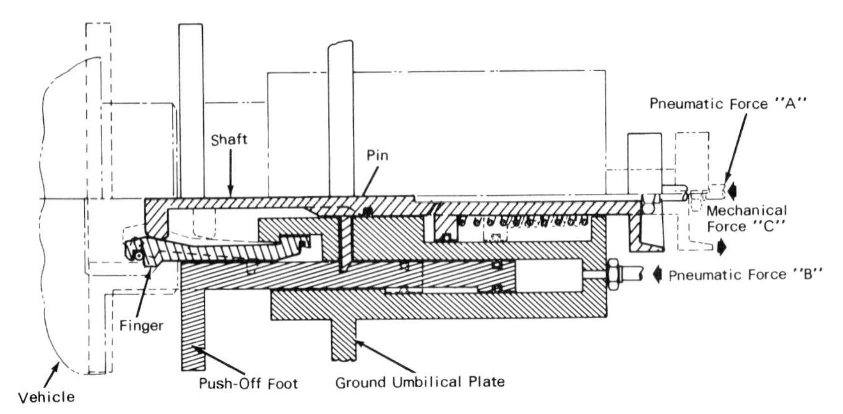

FIG. 14-17 Lock-disconnect mechanism. A unique mechanism locks and unlocks through an internal collet device that is locked and released by the action of a single reciprocating shaft. The appreciable reduction in the number of parts needed to operate the mechanism, compared with conventional umbilical systems, results in a high reliability.

The mechanism, in the locked position, joins the vehicle and the ground umbilical plate. Prior to disconnect, a pneumatic force *B* is applied to preload the push-off foot; the preload is withheld from the vehicle by a series of pins that engage the foot shaft. To unlock the mechanism, either a pneumatic force *A* or a mechanical force *C* is applied to the actuator shaft. As the shaft is forced to the right, the fingers collapse toward the centerline of the mechanism, allowing the ground umbilical plate to be unlocked from the vehicle. Once the device is unlocked, the pins ride down the ramp of the actuator shaft until they reach a predetermined point. The push-off foot is then released and the preload forces it against the vehicle, separating the vehicle and the ground umbilical plate. [C. E. Beaver of the Boeing Co., under contract to Marshall Space Flight Center (MFS-2147).]

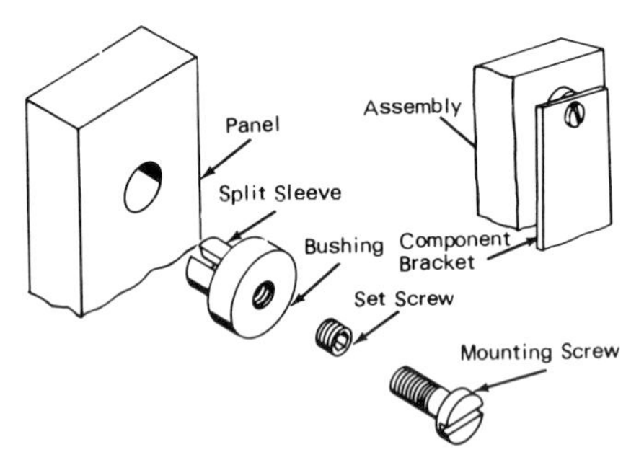

FIG. 14-18 Expandable insert. An expandable self-locking adapter with a female thread that mates with a mounting screw provides a means for securing components to panels where only one side is accessible. The device could be used in electronic and mechanical assembly work, as well as for attaching various household appliances to solid panels.

The adapter, made of metallic or nonmetallic material, has a bushing and a split sleeve which fits inside a mounting hole in the panel. The sleeve is internally threaded to receive a short allen-head setscrew, and the bushing is threaded to receive a mounting screw.

The adapter is securely fastened to the panel by inserting the sleeve into a mounting hole and tightening the set screw with an allen key to expand the split end of the sleeve against the surface of the mounting hole. A mounting screw may then be inserted in the threaded bushing and tightened with a screwdriver. [North American Rockwell Corp., under contract to Manned Spacecraft Center (MSC-301).]

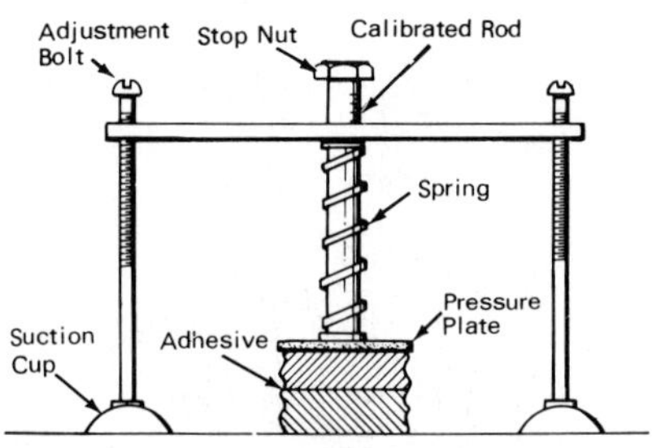

FIG. 14-19 Calibrated clamp. A spring-loaded clamp, with two adjustable legs that terminate in suction cups, can be used to hold materials together or to attach a workpiece to a surface during bonding, machining, welding, and other similar operations.

The two threaded bolts connected to the crossbar are fitted with suction cups in a swivel connection, facilitating attachment to a curved surface. The pressure plate is attached to the spring-loaded rod passing through a hole in the center of the crossbar.

When the device is used to clamp materials together, the suction cups are placed on a supporting surface and fastened, if necessary, with a nonhardening adhesive. The screw bolts are then adjusted to apply the desired pressure to the plate. The spring-loaded rod may be provided with calibrated markings to indicate the applied pressure. [North American Rockwell Corp., under contract to Manned Spacecraft Center (MSC-298).]

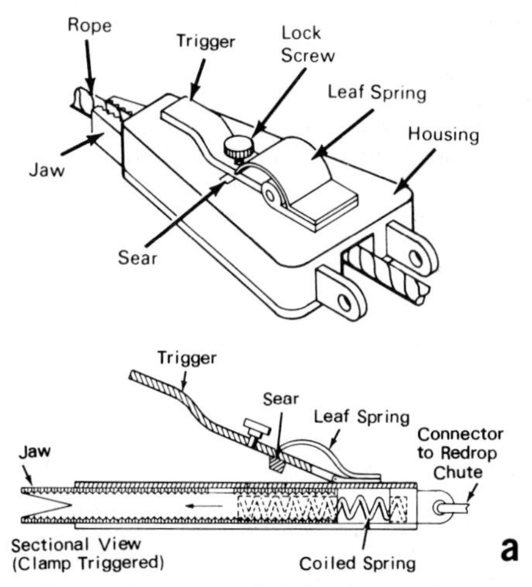

FIG. 14-20 Quick-attach clamp. An improved slideable-jaw clamp can be easily and quickly attached to moving lines such as cables and ropes. The improved clamp would be particularly useful in attaching a redrop parachute to an aerial recovery package.

Although prior clamps are quite efficient, there are many situations where the jaws get in the way when the clamps are being attached. The jaws of such clamps must generally be held apart manually so that the line can be placed between them.

The new clamp (see Fig. 14-20a) has trigger-operated jaws that can be easily actuated to attach a redrop parachute to a moving tow cable. The trigger mechanism maintains the jaws in a retracted position in the housing until they are released for clamping. A sear on the trigger engages a groove in each jaw to hold the jaws retracted. A leaf spring keeps the sear in the grooves before the device is triggered. Two coiled springs force each jaw into clamping engagement when the sear is raised by trigger action. (A removable lock screw is provided to prevent accidental tripping of the trigger.)

With the jaws retracted, the clamp is placed in close proximity to a moving cable. When the clamp is correctly positioned, the trigger is raised to release the sear. The coil springs then immediately snap the jaws into engagement with the moving (or stationary) cable. [A. E. Vano, Flight Research Center, (XFR-05421).]

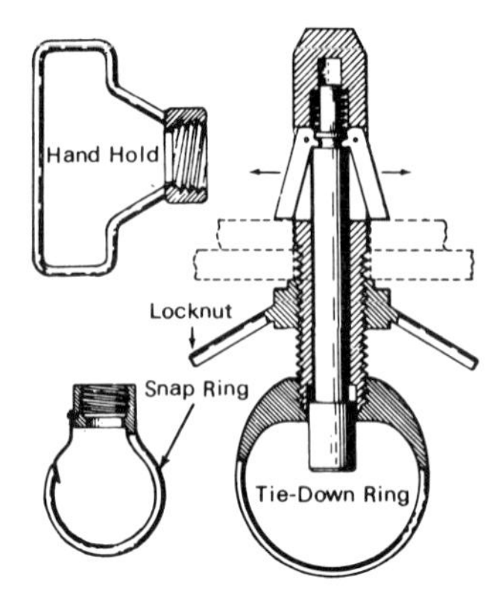

FIG. 14-21 Togglebolt fastener. A new fastener has a cylindrical body with a tapered head for easy insertion into a hole, and a threaded back end for receiving the lock nut and desired attachments (such as tie-down ring, hand hold, plain retainer nut, and snap ring). Slots in the cylindrical body act as receptacles for two or more toggle wings (or detents) which are held in place by pins. The toggle wings are extended, except when the spring-loaded central actuator pin is depressed.

Possible uses of the device include fastening items to a wall or deck where simultaneous access to both sides of the wall or deck is not possible, or fastening two surfaces of opposite curvature; i.e., a concave surface to a convex surface. The device would also be useful in the construction of underwater or space structures that require the use of nuts and bolts.

To insert the fastener into a hole, the fastener is merely pushed into the hole, causing the toggle wings to retract. When the fastener is in place, the actuator pin is released and the toggle wings expand, thereby preventing the fastener from being removed. The fastener is secured by screwing the locknut down firmly, thus exerting a compressive force between the locknut and the toggle wings. With the threaded portion of the cylindrical body functioning as a stud, a suitable attachment may be screwed on.

To remove the fastener from the hole, the locknut is loosened and the exposed portion of the central actuator pin is depressed, causing the toggle wings to retract. [C. C. Kubokawa, Ames Research Center (ARC-10140).]

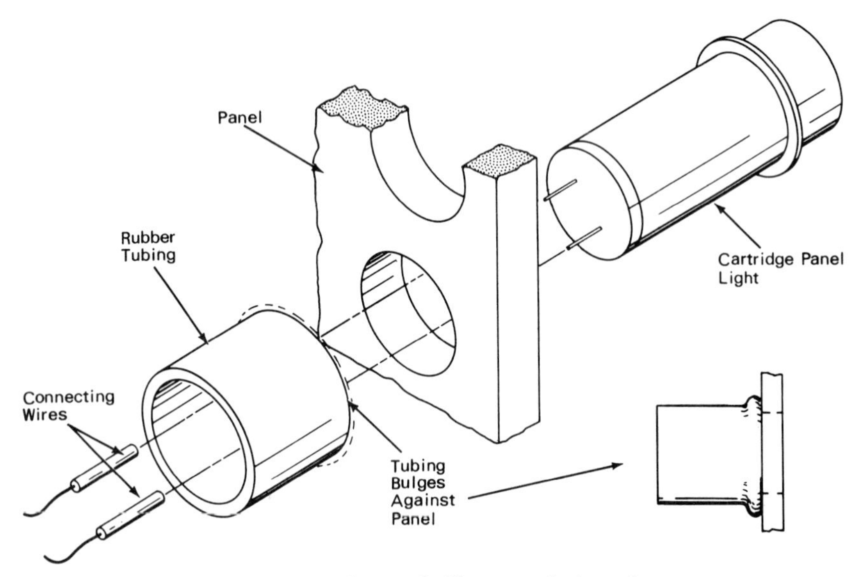

FIG. 14-22 Mounting panel lamps. The use of rubber surgical tubing allows cartridge-type lamps to be mounted in close proximity to other lamps and to be easily removed without damage. Such

lamps can be mounted on digital displays, test control boards, and other display panels using this technique.

A hole large enough to receive the cartridge lamp is drilled in the panel. The lamp is then inserted into the hole. While pressure is applied to the outer portion of the lamp, a short piece of surgical tubing is slid over the part that extends through the hole. A slight bulging of the tubing at the panel secures the lamp during vibration.

The use of an insertion tool, a piece of metal tubing with an inside diameter slightly larger than the outside diameter of the lamp, is helpful in seating the rubber tubing. The lamp can be easily removed by withdrawing the rubber tubing and sliding the lamp from the panel. [R. D. Banta of the Boeing Co., under contract to Kennedy Space Center (KSC-10401).]

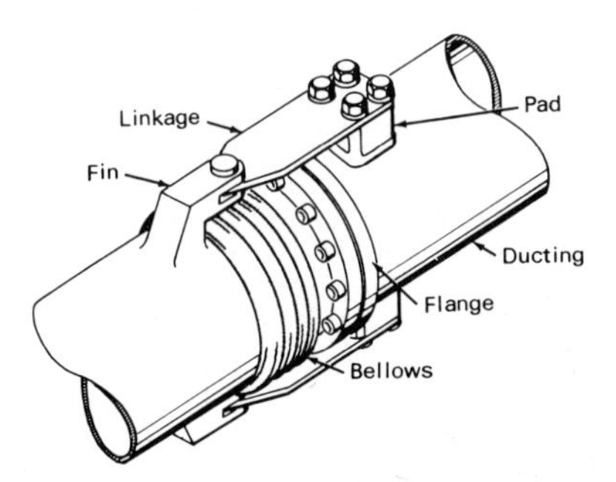

FIG. 14-23 External linkage tie. A proposed concept provides for an external linkage tie to reduce flange thickness and increase seal efficiency in high-pressure ducting and piping systems. Presently, the pressure-separating load is transmitted directly to the flange, which has to be made extra thick to carry the load. The linkage tie would transmit the pressure-separating load to the tube wall behind the flange.

This design may be implemented by extending the linkage tie across the flange and bolting it to a pad directly behind the flanged joint. Since the pressure-separating load would be transmitted to the tube wall behind the flange, the flange could be designed to support only the seal.

A linkage tie based on this concept would also allow the pressure load, caused by the bellows, to put the flanged joint in compression, thereby increasing the efficiency of the seal. [R. O. Pfleger of North American Rockwell Corp., under contract to Marshall Space Flight Center (MFS-823).]

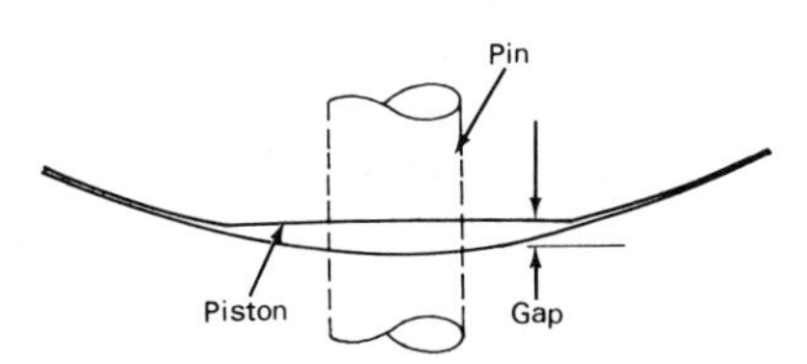

FIG. 14-24 Controlled shear-pin gap. A simple method provides a means for better prediction of shear values by eliminating the side effects of drag and friction. An intentionally controlled gap is made at the shear surface by milling a flat on the piston where the shear pin is installed.

A flat is machined on the piston to establish a gap at the point of shear-pin installation as shown. Such pins will shear at more consistent values when there is a small controlled gap between the shearing surfaces. This method should interest industries whose operations require close control of shear forces. [H. Schmidt of North American Rockwell Corp., under contract to Manned Spacecraft Center (MSC-15014).]

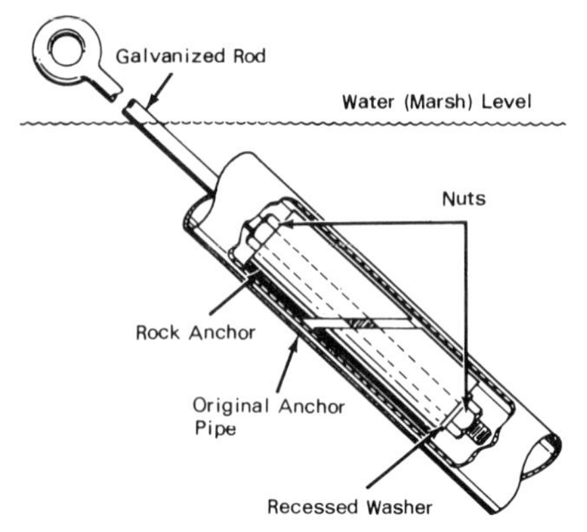

FIG. 14-25 Restoring swamp anchors. Swamp anchors, consisting of steel plates secured at the ends of 5 cm (2 inch) diameter pipes, are screwed into marsh land at depths of 7.5 to 27 m (25 to 90 ft) to hold guys. In time, the exposed portion of pipe at the surface becomes corroded and fails. Previously, a failed swamp anchor was replaced. To eliminate complete replacement, a crushed-rock anchor is installed within the upper portion of the pipe that remains attached to the original swamp anchor.

The rock anchor is fastened to the threaded end of a 1.8 m (6 ft) long galvanized rod, and is inserted into the exposed open end of the pipe secured to the original swamp anchor. The galvanized rod is then rotated to force a recessed washer against the crushed rock, causing it to expand against the inner walls of the pipe. When sufficient compression is achieved, the guy is attached to the eye at the free end of the galvanized rod. Tests indicate that the restored swamp anchor can withstand 1,360 kg (3,000 lb) of tension without failure. [J. W. McAllister, Wallops Station (WLP-10004).]

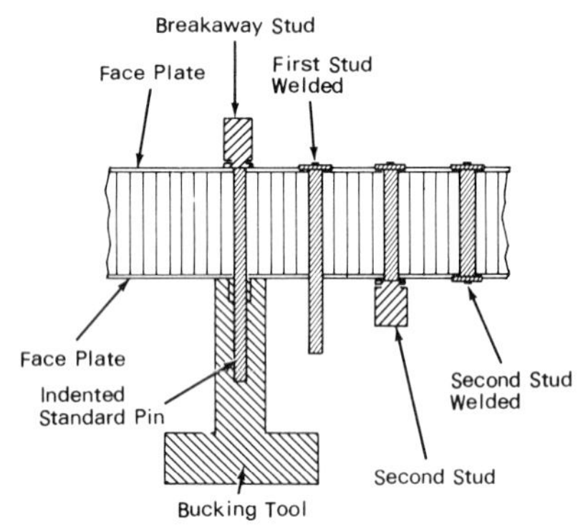

FIG. 14-26 Repairing honeycomb panels. An easy-to-use technique provides a way to repair honeycomb panels by drilling holes and welding breakaway studs to both facing plates. Damaged panels can often be repaired without the use of doublers and with greater strength than doublers can provide. Since this technique requires minimal welding heat, it greatly reduces the distortion of highly stressed panels and makes possible the repair of panels that otherwise could not be repaired.

With a no. 30 drill, holes are drilled at a $\pi/2$-rad (90-degree) angle through both facing plates of a 2.54 cm (1 inch) thick panel and the burrs are then removed. A 0.32 cm ($^1/_8$ inch) diameter pin with an indented end is held in a bucking tool and placed through the holes. The pin is adjusted so that its exposed end is flush with the outer surface of the far plate when the bucking tool is held hard against the opposite plate. With a contact gun, the projection of a breakaway stud is driven into the dent in the end of the pin; the stud is welded to both the face plate and the pin while the bucking tool is held hard against the opposite plate. The body of the stud is then twisted off, using a circular motion, before the bucking tool is removed, and the pin is torque-tested to 1.13 N-m (10 lb-in.).

The other end of the pin, cut flush with the face sheet, is indented with a Starrett no. 18-A automatic center punch, and another breakaway stud is inserted in the dent. Again, the stud is welded to the pin and face sheet before its body is twisted off. [D. F. Bruce of North American Rockwell Corp., under contract to Manned Spacecraft Center (MSC-15046).]

Section 15

Wire Rope and Cable Fastening

edited by

ROBERT O. PARMLEY, P.E.

Consulting Engineer, Ladysmith, Wisconsin

15-1 Wire Rope Wire rope, or cable, is a buildup of strands of wires laid together in the direction opposite the twist which forms the rope. The numbers of wires generally used are: 4, 7, 12, 19, 25, and 37. Standard wire rope is made from 6 strands encompassing a sisal core. Figure 15-1 shows cross sections of typical wire-rope designs.

15-2 Terminology To discuss any specific wire rope, it is necessary to be able to describe it in detail. Example: 200 feet, $^3/_4$ inch 6 × 25 filler wire, preformed, improved plow steel, regular lay, fiber core. This description is broken down into:

Length = 200 feet
Diameter = $^3/_4$ inch
Construction = 6 × 25 Filler Wire
Type = preformed
Grade = improved plow steel
Lay = regular
Core = fiber

The construction, 6 × 25 filler wire, describes the structure of the rope; 6 is the number of strands and 25 is the total number of wires per strand. See Fig. 15-2.

15-3 Breakage and Fatigue Wire rope wears out with repeated usage, improper alignment, undue corrosion, overloading, and whipping of the line. For these and other reasons it becomes necessary to splice the rope.

15-4 Thimble and Loop Splicing A rigger's vise should be used. Other tools required: marlin spike, pliers, a length of pipe, two wooden mallets, serving wire, rope, knife, and a pair of cutters. The extra length to be allowed for splicing a thimble of proper size into a rope varies with the rope diameter as outlined in Table 15-1.

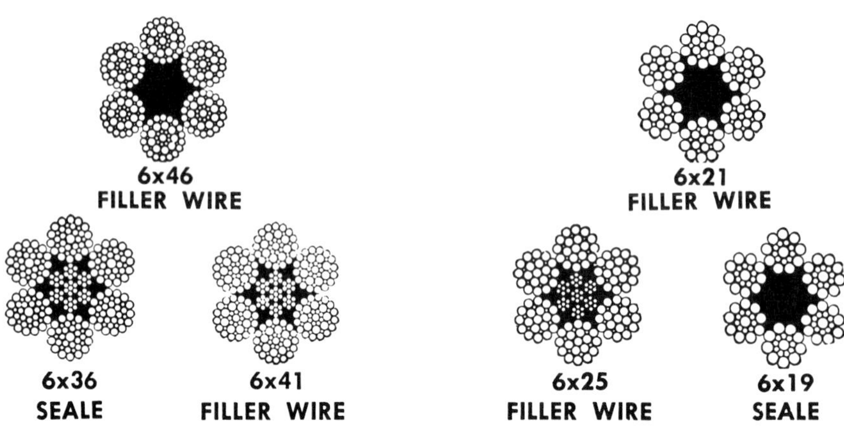

FIG. 15-1 Cross-sectional views of typical wire rope.

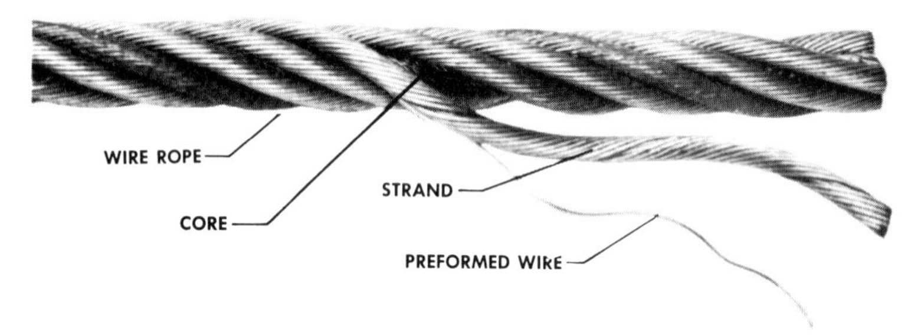

FIG. 15-2 Wire rope terminology. (*Bridon American Corp.*)

TABLE 15-1 Extra Rope to be Allowed for Splicing

Rope diameter, in.	Extra length allowance	Rope diameter, in.	Extra length allowance
$^1/_4$-$^3/_8$	1 ft	$1^1/_4$	3 ft 6 in.
$^1/_2$	1 ft 6 in.	$1^1/_2$	4 ft
$^5/_8$-$^3/_4$	2 ft	$1^3/_4$	4 ft 6 in.
$^7/_8$-1	2 ft 6 in.	$1^7/_8$	5 ft
$1^1/_8$	3 ft	2	5 ft 6 in.

SOURCE: Bridon American Corp.

From this table, if a thimble wire is to be spliced into a $^3/_4$-inch rope, and the finished length is to be 50 ft, 52 ft of rope will be used. Procedure:

1. At approximately the length from end of rope indicated in the table, put a bend in the wire rope, lay a thimble of the correct size into the bed, and clamp securely in rigger's vise. (See Fig. 15-3.)

2. Remove seizing from non-preformed rope so it will untwist. Unlay rope from end to length allowed for splicing to the thimble. Seize each strand to prevent raveling during splicing. If splicing preformed wire rope, straighten loose strands.

3. Cut off the hemp center close to the thimble. If rope contains a wire stand or independent wire rope center, this should be cut off short of the thimble. (Not as close as a hemp center.)

4. Bend two strands to the left and four to the right, placing them in position to make the four-tuck splice which is used in this operation. (To facilitate the remaining operations, slightly untwist or loosen the lay of rope close to the thimble. To do this, wrap a double piece of hemp rope 5 or 6 times around the rope approximately 2 feet from the point of the thimble-wrapping in the same direction as the lay of the rope and spirally toward the thimble. Insert a piece of pipe in the loop and rotate in the same direction as the rope lay to loosen the lay.)

5. Insert a marlin spike under the two strands closest to the point of the thimble. See Fig. 15-4. Rotate the marlin spike a half turn in the direction of the splicer. Insert strand 1 (see Fig. 15-5) through opening. By rotating the spike to the point of the thimble, give one tuck to strand 1.

FIG. 15-3 Step 1 in splicing procedure. *(Bridon American Corp.)*

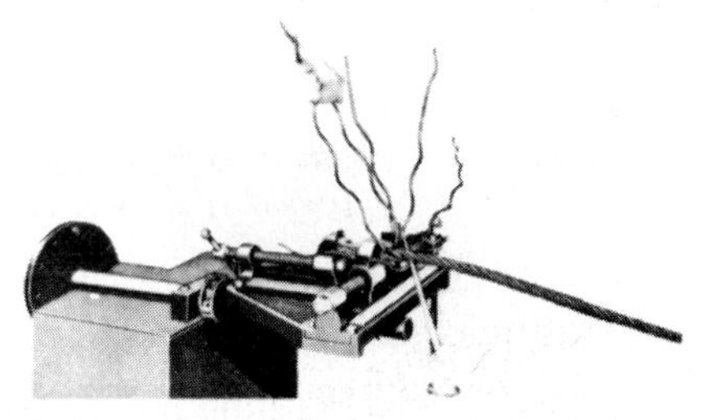

FIG. 15-4 Insert marlin spike—splicing procedure. *(Bridon American Corp.)*

FIG. 15-5 Insert strand 1—splicing procedure. *(Bridon American Corp.)*

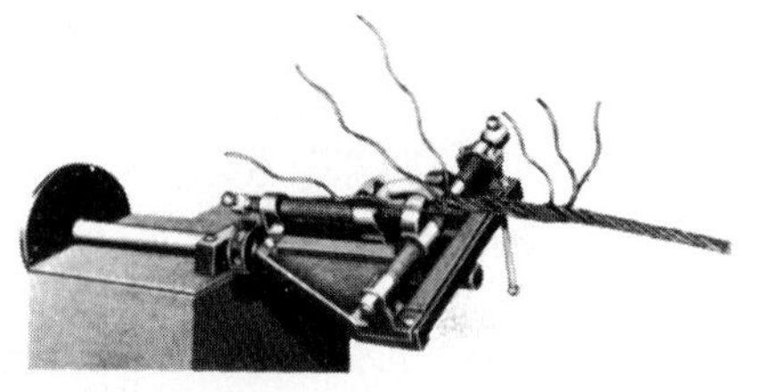

FIG. 15-6 Repeat with strands 4 and 3—splicing procedure. *(Bridon American Corp.)*

6. Tuck strand 2. Use same procedure as used on first strand. The end is tucked through the next single strand in the long end of the rope.

7. Tuck strand 6 under two strands which will be the fifth and sixth strands from the point of the thimble on the long end of the rope.

8. Strand 5 is inserted in upward direction through an opening made by inserting the marlin spike under the next single strand—the fifth from the point of the thimble. Rotate the spike a half turn, pulling the strand up through the opening until the last or fourth tuck is made by inserting the strand down through the opening.

9. Repeat with strands 4 and 3, taking the next laid strands as they come in order on the long end of the rope. (See Fig. 15-6.)

10. Complete strands 2, 1, and 6 in order shown by giving each of them three more tucks.

11. Cut off projecting ends. Hammer out inequalities with wooden mallet. Figure 15-7 shows completed splice.

Approximate strengths of hand splices are shown in Table 15-2.

15-5 Spliced Eye Attachments It is recommended that all eye attachments contain a thimble. Thimbles add substantial reinforcement at critical points in the wire-rope eye. Lacking such reinforcement, ropes kink in the top of the eye and flatten out of shape in other areas of the eye. Strands become displaced.

15-6 Socketing Drop-Forged Sockets on Wire Rope The method described here is recommended by the United States Bureau of Mines in Bulletin 75. It is the most satisfactory method in use today for attaching sockets with molten metal.

1. The wire rope should be securely seized or clamped at the end before cutting. Measure from the end of the rope a length equal to the length of the socket basket. Seize or clamp at this point. Use as many seizings as necessary to prevent the rope from unlaying.

2. Cut rope. Remove end seizing. Cut out hemp core back to seizing. If core is an IWRC, do not cut.

3. Separate wire in strands. "Brush out" the wires in strands by untwisting them. Partially straighten the wires as is necessary. Clean all wires carefully with kerosene, gasoline, or naphtha from the ends to as near the first seizing as possible.

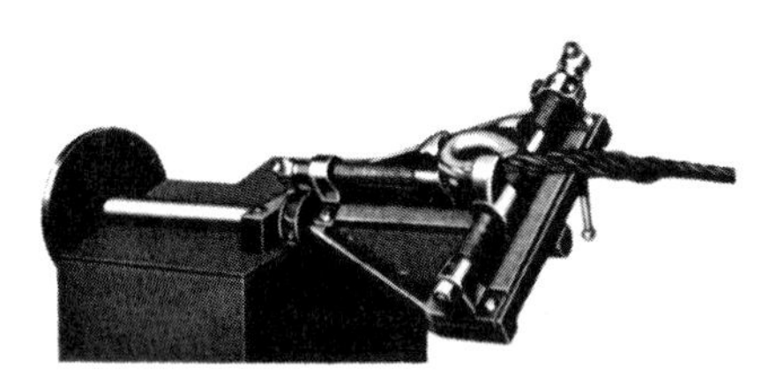

FIG. 15-7 Completed splice. *(Bridon American Corp.)*

TABLE 15-2 Approximate Strength of Hand Splices

Rope dia., in.	Efficiency factor
1/4	0.90
5/16	0.89
3/8	0.88
7/16	0.87
1/2	0.86
9/16	0.85
5/8	0.84
3/4	0.82
7/8 to 2 1/2	0.80

SOURCE: Bridon American Corp.

FIG. 15-8 Dipping wire-rope end into solution. *(Bridon American Corp.)*

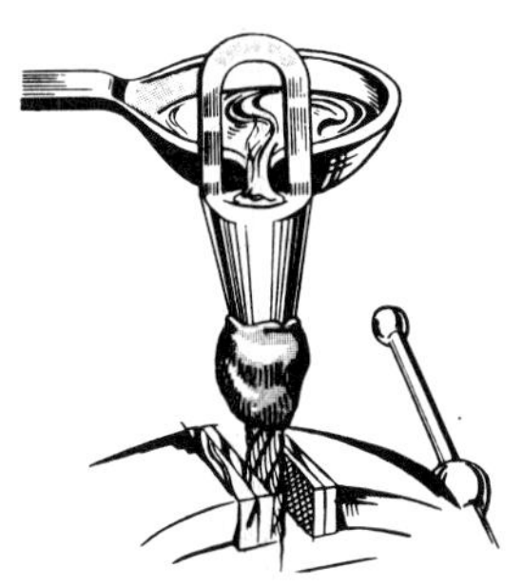

FIG. 15-10 Socket during pouring operation. *(Bridon American Corp.)*

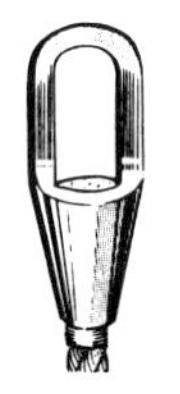

FIG. 15-9 Socket placed in position. *(Bridon American Corp.)*

4. Dip wires to three quarters of the distance to the first seizing in a solution of one half commercial muriatic acid and one half water (use no stronger solution and take extreme care that acid does not touch any other part of the rope) and keep wires submerged long enough to be thoroughly cleaned (see Fig. 15-8). The acid bath must be kept clean. Oil or grease on the surface of the bath from previous dippings should be skimmed off. Thorough cleaning is essential to successful socketing. On removing brush from acid, keep it pointed downward and shake off excess acid thoroughly. Avoid handling brush in an upright position until excess acid has been shaken off.

5. Seize end so that socket can be slipped over wires.

6. Clean basket of socket by swabbing with muriatic acid solution used in step 4.

7. Slip socket over ends of wire, remove end seizing, distribute all wires evenly in basket and flush with top of basket. Be sure socket is in line with axis of rope. Seal the base of the socket around rope with fire clay or asbestos. See Fig. 15-9.

8. Preheat socket basket to expel any moisture and to prevent molten zinc from congealing before it has completely filled the lower end of the basket. Socket should not be heated to a point where it would change color.

9. Heat zinc to approximately 830°F (at this point zinc is fluid for pouring and has no lumps). Zinc must not be too hot or it will anneal wires. Pour zinc into top of basket around wires until basket is full (see Fig. 15-10). Tap sides of socket lightly while pouring. Do not use lead, babbitt, solder or other soft metals.

10 Remove all seizings except one nearest socket.

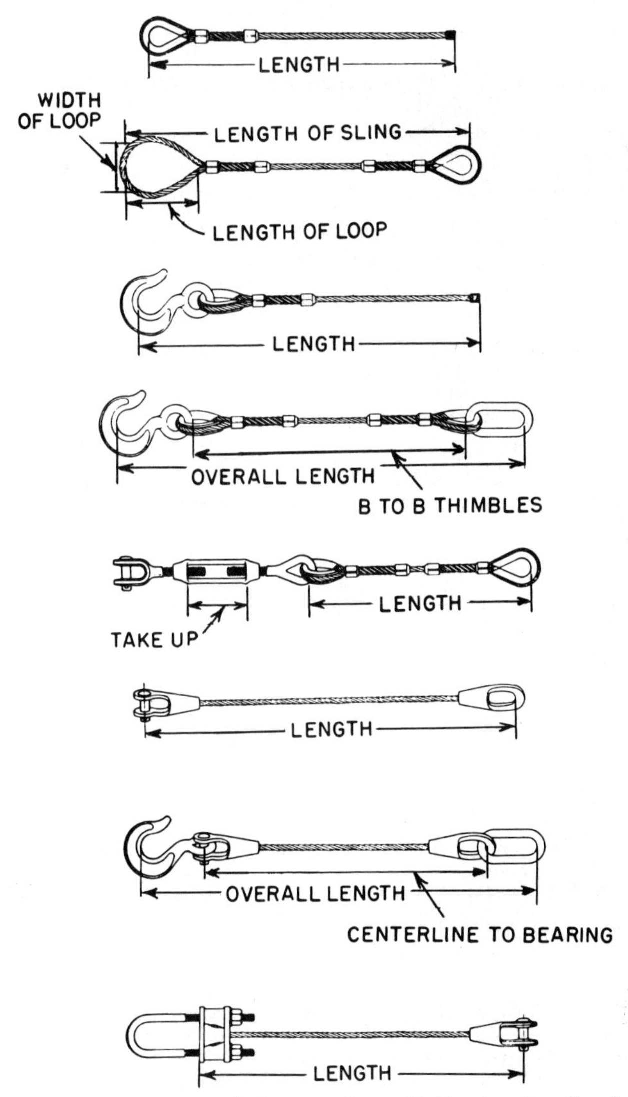

FIG. 15-11 Measurement of wire-rope slings. *(Bridon American Corp.)*

15-7 Wire Rope Slings Wire-rope assemblies wear out under service conditions. The ultimate service and life of the assemblies depend largely on the type of equipment and conditions under which it operates. It should be recognized that in any wire-rope assembly subjected to constant whipping action, deterioration from fatigue will result at the point of vibration arrestment.

Factors requiring the discarding of any wire-rope sling are:

1. Severe corrosion.
2. Localized wear to reduce outer wire diameter more than 50 percent.
3. Damage or displacement of end fittings—hooks, rings, links, or collars—by overload or misapplication.
4. Deformed rope structure caused by kinks, shearing, or twisting.
5. One completely severed component rope in sling body of six- or eight-part braided slings; more than four severed strands in any one component rope of cable-laid slings; more than 15 broken wires in any one strand of strand-laid slings.

Figure 15-11 shows the proper technique to be used when measuring slings with fittings, loops, etc.

15-8 Boom-Cable Assemblies-Terminals Table 15-3 outlines terminal dimensions and minimum breaking strengths of complete boom-cable assemblies based upon the use of 6 × 19 IPS-IWRC wire rope and terminals.

TABLE 15-3 Terminal Dimensions and Minimum Breaking Strengths

Rope diameter, in.	Diameter, pin hole, in.	Maximum thickness at pin hole, in.	Weight per ending—complete, lb	Pin diameter, in.	Openings between jaws, in.	Weight per ending—complete, lb	Minimum breaking strength complete assembly, tons
1/2	1 1/16	15/16	4	1	1	6	11.5
5/8	1 1/4	1 1/8	5	1 3/16	1 1/4	8	17.9
3/4	1 1/2	1 3/8	6	1 3/8	1 1/2	9.5	25.6
7/8	1 3/4	1 5/8	9	1 5/8	1 3/4	16.5	34.6
1	2 1/8	1 7/8	14	2	2	24.5	44.9
1 1/8	2 3/8	2 1/8	21.5	2 1/4	2 1/4	38	56.5
1 1/4	2 5/8	2 5/16	29	2 1/2	2 1/2	52	69.4
1 3/8	2 5/8	2 5/16	34	2 1/2	2 1/2	57	83.5
1 1/2	2 7/8	2 3/4	58	2 3/4	3	85	98.9
1 3/4	3 5/8	3 1/4	90	3 1/2	3 1/2	156	133.0
2	3 7/8	3 3/4	128	3 3/4	4	206	172.0

SOURCE: Bridon American Corp.

15-9 Standard Swaged Terminals Figure 15-12 illustrates a few of the standard swaged fittings available. Swaged fittings are permanently cold-fused on preformed wire rope, with terminal metal mechanically flowed around and between wires and strands, forming an attachment as strong as the rope itself.

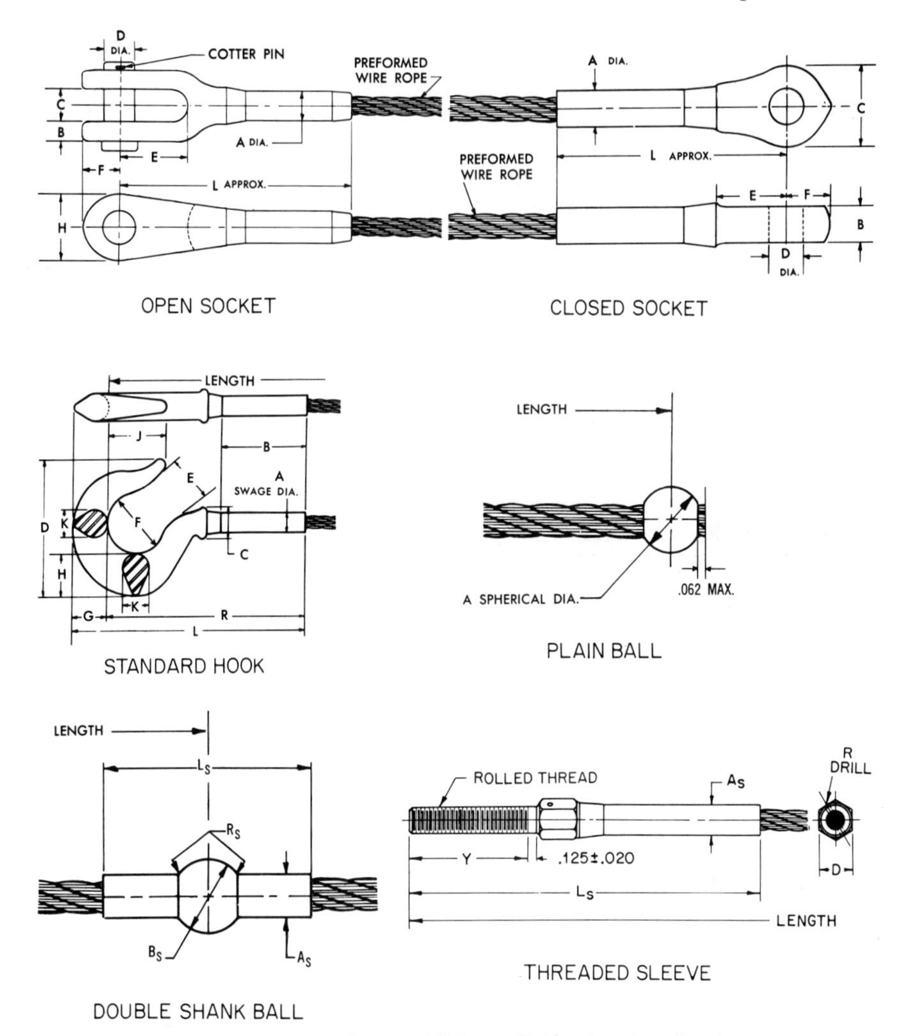

FIG. 15-12 Typical swaged fittings. (*Bridon American Corp.*)

ACKNOWLEDGMENT

The editor wishes to express his thanks for the material supplied by the Bridon American Corp., Wire Rope Div., Exeter, Pa. Their fine assistance and excellent data made this section possible. The bulk of the material used was extracted from their manual, "Here's How To Keep Your Wire Rope Working."

Section 16
Injected Metal Assembly

LAWRENCE B. CURTIS
RONALD J. BORG
Fishertech®, Division of Fisher Gauge, Limited, Peterborough, Canada

16-1 Introduction "Injected metal assembly" (IMA) is a method of joining two or more components by injecting a molten alloy under pressure between these components. It is basically an adaptation of die-casting technology to the joining and fastening process. Die casting is traditionally referred to as "the shortest distance between a design and the finished product." Injected metal assembly applies this adage to joining and fastening by using a low-cost die-casting alloy as both the assembly/locking medium and a production method for additional parts.

Three words can best describe the joining method—"combine," "eliminate," and "substitute." Several operations such as machining, inspection, and the assembly function can be *combined.* Material waste can be *eliminated* as almost all of the molten alloy is used and such surplus alloy as sprues and runners can be recycled for future re-use. Integral injected parts such as cams, pinions, ratchets, etc., can be *substituted* for machined parts with obvious economies as the manufacture and inventory stocking of such are not required. In addition, pinions and the like are not subject to tolerance differences as, being die-cast, they are consistently repeatable.

Injected metal assembly permits the relaxing of some tolerances in the joining area because variants are compensated for by the molten injected alloy.

The process has several distinct areas of application. These are:

- Disks and shafts with such assemblies as gears and shafts, plastic disks and shafts, etc.
- Terminations molded onto wires and cables.
- Abrasives in which small grinding wheels are assembled to mandrels, or alloy hubs are molded into larger grinding wheels.
- Electric motor rotors assembled to shafts. Many types have been assembled including stepper, ac induction, and dc armatures.
- Liquid riveting.

A large variety of applications have been successful and others are constantly being developed as the possibilities stimulate the imagination of creative designers. See Fig. 16-1.

16-2 What Is Injected Metal Assembly? It is as its name implies, an assembly of two or more components joined together through the injection of molten alloy. As the alloy "freezes," it shrinks a minute amount and, in so doing, locks the components together. See Fig. 16-2. The first injected metal assembly machine was developed for the watthour meter industry. A typical meter is shown in Fig. 16-3. The industry's prime problem was the assembly of the damping disk to its shaft. The disk is shown immediately above the word "SANGAMO" in Fig. 16-3. Conventional staking methods required a staking bushing and resulted in a considerable amount of scrap. In addition, the assembly was not always as concentric or perpendicular as required. The location of the disk on the shaft was determined by the center hole, and the whole staking operation resulted in a considerable amount of warpage and wobble. Canadian General Electric approached the president of Fisher Gauge, a small tool-making firm to design and build a machine to overcome these problems. The concept was to hold components in precise tooling and injecting a molten metal, lead alloy in this first application, between the components. The successful application of this concept resulted in a simple machine and tooling being developed in a very short time which produced in-tolerance assemblies.

FIG. 16-1 A varied but typical selection of assemblies formed by the joining of two or more components of similar or dissimilar materials. Components are securely and permanently fastened by the injection of molten alloy at the components' intersection.

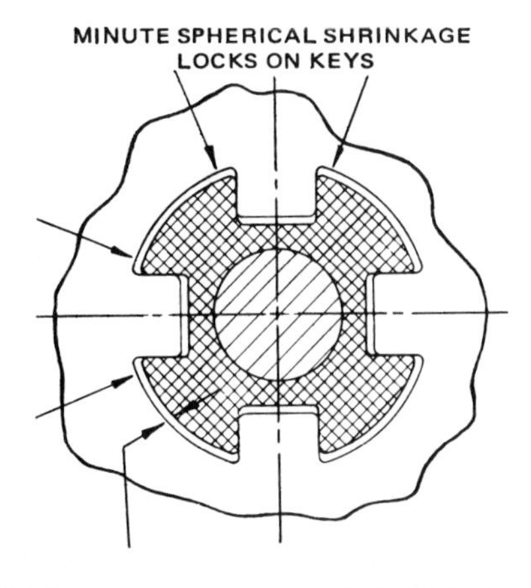

FIG. 16-2 This cross-sectional sketch of an injected metal hub illustrates the principle of spherical shrinkage. The usual injected metal is zinc alloy No. 3, which shrinks about 0.70 percent during cooling, thus causing a mechanical lock onto the shaft and the keys.

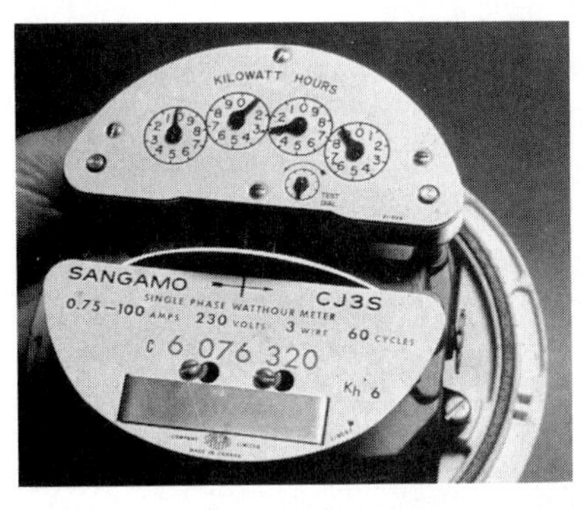

FIG. 16-3 The first manufacturing problem solved by the use of injected metal assembly was the accurate joining of a damping disk to its shaft in an electrical watthour meter. Using the method solved problems of concentricity, perpendicularity, warpage, wobble, and balance.

Shown in Fig. 16-4 are the meter components—the assembled damping disk and aluminum shaft as well as the bearing pin assemblies.

FIG. 16-4 Different watthour meter disks and shafts produced by using injected metal assembly.

This first machine, shown in Fig. 16-5, was built in 1945 and is still in daily use by the original customer over 40 years later.

As the success of this first application became known, the technique was applied to many other components in the meter, such as the gear train. These are shown in Fig. 16-6. All meter manufacturers in North and South America, Europe, and Japan soon converted to this joining process.

In the years since then, injected metal assembly machines have assembled parts with applications in such varied product areas as automobiles, telephones, home appliances, and fractional horsepower electric motors, among many others.

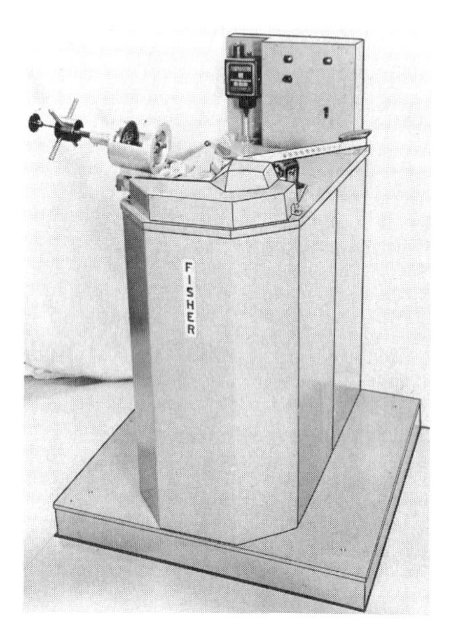

FIG. 16-5 The original 1945 injected metal assembly machine.

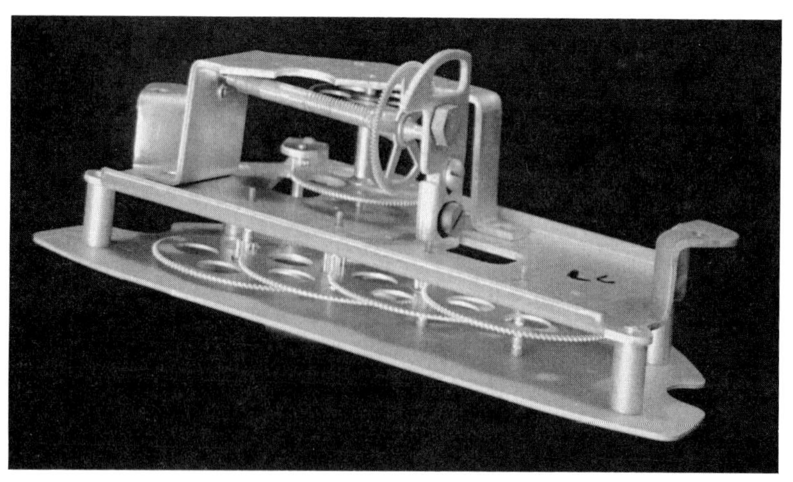

FIG. 16-6 The assembled gear train of the watthour meter. Most of the components are joined into assemblies by use of the IMA technique.

16-3 Process Description Components of similar or dissimilar materials are located and securely held in a fixture (assembly tooling) containing a mold cavity. Molten alloy, usually zinc, is retained in an electrically heated melting pot. A pump consisting of a gooseneck with a closely fitting plunger is immersed in the molten alloy. The components to be assembled are located in the assembly tooling and a "parts-in-place" switch signals a program unit that the components are in the correct location in the closed tooling. The operations to produce an assembly are quite simple and consist of three steps:

In step 1, the parts are placed in the tool. The tool locates parts by their functional features—that is, a gear by its teeth, and the shaft at the center of the gear and at the exact axial position needed by the gear on its shaft. The tool is closed either manually or pneumatically.

Step 2 then occurs. The alloy from the melting pot is injected under pressure into the cavity in the tooling. This alloy surrounds the components to be assembled at their intersection. Almost immediately, the alloy solidifies, cools, and shrinks, thereby locking the components into the required relationship.

In step 3, the movable half of the assembly tooling rotates to shear the sprue which connects the gooseneck nozzle to the cavity in the tooling.

To take full advantage of IMA, locking configurations (such as keys, groves, knurls, etc.) are usually provided in the parts to be assembled. This provides the torque or push-out strength required. In most cases it is advantageous to have relatively large clearance between pins or shafts and the holes in mating parts of the components to be assembled. This permits the injected alloy to flow through the gap and around the components. This then is the source of strength. Although hole tolerances are relaxed, accuracy and repeatability are maintained because parts are located by functional surfaces such as outside diameters, as previously mentioned.

Components can often be eliminated. Cams, ratchets, pinions, keys, pins, cups, etc., can be formed by the injected alloy when assembling parts, thus reducing the number of assembly operations and parts to be stocked. The gear and shaft assembly shown in Fig. 16-7 illustrates the recommended locking configuration on both components. Four lugs or keys can be seen on the gear center hole and in this application a shoulder on the shaft is used to meet the required push-out strength. However, with most gear and shaft assemblies a flush or recessed knurl is used as it best resists the tortional stress normally encountered.

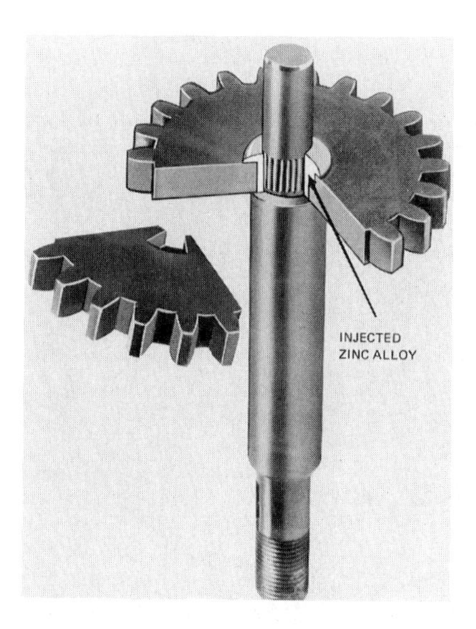

FIG. 16-7 A drive gear is joined to its shaft by means of injected metal assembly. During a torque test, the shaft sheared at 26 ft-lb, while the injected metal hub showed no sign of failure.

16-4 Strength and Torque Properties It is a common misconception that zinc alloy is a weak metal. At many trade shows the assembly of the gear and shaft shown in Fig. 16-8 has demonstrated that a zinc alloy joint can be stronger than the parts being joined. After being joined by means of IMA, the assembly is placed in a torquing device and an attempt is made to twist the 5/16-inch- (7.7-mm) diameter steel shaft out of the gear. In every case the shaft shears before the injected alloy hub shows any sign of failure.

Shown in Fig. 16-9 is an example of high-torque requirement. The specifications called for a torque test of 108 ft-lb (144 N-m). This particular assembly involved the joining of

FIG. 16-8 The joining of gear and shaft by means of IMA demonstrates the strength of the zinc alloy hub. The shaft is torque-loaded, and in all cases the shaft shears before the joint fails.

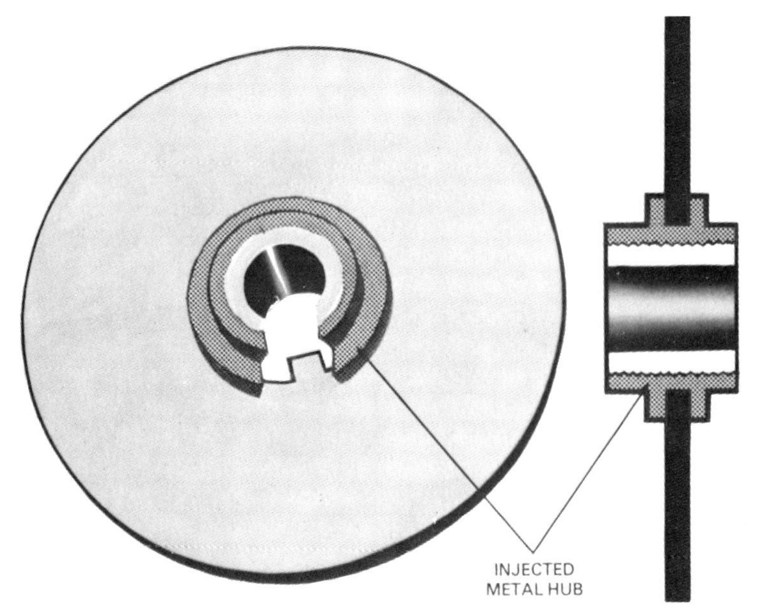

FIG. 16-9 This steel hub and plate was required to withstand a torque of 108 ft-lb. The hub joint remained intact when a high-tensile steel bolt used in the test failed at 160 ft-lb.

a splined center hub to a clutch disk. On a torque test, a 0.625-inch- (16,000-mm-) diameter high-tensile steel shaft sheared at 160 ft-lb (215 N-m) with no deterioration of the actual hub.

In the left-hand view of Fig. 16-9, the locking configuration on the gear and the knurl on the steel shaft is shown. The center view shows the two parts assembled.

16-5 Injected Metals The alloys that can be used with IMA are restricted to those used in hot-chamber die-casting operations. Thus zinc alloys, and occasionally lead alloys, are by far the most commonly used in IMA operations. These alloys are both low in cost and easily obtained from reputable metals suppliers. Zinc Alloy No. 3 (SAE 903) and No. 5 (SAE 925) are recommended and should have alloy certifications from the supplier.

If a high-strength joint is required, one of the zinc alloys is used. If a lower strength is acceptable and a good bearing surface is required, a lead alloy is of advantage.

Figure 16-10 charts the equivalents, or acceptable equivalents, of specifications from certain countries for zinc alloys.

	ZINC DIE CASTING ALLOY (NO. 3)	ZINC DIE CASTING ALLOY (NO. 5)
UNITED KINGDOM (BS)	BS 1004 ALLOY A	BS 1004 ALLOY B
FRANCE (AFNOR)	AFNOR A-55010 Z-A4G	AFNOR A-55010 Z-A4 U1 G
GERMANY (DIN)	GD-ZnAl4 Z-400	GD-ZnAl4Cu1 Z-410
AUSTRALIA	H64-1964 ALLOY A	H64-1964 ALLOY B
JAPAN	ZDC 2	EDC 1
UNITED STATES & CANADA	SAE 903 ASTM AG40A	SAE 925 ASTM AC41A

FIG. 16-10 Actual or acceptable equivalents from certain countries for zinc alloys used for joining by means of the injected metal assembly technique.

16-6 Applications—General There are a number of distinct areas of IMA applications. These include disks and gears to shafts, abrasive wheels to mandrels, terminations to wires and cables, electric motor rotors to shafts, and joining by liquid rivets. The following applications deal with disk and gear applications.

Shown in Fig. 16-11 is the gear train assembly used in the register of all watthour meters, which is an application in which the change in assembly operation to IMA resulted in a cost saving of over 80 percent.

Figure 16-12 demonstrates how for three gear and shaft examples savings in time, material, and money were achieved. The manufacturer had previously to make and stock four different shafts, all of which were machined from pinion stock. The standardization achieved by the IMA process allowed the manufacturer to convert to wire stock for shafts, and, by simply incorporating different stops into the assembly tool, the position

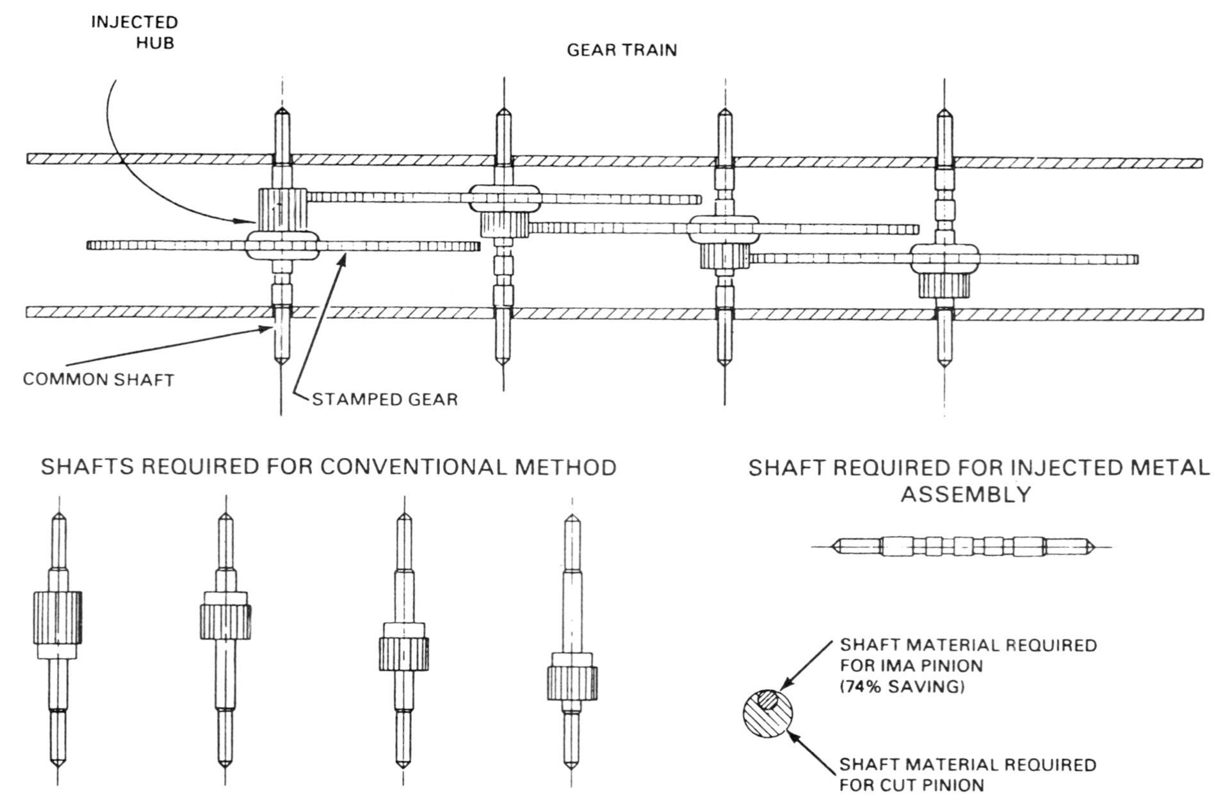

FIG. 16-11 Watthour meter gear train assembly. By changing to IMA, a cost saving of over 80 percent was achieved.

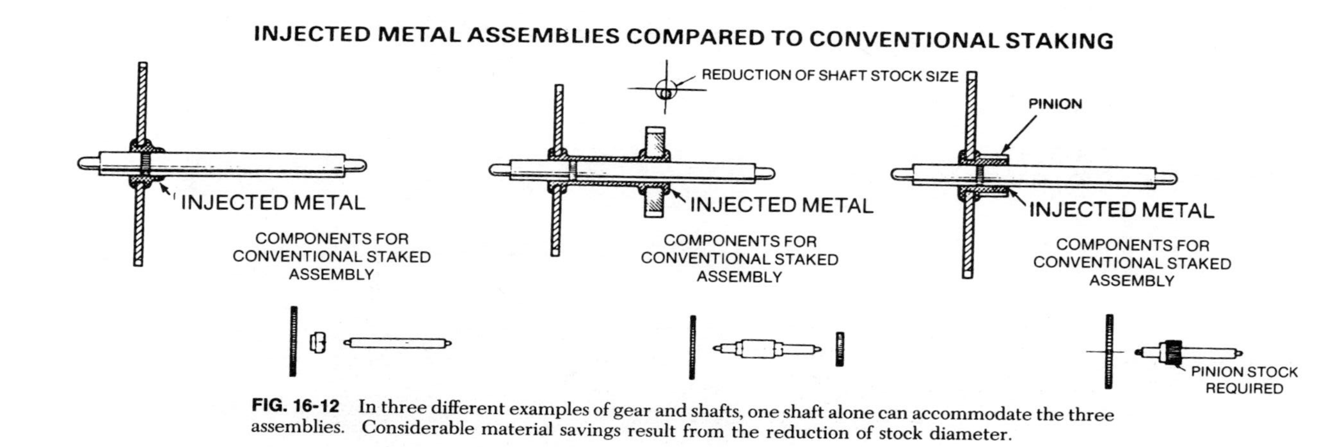

FIG. 16-12 In three different examples of gear and shafts, one shaft alone can accommodate the three assemblies. Considerable material savings result from the reduction of stock diameter.

of the gear on the shaft was controlled. A pinion is required as part of the assembly. In the conventional technique, expensive pinion stock must be used and most of this is wasted in machining it to journal diameter. Alternatively, the IMA technique permits the use of inexpensive wire stock because the required pinion is produced as part of the injected metal hub. Also, the tolerance of the center holes in the gears can be relaxed.

Figure 16-12 compares the advantages of the IMA technique with assembling and joining by conventional methods. In the lower left-hand corner it shows that a staking bushing must be manufactured and used as part of the assembly process. A bushing is a common requirement with a very thin gear.

The conventional method requires that close tolerances must be maintained on the outside diameter (OD) of the shaft, the inside diameter (ID) and OD of the bushing, and the ID of the gear to ensure that the run-out of the assembly does not exceed those precise tolerances.

Using the IMA technique, these two gears are located by their outside diameters. Note on the illustration a large, open-tolerance, clearance hole is provided in the center. Wire stock can be used for shaft material and the three parts joined in one step by injected metal to completely fill the void in the assembly tooling.

16-7 Application—Fan Blade When a hub and fan blade are assembly-produced in the conventional manner, the hub is a machined bushing which is pressed, staked, or welded to the fan blade center hole. This requires close dimensional and concentricity tolerances on the ID and OD of the fan disk. Using IMA (Fig. 16-13) the blade is located by its OD and the hub is injected. Concentricity is readily controlled in the tooling, as is the hub size. Virtually all shapes of fans and impellers can use this process.

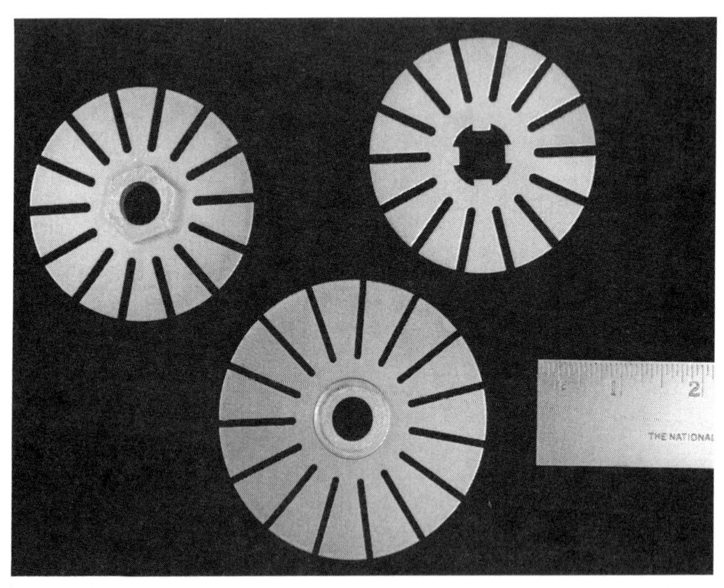

FIG. 16-13 Fan blade hubs are injected with the required shape and ID onto keyed fan blank.

16-8 Application—Miniature Tape Recorder In this application three pins are assembled to the chassis plate of a small tape recorder. In the upper center portion of the top photo in Fig. 16-14, a small slot can be seen. The conventional method of assembly is to produce three staking bushings and to stake the pins and bushings to the plate. It is then necessary to mill a flat on the upper bushing to provide clearance for the small slot.

The bushing is eliminated by IMA. The pins are placed in the tooling together with the plate. In one shot, all three bushings are produced, one of which has a flat to clear the slot. Runners required to join the bushings are easily removed after assembly.

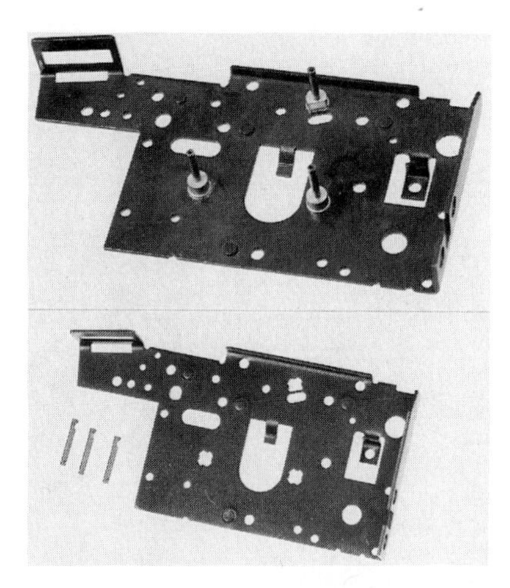

FIG. 16-14 Before (bottom) and after (top) photos of a steel plate for a small pocket dictating machine. Three steel pins shown in the bottom photo are manually inserted into bushings in the assembly tooling. Pins and plate are located in the required positions and zinc alloy is injected to form bosses to fasten the pins. A considerable saving was achieved over a conventional fastening method.

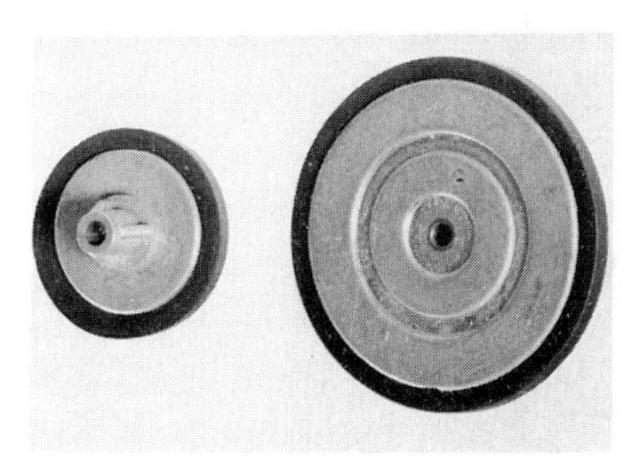

FIG. 16-15 The two-tired wheels shown drive the cassette of a tape recorder. Each tire and central bushing are located in the assembly tooling. Molten zinc alloy is injected between the bushing and the tire to form a wheel. A sprocket is formed simultaneously on one of the wheels.

The wheels shown in Fig. 16-15 are for a tape recorder, and they drive the cassette. The rubber tire and central bushings are placed in the tooling and the wheel injected between the bushing and the tire for each size of wheel. A sprocket is produced simultaneously with the smaller wheel. This method eliminates the need to ream the bushing after a conventional staking operation.

16-9 Terminations onto Wire and Cable Injecting a zinc alloy onto the ends of wire or cable can produce assemblies that can be as strong as and more economical than those produced by using machined fittings. The IMA method eliminates the manufacture and inventory of terminations and their joining onto a wire or cable by crimping, swaging, brazing, and soldering.

The wire or cable may be bare or covered, and its construction may be single wire, strand, cable, or rope. The largest diameter of cable successfully terminated by injected metal assembly equipment thus far is 0.3125 inch (9.50 mm) and the smallest diameter is 0.010 inch (0.25 mm). See Fig. 16-16.

FIG. 16-16 A typical selection of zinc alloy terminations injected onto wire and cable ends. Even full threads can be die-cast onto a cable with no flash or secondary operation.

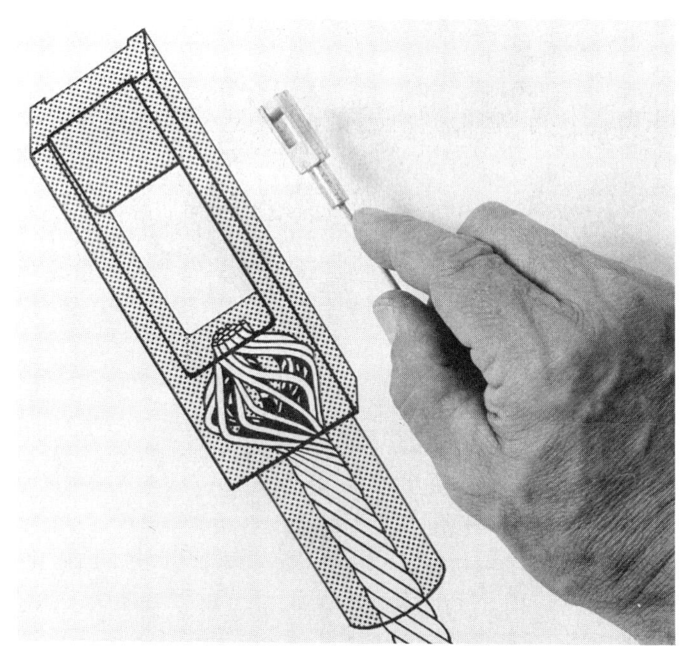

FIG. 16-17 Pull tests repeatably show that, when a cable is "upset" or "birdcaged," a termination will not pull off the cable before the cable itself breaks.

To produce an IMA termination, it is necessary only to place the end of the cable or wire to be terminated in open tooling, the cavity or mold of which has a configuration of the required termination shape. The tooling is closed and a start button depressed, and the assembly machine automatically injects molten zinc alloy under pressure around the end of the wire or cable. The molten metal fills the voids and crevices between cable wires. There is little or no possibility of heat damage to plastic or similar materials used to cover cables, as the total time of injection and solidification is about 150 milliseconds. The cable acts as a heat sink, and the tooling is water-cooled. For solid wire, the metal termination surrounds a premade deformity such as a small right-angled bend. As the termination solidifies and cools, it shrinks onto the wire to form a permanent mechanical lock.

Pull tests have proved that when a cable is "upset" or "birdcaged," as shown in Fig. 16-17, the cable breaks before the termination pulls off. If a termination is to be lightly loaded, the upsetting operation may not be necessary and the shrinkage onto the cable may provide sufficient grip.

A cable upsetting tool can be easily made from standard materials and components. Figure 16-18 shows a termination ground in half and acid-etched to demonstrate how the zinc alloy has flowed between the wires of the cable. A typical pull test made was with 0.010-inch- (0.25-mm-) diameter, 1 × 19 stainless steel, semiflexible aircraft cable. The cable broke under tension at about 1200 lb (545 kg) and the termination at each end of the cable showed no evidence of failure or pulling off the cable.

The control cable assembly shown in Fig. 16-19 has each end of precut lengths of steel-wire-reinforced plastic conduit or "outer" bared to reveal the steel wire. One end

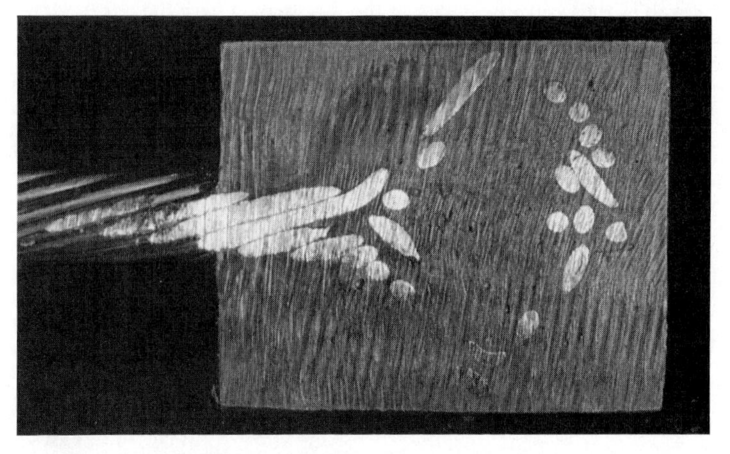

FIG. 16-18 A termination ground in half and acid-etched to show how the zinc alloy has flowed between the strands of the "birdcaged" cable.

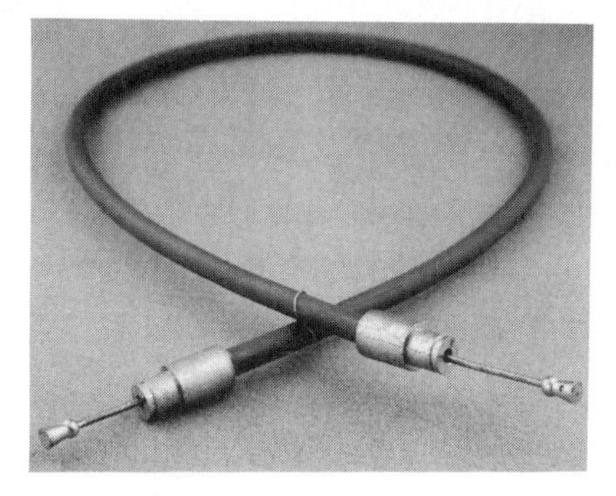

FIG. 16-19 Ferrules and ballends are molded by injected metal assembly onto the ends of steel-wire-reinforced plastic conduit and steel cable. A 10-minute tool change permits the switch from ferrules to ballends. The cable assembly shown is used to control the seat back release mechanism on passenger aircraft.

of the conduit is placed in the assembly tooling, and molten zinc alloy is injected into a ferrule-shaped cavity in the tooling. The zinc flows under pressure into the crevices between the steel wires as the tooling cavity forms the ferrule. This operation is repeated for the other end of the conduit. When a production quantity of conduits has been terminated, a 10-minute tool change can be made and small ball ends are terminated onto the inner cable. One end of the wire is upset to give maximum tension strength and terminated, and the other end is threaded through the conduit. The termination is repeated for the other end of the cable. Each ball end is specified to withstand a pull-off of 75 lb (34 kg) before failure. Pull-off tests on the conduit ferrules have been equally successful.

The production rate for this and similar applications is about 500 terminal injections per hour. Cable assemblies used with IMA equipment are used internationally on motorcycles, autos, snowmobiles, bicycles, trucks, and motorized garden equipment.

A development of this application of IMA enables shapes to be injected along a cable or conduit at predetermined intervals. An example of this is shown on Fig. 16-20. One injected combination shape forms each end of a wound wire conduit. The conduit is sheared between the rectangular flag-like shape and the small grommet-like shape joined to it. Each of these parts forms an end of a conduit.

Similarly, cable terminations are injected at predetermined intervals along a cable which is then cut at desired lengths. The cable is threaded into the conduit and a termination is injected onto the other end. The assembly shown in Fig. 16-20 is used in a dictating machine.

Zinc and lead alloy terminates have been used for electrical connections. Good conductivity is created between the injected termination and the wires of the conductor cable because of the intimate contact of the termination with the wires.

Figure 16-21 shows a termination at one end of the cable of the pulley system which tensions the bow string of a compound bow. This type of bow was first introduced in 1966 and has revolutionized the sport of archery. It is reported that nine out of every ten bows sold today are of this type.

The cable anchor shown in Fig. 16-21 and eyelet zinc alloy injections are formed by IMA onto nylon-coated aircraft specification cable. As the bow string specification calls for up to 70 lb draw weight, the termination must test to a higher weight. To ensure adequate strength, the cable ends are "upset" to give the necessary resistance to pull-off.

The versatility of IMA is demonstrated by the assembly of an electric cord to a thin 0.008-inch (0.20-mm) pad type of heater element. The pad is a lamination of Mylar and aluminum foil. This particular application is used on an electrical heater for a water bed. It has received UL approval and has proved to be durable and efficient. See Fig. 16-22.

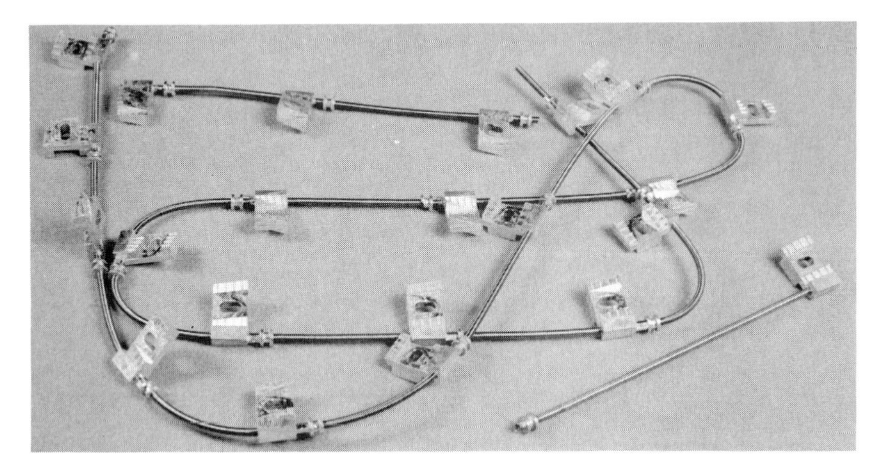

FIG. 16-20 Flag-like terminations are injected onto a cable or conduit at predetermined intervals. Each flag is cut to form each end of a conduit. Similarly, a termination is formed by IMA which is inserted into the conduit, and another termination is formed at the other end of the cable.

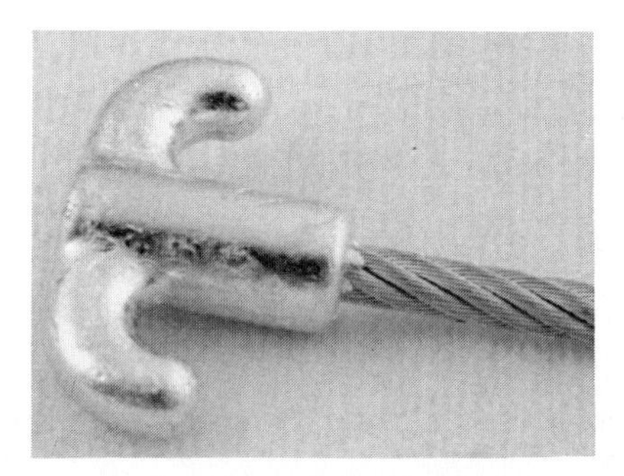

FIG. 16-21 Shows an "anchor" termination made at one end of a nylon-covered steel cable. The anchor is used at one end of the pulley system which tensions the bow string of a compound bow.

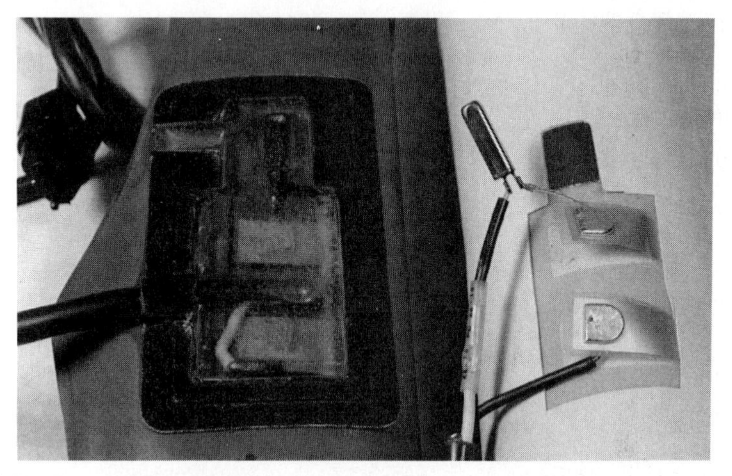

FIG. 16-22 Two small zinc alloy pads enclose bared electric wire ends and fasten them securely to the Mylar and aluminum foil laminate heater element. Two corresponding pads are formed on the reverse side of this water bed heater.

Two small pads of zinc alloy are formed by the injection, which encloses the bared wire ends and securely fastens them to the laminate. Corresponding pads are formed simultaneously on the reverse side of the laminate. Electrical conductivity is excellent.

One of the earliest applications of cable assemblies produced by IMA was the steering mechanism cable used on the motor and steering controls for outboard motor boats. This cable application is currently in production. Another early cable termination application using IMA was for 10-speed bicycles. Both the brake and the derailleur gear mechanism cables were changed to cables terminated by IMA.

The initial sale of IMA cable equipment for bicycles was to a well-known and highly respected British bicycle manufacturer. This adoption of the injected metal method was later followed by the installation of several assembly machines in Japan for similar use. The technique of die casting zinc alloy terminations is considered the international standard today for 10-speed cycles. There is virtually no limit to the shape of terminations that can be injected. It is not difficult to inject an externally threaded termination with a full diameter or with flats. If an internal thread is required, a secondary operation of tapping in a blind cored hole is required.

16-10 Abrasive Points In the assembly of an abrasive point (small grinding wheel) to a steel mandrel, the components are held in the required relationship by tooling while zinc is injected into the void between the abrasive wheel and a knurled mandrel. The molten zinc alloy penetrates the porous abrasive and solidifies, securely joining the abrasive wheel to the mandrel. The assembly can be handled immediately. This penetration and shrinkage onto the knurled mandrel results in a strong mechanical lock. Virtually any shape of abrasive point can have its shaft assembled to it by the IMA method.

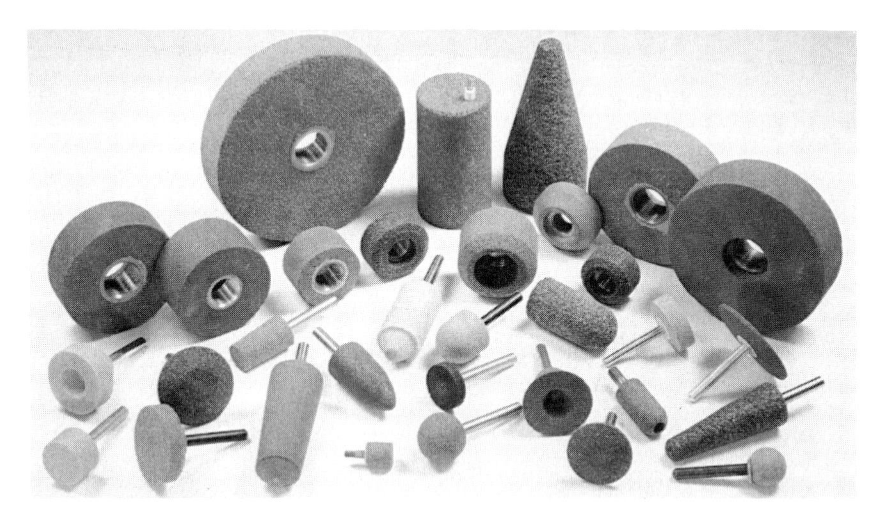

FIG. 16-23 A typical selection of small abrasive points and wheels that are accurately joined to their shafts or hubs by injected metal assembly.

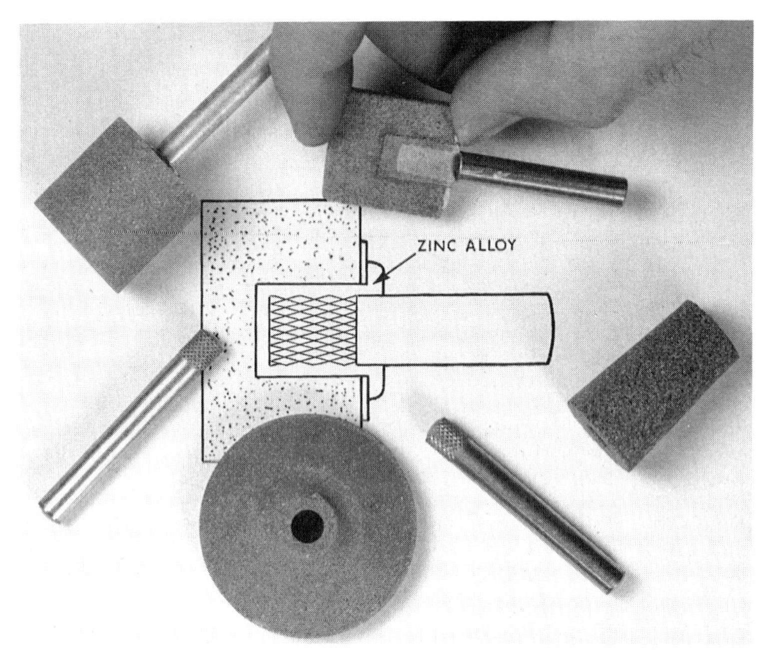

FIG. 16-24 Before and after shots of abrasives and knurled mandrels. The cut-away at top center shows how the molten zinc penetrates the abrasive grit and shrinks onto the mandrel to form a secure and locked joint.

Production rates are good, being between 400 and 600 per hour using low-cost zinc alloy. No cure time, which is necessary with the conventional cementing methods, is needed. Concentricities between 0.003 and 0.004 inch (0.0075 and 0.010 mm) TIR are normal. For most grinding wheel manufacturers, this makes the dressing of wheels unnecessary. See Figs. 16-23 and 24.

Much larger types of grinding wheels can have their center holes lined with lead alloy. The assembly machine injects an alloy hub of the required internal diameter into the grinding wheel. This IMA method of adding the lining to the wheel increases production rates from 20 to 30 pieces per hour by the hand-pour method to about 300 pieces per hour. The capacity is from 3- to 16-inch- (75- to 400-mm-) diameter wheels, a thickness from 0.375 to 2.5 inches (9.5 to 63 mm), and a hole diameter from 0.375 to 2 inches (9.5 to 50 mm). Intermediate sizes of grinding wheels can be accommodated. A hub, either smooth bore or internally threaded, can be injected, giving a high accuracy with a reasonably high production rate.

Although abrasive wheels assembled by IMA do not require dressing before use, some users dress to ensure concentricity and perpendicularity.

Most smaller abrasive points have a small paper disk on the mandrel side of the abrasive wheel to protect the tooling from wear by the wheel, and that disk provides a place for manufacturer's identification. The injection is so fast and the heat dissipates so rapidly that the paper does not char.

16-11 Applications—Sensitive Materials In the application shown in Fig. 16-25, a thin glass plate is joined to an aluminum shaft. A center hub is formed by injected metal assembly for the damping disk and shaft of a polyphased watthour meter. Concentricity and perpendicularity are important and are achieved by injecting metal, in this case a lead alloy, into a cavity in the assembly tooling at the intersection of the two components. This forms the hub that secures the components. An oversize hole is provided in the glass plate to permit a free flow of the molten metal between the components. This is a good example of a brittle material being securely joined unharmed by any heat from the molten metal when using IMA.

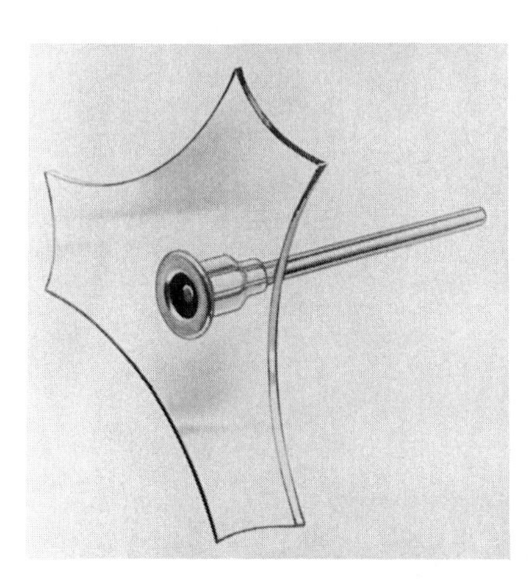

FIG. 16-25 A glass rotor plate for a European watthour meter joined to its aluminum shaft by IMA. An oversize hole is made in the plate.

A timing disk used in electronic equipment is illustrated in Fig. 16-26 to show that heat-sensitive materials can be used without damage when using IMA. Accurate orientation of the cam disk and shaft to the printed circuit is achieved by the tooling as well as the concentricity and perpendicularity of the assembly.

A plastic camming gear and disk are used in a household appliance programmer. The cam orientation is of vital importance, as is the concentricity between the disk OD and the shaft. This is achieved once again by locating the parts in the assembly tool and injecting molten alloy into the tool cavity and the void between the parts. The rapid cooling of the alloy precludes any damage to the heat-sensitive part.

FIG. 16-26 A timing disk shows that heat-sensitive material such as that used in a printed circuit sustains no damage when IMA is employed. The tooling automatically orients the cam disk to the printed circuit. Note the keys on the two disks in the lower half of the photo.

Materials such as ceramics, plastics—even soft rubbers—almost all metals, and even paper products can be used without damage. The speed of the IMA injection, about 15 milliseconds in most cases, and the water cooling of the tooling does not permit heat damage to the parts being joined.

16-12 Unusual Applications In this application, prior to the conversion to IMA, a machined steel pillar was joined to a stamped steel plate to form a mount for an automobile rear view mirror. A considerable cost saving was realized with injected metal assembly by injecting zinc alloy to form the ball and pillar while simultaneously locking the part being die-cast to the stamping. This eliminated the machining and stocking of two parts, the pillar and ball, as well as a staking bushing. See Fig. 16-27.

FIG. 16-27 Only one part is required for this assembly—the plate. IMA injects the ball and pillar onto the plate as they are formed by IMA.

Figure 16-28 shows a different approach to the use of injected metal assembly in that the principle of insert die casting is applied. Three steel shafts or pins are assembled to an injected zinc alloy frame for an electric shaver. The three pins are manually loaded into the assembly tooling, and the complete frame is injected around the pins. No secondary operations are required, which is normal with IMA. The very accurate and repeatable location of the pins and their relationship to each other ensures precision of the end product.

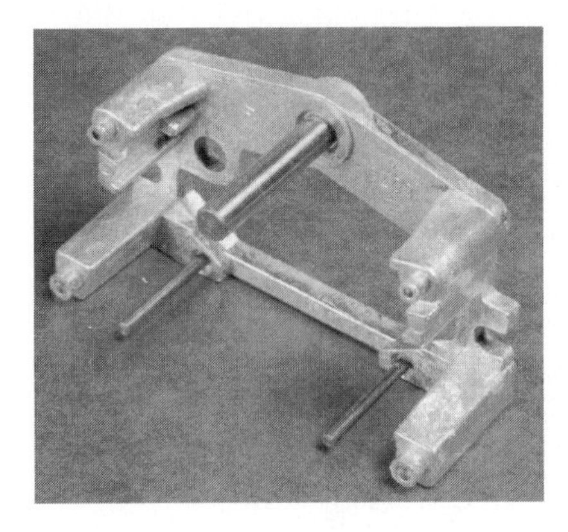

FIG. 16-28 Three steel pins are held in the assembly tooling. Molten zinc is injected to form the frame, ie., an insert die casting. The tooling accurately controls the location of the pins in the frame and the relation of the three pins to each other.

16-13 Small Electric Motor Rotors The assembly of fractional horsepower electric motor rotors to their shafts is one of IMA's most recent and popular applications. Both stepper and induction-type motor rotors are being successfully assembled by IMA. See Fig. 16-29.

FIG. 16-29 Typical small electric motor rotors that have been joined to their shafts by injected metal assembly equipment. The pole piece tooth offset on the stepper motors is accurately achieved as each lamination stack or pole piece is mechanically oriented in the assembly tooling.

The "pushout" or "pull-out" load is often questioned regarding this method of assembling small motor rotors. Obviously one load figure cannot be standardized for all applications. A push-out test on one specific gear and shaft was performed. The steel shaft had a diameter of 0.3125 inch (8.0 mm), and the zinc hub had a length of 0.5 inch (12.7 mm). The push-out measured approximately 2000 lb (8900 N).

When using injected metal assembly to assemble stepper and squirrel-cage-type motor rotors, there are several time- and cost-saving advantages. The center hole of the rotor is deliberately made oversized to easily accommodate the largest of several diameters of shaft that may be required. This hole need not be accurately sized or round, or even concentric with the OD of the rotor. In fact, virtually any convenient shape is satisfactory. However, the best locking configuration is for the holes to have keys to assist in the locking action and the shafts to be knurled or have similar locking means within the joint area, although in very-low-torque applications there is no need for the shafts to be knurled or the rotor center holes to have keys. In those cases, the shrinkage of the injected metal is sufficient to give an adequate lock between the rotor and shaft. Figure 16-30 shows a stepper motor rotor before and after being assembled by IMA. The shaft is shown with a blind hole onto which the injected alloy penetrates and shrinks during cooling to secure the hub to the shaft.

FIG. 16-30 A stepper motor rotor before and after its joining to a shaft by IMA. Note the oversized center holes in the pole pieces and the magnet. The cross-drilling on the shaft assists in the secure fastening to the keyed pole pieces.

The diametric difference between the shaft and the rotor hole should be a minimum of 0.060 inch (0.75 mm) around the shaft to permit free flow of molten alloy to fill the available space in the assembly tooling cavity.

A few applications of IMA to small motor rotor assemblies are illustrated in Figs. 16-31 through 16-36 and described in the following paragraphs.

The small motor armature shown in Fig. 16-31 had been a manufacturing problem until it was assembled by IMA. The problem was that a screw-machined, cup-shaped bushing had a high cost and required hand assembly. Injected metal assembly forms the bushing as part of the hub which secures the laminations simultaneously as it joins the shaft to the armature. The laminations are automatically positioned in the assembly tooling. In addition to saving time and labor, IMA reduces the cost of assembling the components by about four-fifths. Extremely close tolerances are repeatedly achieved and concentricity is greatly improved. Concentricity in this and all other motor rotor applications is held within 0.002 to 0.004 inch (0.05 to 0.10 mm) TIR.

The motor assembly shown in Fig. 16-32 is used as a cooling fan motor in office equipment. A rotor shaft is joined by IMA to a metal spider while being held concentric to and parallel with the internal diameter of the stator laminations. This ensures a concentric and uniform air gap. Secure mechanical locking is achieved by the usual cooling and shrinking of the alloy onto four keys in the rear bracket and onto shaft knurling.

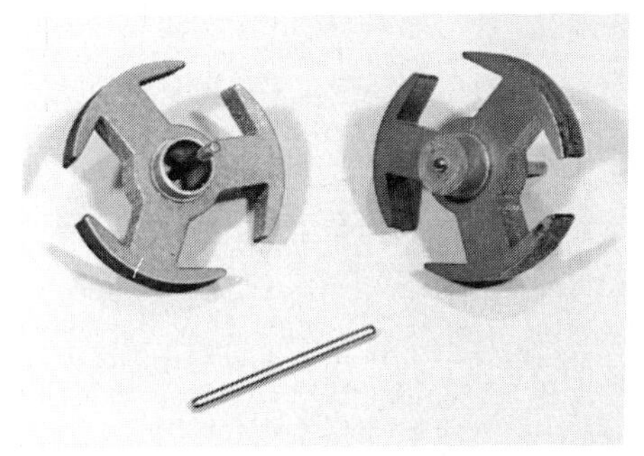

FIG. 16-31 The high unit cost of a screw-machined bushing for a small dc motor rotor was reduced by four-fifths when IMA was used. In addition, close tolerances were repeatably achieved.

FIG. 16-32 A double use of IMA on one product. A rotor shaft is joined to a housing back plate, and a rotor with a knurled end is assembled to a polycarbonate fan by injecting molten zinc between the fan and rotor.

Figure 16-33 shows the shaft for a unit bearing motor assembled to the back plate by IMA. This was one of the earliest applications of this joining method to small electric motors.

Most squirrel cage motors are cast in aluminum alloy using conventional cold-chamber die-casting equipment. This achieves an efficient motor at the usual operating speeds. However, when the noise level in such office equipment applications as a fan must be reduced, a low-speed motor is necessary. This requirement makes zinc alloy die-cast rotors feasible. The size of the end rings and inductor slots will vary with the type of injected metal used. With the IMA method of production, the rotor end rings and slots can be injected simultaneously with the assembly of the shaft to the rotor. The results

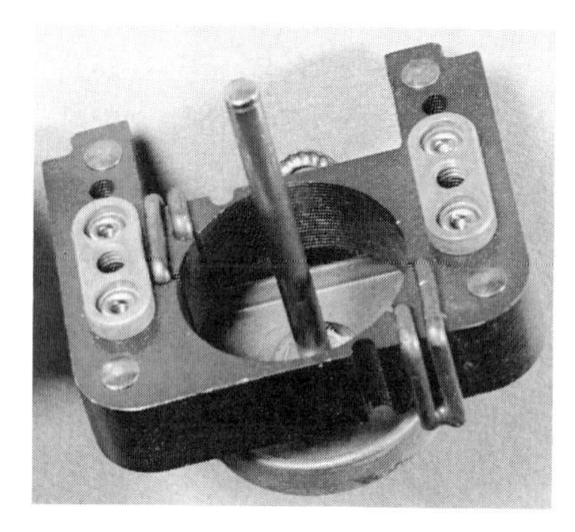

FIG. 16-33 A very early application of IMA to small electric motors. A unit bearing motor shaft is accurately joined to a motor back plate.

are savings in materials used, a reduction in the number of parts and manufacturing operations needed, and an overall reduction in costs.

Figure 16-34 illustrates how a four-part rotor assembly is reduced to three parts. The figure shows a shaft, a copper contact pin, and a pinion assembled to a ceramic magnet. The previous method of assembly required the pinion to be brazed to a hub, which in turn was cemented to the magnet. The primary advantages with IMA are the cost saving through the pinion being injected and the time saved in assembling the hub to the rotor. Cementing requires a considerable time for curing whereas IMA is virtually instantaneous.

Another rotor application is assembly of a synchronous rotor disk to a shaft. The previously used method produced a hobbed shaft with the spline or pinion as part of the shaft. The shaft was then staked to the cup. This method resulted in considerable losses due to excessive eccentricity and wobble imparted by the staking operation. Using the

FIG. 16-34 A steel pin and an offset copper pin are simultaneously assembled by IMA to a ceramic magnet rotor for a battery-operated dc motor.

IMA technique, a large clearance hole is provided in the cup, piano wire is used for the shaft, and a hub and pinion are injected to produce the assembly. See Fig. 16-34. Considerable savings are achieved and an added bonus is that the assembly tool can be used as a checking fixture. If the rotor is out of round, too large or too small in diameter, or too thick, it will not be accepted by the assembly tool. This eliminates costly secondary inspection. In addition, starting torque is reduced because of the high finish and small diameter of the wire compared with that required for the shaft as previously machined.

In stepper motor rotors, use of IMA allows the teeth of one pole piece to be automatically aligned with a tooth or tooth space of an adjacent pole piece. In many applications,

FIG. 16-35 Shafts are assembled to aluminum cups by injected metal assembly for car speedometers. This method of joining makes the traditional balancing operation unnecessary.

FIG. 16-36 All four of these induction motor rotors were joined to their shafts in the same tooling. Interchangeable inserts enabled different lengths of rotor and different shaft diameters to be accommodated in the assembly. The assembly tooling, if required, could assemble rotors of different diameters and shafts of different lengths.

a permanent magnet is located between the pole pieces and secured during the assembly operation.

The tooling fixture can also incorporate interchangeable stops for different shaft positions relative to the rotor. Interchangeable bushings can also accommodate different shaft diameters, and interchangeable locators can be used for different rotor outside diameters. Tooling can be designed to suit a variety of stack heights. All four motor rotors shown in Fig. 16-36 are assembled in the same assembly tooling.

Interchangeable inserts enable different stack heights caused by variances in lamination thickness to be accommodated. These inserts are spring- or air-operated clamp bushings in the tooling. This eliminates the shims or spacers generally required to accurately locate the rotor in the motor housing.

16-14 Benefits of IMA with Motor Rotors It is of interest to compare the IMA technique with the more conventional rotor assembly methods. With the older methods, many dimensions are critical. Achieving the acceptable precision requires assembly by time-consuming and sometimes expensive means. The center hole of the lamination stack must be produced with very precise tolerances and must be concentric with the outside diameter. A shrink fit is achieved during cooling by inserting the shaft through the hole of a heated rotor. This is a slow procedure because of the heating involved and the delays caused by the need to wait for the parts to cool to a handling temperature.

Also with the shrink-fit method, different center hole sizes in lamination stacks and pole pieces must be produced, which requires a very large parts inventory. Usually, two or more spacers and/or shims are required by the older methods to achieve the fixed dimensions from the end faces of the rotor to the end of the shaft as well as the shoulder-to-shoulder dimension. This is to ensure that the rotor will be in the correct relationship with the stator when the motor is assembled, and also to minimize end play. See Fig. 16-37.

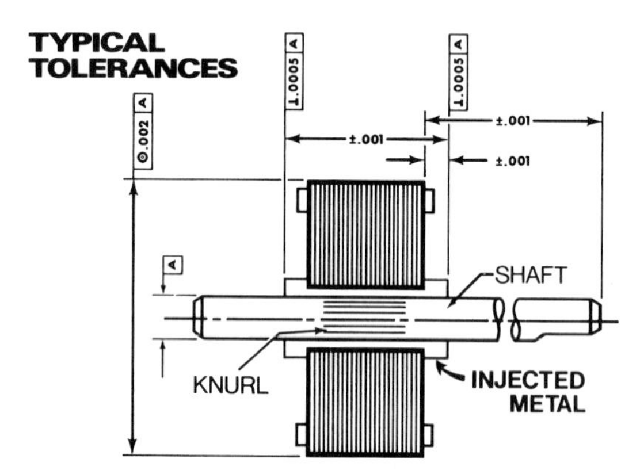

FIG. 16-37 Assembly tooling by the IMA process holds all fixed dimensions for a small induction motor rotor. Varying thicknesses of rotor laminations are automatically compensated for.

As the lamination stack height can vary plus or minus one lamination, such spacers and shims are critical. Once the rotor is secured to the shaft, the outside diameter of the rotor is cylindrically ground to obtain concentricity and OD size. Before the rotor can be assembled into its frame, other secondary operations such as cleaning and deburring may be necessary. It is also common practice to straighten the shaft, which may have been distorted by the heat-shrinking operation.

Besides shrink fitting, other conventional methods of assembly include press fitting, staking, and cementing. All of these are slow and require the trial-and-error process of adding shims and/or spacers.

There are several advantages when assembling by IMA. One is that the lamination stack is centerless-ground to its finished size prior to assembly. This is an inexpensive

FIG. 16-38 Three different variants of a stepper motor rotor produced in the same assembly tooling having interchangeable inserts. These rotors are produced on fully automated assembly equipment working on a fully robotic production line.

method. The same lamination height and length can be used for different shaft diameters, as the molten alloy fills different sized gaps between shafts and rotor center holes. No shims or spacers are required, as this is compensated for in the tooling. One assembly tool using interchangeable stops, locators, and bushings can accommodate different shaft diameters, lengths, and positions, and different stack heights and diameters. There is rarely any need to grind the rotor to size after assembly since the precision of the assembly tooling holds the original diametral dimension created by the centerless grinding. Stages and variants of small stepper motors produced by IMA are shown in Fig. 16-38.

16-15 Riveting with Molten Metal If a customer's parts come from more than one supplier, the accumulation of tolerances can build in stress when riveting conventionally. In addition, holes can be misaligned, preventing the insertion of rivets. This can be a major inspection problem.

However, exact hole location, hole size tolerance, and alignment are not problems with IMA since the technique injects a "liquid rivet" which can compensate for misalignment of holes even if they are off by one-quarter of their diameter. In addition, the riveting holes in the various parts to be riveted need not be exactly the same size.

The components are located and securely held in assembly tooling. A "liquid rivet" of zinc alloy is injected, completely filling the rivet holes. A head is formed at each end by small, shaped cavities in the tooling. A few milliseconds after injection, the riveted assembly is ready to be installed in the end product. A production rate of between 300 and 1000 per hour, depending upon the complexity of the assembly and the number of parts to be loaded by the operator, can be maintained with very accurate assemblies. In some applications, the assembly tooling acts as an inspection fixture, thus eliminating the inspection phase.

Not only can most metals be assembled or riveted by IMA, but also all materials that can be conventionally riveted and even some materials that cannot. Paper products, soft plastics, and brittle materials such as ceramics can be accommodated without damage or rejection.

In the application shown in Fig. 16-39, the gap shown in the lower left-hand corner is

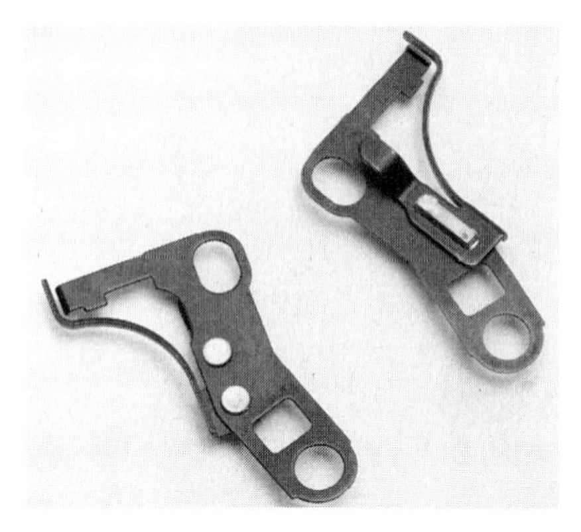

FIG. 16-39 The gap between the spring and plate is critical in this part for a sewing machine. In hard riveting, the location and size of rivet holes controlled the gap and any deviation in those dimensions for holes meant the gap had to be manually adjusted after joining. With IMA, hole deviation is immaterial as the "liquid rivet" finds its own way since parts are accurately located and held by the assembly tooling.

a critical dimension. In the hard-rivet method, the location of the gap and its dimension are controlled by the location and size of the rivet holes; with IMA, on the other hand, the parts are located in the tooling to control the gap, and the rivet holes, regardless of their location and possible oversize, are simply filled by the molten injected metal. The saving is obvious since an adjustment of the gap is not required after assembly and part sizes and locations do not not need to be as accurate.

Figure 16-40 illustrates how a small subassembly—a "hammer" for a computer printer—was redesigned to use IMA. The parts needed were reduced from 18 to four and the number of assembly operations required was reduced from five to one. The figure shows the original design at the left and the one-shot IMA assembly at the bottom and right. The four IMA components are a steel stamping, an armature with 20 laminations, a steel pin, and a tubular rubber stop. The four parts are loaded into assembly tooling. A single runner channels the alloy to perform the following functions:

■ Molten alloy flows through holes pierced in the lamination stack to form rivets holding the stack securely together. Since the alloy is a liquid when injected, neither the accuracy of hole alignment or size is a critical factor.

■ Alloy flows along one arm of the steel stamping to form a hub to secure the steel pin. It also flows along the other arm to form a headed pin which secures the rubber stop already in place in the tooling. The alloy cools so rapidly that the rubber is unaffected by the heat.

An output of about 100 assemblies per hour is achieved by using the IMA process.

Operations required for the original design and now eliminated included tapping holes for fasteners, peening rivets and bushings, and assembly and tightening of screws, nuts, and washers. In addition, a small drilled and tapped plate needed to clamp the lamination stack is eliminated.

The single zinc alloy runner used in the IMA method is easily removed from the back side of the plate. See Fig. 16-41.

The camera manufacturing industry also uses the IMA "liquid-rivet" technique to advantage. The critical factor in this application is the location of the aperture in the plate in relation to the mounting holes. See Fig. 16-42. The control of this location is achieved by the assembly tool. Again, the parts are located in the tool by their critical features and two "liquid rivets" are injected to securely join the parts. The size and location of the rivet holes in the parts are not critical as the liquid rivets fill all the available space. Additional rivets are injected simultaneously. Two rivets have hinge block heads and five are studs of different sizes. As the injected metal cools and shrinks, it permanently joins the parts together. This method of manufacture uses only three parts. The previously used method required the making, stocking, and handling of eight parts.

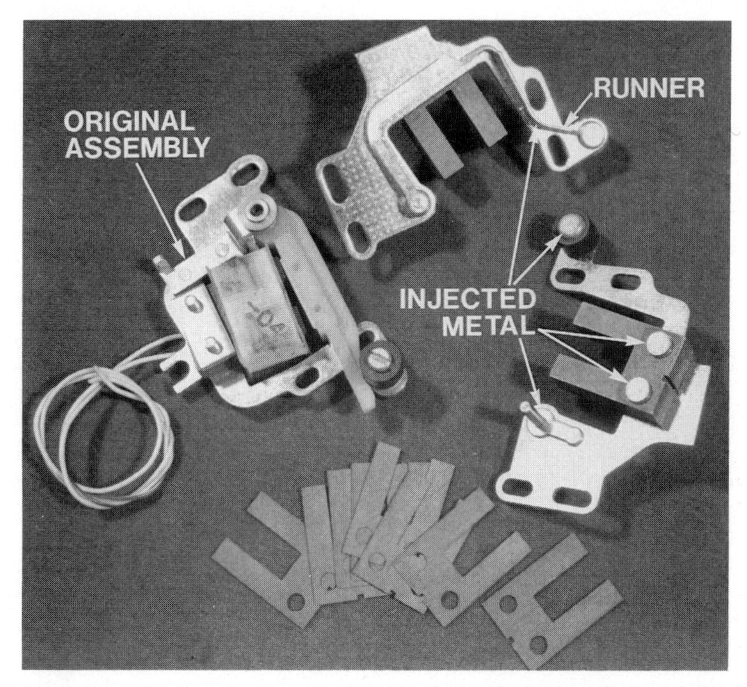

FIG. 16-40 Subassembly of a computer printer "hammer." At left is the original subassembly method, comprising 18 parts plus a number of assembly operations. At right and bottom is the subassembly revamped to use IMA. Because of IMA, the assembly requires only four parts and one operation—placing the four parts in the assembly tooling, closing the tool, and pressing a button. With IMA, the work is done by a "liquid rivet."

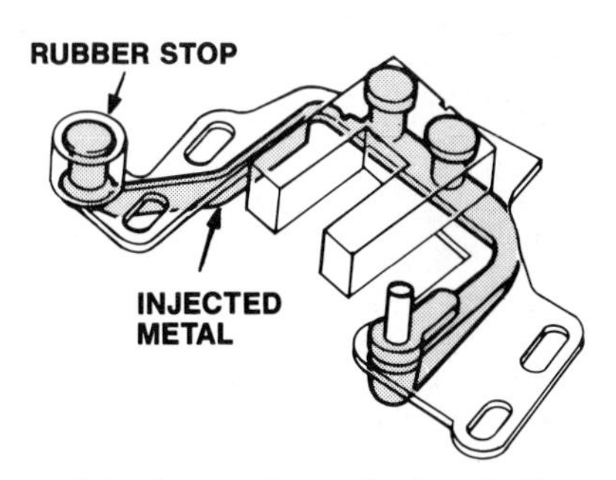

FIG. 16-41 The shaded parts of the "hammer" assembly show the "as cast" features of the joining of the four parts by a "liquid rivet."

16-16 Equipment One of the small IMA assembly machines is shown in Fig. 16-43; it is semiautomatic and will accept a large variety of types of assembly tooling and the operating heads that hold and control the tooling, and can produce quite a large variety of assemblies. This particular model has an injection capacity of 0.3 cubic inch (50 mm^3). A very similar, slightly larger, machine has an injection capacity of up to 0.60 cubic inch (100 mm^3). The largest-capacity assembly machine available at the time of this book's publication has an injection volume of 1.0 cubic inch (160 mm^3).

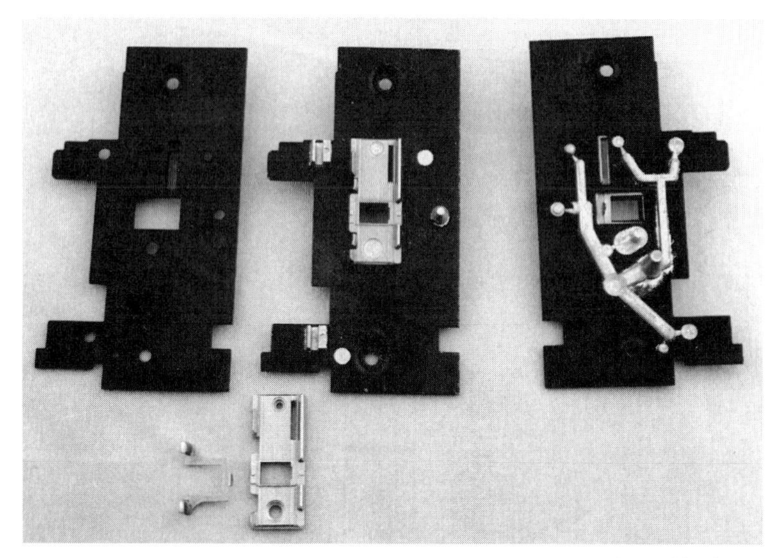

FIG. 16-42 In this aperture plate in a camera, all parts are located in the assembly tool by their critical features and securely joined by the "liquid rivet," the use of which allows for the simultaneous joining of seven additional parts into place on the plate.

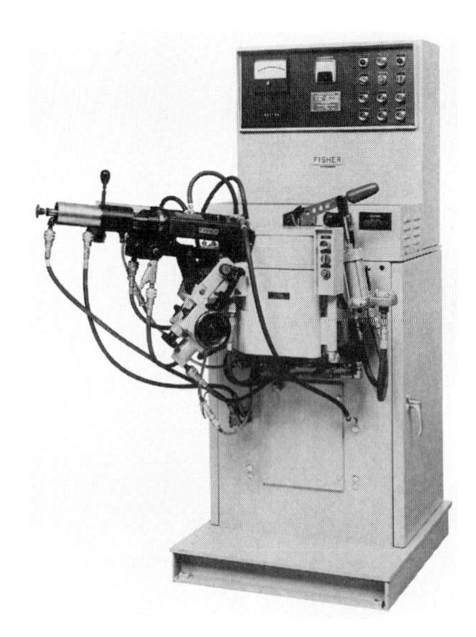

FIG. 16-43 A semiautomatic injected metal assembly machine. The illustration shows its operating head, which holds the assembly tooling.

Fully automatic IMA systems have been designed and produced that feed and load components into assembly tooling. This particular system uses one or more feeders and a programmable logic controller. It has a high production rate of approximately 1000 assemblies per hour and is ideal for long production runs.

The completely automated system shown in Fig. 16-44 is also controlled by a programmable logic controller. Parts are loaded into the tooling by pneumatically operated "pick-and-place" loaders, which can be seen in the illustration. The completed assembly

FIG. 16-44 Fully automated injected metal assembly equipment is shown with its safety cages removed. Parts to be assembled are loaded and unloaded automatically by pneumatic pick-and-place units. Zinc ingots are also automatically fed to the equipment "on demand" from a programmable controller.

is caught and gripped by a pick-and-place unloader on ejection from the assembly tooling. The quantity and type of loaders required are determined by the assembly to be produced. The present system can accommodate up to five pick-and-place units. The relationship between the operating head and grippers of the loading and unloading units can also be seen in the illustration.

Such equipment can operate without interruption for hours at a time at a production rate that has been shown to be more than adequate to keep other robotic equipment supplied with assemblies. One attendant can handle several systems or perform other functions in the area.

With fully automatic equipment, an attendant monitoring several machines has only to add an alloy ingot when necessary and remove the filled boxes of completed assemblies.

With the automated equipment shown in Fig. 16-44, spherical ingots of alloy automatically replenish the melting pot on demand from the pot. Each completed assembly is placed on a conveyor by the unloader and transferred to the station where needed.

Protection of the attendant is vitally important. All pinch points and moving components are covered or protectively screened to prevent access by personnel during equipment operation. In most cases, safety switches are actuated to prevent the system from operating when the screens are removed for any reason.

16-17 Advantages of Injected Metal Assembly The following list summarizes the many advantages of injected metal assembly:

- Dissimilar metals and materials can be joined.
- IMA is a rapid process that contributes to reasonably good production rates.
- Integral injected metal parts can be substituted for machined shapes.
- Dimensional repeatability of assembly can be considered to be absolute.
- Quality of the total assembly is consistent.
- The assembly tooling is often used as a component checking fixture.

- A component's working surfaces are used for location to achieve accuracy in assembly.
- A component's tolerances can usually be opened or reduced.
- Unskilled personnel can be rapidly trained to achieve optimum performance.
- The process is clean, nontoxic, and odor-free.
- The joining alloy is low in cost compared with many other joining or fastening media.
- No cleaning or chemical preparation is required.
- Assemblies can be used immediately with no secondary operations.
- Assemblies can be handled immediately after IMA.
- Installation of assembly equipment into the production line is usually possible.

The quality of assemblies produced by IMA has been proved many times to be superior to that of assemblies produced by other methods, and IMA results in fewer rejected parts than other methods. Because with IMA components are located and held in the assembly tooling in the exact orientation they must have in the completed assembly, the alloy injection secures the components in that exact orientation.

New applications for IMA are being constantly suggested by users in the manufacturing industry. There seems to be no end to the imagination of designers once they appreciate the wide horizons opened up by using this method of assembly.

Section 17
Sheet Metal Fastening

ROBERT O. PARMLEY, P.E.
President
Morgan & Parmley, Ltd., Ladysmith, Wisconsin

17-1 General A major portion of sheet metal work is in the heating, ventilating, and air conditioning (HVAC) industry. Fabrication and material for ductwork, equipment housings, plenums, and related air system components represent from 20 to 40 percent of the total cost of a heating, ventilating, and/or air conditioning system.

Construction of ductwork in HVAC systems must conform to the latest standards of the American Society of Heating, Refrigerating, and Air Conditioning Engineers (ASHRAE); the Sheet Metal and Air Conditioning Contractors National Association (SMACNA); the National Fire Protection Association (NFPA); and applicable state and local codes.

Most sheet metal work in the HVAC industry is custom-made following A/E designs and requires numerous connections. This section is devoted to the presentation of typical or standard sheet metal fastening and joining. However, no attempt is made to include all types of connections in this section, nor should such inclusion be inferred. Adhesive bonding, for one, is not addressed. The reader is referred to sec. 12. The editor, therefore, recommends that publications from the previously mentioned organizations be consulted if further technical data are required.

17-2 Materials The materials most commonly used in the fabrication of sheet metal duct construction are:

1. *Galvanized steel sheets.* Galvanized steel is the most common material used for duct construction. It has a galvanized coating on both sides of the sheet.
2. *Black-steel sheets.* Black steel (hot- or cold-rolled) is generally used to construct hoods and ducts conveying hot air or gases.
3. *Copper and stainless steel sheets.* These metal sheets have high resistance to moisture. They are used in duct work exposed to damp conditions. Stainless steel is used more than copper because of its lower cost and higher strength.
4. *Aluminum sheets.* Aluminum sheets are lighter than other materials, plus they have a high resistance to corrosion. Aluminum is also suitable for use in extreme-humidity environments.
5. *Precoated steel sheet.* Precoated steel sheets, available in a wide range of coating types, colors, and textures, combine the durability and strength of steel with the desired decorative appearance. The precoated sheets still retain good fabricability and joinability.

17-3 Transverse Designs The most common sheet metal connection is the transverse joint. Figure 17-1 illustrates 24 transverse (girth) joints for rectangular ducts. The precise type, size, location, and material used in joint or seam connections are in most instances left open to the prudent judgment of the designer. Figure 17-2 depicts eight standard longitudinal seam joints.

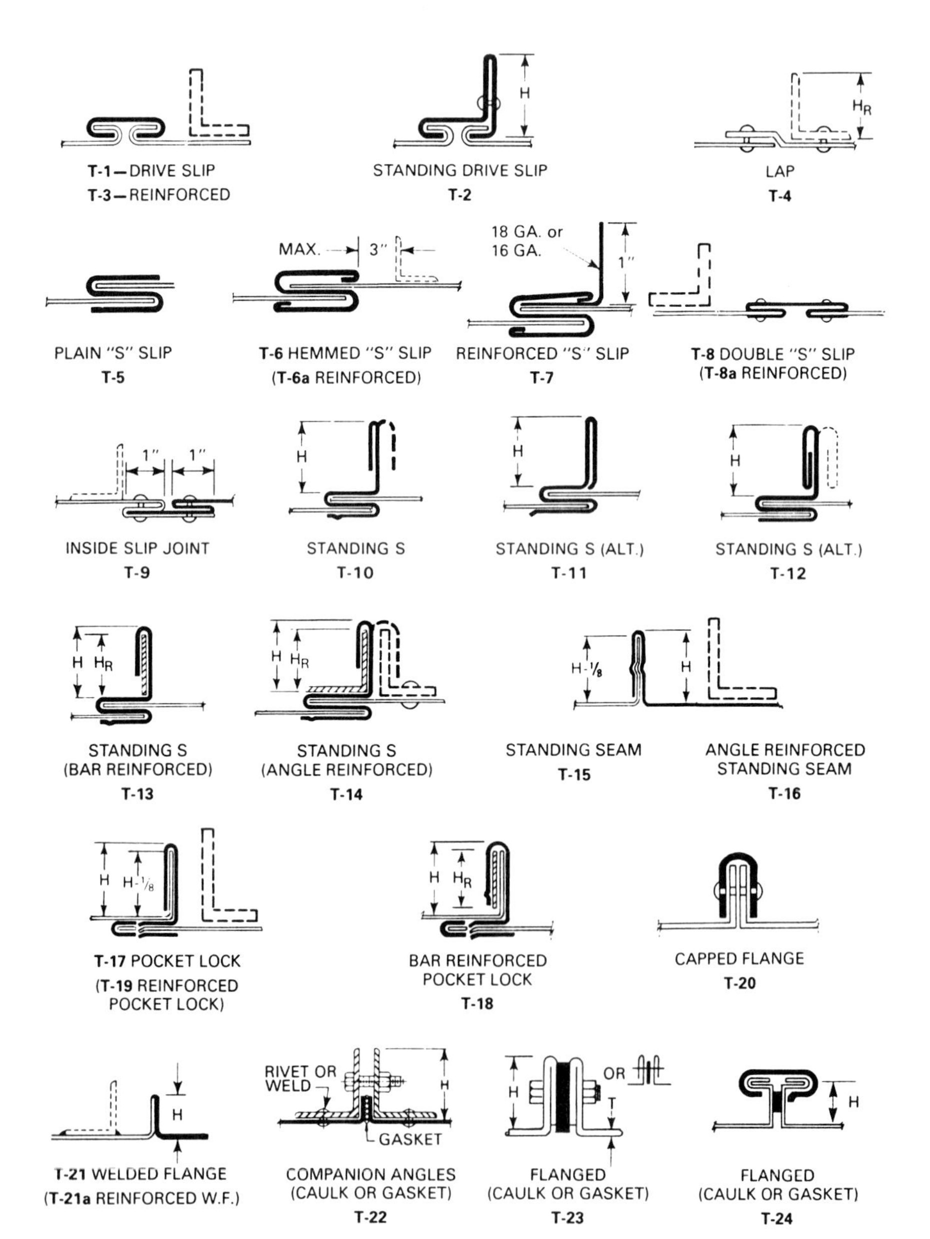

FIG. 17-1 Transverse (girth) joints. *(Courtesy Sheet Metal and Air Conditioning Contractors National Association, Vienna, VA.)*

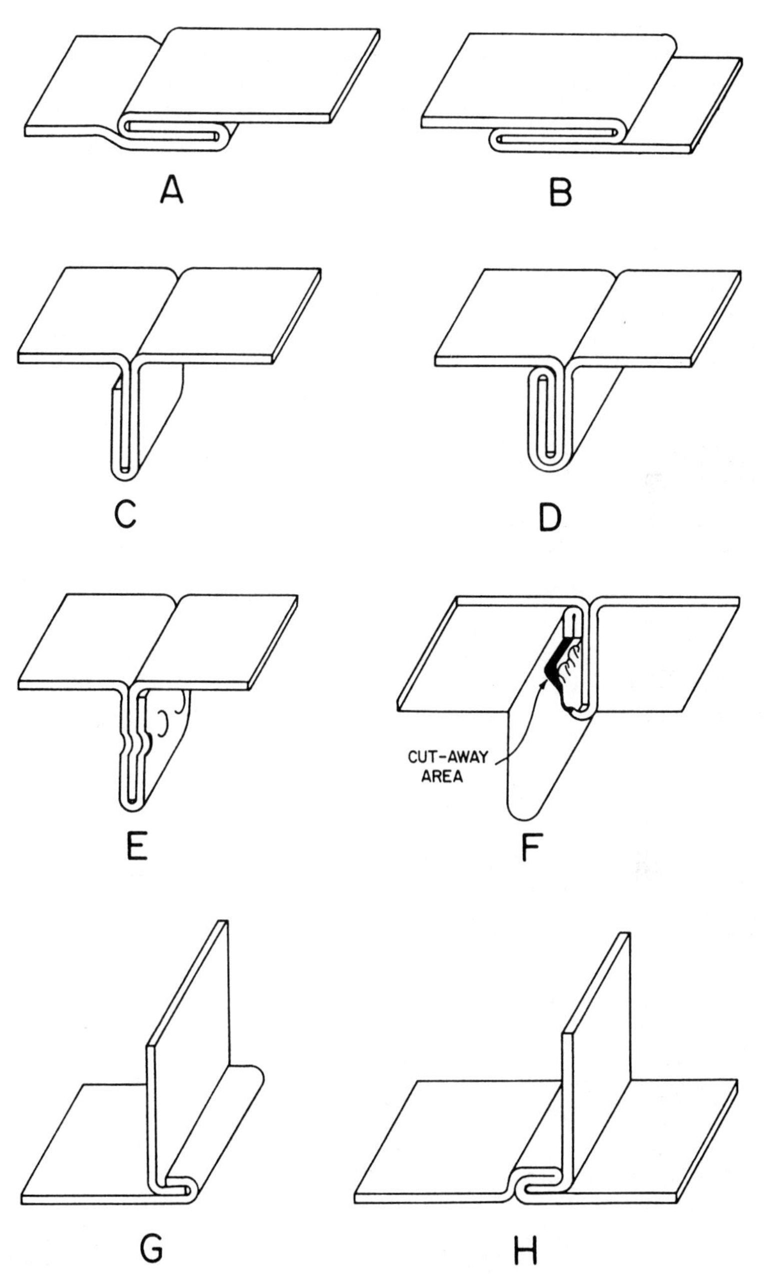

FIG. 17-2 Standard longitudinal seam joints. *(A)* Grooved single-lock seam; *(B)* single-lock seam; *(C)* standing seam; *(D)* double-lock standing seam; *(E)* dimpled standing seam; *(F)* self-locking standing seam; *(G)* folded-over seam; *(H)* modified folded-over seam.

17-4 Corner Closures Figures 17-3 through 17-8 illustrate the typical designs of corner closures for slips and drives, pocket locks, flanges, standing seams, and inside slip joints.

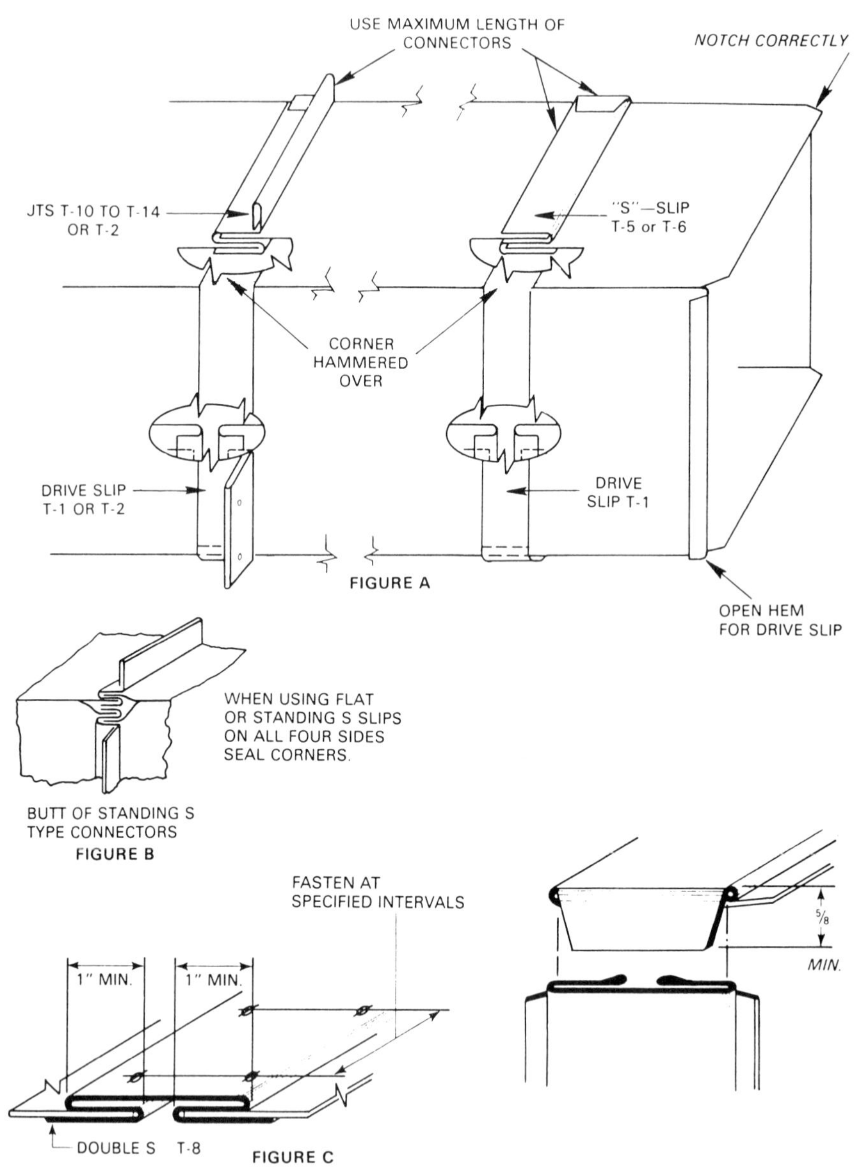

FIG. 17-3 Corner closures—slips and drives. *(Courtesy Sheet Metal and Air Conditioning Contractors National Association, Vienna, VA.)*

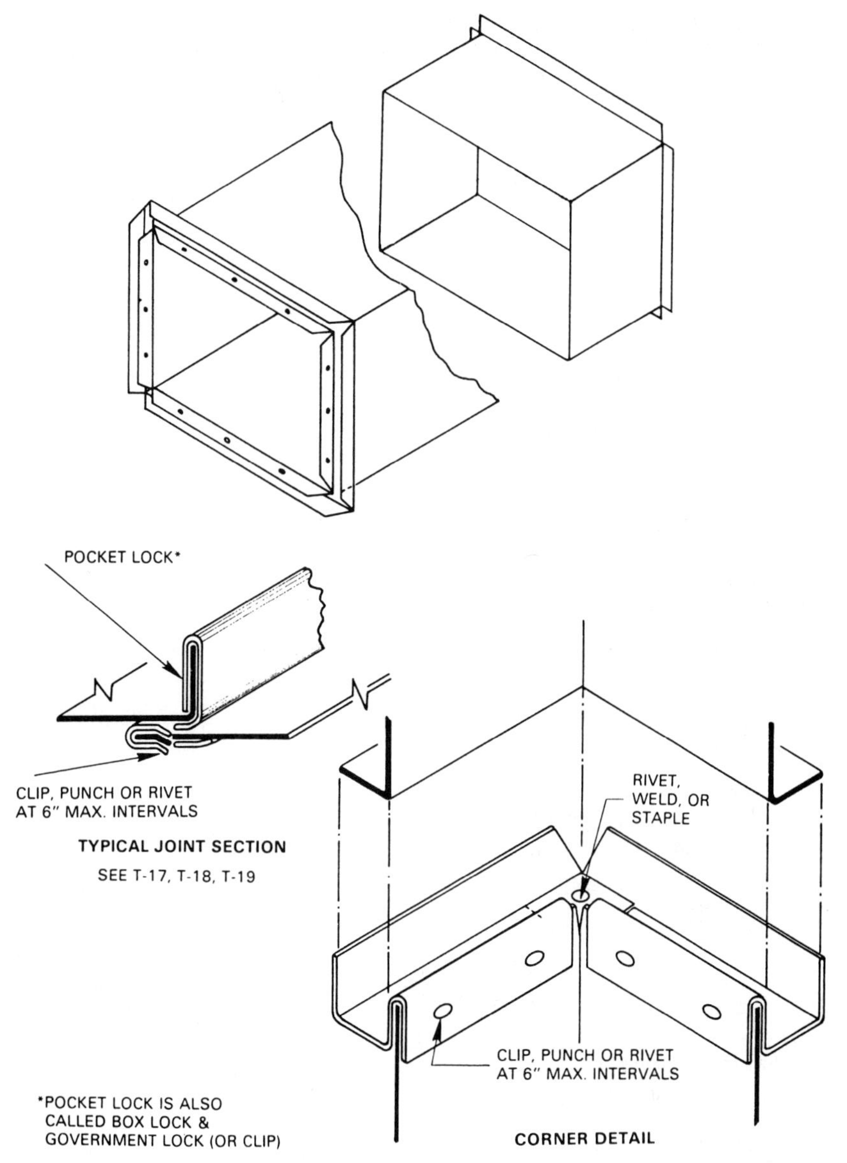

FIG. 17-4 Corner closures—pocket locks. *(Courtesy Sheet Metal and Air Conditioning Contractors National Association, Vienna, VA.)*

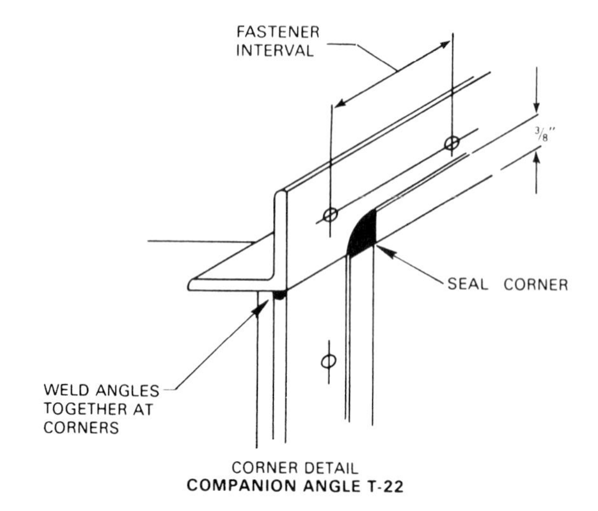

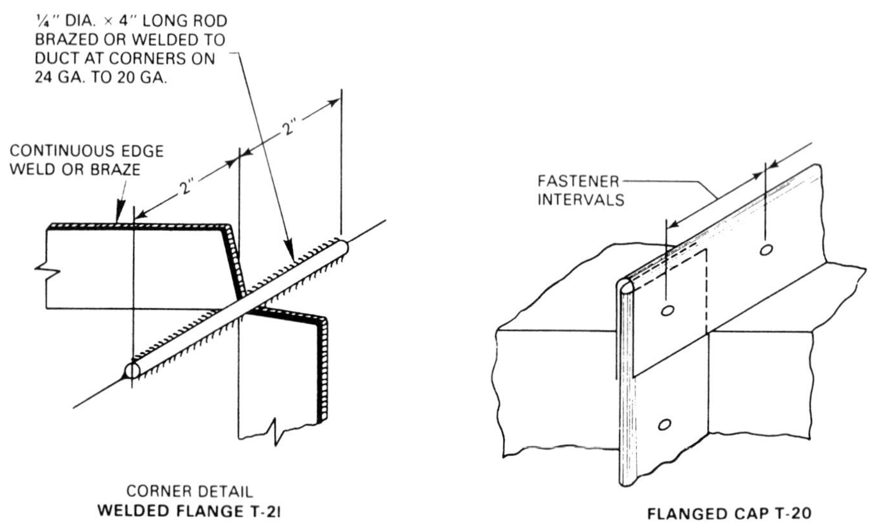

FIG. 17-5 Corner closures—flanges. *(Courtesy Sheet Metal and Air Conditioning Contractors National Association, Vienna, VA.)*

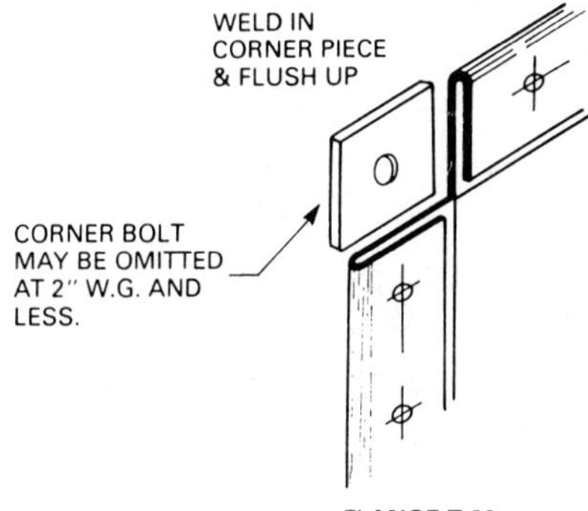

FLANGE T-23

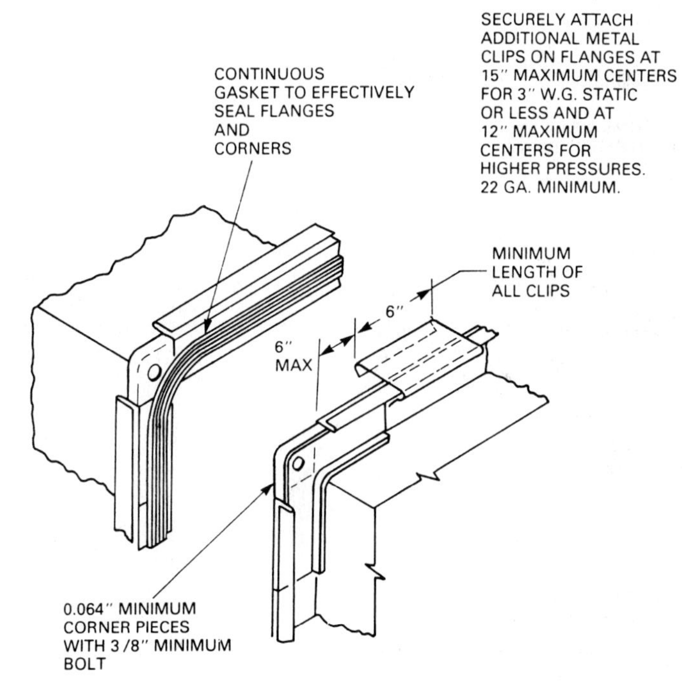

FLANGE T-24

FIG. 17-6 Corner closures—flanges. *(Courtesy Sheet Metal and Air Conditioning Contractors National Association, Vienna, VA.)*

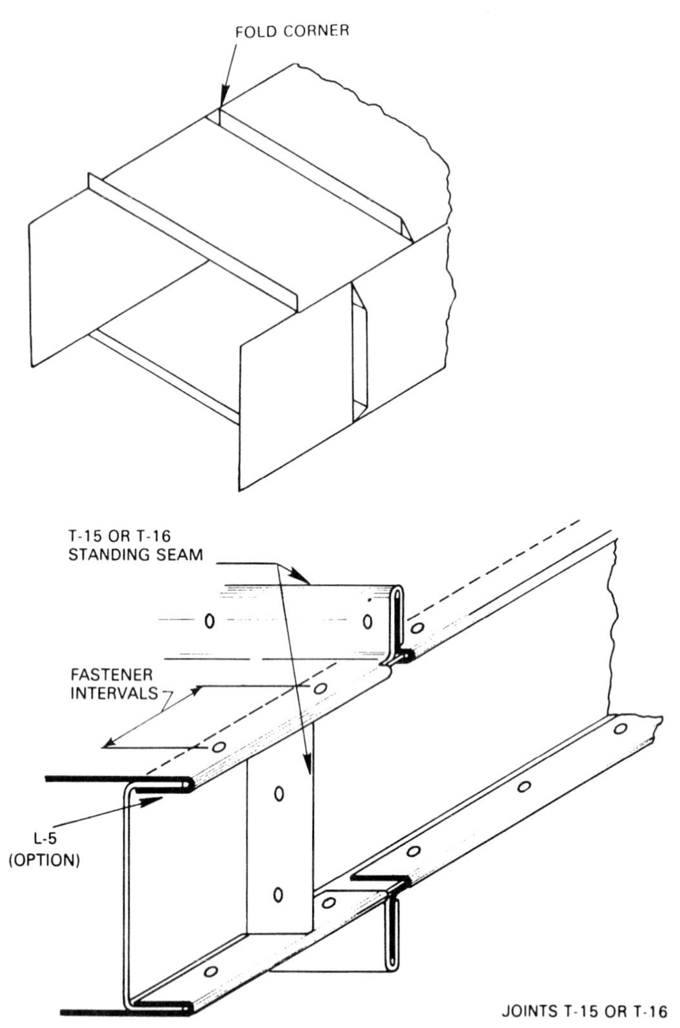

FIG. 17-7 Corner closures—standing seams. *(Courtesy Sheet Metal and Air Conditioning Contractors National Association, Vienna, VA.)*

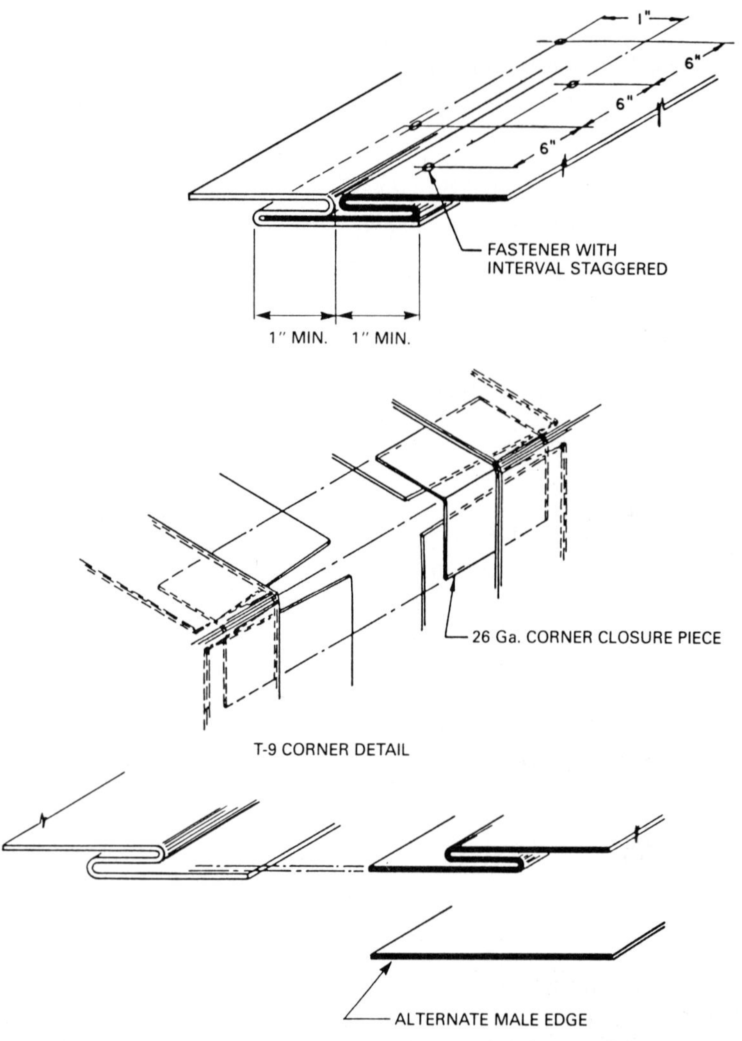

FIG. 17-8 Inside slip joint corner closure. *(Courtesy Sheet Metal and Air Conditioning Contractors National Association, Vienna, VA.)*

17-5 Circular Connections Round duct fittings are shown in Fig. 17-9. Spiral, lap, rivet, groove, butt, pipe lock, flat lock, and snaplock seams are illustrated.

17-6 Locking Tabs and Lanced Parts Under special conditions, sheet metal may be connected by using locking tabs or lanced segments. Figure 17-10 pictures seven standard types in use today.

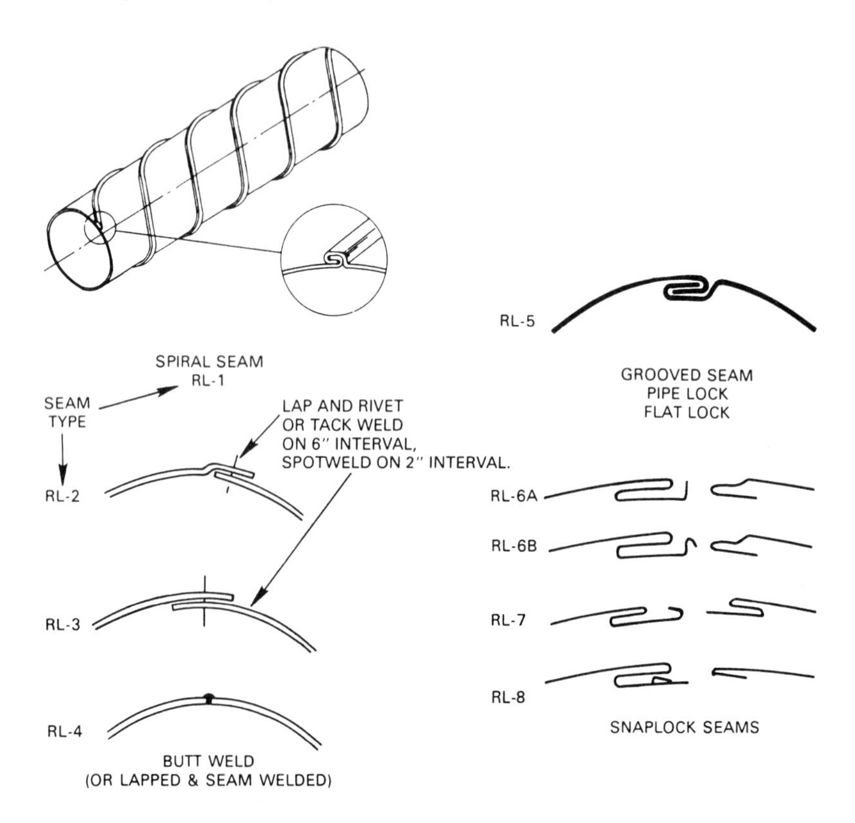

PRESSURE CLASS	SEAM TYPE PERMITTED
Positive	
To +10" w.g.	RL-1, 4, 5 (2*, 3*)
To +3" w.g.	RL-1, 2, 3, 4, 5
To +2" w.g.	ALL
NEGATIVE	
To −2" w.g.	RL-1, 2, 3, 4, 5
To −1" w.g.	ALL

*ACCEPTABLE IF SPOTWELDED ON 1" INTERVALS.

FIG. 17-9 Seams—round duct and fittings. *(Courtesy Sheet Metal and Air Conditioning Contractors National Association, Vienna, VA.)*

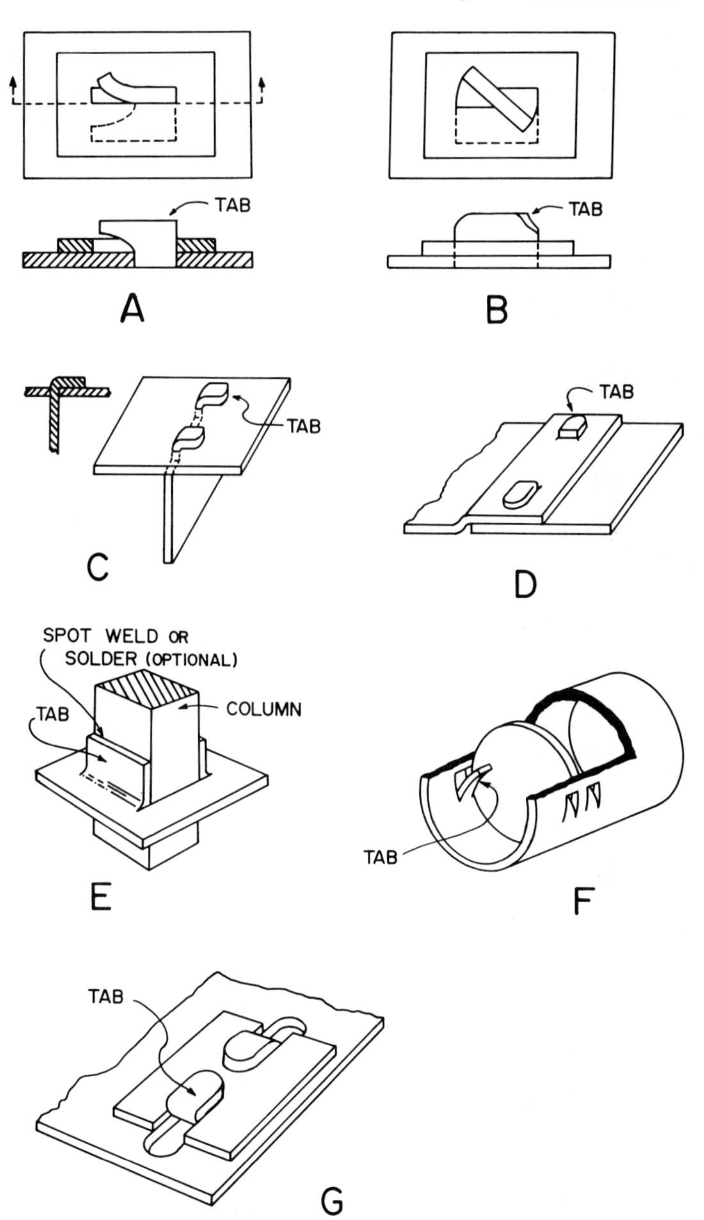

FIG. 17-10 Locking tabs and lanced parts. *(A)* Single-twisted tab; *(B)* double-twisted tab; *(C)* folded tabs; *(D)* opposite folded tabs; *(E)* lanced flaps; *(F)* locator tabs; *(G)* alignment and bent-over tabs.

17-7 Interlocking Fastening Interlocking by folding, forming, or bending tabs is employed on circular configurations, as shown in Fig. 17-11.

17-8 Threaded Fasteners Figures 17-12 and 17-13 reveal only a random sampling of the thousands of threaded fastening components in use.

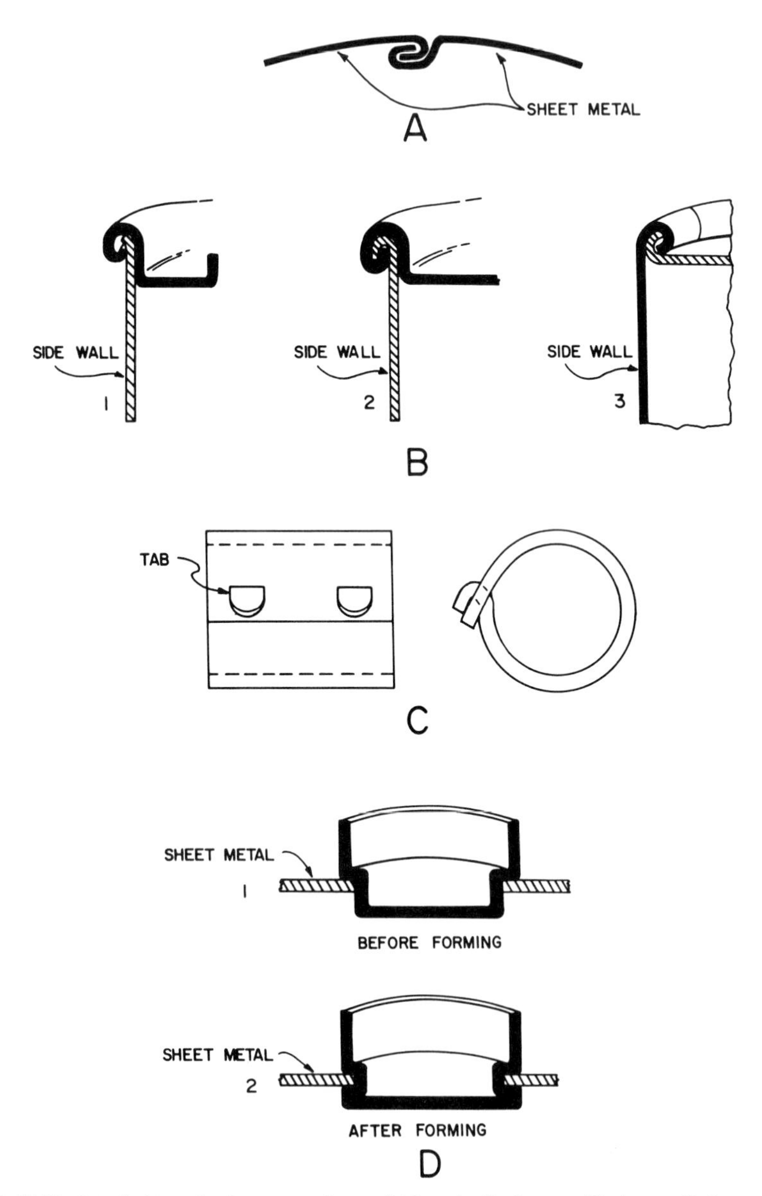

FIG. 17-11 Interlocking circular connections. *(A)* Longitudinal seam; *(B)* three rolled seams; *(C)* bent-over tabs; *(D)* peened-riveted.

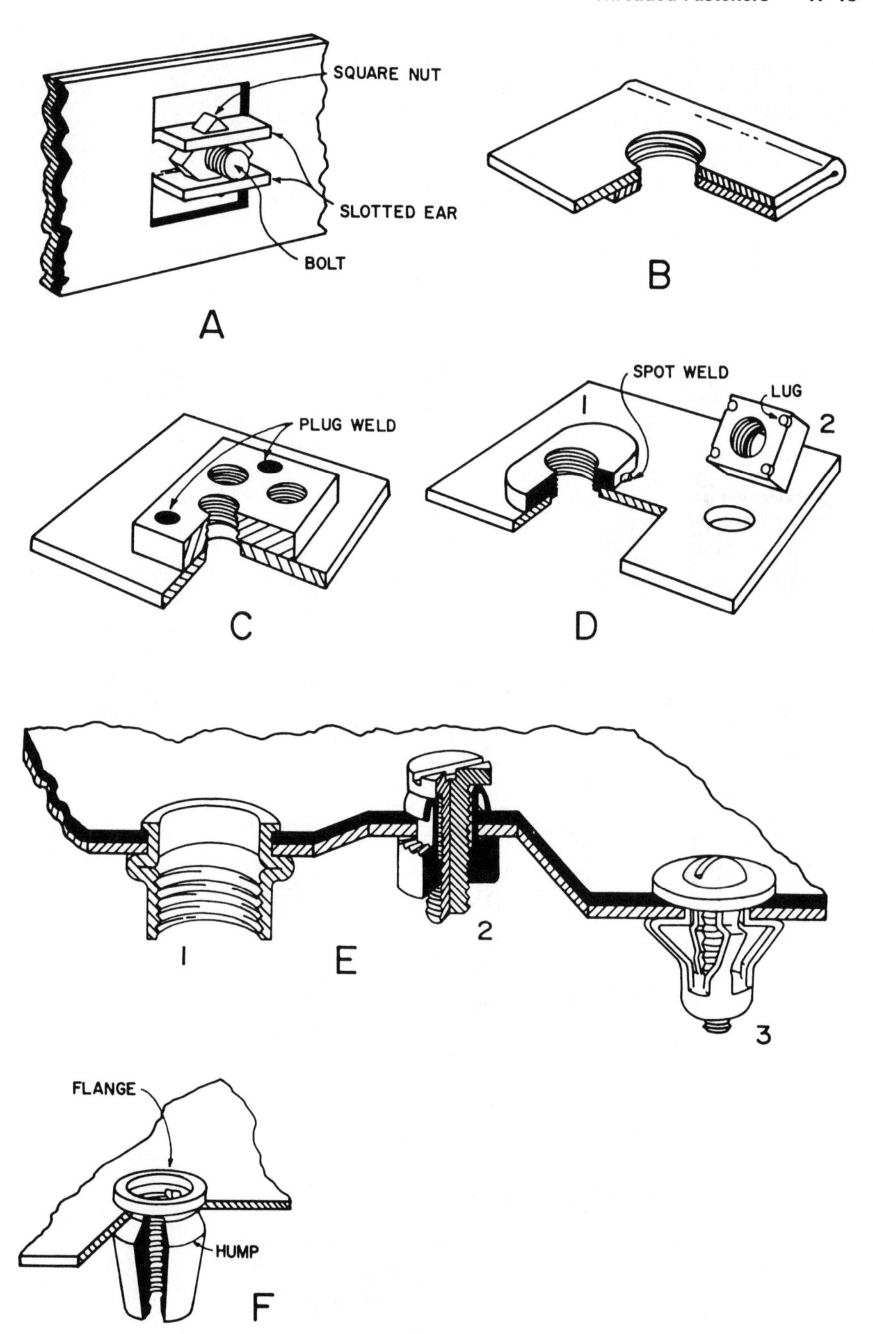

FIG. 17-12 Threaded fastening. *(A)* Captive nut; *(B)* bent-over sheet with threaded hole; *(C)* thread patch; *(D)* threaded lugs; *(E)* three double-duty nuts/blind fasteners; *(F)* threaded plastic grommet.

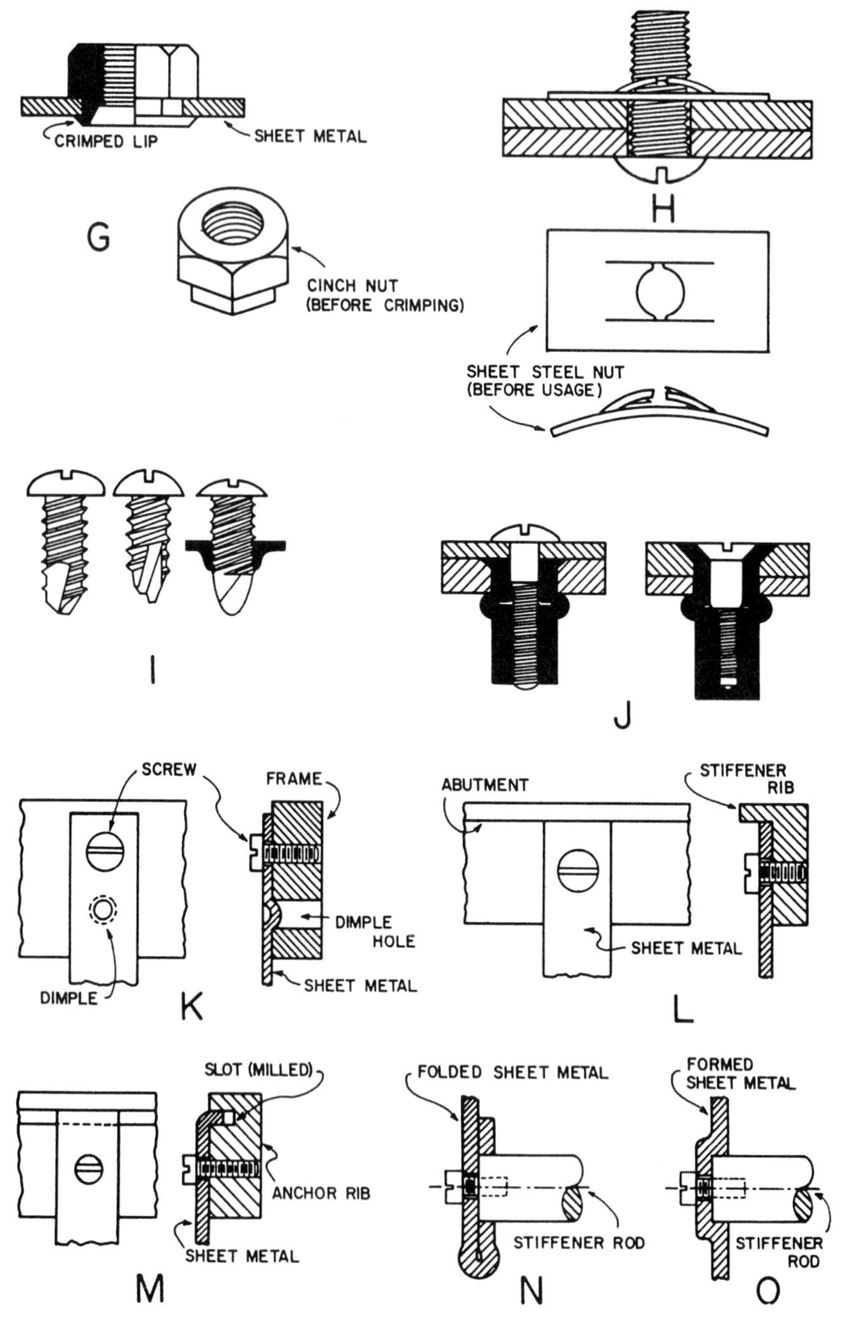

FIG. 17-13 Threaded fastening (cont.). *(G)* Cinch nut; *(H)* sheet steel nut; *(I)* self-drilling screws; *(J)* blind inserts; *(K)* screw with aligning projection; *(L)* screw with aligning abutment; *(M)* screw with aligning slot; *(N)* folded double sheet metal end connection; *(O)* sheet metal embossed connection.

17-9 Welded and Soldered Fastening The varieties of spot welding and soldering connections are endless. Figure 17-14 shows only five basic types. Refer to sec. 6 of this handbook for a detailed study of welding technology.

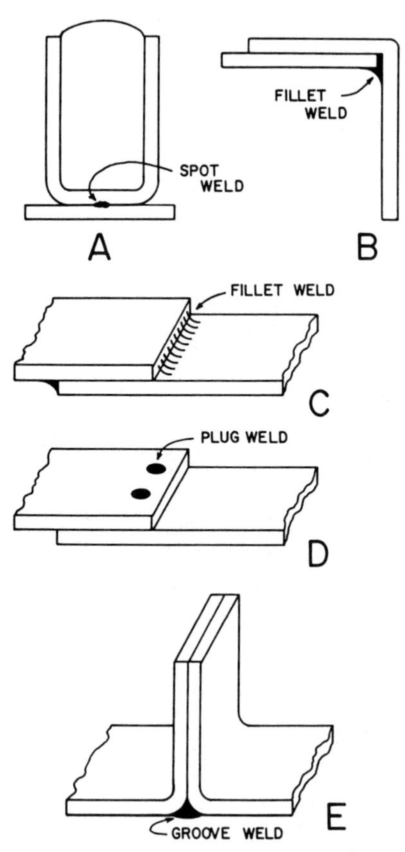

FIG. 17-14 Welded and soldered joints. *(A)* Spot-welded shell; *(B)* corner-reinforced; *(C)* typical lap joint with fillet weld; *(D)* typical lap joint with plug weld; *(E)* groove weld for simple standing rib.

17-10 Riveted Joining Sheet metal is often joined by riveting and is only briefly mentioned here to direct the reader to sec. 13 of this handbook for a detailed presentation of rivet technology.

17-11 Branch Connections Figure 17-15 illustrates two typical branch connections on a rectangular duct. Figures 17-16 through 17-18 picture some standard circular branch connections.

17-12 Liner Fasteners Standard fastening of duct liners is shown in Fig. 17-19.

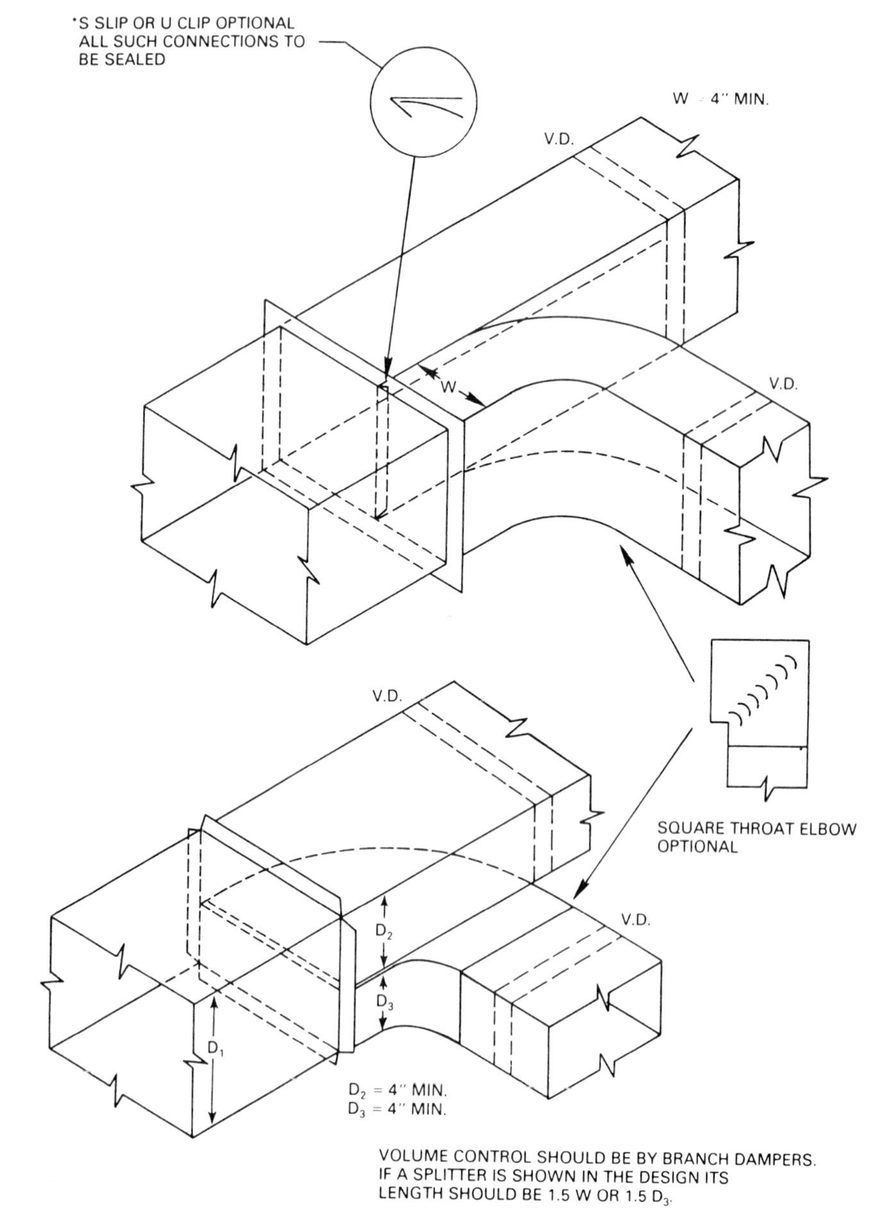

FIG. 17-15 Parallel flow branches. *(Courtesy Sheet Metal and Air Conditioning Contractors National Association, Vienna, VA.)*

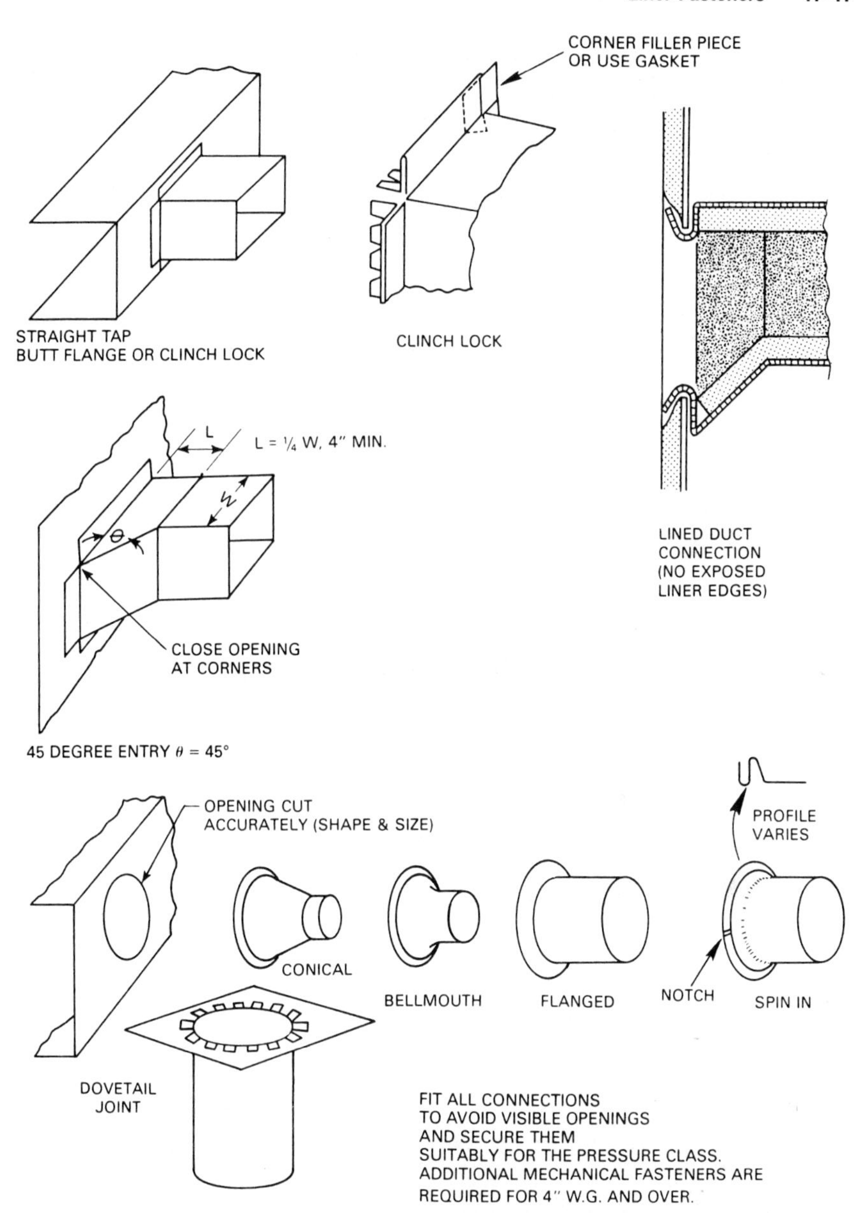

FIG. 17-16 Branch connections. *(Courtesy Sheet Metal and Air Conditioning Contractors National Association, Vienna, VA.)*

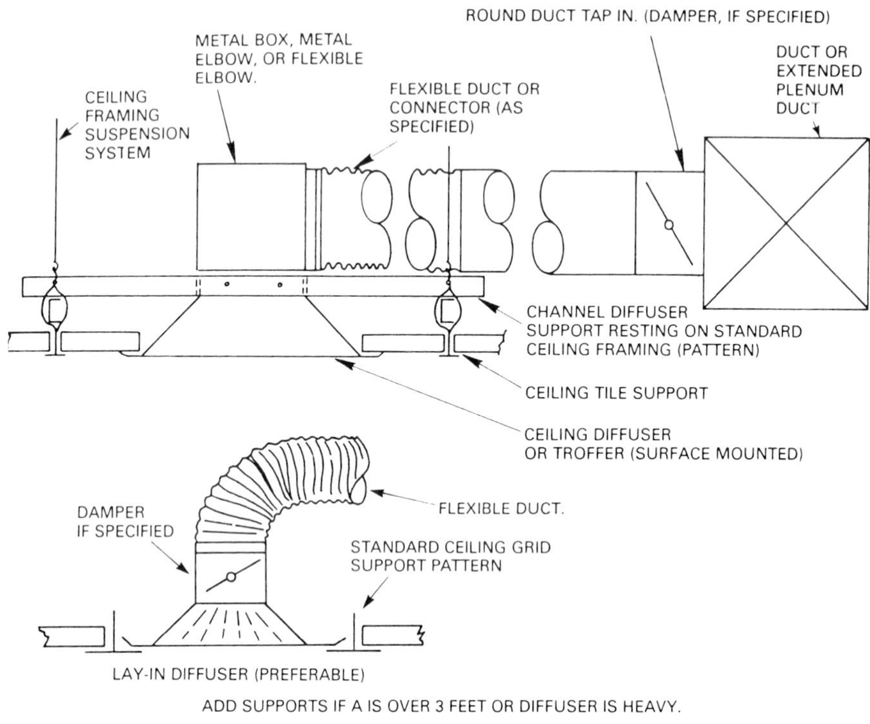

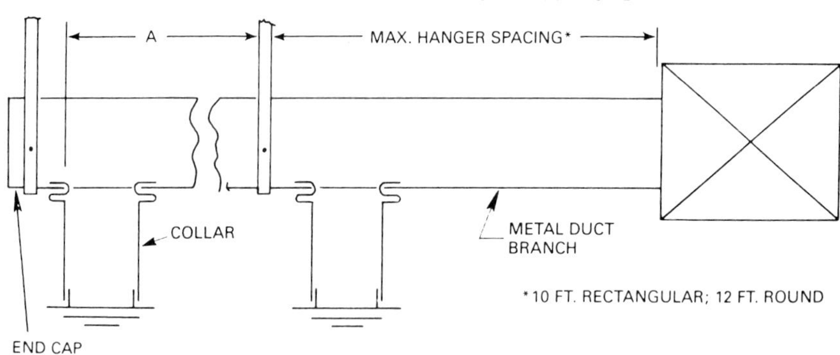

FIG. 17-17 Ceiling diffuser branch ducts. *(Courtesy Sheet Metal and Air Conditioning Contractors National Association, Vienna, VA.)*

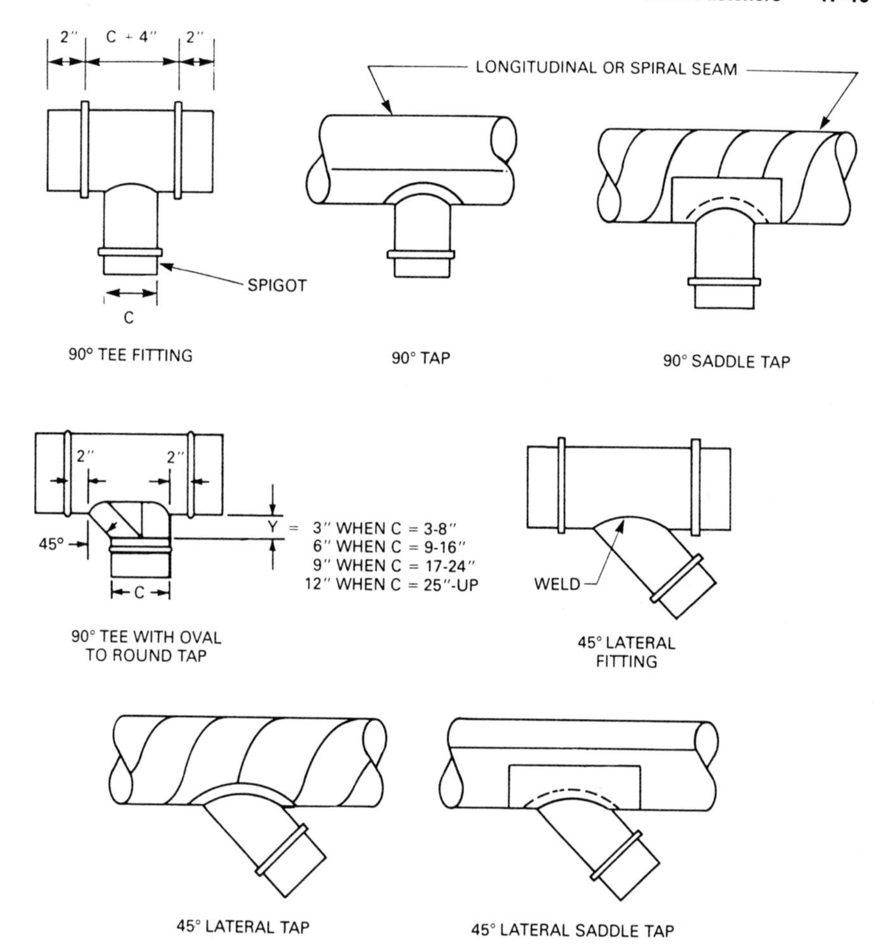

TAPS MAY HAVE SCREWS.
STANDARD SPIGOT LENGTH IS 2″

FIG. 17-18 90° tees and laterals. *(Courtesy Sheet Metal and Air Conditioning Contractors National Association, Vienna, VA.)*

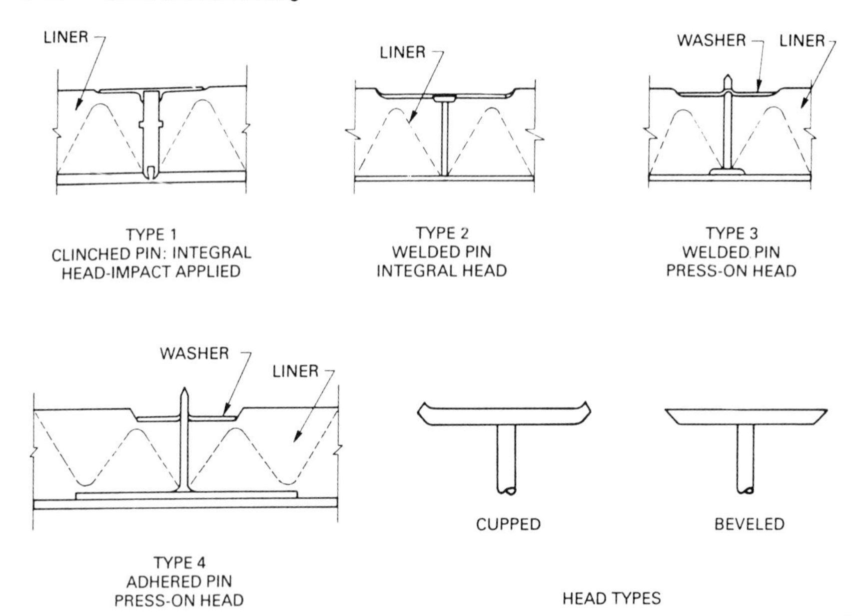

FIG. 17-19 Liner fasteners. *(Courtesy Sheet Metal and Air Conditioning Contractors National Association, Vienna, VA.)*

Section 18
Retaining Compounds*

GIRARD S. HAVILAND
Consultant, Naples, Maine

18-1 General Problem of Inner Space and Hub Stress Whenever two metal surfaces are brought together as an assembly, they must be clamped firmly to produce high friction forces. Even on heavy-force FN-4 fits, the touching is confined to high spots that are limited to 20 to 30 percent of the total surface available (Fig. 18-1). The rest of the area is a gap of "inner space." Often, to get more contact and higher retaining forces, designers make the hub stresses as high as possible, even to the point of yielding the assembled parts, and finishes and tolerances are reduced to an uneconomical minimum. Instead of increasing the pressure, substantial increases in friction can be provided by filling 100 percent of the available space with a strong machinery adhesive. A combination of a light locational fit with high disassembly friction can solve many fitting problems (Fig. 18-2).

*Reprinted from Girard S. Haviland, *Machinery Adhesives for Locking, Retaining and Sealing*, chap. 6, Marcel Dekker, Inc., New York, © 1986, by courtesy of Marcel Dekker, Inc. Various segments have been edited to comply with this handbook's format, and some illustrations have been revised.

The editor wishes to thank the publisher and the author for permission to use this material and to acknowledge the Loctite Corporation, originator of this technology, and the people who share its history.

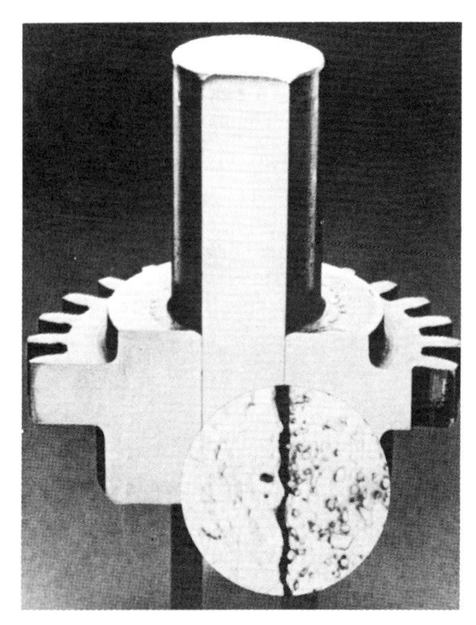

FIG. 18-1 Cutaway of a gear mounted with an FN-4 press fit. Insert 1000:1.

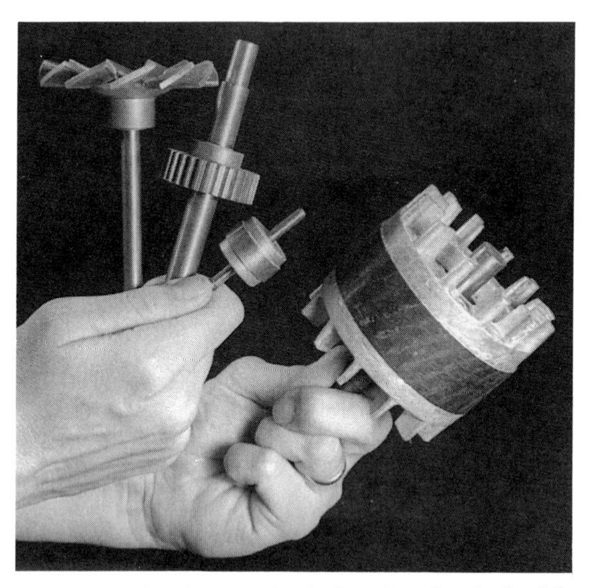

FIG. 18-2 Machined parts with cylindrical, keyed, and splined fits.

18-2 Generic Design Benefits Certain generic design situations are benefited by increasing the friction in force fits with the commonly used fitting materials Grades K, L, S, T, and U. The designer can solve problems of assembly, strength, and cost by using adhesives to assist the fitting process. (Refer to Table 2.1, original publication.)

1. *Eliminating extra bulk maintained only to achieve high friction force.* Thin-walled formed parts are used to reduce weight and cost but are often difficult to press together because of their shape and fragility. In Fig. 18-3, for instance, the parts of an intake manifold for an air compressor are smooth, curved, and difficult to hold. Any high force causes cocking or damaging distortion. Even if assembly is somehow accomplished, the tube compresses easily enough so that the assembly does not stay tight. An adhesive slip fit goes together by hand, is strong, and provides leakproof security.

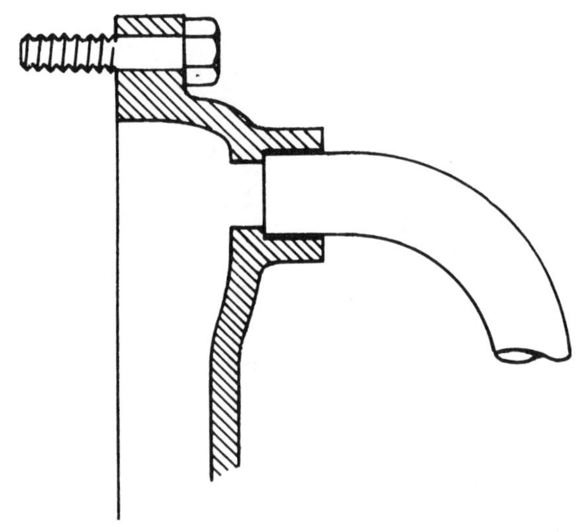

FIG. 18-3 Close elbow assembled in a manifold with adhesive fit.

Slimming of very strong parts is possible when the effective friction between parts can be increased as much as four times. For example, assembling the differential gear in an automobile with a heavy shrink fit and normal friction produced unsatisfactory strength and excessive distortion. A lighter shrink fit with a friction-improver similar to Grade U quadrupled the strength and made regrinding of the gear form unnecessary (Figs. 18-4*a* and *b* and 18-5*a* and *b*).

(a)

(b)

FIG. 18-4 *(a)* Differential ring gear assembled with a light shrink fit and a friction improver similar to Grade U. Interference between the hardened steel ring gear and carrier is 0.04 to 0.12 mm on a diameter of 140 mm (24 mm wide). Separation force was increased from 5 to 30 tons by using the friction-improver. *(b)* Automatic application and assembly station for Renault ring gears. *(Courtesy Renault Cie.)*

(a)

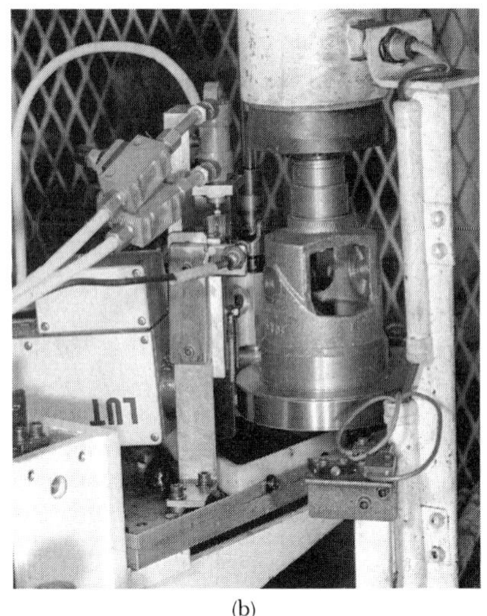

(b)

FIG. 18-5 *(a)* Teflon coating wheel for applying adhesive, Renault. *(b)* Carrier being coated to assembly, Renault.

2. *Augmenting or replacing press fits and reducing cost without changing the design.* This is a benefit that should be considered whenever reasonable tolerances cannot produce the friction or holding necessary without damaging distortion. As we will see, this may be true for any press fit with diameters under 1 in. (25 mm). Distortion is caused by heavy residual stress. The press of a perfectly round bearing into a nonsymmetrical housing can destroy the bearing since the housing imposes its lack of symmetry onto the compressed bearing. Or a slim shaft pushed into a rigid armature will always bend if the fit is tight enough to hold. Restraightening of shafts and repairing galled assemblies is avoided with an adhesive fit.

The tolerances of force fits all lie on the high-cost end of the machining-cost curve, as illustrated in Fig. 18-6. For instance, USA Standard Force and Shrink Fit tables (USAS B4.1-1967) show that a 1-in. shaft has a tolerance minimum of 0.0009 in. and a maximum

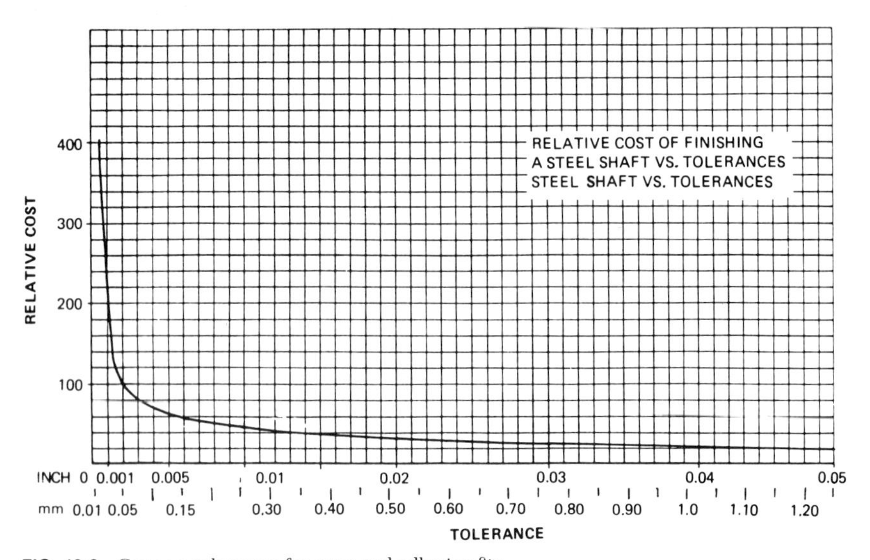

FIG. 18-6 Costs vs. tolerances for press and adhesive fits.

of 0.0033 in., which are on the expensive left-hand side of the cost curve. Friction-improvers and retaining compounds allow a slight to major easing of tolerances and costs.

3. *Combining materials to get the best qualities of each.* Thin brake-cylinder liners are mass-produced from high-wear, corrosion-resistant steel and slipped without distortion into cast housings with an adhesive. Likewise, bronze bushings are mounted in steel structures for medium-speed shaft bearings, and hard drill bushings are reliably assembled in large fixtures of soft, easy-to-cast metal alloys or glass fiber lay-ups.

4. *Replacing O rings used for static seals.* Cured liquid adhesives are superb sealers and can replace static seals under conditions where disassembly is not frequent. Lower-strength materials such as Grade W, X, Y, or Z should be considered so that covers, plugs, or flanges can be easily separated when necessary.

5. *Helping compensate for differential thermal expansion.* This capability of adhesives is one that can save an application that is borderline with a press fit.

6. *Eliminating backlash.* Where keys and splines are used for torsional driving without longitudinal sliding, a fit rigidized with an adhesive can improve fatigue life substantially. The micro shifting of keys and splines usually causes fretting and fretting corrosion, which is the stress raiser that starts fatigue cracks. Reversing loads produce shock and vibration, which can cause keyway wallowing and fatigue failure. Impacting can easily double the experienced stress over the normal steady-state load. An example of this was reproduced in the author's laboratory (Fig. 18-7).

A snugly fitted Woodruff key in a 1-in. (25-mm) shaft was run against a key fitted with a machinery adhesive. The ¼ × 1⅜-in. keys were fitted according to SAE J502 standard, which permitted as much as 0.0008-in. (0.02-mm) clearance but, for this test, were held to a slight interference fit.

After 12 million reversing cycles on a Sonntag fatigue tester of 150 lb-ft (200 N-m), representing 40 percent of the shaft torsional shear strength, the untreated keyway loosened approximately 0.002 in. (0.05 mm) with serious fretting corrosion and shaft fatigue cracks at the slot. The adhesively fitted key developed no backlash and showed no signs of failure.

A practical example of this same effect, shown in Fig. 18-8, is the spline-mounted gear on a crane/shovel platform, which is used to rotate a crane on its base. Shaft failures were

FIG. 18-7 Fatigue crack produced in a shaft when coupled with a tightly fitted key (left) and no failure when assembled with adhesive (right).

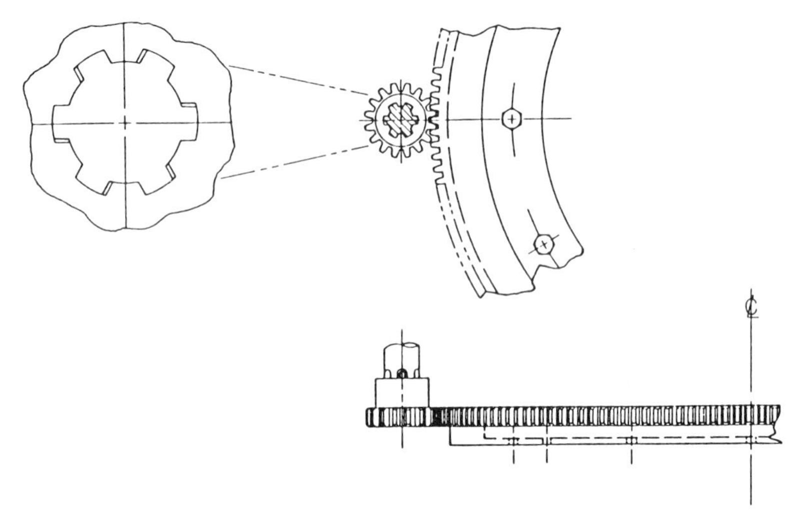

FIG. 18-8 Movement of splined sprocket caused fatigue failure of the shaft in crane turret.

common until the backlash was eliminated with an adhesive in the spline. No other changes were needed to solve the shaft fatigue failure problem.

7. *Making accurate, rigid assemblies.* In the making of sine tables and jig borer vises, it is necessary to hold accurately ground support rods in bored supports without distorting the table or support rods. Sine tables are made to accuracies measured in millionths of an inch. Press fits and set screw pressure both cause more distortion than allowed. Rods are therefore slipped into place, indicated, and adjusted for accuracy, and then a thin machinery adhesive is applied to cast the rod into rigid alignment (Fig. 18-9).

FIG. 18-9 Moore sine table and vise. *(Courtesy Moore Special Tool Co.)*

8. Increasing the holding capability of heavy press fits. Under the extreme pressure and energy of a heavy press fit, most machinery adhesives will cure faster than the fit can be made up. The cured material creates a high friction for the assembly but seems to prevent galling. If the parts will support the high loads, then this is a good way to improve the load capability of the press. Fatigue limits of the assembly can be twice those of a plain press. Holding capability can also be assisted by capillary penetration with Grade R after assembly (sometimes called "wicking action"). It will achieve surprising penetration on clean parts, and the ultimate strength will achieve five figures on the heaviest presses.

DESIGN CALCULATIONS

18-3 Press Fits Tables 18-1 and 18-2 show the results of calculating the holding capability of force fits on shafts of different sizes. The calculation of pressure between hub and shaft is well known from machine design books (see Ref. 1) as derived from thick-walled cylinder formulas (see sec. 18-22). To save calculation time, the results of these calculations are plotted in Figs. 18-10 through 18-12.

Where the hub pressures are known for the interferences and proportions in question, it is possible to compute the assembly (or disassembly) force or the torque capability from the following formulas:

$$F = \pi\, b\, L\, f\, P \text{ (area} \times \text{friction coefficient} \times \text{pressure)} \qquad (18\text{-}1)$$

$$T = \pi\, b^2 L\, f\, P\ /2 \text{ (force} \times \text{radius)} \qquad (18\text{-}2)$$

where F = force, lb (N); T = torque, in.-lb. (N-m); L = length of contact, in. (m); b = shaft diameter, in. (m); f = coefficient of friction; and P = radial hub pressure at the interface of hub and shaft, lb/in.^{2} (Pa) (from Fig. 18-10 or 18-11).

In Table 18-2 note that the shafts under 1 in. (25 mm) in diameter have hub pressures that are excessive (and off the graphs in Figs. 18-10 and 18-11) at reasonable tolerances. Considerable micro and macro yielding takes place, so we calculated the force fit by assuming a maximum hub pressure of 14,500 psi. Practice shows that press fits in the smaller hubs do indeed cause problems of shearing, galling, and shaft bending especially if the hubs are relatively thick and stiff (small b/c). Thin hubs (large b/c) are expanded beyond their yield. Under the conditions producing 14,500 psi hub pressure (e = 0.0017 b/c = 0.65), the tangential tensile stress in the hub is 22,000 psi (Fig. 18-12 extrapolated). Figure 18-13 shows a graphical comparison between hub pressure and tensile stresses in a typical hub.

TABLE 18-1 Limits of Interference and Part Tolerances for Press and Shrink Fits

Nominal diameter, in. (mm)	Tolerance, 1/1000 in. (1/1000 mm)		Interference		
	Shaft	Hole	Locational fit FN-2, 1/1000 in.	Force fit FN-4, 1/1000 in.	FN-4 equivalent (ANSI B4.2) s7-h6, (mm)
0.1	0.25	0.4	0.2–0.85	0.3–0.95	–
(2.5)	(6)	(10)	–	–	(0.008–0.024)
0.2	0.3	0.5	0.2–1.0	0.4–1.2	–
(5)	(8)	(12)	–	–	(0.007–0.027)
0.5	0.4	0.7	0.5–1.6	0.7–1.8	–
(13)	(11)	(18)	–	–	(0.010–0.039)
1.0	0.5	0.8	0.6–1.9	1.0–2.3	–
(25)	(13)	(21)	–	–	(0.014–0.048)
4.0	0.9	1.4	1.6–3.9	4.6–6.9	–
(100)	(22)	(35)	–	–	(0.036–0.093)
10	1.2	2.0	4.0–7.2	10.0–13.2	–
(250)	(29)	(46)	–	–	(0.094–0.169)

TABLE 18-2 Limits of Interference and Part Tolerances for Press and Shrink Fits with Assembly Forces

Nominal diameter, in. (mm)	Allowance e 1/1000, in./in.	Hub[a] pressure, lb/in.^{2}	Ultimate strength	
			FN-4 force[b]	Adhesive slip fit[c]
0.1	6.5	EY @ 14,500	100 lb[d]	160 lb
(2.5)		EY @ 14,500	445 N[d]	700 N
0.2	4.0	EY @ 14,500	410 lb[d]	660 lb
(5)		EY @ 14,500	1.8 KN[d]	3 KN
0.5	2.2	14,500	2600 lb	4100 lb
(13)			12 KN	18 KN
1.0	1.3	11,500	8100 lb	16,500 lb
(25)			36 KN	73 KN
4.0	0.6	5100	58,000 lb	158,000 lb
(100)		260 KN	240 KN	
10	0.32	2500	177,000 lb	330,000 lb
(250)			790 KN	1500 KN

[a]Hub pressure read from Fig. 18-10 for allowance e of FN-4 fits. EY = "exceeds yield" and is off the graph; e was extrapolated for 14,500 psi (approximate hub pressure at tensile yield).

[b]Force required to assemble = $\pi\, dlfp$, where d = shaft diameter; l = length of fit, assumed to be $1.5d$; f = coefficient friction and assumed to be 0.15; and p = unit pressure from Fig. 18-10 at d/D = 0.65.

[c]3500 psi up to 1-in. shaft, 2100 psi for 4-in. shaft, and 750 psi for 10-in. shaft with 75-in.^{2} and 470-in.^{2} bond areas, respectively.

[d]At yield.

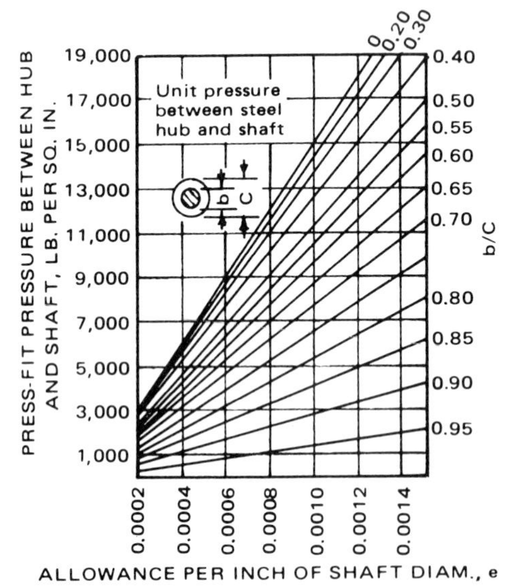

FIG. 18-10 Press-fit pressures between steel hub and shaft. *(From Baumeister, Avallone, and Baumeister, Marks' Standard Handbook for Mechanical Engineers, © 1978. Courtesy McGraw-Hill, New York.)*

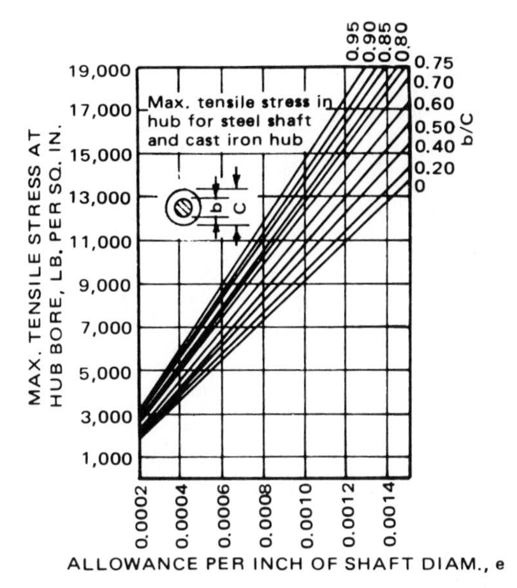

FIG. 18-11 Press-fit pressure between cast iron hub and steel shaft. *(From Baumeister, Avallone, and Baumeister, Marks' Standard Handbook for Mechanical Engineers, © 1978. Courtesy McGraw-Hill, New York.)*

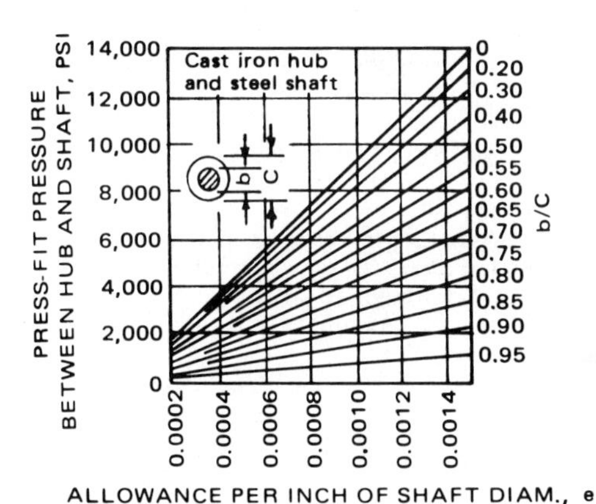

FIG. 18-12 Tensile stress in cast iron hub vs. interference allowance. *(From Baumeister, Avallone, and Baumeister, Marks' Standard Handbook for Mechanical Engineers, © 1978. Courtesy McGraw-Hill, New York.)*

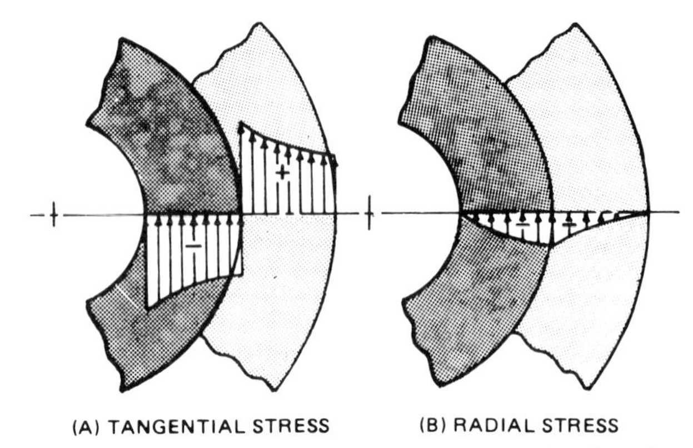

FIG. 18-13 *(a)* Distribution of tangential stresses in shrink-fitted members. *(b)* Distribution of radial stresses in shrink-fitted members. *(From J. E. Shigley, Mechanical Engineering Design, p. 67, © 1977. Courtesy McGraw-Hill, New York.)*

The most variable part of this calculation is the estimate of the coefficient of friction, which from engineering handbooks may vary from 0.10 to 0.33 on steel into cast iron (a 100 percent variation based on the average).

18-4 Adhesive Fits For adhesive fits, the ultimate strength or disassembly force has been calculated in the last column of Table 18-2 for a modest-strength machinery adhesive by the following formula:

$$F = s \times A \text{ (shear stress} \times \text{area)} \tag{18-3}$$

Over 1 in. in diameter, the rated shear stress has been reduced 40 and 80 percent on 4- and 10-in. shafts, respectively, to accommodate the effect of large areas. Studies have shown that the effective average shear strength of bond areas over 10 in.^{2} (6500 mm^2) is inversely related to the area. The reasons for this are purely conjectural, but are probably related to lack of complete fill and stress concentrations more severe than on smaller parts. Under 10 in.^{2}, increasing the length of engagement increases not only the area and disassembly force but also the ultimate stress or strength (Fig. 18-14). This means that on small shafts increasing the length 50 percent can increase the strength 60 to 65 percent. Rated shear stresses were obtained on 0.5-in. shafts with a 0.87 engagement ratio and an area of 0.69 in^2.

In a similar manner the strength is inversely proportional to gaps over 0.003 in. (0.08 mm). The shear stress used in Eq. (18-3) must be reduced for the size, gap, and fatigue effects, as shown in Figs. 18-14 through 18-16, as appropriate. Steel and cast iron will show the same adhesive properties. Aluminum will be about one-third the value of steel. See Figs. 18-14 through 18-16.

In summary, then, the acceptable service stress is expressed:

$$s_s = \text{ultimate rated stress} \times \text{service factor/safety factor} \tag{18-4}$$

where the service factor $= f_m \times f_g \times f_s \times f_t \times f_f \times f_c \times f_h$

Typical Example. Shear rating as obtained on soft steel pin and collar, Grade U = 4500 psi:

f_m,	material factor, steel on steel	= 1
f_g,	gap factor, Fig. 18-15 maximum, diametral clear 0.003 in.	= 1
f_s,	size and engagement ratio, Fig. 18-14, $L/D = 1$ and $D = 4$ in.	= 0.6
f_t,	temperature and environmental factor, 100°F	= 1.0

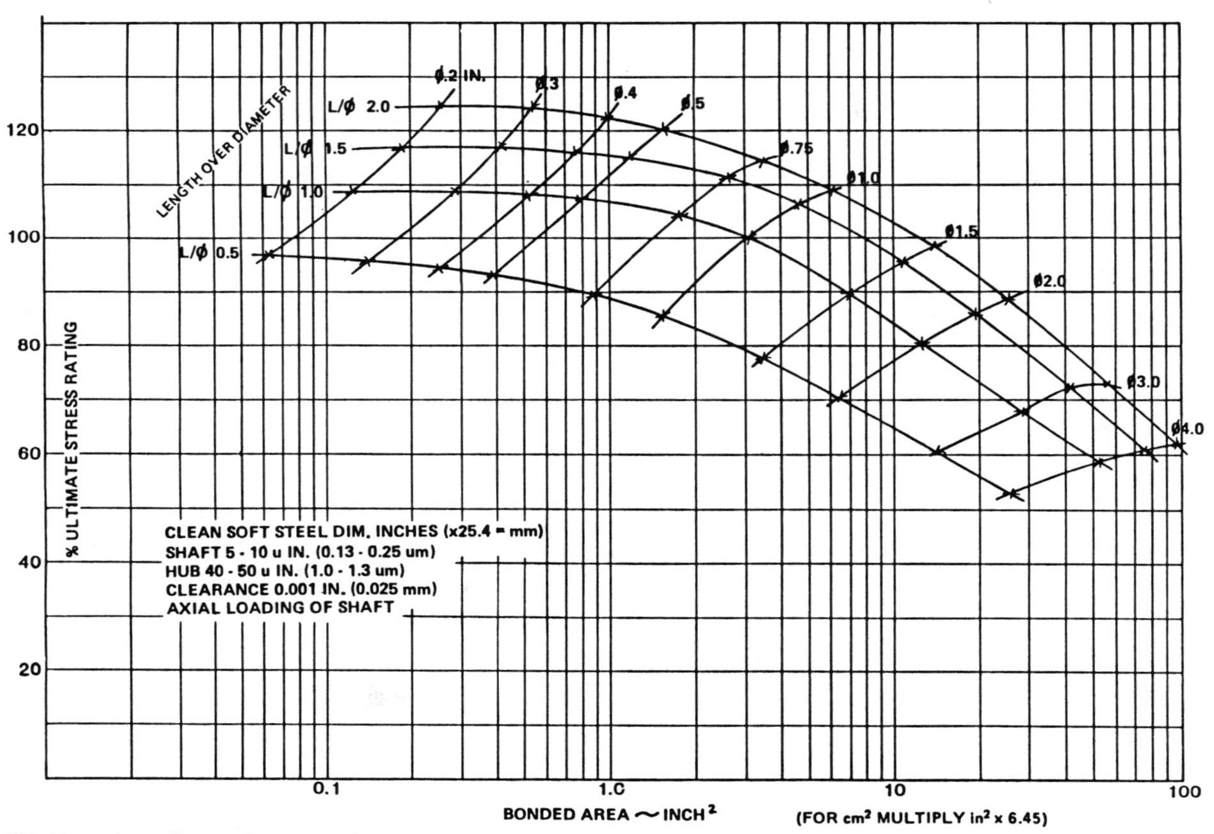

FIG. 18-14 Size effect on shear strength, strength vs. area.

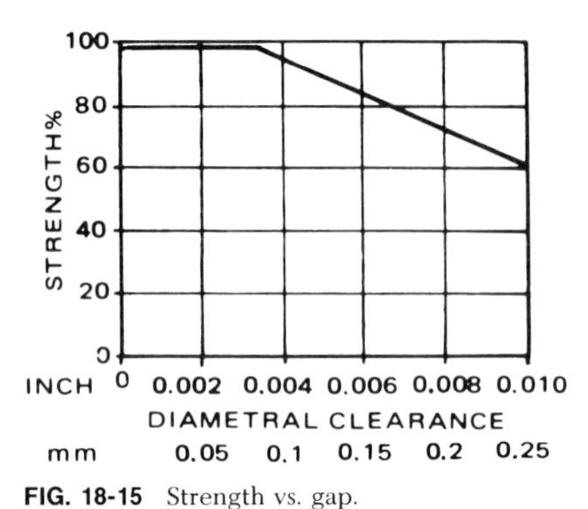

FIG. 18-15 Strength vs. gap.

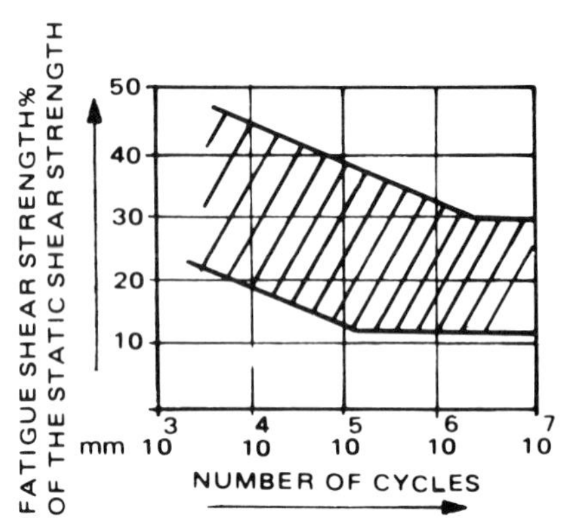

FIG. 18-16 Fatigue strength vs. cycles.

f_f,	fatigue rating, Fig. 18-16, 10^7 cycles	= 0.2
	Safety factor, users practice	= 2
f_c,	chemical resistance factor, air	= 1
f_h,	heat aging factor, Fig. 18-16, none	= 1
	Service factor by multiplication	= 0.06
	Safety factor, users practice	= 2

Acceptable repetitive shear stress:

$$4500 \times 1 \times 1 \times 0.6 \times 1 \times 0.2 \; / \; 2 = 270 \text{ psi}$$

If we had used our rule of thumb of 10 percent of the ultimate shear rating, the result would have been 450 psi. When possible, the ultimate shear strength should be measured under prescribed conditions so that the interactions of factors for material, size, gap, and temperature are exact instead of estimated. Obviously the general factors are conservative in their values. Further evidence of their conservative nature is that endurance limits on 25-mm shafts are at least 38 percent of the ultimate strength. This means that, with a safety factor of 2 (as in our 10 percent assumption), a good design stress is 19 percent of the ultimate, and, in our example, 855 lb/in.^{2}. Because all data are average, the author would prefer to start with 10 percent of ultimate as a design stress, which has a built-in average safety factor near 4. Functional tests would allow reevaluation.

Surface Finish and Direction of Load. Since the stress ratings are done on pins and collars or nuts and bolts, all with surface finishes less than 64 rms μin. (1.6 μm), it is possible to increase the stress rating 60 percent by increasing the roughness to 250 μin. (6.4 μm). The lay of the finish must be perpendicular to the load, or multidirectional. For instance, a pure thrust load is resisted best by circumferential machining marks. A torque load placed on the same parts would show a 40 percent reduction from the thrust load, coming back close to the original rated load.

The author prefers not to use increased directional roughness of machining marks at the design stage as a means of improving strength. The reasons are the uncertainty of the pureness of the direction of most applications and the existence of two surfaces in every assembly, the less rough of which determines the strength. An exception to this practice is to increase roughness by a multidirectional process such as sand blasting or shot peening. Even then, the less adhesive part is the one that should be improved. The results must be confirmed by test.

18-5 Augmented Press Fits When one steps back and takes the long view of how to assemble cylindrical parts, it becomes evident that a combination of light or location press and an adhesive for friction improvement can give benefits available to neither alone. These are:

Good location without high residual stress
High ultimate strength exceeding that of either single method
Fatigue strength exceeding that of either single method
Elimination of fretting, corrosion, and leaking
Relaxed manufacturing tolerances and eliminated rework
Simplified assembly tools and fixtures
A minimum disassembly strength, regardless of tolerances, equal to the adhesive strength

These benefits of adhesives as fitting and friction enhancers have been experienced on fractional horsepower motor shafts as small as 0.1 in. (2.5 mm) with pure adhesive fits, on railroad wheels 5 to 8 in. in diameter (130 to 200 mm) where adhesive was added to a press, and on a 36 in. in diameter (900 mm) rock crusher tapered shaft 96 in. long (2400 mm) fitted to a chilled cast iron sleeve or roller with a shrink fit.

18-6 Adhesive Shrink Fits At the time of this writing, experimental work is being done with various types of fits, with and without adhesives.* The results, which are not quite complete, support the following generalizations:

1. Press fits in steel 0.75 in. (18 mm) and under are very difficult to control and usually result in galling or bending of parts. A slip or locational fit (0.000 to 0.0014-in. interference on 0.5-in. shafts) with an adhesive is the best way to achieve a secure assembly (Fig. 18-17).

2. Press fits 1 in. (25 mm) and over are very beneficially made with an adhesive, although at 3 in. (75 mm) in diameter the ultimate strength can be substantially increased (three to four times) by changing to a shrink fit. At the larger diameters the hub can be expanded enough with heat to give a slip fit at temperature. An adhesive introduced on the cold shaft at this point starts to cure as the parts equilibrate, giving an adhesive layer that is under compressive stress and very strong (Fig. 18-18).

OTHER FACTORS

18-7 Fatigue Considerations It is not enough to stop our calculations at this point and say that the adhesive fit exceeds heavy force fits in force and torque capability.

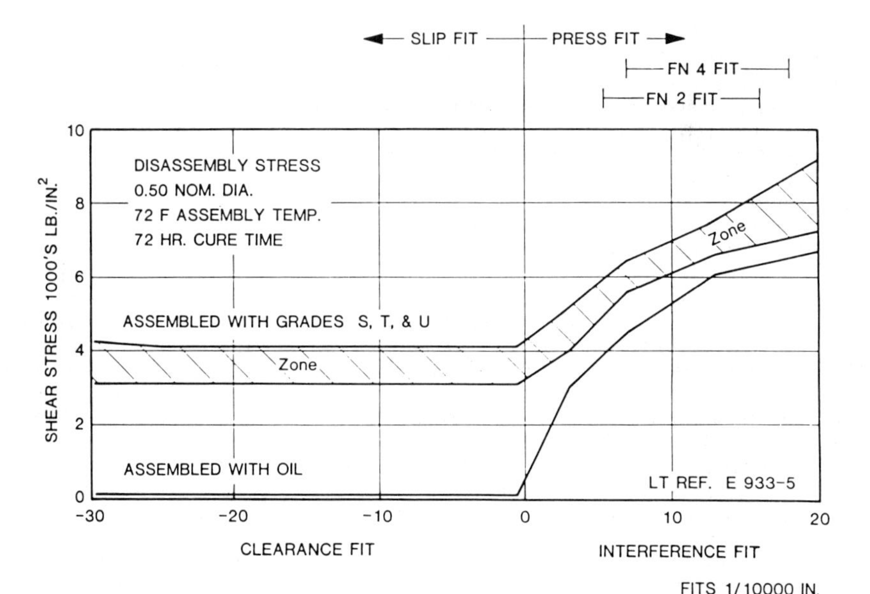

FIG. 18-17 Shear stress vs. press-fit interference for Grade S, T, and U, and without adhesive.

*All machinery and mechanical engineering handbooks and mechanical standards of ISO and ANSI list recommended standards for fits of various degrees; they are not repeated in this book.

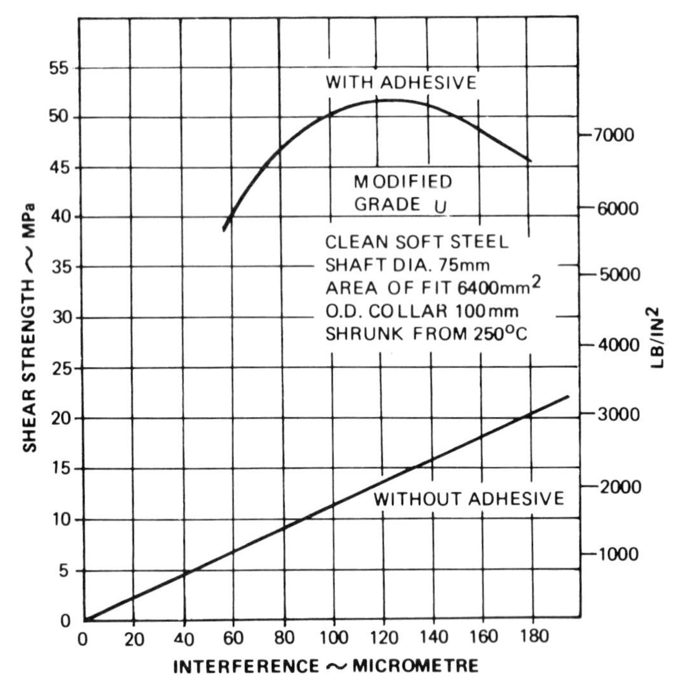

FIG. 18-18 Shear stress vs. shrink-fit interference for modified Grade U and without adhesive.

Although this is true for ultimate strength, fatigue strength is usually the design criterion in shaft and hub applications. Since the force fit is maintained by residual stress in the shaft and hub, any applied stresses of torsion or bending must be superimposed on the residual stress (see Ref. 1). Loss of stress from elevated temperatures or creep can be serious. In small shafts, as we have just seen, the residual stresses are equal to yield, so very careful calculation or tests must be conducted to determine the fatigue life of the assembly and the position of maximum tensile stresses.

Complicating the problems with the press fit may be the stress concentration where the shaft leaves the hub. Although the theoretical value seldom exceeds 2, concentration is dependent on the contact pressure and design of the hub (see Ref. 1). This concentration of stress often means that failure is incipient at the junction where the shaft enters the hub. This is where slight movement occurs during bending or torsion of the shaft, resulting in displacement of the connecting peaks between the hub and shaft. Oxides are formed by the high pressure on microscopic areas, and fretting results with its characteristic brownish-black color. Fretting corrosion can raise the microscopic stress, causing cracking and progressive fatigue failure.

Fatigue calculation of an adhesive fit is less complicated than that of a press fit (Fig. 18-16). It is just a matter of picking the proper reduced stress. Machinery adhesives *on the average* (dangerous, those averages!) will show fatigue strength of 500 psi when tested on relatively stiff shafts and hubs. There are few or no residual stresses, so the calculated stress under functional loads can be used. The rule of thumb used successfully for many years is 10 percent of the ultimate strength, which allows for a 2:1 safety factor and some stress concentration at the shaft hub entry line. Obviously, this represents a considerable downgrading for a material which, at best, is only 10 percent as strong in tension as the joined materials. However, as we saw on shafts of 1 in. and under, press fits have serious problems of exceeding yield by bending, galling, and expanding to achieve less ultimate grip than the adhesive. Where allowed, extra length can be used to improve an adhesive fit where it is not practical with a press fit because of shaft bending.

Field and laboratory experience indicates that the combination of light press plus an adhesive exceeds the fatigue of either a pure press or pure adhesive. Since all fatigue failures occur in the tensile mode, the compression of the joint in a *modest* way can decrease the possibility that the joint will "see" tensile stresses large enough to be harmful.

The laboratory results of a fatigue test done by Steyr-Daimler-Puch in Germany showed that our rules of thumb are most conservative. Figure 18-19 indicates that fatigue or endurance limits are doubled by adding an adhesive to a press fit but, on shrink fits 25 mm in diameter the endurance is the same with or without adhesive. This would indicate that the ultimate strength cannot be used indiscriminately for the basis of fatigue. In this example, fatigue limits of adhesive fits varied from 38 to 50 percent of the ultimate. It is desirable, as always, to test actual assemblies to determine the effect of residual stress from the fitting process (Fig. 18-19).

18-8 Compressive Strength The compressive strength of machinery adhesives in thin films far exceeds what one would expect from a tensile strength of 3000 to 6000 lb/in.^{2} (21 to 41 MPa). In films of 0.003 in. (0.08 mm), loads of 45,000 lb/in.^{2} (310 MPa) can be supported. At this load, soft steel plates start to yield, and the film will indent the plate permanently. Assemblies that use the materials in compression turn out to be very successful (like a rock crusher shaft in a chilled iron roller). A light press fit plus an adhesive keeps the shaft and adhesive in compression in spite of applied loads.

BUSHINGS

18-9 Bushing Mounting Advantages Oil-impregnated, sintered bronze bushings can be installed with a slip fit and a machinery adhesive more effectively than they can with a press fit. This procedure exhibits all of the advantages that one can imagine over trying to press-fit a slippery, compressible sleeve into a housing that has a different coefficient of expansion:

Holding force is 1.5 to 2.5 times the force of a press fit.

"Close-in" of the bushing bore is eliminated, allowing more precise shaft clearance. This means that bushings can be ordered with the inside diameter (ID) to size without the need to perform the extra sizing operation after the parts have been assembled (Fig. 18-20).

The relaxation of housing machining tolerances is possible, with resulting cost reduction (Fig. 18-20).

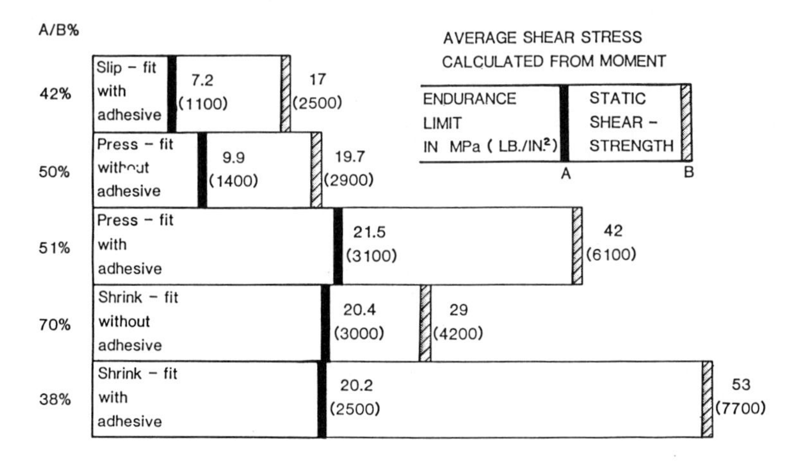

FIG. 18-19 Endurance limit and ultimate strength for press and shrink fits with and without adhesive.

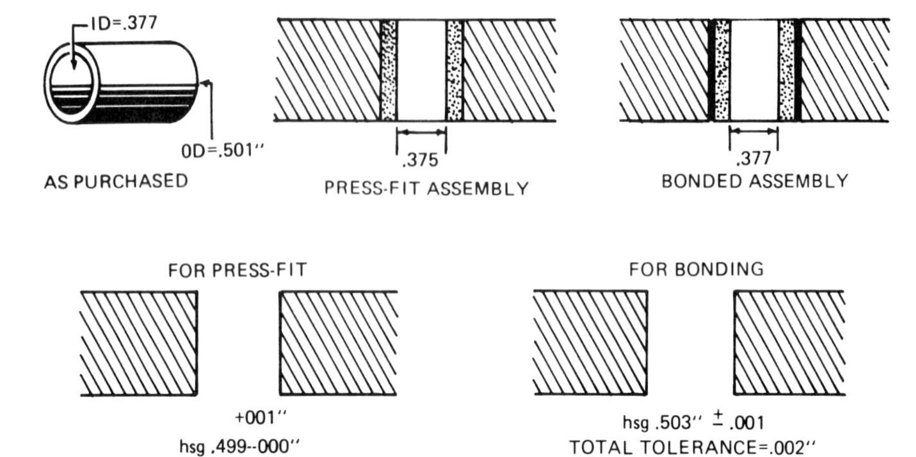

FIG. 18-20 Typical fitting tolerances for press and adhesion fits.

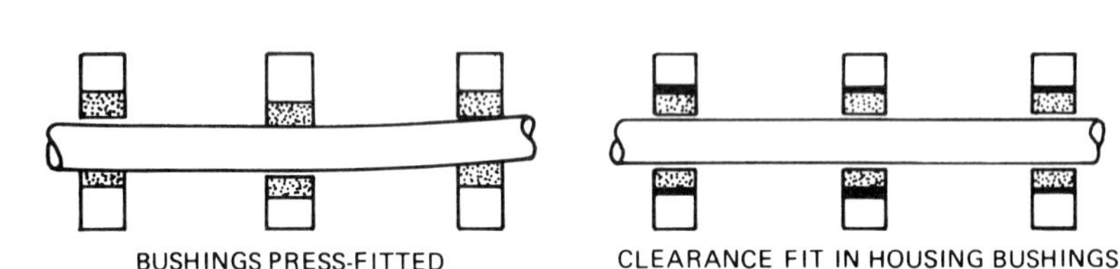

FIG. 18-21 Alignment of multiple bushings from the shaft.

Bushings can be slipped into place without the need for an arbor press.

Improved alignment may be obtained when bushings have a slight clearance fit in the housings and can take their alignment from the shaft. In small motors and sewing machines, this has resulted in freer running and quieter bearings (Fig. 18-21).

18-10 Dimensions and Tolerances Oil-impregnated bushings are available in a broad range of diameters. Because they are formed in a die, it is possible to hold dimensions closely. See Table 18-3.

TABLE 18-3 Typical Bushing Sizes and Tolerances

Inside or outside diameter, in.	Total clearance, in.	Direction of tolerance, in.	
Below 0.75	0.001	+0.000	−0.001
0.75–1.5	0.001	+0.000	−0.001
1.5–2.5	0.0015	+0.000	−0.0015
2.5–3	0.002	+0.000	−0.002
3–4	0.003	+0.000	−0.003
4–5	0.004	+0.000	−0.004
5–6	0.005	+0.000	−0.005

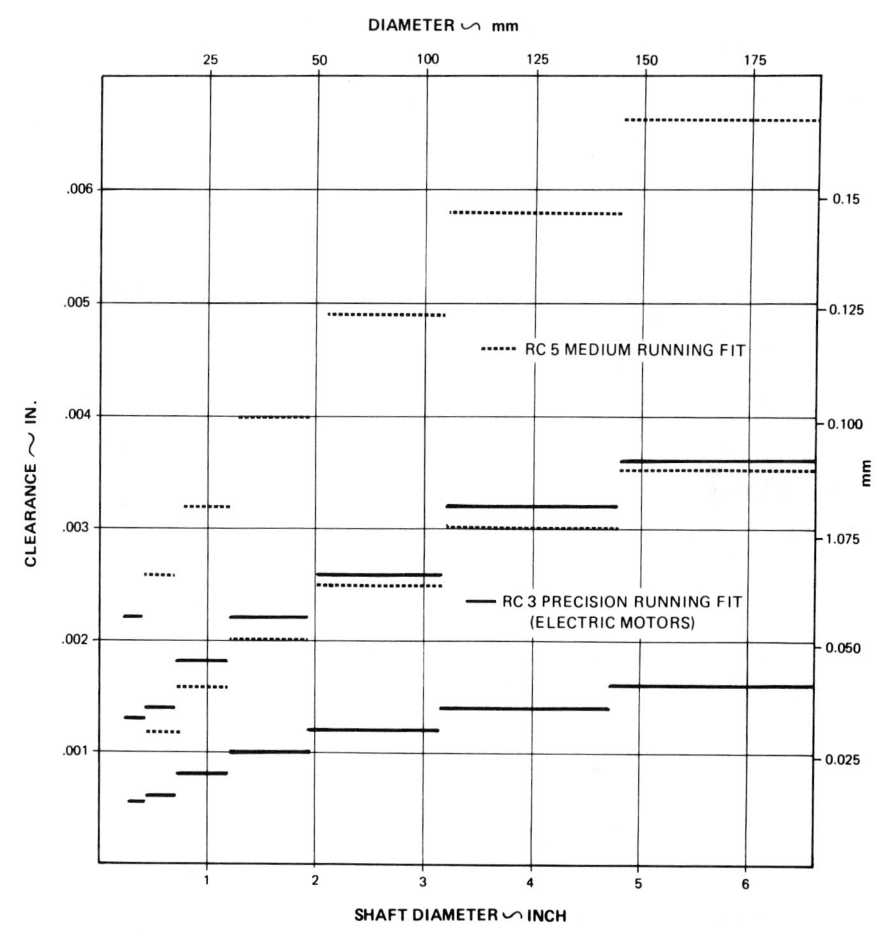

FIG. 18-22 Recommended shaft clearances vs. shaft diameter.

This kind of dimensioning is desirable for controlling critical clearances with the shaft. Figure 18-22 indicates recommended clearances for various applications. For a 1-in. shaft, a maximum diametrical clearance of 0.0015 in. is specified. For smaller, more precise fitting, as with subfractional horsepower motors having 0.25-in. shaft diameters, a diametrical clearance of 0.00075 in. is indicated.

Such clearances require precision grinding of shafts and very accurate control of the ID of the bushing (Figs. 18-20 through 18-22).

18-11 Interference Fitting Practice To obtain a press fit, the interference allowance between bushing outside diameter (OD) and housing or hub ID usually falls in the range of 0.0005 to 0.002 in. per in. (or mm per mm) of shaft diameter. For small bushing sizes, it may approach 0.004 in. per in. (1 mm per mm) in order to achieve reasonable tolerances. One manufacturer specifies a press fit of 0.002 to 0.004 in. (0.05 to 0.1 mm) on a diameter generally for all sizes.

An interference fit must produce sufficient pressure to hold the bushing in position under all operating conditions. Everything done to make a bushing perform satisfactorily as a low-friction bearing makes it difficult to hold with hub pressure and friction. It must be oily, slippery, and nongalling. Stress produced by the interference fit should not exceed the yield point of the bushing or housing material but it often does (see Table

18-2). Excessive stresses are unavoidable if the press fit alone is to be relied on to hold the bushing from moving under functional loads.

Press Fitting Reduces Size of Bushing Bore. The bore of the bushing closes in approximately in proportion to the interference of the OD with the mating hub when press fitting unless a special installation mandrel is employed.

Sleeve bushings with a 0.501-in. OD when fitted into a 0.498-in. hole close in an average of 0.0024 in. or 80 percent of the interference on the OD (Fig. 18-20).

If the bore is not resized, the reduction of bore plus assimilation of the housing tolerance results in a shaft-bearing fit that averages 80 percent looser than the original bushing size allows.

Maintaining the Bushing ID. Resizing of the bore used to be a common practice for restoring the running fit between the bore and shaft. Machining requires accurate positioning of the assembly and extra material on the bore to assure cleanup. Sizing often alters the porosity of the bushing by smearing the pores closed and usually increases the surface roughness. This limits the flow of oil and generally degrades performance of the bearing. The bushing ID is unchanged with an adhesive fit.

Flanged Bushings. When press fitting flanged bushings, there is a greater resistance to close-in at the more massive flange end, with a resulting tapered condition. Tests have shown that flanged bushings maintain the free bore size at the flange end, but collapse at the other end with a loss in parallelism of 0.001 in. (0.025 mm) (Fig. 18-24).

As shown in Fig. 18-22 for a shaft 0.3 in. in diameter, clearance with bushing should be 0.001 in. Thus, loss in parallelism of 0.001 in. produces either line-to-line fit at the unflanged end or double the recommended clearance at the flanged end.

18-12 Adhesive Mounting Method By employing an adhesive to bond the bushing in place, it is possible to open up the hole tolerance in the hub and slip-fit the bushing. In this way, the original bushing ID is maintained. There is no close-in to make it difficult to maintain specified clearances with shafts. If one changes a design from press fit to adhesive fit, the shaft diameter may have to be increased slightly or the bushings brought to finished size in order to gain the improved shaft-bearing fit.

18-13 Holding Capability The use of adhesive for retaining, in addition to better dimensional control, also provides superior holding capability. Press-fitted bushings of 0.5 in. with 0.003-in. interference develop approximately 440 lb average push-out, whereas those treated with Grade S develop 1200 lb.

The presence of oil on the bushing exuding from the capillaries does not destroy the adhesive's effectiveness. If bushings are heavily flooded with oil, the excess should be wiped off. The porous texture mechanically augments strength, compensating for light oil films. Cure on copper-containing bearings is very rapid; however, if immediate testing is to take place, the hub can be primed with an activator. Some assemblers use activated solvent to quickly wash excess oil off the bushing just before assembly. In most cases this step is not necessary with the adhesives listed in this handbook, which are oil-tolerant and fast-curing.

18-14 Heat Transfer As previously explained, press-fitted bushings contact only 20 to 30 percent metal-to-metal with their housings. The remaining section of the circumference is separated from the bore by an air gap, which is an extremely poor conductor of heat. Dead air space is an excellent insulator, with a thermal conductivity coefficient k of 0.015 Btu/hr-ft-°F. If oil fills the space, the conductivity is 0.10.

By contrast, with an adhesive the bushing is surrounded by a hard plastic film completely filling the area. The conductivity of the plastic is 0.08 Btu/hr-ft-°F (0.144 W/m/°C) or about five times as great as that of air and similar to that of oil, although not as high as that of metal.

Tests of assembled parts have indicated that the heat of friction at the bushing was readily dissipated into the housing through the adhesive bond. A slip fit with adhesive transferred heat 93 percent as efficiently as a press fit, and a light press plus an adhesive was 100 percent as effective.

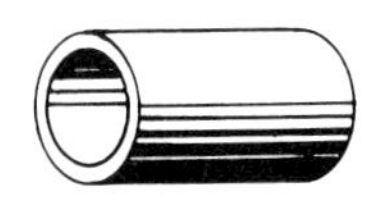

FIG. 18-23 Sleeve bushing.

FIG. 18-24 Flange bushing.

18-15 Bushing Lubrication Bushing lubricants are generally compatible with machinery adhesives and do not impede their cure or effectiveness. It is not necessary to wipe off the bushing OD to ensure satisfactory bonding. However, a few bushing samples should be tested to develop the best technique and establish that the lubricant will not disrupt processing.

If the bushing oil supply is to be replenished from grooves in the casting or from supply passages, their characteristic size and air content usually prevent blocking by cured anaerobic adhesive. The oil seems to find its way into the pores.

18-16 Dry Sintered Bushings While powdered metal bushings are usually furnished impregnated with lubricant, occasionally dry bushings are encountered. The porosity of these parts takes up from 10 to 35 percent of the total volume. If adhesive is applied to dry bushings of this type, the pores will act as a sponge to soak up the liquid adhesive. The effect will be to draw the adhesive out of the joint, destroying its ability to form a strong bond. Adhesively fitted bushings should be either impregnated with oil before mounting or secured with nothing thinner than Grade T or Z.

Impregnation of dry bushings can be carried out by heating them, while they are submerged in oil, to 250°F (120°C) and allowing them to cool to room temperature while still submerged. Heating expands the air out of the pores and cooling pulls the oil in.

18-17 Mounting Other Bushing Types Machinery adhesives are also effective in fitting most hardened plain or flange bushings as well as nonmetallic Teflon or nylon-filled bushings. Plastic parts require Primer N treatment to ensure cure.

Adhesive methods are particularly valuable for securing multiple drill bushings or bearings into broad thin frames or plates, such as might be found in drill fixtures or printing frames. Often the cumulative effect of several press fits will distort the frame beyond usefulness. The stress-free assembly with adhesive retains all the original flatness and hole spacing. Location must be maintained by the fit or external fixtures since the liquid adhesives have no centering ability. Nor will the adhesive fit push the bushing off center as is the tendency with such mechanical fastenings as set screws and tapered keys (Figs. 18-25 through 18-27).

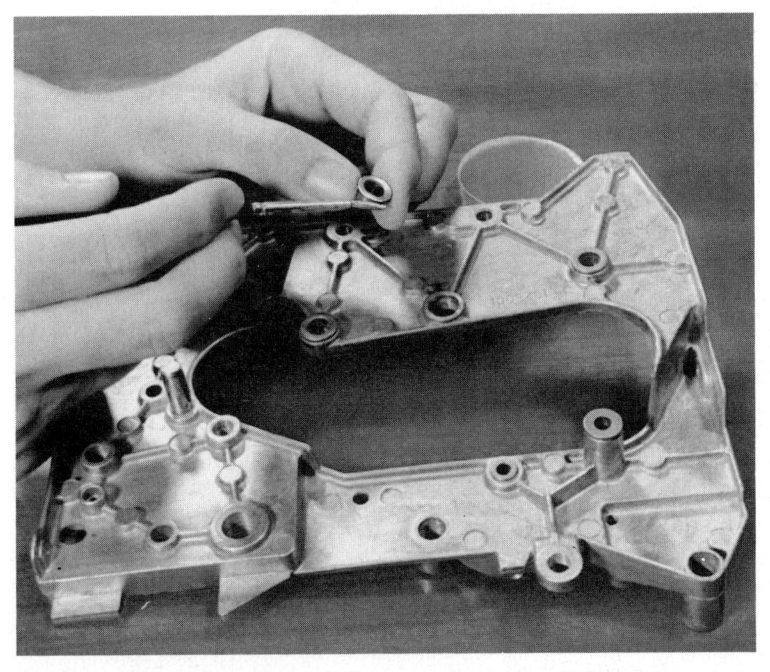

FIG. 18-25 Typical office machine side frame assembled with multiple bushings and rods without distortion of the frame.

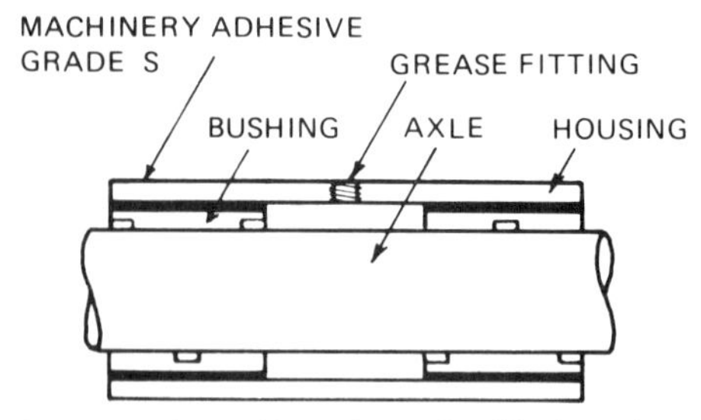

FIG. 18-26 The problem of press-fitted bushings being forced out of their housing by grease-gun pressure was eliminated by the use of a machinery adhesive. The method was developed by Parish Steel Division of Dana Corporation for gear-lever axles on heavy-duty trucks. Oil-impregnated bushings could not be press-fitted tightly enough to resist the 1700-lb./in.^{2} (12-MPa) pressure of grease guns. The friction-improved method required no basic changes in normal assembly procedure.

FIG. 18-27 Straight-sleeve ball bearing bushing diesets, supplied by Danly Machine Specialties, Inc., use machinery adhesive to avoid close-in, which occurs as a result of a press fit. They are supplied with a wring fit and are retained with the resin. When so installed, the bushing bore does not require honing.

BALL AND ROLLER BEARINGS

18-18 Traditional Bearing Mounting Practice There exists some controversy about the mounting of bearings using an assist from machinery adhesives. As in most persistent disagreements, the problem is being viewed from more than one direction, somewhat like the blind men trying to describe an elephant by feeling different parts. We shall attempt to look at all sides of the elephant and give the synopsis of 20 years of successful practice.

Before the invention of machinery adhesives, conventional mounting methods used press fits and lock nuts or caps to retain the bearing races. Tolerances on the bearing diameters and fits of the balls into the races were all standardized around press fits and the resulting change of fits as the races were either expanded or shrunk in the process. Reducing tolerances was the standard way to solve loosening and inaccuracy problems. This system has been very successful but, as with any close tolerance assembly, there are cost penalties and failures. Of the eight most commonly encountered failures (see Ref. 2), four are a result of the mounting process:

1. Bearing overload caused by heavy press fits or out-of-round expansion of housings. Out-of-round can be caused by a very slight galling or pickup as a heavy press is consummated. This often remains undetected until the bearing fails prematurely in service.

2. Brinelling caused by improper pressing of the bearing. The press-fitting load should not be transmitted through the balls; however, contingencies often make this expedient. A heavy press then destroys the bearing.

3. Misalignment caused by bent shafts or crooked or scored bores. Heavy press-fitting practices are not compatible with precision bearings.

4. Cam or inner race fastening methods that cock the bearing. Cam rings and set screws are often used for fastening the inner race. An undersized shaft or overtightening to prevent slip can apply eccentric loads on the bearing.

The other four causes of bearing failure are related not to mounting technique but to environmental effects: contamination, false brinelling caused by vibration when the spindle is stationary, thrust overload, and electric arcing caused by static electricity or faulty wiring.

The Case for Keeping the Press Fit. Bearings are used to locate pulleys, rollers, gears, and cutting tools in a precise position while allowing the rotation of parts or a spindle around a predetermined axis. They must provide extreme stiffness in both axial and radial directions while allowing almost frictionless rotation. In most applications the inner member is a rotating spindle. To this is attached by a press fit the inner race of the ball or roller bearing (Fig. 18-28).

There are four reasons given by the bearing manufacturers or users for maintaining the press fit of the inner race.

1. Concentricity with the spindle is most easily attained by a snug fit. A heavy press is not necessary but some press is. No centralizing is achieved with a liquid adhesive. This is a most persuasive argument which is ignored only during some repair operations, when the cost of machine down-time makes it worth extra effort to align the bearing on a worn shaft with shims or fixtures rather than by plating and regrinding to restore the press.

2. The inner ring can rotate under load. The effect of load on a loose race is a condition shown in Fig. 18-29. Load on the inner ring leaves space at the bottom of the ring. If this shaft were 1.000 in. in diameter and the clearance under load 0.001, the two surfaces would act like gears (as in Fig. 18-29) where the shaft has 1000 teeth and the ring 1001 teeth. As the shaft rotates one revolution, or 1000 teeth spaces, the ring rotates 1000 positions less one. In other words, the ring creeps backwards. This action creates friction, fretting corrosion, wear, and noisy operation. The firm support of a press fit preloads the mating surfaces and prevents rotation in most cases. However, it must be remembered from previous discussions of press fits that a heavy press produces only 30 percent contact of the touching parts, so it is not difficult to see why loosening can occur under repeated loads. If this were the only reason given for a press fit, then machinery adhesives would always be used to avoid one of the four common causes of failure. A

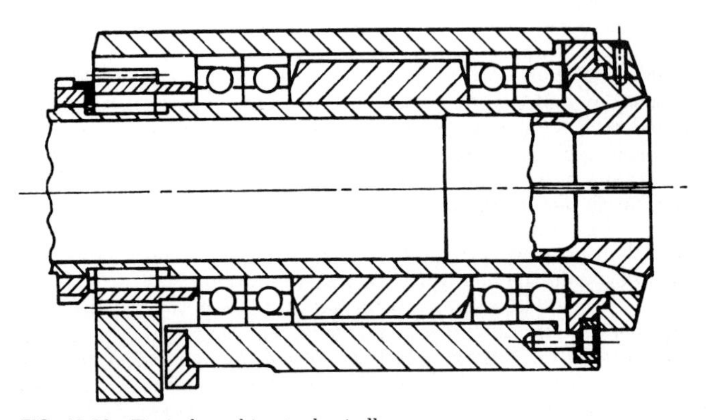

FIG. 18-28 Typical machine tool spindle.

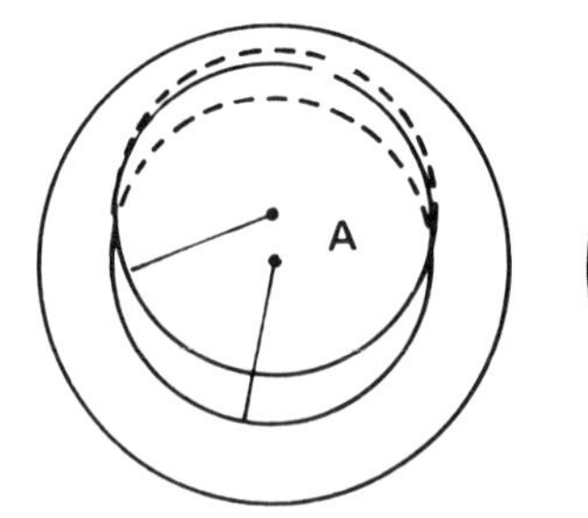

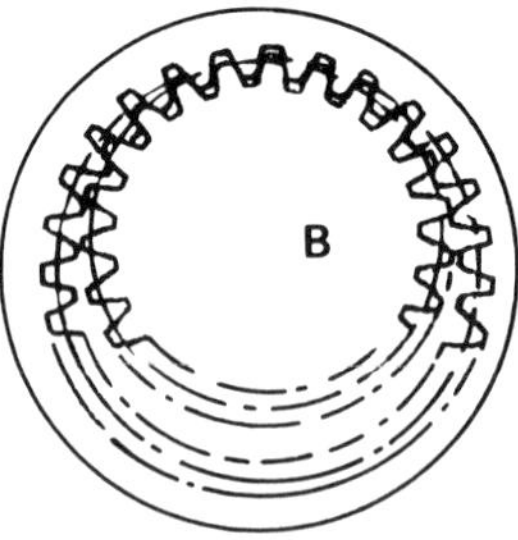

FIG. 18-29 Effect of load on loosely fitted shaft leaves space below shaft. Rotation under load creates gear effect to cause inner ring to rotate backward one tooth per revolution with respect to the shaft.

light press and a machinery adhesive friction improver will hold at least twice as well as a very heavy press.

3. On very heavily loaded inner rings with a premium on space to locate the bearing, such as in a steel rolling mill with tapered roller bearings, the inner race is relatively fragile and must be "rounded up" by the fit. This and drawn cup bearings are the only ones where the bearing roundness and quality do not far surpass the quality of roundness achieved by typical spindle manufacture.

4. The last reason given for the press fit on the inner ring is the success of years of tradition and the dimensional standards that exist. It is stated that on some ball bearings the fit between the balls and the two races is partially controlled by the expansion of the inner race during the press. A press fit is not a reliable method of reducing the clearance in a bearing since it does so only on an average or mean basis and not throughout the spread of tolerances. The best way to reduce clearances is in the manufacture of the bearing. The Anti-Friction Bearing Manufacturers Association (AFBMA) suggests four different fits, given in the following table.

18- to 24-mm diameter	Maximum clearance 0.0000 in.	AFBMA No.
Snug	4	2
Standard	8	0
Loose	11	3
Extra loose	14	4

The Case for Slip Fits. According to some manufacturers, the outer race, under heavy side loading, should intentionally be left loose ("thumb fit") or very lightly pressed in order to allow rotation similar to that shown in Fig. 18-29. The reason is that if this outer ring is kept stationary, all the repetitive ball-rolling load is taken by one side of the ring and fatigue life will be limited. In steel mills where roller bearings are used, outer cups are sometimes press-fitted for rigidity but the races are marked so that they can be reassembled in a different position each time the rolls are disassembled for dressing, which is quite often. Obviously, these procedures of under-load rotation or assembly rotation are practical only under very special circumstances.

Although the above reasoning may be sensible for large roller bearings, most ball bearings are in the range of 18 to 50 mm. Here the most sensitive fatigue member is the inner race because of its convex curvature in contact with the balls. In addition, the control of outer ring rotation by relative looseness is not reliable and can lead to severe fretting of the shaft or bearing (Fig. 18-30). Most users will press the outer ring to one degree or another, with the following important exception.

Longitudinal Slip. Most spindles incorporate at least two bearings separated by enough space to give bending stiffness. The front bearing is as rigidly fixed as is possible both radially and axially. The rear bearing is usually floated with a light slip fit so that

FIG. 18-30 Fretting failure, the result of mounting bearings loosely.

the two bearings do not thrust-load each other as they are assembled or when temperature changes occur.

Example. A shaft 24 in. long operating at 100°F higher temperature than that of the housing will expand 0.000006 in./in./°F. At a strain of 0.0006 in./in. and a modulus of 30×10^6 lb/in.^{2}, this translates into a theoretical stress of 18000 lb/in.^{2} ($s = eE$). With an area of 0.785 in.^{2}, the force is 14000 lb. A typical 1-in. radial bearing is rated at 900 lb thrust.

A temperature differential is quite normal for operating spindles and is especially severe in electric motors where electrical loads end up as heat. One bearing is always floated and the most practical place is the largest diameter of the rear bearing.

Floating Inner Rings for Stationary Spindles or Axles. Where wheels revolve around stationary axles, such as in automobile nondriving wheels or in idler pulleys, the inner ring is floated and the rotating outer ring is assembled with a press. The generalization can be made that the rotating ring is always secured with a press and the stationary ring may or may not be, depending on the requirements for rigidity and axial sliding.

Precision and Light Multidirectional Loading. Machine tool spindles are often loaded in many radial and axial directions. This requires that the races be as rigidly and concentrically mounted as possible. Drill presses, jig borers, portable tools, and internal high-speed grinders all rely on the rigid and concentric mounting of the front bearing(s) to achieve repeatable positioning and to perform without chattering. In most instances, the front bearings will be axially preloaded to remove all slack from the balls and races. This is done either with spring loads or with pairs of preloaded bearings that are manufactured as pairs with built-in preload (Fig. 18-31).

In cases like this, both the inner and outer race of the front bearing must be very rigidly mounted but press fits must be light so that the preloaded bearings are not overstressed. Since the light fit is satisfactory for the light loads encountered, adhesives or friction improvers are not used since spindle rebuilders say "they give an impression of sloppy fits" and are used only when press failures occur and emergency repairs are made. Transitional fits are used according to the requirements for concentricity; thus some degree of interference is achieved on the average.

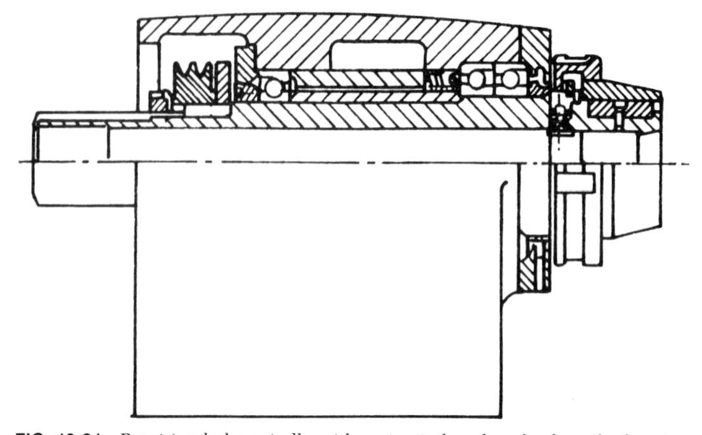

FIG. 18-31 Precision lathe spindle with spring-induced preload on the bearings.

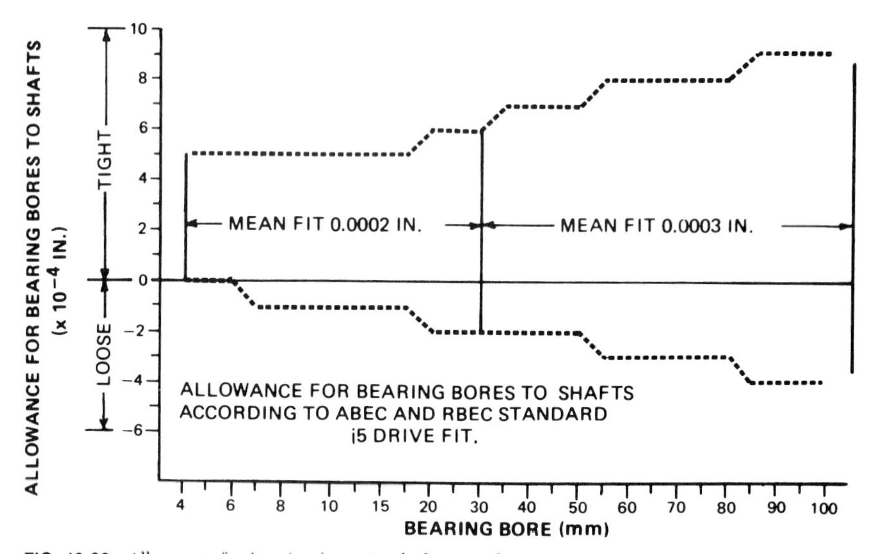

FIG. 18-32 Allowance for bearing bores to shafts according to ABEC and RBEC Standard j5 drive fit.

Shaft and Housing Fits. Realizing the need for different types of interference and clearance fits of bearings, the International Standards Association (ISA) has prepared standards for them. The internal clearances of the bearings are designed with specific amounts of interference or looseness depending on application conditions:

Bearing size and type
Amount of load
Type of load, radial, or thrust
Operating conditions
Possible rotation of load

A graphical representation of the tolerances and allowances for a j5 drive fit is shown in Fig. 18-32. Note that with a mean fit of 0.0002 in. (0.005 mm) the extremes are from loose 0.0002 to tight 0.0006 in. (0.005 to 0.013 mm). The extremes can be very troublesome. When the entire load environment is known and all of the tolerances are carefully controlled, the bearing should operate satisfactorily.

Machinery adhesives as friction improvers can enter this scenario with beneficial results by allowing lighter fits and making loose fits work, provided concentricity requirements are met.

18-19 Generic and Traditional Places for Machinery Adhesives for Ball Bearing Mounting. In the 20 years that machinery adhesives have been used for securing of bearing races, the uses can all be categorized as augmenting or assisting the interference fit. In this role they prevent movement and fretting, improve alignment, lower residual stresses, lengthen life, and solve thermal problems. Friction improvers traditionally have been used for:

1. Augmenting press fits in very heavy-duty applications where heavy presses will yield the housing, bend the shaft, or slip under load. Applications: Bearings for differential gears, truck and automotive transmissions, heavy-duty pumps, and railroad ballast tampers.
2. Augmenting press fits into low-modulus, high-thermal-expansion materials such as aluminum and engineering plastics. Applications: Bearings for automobile transmissions, portable tools, automotive pumps, and aircraft control rod-ends.
3. Augmenting press fits where there is no room for conventional retainers such as nuts, snap-rings, or caps. Applications: Bearings in conveyer rollers, idler pulleys, tension arms, and textile spindles.
4. Providing alignment and security where long or fragile shafts cannot be pressed without disturbing accuracy. Applications: Bearings for typewriter platens, mail-sorting machinery, tape drives, and office machinery in general.
5. Providing fast and ostensibly temporary repair of worn shafts and housings where restoration of press fits may cost thousands of dollars in down-time and repair costs. Machinery adhesives return the machine to duty in hours for pennies. Applications: Bearings for paper driers, steel rolling mills, farm equipment, mining equipment, and oil drilling machinery.

PRESS FITS

18-20 Augmenting Heavy-Duty Fits. As was seen with press fits in general, the adhesive assist can increase the holding ability two to six times. It is not necessary to use the strongest material. In order of increasing strength, Grades M, N, S, Z, and U are appropriate for securing a bearing. Grade S is the best to start with because it is thin enough to wick into the surface finish. It can even be applied after assembly, although Grade P will be faster-wicking.

With all of these fast materials there may be some curing during the press. Grade S will repair itself if cure starts during the press. As long as the press is not stopped partway, the partial cure does not change the coefficient of friction and final strength is not affected enough to be ineffective. Grade Z is so thick that it stays where it is put and can be applied to small areas for limiting the total strength. One might call it a "chemical" set screw.

As with any heavy press fit, the design must allow access for removal. A screw or hydraulic jack is the best way to disassemble a bonded bearing. Hammers and heat are destructive.

18-21 Augmenting Press Fits in Aluminum and Plastic. Demands for lighter weight and greater efficiency in vehicles have led product designers to make greater use of aluminum and reinforced engineering plastics. This section will concentrate on the properties of steel in aluminum fits for simplicity's sake. Everything said for steel and aluminum is germane to any combination of parts with differing moduli and coefficients of thermal expansion.

The problem is threefold: Press fits reach the yield of the low-modulus material before adequate hub pressure is attained; change in temperature varies the hub pressure; and the coefficient of friction between dissimilar materials often is rather low. Aluminum, with all of its strength and lightness, shares these drawbacks when combined with steel.

Historical methods for dealing with this problem included the use of heavy shrink fits, cast-in-place steel sleeves, adhesives, and, most commonly, mechanical retention rings. We will discuss the designing of press and shrink fits in combination with adhesive augmentation. A beneficial synergism results when using the two methods. They can achieve strength levels unattainable with severe shrink fits, at the same time reducing the stress levels in the aluminum (Fig. 18-33).

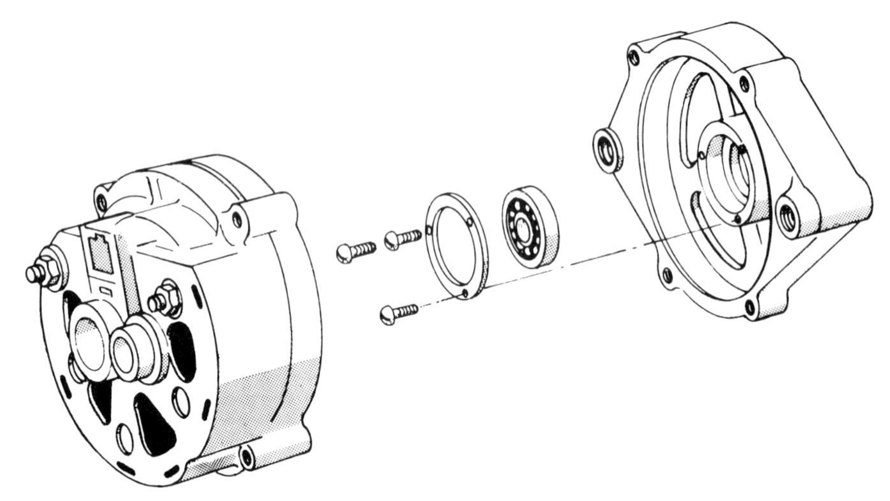

FIG. 18-33 Steel ball bearing mounted in an aluminum housing with a retaining ring.

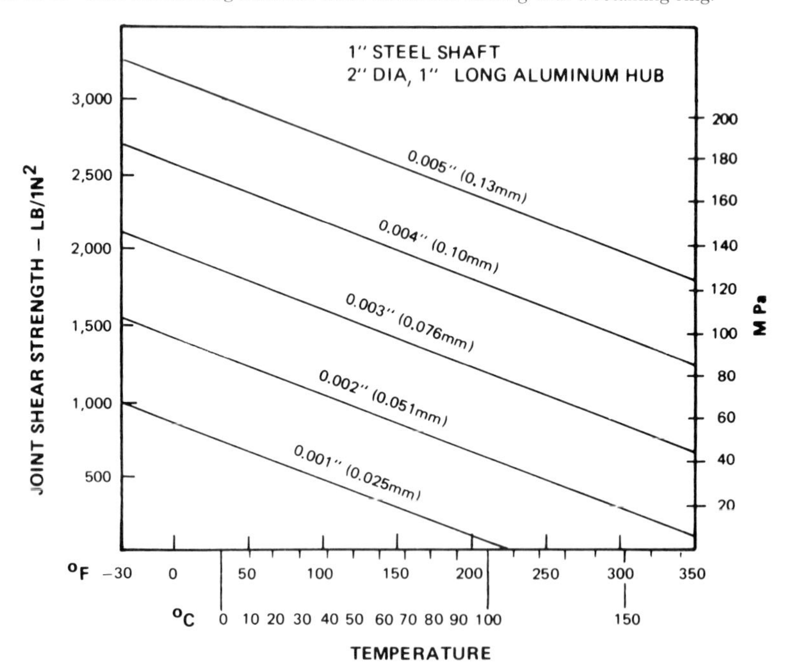

FIG. 18-34 Calculated joint shear strength vs. temperature for steel/aluminum interference fits.

18-22 Performance of Press Fits. Steel into aluminum press fits are shown as calculated in Fig. 18-34 for temperatures ranging from −30 to 350°F (−20 to 177°C). This family of curves was generated for a 1-in. steel shaft pressed into a 2-in. aluminum hub. Interference allowances shown on the curves are for a temperature of 70°F (22°C) (design equations are in the Appendix). Later tests of actual press fits showed about 25 percent lower results, probably because the coefficient of friction used for the calculation (0.12) was too high and because a few "tenths of a thou" error in measurement can have a substantial effect. This disparity between calculated and actual results highlights the problem of predicting press-fit performance from pressure and an assumed coefficient of friction. Small changes in either the fit or coefficient are directly reflected in performance.

Limiting hub stresses: Using the information in Fig. 18-34, it would seem possible to select an interference value that would provide any desired level of joint strength needed. Hub stresses, however, severely limit the choice of press fits.

Figure 18-35 shows the maximum hub stresses that would occur for various temperatures and levels of interferences shown in Fig. 18-34. The hub stress value serves as the limit to the amount of interference that can be used. The yield strength of a typical aluminum casting is shown as the limiting stress in Fig. 18-35. This limit is 24,000 lb/in.^{2} (165 MPa) tangential shear stress, which translates back to a joint shear strength at a coefficient of friction of 0.12 to 1700 lb/in.^{2} (12 MPa) [Eq. (18-9)]. No matter what we do with the press fit, we are always limited to a maximum shear strength of 1700 lb/in.^{2} (12 MPa) or less even with a hub ratio c/b of 0.5, which is very heavy. If we increase the fit beyond this, the hub yields or may split.

We can now replot Fig. 18-34 in a more realistic fashion, as shown in Fig. 18-36.

Any interference over about 0.002 in. (0.051 mm) when carried down to −30°F (−40°C) will yield and, if returned to higher temperatures, goes down the maximum strength line, where at 350°F (180°C) its strength is reduced to about 400 lb/in.^{2} (3 MPa). If we had corrected the curves according to the actual values, which were 25 percent lower, the 350°F (180°C) fit would be zero.

Hub yielding and low friction should be enough evidence to show why press fits are not practical, and the consideration of tolerances will show that the theoretical fits cannot be controlled. A tolerance of ±0.0005 in. (±0.01 mm) on the shaft and ±0.001 in. (0.03 mm) on the hub with a nominal interference of 0.002 in. (0.05 mm) means that the actual interference would range from 0.001 to 0.003 in. (0.025 to 0.08 mm). Some of these parts will loosen at 150°F (76°C).

Functional loads have not been considered. They, of course, must be added to the fitting stresses, the total of which must remain below the yield.

18-23 Use of Adhesives to Alter the Coefficient of Friction. As we have seen, the retention strength is highly dependent on the coefficient of friction, which is notoriously unpredictable. It may change with surface finish, oiliness, temperature, and pressure. For aluminum to steel the frictional coefficient is not only variable but low. Values of 0.10 to 0.15, or 50 percent variation, are often used for design. The use of adhesives to

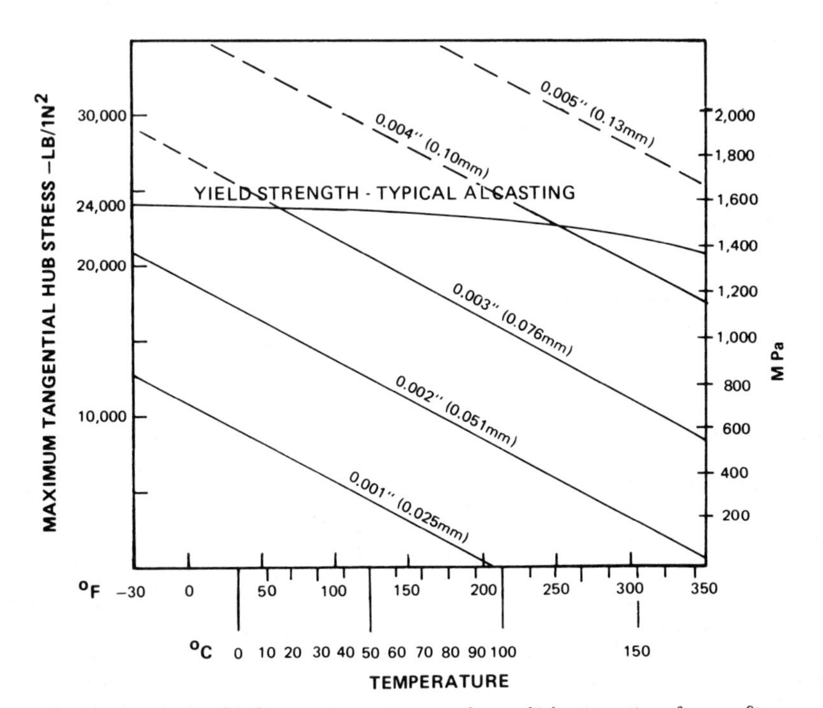

FIG. 18-35 Calculated hub stress vs. temperature for steel/aluminum interference fits.

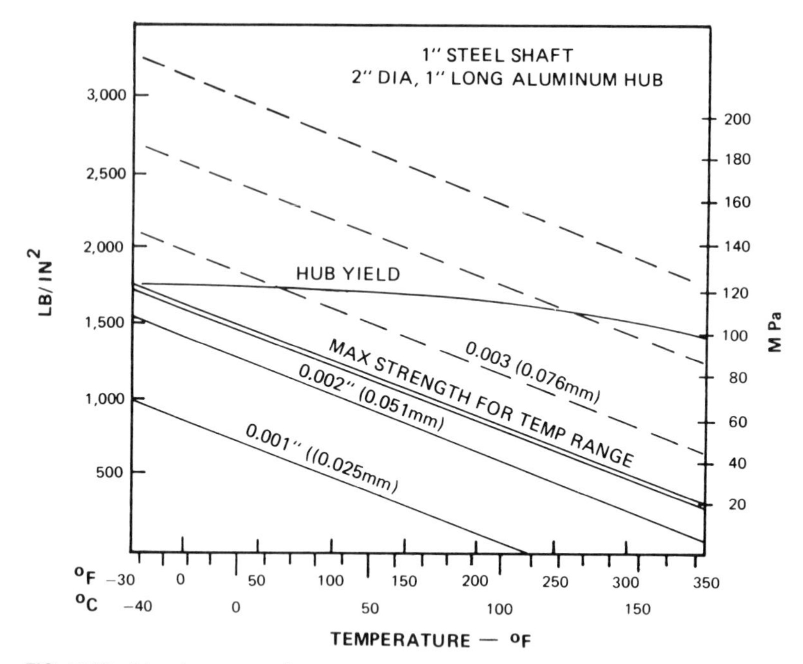

FIG. 18-36 Joint shear strength vs. temperature for steel/aluminum interference fits.

improve the friction at the interface makes sense because we have 70 percent of the surface as unused "inner space" until the adhesive is introduced. Filling the inner space with a strong adhesive did improve the strength by a factor of 2 (Fig. 18-37).

Three adhesive techniques are described below.

1. Adhesive with clearance: This is the conventional bonding method using a 0.002-in. (0.051-mm) clearance with machinery adhesive Grade U. Both components were solvent-vapor-cleaned prior to assembly. One-inch shafts in 2-in.-diameter by 1-in.-long hubs were used throughout. From prior experience we know that clearances could have varied between 0.001 in. (0.025 mm) and 0.005 in. (0.13 mm) without affecting the strength. The shear strength at room temperature (at which it was cured) was 1400 lb/in.^{2} (10 MPa). As the parts were heated, the aluminum grew away from the shaft and the adhesive. Even though the coefficient of thermal expansion of the adhesive exceeds that of the aluminum, the adhesive film is so thin that its total expansion is relatively low and the film begins to fail in tension. Performance was equal to or better than that of the press fit.

2. Adhesive with interference: Grade U was used to augment a 0.002-in. (0.051-mm) press fit with exactly the same preparation as above. Strengths for this method were more than double those of the press fit. The interference fit prevents separation of the surfaces so the adhesive is always effective. The adhesive bond adds joint strength that does not decrease as quickly as frictional forces do when the hub pressure drops. In assembling the parts, about 50 percent more assembly force was required than with an ordinary press fit. The adhesive is not a very good lubricant and may be partially curing during assembly because localized pressure and heat are produced. The hub stress produced is the same as that of a press fit without adhesive.

3. Adhesive shrink fit: Ordinary adhesive technique was used with the usual 0.002-in. (0.051-mm) clearance but this time the hub and shaft were heated to 250°F (120°C), increasing the clearance at temperature to 0.0026 in. (0.066 mm). The parts were quickly coated with Grade U and slipped together. As the assembly cooled, the hub attempted to return to its original 70°F (22°C) dimension but was prevented from doing so by the rapidly curing adhesive, which acted as a shim. This method bears a resem-

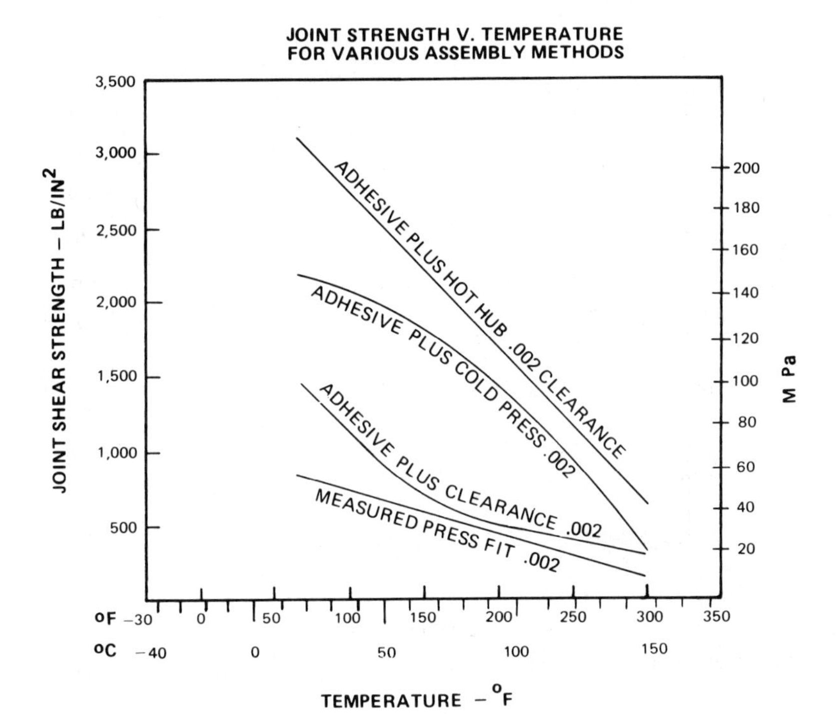

FIG. 18-37 Comparison of assembly methods—joint strength vs. temperature.

blance to a shrink fit, but no metal-to-metal contact occurs. The adhesive is maintained in its toughest compressive mode throughout the temperature range. Fill is better because material is being squeezed out during hub contraction and is not scraped off during assembly as in the press fit.

The adhesive shrink-fitting technique not only is good for bearings in aluminum but has been used successfully to assemble hardened differential gears onto a carrier for an automobile. The strength obtained was two to six times the strength of the maximum press fit. The hub stresses are less than those produced by a modest press fit, as shown in Fig. 18-38.

18-24 Augmenting Press Fits with Space Limitations. Certain types of assembly do not have room for retaining nuts, rings, or heavy press fits. Conveyor rollers, idler and guide pulleys, tension arms, and textile spindles fit into this category. Aircraft ball rod-ends not only have this space limitation but the fit is usually into aluminum or magnesium (Fig. 18-39).

18-25 Preventing Damage and Providing External Alignment. Often the stresses produced by the press fit will damage the shaft or the housing. One of the primary reasons for the adhesive assembly of shafts in small motor armatures is the extra accuracy achieved because the shafts are not bent by the press. Lamination stacks are very difficult to produce with round, straight holes. Shafts pushed into crooked holes will always bend. Before the development of adhesive techniques every rotor-shaft assembly had to go through a straightening operation that cost as much in labor as the assembly did (Fig. 18-40).

The same distortion can ruin the accuracy of bearing support panels in office equipment. Two or three bearings pressed into a large flat plate or casting can dysfunctionally distort the panel. In machinist's terms it "warps like a potato chip" (Fig. 18-25).

In addition to maintaining accuracy, an adhesive can be used to obtain fixture accuracy with ordinary parts. This technique is used to assemble long typewriter platens or rolls. The rolls are thin tubes with precision bearings cast into the bore with adhesive. The

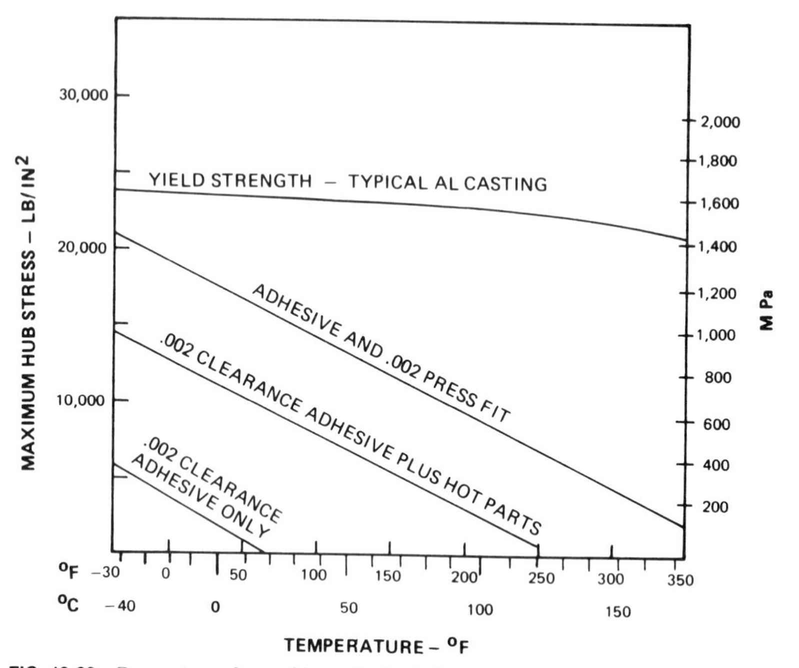

FIG. 18-38 Comparison of assembly methods—hub stress vs. temperature.

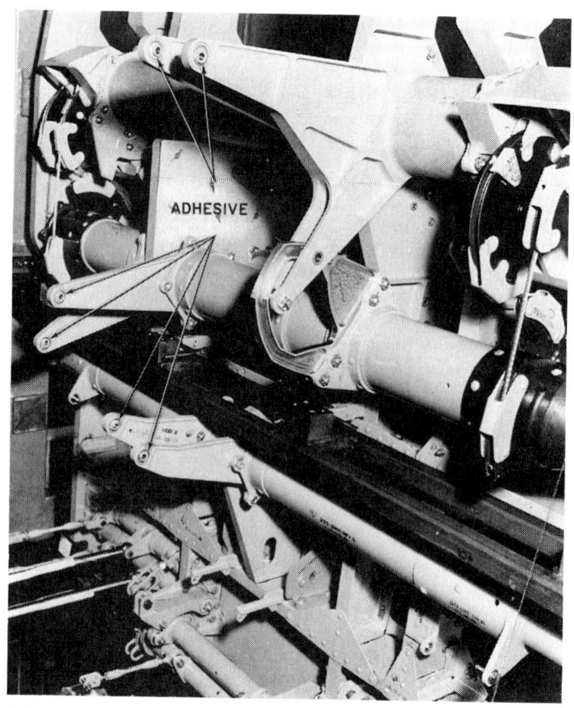

FIG. 18-39 Ball rod-end for aircraft, bonded and staked.

FIG. 18-40 Fractional HP electric motor shaft adhesively assembled without distortion.

0.5-in. (13-mm) shaft can be up to 40 in. (1000 mm) long with bearings spaced intermittently along it for support. The whole device is slip-fitted with adhesive, placed into V blocks to locate centerlines accurately, then induction-heated at the adhesive joints. The cast-in-place accuracy of these long assemblies is a phenomenal maximum of 0.002-in. (0.051-mm) total indicator runout, or about the equivalent of the shaft's original straightness. The key to this type of assembly is to have the fixture compensate for part inaccuracies and the adhesive to cast this precision permanently in place.

18-26 Providing Fast Repair. The greatest number of adhesive-bearing applications has occurred in the field where the bearing or the press has failed. Seldom is it possible to get a replacement for anything other than the bearing; so damaged shafts and housings must be used until major repairs can be scheduled. Tales abound about the money and mission saved by replacing a bearing with shims and adhesive . . . like the front-wheel bearing on a town snow plow, fixed in a blizzard using a torch to warm the adhesive; or the pulp drier that would not keep a bearing until it was adhesively mounted; or the railroad track ballast tamper that had "walking" bearings, fixed on the job. These applications serve to remind us that all techniques of bearing mounting have their strengths and limitations and are worth careful consideration by their implementors.

18-27 Mounting of Spherical Plain or Roller Bearings Spherically seated bearings will bind up if mounted with a heavy press fit and are a natural for a friction-augmented light press fit (Fig. 18-41). The trunion bearings in the USA M-60 tank on which the 155-mm cannon is pivoted are secured with machinery adhesive (Fig. 18-42). All of the shock of firing is absorbed through the bearings and transmitted to the turret, which is secured with bolts locked with Grade K. Any looseness would wreak havoc with the gunner's aim.

18-28 Limitations Limitations are as follows:

Bearing float. As mentioned previously, the rear bearing of a spindle usually has to float to accommodate differential thermal expansion between the spindle and housing. As yet there is no way that machinery adhesives can improve the fit without increasing the friction at the same time, so they must be excluded from this area.

Needle bearings. Drawn cup needle bearings are not round before mounting. They conform to the housing bore into which they are pressed. Looser fits than those recommended by the manufacturer should not be used, but friction improvement with a machinery adhesive is beneficial.

Excess material. Care should be exercised in the application of adhesive to the outer races. Introduction of the adhesive into the ball raceway, where there is potential for

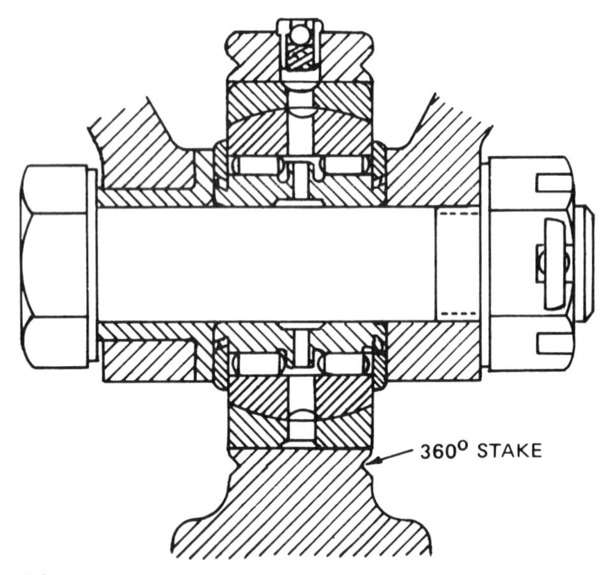

FIG. 18-41 Cutaway of a spherically mounted roller bearing.

FIG. 18-42 Trunion bearings for the M-60 tank.

curing even with only line contact, should be avoided. Usually the lubricant in which the bearing is packed will resist entry of excess adhesive. Buna N seals can be attacked by the adhesive, and sustained contact should be avoided.

APPENDIX: Design Calculations for Interference Fits

While the design procedures for interference fits are well known, it is worthwhile to review them here to see the effect of dissimilar materials and varying temperatures. (See Fig. 18-43.)

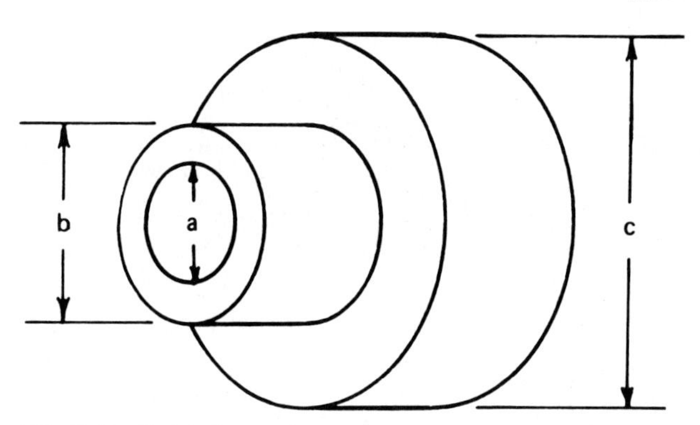

FIG. 18-43 Shaft-hub press fit.

Hub Pressure. The key design parameter for press fits is the radial contact pressure, calculated as follows:

$$P = \frac{\delta}{b\,(A + B)} \qquad (18\text{-}5)$$

where Factor A for Shaft A =

$$\left(\frac{(b^2 + a^2)}{(b^2 - a^2)} - \mu_s\right) \times \frac{1}{E_S}$$

and Factor B for the Hub B =

$$\left(\frac{(c^2 + b^2)}{(c^2 - b^2)} + \mu_h\right) \times \frac{1}{E_h}$$

and P = contact pressure, lb/in^2 (N/m^2 or Pa); a = shaft ID, in. (m); b = shaft OD, in. (m); c = hub OD, in. (m); δ = shaft-to-hub interference, in. (m); E = modulus of elasticity, psi (Pa); and μ = Poisson's ratio. Subscripts s and h refer to shaft and hub, respectively.

Joint strength, axial capacity, and torque capacity can then be calculated as follows (Bibliography, entry 3): Joint strength, lb/in^2 (Pa):

$$S = Pf \qquad (18\text{-}6)$$

Torque capacity, in-lb (N-m):

$$T = \frac{Pf\pi b^2 L}{2} \quad \text{or} \quad \frac{S\pi b^2 L}{2} \qquad (18\text{-}7)$$

Axial capacity, lb (N):

$$F = Pf\pi bL \quad \text{or} \quad S\pi bL \qquad (18\text{-}8)$$

where f = coefficient of friction; L = engagement length, in. (mm).

The interference causes a tangential stress in the hub, which is at a maximum at the hub bore. This stress can be calculated as follows:
Tangential hub stress, lb/in^2:

$$S = \frac{P\,(c^2 + b^2)}{c^2 - b^2} \qquad (18\text{-}9)$$

Temperature Effects. When materials are assembled that have different coefficients of thermal expansion, consideration must be given to the change in interference when the assembly is heated. If the shaft has a lower coefficient of thermal expansion than the hub, the interference decreases with temperature. If the materials were reversed, the interference would increase when heated. As the temperature changes, the values of radial contact pressure, torque capacity, axial capacity, and hub stress all change because of their dependence on interference.

The change in interference can be calculated as follows:

$$\delta_1 = \delta_0 - b \times (\alpha_h - \alpha_s) \times (T_1 - T_0) \tag{18-10}$$

where δ_0 = initial interference at temperature T_0, in. (m); δ_1 = interference at service temperature T_1, in. (m); b = shaft diameter, in. (m); δ_h = coefficient of thermal expansion of the hub (per °F or °C consistent with $T_1 + T_2$); and α_s = coefficient of thermal expansion of the shaft.

REFERENCES

1. J. E. Shigley, *Mechanical Engineering Design,* McGraw-Hill, New York, 1977, Sec. 2–16.

2. *How to Prevent Ball Bearing Failures,* Fafnir Bearing Division of Textron, Inc., New Britain, CT, Form 493.

BIBLIOGRAPHY

1. J. L. Sullivan, "Guarding Against Fatigue Failures in Press Fitted Shafts," *Machine Design,* June 9, 1977.

2. E. Kerekes, R. H. Krieble, R. Wittemann, and R. Nystrom, *A Comparison of Holding Power between Press Fitted and Retaining Compound Bonded Metal Assemblies,* ASME Paper 64-Prod-24, American Society of Mechanical Engineers, New York, 1964.

3. R. Thompson, *Improved Methods for Fastening Steel Parts in Aluminum Housings,* SAE Paper 790503 1979, Society of Automotive Engineers, Warrendale, PA, 1979.

Section 19
Rope Splicing and Tying*

edited by

ROBERT O. PARMLEY, P.E.

President and Consulting Engineer, Morgan & Parmley, Ltd., Ladysmith, Wisconsin

*The editor-in-chief wishes to thank the Columbian Rope Company for permission to use the following material, which was extracted from its publication, *Rope Knowledge for Riggers.* The text was edited slightly to conform to the style used in this handbook.

This section provides the basic essentials of rope splicing and tying techniques which pertain to three- and eight-strand rope.

19-1 Splice Strength The knotting process creates a shearing effect on rope fibers and lowers their resistance to strain.

In splicing, the fibers are not subject to this weakening effect. Thus, a good splice has up to 95 percent of the breaking strength of the original rope, while a typical knot may weaken the rope as much as 50 percent. See Table 19-1 for retained rope strengths for splices.

TABLE 19-1 Rope Strength Retained in Splice

Type of splicing	Strength retained for three-strand, %	Strength retained for eight-strand, %
Eye	100	100
Short	95–100	95–100
Long	85–90	70–80
Long blind	50–60	

SPLICING THREE-STRAND ROPE

19-2 Long Splice This splice is good for any pulley work except power transmission.

1. Unlay the end of each rope a minimum of 30 turns for Manila, 6 to 10 turns more for nylon. Lash securely with twine as shown in Fig. 19-1 to prevent ropes from coming apart further. Place ropes together, alternating the strands from each. Note how strands are numbered to show their relative positions throughout the long splice procedure.

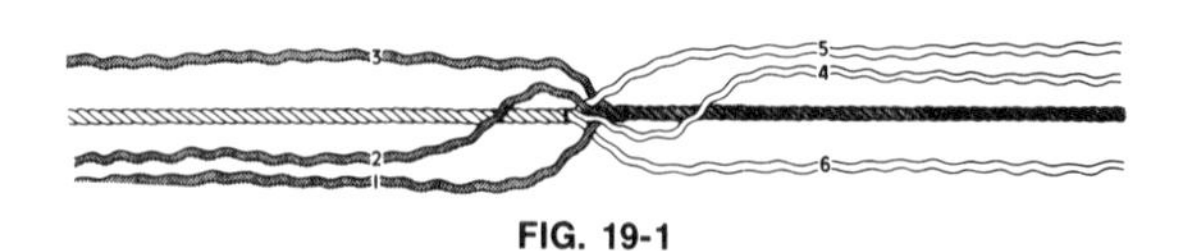

FIG. 19-1

2. Take lashing off one side. Unlay one strand (2) a minimum of 25 turns and replace it with a strand from the other side (5) as it is being unlaid. Lash securely as shown in Fig. 19-2 to hold strands in place. Be sure to place lashing at the "marriage point" to hold strands securely.

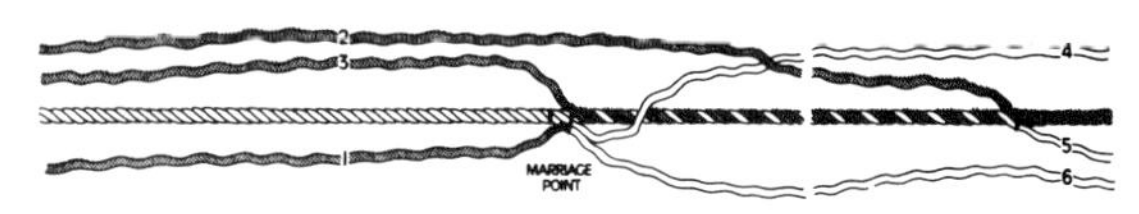

FIG. 19-2

3. Step 3 is like step 2, except in the opposite direction. One strand (6) is replaced with another (3). Each point is securely lashed. This leaves strands 1 and 4 at the "marriage point."

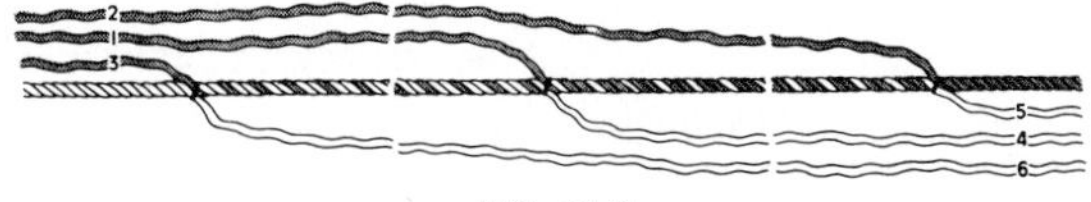

FIG. 19-3

4. Remove lashings and tie each pair of opposing strands—(2) and (5), (6) and (3), (1) and (4)—with an overhand knot. Be sure knot is tied in the direction of strand twist. See Fig. 19-4.

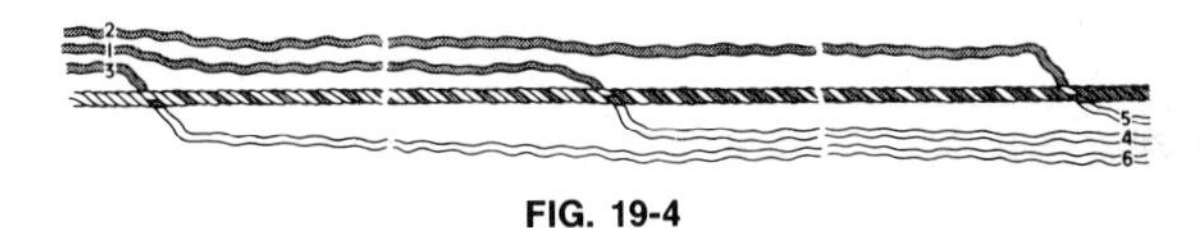

FIG. 19-4

5. Tuck each strand four times for synthetic fiber ropes, or three times for Manila. These tucks should be at right angles to the direction of the twist in the rope. See Fig. 19-5.

FIG. 19-5

6. The splice may be tapered by adding an additional ⅔- and ⅓-strand tuck at each strand junction. Now roll and pound well. Finally, cut the strands off close to the rope. See Fig. 19-6.

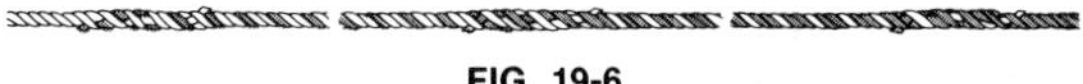

FIG. 19-6

19-3 Short Splice This method is very satisfactory when only a small amount of rope can be spared for making a splice or when the usage will permit an increase of about 50 percent in rope diameter.

1. Unlay ends of ropes to be spliced six or eight turns, depending on the lay of the rope. The softer the rope, the longer the splice should be. Lash as shown in Fig. 19-7 and whip ends of the strands to keep them from coming apart. Bring the two pieces of rope together so that each strand of one part alternates with a strand of the other.

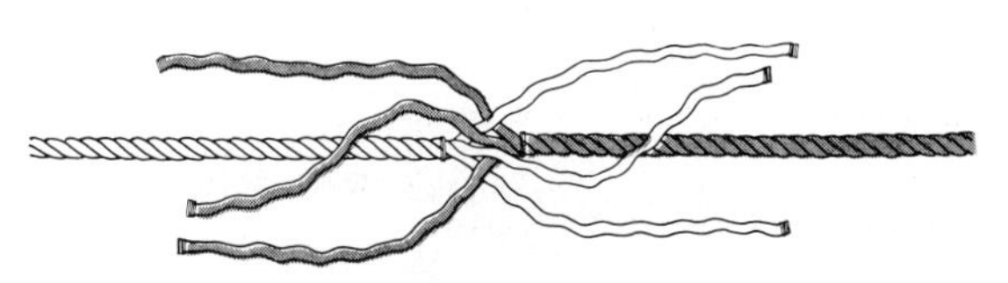

FIG. 19-7

2. Cut the lashing on left side and start tucking strands from the right side over and under next adjacent strands on the left side, as shown in Fig. 19-8.

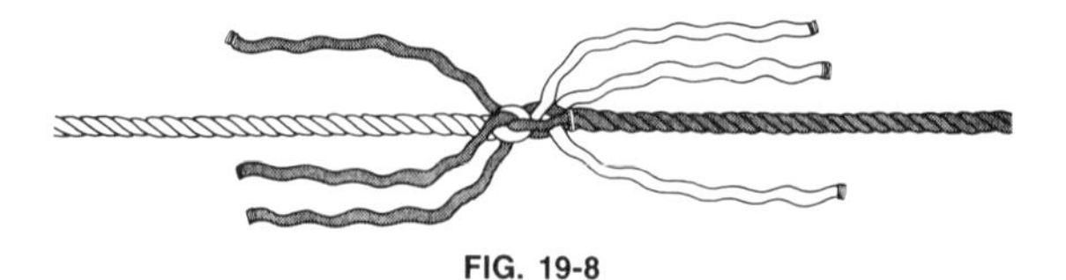

FIG. 19-8

3. Remove lashing from right side and begin tucking the strands from the left side by bringing a strand up over the nearest strand on the right side and down under the next. The tucking should be done at right angles to the direction of the twist in the rope. Pull all strands with a sharp yank on left and right. This tightens up the first tuck. See Fig. 19-9.

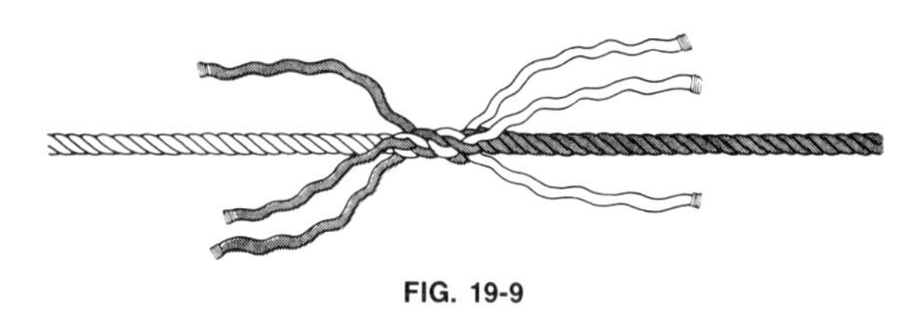

FIG. 19-9

4. Tuck each strand, from left and right sides, at least two more tucks for Manila and three more for synthetic rope. See Fig. 19-10.

FIG. 19-10

5. The splice may be tapered by adding one $2/3$- and one $1/3$-strand tuck. Now cut off all ends and roll the splice beneath your foot or between two boards to give a smooth appearance. See Fig. 19-11.

FIG. 19-11

19-4 Eye Splice or Side Splice The eye splice (sometimes called the "side splice") is used for forming an eye or loop in the end of a rope by splicing the end into the side.

Untwist the strands of the rope end four to six turns. Select as No. 1 the strand that is on top of the rope and in the middle between the other two strands. Raise a single strand on the top of the solid rope and pass strand 1 under this single strand diagonally to the right, as in Fig. 19-12.

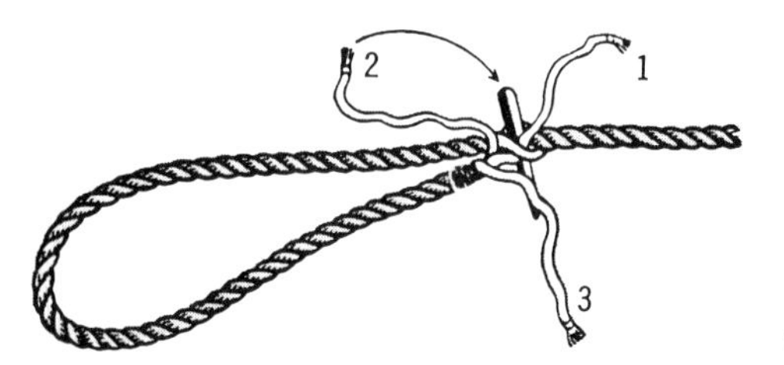

FIG. 19-12

Insert the marlinspike as shown in Fig. 19-12, so that it forces out from the main rope a single strand and so that the end of the marlinspike comes out where strand 1 went in. The marlinspike must *not* enter where strand 1 comes out. Tuck strand 2 so that it passes through the rope in the same direction as the marlinspike did.

Next, insert the marlinspike as shown in Fig. 19-13, starting it where strand 1 comes out and bringing it out where strand 2 goes in. Turn the two ropes over, remove the marlinspike and push strand 3 through. Pull this strand up snugly and the others also. It will now be seen that all three strands come out of the main rope in the same place and that each is separated from the others by a strand of the main rope. Proceed to splice the ends into the solid rope as shown in Figs. 19-14 to 19-17 inclusive. This is in the same manner as given for the short splice with the exception that now we have but three strands and need to work only in one direction. The splice may be finished by dividing each strand as in the short splice and rolling the finished splice under foot.

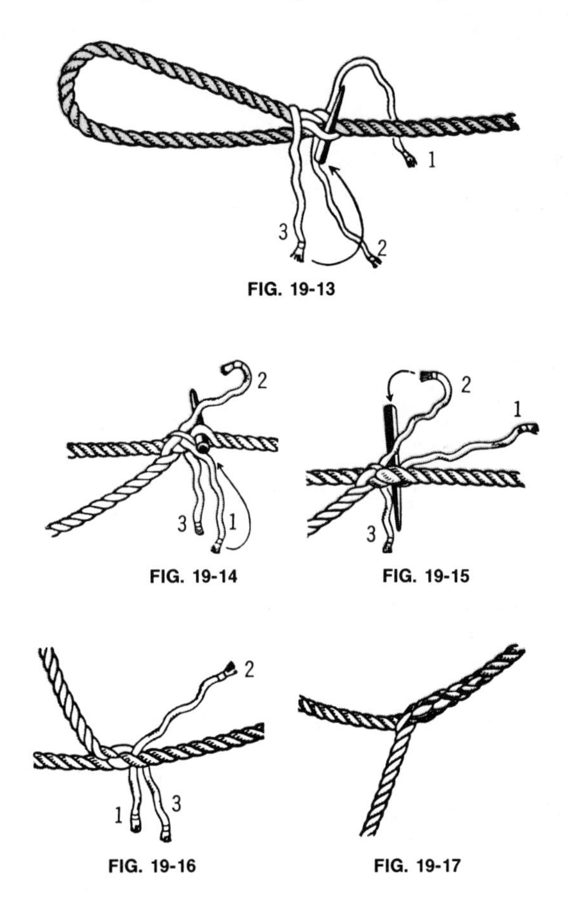

FIG. 19-13

FIG. 19-14

FIG. 19-15

FIG. 19-16

FIG. 19-17

19-5 Splicing Synthetic Fiber Rope Eye splices and short splices in nylon should be given an additional full tuck. When making a long splice in nylon, it is well to unlay one strand 6 to 10 additional turns (more than Manila) before locking the ropes together, to prevent slippage. When these ropes are cut, the ends should be seized to prevent fraying.

19-6 Splicing Ski-Tow Rope Use only the long blind splice, which does not increase rope diameter. The splice should be at least 15 feet long and can be adapted from the four-strand transmission rope splice shown in Fig. 19-7.

To eliminate yarn unraveling and to facilitate tucking and splice finishing, tape the ends of each strand as soon as the cut is made.

SPLICING FOUR-STRAND ROPE

19-7 Transmission Splice Here, we are working with a *four*-strand rope, with a heart. A 16-foot splice is about the proper length for a ¾-inch-diameter rope. Larger ropes should be given proportionately longer splices, up to 42 feet for a 2-inch-diameter rope. *This example* is for a 1¼-inch rope.

First, mark the rope to length, allowing 12 to 15 feet on each end. This will make a 20- to 26-foot splice. Unlay each end back to the markers, then twist the strands together (Fig. 19-18). Cut marker Y, unlay strand 1 5 feet, lay strand A in its place; unlay strand 3 two feet, lay strand C in its place; cut hearts just as they meet. Then, cut marker M, unlay strand D 5 feet, lay strand 4 in its place; unlay strand B two feet, lay strand 2 in its place; cut off all strands about 18 inches long for convenience in manipulation. The rope is now as in Fig 19-19. As the strands are laid in, great care should be used to maintain the original twist in each. Each pair of strands is now successively subjected to the following operations:

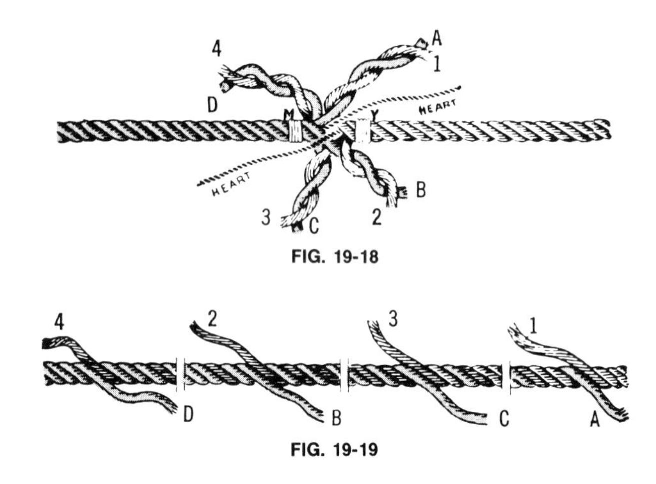

FIG. 19-18

FIG. 19-19

Take strands B and 2, for example: Unlay them each three turns, split and whip one end of each as B and 2^1 in Fig. 19-20. Lay back the split strands B^1 and 2^1 and tie in a simple knot (Fig. 19-21). With a fid or marlinspike raise B^1 and tuck 2^1 around it until it reaches B. Raise 2^1 and tuck B^1 around it until it reaches 2.

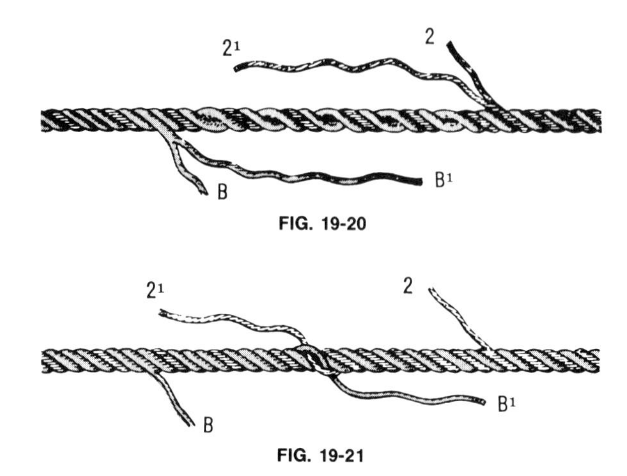

FIG. 19-20

FIG. 19-21

Split B and pass 2[1] through it and through center of rope; split 2 and pass B[1] through it and through center of rope. Do not put much twist into the part which is being tucked around; try to lay a smooth even course of about the same angle as the original strands. These half strands should not pass around more than four times before being drawn through the rope. Figure 19-22 shows B[1] raised with 2[1] being tucked around it, and end B[1] passed through 2 and through the rope ready to be drawn tight.

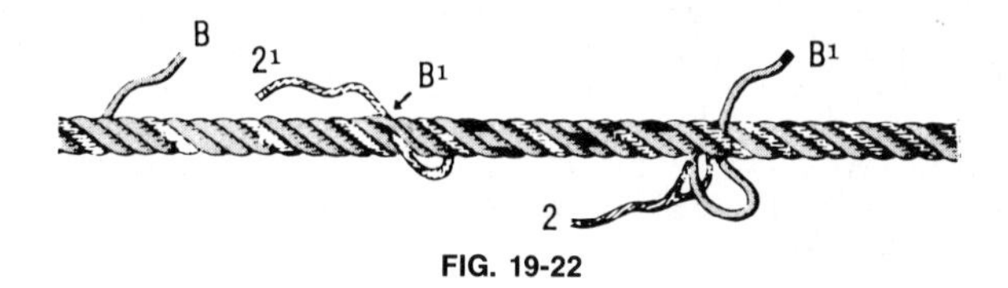

FIG. 19-22

This operation is called "tucking," or "locking the strands." If these tucks or locks are not made small and firm, they will wear more rapidly than the rest of the rope and the splice will fail. When the strand is split, preparatory to making this lock, it is common practice to make the parts used for tucking two or three yarns smaller. In this manner, the lock is made small and wear reduced.

Cut off ends of the strands, leaving 1 or 2 inches projecting, so that as the working tension is put upon the rope, the yarns may draw in somewhat without being loosened. Figure 19-23 shows the completed section of splice.

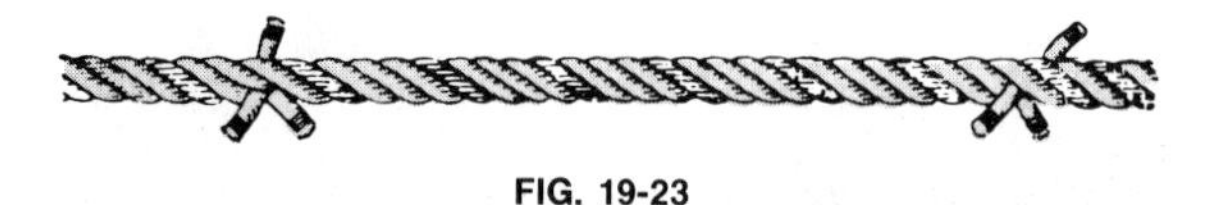

FIG. 19-23

SPLICING EIGHT-STRAND ROPE

You will need:

- A splicing fid
- A sharp knife
- A marking pen or colored chalk
- A roll of plastic or masking tape
- A supply of light, strong string

Plaited rope is made of eight strands grouped in four pairs; two pairs turn to the left and two pairs turn to the right. In the illustrations, for the sake of simplification, we have the two pairs turning to the left in white and the two pairs turning to the right in a dark tone. From here on, we refer to them as the white and dark pairs. Note that the dark pairs are diametrically opposite one another but at a 90° angle to the white pairs and vice versa.

19-8 Long Splice Follow these steps:

1. As shown in Fig. 19-24, lay the two rope ends to be spliced side by side on a flat surface. Taking one rope at a time, carefully determine the two pairs of strands going to the right. If your rope is all white in color, use the marking pen or chalk to clearly mark these two pairs of strands from the end, back along the rope for a distance of 30 pics (Fig. 19-24). From here on we will refer to these as the dark strands, and the remaining two strands (which appear to move to the left) as white.

2. Now make certain that a pair of dark strands are running along the top of each rope. Starting from the ends, count back to the ninth crown (or 9 full pics). Mark this point clearly all around the rope. Repeat this for three counts of six each and clearly mark. This will be the ¾, ½, ¼, and center marks as shown in Fig. 19-24.

3. At the center mark of end A (Fig. 19-24), *securely* tie the string around the rope, over the crown of the dark strands. With end B, tie the string between the center-mark

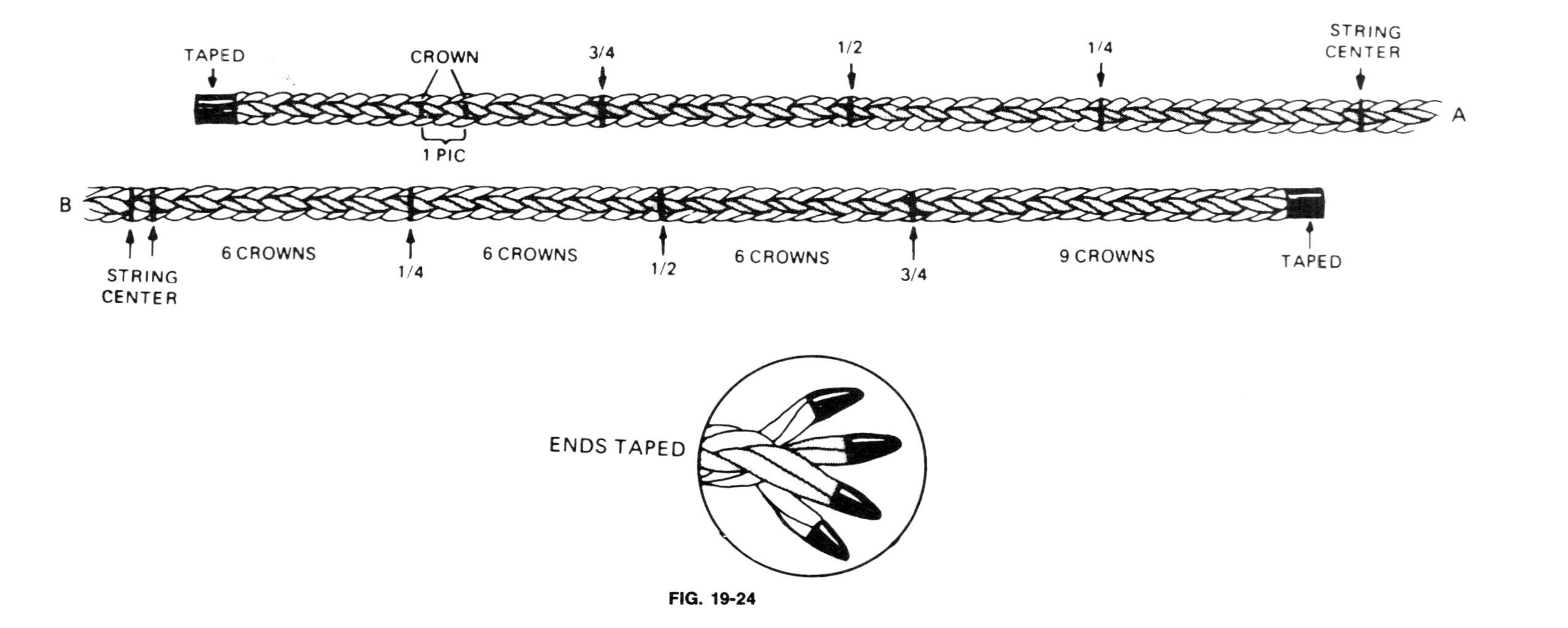
TAPED
CROWN
3/4
1/2
1/4
STRING CENTER
A
1 PIC
B
STRING CENTER
6 CROWNS
1/4
6 CROWNS
1/2
6 CROWNS
3/4
9 CROWNS
TAPED
ENDS TAPED

FIG. 19-24

dark strands and the next pair so that the string passes over the crown of the white strands.

4. Now carefully match your work with Fig. 19-24.

5. Now taking one end at a time, unlay the rope back a short distance. Taking each pair of strands, one at a time, tape them together at the end. Try to work this tape so that it is pointed or conical (see insert in Fig. 19-24); this will help when the splice begins. Now that all four pairs of strands of each rope have been taped together, carefully unlay the ropes back to the strings. Position the strands as shown in Fig. 25. You should now be ready to start your splice.

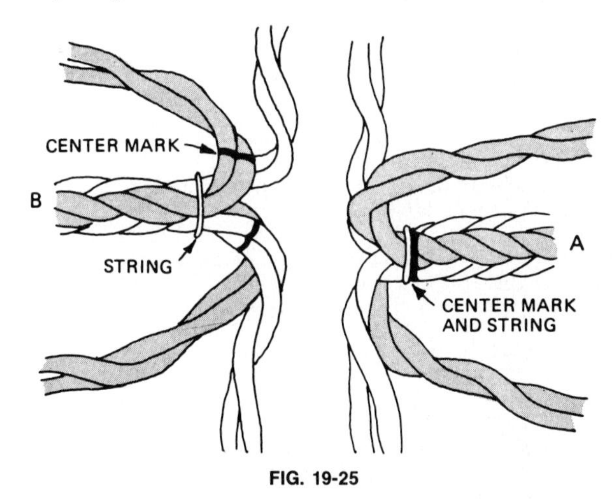

FIG. 19-25

6. Marry the bottom dark pairs, by inserting the pair from the right between the two strands of the pair from the left at the center mark. *Do not pull tight.*

7. Now marry the white strands, on the side away from you, in exactly the same manner. The two white pairs, on the side toward you, should be reversed in that the pair from the left should be inserted between the two strands of the pair from the right at the center mark.

8. To complete this initial step marry the top pair of dark strands by inserting the pair from the left through the pair from the right at the center mark.

9. Now carefully check your work against Fig. 19-26. If it does not conform, recheck the last three steps. Now cut and remove both strings. Taking four pairs of strands in each hand, pull the marriages up tight so that all of the center marks are together.

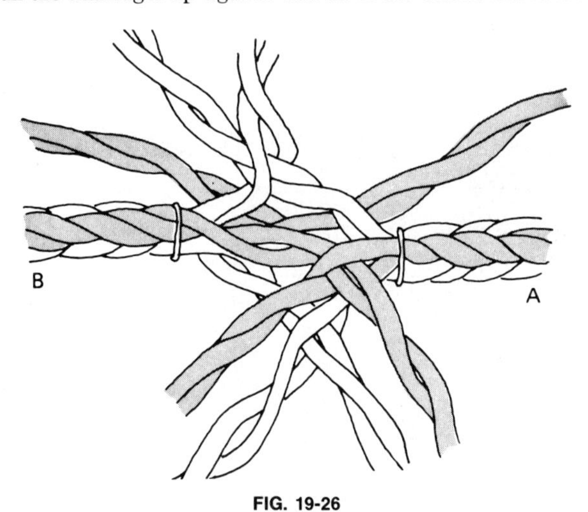

FIG. 19-26

10. Now tie off each of the four marriages individually as pictured in Fig. 19-27. This will take some care to keep them from loosening. The center marks must stay together for a successful splice.

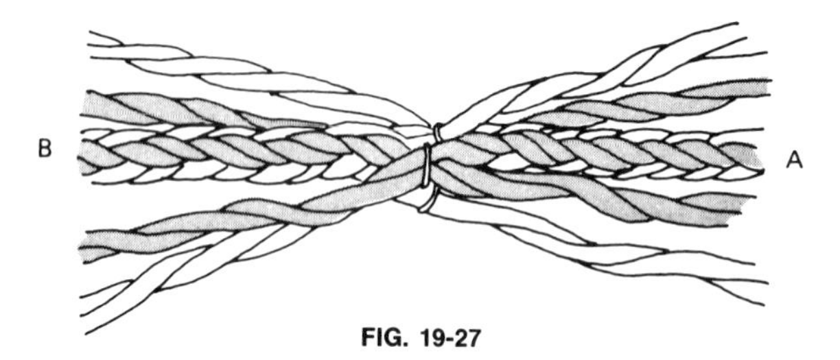

FIG. 19-27

11. Start splicing with the two top (dark) pairs of strands. First cut off the outside pair at the marriage (Fig. 19-28). This should be the pair coming from the right. Now cut the string. Pull the cut ends back from under the white pair. Insert the uncut dark strands (that are coming from the left) in their place. *Make certain now,* and throughout the remainder of the splice, *that the inserted strands are layed in parallel and not twisted on one another.* (See Fig. 19-29.) Continue removing the cut dark strands 1 pic at a time and inserting the dark pair from the left to the ¾ mark on the *strands being inserted.*

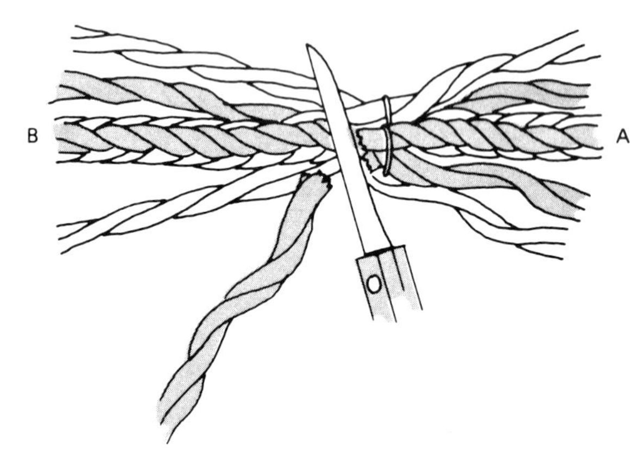

FIG. 19-28

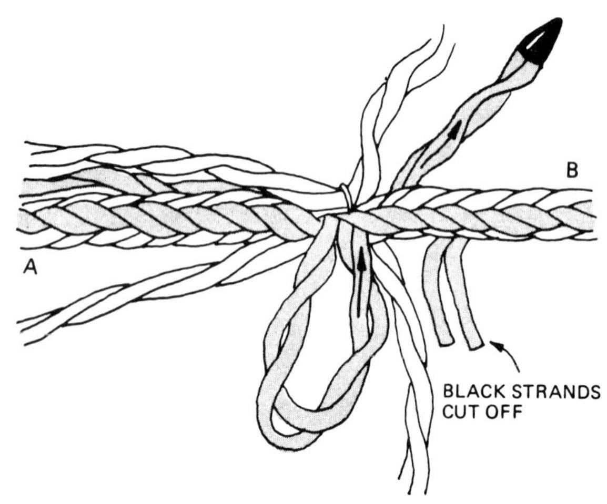

FIG. 19-29

12. Having reached the ¾ mark, cut the tape holding the two strands from the left. Split the pair into two separate strands. Choose one and its exact counterpart from the right. Remove the single strand from the *right* from under the white strands in the same manner as you previously did with the pair. Insert the single strand from the left in its place. Continue removing and inserting for a distance of 6 more pics (Fig. 19-30).

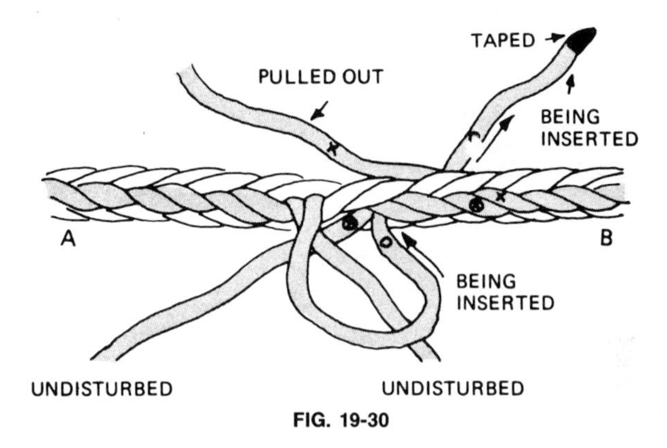

FIG. 19-30

13. Now return to the center marriage point. Choose the white strand marriage on the side away from you. This time cut off the two strands coming from the left. Cut the string. Then moving in the opposite direction duplicate your actions in step 11 to the ¾ mark and step 12 for 6 more pics.

14. From here on it is relatively simple. Return to the center marriage and repeat the above procedure with the remaining pairs of white strands, working back in the original direction as in step 11. *However,* this time go only to the ¼ mark with two strands and to the ½ mark with the single strand.

15. With the remaining pair of dark strands, work to the left to the ¼ mark with two strands and to the ½ mark with a single strand. Your splice should now look like Fig. 19-31.

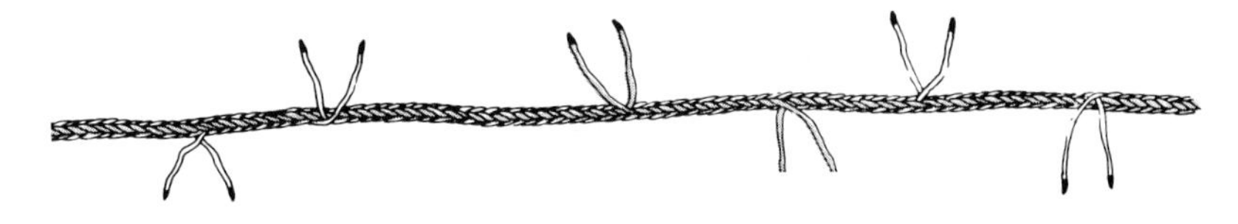

FIG. 19-31

16. The ends now remaining should be cut off about 4 pics long. Then tape the ends in a conical manner as shown in Fig. 19-24. Working in the direction that each has been spliced to this point, tuck each end, in turn, up the center axis off the rope for a distance of about 3 pics (see Fig. 19-32). Whatever short ends remain should now be cut off flush. Your finished splice should now look like Fig. 19-33.

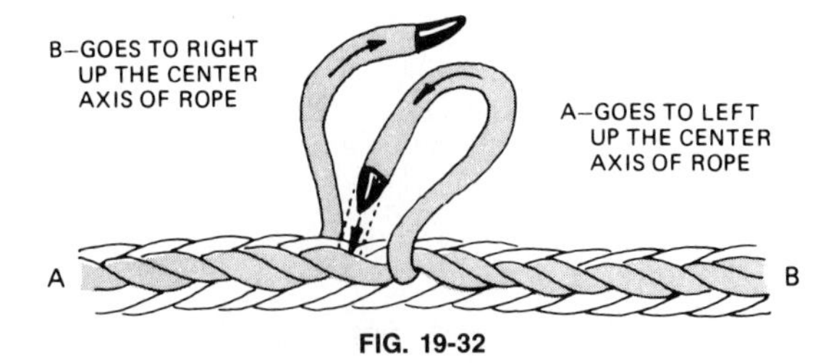

FIG. 19-32

FIG. 19-33

19-9 Short Splice Follow these steps:

1. Lay the two rope ends that are to be spliced side by side on a flat surface. Tape the ends as shown in Fig. 19-34.

FIG. 19-34

2. Determine which pairs of strands go to the right and which go to the left. If the rope which you are splicing is all one color, mark all the pairs which are going to the right with your marker, from the ends back along the rope for a distance of 11 to 12 pics (Fig. 19-34). From here on we will refer to these pairs as dark (going to right) and light (going to left).
3. Count back from the ends of the rope 10 crowns (Fig. 13-34). At this point mark around the entire rope and tie a string securely around the ropes at this point so that it passes directly over the pairs of dark strands on one rope and over the light strands of the second (Fig. 13-34).
4. Unlay the four pairs of strands of both ropes back for a distance of 10 pics and tape the ends, in pairs, as shown in Fig. 19-35. Now, complete the unlaying of the rope back to the strings (Fig. 19-35).

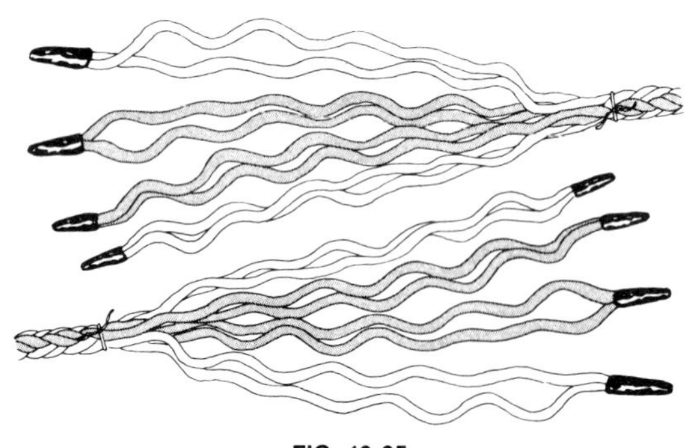

FIG. 19-35

5. Your work should now appear as shown in Fig. 19-35. The undisturbed portion of the ropes above the strings should be laying with the dark strands on top. Throughout the splice these pairs should run parallel and not be allowed to twist over one another.
6. Start to marry the two ropes at the strings in the following manner (Fig. 19-36):

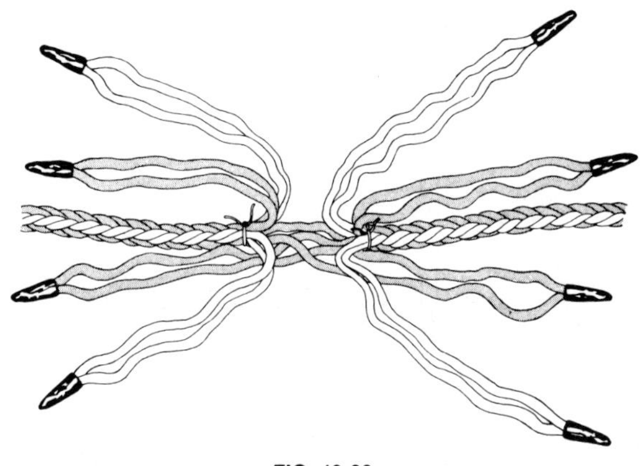

FIG. 19-36

a. Bend the two top (dark) pairs back along the rope.

b. Spread the white strands out at 45° angles to the axis of the rope.

c. Marry the bottom (dark) pairs by inserting the pair from the right between the pair from the left, at the strings.

d. Now marry the white pairs. The right-hand pair between the two strands from the left on one side and the left-hand pair between the strands from the right on the other.

e. Now marry the top dark strands. The pair from the left between the pair from the right. Your splice should now look like Fig. 19-37.

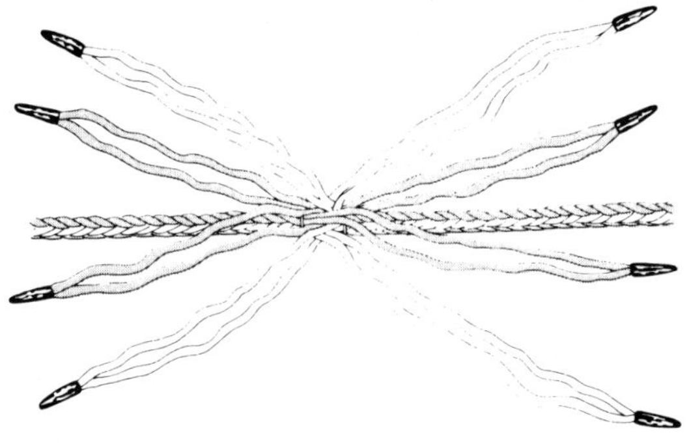

FIG. 19-37

7. Cut off both strings. Using both hands, first grasp the four pairs on the right hand and then the four pairs on the left. Now pull the marriage up snug and tie a string securely around the entire marriage point as illustrated in Fig. 19-38. Cut this string only when it is hopelessly in the way during the next step.

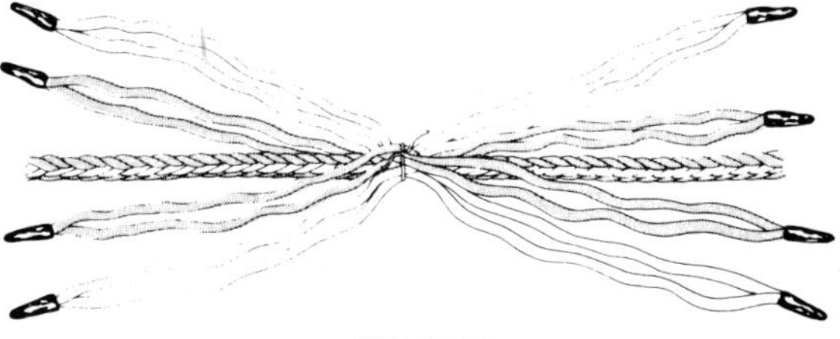

FIG. 19-38

8. Using the fid to make space, start splicing by inserting the white pairs from the left under the dark pairs from the right. Next the white pairs from the right are inserted under the dark pairs from the left. Now complete this first tuck by following the same sequence, as above, with the four pairs of dark strands going under the white strands on the opposite side of the marriage. Your splice should look like Fig. 19-39.

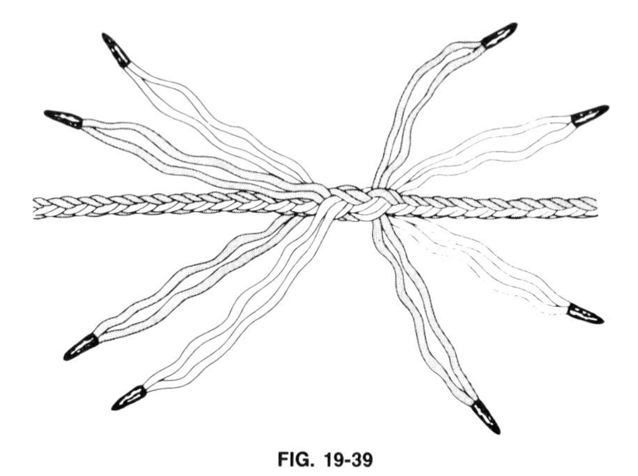

FIG. 19-39

9. Continue splicing pairs as in step 8 for two (2) additional complete tucks. Your work should now look like Fig. 19-40.

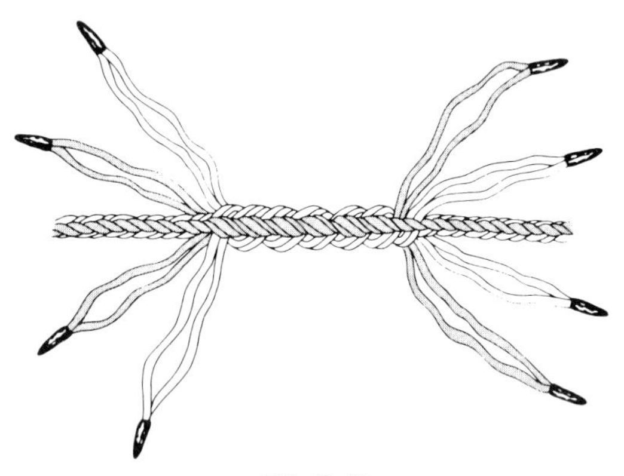

FIG. 19-40

10. Now split the pairs and, using only one (1) strand of each pair, make two additional tucks, as illustrated in Fig. 19-41.

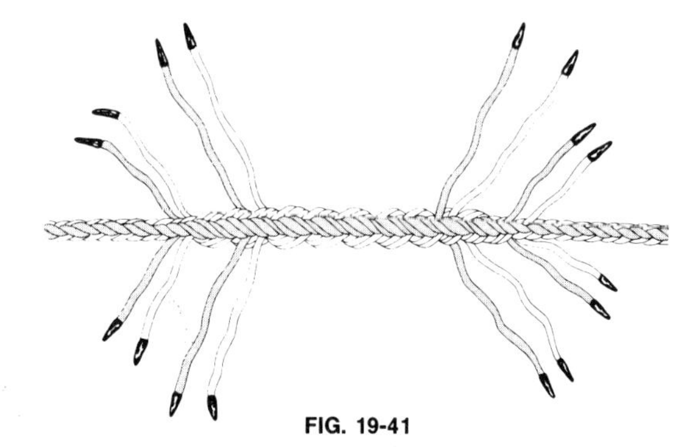

FIG. 19-41

11. Tape and cut the ends off about 2 pics long. Now complete the taping as shown in Fig. 19-42.

FIG. 19-42

This is your finished splice. It should never be used when the spliced work must pass through a cleat or over a sheave. No matter how poor your first attempt may be, you will find that it has a high degree of efficiency in this splice (90 to 95 percent).

19-10 Temporary Short Splice Follow these steps:

1. Lay the two ends of the ropes to be spliced side by side in opposite directions.
2. Starting at a point about 12 pics from the end of the first rope and using a fid or other pointed tool, separate the strands so that there is an opening formed by one pair of right turning strands and one pair of left turning strands on either side of the opening. Tuck the end of the second rope through this opening and pull through about 12 pics.
3. Count down 3 pics and repeat the tuck in the opposite direction. Continue for a total of four tucks.
4. Now start with the end of the first rope and tuck this into standing part of the second rope for four tucks. These first four steps are shown in Fig. 19-43.

FIG. 19-43

5. Pull these loops down until your splice appears as in Fig. 19-44.

FIG. 19-44

6. Cut off excess and tape the ends to the standing parts.

19-11 Eye Splice Preparation: If the rope which you are about to splice is all of one color, it will be simpler if you mark those pairs which turn to the right so that they will conform with the dark pairs in the illustrations. Count back a distance of about 10 pics (see Fig. 19-45) from the end, and tie a string *securely* around the rope so it passes directly over the center of both pairs of dark strands. Place the knot so that it is directly on top of one of these pairs. It is important that this be tied securely to prevent slipping. Now, unlay the pairs of strands back to the string. Making sure not to mix or twist them, tape the ends of the pairs together as seen in Fig. 19-45.

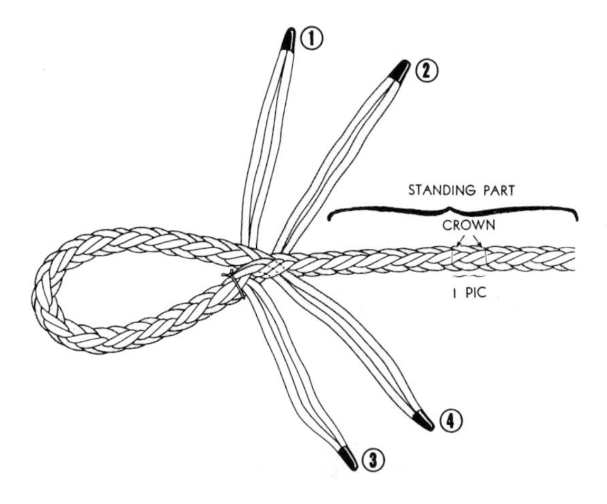

FIG. 19-45

So far, so good—you are now ready to start the splice.

1. Hold or lay the rope so that the pairs of white strands are on top and bottom with a knot to the right as you look toward the end.

2. Bend the rope over to the desired eye in such a way as to keep the knot inside the loop, as shown in Fig. 19-45.

3. Using the fid to make clearance and starting with the dark pairs, tuck them under the diametrically opposite white pairs as shown in Fig. 19-45. *Make sure that you do not disturb the lay of the pairs.* Do not twist them so that the individual strands cross over one another in the pair.

4. Now turn the eye over, and tuck the white pairs under the diametrically opposite dark pairs as shown in Fig. 19-46. (Note that in Fig. 19-46 the splice is turned over from Fig. 19-45.) The white pairs to be tucked should follow the white pairs of the standing part, and the black to be tucked should follow the black pairs of the standing part. The ends in the drawing have been numbered to help you follow their progress.

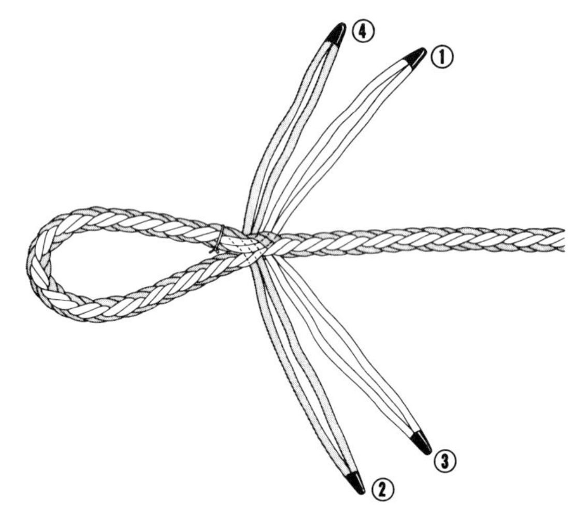

FIG. 19-46

5. Now, you have your eye with the first full tuck complete; pull all four ends down firmly. Starting with the dark pairs, take another full tuck (a full tuck means inserting all four pairs; by starting with dark pairs, you avoid having to go under two pairs at once). Your splice should now look like Fig. 19-47, which now lays on the same side as Fig. 19-45. From here on, you should have no difficulty completing the splice.

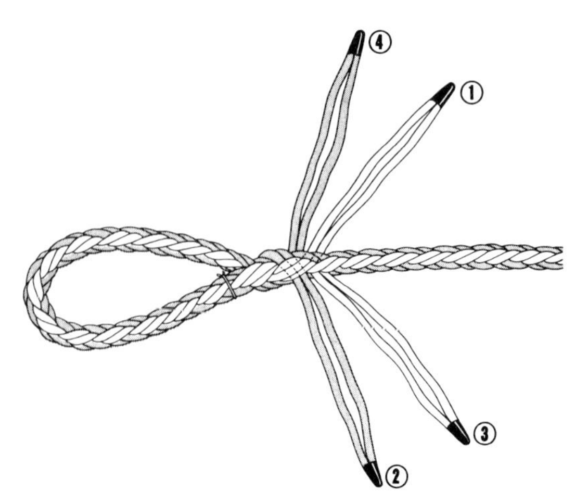

FIG. 19-47

6. Now starting with the dark pairs, take at least one more full tuck. With a very soft rope, it may be necessary to take a fourth or fifth full tuck.

7. Having completed the third full tuck (or fourth or fifth if necessary), select the strand closest to the eye in each pair. Tape this strand close to where it emerges from the tuck and then cut off as shown in Fig. 19-48.

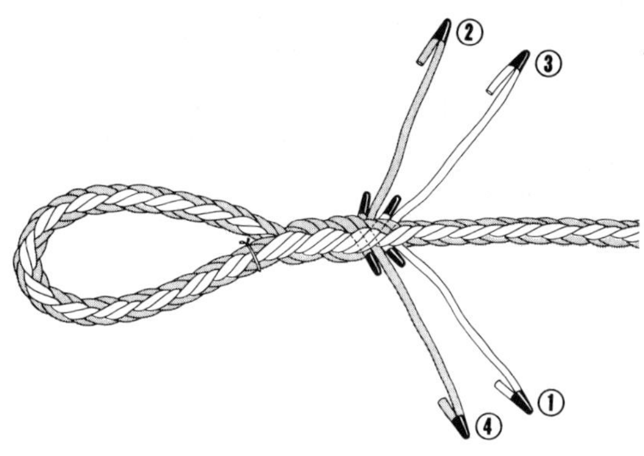

FIG. 19-48

8. Now, splice your remaining single strands just as before for another full tuck. Your splice should now appear as Fig. 19-49, which you will note lays on the side opposite that shown in Fig. 19-50.

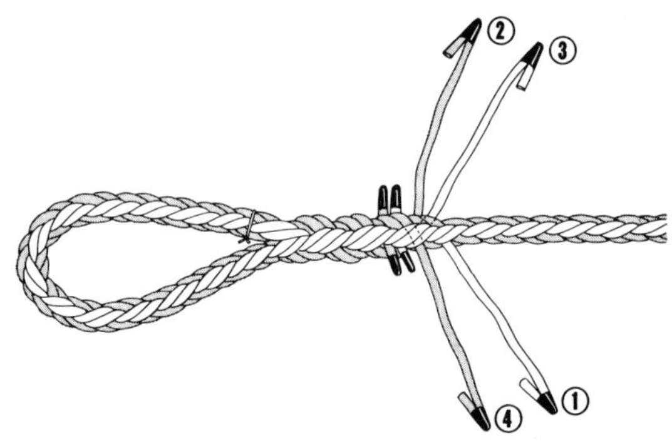

FIG. 19-49

FIG. 19-50

9. Tape first and then cut off the four single strands, as shown in Fig. 19-50. The eight ends may be heated and fused so they will not fray; however, take great *caution* to be certain that you fuse only the ends and do not damage the strands.

A more professional appearance may be achieved by cutting the ends off flush and then taping or whipping the entire splice.

19-12 Temporary Eye Splice It is possible to make a quick splice which will be every bit as efficient as the regular eye splice as depicted in Sec. 19-11. This splice is not designed for permanent use, although under certain circumstances, it would probably give long and lasting utility.

To construct this quick splice, perform the following steps:

1. Determine the size of the eye desired and form a loop as shown in Fig. 19-51. Be sure to allow 10 to 12 inches of excess to be spliced.
2. Using a fid or other pointed tool, separate the strands so that there is an opening formed by one pair of right turning strands and one pair of left turning strands on either side of the opening. Tuck the excess rope to the point where you have formed the size of eye desired.
3. Counting down 3 pics for the top pair of strands, repeat the process in the opposite direction from step 2.
4. Repeat this two more times, as shown in Fig. 19-51.
5. Pull these loops down until your splice appears as Fig. 19-52.

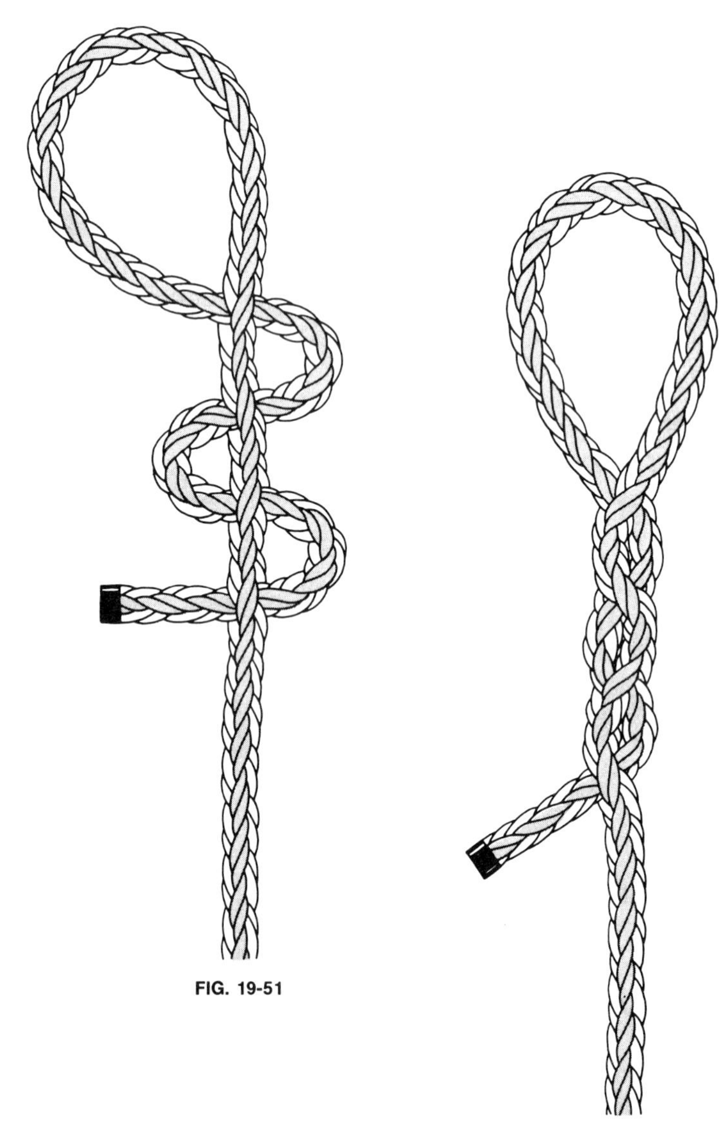

FIG. 19-51

FIG. 19-52

6. Cut off the excess to a distance no closer than 3 inches from the final tuck.

While we consider this type of splice to be temporary in nature, it will afford you a good splice; good enough to be used under most circumstances as a temporary measure.

KNOTS, BENDS, AND HITCHES

In knotting, a rope has three parts used in handling and tying it:

1. The "end" is the end of the rope with which you are working when you tie a knot. See Fig. 19-53.

FIG. 19-53

2. The "standing part" is the inactive length of the rope. See Fig. 19-53.
3. The "bight" is the central part of the rope between the working end and the standing part. See Fig. 19-53.

An "overhand loop" is made by crossing the end *over* the standing part. See Fig. 19-54.

FIG. 19-54

An "underhand loop" is made by crossing the end *under* the standing part. See Fig. 19-55.

FIG. 19-55

A "turn" is taken by looping the rope around an object—often another section of itself. See Fig. 19-56.

FIG. 19-56

A "round turn" is taken by looping the rope *twice* around an object, as shown in Fig. 19-57.

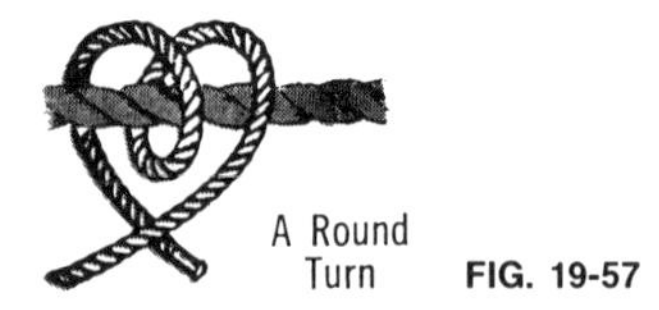

FIG. 19-57

"Over-and-Under" Sequence In tying a knot, whenever two sections of the rope cross each other, one must go *over* and the other *under.* Be careful to follow this "over-and-under" arrangement exactly—otherwise you get either an entirely different knot or no knot at all.

"Drawing Up" Once formed, a knot must be "drawn up" or tightened, *slowly and evenly* to make sure that all sections of the knot arrangement keep their place and their shape. Quick or careless tightening may result in a useless tangle.

19-13 Weakening Effect of Knots The rigger should bear in mind that the sharp bends necessary to form a knot weaken the fibers of the rope, particularly fibers on the outside of the bend which are apt to strain and break. When they break, the entire strain is thrown on the other fibers, which in turn break, and before long the entire rope has parted. Therefore, you can readily see that the knot which requires the least abrupt bends will weaken the rope least. Use knots only where necessary to make a temporary fastening. If a rope breaks, do not knot the ends together; cut out the bad section and splice the rope into one piece.

The percentages of efficiency shown in Table 19-2 are average values and will vary under various conditions of test and stress. However, they serve as a standard to base your calculations upon, and show how various knots reduce the actual strain that may be safely applied to any rope.

TABLE 19-2

Type of knot, bend, or hitch	Percent of retained strength
Anchor bend	
Over ⅝-in.-diam.	55–65
Over 4-in.-diam.	80–90
Two half-hitches	
Over ⅝-in.-diam.	60–70*
Over 4-in.-diam.	65–75*
Square knot	43–47**
Sheet bend	48–58*
Fisherman's knot	50–58
Carrick bend	55–60
Bowline	65–75

*Smaller sizes of nylon are liable to slip without breaking.

**Both nylon and P/D combination ropes in smaller sizes are liable to slip.

19-14 Whipping A good rope deserves good care. And in order to prevent unraveling in the rope, the ends always should be bound or "whipped." In splicing, the separate strands of the rope may also be whipped for convenient handling while splicing.

To make the whipping, a fine yarn, marlin or spun-yarn is generally used. The most common whipping is made by placing the end of the yarn along the end of the rope, and then laying a loop along the rope (see Fig. 19-58). You then wind the yarn tightly around both the loop and the rope, being sure to leave a small portion of the loop uncovered. Wind the whipping for a distance roughly equal to the diameter of the rope.

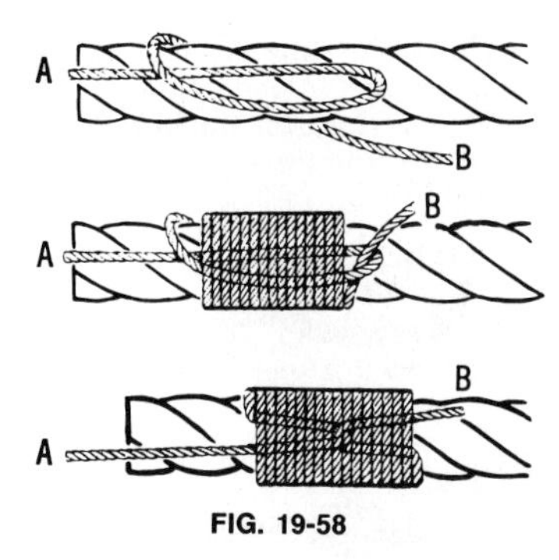

FIG. 19-58

The whipping is finished by putting the winding end B through the uncovered end of the loop—then pulling end A tight, until the loop is drawn back out of sight. Both ends of the whipping are then cut short for a neat finish.

Synthetic rope ends can be secured by wrapping with plastic tape or holding them in a flame.

19-15 Crown Knot and Slip Knot A crown knot is a quick way to prevent the unlaying of a three-strand rope. Use it until you have time to whip.

The slip knot is the most simple slip loop to tie. Starting with the position shown in Fig. 19-60*a*, the end is held in the left hand and the loop is formed by twirling the rope to the right between the thumb and the fingers of the right hand. Note how the loop is in the right hand and the end in the left in Fig. 19-60*b*.

FIG. 19-59

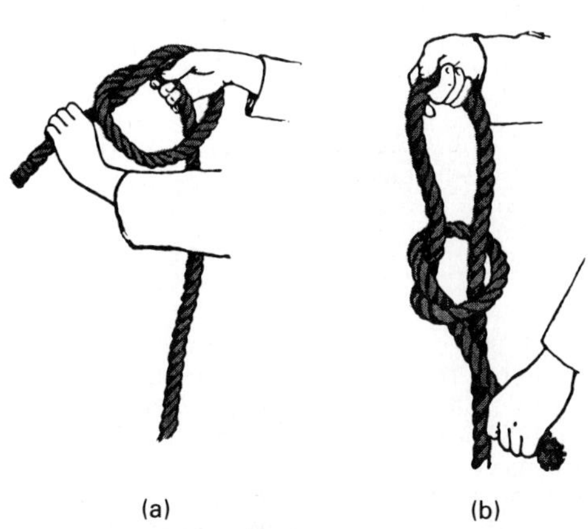

(a) (b)

FIG. 19-60

19-16 Overhand Knot This is the simplest and smallest of all knot forms and the beginning of many more difficult ones. In general, use it only on small cord and twine, since it jams and is hard to untie, often injuring the fiber. To tie: Make an overhand loop. Pass the end *under* and up *through* the loop. Draw up tight. See Fig. 19-61.

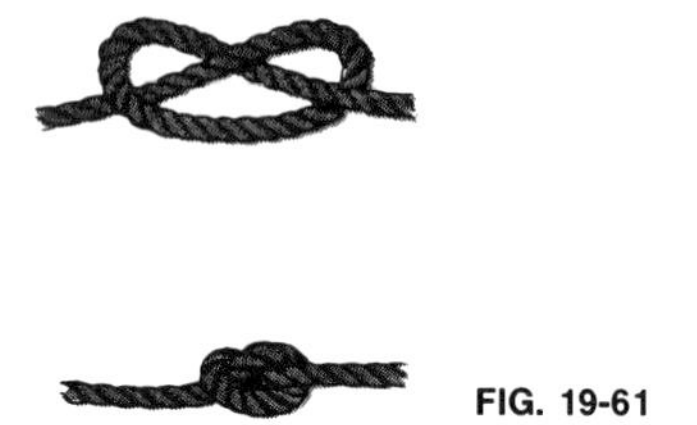

FIG. 19-61

19-17 Figure Eight Knot This is much easier to untie than the overhand knot—is larger, stronger and does not injure rope fibers. It is the best knot to use to keep the end of a rope or "fall" from running out of a tackle or pulley. To tie: Make an underhand loop. Bring the end around and *over* the standing part. Pass the end *under*, and then up *through* the loop. Draw up tight. See Fig. 19-62.

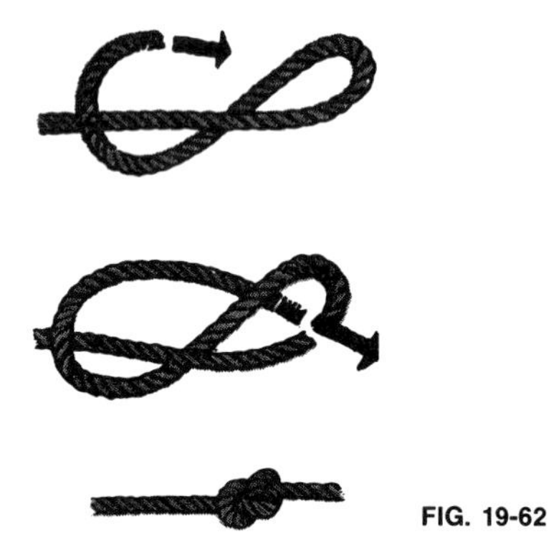

FIG. 19-62

19-18 Stevedore's Knot The stevedore's knot is tied the same as the figure eight knot except that two turns are taken around the rope instead of one. By inserting a small stick or shackle, the knot can be easily untied. See Fig. 19-63.

FIG. 19-63

19-19 Surgeon's Knot This is often used for twine—chiefly to keep the first tie from slipping before the knot is completed. To tie: With one end, take three turns about the other end. Bring both ends up. Pass one end over and under the other end. Draw up tight. See Fig. 19-64.

FIG. 19-64

19-20 Square or Reef Knot This is used at sea in reefing and furling sails, and ashore as the universal package knot for parcels and bundles. Though often used, it is a dangerous knot for tying two ropes together, since it unties easily when either free end is jerked. To tie: Pass the left end *over* and *under* the right end. Curve what is now the left end toward the right. Cross what is now the right end *over* and *under* the left. Draw up tight. See Fig. 19-65.

FIG. 19-65

19-21 Granny Knot Remember that the square knot presents two ends lying together *under* one loop and *over* the opposite loop—while the granny presents one end under and one over on *both* loops. *Watch out for the thief knot,* too. This has the two loose ends coming out of *opposite* sides—instead of from the same side as in the true square knot. See Fig. 19-66.

FIG. 19-66

19-22 Bowline This is used for mooring, hitching, lifting, and joining. Sometimes called the "king of knots," the bowline never jams or slips if properly tied. Generally tied in the hand, it can also be used as a hitch and tied directly around the object.

To tie: Make an overhand loop with the end held toward you. Pass the end up through the loop, then up behind the standing part—then down through the loop again. Draw up tight. See Fig. 19-67.

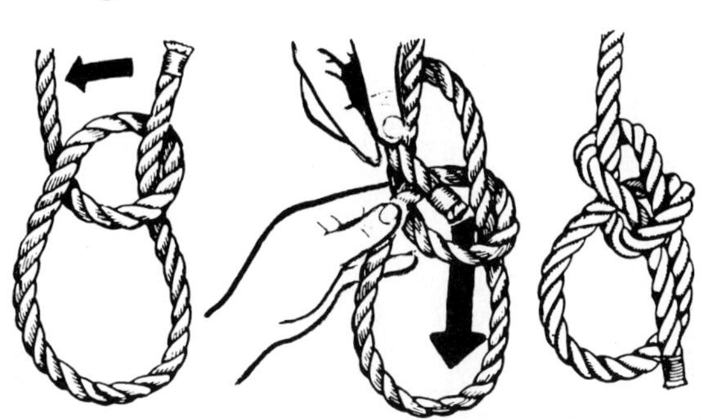

FIG. 19-67

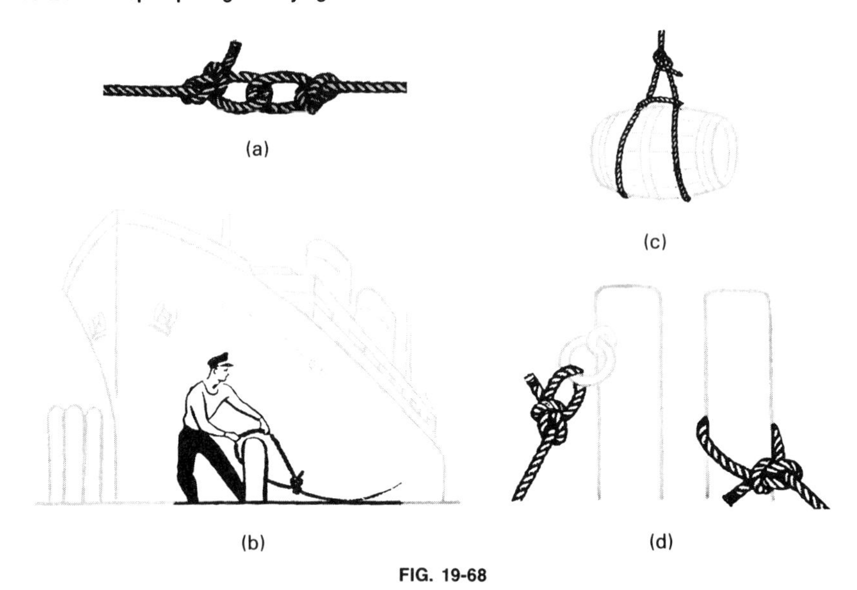

FIG. 19-68

Some uses of the bowline include: Two interlocking bowlines can be used to join two ropes together (Fig. 19-68*a*); the bowline is tied in a hawser and thrown over a post when mooring (Fig. 19-68*b*); a bowline is tied in the end of a rope for hoisting (Fig. 19-68*c*); a bowline can be tied as a hitch directly around a post or a ring when mooring a boat or hitching a horse (Fig. 19-68*d*).

19-23 Double Bowline To tie: Make an overhand loop with the end held toward you, exactly as in the ordinary bowline. The difference is that you pass the end through the loop *twice*—making *two* lower loops, A and B (see Fig. 19-69).

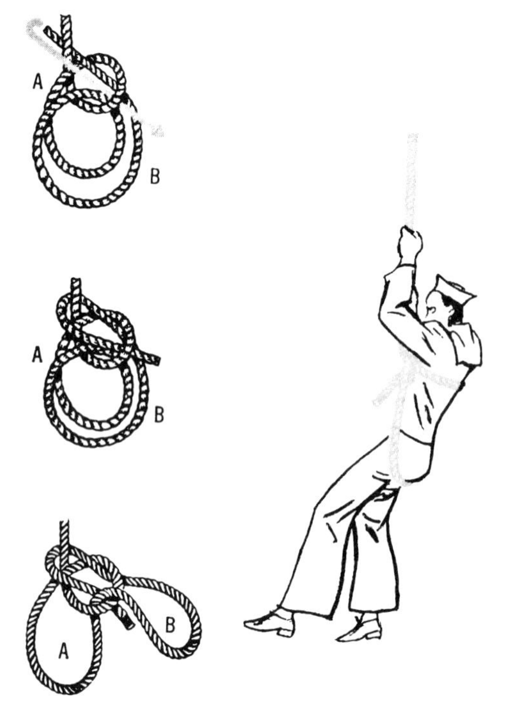

FIG. 19-69

The end is then passed *behind* the standing part and down through the first loop again as in the ordinary bowline. Pull tight. Outside loop B goes under the arms—inside loop A forms the seat.

19-24 Bowline on Bight To tie: Double the rope, make overhand loop C, and draw loop-end D up through it (see Fig. 19-70). Then pass loop end D toward you, down and over loop section A. Bring up in back until D lies behind loop A. Draw loop D tight by a slow even pull on the upper right side of loop A. To use, one leg is put through each loop and loop D passed under the arms before being drawn tight.

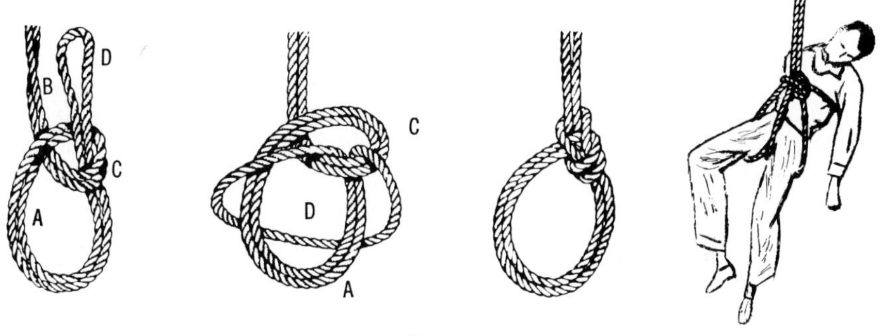

FIG. 19-70

19-25 Spanish Bowline Form three loops, as shown in Fig. 19-71, in any central section of the rope. Turn the large center loop A-B down. Enlarge it so that it encircles the smaller loops C and D. Put your hands through each of the small loops C and D, grasp each side of large loop A and B . . . and pull up through to complete the knot.

This is used to lift an injured person, like the bowline or bight, above, or to sling a ladder as shown in Fig. 19-71.

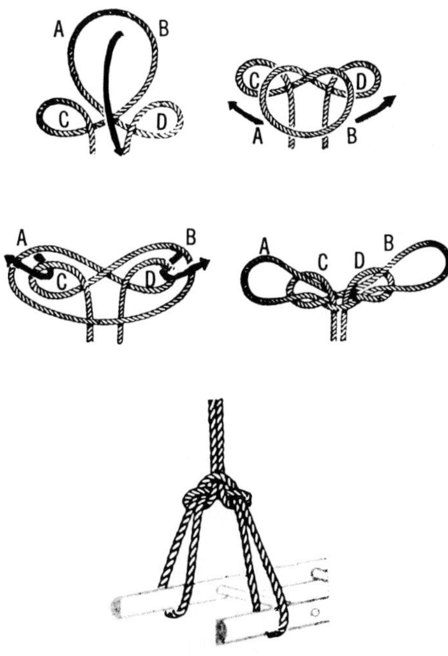

FIG. 19-71

19-26 Man Harness This is used for hauling and mountain climbing. The knot should be tied large enough to go around a person's shoulder, leaving both hands free. It is used in tow and climbing ropes as sketched in Fig. 19-72. To tie: Make a loop in the rope and fold it forward and slightly to the right to get a loop shaped like the one in the first sketch. Then take C up and under A and over B, as shown in the second sketch. A good hard yank on C, to the left, will finish the knot (third sketch).

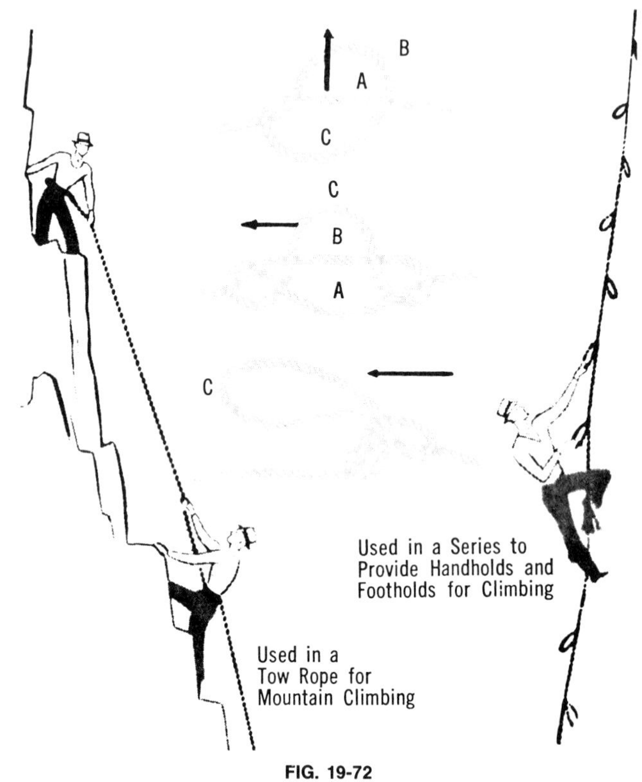

FIG. 19-72

19-27 Half-Hitch A basic knot form, the half-hitch is generally used for fastening to an object for a right-angle pull. To tie: Pass the end of the rope around the object and tie an overhand knot to the standing part. Figure 19-73 shows the half-hitch tied with the end pulled close around the standing part. This is the first step in tying many more complicated hitches, such as the thoroughly reliable timber hitch shown in Fig. 19-86. But for use by itself, it is unsafe tied in this way, since it quickly slips untied. Figure 19-74 shows the half-hitch tied with the end nipped under the turn of the rope some distance *away* from the standing part. This method is fairly reliable for temporary use if the pull is steady and the arrangement is not disturbed.

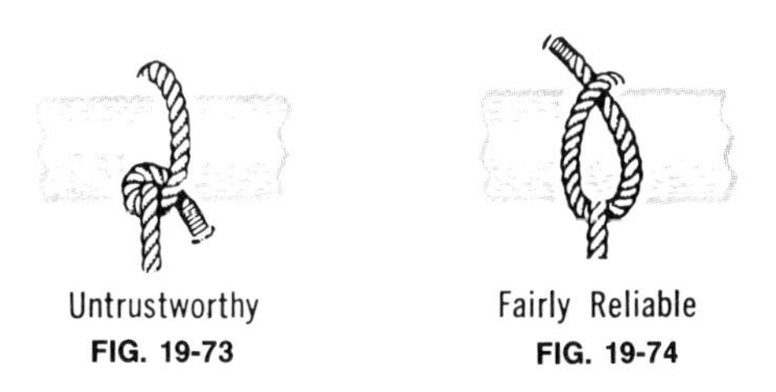

FIG. 19-73 FIG. 19-74

19-28 Two Half-Hitches Quite secure and used for mooring, just as its name states, two half-hitches is simply a half-hitch tied twice, in the manner shown in Fig. 19-75.

For extra security use two round turns in mooring. It is quickly tied and reliable, and can be put to almost any general use. It is the usual method of tying a line to a ring hook in handline and pole fishing.

FIG. 19-75

19-29 Boatswain's Hitch This is a simple hitch used by most workers who have to go aloft—riggers, painters, steeplejacks, etc. To tie, a bight is pulled forward under the eye support for the chair, ladder, or plank. Then a half-turn forms the single hitch required on the hook of the block. See Fig. 19-76.

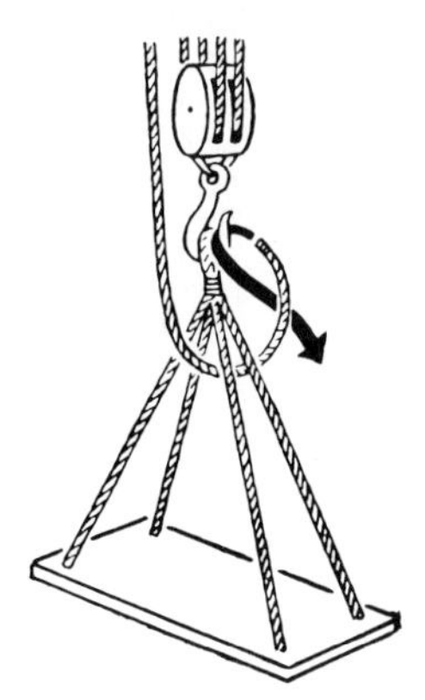

FIG. 19-76

19-30 Midshipman's Hitch To tie: Take a half-hitch around the standing part. Then, take an extra turn around that half-hitch, so that it is wedged inside it when tightened, coming up through the loop again (Fig. 19-77). Finish with a half-hitch around the standing part above the loop (Fig. 19-78). Tied this way, the midshipman's hitch is adjustable—it can be slid to any position on the standing part where it will hold under strain.

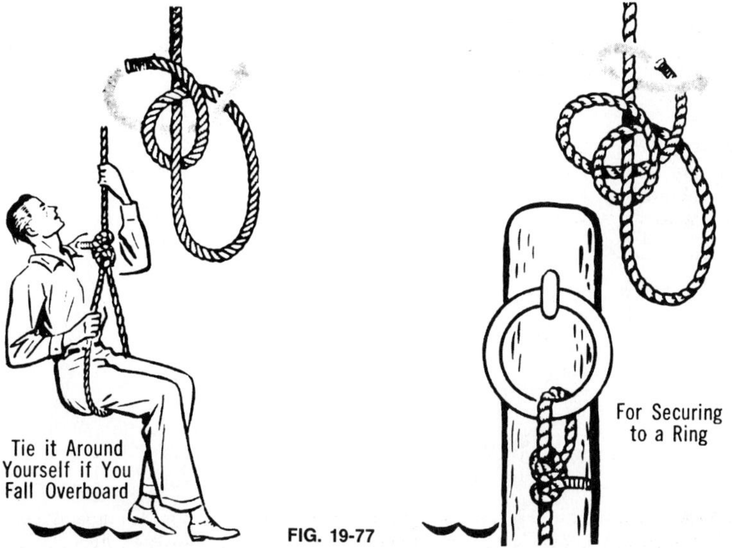
Tie it Around
Yourself if You
Fall Overboard
For Securing
to a Ring

FIG. 19-77

FIG. 19-78

When one line is to be fastened to a working line for the purpose of temporarily taking a load or strain off part of that working line, the midshipman's hitch can be used as the stopper (or fastening) knot.

If you ever fall overboard, tie this knot in the line thrown to you by first passing the end under your legs, thus forming the loop, and then take the hitch around the standing part

19-31 Scaffold Hitch Many occasions arise for the need of a single board scaffold hung by a single rope at each end. For safety, a scaffold of this kind must be hung so that it will not turn.

Lay the short end (A) of the rope over the top of the plank (Fig. 19-79) leaving enough hanging down to the left to tie to the long rope, as shown in Fig. 19-83. Wrap the long end (B) loosely twice around the plank, letting it hang down to the right as shown in Fig. 19-79. Now, carry rope 1 over rope 2 and place it next to rope 3 (Fig. 19-80). Pick up rope 2 (Fig. 19-81) and carry it over 1 and 3 and on over the end of the plank. Take up the slack by pulling rope A to the left and rope B to the right. Draw ropes A and B above the plank (Fig. 19-82) and join the short end (A) to the long rope B by an overhand bowline (Fig. 19-83). Pull the bowline tight, at the same time adjusting the lengths of the two ropes so that they hold the plank level. Attach a second rope to the other end of the plank in the same way and the scaffold is now ready for safe use.

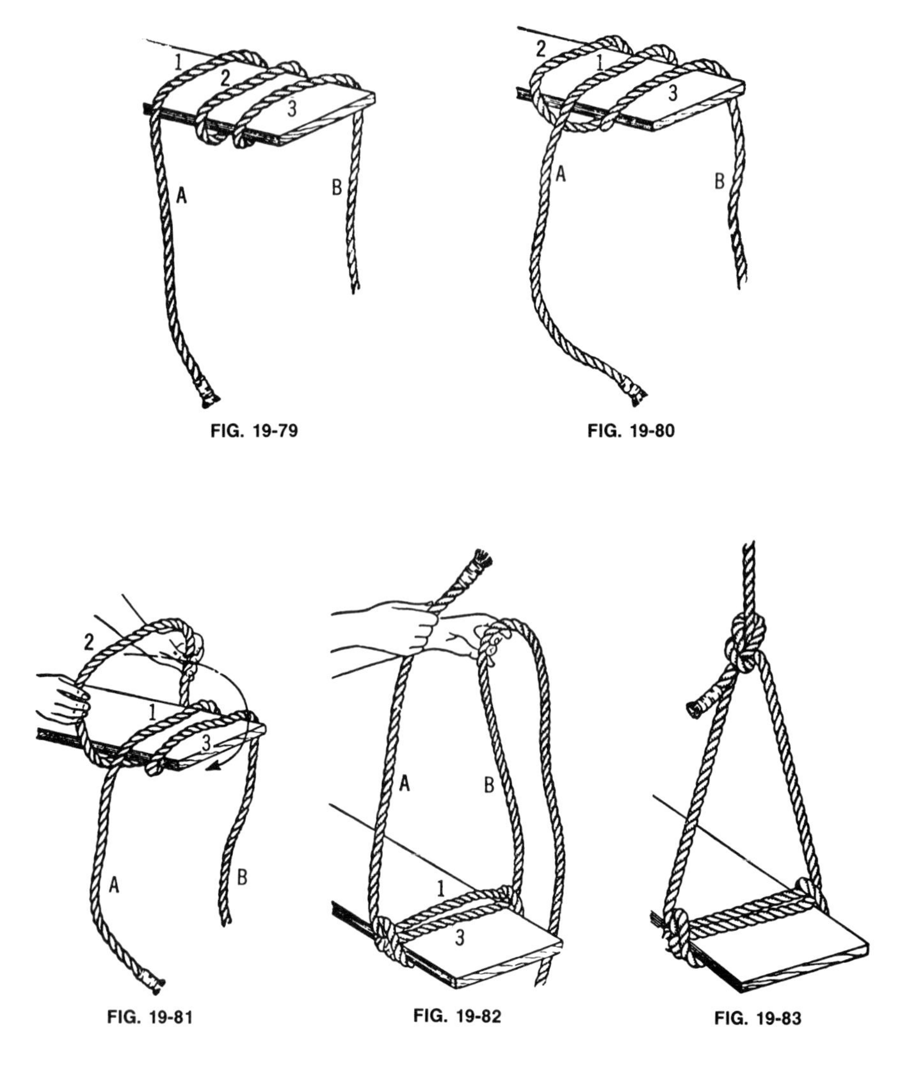

19-32 Clove Hitch This is the "general utility" hitch. It is a quick, simple method of fastening a rope around a post, spar, or stake. It is sometimes called the "builder's hitch" because of its extensive use in fastening staging to upright posts. It can be tied in the middle or end of a rope. But since it has a tendency to slip when used at the end of a rope, the end should be half-hitched to the standing part for greater security.

To tie (see Fig. 19-84): Make a turn with the rope around the object and over itself. Take a second turn around the object. Pull the end up under the second turn so it lies between the rope and the object. Tighten by pulling on both ends.

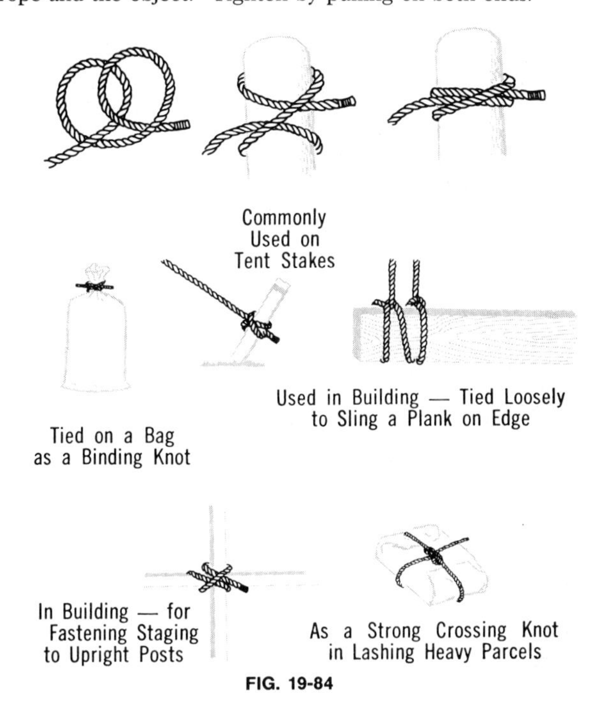
Commonly Used on Tent Stakes
Used in Building — Tied Loosely to Sling a Plank on Edge
Tied on a Bag as a Binding Knot
In Building — for Fastening Staging to Upright Posts
As a Strong Crossing Knot in Lashing Heavy Parcels

FIG. 19-84

19-33 Taut-Line Hitch Many times it is necessary to attach a rope to a second rope that is supporting a load and cannot be bent, i.e., a strand may break and it becomes necessary to attach a new rope above the break, to support the load while the break is repaired.

Wrap the new rope two full turns around the taut one, progressing away from the load or in the direction of the pull. Pass the end over the wrapping and toward the load; draw it firmly, and take one or two half-hitches about the taut rope between the wrapping and the load. The hitch will not hold unless the wrapping and the half-hitch are pulled up securely in the first place and are tightened as the strain is put on the new rope. See Fig. 19-85.

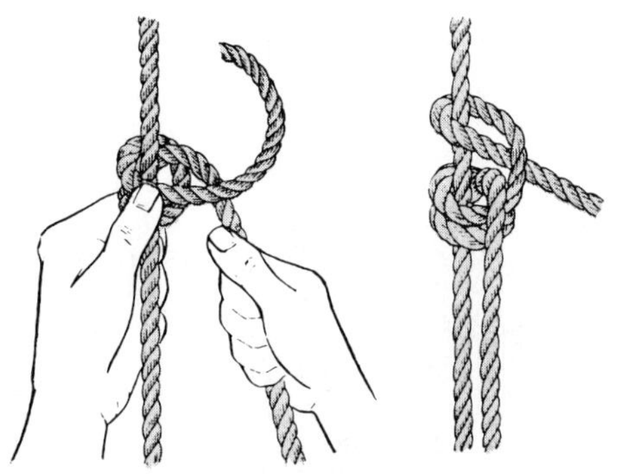

FIG. 19-85

With this taut-line hitch above the break the rope may be spliced or the strand repaired with no danger of the load slipping.

19-34 Timber Hitch This is used for towing and hoisting. It is a simple, convenient hitch which does not jam, and comes undone readily when the pull ceases. It is used mainly to tow or hoist cylindrical objects, such as logs, spars, etc., and also in handling cargo, to hoist small crates and bales. It is used with one or more half-hitches made with the standing part for towing a spar or hoisting a timber on end.

To tie: Pass a rope around the object and take a turn with the end around the standing part. Then, as shown in Fig. 19-86, twist or turn the end back on itself. Three turns back are sufficient and they should follow the lay of the rope.

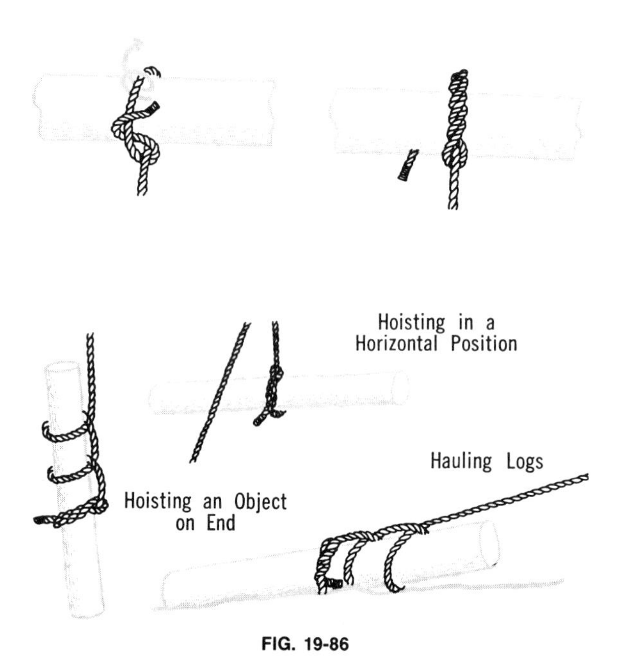

FIG. 19-86

19-35 Well Pipe Hitch Well pipe hitches are used to secure a rope to a pipe or similar object in such a manner that the rope will not slip along the pipe. See Fig. 19-87.

Place the rope along the pipe with the long end running in the opposite direction from which the pulling is to be done. Give the short end (A) two turns about the pipe as shown in Fig. 19-87. Now place the long end of the rope in the direction of the pull. Give both the long and short ends a half-hitch about the pipe as shown, and the hitch is complete.

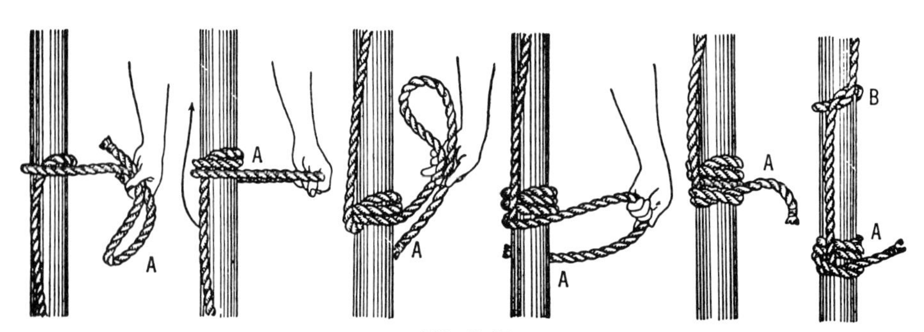

FIG. 19-87

19-36 Single Blackwall Hitch Frequently, it is necessary to attach a rope to a hook. A quick and secure, temporary fastening is the single blackwall hitch. The double blackwall hitch, however, is much safer.

To tie the single blackwall hitch, form a bight (C) in the rope, placing the short end (A) in back of the standing part (B) as shown in Fig. 19-88. Bring this bight up around the hook, as shown in Fig. 19-89. Draw up the bight by pulling on the long end while holding the short end, and then slide the bight down into position (Fig. 19-90). A loop is thus formed around the shank of the hook with the free end (A) pressed against the hook by the tension in the main rope (B).

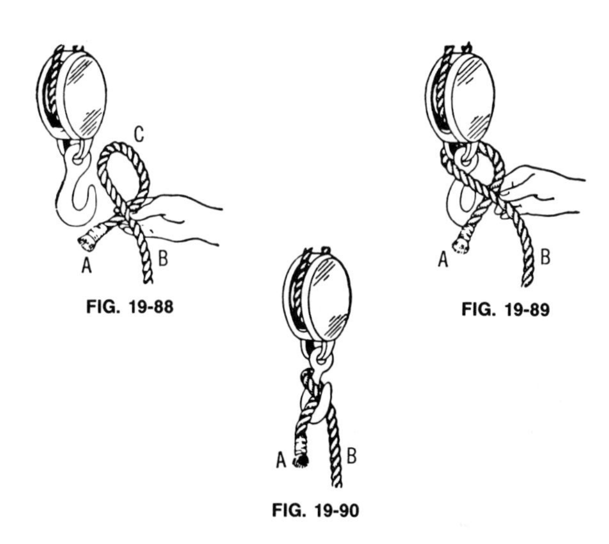

FIG. 19-88

FIG. 19-89

FIG. 19-90

19-37 Double Blackwall Hitch Form a loop (C) in the rope, placing the short end (A) in front of the standing part (B) as shown in Fig. 19-91. Bring the loop up around the hook as shown in Fig. 19-92. Bring the free end (A) from the left, and lay it in the hook, as indicated by the arrow (Fig. 19-92) so that it points to the right as in Fig. 19-93. Now, bring the long end (B) to the right around the back of the hook and lay it in the hook, across end A (Fig. 19-90). Note that the long end (B) crosses over the short end twice, once at the back and again in the hook.

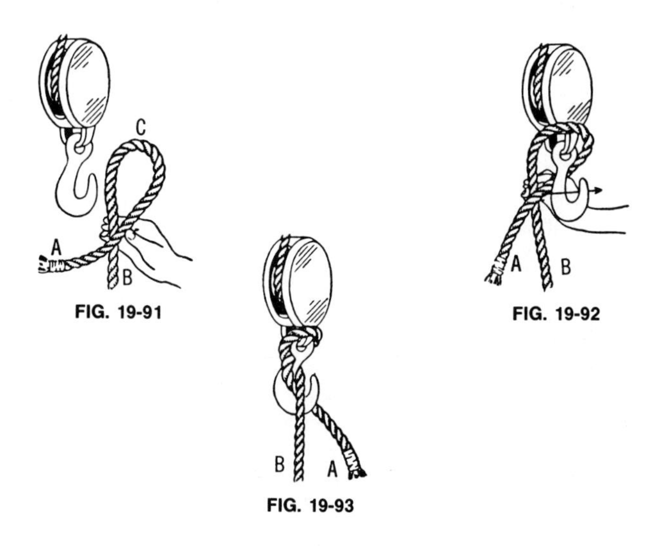

FIG. 19-91

FIG. 19-92

FIG. 19-93

19-38 Catspaw This is a hook hitch for heavy loads. It is generally tied in a "sling" or continuous wreath of rope for heavy hoisting. It does not jam and unties by itself when removed from the hook. See Fig. 19-94. It is tied by grasping two bights held well apart and twisting each of them away from you. The two loops thus formed are then brought together and placed over the hook.

FIG. 19-94

19-39 Sheet Bend or Weaver's Knot Used to join light and medium ropes of equal or unequal size, this is the common utility bend used aboard ship, and unties easily without injuring rope fibers. While it can be tied in larger ropes, such as hawsers and cables, the carrick bend is preferable. Remember that its construction is like that of the bowline—only instead of an end being tied to its own bight, one end is tied to the bight in the end of another rope—and you should find it most easy to tie.

To tie: Make an overhand loop with the end of one rope. Pass the end of the other rope through the loop thus formed, then up behind its standing part—then down through the loop again. Draw up tight. See Fig. 19-95.

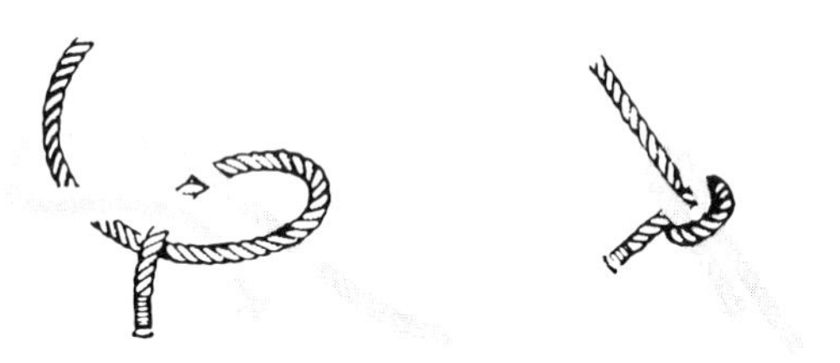

FIG. 19-95

19-40 Carrick Bend Used for heavy ropes, hawsers, and cables, the carrick bend is one of the strongest of knots. It cannot jam and it unties easily. Under strain it always draws up tight correctly—which is important, because very heavy ropes usually cannot be fully tightened by hand. However (although the precaution often is overlooked), for maximum security the ends always should be seized to the standing part.

To tie (see Fig. 19-96): With one rope-end form an underhand loop—with both the free end and standing part pointing away from you. Start the second rope end beneath both sides of the loop. Cross it over the standing part of the first rope. Then under the free end of the first rope. Then over the left side of the loop. Cross it under itself—and let the second free end lie over the right side of the loop. Finish by seizing each end to the standing part.

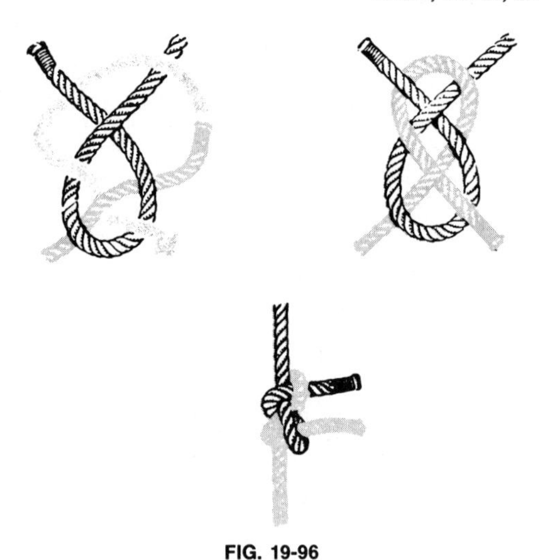

FIG. 19-96

19-41 Sheepshank The sheepshank is intended to shorten a rope for temporary use only. Carefully tied and drawn up tight, it is fairly reliable under a steady pull. But to make it secure for any length of time, both loops should be "stopped" to the standing parts. (An ordinary "stopping" is made with marlin or spun yarn. It consists of several round turns about the parts to be fastened together—the ends finished off with a square knot.)

To tie (see Fig. 19-97): Form an S loop as shown in the top diagram. Then with one free end of the rope make a half-hitch and slip it over one of the loops. Tighten. Repeat the procedure with the other loop.

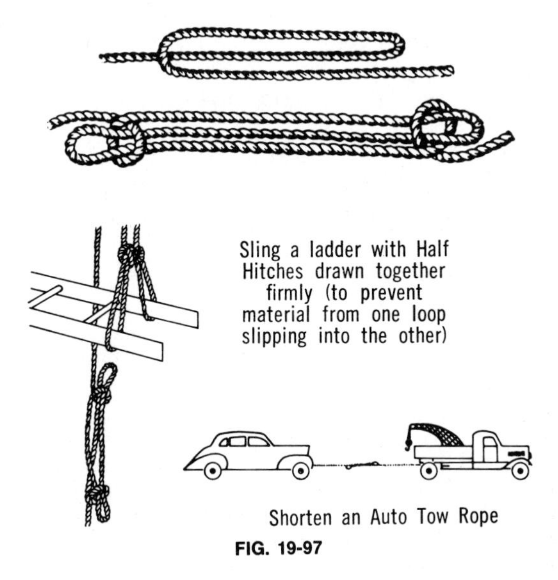

Sling a ladder with Half Hitches drawn together firmly (to prevent material from one loop slipping into the other)
Shorten an Auto Tow Rope

FIG. 19-97

19-42 Fisherman's Knot This is used to join fishlines, small rope, and twine. It is very strong and in common use by anglers. It is also a very handy knot to know in case you are short of twine and must join two lengths together for tying up a package.

To tie (see Fig. 19-98): Lay the two ends together—each pointing in the opposite direction. Then tie an overhand knot in the end of each—*around* the standing part of the other. When drawn tight the two knots slide together, will not slip.

FIG. 19-98

19-43 How to Make Fast Many methods of making fast are used but the correct one is usually the easiest. One easy but effective method is illustrated in Fig. 19-99.

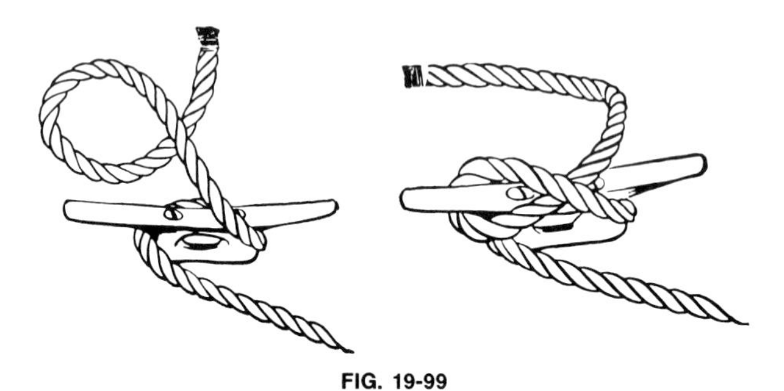

FIG. 19-99

To make fast: Loop the running part around the cleat's far side, away from the direction of the strain. Then take a turn around the stem with the running part and up and over the center (any additional turn would jam the line).

Add several more figure eights or slip a half-hitch over a horn of the cleat immediately if there is little strain.

Your line is now made fast, yet ready for prompt castoff with no part under tension binding loops.

This method makes it easy to cast off without having to take up the slack in the standing part, and ensures against accidents that occur when lines can not be freed quickly.

Section 20

Shafts and Couplings*

ROBERT O. PARMLEY, P.E.

President, Morgan & Parmley, Ltd.,
Ladysmith, Wisconsin

* Reprinted, with permission, from Mechanical Components Handbook © 1985 R. Parmley

SHAFTS

A rotating bar, usually cylindrical in shape, which transmits power is called a shaft. Power is delivered to the shaft through the action of an outside tangential force, resulting in a torsional action set up in the shaft. The resultant torque allows the power to be distributed to other machines or to various components connected to the shaft.

20-1 Usage and Classification Shafts and shafting may be classified according to their general usage. The following categories are presented here for discussion only and are basic in nature.

Engine Shafts An engine shaft may be described as a shaft directly connected to the power delivery of a motor.

Generator Shafts Generator shafts, along with engine shafts and turbine shafts, are called prime movers. There is a wide range of shaft diameters, depending on power transmission required.

Turbine Shafts Also prime movers, turbine shafts have a tremendous range of diameter size.

Machine Shafts General category of shafts. Variation in sizes of stock diameters ranges from $\frac{1}{2}$ to $2\frac{1}{2}$ in (increments of $\frac{1}{16}$), $2\frac{1}{2}$ to 4 in (increments of $\frac{1}{8}$ in), 4 to 6 in (increments of $\frac{1}{4}$ in).

Line Shafts Line shafting is a term employed to describe long and continuous "lines of shafting," generally seen in factories, paper or steel mills, and shops where power distribution over an extended distance is required. Stock lengths of line shafting generally are 12 ft, 20 ft, and 24 ft.

Jackshafts Jackshafts are used where a shaft is connected directly to a source of power from which other shafts are driven.

Countershafts Countershafts are placed between a line shaft and a machine. The countershaft receives power from a line shaft and transmits it to the drive shaft.

20-2 Torsional Stress A shaft is said to be under torsional stress when one end is securely held and a twisting force acts at the opposite end. Figure 20-1 illustrates this action. Note that the only deformation in the shaft is the rotation of the cross sections with respect to each other, as shown by angle ø.

Shafts which are subjected to torsional force only, or those with a minimal bending moment that can be disregarded, may use the following formula to obtain torque in inch-pounds, where horsepower P and rotational speed N in revolutions per minute are known.

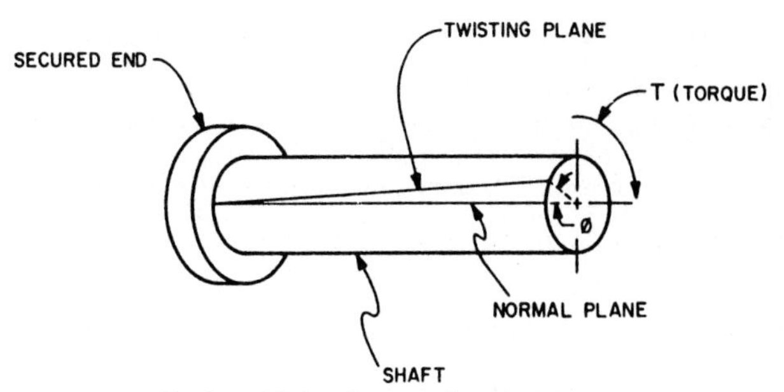

FIG. 20-1 **Shaft subjected to torsional stress.**

Shafts which are subjected to torsional force only, or those with a minimal bending moment that can be disregarded, may use the following formula to obtain torque in inch-pounds, where horsepower P and rotational speed N in revolutions per minute are known.

$$T = \frac{12 \times 33{,}000\ P}{2\pi N} \tag{20-1}$$

20-3 Twisting Moment Twisting moment T is equal to the product of the resultant P_r of the twisting forces multiplied by its distance from the axis R. See Fig. 20-2.

$$T = P_r \times R \tag{20-2}$$

20-4 Resisting Moment Resisting moment T_r equals the sum of the moments of the unit shearing stresses acting along the cross section of the shaft. This moment is the force which "resists" the twisting force exerted to rotate the shaft.

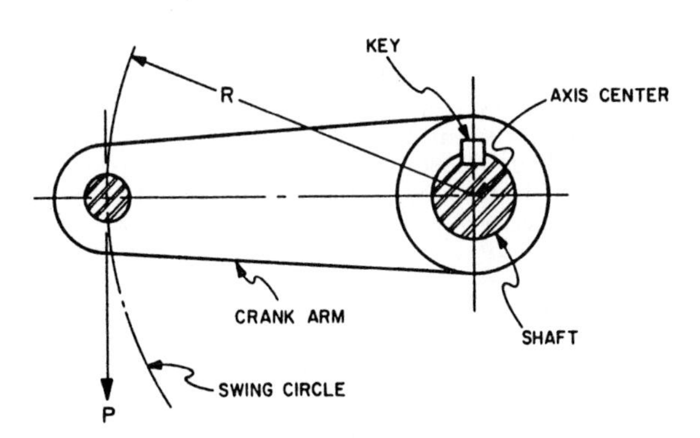

FIG. 20-2 **Typical crank arm forces.**

20-5 Torsion Formula for Round Shafts Torsion formulas apply to solid or hollow circular shafts, and only when the applied force is perpendicular to the shaft's axis, if the shearing proportional limit (of the material) is not exceeded.

Conditions of equilibrium, therefore, require the "twisting" moment to be opposed by an equal "resisting" moment. The following formulas may be used to solve the allowable unit shearing stress τ if twisting moment T, diameter of solid shaft D, outside diameter of hollow shaft d, and inside diameter of hollow shaft d_1 are known.

Solid round shafts:

$$\tau = \frac{16T}{\pi D^3} \tag{20-3}$$

Hollow round shafts:

$$\tau = \frac{16Td}{\pi(d^4 - d_1^4)} \tag{20-4}$$

20-6 Shear Stress In terms of horsepower, for shafts used in the transmission of power, shearing stress may be calculated as follows, where P = horsepower to be transmitted, N = rotational speed in revolutions per minute, and the shaft diameters are those described previously. Maximum unit shearing stress τ is in pounds per square inch.

Solid round shafts:

$$\tau = \frac{321{,}000P}{ND^3} \tag{20-5}$$

Hollow round shafts:

$$\tau = \frac{321{,}000Pd}{N(d^4 - d_1^4)} \tag{20-6}$$

The foregoing formulas do not consider any loads other than torsion. Weight of shaft and pulleys or belt tensions are not included.

20-7 Critical Speeds of Shafts Shafts in rotation become very unstable at certain speeds, and damaging vibrations are likely to occur. The revolution at which this mechanical phenomenon takes place is called the "critical speed."

Vibration problems may occur at a "fundamental" critical speed. The following formula is used for finding this speed for a shaft on two supports, where W_1, W_2, etc. = weights of rotating components; y_1, y_2, etc. = respective static deflection of the weights; g = gravitational constant, 386 in/s^2.

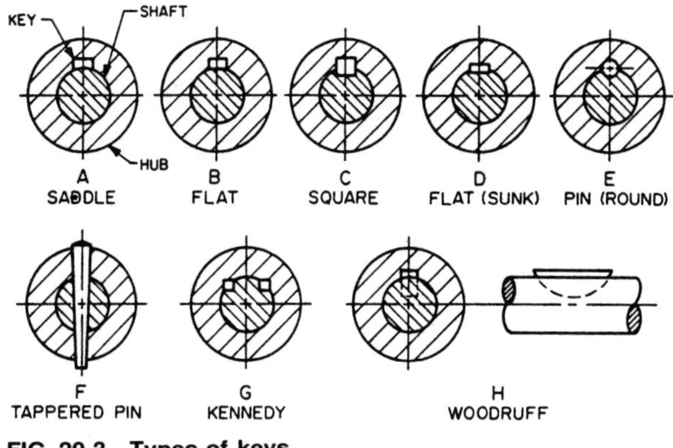

FIG. 20-3 Types of keys.

$$f = \frac{1}{2\pi}\sqrt{\frac{g\,(W_1 y_1 + W_2 y_2 + \cdots}{W_1 y_1^2 + W_2 y_2^2 + \cdots}} \quad \text{cycles/s} \tag{20-7}$$

A thorough discussion of this phenomenon is beyond the scope of this book. Readers should consult the many volumes devoted to vibration theory for an in-depth technical presentation.

20-8 Fasteners for Torque Transmission *Keys.* Basically keys are wedge-like steel fasteners that are positioned in a gear, sprocket, pulley, or coupling and then secured to a shaft for the transmission of power. The key is the most effective and therefore the most common fastener used for this purpose.

Figure 20-3 illustrates several standard key designs, including round and tapered pins. The saddle key (*a*) is hollowed to fit the shaft, without a keyway cut into the shaft. The flat key (*b*) is positioned on a planed surface of the .shaft to give more frictional resistance. Both of these keys can transmit light loads. Square (*c*) and flat-sunk (*d*) keys fit in mating keyways, half in the shaft and half into the hub. This positive holding power provides maximum torque transfer. Round (*e*) and tapered (*f*) pins are also an excellent method of keying hubs to shafts. Kennedy (*g*) and Woodruff (*h*) keys are widely used. Figure 20-4 pictures feather keys, which are used to prevent hubs from rotating on a shaft, but will permit the component part to move along the shaft's axis. Figure 20-4*a* shows a key which is relatively long for axial movement and is secured in position on the shaft with two flat fillister-head matching screws. Figure 20-4*b* is held to the hub and moves freely with the hub along the shaft's keyseat.

A more in-depth presentation of keys can be found in Sec. 10, "Locking Components."

Set screws may be used for light applications. A headless screw with a hexagon socket head and a conical tip should be used. Figure 20-5 illustrates both a "good" design and a "bad" design. The set screw must be threaded into the hub and tightened on the shaft to provide a positive anchor.

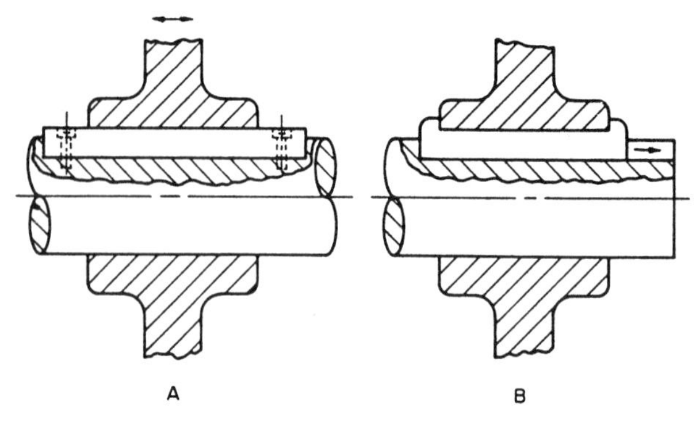

FIG. 20-4 Feather keys.

Pins. Round and taper pins were briefly discussed previously, but mention should be made of the groove, spring, spiral, and shear pins. The groove pin has one or more longitudinal grooves, known as flutes, over a portion of its length. The farther you insert this pin, the tighter it becomes. The spring or slotted tubular pin is a hollow tube with a full-length slot and tapered ends. This slot allows the pin's diameter to be reduced somewhat when the pin is inserted, thus providing easy adaptation to irregular holes. Spirally coiled pins are very similar in application to spring pins. They are fabricated from a sheet of metal wrapped twice around itself, forming a spiral effect. Shear pins, of course, are used as a weak link. They are designed to fail when a predetermined force is encountered.

Pins are discussed in Sec. 2, "Standard Pins" and Sec. 24, "Innovative Connections," which has some sample applications for pins.

20-9 Splines Spline shafts are often used instead of keys to transmit power from hub to shaft or from shaft to hub. Splines may be either square or involute.

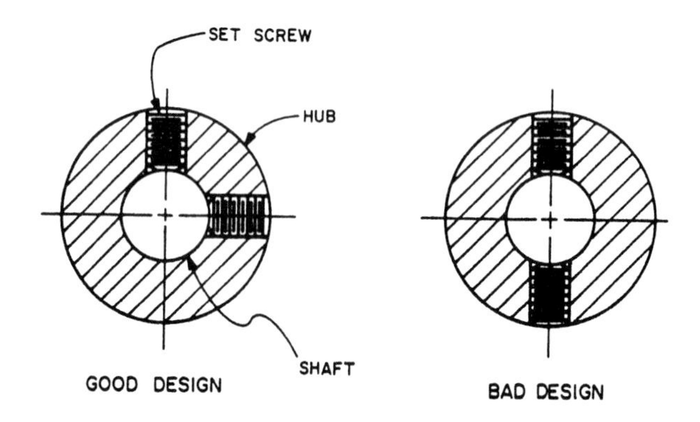

FIG. 20-5 Use of set screws.

One may think of splines as a series of teeth, cut longitudinally into the external circumference of a shaft, that match or mate with a similar series of keyways cut into the hub of a mounted component. Splines are extremely effective when a "sliding" connection is necessary, such as for a PTO (power take-off) on agricultural equipment.

Square or parallel-side splines are employed as multispline shaft fittings in series of 4, 6, 10, or 16. Refer to Sec. 10 for a typical square design pattern.

Splines are especially successful when heavy torque loads and/or reversing loads are transmitted. Torque capacity (in inch-pounds) of spline fittings may be calculated by the following formula:

$$T = 1000NrhL \quad \text{in} \bullet \text{lb} \tag{20-8}$$

where N = number of splines
r = means radial distance from center of shaft/hub to center of spline
h = depth of spline
L = length of spline bearing surface

This gives torque based on spline side pressure of 1000 lb/in^2. Involute splines are similar in design to gear teeth, but modified from the standard profile. This involute contour provides greater strength and is easier to fabricate. Figure 20-6 shows five typical involute spline shapes.

Section 24, "Innovative Connections," illustrates several spline connection designs which will be of interest to the reader.

SHAFT COUPLINGS

In machine design, it often becomes necessary to fasten or join the ends of two shafts axially so that they will act as a single unit to transmit power. When this parameter is required, shaft couplings are called into use. Shaft couplings are grouped into two

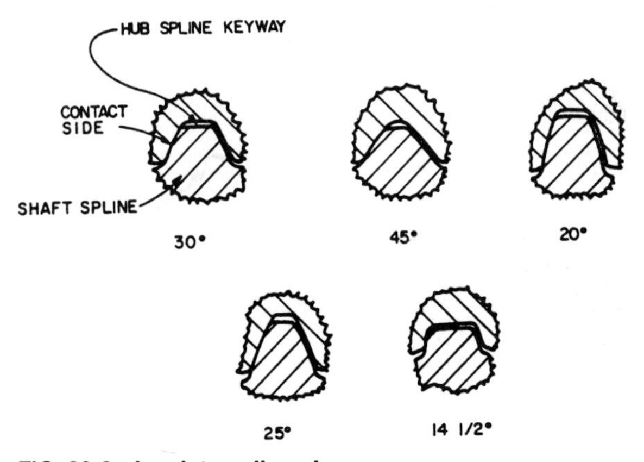

FIG. 20-6 Involute spline shapes.

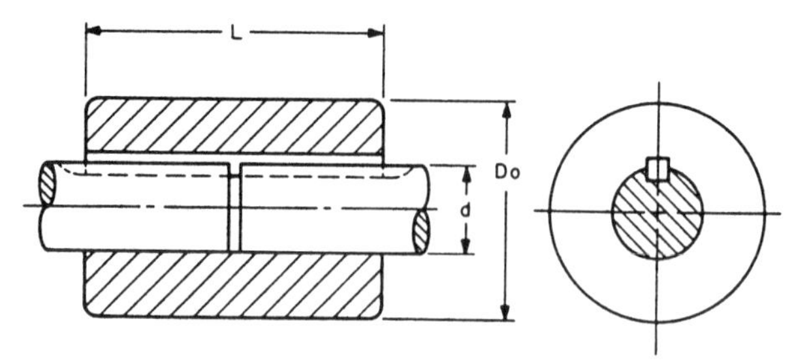

FIG. 20-7 Sleeve coupling.

general classifications: rigid (or solid) and flexible. A rigid coupling will not provide for shaft misalignment or reduce vibration or shock from one shaft to the other. However, flexible shaft couplings provide connection of misaligned shafts and can reduce shock and/or vibration to a degree.

Section 24, "Innovative Connections," pictorially presents several unusual designs of shaft connections.

20-10 Sleeve Coupling Sleeve coupling, as illustrated in Fig. 20-7, consists of a simple hollow cylinder which is slipped over the ends of two shafts fastened into place with a key positioned into mating keyways. This is the simplest rigid coupling in use today. Note that there are no projecting parts, so that it is very safe. Additionally, this coupling is inexpensive to fabricate.

Figure 20-8 pictures two styles of sleeve couplings using standard set screws to anchor the coupling to each shaft end. One design is used for shafts of equal diameters. The other design connects two shafts of unequal diameters.

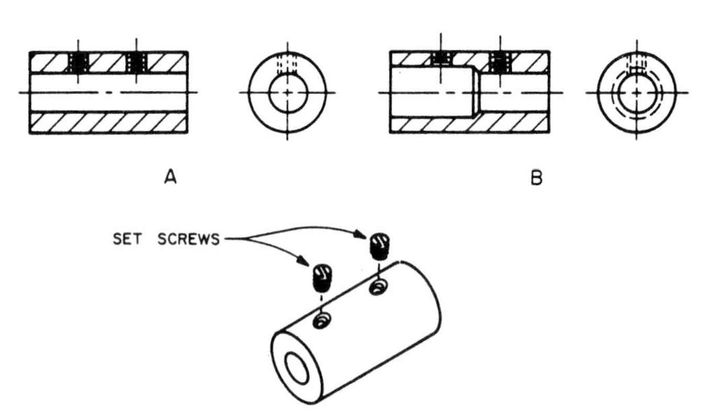

FIG. 20-8 Sleeve shaft coupling.

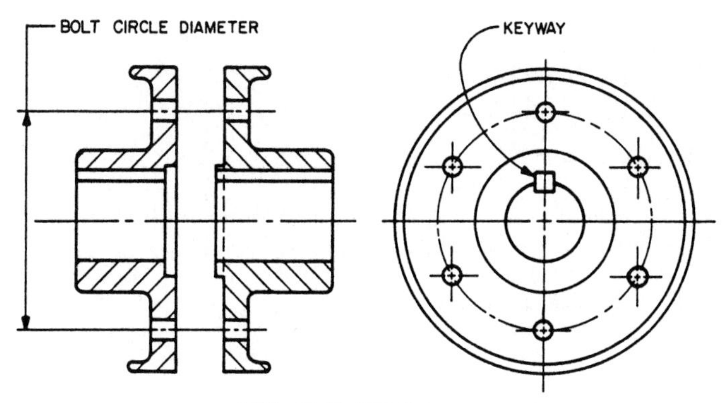

FIG. 20-9 Solid coupling.

20-11 Solid Coupling The solid coupling shown in Fig. 20-9 is a tough, inexpensive, and positive shaft connector. When heavy torque transmission is required, a rigid coupling of this design is an excellent selection.

20-12 Clamp or Compression Coupling The rigid coupling shown in Fig. 20-10 has evolved from the basic sleeve coupling. This clamp or compression coupling simply splits into halves, which have recesses for through bolts that secure or clamp the mating parts together, producing a compression effect on the two connecting shafts. This coupling may be used for transmission of large torques because of its positive grip from frictional contact.

20-13 Flange Coupling Flange couplings are rigid shaft connectors, also known as solid couplings. Figure 20-11 illustrates a typical design. This rigid coupling consists of two components, which are connected to the two shafts with keys. The hub halves are fastened together with a series of bolts arranged in an even pattern concentrically

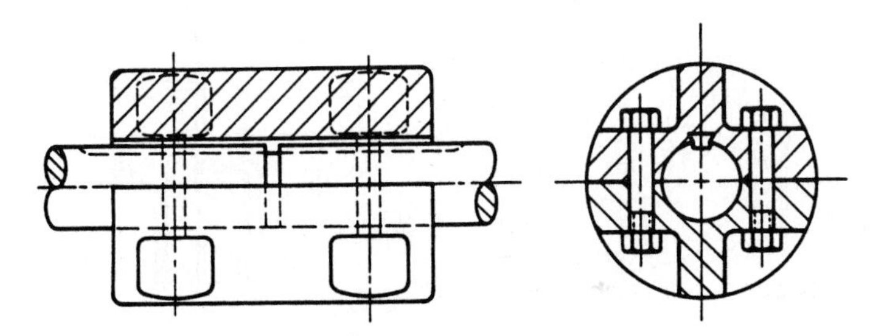

FIG. 20-10 Clamp or compression coupling.

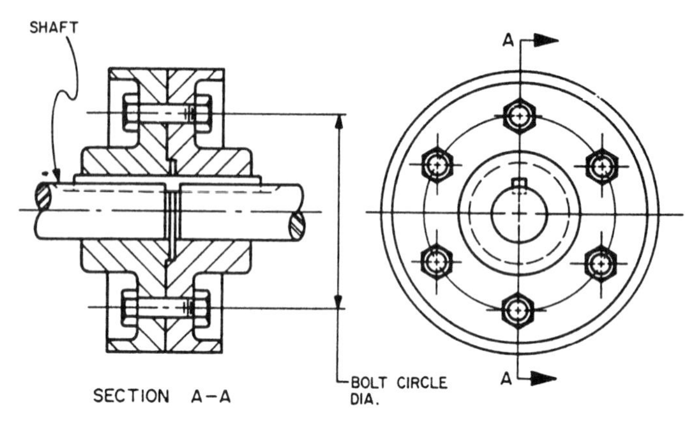

FIG. 20-11 Flange coupling.

about the center of the shaft. A flange on the outside circumference of the hub provides a safety guard for the bolt heads and nuts, while adding strength to the total assembly.

20-14 Flexible Coupling Flexible couplings connect two shafts which have some nonalignment between them. The couplings also absorb some shock and vibration which may be transmitted from one shaft to the other.

There are a wide variety of flexible-coupling designs. Figure 20-12 pictures a two-part cast-iron coupling which is fastened onto the shafts by keys and set screws. The halves have lugs, which are cast on an integral part of each hub half. The lugs fit into

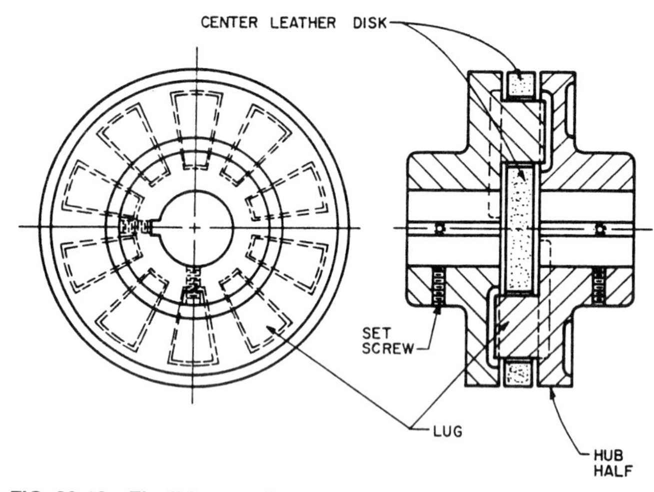

FIG. 20-12 Flexible coupling.

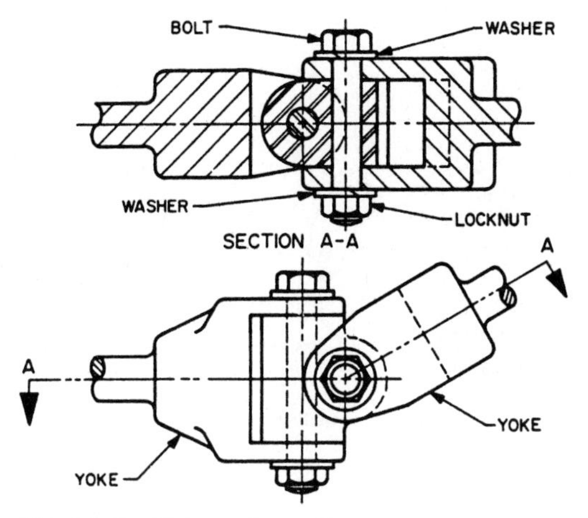

FIG. 20-13 Universal coupling.

entry pockets in a disk made of leather plies which are stitched and cemented together. The center leather laminated disk provides flexibility in all directions. Rotation speed, either slow or fast, will not affect the efficiency of the coupling.

20-15 Universal Coupling If two shafts are not lined up but have intersecting centerlines or axes, a positive connection can be made with a universal coupling. Figure 20-13 details a typical universal coupling.

Note that the bolts are at right angles to each other. This makes possible the peculiar action of the universal coupling. Either yoke can be rotated about the axis of each bolt so that adjustment to the angle between connected shafts can be made. A good rule of thumb is not to exceed 15° of adjustment per coupling.

20-16 Multijawed Coupling This rigid-type shaft coupling is a special design. The coupling consists of two halves, each of which has a series of mating teeth which lock together, forming a positive jawlike connection. Set screws secure the hubs onto the respective shafts. This style of coupling is strong and yet easily dismantled. See Fig. 20-14.

20-17 Spider-Type Coupling The spider-type or Oldham coupling is a form of flexible coupling that was designed for connection of two shafts which are parallel but not in line. The two end hubs, which are connected to the two respective shafts, have grooved faces which mate with the two tongues of the center disk. This configuration and slot adjustment allow for misalignment of shafts. Figure 20-15 shows an assembled spider-type coupling.

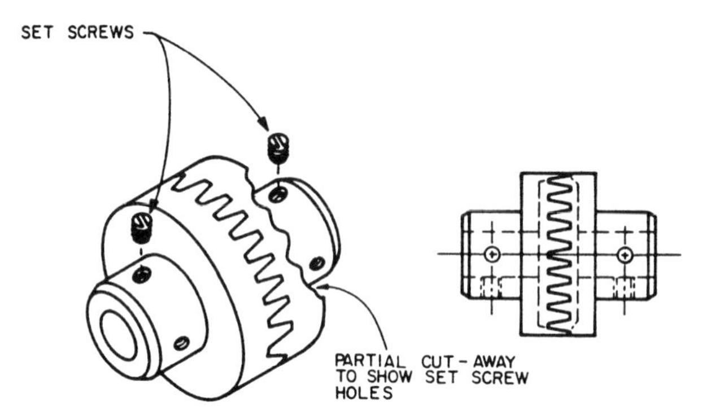

FIG. 20-14 Multijawed coupling.

20-18 Bellows Coupling Two styles of bellows couplings are illustrated in Fig. 20-16. These couplings are used in applications involving large amounts of shaft misalignment, usually combined with low radial loads. Maximum permissible angular misalignment varies between 5° and 10°, depending on manufacturer's recommendation. Follow manufacturer's guidelines for maximum allowable torque. Generally, these couplings are used in small, light-duty equipment.

20-19 Helical Coupling These couplings, also, are employed to minimize the forces acting on shafts and bearings as a result of angular and/or parallel misalignment.

These couplings are used when motion must be transmitted from shaft to shaft with constant velocity and zero backlash.

The helical coupling achieves these parameters by virtue of its patented design, which consists of a one-piece construction with a machined helical groove circling its

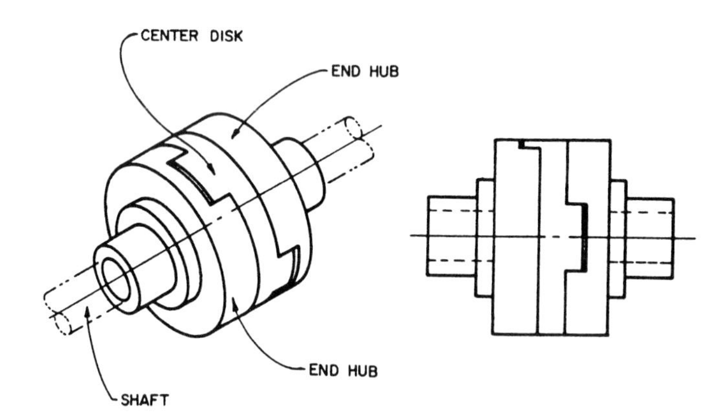

FIG. 20-15 Spider-type coupling.

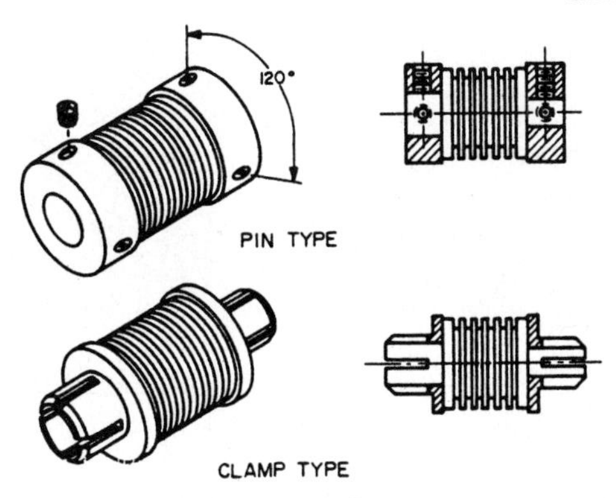

FIG. 20-16 **Bellows couplings.**

exterior diameter. Removal of this coil or helical strip results in a flexible unit with considerable torsional strength. See Fig. 20-17, which pictures both the pin- and clamp-type designs.

20-20 Offset Extension Coupling Figure 20-18 depicts an offset extension shaft coupling. This coupling is used to connect or join parallel drive shafts that are offset ±30° in any direction, with separations generally greater than 3 in. Shafts are secured to the coupling with set screws.

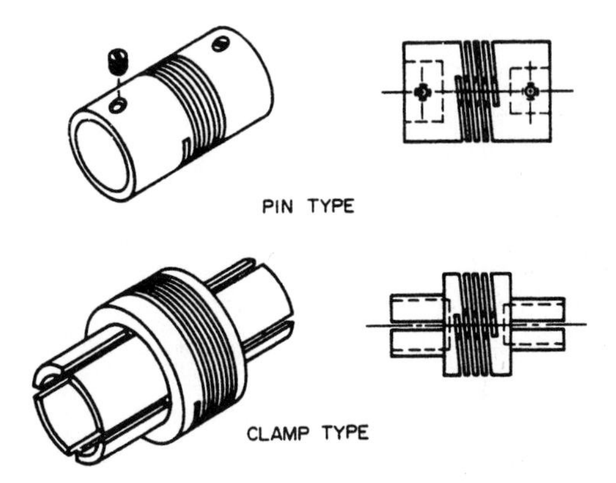

FIG. 20-17 **Helical couplings.**

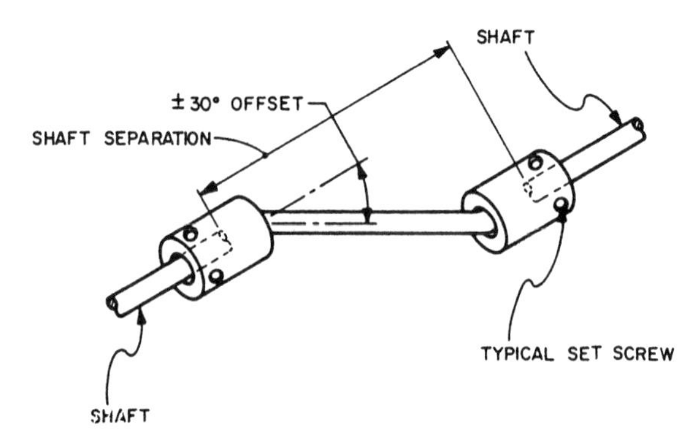

FIG. 20-18 Offset extension shaft coupling.

REFERENCES

Master Catalog 82, Sterling Instrument Division of Designatronics, Inc., New Hyde Park, N.Y.

Levinson, Irving J.: *Machine Design*, Reston Publishing Co., Reston, Va., 1978.

Parmley, R. O.: *Standard Handbook of Fastening and Joining*, McGraw-Hill, New York, 1977.

Spotts, M. F.: *Design of Machine Elements*, 5th ed., Prentice-Hall Englewood Cliffs, N.J., 1978.

Winston, Stanton E.: *Machine Design*, American Technical Society, Chicago, 1956.

Carmichael, Colin, ed.: *Kent's Mechanical Engineer's Handbook*, 12th ed., Wiley, New York, 1958.

Section 21

Seals and Packings

DR. LESLIE A. HORVE, P.E.*

Vice President of Technology, Chicago Rawhide Manufacturing Company, Elgin, Illinois

* This section originally published in Mechanical Components Handbook edited by Robert O. Parmley; Reprinted by permission.

INTRODUCTION

21-1 Seal Requirements Modern machinery is highly dependent upon seals to retain liquids, solids, or gases. A secondary but equally important function of sealing devices is to prevent foreign particles from entering the sealed cavity.

There are many seal designs and sealing principles used in industry. Selecting a sealing principle and ultimately a seal design for a given application is not an easy task. The application must be studied to correctly identify the nature of the sealing problem. The application conditions must be considered to further refine the selection (Fig. 21-1). Some important conditions include system pressure, temperature, medium to be sealed, medium to be excluded, static or dynamic conditions, and type of motion if dynamic. Economics is also a factor. In general, the complexity of seal designs and the cost of sealing increases as the tolerance for leakage decreases.

The most frequently used seals are packings, which are usually the simplest and least expensive solution to sealing problems. Packings are shown as the bottom of the sealing pyramid in Fig. 21-2. The complexity and cost of sealing devices increase as one moves up the pyramid. The horizontal axis represents quantities of seal types used.

21-2 Allowable Leakage Rates Many sources will argue that a condition of zero leakage does not exist. Virtually leakproof static seals can be obtained by plastically deforming metal. The leakage obtained with this gasket type is less than 10^{-8} cm^3/s (atm.) gaseous nitrogen at 300 lb/in^2 (gage) at ambient temperatures. Leakage of fluids is typically measured in terms of drops (approximately 0.05 cm^3) per day. Different applications will tolerate different leakage rates. Allowable leakage rates range from zero to many cubic centimeters per hour.

It has always been necessary to minimize the leakage of expensive, toxic, corrosive,

Seal Variables	Effects Causing Static or Dynamic Leakage	Application Variables
Material ⇒	Thermal expansion & contraction	Static or dynamic
Design	Seal material degradation	Sealed medium properties
	Corrosion	Type of shaft speed & motion
	Fatigue ⇐	Shaft eccentricity
	Vibration	Shaft diameter
	Wear	Shaft finish
	Shaft & bore defects	Shaft material
	Lubricant breakdown	Housing
	Case leakage	Finish
	Misalignment	Eccentricity
		Interference tolerance
		Operation
		Cycles (pressure, temperature)
		Run time
		Ambient
		Temperature & range
		Ozone
		Dust/mud
		Pressure & range

FIG. 21-1 Application conditions.

or noxious gases and liquids. Pressures to reduce energy losses, conserve fuels and lubricants, and protect the environment have increased. It has been estimated that 100 million gallons of fluid are lost to the environment each year from hydraulic machines and systems alone. Additional losses occur from engines, transmissions, and gearboxes.

It is possible to reduce and even stop external leakage. Aircraft hydraulic systems are expensive, but they are virtually leak-free. Leakage in industrial machinery and vehicles has been substantially reduced.

These reductions are due in part to the availability of better seal designs, materials, and technology. They are also due to the great care seal users take in selecting the proper sealing concept for their application.

21-3 Seal Classification Seals are classified into two major categories, static and dynamic. Static seals are used in applications in which there is no movement between mating surfaces. Deformable gaskets and sealants are used extensively in static applications. Dynamic seals are used in applications in which shaft rotation, reciprocation, and oscillation are present. A wide variety of seal designs can be used in dynamic applications. A general classification chart for seals appears in Fig. 21-3.

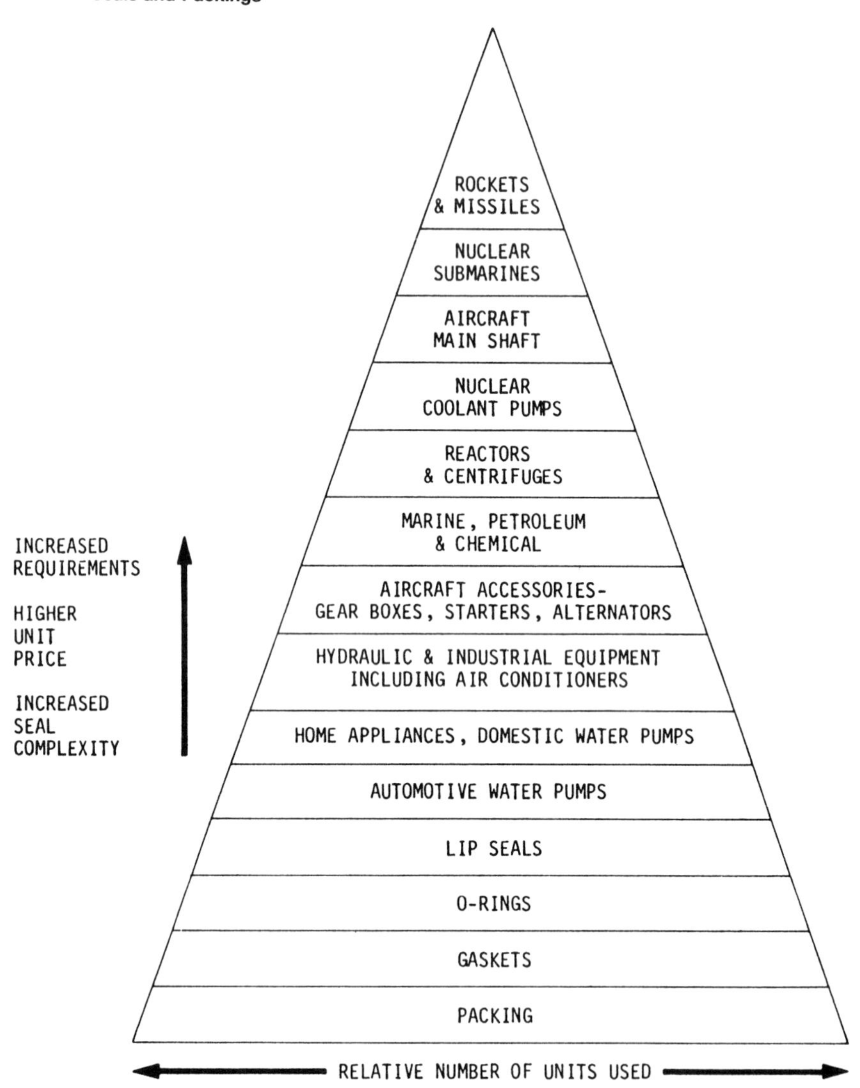

FIG. 21-2 Sealing pyramid.

21-4 Sealing History The earliest seal consisted of leather straps arranged to form a labyrinth at the end of a wheel axle to hold grease or animal fat in place. The industrial revolution of the eighteenth century created demands for more sophisticated lubrication and sealing systems. Compression-packed stuffing boxes were used to seal both reciprocating and rotating shafts. Early packings were made from available fibers, such as flax, hemp, cotton, and wool. Animal fats were used as lubricants and fillers. Then, in the early 1900s, asbestos increased the temperature range and chemical resistance of packings. Synthetic fibers later became available; these are now replacing asbestos because of its recently recognized carcinogenic characteristics. Lubricants and

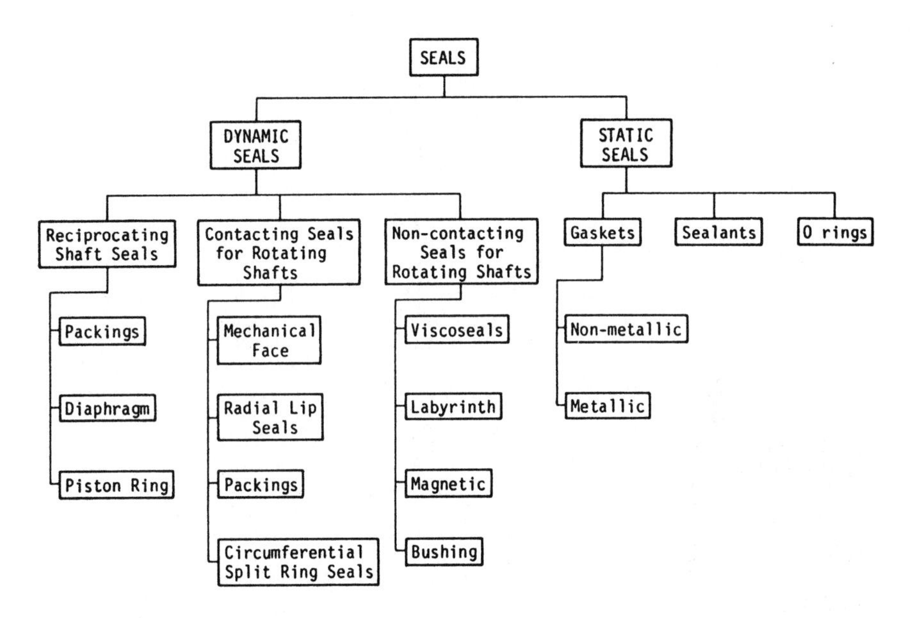

FIG. 21-3 General classification chart.

fillers were improved with petroleum-base lubricants, fish oils, waxes, and synthetic lubricants such as mica, graphite, polytetrafluoroethylene (PTFE), and molydisulfide-base materials. Various fabrication techniques using wire or metal foils all serve to extend the capability of compression packing.

As the speeds of rotating shafts increased, it became necessary to consider an alternative to packings. In the mid-1920s, radial lip seals made of leather were used in automotive applications to follow the eccentricities of high-speed rotating shafts. These assembled seals had advantages that could be readily appreciated: They were compact, self-contained units, and they were easily press fitted into a housing bore.

The mechanical face seal was also developed in the 1920s to provide low leakage rates for the mechanical refrigeration industry. The early refrigeration compressors used a variety of gases, such as ammonia, sulfur dioxide, and methyl chloride. These compressors used stuffing boxes, were difficult to adjust, and permitted significant leakage that drained off the refrigerant gas supply. In addition, the refrigerant gases used at that time were obnoxious or toxic to the personnel in the area. Mechanical seals replaced the packings and lip seals that had been used, solving these problems.

As World War II approached, rotating machinery turned rapidly to antifriction bearings. Shaft speeds increased, and operating temperatures rose. The existing seal designs and materials were no longer adequate. The newly developed oil-resistant synthetic elastomers replaced leather in radial lip seal designs for rotating shaft applications. Upper temperature limitations were extended from 200°F to 300°F. High-pressure hydraulic systems were used in more and more applications. A host of packings made from synthetic elastomers were introduced. These packings (V rings, U rings, O rings, chevrons, and so on) helped to extend the operating pressure range of hydraulic systems while reducing leakage. The jet and space age that started in the 1950s spurred the development of new materials and sealing concepts. Sophisticated labyrinths and bushings were developed for jet engines. All-metallic mechanical seals and new elastomeric materials extended the available temperature ranges for new applications.

Bonding cements were improved, and the assembled rotary lip seal was replaced by simpler designs with the elastomer bonded directly to the metal case. Internal leakage through the assembled components was eliminated. The zero leakage requirements of the space and nuclear applications of the 1970s and 1980s have led to the development of magnetic sealing systems. New sealing systems that will economically ensure that there is little or no leakage must be developed to protect the environment and sophisticated equipment from the inadvertent leakage of harsh chemicals and lubricants.

STATIC SEALS—GASKETS

21-5 Gasket Applications Gaskets are used to develop and maintain a barrier between mating surfaces of mechanical assemblies when the surfaces do not move relative to each other. The barrier is designed to retain internal pressure, prevent liquids and gases from escaping from the assembly, and prevent contaminants from entering the assembly. It is essential that joint design and gasket design be considered together. The gland-type joint captures the gasket (Fig. 21-4) and locks it in place. It is more expensive to machine than the glandless type (Fig. 21-5). Application parameters (liquid or gas to be sealed, internal pressure, temperature, exterior contaminants, types of surfaces to be jointed, surface roughness, and others) must be considered carefully before the gasket type and material are selected. The success of the gasket is dependent on the flange pressure developed at the joint. Flange pressure is defined as the effective compressive load per inch or centimeter of gasket area, expressed in pounds per square inch or pascals. Flange pressure compresses the gasket material and causes the material to conform to surface irregularities in the flange. Flange pressure is developed by tightening bolts that hold the assembly together. About half of the initial tightening torque on a bolt is used to overcome collar friction, and 40 percent is required to overcome mating screw-thread friction. The remaining 10 percent creates the gasket load in flanged joints. The flange pressure can be estimated using Eq. (21-1).

$$F_p = \frac{N_B T}{A_r K D} = \frac{N_B T}{A_r} \qquad (21\text{-}1)$$

where F_p = flange pressure, lb/in² (bars)
N_B = number of bolts in the assembly
T = initial bolt torque, lb•in (N•cm)
K = torque friction coefficient
A_r = gasket area, in² (cm²)
D = nominal bolt diameter, in (cm)
P = bolt clamp load developed by tightening, lb (N)

The effect of load loss due to bolt stretching and other factors must be considered. For many applications a rigid tightening sequence is necessary to ensure that the gasket loading is uniformly applied. The amount of initial bolt torque T required to give a desired bolt clamp load P for dry and lubricated conditions is given in Table 21-1 for various bolt sizes and grades.

21-6 Nonmetallic Gaskets The success of nonmetallic gaskets in a given application is highly dependent on the material selected. The permeability, deformation and flow characteristics under load, resistance to fluids, and cost of the material must be considered when the selection is made. The extent to which gaskets must be compressed to obtain a seal depends on the finish of the faces and the internal pressure. Materials with high compression are generally used for low-pressure applications where a good face finish is not available. Low-compression materials seal high-pressure applications. Materials used for gaskets include Teflon, rubber, asbestos, paper, cork, and

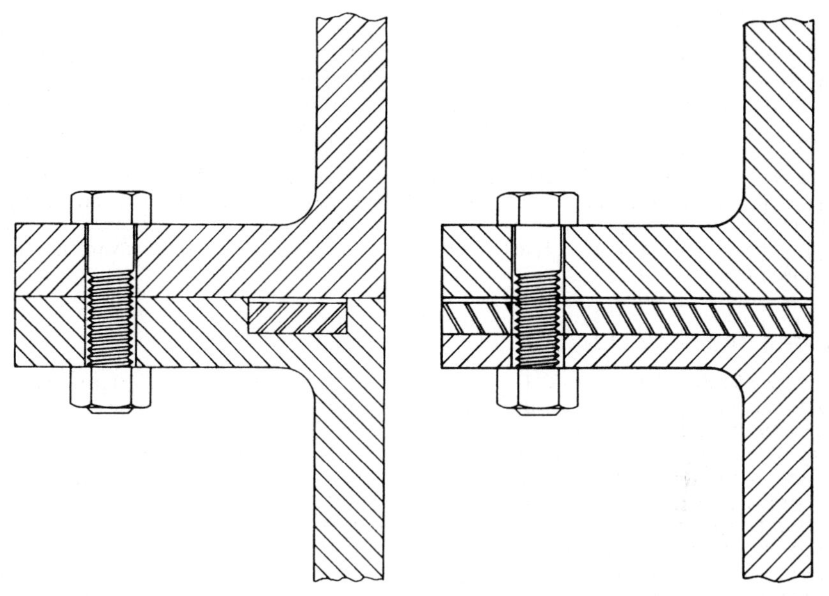

FIG. 21-4 Gland-type joint.

FIG. 21-5 Glandless joint.

various combinations of materials. The amount of compression expected for these materials as a result of various flange pressures is given in Fig. 21-6.

Paper gaskets can be made from organic or mineral (asbestos) fibers. Untreated paper gaskets are highly permeable and are not typically used to contain fluids. They are used to prevent contaminants such as dust or water splash from entering a cavity. The permeability is reduced by saturating the paper with latex or glue. Saturated papers are used to hold oils and gasoline at moderate pressures (0 to 20 lb/in^2) at temperatures less than 160°F. About 500 to 1000 lb/in^2 is required to seat the flange to the paper gasket.

Elastomeric materials can be compounded to produce a great variety of rubber gaskets. Since rubber is incompressible, the designer must provide room for flow or deformation when the rubber sheet is loaded. Rubber gaskets can be molded to virtually any shape to meet special application needs. The material can be stretched over projections during assembly. The hardness and compression set of the rubber materials can be adjusted by compounding. Rubber materials for gaskets are usually compounded to give low compression. There is a wide variety of elastomeric materials to choose from. The desired temperature range and the fluid in the application must be considered when selecting a rubber material (Table 21-2).

Asbestos fibers are low in strength and high in porosity. Rubber and plastic material are usually mixed with asbestos to make a composite material. Rubber-asbestos sheets are used where bolting pressures are high. The material is compacted under high pressure and becomes impermeable. Compressed asbestos requires relatively high loads to achieve intimate contact with the flange faces. It is used in relatively heavy construction where rigid flanges with adequate bolting capacity are used.

Compressed asbestos has little elasticity. After the bolts have pulled up to the point where the gasket is well seated and has conformed to the irregularities of the joint, it will yield little more than the metal components themselves.

TABLE 21-1 Clamp Loading Chart

	Tensile		SAE Grade & Bolts					
						Tightening torque		
Size	Bolt diam. D, in	Stress area, A_s, in	Tensile strength, min lb/in²	Proof load, lb/in²	Clamp† load P, lb	Dry $K = 0.20$	Lub. $K = 0.15$	Tensile strength, min lb/in²
						lb·in	lb·in	
4-40	0.1120	0.00804	74,000	55,000	240	5	4	120,000
4-48	0.1120	0.00661			280	6	5	
6-32	0.1380	0.00909			380	10	8	
6-40	0.1380	0.01015			420	12	9	
8-32	0.1640	0.01400			580	19	14	
8-36	0.1640	0.01474			600	20	15	
10-24	0.1900	0.01750			720	27	21	
10-32	0.1900	0.02000			820	31	23	
1/4-20	0.2500	0.0318			1,320	66	49	
1/4- 8	0.2500	0.0364			1,500	76	56	
						lb·ft	lb·ft	
5/16-18	0.3125	0.0524			2,160	11	8	
5/16-24	0.3125	0.0508			2,400	12	9	
3/8-16	0.3750	0.0775			3,200	20	15	
3/8-24	0.3750	0.0878			3,620	23	17	
7/16-14	0.4375	0.1063			4,380	30	24	
7/16-20	0.4375	0.1187			4,900	35	25	
1/2-13	0.5000	0.1419			5,840	50	35	
1/2-20	0.5000	0.1599			6,600	55	40	
9/16-12	0.5625	0.1820			7,500	70	55	
9/16-18	0.5625	0.2030			8,400	80	60	
5/8-11	0.6250	0.2260			9,300	100	75	
5/8-18	0.6250	0.2560			10,600	110	85	
3/4-10	0.7500	0.3340			13,800	175	130	
3/4-16	0.7500	0.3730			15,400	195	145	
7/8- 9	0.8750	0.4620	60,000	33,000	11,400	165	125	
7/8-14	0.8750	0.5090			12,600	185	140	
1- 8	1.0000	0.6060			15,000	250	190	
1-12	1.0000	0.6630			16,400	270	200	
1 1/8- 7	1.1250	0.7630			18,900	350	270	105,000
1 1/8-12	1.1250	0.8560			21,200	400	300	
1 1/4- 7	1.2500	0.9690			24,000	500	380	
1 1/4-12	1.2500	1.0730			26,600	550	420	
1 3/8- 6	1.3750	1.1550			28,600	660	490	
1 3/8-12	1.3750	1.3150			32,500	740	560	
1 1/2- 6	1.5000	1.4050			34,800	870	650	
1 1/2-12	1.5000	1.5800			39,100	980	730	

Notes:

* Torque-tightening values are calculated from the formula $T = KDP$, where T = torque tightening, lb·in; K = torque-friction coefficient; D = nominal bolt diameter, in; and P = bolt clamping load developed by tightening, lb.

† Clamp load is also known as preload or initial load in tension on bolt. Clamp load (lb) is calculated by arbitrarily assuming usable bolt strength is 75% of bolt-proof load (lb/in²) times tensile stress area (in²) of threaded

	SAE Grade 5 Bolts			SAE Grade 7‡			SAE Grade 8¶		
		Tightening torque			Tightening torque			Tightening torque	
Proof load, lb/in^3	Clamps† load P, lb	Dry $K = 0.20$	Lub. $K = 0.15$	Clamps† load P, lb	Dry $K = 0.20$	Lub. $K = 0.15$	Clamp† load P, lb	Dry $K = 0.20$	Lub. $K = 0.15$
		lb·in	lb·in		lb·in	lb·in		lb·in	lb·in
85,000	380	8	6	480	11	8	540	12	9
	420	9	7	520	12	9	600	13	10
	580	16	12	720	20	15	820	23	17
	640	18	13	800	22	17	920	25	19
	900	30	22	1,100	36	27	1,260	41	31
	940	31	23	1,160	38	29	1,320	43	32
	1,120	43	32	1,380	52	39	1,580	60	45
	1,285	49	36	1,580	60	45	1,800	68	51
	2,020	96	75	2,500	120	96	2,860	144	108
	2,320	120	86	2,860	144	108	3,280	168	120
		lb·ft	lb·ft		lb·ft	lb·ft		lb·ft	lb·ft
	3,340	17	13	4,120	21	16	4,720	25	18
	3,700	19	14	4,560	24	18	5,220	25	20
	4,940	30	23	6,100	40	30	7,000	45	35
	5,600	35	25	6,900	45	30	7,900	50	35
	6,800	50	35	8,400	60	45	9,550	70	55
	7,550	55	40	9,350	70	50	10,700	80	60
	9,050	75	55	11,200	95	70	12,750	110	80
	10,700	90	65	12,600	100	80	14,400	120	90
	11,600	110	80	14,350	135	100	16,400	150	110
	12,950	120	90	16,000	150	110	18,250	170	130
	14,400	150	110	17,800	190	140	20,350	220	170
	16,300	170	130	20,150	210	160	23,000	240	180
	21,300	260	200	26,300	320	240	30,100	380	280
	23,800	300	220	29,400	360	280	33,600	420	320
	29,400	430	320	36,400	520	400	41,600	600	400
	32,400	470	350	40,100	580	440	45,800	660	500
	38,600	640	480	47,700	800	600	54,500	900	680
	42,200	700	530	52,200	860	660	59,700	1000	740
74,000	42,300	800	600	60,100	1120	840	68,700	1280	960
	47,500	880	660	67,400	1260	940	77,000	1440	1080
	53,800	1120	840	76,300	1580	1100	87,200	1820	1360
	59,600	1240	920	84,500	1760	1320	96,800	2000	1500
	64,100	1460	1100	91,000	2080	1560	104,000	2380	1780
	73,000	1680	1260	104,000	2380	1780	118,400	2720	2040
	78,000	1940	1460	111,000	2780	2080	126,500	3160	2360
	87,700	2200	1640	124,005	3100	2320	142,200	3560	2660

section of each bolt size. Higher or lower values of clamp load can be used depending on the application requirements and the judgment of the designer.

‡ Tensile strength (min lb/in^2) of all Grade 7 bolts is 133,000. Proof load is 105,000 lb/in^2.

¶ Tensile strength (min lb/in^2) of all Grade 8 bolts is 150,000 lb/in^2. Proof load is 120,000 lb/in^2.

(Reprinted by permission of the copyright owner *Machine Design*.)

Courtesy: Armstrong Gasket Corporation.

Cork compositions combine granualted cork with a binder to form a widely used gasket material. It is used to mate irregular surfaces where temperatures are moderate (less than 160°F) and pressures are low (less than 30 lb/in²). Cork compositions have high compression under load and are used as shock absorbers for joining glass to metal. They are unaffected by oils and aromatic solvents, but are not recommended for use with acid or alkaline solutions. Cork has a tendency to dry, which can cause shrinking and hardening.

When rubber is mixed with cork and vulcanized, the compression and flow characteristics of the resulting gasket can be controlled. Gaskets which are nearly as compressible as cork or almost as incompressible as rubber can be made. The temperature limitations of the material depend on the elastomer. The material die-cuts well and is used in a wide variety of applications. It is not used for gasket steam lines or high-temperature combusion chambers.

PTFE materials are used in applications where broad temperature ranges, resistance to harsh fluids, and chemical inertness are required. The gaskets can be cut from sintered billets, die-cut from sheets, or molded from powder.

Laminated gaskets are often made by combining two or more materials. The usual reason for laminating is to combine the properties of a strong, incompressible material with those of a weak, highly compressible one. Cork or asbestos is sometimes attached to both sides of a steel sheet to form a highly compressible, strong gasket.

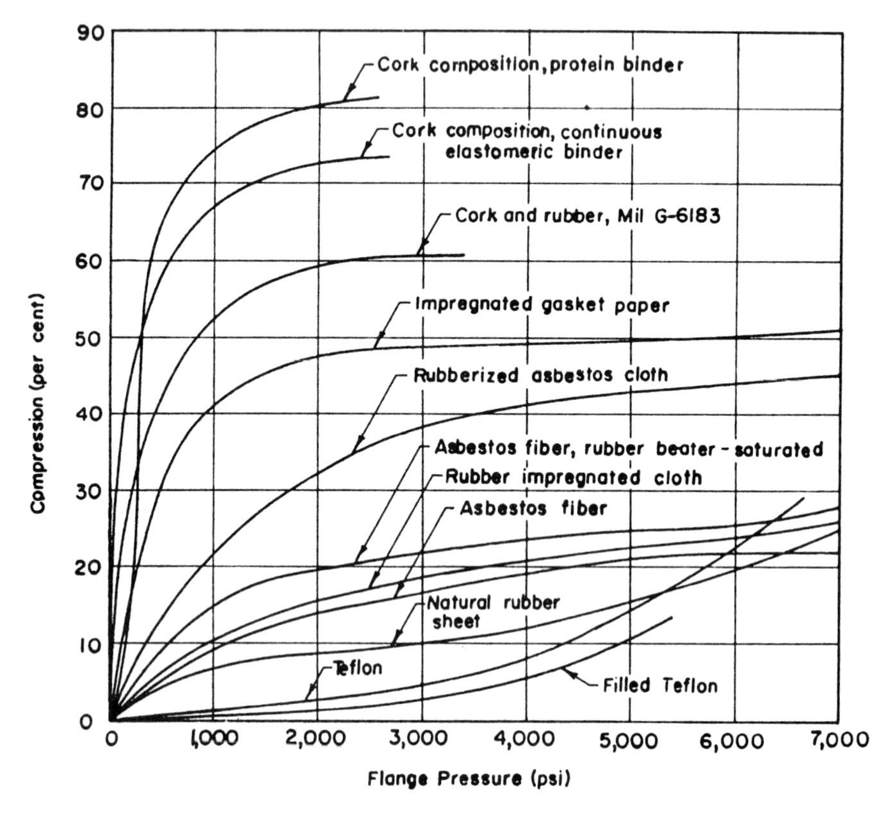

FIG. 21-6 Percent compression vs. flange pressure. (*Courtesy: Machine Design.*)

The properties of nonmetallic gasket materials are given in Table 21-3. Treatments and coatings that are sometimes applied to gaskets are given in Table 21-4.

21-7 Metallic Gaskets Gaskets made from soft materials, such as cork, rubber, and asbestos, are used to seal low-pressure applications. High-pressure gases at high temperatures are usually contained with metallic gaskets. If the product of the internal pressure in pounds per square inch and the operating temperature in degrees Fahrenheit exceeds 250,000, only metallic gaskets can be used. Sealing the head of internal combustion engines is an example of a typical application for metallic gaskets.

The material selected for metallic gaskets depends on operation conditions. Soft metals, such as lead, brass, and copper, are selected if the mating surfaces are rough. Stainless steels are used to prevent corrosion due to oxidation. Aluminum is selected for its light weight and good corrosion resistance. Galvanic corrosion may occur when two dissimilar metals contact an electrolyte. Temperature of the application is also important (Table 21-5) and will dictate what material must be used.

There are many types of metallic gasket designs. Corrugated gaskets (Fig. 21-7) are made from thin metallic sheets (0.010 to 0.031 in) and are used for smooth-faced, low-pressure (500 lb/in^2) applications such as valve bonnets and fuel and combustion lines. In some cases, the corrugated material is coated with a sealing compound to extend the pressure limit to 1000 lb/in^2. Asbestos cord is cemented in the corrugations if the surfaces are uneven. A soft compression filler (usually asbestos) can be partially or totally encased in a metal jacket (Fig. 21-8) to provide more compression than corrugated types. These are used to compensate for flange irregularities in applications up to 1200°F. Spiral-wound gaskets (Fig. 21-9) have the best resilience of all metal asbestos-type gaskets. They are made by winding a preformed V-shaped strip of metal into a spiral. The metallic layers are separated by an asbestos filler. Flat metallic gaskets

FIG. 21-7 Corrugated gasket.

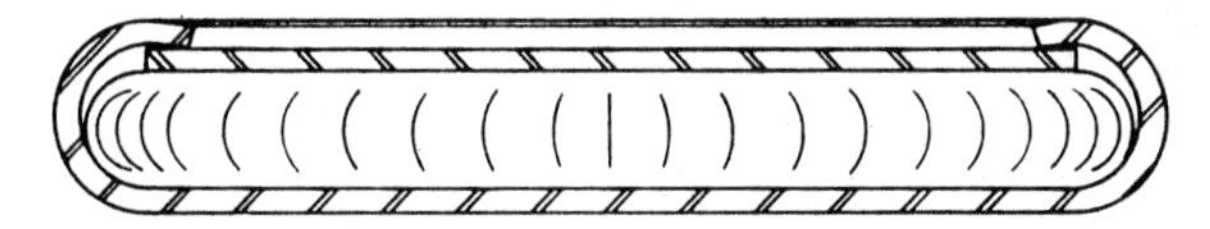

FIG. 21-8 Metal-jacketed soft-filler gasket.

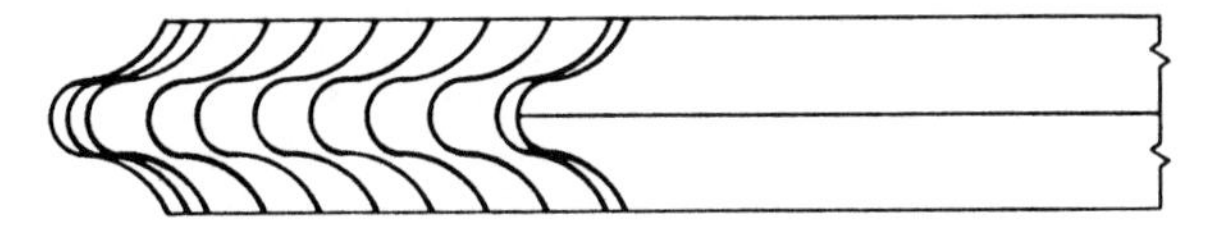

FIG. 21-9 Spiral-wound gasket.

FIG. 21-10 Serrated flat metallic gasket.

TABLE 21-2 Properties of Elastomers Commonly Used for Gasket Materials

Properties	Styrene-butadiene	Ethylene propylene	Polychloroprene	Silicone	Nitrile-butadiene	Chlorosulfonated polyethylene	Fluorocarbon	Polyacrylate
Useful temperature range	−70 to +250°F	−65 to +350°F	−50 to +250°F	−120 to +600°F	−65 to +300°F	−60 to +250°F	−40 to +600°F	−30 to +400°F
Water	E	T	T	T	E	E	E	F
Acid	G	E	G	G	G	E	E	P
Alkali	E	G	E	F	G	E	G	P
Gasoline	P	P	F	P	E	F	E	G
Petroleum oil	P	P	G	P	E	G	G	E
Animal and vegetable oil	P-G	G	G	G	E	F	G	E
Hydrocarbon solvents	P	P	G (except aromatics)	P	G-E	F	E	G
Oxygenated solvents	F	G	P	P	P	F	F	P
Ozone	P	E	G	E	P	E	E	G

Key: E = Excellent, G = Good, F = Fair, P = Poor.

of various shapes and sizes are relatively inexpensive to produce and perform satisfactorily in many types of applications. Flat gaskets seal because the compression loads of the joining bolts exceed the tensile strength of the gasket metal and flow the material into the flange surface irregularities. Some flat gaskets have serrated grooved surfaces to promote deformation (Fig. 21-10). Flat plain metal gaskets should be used with flanges that are grooved or prepared with a surface finish of 80 μin rms.

Solid metal rings with various cross sections are used to provide gas-tight seals at relatively low flange pressure. These gaskets usually seal by line contact or wedging action that causes surface flow. Some designs are pressure-activated. Higher pressures generate greater closing force. Design and uses of solid metal ring gaskets are shown in Fig. 21-11.

Metallic rings of various cross sections (Fig. 21-12) can be designed to expand when internal pressures increase. The rings are usually coated with a soft metallic or plastic material that provides the actual sealing. A typical application is shown in Fig. 21-13. Hollow metallic O rings made from tubing are compressed between parallel faces to form a seal. Vent holes are often provided at the ID for pressure activation (Fig. 21-14).

21-8 Formable Gaskets—Sealants Sealants are used to form gaskets for applications that have relatively low pressure and temperature requirements. The high-

TABLE 21-3 Gasket Materials

Material	Properties	General usage
Untreated paper	Highly permeable, low cost, noncorrosive	To exclude dust and water splash
Treated paper	General purpose, up to 160°F at 0 to 20 lb/in^2	Seal oil, gasoline
Rubber	Incompressible, impermeable, can vary in hardness, can be stretched, can be molded to special shapes	Used when stretching is required for installation. Seal alkalis, hot water, and some acids
Rubber-asbestos	Tough, durable, dimensionally stable, resistant to steam and hot water	Water and steam fittings to 500°F
Cork-compositions	Lightweight, inert, high compression, will not deteriorate, low cost, excellent oil and solvent resistance, high friction, poor resistance to acids and alkalis	Mates irregular surfaces up to 160°F. Used to join glass and ceramic to metal
Cork-rubber	Can compound material to give various levels of compression, high coefficient of friction	General-purpose gaskets. Temperature limits depend upon elastomer
PTFE	Chemically inert, wide temperature range, resistant to most solvents. Expensive	Used in special applications where heat and fluid resistance is required

TABLE 21-4 Treatments and Coatings

Description of treatment	Method of application	Purpose
Synthetic rubber–neoprene	Dipped or sprayed on exterior of cork-composition gaskets	Resists oil penetration
Fungicides (betanaphthol, pentachlorophenol, salicylailide, copper/mercury compounds)	Compounded into materials, dusted onto materials	Resist mold growth
Reflective coating	Dip or spray aluminum paint or lacquer on exterior or gasket	Reflects heat
Graphite	Applied to exterior of gasket. Dusted as a dry flake	Prevents adhesion of gasket to metal surfaces
Adhesives	Applied to exterior of gasket	Bond gasket to flange

viscosity liquids flow over the flange surfaces, filling voids and leveling the contact areas. These properties enable sealants to tolerate surface irregularities that cannot be overcome with preformed gaskets. Formable gaskets can be applied directly to intricate and complex seal-area geometries with relatively simple automatic metering equipment. Relatively low clamping forces are required to seal the joints. Metal-to-metal contact takes place at the high spots, and the sealant fills the void area with a continuous film.

TABLE 21-5 Temperature Limitations for Metallic Gasket Materials

Material	Maximum temperature, °F
Lead	212
Common brasses	500
Copper	600
Aluminum	800
Stainless steel	1600
Soft iron, low-carbon steel	1000
Titanium	1000
Nickel	1400
Monel	1500
Inconel	2000
Hastelloy	2000

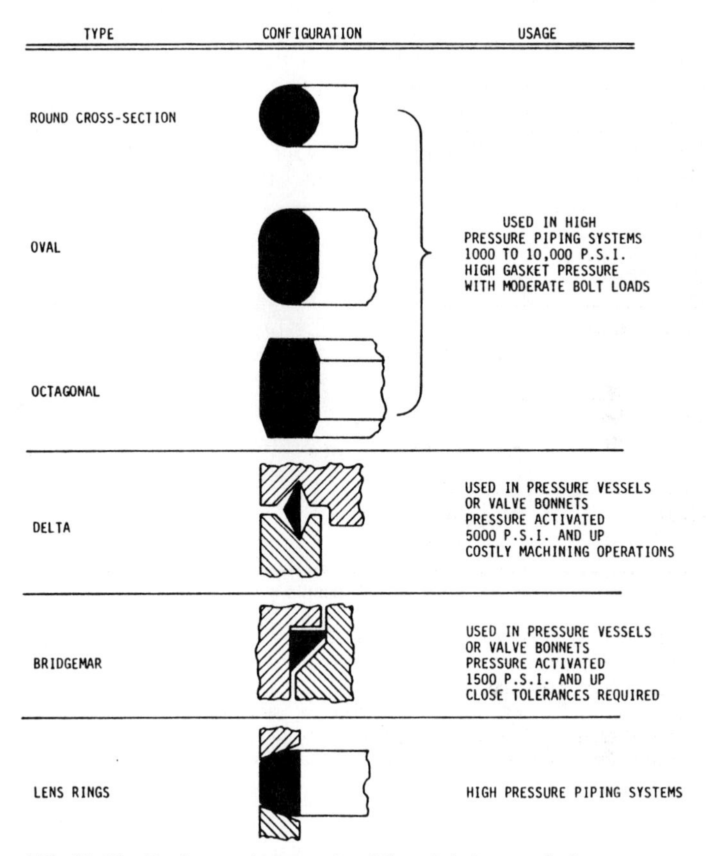

FIG. 21-11 Design and uses of solid-metal ring gaskets.

Sealants can be classified into hardening and nonhardening types. The hardening sealants require cure times that range from 15 min to 2 weeks, depending on sealant type, humidity, and temperature. Some sealants will cure to a rigid state, and others will remain flexible. Nonhardening sealants do not cure. There is no chemical change after application. Solvents may be added to the nonhardening sealants to improve handling. The solvent will evaporate after application to change the consistency of the gasket.

The mechanical properties of sealants are sensitive to composition; minor changes in composition can result in major property alterations. Other factors that must be considered include chemical compatibility, permeability, weather resistance, abrasion resistance, adhesion, electrical characteristics, color, resistance to heat, toxicity, flammability, and method of application.

Sealants are used for static pressure applications. Pipe threads, flange joints, tanks, and pan gaskets are examples of applications for sealants.

Characteristics of typical sealants are given in Table 21-6.

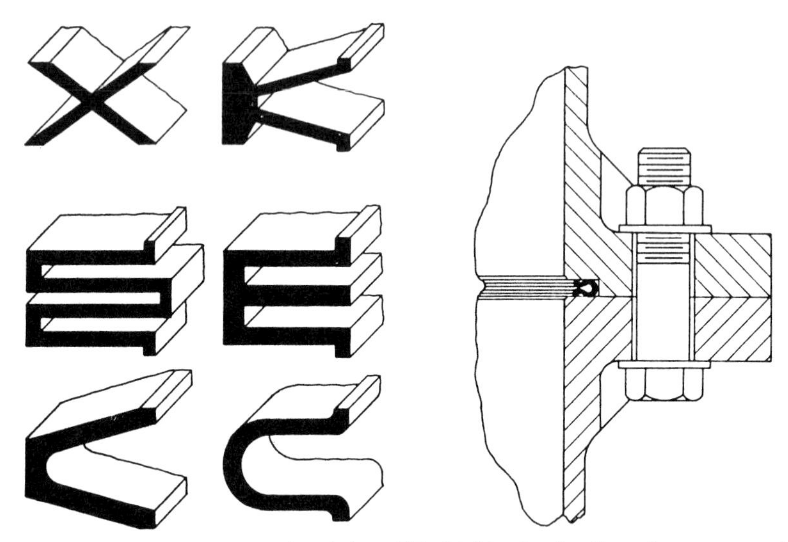

FIG. 21-12 Pressure-actuated metal rings.

FIG. 21-13 Application of pressure-actuated metallic gasket.

STATIC SEALS—ELASTOMERIC

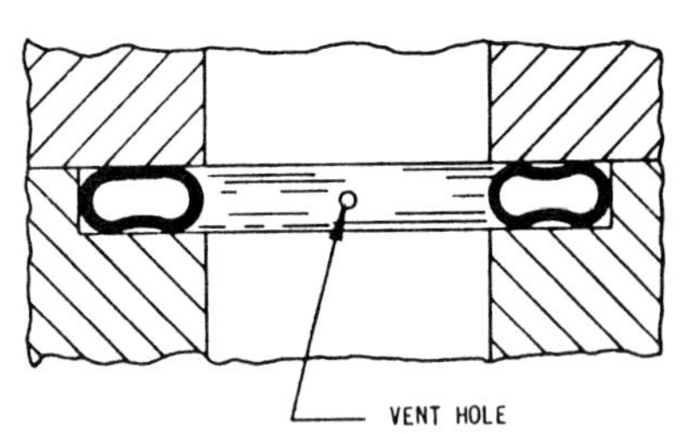

FIG. 21-14 Hollow metallic O ring.

21-9 Elastomeric State Seals A wide variety of shapes can be molded to make an elastomeric static seal. Some special shapes are designed to expand under pressure to increase the rubber-to-metal contact (Figs. 21-15 and 21-16). Cross sections of other seal types include square lathe cut, rectangular lathe cut, oval octagon, pyramid, round, and others (Fig. 21-17). These types of seals are typically installed in machined grooves in one of the surfaces to be sealed. When the surfaces are brought together to form a gland, the rubber ring is squeezed and deformed. This deformation creates a pressure that effectively blocks the leak path in the gland and prevents fluid leakage. The most common shapes are square-cut, rectangular-cut, and round. The round cross section is better known as the O ring.

The usual method of manufacturing rubber rings with square or rectangular cross sections (Fig. 21-18) is to cut them on a lathe to the desired size from vulcanized rubber billets. These rings are recommended only for static applications up to 1500 lb/in^2 fluid pressure. Typical applications for these rings include axial squeeze to form gaskets (Fig. 21-19) and radial squeeze (Fig. 21-20) to seal cylinder walls. The detail of the groove for the lathe-cut rings is given in Figure 21-21.

TABLE 21-6 Sealant Properties

Sealant base	Sealant form	Curing time and temperature range*	Pot life,† h	Relative cost/gal	Uses
Epoxy	One-part liquid	1–16 h @ 180–350°F	½–8		Potting electrical connectors, encapsulating miniature components, coating circuit boards, and cable splicing. Caulking and pipe sealing.
	Two-part liquid	1–8 h @ 75°F; some formulations require 4 to 6 days	½–3	7–12	
	Powder	½–3 h @ 150°F 24 h @ 250°F 2 h @ 320°F 5 min @ 500°F	Unlimited		Used as a sealer and abrasion-resistant coating for concrete
Modified	Two-part liquid modified with polysulfide polymer	½–8 h @ 75°F	⅙–2		
Polyester	Two-part liquid	8–24 h @ 75°F. Less time at higher temperature	⅙–4	6–12	Potting, molding and encapsulating. Gasket and pipe thread sealants
	One-part liquid	¼–24 h @ 75°F	Unlimited if kept in contact with oxygen		
Polysulfide	One-part liquid	14–21 days @ 75°F	1 year	17–25	General construction-type sealing, caulking, and glazing. Sealing between dissimilar metals. Deck caulking and sealing of refracting mirrors
Polyurethane	Two-part	16–24 h @ 75°F	3–6	12–15	Potting and molding of electrical connectors; encapsulating of hydrophones, transducers, and circuit boards. Caulking where compatibility with LOX is required

TABLE 21-6 Sealant Properties (*Continued*)

Sealant base	Sealant form	Curing time and temperature range*	Pot life,† h	Relative cost/gal	Uses
Polyurethane/silicone	One-part	14–21 days @ 75°F	3–9 months	12–18	General construction-type sealing, caulking and glazing
	Two-part	5–10 h @ 180°F	¼–10	45–100	Potting and molding of electrical connectors. Potting firewall connectors and coating of umbilical cables. Sealing of heat shields
Silicone	One-part	1–14 days @ 75°F	6 months	22–35	General construction-type sealing, caulking, and glazing. Potting and molding
Acrylic	Made in one-part release and two-part	Varies, depending on formulation and curing system	—	8–10	Same general uses as nonhardening formulations. Hardening-type materials should be used where pressure limits exceed the limitations of the nonhardening formulations
Oleo-resin	One-part putty or paste	14 days or more @ 75°F to reach specific hardness	Very long	2–6	
Asphalt and bituminous	One-part putty or paste	Several weeks @ 75°F	Very long	1–4	
	Two-part flow-type paste	1–8 h @ 75°F	¼–¼		
		Hardening sealants—nonrigid types			
Polysulfide	Two-part liquid	16–72 h @ 75°F	1–6	10–22	Sealing integral fuel tanks and pressure tanks. Sealing faying surfaces and channel. Potting, molding, and sealing of Plexiglas. General construction sealing and caulking of metal, wood, masonry joints, and glass

Acrylic-Viton	One-part	21 days @ 75°F	3–9 months	8–10	General construction-type sealing, caulking and glazing
	Two-part	24 h @ 75°F	3–4	80–115	For high-temperature service where fuel and oil resistance is required. Sealing fuel tanks, channels, and faying surfaces
Solvent-Release Systems					
Neoprene	—			8–12	Sealing between dissimilar metals. Caulking and general sealing.
Hypalon	—	7–14 days @ 75°F	6–12 months	8–12	Similar in use to Neoprene.
Butyls	—			5–10	Glazing and caulking of metal, glass, and masonry-type joints
Modified epoxy	Two-part	2 h @ 75°F	½–1	7–12	Potting, molding, encapsulating; sealing transformers; high-voltage splicing; capacitor sealing. General construction caulking

TABLE 21-6 Sealant Properties (*Continued*)

Sealant base	Sealant form	Curing time and temperature range*	Pot life,† h	Relative cost/gal	Uses
Nonhardening sealants					
Oleo-resin	One part mastic, putty, or paste	—	—	2–6	Sealing of concrete joints, masonry copings, and tile. Glazing. Sealing of cable pressure splices, electrical conduit, and glass-to-metal meter cases
Asphalt and bituminous	One part mastic, putty, or paste	—	—	1–4	Sealing faying surface metal joints, silos, and air conditioners. Caulking for expansion and contraction joints
Butyl	One part mastic, putty, or paste	—	—	5–10	Caulking expansion and contraction joints. Metal-to-glass seals and metal-to-metal sealing to separate dissimilar materials. Sealing electrical conduit
Acrylic	One part mastic, putty, or paste	—	—	8–10	Pipe joints, glazing, masonry, and metal caulking. Special compounds used as liquid gaskets and pipe dope
Polybutene	One part mastic, putty, or paste	—	—	3–7	General construction-type caulking and glazing. Seal between dissimilar metals

* Ranges given are representative. Formulation and addition of solvents or thinners can alter these values.

† Values given are for standard conditions—75°F and 50 percent relative humidity. Times represent limits obtainable by formulation or addition of solvents. Pot life for specific formulations varies with temperature and humidity.

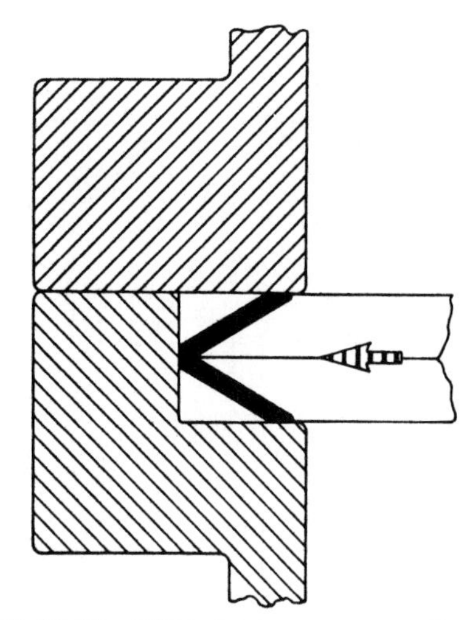

FIG. 21-15 Internal-pressure expanding V ring.

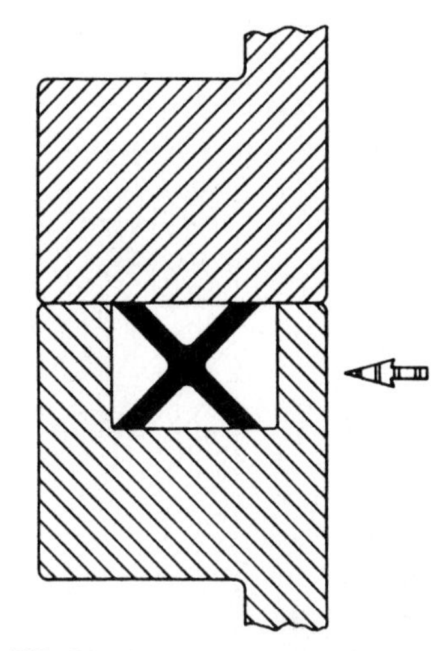

FIG. 21-16 Quad lip seal expands as pressure increases.

The O-ring design (Fig. 21-22) is perhaps the most widely used, since its round cross section readily fits many applications. The depth of the gland must be less than the cross-section diameter of the O ring. The microfinish of the gland should be 32 to 63 μin.

Three O-ring gland shapes suggested for many applications are the rectangle, the vee, and the undercut, Fig. 21-23. The rectangular groove may be used for every type of application, including high-pressure static situations. Its sides slope toward 7° with edges broken to about 0.005-in radius.

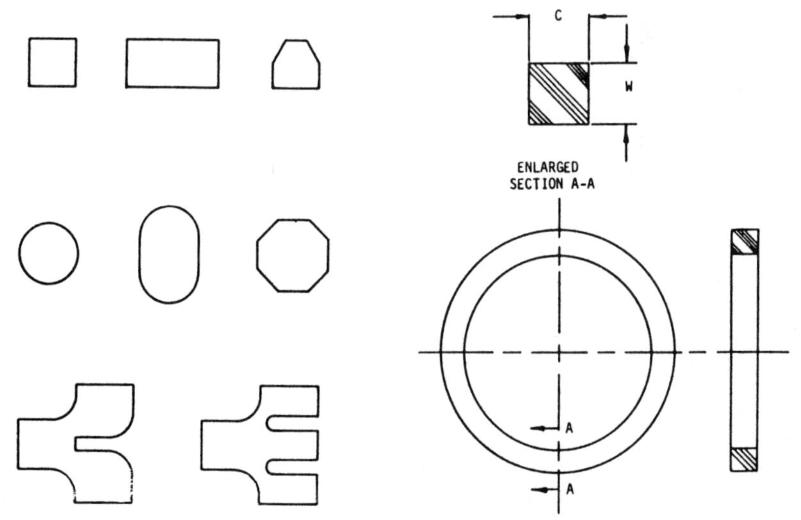

FIG. 21-17 Various cross sections for elastomeric static seals.

FIG. 21-18 Rectangular lathe-cut rings.

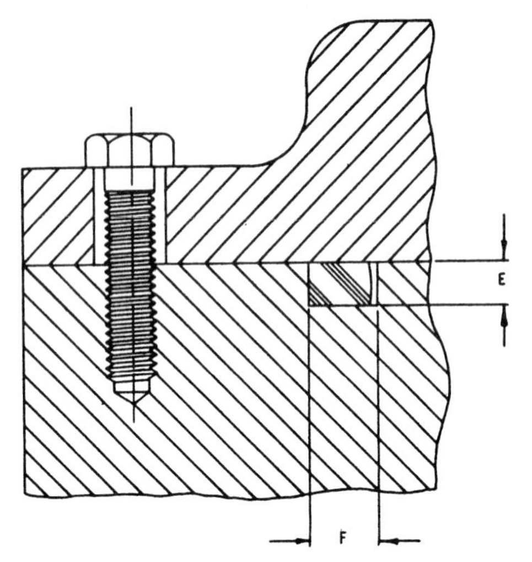

FIG. 21-19 Axial-squeeze application.

The vee groove is suitable for low-pressure applications and low-temperature use, where it resists cold by increasing squeeze on the seal. This gland is the simplest to machine in either male or female configuration, and, therefore, is least expensive. However, a seal in a vee gland generates the highest friction of all three shapes.

The undercut or dovetailed groove prevents extrusion, makes a tighter dynamic seal, and reduces starting and operating friction. Unfortunately, it is the most difficult to machine and, therefore, most expensive.

The minimum squeeze must be 0.006 in, regardless of O-ring cross-section diameter. This will account for differences in the shrinkage of rubber and metal as temperatures

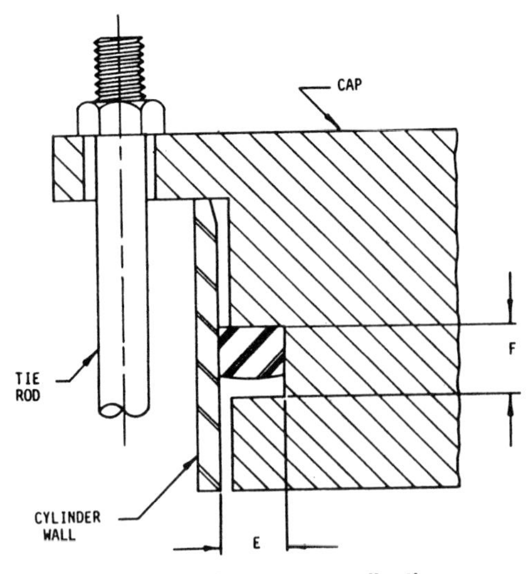

FIG. 21-20 Radial-squeeze application.

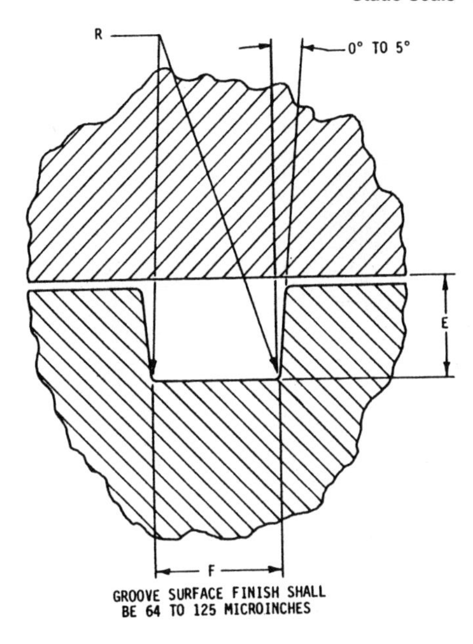

FIG. 21-21 Groove detail for lathe-cut rectangular and square rings.

vary. The maximum recommended squeeze is 35 percent of the O-ring cross-section diameter.

Selection of materials for elastomeric static seals depends on the operating temperature, the internal pressure, and the fluid to be sealed. Table 21-7 provides a guide for material selection.

Static seal leakage can be caused by a variety of reasons ranging from installation damage to extrusion. Extrusion and "nibbling" results when pulsating pressure forces the rubber into the clearance between the metal faces (Fig. 21-24). Increasing the hardness of the elastomer will increase resistance to extrusion and larger clearances between the metal surfaces can be tolerated (Fig. 21-25).

Table 21-8 can be used as a guide when analyzing static seal leaks in the applications.

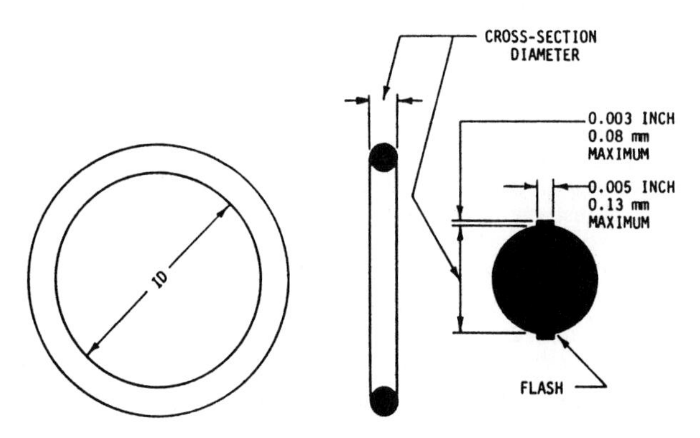

FIG. 21-22 O-ring details.

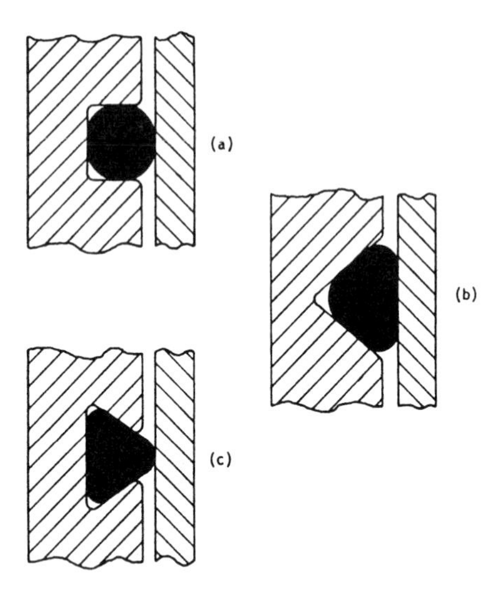

FIG. 21-23 Common O-ring gland shapes.

DYNAMIC SEALS—PACKINGS

21-10 Packing Classification and Usage Packings are used in reciprocating and rotating applications to seal fluids and exclude contaminants. Packings are usually divided into three categories: compression, molded (or automatic), and floating.

Compression packing is used primarily in rotary shaft applications. It is squeeed between the throat of a box and a gland to effect a seal. The packing flows outward to seal against the bore of the box and inward to seal against the moving shaft or rod. Compression packing requires periodic tightening to compensate for wear and loss of volume.

The automatic or molded packing relies on operating pressures to create a seal; therefore, very little gland adjustment is required. These designs use flexible lips to seal against one or both surfaces in a stuffing box. One lip seals against the stationary bore and the other lip against the moving part. In some cases, O rings, quad rings, and other molded static seals are used in dynamic applications. These seals are called "squeeze" seals and depend on internal pressure to effect a seal. More molded packings are used in reciprocating applications. Floating packings include piston rings and segmental rod packings that fit into grooves. These rings are used in rotary and reciprocating motion.

21-11 Compression Packings Compression packings create a seal by being squeezed between the throat of a stuffing box and its adjustable gland (Fig. 21-26). The squeeze forces push the material against the throat of the box and the reciprocating or rotating shaft. When leakage occurs, the gland is tightened further.

Materials are extremely important in selecting the proper packing for an application. The packing must be able to withstand the conditions outlined in Table 21-9. Packing material is made of fabric or metal. The packing cross section can be round, square, or rectangular.

Fabric packings are made from plant fibers, mineral fibers, animal products, or artificial fibers. Plant fibers include flax, jute, ramie, and cotton. Cotton is usually used in cloth form, and the other materials are braided to form high-strength, water-resistant packings. All plant fibers are restricted to low-temperature applications. Asbestos is

TABLE 21-7 Elastomeric Capabilities Guide

	Nitrile (Buna-N)	Ethylene-propylene	(Chloroprene) neoprene	Fluorocarbon (Viton, Fluorel)	Silicone	Fluoro-silicone
			General			
Hardness range, A scale	40–90	50–90	40–80	70–90	40–80	60–80
Relative static ring cost	Low	Low	Low/Moderate	Moderate/High	Moderate	High
Continuous high-temperature limit	257°F	302°F	284°F	437°F	482°F	347°F
	125°C	150°C	140°C	225°C	250°C	175°C
Low-temperature capability	−67°F	−67°F	−67°F	−40°F	−103°F	−85°F
	−55°C	−55°C	−55°C	−40°C	−75°C	−65°C
Compression set resistance	Very good	Very good	Good	Very good	Excellent	Very good
			Fluid compatibility summary			
Acid, inorganic	Fair	Good	Fair/good	Excellent	Good	Good
Acid, organic	Good	Very good	Good	Good	Excellent	Good
Aging (oxygen, ozone, weather)	Fair/poor	Very good	Good	Very good	Excellent	Excellent
Air	Fair	Very good	Good	Very good	Excellent	Very good
Alcohols	Very good	Excellent	Very good	Fair	Very good	Very good
Aldehydes	Fair/poor	Very good	Fair/poor	Poor	Good	Poor
Alkalis	Fair/good	Excellent	Good	Good	Very good	Good
Amines	Poor	Very good	Very good	Poor	Good	Poor
Animal oils	Excellent	Good	Good	Very good	Good	Excellent
Esters, alkyl phosphate (Skydrol)	Poor	Excellent	Poor	Poor	Good	Fair/poor
Esters, aryl phosphate	Fair/poor	Excellent	Fair/poor	Excellent	Good	Very good
Esters, silicate	Good	Poor	Fair	Excellent	Poor	Very good
Ethers	Poor	Fair	Poor	Poor	Poor	Fair
Hydrocarbon fuels, aliphatic	Excellent	Poor	Fair	Excellent	Fair	Excellent
Hydrocarbon fuels, aromatic	Good	Poor	Fair/poor	Excellent	Poor	Very good
Hydrocarbons, halogenated	Fair/poor	Poor	Poor	Excellent	Poor	Very good
Hydrocarbon oils, high aniline	Excellent	Poor	Good	Excellent	Very good	Excellent
Hydrocarbon oils, low aniline	Very good	Poor	Fair/poor	Excellent	Fair	Very good
Impermeability to gases	Good	Good	Good	Very good	Poor	Poor
Ketones	Poor	Excellent	Poor	Poor	Poor	Fair/poor
Silicone oils	Excellent	Excellent	Excellent	Excellent	Good	Excellent
Vegetable oils	Excellent	Good	Good	Excellent	Excellent	Excellent
Water/steam	Good	Excellent	Fair	Fair	Fair	Fair

TABLE 21-7 Elastomeric Capabilities Guide (*Continued*)

Styrene-butadiene (SBR)	Poly-acrylate	Poly-urethane	Butyl	Polysulfide (Thiokol)	Chloro-sulfonated poly-ethylene (Hypalon)	Epichloro-hydrin (Hydrin)	Phospho-nitrilic fluoro-elastomer (PNF)
			General				
40–80	70–90	60–90	50–70	50–80	50–90	50–90	50–90
Low	Moderate	Moderate	Moderate	Moderate	Moderate	Moderate	High
212°F	347°F	212°F	212°F	212°F	257°F	257°F	347°F
100°C	175°C	100°C	100°C	100°C	125°C	125°C	175°C
−67°F	−4°F	−67°F	−67°F	−67°C	−67°C	−67°C	−85°F
−55°C	−20°C	−55°C	−55°C	−55°C	−55°C	−55°C	−65°C
Good	Fair	Fair	Fair/good	Fair	Fair/poor	Fair/good	Good
			Fluid compatibility summary				
Fair/good	Poor	Poor	Good	Poor	Excellent	Fair	Poor
Good	Poor	Poor	Very good	Good	Good	Fair	Fair
Poor	Excellent	Excellent	Very good	Excellent	Very good	Very good	Excellent
Fair	Very good	Good	Good	Good	Excellent	Good	Excellent
Very good	Poor	Poor	Very good	Fair/good	Very good	Good	Fair
Fair/poor	Poor	Poor	Good	Fair/good	Fair/good	Poor	Poor
Fair/good	Poor	Fair/good	Excellent	Poor	Excellent	Fair	Good
Fair	Poor	Poor	Good	Poor	Poor	Poor	Good
Poor	Excellent	Good	Good	Poor	Good	Good	Fair
Poor	Poor	Poor	Very good	Poor	Poor	Poor	Poor
Poor	Poor	Poor	Excellent	Good	Fair	Poor	Excellent
Poor	Fair/poor	Poor	Poor	Fair/poor	Fair	Good	Excellent
Poor	Fair/poor	Fair	Fair/poor	Good	Poor	Good	Poor
Poor	Very good	Good	Poor	Excellent	Fair	Very good	Excellent
Poor	Poor	Fair/poor	Poor	Good	Fair/poor	Very good	Excellent
Poor	Fair/good	Fair	Poor	Good	Fair	Excellent	Fair
Poor	Excellent	Excellent	Poor	Very good	Excellent	Excellent	Excellent
Poor	Excellent	Very good	Poor	Good	Very good	Excellent	Excellent
Fair/good	Very good	Fair	Excellent	Very good	Very good	Excellent	Fair
Poor	Poor	Poor	Excellent	Good	Fair	Fair	Poor
Excellent	Excellent	Excellent	Excellent	Excellent	Excellent	Excellent	Excellent
Poor	Good	Fair	Good	Poor	Good	Excellent	Fair
Fair	Poor	Poor	Excellent	Fair	Fair	Good	Fair

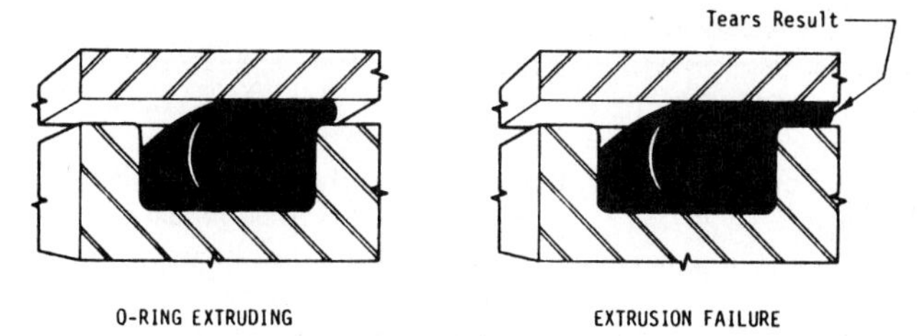

FIG. 21-24 Tears can result from nibbling as a result of pressure pulses.

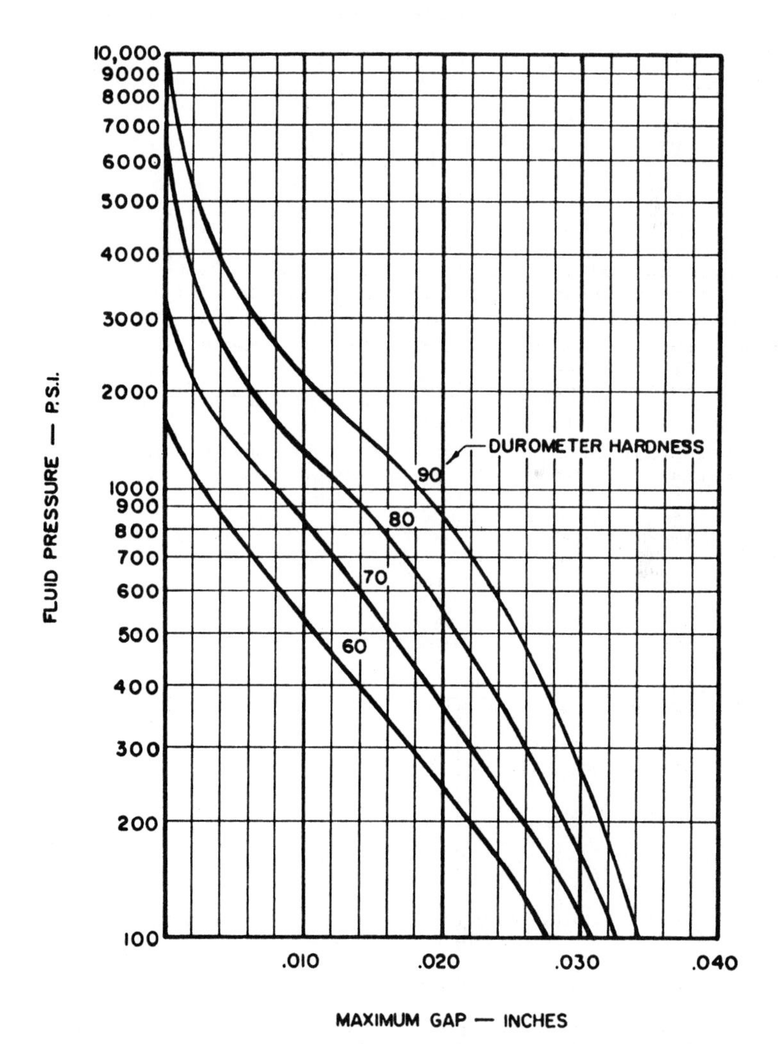

FIG. 21-25 Plot of maximum clearance vs. system fluid pressure for various durometer hardness. (*Courtesy: Hydraulics & Pneumatics.*)

TABLE 21-8 Static-Seal Leakage Analysis

Possible source of trouble	Suggested remedy
Seal has extruded or been nibbled to death	Replace seal and check the following: • Sealing surfaces must be flat within 0.005 in; replace part if out of limits • Initial bolt torque may have been too low; check manual for proper torque • Pressure pulses may be too high; check for proper relief valve setting • If normal operating pressures exceed 1500 lb/in^2, backup rings are required
Seal is badly worn	Replace seal and check the following: • Sealing surface too rough; polish to 16 μin if possible, or replace part • Undertorqued bolts permit movement; check manual for correct setting • Seal material or durometer may be wrong; check manual if in doubt
Seal is hard or has taken excessive permanent set	Replace seal and check the following: • Determine what normal operating temperature is, check system temperature • Check manual to determine that correct seal material is in use
Sealing surfaces are scratched, gouged, or have spiral tool	Replace faulty parts if marks cannot be polished out
Seal has been pinched or cut on assembly	Use petrolatum to hold seal in place with blind assembly; use protective shim if seal must pass over sharp threads
Seal leaks for no apparent reason	Check seal size and parts size; get correct replacement parts

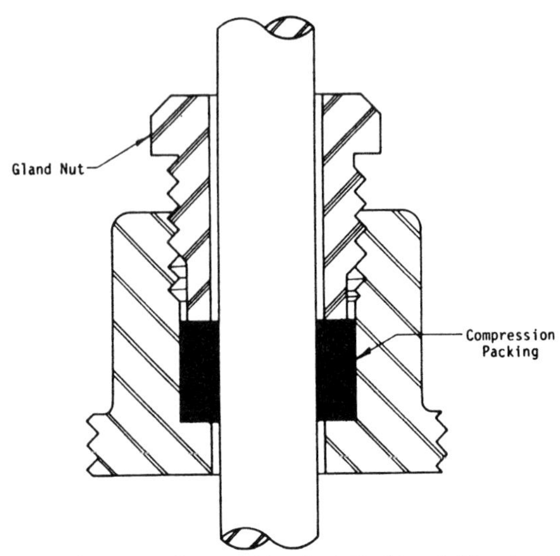

FIG. 21-26 Compression packing installed in stuffing box.

TABLE 21-9 Packing Resistant Conditions

Must withstand the temperature of the applications.
Must be plastic to conform to bore and shaft under moderate pressure.
Must resist application fluid.
Must be elastic to absorb shaft vibrations.
Must be noncorrosive.
Must lose volume slowly to minimize adjustment.

the most versatile mineral fiber. It is resistant to strong mineral acids. Asbestos loses strength at about 800°F.

Leather strips are used to make braided packings. Leather is restricted to applications of 200°F or less. Artificial fibers are used in applications to seal corrosive liquids at temperatures up to 500°F. PTFE and graphite are used to make yarns for packings. The simplest yarn-type packing consists of strands of material twisted together to form a rope (Fig. 21-27). Braided packings can be obtained in round, square, or interlocked configurations (Fig. 21-28). The packings are impregnated with mineral oil, grease, or graphite. This lubricates the moving parts of the application to reduce wear, retain packing flexibility, and help effect a seal. Cloth packings are made from sheets of asbestos or cotton duck that are rolled, folded, or laminated with rubber to form the desired cross section (Fig. 21-29).

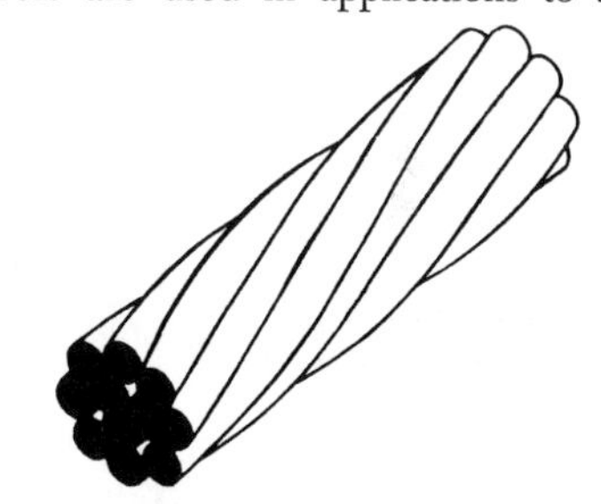

FIG. 21-27 Twisted-yarn fabric packing.

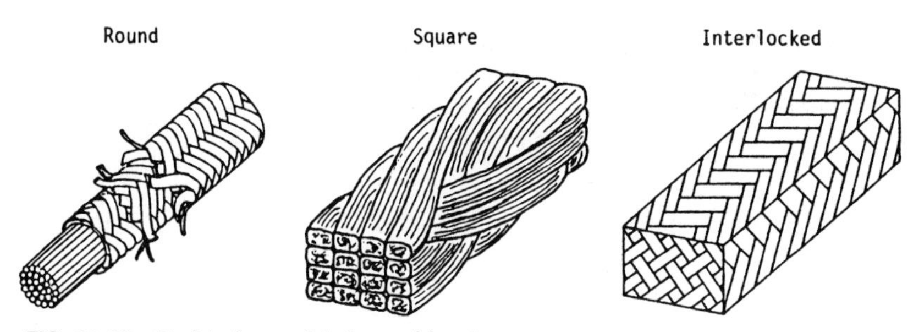

FIG. 21-28 Braided-yarn fabric packing types.

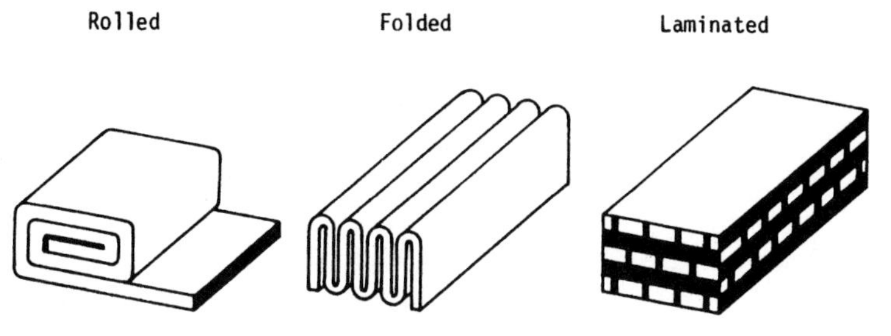

FIG. 21-29 Cloth fabric packing types.

TABLE 21-10 Compression Packings for Various Service Conditions

Fluid medium	Service condition			
	Reciprocating shafts	Rotating shafts	Pistons or cylinders	Valve stems
Acids and caustics	Asbestos (blue) Metallic Semimetallic TFE fluorocarbon resins and yarns	Asbestos (blue) Semimetallic TFE fluorocarbon resins and yarns	TFE fluorocarbon resins	Asbestos (blue) Semimetallic TFE fluorocarbon resins and yarns Graphite yarn
Air	Asbestos Metallic Semimetallic	Asbestos Semimetallic	Leather Metallic	Asbestos Semimetallic
Ammonia	Duck and rubber Metallic Semimetallic	Asbestos Semimetallic	Duck and rubber	Asbestos Duck and rubber Semimetallic
Gas	Asbestos Metallic Semimetallic	Asbestos Semimetallic	Leather Metallic	Asbestos Semimetallic
Cold gasoline and oils	Asbestos Semimetallic	Asbestos Semimetallic	Leather	Asbestos Semimetallic

Low-pressure steam	Asbestos	Asbestos	Duck and rubber	Asbestos
	Duck and rubber	Metallic	Metallic	Duck and rubber
	Metallic	Semimetallic		Semimetallic
	Semimetallic			
High-pressure steam	Asbestos	Asbestos	Metallic	Asbestos
	Metallic	Metallic		Metallic
	Semimetallic	Semimetallic		Semimetallic
Cold water	Duck and rubber	Asbestos	Duck and rubber	Asbestos
	Flax, jute, or ramie	Flax, jute, or ramie		Flax or cotton
	Leather	Semimetallic	Semimetallic	
	Semimetallic			
Hot water	Duck and rubber	Asbestos	Duck and rubber	Asbestos
	Leather	Semimetallic		Duck and rubber
	Semimetallic			Semimetallic

Courtesy: *Machine Design*.

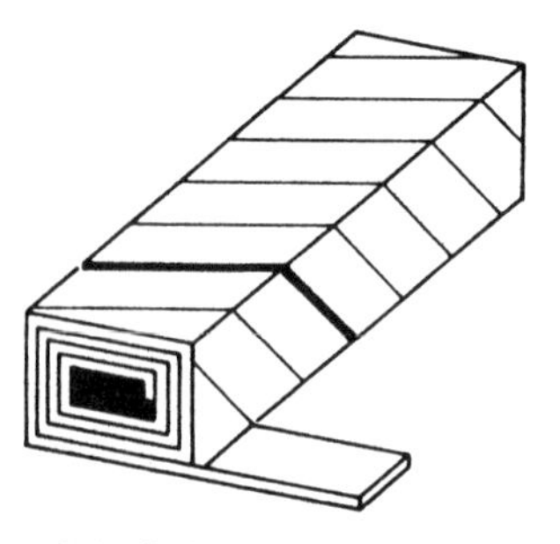

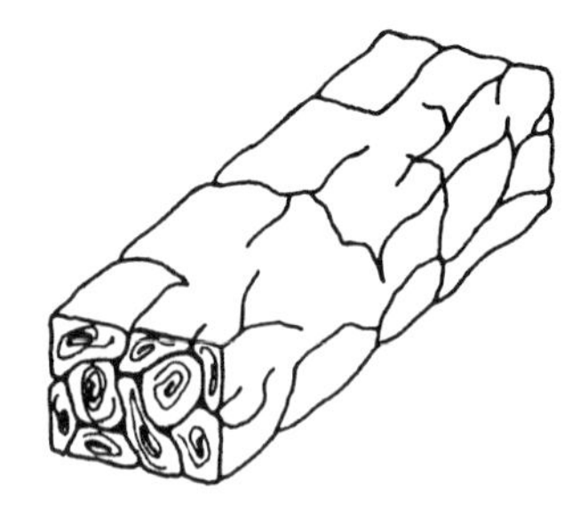

FIG. 21-30 Metallic packings.

Metallic packings are used in high temperature applications. Lead is used in applications up to 450°F. Soft pure copper is flexible; it is used in hot oil, tar, or asphalt pump applications. Aluminum foil is more flexible than copper and is resistant to sour crude oil. Shafts for copper and aluminum packings must be hardened to 500 Bhn; copper and aluminum can handle 1000°F application temperatures. Pure nickel is used to handle steam or caustic alkali materials at temperatures of up to 1500°F. Metallic packing designs are shown in Fig. 21-30. Metal is sometimes combined with other materials to form semimetallic designs (Fig. 21-31).

The types of packings that should be used for various service and fluid conditions are summarized in Table 21-10.

21-12 Molded (Automatic) Packing Molded packings are sometimes referred to as automatic, hydraulic, or mechanical packings. They rely on the fluid pressure of the application to press the packing material against the wear surfaces. Lip-type packings are used primarily for reciprocating shafts. V- and U-ring packings can be used for both high- (50,000 lb/in^2) and low-pressure applications. Multiple V rings are often installed in sets; they are primarily packed on the outside of a reciprocating rod (Fig. 21-32). U rings are usually used to seal a piston (Fig. 21-33). Cup packings have a single lip and are used to seal pistons (Fig. 21-34). Flanged packings seal on the ID (Fig. 21-35) and are used for low-pressure, outside packed installation where there is not enough room for V or U rings.

Exclusion seals are used in conjunction with packings to keep out solid and liquid

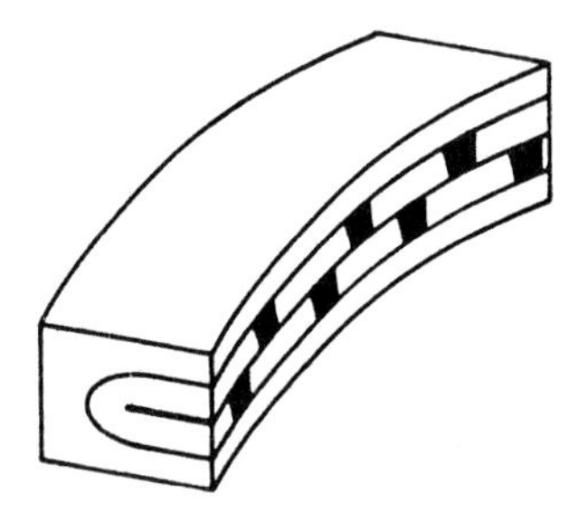

Braided Core

FIG. 21-31 Semimetallic packings.

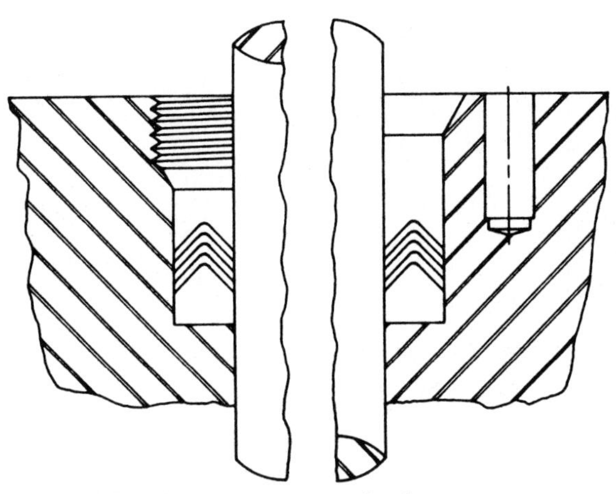

FIG. 21-32 V-ring packing in a gland.

contaminants. Rod wipers (Fig. 21-36) are usually molded from tough materials that resist abrasion. Carboxylated nitriles and polyurethanes are usually used. They are hard (85 to 95 Shore A) with high (6000 lb/in^2) tensile strength. V- and U-cup packings are sometimes combined with a rod wiper to form a utilized seal (Figs. 21-37 and 21-38). Wiper scraper seals (Fig. 21-39) have a metal ring made from copper, aluminum, bronze, or brass that scrapes debris from the shaft. The metal used must have low corrosion and low friction and be soft enough to resist scoring the shaft. Boots (Fig. 21-40) are used to prevent contamination from getting to the packings. Many shapes of boots are available. Specialized boots for hydraulic and pneumatic cylinders accommodate various stroke lengths.

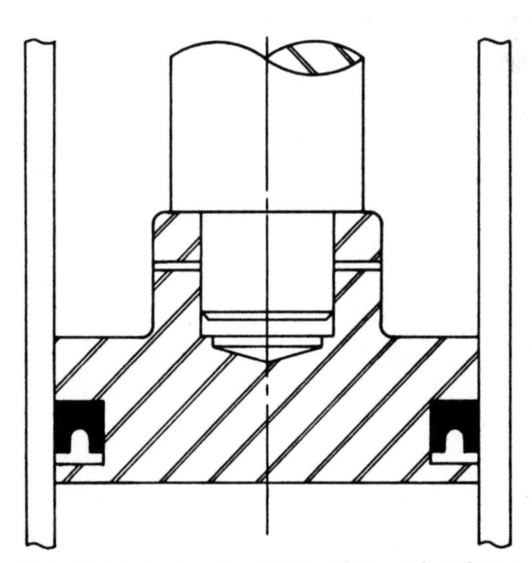

FIG. 21-33 U-ring packing used to seal a piston.

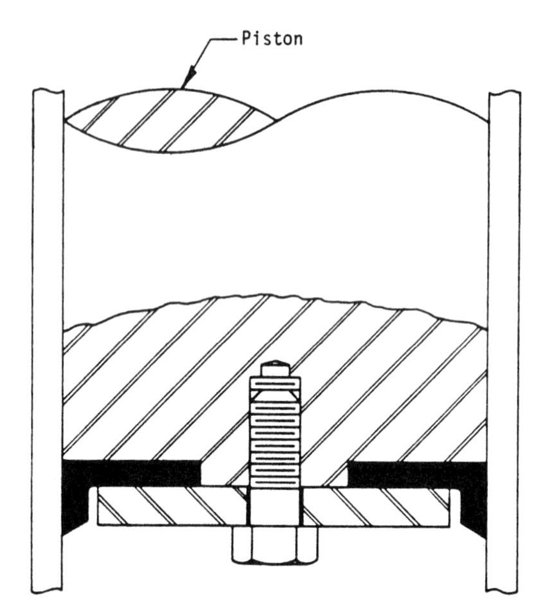

FIG. 21-34 Cup packing.

Packings are made from rubber, fabric-reinforced rubber, and leather. Leather is usually impregnated with a synthetic rubber such as polyurethane to fill voids between the fibers. Leather is quite flexible at temperatures down to −65°F. It deteriorates rapidly at temperatures above 200°F. Leather should not be used for pressurized steam or for strong acids and alkalis. In general, leather can be used if the pH of the liquid lies between 3 and 8.5. Leather will conform to rough surfaces up to 60 μin rms.

Packings are also made from synthetic rubber. The upper pressure limit of homogeneous rubber packings is 5000 lb/in^2. This limit can be extended to 8000 lb/in^2 if the material is reinforced with fabric. Cotton duck is commonly used to reinforce

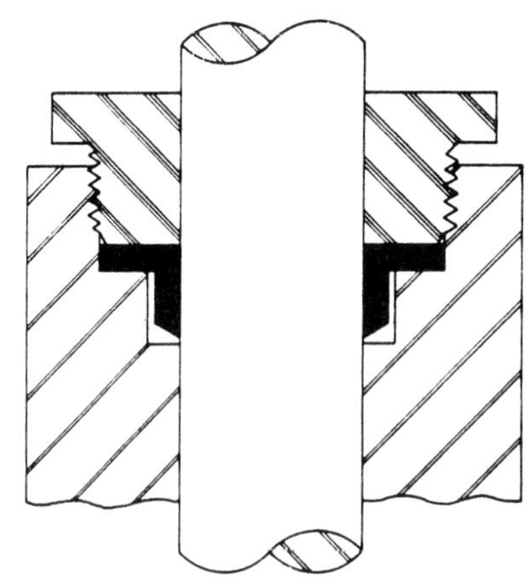

FIG. 21-35 Flange packing.

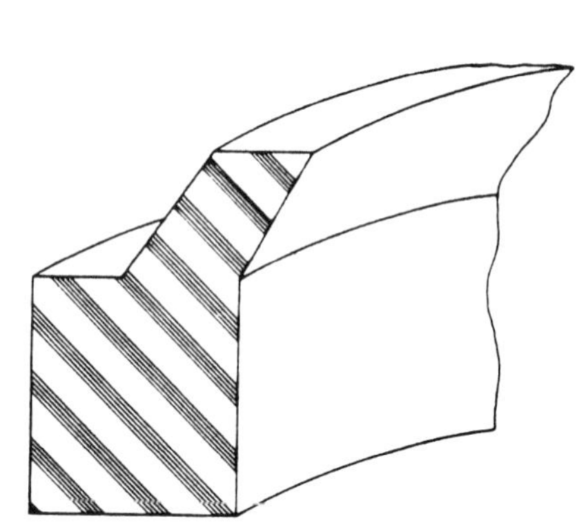

FIG. 21-36 Rod wiper.

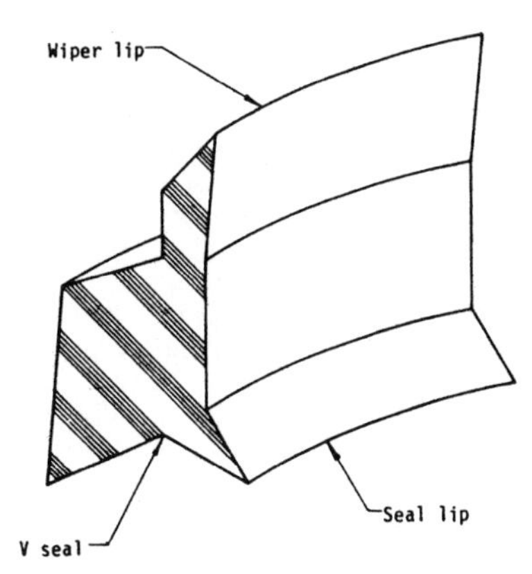

FIG. 21-37 Rod wiper combined with V packing.

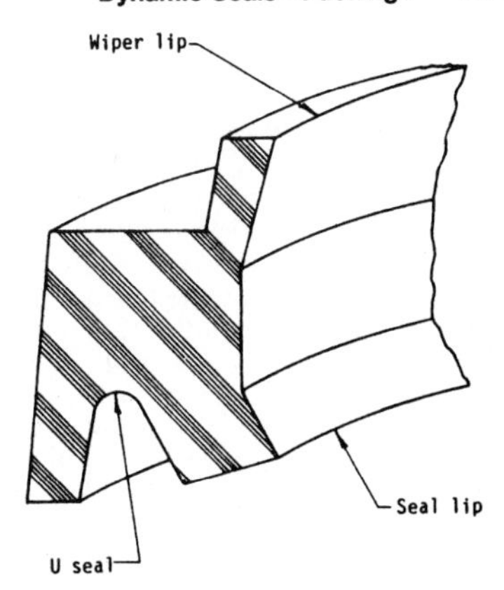

FIG. 21-38 Rod wiper combined with U packing.

rubber when the temperature is less than 250°F. Asbestos is used for temperatures above 250°F. Nylon is used when strength and flexibility are required.

The synthetic rubber used depends on the type of fluid and the temperature. The most common base polymers are polychloroprene, Buna-N, Buna-S, butyl, and fluorelastomer. Polychloroprene and Buna-N are used for oil service, Buna-S for water, and butyl for phosphate esters. Fluoroelastomer is used for high-temperature applications.

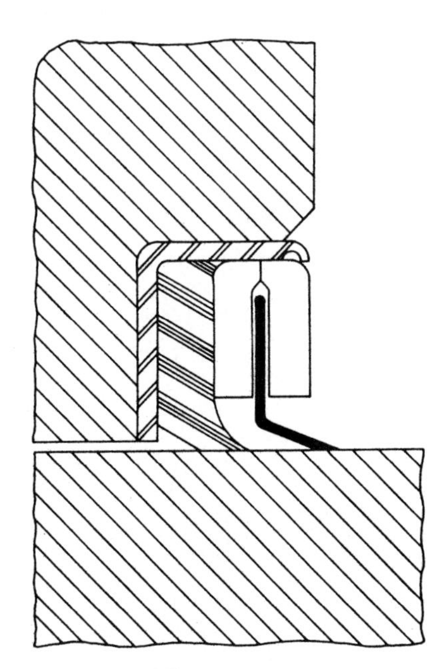

FIG. 21-39 Wiper scraper.

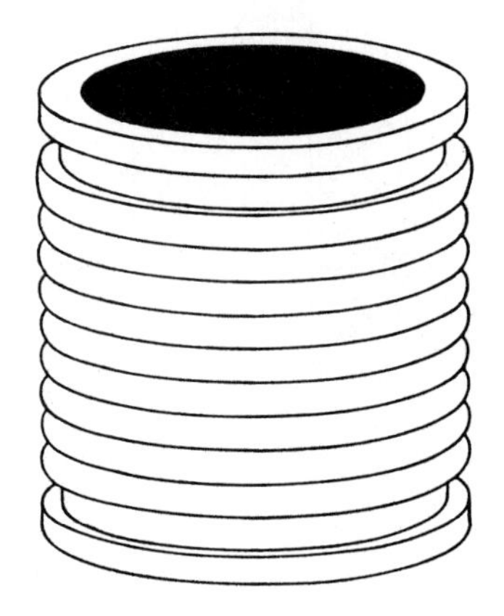

FIG. 21-40 Boots.

Metal surface finish for fabric-reinforced rubber packings should be a maximum of 32 μin rms with a preferred value of 16. On a rough surface, the packing fabric abrades quickly, causing early failure. Homogeneous rubber seals require a surface finish between 8 and 16 μin rms. PTFE packings have very little flexibility, but they are resistant to practically all chemicals and solvents. They are used to seal very corrosive fluids at pressures up to 3000 lb/in^2 and temperatures of 300°F or less.

Table 21-11 can be used as a guide for selecting molded packing materials.

21-13 Floating Packings (Split-Ring Seals) Expanding split-piston-ring seals are used to seal gases in internal combustion engines and fluids in pumps. Contracting split-ring or rod seals are used in linear actuators whenever space, high temperature, or excessive pressure prohibit the use of packings. In all applications, the application pressure forces the rings against the surfaces that require sealing. Expanding split-ring seals must mate on the ID of a cylindrical bore and the side of a piston (Fig. 21-41).

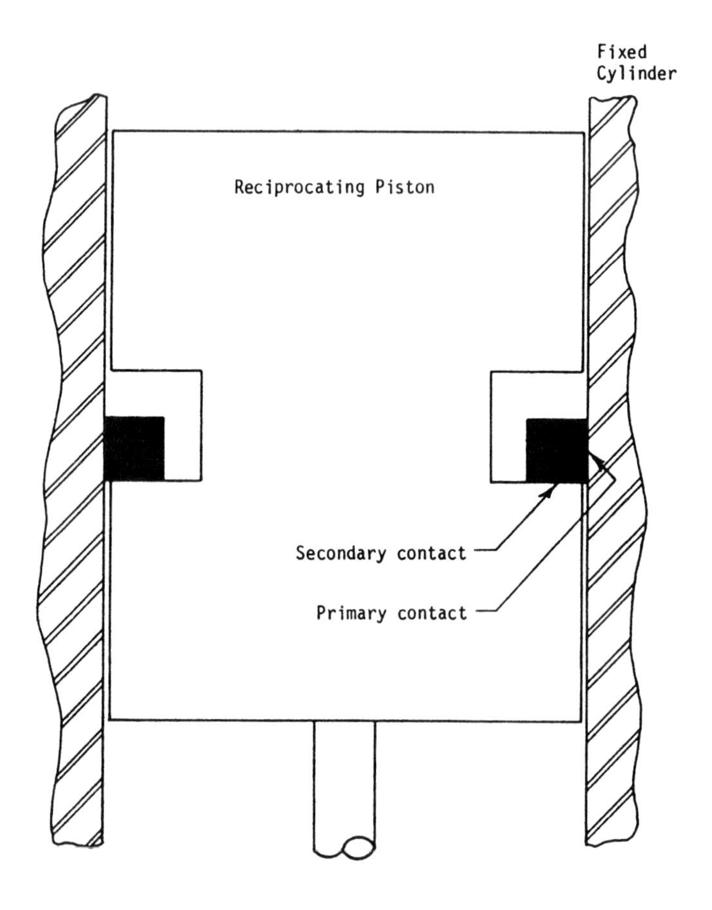

FIG. 21-41 Expanding split ring.

TABLE 21-11 Guide for Selection of Molded Packing Material

Condition	Leather	Homogeneous rubber	Rubber fabric-reinforced
Oil	Good	Good	Good
Air	Good	Good	Good
Water	Good	Good	Good
Steam	Not recommended	Good	Good
Solvents	Not recommended	Good	Good
Acids	Not recommended	Good	Good
Alkalis	Not recommended	Good	Good
Fire-resistant fluids:			
Phosphate ester	Wax or polysulfide impregnation	Butyl base polymer	Butyl base polymer
Water-glycol	Not recommended	Buna-N base polymer	Buna-N base polymer
Water-oil emulsion	Wax, polyurethane, or polysulfide impregnation	Buna-N base polymer	Buna-N base polymer
Temperature range	−65 to +180°F	−65 to +400°F	−40 to +400°F*
Types of metal	Ferrous and nonferrous	Chrome-plated steel and nonferrous alloys with hard, smooth surfaces	Chrome-plated steel and nonferrous alloys with hard, smooth surfaces
Metal finish, rms (max)	60	16	32
Clearances	Medium	Very close	Close
Extrusions or cold flow	Good	Poor	Fair
Friction coefficient	Low	Medium and high	Medium
Resistance to abrasion	Good	Fair	Fair
Maximum pressure, lb/in^2	125,000	5000	8000
Concentricity	Medium	Very close	Close
Side loads	Fair	Poor	Fair
High shock loads	Good	Poor to fair	Fair

* Depending on specific formulation or combination of materials.
Courtesy: *Machine Design*.

Contracting-type seals must contact the side surface of a fixed housing and the OD of a reciprocating rod (Fig. 21-42). The simplest and least expensive split ring has a straight-cut joint (Figure 21-43). It is used as a low-pressure piston seal where joint leakage is not critical. Step joints (Fig. 21-44) of various types are used to reduce joint leakage. Rings are balanced when hydraulic fluid pressure is high. This is accomplished by machining small-circumference grooves in the wear surface of the ring (Fig. 21-45). System fluid flows into the groove and forms a very thin dam, which acts as a pressure reducer. Balancing usually increases leakage. Multiring systems are used to seal high internal pressure. Table 21-12 gives the number of rings required for various pressure levels.

Metal rings are usually used in lubricated applications. If the application is not lubricated, PTFE rings impregnated with carbon-graphite or molybdenum disulfide can be used. Table 21-13 gives the temperature limitations for materials used in ring applications. The limitations of various plating treatments appear in Table 21-14.

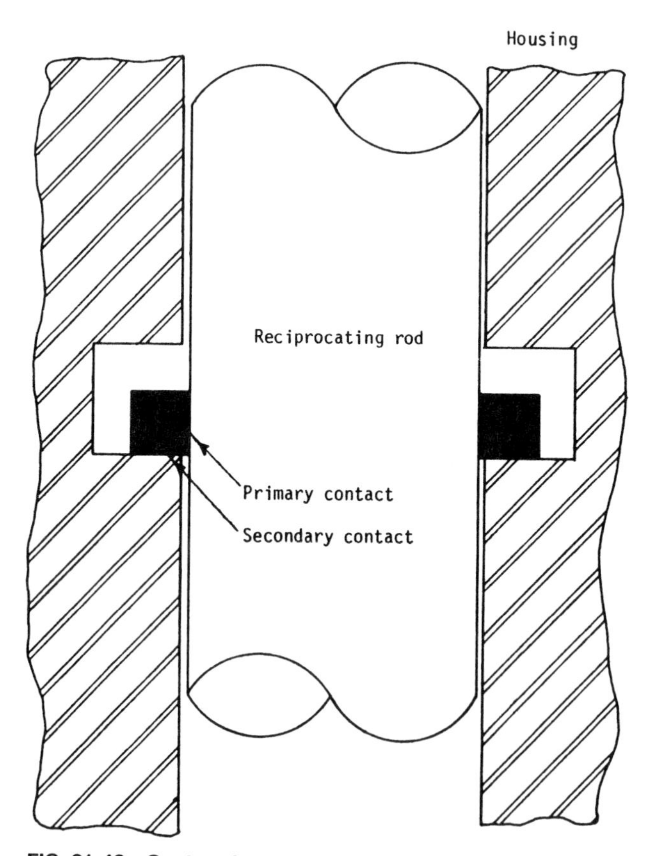

FIG. 21-42 Contracting split ring.

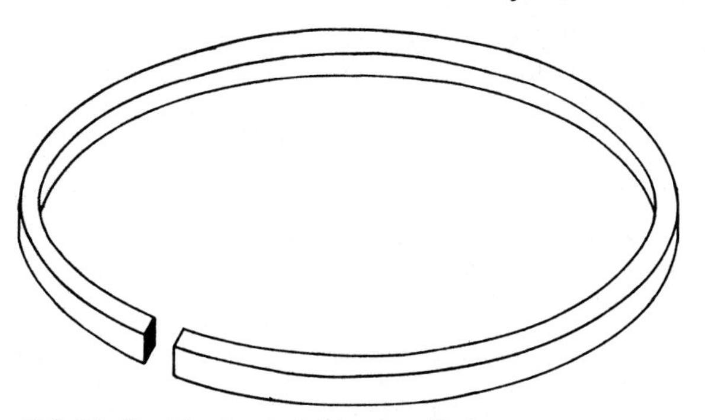

FIG. 21-43 Simple straight-cut split ring.

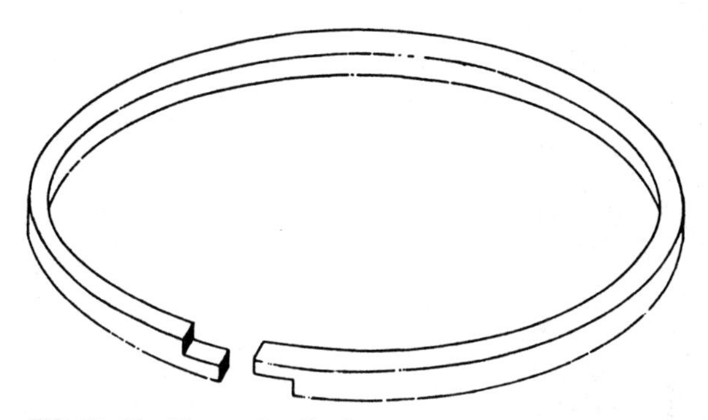

FIG. 21-44 Step-cut split ring.

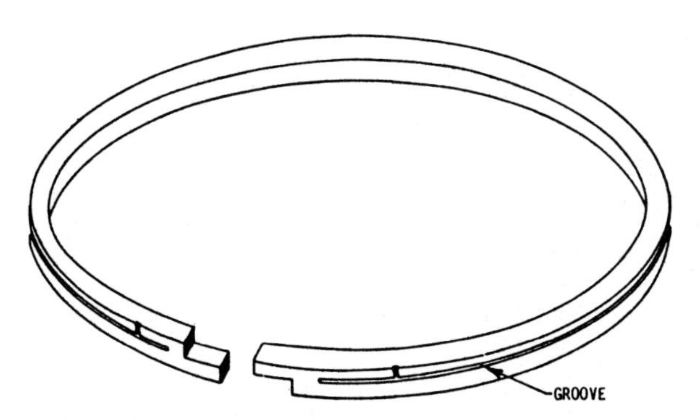

FIG. 21-45 Balanced ring.

TABLE 21-12 Ring Pressure Limitations

Pressure, lb/in^2	No. of rings required
Up to 300	2
300 to 900	3
900 to 1500	4 plain face
	5 balanced
1500 to 3000	5 plain face
	6 balanced
3000 and up	6 balanced

TABLE 21-13 Material Temperature Limitations

Material	Temperature, °F
Low-alloy gray irons	650
Malleable iron	700
Ductile iron	700
Ni-Resist	800
Ductile Ni-Resist	1000
410 stainless steel	900
17-4 pH stainless steel	900
Bronze	500
Tool steel, Tc 62-65	900
Carbon (high temperature)	950
K-30 (filled Teflons)	450 to 500
S-Monel	950
Polyimide	750

Courtesy: *Machine Design*.

TABLE 21-14 Plating Temperature Limitations

Surface treatment	Temperature, °F
Chromium plate	500
Tin plate	700
Silver plate	600
Cadmium-nickel plate	1000
Flame plate	1000–1600
Flame plate LC-1A	1600
Flame plate LA-2	1600

Courtesy: *Machine Design*.

21-14 Diaphragm Seals Diaphragms are membranes that are used to prevent movement of fluid or contamination from one chamber to another (Fig. 21-46). There are static diaphragms that merely separate fluids. These diaphragms are subject to very little displacement. Dynamic diaphragms are attached to the stationary and moving members and usually transmit force or pressure. Dynamic diaphragms can be flat or rolling. A flat diaphragm has no convolutions or convolutions that are less than 180°. Most heavy-duty flat diaphragms have molded convolutions to allow flexibility (Fig. 21-47). They are attached to the stationary housing at the edges. Plates are used to attach the diaphragm to the pushrod. Springs are used to return the diaphragm to the neutral position; they ride in the central disk. Positive mechanical stops are used to prevent overstroking the diaphragm. Diaphragm motion should be less than 90 percent of the maximum possible stroke. The diaphragm must be thin enough to prevent wrinkling.

Rolling diaphragms are used for medium- and high-pressure long-stroke applications (Fig. 21-48). As the piston moves down, the diaphragm rolls off the piston sidewall onto the cylinder sidewall. The diaphragms are usually from 0.010 to 0.035 in thick. The pressure is supported by the cylinder head, and it holds the diaphragm against the piston and cylinder walls. Rolling diaphragms are usually molded in the shape of a top hat, and the convolute is generated by inversion during assembly. In some installations, the resiliency of the material may cause it to revert to its original position (Fig. 21-49). Sidewall scrubbing and high wear will result. This effect is prevented by using retained plates with curved lips. Rolling diaphragms cannot be used in applications where the low- and high-pressure chambers can be reversed. Pressure reversal

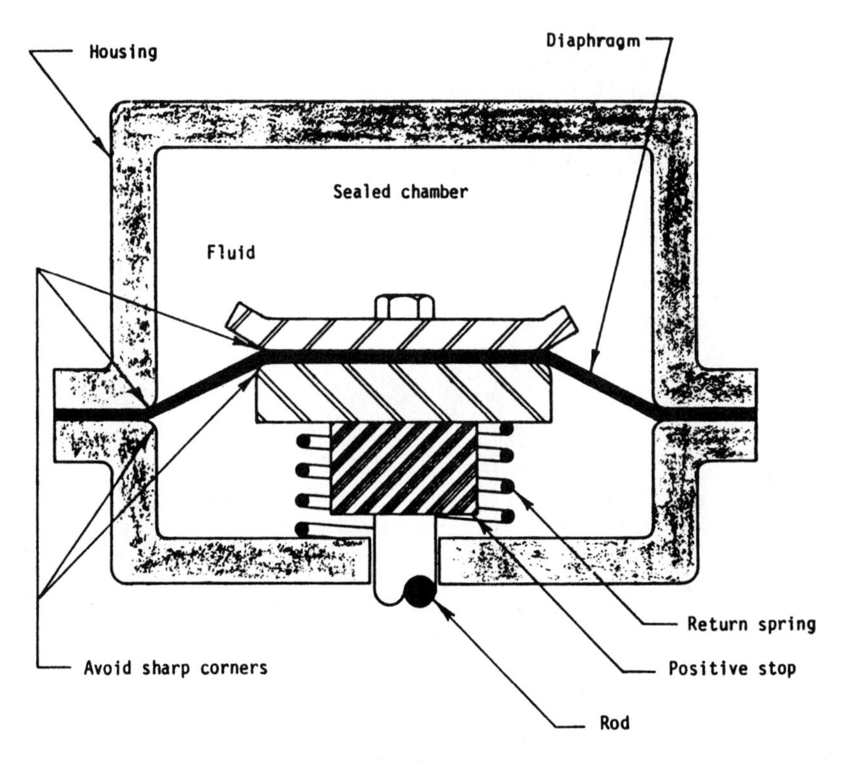

FIG. 21-46 Typical diaphragm application.

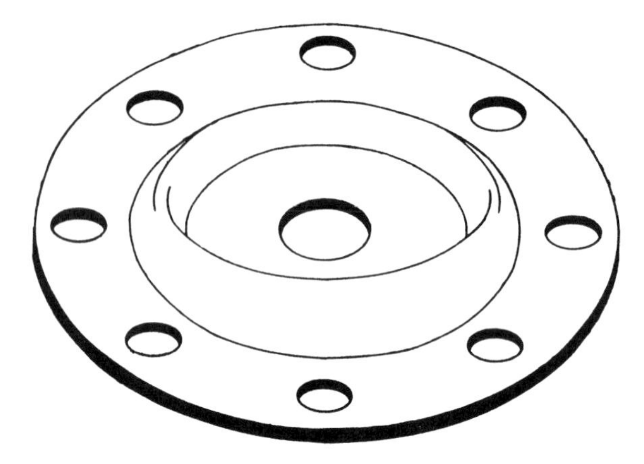

FIG. 21-47 Flat convoluted diaphragm.

will cause the diaphragm to wrinkle and scuff against the sidewall. High wear and premature failure can result.

Materials used to make diaphragms must have a high burst strength. This is the pressure that will rupture the material. Material modulus and tensile strength must be high. Blends of fabrics and elastomers are used in diaphragms. Cottons and nylons are used in flat diaphragms at temperatures less than 250°F. Nylon resists abrasion and fatigue, but will creep under pressure and take a set. Davron can be used at temperatures up to 350°F, and Nomex is used for prolonged exposures at 500°F. All fabrics must be impregnated with elastomers to render the diaphragm impermeable. Elastomeric properties appear in Table 21-15.

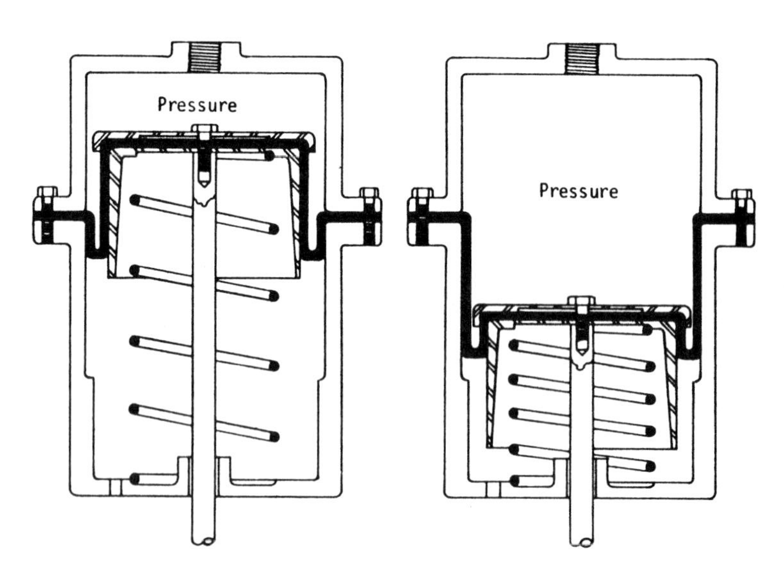

FIG. 21-48 Rolling diaphragm.

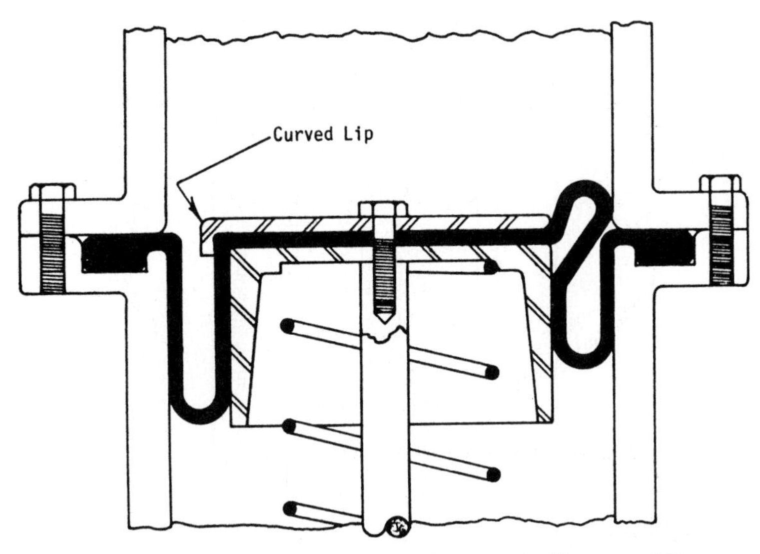

FIG. 21-49 Rolling diaphragm inversion is prevented by curved lip.

DYNAMIC SEALS—CONTACTING TYPE

21-15 Mechanical Face Seals Mechanical face seals are used in applications where low leakage rates and long life are essential. Standard face-seal designs are ideal for high pressure [up to 300 lb/in^2 (absolute)], high shaft speeds (up to 50,000 r/min), and broad temperature ranges (−425 to 1200°F). Special design arrangements are required for conditions that exceed these limits. In some cases, tandem seals are required to handle large pressure differentials.

Mechanical seals employ two wearing faces. The seal head is usually spring-loaded and is pressed against a mating ring or seal. Rotating springloaded heads (Fig. 21-50) made from graphite or plastic are used with shafts that are machined to close tolerances from high-quality materials. The stationary ring, which is mounted in the housing, is made of ceramic or metal. If comparatively high speeds (5000 r/min or more) are encountered, stationary heads and rotating seals (Fig. 21-51) give the best results. The stationary seal head requires less critical dynamic balancing. Close bore tolerances with a high-grade finish must be maintained if a stationary head is used. The method of sealing depends on the direction of the pressure. The configurations shown in Figs. 21-50 and 21-51 must be adjusted to ensure that internal pressure will keep the seal faces closed.

It is necessary to provide a static seal between the wear members and the shaft or housing the member is attached to. This prevents leakage between the members. Elastomeric elements are typically used in pusher-type seals to prevent leakage between the rotating shaft and the rotating head. As the sealing face wears, the static seals are pushed along the shaft by the spring. These elements include the O ring, V ring, U cup, and wedge (Fig. 21-52).

Some seals, known as nonpusher types, use a bellows-shaped element to form a static seal between itself and the shaft. The bellows arrangement is used to compensate for axial movement. Elastomeric bellows (Fig. 21-53) can be used. The upper temperature limitation of the seal is determined by the material used for the static seal or the bellows. In general, most synthetics are limited to −40 to 225°F. TFE fluorocarbon can be used from −400 to 500°F. Metal bellows are used to extend the upper temperature limit to 1200°F. Metal bellows can be produced from tubing by

TABLE 21-15 Elastomeric Properties

Elastomeric type	Air permeability rating	Operating temperature limits, °C	Properties
Silicone	170 to 260	−80 to 260	General-purpose, low-temperature resistant; high permeability
Fluorosilicone	50	−60 to 230	Oil-resistant, high temperature
Nitrile	0.25 to 1.00	−40 to 120	Oil-resistant, low cost, attacked by ozone
Neoprene	1.40	−35 to 120	Weather-resistant; fair oil-resistance
Ethylene propylene	9.60	−40 to 150	Steam-, ozone-, acid-, and alkali-resistant
Fluorocarbon	0.32	−20 to 280	Oil-, fuel-, and chemical-resistant
Polyacrylate	1.50	−30 to 175	Hot-oil- and ozone-resistant
Epichlorohydrin	0.15 to 0.70	−40 to 150	Low permeability, oil-resistant

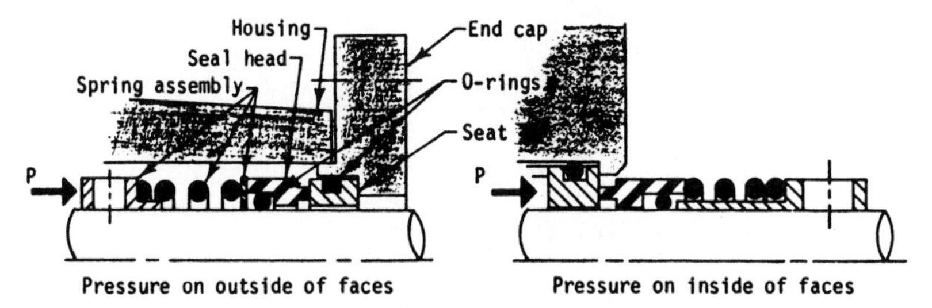

FIG. 21-50 Rotating seal heads.

hydraulic forming in dies or by roll forming (Fig. 21-54). A more expensive method of making the bellows is to weld disks together (Fig. 21-55). The welded bellows requires less axial space than the formed metal bellows.

Metal bellows seals are suited for use in harsh environments that would destroy elastomers. They are used with liquid oxygen, fluorine, acids, and a wide variety of caustic chemicals.

Face Loading. The sealing faces are kept closed under all operating conditions by providing an axial load. The load must be high enough to overcome friction, but not high enough to accelerate seal wear and shorten life.

Belleville, finger, slotted, and wavy washers provide a loading force in a small amount of axial space. These spring types must be heat-tempered to be effective. This requirement limits the selection to materials that are not as corrosion-resistant as stainless steel.

A single spring with a heavy coil that surrounds the shaft (Figs. 21-50 and 21-51) can withstand a relatively high degree of corrosion. This design generally requires a great deal of axial space. Centrifugal forces will attempt to unwind the space. It is generally difficult to provide uniform face loadings with single springs.

Seals with multiple springs require less axial space than single-coil springs. The same springs can be used in various combinations for seals of many sizes (Fig. 21-56).

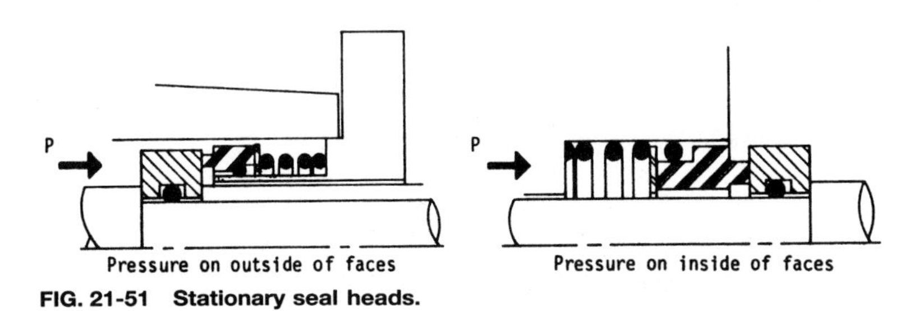

FIG. 21-51 Stationary seal heads.

FIG. 21-52 Pusher types.

FIG. 21-53 Elastomeric bellows.

Multiple springs resist unwinding from centrifugal force. A small amount of axial movement can generate the desired force changes.

The small cross section of the wire used to make the springs will corrode rapidly unless corrosion-resistance materials are used.

Rubber elements are sometimes used to load the faces of seals that are used in relatively low-pressure, low-speed applications. These seal types (Fig. 21-57) are used for track and idler wheels on earth-moving equipment.

Metal bellows that are used to replace static seal elements are also used to supply the force to load the sealing faces. In addition to extending the upper temperature limits of the seal, the metal bellows serves a dual purpose and reduces the number of components in the seal (Table 21-16).

Seal Balancing. As the internal pressure increases, the force pushing the wear faces together can increase dramatically if the seal is unbalanced (Fig.

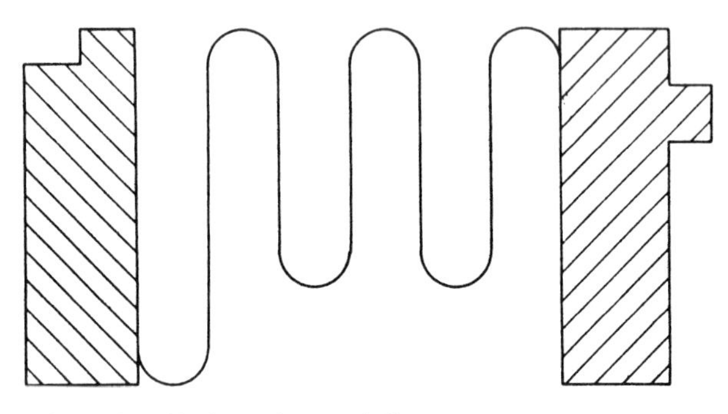

FIG. 21-54 U-shaped convolutions.

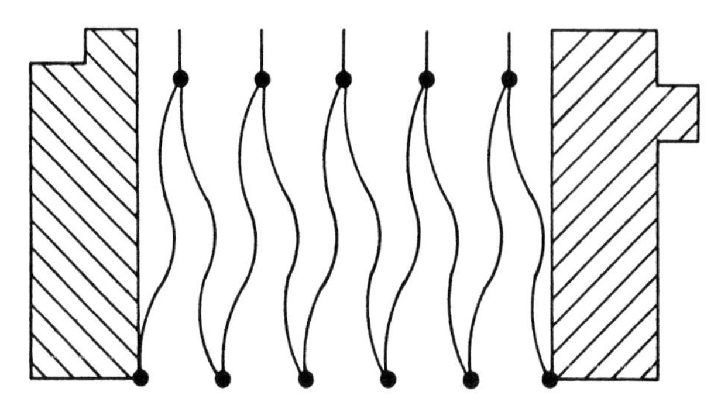

FIG. 21-55 Bellows welded from disks.

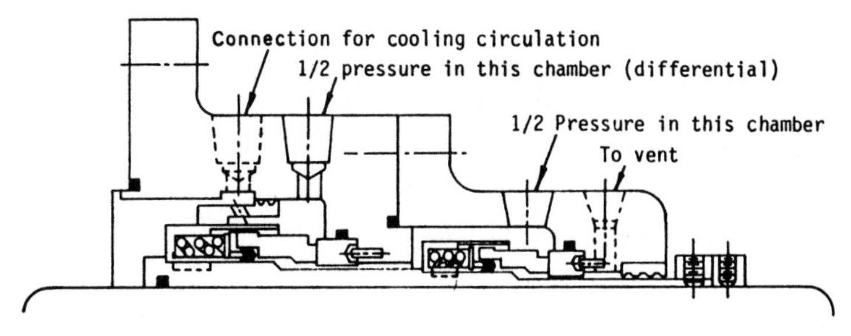

FIG. 21-56 Tandem pusher seal arrangement with V-ring static seals and multiple springs.

21-58). The wear on the face surfaces will increase as the hydraulic pressure increases. The hydraulic pressure can be balanced by designing a step in the seal head. Any portion of the face load can be cancelled by changing the step dimensions (Fig. 21-59). The balance ratio is defined as the ratio of the amount of the face area above the balance line to the amount of face area below. A 55/45 percent ratio is usual with poor lubricants, and an 85/15 percent ratio is usual with good ones. Completely balanced seals (Fig. 21-60) depend only on the spring force to load the faces. The load is independent of internal pressure.

PV Relationships. An important criterion in determining the limitations of seal-face materials is the *PV* factor. The factor is the product of the unit pressure acting on the seal-face junction (in pounds per square inch) and the rubbing velocity (in feet per minute). Unit pressure results from spring, bellows, or diaphragm tension and the unbalanced portion of the hydraulic load.

For any given combination of seal-face materials, the limiting *PV* value is set by factors such as the tenacity of the film on the face surfaces, rate of heat conduction away from the heat-sensitive elements of the seal, and quality of the surfaces.

At high rubbing speeds, the quality of the seal-face surfaces becomes very important. Surface quality is evaluated by roughness, expressed as an rms value, and by flatness, usually given in light bands. As speeds increase, the degree to which planes of the mating surfaces approach the true normal to the shaft axis becomes very important.

The factors that are most significant to seal performance are leakage, power consumption, and life or wear. The wear rate, and thus life, is highly dependent on the maintenance of a lubricant film between the seal faces. Theoretical analyses that examine the factors that influence leakage, power, and life generally define the pressure

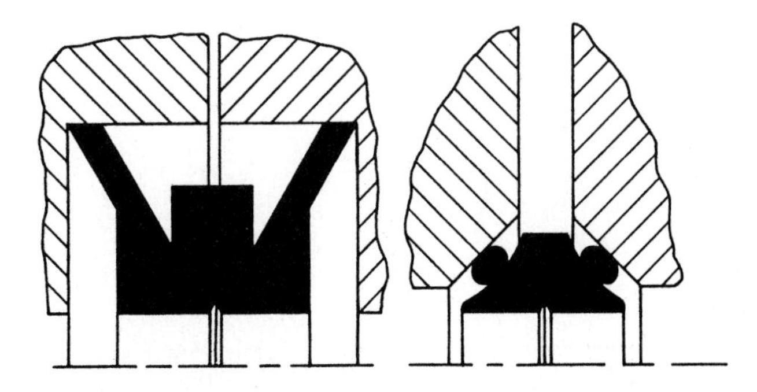

FIG. 21-57 Rubber-loaded faces.

TABLE 21-16 Common Construction Materials for Mechanical Seals

Primary ring	Mating ring	Secondary seal	Spring-housing components
Carbon	Ceramic	Nitrile Ethylene propylene	18-8 stainless
Tungsten carbide	Cast iron	Fluorocarbon resin	316 stainless
Bronze	Cobalt-base alloy		Nickel-base alloys
Silicon carbide	Silicon carbide Tungsten carbide	Fluoroelastomer Perfluoroelastomer	Titanium

distribution across the seal face. This pressure distribution is used to predict the volumetric flow rate and the torque. Most face-seal designs have the high-pressure side of the seal on the outer diameter of the interface. Centrifugal force opposes any leakage that will flow radially inward. The radial pressure distribution for laminar flow between the smooth parallel surfaces defined by the geometry of Fig. 21-61 is given by Eq. (21-2). The amount of seal leakage appears in Eq. (21-3), the theoretical condition for zero leakage is given by Eq. (21-4), and the power consumed is given by Eq. (21-5).

$$p - p_1 = \frac{3\rho\omega^2}{20g}(r^2 - R_1^2) - \frac{6\nu}{\pi h^3}\ln\frac{r}{R} \tag{21-2}$$

$$Q = \frac{\pi h^3}{6\nu \ln(R^2/R^1)}\left[\frac{3\rho\omega^2}{20g}(R_2^2 - R_1^2) - p_2 - p_1\right] \tag{21-3}$$

$$p_2 - p_1 = \frac{3}{20}\rho\omega^2(R_2^2 - R_1^2) \tag{21-4}$$

$$P = \frac{\pi\nu\omega^2}{13{,}200h}(R_2^4 - R_1^4) \tag{21-5}$$

where p = pressure at radial position r, lb/in²
p_1 = pressure at seal ID, lb/in²
p_2 = internal hydraulic pressure, lb/in²
r = radial position, in
R_1 = ID of rotating member, in
R_2 = OD of rotating member, in
h = thickness of fluid between members, in
Q = leakage rate, in³/s
ν = kinematic viscosity, lb•s/in²
ρ = fluid density, lb/in³
ω = rotational speed, rad/s
P = power loss, hp

Troubleshooting Face Seals. Leaking face seals often show telltale contact patterns that help identify the cause of failure. Table 21-17 lists things to look for.

21-16 Radial Lip Seals Radial lip seals are used primarily to retain lubricants in equipment with rotating, reciprocating, or oscillating shafts. The secondary purpose is to exclude foreign matter. Application conditions may vary from high-speed rotation

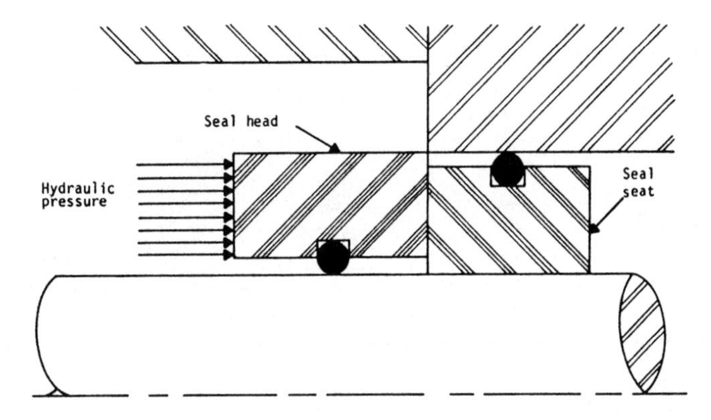

FIG. 21-58 Unbalanced seal. Total pressure pushes head and seat together; heavy wear can result.

up to 7500 r/min in a clean environment to low-speed shaft reciprocation in very dusty environments. The seal must operate under conditions of extreme cold (−50°F). A typical seal design with standard terminology is shown in Fig. 21-62. Sealing is dependent on maintaining an interface between the shaft and the elastomeric member, and between the bore and the seal outside diameter. The basic advantages of radial lip seals include low cost for effective sealing, small space requirements, easy installation, and an ability to seal multivariable applications.

Seal Types. Non-spring-loaded seals (Fig. 21-63*a*) are used to retain highly viscous materials like grease at shaft speeds less than 2000 ft/min (609 m/min). If dust exclusion is desired, a secondary lip can be added (Fig. 21-63*b*). Typical applications include conveyor rollers, vehicle wheels, and so on. For maximum effectiveness as a dirt excluder, the seals are installed with the primary lip facing away from the bearings. Heavy-duty multilip seals (Fig. 21-63*c*) provide good lubricant retention under severe dust conditions, such as in disk harrows. The shaft speed is usually limited to 500 r/min. External multilip seals (fig. 21-63*d*) are used with a fixed shaft and rotating bore. The bore or a wear ring at the bore is used as a sealing surface.

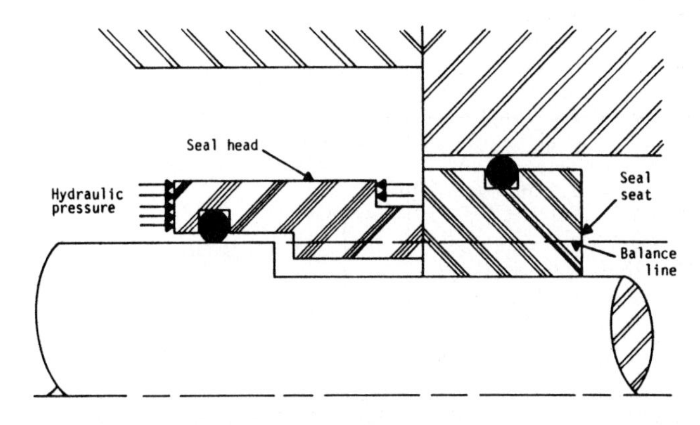

FIG. 21-59 Partially balanced seal. Part of pressure is relieved by step in head.

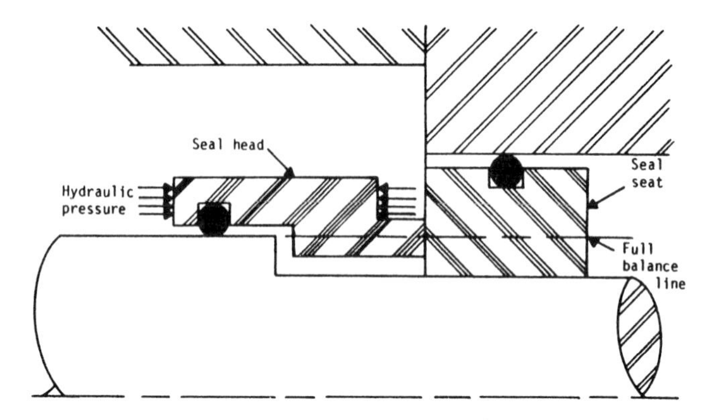

FIG. 21-60 Completely balanced seal. Force between head and seat is independent of pressure.

Spring-loaded seals are used to retain low-viscosity lubricants such as oils at speeds up to 3600 ft/min (1098 m/min). The single-lip seal is the most economical general-purpose oil seal (Fig. 21-64*a*).

Typical applications include engines, drive axles, transmission pumps, electric motors, and speed reducers. If medium-duty dirt exclusion is required, a dual-lip seal is used (Fig. 21-64*b*). These seals are commonly used in automotive, farm, and industrial applications where they are protected from external abuse during operation.

Dual-case seals are also available (Fig. 21-64*c* and *d*) for heavy-duty applications. The secondary case is added to provide additional strength and protection during assembly or operation. These seals are commonly used in construction equipment, farm, and industrial applications. External spring-loaded seals are used in applications with rotating housing and stationary shafts (Fig. 21-65).

In general, pressure reduces seal life. For conventional seal designs, the pressure limitations are found in Table 21-18.

If the pressure requirement is greater, special designs must be considered. These unique designs are usually developed for specific applications. A seal designed for

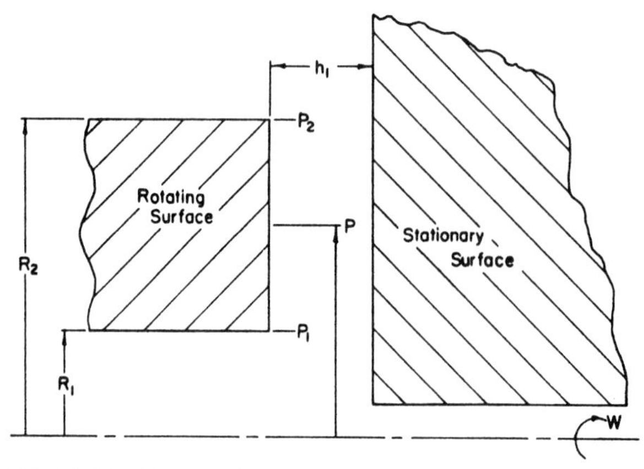

FIG. 21-61 Face seal geometry.

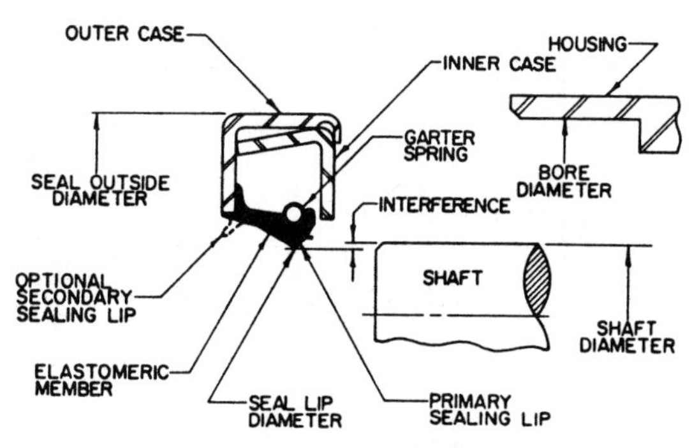

FIG. 21-62 Radial lip seal terminology.

intermittent operation at 1200 to 1500 lb/in^2 (8280 to 10,350 kPa) is shown in Fig. 21-66.

In some cases, a rubber OD seal (Fig. 21-67) is preferred. The rubber OD is used when bore finish recommendations are not followed. If the bore material is different from the seal case material, rubber ODs can be used to eliminate bore leakage resulting from differential thermal expansion.

For applications where the total eccentricity exceeds the recommended levels, special seal designs are used (Fig. 21-68). The elastomer member is convoluted to provide added flexibility, and the ability to follow the shaft is improved. Eccentricity motion as great as 0.060 in (1.524 mm) can easily be handled with these designs. The complexity of the seal design results in a cost penalty.

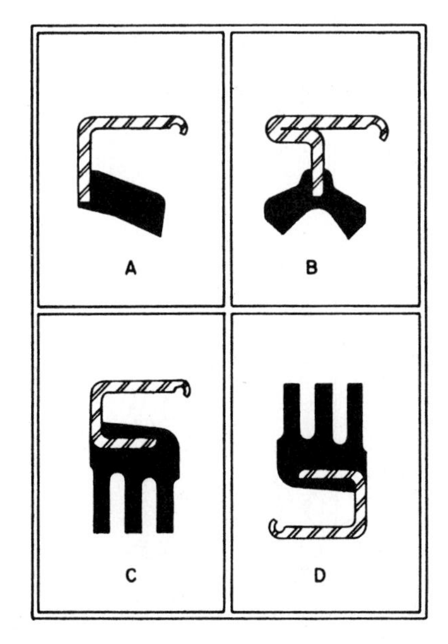

FIG. 21-63 Non-spring-loaded seals.

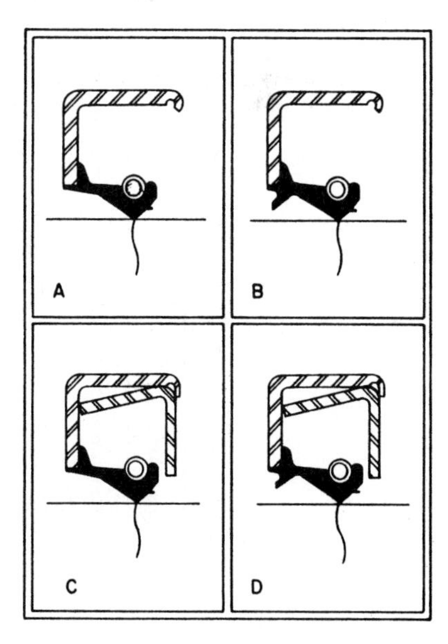

FIG. 21-64 Spring-loaded seals.

TABLE 21-17 Troubleshooting for Face Seals

Symptoms	Appearance	Causes
Seal leaks steadily despite apparent full contact of mating rings	Full 360° contact patterns on mating ring; little or no measurable wear	Secondary seal leakage caused by: • Nicked, scratched, or porous seal surfaces • O-ring compression set • Chemical attack
Steady leakage at low pressure; little or no leakage at high pressure	Heavy contact on mating ring OD; fades to no visible contact at ID. Possible edge chipping on primary ring OD	Deflection of primary ring from overpressurization. Seal faces not flat because of improper lapping
Seal leaks steadily when shaft is rotating; little or no leakage when shaft is stationary	Heavy contact pattern on mating ring ID; fades to no visible contact at OD. Possible edge chipping on primary-ring ID	Thermal distortion of seal faces. Seal faces not flat because of improper lapping
Seal leaks steadily whether shaft is stationary or rotating	Two large contact spots; pattern fades away between spots. Contact through about 270°; pattern fades away at low spot. Contact spots at each bolt location	Mechanical distortion caused by: • Overtorqued bolts • Out-of-square clamping parts • Out-of-flat stuffing box faces • Nicked or burred gland surface • Hard gasket

Seal leaks steadily whether shaft is stationary or rotating. Noise from flashing or face popping	High wear or thermal distress on mating ring. High wear and carbon deposits on primary ring. Possible edge chipping on primary ring. Thermal distress over one-third of mating ring, located 180° from inlet of seal flush. High wear and possible carbon deposits on primary ring. Thermal distress at 2 to 6 locations on mating ring. High wear and possible carbon deposits on primary ring	Sealed liquid vaporizing at seal interface, caused by: • Low suction or stuffing box pressure • Improper running clearance between shaft and primary ring • Insufficient cooling • Improper bushing clearance • Circumferential flush groove in gland plate missing or blocked
Seal leaks steadily whether shaft is stationary or rotating	High wear and grooving on mating ring	Poor lubrication from sealed fluid. Abrasives in fluid
Steady leakage when shaft is rotating; no leakage when shaft is stationary	Contact pattern on mating ring is slightly larger than primary ring width. Possible high spot opposite drive pin hole	Out-of-square mating caused by: • Nicked or burred gland surfaces • Improper drive pin extension • Misaligned shaft • Piping strain on pump casing • Bearing failure • Shaft whirl
Damaged mating ring; seal leaks whether shaft is stationary or rotating	Eccentric contact pattern, although width equals that of primary ring. Possible cracks on mating ring	Misaligned mating ring caused by: • Improper clearance between gland plate and stuffing box • Lack of concentricity between shaft OD and stuffing box ID

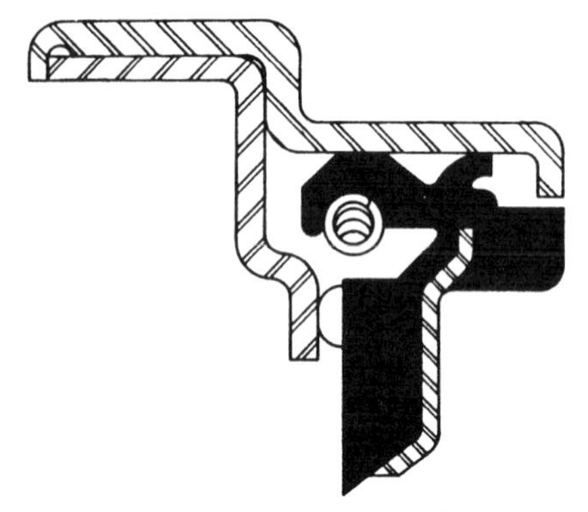

FIG. 21-65 Seal for rotating OD stationary shaft.

TABLE 21-18 Operating Pressure Limits

Shaft speed		Maximum pressure permissible	
ft/min	m/min	lb/in²	kPa
0–1000	0–304.8	7	48.2
1000–2000	305–609	5	34.5
2000–3600	610–1098	3	20.7

FIG. 21-66 High-pressure seal.

FIG. 21-67 Rubber OD seal.

FIG. 21-68 Seal for high shaft runout.

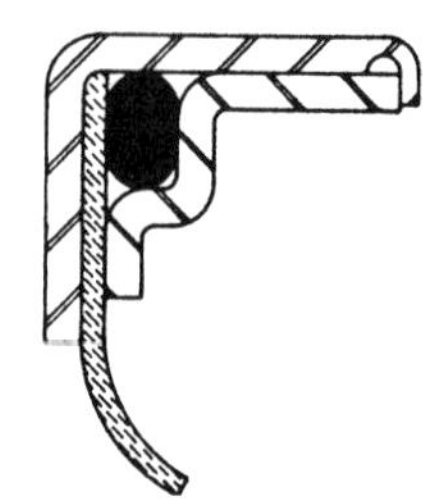

FIG. 21-69 Assembled seal.

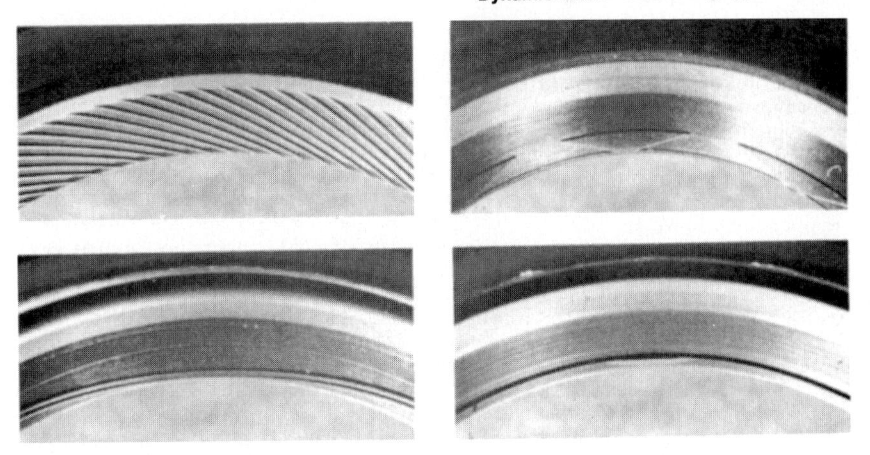

FIG. 21-70 **(*a*) Helix seal—unidirectional. (*b*) Triangular seal—birotational. (*c*) Sine wave—birotational. (*d*) Combination pad and rib—birotational.**

Some seals are made with lip materials that cannot be bonded to the case. The lip material is tightly clinched between metal stampings to prevent the packing from rotating (Fig. 21-69). An internal seal must be provided to prevent fluid from leaking through the seal. The cost of assembled seals is typically more than that of bonded ones.

Some seals have ribs molded on the air side on the rubber element. Any leakage that seeps past the lip is pumped back into the sump. These seals typically run hotter than plain lip seals because there is an increased amount of rubber contacting the shaft. They do not function well in dusty environments, since the ribs will sweep dust into the sump. A helix seal is shown in Fig. 21-70*a*. It will function only if the shaft rotates in a single direction. Seals for birotational operation have triangular pads (Fig. 21-70*b*), sine waves (Fig. 21-70*c*), and combinations of ribs and pads (Fig. 21-70*d*).

FIG. 21-71 **Wave seal.**

One seal type has the entire element formed in a sine-wave configuration. This provides a mild hydrodynamic (Fig. 21-71) action without ribs to create heat and pump dust.

Shaft Seal Materials. Nitrile is a copolymer of butadiene and acrylonitrile. It is the most extensively used shaft seal material because of its good oil resistance, wear resistance, good low-temperature properties, and low cost. The upper temperature limit is approximately 225°F (107°C). High-temperature performance can be improved by increasing the percentage of acrylonitrile in the material. Unfortunately, the low-temperature properties suffer. Carboxylated nitriles are more expensive than standard nitriles and provide better wear resistance at high temperatures. The oil resistance is identical to that of standard nitriles, and low-temperature flexibility is lower. Nitriles are particularly suited for sealing hydrocarbon fluids at moderate temperatures. If low-molecular-weight fluids like gasoline or highly aromatic fluids have to be sealed, some sacrifice in low-temperature properties must be made. Nitriles can be used in EP-type oils at low or moderate temperatures. When sump temperatures get higher than 215°F (102°C), the additives in the oil (i.e., chlorinated paraffins and sulfonated olefins) will harden and crack the material. Nitriles are not recommended for use with low-molecular-weight aromatics such as benzene or polar aliphatics.

The polyacrylic polymers offer improvements over nitrile in high-temperature applications. Polyacrylic compounds function well at 275°F (135°C) in engine or transmission fluids. They resist EP-type additives and should replace nitriles in those applications where hardening and cracking of the nitrile is a problem. Polyacrylate is an ester and is vulnerable to water, acids, bases, and all types of polar solvents like ketones, esters, and so on. The major limitations of polyacrylates have been wear resistance and low-temperature properties. Many polyacrylate compounds will become brittle around 0 to 5°F (−18 to −15°C). New polyacrylate polymers are capable of handling temperatures down to −40°F (−40°C). When selecting a polyacrylate compound for an application, extreme care must be exercised if cold-temperature properties are important.

Silicone is an expensive material that will operate well from −100° to 300°F (−73 to 149°C) in typical engine and transmission oils. Silicones function in water, inorganic acids and bases, diesters, and non-petroleum-based brake fluids. They are not recommended for use in chlorinated or aromatic solvents, gasoline, and other such applications. They are also adversely affected by some EP additives, such as lead naphthanate and zinc dithiocarbamate. Oxidized lubricating oil will also cause the material to revert. Although silicones show fair wear resistance during normal operation, their dry-running characteristics are quite poor. It is recommended that silicone seals be presoaked in oil before installation to prevent dry running during initial break-in. Silicones also have poor tear characteristics, and special care must be exercised during installation to prevent lip damage.

There are two types of fluoroelastomers, which are quite different chemically. Fluorocarbons are used more extensively than fluorosilicone. Both materials are expensive and are applied in critical applications (high temperature, high speed). Fluorocarbons have excellent chemical and high-temperature resistance. The material is very tough and is a popular choice in heavy-duty applications. At low temperatures, the material becomes very stiff, and the sealing lip cannot follow the shaft. Leakage may occur. Fluorosilicone combines the chemical and high-temperature resistance of fluorocarbon with the excellent low-temperature behavior of silicone. Unfortunately, it also has the poor tear strength and wear characteristics of silicone.

Tetrafluoroethylene (TFE) has been used as a sealing material for years, but only within recent times has it been seriously considered for typical rotational shaft seal applications. TFE is resistant to virtually all fluids and has a temperature range from −120 to +400°F (−84 to +204°C). TFE is not an elastomeric material and will not follow shaft motion as well as other seal materials. At temperatures of 100°F (38°C) and below, the TFE material is stiff. The material cannot be easily bonded; thus most TFE seals are of the costly assembled type. The material has a high cost and is easily damaged. Extreme care must be exercised during installation.

Several other materials are used in specialized shaft seal applications. Urethanes are tough materials that are used when the utmost in wear resistance at lower temperatures is required. Butyl and ethylene propylene diene monomer (EPDM) are used to seal non-petroleum-based brake fluids and polar solvents. Epichlorohydrin is an oil-resistant material that lies between nitrile and polyacrylate in cost and properties. In general, these materials represent a small fraction of the shaft seal population because of processing difficulties that create cost penalties.

The chemical resistance of shaft seal materials to various fluids appears in Table 21-19. Even though chemical resistance is of extreme importance, other factors, such as cost, temperature range, and wear resistance, must be considered before the material is selected. These factors appear in Table 21-20. The final material choice is often a compromise that will satisfy several of the most important application parameters.

21-17 Sealing-System Requirements and Recommendations The seal is only part of the sealing system. The efficiency and operating life is often affected by the condition of the two mating surfaces between which it is installed—the bore and the shaft. Table 21-21 provides the recommended bore sizes for a given shaft size.

Under normal conditions, the portion of the shaft contacted by the seal should be hardened to Rockwell C30 minimum to minimize shaft scoring. There is no conclusive evidence that additional hardening will increase the wear resistance of the shaft.

Two types of eccentricity affect seal performance. (1) Shaft-to-bore misalignment (STBM). This is the amount (in inches) that the center of the shaft is offset with respect to the center of the bore. STBM usually exists to some degree as a result of normal machining and assembly inaccuracies. To ensure proper seal performance, the STBM should be as small as possible. Too much STBM increases friction and creates abnormal wear on one side of the seal. The STBM is measured by attaching a dial indicator to the shaft and indicated off the seal bore as the shaft is slowly rotated. (2) Dynamic runout (DRO). This is the amount (in inches) that the sealing surface of the shaft does not rotate about the true center. DRO is caused by misalignment, bending of the shaft, lack of shaft balance, shaft lobing, and other manufacturing inaccuracies. It is measured by the total movement of an indicator held against the side of the shaft while the shaft is slowly rotated.

Sealing performance in eccentric conditions depends largely on the flexibility of the sealing element. As a general rule, typical spring-loaded seals will operate satisfactorily if the total eccentricity (combined indicator readings) does not exceed the maximums shown in Table 21-22.

Steel or stainless steel shafts are recommended for best sealing performance. Nickel-plated surfaces are acceptable. Brass, bronze, aluminum alloys, zinc, magnesium, and similar materials should not be used except under unusual circumstances. The level of interference between the sealing lip and the shaft surface must be precisely controlled to ensure optimum seal performance. The shaft diameter should be held within the tolerances shown in Table 21-23 unless the sealing requirements are not critical.

The corner of the shaft can damage the sealing lip during installation if it is not properly machined. A burr-free chamfer or radius is recommended. (Fig. 21-72). The shaft finish is critical to the proper functioning of the lip seal. The shaft surface must be smooth enough to provide continuous contact between the sealing lip and the shaft surface. It must also be rough enough to provide pockets for lubricant. This lubricant reduces the coefficient of friction between the seal lip and the shaft and provides coolant for the interface. The shaft surface must not cause excessive seal wear or be uneconomical to produce. Machine lead must be avoided, since the spiral marks on the shaft may cause lip damage and pump oil out of the sump.

In many applications, wear sleeves or rings of mild steel are pressed over shafts of cast iron or other soft materials. They permit easy replacement of the sealing surface and should be replaced whenever the seal is changed.

The best method of providing all requirements is to plunge-grind the shaft to a

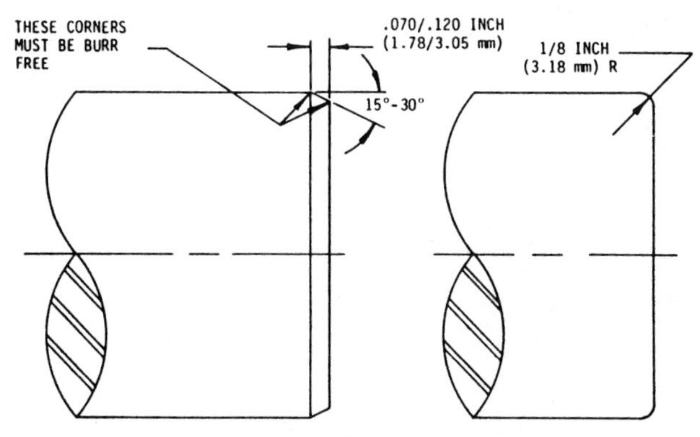

FIG. 21-72 Shaft lead corners.

TABLE 21-19 Seal Element Selection Chart
Chemical resistance to various fluid media

Medium	Nitrile	Polyacrylate	Silicone	Fluoroelastomer		Fluorocarbon TFE
				Fluorosilicone	Fluorocarbon	
Engine oil	Good	Good	Good	Good	Good	Good
ATF-A	Fair	Good	Good	Good	Good	Good
Grease	Good	Fair	Fair	Fair	Good	Good
EP lube	Fair/poor	Good	Poor	Fair	Good	Poor
SAE90	Good	Good	Good	Good	Good	Good
Fuel oil	Good	Fair	Poor	Good	Good	Good
Kerosene	Good	Fair	Poor	Good	Good	Good
Gasoline	Good	Fair	Poor	Good	Good	Good
Petroleum-based hydraulic oil	Good	Good	Good	Good	Good	Good
Brake fluid	Poor	Poor	Poor	Fair	Fair	Good
Skydrol 500	Poor	Poor	Good	Poor	Poor	Good
MIL-L-7808	Fair	Poor	Good	Good	Good	Good
MIL-L-23699	Fair	Poor	Good	Good	Good	Good
MIL-L-6082-A	Good	Good	Good	Good	Good	Good
MIL-L-5606	Good	Good	Poor	Good	Good	Good
MIL-G-10924	Good	Good	Poor	Good	Good	Good
Butane	Good	Good	Fair	Good	Good	Good
Ketone	Poor	Poor	Poor	Poor	Poor	Good
Ammonium gas, cold	Good	Poor	Fair	Poor	Poor	Good
Fresh water	Good	Poor	Good	Good	Good	Good
Salt water	Good	Poor	Good	Good	Good	Good

TABLE 21-20 Seal Element Selection Chart

Material	Approximate relative cost (cpd.)	Approximate relative cost (seals)	Sump oil temperature		Advantages	Disadvantages
			°F	°C		
Nitrile	100	100	−50 to +225	−45 to +107	Low cost, low swell. Good wear and oil resistance at moderate temperatures	Poor resistance to EP additives. Poor high-temperature resistance
Polyacrylic	250	115	−40 to +275	−40 to +135	Good oil resistance. Low swell. Generally resistant to EP additives	Fair wear properties. Poor dry running. Poor water resistance
Silicone	625	130	−30 to +300	−34 to +149	Very broad temperature range.	Poor dry running properties. Poor resistance to oxidized oil and some EP additives. Poor tear characteristics

TABLE 21-20 Seal Element Selection Chart (*Continued*)

Fluoroelastomer (fluorocarbon)	2250	200	−40 to +350	−40 to +177	Excellent oil and chemical resistance. Good wear properties. Low swell	Becomes stiff at low temperatures. Poor followability at low temperatures
Fluoroelastomer	2250	200	−80 to +300	−62 to +149	Excellent oil and chemical resistance. Good low-temperature properties	Poor tear strength. Poor dry running. Poor wear resistance
Fluorocarbon (TFE)		300	−140 to +400	−95 to +204	Excellent oil and chemical resistance. Excellent temperature range	Easily damaged. Becomes stiff at low temperature. High cost. Poor followability at low temperature

TABLE 21-21 RMA Oil Seal Standard Sizes (in)
Single- and dual-lip spring-loaded bonded seals)

Standard bore sizes		Standard shaft sizes																		
		in																		
		0.500	0.625	0.750	0.875	1.000	1.125	1.250	1.375	1.500	1.625	1.750	1.875	2.000	2.125	2.250	2.375	2.500	2.625	2.750
		mm																		
in	mm	12.7	15.9	19.05	22.2	25.4	28.6	31.8	34.9	38.1	41.3	44.5	47.6	50.8	54.0	57.2	60.3	63.5	66.7	69.9
0.999	25.4	×	—	—	—	—	—	—	—	—	—	—	—	—	—	—	—	—	—	—
—																				
1.124	28.6	×	×	—	—	—	—	—	—	—	—	—	—	—	—	—	—	—	—	—
1.250	31.8	×	×	×	—	—	—	—	—	—	—	—	—	—	—	—	—	—	—	—
1.375	35.0	—	×	×	×	—	—	—	—	—	—	—	—	—	—	—	—	—	—	—
1.499	38.1	—	×	×	×	×	—	—	—	—	—	—	—	—	—	—	—	—	—	—
1.624	41.3	—	—	×	×	×	×	—	—	—	—	—	—	—	—	—	—	—	—	—
1.752	44.5	—	—	—	×	×	×	×	—	—	—	—	—	—	—	—	—	—	—	—
1.874	47.6	—	—	—	—	×	×	×	—	—	—	—	—	—	—	—	—	—	—	—
2.000	50.8	—	—	—	—	—	×	×	×	—	—	—	—	—	—	—	—	—	—	—
2.125	54.0	—	—	—	—	—	—	×	×	×	—	—	—	—	—	—	—	—	—	—
2.250	57.2	—	—	—	—	—	—	—	×	×	×	—	—	—	—	—	—	—	—	—
2.374	60.3	—	—	—	—	—	—	—	×	×	×	×	—	—	—	—	—	—	—	—
2.502	63.6	—	—	—	—	—	—	—	—	×	×	×	—	—	—	—	—	—	—	—
2.623	66.6	—	—	—	—	—	—	—	—	—	×	×	×	×	—	—	—	—	—	—
2.750	69.9	—	—	—	—	—	—	—	—	—	—	×	×	×	×	—	—	—	—	—
2.875	73.0	—	—	—	—	—	—	—	—	—	—	—	×	×	×	—	—	—	—	—
3.000	76.2	—	—	—	—	—	—	—	—	—	—	—	×	×	×	×	—	—	—	—
3.125	79.4	—	—	—	—	—	—	—	—	—	—	—	×	×	×	×	×	—	—	—
3.251	82.6	—	—	—	—	—	—	—	—	—	—	—	—	—	×	×	×	×	—	—
3.371	85.6	—	—	—	—	—	—	—	—	—	—	—	—	—	—	×	×	×	×	—
3.500	88.9	—	—	—	—	—	—	—	—	—	—	—	—	—	—	—	×	×	×	×
3.623	92.0	—	—	—	—	—	—	—	—	—	—	—	—	—	—	—	—	×	×	×

TABLE 21-21 RMA Oil Seal Standard Sizes (in) (*Continued*)

Standard bore sizes		Standard shaft sizes																			
		in																			
		2.625	2.750	2.875	3.000	3.125	3.250	3.375	3.500	3.625	3.750	3.875	4.000	4.250	4.500	4.750	5.000	5.250	5.500	5.750	6.000
		mm																			
in	mm	66.7	69.9	73.0	76.2	79.4	82.6	85.7	88.9	92.0	95.3	98.4	101.6	107.9	114.3	120.0	127.0	133.4	139.7	146.0	152.4
3.623	92.0	×	×	×	—	—	—	—	—	—	—	—	—	—	—	—	—	—	—	—	—
3.751	95.3	×	×	×	×	—	—	—	—	—	—	—	—	—	—	—	—	—	—	—	—
3.875	98.4	—	×	×	×	—	—	—	—	—	—	—	—	—	—	—	—	—	—	—	—
4.003	101.7	—	—	×	×	—	—	—	—	—	—	—	—	—	—	—	—	—	—	—	—
4.125	104.8	—	—	—	×	×	—	—	—	—	—	—	—	—	—	—	—	—	—	—	—
4.249	107.9	—	—	—	—	×	×	×	—	—	—	—	—	—	—	—	—	—	—	—	—
4.376	111.2	—	—	—	—	×	×	×	×	—	—	—	—	—	—	—	—	—	—	—	—
4.500	114.3	—	—	—	—	×	×	×	×	—	—	—	—	—	—	—	—	—	—	—	—
4.626	117.5	—	—	—	—	—	×	×	×	×	×	—	—	—	—	—	—	—	—	—	—
4.751	120.7	—	—	—	—	—	—	—	×	×	×	—	—	—	—	—	—	—	—	—	—
4.876	123.9	—	—	—	—	—	—	—	—	×	×	×	—	—	—	—	—	—	—	—	—
4.999	127.0	—	—	—	—	—	—	—	—	×	×	×	×	—	—	—	—	—	—	—	—
5.125	130.2	—	—	—	—	—	—	—	—	—	—	×	×	—	—	—	—	—	—	—	—
5.251	133.4	—	—	—	—	—	—	—	—	—	—	×	×	×	—	—	—	—	—	—	—
5.375	136.5	—	—	—	—	—	—	—	—	—	—	—	×	×	—	—	—	—	—	—	—
5.501	139.7	—	—	—	—	—	—	—	—	—	—	—	—	×	×	—	—	—	—	—	—
5.625	142.9	—	—	—	—	—	—	—	—	—	—	—	—	×	×	—	—	—	—	—	—
5.751	146.1		—	—	—	—		—			—	—	—	—	×	×	—	—	—	—	—
6.000	152.4	—	—	—	—	—	—	—	—	—	—	—	—	—	—	×	×	—	—	—	—
6.250	158.8	—	—	—	—	—	—	—	—	—	—	—	—	—	—	—	×	×	—	—	—
6.375	161.9	—	—	—	—	—	—	—	—	—	—	—	—	—	—	—	×	×	—	—	—
6.500	165.1	—	—	—	—	—	—	—	—	—	—	—	—	—	—	—	—	×	×	—	—
6.625	168.3	—	—	—	—	—	—	—	—	—	—	—	—	—	—	—	—	×	×	—	—
6.750	171.5	—	—	—	—	—	—	—	—	—	—	—			—	—	—	—	×	×	
6.875	174.6	—	—	—	—	—	—		—	—	—	—	—	—	—	—		—	×	×	—
7.000	177.8	—	—	—	—	—	—	—	—	—	—	—	—	—	—	—	—	—	—	×	—
7.125	180.9	—	—	—	—	—	—	—	—	—	—	—	—	—	—	—	—	—	—	×	×
7.500	190.5	—	—	—	—	—	—	—	—	—	—	—	—	—	—	—	—	—	—	—	×

Courtesy: Rubber Manufacturers Association (RMA).

TABLE 21-22 Shaft Eccentricity Chart

Maximum total eccentricity (STBM plus DRO)		Shaft speed, r/min
In	mm	
0.025	0.635	100
0.020	0.508	200
0.018	0.457	500
0.015	0.381	1000
0.013	0.330	1500
0.010	0.254	2000
0.009	0.229	2500
0.008	0.203	3000
0.007	0.178	4000

TABLE 21-23 Shaft Diameter Tolerance Chart

Nominal shaft diameter		Shaft tolerance	
in	mm	in	mm
Up to and including 4.000	Up to and including 101.6	±0.003	±0.08
4.001–6.000	101.63–152.4	±0.004	±0.10
6.001 and up	152.43–254	±0.005	±0.13

finish of 10 to 20 μin (0.25 to 0.50 μm) with no machine lead. Shafts should be ground with mixed-number rotational speed ratios. The grinding wheel should be allowed to spark out to prevent machine lead.

Ferrous materials are most commonly used in the housings that form seal bores. The conventional seal case material is steel; thus thermal expansion is not usually a problem unless other materials (such as aluminum) are used for the housing bore.

Differential thermal expansion rates can cause a sealing problem at the bore, particularly for large-diameter applications. When this problem exists, it can be solved by using special seals with the case material identical to the bore. This usually results in a cost penalty. A steel mating ring pressed into the bore is an alternative solution. In some cases, rubber OD seals are recommended.

To ensure proper sealing at the bore, the proper press fit between the seal OD and the bore must be maintained. The tolerances and press-fit requirements for ferrous bores appear in Table 21-24.

If the bore is stepped, the depth of the bore should exceed the seal width by at least $^1/_{64}$ in (0.397 mm). No specific bore hardness is recommended. It need only be high enough to maintain interference with the seal OD.

Whenever lubricant pressure is present at the outside diameter of the seal, a bore finish of 125 μin (3.15 μm) or less is recommended to prevent bore leakage. If this is not possible, coatings which will effectively fill minor bore imperfections can be applied to the seal OD. If the bore roughness is extreme and tool marks exist, rubber OD seals are sometimes used.

The leading or entering edge of the bore should be chamfered, as shown in Fig. 21-73, to prevent seal damage during installation. The inside corner of the bore should have a maximum radius of $^3/_{64}$ in (1.190 mm).

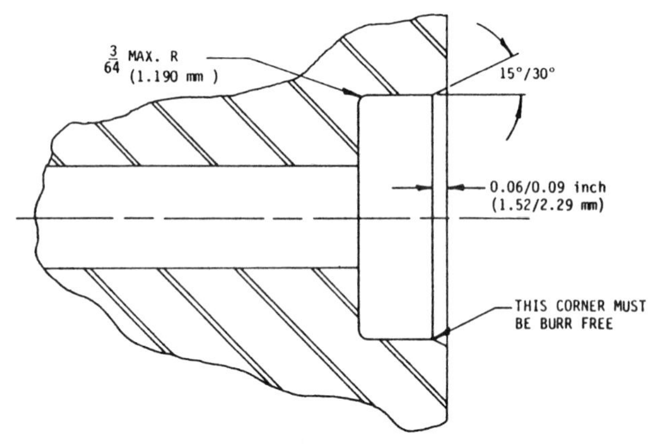

FIG. 21-73 Recommended bore lead corner.

Preventing Sealing-System Failures. Sealing failures can result when one or more deviations from accepted practice are permitted. In many cases, the seal is blamed for leakage when some other component in the sealing system is the culprit. It is difficult to predict or obtain the optimum life of a lip seal, but it is necessary to define and pinpoint reasons for early or premature failure.

Improper installation is one of the main causes of premature leakage. An elastomeric lip seal is a precision mechanical component; it must be assembled properly if the reliability expected for the application is to be obtained. In high-volume production applications, to assemble the seal is desirable. These tools will minimize assembly variables that may affect sealing efficiency.

Installation Procedure. The following installation procedure should be followed to ensure proper seal life.

1. Check dimensions. If the shaft and bore dimensions do not match those specified for the selected seal, leakage is more likely to occur. The seal should be replaced with one that has the proper shaft size and outside diameter.

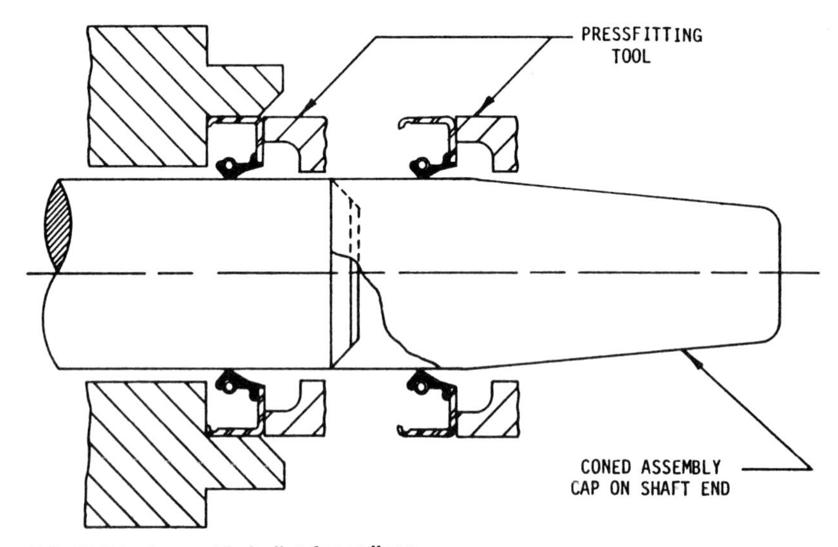

FIG. 21-74 Assembly bullet for splines.

TABLE 21-24 RMA Oil Seal Standard Tolerance
Metal outside diameter seals

Bore diameter		Bore tolerance		Nominal press fit		OD* tolerance		Out of round	
in	mm	in	mm	in	mm	in	mm	in	mm
Up to 1.000	Up to 25.4	±0.001	±0.025	0.004	0.102	±0.002	±0.051	0.005	0.127
1.001– 3.000	25.43– 76.2	±0.001	±0.025	0.004	0.102	±0.002	±0.051	0.006	0.152
3.001– 4.000	76.23–101.6	±0.0015	±0.038	0.005	0.127	±0.002	±0.051	0.007	0.178
4.001– 6.000	101.62–152.4	±0.0015	±0.038	0.005	0.127	+0.003/−0.002	+0.076/−0.051	0.009	0.229
6.001– 8.000	152.42–203.2	±0.002	±0.051	0.006	0.152	+0.003/−0.002	+0.076/−0.051	0.012	0.306
8.001– 9.000	203.23–228.6	±0.002	±0.051	0.007	0.178	+0.004/−0.002	+0.102/−0.051	0.015	0.381
9.001–10.000	228.63–254	±0.002	±0.051	0.008	0.203	+0.004/−0.002	+0.102/−0.051	0.015	0.381
10.001–20.000	254.03–508	+0.002/−0.004	+0.051/−0.102	0.008	0.203	+0.006/−0.002	+0.152/−0.051	0.002	0.051
20.001–40.000	508.03–1016	+0.002/−0.006	+0.051/−0.152	0.008	0.203	+0.008/−0.002	+0.203/−0.051	in/in of seal OD	mm/mm of seal OD
40.001–60.000	1018.03–1524	+0.002/−0.010	+0.051/−0.254	0.008	0.203	+0.010/0.002	+0.254/−0.051		

* The average of a minimum of three measurements to be taken at equally spaced positions.

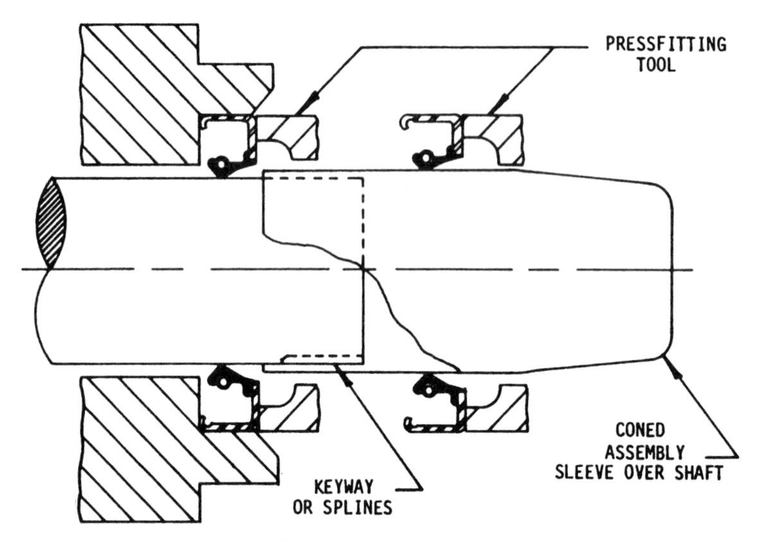

FIG. 21-75 Assembly bullet.

2. Check seal. The seal should be examined for damage that may have occurred prior to installation. A sealing lip that is turned back or nicked will leak. The seal outside diameter should be checked to ensure that there are no dents, scores, or cuts. If there is any damage, the seal should be replaced. Never reuse a seal. Always use a new seal in the application.

3. Check housing bore. The entering edge must be deburred to prevent damage to the seal outside diameter. A chamfer or rounded corner should be provided whenever possible.

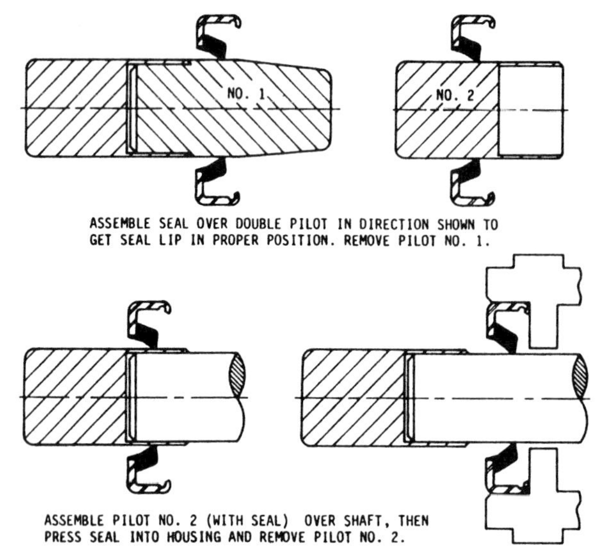

NO. 1
NO. 2
ASSEMBLE SEAL OVER DOUBLE PILOT IN DIRECTION SHOWN TO
GET SEAL LIP IN PROPER POSITION. REMOVE PILOT NO. 1.
ASSEMBLE PILOT NO. 2 (WITH SEAL) OVER SHAFT, THEN
PRESS SEAL INTO HOUSING AND REMOVE PILOT NO. 2.

FIG. 21-76 Piloting tools.

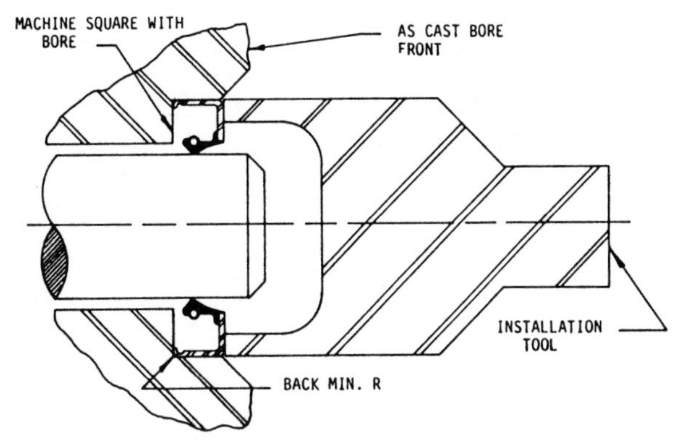

FIG. 21-77 Bottom seal in bore.

4. Check shaft. Remove surface nicks, burrs, and grooves that may damage the sealing lip. Examine the shaft for machine lead. Remove burrs or sharp edges from the shaft end. The shaft end should be chamfered. If shaft burrs cannot be removed, a coned assembly tool must be used (Fig. 21-74).
5. Check splines and keyways. When installing a seal over splines and keyways, an assembly sleeve should be used (Fig. 21-75).
6. Check seal lip direction. Make sure that the new seal faces the same direction as the original. Generally, the lip faces the lubricant or fluid to be sealed. When the shaft is installed against the lip, use piloting tools shown in Fig. 21-76.
7. Prelubricate the sealing element. Carefully wipe the seal lip with lubricant immediatey before installation.
8. Use correct installation tools. Press-fitting tools should have an outside diameter 0.010 in smaller than the bore diameter. Apply pressure only at the seal outside diameter to prevent seal distortion.
9. Use proper driving force. An arbor press is recommended, but a soft-faced mallet can be used with the installation tool. Never hammer directly on the seal face.
10. Bottom out the tool or seal. To avoid cocking the seal in the bore, the installation tool must be designed to bottom the seal in the bore (Fig. 21-77), against the

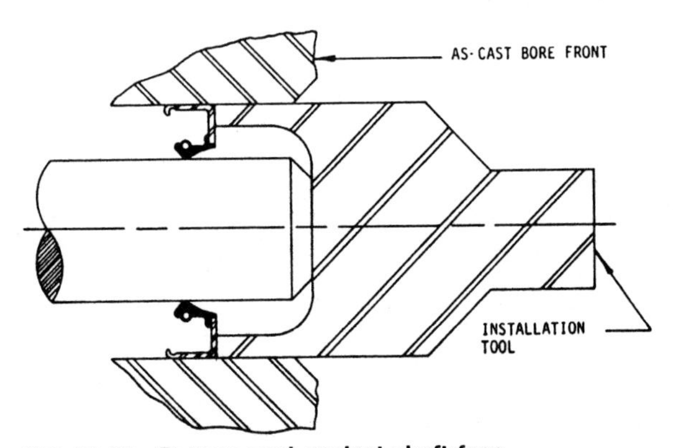

FIG. 21-78 Bottom seal against shaft face.

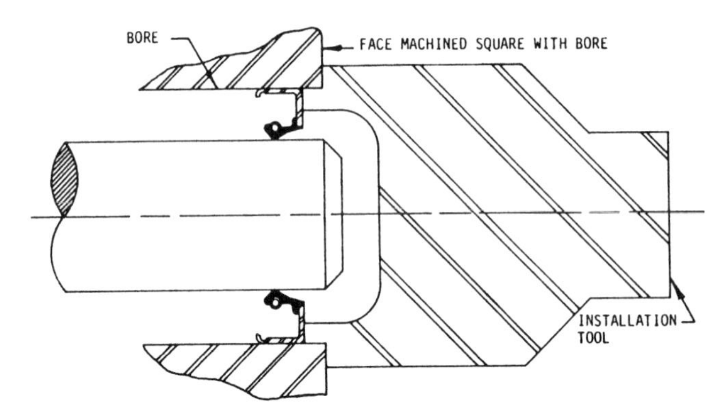

FIG. 21-79 Bottom seal against bore face.

shaft (Fig. 21-78), or against the bore face (Fig. 21-79). If the shaft has been used before, the seal should be positioned to prevent the seal lip from running in the old wear track.

11. Check for parts interference. After the installation is complete, check for other machine parts that may rub against the seal and cause friction and heat.

After Installation. Perfect installations of seals are sometimes ruined by mishandling of the sealing region during normal maintenance. When machinery is painted, the seal area should be masked to avoid getting paint on the lip or the shaft where the lip rides. If the sump is vented to prevent pressure buildup, mask the vents to prevent clogging. If the paint is to be baked or the mechanism exposed to other outside heat sources, care must be taken to keep the seal temperature below material limitations. When cleaning or testing, do not allow cleaning fluids to contact seals. When testing or breaking in machinery, do not subject seals to conditions that exceed design recommendations. Seal damage may not be evident until much later.

The most important factor in the performance of a lip seal is its ability to maintain a radial force between the sealing element and the shaft. Anything that affects this sealing force will shorten seal life and reduce effectiveness. Excessive heat or lubricant additives may harden or soften the rubber lip and affect seal performance. Rapid wear of the sealing lip can be caused by shaft irregularities, lack of lubrication, and excessive temperatures. Typical radial-seal leakage problems, possible causes, and suggested cures appear in Table 21-25. This table is provided as a guide only and cannot be considered all-inclusive.

DYNAMIC SEALS—NONCONTACTING SEALS FOR ROTATING SHAFTS

21-18 Introduction to Noncontacting Seals Noncontacting seals have a small clearance gap between the rotating shaft and the stationary housing. Direct contact of the mating parts does not occur; therefore, frictional wear is eliminated. The seals are durable, reliable, simple in design, and easy to maintain. The design principle allows for some leakage to occur. The pressure differential between the internal sump and the external atmosphere is maintained by throttling the leaking fluid with the seal. The leakage can be minimized by controlling the seal design and the magnitude of the clearance gap. A bushing seal has a constant clearance gap. Leakage is minimized by reducing the gap clearance or increasing the bushing length. A labyrinth seal consists of a series of small bushings joined together. Dynamic leakage is virtually eliminated when screw threads on the shaft or housing pump the fluid back into the sump. Some noncontacting seals use magnetic components. A magnetic fluid is placed and held in the clearance gap. The ferrofluidic seal effectively eliminates leakage.

21-19 Bushings *Fixed-Bushing Seals.* A fixed-bushing seal consists of a close-fitting stationary sleeve or pipe that fits within a housing. The shaft rotates within the bushing (Fig. 21-80).

Fluid seeps through the narrow gap that is formed. The leakage rate of an incompressible fluid flowing in the laminar region through a perfectly concentric and aligned bushing with a small clearance can be estimated with Eq. (21-6). If the center of the bushing does not coincide with the housing center (Fig. 21-81), then Eq. (21-7) must be used to account for eccentricity.

$$Q_L = \frac{\pi R h^3 g}{6\nu\rho} \frac{\Delta P}{L} \tag{21-6}$$

$$Q_{Le} = Q_L(1 + {}^3\!/_2 N^2) \tag{21-7}$$

where Q_L = laminar volumetric flow rate cm³/s
R = mean radius of annulus, cm
h = mean radial clearance, cm
v = kinematic viscosity, cm²/s
ρ = fluid density, g/cm³
p = pressure drop, g/cm²
N = eccentricity/radial clearance = e/h
g = gravitation constant, 980 cm/s²

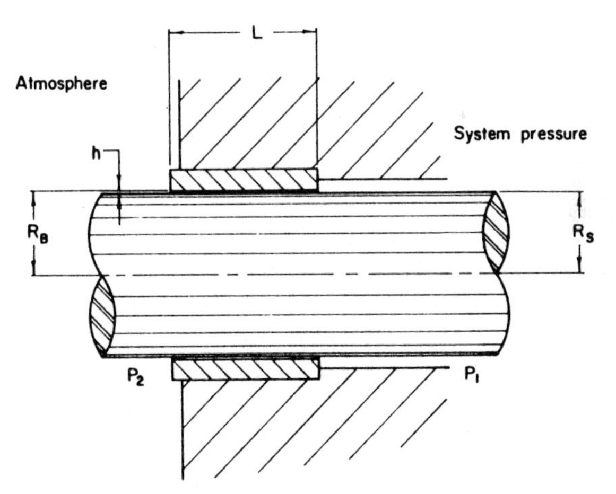

FIG. 21-80 Fixed bushing seal.

If the flow rate becomes large, turbulence will result. Equation (21-8) will then apply for the concentric bushing, and Eq. (21-9) must be used for the eccentric case.

$$Q_T = 2\pi R \left(\frac{1}{0.0665} \frac{h^3 g P}{v\, {}^1\!/_4 \rho L} \right)^{4/7} \tag{21-8}$$

$$Q_{Te} = 1.315 Q_T \tag{21-9}$$

TABLE 21-25 Troubleshooting Lip Seals

Symptom	Possible cause	Corrective action
1. Early lip leakage	(*a*) Nicks, tears, or cuts in seal lip	Examine shaft. Eliminate burrs and sharp edges. Use correct mounting tools to protect seal lip from splines, keyways, or sharp shoulders. Handle seals with care. Keep seals packaged in storage and in transit
	(*b*) Rough shaft	Finish shaft to 10–20 μin rms or smoother
	(*c*) Scratches or nicks on surface	Protect shaft after finishing
	(*d*) Lead on shaft	Plunge-grind shaft surface
	(*e*) Excessive shaft whip or runout	Locate seal close to bearings. Ensure good, accurate machining practices
	(*f*) Cocked seal	Use correct mounting tools and procedures
	(*g*) Paint on shaft or seal element	Mask seal and adjacent shaft before painting
	(*h*) Turned-under lip	Check shaft chamfer for roughness. Machine chamfer to 32 μin or smoother, blend into shaft surface. Check shaft chamfer for steepness. Use recommended lead chamfer. Use correct mounting tools and procedures.

	(*i*) Damaged or "popped out" spring	Use correct mounting tools and procedures to apply press-fit force uniformly. Protect seals in storage and transit
	(*j*) Damaged or distorted case	Use correct mounting tools and procedures to apply press-fit force uniformly. Protect seals in storage and transit
	(*k*) OD sealant on shaft or lip element	Use care in applying OD sealant. Purchase precoated seals
2. Lip leakage, intermediate life	(*a*) Excessive lip wear	Check seal cavity for excessive pressure. Provide vents to reduce pressure. Provide proper lubrication for seal. Check shaft finish. Make sure finish is 10–20 μin
	(*b*) Element hardening and cracking	Reduce sump temperature, if possible. Upgrade seal material. Provide proper lubrication for seal. Change oil frequently. Change seal design.
	(*c*) Element corrosion and reversion	Check material-lubricant compatibility. Change material
	(*d*) Excessive shaft wear	Check shaft hardness. Harden to Rockwell C30 minimum. Change oil frequently to remove contaminants. Use dust lip in dirty atmosphere
3. OD leakage	(*a*) Scored seal OD	Check housing machining. Use 125 μin rms. Check edges on housing bore. Use recommended chamfer. Remove burrs.
	(*b*) Damaged seal case	Use correct mounting tools and procedure to apply press fit uniformly. Protect seals in storage and transit

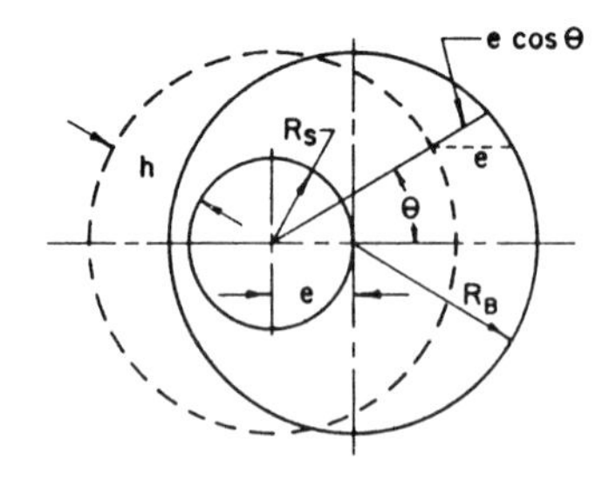

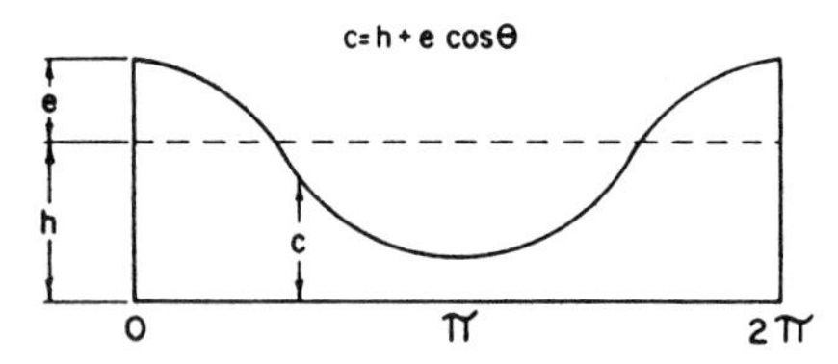

FIG. 21-81 Eccentric fixed bushing.

The leakage rate is reduced as the radial clearance h and the pressure drop from sump to atmosphere are reduced. Leakage rate is also reduced as the bushing length and the fluid kinematic viscosity and density increase.

Leakage rates can be estimated from Table 21-26 and Figs. 21-82 and 21-83.

The geometric and fluid factor F, Eq. (21-10), is calculated and the Reynolds number, Re, Eq. (21-11), is determined from Fig. 21-83. The leakage flow velocity V is calculated with Eq. (21-12), and the volumetric flow rate is obtained from Eq. (21-13).

$$F = \frac{h^3 g \Delta p}{V^2 \rho L} \qquad (21\text{-}10)$$

$$\text{Re} = \frac{Vh}{\nu} \qquad (21\text{-}11)$$

$$V = \frac{\nu e}{h} \qquad \text{cm/s} \qquad (21\text{-}12)$$

$$Q = 2\pi RhV \qquad \text{cm}^3/2 \qquad (21\text{-}13)$$

TABLE 21-26 Leakage Rates

Liquid	°C	Specific weight, g/cm³
Alcohol	20	0.800
Fresh water	15	1.000
Sea water	15	1.024
Lubricating oil	15	0.881–10.946
Glycerin	0	1.260
Fuel oil	15	0.897–1.026

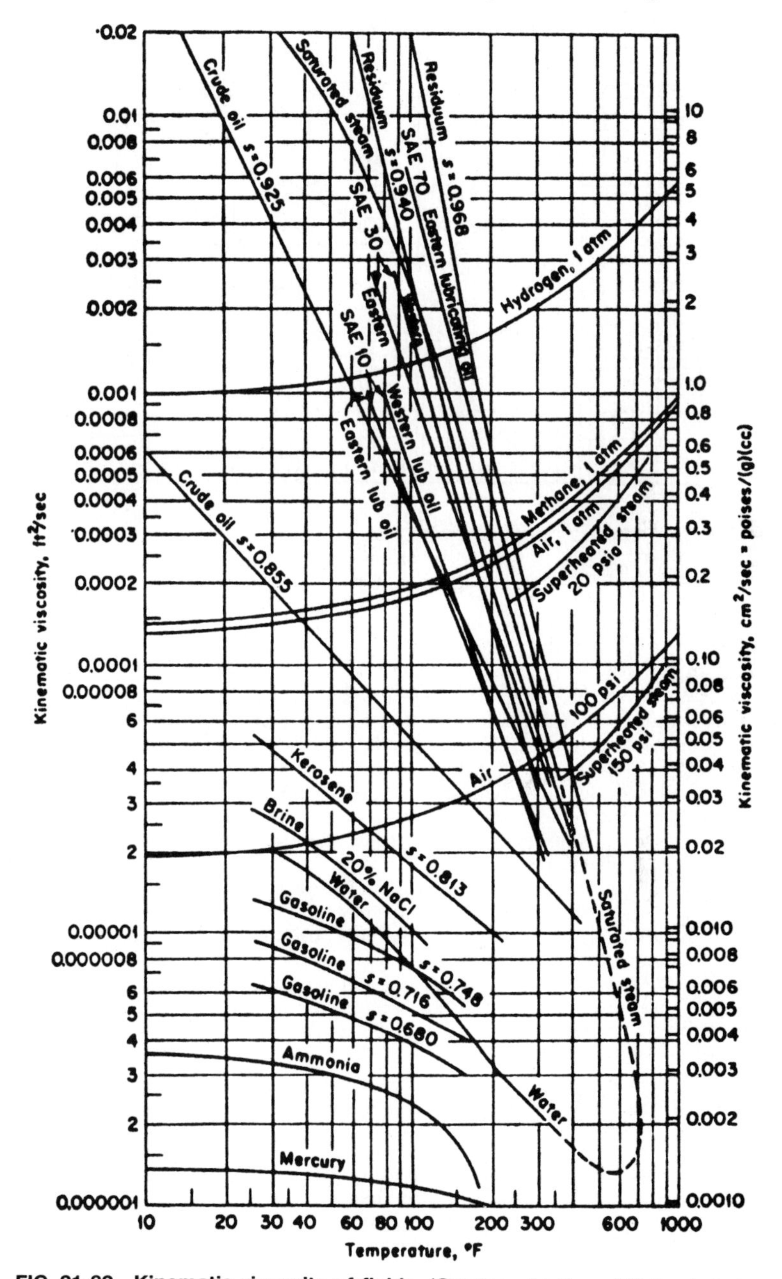

FIG. 21-82 **Kinematic viscosity of fluids.** (*Courtesy: McGraw-Hill Book Co.*)

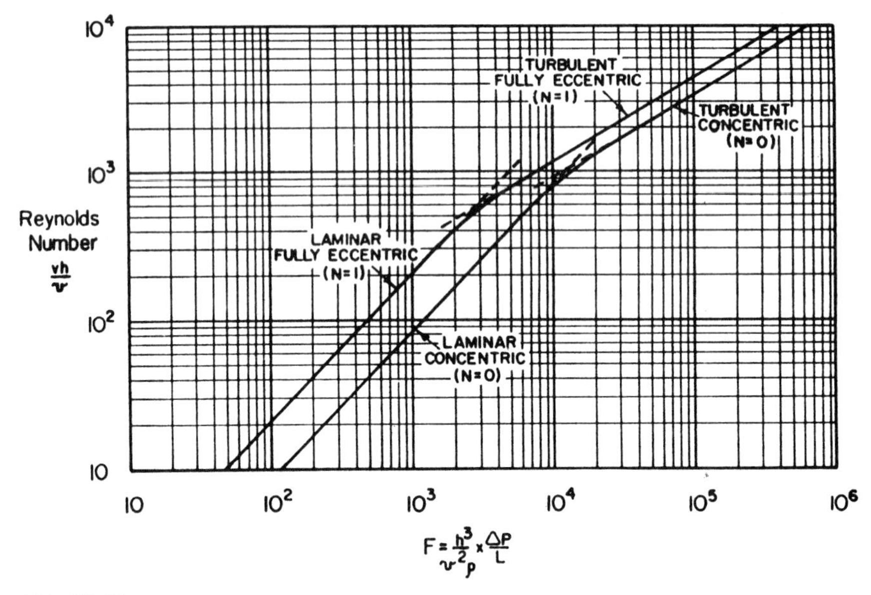

FIG. 21-83

In most cases, the fixed-bushing seal is used for sealing liquids. The viscosity of gases is too low to provide enough friction to keep leakage within tolerable limits.

Since fixed-bushing seals are firmly attached to the housing, the clearance must be large enough to prevent rubbing during operation. Rubbing can occur as a result of vibration, shaft bowing, shaft deflection, thermal growth, shaft-to-bore misalignment, shaft eccentricity, and bore eccentricity. With inadequate clearance, rubbing can induce excessive bearing loads, excessive heat, serious vibration, and seizure. Serious damage to machinery can result.

If large clearances and thus high leakage rates can be tolerated, fixed-bushing seals offer a simple, low-cost, easy to install and maintain solution to sealing problems.

Floating-Bushing and Ring Seals. Operating clearances with fixed bushings may often be too large to obtain desired leakage rates. Floating bushings with smaller clearances can be used to seal gases and reduce leakage (Fig. 21-84). Rotation is prevented with a dowel pin, and the bushing is free to move radially. A helical spring is used to provide a positive axial force to load the bushing face against the housing face to prevent static leakage.

In some cases, multiple bushings or rings are used (Fig. 21-85). The length of each ring in the tandem arrangement is less than the length that would be required if a single bushing was used. This design allows larger shaft misalignments without affecting seal performance.

Instead of using solid rings, it is also customary to design a floating-ring seal as a segmented archbound ring, where the segments are held together by a garter spring (Fig. 21-86). The garter spring provides the contact force required to establish a seal with the shaft surface. The metal retainer is not always necessary.

In some cases, the bushing seal can be balanced by providing a step in the sleeve (Fig. 21-87). The step size can be adjusted to provide the desired sealing force at the bushing/housing interface.

The calculation of leakage rates for gases must compensate for density changes in the clearance space. If the flow is steady, laminar, and isothermal and the bushing and shaft are concentric, Eq. (21-14) can be used to estimate leakage rates. If the ring and shaft are eccentric, Eq. (21-15) must be used.

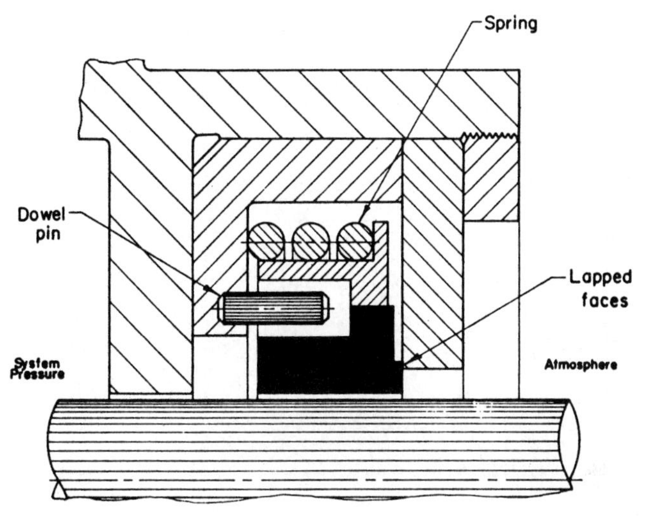

FIG. 21-84 Floating bushing seal.

$$G_L = \frac{\pi R h^3 g}{12 \nu L}\left(\frac{P_i^2 - P_o^2}{P_s}\right)\frac{T_s}{T_i} \qquad (21\text{-}14)$$

$$G_{Le} = G_L(1 + \tfrac{3}{2}N^2) \qquad (21\text{-}15)$$

where G_L = leakage rate of laminar compressible concentric fluid, g/s
G_{Le} = leakage rate of laminar compressible eccentric fluid, g/s
P_i = upstream pressure, g/cm•s^2
P_o = downstream pressure, g/cm•s^2
T_s = standard temperature, K
T_i = upstream temperature, K

Calculations become more complicated if the flow is turbulent. The concentric leakage rate is given by Eq. (21-16). The fully eccentric case is approximately 1.5 times more than the concentric case (see Table 21-27).

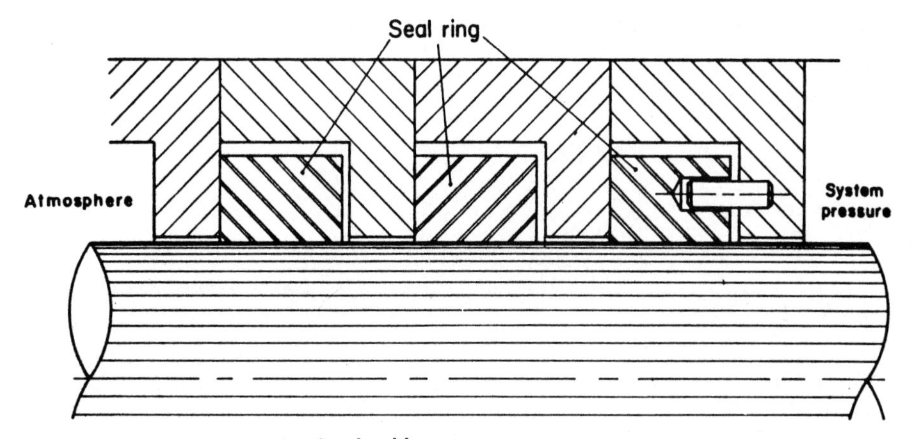

FIG. 21-85 Multistage floating bushing.

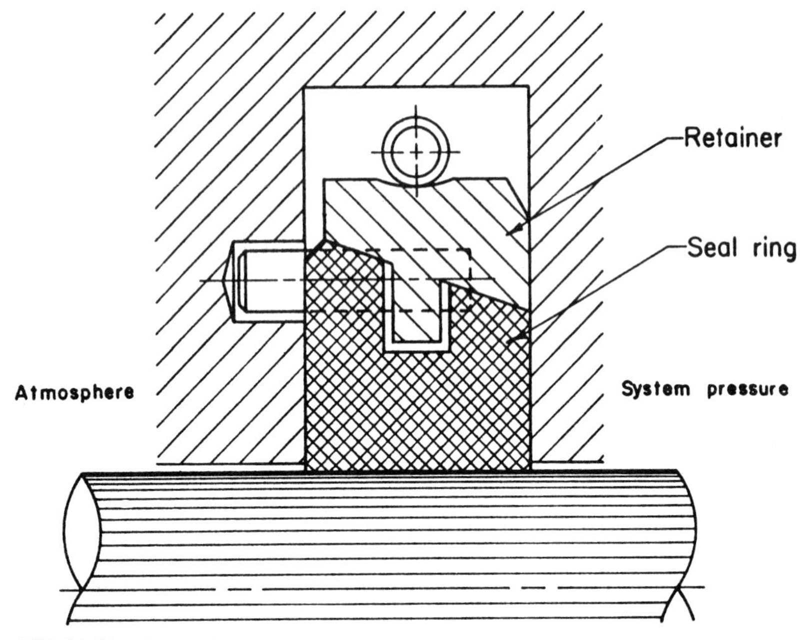

FIG. 21-86 Segmented floating ring seal.

$$G_T = 2\pi Rh\alpha \sqrt{gP_1\rho_i} \tag{21-16}$$

where ρ_i = gas density at the upstream condition, g/cm³
α = flow resistance coefficient

The flow resistance coefficient α is a function of seal geometry, pressure ratio, and shaft speed. The following procedure is used to determine α. The velocity ratio β is calculated from Eqs. (21-17) through (21-19). The multiring factor m is found in Fig. 21-88. The ring resistance factor R_r is determined with Eq. (21-20), and the pressure ratio γ is defined by Eq. (21-21). The values of γ, R_r, and β are used with Fig. 21-89 to determine the flow resistance coefficient. The highest value of β given on the chart is 0.1. If the calculated value of β exceeds 0.1, the data presented for $\beta = 0.1$ should provide a satisfactory approximation for α.

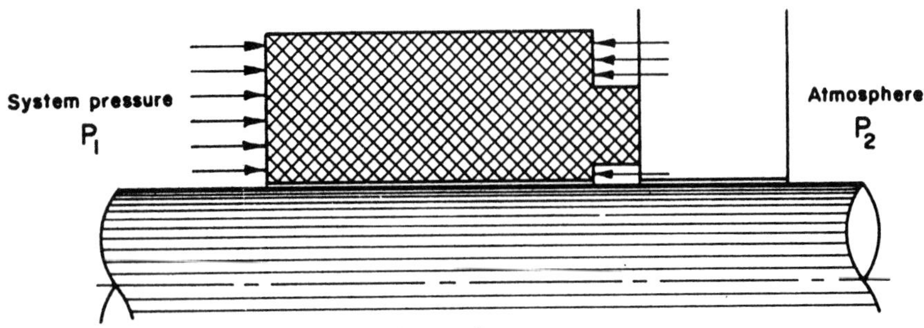

FIG. 21-87 Balanced floating bushing seal.

$$V_c = 2g \frac{k}{kT_i} \frac{P_i}{\rho_i} \tag{21-17}$$

$$V_s = 2\pi RN \tag{21-18}$$

$$\beta = \frac{V_s}{V_c} \tag{21-19}$$

$$R_r = m\left(\frac{w}{h}\right) \tag{21-20}$$

$$\gamma = \frac{P_o}{P_i} \tag{21-21}$$

where w = ring width, in
k = exponent of adiabatic expansion
V_c = critical fluid velocity, cm/s
V_s = linear shaft velocity, cm/s
N = shaft rotational speed, r/s
γ = fluid resistance coefficient (0.5 for steam and 0.02 for air)

For fixed bushings, the best material is an antifriction type, such as carbon-graphite, molybdenum disulfide, or a combination of both with PTFE resins or soft metals used as liners. Where temperatures permit, babbitt is a good candidate. For higher temperatures, bronzes and aluminum alloys are well suited. Any material chosen must be compatible with the system fluid.

Floating-bushing and ring seals are subject to the same basic requirements. Differences in thermal expansion are essential and should be carefully evaluated. With carbon-graphite as a major candidate, it is important to consider the low coefficient of expansion compared to the steel of the shaft. By shrinking it into a metallic retainer, the problem can be minimized. Plasma coatings have proven favorable for high-temperature applications.

The system fluid is an important factor and must be carefully evaluated. It must be clean and should not tend to polymerize or crystallize. When water is used, it

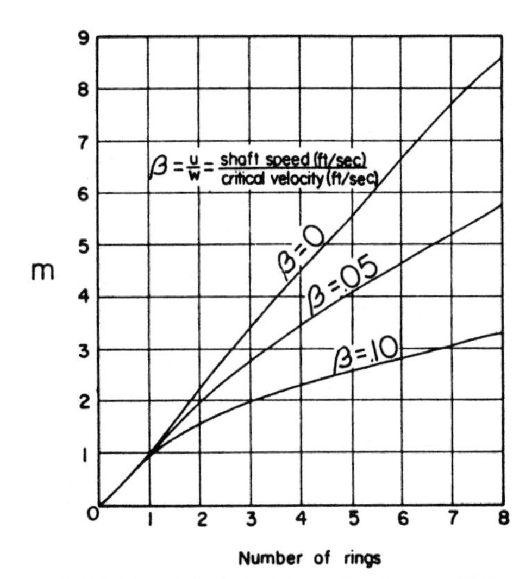

FIG. 21-88 Determination of m for multigland seal rings.

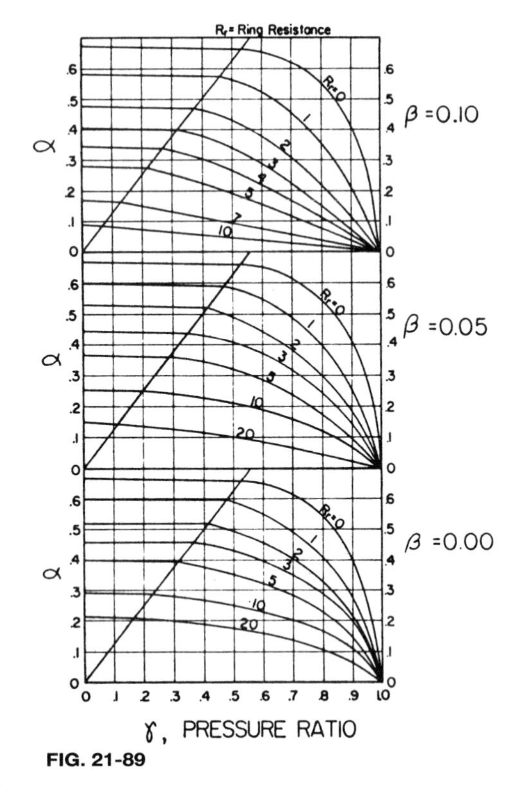

FIG. 21-89

should be "soft." High surface velocities may produce erosion effects along the bushing surface. A materials selection chart appears in Table 21-28.

21-20 Labyrinth Seals Labyrinth seals consist of a series of circumferential strips of material that may extend from the shaft or the bore to form a series of annular orifices. Labyrinths are used primarily to seal gases in compressors and steam turbines. They have leakage rates that are higher than those of bushings or face seals, but they are simple, reliable, and maintenance-free. There are a multitude of labyrinth designs.

TABLE 21-27 Bushing Equations

	Incompressible	Compressible
Laminar-flow leakage rate (concentric)	$Q_L = \dfrac{\pi R h^3 g \Delta P}{6\nu\rho L}$ in³/s	$G_L = \dfrac{\pi R h^3 g}{12\nu L}\left(\dfrac{P_i^2 - P_o^2}{P_s}\right)\dfrac{T_s}{T_i}$
Laminar-flow leakage rate (eccentric)	$Q_e = Q_L(1 + 3/2 N^2)$	$G_{Le} = G_L(1 + 3/2 N^2)$
Turbulent-flow leakage rate (concentric)	$Q_T = 2\pi R\left(\dfrac{1 h^3 g \Delta P}{0.0665\nu^{1/4}\rho L}\right)^{4/7}$	$G_T = 2\pi R h \alpha\sqrt{g P_1 \rho_i}$
Turbulent-flow leakage rate (eccentric)	$Q_{Te} \approx 1.315 Q_T$	$G_{Te} \approx 1.5 G_T$

TABLE 21-28 Floating-Bushing, Ring-Seal Materials, and Environment Combinations

Environment	Seal ring material	Shaft material
Oil	Babbitt	Hardened steel
	Bronzes	Shafting
	Aluminum	Chrome plate
	Carbon graphite	Nitrided steels
Water	Bronzes	440-C
	416 stainless steel	Chrome-plated
	Stellite	
	Carbon graphite	
	Ceramics	Chrome plate
Gas (Air, CO_2), H_2, He, N_2, O_2)	Carbon graphite	Tool steels, hardened
		Tungsten carbide plate
		Ceramic plate
		Chrome carbide
		Chrome plate
		Stainless steel (300)
		Stainless steels (400)
		Stainless steels ≈ 50 Rockwell C

The simplest is the straight-through design (Fig. 21-90). High fluid velocities are generated at the throat of each constriction, and the kinetic energy is dissipated by turbulence in the chamber beyond each throat. There is some velocity carryover with a straight labyrinth. This results in efficiency losses. The labyrinth can be stepped (Fig. 21-91) or staggered (Fig. 21-92) to cause the expanding jet of gas to impinge upon a solid transversal surface. This minimizes the velocity carryover losses. The housing of the staggered labyrinth must be split to allow for assembly. Combinations of labyrinth

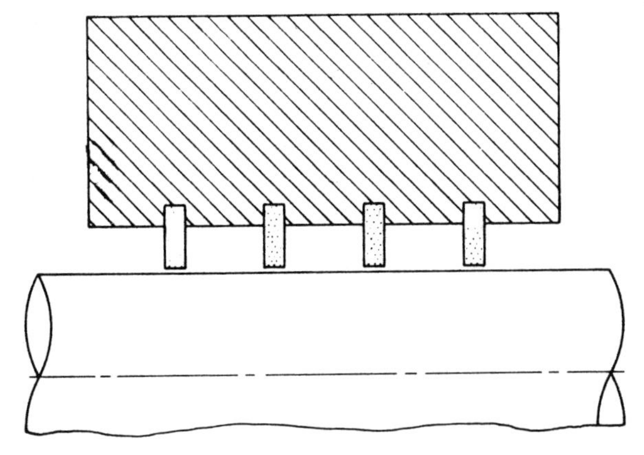

FIG. 21-90 Straight labyrinth.

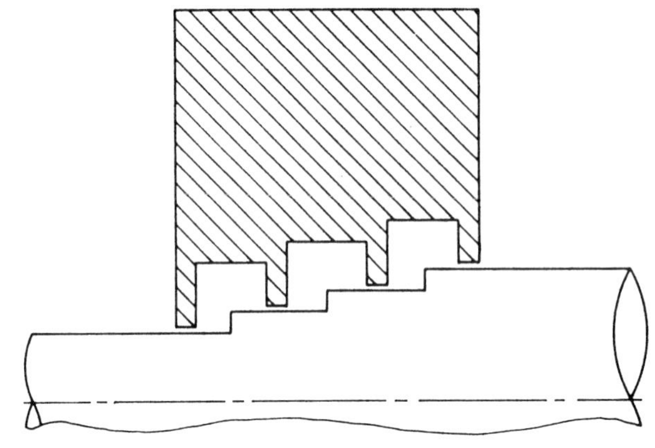

FIG. 21-91 Stepped labyrinth.

types are also used (Fig. 21-93). The combination labyrinth shown uses a buffered inlet to prevent flow of the internal gas to the atmosphere. The barrier fluid is maintained at a pressure higher than that of the process gas. Another variation of the buffered system (Fig. 21-94) uses an aspirating port at the outlet. The outlet pressure is generally equal to or less than the pressure at either end.

The leakage of steam through a labyrinth may be approximated by Eq. (21-22).

$$G = 25KA \sqrt{\frac{P_i}{V_i}\left[1 - \left(\frac{P_o}{P_i}\right)^2\right] \bigg/ \left(N - \ln \frac{P_o}{P_i}\right)} \qquad (21\text{-}22)$$

where G = flow rate of steam, lb/h
A = area through packing clearing space, in^2
P_i = upstream pressure, lb/in^2 (absolute)
V_i = initial specific volume of steam, ft^3/lb
P_o = final pressure, lb/in^2 (absolute)
N = number of stages in labyrinth
K = experimentally determined flow coefficient

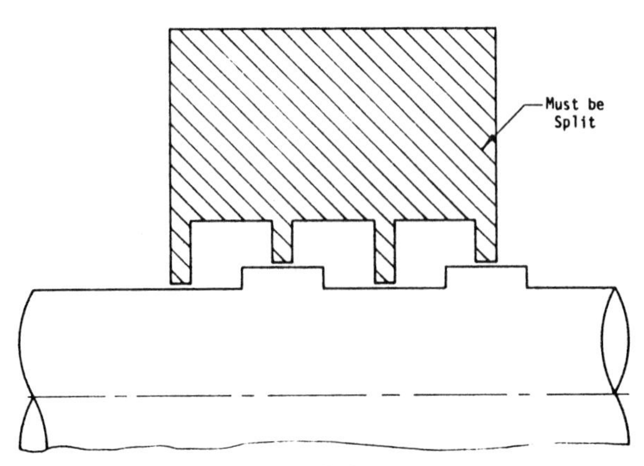

FIG. 21-92 Staggered labyrinth.

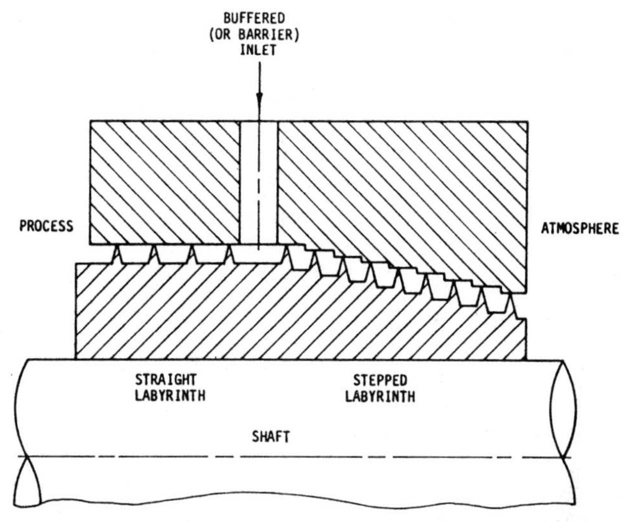

FIG. 21-93 Combination labyrinth.

The coefficient K varies for different types of seal teeth. For simple tooth forms—straight through the inclined teeth—K varies from 60 to 120, with the simplest form yielding the highest coefficient. For staggered tooth forms, K will vary from 30 to 65, with deep high-low forms giving the lowest coefficient.

When designing a labyrinth system, the fluid density, fluid temperature, flow direction, inlet pressure, outlet pressure, amount of leakage that can be tolerated, and available space must all be considered. The simplest forms should be considered first.

The radial clearance at which a labyrinth can operate is probably the most important single factor and the one that most directly affects the effectiveness of the seal; therefore, the clearance selected is usually based on allowing a very slight rub (interference) under the worst conditions. If possible, line-to-line is recommended. In figuring the minimum clearance, machining tolerances on parts together with the fits with control concentricity must be taken into account for static assembly conditions. In addition, however, the effect of differential expansion and of deflection of the parts as a result of the forces during operation must be considered. One must remember that no rotating shaft is free from wobble or movement in some orbit, and that very

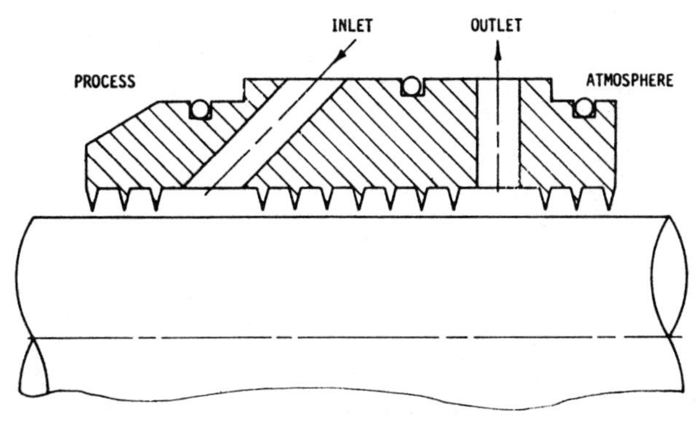

FIG. 21-94 Buffered-educted labyrinth.

often at least one critical speed must be gone through. Prior experience usually determines the amount of clearance which a particular machine will produce in rotating about a center other than the shaft center. One must also consider that the amount of clearance required increases as the span between bearings increases. There is really no merit in providing a very small static clearance when it will be worn to a relatively large value on the first run. In some instances, the labyrinth is designed to "run in." Experience dictates this allowance, but in any case it will be a function of the rubbing speed and the materials used. As an example, aluminum or bronze seal strips or knives running against a steel surface will usually tolerate several thousandths of an inch wear without causing appreciable damage either to the shaft or to the strips or knives.

After the clearance is established, the number of strips or knives must be considered. This selection will be influenced by the allowable leakage and space considerations. Concurrently, the proportions of the strips of knives must be considered. No hard-and-fast rule concerning knife form, knife pitch, or knife depth can be given. As a general guide, one might say that the tip of the knife or strip usually has a width of 0.005 to 0.015 in and that the included angle on the knife or strip is 8° to 12°; the pitch will be in the range of 0.150 to 0.220 in, and the depth of the knife or strip will approximate the pitch. In aircraft service, the number of throttlings for a single pass usually does not exceed four. If more throttlings are desired, this is accomplished in two or more tiers. In steam turbine applications, the end seals may have as many as 30 throttlings at lesser pitch.

The materials used in the rubbing surfaces are a very important consideration. The material combination is chosen on the basis of rubbing qualities and the ability to stand up under the particular atmospheres and temperature existing in the service. With reference to wearing qualities, it is preferable that the end of the tooth remain narrow and clean-cut so as to obtain the minimum flow coefficient. Materials that crumble and break away cleanly are desirable. Materials that mushroom are not desirable, since the flow coefficient, and consequently the leakage, increases because of the rounding at the orifices. Materials that ball up as a result of melting and air hardening are undesirable because they will tear or gouge out a larger clearance than would be expected from normal wear, and thus will cause large leakage. For low-temperature work a soft brass may work well. In steam, which can be considered a protective atmosphere, a leaded nickel-tin bronze is very satisfactory and may be used at temperatures up to 950°F. In the presence of oxygen or air, however, it will oxidize rapidly, and a limiting temperature of 400 to 450°F is indicated. 25 Chrome-20 nickel stainless steel appears to work reasonably well, although it does have a high coefficient of friction. Some of the aluminum bronzes seem to have both wear resistance and oxidation resistance at high temperature. Aluminum is a likely candidate for some applications, but it should not be anodized, since aluminum oxide is formed, and this may score adjacent part.

Stationary seal elements that have a greater coefficient of expansion than the casing (e.g., rolled in strips), should be split into segments with joint clearance to allow for minimum radial clearance under all conditions. A method of reducing clearance is to flexibly mount a labyrinth sleeve to permit radial motion relative to the fixed housing. Care must be exercised and pressure balancing considered to assure minimum frictional restraint due to axial forces induced by high pressure. Another method is to use soft or abradable material. In this category, aircraft use silver plate, steel honeycomb, fiber metal, and abradable plasma sprays. Industrially, carbon-graphite labyrinths have found some acceptance. Generally, in steam turbines the clearances are adjusted so that rubbing is not problematical.

21-21 Visco Seals Visco seals have screw threads machined on the shaft (Fig. 21-95) or into the housing (Fig. 21-96). They are used primarily to screw liquids from the outside toward the interior of a sump. This pumping prevents fluid leakage.

Visco seals are designed to function properly only when the shaft operates at a certain minimum rotational speed. Thus, for very low or zero shaft rotation, a secondary sealing device must be provided. Several designs for industrial visco seals utilize the

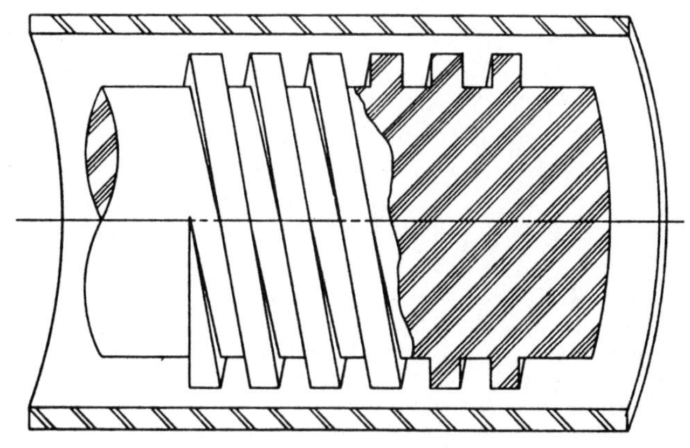

FIG. 21-95 Screw threads machined on shaft.

centrifugal force during rotation to keep built-in lips from establishing contact with the shaft, but once the shaft stops rotating, the lips contact the shaft and establish the necessary seal. This provides a satisfactory seal as long as temperature is not a problem for elastomeric lip materials.

There is no contact between the rotating and the stationary components of the seal; therefore, visco seals have long reliable life.

The key design parameters appear in Fig. 21-97. The optimum sealing performance for laminar flow occurs when the helix angle α is 15 to 20°, γ is 0.5, and β is 3.6 to 4.1. The parameter γ is defined to be the ratio of groove width b to land and groove width at b. The parameter β is the ratio of groove depth h and radial clearance c to the radial clearance, as shown in Eqs. (21-23) and (21-24).

$$\gamma = \frac{b}{a + b} \tag{21-23}$$

$$\beta = \frac{h + c}{c} \tag{21-24}$$

These rules do not apply if the flow becomes turbulent. Helix angles of 10 to 15°, a γ of 0.62, and a β of 4.1 to 6.5 give good results in the turbulent region.

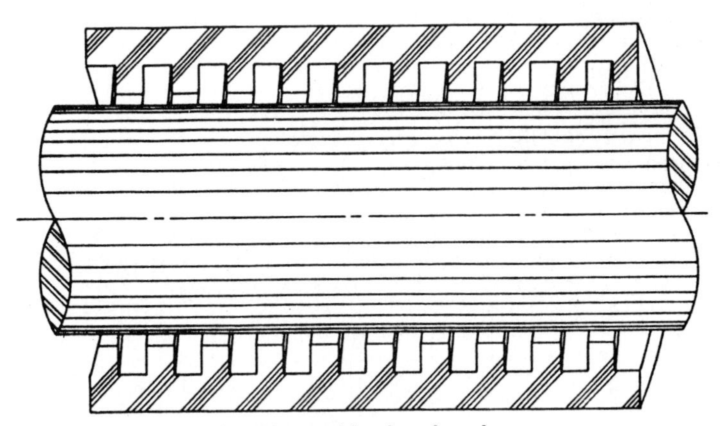

FIG. 21-96 Screw threads machined on housing.

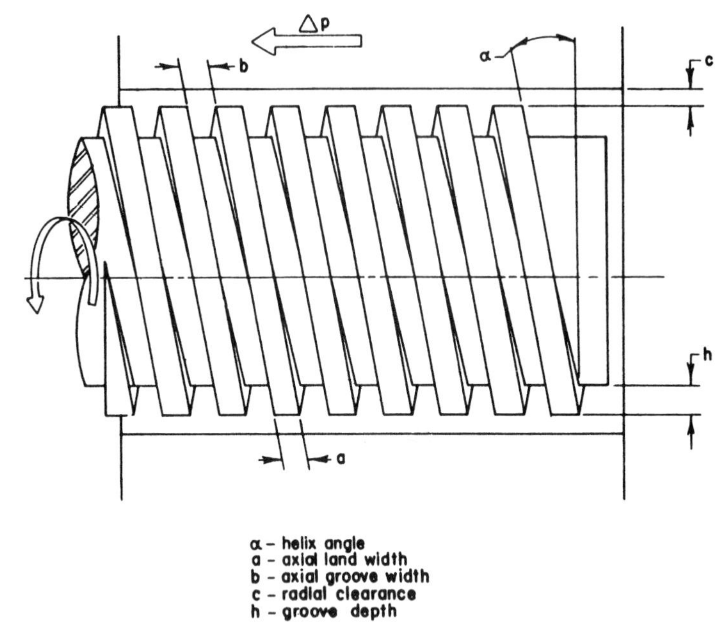

FIG. 21-97 Geometrical factors.

21-22 Ferrofluidic Seals Ferrofluidic seals use a magnetic fluid to seal the clearance gap in a labyrinth-type seal. This magnetic fluid prevents any leakage from occurring and is thus ideally suited for applications where any leakage is intolerable.

A ferrofluid consists of a carrier liquid that contains ultramicroscopic particles of a magnetic solid, such as magnetite. They are colloidally suspended, then stabilized by physiochemical means. To prevent flocculation even under the influence of a magnetic field, the particles are coated, and random collisions (brownian motion) with the molecules of the carrier liquid keep the particles in colloidal suspension for an indefinite period of time.

At the present time, all commercially available ferrofluids are electrically nonconductive in carrier liquids, such as fluorocarbons, hydrocarbons, polyphenyl ether, and

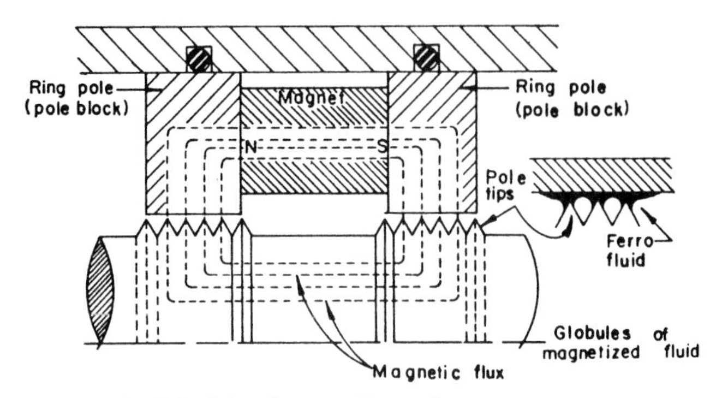

FIG. 21-98 Principle of magnetic seal.

TABLE 21-29 Summary of Physical Properties of Ferrofluids

Carrier fluid	Saturation magnetization, gauss	Viscosity, cP, 27°C	Evaporation rate, g/s·cm^2, 240°C	Density, g/mL	Thermal conductivity, mW/m·K	Initial susceptibility	Pour point, °C	Permeability	Electrical resistivity, Ω·cm
Ester	450	450	1.39×10^{-6}	1.490	209	2.10	−28	1.0–1.4	1.60×10^{9}
Ester	450	450	1.07×10^{-5}	1.410	168	2.20	−44	1.0–1.4	0.27×10^{9}
Synthetic petroleum	300	120	3.69×10^{-6}	1.195	170	1.49	−54	1.0–1.3	0.94×10^{9}
Petroleum	200	200	3.17×10^{-6}	1.080	—	1.14	−51	1.0–1.2	1.32×10^{9}
Ester	300	75	5.95×10^{-6}	1.258	185	1.37	−54	1.0–1.3	11.30×10^{7}
Ester	450	100	1.27×10^{-5}	1.440	186	2.06	−63	1.0–1.4	9.30×10^{7}
Fluorocarbon	300	3500	2.24×10^{-6}	2.245	117	1.36	−27	1.0–1.2	13.20×10^{9}
Polyphenyl ether	450	4500	9.92×10^{-7}	1.665	—	2.13	−12	1.0–1.4	0.18×10^{9}
Synthetic petroleum	200	1000	6.94×10^{-6}	1.125	158	0.67	−38	1.0–1.1	9.90×10^{9}

Source: Ferrofluidics Corporation, Nashua, N.H.

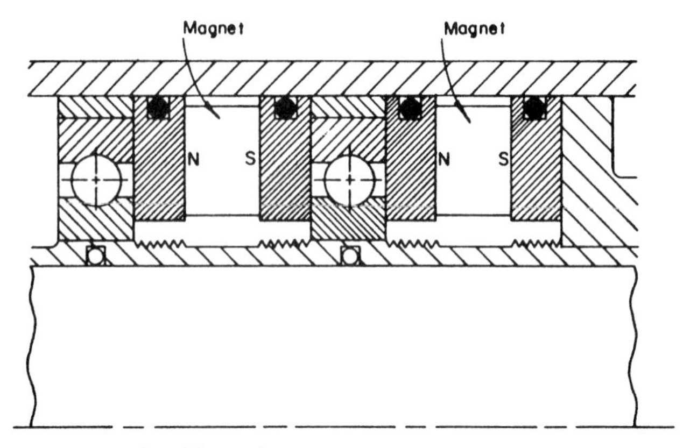

FIG. 21-99 Double seal arrangement.

TABLE 21-30 Noncontacting Seals

Problem	Causes	Corrective action
Excessive leakage (all seals leak at all times)	Excessive radial clearance	Decrease radial clearance
	Excessive pressure	Improve or add vent
	Excessive lubricant fill	Decrease fill or increase cavity
	Excessive temperature	Decrease fill or increase cavity
	Excessive vibration	Increase lube cavity
	Excessive end play	Increase lube cavity
Excessive leakage (some seals leak at all times)	Excessive radial clearance	Improve quality control
	Radial contact	Decrease misalignment
	Bypass leakage	Fill scratches and voids
Excessive leakage (all seals leak after a period of time)	Loss of lubricant viscosity	Improve lubricant
	Excessive relubrication	Decrease relubrication
	Contaminant ingress	Add contaminant shield
Water ingress	Static leakage	Add static seal
	Partial vacuum	Improve or add vent
	No lube leakage	Increase lube fill
	Excessive wet environment	Add water shield
Dirt ingress	Excessive radial clearance	Decrease radial clearance
	Excessively dirty environment	Add dirt seal
	Partial vacuum	Improve or add vent
	No lube leakage	Increase lube fill

aqueous solutions. Chemically and mechanically, the magnetic fluid offers the same characteristics as those provided by the carrier liquid in which the magnetic particles are colloidally suspended. Physical properties of typical ferrofluids appear in Table 21-29.

The operating principle of magnetic seals is shown in Fig. 21-98. A ring magnet is placed over the shaft contacting a ring pole on either side. With this arrangement, a magnetic field is produced; it is enforced in its effect by placing thread-type serrations either on the shaft surface or on the ID of the ring poles. The magnetic flux path is complete though the clearance gap between the ring poles and the shaft is filled. The fluid bridges the passageway, blocking any trace of leakage flow. The shaft is free to rotate without frictional disturbances by the ferrofluid. Without solid mechanical·contact, no wear is possible, and the friction generated in the fluid film is negligible. Even at extreme rotational shaft speeds, maintenance is not required.

Manufacturers claim successful operations at speeds of the order of 10,000 r/min in gaseous atmosphere under pressure or vacuum without leakage. Clearance gaps are typically 0.002 to 0.005 in. Temperatures must be maintained at 225°F or below to prevent sufficient carrier fluid evaporation. A typical double-seal arrangement is shown in Fig. 21-99. At the present time, the magnetic seal is applied to equipment sealing gases only.

21-23 Troubleshooting Noncontacting Seals Table 21-30 may be helpful when correcting leakage that occurs with noncontacting seals.

Section 22

Self-Clinching Fasteners

JOSEPH F. LOPES

Vice President, Engineering
Penn Engineering & Manufacturing Corp.
Danboro, Pennsylvania

INTRODUCTION

The original self-clinching fastener was introduced to sheet-metal fabricators more than 50 years ago by K. A. Swanstrom. It was designed as a device to provide load-carrying threads in metal sheets too thin to tap. As the need to hold together ultra-thin and ultra-light metals grew among existing and emerging industries, the original self-clinching fastener was modified to accommodate new applications, making a wide variety of thin-metal designs possible for the first time. Today, self-clinching nuts, studs, spacers, standoffs, and similar hardware are specified worldwide in a variety of manufacturing industries to assemble electronic components, business machines, automotive and aerospace equipment, telecommunications systems, and other precision products. Self-clinching fasteners deliver universal economies during the assembly process by reducing time, labor, parts, and costs. Once installed, these fasteners meet high performance standards and can even allow for easier disassembly of components requiring repair or service.

22-1 Purpose and Function By general definition, a self-clinching fastener is any device, usually threaded, which, when pressed into place in a properly sized drilled or punched hole, displaces the host material around the mounting hole. (Sample fasteners are shown in Fig. 22-1.) This pressing or squeezing process causes the displaced panel material to cold flow into a specially designed annular recess in the shank or pilot of the fastener, thereby locking the fastener in place. A serrated clinching ring, knurl, ribs, or hexagonal head prevents the fastener from rotating in the metal when tightening torque is applied to the mating screw or nut. The result is that self-clinching fasteners become a permanent and integral part of the panel, chassis, bracket, PC board, or other item in which they are installed.

Design engineers who specify self-clinching fasteners have a very specialized use in mind, one that allows very little margin for error in fastener performance. These fasteners are used primarily where good pushout and torque loads are required in sheet metal and panel materials that are too thin to provide secure fastening by any other method.

As a rule, a self-clinching fastener should be selected whenever a component must be replaced readily and where "loose" nuts and hardware would be inaccessible. If the attaching nuts and screws cannot be reached after a component is assembled, self-clinching fasteners can be installed during the initial fabrication process, thereby simplifying and expediting the component mounting and assembly operations, including those performed in the field.

FIG. 22-1 Typically used self-clinching fasteners. *(Penn Engineering & Manufacturing Corp.)*

22-2 Advantages Self-clinching fasteners are recognized for their production economies, inherent design benefits, and in-use performance. These fasteners usually require fewer assembly operations than many other types of fastener hardware, such as caged or anchor nuts. They also eliminate the need to stock or handle additional hardware, such as washers, lock washers, loose nuts, or bolts during the component assembly or disassembly operations. In addition, the number of steps required during the final assembly can be reduced, because the fastener is installed during the fabrication process.

These fasteners can be installed quickly and easily by using a parallel squeezing force. The installation process requires no special preparation of the mounting hole, such as chamfering or deburring. They also provide a secure fastening method with strong threads in many sheet and panel materials as thin as 0.020 in (0.51 mm).

The installed internally threaded fasteners do not require any retapping, and their holding power usually exceeds that of sheet metal screws. Self-clinching fasteners are designed for those applications where high pushout and torque-out resistance forces are required. They can even enhance end-product appearance, because their compact design allows them to be installed flush to the panel. In many applications, they can support the switch to a thinner sheet metal or panel material and thereby generate a reduction in total installed costs over those employing traditional fastener designs.

22-3 Materials Self-clinching fasteners are manufactured from a variety of materials, such as low- and medium-carbon steels, martensitic and austenitic stainless steels, aluminum, and phosphor bronze, which are just some of the more popular materials currently used.

When one is designing self-clinching fasteners for a particular assembly application, several factors must be considered. First, the fastener material must be compatible with the panel or sheet material. Second, the fastener hardness must be greater than the hardness of the softer panel material into which it is installed. A general rule of thumb is that the fastener should be at least 20 points HRB greater than the panel or sheet material hardness. Punched holes will produce an increased hardness around the mounting hole. This hardness increase is caused by the cold working of a localized layer of material generated during the hole-piercing operation.

The increased hardness of the mounting hole in steel and aluminum panels does not generally affect its ability to flow into the annular recess of the fastener. However, the effect of piercing holes in already cold-worked stainless steel sheets can be significant and should be considered when one is designing austenitic stainless steel fasteners in 300 series stainless steel sheets.

PRIMARY FASTENER TYPES

Self-clinching fasteners traditionally fall into one of several primary categories: nuts, studs, and spacers and standoffs.

22-4 Nuts These self-clinching fasteners (Fig. 22-2), which feature thread strengths greater than those of mild screws, are commonly used wherever strong internal threads are needed for component attachment or fabrication assembly.

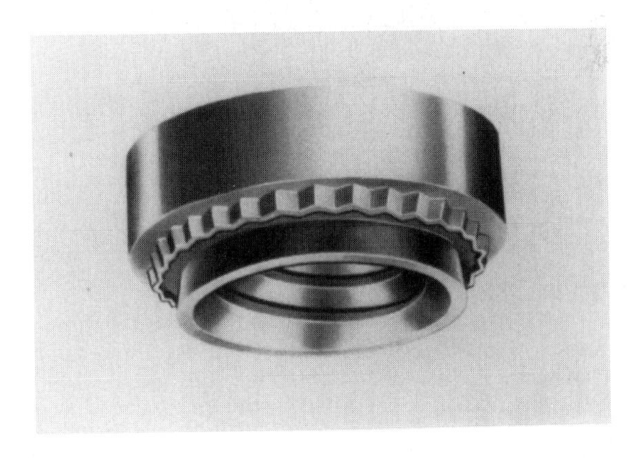

FIG. 22-2 Self-clinching nut. *(Penn Engineering & Manufacturing Corp.)*

During installation, a clinching ring locks the displaced metal behind the fastener's tapered shank, resulting in a high pushout resistance. High torque-out resistance is achieved when the knurled platform is embedded in the sheet metal. Proper installation forces will not distort or damage the threads, because the recommended shank length is always less than the minimum sheet thickness. The clinching action of these nuts occurs on the fastener side of the thin sheet, with the reverse side remaining flush and smooth.

22-5 Studs These externally threaded self-clinching fasteners (Fig. 22-3) are used where the attachment must be positioned before being fastened. Flush-head studs are usually specified, but variations have been designed for high-torque, thin-sheet, or electrical applications. Manufactured from a variety of materials, self-clinching studs are offered in a wide range of thread sizes. Studs are also available without threads for use as permanently mounted guide pins or pivots.

FIG. 22-3 Self-clinching stud. *(Penn Engineering & Manufacturing Corp.)*

22-6 Spacers and Standoffs These fasteners (Fig. 22-4) are used where it is necessary to stack or space components away from the panel. Through-threaded or blind types generally are standard. Their material can be steel, stainless steel, or aluminum. Common thread sizes for through-hole and blind-type standoffs range from 4-40 through 10-32 and M3 through M5. Unthreaded standoffs are designed for spacing multipanel assemblies.

Self-clinching standoffs are installed with their heads flush with one surface of the thin mounting sheet. When blind-threaded types are used, outer panel surfaces are both smooth and closed. (See Fig. 22-5.)

Variations of standoffs have been developed over the years to meet emerging applications, primarily in the electronics industry. These types include standoffs with concealed heads, others that allow boards to snap into place for easier board assembly and removal, and those designed specifically for use in printed-circuit boards.

FASTENER VARIATIONS

Since the original S nut was developed in 1943, a wide range of self-clinching fastener variations have been engineered for specialized assembly applications. Each of these different types offers a particular advantage, whether in terms of meeting a design challenge, performance requirement, or even enhancing component appearance. These variations, each addressing a unique design need, share the common benefits of simple

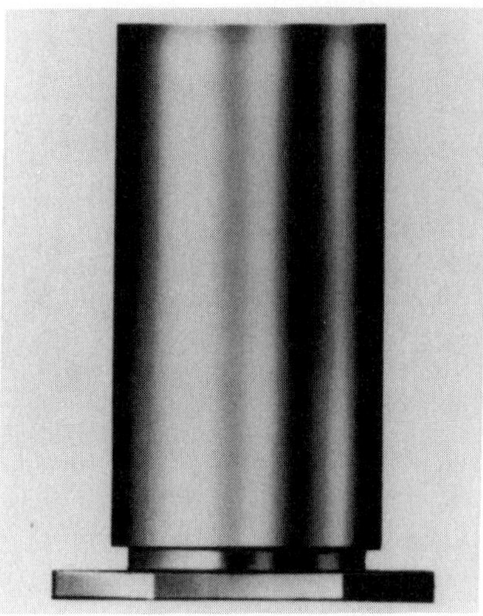

FIG. 22-4 Self-clinching standoff. (*Penn Engineering & Manufacturing Corp.*)

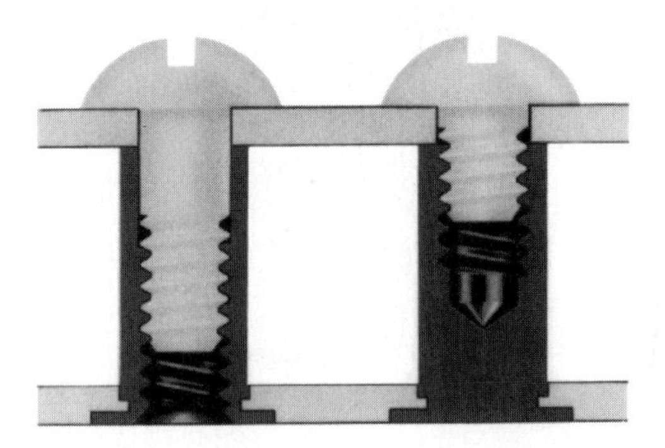

FIG. 22-5 Installed standoffs with through-hole threads (left) and blind threads (right). (*Penn Engineering & Manufacturing Corp.*)

installation, holding power, and assembly economies, which underlie self-clinching fastener technology.

22-7 Flush Fasteners When installed, these fasteners (Fig. 22-6) are completely flush with the metal sheet. They are designed for applications where a thin panel or sheet requires load-bearing threads but still must remain smooth, with no protrusions on either surface. This can enhance the functional and cosmetic qualities of an entire assembly.

Self-clinching stainless steel flush nuts can be installed in metal sheets as thin as 0.060 in (1.5 mm) even before bending and forming operations. This can provide strong threads in places which would be inaccessible for installation after the chassis is formed. Installation in these instances is typically performed into a flat sheet so that the fasteners will not interfere with the subsequent bending and forming operations.

The fastener's hexagonal head along with the self-clinching design delivers high axial and torsional strength.

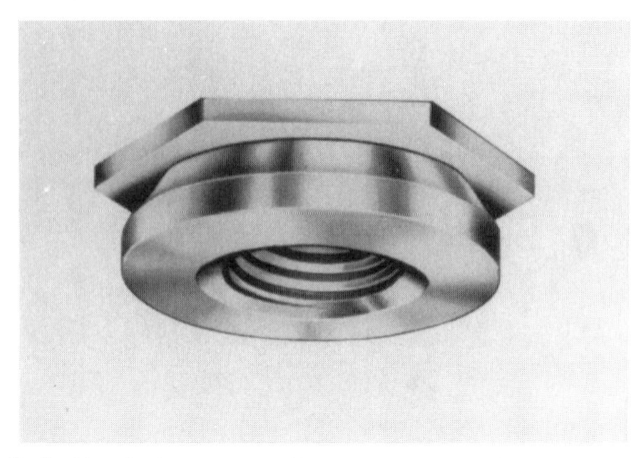

FIG. 22-6 Self-clinching flush fastener. *(Penn Engineering & Manufacturing Corp.)*

22-8 Floating Nuts This type of self-clinching fastener (Fig. 22-7) was developed to compensate for mating hole misalignment by having a floated threaded element. These fasteners provide load-bearing threads in sheets as thin as 0.040 in (1 mm) and permit up to 0.030 in (0.8-mm) adjustment. They are made with locking and nonlocking thread styles. The locking thread types should use the same design criteria as stated in Paragraph 22-10 (Self-Locking Fasteners).

FIG. 22-7 Self-clinching floating nut. *(Penn Engineering & Manufacturing Corp.)*

22-9 Spring-Loaded Panel Fasteners Completely preassembled, spring-loaded panel fastener assemblies (Fig. 22-8) are generally used on enclosures where the screw must remain with the door or panel. These fasteners offer the advantages of ease of assembly and quick panel removal without loose hardware. Related hardware includes flush-mounted panel screw components and plunger assemblies.

Self-clinching panel fastener assemblies are manufactured in a range of thread sizes and assorted screw lengths to satisfy many application demands. Separate spring-loaded assemblies have been created specifically for mounting printed-circuit boards.

Flush-mounted panel screw components are specially designed so that the tightened screw is completely flush with the top of the sheet. This fastener's separate components

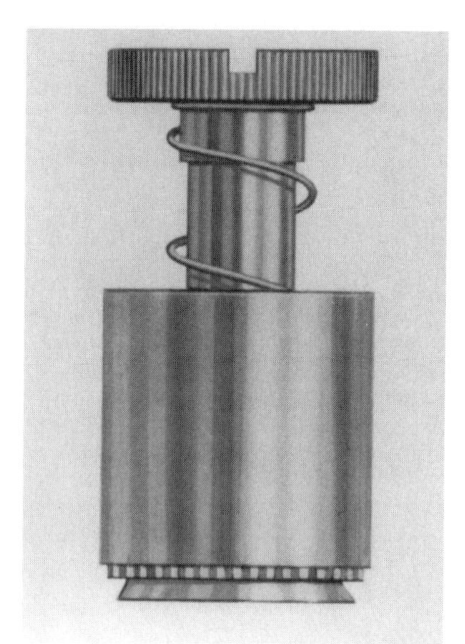

FIG. 22-8 Spring-loaded panel fastener. *(Penn Engineering & Manufacturing Corp.)*

consist of a screw, retainer, and receptacle nut. When installed together, they provide a secure, flush-mounted, captive screw assembly in sheets as thin as 0.125 in (3.2 mm).

A plunger assembly is a retractable, spring-loaded, locating pin for applications including drawer and glide stops, indexing pins, and latch detents. When the plunger is retracted, the reverse side of the sheet is completely flush. A quick-turn, lock-out feature keeps the plunger in a hands-free, retracted position until released.

All screws and plunger ends are designed for easy entry into receiving hardware. Clearance with panel fastener retainers permits maneuverability to locate and seat mating threads.

22-10 Self-Locking Fasteners These fasteners (Fig. 22-9) provide a prevailing torque locking feature to restrict the rotation of the mating screw under adverse conditions of vibration. The locking feature usually incorporates a self-contained all-metal or nylon type feature. The self-locking feature provides a reusable and prevailing torque thread lock which resists the torque-out of the mating fastener after installation.

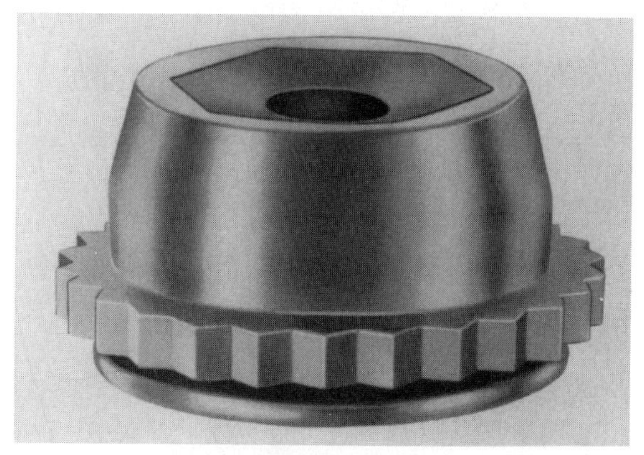

FIG. 22-9 Self-locking fastener. *(Penn Engineering & Manufacturing Corp.)*

These nuts should be designed so that the spin-out or torque-out of the nut's clinching feature greatly exceeds the torque-out requirements of the self-locking feature.

The installed nut's knurled collar is embedded in the panel and its undercut cavity beneath the collar is filled with displaced panel material, thereby capturing the fastener in the sheet or panel material.

These nuts should be coated with a lubricant that will prevent galling or nut-bolt seizure. The lubricant will enhance the reusability qualities of the nut-bolt system. Specifications on locking nut performance requirements can be found in MIL-N-25027 or IFI 100/107 standards. These are the most popular standards applied to prevailing locking torque nuts in the United States.

22-11 Nonthreaded Standoffs These variations of self-clinching standoffs (Fig. 22-10) are designed so that a printed-circuit board or panel can be slipped quickly into place and then removed from an assembly. They can be used for spacing or hanging replaceable components and allow for assembly or removal of components without the need for screws or additional hardware.

Two types of nonthreaded self-clinching standoffs are used most often and can be distinguished in the manner in which board or panel assembly and removal is achieved. One fastener design allows board or panel removal by sliding the component sideways and lifting it off. An alternative design offers the capability to snap the board into place on the fastener.

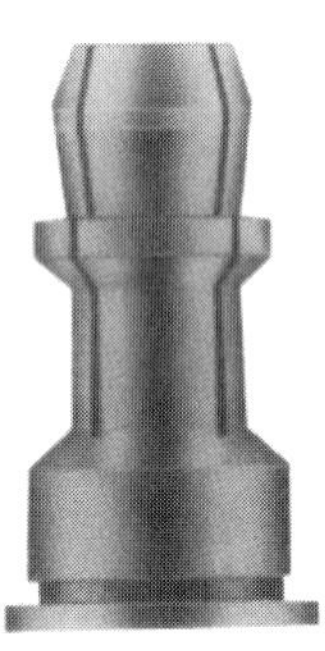

FIG. 22-10 Nonthreaded standoff. *(Penn Engineering & Manufacturing Corp.)*

22-12 Concealed-Head Studs and Standoffs The concealed-head feature of these self-clinching fasteners (Fig. 22-11) allows the side of the panel opposite the installation to remain smooth for applications where surfaces must remain unmarred. The fastener head is locked securely in a blind, milled hole and is able to handle substantial loads.

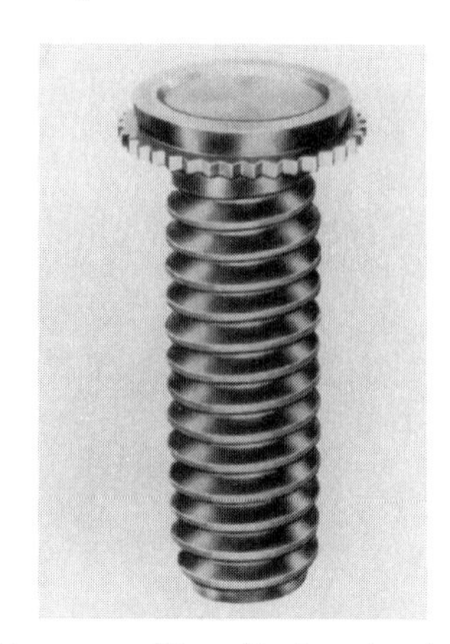

FIG. 22-11 Concealed-head fastener. *(Penn Engineering & Manufacturing Corp.)*

Concealed-head aluminum or stainless steel studs and stainless steel standoffs can be mounted in thin metal sheets and are installed in the milled blind hole to the recommended minimum depth.

22-13 Blind Fasteners Self-clinching blind fasteners (Fig. 22-12) have a closed end that limits screw penetration and are useful for protecting internal components from damage by inadvertent insertion of extra-long screws. Barriers protect the threads from damage and foreign matter. These fasteners provide permanently mounted blind threads in thin metal sheets. The shanks of these fasteners can act as their own pilots. The threads are also protected from damage and foreign matter after installation.

22-14 Miniature Fasteners The primary advantage of miniature self-clinching fasteners (Fig. 22-13) is their design to fit into a minimal space, while providing strong, reusable threads. Both nonlocking and self-locking types have been developed. Normal thread sizes can range from 0-80 through $\frac{1}{4}$-28 and M2 through M6.

FIG. 22-12 Self-clinching blind fastener. *(Penn Engineering & Manufacturing Corp.)*

FIG. 22-13 Self-clinching miniature fastener. *(Penn Engineering & Manufacturing Corp.)*

These fasteners provide immediate visual indication when proper installation has been accomplished. A strong, knurled collar, which is completely embedded in the sheet, guarantees against rotation of the fastener in the sheet. The self-locking type of fasteners should be designed so that the spin resistance of the knurl greatly exceeds the torque that can be exerted by the self-locking feature. When the collar is embedded in the sheet, the undercut cavity beneath the collar is filled with displaced material, thereby developing pushout resistance.

These nuts should be supplied with a thread lubricant to provide a smooth, nongalling prevailing torque performance necessary for reliable locking and reusability.

Screws for use with self-locking fasteners should be class 3A/4h fit or no smaller than class 2A/6g. Screws should be long enough that at least two threads project through the fasteners when tightened.

22-15 Custom Designs Standard self-clinching fasteners have resulted over the years from research and development efforts and in response to evolving technical needs. Equally important to product development has been the custom design process among a wide range of industries. Nowhere is this custom design process more evident than in the electronics industry.

As an example, the attachment of connectors traditionally has required excessive loose hardware and lengthy assembly time with as many as 10 pieces of loose fastening hardware per connector. In some cases, special front- and rear-mating shells have been necessarily used by connector manufacturers to accommodate standard fastening hardware. In others, cable houses and OEMs have become frustrated with increased instances of insecure mounting and unreliable attachment and contact of components. These concerns and the need to ensure component integrity have led to a variety of custom self-clinching fasteners designed exclusively to satisfy common connector attachment applications.

Some of these connector applications include D-subconnector-to-panel mounting, slide latch attachments, and cable-to-cable mounting. All custom self-clinching fasteners for these applications have been developed to provide clear advantages over traditional assembly methods. When installed, the fasteners are held securely in place, preventing them from dropping into delicate electronic circuitry. They also ensure proper spacing to maintain proper contact of connectors. Fewer parts translate to a reduction in handling, inventory, and assembly costs. (See Fig. 22-14 for a comparison schematic.)

Applying the basic, proven principles of self-clinching fastener technology, fastener designers in the years ahead can be expected to originate even more application-specific, custom parts as assembly needs and processes change around the world.

FASTENER INSTALLATION

22-16 Requirements Self-clinching fasteners can be installed into properly sized mounting holes with any parallel-acting press that can be adjusted to predetermined forces. Mounting holes should be punched, drilled, or cast to specified dimensions. There should be no deburring of the hole on either side of the sheet before installation, since deburring will remove metal required for clinching the fastener into the sheet.

Host panel hardness and thickness should be evaluated to ensure fastener compatibility and performance in service. The panel must be made of a ductile material softer than the fastener which is going into it. The panel also must meet the minimum sheet thickness required by the particular fastener. Some self-clinching fasteners can be installed into sheets as thin as 0.020 in (0.51 mm), but generally 0.030 in (0.76 mm) or 0.040 in (1 mm) is the minimum sheet thickness necessary. While there is no specified maximum thickness for sheets, a few fastener types do specify a thickness range, which includes a maximum, as a result of their special design and function.

Access to both sides of the sheet is typically required to properly install self-clinching fasteners. However, there are some ¼-in (M6) or larger nuts which can be drawn in from one side by using special tooling and an impact-torque wrench.

22-17 Procedure Installation of self-clinching fasteners can be accomplished in three basic assembly steps.

First, the shank or pilot of the fastener should be inserted squarely into the previously prepared mounting hole. Next, force is applied, using a press until the head of the nut contacts the sheet. (Some types of fasteners will be fully installed when the head is flush within the sheet.) The process is complete after the mating piece is inserted into the side opposite the head of the fastener. (Figures 22-15, 22-16 and 22-17 illustrate the self-clinching process for typical nuts, studs, and standoffs.)

22-18 Special Considerations Mounting holes should not be chamfered or have broken edges in excess of 0.005 in (0.127 mm). Hole tolerances of +0.003-.000 in (+0.08 mm) must generally be held. The fastener should be installed on the punch side if the sheet is 0.090 in (2.29 mm). or thicker because of the blowout condition generated on the die of the sheet during the piercing operation. Holes pierced in

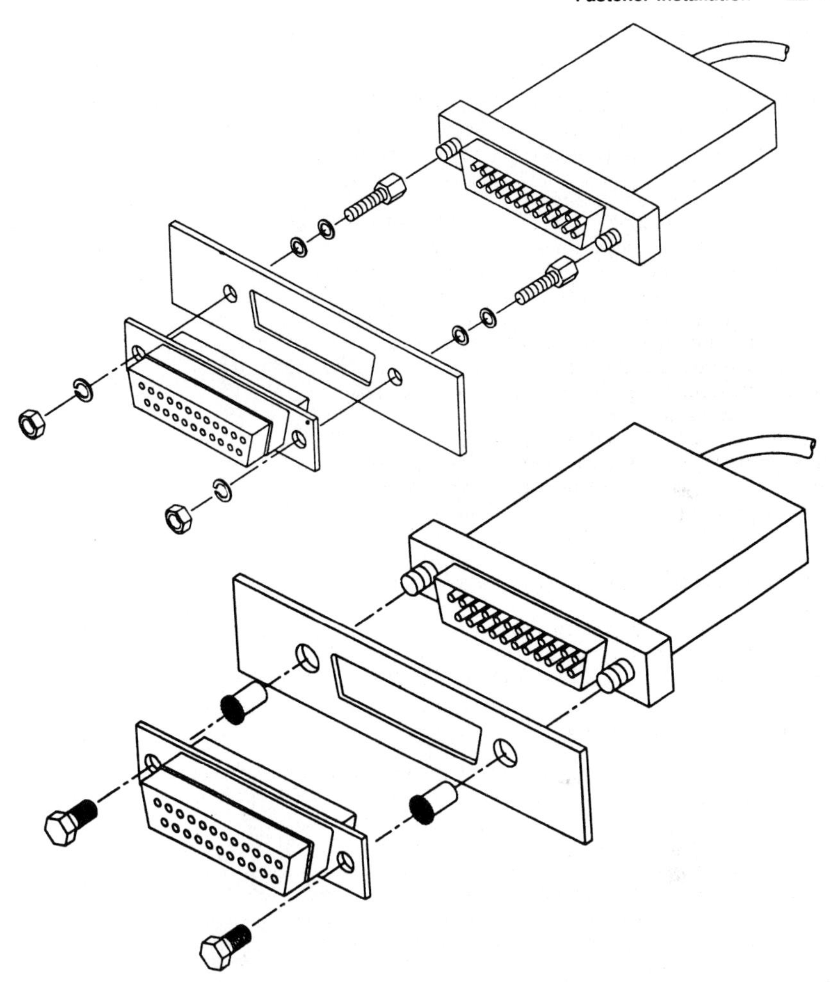

FIG. 22-14 Use of self-clinching fasteners (bottom) allows for use of fewer parts compared with conventional attachment methods (top). *(Penn Engineering & Manufacturing Corp.)*

sheets thinner than 0.090 in (2.29 mm) may have fasteners inserted on either side. In all cases, the manufacturer's minimum centerline of hole to edge of sheet distance should be observed.

During installation, the fastener must be squeezed into place between parallel surfaces. A hammer blow or impact does not allow sufficient time for the sheet material to flow into the recesses of the shank and undercut. Sufficient squeezing force should be applied to totally embed the clinching ring around the entire circumference and to bring the shoulder squarely in contact with the sheet. Oversqueezing can crush the fastener head, distort threads, and buckle the sheet.

When installation is done with the recommended squeezing force (depending on the size of the fastener and hardness of the sheet metal), there is little or no distortion of the sheet or damage to the finished surface. Fasteners generally should be installed after plating, finishing, or anodizing, although with a proper plating process, they can be installed before plating.

Installation results in a flush surface on one side of the panel. Conversely, staked or crimped fasteners require special counterboring to obtain a one-sided flush surface.

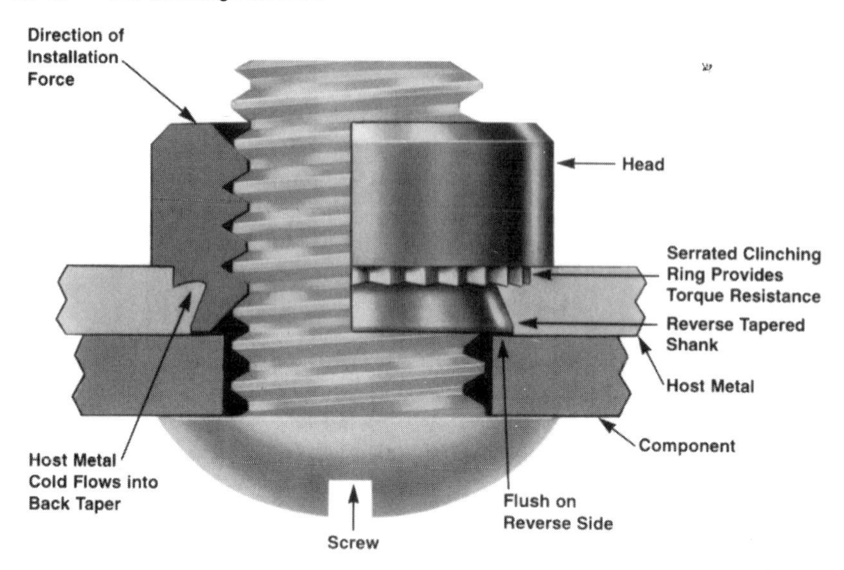

FIG. 22-15 Typical installed self-clinching nut. *(Penn Engineering & Manufacturing Corp.)*

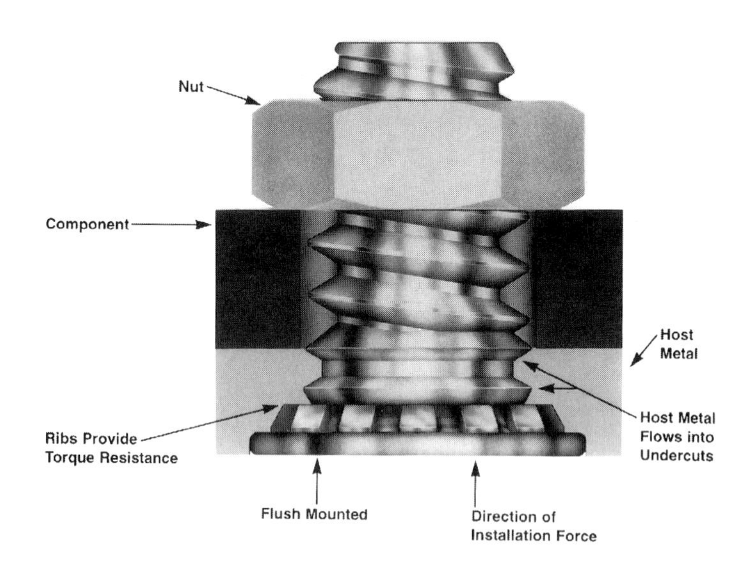

FIG. 22-16 Typical installed self-clinching stud. *(Penn Engineering & Manufacturing Corp.)*

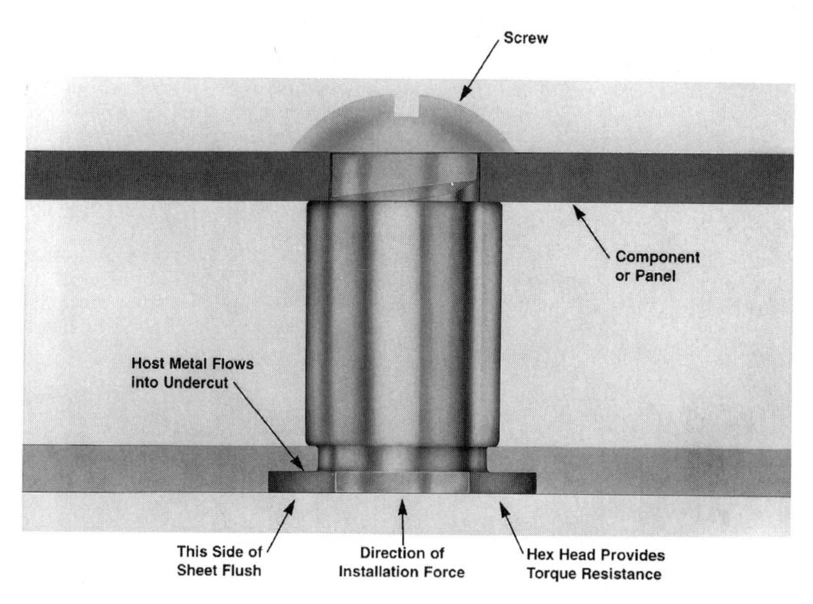

FIG. 22-17 Typical installed self-clinching standoff. *(Penn Engineering & Manufacturing Corp.)*

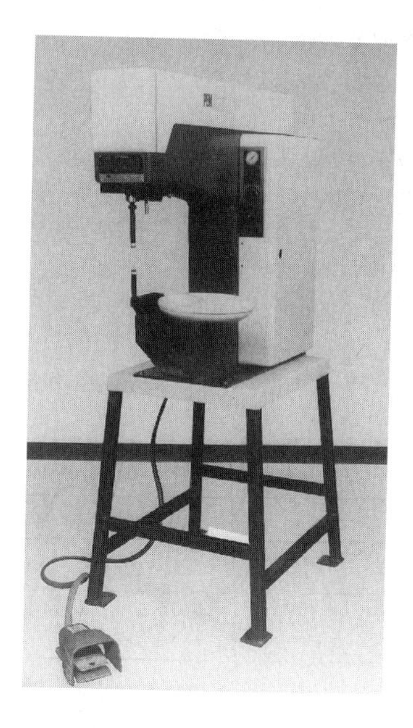

FIG. 22-18 Stand-alone manual press system for installing self-clinching fasteners. *(Penn Engineering & Manufacturing Corp.)*

22-19 Tooling Since all self-clinching fasteners must be squeezed into place, any press or vise that provides the necessary parallel force may be used to install them. However, assembly considerations and production needs have led to the development of specialized precision fastener-installation presses with particular features and benefits for applications ranging from high-volume to prototype work.

The larger stand-alone manual press systems, which are equipped with an adjustable stroke length to speed operation, can deliver a ram force of 500 to 12,000 lb. (2.2 to 53.3 kN). (See Fig. 22-18 for an example.) Rapid tooling changeover offers the capability to accomplish short production runs efficiently. These presses are pneumatically powered.

Bench-mounted presses (such as in Fig. 22-19) are available which can develop a squeezing force of 8000 lb (35.6 kN).

Ram forces of up to 6000 lb (26.7 kN) can be achieved with portable hand presses (see Fig. 22-20), which typically weigh only 10 lb. (4.6 kg). These hand presses install self-clinching nuts and studs in unified and metric sizes and provide appropriate punches and anvils, which can be snapped into place. Hand presses also can be bench-mounted.

FIG. 22-19 Bench-mounted press for fastener installation. *(Penn Engineering & Manufacturing Corp.)*

22-20 Automated Assembly An automated press should be considered for high-volume fastener installations. Some automated presses are specifically designed to feed self-clinching fasteners automatically into punched or drilled holes in sheet metal, seating them correctly with a parallel squeezing force. Feeding rates are typically five or six times faster than manual insertions, and squeezing action is adjustable to compensate for variations in thickness and hardness of the sheet and the height of the fasteners.

Such presses can be placed at the most advantageous location in an automated production line. No special venting is needed. There is little need for parts and inventories of expendables, as there is with welding procedures. Depending on the complexity of panel configurations and how much handling is required, an operator can install up to 20 fasteners per minute.

One of the recent advances in automated tooling is a system that operates in tandem with turret punch presses. A turret nut tool, which is fully loaded with prepackaged "sticks" of fasteners and dropped into one of the $1\frac{1}{4}$-in stations of a turret press, can achieve an installation rate of 5400 self-clinching nuts in 1 h.

When correctly programmed, the press will punch the proper holes in the flat sheet, then rotate the turret to the nut tool position, and install the fasteners in their respective mounting holes. The insertions are done with CNC precision so that each nut is installed correctly and reliably. The integrity of each fastener is also maintained. Each nut will exhibit the same pushout and torque-out values that are achieved with traditional installation methods.

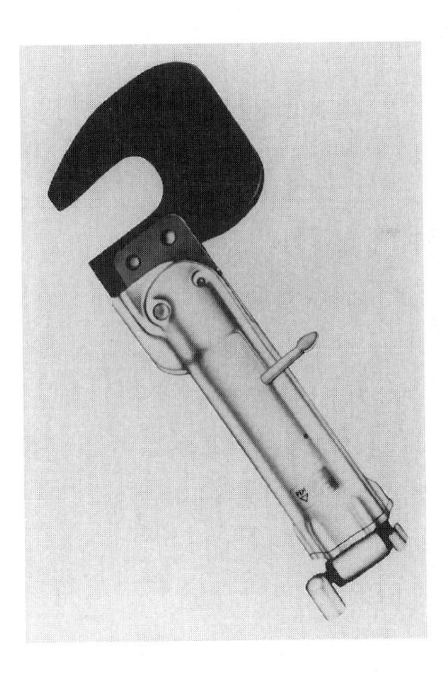

FIG. 22-20 Portable hand press for installing self-clinching fasteners. *(Penn Engineering & Manufacturing Corp.)*

22-21 Troubleshooting A review of some of the potential problems that could be encountered with self-clinching fasteners upon installation underscores the importance of factors such as proper squeezing force and assembly procedure.

- If a fastener is not seated squarely and exhibits reduced holding power, then the punch and anvil faces may not be parallel and the panel may be cocked during installation. The solution is to ensure that punch and anvil are flat, parallel, and hard, and that large panels are held perpendicular to the punch and anvil.

Thread Size		Panel Material											
		5052-H34 Aluminum .040" (1.02 mm) thick						Cold-rolled Steel .040" (1.02 mm) thick					
		Installation		Pushout		Torque-out		Installation		Pushout		Torque-out	
		(lbs.)	(kN)	(lbs.)	(N)	(in. lbs.)	(N•m)	(lbs.)	(kN)	(lbs.)	(N)	(in. lbs.)	(N•m)
2-56	(M2)	1500-200	6.7-8.9	89	400	10.5	1.13	2500-3500	11.2-15.6	125	550	15	1.7
4-40	(M3)	1500-2000	6.7-8.9	89	400	10.5	1.13	2500-3500	11.2-15.6	125	550	15	1.7
6-32	(M3.5)	2500-3000	11.2-13.5	105	400	17	1.92	3000-6000	13.4-26.7	130	570	20	2.3
8-32	(M4)	2500-3000	11.2-13.4	105	470	23	2.6	4000-6000	18-27	145	645	35	4
10-32	(M5)	2500-3500	11.2-15.6	110	480	32	3.6	4000-9000	18-38	180	800	40	4.5
1/4"	(M6)	4000-7000	18-32	360	1580	125	14.1	6000-8000	27-36	400	1760	150	17
5/16"	(M8)	4000-7000	18-32	380	1570	210	23.7	6000-8000	27-36	420	1870	230	26

Torque-through generally applies to self-clinching studs and is not shown in this chart

FIG. 22-21 General pushout and torque-out values for self-clinching nuts (unified and metric data included). *(Penn Engineering & Manufacturing Corp.)*

• If a fastener has poor holding power and falls out of the panel, then inadequate installation force may have been applied or the panel may be too hard for the fastener material. The solution is to seat the fastener against the shoulder by applying more force or changing the shut height of the installation press.

• If the internal threads are tight and the sheet buckles, the possible cause may be oversqueezing of the fastener. The installation force should be reduced.

• If a fastener exhibits poor holding power and the nut is off-center of the hole, the mounting hole is probably oversized and the nut is cocked in the hole. This produces a shearing action on the side of the hole. The solution is to ensure that the hole is punched or drilled to specified dimensions and to check that the shank of the nut is squarely in the hole before squeezing.

• If the internal threads are tight or cracked, the shank length of the fastener may be extending through the sheet. If this is the case, a different fastener should be selected with proper shank length for sheet thickness.

• If a panel buckles badly with a stud in 0.040- to 0.059-in (1- to 1.5-mm) material, the probable cause is lack of countersink in the anvil. This can be corrected by providing countersink in the anvil to specified dimensions.

• If standoffs or studs hold poorly in a panel, the hole in the anvil may be too large or chamfered. Use of an anvil with a hole matching the manufacturer's specifications can remedy this situation.

• If the head of a flush-head stud or standoff cups during installation, the punch diameter may be too small or not hard and flat. The punch must be larger than the head of the stud or standoff and preferably equal to the anvil diameter.

• If the fastener exhibits poor holding power, the die side of the panel may have a hole too large for the mating fastener. This blowout condition is prevalent with pierced or punched holes. The hole size tends to increase on the die side and increases as the sheet material thickness increases. There are other factors to consider as this blowout condition occurs: Evaluate the condition of the punches, since dull punches will increase the blowout condition, and evaluate the sheet hardness to determine whether the hardness is uniform within the entire sheet and is within the supplier's recommended hardness range. This blowout condition can be remedied by punching the hole on the opposite side of the panel and by ensuring that the installation takes place only on the punched side of the sheet or panel material.

• If the edge of a panel bulges, the mounting hole likely violates the specified minimum edge distance, or the nut is oversqueezed. This problem can be corrected by putting the panel or bracket in a restraining fixture during installation or moving the mounting hole away from the edge. If possible, the installation force should be reduced.

PERFORMANCE CONSIDERATIONS

The reliability of a self-clinching fastener in service depends on many factors. Some of the most crucial are reviewed in this section.

22-22 Design Integrity There are both subjective and objective areas to examine with respect to design integrity. One subjective area might be whether the supplier originated or copied the particular fastener design specified. Objective data can also be gathered to help determine whether a self-clinching fastener will perform as desired. Areas to evaluate with respect to design include a fastener's thread strength to ensure it matches the thread strength of the mating part and the ratio of raw material used with respect to performance, weight, and strength.

22-23 Dimensional Tolerances A self-clinching fastener requires very tight tolerances to maximize its performance. Slight variations in the clinching feature can affect the torque-out and pushout performance by 20 percent. Simple visual inspection may not always prove revealing. However, there may be obvious variations from part to part, such as ragged surfaces.

These are some dimensional areas for consideration: Is the clinching profile consistent (the knurl, undercut, and shank)? Is the overall outer envelope enough for automated insertion? What is the variation in threads? In all dimensional tests to which a supplier's parts are subjected, the smaller the variations, the better the performance

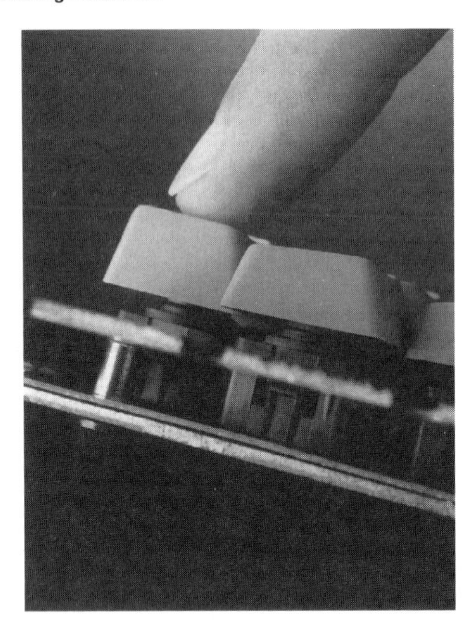

FIG. 22-22 Application for self-clinching threaded standoffs. A computer keyboard manufacturer utilizes stainless steel standoffs to attach the keyboard frame to the printed-circuit board assembly. The standoffs (one in each of four corners) help maintain the spacing between the layers of assembly by permitting stacking and ensure reliable contact and operation. The standoffs are pressed into round holes in the $1\frac{1}{16}$-in-thick metal frame, where they become an integral part of the host material. The hexagonal head of each standoff is completely flush with the sheet surface, making the head virtually impossible to see. This promotes a desired attractive panel appearance after the finishing operation. (*Penn Engineering & Manufacturing Corp.*)

values. Rigorous tests should be conducted before assembly gets underway to ensure that dimensional tolerances are as specified.

22-24 Thread Fit A part may be specified because it meets one or more government specifications for thread tolerances. If equivalents are considered, they should be checked to confirm that they meet specifications. Then, all should be tested. For example, ANSI B1.1 or ASME B1.13M covers the major diameter, minor diameter, and pitch diameter of nuts and studs. Federal Standard H-28 lists all dimensions and gaging standards.

22-25 Heat Treatment This is a critical quality area. Improper heat treatment can cause a fastener to fail during or after installation. Improper tempering can cause fastener brittleness, causing the fastener to crack. Inadequate treatment can cause fasteners to be so soft that they are literally crushed during installation.

22-26 Plating Plating standards (such as ASTM-B-633 for zinc) set limits for preparation of the metal, plating thickness, adhesion, rust corrosion protection, hours of salt spray testing, and other operations. A poorly plated part will diminish the performance and appearance of a final product.

22-27 Mechanical Properties One of several tests applicable to a self-clinching fastener to determine its reliability in service is *torque-out*, which determines the fastener's ability to resist rotation within a panel. This test often is done at the head of the fastener, often with values exceeding the ultimate torsional strength of the mating screw or nut. Torque-out is usually expressed in inch-pounds or newton-meters. Typical torque-out values are generally quite high compared to the rotational force

FIG. 22-23 Application for self-clinching panel fasteners. A manufacturer of rugged enclosures for military and industrial applications utilizes 12 panel fastener assemblies (six on the front panel and six on the rear) in each unit to secure the panels and provide easy front-panel access to boards and rear access to cables during service. With their captive-screw design, the panel fasteners enable rapid panel removal and replacement without loose hardware. No special tooling is required for access. The heavy-duty enclosures are designed to protect 6U systems from severe shock and vibration, and the fasteners have demonstrated consistent holding power. The manufacturer opted for black panel fasteners for aesthetics. (*Penn Engineering & Manufacturing Corp.*)

that will be put on them. In fact, for most qualify self-clinching nuts, the screw will fail before the nut rotates in the material.

A second performance reliability measure is *torque-through*, which is the resistance of the fastener to pulling through the metal sheet when a clamping torque is applied. Torque-through generally applies to self-clinching studs and is expressed in inch-pounds or newton-meters.

Another performance measure is the pushout values. These indicate the axial resistance of a fastener to being removed from the sheet opposite to the direction from which it was installed and should be roughly 5 to 10 percent of the force used to install the fastener. Pushout is expressed in pounds or newtons.

(General pushout and torque-out values are shown in Fig. 22-21.)

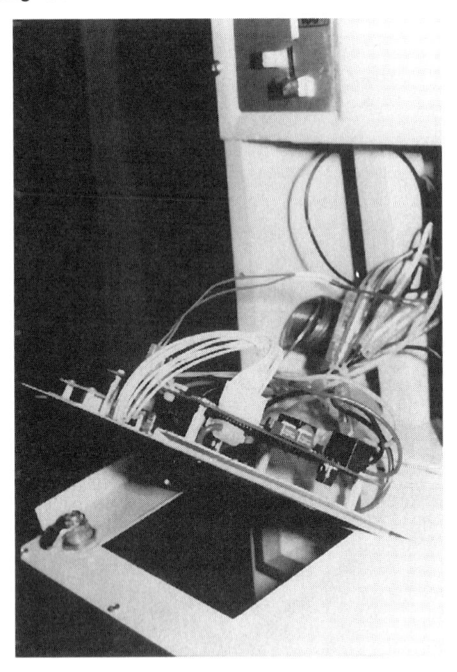

FIG. 22-24 Application for self-clinching snap-fit standoffs. A manufacturer of commercial clothes dryers reports a fivefold improvement in production times for control panel assemblies by switching to snap-fit self-clinching standoffs. Production improvements were achieved by eliminating unnecessary fastening components (six per assembly) and eliminating process steps (welding of three components). The standoffs are squeezed into prepunched holes in 18-gage (0.048-in) sheet metal. Components then snap into place onto the fasteners, which feature a spring action to hold subassemblies securely and become a permanent part of the installation. The snap capability allows quick removal of the panel to facilitate in-field service. (*Penn Engineering & Manufacturing Corp.*)

FIG. 22-25 Application for self clinching nuts. Dozens of self-clinching stainless steel nuts (shown on the underside of the panel) secure all electrical components, including circuit boards, in a fully automated, large engineering-document folder. The use of self-clinching fasteners eliminates the need for additional hardware, such as lockwashers and nuts, which previously had proved cumbersome in the assembly process and often were prone to loosening during shipping or while the machine was operating. (*Penn Engineering & Manufacturing Corp.*)

APPLICATIONS

22-28 Noteworthy Examples Figures 22-22 to 22-25 illustrate the use of typical self-clinching fasteners in actual applications.

GLOSSARY OF SELF-CLINCHING TERMS

Following is a review of some of the commonly used terms relating to the self-clinching fastening process.

Anvil. An insert, either solid or hollow, which is used on the underside of a panel to resist the installation force.

Blind hole. A hole, usually threaded, which is open from only one end.

Chamfer. A beveled edge or corner.

Cold flow. The movement of a ductile material under pressure.

Floating. The ability of a fastener to move in a direction parallel to the mounting panel and allow for mating hole misalignment.

Flush. The ability of a fastener to be contained completely within the thickness of a panel. This also refers to the absence of a protrusion above the surface of the panel.

Head. The portion of a fastener which forms its largest diameter.

Installation force. A term for force, expressed in pounds, tons, or newtons, applied axially to a self-clinching fastener to achieve proper installation.

Interference fit. The insertion of one member into another whose diameter is slightly smaller than the part being inserted.

Knurled clinching ring. The displacer portion of a fastener which has corrugations and is used to develop torque resistance when installed in sheet metal.

Minimum distance. The minimum distance from the center of a fastener mounting hole to the nearest edge of a panel which will keep the edge from deforming. This distance may be reduced by suitable fixturing or increasing thickness of panel material.

Minimum sheet thickness. The thinnest section of a panel, usually measured in thousandths of an inch or millimeters, into which a fastener may be properly installed. The same fastener may be installed in panels having any thickness greater than minimum.

Mounting hole. A properly sized round opening in a panel to receive the shank of a self-clinching fastener.

Positive stop. A visual indication that the proper depth of penetration of the knurled ring has occurred or when the "head" is in contact with the top surface of the panel. A synonym for positive stop is "shoulder."

Pull through. The resistance of a fastener to a force applied in the same direction to which it was installed.

Pushout. The force required to remove a fastener from a panel in a direction opposite to the way from which it was installed. Pushout is expressed in pounds or newtons.

Rockwell hardness. A relative measure of hardness. Rockwell C Scale is used for hard materials and Rockwell B is used for softer materials, such as sheet metal.

Self-clinching. The method by which a fastener is securely attached to a sheet of ductile material by causing the material to cold flow under pressure into an annular recess of the fastener, thereby securely locking it in place.

Shank. The portion of a fastener which is slightly smaller than the fastener's mounting hole and provides a positive location for the fastener in the hole. A shank also incorporates an annular grove which becomes filled with panel material as the fastener is installed. The retention of this material provides pushout resistance.

Shank length. The actual length of that portion of a fastener which is embedded in the panel material.

Shoulder. The surface area of a fastener which contacts the top surface of the sheet material.

Thread class. A measure of clearance or fit between the screw and the nut taken at the pitch diameter. Class 2B for nuts and Class 2A for screws or studs denote a medium clearance fit. Classes 3A and 3B denotes close fits.

Throughhole. A hole, threaded or not threaded, which transverses the entire length of a part and is usable from either end.

Tolerance. The absolute amount of maximum or minimum dimensional deviation allowed that will not affect the performance of a mechanical part.

Torque-out. The torsional holding power of a self-clinching fastener in a sheet. Torque-out testing is conducted by applying pure torsion (no axial load) to the self-clinching fastener. Torque-out is expressed in inch-pounds or newton-meters.

Undercut. The reduced diameter of a fastener which receives the sheet material when a fastener is installed. Depending on the type of fastener, the undercut may be rectangular or back-tapered.

Section 23

Robotic Assembly

CHRISTOPHER T. ANDERSON

Product Manager
Motoman Inc.

23-1 Automated Operations Process automation is high technology with robots, CNC equipment, programmable logic controllers (PLCs), and cameras for vision guidance. Very seldom is this really the case. Automation can be a self-feeding screwdriver positioned over an assembly operation. Automation can be a welding torch on an air cylinder making a single linear weld. And automation can be robots programmed to carry out a task. While this book covers a variety of fastening and joining processes, this chapter will cover some aspects of automating them.

There are different degrees of automation, and the decision to automate will have to take into account many considerations. Motion and process technologies are becoming more "off the shelf" for easy integration into systems. At the same time, manually operated equipment is becoming more sophisticated with improved ergonomics and efficiency for operators. In general, automation is suited for batch or fixed production where little to no decision making is required to assemble the part. Human operators have high dexterity and can adapt to changing situations on an assembly line. The best way to apply automated assembly operations is to design the part for automated assembly. Automation can be applied in stages of assembly; i.e., an automatic fixture can be used to locate and drill holes through an assembly while the rivets are applied manually.

Automation covers a spectrum from simple dedicated devices, such as a welding torch on a cylinder, to reprogrammable robotic systems. Dedicated equipment can be the cheapest solution for fixed production with high part volumes. However, a dedicated machine which performs several operations can be very expensive. Robotic systems are good solutions where a wide range of motion is required, as they can be easily reapplied on future model changes or products. Manufacturers offer integrated robot work cells, as shown in Fig. 23-1. These "packaged" cells are very economical and are available in a variety of configurations.

23-2 Reasons to Automate

Productivity. Automation equipment is justified by a gain in productivity. Automated equipment will often reduce the cycle time required to produce a part. Even if the part can be produced as quick manually, automation can still improve throughput with consistent performance over long periods. In printed-circuit board assembly, automated equipment can precisely locate elements several times faster than human operators. Arc welding robots weld at approximately the same speed as manual welders; however, robots are welding 90 percent of the time versus 50 percent for the human welder.

FIG. 23-1 ArcWorld™ packaged robotic welding work cells are an example of economical integrated systems. (*Courtesy: Motoman, Inc.*)

The improved productivity is translated into lower part costs. A robotic welding cell with a single operator can produce the equivalent of several manual welders. This output is normally enough to justify the purchase of equipment. There are usually intangibles that go along with automating. One manufacturer noted a savings of $30,000 per year for welding wire after installing a single robot. The robot was applying a uniform weld size while manual operators were applying excessive weld.

The same manufacturer was using the robot in a critical welding application. Manual welders were failing 20 percent of the time in a pressurized leak check. The robot installation brought the repair rate down to 5 percent or less. Another factor in this manufacturer's purchase decision was the difficulty of finding and keeping trained welders. The robot's consistent weld size also reduced the need for finishing operations such as grinding and polishing. In another installation, a robot was welding an assembly with threaded inserts that had been welded by hand. The system was justified based on the welding operation, but a 30 percent reduction in subsequent assembly time was realized when hole patterns matched up consistently.

Quality. Automation can reduce costs and improve quality by controlling process parameters. These uniform process parameters can be a pitfall to manufacturers who have counted on manual operators to compensate for poor fit-up. Welding, in particular, is a process which normally takes place near the end of the line. Weld joints need to be located within one-half of the diameter of the welding wire.

The consistent fit-up and uniform process parameters will yield improved cosmetics on the part. A commercial food equipment manufacturer relied heavily on the improved arc weld cosmetics to justify the purchase of a robot. Another factor was substantial warranty claims due to leaks caused by the manual operator's lack of process control.

With consistent part fit-up and uniform process parameters, the implementation of statistical process control (SPC) is meaningful for automation. SPC is a quality control measure whereby a process is certified and then random samples are taken to verify the process is still in control. With automation there is a better chance that all the screws or welds will be in place on an assembly. When a robotic welding cell is first implemented, the parts are normally checked for strength and proper dimensions. After certification, the operator visually inspects the parts, and only random parts are tested.

For more critical applications, continuous quality control is provided for monitoring the process. In a welding cell, process parameters such as amperage, voltage, wire feed speed, and gas flow can be continuously monitored. After the cell is certified, alarm limits can be established for these parameters and suspected defects identified. A manufacturer of air bag canisters is using process monitoring to record each weld, and these are stored for the serialized canister.

Safety. During the 1990s the government and other industries became more aware of work-related injuries. Industry has focused on ergonomics to improve productivity and reduce lost days and injury claims, while government has increased regulations in the area of ergonomics. Fastening methods often involve impact tools, torque-driving equipment, or high-temperature processes. Automation can reduce work-related injuries by taking fatiguing processes out of the hands of operators. Robots have been applied to repetitive, tedious applications such as repositioning heavy parts on an assembly line or stacking products coming off a production line (Fig. 23-2).

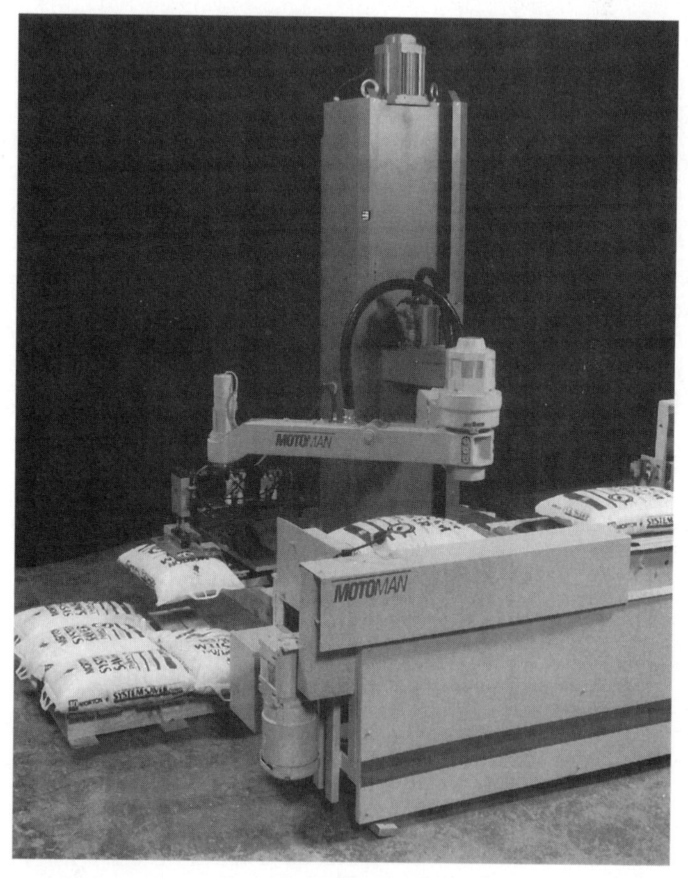

FIG. 23-2 Automation can reduce work-related injuries from poor ergonomic conditions. (*Courtesy: Motoman, Inc.*)

Automated equipment can also remove an operator from potentially hazardous environments. Welders are exposed to high temperatures and fumes from the melting metal. The hazards can be reduced by moving the operator away from the process and applying the weld automatically (Fig. 23-3). Fume extraction equipment can be placed directly over a welding cell to improve the overall plant environment. Adhesives are another source of fume.

FIG. 23-3 Robotic work cell isolates the operator from welding hazards. (*Courtesy: Motoman, Inc.*)

Automated equipment needs to be properly safeguarded. This may include machine guards for fixed equipment or barriers and interlocks for robotic cells. The level of guarding is normally specified by OSHA regulations and applicable ANSI standards. Safeguards should be addressed with the manufacturer when one is considering the purchase of automated equipment. Operators tasks associated with the automated equipment need to be evaluated. The automated equipment should not create a worse ergonomic situation than the original manual process.

23-3 Dedicated and Robotic Systems

Dedicated Equipment. Dedicated equipment is designed for a specific task or application and does not normally provide much flexibility. Dedicated machines can range from simple devices with relay logic to sophisticated cells with PLCs, programmable motion, and hundreds of sensors. The machine may require a single operator for part loading and unloading, or may utilize self-feeding devices to reduce manual intervention, or may be set up in a line with several operators working on various stages of production.

Dedicated equipment is normally justified by a large-volume part run. The equipment can be designed to perform multiple operations in a single station. A simple example is a weld lathe with a fixed torch and a rotating part beneath it. A manufacturer of fire extinguishers utilized a weld lathe with three fixed torches to weld the circumference of a three-stage design. An operator loaded and unloaded parts from the fixturing and initiated the weld cycle. This fixture beat robotics in both cost and cycle time. The same principle applies to fixtures with automatic screwdrivers for fixtured parts. The operator simply loads parts into the fixture, and self-driving heads with part feeders can apply one or several screws.

The fire extinguisher manufacturing process is a good example of the potential of dedicated equipment, but also has its limitations. The fixture is limited to cylindrical parts. If the part changes in length or circumference, the fixturing can probably be adapted. The torches can be moved to different positions along the cylinder, or the number of torches could be changed. Fire extinguishers will always be cylinders, but most manufacturers deal with greater product variation.

Many bicycle manufacturers use this weld lathe approach with flexible torch positioning. The compound miter joints in bicycle frames create a complex geometry when the part is rotated. Machines, called *cabinet welders*, used cams to position the weld torch while the part spun underneath it. Today these cabinet welders incorporate programmable motion devices with servomotor controls. This programmability increases the flexibility of the device.

Sophisticated equipment is used in electronic assembly. Circuit boards are sent down lines that include several stations with multiple operations being performed at each station. Feeding devices are used for electronic components. Robots or linear devices are used for component insertion. Soldering is done automatically. This assembly equipment may be designed to produce a few board designs with minimal cycle time, or have high flexibility to build dozens of board combinations.

Robotic Systems. The Robotic Industries Association (RIA) defines a *robot* as "a reprogrammable multi-functional manipulator designed to move material, parts, tools, or specialized devices, through variable programmed motions for the performance of a variety of tasks." Robots are available in a variety of sizes and configurations for all types of applications. Robots are categorized by the number and configuration of moving axes and the rated payload capacity. Most fastening applications will employ articulated-arm robots with up to six axes of motion. The payloads can range from 6 kg (13.2 lb) for arc welding or sealing to 100 kg (220 lb) or more for spot-welding applications.

The advantage of robotics is its high degree of flexibility. It is very easy to change a robot program, should there be part changes. This includes not only part redesign but also variations in press runs on stamped parts. Robots can also be reapplied. A robot that was used for arc welding could be used for plasma cutting. Some manufacturers have used automatic tool changers to have the robot perform plasma cutting and then welding on the same part.

As with computers, the price of robots has come down over the years while performance has improved. These factors have helped expand the use of robotics. Robots are integrated into cells as single stations or in a line of multiple units. They are used with the following application variations: robot manipulating process to part (faster cycle times), part positioner with robot (higher flexibility), robot manipulating part to process (multiprocess capability).

The most common use of robots is to mount a process tool on the end of the arm and move the process around the part (Fig. 23-4). The payload capacity of the robot is determined by the tool, and the robot reach is determined by the size of the part. Applications such as arc welding, sealing, and spot welding can benefit from a three-axis wrist which allows variation in tool angles. For arc welding and sealing, the robot controls the process along a three-dimensional path. For spot welding, the robot merely repositions the tool to stationary positions and waits for the process to take place. The speed of the robot allows the tool to be repositioned quickly and reduces the cycle time for producing the part.

Part positioners can be combined with robots to increase their flexibility (Fig. 23-5). Part positioners can have independent indexing motion or can be integrated with the robot's motion controller. Indexing equipment has its own control that turns the

FIG. 23-4 Robot equipped with gas metal arc welding torch. The robot moves the torch along the part path to complete the weld. (*Courtesy: Motoman, Inc.*)

FIG. 23-5 Robotic work cell with coordinated part positioners provides a high degree of flexibility. The robot and positioner can move simultaneously to weld parts while they are turning. (*Courtesy: Motoman, Inc.*)

device to fixed stops when it gets a signal from the robot. This type of device would allow a robot to weld the top side of a part and then index it 180° to weld the back side. Positioners that are integrated with the robot controller would allow a turning part to be welded all the way around. The positioner also allows part to be turned at any angle. Robots can be placed on tracks to increase the size of their working area (Fig. 23-6).

A cabinet welder is an example of a dedicated machine used to produce bicycle frames. Cabinet welders are dedicated to a particular weld joint. Therefore, the bike frame is produced by several machines with work being transferred between stations. Robots are gaining popularity in this industry with the added flexibility in torch positioning for the most desirable weld angle. The robot cell can produce the entire frame and reduces work in process and material handling.

FIG. 23-6 A track increases the reach of the robot, and part positioners turn parts within the robot's envelope. One positioner can be loaded while the robot welds large parts on the other positioner. *Note: Point-of-operation safeguards are not shown.* (*Courtesy: Motoman, Inc.*)

In some cases robots are used to carry parts to the process. This can create a longer cycle time for the robot to pick up, move, and place the parts, but it will relieve operators from the task. The robot can then move the part to several operations and unload it when it is completed. These operations might include stationary arc welding torches, fixed-pedestal spot welding machines, projection welding machines with nut feeders, or finishing equipment. Several robots can be set in a line and can pass the part down the line as they complete their operations. Operators can inspect and add components during the process, and finished parts can be placed on conveyors for bins or assembly operations.

An automotive parts supplier used several robots in a line to produce a suspension swing arm. An operator loaded piece parts into one gripper while the robot was working with the other. The robot moves the gripper to a fixed spot welder for several spot welds and then moves it under an arc welding torch. At the end of its cycle, the robot places the part on a stand for the next operation and returns the empty gripper to the operator.

23-4 Automated Threaded Fasteners

Screws and Bolts. Screws and bolts can be driven automatically. The application may call for self-starting pointed screws or bolts. The screw needs to be delivered to the driving head in the proper orientation. Bowl feeders use vibration to feed and orient a screw from a hopper. The fastener is then loaded into the head by a pick-and-place unit or by a blow feeder, which uses compressed air to push the oriented fastener through flexible air line to the driving head.

The drive for the screw may use electric or pneumatic motors. The driving head incorporates linear travel to drive the screw. The depth of travel can be adjusted by mechanical stops or electric switches. The driver may be equipped with a clutch so that the fastener is always tightened to the proper torque.

These driving heads are easy to incorporate into a stationary machine. A simple PLC can be used to control several heads with feeders. Parts are often loaded on an indexing dial table and are turned under the fixed heads. The same principle is used to drive rivets or stake assembled parts. Multiple stations can be used to maximize throughput. There is a wide range in costs for this equipment, and it is normally limited to smaller parts.

A driving head can be fitted to the end of a robot. The robot normally moves to a stationary position and lets the linear action of the head drive the screw. The robot needs to be sized to carry the tool and resist the torque generated by the driver. The robot can move to a bowl feeder to load a screw, but it is more productive to use a blow feeder to deliver a screw while the robot is moving to the next location.

Small robots are used for printed-circuit board assembly, but they can be used on a larger scale, too. An example is a manufacturer of home satellite dishes who wanted to screw expanded metal panels to a welded frame. The application required special ⅜-in self-tapping screws to be blow-fed from a bowl feeder. The driving head used a compliant mount to the robot. If the screw started on top of the expanded metal, the compliant head allowed the screw to "skip" and drive into the frame. The robot's reach could cover the large area of the dish, and the wrist allowed the tool to stay perpendicular along the parabolic curve.

Two robots are working to assemble another robot in Fig. 23-7. One robot positions an upper-arm assembly while another robot inserts bearings and bolts on a cover plate. Automated assembly of the arms increased the mean time before failure from 15,000 to 20,000 h.

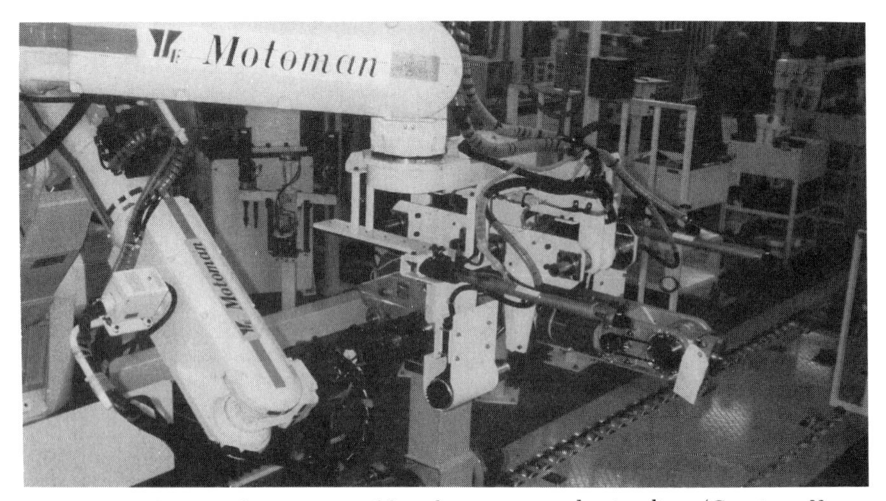

FIG. 23-7 Robots working to assemble robots on a production line. (*Courtesy: Yaskawa Electric, Kitakyushu, Japan*)

Welded Fasteners, Stud and Projection Welding. Threaded studs allow components to be bolted in place while trim studs have a unique profile for retaining molding. Like screws, these studs can be bowl-fed for orientation and then delivered to the gun mechanically or by blow feeder. Studs are applied with a spring-loaded gun and a capacitive discharge power supply. The gun is placed against the workpiece, and the stud is retracted, creating an arc. The spring then drives the stud into the molten area where it solidifies. The weld is completed in seconds.

Stud welders have been incorporated into turret presses. For electrical enclosures, the welded studs and nuts can be added at the same time as the blank is being produced. The blank is then folded into a box with the studs in the proper location.

Robots can apply studs on multiple planes instead of a flat surface. The robot needs to be sized to counter the spring force in the gun. Robots can be used to apply trim studs in the rear opening of an automobile to retain the window molding. The robot positions the gun against the workpiece and waits for the welder to indicate that the cycle is complete.

Studs have a straight shaft which is welded to the workpiece. Nuts and threaded fasteners with a base are projection-welded to the workpiece. As in spot welding, electrodes squeeze the parts together while high current creates a melt zone where the two pieces contact. Fasteners are carefully designed so that the projections provide a smooth transition of weld current and produce a properly shaped weld nugget.

Stud welding guns are compact and can be hand-held. The forces exerted for project welding requires pedestal-mounted welders with powered clamping of the electrodes. The high welding current also requires heavy transformers and copper electrodes. Parts are often manually positioned on the "ped welder" electrodes. Automatic feeders assist the operator by placing the nut on a pin in the electrode. The operator initiates a cycle, causing the electrodes to clamp and to apply the weld current. Then a squeeze time allows the weld nugget to solidify. The weld is completed in a couple of seconds.

The appliance industry has incorporated projection welding in dedicated equipment. Blanks are fed into a fixture with multiple welding heads. The fixture closes as a press does, with electrodes contacting the locations of the nuts. The welding current is applied, and in the case of multiple nuts, it has to be sequenced to fire at different times. After the last squeeze time, the blank is unclamped and fed through.

Robots have been used in the automotive industry to position parts in pedestal welders with nut feeders. The robots can pick the parts and then sequence them through the welder, applying nuts in a variety of orientations. Different equipment or feeders can be used to apply different size nuts or to perform spot welds on the assembly in the same cell.

A stamping plant was producing a gas filler well that fits in the side of an automobile. The stamping required several nuts for mounting to the vehicle and supporting the door to the compartment. Manual operators could produce faster than a robot, but a robot provides improved efficiency and can produce consistently throughout the day. The robot improved quality by not forgetting to place a nut on the component.

An example of an automated welding line is shown in Fig. 23-8. The operator to the right loads stampings into a dedicated spot welding fixture. Several spot welds are applied to join the stampings into a subassembly. A single robot with a gripper is used to move the subassembly to two stationary nut welders to projection-weld 6 and 8-mm nuts in various locations. This subassembly is placed on a conveyor which travels to a second operator. The second operator added brackets with nuts, which she or he had manually welded, into a fixture with the subassembly. The first robot picks up the part and indexes it into a stationary welder to apply several spot welds. The assembly is then placed in a nest, and the robot returns to pick up another part from the fixture. The second robot picks the assembly out of the next and passes it down to the next robot after applying more spot welds. The final assembly exits the cell on a conveyor and is loaded with other parts into robotic arc welding cells, which are not shown.

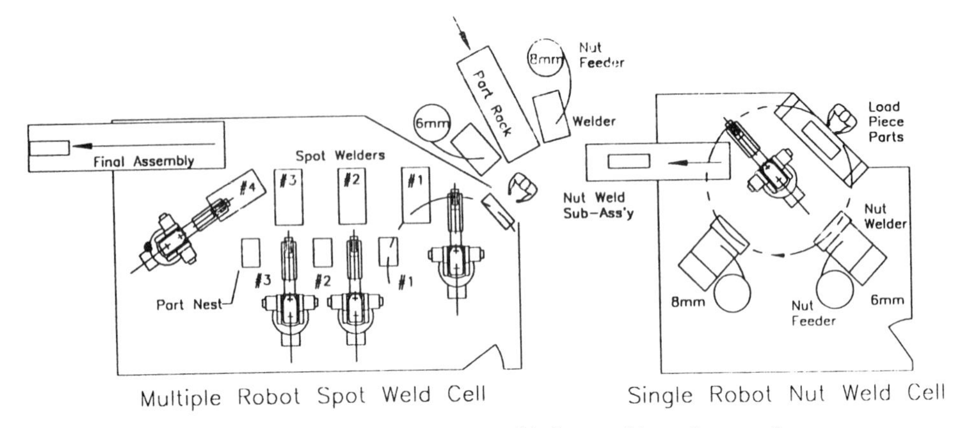

FIG. 23-8 Automated line to produce spot welded assembly with tapped nuts.

23-5 Arc Welding Gas metal arc welding (GMAW) is used extensively for automated welding. A welding wire is continuously fed into the workpiece. A welding power supply provides the energy to melt the wire and surrounding base metal. The faster the wire is fed into the workplace, the more energy is supplied to melt the wire, and the more base material is melted. A gas flow surrounds the arc to shield it and the melted base material from the atmosphere.

Due to hazardous conditions and operator inefficiencies, automated welding processes are more attractive than manual welding. The melting metal produces a fume which rises directly over the weld. Manual welders are in close proximity to this fume, but the use of robots removes the operator from direct exposure. The heat and sparks generated by welding can be hazardous and uncomfortable for welders. Welders have to wear hoods with dark lenses to protect them from intense ultraviolet radiation. Many companies figure their welders are only 50 percent efficient due to fatigue and time spent repositioning between weld joints.

A typical robotic welding system includes a welding power supply, wire feeder, welding torch, and cabling to integrate the equipment into the robot controller (Fig. 23-9). The robot program not only starts and stops the welding process but also sets the desired wire feed rate and arc voltage. The robot controller includes software which will detect and stop due to arc faults. The robot will deposit the exact amount of weld required, will not skip welds, and will move quickly between welds, improving the efficiency.

Arc welding robots have been applied in industries including automotive, construction equipment, farm machinery, furniture, sporting goods, recreational vehicles, and many more. Robots are welding in companies with a few employees to Fortune 500 companies. Most companies justify robots on the basis of improved productivity. Robots often accrue intangible benefits. A manufacturer that had several manual welders implemented a robot system to do the same task. Repair rates were cut in half, and scrap was eliminated. A savings of $30,000 in weld wire was realized because the robot applied the proper size weld. Another company that makes food equipment used a robot to produce a stainless steel machine frame. The frame was constructed of square tubing and included some threaded bosses with very tight tolerances. Not only did the robot achieve the targeted cycle times, but also the assembly time was reduced by one-third due to the improved alignment of the frame.

Welding robots are typically integrated into work cells. A part positioner not only adds flexibility, but also serves as a safeguard to keep the operator out of the robot envelope. A barrier around the robot safeguards plant personnel. The work cell should have protective curtain to block the harmful ultraviolet light from the arc. This curtain

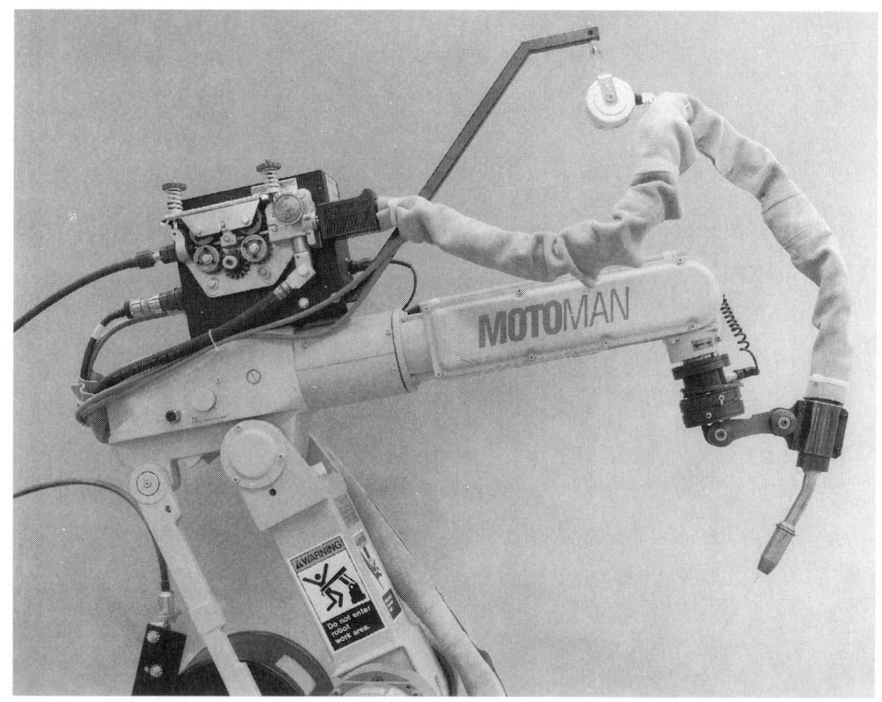

FIG. 23-9 Welding robot with wire feeder, welding torch, and impact-sensing torch mount. (*Courtesy: Motoman, Inc.*)

also contains the welding sparks. The plant may require ventilation installed over the robot to remove fume. The work cell includes controls for an operator interface (Figs. 23-10 and 23-1).

In addition to GMAW, or MIG welding, the gas tungsten arc welding (GTAW) process can be automated. The GTAW process is normally used for improved cosmetics, but at a price of one-third the travel speed of GMAW. High-frequency voltage is required to start the arc between the tungsten electrode and the workpiece. This can interfere with the robot's computer control if it is not properly shielded. Plasma arc welding (PAW) is similar to GTAW but uses a pilot arc which minimizes electrical noise. Either process can be used to weld with or without the addition of filler material (Fig. 23-11). The use of filler material is based primarily on the joint design.

Automated systems can perform cutting operations in addition to welding. The plasma arc cutting (PAC) process can be used at travel speeds of 100 to 200 ipm (inches per minute) on gage thicknesses. There are *XY* tables which move the torch to cut flat sheet. They usually include software for cutting shapes and nesting material on the sheet. Robots can articulate the torch and cut structural shapes in multiple planes. Robot cells can be equipped to change tools to cut and weld on the same part.

23-6 Resistance Welding Resistance spot welding is performed with high electric currents passing through overlapped sheet metal clamped between copper electrodes. The high current causes metal to melt in the area between the two sheets due to the high electrical resistance at that point. The weld current is applied long enough to form a weld nugget between the surfaces but not melt completely through both thicknesses. The copper electrodes stay clamped after the current is off, to allow the molten nugget to solidify.

All spot welding equipment requires a weld controller. This controller actuates electrode clamping, sets the value and duration of weld current, and keeps the electrodes clamped for the specified *squeeze time*. The weld controller may have additional

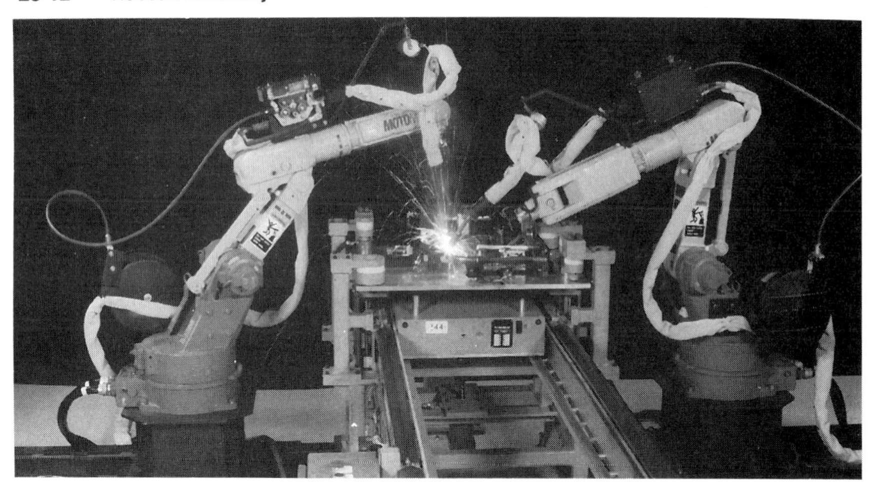

FIG. 23-10 Welding cell conveys fixture pallets into lift/locate station with two robots. The welded parts are conveyed to the operator who reloads them outside the robot envelope. *Note: Point-of-operation safeguards are not shown.* (*Courtesy: Motoman, Inc.*)

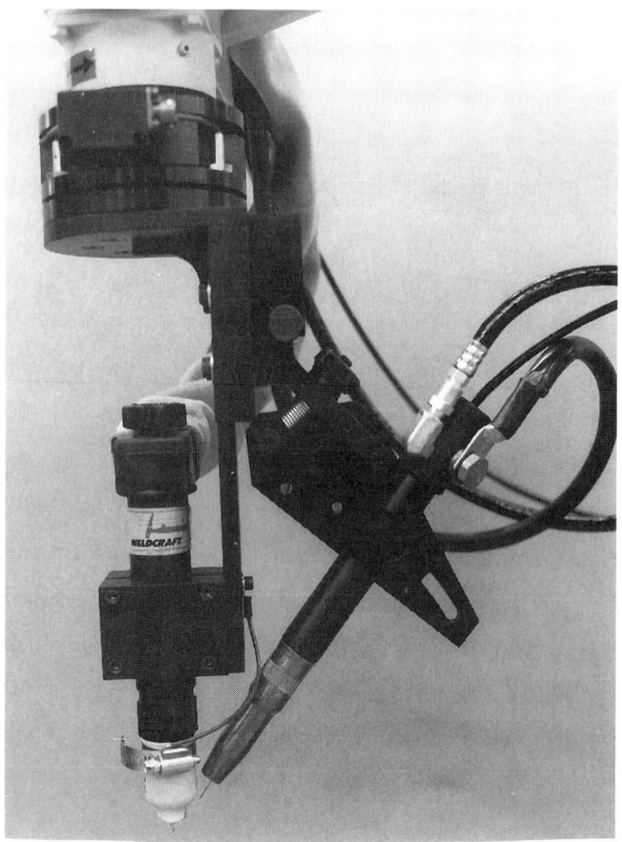

FIG. 23-11 Gas tungsten arc welding torch with "hot" wire feed. (*Courtesy: Motoman, Inc.*)

capabilities such as weld preheat and postheat, power line compensation, multiple weld programs with remote selection, or adaptive feedback to ensure a good weld. Resistance welding uses alternating current, and sequencing in the controller is set in *cycles*, which is $\frac{1}{60}$ s, (based on a 60-Hz line frequency used in the United States). Most sheet-metal spot welds are completed in 1 to 2 s.

Spot welding systems can be configured as dedicated equipment, can incorporate robotics, or can use a combination of both. Dedicated equipment is configured similar to a press; the parts are nested on tooling, and several copper electrodes are on a ram that comes down to clamp. An operator loads the parts and initiates the sequence; then the weld controller clamps, welds, and unclamps the parts. The advantage of dedicated equipment is that multiple spot welds can be performed in one operation in a few seconds. The disadvantages are the high equipment cost and the difficulty in adapting it to part changes. In fixtures where several welds are being made, the controller sequences the welds because performing all the welds simultaneously would cause excessive voltage drop in the plant power line.

Resistance welding requires a transformer to step the power line voltage down to a low voltage and high current. Heavy cabling is required to carry the high current from the secondary side of the transformer to the electrodes. Spot welding guns were designed for robotics with a transformer connected directly to the electrodes. These "transguns" weigh over 100 lb, but the need for heavy cabling is eliminated and flexibility for positioning the gun is increased (Fig, 23-12). Transguns can also be incorporated into dedicated fixtures.

Spot welding robots are designed to carry the heavy payload of a transgun and have about a 6-ft reach. Robots are also used to reposition parts between the electrodes of a fixed-pedestal welder (Fig. 23-13). The robot signals the weld controller to initiate the weld sequence and then waits for a weld-complete signal before repositioning.

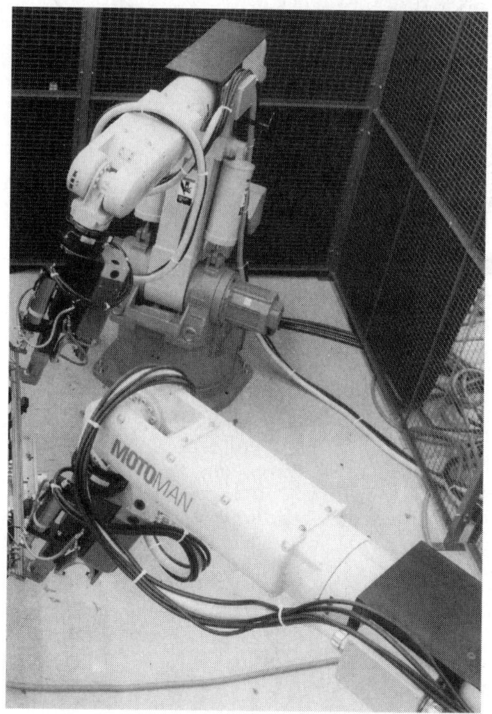

FIG. 23-12 Robots with self-equalizing transguns perform resistance welds on part. (*Courtesy: Motoman, Inc.*)

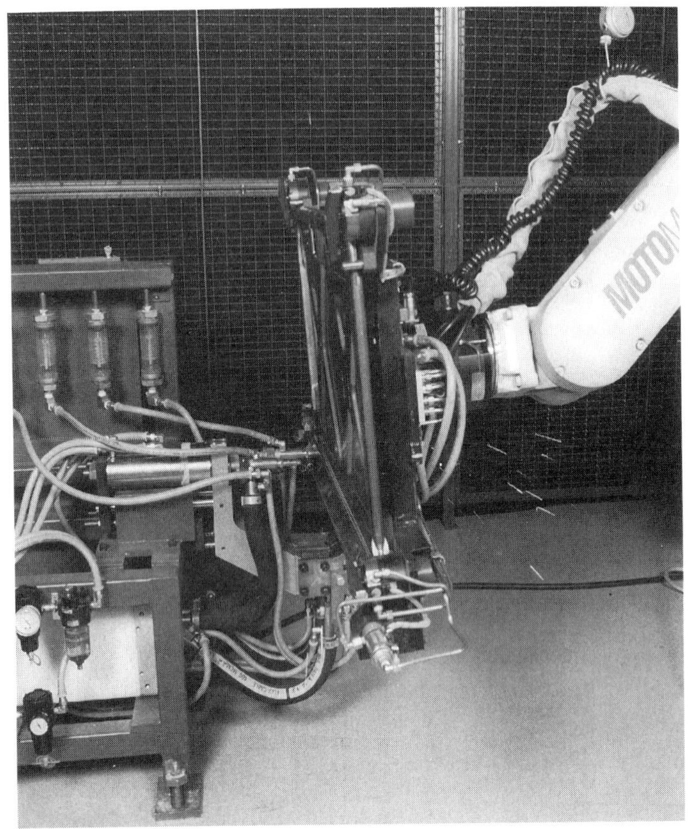

FIG. 23-13 A robot picks up manually loaded fixture and moves it through a stationary welder with multiple-spot-weld capability. (*Courtesy: Motoman, Inc.*)

The robot can select different weld settings from controllers with programmable schedules. When used with robotics, the transgun or pedestal welder should have self-equalizing clamping between the electrodes. The self-equalizing design allows the electrodes to pivot so the high clamping forces are not transferred back to the robot arm.

A classic example of spot welding automation is in the production of car bodies. Dedicated fixtures are used to position the different body panels and spot weld them together. These fixtures are very expensive due to their large size, high mechanical complexity, and critical manufacturing tolerances. The panels are held together with just enough spot welds to keep the body dimensionally stable. The assembly, or "body in white," is then indexed through a line of welding robots. The line has several stations with numerous robots on either side to apply hundreds of welds to complete the body. The robots move transguns around the body and even inside openings to access all the weld locations. The flexibility of the robots allows them to reach a wide area with a variety of gun angles. The robot can easily be reprogrammed for a design or model change while the dedicated fixture will need to be retooled.

23-7 Adhesive Bonding Automated equipment for dispensing adhesives will depend on whether the material is a one- or two-part epoxy and if it is heated before applying. A dispensing system consists of a pump to deliver the material, a valve to control the flow of material, and a head to apply the material.

Adhesives are supplied in drums, and the pumps are designed to mount directly on the drum. A two-part epoxy would require two pumps to deliver each component to the valves. Valves are used to turn on and turn off the flow of adhesives and mount on the end of the robot as close to the point of delivery as possible. This prevents excess material from hardening in the head and restricting flow. Some heads have controlled variable valves so the robot can regulate the rate of flow in addition to turning on and off. The dispensing head can be as simple as a metal tube or may be designed with orifices that fan or swirl the exiting fluid to create a wide bead. For two-part epoxies, the head will incorporate a mixing chamber to swirl the two fluids together in proper proportions before exiting the head.

The main advantage of the robotic dispensing is the uniformity of material that is deposited. A consistent bead size will improve quality and reduce material costs. Robotic work cells are often part of an automation line that uses equipment to control assembly and the curing process. Adhesives need to be controlled from application through curing to ensure a proper bond has occurred.

Robots are also used to apply sealant material which is similar to adhesives (Fig. 23-14). Robots with dispensing heads apply sealant around front and rear automobile windshields. Bigger robots are then used to insert the glass in cars on the assembly line. A single robot cell is used to apply a bead of sealant in a channel around the perimeter of a plastic tail light. The robot must follow a complicated three-dimensional path and keep the head perpendicular to the channel. A head with variable flow is used to reduce the material flow as robot travel speed is reduced through corners. The cured bead will form a gasket when the tail light is mounted.

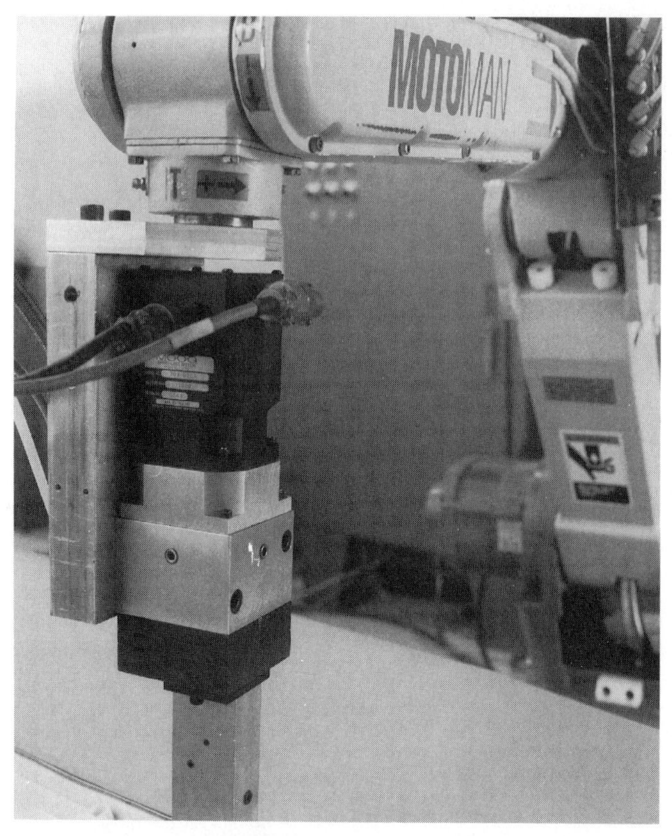

FIG. 23-14 A robot applies a bead of sealant with a programmable flow head. The bead will form a gasket when the part is installed. (*Courtesy: Motoman, Inc.*)

Section 24
Innovative Connections

ROBERT O. PARMLEY, P.E.

President & Principal Engineer
Morgan & Parmley, Ltd.
Ladysmith, Wisconsin

This section has been included for the single purpose of illustrating that individual standard fasteners and components have wide potential uses, far beyond their original concept. All designers must be aware of this fact so that they can use stock items to ensure cost-effective products.

The following illustrated examples depict a broad range of some fastening and sealing components, each with a multitude of uses.

All of the material in this section was written by the Editor-in-Chief and most of the following pages originally appeared in *Product Engineering*. The remaining articles were extracted from *Assembly Engineering* and *Machine Design*. Written permission has been granted for their use, and proper credit is given each publication.

24-1 Retaining Rings

The retaining ring comes in a variety of sizes, shapes, and designs. This versatile fastener functions both as a shoulder and as a locking device, thus reducing machine and assembly complexity.

Figures 24-1 to 24-5 illustrate the versatility of this mechanical component, not only as a fastener, but also as a key machine component.

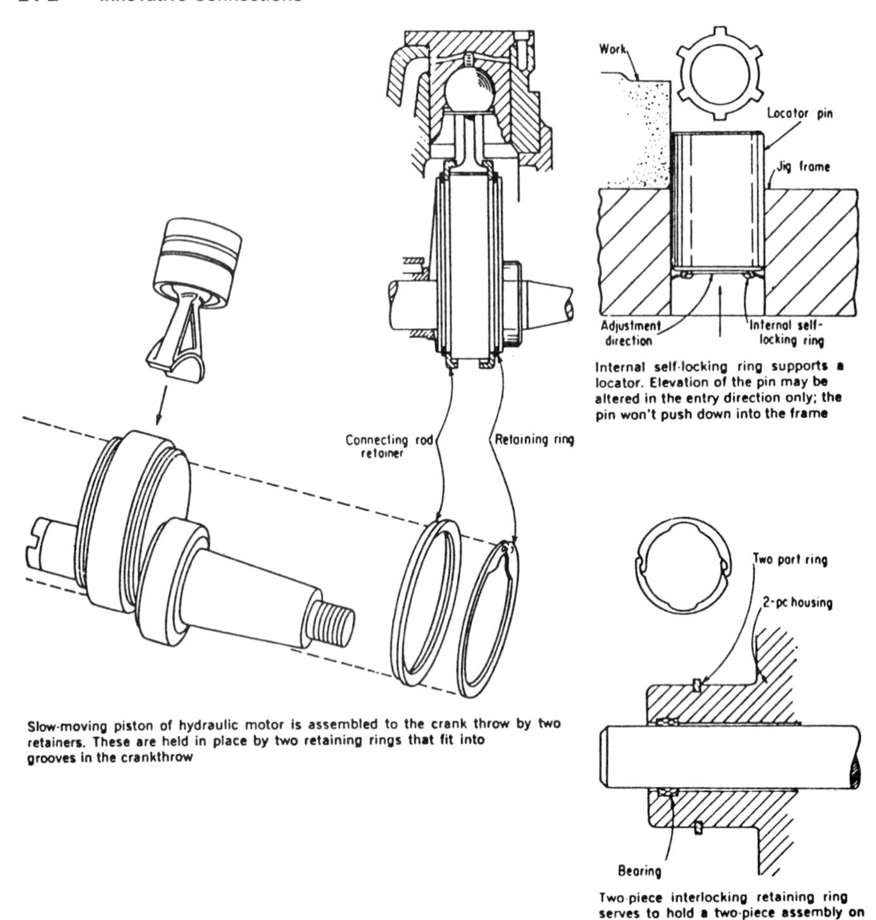
Work
Locator pin
Jig frame
Adjustment direction
Internal self-locking ring
Internal self-locking ring supports a locator. Elevation of the pin may be altered in the entry direction only; the pin won't push down into the frame
Connecting rod retainer
Retaining ring
Slow-moving piston of hydraulic motor is assembled to the crank throw by two retainers. These are held in place by two retaining rings that fit into grooves in the crankthrow
Two part ring
2-pc housing
Bearing
Two-piece interlocking retaining ring serves to hold a two-piece assembly on a rotating shaft, and is more simple than a threaded cap, a couple of capscrews or other means of assembly

FIG. 24-1 Retaining rings aid assembly—1. By functioning as both a shoulder and as a locking device, these versatile fasteners reduce machining and the number and complexity of parts in an assembly. (*From R. O. Parmley, "Retaining Rings Aid Assembly, 1," in American Machinist, June 3, 1968. Reprinted with permission. American Machinist. A Penton Publication. Prepared with the cooperation of the Truarc Retaining Rings Division of Waldes Kohinoor, Inc., Long Island City, N.Y.*)

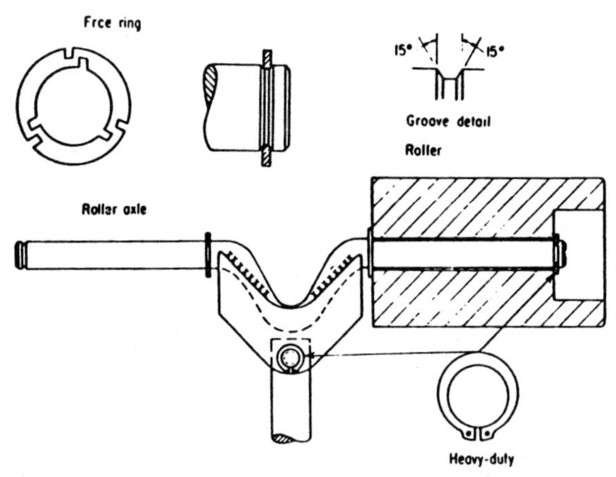

Two types of rings may be used on one assembly. Here permanent-shoulder rings provide a uniform axle step for each roller, without spotwelding or the like. Heavy-duty rings keep the rollers in place

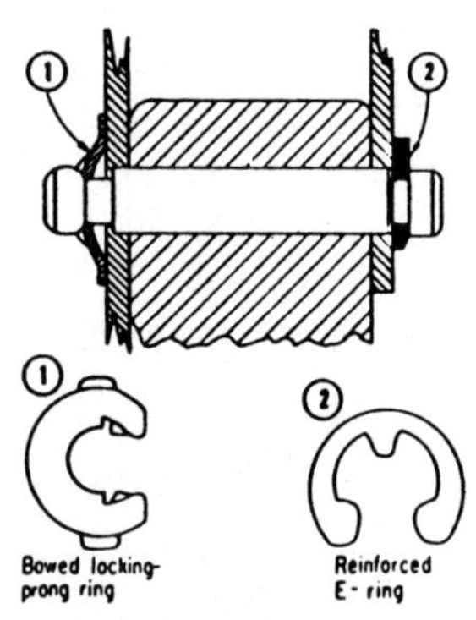

Snug assembly of side members to a casting with cored hole is secured with two rings: 1—spring-like ring has high thrust capacity, eliminates springs, bow washers, etc; 2—reinforced E-ring acts as a retaining shoulder or head. Each ring can be dismantled with a screwdriver

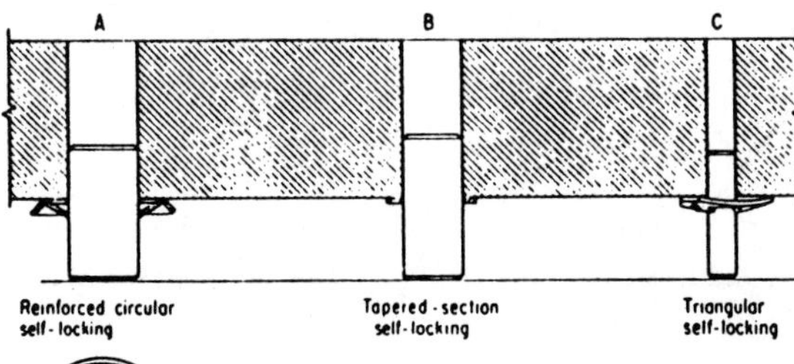

These three examples show self-locking retaining rings used as adjustable stops on support members (pins made to commercial tolerances): A—external ring provides positive grip, and arched rim adds strength; B—ring is adjustable in both directions, but frictional resistance is considerable, and C—triangular ring with dished body and three prongs will resist extreme thrust. Both A and C have one-direction adjustment only

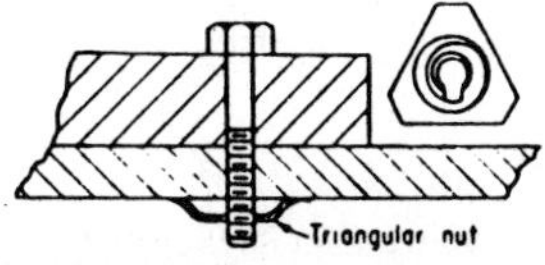

Triangular retaining nut eliminates the need for tapping mounting holes and using a large nut and washer. Secure mounting of small motors and devices can be obtained in this manner

FIG. 24-1 (*continued*)

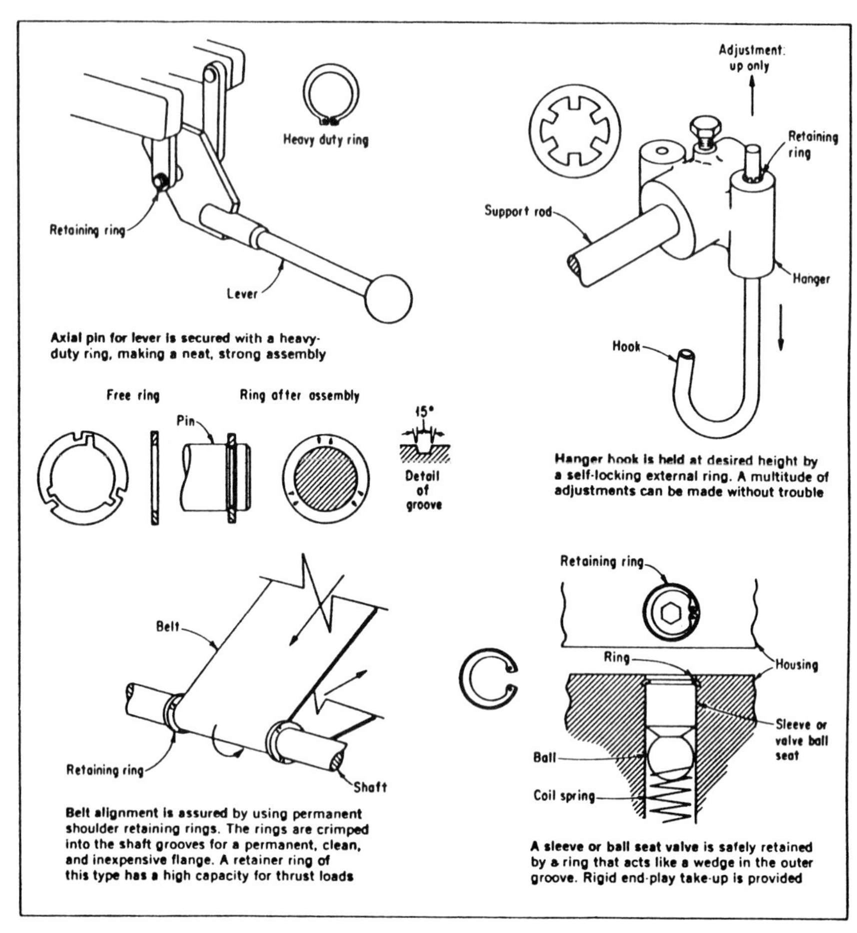
Heavy duty ring
Retaining ring
Lever
Axial pin for lever is secured with a heavy-duty ring, making a neat, strong assembly
Adjustment: up only
Retaining ring
Support rod
Hanger
Hook
Hanger hook is held at desired height by a self-locking external ring. A multitude of adjustments can be made without trouble
Free ring
Ring after assembly
Pin
15°
Detail of groove
Belt
Retaining ring
Shaft
Belt alignment is assured by using permanent shoulder retaining rings. The rings are crimped into the shaft grooves for a permanent, clean, and inexpensive flange. A retainer ring of this type has a high capacity for thrust loads
Retaining ring
Ring
Housing
Sleeve or valve ball seat
Ball
Coil spring
A sleeve or ball seat valve is safely retained by a ring that acts like a wedge in the outer groove. Rigid end-play take-up is provided

FIG. 24-2 Retaining rings aid assembly—2. Here are eight more thought-provoking uses for retaining rings, to be added to the previous assortment of ideas. (*From R. O. Parmley, "Retaining Rings Aid Assembly," in American Machinist, Feb. 22, 1971. Reprinted with permission, American Machinist. A Penton Publication.*)

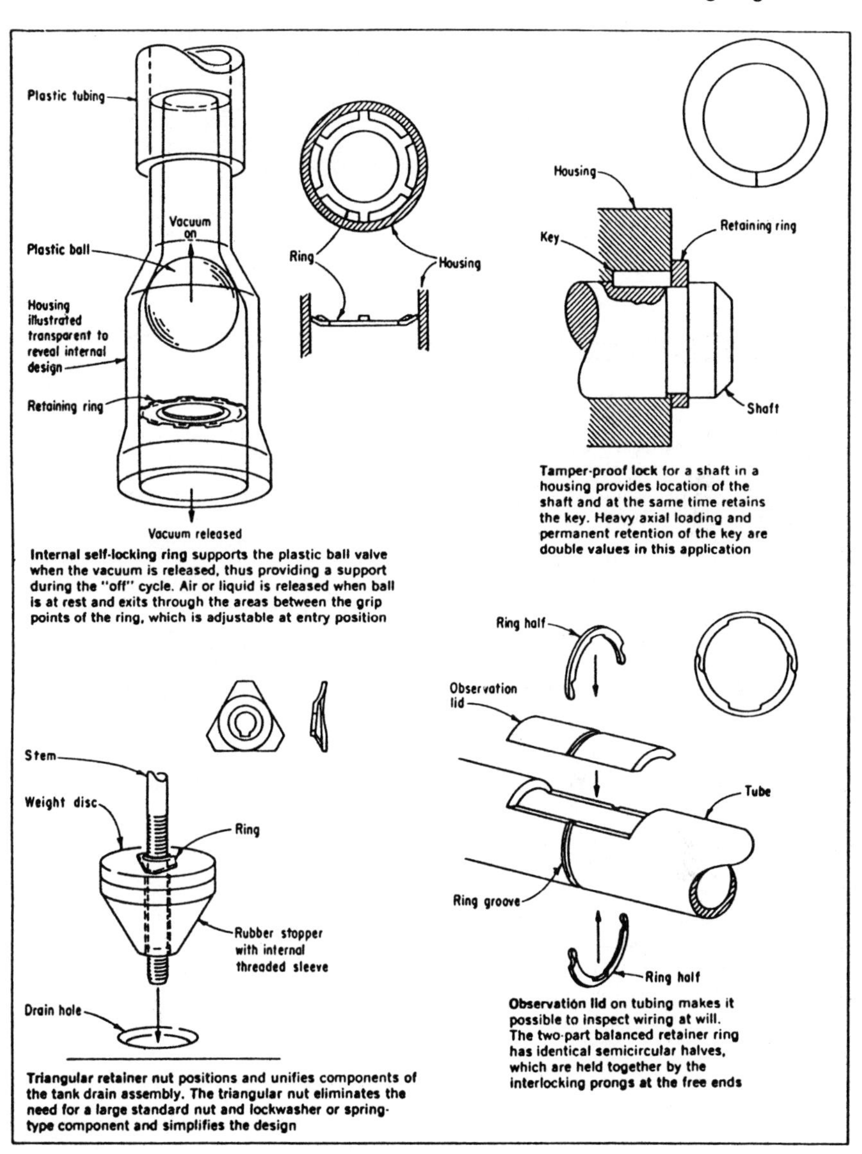

Internal self-locking ring supports the plastic ball valve when the vacuum is released, thus providing a support during the "off" cycle. Air or liquid is released when ball is at rest and exits through the areas between the grip points of the ring, which is adjustable at entry position

Tamper-proof lock for a shaft in a housing provides location of the shaft and at the same time retains the key. Heavy axial loading and permanent retention of the key are double values in this application

Triangular retainer nut positions and unifies components of the tank drain assembly. The triangular nut eliminates the need for a large standard nut and lockwasher or spring-type component and simplifies the design

Observation lid on tubing makes it possible to inspect wiring at will. The two-part balanced retainer ring has identical semicircular halves, which are held together by the interlocking prongs at the free ends

FIG. 24-2 (*continued*)

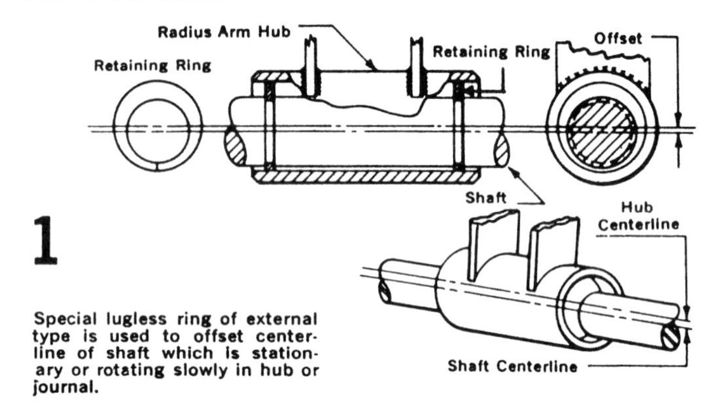
Radius Arm Hub
Retaining Ring
Retaining Ring
Offset
Shaft
Hub Centerline
1
Special lugless ring of external type is used to offset centerline of shaft which is stationary or rotating slowly in hub or journal.
Shaft Centerline

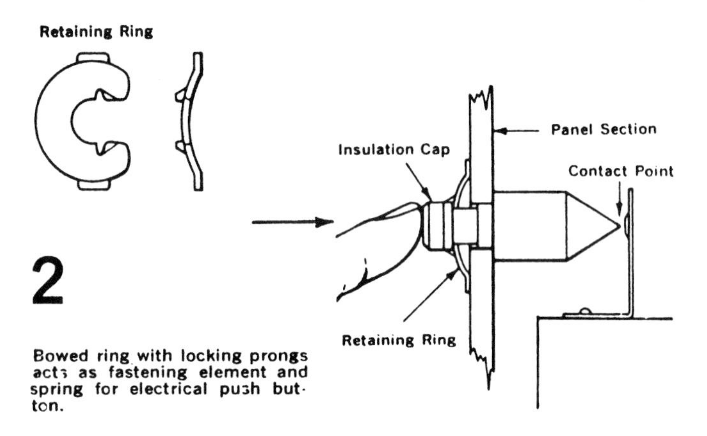
Retaining Ring
Panel Section
Insulation Cap
Contact Point
2
Retaining Ring
Bowed ring with locking prongs acts as fastening element and spring for electrical push button.

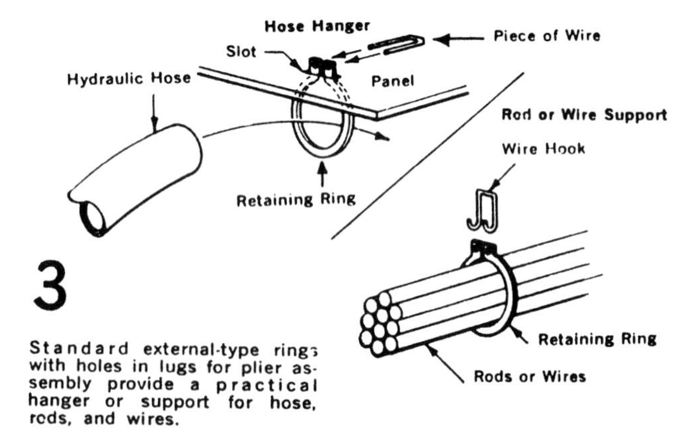
Hose Hanger
Piece of Wire
Slot
Hydraulic Hose
Panel
Rod or Wire Support
Wire Hook
Retaining Ring
3
Retaining Ring
Rods or Wires
Standard external-type rings with holes in lugs for plier assembly provide a practical hanger or support for hose, rods, and wires.

FIG. 24-3 Multiple-purpose retaining ring. A roundup of 10 unusual ways for putting retaining rings to work in assembly jobs. Examples shown are based on stamped ring designs produced by Truarc Retaining Rings Division of Waldes Kohinoor, Inc. (*From R. O. Parmley, "The Multiple Purpose Retaining Ring," in Assembly Engineering, July 1966, vol. 9, no. 7. Excerpted from Assembly Engineering. By permission of the publisher ©1966 by Hitchcock Publishing Co. All Rights Reserved.*)

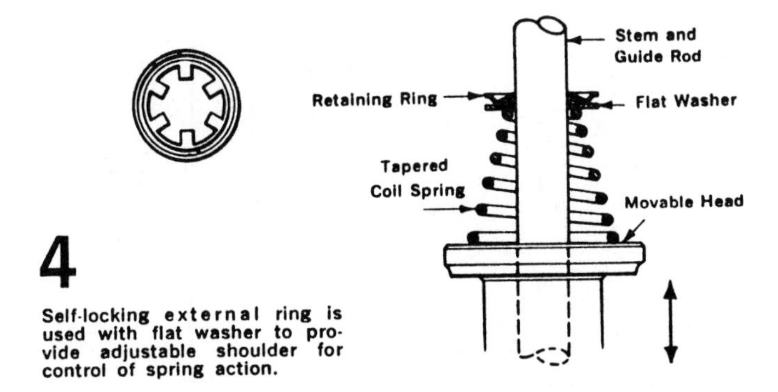

Self-locking external ring is used with flat washer to provide adjustable shoulder for control of spring action.

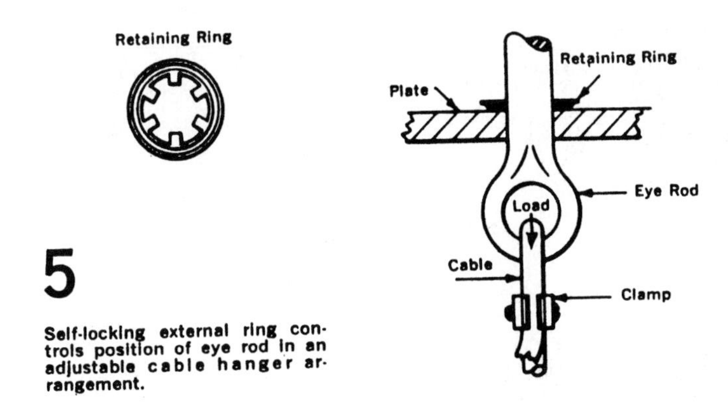

Self-locking external ring controls position of eye rod in an adjustable cable hanger arrangement.

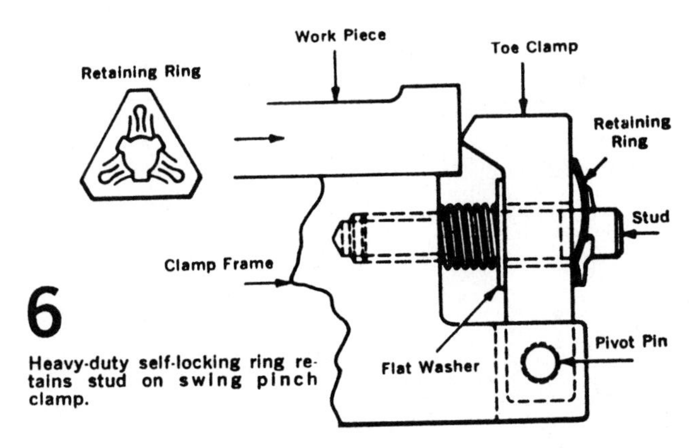

Heavy-duty self-locking ring retains stud on swing pinch clamp.

FIG. 24-3 (*continued*)

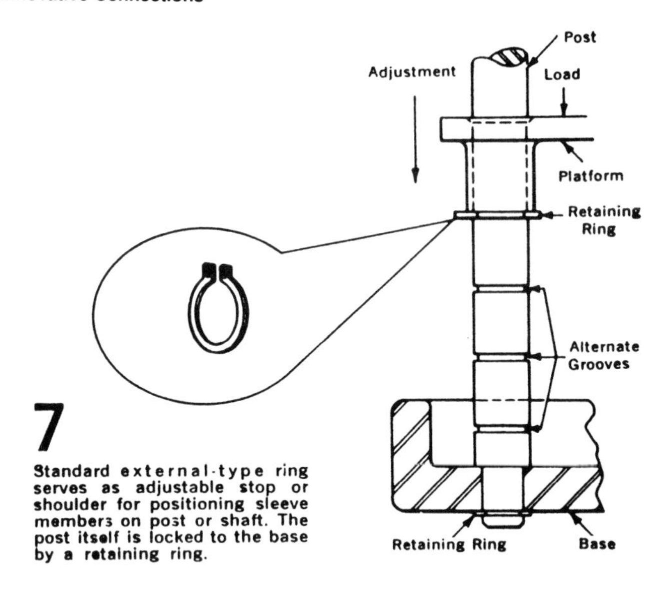

7

Standard external-type ring serves as adjustable stop or shoulder for positioning sleeve members on post or shaft. The post itself is locked to the base by a retaining ring.

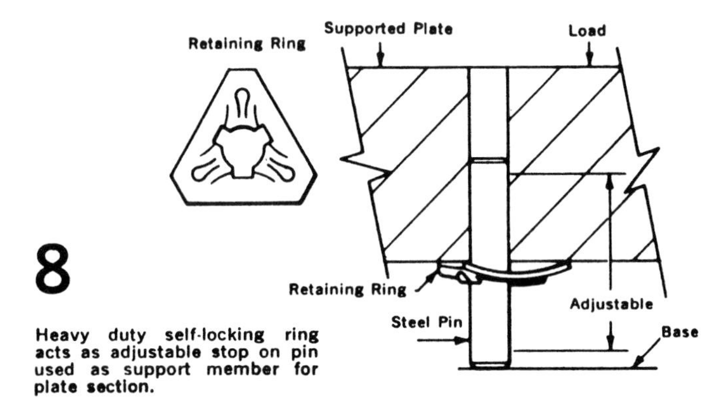

8

Heavy duty self-locking ring acts as adjustable stop on pin used as support member for plate section.

FIG. 24-3 (*continued*)

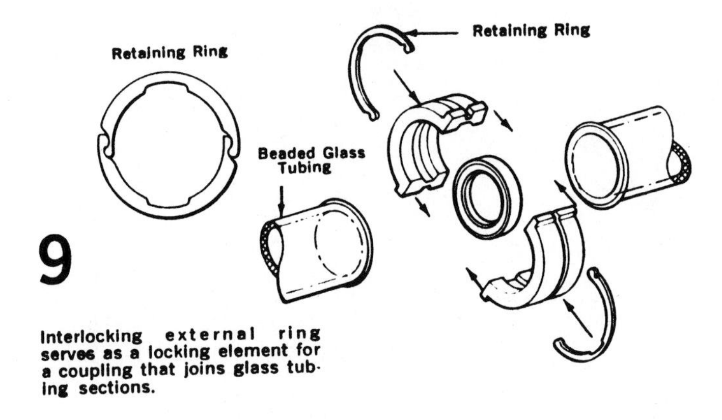

Interlocking external ring serves as a locking element for a coupling that joins glass tubing sections.

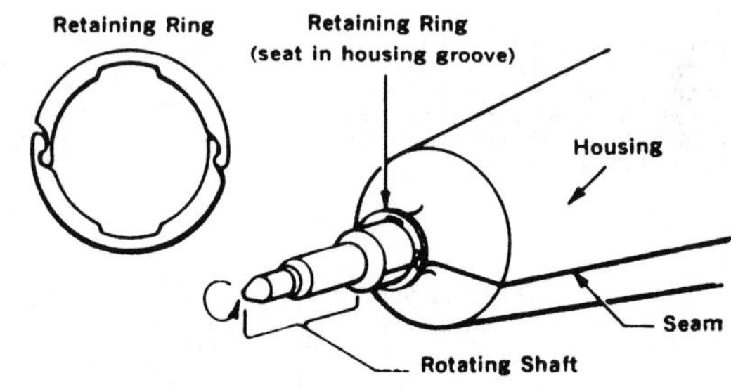

Interlocking external ring locks two-piece housing that fits around a rotating shaft.

FIG. 24-3 (*continued*)

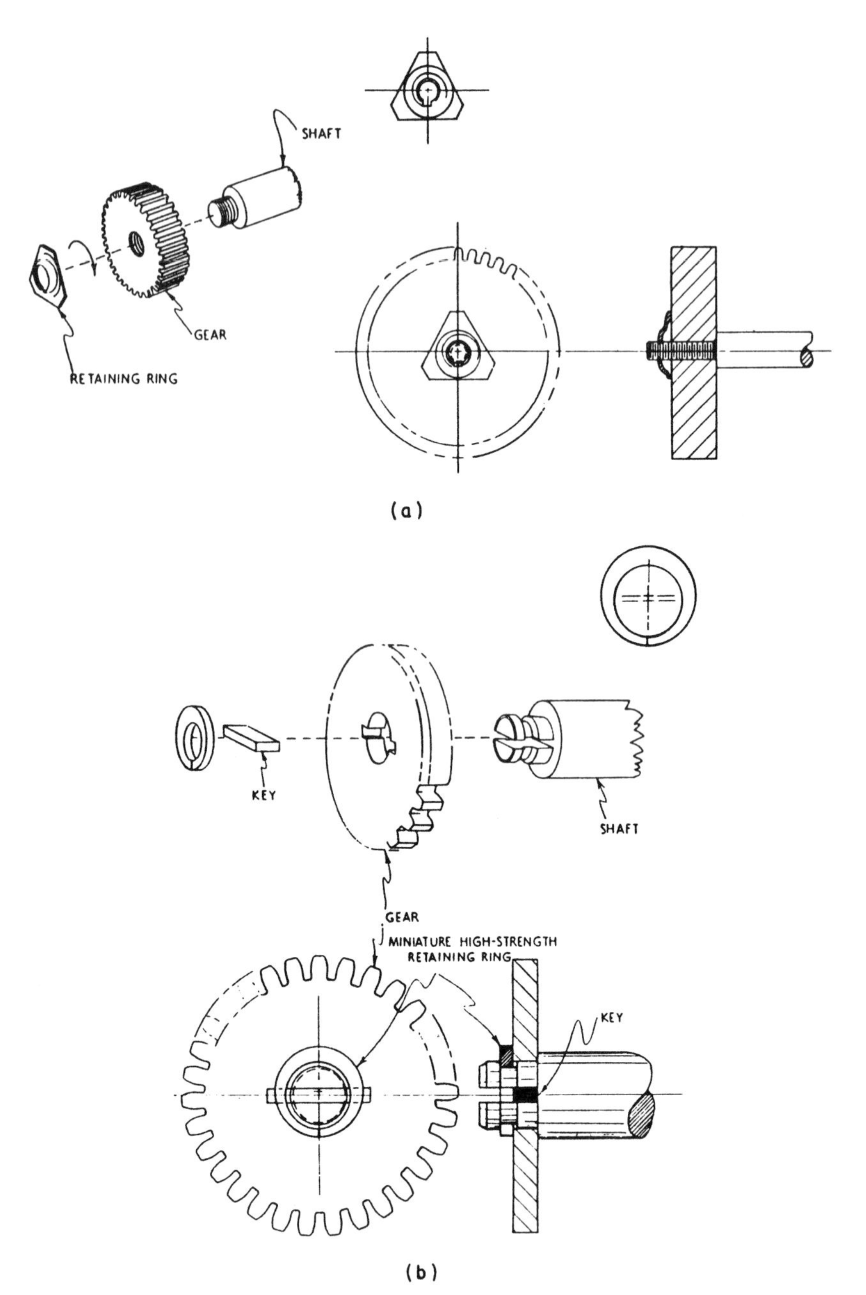

FIG. 24-4 Versatile retaining ring. A design roundup of some unusual applications for retaining rings. (*a*) The assembly of a hubless gear and threaded shaft may be accomplished by using a triangular nut retaining component which eliminates the need for a large standard nut and lock washer or other spring-type part. The dished body of the triangular nut flattens under torque to lock the gear to the shaft. (*b*) This heavy-duty hubless gear and shaft are designed for high torque and end thrusts. The retaining ring seated in a square groove and the key in a slot provide a tamper-proof lock. This design is recommended for permanent assemblies in which the ring may be subjected to heavy loads from either or both axial directions. An angled groove can be provided which has one wall cut at a 40° angle to the shaft axis. This will permit the ring to

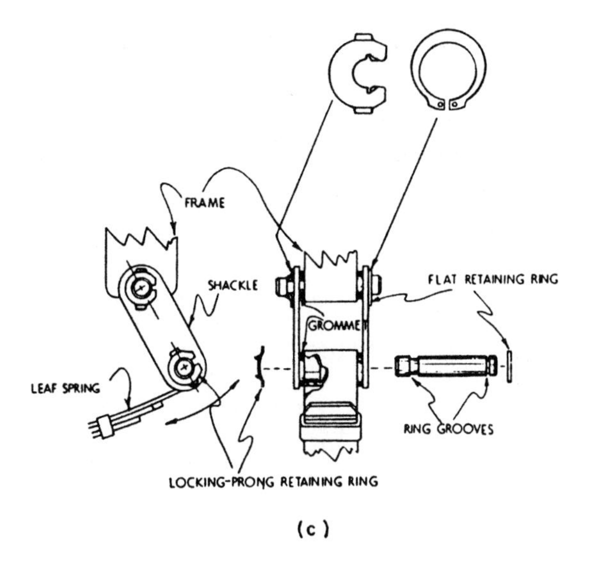

(c)

"A"
"A"
FLOAT
"SEC. A-A"
FLOAT ACTION
FLOAT GUIDE ROD
SELF-LOCKING RETAINING RING (ADJUSTABLE)

(d)

FIG. 24-4 (*continued*) be removed without damage. (*c*) Two different types of retaining rings are used in this application involving a leaf spring and shackle assembly. A locking-prong retaining ring is bowed for tension while the prongs act as fastening elements to secure the pivot bolt. A flat or standard external ring is used as a flange or bolt head. (*d*) The self-locking retaining rings used in this application provide stops for a float. The rings are adjustable on the guide rod and yet the friction force produced by the heavy spring pressure makes axial displacement from the lightweight hollow float impossible. (*e*) Retaining rings provide a uniform circular shoulder for small-diameter parts such as the pipe nipple shown here. In this case the retaining ring shoulder is used as a stop for the plastic tube. The wall thickness of the nipple should be at least 3 times as thick as the depth of the groove. When assembling the ring in the groove, the nipple should be supported by inserting a mandrel of rod. (*f*) This

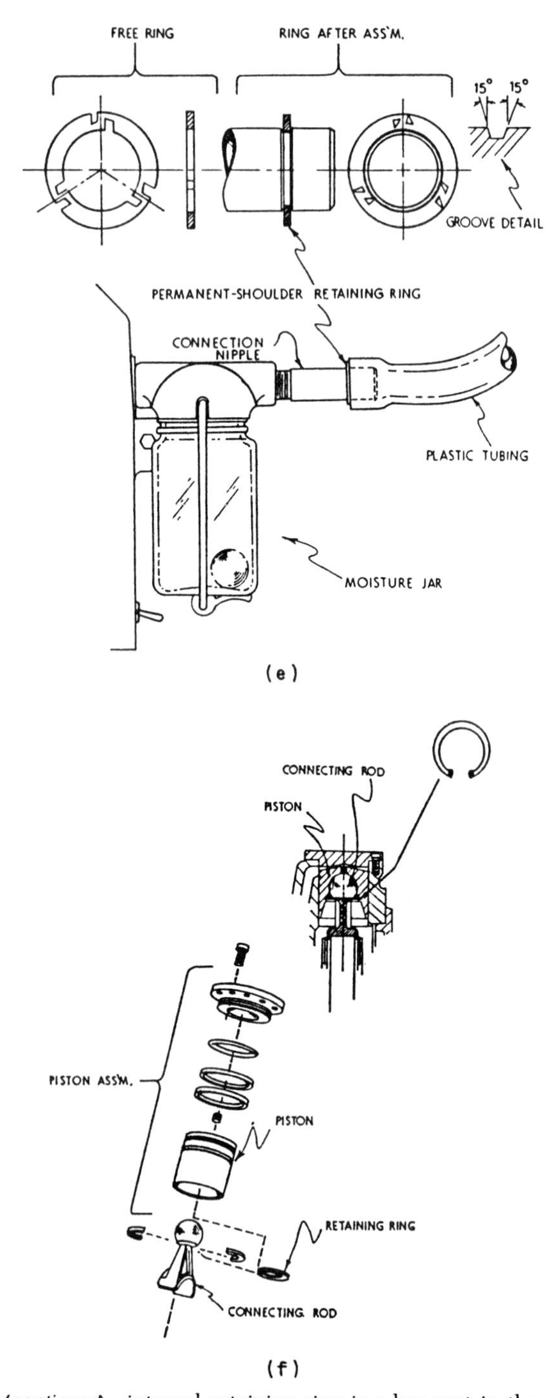

FIG. 24-4 (*continued*) internal retaining ring is a key part in the assembly of a connecting rod and piston for a hydraulic motor. The ring's lug holes make rapid assembly and disassembly possible when the proper pliers are used. The piston assembly in this case is slow moving and is not subject to heavy cycle loading. (g) Internal self-locking rings can act as a support carrier when the ID of a sleeve or housing cylinder is too large to center and stabilize small rods or conduit. The rings are adjustable in the

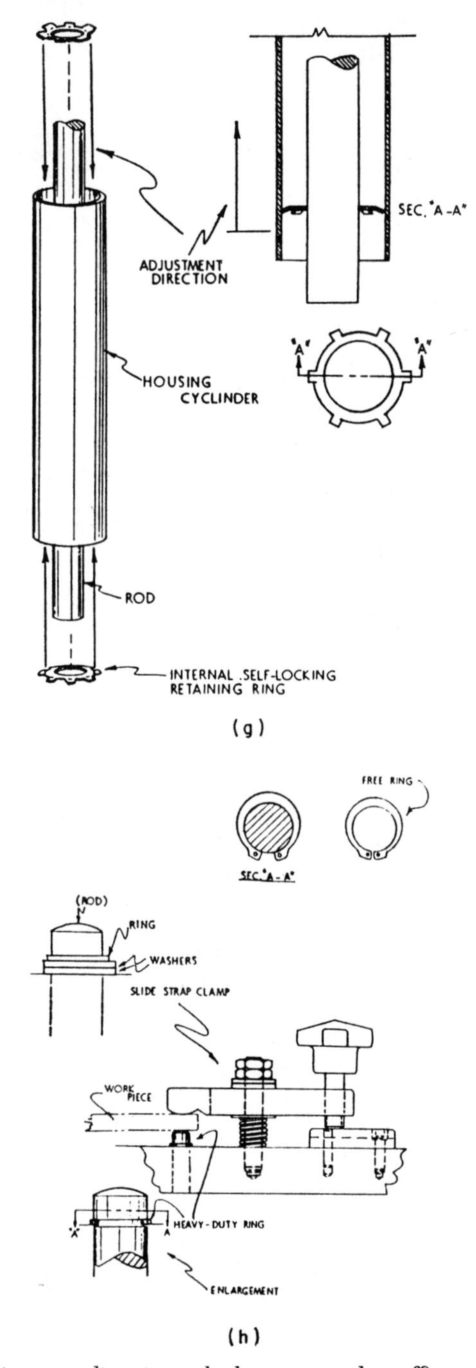

FIG. 24-4 (*continued*) entry direction only, however, and a sufficient number should be used to secure the rod. (*h*) The heavy-duty external retaining ring shown here controls the elevation or position of a support post in a holding clamp. This type of ring is ideal for heavy-duty applications where extreme loading conditions are encountered. By adding washers under the ring the elevation of the support post can be adjusted as required. (*From R. O. Parmley, "The Versatile Retaining Ring," in Assembly Engineering, February 1968, vol. 2, no. 2. Excerpted from Assembly Engineering. By permission of the publisher ©1968 by Hitchcock Publishing Co. All rights reserved.*)

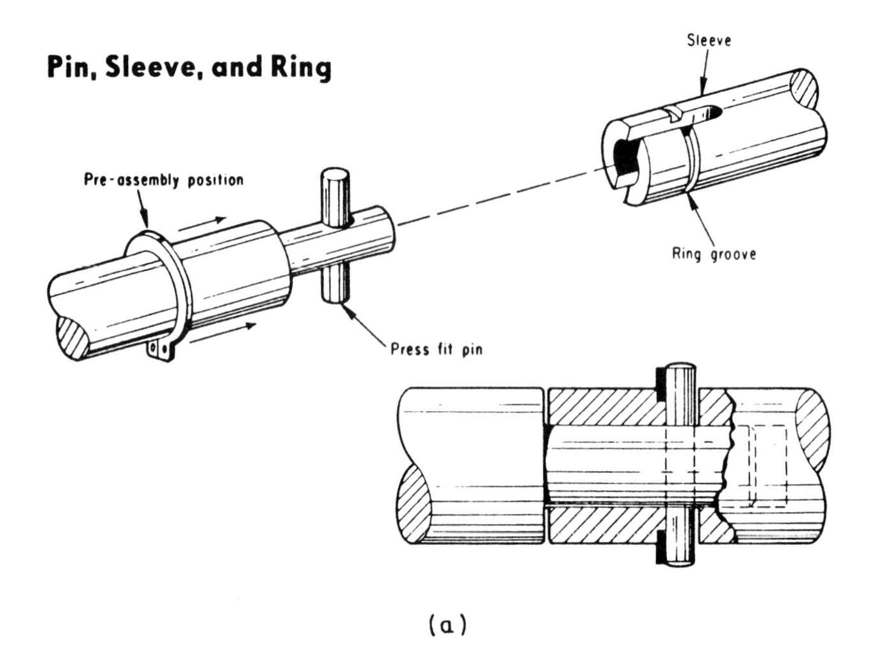

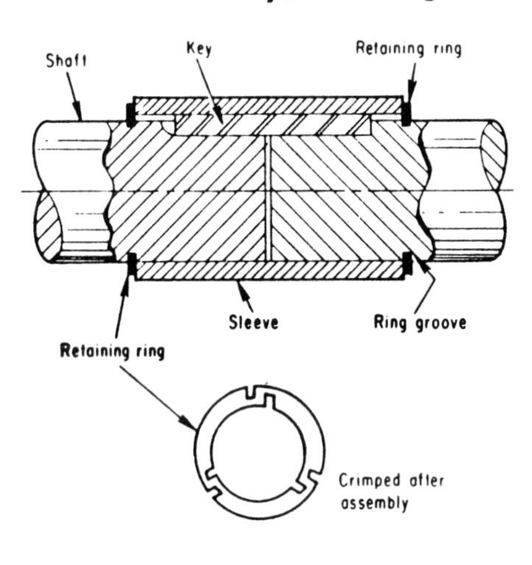

FIG. 24-5 Coupling shafts with retaining rings. These simple fasteners can provide an original way around certain design snags. For example, here are eight ways they are used to solve shaft-coupling problems. (*a*) This inexpensive connection is for light torques and moderate loads where accurate positioning is not required. A heavy-duty ring is used to resist high-impact and thrust loads. (*b*) Crimping the retaining ring into the groove produces a permanent, simple, and clean connection. This method is used to avoid machining shoulders in expensive materials, and to permit use of smaller-diameter shafts. When the ring is compressed into a V-shaped groove on the shaft, the notches permanently deform into small triangles, causing a reduction of the inner and outer diameters of the ring. Thus, the fastener tightly grips the groove, and provides a 360° shoulder around the shaft. Good torsional strength and high thrust-load capacity are provided by this connection. (*c*) A balanced two-part ring provides

Two-Shaft Splice

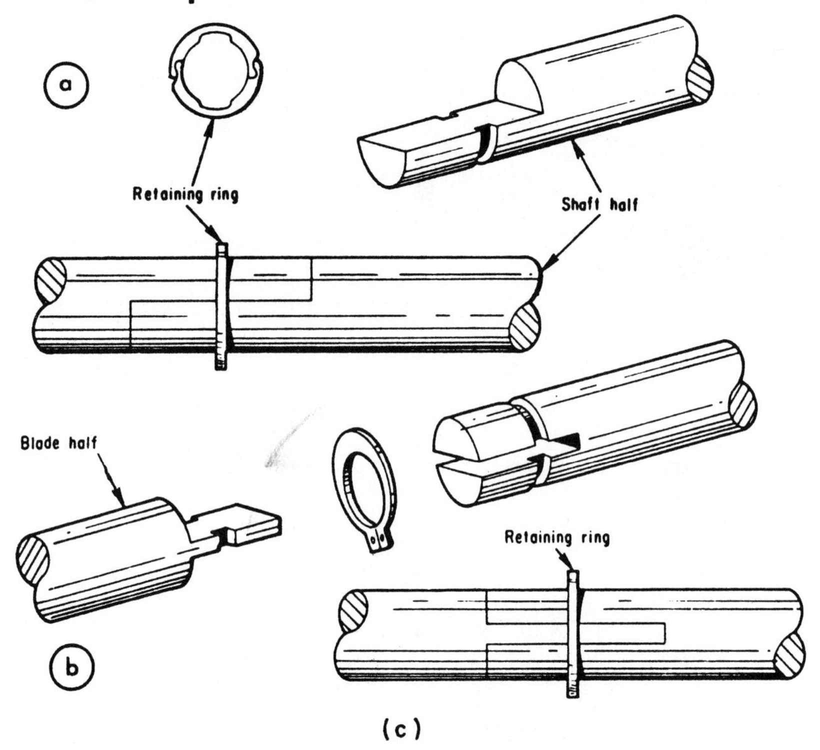

End-Flange Connection

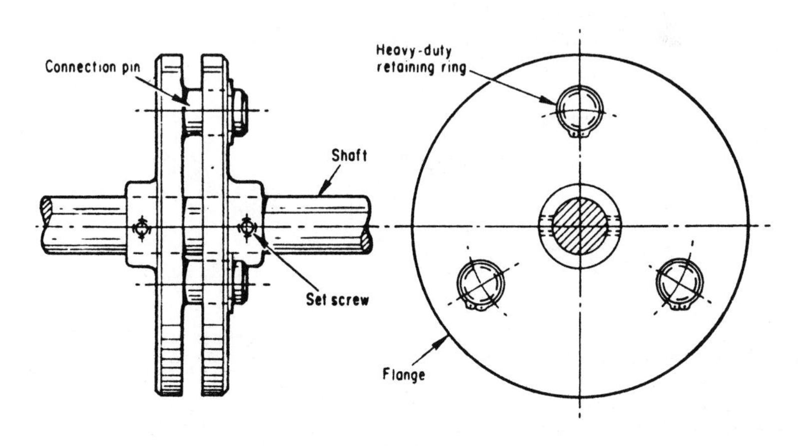

(d)

FIG. 24-5 (*continued*) an attractive appearance in addition to withstanding high rotational speeds and heavy thrust loads. 1. The one-piece ring, 2, secures the shafts in a high-torque capacity design. (*d*) This assembly for heavy-duty service requires minimum machining. Ring thickness should be substantial, and extra ring-section height is desirable. (*e*) For a connection that requires axial shaft adjustment, the self-locking ring requires no groove ⓐ An alternative solution ⓑ uses an inverted-lug ring seated in an internal groove. Extra ring-section height provides a good shoulder. The ring is uniformly

Collar, Rings, and Threaded Shaft

a

Collar

Shaft

Adjustment

Tapered-section, self-locking ring

b

Shaft

Ring-groove clearance

Equals ring-groove clearance for assembly

(e)

Coupler and Ring

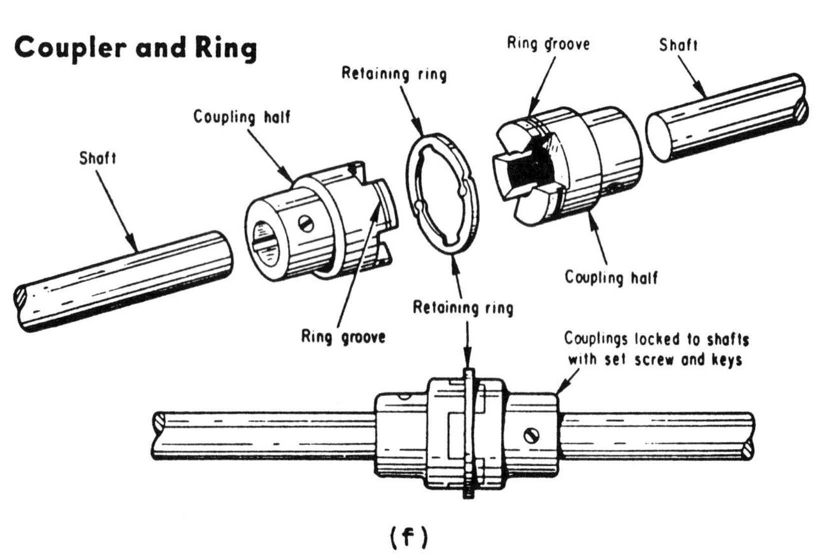

(f)

FIG. 24-5 (*continued*) concentric with housing and shaft. (*f*) Where attractive appearance is desired in a dependable locking device, this connector and ring can be used. (*g*) A slotted sleeve with tapered threads connects shafts which cannot be machined. Prongs on the retaining ring provide positive shaft gripping to stop collar movement. The arched rim adds extra strength. (*h*) An alternative solution for coupling

Slotted Sleeve With Tapered Threads

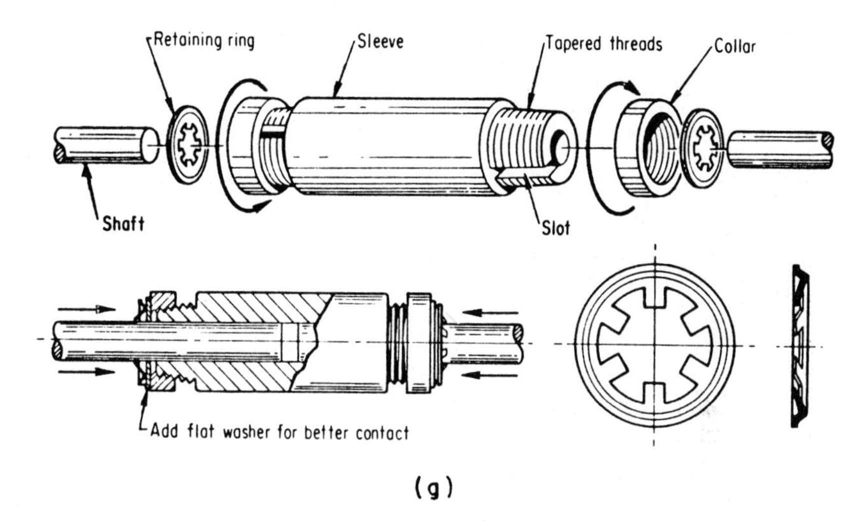

Bossed Coupling and Rings

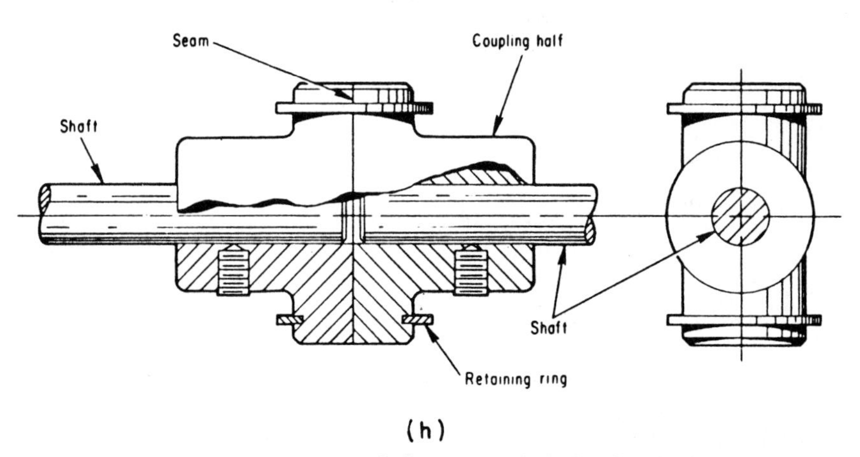

FIG. 24-5 (*continued*) concentric with housing and shaft. (*f*) Where attractive appearance is desired in a dependable locking device, this connector and ring can be used. (*g*) A slotted sleeve with tapered threads connects shafts which cannot be machined. Prongs on the retaining ring provide positive shaft gripping to stop collar movement. The arched rim adds extra strength. (*h*) An alternative solution for coupling

24-2 Pins

Space does not allow full coverage of all of today's uses of stock pins. However, to show a small segment of the versatility of pins in general, several have been selected.

Slotted or split-spring pins are widely used today. Figures 24-6 and 24-7 present a clear picture of their universal potential in machine design. See Sec. 2.

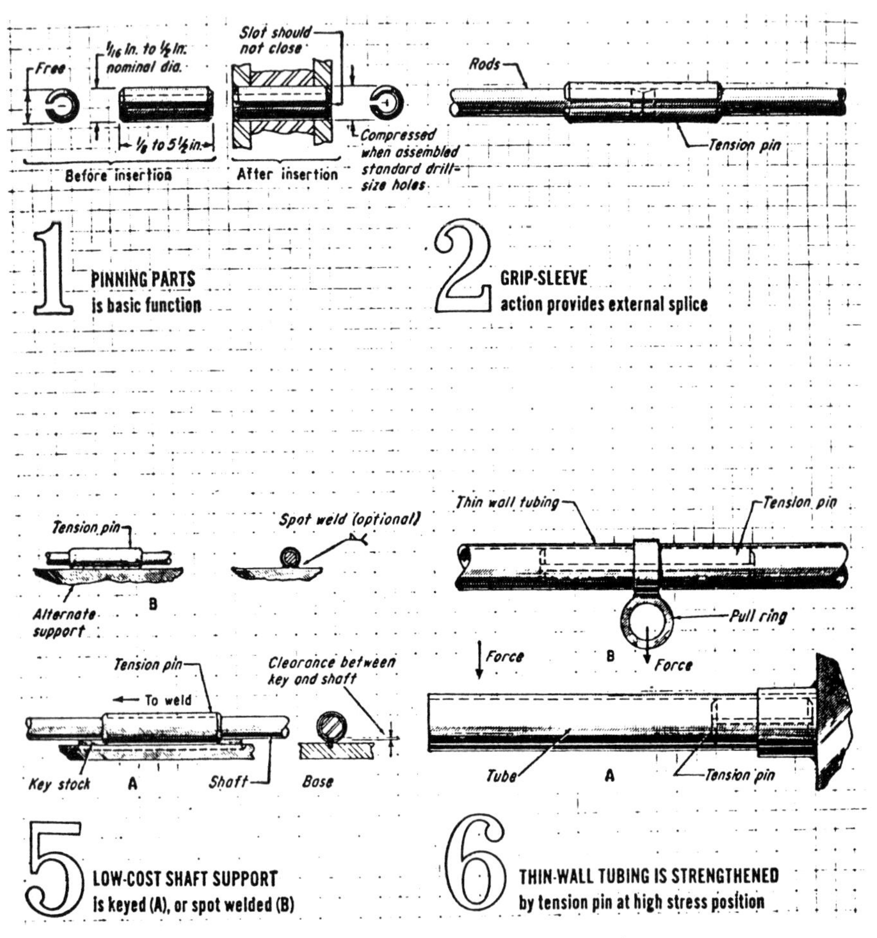

FIG. 24-6 Slotted spring pins. Assembled under pressure, these fasteners provide powerful gripping action to locate parts and hold them together. (*From R. O. Parmley, "Slotted Spring Pins Find Many Jobs," in Product Engineering, Aug. 31, 1964. Reprinted by permission, Morgan-Grampian Publishing Co.*)

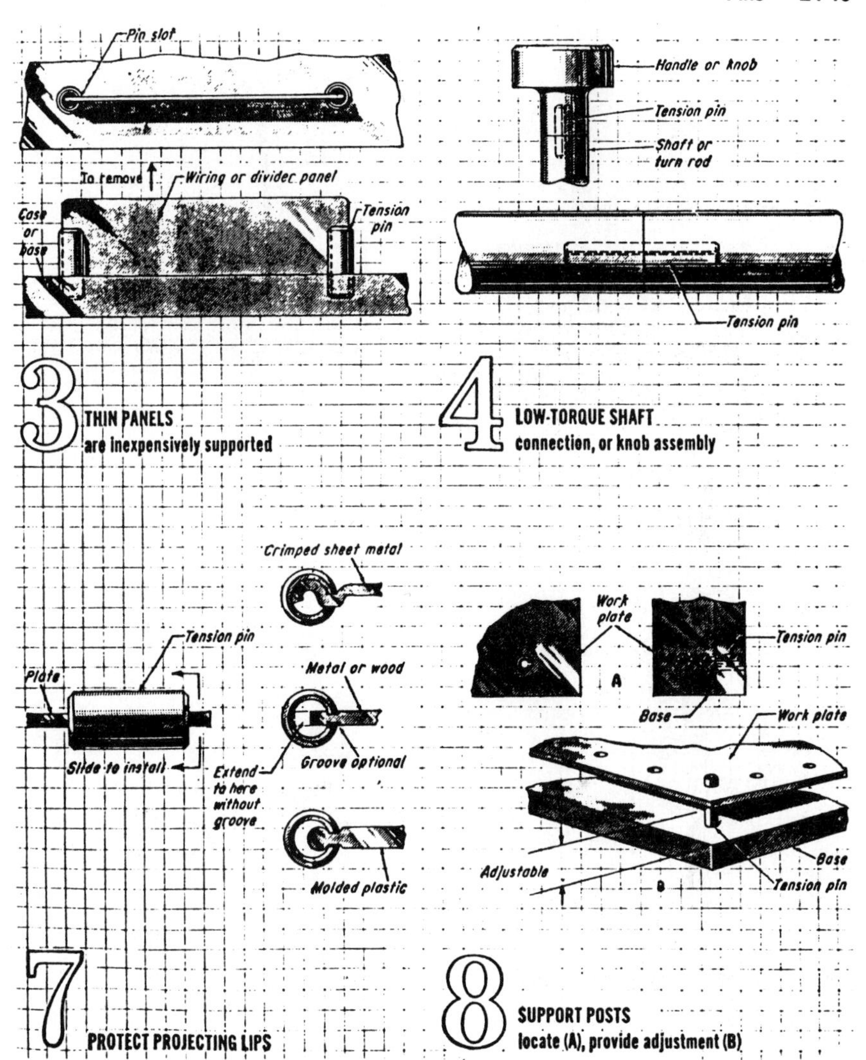

FIG. 24-6 (*continued*)

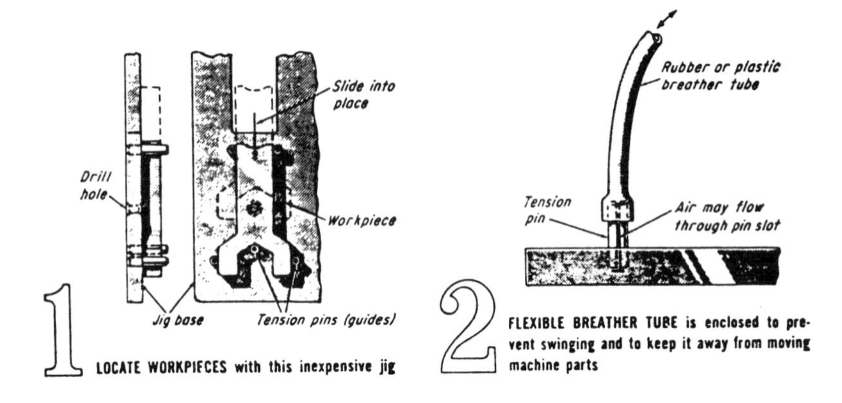

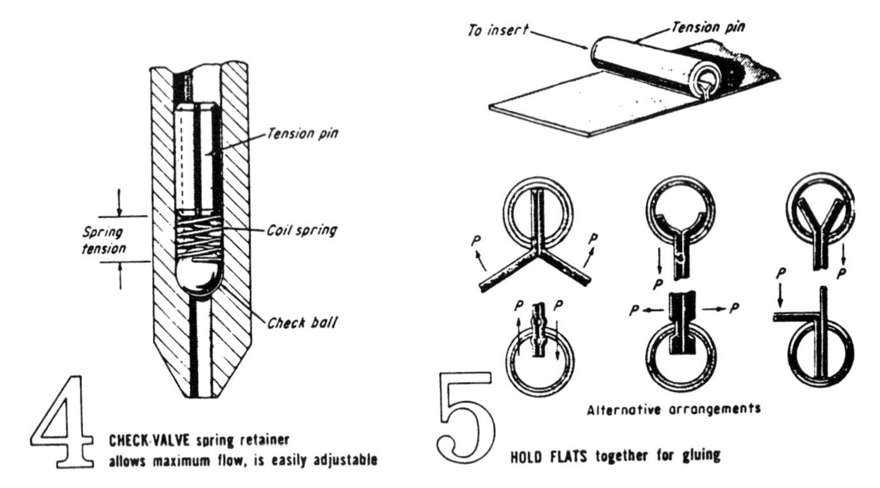

FIG. 24-7 More spring pin applications. Some additional ways that these fasteners, assembled under pressure, can grip and locate parts. They can even valve fluids. (*From R. O. Parmley, "8 More Spring-Pin Applications," in Product Engineering, Sept. 28, 1964. Reprinted by permission, Morgan-Grampian Publishing Co.*)

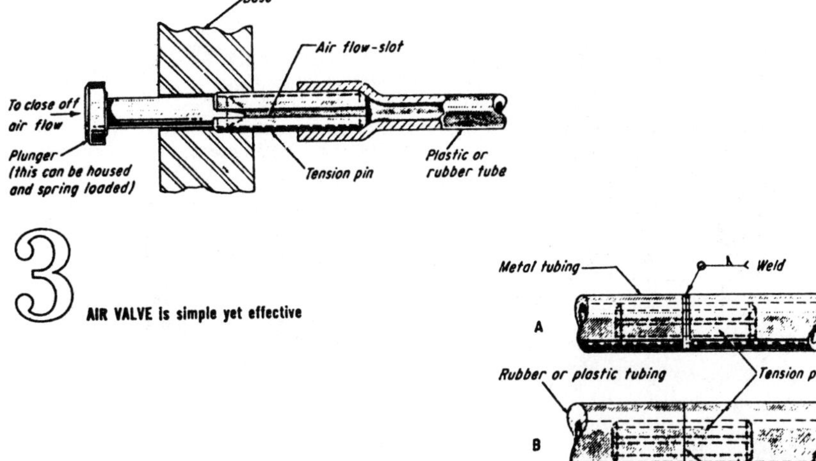

3 **AIR VALVE** is simple yet effective

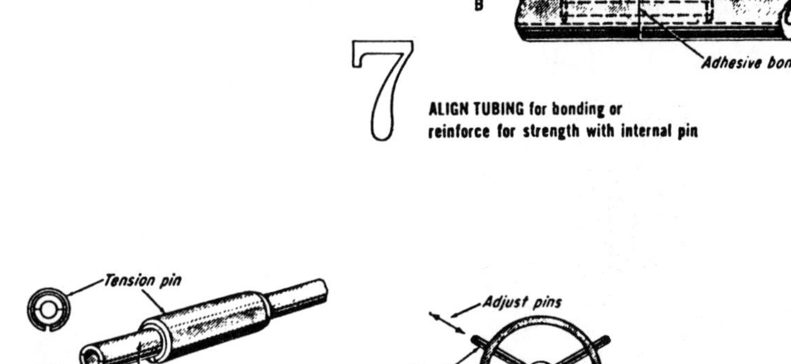

7 **ALIGN TUBING** for bonding or reinforce for strength with internal pin

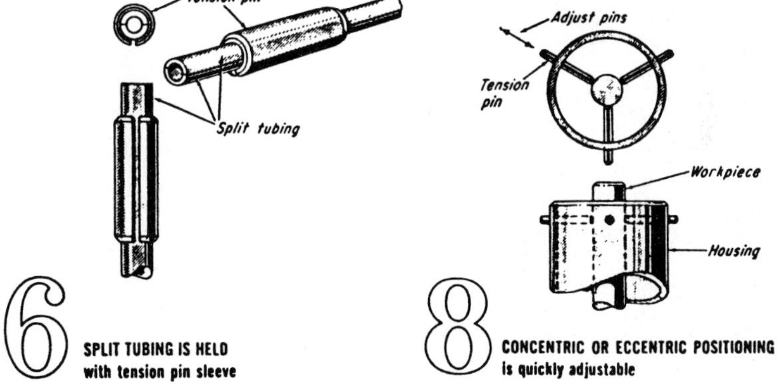

6 **SPLIT TUBING IS HELD** with tension pin sleeve

8 **CONCENTRIC OR ECCENTRIC POSITIONING** is quickly adjustable

FIG. 24-7 (*continued*)

24-3 Washers

Washers have many designs, the most common being flat and plain. Figures 24-8 to 24-10 illustrate examples of uses for flat steel and flat rubber washers.

The serrated flat washer is also a stock item with a large scope for potential usage. See Fig. 24-11.

Cupped washers are not as common, but have as much potential as their more common cousins. Figure 24-12 pictures some thought-provoking designs. Dished washers used in innovative designs are illustrated in Figs. 24-13 and 24-14.

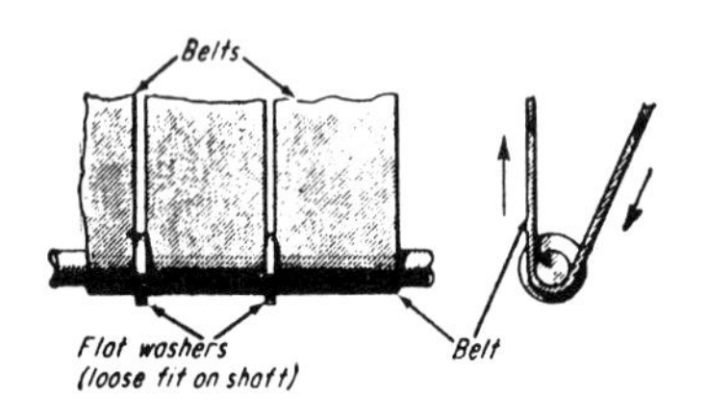

1 Are your belts overlapping? A flat washer, loosely fitted to the shaft, separates them.

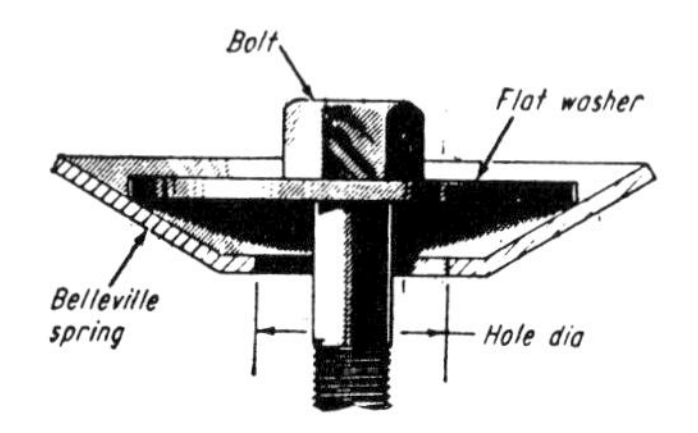

3 Need to hold odd-shaped parts? A flat washer and Belleville spring make a simple anchor.

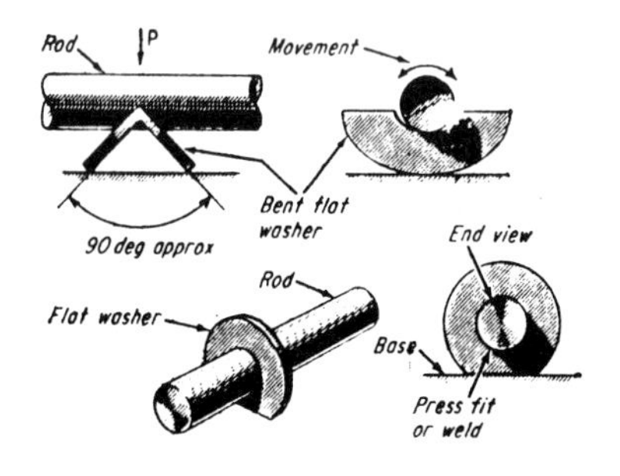

2 How about a rod support? A bent washer permits some rocking; if welded, the support is stable.

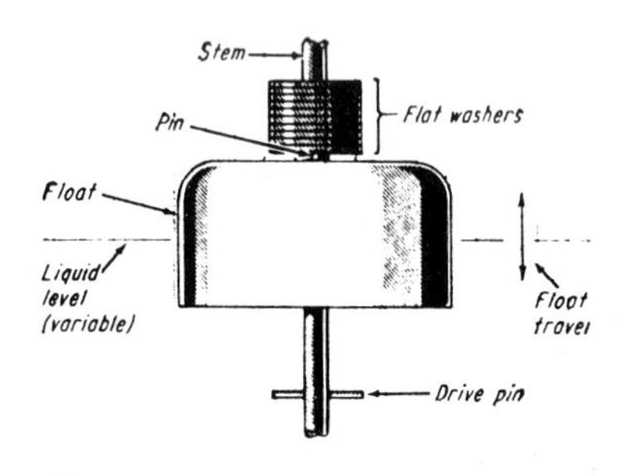

4 Got a weight problem? Adding or subtracting flat washers can easily control float action.

FIG. 24-8 Ideas for flat washers. You can do more with flat washers than you may think. Here are 10 ideas that may save the day next time you need a simple, quick, inexpensive design. (*From R. O. Parmley, "10 Ideas for Flat Washers," in Product Engineering, June 21, 1965. Reprinted by permission, Morgan-Grampian Publishing Co.*)

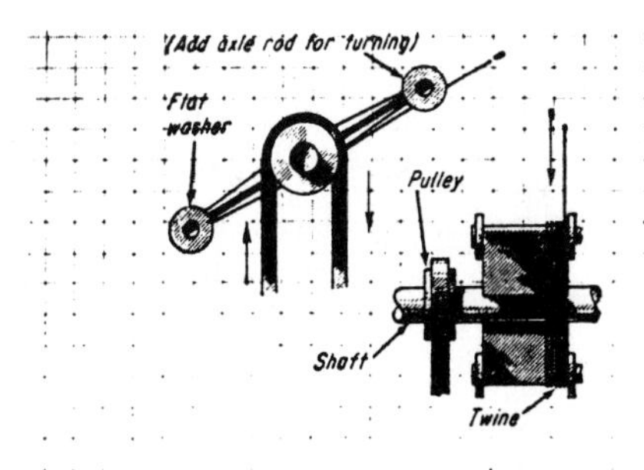

5 How about some simple flanges? Here the washers guide the twine and keep it under control.

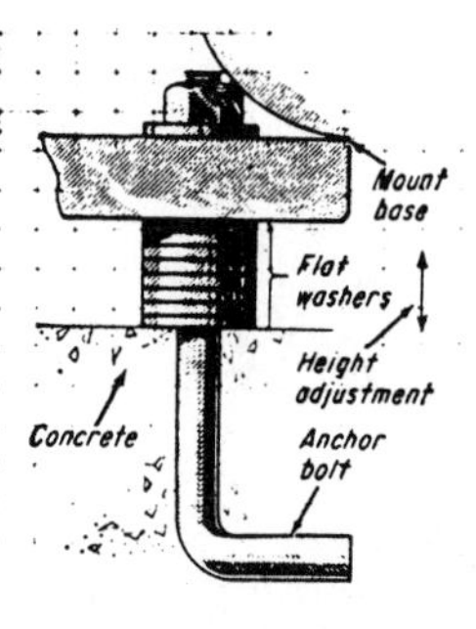

8 Does your floor tilt? Stacked washers can level machines, or give a stable height adjustment.

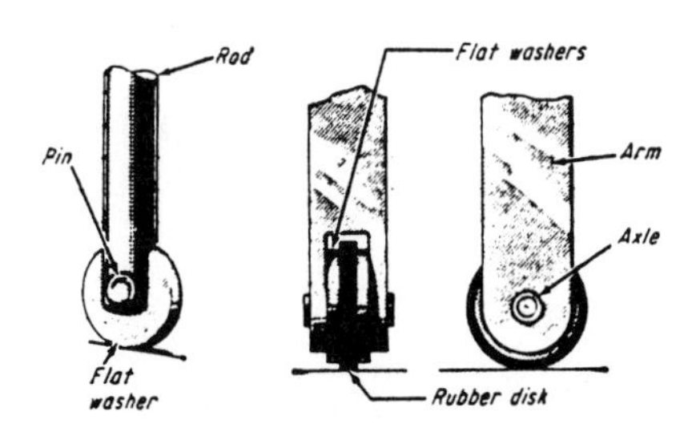

6 Need some wheels? Here flat washer is the wheel. A rubber disk quiets the assembly.

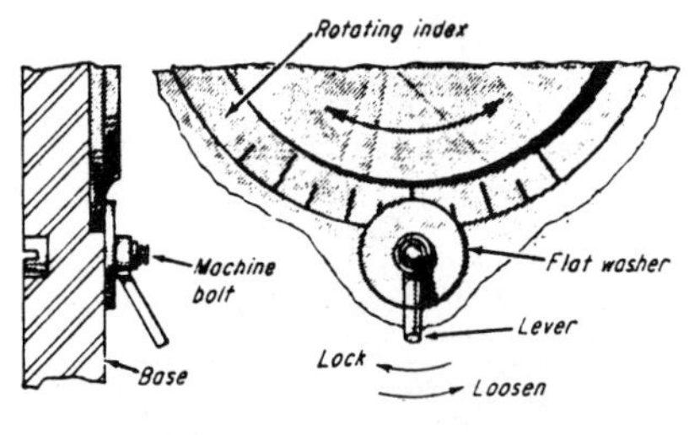

9 Here's a simple lock. A machine bolt, a washer, and a wing or lever nut make a strong clamp.

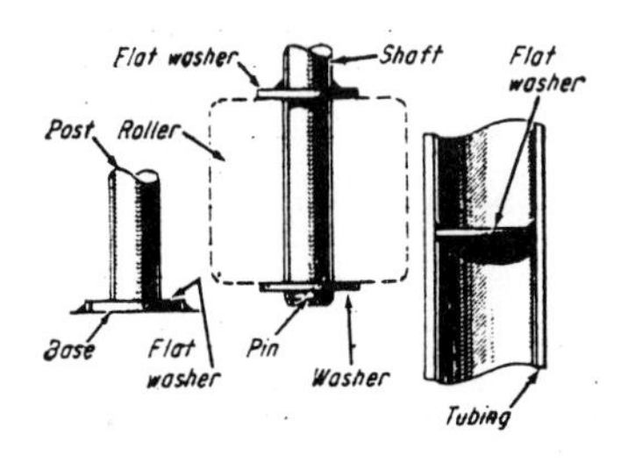

7 Want to avoid machining? Washers can make anchors, stop shoulders, even reduce tubing ID.

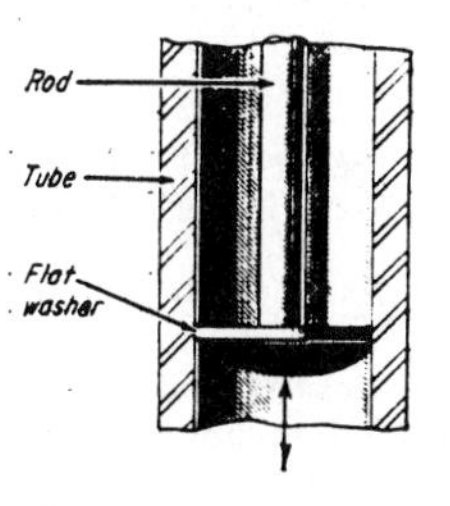

10 Need a piston in a hurry? For light service, tube, rod, and washer will be adequate.

FIG. 24-8 (*continued*)

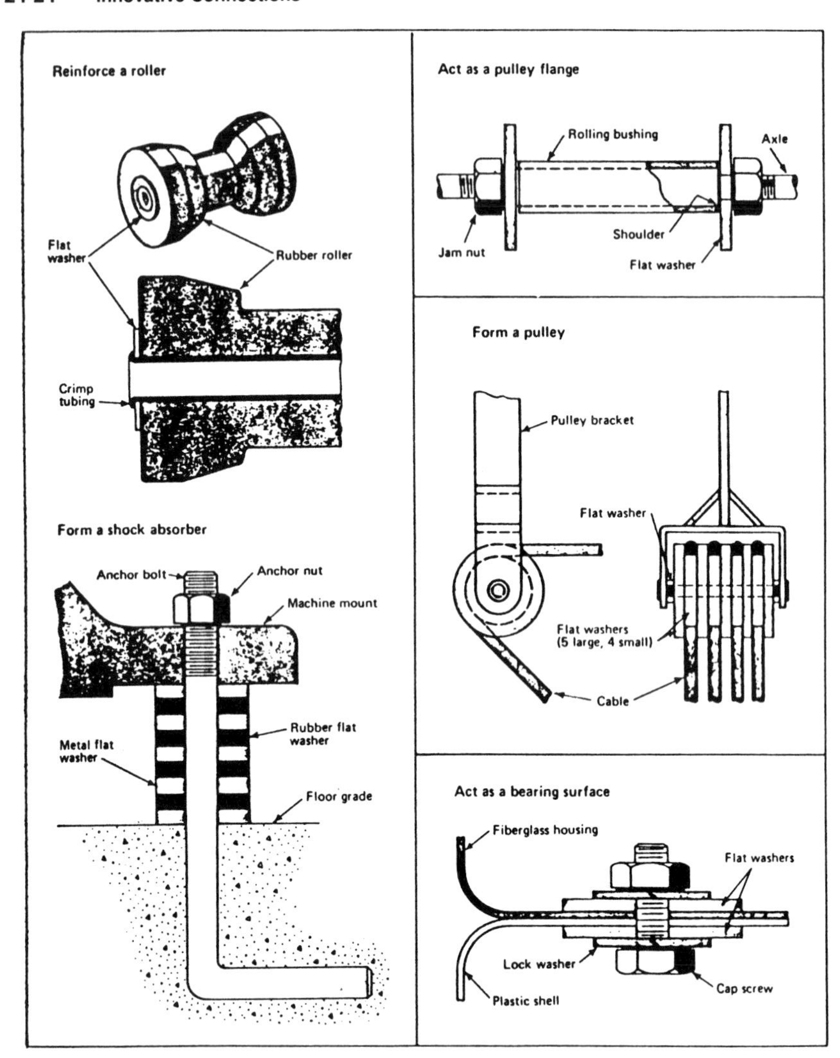

FIG. 24-9 Versatile flat washers. Washers are usually thought of as bearing surfaces placed under bolt heads. However, they can be used in a variety of ways that could simplify design or be an immediate fix until a design part is available. (*From R. O. Parmley, "Versatile Flat Washers," in Product Engineering, January 1973. Reprinted by permission, Morgan-Grampion Publishing Co.*)

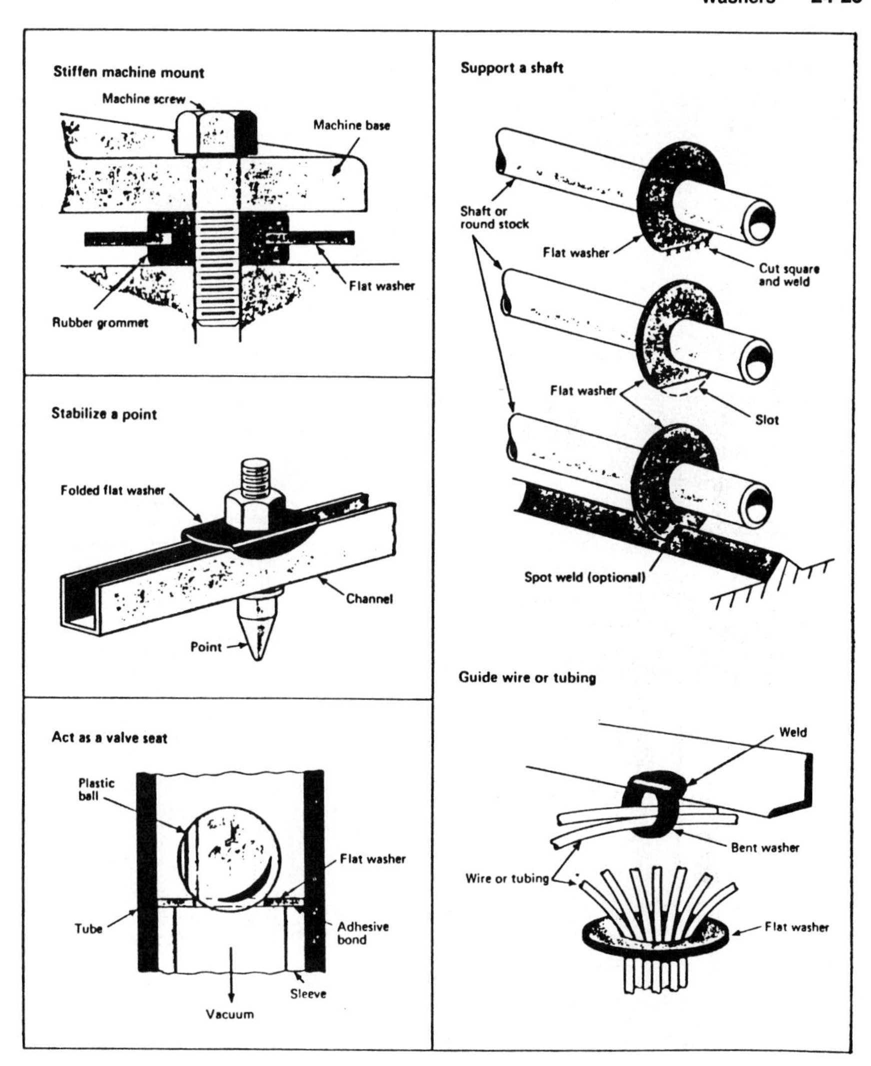

FIG. 24-9 (*continued*)

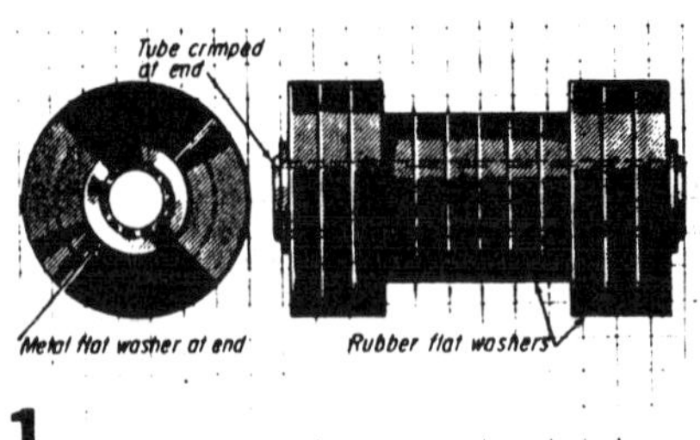

1 Step roller

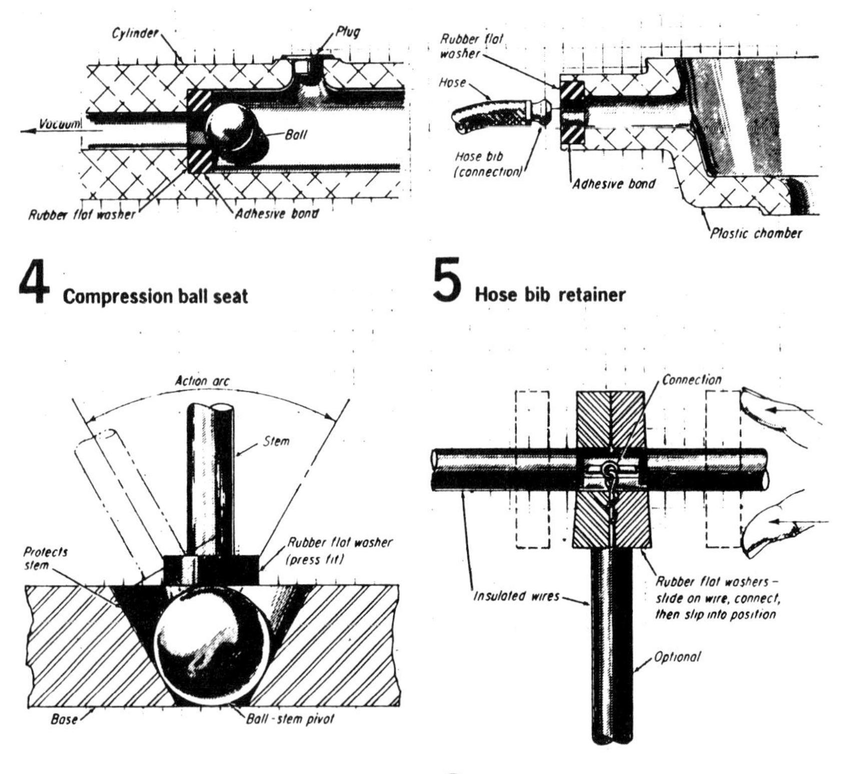

4 Compression ball seat

5 Hose bib retainer

8 Protective bumper

9 Expansion isolator

FIG. 24-10 Jobs for flat rubber washers. Rubber washers are far more versatile than you realize. Here are some odd jobs they can do that may make your next design job easier. (*From R. O. Parmley, "Jobs for Flat Washers," in Product Engineering, Nov. 22, 1965. Reprinted by permission, Morgan-Grampian Publishing Co.*)

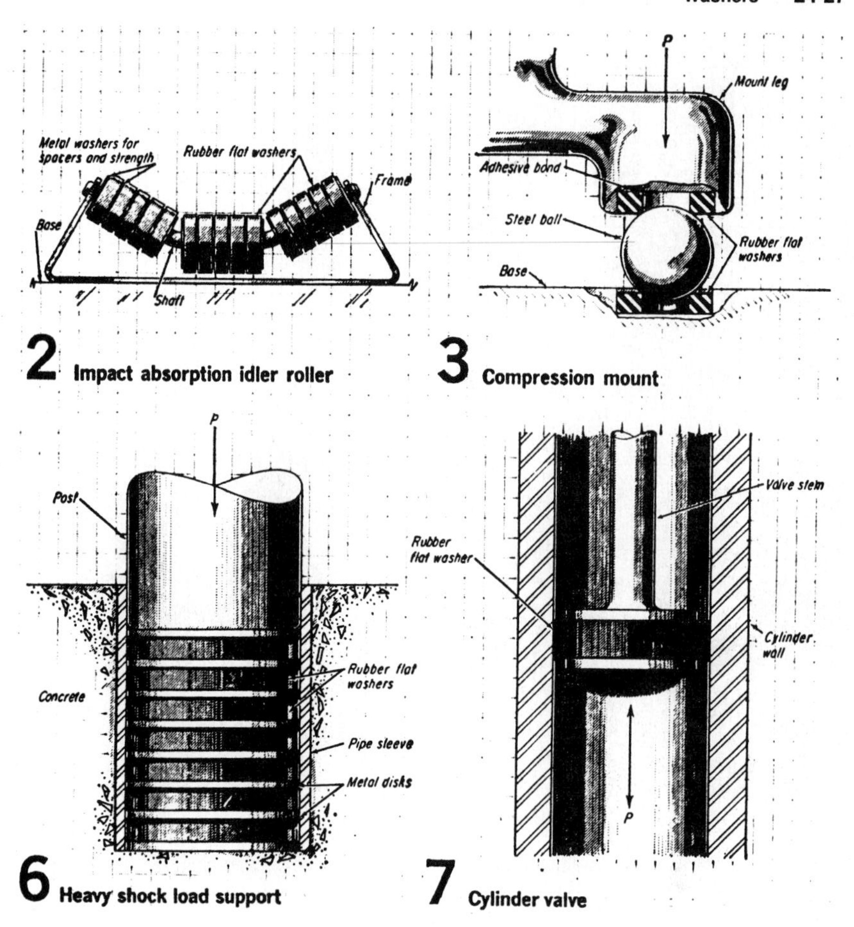

FIG. 24-10 (*continued*)

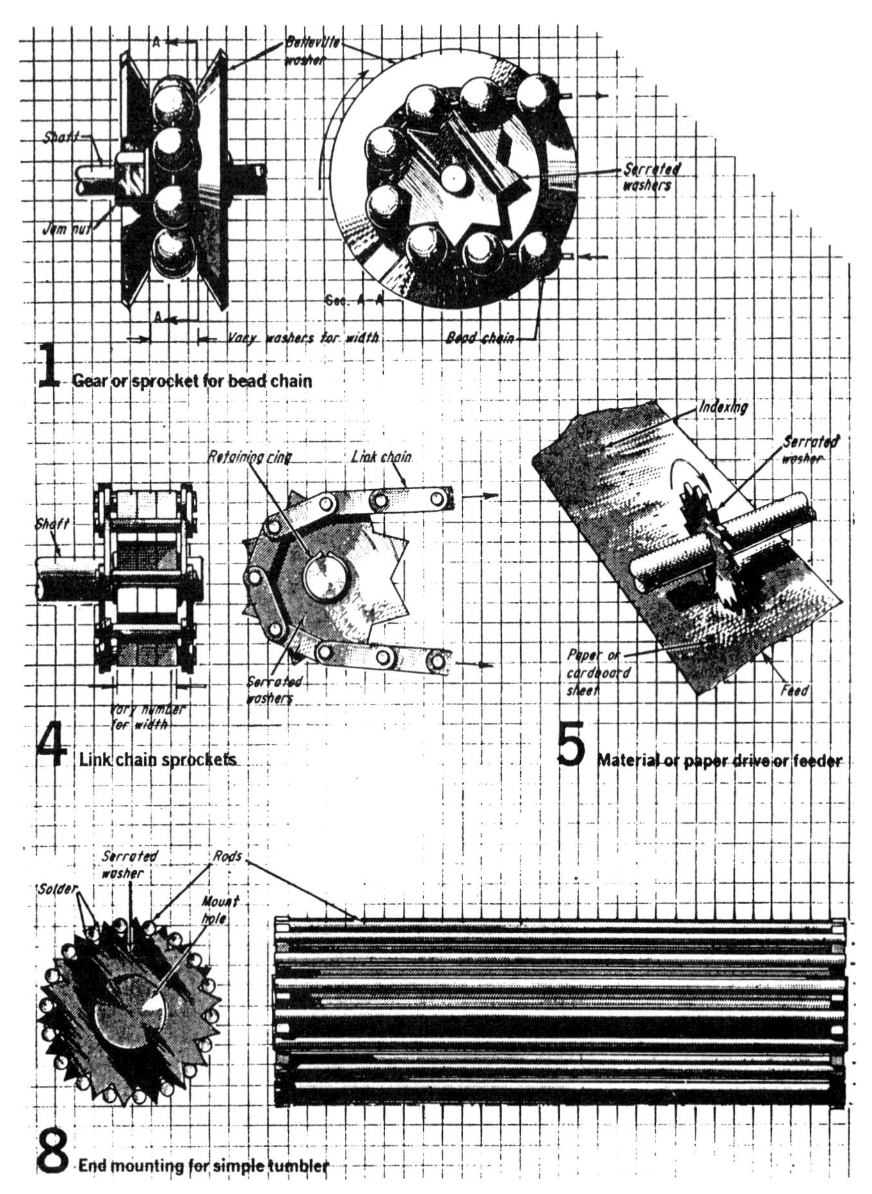

FIG. 24-11 Serrated washers are a stock item and come in a wide variety of sizes. With little thought they can do a variety of jobs. Here are just eight. (*From R. O. Parmley, "Take Another Look at Serrated Washers," in Product Engineering, Sept. 27, 1965. Reprinted by permission, Morgan-Grampian Publishing Co.*)

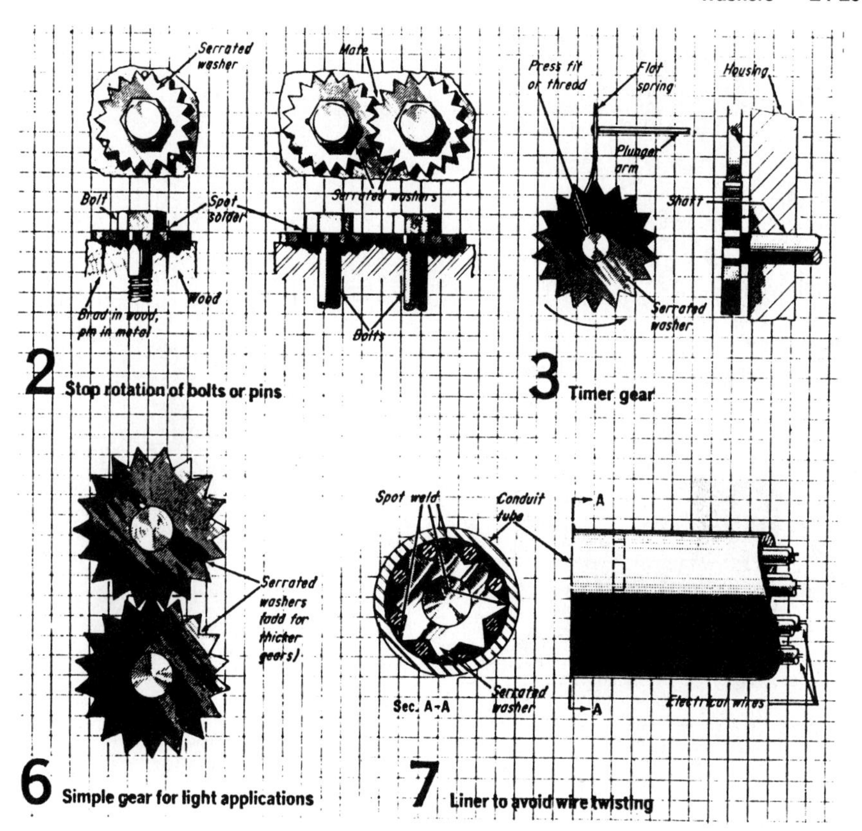

FIG. 24-11 (*continued*)

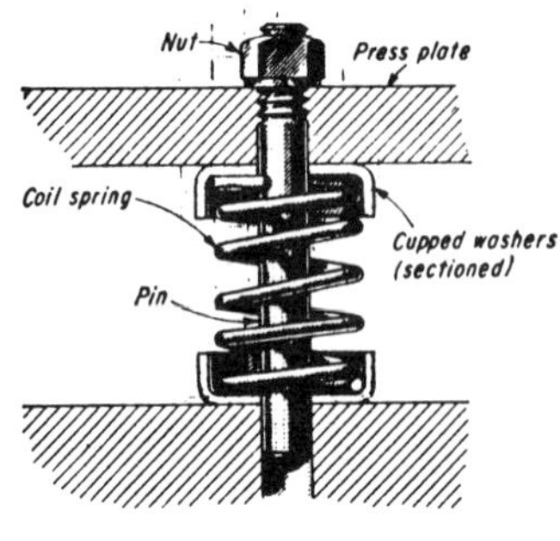

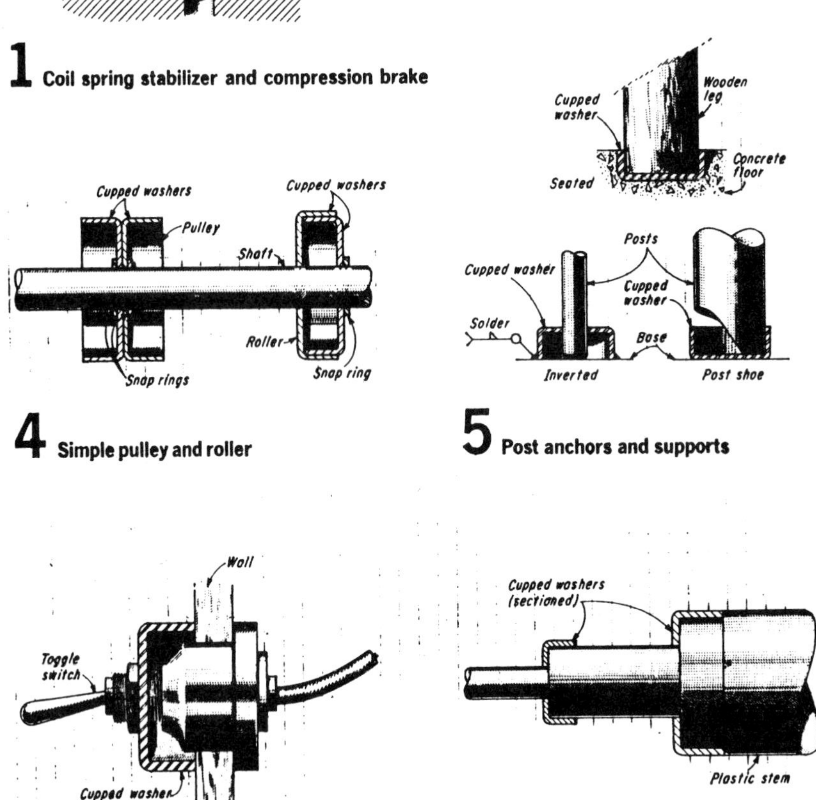

FIG. 24-12 Creative ideas for cupped washers. A standard "off-the-shelf" item with more uses than many ever considered. (*From R. O. Parmley, "Creative Ideas for Cupped Washers," in Product Engineering, Dec. 20, 1965. Reprinted by permission, Morgan-Crampian Publishing Co.*)

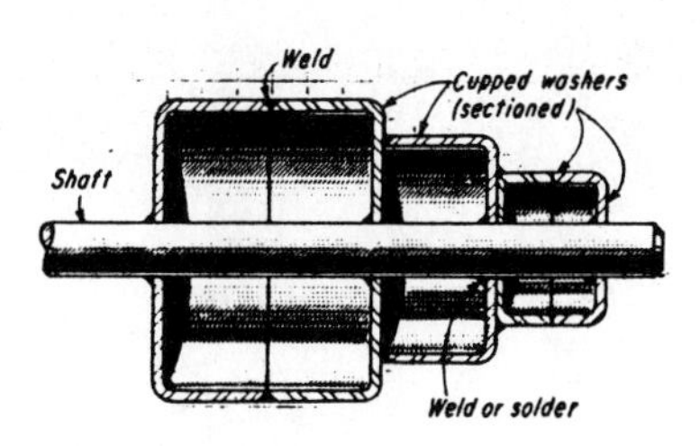

2 Simple step pulley

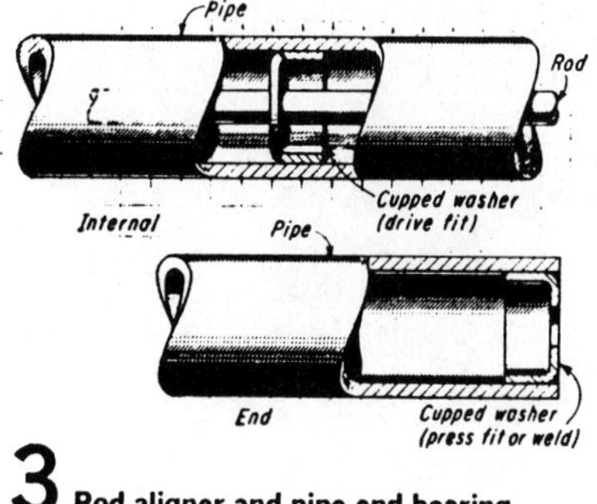

3 Rod aligner and pipe-end bearing

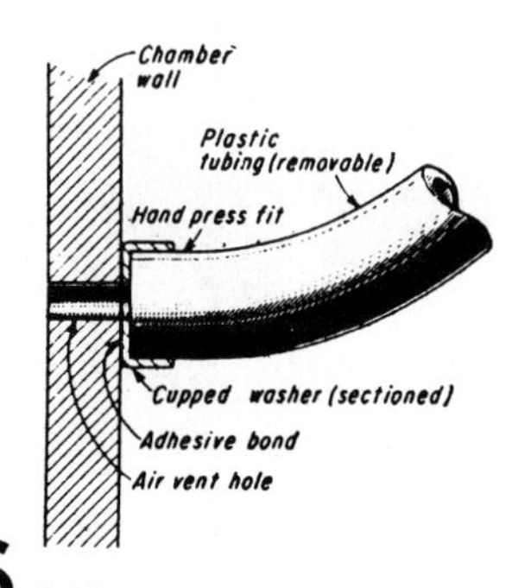

6 Tubing connector

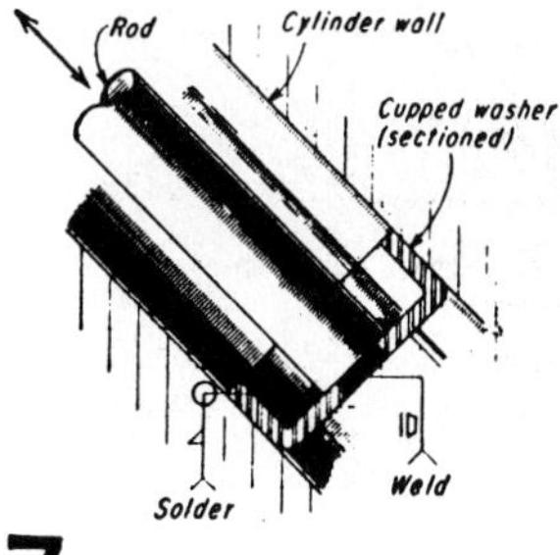

7 Simple piston for cylinder

FIG. 24-12 (*continued*)

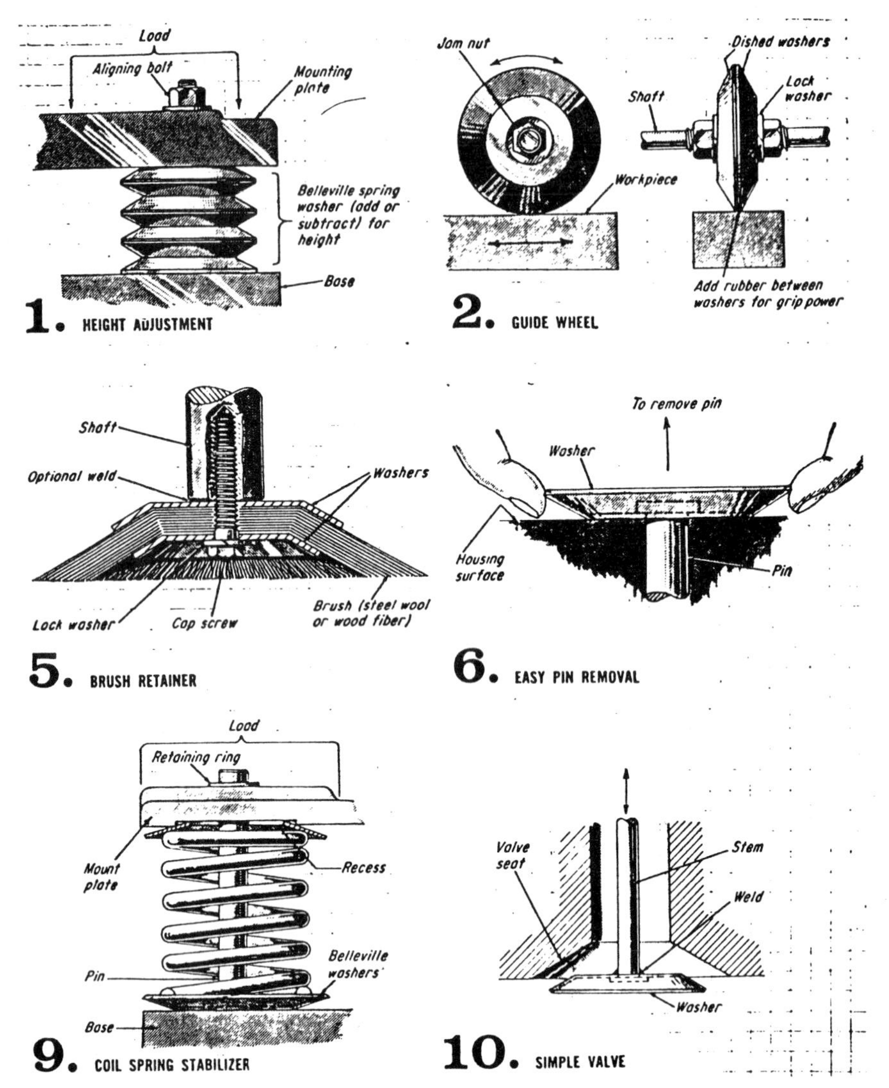

FIG. 24-13 Dished washer designs. Let these ideas spur your own design creativity. Sometimes Belleville washers will suit; otherwise you can easily dish your own. (*From R. O. Parmley, "Dished Washers Get Busy," in Product Engineering, Dec. 7, 1964. Reprinted by permission, Morgan-Grampian Publishing Co.*)

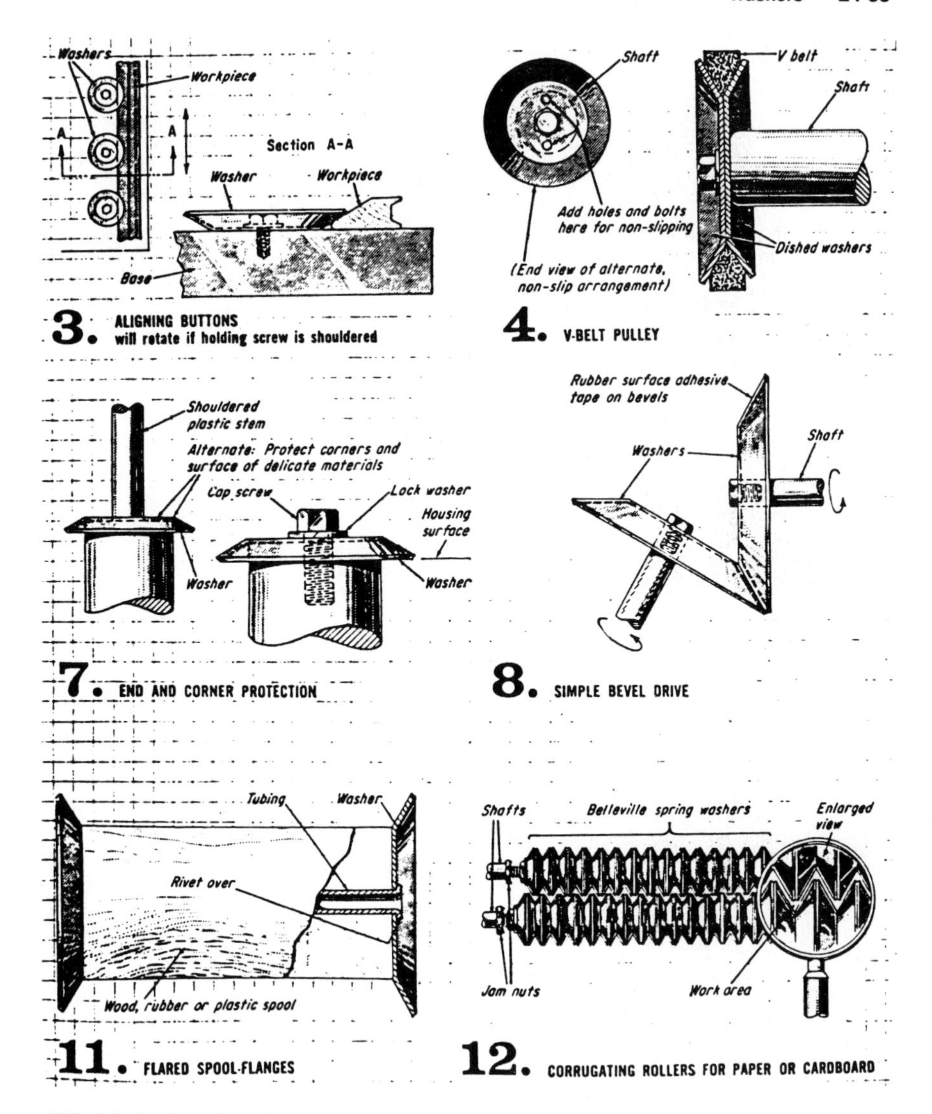

FIG. 24-13 (*continued*)

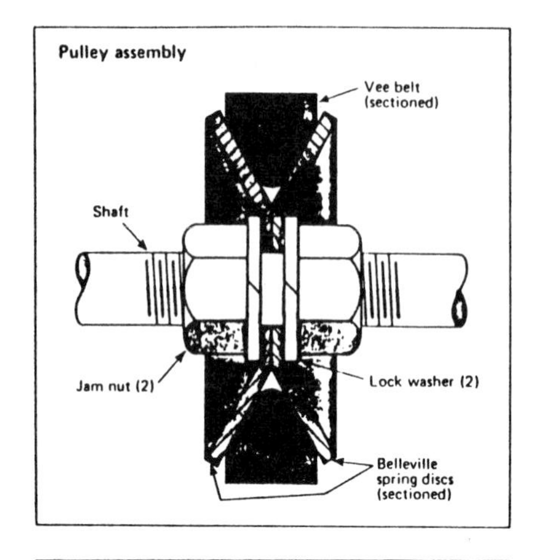

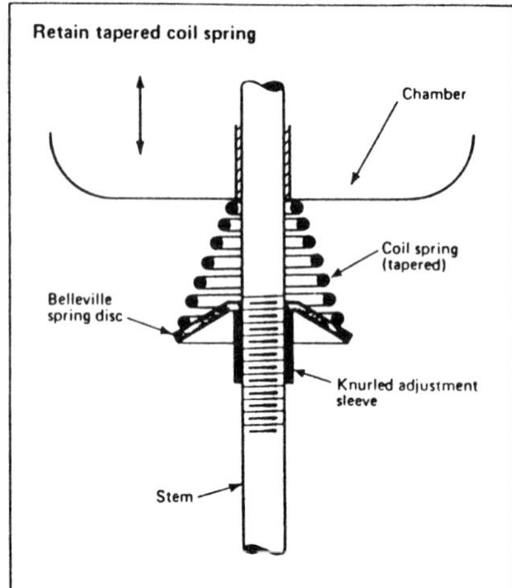

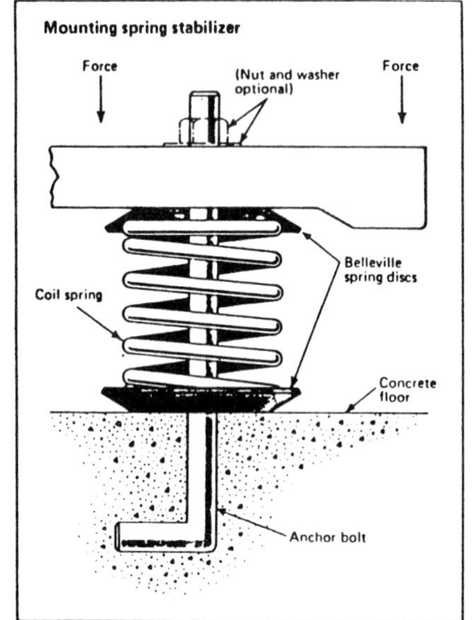

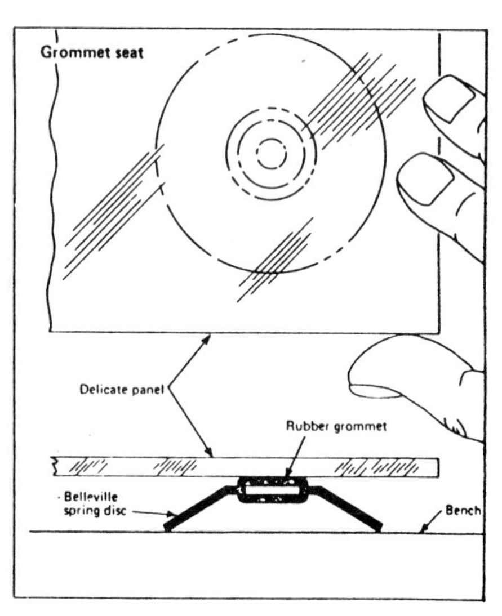

FIG. 24-14 Disk springs solve design problems. Disk springs or Belleville springs are a versatile component that offer a wide range of applications. There are many places where the component can be used and its availability as a stock item should be considered when confronted with a design problem that requires a fast solution. *(From R. O. Parmley, "Disk Springs Are an Easy Way to Solve Some Design Problems," in Product Engineering, February 1973. Reprinted by permission, Morgan-Grampian Publishing Co.)*

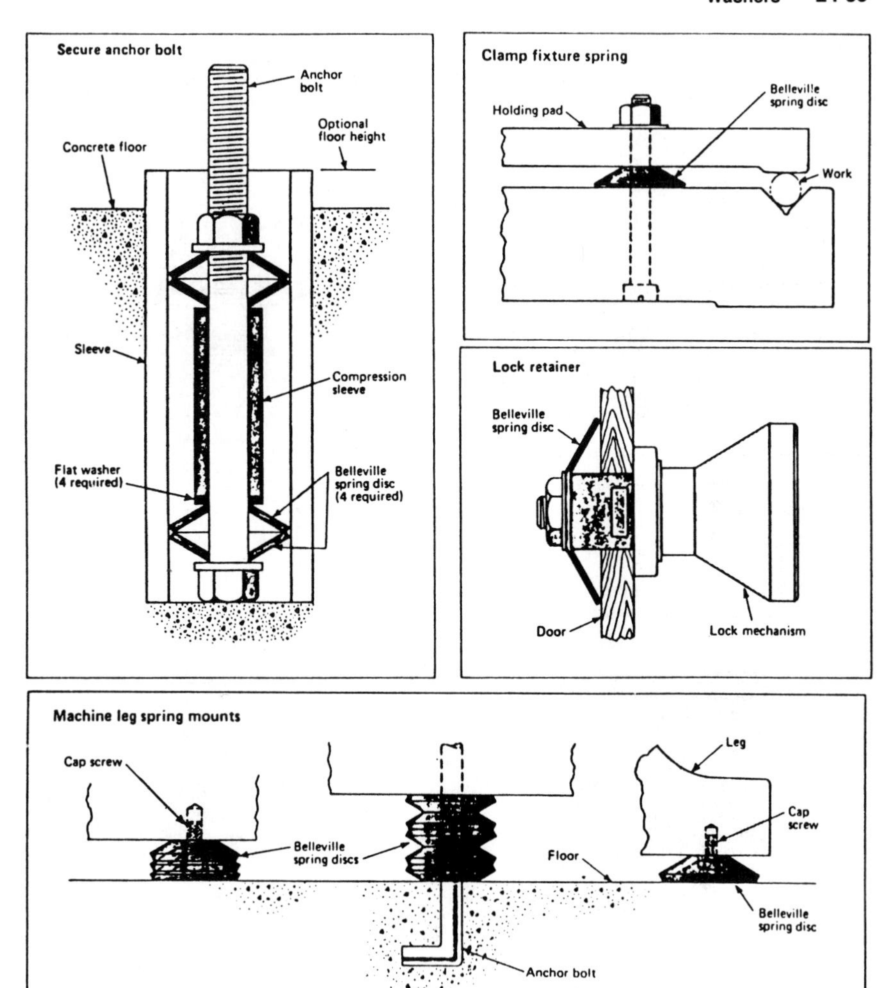

FIG. 24-14 (*continued*)

24-4 Flanged Bushings

Steel flanged bushings (journal bushings) and flanged rubber bushings are widely known and used. Figures 24-15 and 24-16 depict some common and unusual uses.

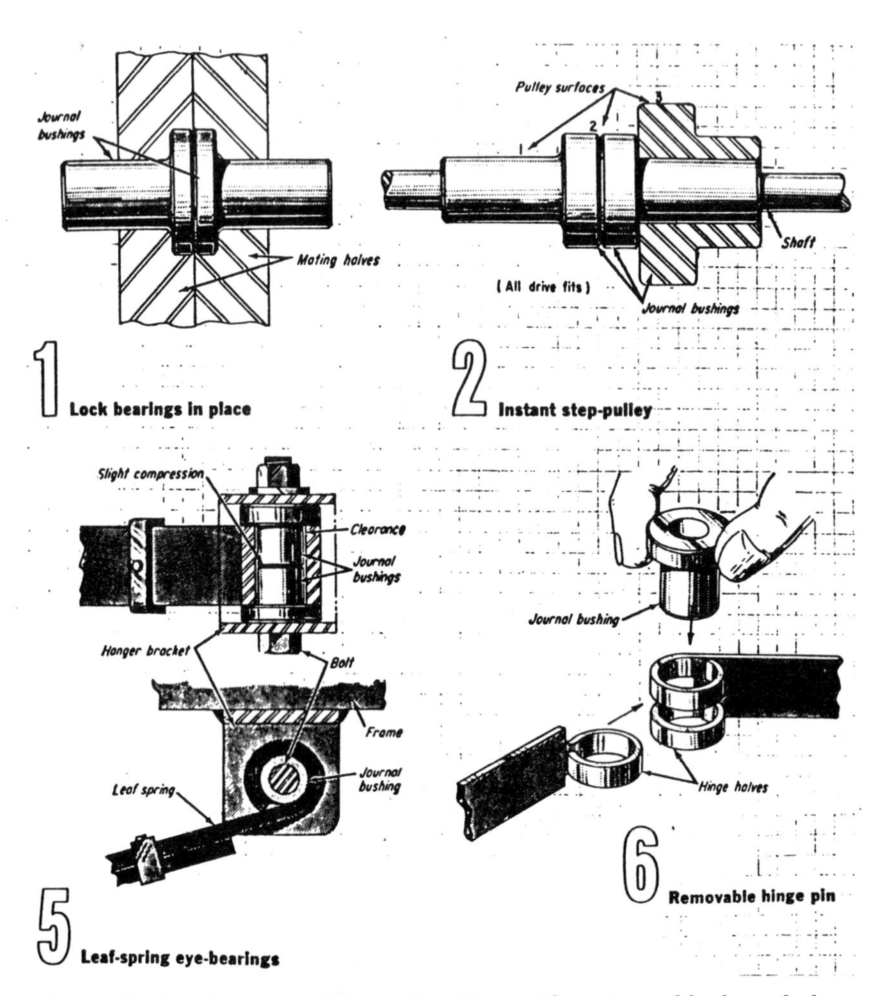

FIG. 24-15 Creative usage of flanged bushings. These sintered bushings find a variety of jobs and are available in 88 sizes, from $\frac{1}{8}$ to $1\frac{5}{8}$ in internal diameter. (*From R. O. Parmley, "Go Creative With Flanged Bushings," in Product Engineering, March 15, 1965. Reprinted by permission, Morgan-Grampian Publishing Co.*)

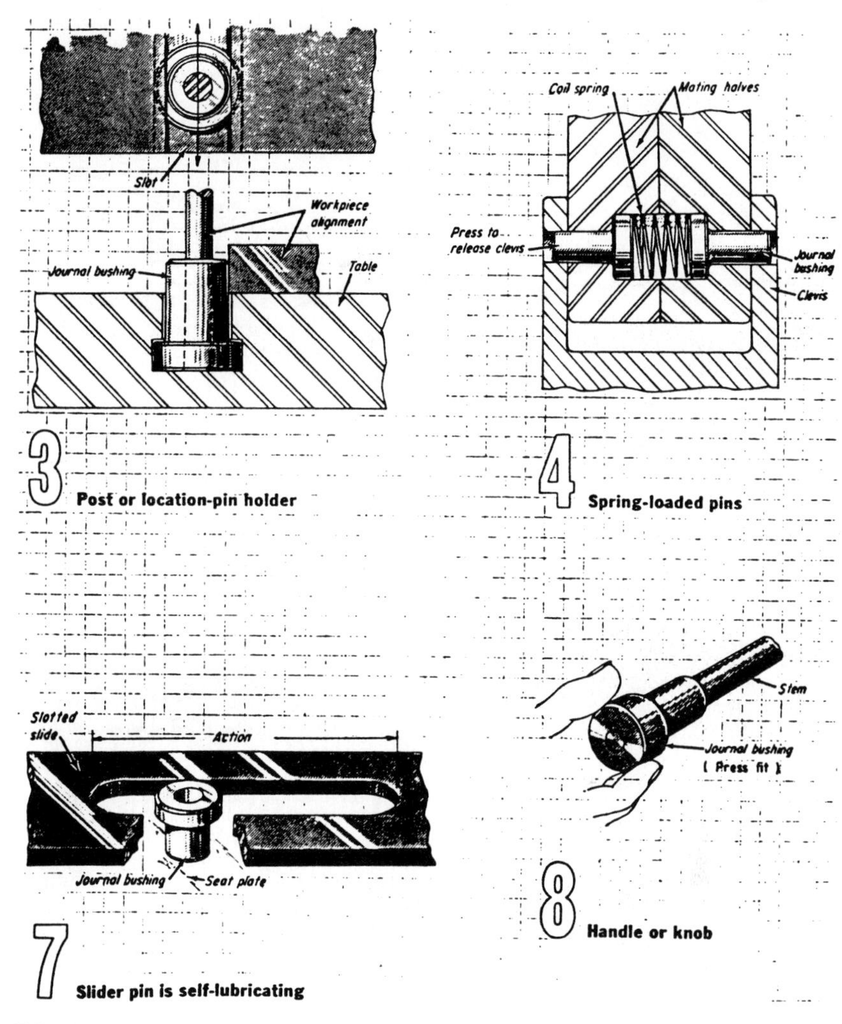

FIG. 24-15 (*continued*)

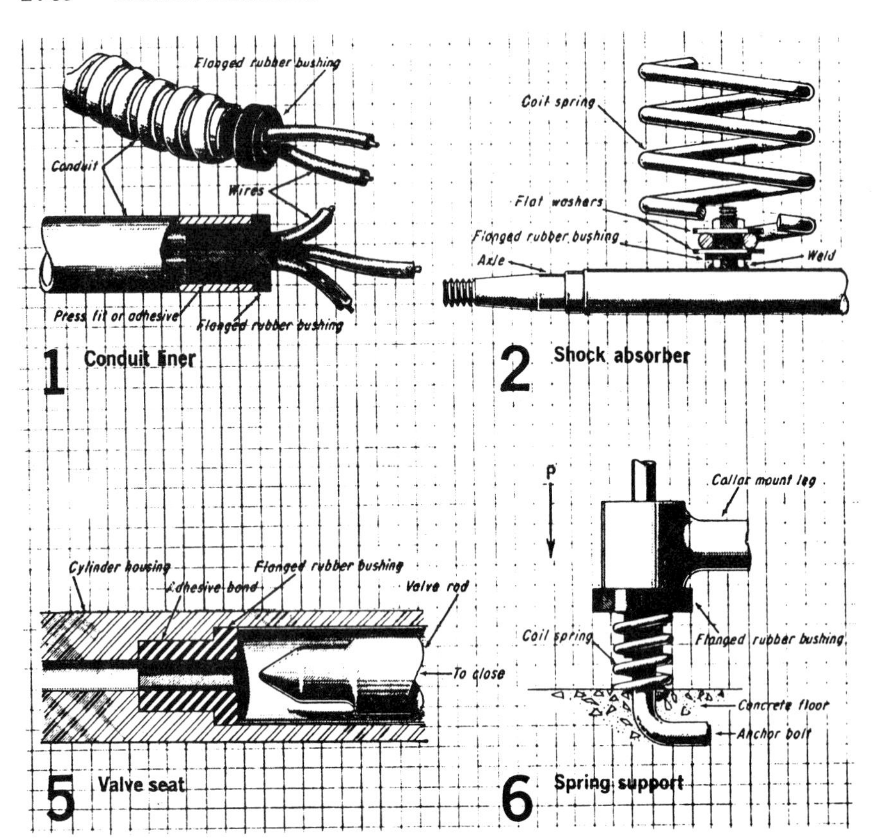

FIG. 24-16 Ideas for flanged rubber bushings are simple, inexpensive, and often overlooked. Check your design for places where rubber bushings may be a solution to a design problem. (*From R. O. Parmley, "Seven Creative Ideas for Flanged Rubber Bushings," Product Engineering, Feb. 14, 1966. Reprinted by permission, Morgan-Grampian Publishing Co.*)

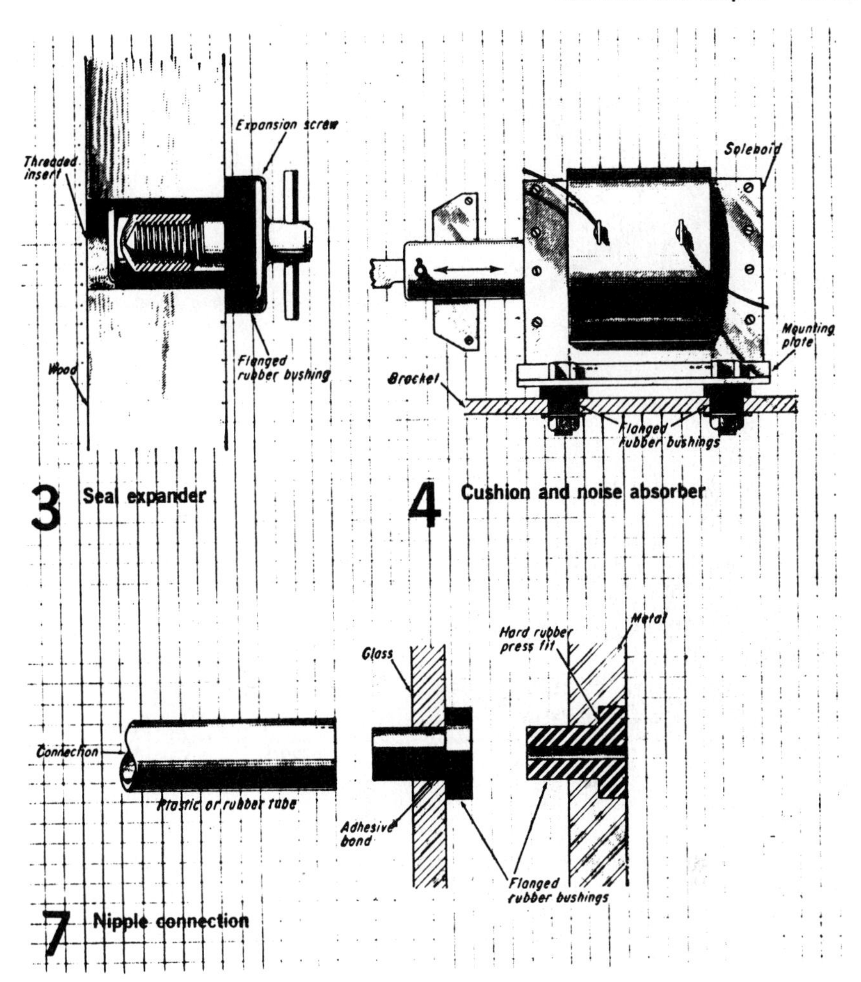

FIG. 24-16 (*continued*)

24-5 Grommets and Bumpers

Rubber grommets and bumpers have a wide range of uses as key components.

Figure 24-17 reveals eight unusual applications of rubber grommets, ranging from seals to shock absorbers.

Rubber mushroom bumpers also serve a broad range of applications, as shown in Fig. 24-18.

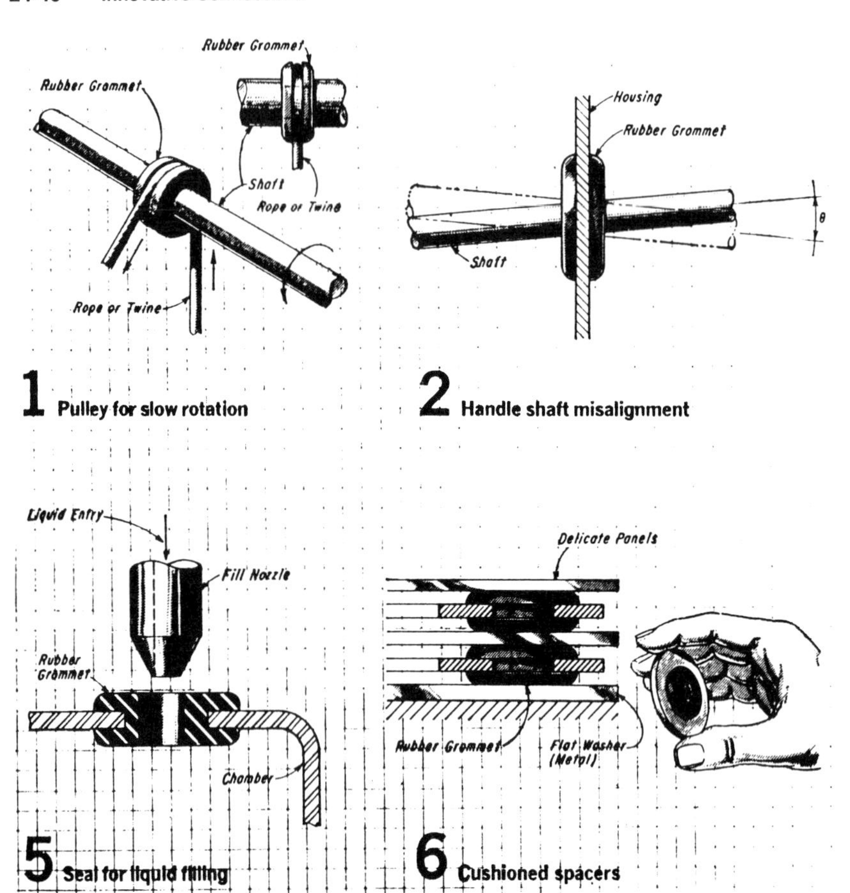

FIG. 24-17 A fresh look at rubber grommets. A small component that is often neglected in the details of a design is shown here in eight unusual applications. (*From R. O. Parmley, "A Fresh Look at Rubber Grommets," Product Engineering, Jan. 3, 1966. Reprinted by permission, Morgan-Grampian Publishing Co.*)

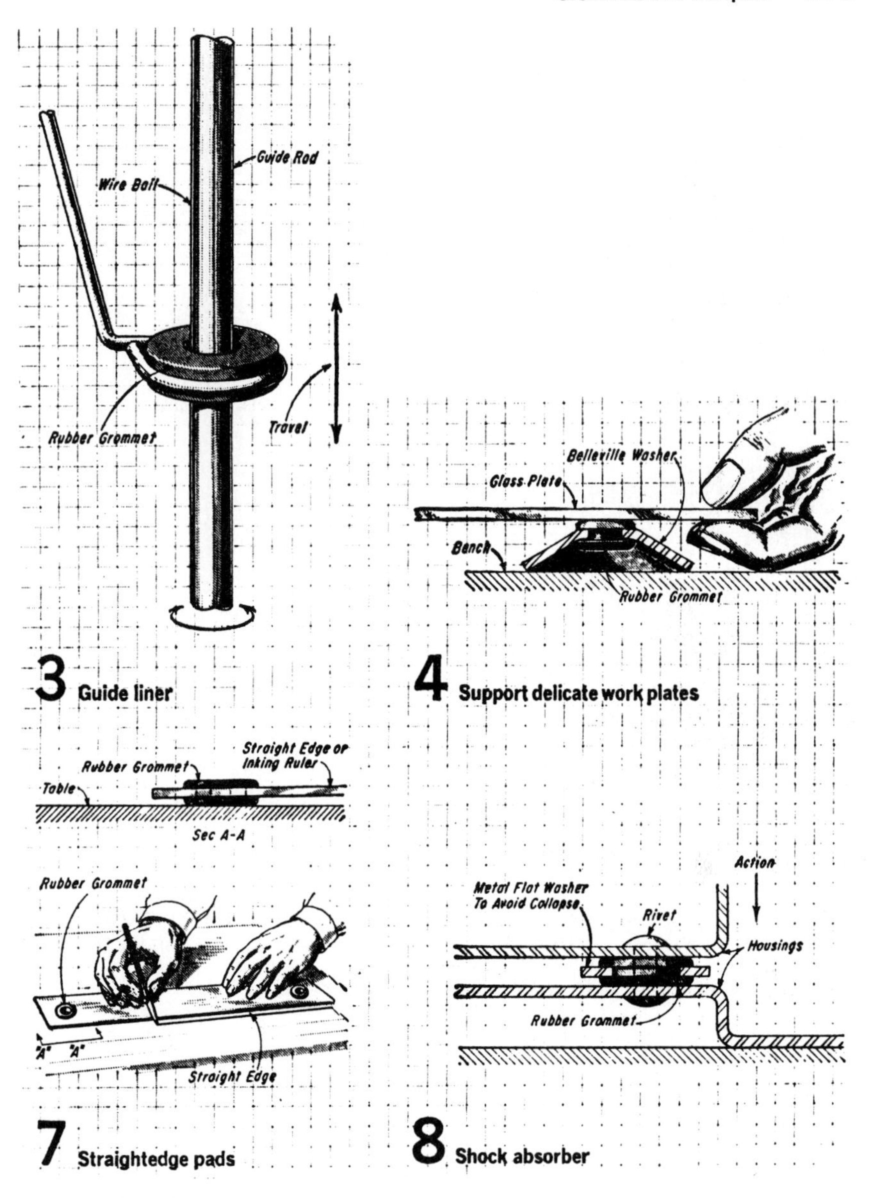

FIG. 24-17 (*continued*)

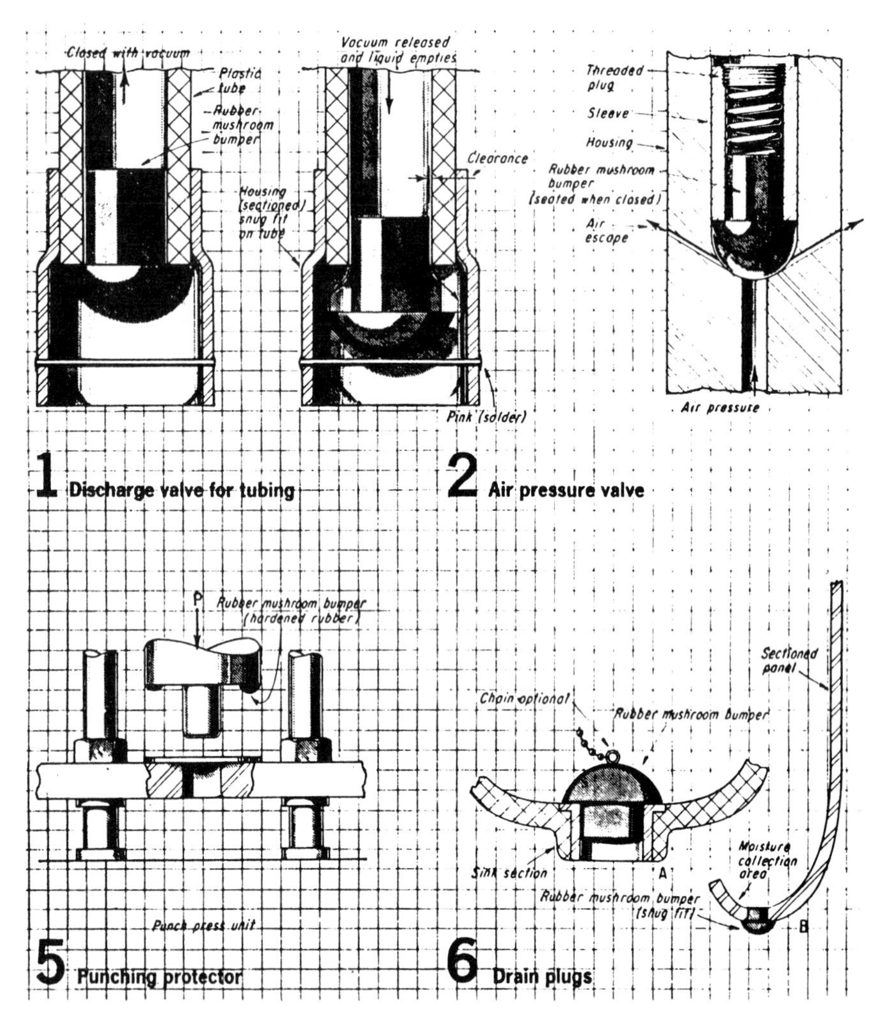

FIG. 24-18 Rubber mushroom bumpers. High energy absorption at low cost is the way mushroom bumpers are usually billed. But they have other uses; here are seven that are rather unconventional. (*From R. O. Parmley, "Odd Jobs for Rubber Mushroom Bumpers," Product Engineering, Jan. 31, 1966. Reprinted by permission, Morgan-Grampian Publishing Co.*)

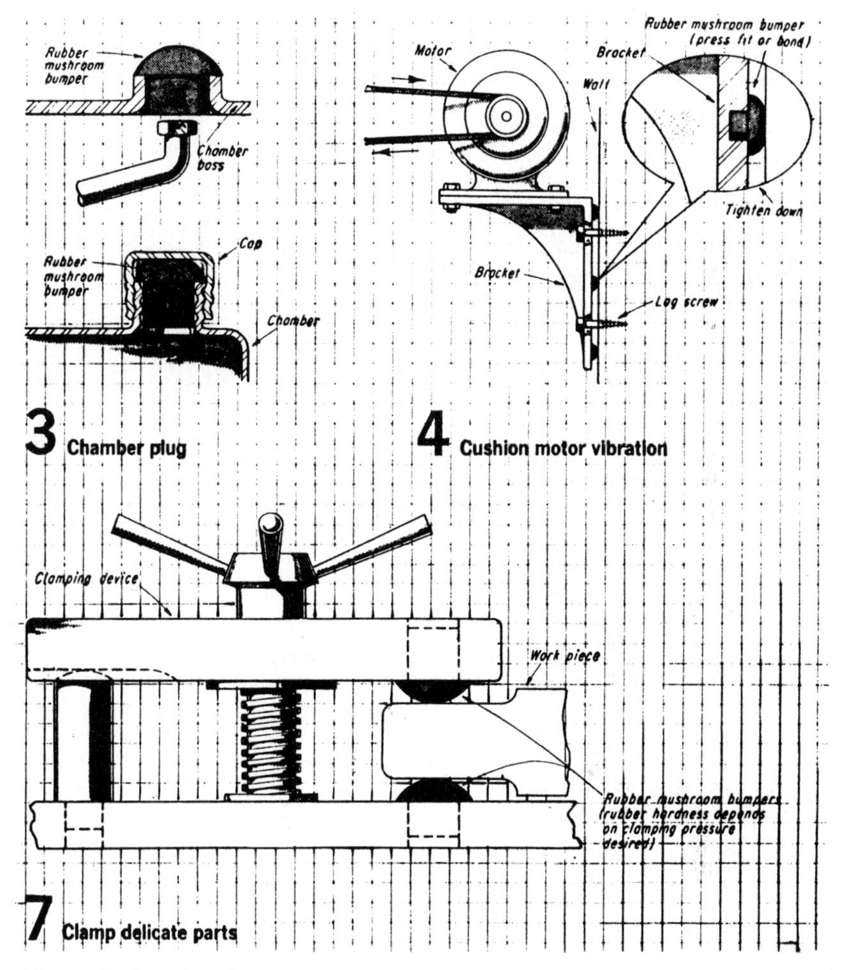

FIG. 24-18 (*continued*)

24-6 O Rings

Rubber O-rings are excellent seals, but they may be used in a wide variety of other applications, as shown in Figs. 24-19 thru 24-23.

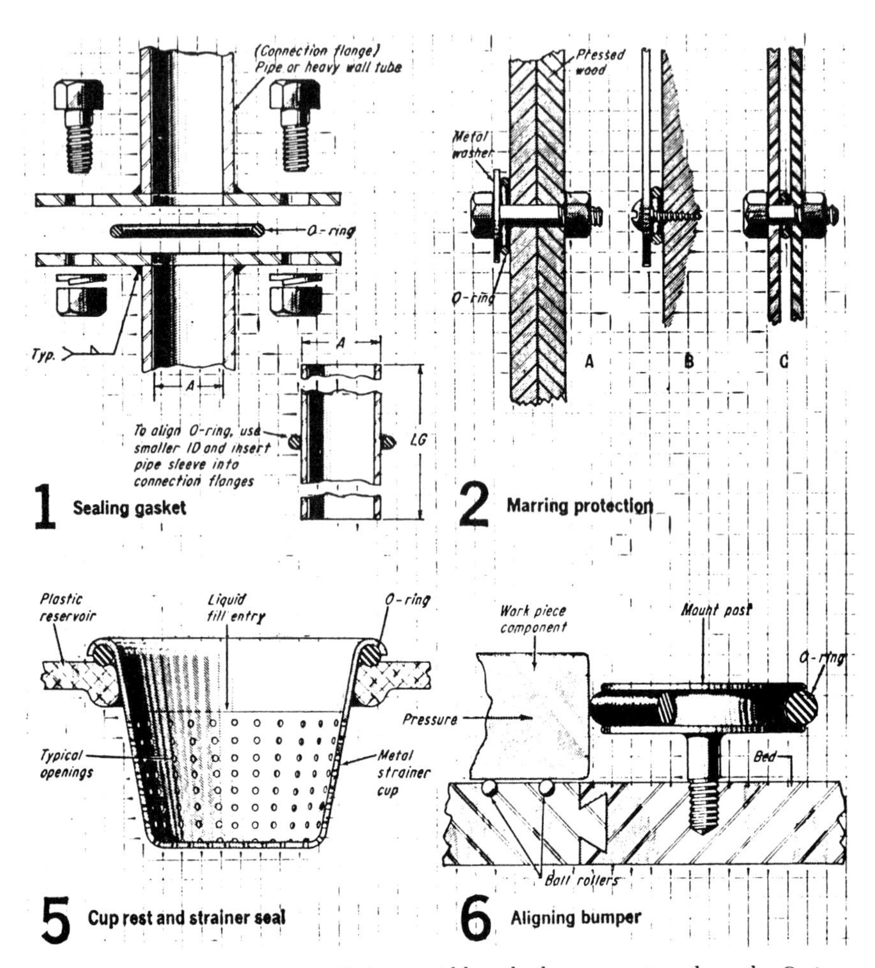

FIG. 24-19 A different look at O-rings. Although they are primarily seals, O-rings can do a variety of other jobs as well as more sophisticated pieces of hardware. (*From R. O. Parmley, "Look At O-Rings Differently," in Product Engineering, Aug. 16, 1965. Reprinted by permission, Morgan-Grampion Publishing Co.*)

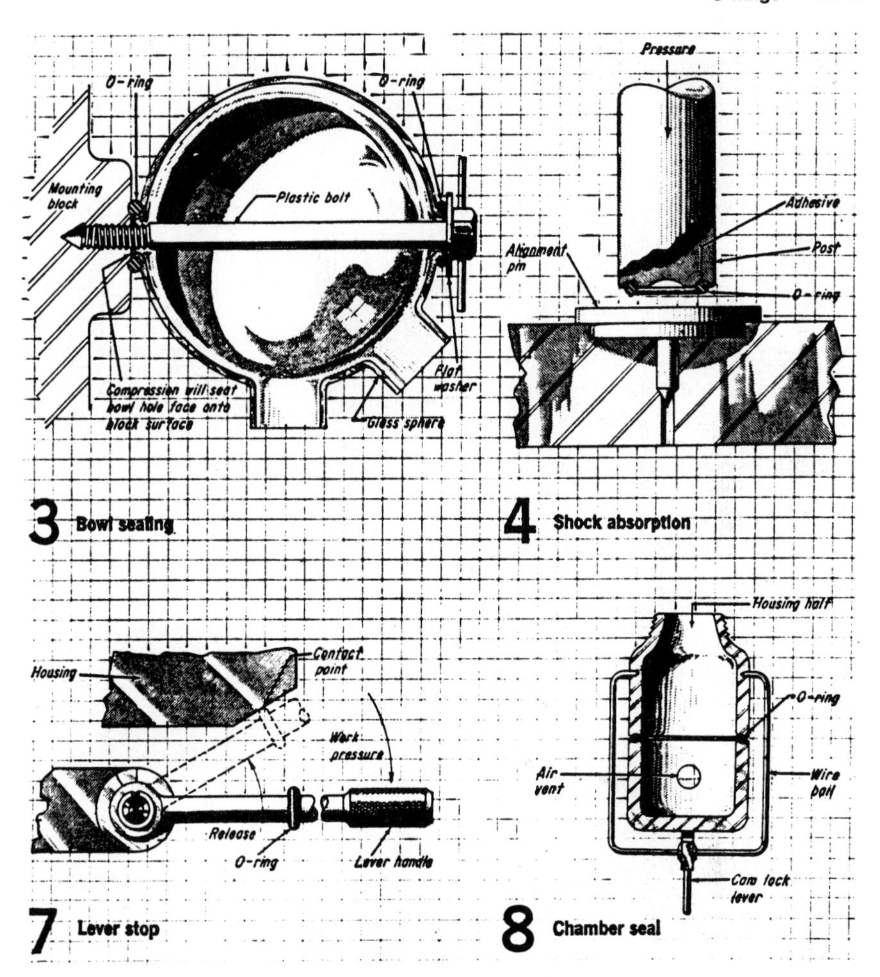

FIG. 24-19 (*continued*)

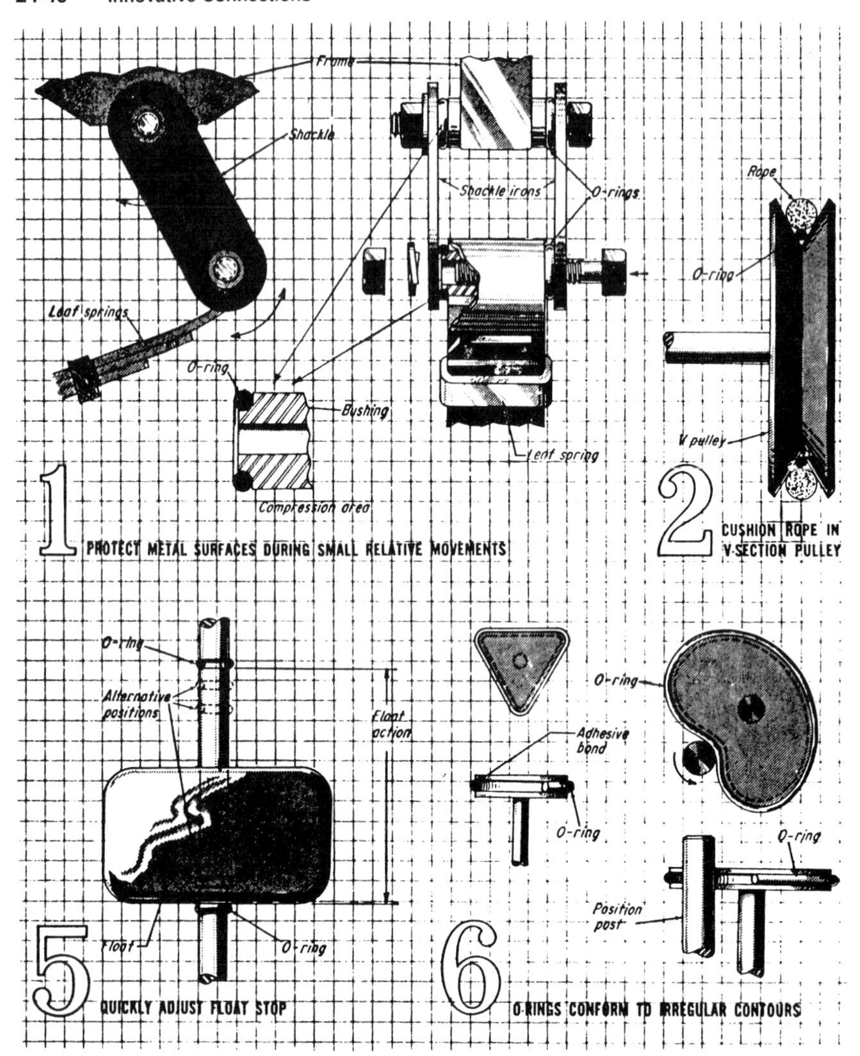

FIG. 24-20 Unusual O-ring applications. Playing many different roles, O-rings can perform as protective devices, hole liners, float stops, and other key design components. *(From R. O. Parmley, "8 Unusual Applications for O-Rings," in Product Engineering, Nov. 25, 1963. Reprinted by permission, Morgan-Grampian Publishing Co.)*

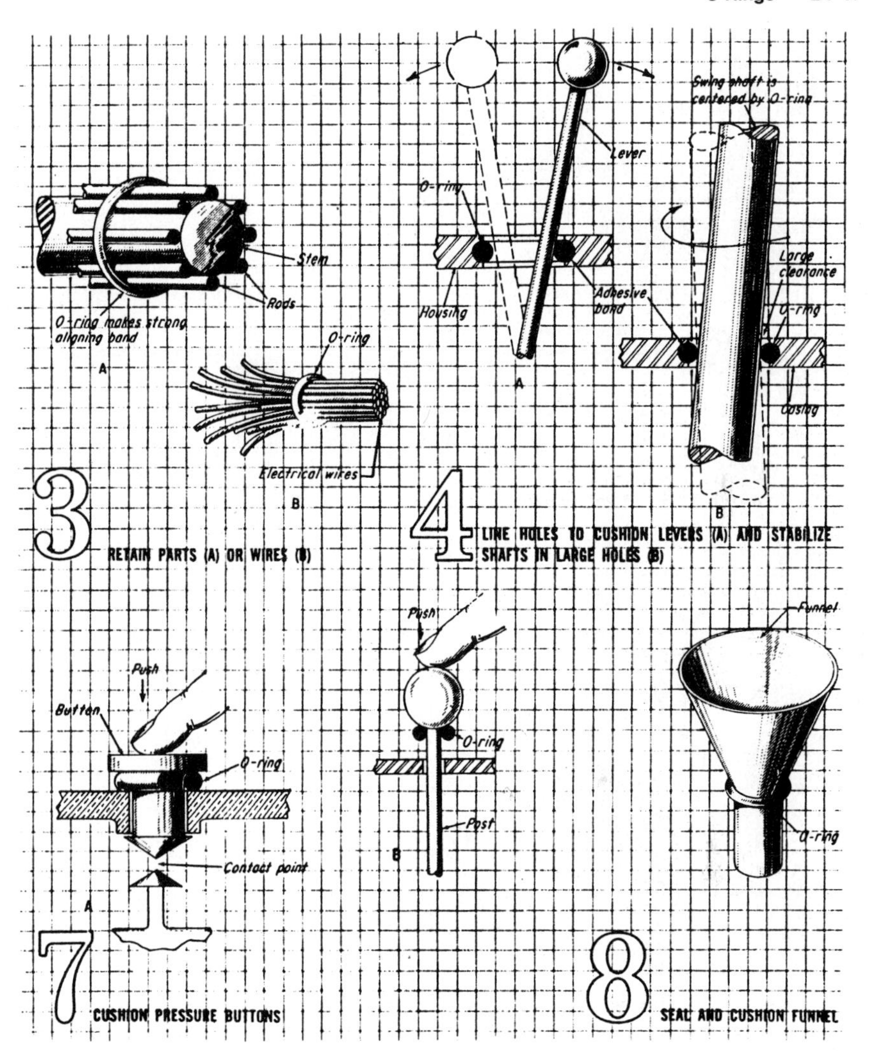

FIG. 24-20 (*continued*)

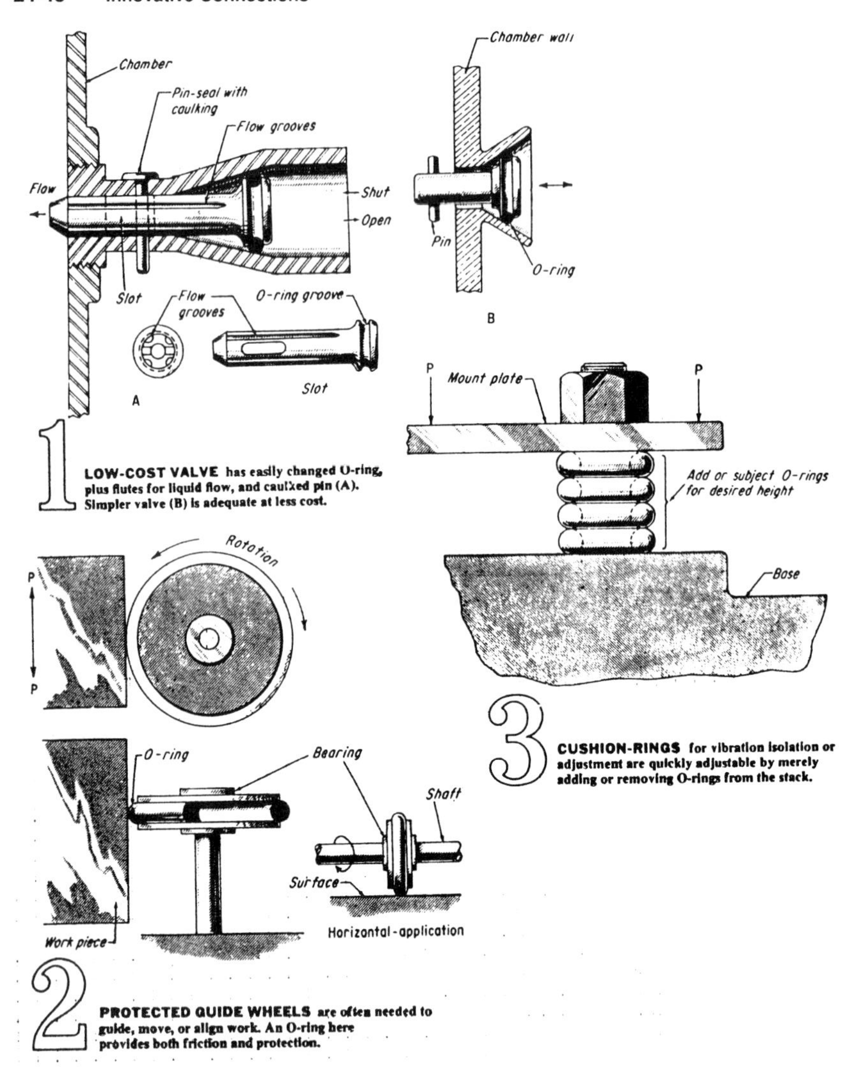

FIG. 24-21 More ways to use O-rings. O-Rings are shown here performing in valves, on guide wheels, and as cushioning components. (*From R. O. Parmley, "7 More Applications for O-Rings," in Product Engineering, Dec. 9, 1963. Reprinted by permission, Morgan-Grampian Publishing Co.*)

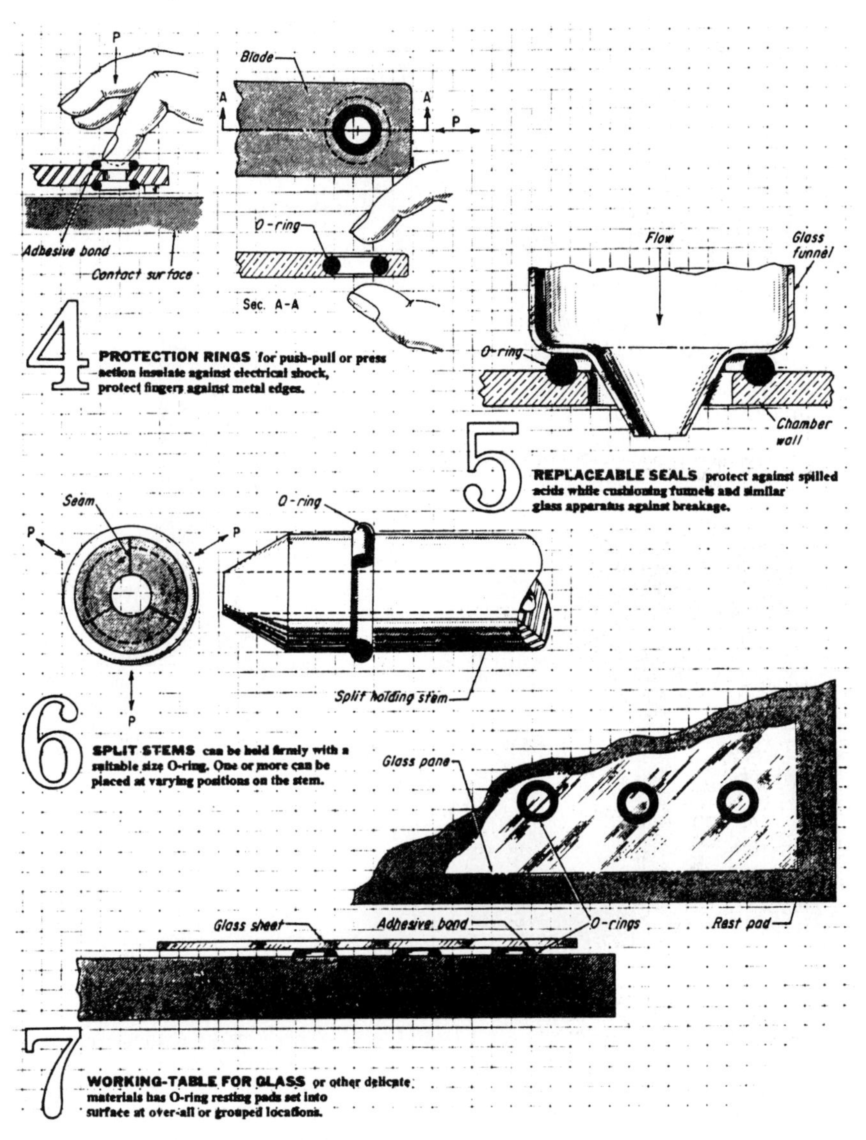
P
Blade
A
A
P
Adhesive bond
Contact surface
O-ring
Sec. A-A
4
PROTECTION RINGS for push-pull or press action insulate against electrical shock, protect fingers against metal edges.
Flow
Glass funnel
O-ring
Chamber wall
5
REPLACEABLE SEALS protect against spilled acids while cushioning funnels and similar glass apparatus against breakage.
Seam
O-ring
P
P
P
Split holding stem
6
SPLIT STEMS can be held firmly with a suitable size O-ring. One or more can be placed at varying positions on the stem.
Glass pane
Glass sheet
Adhesive bond
O-rings
Rest pad
7
WORKING-TABLE FOR GLASS or other delicate materials has O-ring resting pads set into surface at over-all or grouped locations.

FIG. 24-21 (*continued*)

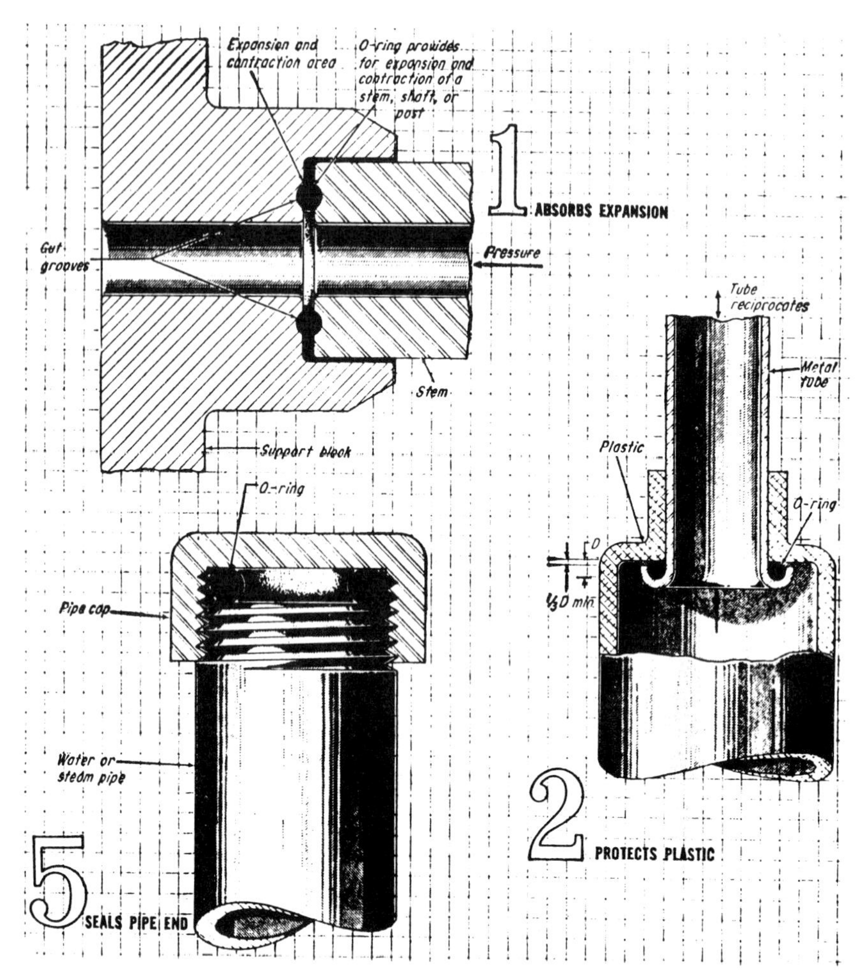

FIG. 24-22 Design problems solved with O-rings. Rubber O-rings provide thermal expansion, protect surfaces, seal pipe ends and connections, and prevent slipping. *(From R. O. Parmley, "O-Rings Solve Design Problems," in Product Engineering, May 11, 1964. Reprinted by permission, Morgan-Grampian Publishing Co.)*

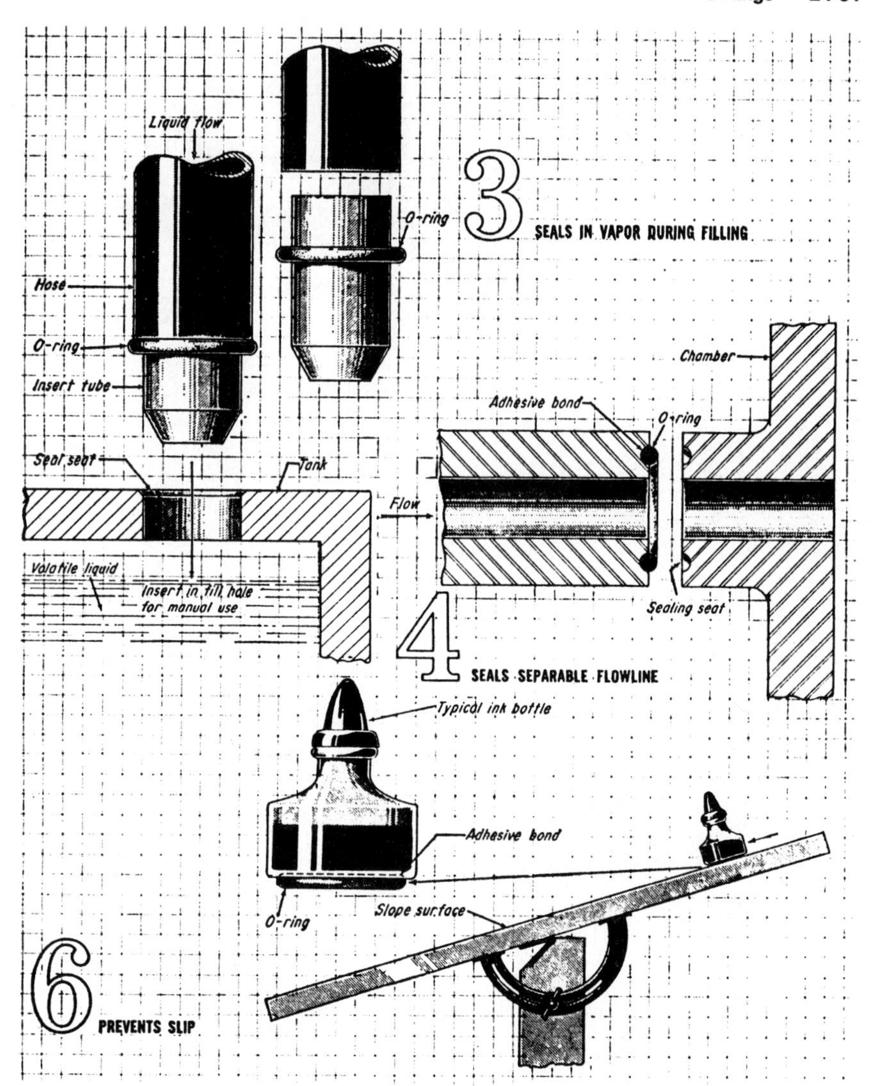

FIG. 24-22 (*continued*)

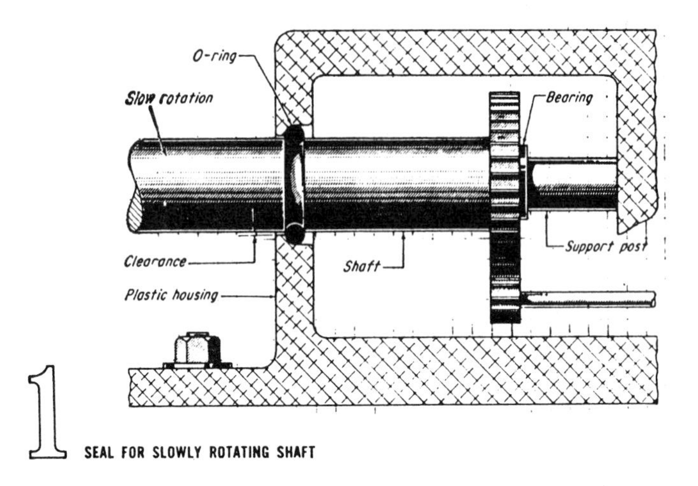

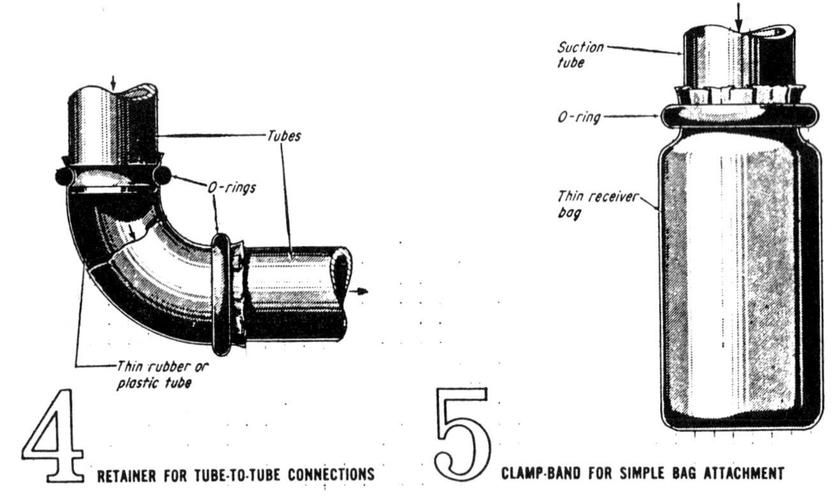

FIG. 24-23 Solve design problems with O-rings. More examples of how rubber O-rings provide seals for shafts, lids, nozzles, and elbows, as well as protect corners and cushion metal surfaces. (*From R. O. Parmley, "O-Rings Solve Design Problems II," in Product Engineering, May 25, 1964. Reprinted by permission, Morgan-Grampian Publishing Co.*)

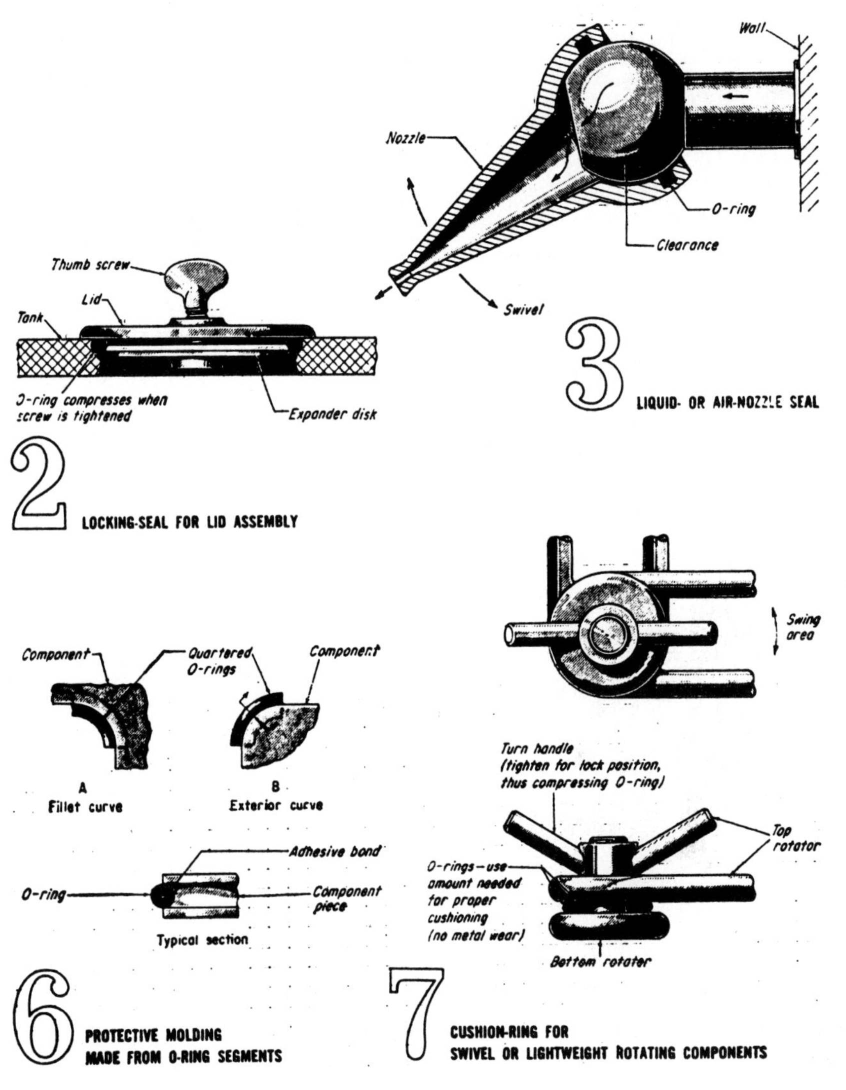

FIG. 24-23 (*continued*)

Section 25

Metrics and Conversion Data*

ROBERT O. PARMLEY, P.E.
President, Morgan & Parmley, Ltd.
Ladysmith, Wisconsin

*This section reprinted with permission from Robert O. Parmley, P.E., *Mechanical Components Handbook,* McGraw-Hill, New York, 1985.

25-1 History Records of measurement date back to approximately 5000 B.C. to ancient civilizations located on the Chaldean plains and areas adjacent to the Nile River. These fledgling attempts at standardization of dimensions involved length units and were directly asscociated with the human body, specifically the thumb, hand, foot, forearm, and pace. Commencing about 2500 B.C., balances were used to weigh gold, as shown on records buried in ancient Egyptian tombs.

The "cubit," a measurement often mentioned in the Bible, is the distance from the outstretched tip of a person's middle finger to the point of the elbow. It had become the principal length unit by 400 B.C. and was somewhat standardized at what is now approximately 460 mm.

The "span" is approximately one-half a cubit; it is measured from the tips or points of the outstretched thumb and little finger.

As civilization progressed, each country introduced its own standard for weights and measures. English history tells us that by A.D. 1500, the English used the following set of conversions:

1 inch = 3 barley corns
1 foot = 12 inches
1 yard = 3 feet
1 span = 9 inches
1 ell = 5 spans
1 pace = 5 feet
1 furlong = 125 paces
1 rod = 5½ yards
1 statute mile = 8 furlongs
1 league = 12 furlongs

25-2 Metric System As world trade became more common, there was an ever-increasing need for a universal standard of measurement based on a logical scientific system.

A commission of French scientists developed the metric system, and France adopted it as their legal system of weights and measures in the year A.D. 1799. Several revision have occurred since that time. In 1960, the present form, known as the "Système International d'Unités" (International system of Units) or SI, was approved.

SI metric has seven basic units of measure and two supplemental units. All other units of measure can be derived from these basic units. (See Table 25-1 and 25-2).

TABLE 25-1 Basic SI Units*

Quantity	Unit
Length	meter (m)
Mass	kilogram (kg)
Time	second (s)
Electric current	ampere (A)
Temperature (thermodynamic)	kelvin (K)
Amount of substance	mole (mol)
Luminous intensity	candela (cd)

* From *Metrication Manual* by Tyler G. Hicks. © 1972 McGraw-Hill Book Company. Used with permission of the publisher.

TABLE 25-2 Prefixes for SI Units*

Multiple and submultiple	Prefix	Symbol
1,000,000,000,000 = 10^{12}	tera	T
1,000,000,000 = 10^{9}	giga	G
1,000,000 = 10^{6}	mega	M
1,000 = 10^{3}	kilo	k
100 = 10^{2}	hecto	h
10 = 10	deka	da
0.1 = 10^{-1}	deci	d
0.01 = 10^{-2}	centi	c
0.001 = 10^{-3}	milli	m
0.000 001 = 10^{-6}	micro	μ
0.000 000 001 = 10^{-9}	nano	n
0.000 000 000 001 = 10^{-12}	pico	p
0.000 000 000 000 001 = 10^{-15}	femto	f
0.000 000 000 000 000 001 = 10^{-18}	atto	a

* From *Metrication Manual* by Tyler G. Hicks. © 1972 McGraw-Hill Book Company. Used with permission of the publisher.

25-3 Simpler Language The metric system is a much simpler language; however, conversion from the common U.S. Customary System is very expensive and time-consuming for American engineering disciplines.

The seven basic SI units previously mentioned are relatively easy to use and convert to, or from, common English units. However, in the case of derived units, those which are derived from one or more basic SI units, there can be some difficulty. Table 25-43 includes the most common derived units used in SI.

25-4 Conversion Factors Table 25-4 gives the definitions of various units of measure that are exact numerical multiples of coherent SI units, and provides multiplication factors for converting numbers and miscellaneous units to corresponding new numbers and SI units.

The first two digits of each numerical entry represent a power of 10. An asterisk follows each number that expresses an exact definition. For example, the entry "−02 2.54*" expresses the fact that 1 inch = 2.54×10^{-2} meter, exactly, by definition. Most of the definitions are extracted from National Bureau of Standards documents. Numbers not followed by an asterisk are only approximate or are the results of physical measurements. The conversion factors are listed alphabetically and by physical quantity.

The listing by physical quantity includes only relationships which are frequently encountered and deliberately omits the many combinations of units which are used for more specialized purposes. Conversion factors for combinations of units are easily generated from numbers given in the alphabetical listing by the technique of direct substitution or by other well-known rules for manipulating units. These units are adequately discussed in many science and engineering textbooks and are not repeated here.

TABLE 25-3 Derived Units of the International System*

Quantity	Name of unit	Unit symbol or abbreviation, where differing from basic form	Unit expressed in terms of basic or supplementary units†
Area	square meter		m^2
Volume	cubic meter		m^3
Frequency	hertz, cycle per second‡	Hz	s^{-1}
Density	kilogram per cubic meter		kg/m^3
Velocity	meter per second		m/s
Angular velocity	radian per second		rad/s
Acceleration	meter per second squared		m/s^2
Angular acceleration	radian per second squared		rad/s^2
Volumetric flow rate	cubic meter per second		m^3/s
Force	newton	N	$kg \cdot m/s^2$
Surface tension	newton per meter, joule per square meter	N/m, J/m^2	kg/s^2
Pressure	newton per square meter, pascal‡	N/m^2, Pa‡	$kg/m \cdot s^2$
Viscosity, dynamic	newton-second per square meter, poiseuille‡	$N \cdot s/m^2$, Pl‡	$kg/m \cdot s$
Viscosity, kinematic	meter squared per second		m^2/s
Work, torque, energy, quantity of heat	joule, newton-meter, watt-second	J, N·m, W·s	$kg \cdot m^2/s^2$

Power, heat flux	watt, joule per second	W, J/s	$kg \cdot m^2/s^3$
Heat flux density	watt per square meter	W/m^2	kg/s^3
Volumetric heat release rate	watt per cubic meter	W/m^3	$kg/m \cdot s^3$
Heat transfer coefficient	watt per square meter-kelvin	$W/m^2 \cdot K$	$kg/s^3 \cdot K$
Heat capacity (specific)	joule per kilogram-kelvin	$J/kg \cdot K$	$m^2/s^2 \cdot K$
Capacity rate	watt per kelvin	W/K	$kg \cdot m^2/s^3 \cdot K$
Thermal conductivity	watt per meter-kelvin	$W/m \cdot deg$, $J \cdot m/s \cdot m^2 \cdot K$	$kg \cdot m/s^3 \cdot K$
Quantity of electricity	coulomb	C	$A \cdot s$
Electromotive force	volt	V,W/A	$kg \cdot m^2/A \cdot s^3$
Electric field strength	volt per meter		V/m
Electric resistance	ohm	Ω, V/A	$kg \cdot m^2/A^2 \cdot s^3$
Electric conductivity	ampere per volt-meter	$A/V \cdot m$	$A^2s^3/kg \cdot m^3$
Electric capacitance	farad	$F, A \cdot s/V$	$A^3s^1/kg \cdot m^2$
Magnetic flux	weber	$Wb, V \cdot s$	$kg \cdot m^2/A \cdot s^2$
Inductance	henry	$H, V \cdot s/A$	$kg \cdot m^2/A^2s^2$
Magnetic permeability	henry per meter	H/m	$kg \cdot m/A^2s^2$
Magnetic flux density	tesla, weber per square meter	$T, Wb/m^2$	$kg/A \cdot s^2$
Magnetic field strength	ampere per meter		A/m
Magnetomotive force	ampere		A
Luminous flux	lumen	lm	cd sr
Luminance	candela per square meter		cd/m^2
Illumination	lux, lumen per square meter	$lx, lm/m^2$	$cd \cdot sr/m^2$

* From *Metrication Manual* by Tyler G. Hicks. © 1972 McGraw-Hill Book Company. Used with permission of the publisher.

† Supplementary units are plane angle, radian (rad), solid angle, steradian (sr).

‡ Not used in all countries.

TABLE 25-4 Conversion Factors as Extracted Multiples of SI Units†

To convert from	To	Multiply by
abampere	ampere	+01 1.00*
abcoulomb	coulomb	+01 1.00*
abfarad	farad	+09 1.00*
abhenry	henry	−09 1.00*
abmho	siemens (mho)	+09 1.00*
abohm	ohm	−09 1.00*
abvolt	volt	−08 1.00*
acre	$meter^2$	+03 4.046 873
ampere (international of 1948)	ampere	−01 9.998 35
angstrom	meter	−10 1.00*
are	$meter^2$	+02 1.00*
astronomical unit	meter	+11 1.495 979
atmosphere	pascal ($newton/meter^2$)	+05 1.013 250*
bar	pascal ($newton/meter^2$)	+05 1.00*
barn	$meter^2$	−28 1.00*
barrel (petroleum, 42 gallons)	$meter^3$	−01 1.589 873
barye	$newton/meter^2$	−01 1.00*
British thermal unit (ISO/TC 12)	joule	+03 1.055 06
British thermal unit (International Steam Table)	joule	+03 1.055 04
British thermal unit (mean)	joule	+03 1.055 87
British thermal unit (thermochemical)	joule	+03 1.054 350 264 488
British thermal unit (39°F)	joule	+03 1.059 67
British thermal unit (60°F)	joule	+03 1.054 68
bushel (U.S.)	$meter^3$	−02 3.523 907 016 688*
cable	meter	+02 2.194 56*
caliber	meter	−04 2.54*
calorie (International Steam Table)	joule	+00 4.1868
calorie (mean)	joule	+00 4.190 02
calorie (thermochemical)	joule	+00 4.184*
calorie (15°C)	joule	+00 4.185 80
calorie (20°C)	joule	+00 4.181 90
calorie (kilogram, International Steam Table)	joule	+03 4.1868
calorie (kilogram, mean)	joule	+03 4.190 02
calorie (kilogram, thermochemical)	joule	+03 4.184*
carat (metric)	kilogram	−04 2.00*
Celsius (temperature)	kelvin	$t_K = t_C = 273.15$
centimeter of mercury (0°C)	$newton/meter^2$	+03 1.333 22
centimeter of water (4°C)	$newton/meter^2$	+01 9.806 38
chain (engineer or ramden)	meter	+01 3.048*
chain (surveyor or gunter)	meter	+01 2.011 68*
circular mil	$meter^2$	−10 5.067 074 8
cord	$meter^3$	+00 3.624 556 3
coulomb (international of 1948)	coulomb	−01 9.998 35
cubit	meter	−01 4.572*

† From *Metrication Manual* by Tyler G. Hicks. © 1972 McGraw-Hill Book Company. Used with permission of the publisher.

TABLE 25-4 (*Continued*)

To convert from	To	Multiply by
cup	meter^3	−04 2.365 882 365*
curie	disintegration/second	+10 3.70*
day (mean solar)	second (mean solar)	+04 8.64*
day (sidereal)	second (mean solar)	+04 8.616 409 0
degree (angle)	radian	−02 1.745 329 251 994 3
denier (international)	kilogram/meter	−07 1.00*
dram (avoirdupois)	kilogram	−03 1.771 845 195 312 5*
dram (troy or apothecary)	kilogram	−03 3.887 934 6*
dram (U.S. fluid)	meter^3	−06 3.696 691 195 312 5*
dyne	newton	−05 1.00*
electron-volt	joule	−19 1.602 10
erg	joule	−07 1.00*
Fahrenheit (temperature)	kelvin	$t_K = (5/9)(t_F + 459.67)$
Fahrenheit (temperature)	Celsius	$t_C = (5/9)(t_F - 32)$
farad (international of 1948)	farad	−01 9.995 05
faraday (based on carbon 12)	coulomb	+04 9.648 70
faraday (chemical)	coulomb	+04 9.649 57
faraday (physical)	coulomb	+04 9.652 19
fathom	meter	+00 1.828 8*
fermi (femtometer)	meter	−15 1.00*
fluid ounce (U.S.)	meter^3	−05 2.957 352 956 25*
foot	meter	−01 3.048*
foot (U.S. survey)	meter	+00 1200/3937*
foot (U.S. survey)	meter	−01 3.048 006 096
foot of water (39.2°F)	newton/meter^2	+03 2.988 98
foot-candle	lumen/meter^2	+01 1.076 391 0
foot-lambert	candela/meter^2	+00 3.426 259
furlong	meter	+02 2.011 68*
gal (galileo)	meter/second^2	−02 1.00*
gallon (U.K. liquid)	meter^3	−03 4.546 087
gallon (U.S. dry)	meter^3	−03 4.404 883 770 86*
gallon (U.S. liquid)	meter^3	−03 3.785 411 784*
gamma	tesla	−09 1.00*
gauss	tesla	−04 1.00*
gilbert	ampere-turn	−01 7.957 747 2
gill (U.K.)	meter^3	−04 1.420 652
gill (U.S.)	meter^3	−04 1.182 941 2
grad	degree (angular)	−01 9.00*
grad	radian	−02 1.570 796 3
grain	kilogram	−05 6.479 891*
gram	kilogram	−03 1.00*
hand	meter	−01 1.016*
hectare	meter^2	+04 1.00*
henry (international of 1948)	henry	+00 1.000 495
hogshead (U.S.)	meter^3	−01 2.384 809 423 92*
horsepower (550 foot lbf/second)	watt	+02 7.456 998 7

TABLE 25-4 (*Continued*)

To convert from	To	Multiply by
horsepower (boiler)	watt	+03 9.809 50
horsepower (electric)	watt	+02 7.46*
horsepower (metric)	watt	+02 7.354 99
horsepower (U.K.)	watt	+02 7.457
horsepower (water)	watt	+02 7.460 43
hour (mean solar)	second (mean solar)	+03 3.60*
hour (sidereal)	second (mean solar)	+03 3.590 170 4
hundredweight (long)	kilogram	+01 5.080 234 544*
hundredweight (short)	kilogram	+01 4.535 923 7*
inch	meter	−02 2.54*
inch of mercury (32°F)	pascal (newton/meter^2)	+03 3.386 38
inch of mercury (60°F)	pascal (newton/meter^2)	+03 3.376 85
inch of water (39.2°F)	pascal (newton/meter^2)	+02 2.490 82
inch of water (60°F)	pascal (newton/meter^2)	+02 2.488 4
joule (international of 1948)	joule	+00 1.000 165
kayser	1/meter	+02 1.00*
kilocalorie (International Steam Table)	joule	+03 4.186 74
kilocalorie (mean)	joule	+03 4.190 02
kilocalorie (thermochemical)	joule	+03 4.184*
kilogram mass	kilogram	+00 1.00*
kilogram force (kgf)	newton	+00 9.806 65*
kilopond force	newton	+00 9.806 65*
kip	newton	+03 4.448 221 615 260 5*
knot (international)	meter/second	−01 5.144 444 444
lambert	candela/meter^2	+04 $1/\pi$*
lambert	candela/meter^2	+03 3.183 098 8
langley	joule/meter^2	+04 4.184*
lbf (pound force, avoirdupois)	newton	+00 4.448 221 615 260 5*
lbm (pound mass, avoirdupois)	kilogram	−01 4.535 923 7*
league (British nautical)	meter	+03 5.559 552*
league (international nautical)	meter	+03 5.556*
league (statute)	meter	+03 4.828 032*
light year	meter	+15 9.460 55
link (engineer or ramden)	meter	−01 3.048*
link (surveyor or gunter)	meter	−01 2.011 68*
liter	meter^3	−03 1.00*
lux	lumen/meter^2	+00 1.00*
maxwell	weber	−08 1.00*
meter	wavelengths Kr 86	+06 1.650 763 73*
micron	meter	−06 1.00*
mil	meter	−05 2.54*
mile (U.S. statute)	meter	+03 1.609 344*

TABLE 25-4 *(Continued)*

To convert from	To	Multiply by
mile (U.K. nautical)	meter	+03 1.853 184*
mile (international nautical)	meter	+03 1.852*
mile (U.S. nautical)	meter	+03 1.852*
millibar	newton/meter^2	+02 1.00*
millimeter of mercury (0°C)	newton/meter^2	+02 1.333 224
minute (angle)	radian	−04 2.908 882 086 66
minute (mean solar)	second (mean solar)	+01 6.00*
minute (sidereal)	second (mean solar)	+01 5.983 617 4
month (mean calendar)	second (mean solar)	+06 2.628*
nautical mile (international)	meter	+03 1.852*
nautical mile (U.S.)	meter	+03 1.852*
nautical mile (U.K.)	meter	+03 1.853 184*
oersted	ampere/meter	+01 7.957 747 2
ohm (international of 1948)	ohm	+00 1.000 495
ounce force (avoirdupois)	newton	−01 2.780 138 5
ounce mass (avoirdupois)	kilogram	−02 2.834 952 312 5*
ounce mass (troy or apothecary)	kilogram	−02 3.110 347 68*
ounce (U.S. fluid)	meter^3	−05 2.957 352 956 25*
pace	meter	−01 7.62*
parsec	meter	+16 3.083 74
pascal	newton/meter^2	+00 1.00*
peck (U.S.)	meter^3	−03 8.809 767 541 72*
pennyweight	kilogram	−03 1.555 173 84*
perch	meter	+00 5.0292*
phot	lumen/meter^2	+04 1.00
pica (printers)	meter	−03 4.217 517 6*
pint (U.S. dry)	meter^3	−04 5.506 104 713 575*
pint (U.S. liquid)	meter^3	−04 4.731 764 73*
point (printers)	meter	−04 3.514 598*
poise	newton-second/meter^2	−01 1.00*
pole	meter	+00 5.0292*
pound force (lbf avoirdupois)	newton	+00 4.448 221 615 260 5*
pound mass (lbm avoirdupois)	kilogram	−01 4.535 923 7*
pound mass (troy or apothecary)	kilogram	−01 3.732 417 216*
poundal	newton	−01 1.382 549 543 76*
quart (U.S. dry)	meter^3	−03 1.101 220 942 715*
quart (U.S. liquid)	meter^3	−04 9.463 529 5
rad (radiation dose absorbed)	joule/kilogram	−02 1.00*
Rankine (temperature)	kelvin	$t_K = (5/9)t_R$
rayleigh (rate of photon emission)	1/second-meter^2	+10 1.00*
rhe	meter^2/newton-second	+01 1.00*
rod	meter	+00 5.0292*
roentgen	coulomb/kilogram	−04 2.579 76*
rutherford	disintegration/second	+06 1.00*

TABLE 25-4 (*Continued*)

To convert from	To	Multiply by
second (angle)	radian	−06 4.848 136 811
second (ephemeris)	second	+00 1.000 000 000
second (mean solar)	second (ephemeris)	Consult American Ephemeris and Nautical Almanac
second (sidereal)	second (mean solar)	−01 9.972 695 7
section	meter^2	+06 2.589 988 110 336*
scruple (apothecary)	kilogram	−03 1.295 978 2*
shake	second	−08 1.00
skein	meter	+02 1.097 28*
slug	kilogram	+01 1.459 390 29
span	meter	−01 2.286*
statampere	ampere	−10 3.335 640
statcoulomb	coulomb	−10 3.335 640
statfarad	farad	−12 1.112 650
stathenry	henry	+11 8.987 554
statmho	mho	−12 1.112 650
statohm	ohm	+11 8.987 554
statute mile (U.S.)	meter	+03 1.609 344*
statvolt	volt	+02 2.997 925
stere	meter^3	+00 1.00*
stilb	candela/meter^2	+04 1.00
stoke	meter^2/second	−04 1.00*
tablespoon	meter^3	−05 1.478 676 478 125*
teaspoon	meter^3	−06 4.928 921 593 75*
ton (assay)	kilogram	−02 2.916 666 6
ton (long)	kilogram	+03 1.016 046 908 8*
ton (metric)	kilogram	+03 1.00*
ton (explosive energy of one ton of TNT)	joule	+09 4.184
ton (register)	meter^3	+00 2.831 684 659 2*
ton (short, 2000 pound)	kilogram	+02 9.071 847 4*
tonne	kilogram	+03 1.00*
torr (0°C)	newton/meter^2	+02 1.333 22
township	meter^2	+07 9.323 957 2
unit pole	weber	−07 1.256 637
volt (international of 1948)	volt	+00 1.000 330
watt (international of 1948)	watt	+00 1.000 165
yard	meter	−01 9.144*
year (calendar)	second (mean solar)	+07 3.1536*
year (sidereal)	second (mean solar)	+07 3.155 815 0
year (tropical)	second (mean solar)	+07 3.155 692 6
year 1900, tropical, Jan., day 0, hour 12	second (ephemeris)	+07 3.155 692 597 47*
year 1900, tropical, Jan., day 0, hour 12	second	+07 3.155 692 597 47

TABLE 25-4 (*Continued*)

To convert from	To	Multiply by
LISTING BY PHYSICAL QUANTITY		
Acceleration		
foot/second^2	meter/second^2	−01 3.048*
free fall, standard	meter/second^2	+00 9.806 65*
gal (galileo)	meter/second^2	−02 1.00*
inch/second^2	meter/second^2	−02 2.54*
Area		
acre	meter^2	+03 4.046 856 422 4*
are	meter^2	+02 1.00*
barn	meter^2	−28 1.00*
circular mil	meter^2	−10 5.067 074 8
foot^2	meter^2	−02 9.290 304*
hectare	meter^2	+04 1.00*
inch^2	meter^2	−04 6.4516*
mile^2 (U.S. statute)	meter^2	+06 2.589 988 110 336*
section	meter^2	+06 2.589 988 110 336*
township	meter^2	+07 9.323 957 2
yard^2	meter^2	−01 8.361 273 6*
Density		
gram/centimeter^3	kilogram/meter^3	+03 1.00*
lbm/inch^3	kilogram/meter^3	+04 2.767 990 5
lbm/foot^3	kilogram/meter^3	+01 1.601 846 3
slug/foot^3	kilogram/meter^3	+02 5.153 79
Energy		
British thermal unit (ISO/TC 12)	joule	+03 1.055 06
British thermal unit (International Steam Table)	joule	+03 1.055 04
British thermal unit (mean)	joule	+03 1.055 87
British thermal unit (thermochemical)	joule	+03 1.054 350 264 488
British thermal unit (39°F)	joule	+03 1.059 67
British thermal unit (60°F)	joule	+03 1.054 68
calorie (International Steam Table)	joule	+00 4.1868
calorie (mean)	joule	+00 4.190 02
calorie (thermochemical)	joule	+00 4.184*
calorie (15°C)	joule	+00 4.185 80
calorie (20°C)	joule	+00 4.181 90
calorie (kilogram, International Steam Table)	joule	+03 4.1868
calorie (kilogram, mean)	joule	+03 4.190 02
calorie (kilogram, thermochemical)	joule	+03 4.184*

TABLE 25-4 *(Continued)*

To convert from	To	Multiply by
electron-volt	joule	−19 1.602 10
erg	joule	−07 1.00*
foot-lbf	joule	+00 1.355 817 9
foot-poundal	joule	−02 4.214 011 0
joule (international of 1948)	joule	+00 1.000 165
kilocalorie (International Steam Table)	joule	+03 4.1868
kilocalorie (mean)	joule	+03 4.190 02
kilocalorie (thermochemical)	joule	+03 4.184*
kilowatt-hour	joule	+06 3.60*
kilowatt-hour (international of 1948)	joule	+06 3.600 59
ton (nuclear equivalent of TNT)	joule	+09 4.20
watt-hour	joule	+03 3.60*
	Energy/area time	
Btu (thermochemical)/foot²-second	watt/meter²	+04 1.134 893 1
Btu (thermochemical)/foot²-minute	watt/meter²	+02 1.891 488 5
Btu (thermochemical)/foot²-hour	watt/meter²	+00 3.152 480 8
Btu (thermochemical)/inch²-second	watt/meter²	+06 1.634 246 2
calorie (thermochemical)/ centimeter²-minute	watt/meter²	+02 6.973 333 3
erg/centimeter²-second	watt/meter²	−03 1.00*
watt/centimeter²	watt/meter²	+04 1.00*
	Force	
dyne	newton	−05 1.00*
kilogram force (kgf)	newton	+00 9.806 65*
kilopond force	newton	+00 9.806 65*
kip	newton	+03 4.448 221 615 260 5*
lbf (pound force, avoirdupois)	newton	+00 4.448 221 615 260 5*
ounce force (avoirdupois)	newton	−01 2.780 138 5
pound force, lbf (avoirdupois)	newton	+00 4.448 221 615 260 5*
poundal	newton	−01 1.382 549 543 76*
	Length	
angstrom	meter	−10 1.00*
astronomical unit	meter	+11 1.495 978 9
cable	meter	+02 2.194 56*
caliber	meter	−04 2.54*
chain (surveyor or gunter)	meter	+01 2.011 68*
chain (engineer or ramden)	meter	+01 3.048*
cubit	meter	−01 4.572*
fathom	meter	+00 1.8288*
fermi (femtometer)	meter	−15 1.00*

TABLE 25-4 (*Continued*)

To convert from	To	Multiply by
foot	meter	−01 3.048*
foot (U.S. survey)	meter	+00 1200/3937*
foot (U.S. survey)	meter	−01 3.048 006 096
furlong	meter	+02 2.011 68*
hand	meter	−01 1.016*
inch	meter	−02 2.54*
league (U.K. nautical)	meter	+03 5.559 552*
league (international nautical)	meter	+03 5.556*
league (statute)	meter	+03 4.828 032*
light year	meter	+15 9.460 55
link (engineer or ramden)	meter	−01 3.048*
link (surveyor or gunter)	meter	−01 2.011 68*
meter	wavelengths Kr 86	+06 1.650 763 73*
micron	meter	−06 1.00*
mil	meter	−05 2.54*
mile (U.S. statute)	meter	+03 1.609 344*
mile (U.K. nautical)	meter	+03 1.853 184*
mile (international nautical)	meter	+03 1.852*
mile (U.S. nautical)	meter	+03 1.852*
nautical mile (U.K.)	meter	+03 1.853 184*
nautical mile (international)	meter	+03 1.852*
nautical mile (U.S.)	meter	+03 1.852*
pace	meter	−01 7.62*
parsec	meter	+16 3.083 74
perch	meter	+00 5.0292*
pica (printers)	meter	−03 4.217 517 6*
point (printers)	meter	−04 3.514 598*
pole	meter	+00 5.0292*
rod	meter	+00 5.0292*
skein	meter	+02 1.097 28*
span	meter	−01 2.286*
statute mile (U.S.)	meter	+03 1.609 344*
yard	meter	−01 9.144*
	Mass	
carat (metric)	kilogram	−04 2.00*
dram (avoirdupois)	kilogram	−03 1.771 845 195 312 5*
dram (troy or apothecary)	kilogram	−03 3.887 934 6*
grain	kilogram	−05 6.479 891*
gram	kilogram	−03 1.00*
hundredweight (long)	kilogram	+01 5.080 234 544*
hundredweight (short)	kilogram	+01 4.535 923 7*
kgf-second^2-meter (mass)	kilogram	+00 9.806 65*
kilogram mass	kilogram	+00 1.00*
lbm (pound mass, avoirdupois)	kilogram	−01 4.535 923 7*
ounce mass (avoirdupois)	kilogram	−02 2.834 952 312 5*
ounce mass (troy or apothecary)	kilogram	−02 3.110 347 68*

TABLE 25-4 (*Continued*)

To convert from	To	Multiply by
pennyweight	kilogram	−03 1.555 173 84*
pound mass, lbm (avoirdupois)	kilogram	−01 4.535 923 7*
pound mass (troy or apothecary)	kilogram	−01 3.732 417 216*
scruple (apothecary)	kilogram	−03 1.295 978 2*
slug	kilogram	+01 1.459 390 29
ton (assay)	kilogram	−02 2.916 666 6
ton (long)	kilogram	+03 1.016 046 908 8*
ton (metric)	kilogram	+03 1.00*
ton (short, 2000 pound)	kilogram	+02 9.071 847 4*
tonne	kilogram	+03 1.00*
	Power	
Btu (thermochemical)/second	watt	+03 1.054 350 264 488
Btu (thermochemical)/minute	watt	+01 1.757 250 4
calorie (thermochemical)/second	watt	+00 4.184*
calorie (thermochemical)/minute	watt	−02 6.973 333 3
foot-lbf/hour	watt	−04 3.766 161 0
foot-lbf/minute	watt	−02 2.259 696 6
foot-lbf/second	watt	+00 1.355 817 9
horsepower (550 foot lbf/second)	watt	+02 7.456 998 7
horsepower (boiler)	watt	+03 9.809 50
horsepower (electric)	watt	+02 7.46*
horsepower (metric)	watt	+02 7.354 99
horsepower (U.K.)	watt	+02 7.457
horsepower (water)	watt	+02 7.460 43
kilocalorie (thermochemical)/ minute	watt	+01 6.973 333 3
kilocalorie (thermochemical)/ second	watt	+03 4.184*
watt (international of 1948)	watt	+00 1.000 165
	Pressure	
atmosphere	newton/meter^2	+05 1.013 25*
bar	newton/meter^2	+05 1.00*
barye	newton/meter^2	−01 1.00*
centimeter of mercury (0°C)	newton/meter^2	+03 1.333 22
centimeter of water (4°C)	newton/meter^2	+01 9.806 38
dyne/centimeter^2	newton/meter^2	−01 1.00*
foot of water (39.2°F)	newton/meter^2	+03 2.988 98
inch of mercury (32°F)	newton/meter^2	+03 3.386 389
inch of mercury (60°F)	newton/meter^2	+03 3.376 85
inch of water (39.2°F)	newton/meter^2	+02 2.490 82
inch of water (60°F)	newton/meter^2	+02 2.4884
kgf centimeter^2	newton/meter^2	+04 9.806 65*
kgf/meter^2	newton/meter^2	+00 9.806 65*
lbf/foot^2	newton/meter^2	+01 4.788 025 8
lbf/inch^2(psi)	newton/meter^2	+03 6.894 757 2
millibar	newton/meter^2	+02 1.00*

TABLE 25-4 (*Continued*)

To convert from	To	Multiply by
millimeter of mercury (0°C)	newton/meter^2	+02 1.333 224
pascal	newton/meter^2	+00 1.00*
psi (lbf/inch^2)	newton/meter^2	+03 6.894 757 2
torr (0°C)	newton/meter^2	+02 1.333 22
	Speed	
foot/hour	meter/second	−05 8.466 666 6
foot/minute	meter/second	−03 5.08*
foot/second	meter/second	−01 3.048*
inch/second	meter/second	−02 2.54*
kilometer/hour	meter/second	−01 2.777 777 8
knot (international)	meter/second	−01 5.144 444 444
mile hour (U.S. statute)	meter/second	−01 4.4704*
mile/minute (U.S. statute)	meter/second	+01 2.682 24*
mile/second (U.S. statute)	meter/second	+03 1.609 344*
	Temperature	
Celsius	kelvin	$t_K = t_C + 273.15$
Fahrenheit	kelvin	$t_K = (5/9)(t_F + 459.67)$
Fahrenheit	Celsius	$t_C = (5/9)(t_F - 32)$
Rankine	kelvin	$t_K = (5/9)t_R$
	Time	
day (mean solar)	second (mean solar)	+04 8.64*
day (sidereal)	second (mean solar)	+04 8.616 409 0
hour (mean solar)	second (mean solar)	+03 3.60*
hour (sidereal)	second (mean solar)	+03 3.590 170 4
minute (mean solar)	second (mean solar)	+01 6.00*
minute (sidereal)	second (mean solar)	+01 5.983 617 4
month (mean calendar)	second (mean solar)	+06 2.628*
second (ephemeris)	second	+00 1.000 000 000
second (mean solar)	second (ephemeris)	Consult American Ephemeris and Nautical Almanac
second (sidereal)	second (mean solar)	−01 9.972 695 7
year (calendar)	second (mean solar)	+07 3.1536*
year (sidereal)	second (mean solar)	+07 3.155 815 0
year (tropical)	second (mean solar)	+07 3.155 692 6
year 1900, tropical, Jan., day 0 hour 12	second (ephemeris)	+07 3.155 692 597 47*
year 1900, tropical, Jan., day 0, hour 12	second	+07 3.155 692 597 47
	Viscosity	
centistoke	meter^2/second	−06 1.00*
stoke	meter^2/second	−04 1.00*

TABLE 25-4 (*Continued*)

To convert from	To	Multiply by
foot^2/second	meter^2/second	−02 9.290 304*
centipoise	newton-second/meter^2	−03 1.00*
lbm/foot-second	newton-second/meter^2	+00 1.488 163 9
lbf-second/foot^2	newton-second/meter^2	+01 4.788 025 8
poise	newton-second/meter^2	−01 1.00*
poundal-second/foot^2	newton-second/meter^2	+00 1.488 163 9
slug/foot-second	newton-second/meter^2	+01 4.788 025 8
rhe	meter^2/newton-second	+01 1.00*
	Volume	
acre-foot	meter^3	+03 1.233 481 9
barrel (petroleum, 42 gallons)	meter^3	−01 1.589 873
board foot	meter^3	−03 2.359 737 216*
bushel (U.S.)	meter^3	−02 3.523 907 016 688*
cord	meter^3	+00 3.624 556 3
cup	meter^3	−04 2.365 882 365*
dram (U.S. fluid)	meter^3	−06 3.696 691 195 312 5*
fluid ounce (U.S.)	meter^3	−05 2.957 352 956 25*
foot^3	meter^3	−02 2.831 684 659 2*
gallon (U.K. liquid)	meter^3	−03 4.546 087
gallon (U.S. dry)	meter^3	−03 4.404 883 770 86*
gallon (U.S. liquid)	meter^3	−03 3.785 411 784*
gill (U.K.)	meter^3	−04 1.420 652
gill (U.S.)	meter^3	−04 1.182 941 2
hogshead (U.S.)	meter^3	−01 2.384 809 423 92*
inch^3	meter^3	−05 1.638 706 4*
liter	meter^3	−03 1.00*
ounce (U.S. fluid)	meter^3	−05 2.957 352 956 25*
peck (U.S.)	meter^3	−03 8.809 767 541 72*
pint (U.S. dry)	meter^3	−04 5.506 104 713 575*
pint (U.S. liquid)	meter^3	−04 4.731 764 73*
quart (U.S. dry)	meter^3	−03 1.101 220 942 715*
quart (U.S. liquid)	meter^3	−04 9.463 529 5
stere	meter^3	+00 1.00*
tablespoon	meter^3	−05 1.478 676 478 125*
teaspoon	meter^3	−06 4.928 921 593 75*
ton (register)	meter^3	+00 2.831 684 659 2*
yard^3	meter^3	−01 7.645 548 579 84*

Index

NOTE: Subjects are followed by double numbers to indicate each page reference. The (first) bold-face number indicates the section; the (second) light-face number is the page number of that section.

About the Editor in Chief

Robert O. Parmley, P.E., CMfgE, CSI, is President and Principal Consulting Engineer of Morgan & Parmley, Ltd., Professional Consulting Engineers, Ladysmith, Wisconsin. He is also a member of the National Society of Professional Engineers, the American Society of Mechanical Engineers, the Construction Specifications Institute, the American Design Drafting Association, the American Society of Heating, Refrigerating, and Air-Conditioning Engineers, and the Society of Manufacturing Engineers, and is listed in the AAES *Who's Who in Engineering*. Mr. Parmley holds a BSME and a MSCE from Columbia Pacific University and is a registered professional engineer in Wisconsin, California, and Canada. He is also a certified manufacturing engineer under SME's national certification program and a certified wastewater plant operator. In a career covering more than three decades, Mr. Parmley has worked on the design and construction supervision of a wide variety of structures, systems, and machines—from dams and bridges to municipal sewage treatment facilities and water projects. The author of over 40 technical articles published in leading professional journals, he is also the Editor-in-Chief of the *Field Engineer's Manual*, Second Edition, the *HVAC Field Manual*, the *Hydraulics Field Manual*, the *HVAC Design Data Sourcebook*, and the *Mechanical Components Handbook*, all published by McGraw-Hill.